PLANT

PLANT

EDITOR-IN-CHIEF JANET MARINELLI

DK

LONDON, NEW YORK, MELBOURNE, MUNICH, AND DELHI

Project editor Louise Abbott, with Annelise Evans
Project art editor Vanessa Hamilton, with Alison Donovan

Editors Pamela Brown, Candida Frith-Macdonald, Linden Hawthorne, Letitia Luff, Constance Novis, Francis Ritter, Martin Walters, Fiona Wild
Art editors Stephen Josland, Elly King, Rachael Smith

Senior managing editor Anna Kruger
Senior managing art editor Lee Griffiths

DTP designer Louise Waller
Picture researcher Samantha Nunn
Additional picture research Celia Dearing
Media resources Lucy Claxton, Richard Dabb, Neale Chamberlain
Illustrator Debbie Maizels
Cartographer Rob Stokes

Indexer Jane Parker
Production controller Mandy Inness

Pages 16–85 produced by **cobaltid**
Designers Paul Reid, Darren Bland, Lloyd Tilbury
Editors Marek Walisiewicz, Kati Dye
www.cobaltid.co.uk

First published in Great Britain in 2004 by
Dorling Kindersley Limited, 80 Strand, London, WC2R 0RL
A Penguin company.

1 3 5 7 9 10 8 6 4 2

A CIP catalogue record for this book is available from the British Library
ISBN 0-7513-4797-3

Colour reproduction by Colourscan, Singapore
Printed and bound in Slovakia by TBB

see our complete catalogue at
www.dk.com

EDITOR-IN-CHIEF
JANET MARINELLI

Director of Publishing at Brooklyn Botanic Garden and editor of the BBG's internationally renowned series of gardening handbooks, Janet Marinelli is a leading advocate of plant conservation, a respected author, journalist, and pioneer in the field of ecological landscaping. Her previous books include *Stalking the Wild Amaranth*, of which it has been said: "not since Voltaire has anyone so eloquently proclaimed the wisdom and the necessity of cultivating our own gardens".

CONTRIBUTORS AND CONSULTANTS

Andrew Byfield
Plantlife International

Dr Steven Clemants
Brooklyn Botanic Garden

Gib Cooper
Director, Bamboo of the Americas (BOTA)

Professor Sir Peter Crane
Director, Royal Botanic Gardens, Kew

Dr Phillip Cribb
Royal Botanic Gardens, Kew

Dr Kingsley Dixon
Kings Park and Botanic Garden, West Perth, Western Australia

Dr John Donaldson
National Botanical Institute, South Africa

Martin Gardner
Royal Botanic Garden Edinburgh

Dr Mary Gibby
Royal Botanic Garden Edinburgh

Dr David R. Given
IUCN; Christchurch Botanic Garden, New Zealand

Carlos Gómez-Hinostrosa
UNAM, Mexico

Dr Christopher Grey-Wilson
The Alpine Garden Society, UK

Madeleine Groves
Royal Botanic Gardens, Kew

Dr Héctor M. Hernández
UNAM, Mexico

Craig Hilton-Taylor
IUCN Red List Programme Officer

Dr Kathryn Kennedy
Executive Director, The Center for Plant Conservation, St Louis

Jeffrey Kent
Kent's Bromeliad Nursery Inc., California

Sabina G. Knees
Royal Botanic Garden Edinburgh

Dr Mike Maunder
Director, Fairchild Tropical Botanic Garden, Miami; IUCN/SSC Plant Conservation Committee

Dr Alan W. Meerow
President, International Bulb Society; Chair, IUCN Bulb Specialist Group

Dr Abisaí J. García Mendoza
UNAM, Mexico

Dr Ulrich Meve
University of Bayreuth

Sara Oldfield
Global Trees Campaign, Fauna & Flora International (FFI)

Dr Sarah Reichard
Center for Urban Horticulture, University of Washington

Christopher S. Robbins
WWF-US and Biota LLC

Dr David L. Roberts
Royal Botanic Gardens, Kew

Professor Gideon F. Smith
National Botanical Institute, South Africa

Dr Wendy Strahm
IUCN Plants Officer, Species Programme, Switzerland

Dr Chris Stapleton
Royal Botanic Gardens, Kew

Dr Wolfgang Stuppy
Royal Botanic Gardens, Kew

Dr Nigel Taylor
Royal Botanic Gardens, Kew

Dr Mark Tebbitt
Brooklyn Botanic Garden

Kerry S. Walter
BG-BASE (UK) Ltd & Royal Botanic Garden Edinburgh

Martin Walters
People and Plants Editor, WWF-UK

Dr Paul Wilkin
Royal Botanic Gardens, Kew

Dr Peter Wyse Jackson
Botanic Gardens Conservation International (BGCI)

Dr Scott Zona
Fairchild Tropical Botanic Garden, Miami

PUBLISHER'S ACKNOWLEDGMENTS

The views expressed in *Plant* are those of the individual authors and of the Editor-in-chief, and should not be interpreted as representing the policies of any of the organizations below. However, this book would not have been possible without the participation and access to information provided by the following:

Botanic Gardens Conservation International (BGCI)
Royal Botanic Gardens, Kew
Royal Botanic Garden Edinburgh
Fauna & Flora International (FFI)
IUCN – The World Conservation Union
TRAFFIC North America
UNEP-World Conservation Monitoring Centre
WWF International (formerly World Wide Fund for Nature)

Further information and contact details for these organizations, for *Plant*'s sponsoring botanic gardens, and for many other bodies working for plant conservation can be found in *Useful Addresses, pp. 473–78*.

Contents

The world of plants

Global habitats

Plant encyclopedia

Invasive plants

PREFACE

It is a scientific fact, but one that is not sufficiently widely acknowledged, that the biological diversity covering the land surface of our planet rests almost exclusively on the ecological foundation provided by plants. Plants are therefore of fundamental importance in the maintenance of the biosphere. Diversity in the world of plants also supports the variety of mammals, birds, amphibians, and other animals that enrich our lives, and contributes to the ecological processes that sustain us.

For many years the recognition that gardens are important safe havens for plants that are threatened in their native habitats has been steadily increasing. The ginkgo, the dawn redwood, the Chilean blue crocus, and the chocolate-scented cosmos are all examples of plants with narrow and highly threatened ranges in the wild, but whose future survival is assured by their popularity in cultivation. Nevertheless, among the millions of people around the world who derive great pleasure from gardening, there are still many who do not fully appreciate the importance of what they can do for plant conservation.

This book is the first authoritative reference for gardeners interested in how their passion for plants can contribute to the survival of species. It highlights those plants that are currently vulnerable and under threat. It takes a continent-by-continent look at hotspots of plant diversity and plants at risk. And it draws attention to those plants that are currently endangered in the wild by overcollection for the horticultural trade. It also explains the vitally important laws and regulations that govern the transfer and exchange of plants worldwide. The aim is to show how gardeners can contribute to conservation, while at the same time encouraging conservation in the wild, and ensuring that their actions do not unwittingly create new ecological problems, such as that of invasive species. Especially important for the goals of global conservation is to help people around the world benefit from the conservation of their local plants.

In these and in many other areas, this book takes a positive and proactive approach. It highlights the key issues and shows how gardeners can help plants and people across the globe. It is a superbly rich and completely indispensable resource for all those who love plants, understand their importance, and are concerned about the variety of plant life and its future.

Janet Marinelli, her contributing authors, and Dorling Kindersley deserve great credit for developing such an important and beautiful book. This magnificent work introduces a new ethic of gardening: one in which gardeners ask not simply what plants can do for them, but where they also pause to consider what they can do for plants and our planet.

Professor Sir Peter R. Crane FRS
Director, Royal Botanic Gardens, Kew

HOW TO USE THIS BOOK

Plant is divided into five sections. *The world of plants* describes how plants evolved and how many of them need our help to survive. In *Global habitats*, links are traced between plants and their native conditions. The *Plant encyclopedia* presents the plant kingdom in 12 parts, from *Trees and shrubs* through to *Carnivorous plants*. *Invasive plants* identifies the species that most threaten biodiversity on Earth. The reference section has useful addresses, conservation maps and information, and a glossary.

THE PLANT ENCYCLOPEDIA

The chapters in this section each list plant entries alphabetically by botanic name. There are two types of entry. Each entry headed by a genus name discusses a genus, or plant group, with particular reference to several species and forms: first mentions of illustrated plants are highlighted in bold. The information, including the map, at the head of the entry applies to all plants in the genus. An entry headed by a full species name contains information about that species alone. The Red List and CITES status of all plants mentioned is given either in the status bar at the top of the entry or on their first mention in the text.

▲ CHAPTER INTRODUCTION

◄ STANDARD ENCYCLOPEDIA ENTRIES

▼ FEATURED ENCYCLOPEDIA ENTRY

Conservation status bar gives details of the status of the species or genus on the Red Lists and CITES appendices (see below)

Botanical family notes where the species or genus belongs in botanical classification

The natural distribution of the species or genus is given in the map and supporting text

Symbols indicate major habitats (see key, below) *in which plants are found*

Summary of the conditions required by the plant

Information about related invasive species

Red List: Endangered*

Deschampsia cespitosa *subsp.* alpina

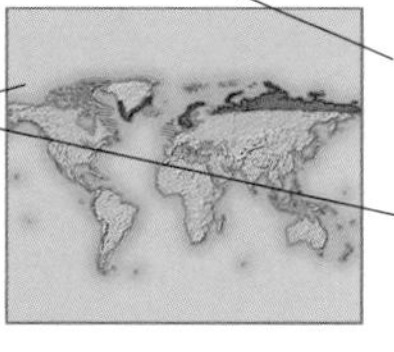

Botanical family
Gramineae

Distribution
Canada: Labrador, Nunavut, Quebec; Greenland; Iceland

Hardiness
Fully hardy

Preferred conditions
Rocky soils, moist sites; tolerates cold conditions.

Alpine (tufted) hairgrass is a plant of alpine and mountain meadows. Although it has quite a wide

germinate, and at least three years must transpire before the plants first flower.

See also Invasive plants, *p.454.*

For the purposes of this book, hardiness is given using the following categories:

Fully hardy: *to −15°C (5°F)*
Half-hardy: *to −0°C (32°F)*
Frost-tender: *to 5°C (41°F)*

THE IUCN RED LISTS

Two sources (*see also p.482*) for the conservation status of plants exist: a printed 1997 IUCN Red List of Threatened Plants (www.unep-wcmc.org/species/plants/red_list.htm), and the IUCN Red List of Threatened Species, updated annually since 2000 (www.iucnredlist.org). Categories used are given below. Plants named as "Unlisted" have not yet been evaluated, but may still be globally threatened.

1997	2000
(Ex) Extinct	(EX) Extinct
(Ex/E) Extinct/Endangered	(EW) Extinct in the Wild
(E) Endangered	(CR) Critically Endangered
(V) Vulnerable	(EN) Endangered
(R) Rare	(VU) Vulnerable
(I) Indeterminate	(LR) Lower Risk
	(DD) Data Deficient

An asterisk (*) denotes listings from the 1997 Red List

CITES

Appendix I: Trade in specimens of these species is permitted only in exceptional circumstances.
Appendix II: Trade in species is controlled in order to avoid utilization incompatible with their survival.
Appendix III: The named species are protected in at least one country, which has asked other CITES Parties for assistance in controlling the trade.
For a full explanation of CITES listings, *see page 483.*

HABITAT SYMBOLS

- Temperate forest
- Coniferous forest
- Tropical and rainforest
- Mountain, highland, scree
- Mediterranean scrub
- Desert and semi-desert
- Open grassland
- Wetland

INVASIVE PLANTS

The section on Invasive Plants (*pp.440–71*) gives an A–Z guide to plants that, to a greater or lesser degree, have had an adverse effect on native plant communities in areas where they were introduced. In this section, the ecological impact of each species is described, warning the gardener of the worst offenders and providing information on those that are only regionally harmful.

Eichhornia crassipes 100

Water hyacinth

While considered to be one of the worst weeds in the world, this aquatic plant is still commonly grown as a pond ornamental. It

Botanical name

Featured in One Hundred of the World's Worst Invasive Alien Species (*see p.442*).

Common name

INVASIVE PLANTS CHAPTER ENTRY

THE WORLD OF PLANTS

WHAT PLANTS DO FOR THE PLANET

Of all the characteristics of life on Earth, none is so dazzling as its sheer diversity – the incredible multitude of life forms, the vast genetic archive they embody, the complex ecological associations they form, and the fascinating behaviours they exhibit. The process that fuels this diversity is one of the great marvels of evolution – photosynthesis. All life depends on the ability of plants to capture the energy in sunlight and convert it to the chemical energy all organisms need to survive.

LAND AND SEA

The diversity of living things is far larger on land than in the oceans. Some 250,000 species of marine organisms have been recorded, compared with more than 1.5 million terrestrial ones. Scientists believe this is because there is relatively little physical variation in marine environments, and because ocean habitats have changed comparatively little over time. However, the diversity of major lineages (phyla, divisions, and classes) is much greater in the sea than on land. Among the most species-rich of these major groups are protists (an extremely diverse group of microbes with a fascinating variety of shapes), molluscs, and crustaceans.

The number of known marine species of the phylum Anthophyta, the flowering plants, is astonishingly low. Only 50 – the seagrasses – live, grow, and flower permanently submerged beneath the waves, compared with at least 270,000 on land. Being less accessible, marine species are also less well known.

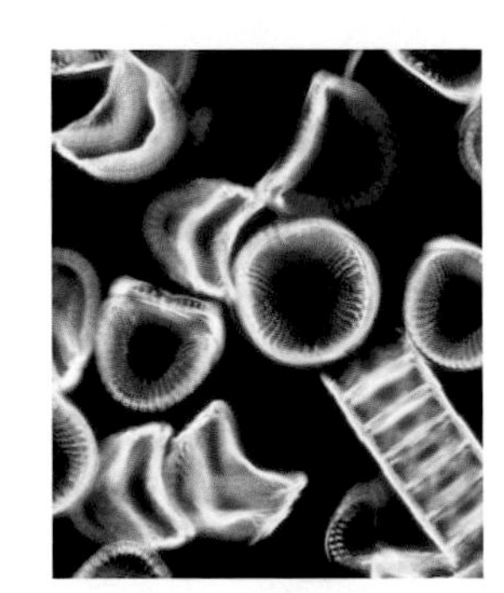

◄ SIMPLE AND COMPLEX *The simplest algae* (left) *are capable of carrying out photosynthesis to nourish life, yet the world is full of elaborate plants with complex flowers* (below). *The diversity of living things has been shaped over the millennia in a multitude of habitats by a range of evolutionary forces.*

THE WATER PLANET

Terrestrial habitats, including freshwater areas, cover just 29 per cent of the Earth's surface. The rest is ocean. For living things, the distinction between terrestrial and marine habitats is fundamental. Land plants and animals inhabit an environment where water loss and dehydration are constant threats, and they have evolved a range of anatomical and physiological solutions to this problem, from water-storage organs to water-conserving metabolisms.

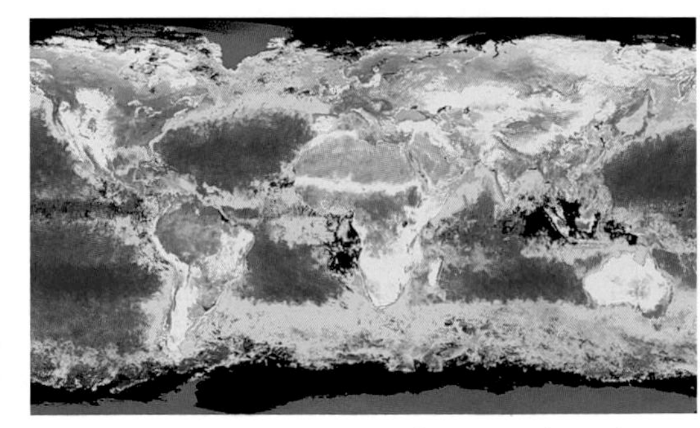

▲ PLANT ABUNDANCE *A satellite image shows the distribution of plants on Earth. Colours represent densities of green plant pigment; they range from red (most dense) to blue in the oceans, and green (most dense) to orange on land.*

UNDERWATER FORESTS

Photosynthesis is the driving force of life in the sea, as on land. Single-celled phytoplankton and picoplankton, or minute unicellular cyanobacteria, are the ocean's primary photosynthesizers. No marine plants are directly analogous to the woody plants that provide structurally rich habitats for communities of organisms on land; the closest equivalents are large brown algae called kelps.

GIANT KELP *Forests of* Macrocystis pyrifera *grow off the Californian coast. Kelps have large stipes and blades – the equivalent of stems and leaves on a tree – which form a canopy layer.*

PLANT SUCCESSION

When fire or another agent of disturbance lays bare a forest, the former mix of shade-loving plants perishes, and the site is recolonized by sun-loving "pioneers". As these grow, the light conditions change and ultimately begin to decline, and shade-tolerant species reappear. This process promotes diversity by populating the landscape with unique associations of plants and animals in various stages of succession. As humans rapidly destroy old-growth forests and other late-succession communities, the species that inhabit them are being lost: some are becoming extinct, and those that can return may take many hundreds of years to mature.

▲ INTREPID PIONEER *Morning glory recolonizes volcanic ash in the crater of the volcano at Krakatoa, Indonesia. Few terrestrial habitats are too hostile to support plant life, and terrain cleared by fires or volcanic activity is soon recolonized. Such disturbances begin the chain of plant succession and keep living communities in a state of flux.*

PARADISE AT RISK *Old-growth forests, such as this in Kauai, Hawaii, are being rapidly destroyed. For every plant species that becomes extinct, as many as 30 other animal and plant species may also decline. Scientists are just beginning to understand how the diversity of living things is essential to the health of our planet.*

LIFE SUPPORT *Plants are major players in the support systems that are vital to the well-being of life on our planet. They are oxygen factories – producing this important gas as a by-product of photosynthesis – making the Earth's atmosphere breathable.*

LIGHT AND LIFE

Green plants, as well as algae and some bacteria, are able to harness sunlight and convert it into sugars – energy-rich organic compounds that are used by the plant to fuel its life support processes, growth, and reproduction. Plant matter is consumed by herbivores, and forms the basis of every food chain, so plants' ability to capture solar energy sustains almost all life on Earth.

The process of trapping and fixing solar energy is called photosynthesis, and it takes place in minute structures called chloroplasts within a plant cell. The first step of the process is the absorption of light by pigment molecules in the chloroplasts. The most significant of these pigments – chlorophyll – absorbs light mostly in the blue and red wavelengths, and reflects green light; this is what makes plants green. Light energizes the chlorophyll molecules, which then pass their chemical energy through a complex chain of intermediates, eventually resulting in the formation of sugars, which can be used immediately by the plant or stored. The raw materials for the process are carbon dioxide, which is taken in from the atmosphere, and water, which is drawn up from the plant's roots. During photosynthesis, water is split chemically, releasing oxygen gas as a by-product. Virtually all the oxygen in the atmosphere is thought to have been generated in this way.

GREEN ENERGY

In most flowering plants, the primary photosynthetic organ is the leaf, although in cacti this function is taken over by the stem. The broad, flat structure of a leaf is perfect for intercepting light, and its surfaces allow carbon dioxide gas to enter while minimizing water loss to prevent dehydration.

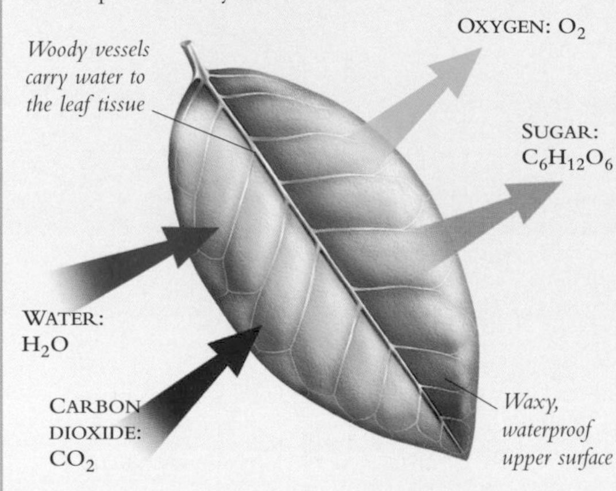

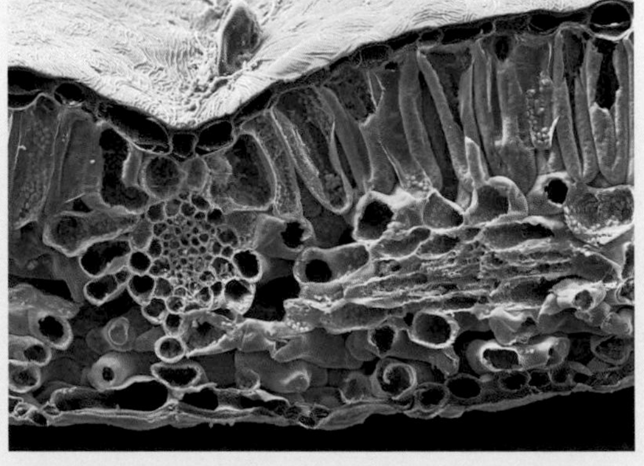

▲ **LEAF SECTION** *An electron microscope reveals the fine detail of leaf anatomy. The highest photosynthetic activity occurs in the palisade layer – the row of elongated cells just below the leaf's waxy upper surface.*

◄ **CHEMICAL BALANCE** *Photosynthesis is a complex biochemical pathway, but even early in the 19th century, scientists knew that the raw materials of water and carbon dioxide were used to produce sugar and oxygen.*

PLANTS AS FOOD

Without the grasses of the North American prairie, there would be no herds of grazing bison. And without the savannas and woodlands of Africa that sustain vast herds of herbivores, there would be no leopards or lions stalking prey. Plants support not only the large, spectacular herbivores, but also the myriad insects and invertebrates that contribute so much to the diversity of life on Earth.

The energy stored in plant bodies passes through ecosystems when animals feed, and organisms are sometimes grouped according to their source of energy. Plants, which use light energy to make their own food, are called primary producers. Herbivores are primary consumers because they feed on living plants. Carnivores and parasites are known as secondary consumers because they feed on the animals that feed on plants, while decomposers, such as fungi and bacteria, break down organic matter stored in the dead bodies of other organisms.

▲ **PRIMARY CONSUMERS** *Plains zebras must eat almost continuously to obtain nutrients from tough grasses. Food passes quickly through their gut where only the most readily digestible fractions are extracted; the rest is excreted.*

FOUNDATION OF LIFE

The total weight of all living organisms in any particular ecosystem is called its biomass. The bulk of the biomass – usually around 90 per cent – is made up of plants, the primary producers. The biomass pyramid illustrates the extent to which plants sustain the various levels of consumers.

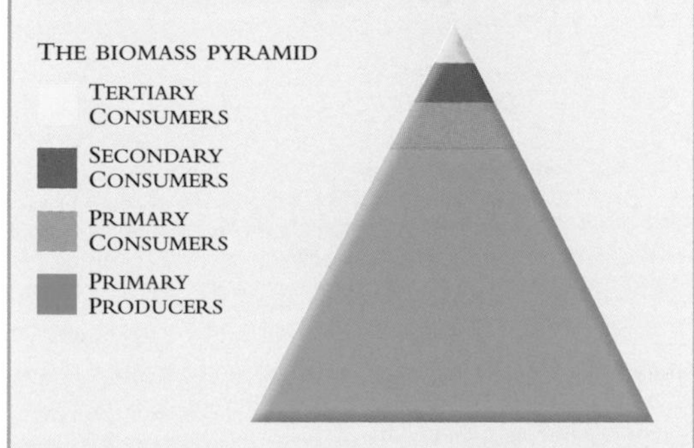

NATURAL RECYCLING

In a typical temperate forest, only about one per cent of plant biomass is eaten by large herbivores. The rest is dealt with by decomposers – bacteria, fungi, insects, and worms – which break down the material and return key nutrients, such as nitrates, to the soil, and carbon dioxide to the atmosphere. The role of these organisms is vital. Not only do they prevent the build-up of vast amounts of dead plant matter, but they also recycle the raw materials of life so that new growth can begin.

The resources available in an ecosystem are limited, so the rate of new plant growth depends partly on how fast minerals can be recycled. In the heat and humidity of tropical rainforests, decomposers work extremely fast, allowing a profusion of plant life despite the typically poor soils.

▲ FOREST RENEWAL *A beech seedling emerges from the leaf litter of a forest floor, taking up nutrients released by decomposers. A single gram of forest soil may contain around five billion decomposing bacteria.*

Carbon cycling

Plants play a fundamental role in maintaining the quality of our atmosphere. Not only do they produce breathable oxygen, but they also take up carbon dioxide – an important "greenhouse" gas (*see p.35*) – as they photosynthesize. Carbon dioxide is released by plants and animals as they respire, so carbon – the element that makes up all living things – constantly cycles between organisms. Huge amounts of carbon are stored in the woody trunks of trees, as dissolved carbon dioxide in the oceans, locked up in limestone rocks, or buried as oil and coal deposits. For the past 10,000 years, the natural processes of photosynthesis, respiration, and carbon storage have essentially been in balance, keeping the concentration of atmospheric carbon dioxide relatively constant. In the last century, however, the amount of carbon dioxide in the atmosphere has been increasing, due mainly to the burning of fossil fuels and the destruction of forests.

STORAGE SYSTEM ► *Every year, 100 billion tonnes of carbon in the atmosphere are bound into organic compounds by photosynthesis. Forests can store carbon for centuries, and their destruction contributes to the greenhouse effect and global climate change.*

Water management

Plants are an integral part of the processes that distribute and purify the water that sustains all life on the planet. They help hold soil in place and renew its fertility, provide natural flood control, detoxify and decompose wastes, purify the air and water, and stabilize the climate.

Through transpiration (*see box*), plants pump moisture from the soil into the air, so helping to cycle water between the atmosphere, the oceans,

▼ COASTAL BUFFERS *Mangroves* (Rhizophora stylosa) *stabilize sediments, recycle nutrients, and provide a protective habitat for many species. They also protect coastal areas from wave damage during storms.*

and the land, and influencing weather patterns. Wetland plants have a great influence on the hydrological cycle on land, because they slow the flow of water, allowing it more time to infiltrate underground. They prevent flooding by capturing and containing large volumes of potentially destructive floodwaters – one hectare (2.5 acres) of wetland can store up to 17 million litres (3 million gallons) of storm water – and at the same time, remove pollutants from the water.

Plant cover protects the land from erosion. When deforestation, overgrazing, and over-cultivation expose the soil to the elements, it is quickly washed into rivers by rainstorms and blown away by the wind. Once the nutrient-rich layer of topsoil is gone, the land becomes barren – a process called desertification (*see p.34*). The presence of plants slows down water flow and binds soil in position – particularly important in countries where high rainfall and mountainous terrain promote erosion. The mudslides and other damage caused in 1998 by Hurricane Mitch, which battered Central America and took thousands of lives, were exacerbated by deforestation in mountainous inland areas.

MOVING WATER

Water that evaporates from the leaves and other above-ground parts of a plant is replaced by water from the roots, a process known as transpiration. About 99 per cent of water taken up by the roots is not used by the plant's tissues, but rather transpires. Water rises to the leaves of the tallest trees by "pulling itself" up through narrow vessels in the trunk. This pull comes from capillary action – the same process by which ink spreads through blotting paper. As long as the water column is unbroken, moisture can rise to the tops of even the tallest trees.

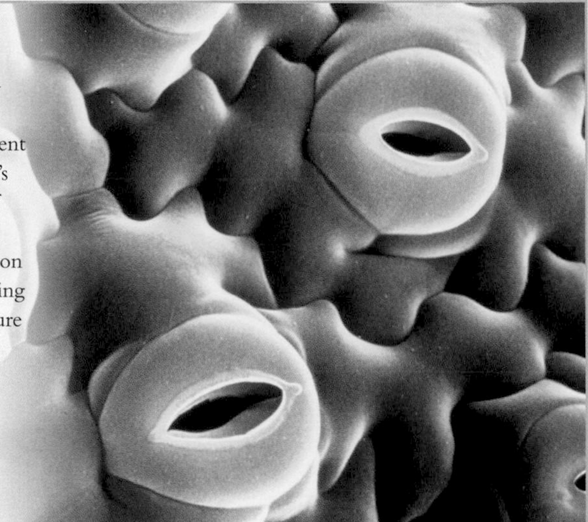

PLANT PORES *Most of the water given off by vascular plants is released through tiny pores, or stomata, in their leaves. Guard cells lining the stomata change the size of the pores depending on the degree of water stress.*

FLOW CONTROL *With its tall canopy trees and vertical layers of understorey vegetation, this forest in tropical Costa Rica functions like a giant sponge, soaking up most of the water from torrential rainstorms before gradually releasing it back into the atmosphere as water vapour.*

WHAT PLANTS DO FOR US

Every day the spectacular diversity of plants on the planet provides the goods essential for human life: the foods that sustain us, the medicines that keep us healthy, the materials with which we build our homes and fashion our clothing, and the fuels that power our energy-hungry societies. Plants also enrich our lives with their beauty, intoxicating fragrances, and fascinating habits, as well as with the raw materials we use to make products as diverse as alcoholic drinks and beads for jewellery.

▲ WATER WORTHY *Wood is an excellent material for boat building, not least because it expands when wet, creating watertight seals between adjacent pieces. It is used widely in vessels today, from rustic canoes to luxury sailing boats.*

FEEDING THE WORLD

For most of human history, we have been a species of hunters and gatherers of wild plants. The domestication of plants is a relatively recent practice, which began about 9,000 years ago in Mesopotamia (parts of present-day Iraq, Turkey, Syria, and Jordan). Of the 7,000 plant species that have, at one time or another, been collected or cultivated for consumption, only about 200 have been domesticated, and today, just 12 crop plants provide some 75 per cent of human calorie intake. These are, in alphabetical order, bananas and plantains, beans, cassava, maize, millet, potatoes, rice, sorghum, soya beans, sugar cane, sweet potatoes, and wheat.

▲ PAPAYA *Grown mainly for its fleshy fruit, the papaya also has medicinal and industrial uses.*

The husbandry, and deliberate selection, of wild plants over many generations has made them progressively easier to grow and more productive. It has also transformed their appearance: by breeding for maximum yield, the useful parts of plants – their seeds, leaves, or tubers, for example – have grown to huge sizes and the plants have then had to change their growth form to support the giant parts.

NATURAL MATERIALS

Wood is probably the most versatile construction material on our planet. It is used to make everything from houses to musical instruments, and despite the advent of plastics, it shows no sign of going out of fashion. Wood is composed of the dead cells of a tree trunk, particularly the water-carrying xylem vessels, which are made up of biological polymers – chiefly cellulose and lignin. Wood of angiosperms, known as hardwood, has dense clusters of heavily lignified, thick-walled fibre cells, which is why it is generally harder and heavier than the wood of conifers, known as softwood. At least a dozen species of trees called ironwoods are so dense and heavy that they actually sink in water. One of these species, the Caribbean lignum vitae (*Guaiacum officinale*) is harder-wearing than iron and was once widely used to make pulley sheaves, bearings, and castors.

▲ WORKABLE MATERIALS *Soft and relatively easy to work with simple tools, wood has always been favoured by builders. Even today, most new homes are of wood-frame construction.*

Equally essential plant products are fibres – the raw materials for the manufacture of paper, textiles, cords, and ropes. In botanical terms, fibres are long, narrow, tapering cells with thick, heavily lignified cell walls. Several types of fibre are used to make textiles. Surface fibres, like cotton and coconut coir, grow on the surface of seeds, leaves, or fruits. Soft or bast fibres, such as linen, jute, hemp, and ramie, are found in the phloem or inner bark of the stems of dicots. Hard or leaf fibres like sisal come from the long, narrow leaves typical of monocots.

STAPLE CROP *More than 400 million tonnes of rice are grown worldwide every year. Asia accounts for 90 per cent of the world's production and consumption because of its favourable hot and humid climate. Much of the rice is grown on the terraced fields of numerous small farms.*

FUEL FOR LIFE

Human beings have an insatiable thirst for energy. Globally, we use the equivalent of 10,000 million tonnes of oil every year to run our cars, heat our homes, and sustain our industries, and most of this energy comes directly or indirectly from plant biomass. Around 40 per cent of the world's people rely on wood harvested from forests as their primary source of energy for cooking and heating. And fossil fuels – coal, oil, and gas – are derived ultimately from the photosynthetic activity of plants that lived hundreds of millions of years ago. Most of our use of plant material as fuel is unsustainable – known coal reserves will last only another 250 years or so – but increasingly, new, clean technologies are enabling us to exploit plant biomass in more responsible ways.

▲ BURNING ISSUES *Fuelwood consumption has increased by 250 per cent since 1960. In less developed countries, the collection of fuelwood and building material from natural forests is a cause of significant deforestation.*

◄ WOOD ENERGY *Charcoal is produced by burning wood at a high temperature in the absence of air – a process known as carbonization. This makes it easier to transport, store, and use.*

STIRRING THE SENSES

Across the globe, plants are celebrated and valued for their aesthetic qualities – colour, form, and scent. The total size of the world market for flowers and potted plants is estimated to be at least $50,000 million – a staggering figure that excludes the huge sums spent by gardeners and by visitors to areas of natural beauty. In the deciduous forests of the eastern USA, plant lovers make pilgrimages to see the fleeting spring blooms of bloodroot, trout lily, and other ephemeral wild flowers; and in the autumn, "leaf peepers" flock to New England to see the forests explode into a fireball of foliage colour.

Plants produce natural perfumes to attract insect pollinators. The distillation of these fragrant compounds is an art that dates back to the ancient world, when scents were valued for incense used in religious ceremonies, and used to anoint the body. Oil of lavender (from *Lavandula angustifolia*) and attar of rose (from *Rosa damascena* and others) are commonly used in modern perfumes.

The beauty of plants is a cause for contemplation as well as celebration. The famed American biologist E.O. Wilson coined the phrase "biophilia" to describe what he believed to be an innate human need for the natural world. Wilson argued that this need is biologically based – integral to our development as individuals and as a species. Environmental destruction, he argued, could have a significant impact on our quality of life, not just materially but psychologically and even spiritually.

PLANTS AS MEDICINE

Used for centuries to treat inflammation and pain, the bark of the willow tree is a rich source of salicin. This naturally occurring compound is a chemical precursor of acetylsalicylic acid, better known to millions of people worldwide as aspirin. Not just aspirin, but about one half of all the pharmaceuticals sold across the world owe their genesis to compounds first extracted from plants. Tens of thousands of other plant species are used globally in traditional medicine, the mainstay of healthcare for 80 per cent of the world's population.

According to the Food and Agriculture Organization (FAO), 13 per cent of all flowering plants, or nearly 53,000 plant species, are used medically. Increasingly, medicinal plants that have their origins in traditional medicine are sold as herbal remedies in western countries and cities. A small selection of these plants is given below:

▲ **HEALING SCIENCE** *Technicians catalogue plants in the herbarium of a pharmaceutical company in San Francisco, USA. The plants may provide a basis for drug development.*

▲ **HERBAL REMEDIES** *Traditional herbalists spend years learning to treat illnesses using native plants. The incursion of western culture may threaten this unique knowledge base.*

- Kava (*Piper methysticum*) is a shrub belonging to the pepper family, native to the South Pacific. It is traditionally prepared by islanders as a ceremonial and social beverage. The root contains high concentrations of kavalactones – chemicals that produce a mild tranquillizing effect. Kava is exported to herbal markets in western Europe, North America, and Asia where the herb is promoted to relieve insomnia and anxiety.
- Cat's claw (*Uncaria tomentosa*) is a liana found in parts of Central and South America, where it climbs to heights of 30m (100ft) in the rainforest. Tribes of the Peruvian Amazon use the bark, which is dried, ground, and taken in the form of tea, to treat a variety of ailments such as inflammation and sexually transmitted diseases. In North America and Europe, herbalists tout the plant's ability to benefit the immune system and aid digestion.
- Rooibos (*Aspalanthus linearis*), or red bush, is a member of the pea family that grows in Cedarberg, South Africa. Natives of the region prepare a tea from its dried, fermented leaves and stem. The plant is a concentrated source of polyphenols – powerful antioxidants linked to purging the body of damaging free radicals.
- Devil's club (*Oplopanax horridus*), aptly named for the spines protruding from its leaves and stems, is a deciduous shrub belonging to the same family as ginseng (*Panax* species). It inhabits old growth forests of the US Pacific Northwest, preferring moist soil and habitat. Native Americans harvest the roots and stem bark to treat a range of health problems, including arthritis, respiratory ailments, and indigestion. Devil's club inhibits bacterial, viral, and fungal activity and contains chemical compounds that regulate blood sugar, making it potentially useful to diabetics.

Contemplative design *The best Japanese gardens inspire an active but tranquil contemplation of nature. The power of these traditional landscape designs stems from their ability to distil the sights, sounds, and fragrances of nature, not from the profusion of showy flowering plants.*

Plants that changed the world

All human societies depend on plants. This simple fact is a daily reality to indigenous peoples, who harvest plants directly for food, shelter, fuel, clothing, and medicines. In the developed world, our total dependence on plant ecosystems is less visible, but no less real.

Over the past 300 years, plants, and the trade in plant products, have shaped human history, dictating patterns of colonization and settlement that are reflected in the wealth and status of nations today. Two examples demonstrate the historic and present influence of plants on society: the colonization of Virginia by the British in the early 1600s would not have been economically viable without the lucrative tobacco trade and the superior tobacco growing fields in North America; and today, production of cocaine from the coca plant in Colombia continues to undermine government authority, causing political instability and violence.

The plants described below have all had a seismic influence on societies around the world, enriching some while casting others into slavery.

◄ **Cinchona** *Quinine, a chemical derived from bark of the South American* Cinchona *tree, gives protection from the parasites that cause malaria. In colonial times, availability of quinine made it possible to colonize areas affected by this deadly disease.*

◄ **Wheat** *Four grasses – wheat, barley, rice, and maize – underpin the success of the world's great civilizations. Their high yield enabled farmers to build up large surpluses of wealth, which liberated other members of the community from toiling in the fields.*

▼ **Rubber** *The trunk of the Brazilian rubber tree* (Hevea brasiliensis) *yields a milky sap that can be hardened and treated to make an elastic solid – rubber. Early industry was entirely dependent on natural rubber until the development of synthetic alternatives.*

◄ **Cocoa** *Chocolate is made from the beans of the cacao tree* Theobroma cacao, *which were first exported from Central America to Europe in the 1500s.*

Pepper ► *In the Middle Ages, pepper* (Piper nigrum) *was so precious that peppercorns were used as a form of payment. Trade in pepper fuelled the European Age of Exploration.*

Cotton ► *Trade in cotton* (Gossypium *species) between the American South and the mills of northern England created and sustained the last great slave workforce. The issue of slavery was among the triggers of the American Civil War of 1861–1865.*

Papyrus ► *Once common in the Nile Valley, the reed* Cyperus papyrus *was cut into thin strips to make papyrus, an ancient precursor of paper. Papyrus was first used about 4000 BC, and became one of Egypt's major exports.*

THE EVOLUTION OF PLANTS

In the words of one botanist, the origin of land plants is "as lost in the mists of time as anything can be". Although microscopic marine life dates back as far as 3.5 billion years, the Earth's land masses remained largely uninhabited until about half a billion years ago. Our view of early land plants has been constructed from fossils left in ancient sedimentary rocks by improbable accidents that preserved pieces of once-living tissue – spores, leafy debris, and wood from prehistoric forests.

COLONIZATION OF LAND

The Ordovician period (490–443 million years ago, or mya) is best known for its trilobites and other marine invertebrates, but there is also evidence that plants began to invade the land at this time – not from the sea, as scientists once thought, but rather from fresh water. There is good evidence that they evolved from green algae. This algal ancestor diverged to produce ancestors to the bryophytes – small plants that lack vascular tissue, such as today's mosses, hornworts, and liverworts – as well as ancestors to the more complex vascular plants. One of the earliest known vascular land plants, *Cooksonia*, dates back to the Silurian age, 428 mya. A small plant, it had green, leafless, bifurcating branches. From this simple beginning, vascular plants have been the basis of terrestrial ecology ever since.

◄ COOKSONIA FOSSIL *This small, simple plant was probably the first to grow upright.*

Once they mastered the rudimentaries of life on land, plants were free to evolve in a virgin landscape. They spread and colonized much of the Earth's surface, encountering virtually no competition, paving the way for the subsequent evolution of land animals.

VASCULAR TISSUE

Provided a plant grows on a moist surface and close to the ground – as bryophytes such as mosses do – a vascular system is not required. "Plumbing" to move water and nutrients is essential only when a plant adopts an upright habit and grows larger, defying gravity and exposing itself to the desiccating effects of the wind. By the end of the Devonian age, plants had xylem – conductive tissue that enables water to reach the tallest parts of the plant – and phloem, for the transport of nutrients.

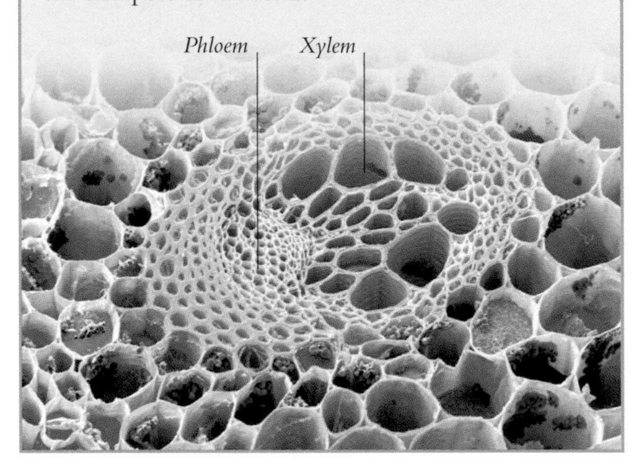

THE FIRST TREES AND FORESTS

During the Devonian age (418–354 mya), with increasing competition for space and light, plants reacted by growing upwards. They were able to do this because of the development of vascular tissue (*see above, right*) – fluid-conducting tubes that allow water and nutrients to be transported throughout the plant. The vascular tissue was also impregnated with an organic substance – lignin – that provided internal and mechanical support. The vertical growth of plants was bolstered by such innovations, and a variety of trees and shrubs of all sizes evolved. Plant communities became vertically layered and more complex. At the same time, plants were becoming more widely distributed, and different kinds of plants evolved in different places. Complex communities evolved in relation to their environments, resulting in specialized plants adapted to different habitats.

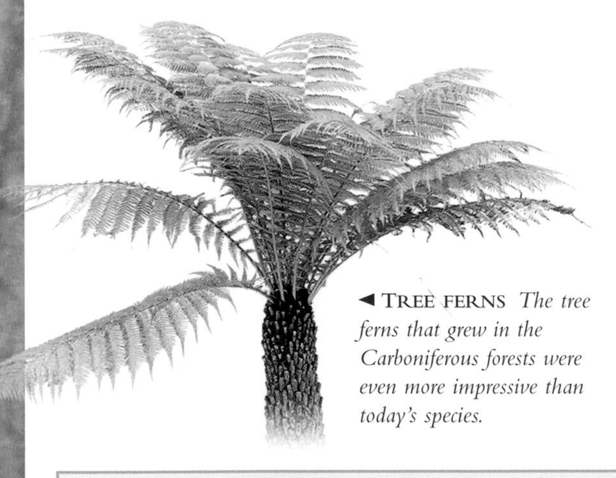

◄ TREE FERNS *The tree ferns that grew in the Carboniferous forests were even more impressive than today's species.*

◄ HORSETAILS *Modern equisetums still closely resemble their ancestors, except in terms of scale – Carboniferous-age* Calamites *towered over the landscape.*

During the Carboniferous age (354–290 mya), size dominated plant evolution. The tropical climate was warm and moist, and great swamp forests grew, creating ideal conditions for the proliferation of pteridophytes – plants that reproduce by spores. The most imposing plants in these lush forests were arborescent lycopods that soared to 40m (130ft) tall. These forest giants were the ancient relatives of the modern quillworts, which generally grow close to the ground. Alongside the lycopods were magnificent tree ferns that grew to a height of 20m (65ft). Huge horsetails, called *Calamites*, also loomed large in Carboniferous plant communities, growing in dense clumps like bamboo.

▲ PRIMEVAL FOREST *The Carboniferous age witnessed great swamp forests, whose dense layers of plant matter were laid down to become today's coal seams. Early reptiles and huge cockroaches and dragonflies roamed these ancient forests.*

EVOLUTIONARY TIMELINE

Life has existed on the Earth for around 3,800 million years. The first living things were single-celled bacteria that lived in the oceans, and by 2,000 mya the first multi-celled organisms had appeared. Early plants – moss-like bryophytes – colonized the land 465 mya, and by the Silurian age, simple vascular plants, such as *Cooksonia*, had developed. It was not until the Cretaceous that the first flowering plants appeared.

PALAEOZOIC ERA						
CAMBRIAN	ORDOVICIAN	SILURIAN	DEVONIAN	CARBONIFEROUS		PERMIAN
				MISSISSIPPIAN	PENNSYLVANIAN	
543 mya	490 mya	443 mya	418 mya	354 mya	323 mya	290 mya

THE RISE OF SEED PLANTS

Most plants of the Carboniferous swamps were dependent on abundant supplies of water. In order for spore-bearing plants to reproduce, sperm cells must swim through water on plant surfaces to reach the eggs, so when the climate became drier and cooler at the end of the Palaeozoic era, most of the plants that flourished in these forests became extinct. But gymnosperms – conifers and other advanced seed-bearing plants – were pre-adapted to drier conditions and survived. The Mesozoic era (252–65 mya) was the heyday of conifer diversity and they became successful worldwide. Many modern conifer families date back to this period, including podocarpus, araucaria, cypress, and pine. Although the flowering plants later ousted them from many of their habitats, the conifers continue to dominate large areas of the Earth's vegetation today.

Other gymnosperms attained prominent positions in the Mesozoic vegetation. Today's cycads seem to have changed very little since the days of the dinosaurs, when they were much more widespread and numerous. It is estimated that during the Triassic and Jurassic periods, about 20 per cent of land plants were cycads. Mesozoic ginkgos were also found virtually worldwide. The mass extinction of the dinosaurs at the end of the Cretaceous period was accompanied by great vegetational change, paving the way for the reign of the flowering plants. Today, about 750 species of gymnosperms survive.

GINKGO BILOBA

◄ MESOZOIC SURVIVOR *The only surviving species from an ancient group of plants,* Ginkgo biloba *– or maidenhair tree – is often referred to as a "living fossil" because the modern ginkgo is almost identical to those found in the fossil record. While Mesozoic ginkgos were distributed almost worldwide,* Ginkgo biloba *today grows wild only in south-east China.*

FOSSILIZED GINKGO LEAF

THE IMPORTANCE OF SEEDS

The development of seeds was one of the most important events in the evolution of land plants, enabling them to diversify and adapt to a range of different environments. Seeds – embryonic plants contained within a hard protective coat – can be dispersed long distances by air or animals and remain viable for long periods of time. Then, when conditions are favourable, the seed can germinate and become a new adult plant. This innovation has allowed plants to inhabit some of the most inhospitable climate areas in the world. Some desert species, for example, survive through long periods of drought as seeds, only springing back to life when moisture becomes available.

CONE AND SEEDS OF A MODERN SILVER FIR

THE AGE OF FLOWERING PLANTS

Darwin called the origin of angiosperms, or flowering plants, an "abominable mystery". Learning how and when they appeared on Earth has been the Holy Grail for botanists for over a century. They have been traced back to the earliest Cretaceous period, about 135 mya. However, these discoveries do not pinpoint the first flowering plant but rather early branches from that unknown ancestor.

Angiosperm means "seed borne in a vessel" while gymnosperm means "naked seed", a reference to the lack of a protective structure enveloping the seed. One reason that flowering plants were able to diversify so dramatically and spread so rapidly during the Cretaceous and the Cenozoic, or modern, era was the evolution of new structures and tissues such as the carpel, a womb-like vessel that encloses angiosperm seeds, and the endosperm, a placenta-like tissue that nourishes the young plant as it develops within the seed. Today, angiosperms dominate terrestrial life on the planet. At an estimated 422,000 species (*see p.28*), they comprise by far the largest group of plants. They grow in a greater range of environments, exhibit a wider range of growth habits, and display more variation in form than any other living group of vascular plants. In size, angiosperms range from tiny duckweeds to eucalyptuses more than 100m (330ft) tall. The explosion of angiosperm diversity has gone hand in hand with the proliferation of insects, birds, and other animals that pollinate their flowers, disperse their fruits and seeds, and eat their leaves.

PAPHIOPEDILUM INSIGNE ► *The orchid family contains some of the most highly specialized and complex adaptations of all flowering plants.*

Mesozoic era				Cenozoic era						
Triassic	Jurassic	Cretaceous		Tertiary					Quaternary	
		Lower	Upper	Palaeocene	Eocene	Oligocene	Miocene	Pliocene	Pleistocene	Holocene
252 mya	199.5 mya	142 mya	99 mya	65 mya	54.8 mya	33.5 mya	24 mya	5 mya	1.8 mya	0.01 mya – 0 mya

PLANT DIVERSITY TODAY

▲ SPECTACULAR FLOWER *Several species of* Rafflesia *are native to the jungles of south-east Asia, and many are rare or threatened.* Rafflesia arnoldii *is the largest, its flower reaching 1m (3ft) in diameter.*

Over the millennia, the world's green cloak of primitive algae, lush ferns, and giant conifers has exploded into the multicoloured floral ebullience of the modern era. Unprecedented numbers of plant species have evolved to inhabit the Earth. Biologists distinguish between the major levels of diversity: not only the diversity of species, but the genetic material within species, as well as the diversity of communities, ecosystems, and landscapes in which plants exist and evolve. The term "biodiversity", made popular by American biologist Edward O. Wilson, refers to this total variability of life.

GENETIC DIVERSITY AND SPECIES RICHNESS

The most obvious measure of plant biodiversity is the raw number of species on the planet. For the past decade, many authorities have put the number of flowering plant species at around 270,000; however, recent analyses have produced substantially higher figures. One study by David Bramwell, Director of the Jardín Botánico Canario Viera y Clavijo in Las Palmas, Canary Islands, suggests a figure of around 422,000. This estimate was made by taking the flora of the largest country in each region and adding to it the number of local endemic species in each of the other countries in the region, then tallying the regional totals. Even this higher number may not be a true reflection of species diversity, because it does not include undiscovered species, which may number around 50,000.

All plants created via cross-pollination, with the random recombination of genes that occurs during sexual reproduction, are unique at the genetic level. The concept of biodiversity also encompasses these genetic differences among members of the same species. The number of genes in *Arabidopsis thaliana*, a small weed used worldwide in laboratory studies of plant genetics, is around 25,500 – similar to the total number in the human genome (about 32,000 genes). The genetic diversity of a species is a measure both of the number of genes possessed by any one individual, and of the number of variants of genes (or alleles) that exist within the species. If each gene has just a small number of alleles, or variant forms, the number of potential genetic combinations is truly enormous. This reservoir of diversity allows species to adapt to changing environmental conditions. It is the material upon which evolution acts, ultimately creating new species.

COMMUNITY DIVERSITY

Different plant species have very different requirements in terms of climate, soil type, aspect, and their essential biological relationships. For this reason, very different plant communities are found in different habitats. Broadleaf deciduous trees, for example, dominate temperate forests, yet cannot function well in more arid temperate ecosystems, such as grasslands and deserts, where herbaceous and succulent species thrive. Maintaining the diversity of ecosystems and their native plant communities is critical to global plant diversity – not just because more ecosystems means more species, but because plant communities are more than just the sum of their parts. Ecological studies demonstrate that plant productivity, nutrient use, and other processes can falter when biodiversity is reduced; one long-term field study found that reductions in the number of species made a grassland more vulnerable to drought. Declines in productivity and stability can be particularly acute when the number of species is low, as it is in artificial ecosystems, such as croplands and timber plantations.

◄ TREE AS HABITAT *A single mature tree may be home to hundreds of other plants and animals.*

UNIQUE LIFE FORMS

One of the most important gauges of biodiversity is endemism – the occurrence of species with narrow geographical ranges. Many species are endemic to a single country; more significantly, some are endemic to a very small area – a mountain slope, desert spring, island, or other isolated habitat. Areas with many endemic species are called centres of endemism.

Discrete areas with complex topography, especially in the tropics, often have high levels of endemism. Globally, habitats with the highest percentage of endemic plants include the moist and the dry forests of Madagascar, New Caledonia, and Hawaii; Brazil's Atlantic coastal moist forest; and the fynbos of South Africa's Cape Floral Kingdom. Generally, but not always, areas with high levels of endemism are also rich in species diversity. Countries or other areas with the greatest endemism and the highest species diversity are called "megadiverse", and are of great conservation value. High degrees of endemism are typically a result of geographical isolation, such as that of oceanic island groups. Endemics can also evolve in biologically isolated areas, such as remote mountain ranges.

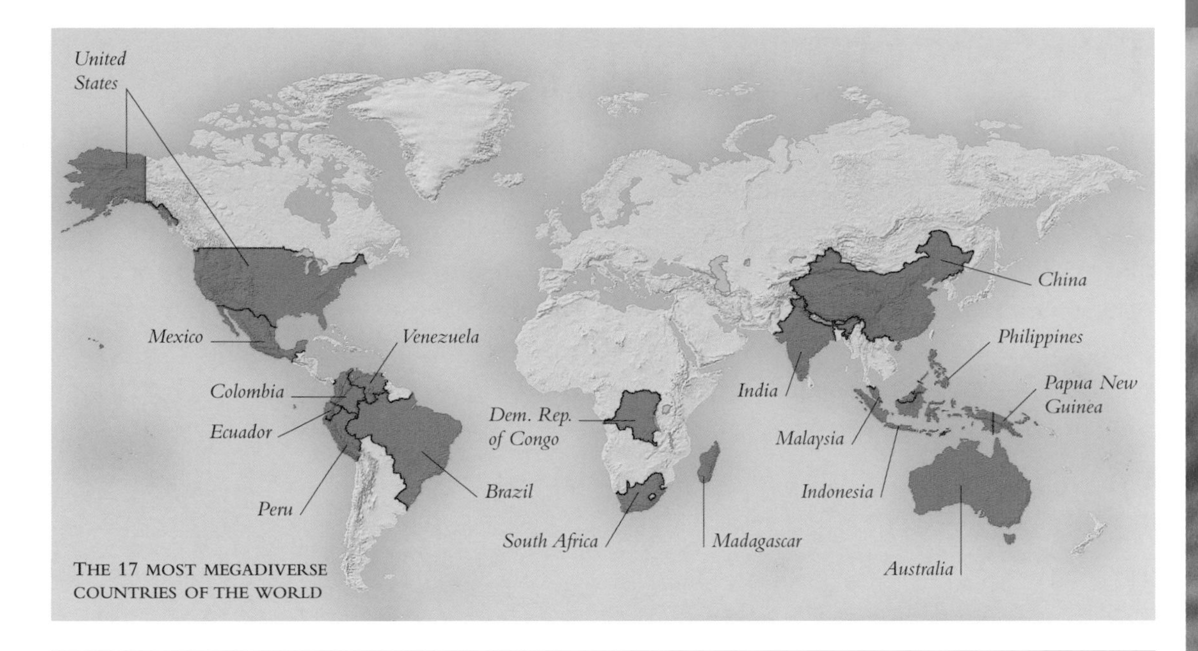

THE 17 MOST MEGADIVERSE COUNTRIES OF THE WORLD

BIODIVERSITY HOTSPOTS

In 1988, British ecologist Norman Myers published an article in which he proposed that international conservation priorities should be set on the basis of "hotspots" – areas with exceptionally high species richness and endemism that are also under threat. Myers first pinpointed ten such hotspots – all tropical forest regions – but his ideas were soon adopted by global conservation groups who identified many other hotspots around the world. The diversity of vascular plants – which are necessary for the survival of most other kinds of organisms – was used as the top criterion for hotspot status.

In 1998, the environmental organization Conservation International published a study called *Megadiversity: Earth's Biologically Wealthiest Nations*, which identified 17 countries that contain more than two-thirds of the world's biodiversity (*see above*); again, the primary criterion for assigning "megadiverse" status was plant endemism. Conservation International later recognized 25 hotspots (*see p.479*), which between them cover less than 2 per cent of the Earth's land area but account for 44 per cent of all vascular plant species. Fifteen of these 25 hotspots are tropical forests, and five are Mediterranean climate zones; almost all tropical islands fall into one of these hotspots. The tropical Andes is the richest hotspot in terms of both plant diversity (with 45,000 species) and plant endemism (with 20,000 species found nowhere else in the world). The World Wildlife Fund has also published a list of worldwide hotspots, called the Global 200 (*see p.480*). From the Everglades, the huge freshwater grassland in Florida, USA, to Australia's Great Barrier Reef, these are ecoregions with rich and unique expressions of biological diversity.

◄ CULTURAL VALUE *Mountain flax* (Phormium cookianum) *is a rare native of New Zealand, where it was valued by the Maoris as a source of fibre.*

BEAR'S BREECHES ► Acanthus spinosus *is native to the Mediterranean basin – one of five similar climate regions worldwide with many endemic plants.*

ISLAND REFUGES

Remote oceanic islands often harbour disproportionately high numbers of plants found nowhere else in the world. These endemic species may have evolved from immigrant plants that arrived on sterile volcanic outcrops after a long sea crossing, diversifying into an astonishing range of species in a process called adaptive radiation. Some islands are sanctuaries for relict plant communities once widespread, but now largely extinct, such as the laurisilva of the Canary Islands (*see p.110*). Island floras are highly vulnerable due to their very restricted geographical ranges.

◄ ISLAND EXTINCTION *Oceanic islands like Hawaii are flashing early warning signs of mass extinction. Many of their endemic plants – including 20 per cent of Hawaiian lobeliads – are thought to have already vanished.*

HAWAIIAN ENDEMISM

The Hawaiian Islands stretch over 2,650km (1,646 miles); their distance from the nearest mainland is 3,400km (2,110 miles). Some estimated percentages of endemic species in selected plant and animal groups:

Group	%
Mosses	46%
Ferns	70%
Flowering plants	91%
Gymnosperms	91%
Arthropods	99%
Birds	81%

CLASSIFYING PLANTS

Naming and cataloguing the vast diversity of life on Earth is one of the oldest preoccupations of natural scientists. Before Darwin, systems of classification either sought to reveal the supposed divine pattern of Creation, or grouped plants and animals by arbitrary, and rather artificial, criteria, giving them long and often convoluted names. Since the "discovery" of evolution, however, biologists have given plants and animals names that reflect their similarities and reveal their evolutionary relationships with other species. In this system of binomial classification, each species is known by an internationally recognized two-word Latin scientific name.

DEFINING TRAITS

Most biologists recognize five principal groups, or kingdoms, of living things – monerans (bacteria), protists (to which algae belong), fungi, plants, and animals. The Plant Kingdom comprises hundreds of thousands of plant species adapted to life on land. Its main divisions are themselves ordered into hierarchies of lower groupings (taxa) that reflect the biological relatedness of their members (*see below*). This relatedness is assessed primarily on the basis of shared characteristics that have been inherited from a common ancestor.

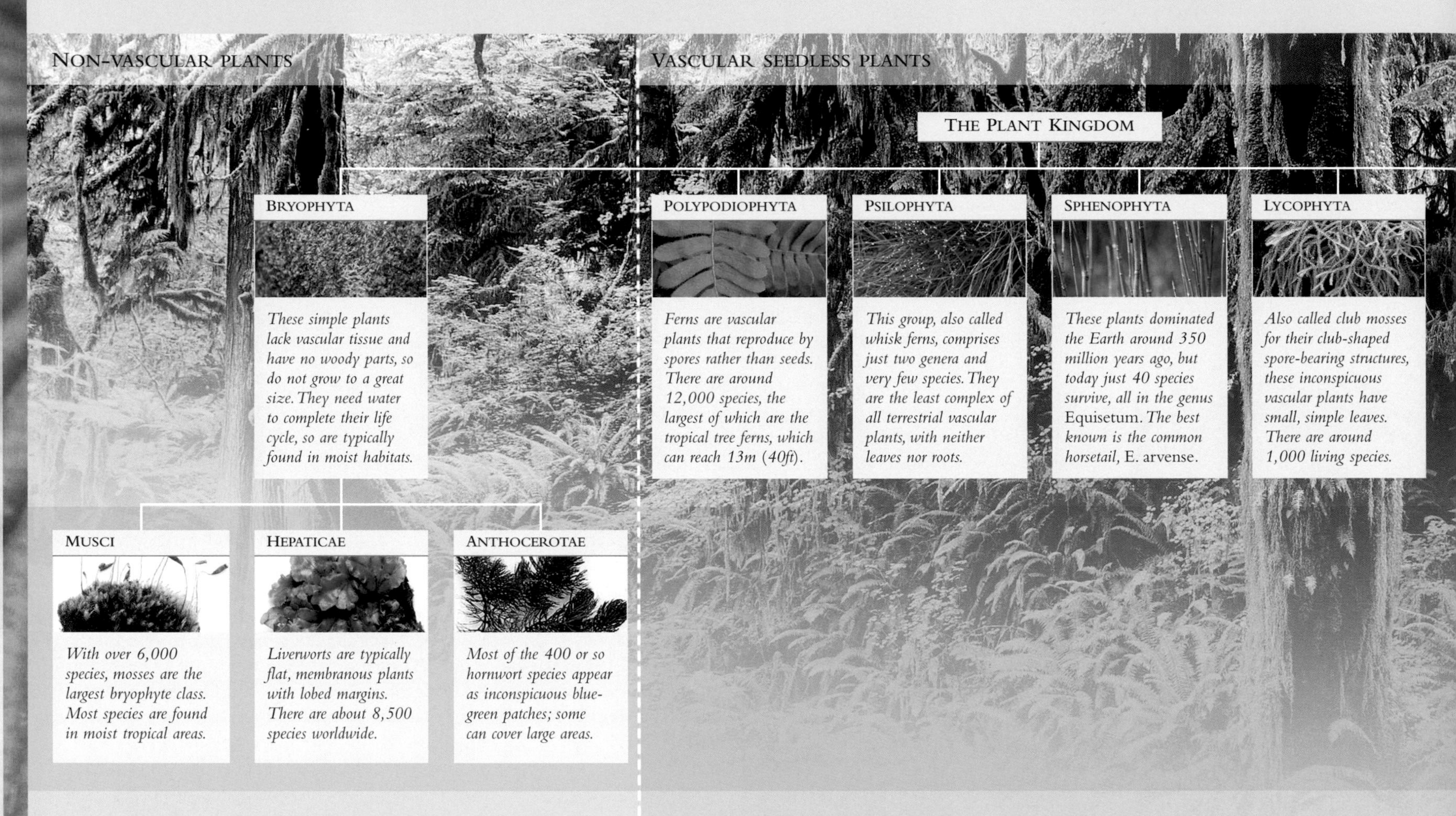

FATHER OF TAXONOMY

The foundations for the binomial system of plant naming still in use today were laid down in the 18th century by the Swedish naturalist Carolus Linnaeus. His taxonomy, which predated the theory of natural selection, was based on the number and arrangement of the reproductive organs – the stamens and pistils – of a plant. Linnaeus created large numbers of genera and assigned each species to a genus. Today, taxonomists group genera into orders, orders into classes, and classes into divisions.

CAROLUS LINNAEUS ►
Linnaeus (b. 1707) studied medicine before developing his system of plant classification.

HOW PLANTS ARE NAMED

Every species has a two-part name made up of its generic (genus) name followed by a specific epithet, or species name. By convention, both are written in italic. For example, *Viola pedata* is the botanical name for the eastern North America native bird's-foot violet, named for its characteristic many-lobed leaves. Some species consist of one or more similar races, called subspecies or varieties. These plants are given an additional third name, also written in italics. For example, *Epimedium grandiflorum* var. *coreanum* is more upright than the species, *Epimedium grandiflorum*. The majority of specific epithets describe some aspect of the plant – its colour, shape, or number of petals, or habitat or season of flowering. Understanding these words, which are typically derived from Latin or Greek, can help to identify a plant or understand an aspect of its biology. Given here is a selection of epithets and their meanings.

COLOUR
- *aeneus:* bronze
- *flavus:* yellow
- *luteus:* deep yellow
- *niger:* black
- *ruber:* red
- *virens:* green

SHAPE AND SIZE
- *angusti-:* narrow
- *brevi-:* short
- *glosso-:* tongue-like
- *hetero-:* different
- *lepto-:* slender
- *lanci-:* lance-shaped
- *mega-:* very large
- *pachy-:* thick

GROWTH HABIT
- *arborescens:* tree-like
- *fruticosus:* shrubby
- *patens:* spreading
- *repens:* creeping
- *caespitosus:* tufted

HABITAT
- *arvensis:* of cultivated land
- *alpinus:* of the mountains
- *aquaticus:* of water
- *fluviatilis:* of rivers
- *lacustris:* of lakes
- *littoralis:* of shores
- *pratensis:* of meadows
- *sylvaticus:* of woods
- *terrestris:* of dry land

GREEN GROUPS

Plants can be classified in ways other than by their evolutionary relationships. Ecologists, for example, have long categorized plants according to the communities in which they naturally occur.

A native of Germany, Alexander von Humboldt (1769–1859) was one of the greatest scientific travellers of all time. He was also one of the first scientists to observe that plants tend to occur in similar groupings, called communities, and that wherever there are similar conditions, such as climate or soil, the same types of plant groups appear. Botanists and plant ecologists have since devised a classification system of principal vegetation types. On the global level, major plant communities are known as biomes (*see also* Global Habitats, *pp.60–85*). For example, regions with mild winters and long, dry summers, wherever they occur in the world, tend to be characterized by vegetation known as Mediterranean scrub. Mediterranean scrub plant communities on the west coast of North America are called California chaparral, while in the Mediterranean region they are called maquis or garrigue.

Some of the categories commonly used by gardeners to group plants reflect these habitats, or biomes – for example, alpines and aquatics. Others are based on botanical groupings, such as conifers, or even individual families – for example, the orchids. Much more generalized garden plant groupings are based on such factors as size, habit, and life cycle, from "trees", "climbers", and "herbaceous perennials" to such artificial horticultural categories as "summer bedding". But while such broad groupings can give some insight into cultivation needs, it is increasingly apparent that recognition of an individual plant's native habitat, be it forest, prairie, marsh, or desert, is the key to its successful care in the garden.

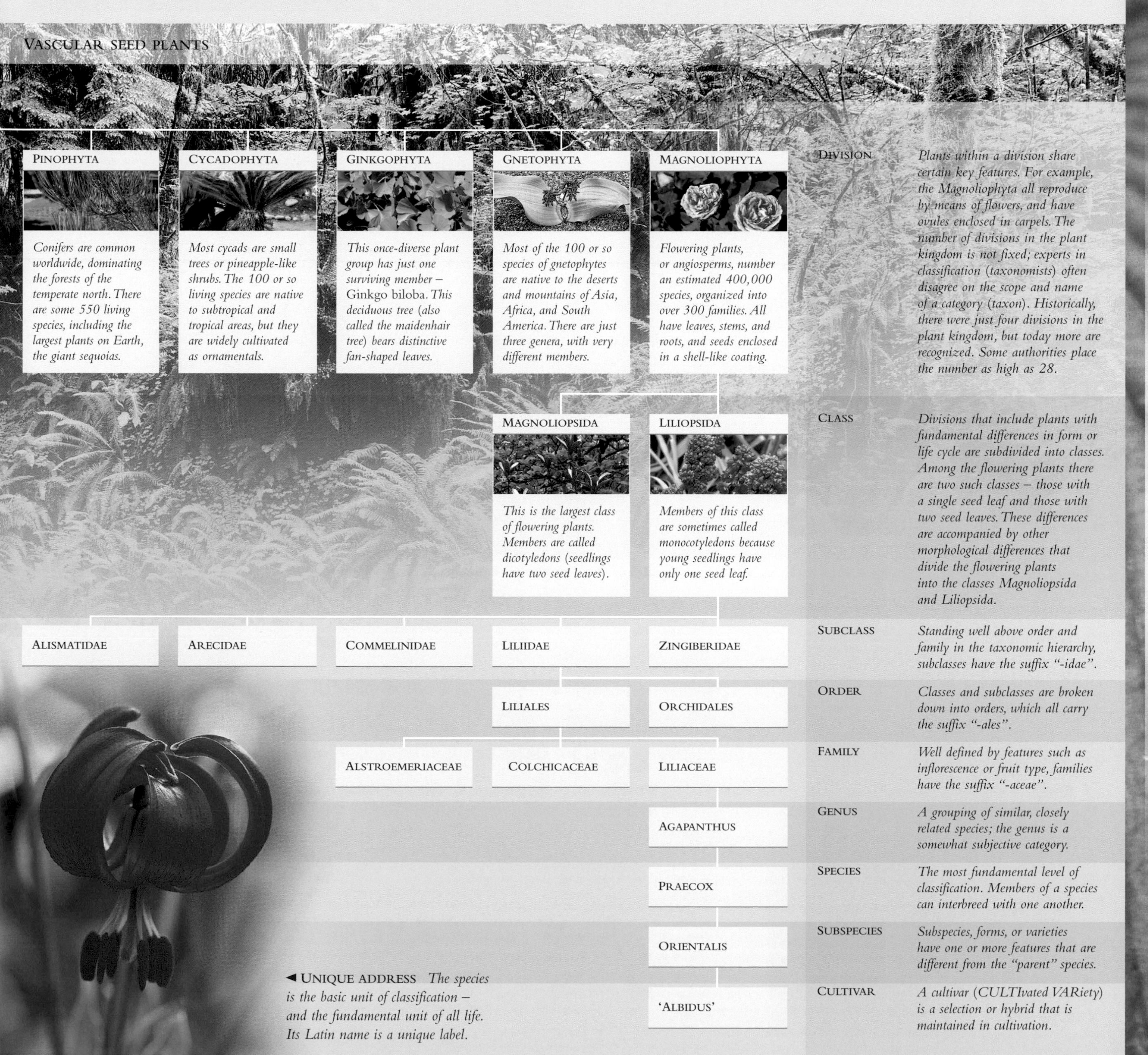

◄ UNIQUE ADDRESS *The species is the basic unit of classification – and the fundamental unit of all life. Its Latin name is a unique label.*

THE EXTINCTION CRISIS

We are poised on the brink of a biological disaster that could rival anything in evolutionary history, including the mass demise of the dinosaurs during the Cretaceous period some 65 million years ago. The fossil record shows that, unlike animals, plants have rarely suffered mass extinctions. However, if current trends continue, a devastating number of the world's species are likely to go extinct in the coming decades – including one-fifth to one-quarter of the planet's flowering plants.

▲ **JURASSIC PLANTS** *Conifers, club mosses, horsetails, ferns, and seed ferns dominated the Jurassic landscape, but were later replaced by flowering plants after a mass extinction episode.*

ACCELERATING LOSSES

Extinction is, and always has been, a natural process. The fortunes of different species are affected by changes in climate, topography, and the living environment, which give one group a competitive advantage over another. Groups that cannot adapt to change fast enough are eliminated, while others survive the evolutionary challenge. Palaeontologists differentiate between mass extinctions – relatively short periods when many groups disappear – and a more gradual "background extinction rate" that characterizes other periods. It is thought that the current extinction rate is at least 1,000 times higher than the background rate, as a result of human activity.

Botanists agree that the number of threatened plants worldwide is increasing at an alarming rate. However, the precise number of plant species at risk of extinction is a matter of speculation. This is unsurprising, given the fact that despite 250 years of cataloguing and describing members of the plant kingdom, there is still no accurate estimate of the total number of flowering plants on our planet, threatened or not.

Estimates of the current number of threatened plants vary widely. More than 34,000 taxa (plant groups) are on the 1997 IUCN Red List of Threatened Plants (*see p.36*), but this figure is just the tip of the iceberg – not enough data on the highly diverse and threatened tropical floras is yet available in order to include many of the plants on the Red List. Peter Raven, head of Missouri Botanical Garden, puts the number of taxa at risk of extinction closer to 100,000 – more than 23 percent of the total global flora.

Human-driven destruction of the Earth's vegetation is not confined to particular areas, or particular groups of plants. It is truly global in scope, with entire habitats under assault. A mass die-off could feasibly occur in a matter of decades, not millennia, as in previous extinction episodes.

IMPERILLED GARDEN PLANTS

Horticultural groups are not immune to extinction. This list, taken from the 1997 Red List, the Cycad Action Plan, and the World List of Threatened Trees, gives the number of threatened species in a selection of garden plant groups, followed by the percentage of that group currently under threat.

- Agave family: 68 species, 17.9%
- Aloe family: 206, 29.4%
- Amaryllis family: 176 species, 25.1%
- Bromeliads: 480 species, 24%
- Cacti: 581 species, 38.7%
- Conifers: 550 species, 55.8%
- Cycads: 149 species, 52%
- Ferns: 683 species, 7.5%
- Grasses and bamboos: 776 species, 9.7%
- Hyacinth family: 134 species, 20.5%
- Iris family: 484 species, 32.3%
- Lily family: 149 species, 32.4%
- Nepenthes family: 19 species, 25.3%
- Orchids: 779 species, 5.9%
- Palms: 869 species, 29%

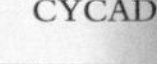

CYCAD

MASS EXTINCTIONS

A mass extinction episode is usually defined as one in which about 75 to 95 per cent of species become extinct in a geologically short period, perhaps as little as a few hundred thousand years. Mass extinctions are typically followed by 5–10 million years of very low diversity, when a small number of species come to dominate, after which biodiversity again begins to increase.

At least five episodes of mass extinction are evident in the fossil record. In each case, despite a huge loss of species, the complexity and diversity of life recovered, and new dominant life forms emerged. For example, there was an explosion in mammal diversity after the disappearance of the reptiles that reigned during the Jurassic and Cretaceous periods. Mammals became dominant.

Severe extinction episodes have played a major role in animal evolution, but the fossil record suggests that only one mass extinction has had a serious impact on plants. This was the extinction at the end of the Permian period, when ferns and other spore-bearing plants were replaced by cycads, ginkgos, conifers, and other seed-bearing plants (*see pp.26–27*). The Cretaceous extinction saw a small loss of plant species – about 15 per cent – but since then, the extinction rate has remained relatively low – at least until recently. Many scientists believe that the spread of human civilization marks the onset of the sixth – and possibly most catastrophic – mass extinction. For thousands of years we have been modifying the Earth's ecosystems, favouring some species while driving others to extinction. In North America, for example, 70 per cent of all large mammals became extinct soon after people arrived on the continent after crossing the Bering land bridge. Although the loss of mammal and bird species has attracted more public attention, the consequences for plants are also dire, and will continue to grow worse.

LEPIDODENDRON ► *These trees grew to 30m (100ft) in the swamps of the Late Carboniferous. They had fragile, slow-growing stems, which some suggest may have contributed to their extinction.*

HUMAN IMPACT

Rapid growth of the human population is placing huge stresses on natural communities of plants and animals. Clearance for agriculture, pollution, overexploitation, and climate change are all contributing to the next extinction and the wide gap in distribution of income is hastening ecological decline. The richest 20 per cent of the world's people take 83 per cent of global income, the poorest just 1.4 per cent. This leads to excess consumption and waste in the most affluent countries, while the poor are forced to cut down forests to support a subsistence lifestyle.

▲ DISAPPEARING WILDLANDS *Habitat loss – especially to agriculture – is identified as a main threat to 85 per cent of all species described in IUCN Red Lists.*

WORLD POPULATION GROWTH

As recently as the beginning of the 19th century, the Earth's human population was less than one billion. In September 1999 it reached six billion, and by the end of this decade, it is expected to grow by another billion – the equivalent of adding another China. Most population growth occurs in the developing regions of Africa, Asia, and Latin America, and the contribution of these regions will rise over the next 25 years.

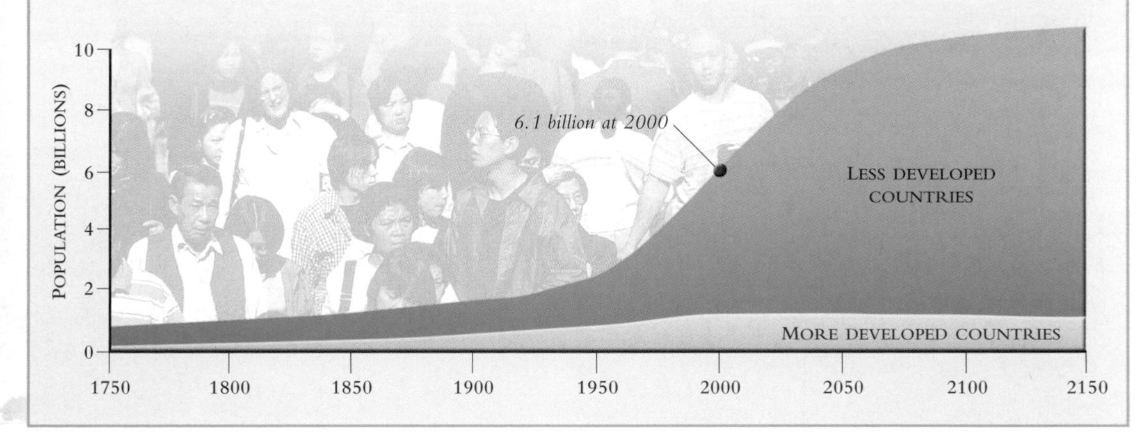

FUTURE SCENARIOS

To minimize the effects of human activity on plant and animal populations, conservation scientists must know where and how to focus their efforts. A project called GLOBIO (the global methodology for mapping human impacts on the biosphere), which was launched in 2001 by the UN Environment Programme, is helping them to do so. GLOBIO maps the present and future impacts of human development on biodiversity. The projections of this project suggest that biodiversity will be threatened in almost 72 per cent of global terrestrial land area by the year 2032, compared with close to 47 per cent today. Losses of biodiversity are likely to be very severe in Southeast Asia, the Congo basin, and parts of the Amazon. In these areas, conversion to agriculture, plantations, and urban areas may increase to 48 per cent of the land area, from 22 per cent today.

SALINIZATION *The replacement of native plants with crops in Western Australia has caused a form of pollution known as salinization. The change in land use causes the water table to rise, forcing salt in the soil to form surface crusts that kill plants and aquatic animals.*

MAJOR THREATS

The loss and fragmentation of natural habitats, the overexploitation of many species, and the biological invasion of introduced plants and animals are among the chief causes of the current extinction crisis. These pressures on the world's flora and fauna are exacerbated by increasing climate change.

The catastrophic loss of plants is not only extinguishing entire species but also causing misfortune for untold numbers of people. As human populations swell, and lands are depleted, subsistence farmers are forced to migrate even deeper into virgin forests. About two-thirds of all terrestrial plant species are found in the tropics, mostly in rainforests – areas that once covered some 16 million square kilometres. Human activity has halved the extent of these forests, and their destruction through logging, clearance for agriculture, and burning continues at an unprecedented rate. Habitat loss in such centres of biodiversity is causing major losses of plant species worldwide. The areas of natural habitat that still remain are often no more than small islands in a sea of human-modified landscapes.

Pollution adds to the litany of assaults on plants and ecosystems. Acid rain, for example, weakens trees by damaging their leaves, limiting the nutrients available to them, and increasing their uptake of toxic substances, such as aluminium, in the soil. Ozone, formed in the troposphere from the products of burning fossil fuels, causes premature loss of foliage.

Desertification – where overcultivation and overgrazing turn dry lands into deserts – threatens up to a third of the planet's land area, and already affects at least one billion people in more than a hundred countries.

FRAGMENTATION ► *Building roads to logging and mining sites opens vast new stretches of virgin rainforest to exploitation. It also carves the forest into ever smaller blocks that can support far fewer species than continuous expanses of habitat. Remaining plant and animal populations become split up and isolated, decreasing gene flow and increasing genetic erosion.*

▲ CHOKED BY INVASIVES *Water hyacinth* (Eichhornia crassipes) *is a South American species that has aggressively invaded waterways in many other parts of the world. It has a devastating effect on native species and on water quality.*

BIOLOGICAL INVADERS

Oceans, mountain ranges, rivers, and deserts are natural barriers that impede the movement of plants and animals. Over millennia, this geographic isolation permits the evolution of unique species and ecosystems. Over the last few centuries, however, trade and travel have broken down these barriers, and people have carried plants and animals beyond their natural ranges. Some non-native plants have become established in new environments and spread unchecked, threatening indigenous plants and animals. These are known as "invasive species".

Invasive plants overtake native species; invasive pathogens can destroy entire species such as the American elm; and invasive animals destroy roots, and eat or trample seedlings.

OVEREXPLOITATION

Helping to drive the extinction crisis is the fact that plants, and the products derived from them, are valuable commodities for human societies. Ironically, the overharvesting of certain tree species for timber and the illegal collection of ornamental and medicinal plants is threatening not only their continued survival, but also the livelihoods and health of the communities that depend on them. Economically valuable Caribbean mahogany has, for example, already been logged out and today is considered commercially extinct throughout Central America and the Caribbean islands. A number of orchids, cacti and succulents, cycads, and carnivorous plants are removed from the wild for collectors willing to pay vast sums for rare and often endangered species. Some, such as the giant pitcher plant (*Nepenthes* species), have already been driven to near extinction.

Demand for medicinal plants, too, is at an all-time high, as the popularity of herbal remedies grows in western countries. Rural harvesters earn income from selling raw plant material but may unknowingly collect plants at an unsustainable rate.

◄ COLLECTOR'S TARGET *Showy orchids may take a decade to flower in the wild, making them highly vulnerable to overcollection.*

HOTHOUSE EARTH

Our planet is warming faster than at any time in the past 10,000 years. Human activities, primarily the burning of fossil fuels and the destruction of forests, have caused heat-trapping gases such as carbon dioxide, methane, nitrous oxide, and chlorofluorocarbons to reach their highest atmospheric concentrations in more than 400,000 years. Many climatologists believe that the increased levels of these "greenhouse" gases are likely to raise world temperatures by 2°C (3.6°F) by 2030, at the same time causing sea levels to rise by 30–50cm (12–20in). Coastal ecosystems and human communities may be inundated as sea levels rise. Many species will not be able to migrate fast enough to keep up with the changing climate, and some, such as alpine plants, will simply run out of headroom when they reach the highest altitudes. Among the other plant species most likely to be affected by global climate change are endemic, or geographically isolated, species; specialized species, which have adapted to a narrow range of conditions; and annuals – because local populations can suffer complete reproductive failure in one year.

Scientists are already observing changes caused by global warming – for example, an advance in the time of leaf emergence and flowering of many northern species due to earlier springs.

EDELWEISS ► *Climate change threatens sensitive alpines, such as the famous edelweiss. If temperatures rise by a mere 2°C (3.6°F), such plants will need to move about 300m (990ft) higher – not possible for species near summits.*

BITTER HARVEST *Logging for the timber trade continues to destroy plants and animals that depend on old-growth forests. Logging continues not only in tropical forests, but also in the old-growth forests of Canada and the USA.*

GLOBAL ACTION

In recent decades, conservation strategies designed to protect plants have been put in place at local, national, and international levels. Much of the current conservation work focuses on ensuring the survival of single species. National laws, such as the US Endangered Species Act, authorize action and appropriate resources to save declining plants. On the global stage, a number of initiatives aim to compile a comprehensive worldwide list of imperilled species, control international trade in endangered species, and promote the sustainable use of plants, particularly food crops, medicinal species, fuel wood, and timber essential to human well-being.

PAPHIOPEDILUM LOWII *Trade in species listed in Appendix I of CITES is prohibited. Included in the Appendix are 298 plant species, mainly cacti and succulents, cycads, carnivorous plants, and orchids, including two genera of tropical slipper orchids,* Paphiopedilum *and* Phragmipedium.

THE IUCN RED LIST OF THREATENED SPECIES

The Red List, produced by IUCN – The World Conservation Union, is the most comprehensive global inventory of the conservation status of plants and animals. It helps conservation agencies to set their priorities at all levels, and establishes a baseline from which to monitor the future status of species. The Red List was first produced as a series of printed books, but now exists as a searchable online database. It is available at www.iucnredlist.org.

The Red List is based on information supplied by IUCN's Species Survival Commission – the world's largest network of species conservation experts. There are 35 Plant Specialist Groups, each focusing on different types of plants, or plants from various geographical areas.

▲ THREAT CATEGORIES *The IUCN Red List assigns each species a threat category* (see p.332), *from EX (Extinct) to LC (Least Concern). The bastard quiver tree* (Aloe pillansii) *of southern Africa is classified as CR – Critically Endangered.*

◄ CRITICAL LIST *The Red List dates back to 1959, when Lt Colonel Leofric Boyle, Chair of the Survival Service (now the Species Survival Commission), created a card index system to log the status of threatened species.*

THE CONVENTION ON BIOLOGICAL DIVERSITY

At the 1992 Earth Summit in Rio de Janeiro, Brazil, world leaders agreed on a strategy for "sustainable development" – meeting human needs, while at the same time maintaining a healthy world for future generations. One of the key agreements adopted at the Earth Summit was the Convention on Biological Diversity (CBD). The CBD has three major goals: to conserve the world's biodiversity; to promote the sustainable use of plants and animals; and to ensure that any benefit that is obtained through the development of this planet's plant life is shared fairly by all the world's people.

In 2002, the parties of the CBD endorsed the Global Strategy for Plant Conservation – the first major global agreement to include clear conservation targets. The 16 targets of this strategy, to be achieved by 2010, include:

- A preliminary assessment of the conservation status of all known plant species.
- Protection of 50 per cent of the most important areas for plant diversity.
- Conservation of 60 per cent of the world's threatened species in their natural habitats.
- Conservation of 60 per cent of threatened plant species in seed or tissue banks, and in cultivation.

▲ TREATY BENEFITS *The succulent* Hoodia *is chewed by the San Bushmen of South Africa to stave off hunger and thirst. Development of the plant's extracts into an anti-obesity drug may pay royalties to the bushmen under the CBD.*

PLANT TRADE ► *Over 28,000 plant species are listed by CITES in one of three Appendices, according to the degree of protection they need. Many commonly grown plants are covered by the convention, and a breach of the rules can be a criminal offence.*

CITES

The Conventional on International Trade in Endangered Species of Wild Fauna and Flora, better known as CITES, is an agreement between 164 governments that came into effect in 1975 (*see p.483*). Its goal is to ensure that global trade in wild plants and animals does not threaten their survival. The international wildlife trade is estimated to be worth billions of dollars annually, and to include hundreds of millions of plant and animal specimens. The trade in plants ranges from whole plants to bulbs, seeds, and other plant parts, as well as a vast array of products derived from them, including foods and medicines. Most of this trade is legal, but a significant portion is not. CITES allows trade in species that can withstand current rates of exploitation, but restricts trade in those threatened with extinction.

GLOBAL CONSERVATION: A GARDENER'S GUIDE

Before buying any plant from an overseas supplier, check to see if it is listed in CITES (www.cites.org). If you wish to import or export a CITES-listed plant, even if you are given a plant as a gift from a friend's garden, you must obtain in advance a permit from the CITES Management Authority in your country. A list of Management Authorities is available on the CITES website. In European Union countries, EU rules have, for the most part, superseded regulations of member nations, and require that a Management Authority see an export permit from the country of export before issuing an import permit. Because the permitting process can be complex, most people place orders through established commercial sources. Permits are not given for Appendix I species, except for plants that have been artificially propagated and for which certified proof of nursery origin has been obtained. Check import permits and documentation of artificial propagation to make sure that the plants have been obtained legally.

The Convention on Biological Diversity does not affect the movement of plants into most countries. It does, however, affect the export from an increasing number of countries around the world that are signatories to the convention. Even if they are granted, export permits will probably contain carefully worded stipulations regarding the future use of the plants or plant parts. These restrictions are designed to ensure that no commercial gain or other advantage is taken from the plants or their products without the specific agreement of the owners of the genetic resource they represent.

PROTECTING NATURAL AREAS

For millennia, people have been protecting plant communities – such as the sacred groves in India and China – for religious or medicinal purposes. However, the modern era of environmental conservation began only in the latter half of the 19th century with the foundation of the first national parks. The protection of the Earth's remaining natural habitats continues to be vital if we are to maintain the ecological and evolutionary processes that promote biodiversity and enhance human life.

LEVELS OF PROTECTION

Protected areas around the world are classified using a system of six categories developed by the World Conservation Union (IUCN). The six categories are based on management goals for the areas and represent a gradation in the extent to which natural systems have been modified and the degree to which human activity is permitted.

Category I is the strictest designation, covering reserves used mainly for scientific research, as well as wilderness areas. Category II areas are national parks, managed primarily to conserve ecosystems and provide opportunities for recreation.

Category III areas – natural monuments – preserve specific natural features. Wildlife refuges and nature reserves that are actively managed to conserve species or habitats are classified as Category IV, while Category V areas include landscapes or seascapes where the interaction of people and nature over time has produced areas with a distinct character and high ecological, aesthetic, and cultural value. Category VI includes natural areas predominantly unmodified by humans, which are managed to ensure the long-term protection of biological diversity while providing a sustainable flow of natural resources.

UNESCO RESERVES

Reserves established under the United Nations Educational, Scientific, and Cultural Organization (UNESCO) Man and the Biosphere Programme show how conservation can work in a landscape that supports economic activity.

Each reserve is organized into three zones: a core area that provides for the long-term protection of the landscapes, ecosystems, and species; a surrounding buffer that can be used for scientific research, ecological restoration, education, and recreation; and an outer transition area that may include human settlements, agriculture and other activities. The goal is sustainable use of natural resources – living on the interest of the land without depleting its capital.

Monuments and natural areas that are of outstanding value to humanity are designated as World Heritage Sites under a UNESCO convention adopted in Paris in 1972. The convention encourages international cooperation in safeguarding these areas, of which there are currently over 750. Most of the sites, which are spread across the globe, are designated for their natural value. They include areas such as the Serengeti National Park in Tanzania and the Galapagos Islands of Ecuador, as well as cultural gems like the Acropolis in Athens.

▲ BIOSPHERE RESERVE *Siberut Island, off the west coast of Sumatera, is one of only four UNESCO Biosphere reserves in Indonesia. Worldwide, there are more than 440 such reserves spread over 97 countries.*

GLOBAL COVER

The number of protected areas has increased rapidly over recent decades. According to a list compiled in 2003 by the United Nations (UN), there are now 102,102 such areas worldwide covering 18.8 million sq km (7.3 million sq miles). This amounts to 12.65 per cent of the Earth's land surface; 11.5 per cent is terrestrial – an area almost as large as the continent of South America. The terrestrial biomes least represented in the UN list occur in the temperate regions of the world. They are temperate grassland (with just 4.59 per cent protected), temperate broadleaf forest (7.61 per cent), and temperate needleleaf forest (8.61 per cent). The most represented terrestrial biomes on the list are island ecosystems (29.73 per cent) and tropical rainforest (23.31 per cent).

Of course, official designation alone does not guarantee that natural areas will be effectively protected. The resources needed to patrol and manage the sites are often inadequate, and poaching and other activities continue to damage some protected areas.

▲ **WETLAND SITES** *The Ramsar Convention, signed by 138 nations in Ramsar, Iran, in 1971, accords protected status to wetlands of international importance, such as the River Kinloch in Scotland.*

WILDLIFE CORRIDORS

Human activity tends to fragment the natural landscape, creating islands of valuable habitat surrounded by a sea of development. A growing number of conservation biologists believe that a network of green corridors should be established to connect these fragments. Creating wildlife corridors allows species to move freely across their traditional range – from large herbivores making their annual migrations to birds flying daily between feeding and roosting sites, dispersing seeds along the way. Corridors connect different populations of a species, promoting gene flow, biodiversity, and continued evolution. They are needed in varying dimensions – from hedgerows that connect small copses and parks, to wide swathes of land that link together the huge territories of large predators.

NATURAL NETWORKS *Intensively farmed landscapes provide few opportunities for wildlife, but preserving strips of woodland along field margins and riverbanks allows small animals to move and plants to disperse their seed widely.*

YELLOWSTONE NATIONAL PARK *The world's first national park was established at Yellowstone in the western USA in 1872. It became one of the first biosphere reserves about 100 years later. Its original aim – to preserve the area's extraordinary volcanic landscape – has evolved to encompass conservation of habitats and threatened species.*

BOTANIC GARDENS

The estimated 2,204 botanic gardens in 153 nations around the globe harbour a significant portion of the Earth's plant diversity. According to a conservative estimate by the organization Botanic Gardens Conservation International (BGCI), at least 6.13 million plants are currently growing in botanic gardens, representing 80,000 to 100,000 species – perhaps a quarter of all vascular plants. Growing in the grounds of these living museums are more than 10,000 rare and endangered plant taxa. The living collections are complemented by 142 million dried specimens housed in botanic garden herbaria.

EARLY GARDENS

Botanic gardens sprang from different origins in different parts of the world. The earliest European gardens were established by universities for the study of medicinal plants, and the oldest of them all – the Botanic Garden of Padua, Italy, established in 1545 – is still in operation. Called the Horto Medicinale, it was founded for the study and cultivation of medicinal plants from Italy, Greece, the Middle East, Arabia, and the coast of Africa. Designated a World Heritage Site by UNESCO, the Padua garden is today involved in the conservation of imperilled plants of the region.

▲ **BOTANIC GARDEN OF PADUA** *The semi-circular plant beds directly in front of the greenhouse contain the only medicinal herbs at the Padua gardens today, although they were once its raison d'être.*

Another recently designated World Heritage Site is the Royal Botanic Gardens, Kew, in west London. Originally a playground for the royal family, the gardens flourished under the patronage of King George III, who appointed renowned plant collector Joseph Banks to direct the gardens in the 1770s. The gardens grew in size and scope under illustrious directors including Joseph Hooker, and today attract well over a million visitors a year. Modern Kew's 120-hectare (300-acre) site, in which 33,000 species are grown, has vastly increased from the original 3.6 hectares (9 acres).

EARLY COLLECTORS ► *Botanic gardens owe their origins to dedicated collectors and gardeners, such as John Tradescant the Elder, who travelled widely in 17th century Europe.*

CHANGING ROLES

In the 19th century, many botanic gardens were in the business of economic botany. The colonial powers of the day used botanic gardens to house, and develop, cash crops from one part of the world before transferring them to another far-flung colony. By the 20th century, the missions of most gardens had changed, focusing more on research, public education, and pleasure. Missouri Botanical Garden, for example, which was inspired by the Royal Botanic Gardens, Kew and the Great Exhibition of 1851 at Crystal Palace, London, stands today as one of St Louis' most popular attractions and one of the world's leading centres of botanical research and science education.

KEBUN RAYA ► *Officially opened as the* Kebun Raya *(Great Garden) in 1817, the Bogor Botanic Garden, Indonesia, became a centre for the development of colonial cash crops, such as tea, tobacco, and oil palm.*

MODERN GARDENS

In the face of species loss and habitat destruction, botanic gardens have today become centres for integrated conservation. Their work ranges from micropropagation of rare and endangered plants for safekeeping in seed banks and in cultivation, to collaboration with governments and other groups to reintroduce the species to the wild and restore their natural habitats.

An exemplary modern garden is the Jardín Botánico Canario Viera y Clavijo, which is devoted to the study and preservation of the plant life of the Canary Islands. Founded in 1952 by Swedish botanist Enrique Sventenius, the garden represents the many vegetation zones native to the islands, including most of the species at risk. Environmental education programmes showcase the islands' botanical treasures.

KEW PALM HOUSE *At 20m (70ft) in height, the Palm House, designed in the 1840s, is just one of the many attractions at Kew. The Royal Botanic Gardens opened to the public in 1841 and is now one of the world's leading botanical research and training centres.*

▲ CANARY ISLANDS GARDEN *Dragon trees* (Dracaena draco) *are endemic to the Canary Islands, but very few grow naturally in the dry bush. They find refuge on the rocky slopes of the Jardín Botánico Canario.*

▲ INTERNATIONAL LINKS *Botanic gardens around the world cooperate to implement measures outlined in the International Agenda for Botanic Gardens in Conservation – a global policy framework coordinated by Botanic Gardens Conservation International.*

GARDEN NETWORKS

Botanic gardens work with one another at regional, national, and international levels to maximize their impact on plant conservation. In the USA, for example, threatened species are protected in the National Collection of Endangered Plants – an initiative of the Center for Plant Conservation (CPC), which is a network of 33 of the nation's botanic gardens.

The Southern African Botanical Diversity Network (SABONET) covers ten countries (Angola, Botswana, Lesotho, Malawi, Mozambique, Namibia, South Africa, Swaziland, Zambia, and Zimbabwe) which are rich in botanical diversity but have limited financial resources. Collaboration through SABONET has allowed these countries, which have a combined area of 6 million sq km (2.3 million sq miles) to train a core of plant specialists, conduct collaborative expeditions, and develop a regional botanic garden conservation strategy.

UNDOING THE DAMAGE

Scientists believe that the best place to safeguard the most imperilled plants is in their natural habitats. But creating and maintaining reserves in which such plants can thrive and reproduce in self-sustaining natural communities involves much more than just putting a fence around a suitable area. Detailed recovery programmes for each threatened species must be developed and implemented; and aggressive habitat restoration and long-term management and monitoring may be needed. Such efforts often involve collaboration among land-management agencies, research institutions, local communities, private conservation groups, and botanic gardens.

AN EXTENDED REACH

According to Botanic Gardens Conservation International (BGCI), more than 400 botanic gardens around the globe manage natural areas within their boundaries. In recent years, many gardens have been active in improving the conservation value of these areas. For example, the Royal Botanical Gardens, Hamilton, Canada includes more than 120 hectares (300 acres) of display gardens and more than 900 hectares (2,200 acres) of natural habitat, one-third of it wetlands. In the past, the wetlands had declined, due partly to the introduction of non-native carp, which destroyed the native aquatic plants that shelter amphibians, mammals, birds, and insects. In 1991, the garden launched Project Paradise to restore the habitat and remove the carp. A "fishway" was created that prevents carp from entering the wetlands from Lake Ontario, but enables native fish to migrate.

▲ RESTORED WETLANDS *Native fish and amphibians have now returned to the wetlands of the Royal Botanical Gardens, Hamilton, along with wild rice and mud plantain – species that had not been recorded at the site for 50 years.*

THE CURTIS PRAIRIE RESTORATION

The world's first ecological restoration project began in 1936 at the Curtis Prairie, a 25-ha (63-acre) tall-grass prairie in the grounds of the University of Wisconsin–Madison Arboretum. At the time, the tall-grass prairie of the American Midwest was already becoming rare; today it is one of the most threatened ecosystems in the USA. Flat topography and fertile soils had encouraged the conversion of prairie lands for agricultural use, and much of the restoration site had been ploughed regularly for a century and planted with corn, oats, and pasture. Restoration began with the planting of native grasses and wild flowers to replace the quackgrass and other non-native plants that dominated the area. Years of intensive management and ecological monitoring have restored the prairie, which now closely resembles natural areas in plant composition, but it will never regain its large herbivores, such as bison and elk, because the area is too small to support them.

▲ REPLANTING *The Curtis Prairie was initially replanted with 42 native species. The work was carried out in 1936 by 200 recruits of the Civilian Conservation Corps.*

▲ WEED CONTROL *The Prairie is burned, section by section, every two or three years. This management regimen helps to control non-native weeds and woody plants.*

▲ RESTORED PRAIRIE *The soils of the Curtis Prairie are beginning to recover their former structure, but prairie soil horizons can take from 500 to 1,400 years to develop fully.*

THE GARDENER'S ROLE

Plant reintroduction is a job for geneticists, botanists, land managers, and others – not gardeners going it alone. For years, Betty Guggolz and her husband Jack monitored two wild populations of yellow larkspur, *Delphinium luteum*, a rare and beautiful wild flower found in the coastal scrub north of San Francisco, USA. Betty, a long-time member of the California Native Plant Society, also cultivated the plant on her family's property, and was eager to create another population of yellow larkspur in the wild. She enlisted the expert help of the University of California Botanical Garden, state officials, and plant geneticists, who determined that her plants had sufficient genetic diversity to be used in future reintroductions without harming the gene pool.

YELLOW LARKSPUR *is a spectacular but rare horticultural subject. One Californian gardener is helping to safeguard the future of the species by donating her plants to officials for reintroduction into the wild.*

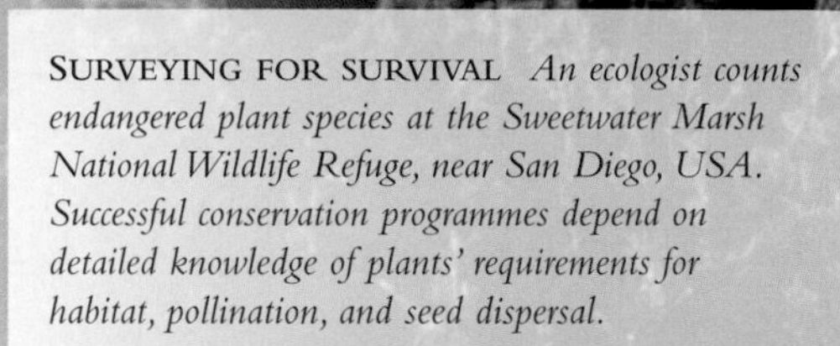

SURVEYING FOR SURVIVAL *An ecologist counts endangered plant species at the Sweetwater Marsh National Wildlife Refuge, near San Diego, USA. Successful conservation programmes depend on detailed knowledge of plants' requirements for habitat, pollination, and seed dispersal.*

RETURNING HOME

Plant reintroduction – establishing new populations of rare or endangered plants within their historical ranges – has become a common strategy for conserving species on the brink of extinction. When humans carry out reintroductions, they are often just taking over the role of natural dispersers, such as birds or mammals, which are no longer present or able to move freely enough to disperse plants to new habitat sites. Reintroduction projects have been carried out successfully worldwide. Critically endangered *Impatiens gordonii* was for many years presumed extinct. Historically, this endemic balsam had been recorded on only two of the 32 islands of the Seychelles. In the 1990s plants were rediscovered on the islands Mahe and then Silhouette. Seed was propagated *in vitro* at the Royal Botanic Gardens, Kew and conservationists in the Seychelles, with the assistance of the Eden Project (*see right*), are establishing new populations of this species.

▲ IMPATIENS GORDONII *This rare balsam is one of 76 species that occur only on the Seychelles. The islands – granitic fragments of the break-up of the ancient supercontinent Gondwanaland – were once thought to be remnants of the Garden of Eden.*

▲ THE EDEN PROJECT *Situated in Cornwall, England, the Eden Project is a botanical and horticultural garden centred on two large glasshouse biomes, which together form the largest greenhouse in the world. Its scientific team collaborates with a number of partners to promote plant conservation.*

SAFEGUARDING THE FUTURE

Not all threatened plants grow within the protected confines of the world's parks and reserves, and even these species and habitats can be destroyed overnight by severe storms and other natural disasters. Moreover, plant populations in protected areas may represent only a fraction of the genetic diversity essential for a species' long-term survival. Seed banking and other scientific conservation tools are used to complement protection in the wild, offering an insurance policy against extinction.

FROZEN GARDENS

During the 1980s when their formal role in conservation began, botanic gardens were envisioned as modern-day arks – living repositories of imperilled plants. When it became apparent that it would sometimes be necessary to cultivate numerous specimens, sometimes hundreds, to preserve the genetic diversity of a single species, gardens began establishing seed banks to complement their plant collections as the last line of defence against extinction. In a seed bank, thousands of seeds from a multitude of populations representing the range of genetic diversity found in a species may be stored – for decades or even centuries. The goal is not only to store the seed long-term in these frozen gardens, but also to use the seed banks to reinforce or re-establish populations in the wild.

The seeds of dryland and temperate plants are generally more amenable to seed banking than seeds of wetland and tropical plants. The maturing seeds are able to "switch off" their metabolism and become dormant as their water content decreases.

DORMANT DIVERSITY ► *To increase their longevity in a bank, seeds are dried, packed in airtight containers, and stored at around −18°C (0.4°F).*

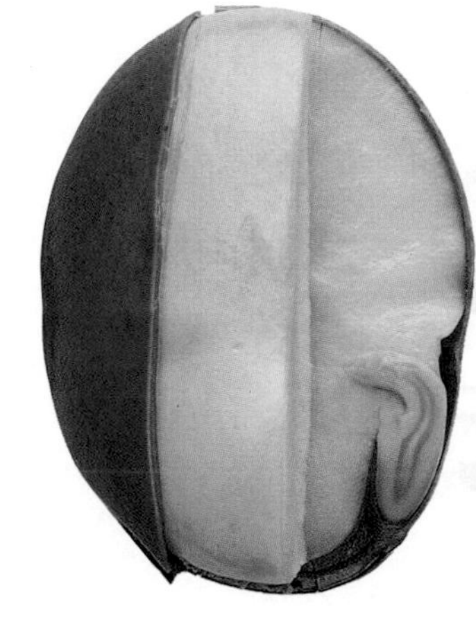

▲ PROTECTIVE STRUCTURE *A seed typically consists of an embryo (the nascent plant), endosperm (the food reserve that fuels germination), and a seed coat or protective covering.*

RECALCITRANT SEEDS

Seeds with a high moisture content cannot be banked in conventional cold storage because they begin germinating as soon as they are shed by the plant. Researchers have stored a small number of these so-called recalcitrant seeds using a newer method – cryopreservation in liquid nitrogen at about −160°C (–256°F). In this extreme cold, metabolism virtually shuts down and the seed can be banked almost indefinitely. The principles involved are the same as those used for the storage of animal semen and human embryos, but the storage of seeds is complicated by their large size – about 7,500,000 cells, compared to semen and embryos, which are made up of fewer than 100 cells.

GLOBAL EXPERTISE

Their scientific collections of living plants and dried herbarium specimens and expertise in seed germination enable botanic gardens to carry out much of the global work in conserving wild plants in seed banks and in cultivation. Over time many older gardens have developed specialist collections of plants, such as orchids from around the world, and now lead conservation efforts for these species. Research at botanic gardens often focuses on identifying and conserving the floras of the local region.

▲ SHARING KNOWLEDGE *A botanist checks specimens at the Vavilov Institute in Russia. Seeds, research data, and conservation technology are shared between research bodies.*

THE MILLENNIUM SEED BANK

One of the most ambitious global plant conservation efforts is the Millennium Seed Bank Project (MSBP) run by7 the Royal Botanic Gardens, Kew. The MSBP has already collected seed from virtually all of the UK's native plants, and now aims to collect ten per cent of the world's seed-bearing plants, principally from drylands, by 2010. Drylands cover a third of the Earth's land surface, including many of the poorest nations, and support about a fifth of its human population. Fortunately, the seeds of most dryland plants are desiccation-tolerant, making seed conservation highly feasible. In accordance with the Convention on Biological Diversity, the MSBP has developed partnerships with dryland nations, based on a legally binding benefit-sharing agreement. Seeds, research data, and conservation technology are shared.

▼ INTERNATIONAL COLLABORATION *The MSBP works closely with countries including Australia, Madagascar, Lebanon, Namibia, Mexico, Saudi Arabia, and the USA.*

ORCHID PROPAGATION *Most orchids form symbiotic associations with fungi to help them gain nutrients. Their seeds have almost no internal food supplies, and are hard to germinate conventionally.* In vitro *propagation in special media achieves high rates of germination, and is essential for orchid conservation.*

▲ **HIGH-TECH CONSERVATION** *Tissue culture is particularly useful in the propagation of orchids and carnivorous plants. Venus flytrap* (Dionaea muscipula) *plants are often available from garden centres in micropropagation tubes complete with instructions for transferring them to compost.*

TEST-TUBE PLANTS

Micropropagation – growing plants from seed or small pieces of tissue under sterile conditions on nutrient-enriched media – has enabled botanic gardens to multiply the stocks of many rare and endangered plants. Today, *in vitro* techniques are widely used not only to germinate seed but also to culture tissue from threatened plants that are unable to reproduce naturally in sufficient numbers, or at all. The plants are grown on a variety of media that contain a carbohydrate source and mineral salts, and sometimes vitamins, growth regulators, and other substances that encourage growth.

Notable successes of micropropagation include the Tunbridge filmy fern, *Hymenophyllum tunbrigense*, which grows mostly on rocks in woodlands in the British Isles. The plant produces green spores that can be stored only for a few days; laboratory techniques have made rapid germination and growth possible, assisting recovery efforts for the species. The stocks of extremely rare species have been successfully increased by tissue culture. The critically endangered Corrigin grevillea, *Grevillea scapigera*, was presumed to be extinct in its native Australian habitat by 1986. The last surviving plants were established in tissue culture at the Royal Botanic Gardens in Sydney and Kings Park and Botanic Gardens in Perth. Hundreds of micropropagated plants have been reintroduced back into the wild in western Australia, vastly improving the species' prospects.

In a novel method of tissue culture called somatic embryogenesis, tiny pieces of somatic tissue (ordinary tissue from a leaf or stem that does not normally undergo cell division) are propagated *in vitro*. The tissue fragments are treated with chemical signals, resulting in the formation of embryo-like structures without the need for sexual reproduction. Thousands of such somatic embryos can be formed from a single gram of tissue. The technique is being used to help save the critically endangered endemic Mauritian bottle palm (*Hyophorbe lagenicaulis*) and, it is hoped, may someday ensure the future of its close relative *Hyophorbe amaricaulis*, of which only one plant remains.

THE GARDENER'S ROLE IN CONSERVATION

As wilderness shrinks and garden acreage increases, our role in the current extinction crisis is growing ever greater. Most gardeners are passionate admirers of plant diversity, and fill their gardens with cultivars derived from species around the globe. Across continents of breathtaking diversity, however, the vast majority of gardeners cultivate a comparatively tiny number of ornamental plants. Yet gardeners can potentially play a major part in efforts to save plants and promote a greater richness of earthly life.

▲ POOLED RESOURCES *Gardeners can play a key role in efforts to save not only native species and plant communities, but also threatened plants from other parts of the world. Private collections of orchids and cycads, for example, are an important gene pool for some threatened species.*

THE DOMESTIC GARDEN AS SANCTUARY

Some conventional horticultural practices have contributed directly to the degradation of natural environments and the erosion of biodiversity. But the past few decades have seen a remarkable series of changes to traditional horticultural techniques and tools. In the 1960s and '70s, integrated pest management, organic gardening, and other alternatives to toxic herbicides and pesticides gained ground. At the same time, recognition of the detrimental effects of the overuse of fertilizers encouraged composting and other techniques that recycle organic matter in the garden, mimicking the nutrient cycle of natural plant communities. More recently, as it has become clearer that plants and animals are rapidly disappearing, interest in planting native species and gardening for wildlife has soared.

We stand at the threshold of a deeper, more coherent kind of ecological gardening, one that aims to support the complex web of biological relationships found in the natural landscape. The new ecological gardens strive to preserve not only individual species but also entire habitats, and to bolster the shrinking gene pools of plants. As the number of such gardens grows, they can help form a network of corridors crossing ecosystems and continents, connecting fragmented nature reserves to allow wildlife to move freely and seeds to disperse.

Ecological gardens include plantings that honour our own species' long and fruitful relationship with the land, whether a formal herb garden full of culinary plants or a patio edged with fragrant flowers. There is a place in biodiverse gardens for healthy, home-grown fruit and vegetables, as well as food and flower cultivars that are becoming horticultural heirlooms as they disappear from the nursery trade.

AT ONE WITH THE LAND *Traditional gardens interact in a limited way with each other and the surrounding wild landscape. Modern gardens created following ecological principles range from beautifully integrated plantings, such as this, to small city borders filled with native flowers to form oases for wildlife.*

EXTINCT IN THE WILD, ALIVE IN THE GARDEN

For a few dozen plants, gardens are the last remaining refuge. "Extinct in the Wild", in the parlance of conservation biologists, they survive only in cultivation, far from their original habitats. Total extinction has been prevented – or at least delayed – by the skills of generations of gardeners. Some of these species can be bought today from well-stocked garden centres. Ironically, collection from the wild for the horticultural trade may well have been what brought them to the brink of extinction in the first place, even though nurseries offering them today may be innocent of such practices.

THE LIVING DEAD

The beautiful Chilean blue crocus, *Tecophilaea cyanocrocus* (*see p.255*) is a cormous perennial, prized for its intense gentian-blue flowers. It was thought to have become extinct in the wild in the 1950s – a victim of overcollecting by bulb dealers and overgrazing by cattle – but was rediscovered in 2001 in mountains south of Santiago, Chile. The species persisted in gardens throughout this period in the form of self-seeding and reproducing populations.

▲ TULIPA SPRENGERI *This species of tulip was first described in 1894, growing in the wild in north-central Anatolia, Turkey. Following its glowing description in European horticultural journals, it was rapidly eradicated by collectors.*

Another victim of overcollection, *Tulipa sprengeri* (*see p.257*), now no longer exists in its native habitat. However, unlike many of the botanicals, or wild tulips, it perseveres well in gardens, propagating itself by seed and by bulb. Cultivated plants are today considered highly desirable and command high prices in the marketplace.

Clearance of land for farming destroyed the forest habitat of *Firmiana major* in central and western Yunnan, China. This tree, with its large, palmate leaves, smooth grey bark, and red flowers, survives only around local temples and in villages where it has been cultivated for centuries. Similarly, the spectacular *Rhododendron kanehirai* (*see p.122*) became extinct in the wild in 1984 when dam construction flooded its only known wild habitat at Feitsui in Taiwan. Fortunately, the shrub, which is virtually unknown in the West, survives in gardens throughout Taiwan.

GENETIC FRAGMENTS

Many plants that have become extinct in the wild survive, and may even thrive, in our gardens. However, these plants often represent just a small fragment of the genetic diversity present in former wild populations of the species. This is because they have typically been derived from just a few collected individuals; and commercial nurseries breed plants not to enhance their genetic diversity but rather to maintain desirable characteristics and ease of propagation and cultivation (*see also pp.56–7*).

For species with low diversity, the prospects for successful reintroduction into the wild are not good, because populations have insufficient genetic variability to cope with natural environmental challenges, or even to breed with one another to form viable offspring. An extreme example is provided by chocolate-scented cosmos, *Cosmos atrosanguineus*. This perennial was last seen in its native Mexican habitat in 1902, but is now found only in gardens. All cultivated plants in the UK and North America appear to represent a single clone, propagated solely from cuttings, and one other genotype is reportedly being registered in New Zealand. This places the plant's future at great risk, because remaining stocks come from a very narrow genetic base.

▲ COSMOS ATROSANGUINEUS *This popular garden plant, with its rich, brown-red flowers, no longer exists in the wild. Plants propagated by tissue culture* (see p.45) *may be able to be reintroduced to their native Mexico, but the genetic diversity of such vegetatively propagated clones will be limited.*

ECOLOGICAL MYSTERIES

Attempts to reintroduce "extinct in the wild" plants to their natural ranges are sometimes hampered by a lack of information on the ecology of their original habitats. For example, the native habitat of *Laelia gouldiana* (*see also p.405*), an orchid with spectacular deep purple flowers, is a mystery. It is believed to have been endemic to Mexico's Sierra Madre Oriental, but repeated searches in the wild have been unsuccessful. Semi-feral specimens have been found planted on stone fences and mesquite trees, but these plants are thought to be divisions of a single clone and do not produce seed.

▲ FRANKLINIA ALATAMAHA *Prospects of reintroduction for this beautiful, white-flowered shrub are poor. Once native to a very restricted and specialized North American habitat, it may now require permanent sanctuary in gardens.*

▲ LAELIA GOULDIANA *This species is widespread in the orchid trade and has been successfully crossed with other related species. Unfortunately, attempts to breed the pure species have failed.* Laelia gouldiana *will survive only if propagated indefinitely by vegetative means, by breeders and other orchid enthusiasts.*

Second chances

Plants that survive only in cultivation have been described as orphan species in a foster habitat. Endangered species will always face uncertain futures in gardens, which are essentially short-lived collections subject to the intuition and skill of the grower, the influences of soil and weather, and the vagaries of horticultural fashion. It is no surprise that many of these orphans become extinct – either by accident or by design.

One such species, which found only temporary sanctuary in the garden before final extinction, was *Streblorrhiza speciosa* (*see p.217*). A member of the pea family originally native to Norfolk Island in the South Pacific, it was eliminated from its native habitat by imported livestock. The plant was introduced to cultivation in 1840, but after a brief period of horticultural enthusiasm, it faded from gardens early in the 20th century.

◄ DEPPEA SPLENDENS *grows in cultivation to a height of 1.5m (5ft) and bears clusters of 5cm (2in) yellow and orange tubular flowers set off by claret calyces that dangle from wiry stems. In its native habitat, it was most likely pollinated by hummingbirds.*

Today's increasingly sophisticated genetic screening technologies and propagation techniques are giving many "extinct in the wild" plants a new lease of life beyond the garden walls. In some cases, concerted efforts are already being made to bring together both local and international conservation authorities and re-establish plants in protected areas in their native ranges. For example, a plan to reintroduce the Chilean blue crocus to its native habitat has been drafted by Royal Botanic Gardens, Kew, Chilean officials, and the Alpine Garden Society in the UK.

An ambitious reintroduction programme is also underway for *Deppea splendens* – a beautiful shrub once found only in a cloud forest canyon on the south slope of Cerro Mozotal in southern Chiapas, Mexico. The first pressed specimens of the plant were collected in 1972, and nine years later seeds were distributed to a small number of botanical gardens in California, USA. The plant's cloud forest site was cleared for corn farming in 1986, and the species continued to survive only in cultivation.

Huntington Botanical Gardens, along with other US botanic gardens and nursery growers, is now working to assemble a collection of genetic material from *Deppea splendens*. This shrub has also been successfully hand-pollinated, paving the way for possible reintroduction to the wild.

In the meantime, the species is being made available to gardeners on a limited basis.

Partnerless plants

Some of the world's rarest plants teeter on the brink of extinction because they lack a compatible partner, without which they can never re-establish a viable, reproducing population in the wild.

Just one plant of the Australian shrub Sargent's snakebush, *Hemiandra rutilans*, is known to exist – a solitary specimen in the care of King's Park and Botanic Garden, Perth, nurtured from a cutting taken from the last known plant in the wild, which died in 1994. In a cruel twist, Sargent's snakebush is self-incompatible: it cannot set seed on its own flowers. Unless another genetic individual can be located to pollinate the plant at King's Park, this charming, scarlet-flowered, grey-leaved shrub will remain on the edge of extinction.

The majestic Wood's cycad, *Encephalartos woodii*, is the most eligible bachelor in the plant world. The species is extinct in the wild, and the 500 or so specimens that exist in cultivation are all clones of a single male discovered in Zululand in 1895. No females have ever been found, so the species cannot reproduce sexually. Its only hope lies in recent research that indicates that, in very rare circumstances, cycads have the potential to change sex. Nearly all the instances had one factor in common – a traumatic incident, such as transplantation damage or exposure to severe cold – just before the sex change. The possibility of inducing such a change has naturally become the focus of intense interest.

▲ ENCEPHALARTOS WOODII *A sex change, whether accidental or induced, seems the only hope of jump-starting the sex life of this majestic cycad, perhaps enabling it to redevelop genetic diversity.*

▲ HEMIANDRA RUTILANS *The original cutting would not thrive on its own roots, and had to be grafted in order to survive. The tentative hold on life of this lone specimen renders even more urgent the search to find another compatible plant in the wild.*

A PLANT BUYER'S GUIDE

The vast majority of gardens feature plants with large, showy flowers, colourful foliage, or other characteristics that reflect generations of expert breeding by horticulturists. However, every year gardeners buy plants that have been dug up from natural areas worldwide, unwittingly contributing to the problems facing some threatened species. By becoming familiar with the species most likely to be collected from the wild and asking appropriate questions when purchasing plants, gardeners can make a major contribution to conservation efforts around the globe.

WILD-COLLECTED PLANTS

Certain groups of plants are particularly threatened by collection from the wild. Since the earliest days of orchid cultivation it has been considered glamorous to grow plants collected from steamy jungles or vapour-shrouded mountain slopes. Today it is still possible to find wild-collected orchids for sale in the tropics, on the internet, and even on the plant lists of reputable nurseries. For decades there has been concern about the effects of wild collection on populations of bulbs, including species of *Galanthus*, *Cyclamen*, and *Sternbergia*. Cactus rustlers routinely poach plants from the Chihuahuan and Sonoran Deserts in Mexico and southwestern USA. The number of bizarre succulents and fascinating carnivorous plants collected from the wild for sale has increased as their popularity has grown. The primeval beauty of tree ferns and rare cycads makes them vulnerable to illegal collection. And exquisite wildflowers, such as North American trilliums, are dug up from their habitats and sold in their native countries and abroad. The consumers of these wild plants are not only specialist collectors but also ordinary gardeners who take advantage of low-priced bulbs, wildflowers, and other plants.

The vast majority of plants collected from the wild are harvested without any management plan or monitoring system. Harvest rates are rarely based on scientific information, and so many populations have been overcollected and have declined or disappeared altogether. Many of the plants that are threatened by overcollection from the wild are difficult to propagate, or take a long time to flower and reach saleable size. Trilliums, which typically take five years or more to flower from seed, are the classic example. Wild-collected specimens, therefore, are often more profitable for suppliers and less expensive for consumers.

Overcollection not only harms the prospects of some species in the wild, but also diminishes the options available to breeders in the future. Much of the selection and breeding of ornamental plants relies on the diversity of gene pools of wild species growing in their native habitats. By purchasing rare plants taken from the wild, gardeners may be depriving generations to come of showy hybrids, disease-resistant varieties, and other garden treasures.

▲ STOLEN ORCHID *It may be tempting to rationalize that "just one plant" cannot possibly make a difference. But it is almost never just one plant, taken by just one gardener or enthusiast. The cumulative impact of constant depredations can have a catastrophic effect on wild populations.*

▲ BULBS UNDER THREAT *Cyclamens, snowdrops, and sternbergias are all listed by CITES, the convention governing trade in endangered species, but illegal trade continues. If suppliers do not explicitly state that theirs are nursery-propagated, ask. Responsible suppliers will be aware of the origins of their plants.*

CONSCIENTIOUS CULTIVATION
Orchid farms such as this one outside Chiang Mai, in Thailand, help take the pressure off wild populations. Supplying plants and flowers for the cut-flower industry is a valuable source of revenue for many developing countries.

THE IMPORTANCE OF PROVENANCE

To help preserve unique gene pools in your area, when purchasing native species look for plants of local provenance, generally defined as propagated from seed collected within 160km (100 miles) or 305m (1,000ft) in elevation. Many species exhibit variation that results from genetic adaptation to different conditions in various parts of their natural range. Plants of one provenance may differ from plants from elsewhere in cold-hardiness, moisture requirements, flower colour, or other characteristics. Some countries, such as the UK, lie at the edge of the range of many native species, and the local genetic makeup (the genotype) of plants may be significantly different from populations in other regions. Not just genetic diversity is at stake – non-local plants may not perform as well in your garden as local genotypes.

Provenance is also an important consideration for gardeners growing rare non-native plants. For conservation purposes it is important to know the origin of a particular plant. Choose reputable, knowledgable suppliers who can furnish these details. Many specialist plant societies sponsor seed banks, making propagated examples of rare plants of known provenance available to members. One interesting example is the International Carnivorous Plant Society.

▲ **GOING NATIVE** *At plant fairs and markets, check out the stalls of local conservation and native plant societies. Many propagate locally growing plant species for sale; buying these will fill your garden with plants of appropriate provenance, as well as supporting valuable work.*

WHAT GARDENERS CAN DO

- Never dig up plants from the wild, and do not purchase plants that have been lifted from the wild. There is an exception to this rule – rescued plants that would otherwise have faced obliteration by development, dug up under license.
- Purchase species only from reputable dealers. If you don't know how to find responsible suppliers, ask for recommendations from plant societies.
- Buy rare species only from suppliers who state explicitly that their plants are nursery-propagated, not collected from the wild. Do not be misled by ambiguous phrases such as "nursery-grown" or "field-grown", which may mean that the plant was dug from the wild and grown on in a nursery for a period of time. The words to look for are "nursery-propagated" or "from cultivated stock".
- Educate yourself about the conservation status and propagation and cultivation requirements of the plants you would like to buy. Be especially cautious if buying any of the plants listed in this book. Be sceptical where plants are difficult to propagate or slow to flower or reach marketable size.
- Think twice about purchasing species being sold at bargain prices. Be particularly cautious about species being sold cheaply on the internet.
- Be aware of local, national, and international laws and regulations that protect wild plants (*see p.37*).
- Support nursery suppliers who are actively engaged in the conservation of rare species. Botanical gardens can be another good source of seeds, bulbs, and plants.

GALANTHUS

REDEFINING THE WEED

Just a century or two ago, gardens were tiny enclaves in a vast wilderness. People struggled daily against the forces of nature, especially weeds – hence the traditional definition of a weed as any plant from outside the garden that ends up inside the garden, where it is not wanted. But today, mere fragments of wilderness are hemmed in by human-dominated land, and our activities, including gardening, threaten native plant communities. Hundreds of plants, including prized ornamentals typically foreign in origin, have jumped the garden gate and become intractable pests in parks, preserves, and other natural areas. These are the true weeds of the modern world.

PLANT INVADERS

People have been rearranging the planet's flora for centuries. Many of the exotic plants we have planted in gardens have been beneficial to us and are ecologically benign. But a small percentage has run rampant. According to a study by Brooklyn Botanic Garden in New York, half of the worst invasive plants causing ecological and economic damage in the USA were brought there for horticultural use. An even larger percentage of invasive woody plants, an estimated 82 per cent, were introduced for landscaping. Studies suggest that a similar percentage of the non-native plants now growing in the wild in the UK, Scandinavia, and Australia were imported intentionally, the majority for gardens. Most of the worst weeds invading natural areas in South Africa were introduced as ornamental plants. Some of the world's worst weeds are aquatic plants, for example the water hyacinth, *Eichhornia crassipes*, a wayward ornamental native to the Amazon basin that is now a scourge of waterways in the tropics and subtropics.

New research suggests that gardeners, along with farmers and foresters, have inadvertently encouraged plant invasion not only by dispersing new species around the globe, but also by giving them loving care. Most introduced plants perish when subject to the vagaries of a new environment. It is extremely difficult for them to become established as reproducing wild populations unless they are nurtured by humans for long enough to enable their repeated attempts at colonization finally to succeed.

▲ JAPANESE KNOTWEED Fallopia japonica *is an imposing perennial that was introduced from Asia to Europe in the mid-19th century as an ornamental and fodder plant. The species has spread and become highly invasive throughout Europe, and also North America and New Zealand.*

RUNNING AMOK

The intense magenta flower spikes of a stand of purple loosestrife, *Lythrum salicaria*, are a sight to behold, silhouetted against a blue summer sky. Gardeners can easily understand the passion of Charles Darwin, one of the species' most passionate admirers, who wrote that he was "almost stark, staring mad over *Lythrum*". As Darwin observed in his Kentish garden, purple loosestrife is a plant well endowed for reproduction. A mature plant may have as many as 30 flowering stems, capable of producing an estimated two to three million seeds a year. The plant also readily reproduces vegetatively via underground stems at a rate of about 30cm (12in) a year. Outside its native Eurasian range, in the absence of the competitors, predators, and pathogens with which it evolved, *Lythrum salicaria* has reproduced explosively, invading wetlands throughout the USA and Canada, replacing native grasses, sedges, and other plants that provide a better source of nutrition for wildlife, as well as threatened wetland orchids.

◄ PURPLE LOOSESTRIFE *The World Conservation Union (IUCN) has this species on its list of the World's 100 Worst Invaders* (see also p.462). *So-called sterile cultivars are actually highly fertile and able to cross freely with other members of the species.*

HELP PREVENT PLANT INVASIONS

Plant selection, the most basic aspect of gardening, has become a significant act with potentially serious consequences for other species. Gardeners can play a key role in controlling invasive species by being careful not to grow invasive or potentially invasive plants.

- Become aware of the invasive species in your area. Unfortunately, in most countries, invasive plants are still commercially available. Fortunately, there is a growing number of international, national, and regional web sites hosted by governments and conservation groups that include lists of the worst invasive species in the area – plants that gardeners definitely should not grow (*see also pp.440–571*). If such species are already growing in your garden, consider removing and destroying them.
- It is much harder to find information on plants that are just beginning to show signs of invasiveness in a region. Yet it is very important for gardeners to be aware of these plants, because stopping invasives at an early stage is critical. To reduce the risk of future invasions, Australia and New Zealand have implemented "clean list" policies: unless a plant has been determined beforehand to be non-invasive, it cannot be imported until an evaluation has been completed. People in other countries must grapple with the problem on their own. There is little gardeners can do to stop an invasive species that has already swamped native plant communities, but they can play a major part in preventing others from gaining a foothold in the wild. Many gardeners, for example, are fond of *Buddleja davidii* for its colourful flowers and appeal to butterflies. However, the species is becoming invasive in Ireland, Scotland, and the Pacific Northwest region of the USA, and is beginning to naturalize in other areas. It is unwise for gardeners in regions with similar climates and growing conditions to grow the shrub.
- Look for non-invasive alternatives to problematic species. Usually there are better-behaved plants with similar showy attributes. For example, in North America a variety of plants with impressive spikes of magenta flowers are being promoted as alternatives to purple loosestrife, including anise hyssop (*Agastache foeniculum*), blazing stars (*Liatris* species), and fireweed (*Epilobium angustifolium*). Local botanical gardens may be able to suggest alternatives. Encourage nursery suppliers to stock these instead of invasive plants.
- If you choose a plant indigenous to your region, you can be almost certain that you will not be aiding and abetting a new invader. The vast majority of invasive plants in an area are not native to the site.
- Most of the showy garden hybrids developed by breeders are safe to grow. Research suggests that only a small percentage of invasive plants are hybrids. The same is not true of showy selections of invasive plants. There is as yet no research indicating that, generally, cultivars of an invasive species are any safer to grow than the species itself.

BUDDLEJA DAVIDII

PESTS AND PATHOGENS

Around the world, native plant communities are threatened not only by introduced plants, but also non-native pests and pathogens. Forests are often the hardest hit. For example, the flowering dogwood (*Cornus florida*), a beautiful understorey tree in the temperate deciduous forests of North America and a popular ornamental, is now under siege by a virulent fungal disease, dogwood anthracnose, which is believed to have arrived on the continent in the mid-1970s. In shaded woodland many dogwoods have perished one to three years after infection, but trees in sunnier sites have often survived. Research on how environmental conditions affect dogwood anthracnose has led to specific cultural recommendations for gardeners, including proper siting, watering, and pruning. A handful of disease-resistant cultivars, such as *Cornus florida* 'Appalachian Spring', have been introduced in the nursery trade. With proper care of mature specimens, gardeners are deterring the pathogen.

CORNUS FLORIDA ► *Planting resistant cultivars may bolster the species' resistance, so that one day it will be better able to defend itself against anthracnose disease.*

CARPOBROTUS EDULIS *The Hottentot fig or freeway iceplant, here gaining ground on native plants in northern Tunisia, was widely introduced for landscaping and soil stabilization purposes and is now causing environmental problems in areas far from its native Cape, such as California and the Pacific Islands.*

NATURAL GARDENING

Theories evolving on the frontiers of science, especially the science of ecology, suggest that how we design gardens has a profound effect on the rest of nature. In an age of extinction, we need to create gardens that act like nature, restoring the structure and functions of native plant communities. Although it has been interrupted on a handful of occasions by episodes of mass extinction, the general increase in the diversity and complexity of species since life began, three-and-a-half billion years ago, is astonishing. And so a garden that acts like nature, at the deepest level, is designed to nurture a greater richness and diversity of life.

▲ IMITATING NATURE *Using grasses and wildflowers native to the midwestern USA, landscape designer Clifford Miller recreated a parcel of tall-grass prairie, one of North America's most critically endangered plant communities.*

IMITATING NATURE

In 1625, during an age of highly formal, symmetrical gardens, Francis Bacon asserted that the ideal garden should include a "heath or wilderness" – one of the first recorded calls for a more natural landscape. One of the first books to offer guidance on creating natural landscapes was *The Wild Garden,* first published in 1870, in which William Robinson railed against what he called the "dark ages of flower gardening" and the Victorian use of tropical species in elaborately patterned "carpet beds". Instead, he advocated a wild garden created by "naturalizing many beautiful plants of many regions of the Earth".

Today, much of the natural landscape has been destroyed by human activity, fragmenting plant communities, threatening species, and eroding the genetic diversity that enables them to adapt and evolve. Two major goals of a garden that acts like nature are to stitch together fragmented islands of wilderness, and to augment the gene pools of shrinking populations of native plants.

▲ RETURN OF THE NATIVE *Gardens created following ecological principles are a critical part of the larger evolving wild landscape. For this reason, native species, combined in plantings inspired by local plant communities, play the leading roles in these designs.*

UNDOING THE DAMAGE

During the last 100 years, much has been learned about the structure and functions of natural landscapes. We are discovering not only how to avoid doing ecological damage, but also how to undo it once it has been done. In places where the natural landscape has been largely destroyed, we are learning how to restore native vegetation and natural processes so that nature can heal itself and get on with evolution.

Because the best way to prevent plant extinction is to preserve their habitats, it is wiser to build in areas where the natural landscape has already been transformed by development than to destroy even more remaining forest, desert, grassland, or shrubland. When it is necessary to build in an undisturbed area, great care should be taken to clear the smallest possible area around the structure and literally cordon off the surrounding native vegetation from construction workers and machinery.

▲ SEASHORE DESIGN *The late film-maker Derek Jarman created his celebrated garden on a windswept Kent beach using native seaside plants, punctuated by flotsam gathered along the shore and transformed into garden art.*

RECYCLING RESOURCES

Gardens that act like nature mimic the major functions of any ecosystem, such as the recycling of nutrients. In nature, almost everything produced in the landscape is returned: in a forest, for example, fallen leaves become food for vast populations of organisms that help it decompose into food for the plant community. Since the beginning of the Industrial Age, gardeners have ignored the natural nutrient cycle, raking up leaves and clippings and sending them off to landfill sites, then buying fertilizers to compensate for the lost nutrients. In an ecological garden, the natural nutrient cycle is left to take its course. Leaf litter may be shredded, but is returned to the woodland garden to revitalize the soil and mulch the plants. For the more highly cultivated parts of the garden, the natural process of decomposition is speeded up in the compost heap.

Ecological gardens require little supplementary irrigation. To preserve aquatic habitats and fresh water supplies, gardeners have been rethinking plant selection – favouring species best adapted to local conditions and climate – and also, with the creation of "rain gardens", rethinking garden design. In a natural landscape, there is generally very little run-off, because the soil and its dense cover of vegetation and leaf litter act as a sponge; but where this has been replaced by impervious surfaces, rainfall can no longer soak into the soil as readily as it once did. Instead, storm water flows off roofs and lawns, down driveways and roads, eroding wetland habitats and carrying contaminants to nearby streams and rivers. Rain gardens reverse this trend: they are created by sculpting appropriate parts of the landscape into swales and wetland plantings to capture damaging storm water, while creating habitat for moisture-loving plants and animals.

▲ HARVESTING WATER *"Rain gardens" such as this one, designed by Edgar David, are designed to protect aquatic habitats from erosion and pollution caused by huge volumes of storm water that flow off residential and commercial properties.*

Rebuilding nature

A native plant community's structure depends primarily on its characteristic type of vegetation. If you live in a forested region and want to design a garden that acts like nature, you should make a woodland planting the core of your landscape. In an age of extinction it is important to incorporate vertical structure when designing a garden. Plant life is organized into vertical layers, which are most obvious in forests but are found in all plant communities. Choosing plants to fill all the various layers promotes biodiversity because, in general, the greater the number of distinctive vertical strata, the more diverse the plant life, and therefore the more habitat niches for a more diverse array of animal life.

For garden designers, orchestrating a long-lasting sequence of colourful blooms has traditionally been a question of aesthetics: of combining flowers in just the right shades of gold and blue, or contrasting different forms and shapes. To promote biodiversity, the pictorial sense should be grounded in the larger natural context. Gardeners rarely consider the many interactions between the changing spectacle of plants and the wildlife that pollinate their flowers, eat their fruits, or disseminate their seeds. Few American horticultural texts note that the eastern box turtle spreads the seeds of the native may apple, or that the native eastern columbine blooms when the ruby-throated hummingbird arrives in spring. However, gardeners should be aware of these relationships, because supporting plant and animal interactions is another way that biodiversity can be promoted in designed landscapes. And although gardeners know from experience that non-native ornamental plants provide food and shelter for many "generalists" – species of birds, bees, and butterflies that are not finicky about what they eat or where they perch or sleep – some creatures have specialized relationships with local plants. For example, wild lupin, *Lupinus perennis*, is the only host plant for caterpillars of the "Karner" melissa blue butterfly, native to northeastern USA.

▼ **Dispersal duet** *Box turtles are the only known vertebrates able to eat the toxic fruits, and thus spread the seeds, of the may apple,* Podophyllum peltatum *– a herb with possible pharmaceutical potential in the treatment of rheumatism and skin cancers.*

Layered planting *The Garden in the Woods, in New England, replicates temperate forest "layering": below the canopy of the tallest trees is an understorey of smaller trees, then sapling trees and shrubs make up the next, shrub layer. Wildflowers, ferns, and mosses create the tapestry known as the ground layer.*

NURTURING WILDLIFE

Gardens have become important habitats for wild flora and fauna. In developed areas, their value as year-round sanctuaries and as stopover points for winged creatures in transit is increasingly significant. Conservation biologists suggest that green corridors – bands of restored native vegetation of different sizes to suit the needs of various species – may be able to link parks and reserves, transforming lonely islands of wilderness into functional networks that preserve the ecological integrity of entire biomes. Gardens could play a vital role in the larger efforts to compensate for habitat fragmentation by restoring habitat and connecting scattered nature preserves.

GREEN CORRIDORS

For the past few decades, ecologists have been questioning why extinction rates have always been higher on islands than on mainlands, and whether a similar fate will befall continental species where native communities of plants and animals are currently being fragmented into island-like bits. Biologists have long suspected that the ever-increasing reach of human society has taken its toll on native plant life, but a lack of detailed historical data left them unable to calculate precisely the extent of the damage. In 1993, three Rutgers University scientists analyzed an unusual series of plant inventories taken over a period of 112 years on Staten Island, a 180-sq km (70-sq mile) borough of New York City, offering the first extensive look at the fate of native species in the wake of urban development. Unlike other parts of the New York metropolitan area, an urban corridor that has the highest population density in the USA, the island has only recently begun to experience intensive suburbanization, and about 10 per cent of the land is protected. Yet its native flora has been drastically reduced – more than 40 per cent of the native species once catalogued are now missing. At the same time, non-native invasive species have become an increasingly dominant part of the island's plant life.

◄ CITY LINKS *Today, domestic gardens comprise the largest area of green space in the typical British city, some 15 per cent of the land area, typically far more than offered by public open spaces such as parks.*

GARDENING FOR POLLINATORS

For millions of years, birds, butterflies, and bees have been the agents of biodiversity, scattering seed and pollen across the land. However, unbridled development, conversion of natural areas to crop monocultures, and overuse of toxic pesticides have decimated some pollinator populations. This in turn has intensified the threat to native plants with which they have co-evolved. Gardeners can not only help restore and reconnect wild populations of indigenous plants, but also nurture the pollinators on which they depend for reproduction. By planting a butterfly garden, for example (*see box, facing page*), it is possible to sustain the entire life-cycle of some species, and to provide food and habitat for long-distance migrants such as the North American monarch butterfly. If neighbouring gardens also provide suitable habitat, creating corridors of nectar and caterpillar plants, butterflies will be more likely to satisfy their needs without falling prey to predators.

BUTTERFLY BORDER ► *In return for planting flowers such as milkweed, coneflowers, and larkspur, gardeners may observe the habits of some of nature's loveliest pollinators.*

BUMBLEBEE FLOWERS

The bumblebee is one of the world's most proficient pollinators – capable of transporting loads equal to 90 per cent of its body weight, often out of the hive before dawn and frequently still at work after sunset. Worldwide, bees are beset by pests, insecticides, and habitat loss, but gardeners can provide refuge. Choose flowers that have evolved characteristics suited to the creatures' girth, behaviour, and other attributes. The flowers must be sturdy to support a bumblebee's weight – some, like the snapdragon, have tailor-made "landing platforms". Many have bizarre wavy lines or leopard-like spots on their petals that signpost the bees towards the nectaries. Many bumblebees have a long tongue that ensures that only they are able to probe the deep recesses of elongated blooms to suck up nectar. Consequently, the petals of bumblebee flowers often form elongated bells, funnels, tubes, or spurs with the nectaries deep inside.

▼ COLOUR CODED *Bees, which can see ultraviolet, are partial to flowers in the blue, purple, and ultraviolet shades that dominate one end of the colour spectrum.*

STOPOVER GARDENS

Every spring and autumn around the globe millions of migratory birds fly great distances to and from the places where they nest and spend the winter. In recent decades, vast stretches of natural habitat along major coastal and inland migration routes have been carved up and developed, contributing to the decline of some songbird species. However, gardeners can re-create a "stopover habitat" by transforming their properties into bed-and-breakfasts for weary migrants. The key to creating a stopover garden is to re-create as many of the different vertical layers of plants found in a forest as possible. Because some species perch or feed in canopy trees while others prefer understorey trees or shrubs, this will provide the greatest number of niches for the most diverse assortment of birds. Along the edges, autumn-fruiting shrubs should be planted. Native trees and shrubs are generally best, because they are genetically programmed to leaf out, bloom, and fruit at just the right time for the migrants with which they have co-evolved. The plants feed the birds, which return the favour by dispersing their seeds.

▲ FAMISHED FRIEND *According to "stopover" ecologists, even small patches of habitat in the midst of urban and agricultural landscapes can mean the difference between successful migration and starvation for many bird species.*

AMAZING MIGRATION *The great seasonal winged migrations include butterflies as well as birds. Monarchs* (see above) *feast on garden flowers during their 2,000-km (3,000-mile) journey to and from their winter roosts in Mexican fir forests and their northernmost breeding range in Canada.*

MAKING A BUTTERFLY GARDEN

A butterfly garden's style is not as important as its content, so nurturing butterflies does not require major changes to existing gardens. Here are a few ways to provide habitat for butterflies in gardens.

- Butterflies avoid shady or windy areas. Locate the plants they love in a sunny, sheltered spot.
- Observe butterflies in nearby areas to see which flowers they prefer.
- Provide caterpillar food plants as well as nectar plants for adults. Because natural predators usually keep caterpillar populations in check, prized ornamental plants rarely suffer severe damage. While native species play an important role as host plants for larvae, most adult butterflies have cosmopolitan tastes, feeding on native wildflowers as well as non-native ornamentals. Old-fashioned, single-flowered cultivars are generally most attractive to butterflies and other pollinators.
- Grow large clumps of the most favoured species and varieties, not just one or two plants.
- Select a variety of plants that provide a continuous sequence of bloom and supply of nectar.
- Provide plants in a range of heights, from the tallest to the smallest herbaceous species, as well as flowering trees and shrubs: butterflies need perches from which to watch for mates and hide from predators.
- Leave an empty, damp spot where butterflies can "puddle". In addition to the nectar they sip from flowers, most butterflies need mineral salts and other compounds they cannot get from plants. Many of these are naturally available in garden soil.
- Avoid using insecticides, even biological control agents such as *Bacillus thuringiensis*, which kill caterpillars indiscriminately.
- Moths are also valuable pollinators. Many of the plants they prefer (*see below*) have equal appeal for busy gardeners relaxing outdoors on summer evenings.

▲ EVENING ATTRACTION *Many butterflies have a predilection for pink and yellow flowers, while moths, their nocturnal cousins, favour bright whites and pastel colours that glow in the dusk. Strong fragrance also allows moths to locate the blooms they prefer.*

PROPAGATING DIVERSITY

Most plants found at nurseries today are cultivars that have been propagated in ways that tightly control or even eliminate genetic variation. This guarantees that gardeners get the plant they have seen on the seed packet or in the nursery catalogue – with giant flowers, for example, or a tall, columnar growth habit. Modern propagation techniques have resulted in the expansion of plant diversity that is so evident in the development of both food and ornamental crops. But the survival of rare or declining species is better served by other propagation methods that promote genetic variation.

The female parts of a flower, or pistil, hold the stigma aloft

Stamens, or male flower parts, are tipped by bean-shaped anthers

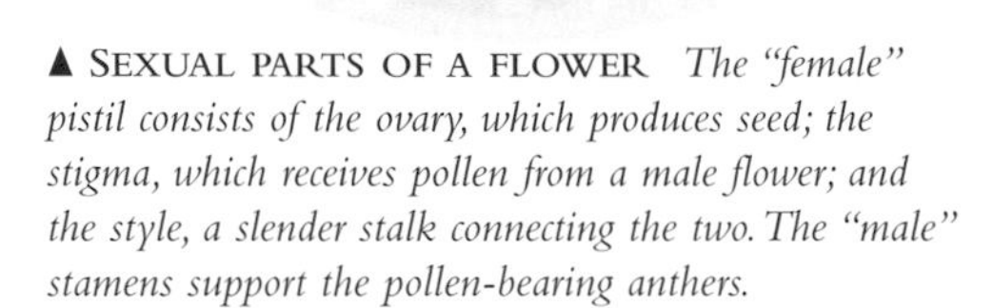

▲ SEXUAL PARTS OF A FLOWER *The "female" pistil consists of the ovary, which produces seed; the stigma, which receives pollen from a male flower; and the style, a slender stalk connecting the two. The "male" stamens support the pollen-bearing anthers.*

THE SEX LIFE OF PLANTS

With the advent of high-tech screening tools, researchers have discovered that plant species generally have high levels of genetic variation. The way most species acquire genetic variation is via sexual reproduction, with its random recombination of genes that occurs with each new generation. Sexual reproduction is the most important method used by many, if not most, species to survive and increase: genetic material from a male and a female unite in a seed to form a new, genetically unique individual. Biologists believe that the genetic variation resulting from sexual reproduction provides species with the wherewithal to overcome ecological pitfalls thrown in their path – a new insect pest or pathogen, a warming planet – what Darwin called natural selection pressures. A few of these diverse individuals, they reason, are likely to have the genetic repertoire needed to cope with changing conditions. However, asexual reproduction (or vegetative propagation) has advantages for plant populations already well adapted to their environments, impelled to spread but under no pressure to evolve. Many kinds of cloning can be seen in the plant world: the formation of "daughter" bulbs and corms, the production of runners – even the growth of tiny plantlets along the edges of leaves.

UP CLOSE AND PERSONAL ► *A coloured scanning electron micrograph (or SEM) of the pistil (female receptor) of a morning glory flower with red pollen grains attached.*

SEXUAL STRATEGIES

In angiosperms, the process of seed production begins in the flower. Before it can produce seeds, the flower must be pollinated. Species that can produce seeds by self-pollination, or "selfing", are common in nature. These plants can reproduce even if pollinators fail to appear. But if a plant pollinates itself instead of receiving pollen from another individual, genetic variation in the seed is reduced. Plants have evolved a marvellous assortment of sexual techniques, and use a variety of ingenious strategies to promote cross-pollination and genetic diversity.

Some plants, such as hollies, are dioecious, meaning that each plant is either male or female. In these species, self-pollination is impossible; gardeners know from experience that a dioecious plant will not flower and fruit unless a member of the opposite sex is in the vicinity. Other plants, like the common cane begonia, are monoecious: individual specimens have separate male and female flowers. Many species are hermaphroditic: their flowers have both male and female organs. However, the stigma and anther may ripen at different times to discourage self-pollination, or the styles may be elongated to prevent the stigmas from being fertilized by physical contact with the male parts.

Plants often entice animals – butterflies, bees, birds, bats, beetles, flies, moths, even small mammals – to transfer pollen from one flower to another. They attract the pollinators with showy petals or appealing scents, and reward them with sweet nectar or protein-rich pollen. Orchids have evolved some of the most ingenious ways to lure potential pollinators, producing flowers that look or smell like female insects so that males will be tempted to copulate with the flower and in the process fertilize it with pollen.

▲ FLIRTY FLOWER *The labellum, or lip, of this orchid,* Ophrys scolopax, *imitates in shape, colour, and smell the female of the long-horned bee* Eucera longicornis, *enticing the males, its preferred pollinating partners, to "mate" with it.*

GARDEN CULTIVARS

A cultivar is a food or ornamental plant that has been selected or bred in cultivation, and when propagated sexually or asexually, retains its desirable characteristics. Cultivars are prized for their valuable or unusual characteristics, and the surest way for growers to guarantee that gardeners will get the desired giant flowers or weeping habit is to produce clones, typically from cuttings. But many other vegetable and border plants are produced commercially by seed. Some seed-produced cultivars are the product of self-pollination. To produce true-breeding lines in a self-pollinated cultivar, plant breeders remove unwanted genetic variation and "fix" desirable characteristics by producing six to ten consecutive generations of self-fertilized plants. Enforced self-pollination of naturally cross-pollinating species through consecutive generations can result in small, weak, and less productive plants. If inbred lines are crossed, however, the resulting plants may be even bigger and stronger than either parent, a phenomenon known as hybrid vigour. Inbred lines are produced primarily for later production of F1 hybrids and "synthetic lines".

▲ SYNTHETIC LINES Phlox drummondii *'Sternenzauber' is a mixture of different colour forms but has the same star-like corolla shape as the parent species; the Million Bells colour series of petunias is another very popular example.*

BIOLOGICAL COSTS OF MODERN PROPAGATION

Cultivars, like corgi dogs and Siamese cats, are domesticated organisms, created when human beings apply artificial selection pressures to control a species' evolutionary course. Unfortunately, those characteristics that appeal to humans may make a plant less fit for life in the wild, and domesticated plants can become utterly dependent upon people for their growth, reproduction, and evolution.

Horticultural fashion often selects traits, such as double flowers or large, round, and flat blooms, that can make pollination impossible, "sterilizing" the plant. For example, in recent decades threatened tropical slipper orchids, such as *Paphiopedilum sukhakulii* (*see also p.410*), have been successfully propagated. However, they have been line-bred for rounder flowers or wider petals and are now quite different from the wild plants from which they were derived.

A few cultivars in a planting can be a good thing. Disease-resistant cultivars of the flowering dogwood, *Cornus florida*, could help save the species (*see p.53*). However, the overuse of clones in horticulture is risky, because a disease or insect attack can wipe them out. A clone can perpetuate itself in nature, sometimes even better than a plant resulting from a sexual union, but only as long as the environment remains reasonably benevolent.

If the environment changes drastically, a clonally reproduced or highly self-fertilizing species may be at a disadvantage because it is less likely to be able to evolve forms better adapted to the new conditions. For example, the large population of English elms in the UK was destroyed in the 1960s and '70s by Dutch elm disease. The trees typically reproduce by root suckers so were represented by just a few genetically different clones.

▲ DOMESTICATION SYNDROME *Gigantism is a very common characteristic of the plants we select and breed: the parts of the plant that we value grow huge, be they dahlia blooms the size of dinner plates or plump, super-sized corn cobs.*

POTTED HISTORY *Since our Neolithic ancestors began domesticating wild plants, 10–15,000 years ago, humans have been encouraging certain traits in them and eliminating others. Only now are we realising the relative short-termism of our approach, and the integral value of variety and diversity in the plant world.*

GLOBAL HABITATS

GLOBAL HABITATS

The millions of dazzling creatures found around the world do not exist as isolated individuals, but rather live together in interdependent communities. Terrestrial habitats typically are classified according to the dominant plants that determine their structure and appearance, whether towering tree, compact shrub, or thick-rooted grass. The major plant communities at the global level are known as biomes. These basic vegetation types have evolved in response to prevailing conditions, especially climatic factors such as precipitation and temperature. Across the planet, wherever conditions are comparable, similar plant communities cloak the ground, creating habitat in turn for an astonishing array of animals, fungi, bacteria, and other organisms.

REGIONAL VARIATIONS

German scientific traveller Alexander von Humboldt (*see also p.31*) was one of the first scientists to make the important observation that the same types of plants grow in similar groupings in far-flung corners of the globe as long as climate and other conditions are similar. In parts of the world where moisture is ample and temperatures are relatively high for all or most of the year – creating conditions most conducive to plant growth – trees tend to be the dominant plants, forming communities of forest or woodland. In areas where water is abundant but temperatures are too low to support sustained plant growth for much of the year, such as the polar regions, trees cannot thrive and tundra is the predominant vegetation form. Grasses and low shrubs typically are the major plant types in dryland regions, forming grassland, savanna, or scrub communities. These global vegetation types, or biomes, rarely have defined boundaries; instead edges tend to be blurred.

At the regional level, biomes can be subdivided into plant provinces known as associations, formed by variations in precipitation patterns, averages and extremes of temperature, the length of the growing season, and the amount of available sunlight. The grasslands of central North America, for instance, are divided into three main associations: lush tall-grass prairie in areas with relatively high moisture levels; sparser short-grass prairie in more arid areas; and mixed-grass prairie in areas with moisture levels between the two.

ELEVATION AND LATITUDE

Within the tropical and temperate zones, precipitation levels and elevation are the major forces that have driven the formation of different habitat types. In the tropics, for example, distinct biomes have evolved along a moisture gradient, producing, from wettest to most arid conditions: rainforest, evergreen seasonal forest, dry forest, thorn woodland, desert scrub, and desert. At temperate latitudes, varying moisture levels, from wettest to driest, have led to: mesophytic (moist) forest, woodland, tallgrass prairie, shortgrass prairie, and desert. Changes of elevation in the tropics, from lowland to mountaintop, have stimulated the development of: rainforest, cloud forest, elfin woodland, and paramo, a cold, damp, high-altitude community of stunted trees. In temperate regions, as elevation increases, mesophytic forests become conifer-dominated forests and eventually tundra.

Within regional vegetation associations, the contour of the land and the character of the soil vary greatly, creating another dimension of plant diversity. On Long Island, off the coast in the north-eastern USA, for instance, some land is blanketed with oaks, tulip trees, hickories, and American beeches – characteristic of the greater Middle Atlantic oak forest association. This is found alongside red maples, tupelos, and sassafras typical of bottomland swamp forests; pitch pines of the island's glacial barrens; and straight-trunked cedars of the Atlantic white cedar swamps that dot Long Island's southern shore.

This variety of vegetation types at global, regional, and local levels has spurred the amazing diversity of associated plant and animal species. For aeons, barriers to dispersal, such as oceans or mountain ranges, separated ecologically similar biomes of the world, increasing the diversity as new species evolved to fill the corresponding niches.

◄ **CONVERGENT EVOLUTION** *Leaf shape, size, and texture is strikingly similar among shrubs of Mediterranean regions around the world, even though their characteristic species are only distantly related – a phenomenon known as convergent evolution. The leaves of the Australian native shrub* Grevillea rosmarinifolia (above left)*, for example, share many features with those of* Lavandula dentata (left)*, a plant native to the Mediterranean basin.*

◄ **SALVIA ALBIMACULATA**
Members of the Salvia *genus are distributed worldwide in temperate and tropical regions, except in very hot and humid habitats. Often aromatic, they have woolly, hairy, or silvery foliage.*

MAJOR BIOMES ►
The lion's share of terrestrial plant diversity in the world is contained within four major biomes. Forests cover much of the tropical, temperate, and northern boreal zones. About 20 per cent of terrestrial habitat is temperate grassland, tropical savanna, and alpine meadow, while a further 20 per cent is desert. Five relatively tiny Mediterranean-climate regions are scattered around the planet, supporting a disproportionate number of endangered plants found nowhere else in the world.

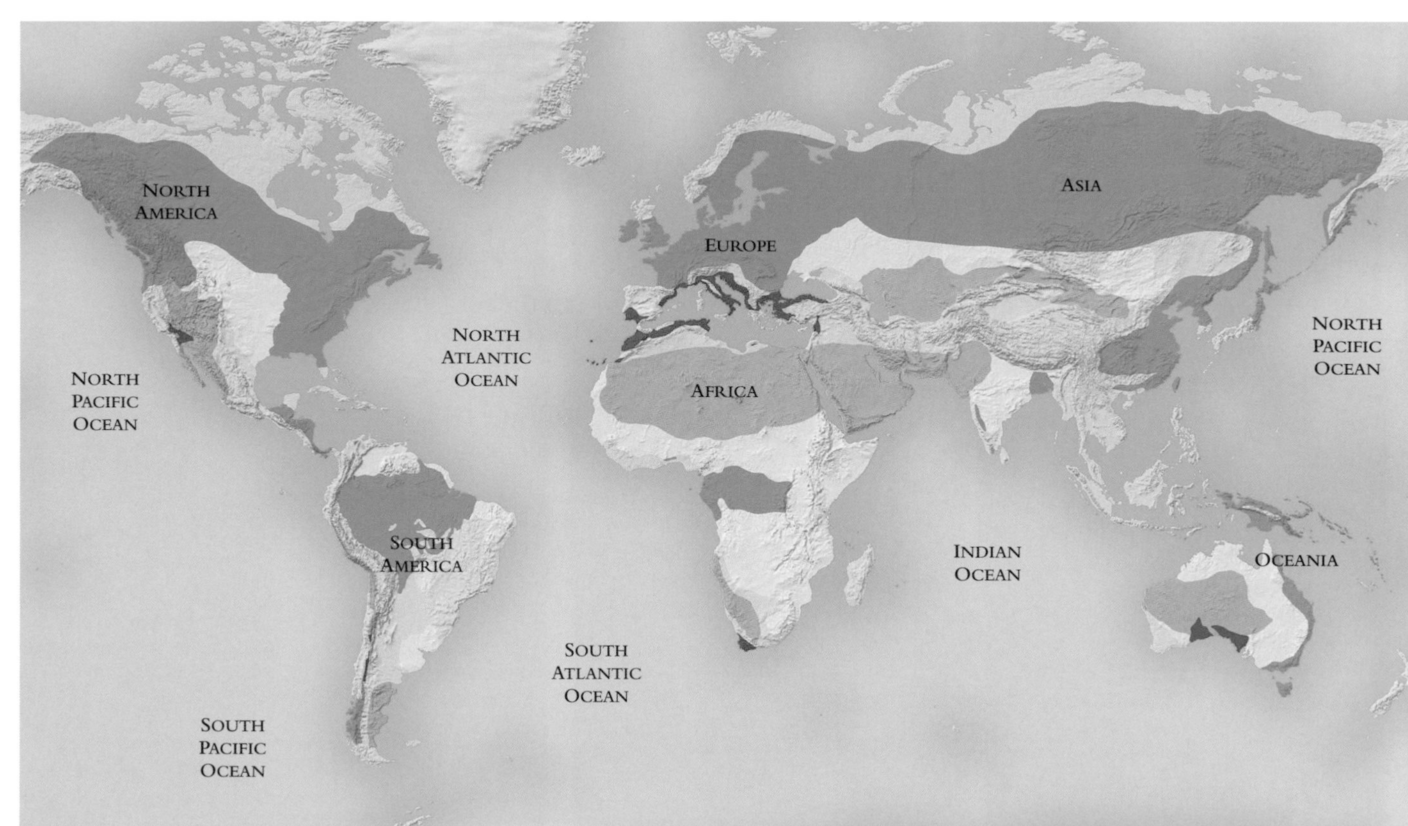

KEY TO HABITAT LOCATIONS:

- FORESTS
- GRASSLANDS/SAVANNA
- MEDITERRANEAN
- DESERTS

FORESTS

Trees dominate forest life. Under their protective canopies, plants with a variety of growth habits have evolved to take advantage of any available space and light: understorey trees and shrubs, lianas, ephemeral wild flowers, and luxuriant ferns and mosses. Forests are grouped into three broad categories based on latitude: tropical, temperate, and boreal.

GRASSLANDS

Vast, grass-dominated plant communities occur in areas where it is too dry or cold to support the enormous biomass of trees. In temperate latitudes, perennial grasses and herbaceous wild flowers are the dominant growth forms. Tropical savannas have a continuous cover of perennial grasses, dotted with tough, drought-resistant trees or shrubs.

MEDITERRANEAN

Evergreen or drought-deciduous shrubs dominate life in the world's five relatively small, and critically endangered, Mediterranean vegetation zones. The flora in each of these distant areas has evolved in response to a characteristic climate of mild, rainy winters and warm, dry summers.

DESERTS

Scorching sun and scant rainfall are the realities of desert life. In this inhospitable habitat, desert denizens can survive because they have evolved an array of strategies to conserve water. Ground-hugging shrubs and stubby, sparse-leaved trees are the dominant growth forms of these arid lands.

FORESTS

Forests are the most biodiverse terrestrial habitats on the planet, and rival the oceans in their influence on the biosphere. They stabilize land, control flows of water, and modulate global as well as local climates. They also supply food, medicine, fibre, and building materials for people. About half of the forest that covered the Earth before the spread of human influence has disappeared: forests once blanketed about half of the planet's land area, but today, they cloak only about one-quarter. Even with these losses, forests still account for more than two-thirds of the leaf area of plants, and contain about 70 per cent of the carbon present in living things.

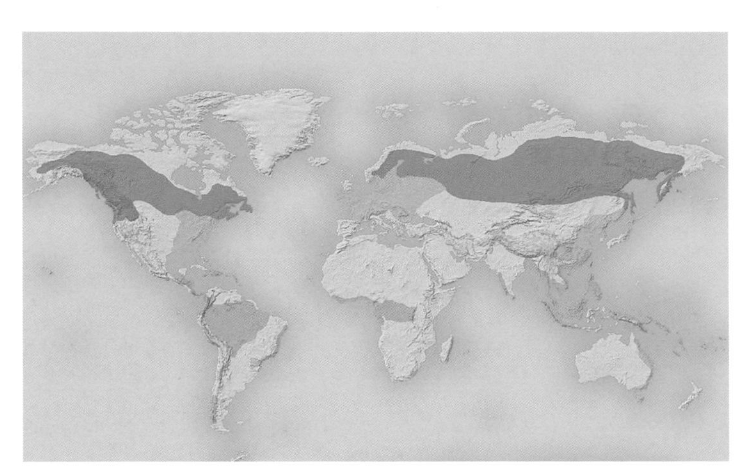

DISTRIBUTION *Forests are found on every continent in the world except Antarctica. They exist in a breathtaking array of forms and types, from hot, dry tropical forests and rainforests near the equator to cold, coniferous boreal forests – or taiga – in the northernmost latitude. In between the two are the temperate forests of eastern North America, central and western Europe, and eastern Asia.*

Boreal forests
Temperate forests
Tropical forests

A forest is an ecosystem in which trees are the dominant life form, covering at least 10 per cent of the land with their canopies. Where the canopy cover is 40 per cent or more, they are known as closed forests. The world's remaining large, intact natural forest ecosystems are known as old-growth or frontier forests. Few of these remain, however, and much of the world's forests, including temperate and subtropical forests as well as tropical deciduous lowland and mountain forests, are highly fragmented and under great pressure from human activities.

Globally, tropical dry forest has suffered the largest losses, having been reduced in area by about 70 per cent. Some 60 per cent of the original temperate broadleaf and mixed forests have been cleared, as well as 30 per cent of temperate needleleaf forests. An estimated 45 per cent of tropical moist forest has been destroyed. Developing countries are continuing to decimate their tropical forests, while deforestation has slowed in temperate regions.

▲ **UNDER THE CANOPY** *The types and density of the trees in a forest define the amount of canopy cover* (above)*, and so determine the amount of light penetration* (below)*, and thus life on the forest floor.*

BEECH FOREST, SNOWDONIA NATIONAL PARK, UK

TYPICAL PLANTS OF FOREST HABITATS

CANOPY TREES *The tallest trees in a forest emerge from the understorey to form a canopy that hogs the light and air.*

LIANA VINE *These woody rainforest vines use the trunks of trees and other vegetation to climb their way up to the light-rich environment near the tree canopy.*

MOUNTAIN DOGWOOD *In temperate zones, understorey trees and shrubs grow in the dappled shade beneath the canopy. They are adapted to thrive in lower light conditions.*

BROMELIADS *Epiphytes, such as this characteristic rainforest plant, are not rooted in the ground but grow on other plants, such as the branches and trunks of trees.*

DECOMPOSERS *Fungi, such as these mushrooms, cannot photosynthesize, but instead obtain their nutrition from dead plant material that builds up on the forest floor.*

TRILLIUM GRANDIFLORUM *A dense carpet of shade-tolerant wild flowers thrives in the rich layer of organic material that covers the floor of a temperate deciduous forest.*

High humidity
The defining feature of a rainforest is that it receives large amounts of rainfall each year. Rainforests in temperate regions experience long, mild, and very wet winters. The summers are dry and cool but foggy, which provides the moisture necessary to sustain plant life.

Giant vegetation
Temperate rainforest habitats are home to some of the largest living things on the planet. The giant conifers that thrive in this habitat are extremely long-lived and can grow to enormous sizes. The Sitka spruce, for example, can grow to over 60m (200ft) tall with a trunk 2.5m (8ft) in diameter.

Forest layers
Plants grow in almost every last inch of space in this fertile habitat. Tree branches reach up towards the sky, covered with epiphytic plants, such as mosses and lichens, which take advantage of their lofty position. Beneath, shade-loving trees fill the understorey, while ferns and small plants carpet the forest floor.

Fertile soil
The forest floor is covered with a thick layer of organic matter, formed when dead plant material is slowly decomposed by bacteria, insects, and fungi. This creates a nutrient-rich soil that sustains the lush forest vegetation.

Temperate rainforest
Olympic National Park in Washington is the wettest place in the continental USA. It is characterized by a dense growth of timber.

TROPICAL FORESTS

Tropical forests teem with life. More species of plants and animals live in tropical rainforests than in all of the rest of the world's biomes combined. An astonishing variety of trees grows in tropical rainforests, as many as 300 species per hectare in the western Amazon basin, the most diverse of all tropical forest regions.

DISTRIBUTION OF TROPICAL FORESTS

The Amazon and Zaire river basins hold the two greatest and most intact lowland rainforest regions in the world. Tropical rainforests are also found on the Atlantic coast of Brazil, in Central America, eastern Mexico, and some Caribbean islands, as well as West Africa and south-east Asia, from eastern India to Thailand, Indonesia, and the Philippines. However, most of these have been fragmented or destroyed.

In areas with high rainfall spread more or less equally throughout the year, the lowland forests are evergreen. Many of these tropical rainforests have canopies 40–50m (130–160ft) tall, and some have emergent trees that rise above the main canopy to heights of 60m (200ft) or more. Climbing vines called lianas hang from the trees, often connecting them at various levels. Epiphytes – plant that do not root in the soil but grow on other plants – cover the trunks and branches of the trees, taking advantage of the higher light levels found in the forest canopy. The rich diversity of tree species in tropical rainforests is mirrored in the diversity of lianas and epiphytes, which include thousands of orchids and bromeliads. These support a wealth of wildlife, including ants, pollinators, and many other species. The lower layers of the forest consist largely of tree seedlings, because light levels are so low that few herbaceous plants grow there. Those that do are found mostly in gaps in the canopy created by treefalls. Vast populations of organisms decompose dead wood and leaf litter on the forest floor, while large cats stalk prey through the undergrowth.

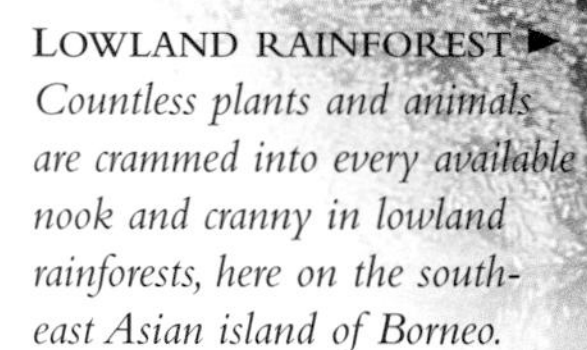

LOWLAND RAINFOREST ► *Countless plants and animals are crammed into every available nook and cranny in lowland rainforests, here on the south-east Asian island of Borneo.*

▼ SUMATRAN TIGER *The Sumatran tiger is found only on the Indonesian island of Sumatera, where it roams the rainforest floor hunting for wild pigs and muntjak deer. Conservationists estimate that there are now fewer than 500 of these majestic animals outside captivity.*

In areas where rainfall is seasonal, tropical forests are deciduous, with trees shedding their leaves during the few dry months each year. In south-east Asia, these are known as monsoon forests, because the trees open their leaves during the onset of the annual monsoon, when moisture-bearing winds blow steadily off the ocean. The best-known tree species in these forests is teak.

▲ EMERALD TREE BOA *This striking, brilliant-green rainforest snake,* Corallus caninus, *spends almost all of its life above the ground in the lush vegetation of the forest canopy. It survives there on a carnivorous diet of small rodents, birds, and bats.*

A HABITAT AT RISK

According to the UN Food and Agriculture Organization (FAO), more than 94 per cent of the forest habitat lost each year during the 1990s was tropical forests. Clearing of forests for agriculture has been the main cause of tropical forest loss. Government resettlement programmes have moved large numbers of poor farmers into once-remote forests. Landless settlers have hacked down and burnt countless thousands of square kilometres of both logged and undisturbed forests, to provide land suitable for cultivation (*see facing page*). These so-called slash-and-burn techniques are destroying large areas of rainforest, leaving behind infertile, barren land.

The spread of cattle ranching, encouraged by tax breaks, has posed a serious threat to the rainforests of Latin America. Huge areas of forest have been felled to produce beef not only for domestic consumption, but also for export to the fast-food industry in North America and Europe. Soil fertility in these deforested areas declines rapidly, however, and the ranchers move on to new pastures when productivity falls.

For tropical countries, timber is an important source of foreign currency. One-fifth of the world's industrial roundwood originates in tropical forests, including such prized species as mahogany and teak. The timber supplies of some countries have already been exhausted, and a few major producers –

▲ SLASH AND BURN *Tropical soils tend to be very poor, so settlers have traditionally returned the nutrients stored in the plants to the soil by cutting down and burning areas of forest. In a few years, however, the nutrients are depleted, and the settlers are forced to move on and repeat the process elsewhere.*

▲ LIFE IN THE CANOPY *Tropical forests harbour a fantastic variety of wildlife, and much of this biodiversity is found high above the ground in the canopy. Conservationists regularly survey the canopy ecosystems, to gain insight into the relative health of the rainforest as a whole.*

Indonesia, Malaysia, Brazil, and India – now dominate production. Logging threatens the biodiversity of tropical forests in a number of ways. Although some timber species are naturally abundant and can survive commercial exploitation, many have suffered extensive population and genetic losses. Most under threat are rare species that are felled because they resemble their commercially valuable relatives. Furthermore, logging operations create access to forest areas that may otherwise have remained remote and isolated. This can pave the way for colonization and conversion of the land for agricultural use. Access to the land may also attract hunters, to satisfy the popular demand for bushmeat, and so threatens forest animals.

Felling trees for fuelwood is a growing threat to tropical forests in developing countries. Fuelwood and charcoal consumption more than doubled between 1961 and 1991, and is still rising.

To many tropical nations, however, forests are not just a source of land, timber, and fuelwood but also a wealth of minerals, including gold. Mining is a relatively minor cause of direct deforestation, but the access roads that are built by the mining companies frequently attract settlers who can then reach previously remote areas of rainforest.

Climate change is affecting some tropical forest habitats, such as the mountain cloud forests, which depend on mists for moisture. Research has shown that the cloud base is moving upwards on tropical mountains as a result of shifts in climate. The forest species are not able to migrate at such a rapid rate, and at some point will be limited by the land area available at higher elevations. Local extinctions of cloud forest amphibians, including the critically endangered golden toad, have been attributed to this temporary or long-term change in the climate.

The destruction of rainforests spells disaster for some plants and animals. Tropical plants and their associated wildlife often have restricted ranges, and species can be lost as rainforests are toppled. With each felled tree, a veritable ecosystem can perish. Fragmented islands of tropical forest left in the wake of development often cannot support populations of vital pollinators, seed dispersers, and other animals, and are vulnerable to invasion by aggressive weeds.

◄ PITCHER PLANT *Some rainforest species have developed unusual ways to supplement their diet. Pitcher plants (*Nepenthes *species), for example, have cup-shaped leaves filled with nectar or collected rain water, which attract and trap small insects.*

▼ BUTTRESSED ROOTS *The layer of organic material on the rainforest floor is extremely thin, so most trees spread their roots widely but not deeply into the ground. To gain better anchorage and support, the trees develop buttressed roots that grow out from the base of the trunk, sometimes as much as 4.5m (15ft) above the ground.*

TYPES OF FOREST

FLOODED FOREST, MALAYSIA
Some areas of tropical forest are subject to annual flooding, when rivers break their banks during the rainy season. This seasonal inundation restores nutrient levels to the forest, helping the trees and other vegetation to grow. Many trees found in this habitat depend on the wildlife that the flood water attracts – such as fruit-eating animals and fish – to disperse their seed.

MONTEVERDE CLOUD FOREST, COSTA RICA
On mountain slopes in the tropics, the hot, sticky humidity of the lowland rainforest is replaced with a cooler dampness. Mists engulf the canopies of cloud forests, and the foliage literally drips with moisture. Lichens drape from tree branches, and bright green mosses cover tree trunks and the forest floor. The constant source of water above ground makes this an ideal habitat for epiphytes.

DRY FOREST, MADAGASCAR
The canopy of tropical dry forests is generally lower than that of rainforests, and there are fewer epiphytes but more lianas and other climbing plants. The areas of dry forest that are still intact are often important refuges for once widespread species, including endemic mammals. For example, the forests of western Madagascar are inhabited by about 40 per cent of the island's endemic lemurs.

TEMPERATE AND BOREAL FORESTS

The forests north and south of the tropics range from lush, temperate rainforests and leafy deciduous forests to the vast coniferous boreal forests of the Earth's northern reaches. Often blanketed by snow, boreal forests have the distinction of being the most extensive of all terrestrial biomes.

TEMPERATE FOREST ► *Beech trees fill this temperate deciduous forest in New Zealand, while a dense understorey of ferns and shrubs creates a rich, green carpet that covers the forest floor.*

DISTRIBUTION OF TEMPERATE AND BOREAL FORESTS

Temperate broadleaf deciduous forests can be found on all of the major land masses in the northern hemisphere, particularly eastern North America, western and central Europe, and northeast Asia. Oaks, maples, and other broadleaf trees are the dominant species in these forests. Growth in this habitat is largely in the spring and summer, while the winter brings a period of dormancy. As spring sunlight streams through the bare tree branches, an explosion of growth begins. Among the first plants to bloom are ephemeral wild flowers like bluebells and trilliums, which take advantage of the temporary sunshine on the forest floor. Leaves soon appear on the trees and by early summer, most of the wild flowers have disappeared, to be replaced by a covering of ferns and mosses.

Temperate forests are less rich in species than their tropical counterparts. The mixed hardwood forests of the southern USA may include as many as 30 canopy and subcanopy tree species. Europe's deciduous forests, which lost species due to glaciation during the Pleistocene period, tend to be less diverse, while the forests of north-east Asia, which once may have been the richest of all, have largely been replaced by farms and artificial forests. The plants found in each of these three regions are remarkably similar, and often represent related species of the same genera, such as oaks (*Quercus*), dogwoods (*Cornus*), and lady's slipper orchids (*Cypripedium*). Bordering the northernmost deciduous forests are mixed forests in which conifers mingle with the broadleaf species.

There is no direct counterpart to the broadleaf deciduous forests of the north in the southern hemisphere. Instead, humid, subtropical climate regions have mixed broadleaf and needleleaf evergreen forests that are home to araucarias (*Araucaria*), podocarps (*Podocarpus*), and southern beech (*Nothofagus*). Also found in these areas are most of the rare, temperate rainforests.

Some temperate wetland areas support the growth of water-loving trees and some woody shrubs. Known as swamp forests, these areas are covered with standing water for at least part of the year. The loose, water-soaked soil does not provide a firm base in which the trees can anchor their roots, and so many develop buttressed roots (*see p.67*) for additional support. The tree species that grow in the temperate swamp forests of North America vary according to the region. In the south, the swamps are home to cypresses (*Taxodium*), while in the northwest, willows (*Salix*) and western hemlock (*Tsuga heterophylla*) are found in these watery habitats.

Temperate and boreal needleleaf forests are composed almost entirely of conifers. Temperate needleleaf forests mostly occupy higher altitude areas in the northern hemisphere. Although temperate and

▼ BOREAL HERBIVORES *Caribou herds make their winter home among the conifers of boreal forests, here in China. These large mammals survive through the winter feeding on the lichens that grow on the trunks of the trees.*

▼ SEASONAL CHANGES *Broadleaf deciduous forests harbour the most dramatically changing plant communities on the planet. The vibrant greens of the spring and summer are replaced in autumn by a fireball of foliage colour, before the plants shed their leaves and settle into winter dormancy.*

▲ SWAMP FOREST *Bald cypresses* (Taxodium distichum) *are so-called because they drop their needles in autumn. These long-lived, water-loving trees grow in watery habitats, such as this swamp forest in Louisiana, USA.*

boreal needleleaf forests typically do not contain a high diversity of tree species, and many of their conifers are threatened with extinction, they do harbour an extraordinary diversity of mosses and lichens, important food sources for the animals they support. Compared to broadleaf deciduous forests, needleleaf forests are structurally simple. Their dense canopies allow very little sunlight to filter through, which reduces the opportunities for other plant species to grow on the forest floor below.

Boreal forests occur in a nearly continuous belt across North America and Eurasia between 50° and 60° north latitude. They are dominated by a few species of the four main genera of conifers: spruce (*Picea*), fir (*Abies*), pine (*Pinus*), and the deciduous larch (*Larix*). These spire-shaped needleleaf trees are well adapted to the short, moist summers and long, cold winters – often with high snowfall – that are characteristic of the far north.

FORESTS UNDER THREAT

The chief threats to the world's temperate and boreal forests are the increasing human population in many areas, the continued loss of forest to agriculture and urban sprawl, and overexploitation for timber. As in tropical regions, their immediate value for timber continues to dominate virtually all decisions regarding the management of temperate forests. Timber harvest and related damage to surrounding forests threaten the health and long-term survival of entire ecosystems as well as individual species. *Fitzroya cupressoides*, a conifer native to Argentina and Chile, has become an endangered species because its durable wood is felled for house construction and roof shingles.

For centuries, temperate forests have been cleared for human habitation and farmland. The temperate broadleaf deciduous forests of China are known primarily from the fossil record; the natural vegetation of the region has been stripped to be replaced by intensive agriculture for at least 4,000 years. The natural forests of Japan have largely been replaced with man-made woodlands. In Europe, most forests were cleared for agricultural use centuries ago, and remnants survive only in some royal hunting preserves. Only in the mountains of South Korea are the forests largely intact.

Almost all the forests of eastern North America are second growth, but in them the world's greatest diversity of temperate broadleaf flora and fauna survives, especially in the Appalachian plateau region of eastern Kentucky and Tennessee and western North Carolina and Virginia. The ecological importance of this area has been recognized by the United Nations, which designated the Great Smoky Mountains National Park in Tennessee as an international biosphere reserve.

◄ LICHENS *Lichens colonize the trunks and branches of trees in habitats that range from the northernmost and coldest boreal forests to hot, sticky, tropical rainforests. These unusual organisms are a symbiotic partnership between an alga and a fungus.*

▼ FOREST FUNGI *In coniferous forests, undergrowth is sparse because very little light reaches the floor, and decomposing pine needles make the soil very acidic. Fungi, however, which do not require light to grow, thrive under such conditions.*

TYPES OF FOREST

TEMPERATE BROADLEAF FOREST, USA
The deciduous forests of the Great Smoky Mountains National Park are world-renowned for their diverse plant and animal life. Characteristic plants in this habitat include sugar maples (Acer saccharum), *which provide a blaze of yellow, scarlet, and orange in autumn. American chestnuts were once the most common tree in the park, but numbers have been decimated by chestnut blight* (see p.70).

TEMPERATE EVERGREEN FOREST, USA
The evergreen forests of northern California are home to some of the tallest trees in the world, including the giant redwoods (Sequoia sempervirens) *that can live for 2,000 years and reach heights of 100m (330ft). Beneath these giants, an understorey of smaller trees, such as hemlocks* (Tsuga *species*) *and Douglas fir* (Pseudotsuga menziesii), *almost creates a second canopy.*

BOREAL FOREST, RUSSIA
The temperature in the boreal forests of the Urals does not climb above freezing for six months of the year. In this extreme climate, the diversity of plant species that can survive is reduced. These needleleaf forests, also known as taiga, are dominated by evergreen conifers, such as white spruce (Picea glauca), *that are adapted to the cold climate and extremely short growing season.*

◄ DICENTRA CUCULLARIA *Many forest species are dependent on their associated wildlife in order to complete their life cycle. This wild flower, found in temperate forests, relies on ants for the dispersal of its seeds, and so any threats to local ant populations also threaten the plant itself.*

DECLINING HABITAT

In regions of temperate forest where extensive clearing has taken place, the areas that remain are extremely fragmented. These isolated patches of forest lose plant and animal species because they are not large enough to support cross-pollination and seed dispersal or the resident animals that require room to roam. For example, forest fragmentation has led to the decline of the spotted owl, because its habitat – the old-growth coniferous forests of the Pacific coast of North America – has been carved into small pieces. Fragmentation also impedes the migration of birds and other species.

Temperate and boreal forests are also plagued by insects and diseases introduced from other regions. One notable example is the American chestnut (*Castanea dentata*), once known as the king of the deciduous forests of eastern North America. This dominant tree species was virtually eliminated by chestnut blight, a disease caused by the fungal pathogen *Endothia parasitica*. Introduced from Asia in 1904, this disease decimated the population of mature chestnuts, magnificent trees that could grow over 2m (6ft) wide and 35m (120ft) tall in less than 40 years.

In and around industrialized regions, temperate and boreal forests have been damaged by airborne pollutants – especially ozone – and by acid rain, which poisons trees, especially those in forests with already acidic soils. Acid rain forms when pollutants created by the burning of fossil fuels in power stations and cars dissolve in atmospheric moisture, creating sulphuric and nitric acids. These return to the earth as rain or fog. The Black Forest of Germany and the Adirondack mountains in New York state are just two of the forest regions that have suffered significant damage from acid rain.

Global warming poses what may be even more intractable problems for the planet's temperate and boreal forests. The swamping of coastal cities by melting polar ice caps and the transformation of the world's major food-producing regions into dustbowls are two of the potential disasters that could be caused by the pumping of massive amounts of carbon dioxide and other greenhouse gases into the atmosphere. Less well-known data suggests that the ranges of many forest plants could shift northward, or upward in mountainous areas, and many species would not survive these rapidly changing conditions.

▼ POLLUTION DAMAGE *The effect of acid rain is all too obvious in this area of forest in Wales. Acid rain poisons the soil and damages the trees, and this fatally reduces their ability to cope with natural stresses, such as drought, disease, or cold weather.*

▲ GLOBAL WARMING *The effects of climate change on boreal forests are highlighted in this area of taiga forest in Canada. As the bitterly cold winters that have shaped this habitat become warmer, deciduous trees, which once would not have been able to survive here, are able to colonize the forests.*

▲ SUDDEN OAK DEATH (SOD) *Oak trees suffering from this devastating disease were first discovered in California in 1995. Caused by the fungus-like pathogen* Phytophthora ramorum, *SOD has reached epidemic proportions in this part of the world. Affected oaks develop brown patches of infection on their trunks before suddenly dying.*

▲ TEMPERATE WOODLAND CLEARANCE *The dense covering of vegetation that a forest provides protects the land from the elements. If the trees are felled – for timber, for example – erosion of the vital soil layer can occur more quickly than new trees can grow, leaving behind barren and infertile land that can no longer support diverse forests.*

GARDENING WITH WOODLAND PLANTS

A typical temperate deciduous forest is made up of four vertical layers: the canopy; an understorey of smaller trees; a layer of saplings and shrubs; and a ground layer of wild flowers, ferns, and mosses. Re-create these layers for maximum diversity and interest in a woodland garden. Like natural forests, woodland gardens are dominated by their tallest trees. The shape, density, and degree of overlap of the crowns of the trees determine the amount of shade, the temperature, and the moisture conditions on the site. Plants in the lower layers must be appropriate for these specific conditions or they may not survive: plants adapted to the deep shade of conifer-dominated forests may perish in sunny spots, while species such as sunflowers, which do best in open gaps in native forests, may perish in deep shade.

Most herbaceous woodland plants benefit from the addition of organic matter to the soil at planting time, while trees and shrubs grow better in the original soil, with no supplementary nutrients. If there is a miracle ingredient for a woodland garden, however, it is mulch. Nature renews the mulch in forests each year as leaves are shed and fall to the forest floor. Woodland gardens similarly benefit from an annual application of leaf mulch, typically in the autumn.

▲ CORNUS MAS *The Cornelian cherry is a vigorous deciduous shrub that grows to 5m (15ft). Profuse clusters of dainty yellow flowers are produced in early spring before the leaves, which are dark green and turn purple in the autumn.*

▲ HAZEL IN THE UNDERSTOREY *Choose plants not only for their looks, but also for their value to wildlife. Even small trees, such as hazel, provide a place where birds and an array of other creatures can feed, breed, and rest.*

GARDEN IN THE WOODS ► *Drifts of phloxes and trilliums bloom in the New England Wild Flower Society's woodland garden in north-eastern USA.*

▲ ERYTHRONIUM CALIFORNICUM *Trout lilies and other beautiful woodland bulbs come into leaf and bloom in spring, before the tree canopy closes overhead and casts dense shade on the forest floor.*

GARDENER'S GUIDE

Woodland gardening involves manipulating the layers of vegetation under the tallest trees' canopies. Choose a mix of understorey trees and shrubs, climbers, ferns, herbaceous wild flowers, and bulbs, like those listed below. Woodland wild flowers are threatened by overcollection; buy rare woodland wild flowers and bulbs only from suppliers that sell plants that have been propagated in a nursery.

IRIS

- *Camellia* (see p.97)
- *Cercis canadensis* (see p.98)
- *Franklinia alatamaha* (see p.108)
- *Magnolia* (see p.113)
- *Arisaema* (see p.169)
- *Kirengeshoma* (see p.184)
- *Shortia galactifolia* (see p.197)
- *Trillium* (see p.200)
- *Dicentra* (see p.174)
- *Adiantum* (see p.312)
- *Passiflora* (see p.200)
- *Erythronium* (see p.218)
- *Lilium* (see p.230)

GRASSLANDS

The world's great grasslands and savannas are shaped by climatic factors that impede the growth of woody plants, such as trees or large shrubs, creating a landscape covered with grasses, sedges, and herbaceous wild flowers. Grasses have basal meristems, or growing tips, that are not easily destroyed by fire or by grazing – the other major influences on grassland ecosystems. The world's major grasslands were once inhabited by thunderous herds of grazing mammals, but they have been hunted almost to extinction. Today, they have largely been replaced by domestic livestock, and vast expanses of grasses and wild flowers have been cleared for the cultivation of crops.

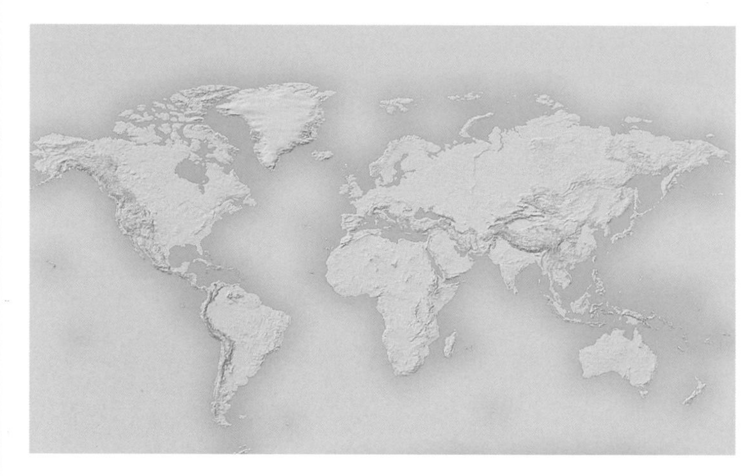

DISTRIBUTION *About 20 per cent of the Earth's land surface (excluding Antarctica) supports grasslands, although these areas now vary greatly in their naturalness. Temperate grasslands make up around one-quarter of the total area, while the rest is savanna. Temperate grasslands are found in semi-arid, continental climates of the middle latitudes, forming the zone between deciduous forests and deserts. In the tropics, savannas are found between the rainforests of the equatorial zone and the deserts of the subtropics.*

Grasslands and savanna

Grasslands occur in semi-arid climate areas, where it is too dry for trees and shrubs to grow, but not so dry that the area becomes desert. Temperate grasslands can be divided into three categories – tall grass, mixed grass, and short grass – depending on the annual rainfall of the area, with short grass characteristic of the drier climates. The most extensive temperate grasslands are the prairies of North America, the pampas of South America, and the Eurasian steppes.

Savannas are temperate or tropical ecosystems in which grasses dominate the ground layer. In the tropics, they range from treeless plains to woodlands with a grassy understorey, and typically have less rainfall than tropical forests, and a distinct dry season.

▲ WATER BIRDS *Most of the rain that falls on the tropical savanna of Australia does so during the rainy season in the summer, when many lakes and rivers break their banks. The flooded plains become a magnet for wildlife, such as this flock of wandering whistling duck* (Dendrocygna arcuata).

▼ GRAZING GIANTS *This bison herd in Yellowstone National Park represents a relic from prehistoric times, when enormous grazing animals, most now extinct, roamed the Earth. Once common on the prairies of North America, bison today survive only in isolated refuges and a few ranches.*

TYPICAL PLANTS OF GRASSLAND HABITATS

BAOBAB TREE *This drought-adapted species found in the African savanna can survive the long dry season by utilizing water reserves it has stored in its thick trunk.*

WHISTLING THORN SHRUB *The ants that find a home in the bulbous galls of this African acacia protect it from grazing animals, by stinging any that venture too close.*

ELEPHANT GRASS *The roots of tall grasses extend deep into the ground. This African grass, which can grow as tall as 4m (12ft), can be invasive outside its natural habitat.*

CONEFLOWER *Non-woody, broad-leaved perennials, known as forbs, provide splashes of colour among the prairie grasses in tall- and mixed-grass prairies.*

THREEAWN AND HAIRY GRAMMA *The grasses that cover the driest prairies grow in low, dense bunches and have shallow roots, to take advantage of the lightest rainfall.*

HIMALAYAN WILD FLOWERS *High-altitude grasslands are a sea of colour in spring, when the herbaceous wild flowers that have lain dormant in the dry season burst back into life.*

Severe weather
Short-grass prairie grasslands are extremely dry. Some areas receive just 25cm (10in) of rainfall in a year – most of which falls within the space of a few months in late spring. Periodically, these areas experience severe drought when no rain falls for several years.

Nutrient-rich soil
The landscape may look barren and the vegetation sparse, but much of the biomass in the prairies is located underground. The roots, bulbs, and rhizomes of prairie plants extend deep into the soil, enriched by the large amounts of organic matter created when the plant tops die back each year and decompose.

Survival strategies
Prairie grasses are adapted to survive the long months of the dry season without water. Some are annuals, and remain as seed through the winter months. Others die back in the winter, and lie dormant underground, sprouting again from the roots when the rain falls in spring.

Plant protection
Many prairie plants have developed mechanisms for avoiding the unwanted attentions of grazing herbivores. Some arm their stems and leaves with thorns or prickles, while locoweeds (Astragalus *species) produce a substance that is highly toxic to grazers.*

Short-grass prairie
The short-grass prairies of North America are a mixture of low-growing grasses, wild flowers, and drought-resistant shrubs.

TYPES OF GRASSLAND

SAVANNA, NORTHERN AUSTRALIA
The tropical savanna of northern Australia experiences extreme seasonal weather changes: the summer months are hot, humid, and very wet, followed by a cooler dry season lasting up to five months, in which very little rain falls and periodic fires sweep the parched landscape. The dense grass of the savanna is scattered with trees, such as Eucalyptus *and desert oak* (Allocasuarina decaisneana).

STEPPE, RUSSIA
The vegetation of the dry, cold steppe grasslands of Russia and Mongolia mainly consists of short grass, dotted with shrubs, such as sagebrush (Artemisia *species*). *The fauna of the Russian steppes once included the wisent – a cousin of the bison, the tarpan or wild horse, and the saiga antelope, but as with much of the grassland around the world, large areas of their habitat has now been cleared for cultivation.*

HIGH-ALTITUDE GRASSLAND, TIBET
Grassland found in the extreme climate of alpine regions – where temperatures remain low all year round and there is little precipitation – supports a wealth of magnificent wild flowers. The relatively undisturbed alpine steppe of the Tibetan plateau, in Qinghai Province, China, is the native habitat of Tibetan antelope and gazelle, wild yak, blue sheep, and the endangered snow leopard.

▲ **OCCASIONAL TREES** *Savanna grasslands, such as these plains in Tanzania, are shaped by a climate consisting of a short wet season, followed by a long period of drought. These conditions are too dry for trees to dominate, but can support scattered trees, such as acacias.*

◄ **ECOTOURISM** *The Masai Mara National Reserve in Kenya, which attracts many thousands of visitors each year, highlights the paradoxical nature of ecotourism. Endangered species, like the black rhino, are being preserved, but the increase in human activity is threatening the habitat itself.*

THREATS TO GRASSLANDS

Temperate grasslands, especially those in humid environments, have some of the world's most fertile soils. In this habitat, a high proportion of plant biomass – in the form of roots and rhizomes – is located underground, and consequently the soils are extraordinarily rich in organic matter. This has been the grassland's blessing and also its curse. The rich soils once supported a lush flora of grasses, sedges, and wild flowers. However, the land has been almost completely converted for agriculture, including vast acreages of wheat in the steppes of central Europe and the northern plains of North America, and of corn in the American Midwest.

Grazing of domestic livestock is the most common human use of grasslands in most arid or semi-arid areas. The westernmost and driest of the North American grasslands, known as the short-grass prairies, were once home to a variety of low grasses, including gramma grasses, buffalo grass, and needle-and-thread, as well as a host of colourful wild flowers, such as blanketflower, Indian paintbrush, evening primroses, and penstemons. These plant communities have been severely overgrazed by cattle and many have turned into deserts. Overgrazing by domestic animals also threatens the diversity of plants and animals of tropical savannas and mountain meadows around the globe.

Subsistence hunting has been a part of grassland ecosystems since the evolution of early hominids. However, where people rely on bushmeat as a major source of protein, populations of several antelope species have declined at alarming rates, and other mammals are also at risk. Large grassland mammals are particularly vulnerable to poaching, a major threat to the Tibetan antelopes and snow leopards of Asia as well as the large game animals of the Serengeti and other east African savannas. According to the World Conservation Monitoring Centre (WCMC), an estimated 200,000 animals in the Serengeti National Park alone are killed by poachers each year. Elephants are killed for their skins and ivory tusks. The black rhinoceros has been hunted to extinction in much of its habitat, killed by trophy hunters and poachers seeking their valuable horns. Only in heavily guarded parks and reserves or in other sanctuaries where poaching has been curtailed do elephant families bathe in water holes while black rhinos wallow in the mud to keep cool.

The cycle of periodic fires in grassland areas has played a critical role in the development and maintenance of this habitat. Wildfires in these arid lands are started spontaneously by lightning strikes. Native peoples would also set fires deliberately to enhance soil fertility for crop planting, and by

▲ **ESSENTIAL FIRE** *The landscape of the African savannas has been shaped by a history of periodic fires. Grasses are adapted to survive fire, because their growing points are under the ground, but fire limits the growth of many shrubs and trees.*

improving the grassland, they also increased the herds of animals on which they depended for food. These natural and planned fire regimes have been interrupted in many regions. In highly populated areas where wildfires have been suppressed, as well as in areas where native peoples no longer burn the landscape, grasslands have changed into forests or other plant communities.

Disturbed grasslands are vulnerable to invasion by non-native plants and animals. The overgrazed short-grass prairies of North America have been invaded by prickly pear cacti and other thorny plants. In the *Eucalyptus* savannas of northern Australia, many exotic plants introduced for pasture or ornamental use have become invasive pests, including noogoora burr (*Xanthium strumarium*), *Parkinsonia aculeata*,

▲ GRASSLAND AGRICULTURE *Very little of the original prairie land in North America exists today, because most of this fertile grassland habitat has been converted for agricultural use, such as beef-cattle farming.*

INVASIVE PLANTS ► *The castor bean* (Ricinus communis) *is native only to Africa, but is cultivated in other parts of the world for the oil in its seeds. It is now an invasive plant in the Australian savanna, having escaped cultivation.*

and the bellyache bush (*Jatropha gossypifolia*). Arnhem Land in northern Australia has particularly suffered from the introduction of non-native wildlife. Mission grass (*Pennisetum polystachion*) and gamba grass (*Andropogon gayanus*) have invaded vast areas of open *Eucalyptus* savanna and feral cats have been implicated in the decline of some of the region's native mammals. Feral water buffalo and pigs have wreaked environmental havoc in the region. The cane toad (*Bufo marinus*), originally introduced into Queensland to control pests affecting sugar cane, poses a particularly serious threat to this habitat. With no natural predators, the cane toad population has increased and spread rapidly, and its voracious appetite for grubs and beetles is eliminating the prey of the native wildlife.

Fragmentation is a growing problem for grassland and savanna ecosystems, even in eastern Africa, where much savanna habitat has been preserved in parks and other protected areas. As human populations expand, viable corridors between these areas, especially the smaller reserves, are being lost, impeding the movements and seasonal migrations of animal species. Tropical savanna regions are also at risk from the felling of trees for fuelwood and charcoal production. The charcoal, made from many of the woodland tree species, is used for cooking as well as for drying tobacco – a form of dryland agriculture that is increasing in eastern Africa.

Ecotourism is a growing threat to the world's remaining wild grasslands. While the attraction of visitors can be an important source of revenue in developing nations and an incentive to preserve plants and animals, ecotourism can also add to the pressure on these habitats. Increasing numbers of tourists are visiting Africa's savannas to see the massive herds of gazelles, wildebeest, zebras, and other grazing mammals, stalked by cheetahs, lions, and leopards. With them, these visitors bring increased pollution and disturbance of the habitat. In addition, in the eastern Himalayan region, some tourists remove native plants, and trekking staff often cut slow-growing shrubs for firewood. In tourist locations around the Tibetan plateau – one of the most impressive alpine landscapes on the planet – snow leopard pelts are displayed for sale.

GRASSLAND GARDENS

Like their wild counterparts, grassland gardens are dominated by sun-loving grasses and herbaceous wild flowers; savannas are also dotted with trees. In native grasslands, different communities of species are found in varying dry to moist and cold to hot conditions. Another set of plants that tolerate some shade thrives under the dappled sunlight found below trees in tropical and temperate savannas. To create complexity and interest in a grassland garden, select a variety of grasses, wild flowers, and, in the case of a savanna garden, trees, that suit the conditions on the site. Removal of existing vegetation, often turf or pasture with persistent perennial weeds, is usually necessary before planting. Once they have been cleared, small sites can be planted out with potted or bare-root specimens, while larger areas are most economically started by seed.

GARDENER'S GUIDE

Grasses form the matrix of the garden. To mingle with them, choose herbaceous wild flowers, including bulbs and orchids, that provide continuous bloom. To discourage irresponsible collection, purchase plants that have been nursery propagated, or those harvested sustainably from the wild as part of an official conservation plan.

Festuca (see p.270)
Allium (see p.224)
Angelicum pachycarpa (see p.167)
Crinum (see p.229)
Echinacea (see pp.176–7)
Lupinus (see pp.186–7)
Dactylorhiza (see p.398)
Plantathera (see p.413)
Ophrys (see p.408)

▲ RUDBECKIA MAXIMA *The cabbage-leaf coneflower produces summer blooms on tall stems, and impressive clumps of large, silvery blue-green leaves.*

PRAIRIE GARDEN ► *This P. Clifford Miller garden in Illinois is planted with native grasses and wild flowers, with aspens for shade.*

WET GRASSLANDS

Marshes, bogs, fens, and salt marshes are some of the most productive ecosystems on the planet. Like swamps, their wooded counterparts, they are sprinkled throughout the world's major biomes. These wetlands are nature's flood-control systems, absorbing the vast volumes of water unleashed by storms. Without wetlands, groundwater would not be recharged, and aquifers would not be replenished. Wetlands also filter sediments and break down pollutants that wash off drier lands.

WETLAND WILDLIFE

Areas with standing or flood water for at least part of the year, and that support the growth of emergent and aquatic plants, are known as wetlands. The plants that grow there have evolved special root systems and other adaptations to enable them to live at least partially submerged in water for extended periods.

When the lush vegetation that grows in wetland areas dies back and decays, it releases nutrients into the water. This nutrient-rich water teems with a vast array of animals that feed on the decaying vegetation, and these animals form the foundation of an important food web that includes insects, fish, turtles, frogs, and other amphibians, and larger animals such as mink and otter. Wetlands are also a crucial habitat for a wide variety of bird species, including waders and wildfowl.

▲ MOIST SAVANNA
Seasonal flooding in tropical savannas forms wetland habitats interspersed with grasslands, patches of trees or open woodlands, and so-called gallery forests that line streams and rivers.

WETLAND GARDENS

Wetland plantings can be modelled on their various wild counterparts. Marshes are usually open and sunny, with shallow water covered by herbaceous vegetation. A true bog is fed by groundwater and characterized by highly acidic soils, poor drainage, and sphagnum mosses. Streamside, or riparian, wetlands accommodate plants along their banks or even in running water. Wetland plants typically are arranged in a series of concentric zones determined by water depth; these can be re-created to increase diversity in a wetland garden. The wet meadow, where wetland grades into upland, is dominated by herbaceous plants and occasionally trees or shrubs. Next is the emergent zone, dominated by soft-stemmed, herbaceous plants, such as cattails, that grow partially in water. Submerged aquatics and floating plants like waterlilies and pondweeds grow in the deeper water.

▲ NUPHAR LUTEA
The yellow pond lily or spatterdock floats its leaves on the surface of still water, and blooms only in full sun.

NATIVE PLANTING ►
This wetland garden in south-west England, created by landscape designer John Brookes, features indigenous white waterlily.

GARDENER'S GUIDE

Selecting plants for the garden can have serious consequences for native wetlands. To protect these vital ecosystems, do not grow invasive or potentially invasive wetland plants. Many specialized wetland denizens, such as Venus flytrap and pitcher plants, are threatened by overcollection in the wild, so purchase only nursery-propagated specimens.

Nuphar (see pp.190–1)
Dionaea (see p.430)
Sarracenia (see p.438)
Drosera (see pp.432–3)
Typha (see p.278)
Illecebrum (see p.183)
Amsonia (see p.166)
Franklinia (see p.108)
Lindera (see p.112)
Liquidambar (see p.112)

POLLUTION ► *Wetland ecosystems are threatened by the enormous volumes of chemicals running off agricultural land and paved surfaces, and by industrial effluent, as shown here pouring into the Silbersee, Germany.*

Marshes are open wetlands dominated by herbaceous plants. Emergent plants, such as cattails, bulrushes, grasses, sedges, and water-tolerant forbs, are best suited to persistent flooding; just enough of the leaves remain above the water to photosynthesize sugars for the entire plant.

Areas of coastal marsh that are frequently or continually flooded by tidal water can be brackish – partially salty – or fully saline. Salt marshes are bursting with wildlife, inluding fish, shellfish, and other marine organisms.

Peatlands form in glacial lake beds in regions where temperatures are too cold for rapid decay of organic matter. The organic matter, called peat, accumulates faster than it decomposes, filling the wetlands over time. Bogs are specialized, acidic peatlands derived mostly from species of sphagnum moss. Fens are peat-forming wetlands that typically receive nutrients from the flow of minerals through surrounding soils and from groundwater.

THREATS TO WETLANDS

Some 84 per cent of the wetlands safeguarded by Ramsar, the international convention on wetlands, have undergone or are threatened by ecological damage – and these are wetland areas that are relatively well protected. Estimates suggest that the world may already have lost half of its wetlands, and the remaining half are under intense pressure from human activities.

Drainage of wetlands for agricultural production has been the major cause of their demise. By 1985, according to Ramsar documents, up to 65 per cent of available wetlands had been drained for intensive crop cultivation in Europe and North America. Loss of wetland habitats in Asia has been under way for thousands of years; vast areas have been converted to ricefields alone. Pollution and damage from the enormous volumes of stormwater that run off paved surfaces threaten wetlands in urbanized regions. Wetland destruction has been less dramatic in tropical and subtropical areas, but is now increasing at a rapid and potentially disastrous pace.

The biodiversity of wetlands around the world is being destroyed or degraded by invasive species. The Australian melaleuca tree (*Melaleuca quinquenervia*), for example, was introduced into Florida in the early 1900s for horticultural use. Over the past 30 to 40 years, this species' prolific seed production, adaptation to fire, tolerance of flooding, and lack of natural competitors have enabled stands of melaleuca to explode in freshwater wetlands in south Florida, including the Everglades. Purple loosestrife (*Lythrum salicaria*), which was introduced for horticultural use in the early 1800s, has since spread throughout the temperate regions of North America, colonizing wet meadows, marshes, river banks, and the shores of lakes and ponds, and disrupting the native plant communities in these habitats.

▲ RAISED BOG *Bogs are poor in nutrients and generally form in basins with no inlet or outlet for surface water. The lack of fresh water causes the acidity to increase to a level that is toxic to all but adapted plant species.*

▼ CARNIVOROUS PLANTS *In nutrient-poor bog environments, some plants supplement their diet by feeding on insects and other small prey. This sundew* (Drosera intermedia) *has trapped a white-legged damselfly.*

TYPES OF WET GRASSLAND

OKAVANGO DELTA, BOTSWANA
One of the largest and most ecologically important wetland areas of the world, the Okavango delta is a maze of floodplains and reed- and Eucalyptus*-lined waterways and rivers. Much of the delta is permanently covered with water, while other areas are periodically swelled by seasonal flooding. Surrounded by the parched plains of the Kalahari desert, these wetlands are a magnet for animals and birds.*

FENLAND, EASTERN ENGLAND
Fens are peatlands, but are less acidic than bogs and have higher nutrient levels. These wetlands harbour a distinct group of calcium-loving plant species called calciphiles, such as grass-of-Parnassus (Parnassia *species), often grown in gardens for their large, solitary, saucer-shaped white to pale yellow flowers. Fens support a rich diversity of insects, which in turn attract insect-eating birds.*

SALT MARSH, NORTH-WEST ENGLAND
Salt marshes are coastal wetlands found in estuaries and backwaters, behind barrier beaches, and along bays and inlets. They are subject to dramatic, twice-daily water-level fluctuations with the high and low tides. Grasses dominate this habitat, but a variety of sedges, glassworts (Salicornia *species), and other forbs are also found. Salt marshes are vital breeding grounds for many species of bird.*

MEDITERRANEAN

There are only five small Mediterranean-climate regions around the globe, but they harbour about 20 per cent of the plant species on the planet. In fact, they are almost as rich in species as tropical rainforests. The most distinctive plant communities that have evolved in the cool, wet winters, warm to hot, dry summers, and poor soils of these regions are low-growing, fire-adapted, and drought-tolerant shrublands. Each of the Mediterranean-climate regions is a great distance from the others and has evolved its own unique assemblage of plant species. Many of these plants are found nowhere else in the world, and are now in great danger of extinction.

DISTRIBUTION *Mediterranean-climate regions cover just one to two per cent of the land surface of the Earth. They occur between about 30° and 40° latitude, on the west coasts of continents and to the east of cold ocean currents that generate winter rainfall. More than two-thirds of the total Mediterranean biome is found in the Mediterranean basin, from the Pyrenees to Crete to the Taurus Mountains, but its unique vegetation is echoed in the Chilean matorral, South African fynbos, Australian mallee, and Californian chaparral.*

Mediterranean regions

MEDITERRANEAN ECOLOGY

The shrubs and trees characteristic of Mediterranean habitat areas are well adapted to this unique climate. Mediterranean species have a relatively short growing season during the cool part of the year when moisture is most abundant. The luxuriant growth of spring is followed by summer drought and dormancy, when the landscape looks dry and barren.

Fire is a major ecological factor in Mediterranean-climate regions, and many plants that grow in this habitat are adapted to fire, with protective bark or the ability to regenerate from base roots or bulbous structures below ground, where they are relatively safe from the heat and the flames. Some species are dependent on fire to release seeds in closed cones or to penetrate special seed coats. Fire also stimulates many bulbous plants to increase or flower.

Many typical members of the Mediterranean shrub flora are aromatic – including the culinary standards sage, rosemary, and thyme. The leaves of these pungent species contain highly flammable, aromatic oils that help to protect the plant against water loss, but also exacerbate the spread of fire.

◄ **LAUREL** *Plants in Mediterranean-climate areas are typically low-growing, evergreen shrubs that have small, leathery (sclerophyllous) leaves with thick cuticles.*

▼ **MAQUIS SHRUBLAND** *There are two main types of shrubland in the Mediterranean region itself: maquis* (below), *with dense communities of short evergreen shrubs and olive and oak trees, and garrigue* (see facing page).

TYPICAL PLANTS OF MEDITERRANEAN HABITATS

OLIVE TREE *The European olive is typical of sclerophyllous (hard-leaved) woody plants, often with silvery foliage, that have adapted to the long, dry summers.*

PROTEA REPENS *The showy flowers of shrubs, such as this South African sugarbush, attract nectar-feeding birds, helping to sustain the rich birdlife of this habitat.*

BANKSIA *Many Mediterranean-climate shrubs, such as this Australian* Banksia, *are fire-adapted, regrowing from the base and often releasing seed in response to fire.*

CISTUS SALVIIFOLIUS *The seeds of this Mediterranean shrub have extremely hard outer coats which must be ruptured by fire before they can germinate.*

YELLOW MARIPOSA *Bulbous plants, such as this California lily, thrive in Mediterranean areas, surviving dry spells and fire as bulbs under the ground.*

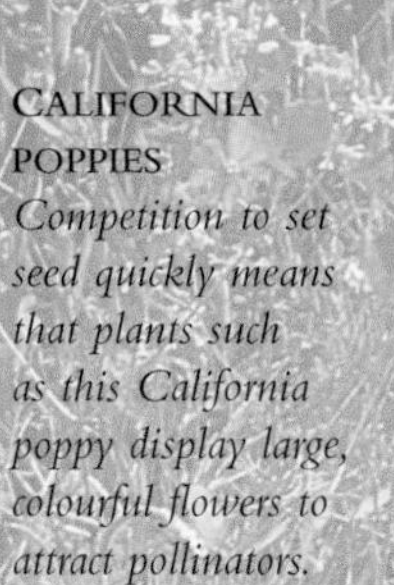

CALIFORNIA POPPIES *Competition to set seed quickly means that plants such as this California poppy display large, colourful flowers to attract pollinators.*

Coastal influence
Plants of Mediterranean regions have evolved to tolerate the desiccating, salt-laden winds that blow in from the sea; strategies include low, ground-hugging growth, wiry stems, and small, leathery leaves.

Sloping sites
Nearly all regions in Mediterranean-climate zones lie between the sea and a range of mountains, the latter precipitating the winter rains that define the climate.

Binding poor soil
Thin, shallow, fast-draining soils over rock are typical, and plants develop wide-ranging, intertwining root systems to anchor themselves to stony hillsides. They protect the land from the constant erosion caused by winds and winter rainfall, and provide themselves with a large area from which to draw water.

Floral richness
Late spring is a time of simultaneous flowering for many plants; the mild wet winters encourage growth to start early in the year. Many are annuals, surviving the summer drought as seeds; perennials aim to complete the energetic process of reproduction by the time the dry season begins.

Garrigue shrubland
Typical garrigue habitat (here in Greece) is an open mixture of heaths and aromatic shrubs, such as lavender and thyme.

MEDITERRANEAN GARDENS

The gardens best suited to Mediterranean-climate regions, like their counterpart plant communities in the wild, tend to be mounding mosaics of drought-tolerant shrubs and small, shrubby trees, with a smattering of bulbs and herbaceous wild flowers. Plant choices for a Mediterranean garden should be based on the particular conditions of the site: some species thrive on hot, dry slopes, while others prefer somewhat moister, cooler exposures. Group together the taller shrubs and trees to create keynotes in the garden. Space them widely enough so that when they are mature they will have enough soil area from which to draw water and nutrients; trees and shrubs are best planted out as bare-root or container-grown specimens. Other species, particularly annual wild flowers, are easily started from seed; sow them in late summer or fall, just before the rains begin.

▲ LAVANDULA *One of the best-loved Mediterranean shrubs, lavenders have long been used in potpourris, flower arranging, and cooking, as well as in the garden.*

ANCIENT DESIGN ► *Olive trees, cypresses, and aromatic shrubs in terraced hillside plantings have been the classic elements of Mediterranean gardens for centuries.*

GARDENER'S GUIDE

Gardeners in the world's Mediterranean regions are blessed with a wealth of beautiful plants. Distinctive flowering shrubs and small trees are the backbone of Mediterranean-climate gardens. Tuck appropriate grasses, aromatic herbs, and bulbs between islands of woody plants, and sow seeds of colourful annuals for even greater diversity and visual impact.

Protea (see p.116)
Banksia (see p.95)
Babiana (see p.225)
Brodiaea (see p.226)
Arctostaphylos densiflora (see p.94)
Eucalyptus rhodantha (see p.106)
Cotoneaster granatensis (see p.99)

▲ URBANIZATION *The growth of sprawling coastal cities is squeezing the ecologically important Mediterranean habitat areas that surround them. The rapid expansion of Cape Town, for example, which lies on the far south-western tip of Africa, is threatening the South African fynbos.*

THREATS TO MEDITERRANEAN CLIMATE REGIONS

The world's Mediterranean-climate regions have not only been favoured by a rich diversity of plant life, but have also been a favourite habitat for humans. As a result, the world's smallest biome, and one of its most biodiverse, is also the most highly altered on the planet. Few undisturbed areas remain.

For millennia, the Mediterranean basin has been the site of intensive human activities. Little of the region remains intact. The vegetation has been cleared for agriculture – the quintessential symbol of the Mediterranean, the olive tree, has been cultivated since about 3,000 BC. In some areas, timber cutting has had a major impact. In the early 20th century, for example, large expanses of marginal land in Spain and Portugal were cleared for agriculture. When crops proved disappointing, the lands were transformed into tree plantations, notably *Eucalyptus*. Today, Portugal has the world's second largest eucalypt plantation. The continued cultivation of these trees to provide timber and pulpwood threatens the long-term productivity of the land. Native plant communities throughout the Mediterranean basin have also been severely overgrazed, have given way to cities, and are under pressure from tourism. In the eastern Mediterranean, bulbs have been dug up for the international horticulture trade, damaging wild populations of snowdrops, tulips, and other species.

Mediterranean climate regions are threatened by rampant urbanization and its associated problems, nowhere more so than in the chaparral region of the California coast. For decades, millions of Americans have been flocking to make their home in this beautiful location. Situated on some of the most prized coastal land on the planet, coastal sage chaparral – a particularly diverse and globally rare habitat type – is threatened by human development. Only an estimated 15 per cent remains intact, and it is now one of the most endangered plant associations in North America. What little coastal sage scrub remains is extremely fragmented and surrounded by

▼ FIRE MANAGEMENT *Mediterranean vegetation depends on regular fires to maintain the ecological balance of the habitat. Many plant communities are at risk because, as human habitation inches ever closer, shrubland fires are extinguished to safeguard homes.*

areas of intensive development, preventing effective dispersal of most species. This habitat is further threatened because fires are often controlled to prevent damage to homes and other property, changing the natural patterns that many native chaparral species need to survive. Suppressing fire can also encourage invasion by non-native plants. At the same time, pet as well as feral cats and dogs take a terrible toll on birds and other wildlife struggling to survive in the few remaining chaparral fragments.

The greatest threat to the Chilean matorral is the burgeoning population of central Chile. With more people have come more roads, and the clearing of large areas to be used for growing crops and grazing livestock. Fire, which is sometimes used to clear the land, has destroyed a large part of the Chilean matorral. As a consequence, hundreds of plant species are now characterized as rare or endangered.

Put thousands of flower shops together and they still would not come close to the diversity of plant life in the fynbos region of South Africa. Like most Mediterranean-climate regions, however, the region has been plagued by invasive non-native plants. These species not only crowd out native species, but also transform natural communities in ways that make it even more difficult for the indigenous species to survive. The typical fynbos community consists of a variety of shrubs in the heath family growing below proteas and other broader-leaved shrubs, but it has been severely affected, for example, by acacias introduced from Australia. These species create large amounts of leaf litter, adding an unnatural amount of organic matter to the region's poor soils. At the same time, acacias fix nitrogen, enriching the soil even further. Although the acacias benefit from the richer soil, native fynbos species are not adapted to using the extra nutrients. This gives the invasive species an even bigger advantage and allows them to out-compete the native plants. Urbanization is also a serious threat to the fynbos, especially in the two metropolitan areas, Cape Town in the west and Port Elizabeth in the east, and along the coast.

Global climate change poses a serious, long-term threat to Mediterranean-climate regions. It is likely to have an adverse effect on the biodiversity of the fynbos, particularly on the numerous endemic species with particular habitat requirements. In addition, global warming could endanger Australia's species-rich Mediterranean shrublands as the desert spreads southwards, reducing the area under the influence of a Mediterranean climate.

DIVERSE BIRDLIFE *Mediterranean climate areas teem with wildlife. Cape sugarbirds* (below) *feed on the nectar of the large, colourful flowers of proteas in the South African fynbos, and play an important role in their pollination. California condors* (right) *were once a spectacular sight over the California chaparral, but are now an endangered species.*

▼ PRESERVING KEY HABITATS *The National Botanical Garden at Kirstenbosch, South Africa, is a guardian of the spectacularly diverse fynbos flora of the Cape region. About 8,500 plant species grow in the fynbos; nearly 70 per cent of these do not grow anywhere else.*

TYPES OF MEDITERRANEAN HABITAT

CHAPARRAL, CALIFORNIA
West of the Sierra Nevada mountains, California is covered by coastal sage chaparral and foothills chaparral. Coastal chaparral is characterized by low, aromatic, and drought-deciduous shrubs such as black, white, and Munz's sage (Salvia mellifera, Salvia apiana, *and* Salvia munzii). *Foothills chaparral is dominated by a variety of woody shrubs such as chamise* (Adenostoma fasciculatum).

MATORRAL, CHILE
Central Chile's Mediterranean vegetation, as that of California, is confined to the coast by high mountains. Called matorral, it includes more deciduous species than California's chaparral, as well as many cacti and other thorny species. One tree, the huge Chilean palm (Jubaea chilensis), *has an extremely restricted range and is threatened with extinction.*

MALLEE SCRUB, AUSTRALIA
Southern and south-western Australia's mallee shrubland is the second most diverse Mediterranean region. The vegetation is dominated by pungently scented Eucalyptus *species, including the red-flowered mallee. Many magnificent wild flowers are also found in the region, including kangaroo paw, which has velvety, tubular flowers that provide nectar for the mouse-sized honey possum.*

DESERTS

Deserts are places of extremes. Most receive very little, if any, rain each year. In some coastal deserts, plants and animals must survive solely on moisture from fog that rolls in from cold offshore ocean currents. Besides being dry, deserts endure intense sunshine, and greater daily temperature swings than any other terrestrial habitat. Rain – when it comes – typically falls in brief but torrential downpours, while strong winds drive sand and grit almost horizontally through the air. Although the vegetation is sparse, deserts are home to some of the world's most bizarre and beautiful plants, which have evolved a host of characteristics to enable them to live in these extreme environments.

DISTRIBUTION *Deserts cover about a fifth of the land surface of the Earth. Nearly 10 million sq km (3.9 million sq miles) is true desert, where rainfall is extremely low, typically less than 20cm (4in) a year. The great deserts of the world are all located in the zones of atmospheric high pressure that flank the tropics. The Sahara, which extends from the Atlantic coast of Africa to the Arabian peninsula, is the largest desert in the world.*

Desert regions

Extensive areas of southern Africa, as well as northern Africa, are covered by desert. Considered the world's oldest desert, the Namib stretches along the coast of Namibia, sloping towards the Atlantic in a sea of giant red sand dunes. The adjacent Succulent Karoo Desert, which extends into South Africa, includes an astonishing variety of succulents and bulbous species, many found nowhere else.

The four major deserts of North America stretch from Mexico north into the south-western USA. The Sonoran Desert is distinguished by its forests of tall cacti and by the ocotillo (*Fouquieria splendens*), a tall spindly shrub with red, cone-shaped flowers that bloom after the first spring rain.

The Atacama and Sechura Deserts, considered the world's driest, form a narrow band that stretches 2,100km (1,300 miles) between the Andes and the Pacific Ocean in Peru and Chile. In some areas of the Atacama, rainfall has never been recorded.

▲ THORNY DEVIL *Water conservation is the key to survival in the harsh desert environment. When rain or dew lands on the back of the thorny devil* (Moloch horridus)*, grooves in its thick skin channel the water to the corners of its mouth.*

Much of the desert is barren, but where isolated mountains or steep coastal slopes intercept the clouds, unique fog-zone plant communities can be found.

No other desert in the world experiences the combination of very cold winters and blistering hot summers characteristic of the Central Asian Deserts, which are dominated by communities of shrubs and semi-shrubs, each adapted to a specific soil type.

▼ SALT DESERT *The dry rock and snow-white salt lakes in the Atacama Desert represent the absolute limits of where life is possible. Salt lakes form when water evaporates faster than it is replaced by rainfall, increasing the salt concentration.*

TYPICAL PLANTS OF DESERT HABITATS

LIVINGSTONIA PALM TREE *Many desert trees, including some palms, have very long roots that probe deep into the earth to tap underground sources of water.*

BARREL CACTUS *The broad stems of cacti are highly adapted for water conservation, with a fleshy central core to store water and thick, waxy skin to prevent water loss.*

AGAVE *The thick, succulent leaves of agaves not only store water, but are sharp and pointed, affording the plant protection from foraging animals.*

EUPHORBIA *Like cacti, the euphorbias are stem succulents: they collect and store water in their stems and roots, enabling them to survive long periods of drought.*

WELWITSCHIA *The extraordinary gymnosperm* Welwitschia mirabilis *is found only in the Namib Desert, and can live for many thousands of years.*

BABIANA SPECIES *Bulbous plants, such as members of the lily family, survive drought as a dormant bulb under the ground, only sprouting when water is available.*

Dry atmosphere
With little rainfall, the desert atmosphere has a very low water content. Dry desert winds intensify the aridity and sweep across the landscape, sometimes building up into major dust and sand storms that can reshape dunes and erode rock.

Specialized plants
Plant life in the desert is not only adapted to survive the harsh climate, but also to afford protection from grazing animals. Some species hold their leaves high out of the reach of herbivores, while others arm themselves with sharp spines or pointed leaves.

Poor soil
Desert soils are coarse and gravelly, because lighter material, such as sand, is picked up and carried away by the wind. They may look barren, but many are full of the seeds of annuals and perennials waiting for the next rainfall before bursting into life.

Little water
With water at a premium, the roots of desert plants compete to soak up as much precipitation as possible, so plants are typically widely spaced apart. Cacti, for example, have extensive but shallow root systems, allowing the capture of water after even the lightest rain shower.

Namib Desert
Vegetation is sparse in the Namib, but includes lichens and drought-tolerant shrubs, such as this large, mature Aloe dichotoma.

TYPES OF DESERT

SONORAN DESERT, SOUTH-WESTERN USA
The slow-growing, many-armed saguaro cactus (Carnegia gigantea) *is endemic to the Sonoran Desert. This desert receives more rainfall than any other desert in North America, and is well-known for its abundance of cacti and other succulents. It is also rich in animal life, including rare desert tortoises and poisonous Gila monsters, which move slowly across the sand.*

EASTERN PAMIR MOUNTAINS, TADZHIKISTAN
The Central Asian Deserts are characteristic of "cold" deserts; after the hot summers, they experience very cold winter temperatures. These regions are extremely dry, but, when it does fall, the main form of precipitation is snow. The gravelly deserts of the Pamir Mountains region are rocky and bleak, and plant life is sparse. These barren lands are home to the endangered snow leopard.

GREAT SANDY DESERT, WESTERN AUSTRALIA
Australia's desert, which includes the Great Sandy, Gibson, and Great Victoria – collectively known as the Great Western Desert – covers 44 per cent of the continent and is the world's second largest after the Sahara. The characteristic red sand plains are dotted with clumps of tough spinifex grass (Triodia *species*)*, saltbush shrubs* (Atriplex *species*)*, and patches of desert oak* (Allocasuarina decaisneana).

THREATS TO DESERT HABITATS

Agriculture and the overgrazing and trampling caused by domestic livestock are the principal threats to the planet's deserts. Vast areas of native vegetation in North America's largest desert, the Chihuahuan Desert, have been replaced by cattle farms and other forms of agriculture. Conversion of lands for cattle, goat, ostrich, and sheep ranching has also damaged fragile areas in the Namib-Karoo-Kaokoveld Deserts ecoregion of southern Africa. The native bastard quiver tree, one of the world's largest aloes, is now considered endangered due to overgrazing by goats, which prevents the species from reproducing. Livestock grazing is particularly destructive because it not only harms sensitive plant communities that grow in these arid conditions, but also vegetation along streamsides and water holes, which are critical habitats for birds and other desert wildlife.

Ironically, overgrazing not only damages deserts, but also expands them. Cultivation of marginal land and overgrazing by an estimated three billion cattle, sheep, and goats are transforming drylands around the globe into deserts. In the Sistan basin in Afghanistan and Iran – a former oasis that as recently as several years ago supported at least a million cattle, sheep, and goats – windblown dust and sand from overgrazed pastures have literally buried more than a hundred villages. Desert margins are particularly at risk, but this process, called desertification, threatens as much as one-third of the Earth's land area in 110 countries.

INTRODUCED WILDLIFE

Invasive non-native plants and feral animals add to the pressure on desert flora and fauna. Introduced plants, such as tamarisk trees and buffel grass, are driving out native species in the Sonoran Desert in Arizona and Baja, Mexico. Several *Tamarix* species, imported into the USA in the early 1800s for horticultural use, escaped cultivation and now grow along the banks of streams in dense, impenetrable tangles up to 9m (30ft) tall. They not only displace native river-bank plants, but also consume more water, drying up desert oases by lowering the water table. Introduced animals, such as cats, foxes, and

▲ **FLOWERING DESERT**
When rain falls in the Chihuahuan Desert (above)*, the barren landscape bursts into bloom. The seeds of ephemeral plants, such as the desert paintbrush,* Castilleja chromosa (left)*, respond rapidly to rainfall, some putting up flower stalks within only a few days.*

camels, are all serious concerns in Australia's Great Sandy-Tanami Desert, a vast wilderness of red sand plains and high dunes. Wild camels roam Australia's arid interior, damaging and degrading the desert vegetation. Imported from Afghanistan and widely used by early explorers and pioneers, they were replaced by motor vehicles in the 1920s but strayed or were set loose. The introduction of predators has seriously threatened some native Australian animals. The brush-tailed bettong (*Bettongia penicillata*), for example, was once widespread but has been hunted by introduced foxes and cats and is now an endangered species.

SPREADING CITIES

Urban sprawl is becoming an increasing threat in some desert areas, including the Sonoran and Chihuahuan Deserts and the Atacama-Sechura Deserts of South America's Pacific coast. In this latter region, the human population is large and growing. Cities are expanding and new roads are encroaching ever further into the desert, posing an increasing threat to its rare and endemic plant species. Two of the largest and fastest-growing American cities, Phoenix and Tucson, are destroying and fragmenting ever-larger areas of the Sonoran and Chihuahuan Deserts. Aggravating the problems near populated areas are off-road vehicles, which tear up dryland vegetation. Tourism is increasing, bringing with it increased pollution and disruption of the habitat.

Irrigation projects that divert water for use by urban populations as well as farms also take their toll on the desert environment. For centuries, rivers that flow through the deserts of Central Asia have fed the

Aral Sea, once the fourth-largest inland body of water in the world. But so much water has been diverted from the Amu Darya and Syr Darya rivers for cotton farms and other kinds of agriculture that the lake is drying up.

Plant poachers

The striking traits that give cacti and other succulents the ability to survive in desert environments also make them irresistible to collectors and attractive to gardeners trying to save water and money by landscaping with desert plants. According to a recent study by TRAFFIC North America, an organization that monitors trade that threatens plants and animals, between January 1998 and June 2001, nearly 100,000 cacti, agaves, yuccas, and other succulents were harvested from mostly wild populations in the state of Texas, USA, or were illegally imported into Texas from Mexico. These plants were destined for consumers in cities such as Phoenix and Tucson. The study concluded that this illegal trade is a potential threat to some cactus species and to their natural populations in the Chihuahuan Desert, home to almost 25 per cent of the cactus species known to science. Private collectors, who represent a small but significant part of the problem, are driven by a desire for the rarer and newly discovered species, particularly from Mexico. Showy species, such as hedgehog cacti (*Echinocactus* species), barrel cacti (*Ferocactus* species), prickly pear cacti (*Opuntia* species), and ocotillo (*Fouquieria splendens*), are most at risk. Poaching of wild plants is also a serious problem in southern Namibia and South Africa's Succulent Karoo. Recognizing the threat unbridled international trade poses for these plants, CITES protection has been extended to many succulents, including cacti, aloes, agaves, euphorbias, and pachypodiums.

▲ CHANGING LANDSCAPE *The effect of desertification is highlighted in this area of the Guadaloupe Mountains, Texas. Overgrazing has transformed the land to the right of the fence to desert, a stark contrast to the ungrazed grassland to the left.*

▼ ARABIAN ORYX *Hunted to extinction in the wild in 1970s, this desert antelope* (Oryx leucoryx)*, which once lived in the Arabian peninsula, has now been successfully reintroduced in Saudi Arabia.*

Desert gardens

Desert gardens are defined by intense sunlight, temperature extremes, and lack of water. It is best to fill them with drought-resistant shrubs, and cacti and succulents, which have adapted leaves or stems, or other adaptations that enable them to conserve water and survive extended periods of drought. Annual wild flowers in the garden, as in the wild, bloom and set seed quickly after rainfall. Microclimates are especially important in arid regions: relatively shady or moist spots, such as along desert washes, support plants which are quite different from those in the sunniest and driest areas. Desert landscapes do not have thick carpets of plants – typically only about 25 per cent of the ground is covered with vegetation – so pay special attention to "negative" or unplanted spaces, choosing an inorganic mulch that enhances the setting.

GARDENER'S GUIDE

Select a mix of desert plants: true xerophytes, such as succulents, that need no supplementary water once established; annual wild flowers that bloom after rainfall; and drought-resistant trees and shrubs. Many desert plant species are threatened by overcollection, so purchase only nursery-propagated plants.

Agave (see pp.330–1)
Aloe (see pp.332–3)
Beschorneria (see p.335)
Ceropegia (see p.335)
Echinocactus (see p.340)
Euphorbia (see p.342)
Lithops (see p.345)
Mammillaria (see p.346)
Yucca (see p.355)

▲ DESERT SPECIMEN Agave victoria-reginae *is a striking specimen plant, but place it away from patios and pathways because it has very sharp spines.*

NATURAL INSPIRATIONS ► *Bold and sculptural desert plants, such as prickly pears, feature in this garden by designer Philip Van Wyck, near Tucson, Arizona, USA.*

PLANT ENCYCLOPEDIA

TREES AND SHRUBS

These broad-leaved, evergreen or deciduous flowering plants have a permanent woody framework. They live for tens to hundreds of years and are indispensable to all air-breathing creatures, producing much of the oxygen needed to sustain life. Their living biomass is the primary engine driving all the major systems that make Earth habitable, and is critical to the carbon, nitrogen, and water cycles. Humans have destroyed half the world's forests, but trees and shrubs still make up over 70 per cent of Earth's plant life.

WHERE AND HOW THEY GROW

A tree has a single stem that branches to form a crown, while a shrub has many stems arising from the base. In practice, this distinction is often blurred: shrubs may develop a single stem, and trees can become multi-stemmed if they repeatedly lose their leading shoot. Both grow in environments from benevolent to extreme, from lush rainforest to arid semi-desert, from tropical to boreal zones.

The greatest biomass of woody plants is found in rainforests (*see pp. 66–67*), and over two-thirds of all rainforest plants are trees. In deciduous forests, light reaches the ground under the trees after leaf fall and shrubs are able to form a layer beneath the canopy. Shrubs are often seen at the woodland or forest margin as well.

Shrubs and trees are also found in scrub and savanna. On mountains, the plants at the tree line are usually conifers, but woody shrubs, such as *Arctostaphylos* and *Rhododendron* species, grow higher up. In parts of the Himalaya, the white-stemmed birch tree, *Betula utilis*, is seen at even higher levels than spruces (*Picea* species) and pines (*Pinus* species).

Annual ring
Ray
Heartwood
Xylem tissue, conducting water and salts
Outer bark
Bark cambium
Phloem, the main food-conducting tissue
Cambium layer between the wood and the bark

▲ **WOODY STEM STRUCTURE** *As stems grow, a substance called lignin builds up in the tissues. This makes the stems woody and strong enough to support the mass of a large plant.*

ENDEMIC FOREST ► *Northern Madeira still retains its ancient laurel forest. This vegetation, found in north Africa and Europe two million years ago, is now unique to Atlantic islands.*

▲ **RACE TO THE LIGHT** *Strong, woody tissues allow trees to grow tall and reach light and air. Their crowns form the upper canopy in forests, as here in Monteverde, Costa Rica.*

◄ **AFTER FIRE** *Some eucalypts have woody tubers (lignotubers) in the ground. These enable the trees to survive fire and drought, and still regrow rapidly afterwards.*

SURVIVAL STRATEGIES

A permanent woody structure equips trees and shrubs for survival. Height is an advantage for trees in seeking light and air, and deep, wide-spreading roots maximize access to groundwater. Evergreen trees and shrubs can function all year round; deciduous types have a food store and plenty of growth buds, so they can lie dormant when conditions are cold or dry, but spring into growth and begin photosynthesis when conditions allow.

Woody plants have evolved to thrive in particular environments. Small, fleshy leaves with a waxy or leathery surface or a covering of fine hairs help plants to conserve moisture, and are commonly found on trees and shrubs in high mountains and arid places. Cucumber trees (*Dendrosicyos socotrana*) and baobabs (*Adansonia* species) store water in their vast and bulbous stems to sustain them during periods of drought. Some plants offer benefits to the creatures they depend on for pollination or seed dispersal; for example, bird-pollinated banksias and proteas are adapted to provide nectar to their pollinators.

Generally, more extreme environments contain more highly specialized plants, which are more vulnerable to environmental change.

ANCIENT WOODLAND *The shallow roots of beech* (Fagus) *prevent a lower layer of shrubs from developing. They also make these trees particularly vulnerable to climate change.*

◄ TREE DWELLER *Many birds live in holes in old tree trunks. They range from small insect-eating species to large predators, such as this tawny owl* (Strix aluco).

▲ LEAF LITTER *Deciduous trees and shrubs shed a huge volume of leaves each autumn. This covering returns nutrients to the soil and supports a host of fungi and micro-organisms.*

SUPPORTING LIFE

Woody plants provide shelter for smaller plants, including mosses and lichens, and support wildlife from microscopic fungi and tiny insects to birds and small mammals. Trees, in particular, are home to an incredible diversity of animal life, especially birds and insects.

On a larger scale, woody plants provide the oxygen that we breathe. They are at the core of the carbon cycle (*see p.20*), and play a critical role in bringing nitrogen, the essential building block of all proteins, into the food web. They are also crucial to the hydrological cycle (*see p.21*), maintaining the Earth's water in a state of dynamic equilibrium between evaporation, precipitation, and storage, and moderating water flow on land.

Their roots stabilize soils, especially on steep slopes, and prevent landslips and erosion. By moderating the pace at which rainwater percolates into the ground, woody plants reduce nutrient leaching and prevent the excessive surface run-off that causes floods. They ensure that underground water escapes into rivers and streams in an even flow, often sustained even during periods without rain. This also results in the steady replenishment of the Earth's groundwater.

▲ RHODODENDRON PONTICUM *This and many other plants reduce competition by allelopathy. Chemicals harmful to other plants are exuded from the roots, or are in the leaves, flowers, or fruits, which fall in a germination-inhibiting carpet around the plant.*

USES AND EXPLOITATION

Since prehistory, woody plants have provided us with shade and shelter; fuel for warmth; berries, fruits, and nuts to eat; and even medicines.

Timber was used in ancient buildings, and is still important in construction today. Wood has furnished our homes from the earliest rough tables, trunks, and stools in native timbers, to highly skilled work in tropical rosewood and mahogany. Since at least 3000BC, charcoal has been used in forging ironwork, producing the agricultural tools that revolutionized farming. Until the advent of steel, timber was the staple material of ship and boat building throughout the world, from English oak to Indian and Burmese teak. This has influenced cycles of deforestation and reafforestation for centuries. In 16th-century Europe, widespread fuel shortages followed felling to rebuild national fleets; concerns for a strategic reserve of timber prompted some of the earliest tree surveys.

There is no doubt of the potential of woody plants to yield new medicines, but no way to assess what will have been lost if the plants become extinct first.

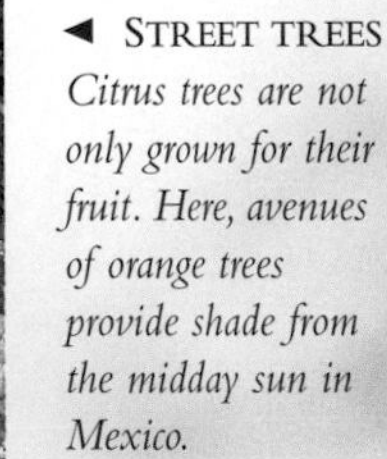

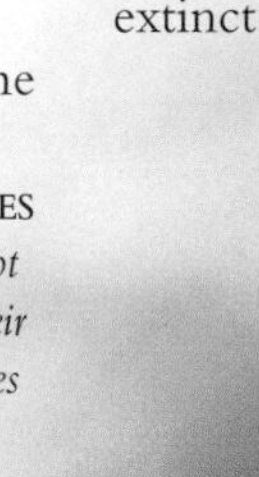

◄ STREET TREES *Citrus trees are not only grown for their fruit. Here, avenues of orange trees provide shade from the midday sun in Mexico.*

LATE BLOOMERS ► *Some woody plants, such as magnolias, take 20 or 30 years to flower and are vulnerable to any sudden changes in their habitat.*

WHAT ARE THE THREATS?

Like other plant groups, woody plants suffer from habitat loss due to building developments, mining, agriculture, or forestry. Non-native species may take up available water, or occupy cleared land and prevent natural regeneration. Suppression of fire near urban areas affects plants like proteas, which have evolved to depend on fire (*see p.78*). Some plants are overcollected for the horticultural or floral trades, or for medicines, dyestuffs, fuelwood, and timber.

If habitats become fragmented (*see p.34*), plants become separated from potential cross-pollinators, leading to inbreeding depression and lack of vigour. This is seen in timber reserves, where logging of the tallest, straightest trees has left lesser plants to reproduce, resulting in inferior offspring.

With climate change (*see p.35*), the boreal tree line is likely to shift, putting pressure on the taiga and tundra; warmer latitudes will suffer more drought. Rising sea levels will affect mangrove swamps; mountain species are likely to be lost as temperatures rise and water levels fall. The severity of these effects will depend on how well woody plants adapt to changed habitats, or how rapidly they are able to find suitable new living spaces.

▲ GOOD WOOD *These carvings for sale in Mombasa, Kenya, are from a People & Plants Good Wood project using sustainable crops of neem* (Azadirachta indica).

SUSTAINABLE USES

The Global Trees Campaign, a partnership between Fauna and Flora International (FFI) and the World Conservation Monitoring Centre (WCMC), operates a worldwide programme of field research. They survey, record, and monitor the status of the world's trees and forests, and make the most up-to-date scientific information available to governments and non-governmental organizations. They also create conservation plans to protect habitats and promote wise use of forest resources, and aim to educate and raise awareness in schools and involve local people in conservation programmes. In another joint venture, FFI and WCMC developed the Soundwood campaign. This brings together educators, scientists, musicians, and the music industry to improve the management of trees used for making musical instruments; they also produce teachers' packs for use in schools. Some of the world's threatened woody plants are protected in state and national conservation areas of various grades, reserves, national parks, and World Heritage Sites, but these often represent fragments of formerly more extensive habitats. The World Land Trust raises money to purchase land, managing it in reserves. They also research sustainable forest use that will benefit surrounding communities. They have helped to purchase and protect 300,000 acres of biologically important and threatened land around the world. This type of direct action may be one solution to habitat fragmentation.

▲ COPPICING *This old practice, here seen on sweet chestnut* (Castanea sativa) *in the UK, ensures a steady supply of wood, as the tree resprouts from the stump.*

GARDENER'S GUIDE

- Some woody plants are directly threatened by over-collection. *Davidia*, for example, has been taken from the wild for horticulture, and *Banksia* species are collected for the floristry trade.
- Trade monitoring of plants that are threatened by over-collection continues under CITES regulation. Gardeners can contribute by insisting on buying plants and floristry products of known, cultivated provenance.
- When buying timber products for the home or garden, first check whether you can find material that has been reused. Reclaimed floorboards or furniture made from recycled wood can be more attractive than new products. The timber used in household products, and the household's contribution to climate change, are more likely to be damaging forests.
- Choose new products that carry a Forest Stewardship Council (FSC) certificate. The FSC is active in 40 countries, and aims to promote sustainable forestry by providing principles and criteria for good management. These include benefits for the environment and indigenous people. The certificates are monitored by independent organizations, such as the Soil Association, and guarantee a "chain of custody". This ensures that the certified timbers do not become mixed with illegally logged material between the forest and the timber yard.
- If you have a charcoal barbecue, purchase charcoal from sustainable sources. Much of the charcoal sold is from tropical wood, and its production often destroys vast swathes of forest. Ask for charcoal that is FSC certified or from coppiced domestic sources.
- Future Forests organize forest planting and help companies to invest in cleaner or renewable energy sources. As part of their Carbon Neutral campaign, they have created a website that calculates your CO_2 emissions and advises you on how to reduce them.

Red List: Endangered*

Abeliophyllum distichum

Botanical family
Oleaceae

Distribution
Central Korea

Hardiness
Fully hardy

Preferred conditions
Full sun. Fertile and well-drained soil. A warm site with some protection from wind.

Commonly known as white forsythia, ***Abeliophyllum distichum*** is the only species in its genus. It is native to central Korea, where it grows on open hillsides in dry scrub vegetation. A pretty, multi-stemmed, deciduous shrub, it grows to 2m (6ft). The leaves are paired and oval in shape. The sweetly scented, four-petalled flowers are white, sometimes with a pink tinge, and are borne in dense clusters in late winter or early spring. The dry, one-seeded fruit (a samara) has an almost circular, wing-like appendage.

Urgent action is needed to prevent the extinction of this critically endangered shrub in the wild, but fortunately it is well established in cultivation. The species was first described in 1919. Seeds and live material were sent to the UK, and the species soon became popular when it became commercially available in 1937.

White forsythia is easy to grow and can be propagated by greenwood or semi-ripe cuttings, or by layering, in summer, or by seed. The shrub grows rapidly and does well trained up a wall.

▲ WINTER APPEARANCE *The flowers open from purple buds, covering the stems in profusion before the leaves appear.*

13*/3 on Red List

Acer

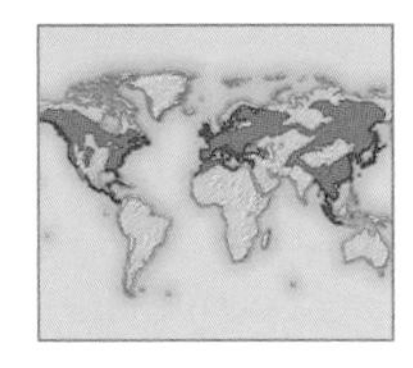

Botanical family
Aceraceae

Distribution
Europe, North Africa, Asia, North and Central America

Hardiness
Fully hardy – half-hardy

Preferred conditions
Full sun or partial shade. Fertile, moist, but well-drained soil. Least hardy are *Acer palmatum* and *Acer japonicum*, which require shelter from cold winds and root protection in freezing conditions.

The genus *Acer* consists of mainly woodland plants that grow as either large trees or understorey shrubs in temperate woods and forests. There are approximately 120 species and many are grown for their attractive foliage. The conservation status of many species in the wild remains poorly known.

Acer duplicatoserratum, Vulnerable in the 2000 Red List, is found in submontane forest on the island of Taiwan. Collection of whole plants for the horticultural trade has been one problem faced by this attractive species, which is is not currently protected in its native habitat. Another threatened species is *Acer pycnanthum,* which grows in Japan and is known in that country as *Hana no ki*. In the wild, this rare maple, Vulnerable in the Red List of 1997, is restricted to the small Tokai Floristic Region in the centre of Honshu island. Some large specimens have been designated "national natural monuments" by the Japanese government and are protected by fencing.

Acer pycnanthum grows in damp locations at an altitude of 400-500m (1,300–16,000ft). The branches are reddish brown to grey-brown and the flowers are red. The tree looks similar to its common North American counterpart, the red maple, ***Acer rubrum***, although it is usually smaller than that species. *Acer pycnanthum* is not as popular in cultivation as the red maple, famous for its brilliant autumn colours.

Several other acers have declined as a direct result of unsustainable use of their timber. They include the Vietnamese *Acer erythranthum* (classed as Lower Risk in the 2000 Red List), and the Himalayan *Acer caesium*, which is found to the west of China in open grazing grounds and as isolated trees in coniferous forest growing at altitudes of 2,400–3,800m (7,800–12,500ft).

See also Invasive plants, *pp.442–43.*

◄ACER RUBRUM *The red maple has a widespread natural distribution in North America. This fast-growing tree is also a familiar sight in European parks and gardens.*

6 on Red List

Adansonia

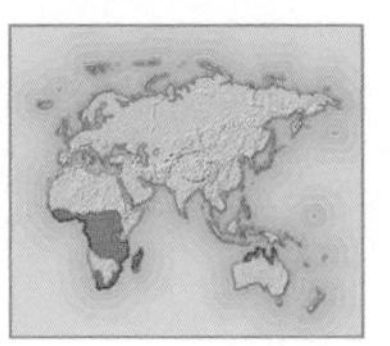

Botanical family
Bombacaceae

Distribution
Africa, Madagascar, Australia

Hardiness
Frost-tender

Preferred conditions
Full sun. Well-drained, sandy, moderately fertile soil. Adansonias require moderate rainfall in summer, less rain in spring and autumn, and dryness during the winter season.

Adansonias are the extraordinary and legendary baobab trees. They are sometimes known as the upside-down tree because the branches, when they are without leaves, resemble roots sticking up into the air. Another common name, the bottle tree, refers to the swollen trunk. In total there are eight species, all growing in tropical, semi-arid regions. One species, ***Adansonia digitata***, is widespread in Africa; six additional species are endemic to Madagascar; and one, ***Adansonia gregorii***, grows in Australia.

Three of the six Madagascan *Adansonia* species are considered to be globally threatened, and the remaining three are close to globally threatened status. The largest Madagascan species is

▲ **ADANSONIA GREGORII,** *like all baobabs, has a succulent and often swollen trunk that stores a large volume of water.*

Adansonia grandidieri, classified as Vulnerable in the Red List of 2000. Mature examples are now mainly found in degraded agricultural land, where regeneration is poor. Unusually, the flowers are pollinated by nocturnal lemurs – other Madagascan adansonias are pollinated by fruit-eating bats. The species is of considerable commercial importance because of the high-quality oil that is extracted from the seed.

Adansonia gregorii, the Australian species of baobab, or baob as it is known in Australia, lives for over 2,000 years. Scattered over a large area of the western Australian outback, trees of this species (unlisted by IUCN) have great cultural importance for the aboriginal people.

PLANTS AND PEOPLE

A LEGENDARY TREE

Adansonia digitata, the African baobab, grows from Sudan southwards to South Africa, and from the Cape Verde Islands in the west to Ethiopia and Madagascar in the east. The name *digitata* derives from the nature of the leaves – appearing in late spring or early summer, each has five to seven leaflets that resemble the fingers (digits) of a hand. The large, white, sweet-scented flowers are pollinated by a range of nocturnal animals.

This tree is a great benefit to humans. Kalahari bushmen exploit the water-storage properties of the trunk by sucking water from it through the hollow stems of grasses. Fibre from the trunk is made into rope and is woven to make cloth; the fruits are edible, as are the young shoots and leaves; and the bark, leaves, and fruit of baobab trees all have roles in African traditional medicine.

ADANSONIA GRANDIDIERI *Near Morondava to the north of Madagascar is the Avenue of Baobabs, a lane of fine specimens of* Adansonia grandidieri. *The trees are not planted but a remnant of what was once an entire forest of this magnificent species.*

Red List: Lower Risk

Aglaia odorata

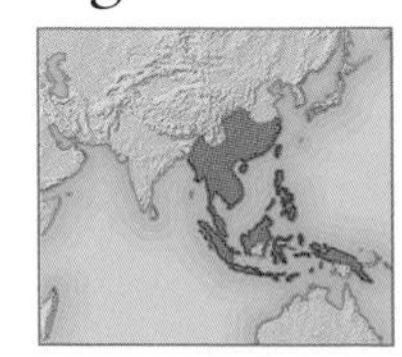

Botanical family
Meliaceae

Distribution
China, south-east Asia, Pacific islands

Hardiness
Half-hardy

Preferred conditions
This rainforest species prefers full sun or semi-shade, and rich, fertile soil, which can be mildly acidic or neutral.

The genus *Aglaia* consists mainly of trees that grow in the tropics. More than 50 species are recorded as globally threatened by IUCN; around 40 additional species are of conservation concern and are close to meeting the criteria for IUCN Red Listing.

One of the latter species is ***Aglaia odorata***, a shrub or small tree commonly known as the peppery orchid tree. The species is found in evergreen forests in parts of Southeast Asia, although its status in the wild remains unclear. It is an important ornamental species, and its wood is particularly good for turnery.

▲ SCENTED FLOWERS *The small, yellow flowers can be very fragrant when grown in habitat, but may be scentless in cultivation.*

The plant produces a highly effective chemical insecticide to safeguard itself from being eaten by insects. This same chemical is thought to inhibit cancer-cell proliferation in humans. The species also yields an essential oil, which is extracted from the seeds.

Red List: Vulnerable

Arbutus canariensis

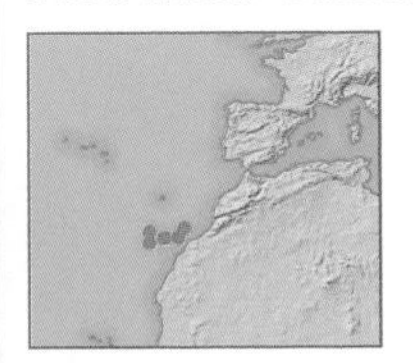

Botanical family
Ericaceae

Distribution
Canary Islands

Hardiness
Fully hardy

Preferred conditions
Full sun, in a site sheltered from cold winds, even when the tree is mature. Fertile, humus-rich, and well-drained soil.

Endemic to the Canary Islands of Tenerife, Gomera, Hierro, and Gran Canaria, ***Arbutus canariensis***, or madroño, is a component of the islands' cloud forests (*see p.110*). It is an attractive small tree that is well suited to cultivation in warm, temperate climates. Growing to 15m (50ft) high, it has brown bark that peels in flakes. White or greenish bell-shaped flowers, often tinged pink, are borne on branching flowerheads. The fruit is edible and similar to that of the strawberry tree, *Arbutus unedo*.

There are about ten populations of the species in the wild, comprising no more than 10,000 individuals. Numbers seem stable, although declining water levels and fires may be affecting some stands. An *ex-situ* collection grows at the Jardín Botánico Viera y Clavijo on Gran Canaria.

▲ LUSCIOUS FRUITS *This species bears fruit reputed to be the Golden Apple of the Hesperides sought by Hercules in Greek myth.*

Red List: Endangered*

Arctostaphylos densiflora

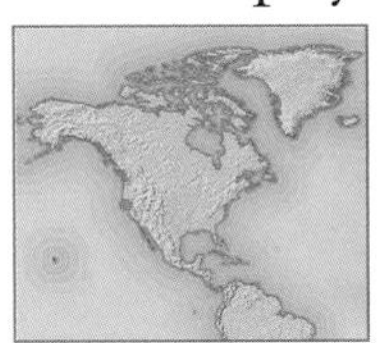

Botanical family
Ericaceae

Distribution
USA: California

Hardiness
Half-hardy

Preferred conditions
Full sun to partial shade, in a dry location. The shrub survives well on little water, requiring spells of dryness between showers in summer.

Known as the Vine Hill manzanita, ***Arctostaphylos densiflora*** is an extremely rare species endemic to Sonoma County, California, where it grows in acidic, sandy soils of open heath in the pine barrens. Most of the natural vegetation in the area of Sonoma County where the Vine Hill manzanita grows has already been cleared in favour of orchards and vineyards.

▼ EVERGREEN ATTRACTION *In the garden, this undemanding, glossy-leaved plant thrives if left alone.*

This low, spreading, evergreen shrub has shiny green leaves, black branches, and small, white-to-pink flowers. Three of the four known populations of the species have become extinct in the wild. The main threats to the surviving individuals include roadside maintenance and a fungal disease that was reported as affecting the plants in 1987. In the 1990s, the California Native Plants Society (CNPS) bought the land on which the last known population grows, and this site is now managed for the benefit of the species.

The Vine Hill manzanita is cultivated at the Regional Parks Botanical Garden, Berkeley, California, a member of the US Center for Plant Conservation. The botanical garden, established in 1940, is a haven for hundreds of California's rare and endangered plants. The centre possesses what is probably the most complete collection of Californian manzanitas in cultivation.

The small, evergreen leaves and dense foliage of this medium-sized Californian native shrub make it a good screening plant for gardens in similar climates. *Arctostaphylos densiflora* cultivars that may be purchased by gardeners include 'Harmony', 'Howard McMinn', and 'Sentinel'.

WILDLIFE

DEPENDENT SPECIES

Manzanitas are part of the chaparral vegetation in California and provide significant benefits to the wildlife community of the chaparral. Flowers of the manzanita species provide nectar for butterflies and native bees in the spring; berries provide food for chipmunks, racoons, scrub jays, and a variety of birds in late summer; and the low-growing habit of the plant makes it a good site for the nests of such birds as California valley quails and wrentits. The latter birds prefer to fly within a shrubby habitat of dense, stiff brush that provides a varied diet of insects, spiders, berries, and small fruits. Wrentits also inhabit groves of ceanothus, chamise, and scrub oak.

WRENTIT (CHAMAEA FASCIATA)

CITES I

Balmea stormiae

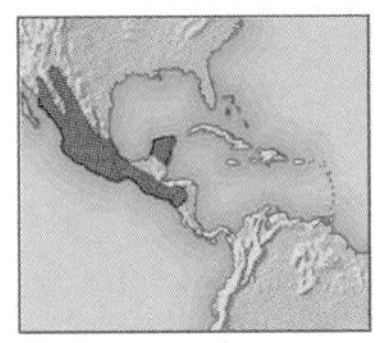

Botanical family
Rubiaceae

Distribution
El Salvador, Guatemala, Honduras, Mexico

Hardiness
Frost-tender

Preferred conditions
Full sun or light, dappled shade. Moist but well-drained, neutral to acid soil. Prefers the same conditions as coffee and cinchonas (quinine tree).

The genus *Balmea* contains only one species, *Balmea stormiae*; the tree's common name is ayuque. It grows in varied conditions, from dry, stony terrain in Mexico to wet mountain forest growing at an altitude of 425–700m (1,400–2,300ft) in Guatemala. The tree may be more common than suspected in botanically unstudied parts of Mexico.

Ayuque has brilliant, scarlet-red flowers that have long been appreciated by local people in the areas where it grows. In markets at Uruapan, Mexico, it is common to see it cut and sold as a Christmas tree. Ironically, overharvesting of balmeas arose only on the enforcement of laws that made it illegal to cut conifer saplings for use at Christmas.

The local popularity of ayuque is unfortunately one of the threats to its survival, and stands are rapidly exploited wherever they are newly discovered. Concerns about the threat of trade in *Balmea stormiae* led to its listing in Appendix I of CITES in 1975. The tree's current conservation status is unclear.

15 on Red List*

Banksia

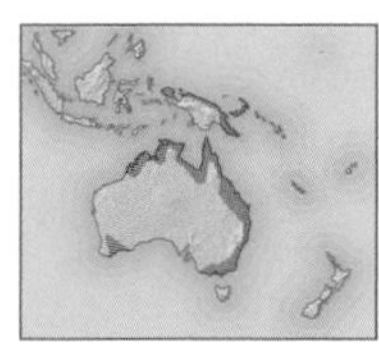

Botanical family
Protoaceae

Distribution
Australia, New Guinea (one species)

Hardiness
Half-hardy – frost-tender

Preferred conditions
Full sun or partial shade. *Banksia brownii* prefers acid-neutral, well-drained soil. *Banksia solandri* requires well-drained soil that is not alkaline.

Named after the naturalist Sir Joseph Banks (*see right*), the genus *Banksia* contains around 70 species, all native to Australia's temperate to tropical scrub and forest, except for one that extends to the island of New Guinea. The greatest diversity of wild species is found in southwest Western Australia. Banksias are evergreen shrubs or small trees with alternate leathery leaves. The flowers are usually borne in dense cylindrical spikes and the fruits, large woody "cones", are used in decorative flower arrangements.

Overcollection of flowers, foliage, and fruits has been one of the threats faced by rare species in this genus. Habitat destruction remains the major threat. Many banksias are cultivated, and the flowers attract nectar-feeding birds.

Brown's banksia, ***Banksia brownii***, is classified as Endangered in the Red List of 1997. It was named after the British botanist, Robert Brown, who voyaged to Australia in 1801 with Captain Matthew Flinders. The ornamental, erect shrub grows to 6m (20ft) in height. Its small, handsome, fern-like leaves have white undersides. The pairs of red to reddish-brown flowers bloom in January and then again between April and August.

Brown's banksia is found only at several widely separated sites in southern Western Australia, in the Stirling Ranges and northeast of Albany. Fortunately it occurs in protected areas, including Stirling Range National Park, where it is protected by law. However, the plant is endangered as a result of infection by the fungus *Phytophthora cinnamomi*, which causes root rot. All of the populations, except one in the Stirling Ranges, are infected and the species faces extinction in the wild. However, the plant is fairly common in cultivation and this garden population offers the possibility of reintroduction into the wild should the fungal disease come under control.

Solander's banksia, ***Banksia solandri***, is classed as Rare in the Red List of 1997. It is named after Daniel Solander, an English botanist who travelled with Joseph Banks on Captain James Cook's first expedition to Australia (1768–1771). The decorative shrub has ornamental, oak-like leaves that grow to 30cm (12in) in length. The leaves are dull green on the upper surface and whitish green on the under surface. The flower spikes are bronze with tightly compressed flowers.

In the wild this shrub is confined to the elevated and rocky slopes of the Stirling Ranges of Western Australia. The climate there is generally cool, due to the high altitude; rainfall averages about 600mm (24in) per annum.

▲ BANKSIA BROWNII *has a tall, narrow, cylindrical flowerhead. In other banksias, the flowerhead may be shorter and wider, or spherical like the flowerhead of an artichoke.*

▲ BANKSIA SOLANDRI, *with its striking bronze flowerheads, is among the* Banksia *species cultivated for their ornamental blooms and foliage by Australian floriculture companies.*

▼ BANKSIA SPECIOSA, *seen here in Cape le Grand National Park on the southern coast of Western Australia, is one of many banksias that flourish in the area's warm, humid conditions.*

PEOPLE

JOSEPH BANKS

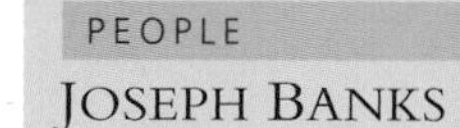

Inheriting great wealth at a young age, the naturalist Joseph Banks financed the botanical side of Captain Cook's first expedition to Australia in 1768. With fellow botanist Daniel Solander, Banks collected and described many hundreds of the strange and exotic plants to be found at Botany Bay and further north on the east coast. Banks remained closely involved in the affairs of the new British colony for the rest of his life, corresponding with the first four governors of New South Wales. He was botanical advisor to Kew Gardens from 1771 and was knighted in 1781. President of the Royal Society from 1778 until his death in 1820, he sponsored many scientists of the age, including Robert Brown, the botanist made famous by his description of Brownian motion.

BANKS IN TIERRA DEL FUEGO

Red List: Vulnerable

Brugmansia aurea

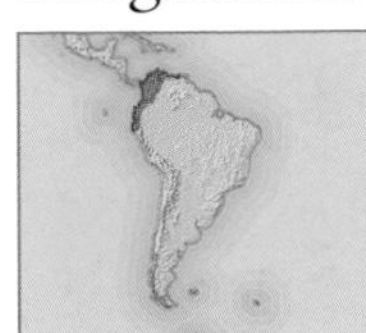

Botanical family
Solanaceae

Distribution
Colombia, Ecuador, Venezuela

Hardiness
Frost-tender

Preferred conditions
Flowers best in full sun and moist, well-fertilized soil. Tolerates light frosts of –5°C (23°F) and even lower if growing in a sheltered site.

Truly an enticing plant, ***Brugmansia aurea*** grows in scrub in the high Andes. It has golden yellow to white, pendulous, trumpet-shaped flowers, 15–30cm (6–12in) long, with a strong, distinct fragrance at dusk. In the wild it grows as a tree to a height of 8m (25ft), with short and densely leafed branches. Trade and overexploitation are potential threats to the species at a national level in Ecuador. The 2000 Red Listing is based on its status in Ecuador only.

Like many other members of the family Solanaceae, *Brugmansia aurea* is toxic and extremely potent. Despite this, native South American Indian people use the plant in witchcraft (*see below*) and medicinally to treat a variety of ailments.

Although tropical in origin, *Brugmansia* species in general are grown as large shrubs. In winter they are either cut back or kept in a cold greenhouse. Breeders have produced very successful hybrids of *Brugmansia aurea* and *Brugmansia versicolor.* Named *Brugmansia* x *candida*, they have beautiful and fragrant flowers.

The common name, angels' trumpets, derives from the golden, trumpet-shaped flowers

▲ IRRESISTIBLE PERFUME *The spectacular flowers are only part of the attraction of this species, since they are pollinated by night-flying butterflies lured by their powerful scent.*

PLANTS AND PEOPLE

POWERFUL MEDICINE

In the Sibundoy Valley of Colombia, *Brugmansia aurea* is employed by medicine men of the Inga and Kamsá Indians in their witchcraft. The local Kamsá name for the plant translates as "intoxicant of the jaguar". The leaves can cause frightening hallucinations and are used by the medicine men in a drink that they believe aids in divination and prophecy. The witch-doctors cultivate their own plants and are the only members of their tribes who are entitled to use the drug.

Red List: Endangered

Caesalpinia echinata

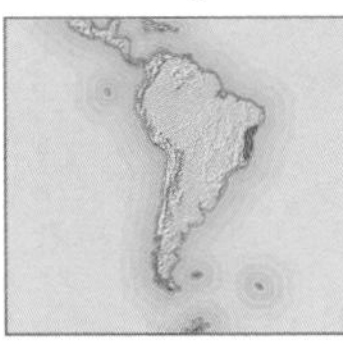

Botanical family
Leguminosae

Distribution
Brazil

Hardiness
Frost-tender

Preferred conditions
The tree grows in a wet, low-altitude environment that receives strong light. It is found in coastal and waterside forest and woodland growing on well-drained sandy or sand-clay soils.

Pau brasil or brazilwood, as *Caesalpinia echinata* is commonly known, is the national tree of Brazil. It is an important flagship species for the tropical rainforest ecosystem on the Atlantic coast of Brazil, which is a global biodiversity hotspot. As a result of clearing for agriculture and plantation crops, mining, and industrial and urban development, only about five per cent of the original extent of Brazil's coastal forest now remains.

The species has suffered from both direct exploitation and habitat loss. Natural stands of *Caesalpinia echinata* were almost completely destroyed by the dye industry, except for restricted populations that remained in a few areas on the coastal plain. These have since been the subject of further exploitation, partly for use in the manufacture of bows for musical instruments. Despite protection by law, illegal cutting continues to take place. IBAMA, the state conservation agency, includes pau brasil on its official list of threatened Brazilian plants.

A conservation action plan for pau brasil was prepared in 1997 as part of the SoundWood Programme of Fauna & Flora International (*see p. 100*). Various components of this are now being implemented and receive developing support from the bow-makers.

Conservation action includes protection of the tree within designated areas, including reserves in Bahia and Pernambuco. The Rio de Janeiro Botanic Garden is currently studying the conservation requirements of the species in its remnant forest habitats within Rio de Janeiro state. The species is also in cultivation and young cultivated trees have been distributed to schools by Fundacion Nacional do Pau-Brasil (FUNBRASIL). The fame of this species, the national tree of Brazil, should guarantee that its future is secured.

RENEWAL

IMPROVING QUALITY

Caesalpinia echinata grows well from seed and various small-scale plantations have been established, but so far there has been some reluctance on the part of bow-makers to buy the small quantities of wood available from the plantations – the quality is believed to be too poor for bow-making. To relieve the pressure on the wild populations of pau brasil, horticulturalists must find ways to raise the quality of trees in cultivation. This will involve painstaking research into better growing conditions, techniques for commercial cultivation, and methods for improving production in the species' natural forest environment.

PEOPLE

RENAISSANCE MAN

Born in 1525 at Arezzo, Italy, Andrea Cesalpino studied philosophy, medicine, and botany at the University of Pisa. His instructor in botany, Luca Ghini, became the director of the second botanical garden to be created in Italy, at Pisa in 1547, to be succeeded by Cesalpino from 1554–58. Cesalpino is known as one of the first botanists to make a herbarium; the three-volume set containing 768 plant varieties is now in Florence's museum of natural history.

In 1583, Cesalpino published his most significant contribution to science. Entitled *De plantis libri XVI*, his treatise included the world's first attempt to describe and classify plants according to their organs of reproduction. For the first time, acute observations of flowers, fruits, and seeds were organized in a systematic way.

Cesalpino, who died in 1603, is also celebrated for his description of the blood circulation, which anticipated the work of William Harvey. He is remembered by the genus *Caesalpinia*, which is derived from the latinization of his name, as well as the plant family Caesalpiniaceae.

17*/11 on Red List

Camellia

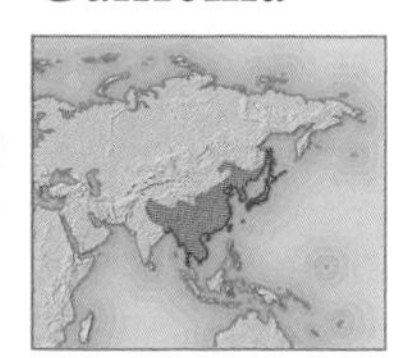

Botanical family
Theaceae

Distribution
India, China, Japan, Indonesia

Hardiness
Fully hardy – frost-tender

Preferred conditions
Light, dapped shade, in a site sheltered from cold, dry winds, early morning sun, and late frosts. The most favoured soil is moist but well drained, humus-rich, and acid (pH 5.5–6.5).

The genus *Camellia* contains over 80 species of evergreen trees and shrubs. With their large, attractive flowers and handsome, usually glossy leaves, camellias include some of the most popular shrubs in cultivation. In the wild, the species are native to woodlands of the eastern Himalayas, China, Japan, and south to Malaysia. Camellias have been cultivated in China and Japan for over 1,000 years. They are grown for their oil-bearing seeds, the tea made from their leaves, and as ornamentals.

◄ CAMELLIA CHRYSANTHA *is considered an outstanding species because of its golden hue and its striking foliage. As many as 200 blooms may appear on a single plant.*

The golden camellia, ***Camellia chrysantha***, Vulnerable in the 2000 Red List, is a shrub or small tree restricted to wet areas of rainforest below 500m (1,600ft) in southwest Guangxi in China, extending into Vietnam. It is one of several Chinese camellias with yellow flowers. The extent and size of the population is uncertain. Populations continue to be threatened by overcollection of seedlings.

Clearance of the golden camellia's monsoon forest habitat is another significant threat for the species, which regenerates poorly in the open. The golden camellia requires the shady, humid environment provided by its forest cover. The species happens to occur in a number of nature reserves where its natural habitat is protected, and Chinese legislation prohibits collection without authorization.

The species can be propagated by seed, grafting cuttings, layering, and tissue culture. The seeds should be collected as soon as they ripen and sown immediately. Plants grown from seed usually flower after 7–8 years. Unfortunately this remarkable species is very tender, but attempts are being made to cross it with hardier species. The golden camellia is surviving well in cultivation and is found in botanic gardens around the world.

Perhaps the rarest species of the genus is *Camellia crapnelliana*, classed as Vulnerable in the 2000 Red List. This is an attractive, evergreen tree, growing to 12m (40ft) in height. The bark of the trunk is brick-red. The leaves are dark green and leathery. Large, white flowers with 6–8 petals, growing to 9cm (4in) in diameter, are borne on the end of branchlets in November and December.

The first known specimen was discovered in 1903 on the south side of Mount Parker, Hong Kong. The single tree was cut and the species thought to be extinct until it was rediscovered at the same site in 1965. Other populations were subsequently found on the Chinese mainland. A species of the broadleaved forest, *Camellia crapnelliana* occurs along the coast in Fujian, southwest Zhejiang, southern Guangdong, and southern Guangxi. The climate in this area is warm and humid. Populations of the species are steadily diminishing with increasing cutting and loss of habitat.

The large flowers and fruits of *Camellia crapnelliana* make it an important oil-bearing and ornamental tree.

Propagation is by seed or cuttings. Seeds collected in autumn tend to deteriorate rapidly as a result of their high oil content and should therefore be sown immediately after collection. Seeds germinate 4–6 weeks after sowing and seedlings should be placed in shade.

CULTIVATION

CAMELLIA HYBRIDS

The people of China and Japan have appreciated camellias for centuries. Europe saw its first camellias in the early 18th century and has intensively cultivated them ever since. At least 5,000 hybrids have been created, many of them from *Camellia japonica*, although a number belong to *Camellia sasanqua* and *Camellia reticulata*.

When, in the 1930s, the breeder J.C. Williams crossed *Camellia japonica* with *Camellia saluenensis*, the resultant hybrid, called *Camellia* x *williamsii*, was to begin a important line of hardy, free-flowering hybrids from those parent species. Further hardy lines were established by crossing *Camellia oleifera* with a number of species.

Camellias need acidic soil and are often best seen in dedicated gardens. In the UK, Caerhays and Trewidden gardens in Cornwall have long-established collections. The American Camellia Society displays at Massee Lane Gardens, Fort Valley, Georgia.

▼ CAMELLIA SINENSIS *This Indian tea plantation consists of thousands of camellia shrubs, which can live for more than 100 years.*

Red List: Lower Risk

Cercidiphyllum japonicum

Botanical family
Cercidiphyllaceae

Distribution
China, Japan

Hardiness
Fully hardy

Preferred conditions
Sun or dappled shade. The tree grows in deep, fertile, humus-rich, moist but well-drained soil, preferably neutral to acid, although autumn colours are less pronounced on lime-rich soil.

The katsura tree, ***Cercidiphyllum japonicum***, is a rare deciduous tree found in temperate beech forests in Japan and in remnant patches of broad-leaved forest in China. It achieves a height of 40m (130ft) in the wild. The young leaves are red as they unfold but soon turn green. The round, toothed leaves are grey-green underneath. Tiny red flowers emerge with the young leaves. In autumn the leaves turn to striking shades of red and yellow.

The species is considered to be a relic from the Tertiary period. The main threat to the species is poor regeneration in its habitat, but felling is also a threat.

▲ TOFFEE-SCENTED TREE *This species has attractive red and yellow tints in autumn, and the foliage develops the smell of toffee.*

The katsura tree is an important timber species in Japan. The species has been cultivated since 1881 in the UK, where it is considered to be fully hardy, although late frosts can be damaging to new growth. The tree is propagated by seed.

Unlisted

Cordyline kaspar

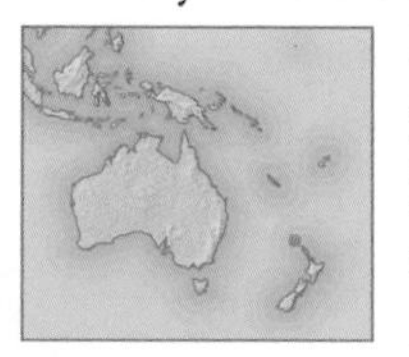

Botanical family
Agavaceae

Distribution
Three Kings Islands, New Zealand

Hardiness
Half-hardy

Preferred conditions
Sun or partial shade. This cordyline prefers deep, fertile, moist but well-drained soil and requires protection from heavy frosts.

Also called the Three Kings cabbage tree, ***Cordyline kaspar*** is endemic to the broadleaved forest and shrubland of New Zealand's Three Kings Islands, a small group of islands lying about 64km (40 miles) northwest of Cape Reinga at New Zealand's northern tip. The tree has a short, stout trunk growing to 2m (6ft), and the branches spread with dense clusters of stiff leaves, which are up to 80cm (32in) long and up to 7cm (3in) across. The flower spikes are scented and produce fruits streaked in blue and white.

The Three Kings Islands have a remarkable endemic flora that includes 12 plant species found nowhere else in the world. Of particular note is South West Island. The second-largest of the group, it is mainly formed of steeply sloping basaltic lava, without beaches and permanent streams. *Cordyline kaspar* is naturally rare in this habitat, but is well-established in cultivated varieties.

▲ GARDEN HYBRID Cordyline kaspar *'Green Goddess' is just one of the popular hybrids derived from the New Zealand species.*

Unlisted

Cercis canadensis

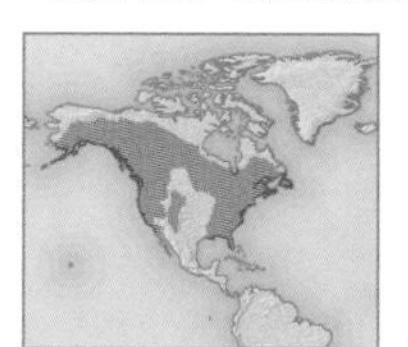

Botanical family
Leguminosae

Distribution
Canada, USA, Mexico

Hardiness
Fully hardy

Preferred conditions
Full sun or dappled shade. Preferred soil is fertile, deep, moist but well-drained, a loam of sand, clay, and organic material. Disturbed plants do not reestablish themselves easily.

The species ***Cercis canadensis***, known as the eastern redbud or Judas tree, is one of six species in the genus *Cercis* and has a relict distribution from the Tertiary era. Its natural distribution extends from Canada through the USA to Mexico. While it is common in the USA, growing in various temperate forest habitats in the wild, it is now recorded as extinct in Canada.

Eastern redbud is conspicuous in the spring because it flowers before the leaves of other trees form. The flowers are usually pink to reddish purple, and rarely white. They are borne on stalks in clusters of 2–8 and are produced from small buds on old twigs, branches, and trunks. Pollination is usually accomplished by bees. The fruits, which are flat, reddish-brown pods containing up to ten brown, bean-like seeds, remain on the tree until after leaf fall. The eastern redbud is extensively planted as a valuable ornamental species.

▼ PINK CLUSTERS *of* Cercis canadensis *flowers vie here with the white blooms of* Cornus florida *at Cumberland Falls, Kentucky, USA.*

◄ CERCIS CANADENSIS 'FOREST PANSY' *is a red-leaved cultivar of the eastern redbud.*

4*/1 on Red List

Corylopsis

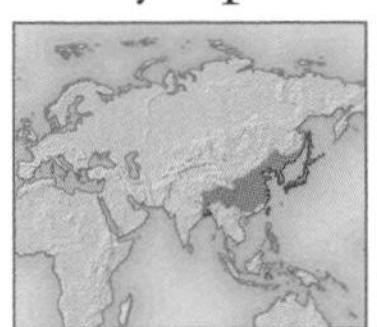

Botanical family
Hamamelidaceae

Distribution
Asia, from Bhutan to Japan

Hardiness
Fully hardy

Preferred conditions
Corylopsis species grow well in light shade, and where their flowers are protected from spring frosts. They will grow in most garden soils, preferring lime-free conditions.

The genus *Corylopsis* is a member of the witchhazel family. There are around 12 species, growing in woodlands and scrub from Bhutan to Japan. All are deciduous shrubs with alternate leaves and woody seed capsules. The catkin-like flowers smell like cowslips and are produced before the leaves in the spring.

Propagation of *Corylopsis* species is by layering shoots in October and removing them from the parent plant after one or two years. Cuttings from lateral shoots can also be taken in late summer.

Fragrant winterhazel, ***Corylopsis glabrescens*** is native to Japan and Korea and is classed as Rare in the Red List of 1997. It is a spreading, multi-stemmed shrub that grows to a height of about 4m (13ft). The leaves, which appear after the fragrant, pale yellow flowers, are alternate, roundish-ovate, and more or less heart-shaped at the base.

Another species, commonly known as starwood in its native Japan and Taiwan, is *Corylopsis pauciflora* (classed as Data Deficient in the 2000 Red List). In Taiwan only a few wild populations remain, restricted to a few small areas of broadleaved submontane forest. The wild populations are not protected and are regenerating poorly. The early flowers are followed by rounded, toothed leaves.

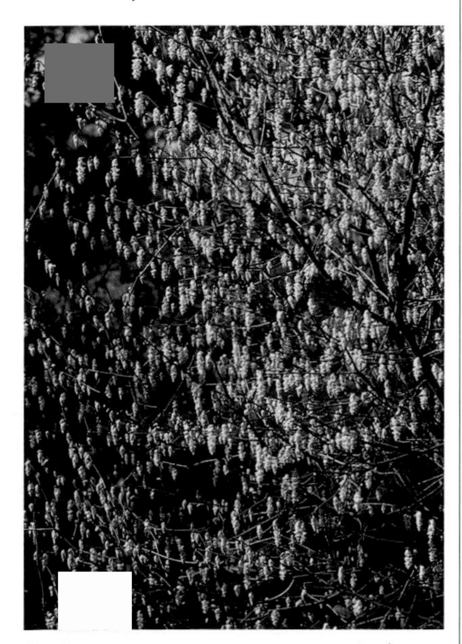

▲ CORYLOPSIS GLABRESCENS *is the hardiest species of its genus. The pale yellow flowers appear in late winter or early spring.*

Red List: Endangered

Corylus chinensis

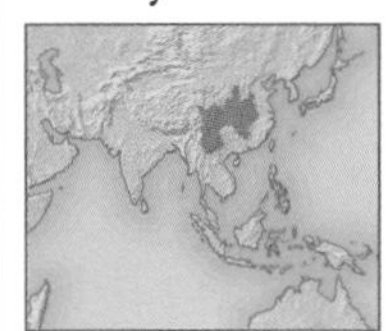

Botanical family
Betulaceae

Distribution
China

Hardiness
Fully hardy

Preferred conditions
Sun or partial shade. Fertile, well-drained soil is preferred, and chalky soils are ideal. This tree has a cool, high-altitude native habitat and grows in fertile soils in a relatively warm climate.

The Chinese hazel, ***Corylus chinensis***, occurs quite widely at elevations of 900–3,500m (2,950–11,500ft) in the broadleaved forests of Henan, Hubei, Hunan, Shaanxi, Sichuan, and Yunnan provinces in China. The species prefers a temperate, cool, and humid climate.

This hazel species is a deciduous tree with spreading branches growing to 20m (65ft) tall with grey-brown, vertically fissured bark and simple elliptical or ovate leaves. The male flowers are in the form of catkins that grow up to 5cm (2in) long.

Trees are fast-growing and, like European and North American hazels, provide edible nuts and good-quality timber. A decline in the number of mature individuals has been reported, caused largely by overexploitation of the nuts and timber.

Corylus chinensis was introduced into cultivation in 1900 by the plant hunter Ernest Wilson (*see p. 113*). The species can be propagated by sowing the seeds as soon as they are ripe, or by layering.

▲ CHINESE HAZEL *While this species is endangered by loss of habitat, another factor in its decline has been poor regeneration in the wake of devastation of the nut crops by birds.*

Red List: Lower Risk

Cotoneaster granatensis

Botanical family
Rosaceae

Distribution
Spain: Sierra Nevada

Hardiness
Fully hardy

Preferred conditions
Full sun, in a site sheltered from cold, dry winds. Preferred soil is moderately fertile and well-drained, although dry soil is tolerated.

Endemic to the scrubby hillsides of the Sierra Nevada region of Spain, *Cotoneaster granatensis* is a shrub that is known locally as *durillo dulce*. It grows naturally on calcareous and ultrabasic rocks in vegetation that also includes *Berberis hispanica*, *Prunus ramburii*, *Amelanchier ovalis,* and *Lonicera arborea*.

Most of the distribution of the shrub is contained within protected sites. For land managers, it makes a useful colonizer and stabilizer of slopes that are prone to erosion. The species has entirely elliptical leaves. The white flowers are hermaphroditic (both sexes are contained in the flower); the fruit is oval and red.

Recent research into ways to promote reforestation of the Sierra Nevada with such species as pine (*Pinus nigra*) and yew (*Taxus baccata*) has indicated that "nurse shrubs", including *Cotoneaster granatensis*, have a vital role in protecting seedlings from dehydration during the very hot summer, and from drying winds and frost in winter. Spiny, fleshy-fruited shrubs may be planted for this purpose, and also to form natural barriers against predation of seedlings and saplings by herbivores.

HABITAT

SIERRA NEVADA, SPAIN

The Spanish Sierra Nevada is the second highest mountain chain in Europe after the Alps. It has the highest concentration of endemic species in the Iberian Peninsula and was named as a UNESCO Biosphere Reserve in 1986. The upper slopes were declared a National Park in 1999.

The scrubby vegetation covering the slopes is known as garrigue and contains many aromatic herbs, including thyme, rosemary, savory, and bay. A recent survey identified 147 threatened plant species in the region, of which 98 are not formally protected. While tourism, overgrazing, and overcollection are problems, the rarer high-altitude species are also threatened by climatic change caused by global warming.

Red List: Endangered*

Cyanea superba

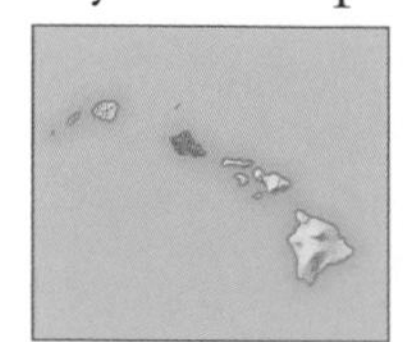

Botanical family
Campanulaceae

Distribution
Hawaii

Hardiness
Frost-tender

Preferred conditions
This species is a forest understorey plant and grows on well-drained, sloping terrain underlain by a rocky substrate.

Known in the Hawaiian Islands as *Oha wai*, *Cyanea superba* is one of the Hawaiian lobelias, which have co-evolved with the endemic finches of the islands. The species has two subspecies, one of which, *Cyanea superba* subsp. *regina*, is confirmed as already extinct.

Cyanea superba subsp. *superba* is now known from two small populations, totalling fewer than ten plants, in tropical forest of the Waianae Mountains of Oahu Island. One population is on federal property in Kahanahaiki Valley, and the other on state land in Pahole Gulch. Habitat degradation caused by the spread of introduced species is the most serious threat to the remaining wild plants. Protective measures and planting are being carried out at Pahole. Propagation has taken place in Hawaii's botanic gardens. The taxon is listed under the US Endangered Species Act.

Red List: Vulnerable*

Darwinia oxylepis

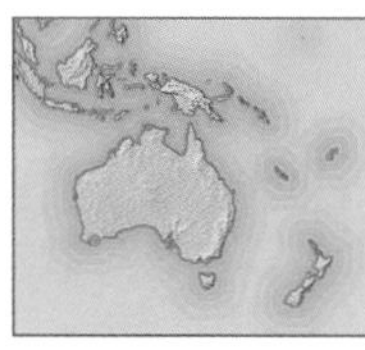

Botanical family
Myrtaceae

Distribution
Western Australia

Hardiness
Frost-tender

Preferred conditions
Full sun. Preferred soil is sandy, humus-rich, and well-drained. Moderate water is needed during the growing season, followed by winter dryness.

There are around 35 species in the genus *Darwinia*, various of which are commonly known as "mountain bells". *Darwinia* is an Australian genus of small shrubs found on heaths, sandy plains, and scrub. They generally have small flowers within conspicuous, bell-shaped modified leaves (bracts) borne at the end of branches. The leaves smell like eucalyptus oil when crushed and have been used as a source of essential oils for perfumery. Over half of the species are in need of conservation attention. They all make attractive garden plants but can be difficult to cultivate. Propagation is usually from cuttings because of difficulties with seed germination.

Darwinia oxylepis is a shrub that occurs naturally in the south of Western Australia. It is small, growing to 1.5m (5ft) in height, and the leaves are needle-like, around 10mm (½in) long with inwardly curving margins. Samples of the plant were first collected by James Drummond in 1848. The species is thought to be pollinated primarily by birds feeding on the nectar.

Wild populations of the shrub are confined to seasonally moist gullies near the lower slopes of mountains in Western Australia's Stirling Range National Park, where they grow in mallee heathland on acid soils. The threats faced by this plant include the "dieback" disease caused by the fungus *Phytophthora cinnamomi*, disturbance from tourist activities, illegal picking, bushfires, and drought.

Conservation measures for this species include the careful monitoring of wild populations to assess the nature and degree of pressures from visitors, and *ex-situ* conservation in botanic gardens.

◄ CURIOUS BLOOMS *What may look like red petals are actually modified leaves that surround tiny, petal-less flowers.*

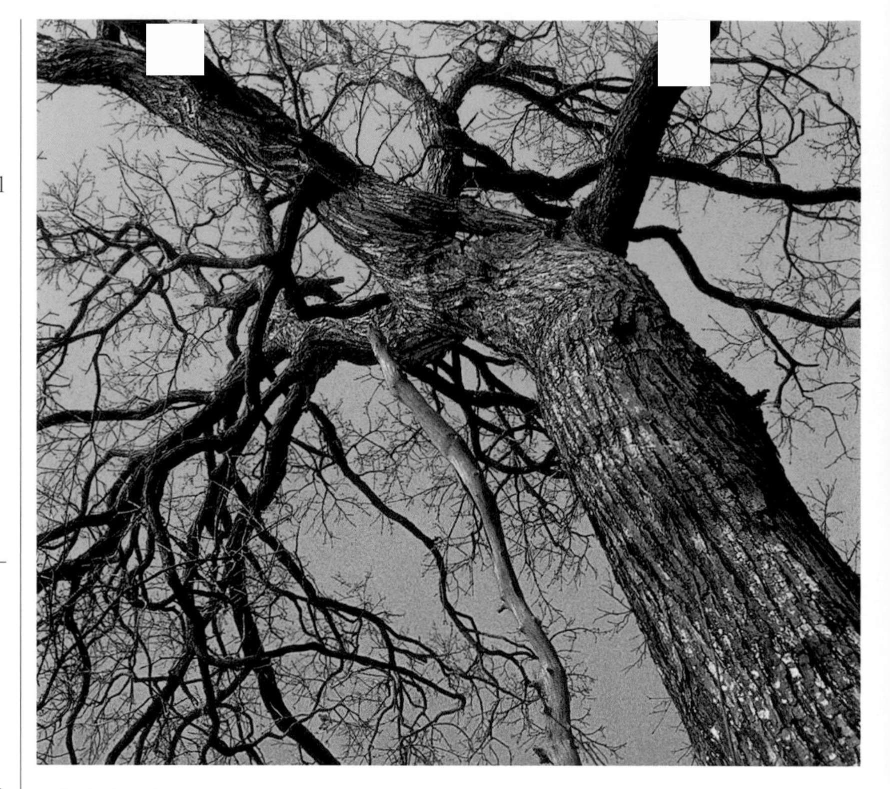

▲ COCOBOLO *Much coveted for its wood, this species grows to only 15–20m (50–65ft) in height and is usually of poor form.*

Red List: Vulnerable

Dalbergia retusa

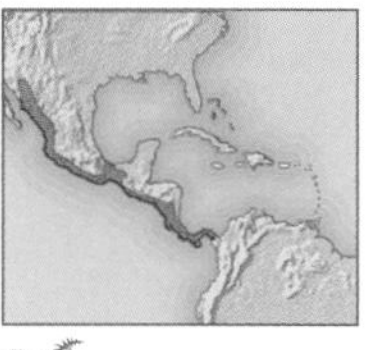

Botanical family
Leguminosae

Distribution
Central America, Mexico

Hardiness
Frost-tender

Preferred conditions
Full sun or light, dappled shade. Preferred soil is fertile and well-drained.

Cocobolo, the common name of the species *Dalbergia retusa*, is a tropical dry forest tree found in the Pacific coastal lowlands of Central America and Mexico. The species has the distinction of producing a highly prized rosewood, also known as cocobolo, which has an unusual, exotic appearance. The wood is orange, purple, and yellow in colour, with purple and black grains. The bright colours fade on exposure to light and the wood darkens to a deep orange-red. Regions where the species was formerly widespread, such as Costa Rica, are now almost completely exhausted. There is a limited occurrence of the species north of the canal in Panama and in Mexico. The tree's habitat has also been exploited for 400 years, and cattle ranching and excessive forest burning continue to reduce numbers of the species.

Cocobolo wood has long been popular, particularly in the USA, for making knife handles. It continues to be used for hairbrushes, chess pieces, and boxes, as well as high-quality electric guitar bodies and other instrument parts.

Cocobolo wood is now so rare that very little of it reaches the world market. Today, most of the wood derives from privately owned fincas where cocobolo trees were planted 80–100 years ago. Small-scale plantations continue to be grown to supply the timber trade.

RENEWAL

MUSIC PROGRAMME

Dalbergia retusa, along with other species associated with the manufacture of musical instruments, is the focus of the SoundWood Programme set up by Fauna & Flora International. This programme aims to conserve globally threatened trees that provide woods used in making clarinets, oboes, violins, violas, harps, and other woodwind and stringed instruments. The ultimate goal is to enable future generations to enjoy the compositions of our musical heritage played on traditional wooden instruments, exactly as their composers intended them to be heard.

Red List: Rare*

Davidia involucrata

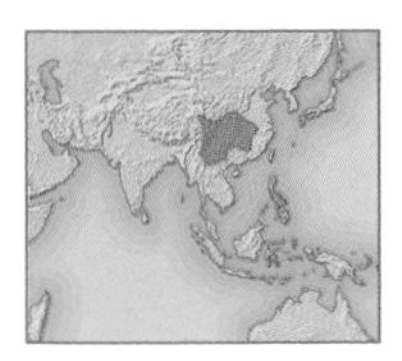

Botanical family
Cornaceae/Davidiaceae

Distribution
China

Hardiness
Fully hardy

Preferred conditions
Davidias prefer sun or partial shade and shelter from the wind. They thrive in fertile soil with ample moisture. Trees are killed by honey fungus.

The ghost tree, also known as the handkerchief or dove tree, ***Davidia involucrata*** is a Chinese relative of the dogwoods and was discovered in 1869 by a French missionary and botanist, Père David. The genus *Davidia* contains only this species, which is native to China and grows in temperate forest. It is now so rare in the wild that the two beautiful specimens at Harvard's Arnold Arboretum in Boston, USA, are regularly visited by Chinese botanists who have never had the opportunity to study the tree in their homeland. The species is of great interest as a relic of Tertiary flora.

There are two varieties, distinguished by differences in the leaf, and both are classed as Rare in the Red List of 1997. *Davidia involucrata* var. *involucrata* occurs naturally in montane or cloud forest with a cool, humid climate, and was introduced into the West in 1904 by Ernest Wilson (*see p. 113*). *Davidia involucrata* var. *vilmoriniana* is the more threatened of the two varieties. Its distribution is similar to that of var. *involucrata*, but with fewer populations. It was first cultivated in the West in 1897. There is a strong demand for seeds and plants of var. *vilmoriniana* in the horticultural trade, and overcollection from wild populations has severely reduced natural regeneration.

Propagation is by seed, cuttings, or layering. Long shoots can be layered in October and detached from the parent plant after two years. Cuttings of side shoots are preferably taken with a heel.

▲ GHOSTLY BRACTS *In spring, gardeners eagerly await the spectacular flowering of this species, when drooping, papery, floral modified leaves complement the ornamental foliage.*

Red List: Vulnerable

Delonix regia

Botanical family
Caesalpiniaceae/Leguminosae

Distribution
Madagascar

Hardiness
Frost-tender

Preferred conditions
Full sun, in a site sheltered from strong winds. The best soil is fertile and moist but well-drained. Leaves fall if temperatures approach freezing.

The flamboyant or flame tree, ***Delonix regia***, is a tree species seen much more often in cultivation than in its native open, dry forest habitat. One of the most attractive ornamental trees of the tropics, it is widely planted as a shade tree in parks, gardens, and avenues throughout Southeast Asia, Australia, and islands of the Pacific. Its flower has been adopted as the national flower of Puerto Rico in the Caribbean, even though its native habitats are in Madagascar.

The flamboyant tree grows to 15m (50ft) in height. When mature, at up to ten years of growth, it has a broad, umbrella-shaped crown. The leaves, which reach a length of 30–50cm (12–20in), are composed of numerous tiny leaflets. The magnificent flowers, which are scarlet with orange-striped petals, are pollinated by birds. Where there is nothing to restrict its reproduction, the species can form dense stands that exclude alternative species.

In Madagascar this species occurs naturally in the seasonal forests in the west and north of the island. The main native populations grow around the northern town of Antseranana. Since this area is threatened by charcoal production, the tree has been given Vulnerable status.

Propagation is from the bean-like seeds, which develop in pods up to 60cm (2ft) long. Germination can be assisted by placing the seeds in boiled water for two minutes, then soaking in cold water for a further 24 hours before planting.

▼ CROWN OF FIRE *From spring to summer the wide-spreading crown of this popular shade tree bears a profusion of flowers and yellow-and-red striped leaves.*

Red List: Vulnerable

Dendrosicyos socotrana

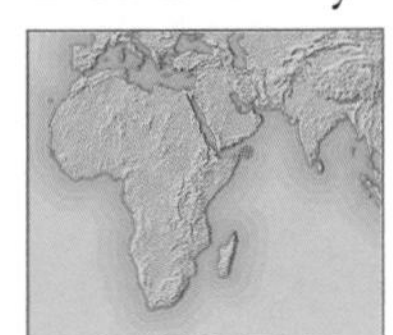

Botanical family
Cucurbitaceae

Distribution
Socotra Island, South Yemen

Hardiness
Frost-tender

Preferred conditions
This plant prefers full sun. The soil in its habitat contains a high proportion of pumice. While its habitat is dry, it benefits from liberal rainfall.

One of the very few species in the cucumber family to grow to the stature of a tree, *Dendrosicyos socotrana* is endemic to South Yemen's island of Socotra. The tree is fairly common on the coastal plains and foothills, mainly on limestone but also in granitic areas. It is most common on slopes near the north coast of the island. The cucumber tree often grows alongside a euphorbia tree species that is also endemic to Socotra.

The Socotran cucumber tree is notable for its characteristic swollen, succulent trunk and grows to 7m (23ft) in height. The trunk bears a small, spreading crown consisting of a few branches with tufts of rounded or heart-shaped leaves at the ends. The leaves are 4–8cm (2–3in) long. The flowers occur in clusters on irregular protuberances that arise from the stem. Male and female flowers appear on separate protuberances; the male flowers are rather larger than the female.

HABITAT

EVOLUTIONARY NOVELTY

Lying 240km (150 miles) to the east of the Horn of Africa, Socotra is the largest island of an archipelago that split away from mainland Africa and Arabia millions of years ago. Its flora evolved in complete isolation – 30 per cent of its plants are endemic to the islands, and ten genera are found nowhere else in the world.

Socotran plants evolved their often swollen and stubby shapes and other characteristics in response to the extremely dry climate and fierce monsoonal winds from Africa. Some shrubs are umbrella-shaped and grow in dense thickets to the same height, giving mutual protection against the wind. Many have waxy surfaces to reflect the sun. One plant, *Adenium socotranum,* prevents itself from overheating by manipulating the flow of its sap.

As a result of overgrazing and increasing human activity in the archipelago, more than 50 of Socotra's endemic plants are now Red Listed. Botanists, calling Socotra "the Galapagos of the Indian Ocean" for its unique biodiversity, number it among the world's top ten endangered islands.

The flora of Socotra is under extreme threat, with over half of the endemic species facing extinction. A general threat to the vegetation, as on so many islands, is overgrazing by introduced livestock. In times of shortage, the whole plant of the cucumber tree is cut and pulped for livestock fodder. Although current levels of this form of use are sustainable, an increase in the numbers of livestock on the island would place the cucumber tree populations under severe pressure.

Dendrosicyos socotrana was previously extremely rare in cultivation, but due to its unusual characteristics it is now displayed in numerous botanical centres. Hand-pollination of the freshly opened female flowers with pollen from fresh male flowers is the most reliable way to encourage fruiting and obtain seed. The seeds are planted in a mix of mainly pumice and watered generously. The seedlings, appearing within two weeks, are grown on in a warm glasshouse.

81*/78 on Red List

Diospyros

Botanical family
Ebenaceae

Distribution
China, Taiwan, Hong Kong

Hardiness
Fully hardy – frost-tender

Preferred conditions
Full sun, in a site sheltered from cold, drying winds and late frosts. Preferred soil is loamy, deep, fertile, and well drained. Although some cultivars of *Diospyros kaki* produce fruit without a male, pollination usually results in larger crops.

Trees and shrubs of the genus *Diospyros*, which includes the true ebony wood species, grow in tropical, subtropical, and warm temperate forests. Also included in the genus are several fruit trees, including *Diospyros kaki,* the edible Chinese persimmon, and *Diospyros vaccinioides*, the small persimmon.

The latter plant, a heavily exploited ornamental species, is just one example of the 78 *Diospyros* species currently considered by IUCN to be threatened. Overcollecting in Taiwan has led to the complete absence of mature trees in the wild. However, a small population may exist in Fengkang. Populations have also been recorded in Chuhai and Huiyang in China, and in Hong Kong. *Diospyros vaccinioides* is also in cultivation in the Heng-Chun Tropical Botanical Garden, located at the southern tip of Taiwan.

On the island of Mauritius, a number of *Diospyros* species are critically endangered. *Diospyros angulata* has been reduced to a single tree. So far, *ex-situ* propagation has not proved successful.

◄ PERSIMMON FRUIT *The persimmon species most grown commercially for its fruit is* Diospyros kaki, *a deciduous tree native to China and Japan. It is now widely grown in temperate, subtropical, and tropical areas.*

▲ FOREST OF EBONY *Many of the* Diospyros *species on the IUCN Red List have long been threatened by unsustainable exploitation of their ebony woods, worth more by weight than any other timber. In Africa, thousands of workers depend on ebony carvings for their livelihood.*

2 on Red List

Dipteronia

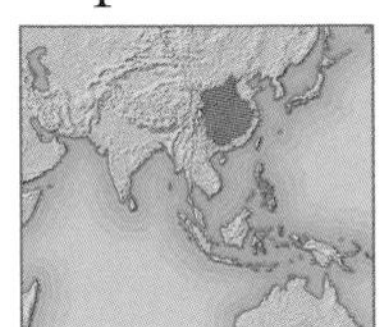

Botanical family
Aceraceae

Distribution
China

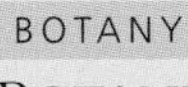

Hardiness
Fully hardy

Preferred conditions
Sun or partial shade. The two *Dipteronia* species grow to their full potential in soil that is loamy, fertile, and moist but well-drained.

Unusual members of the maple family, the two species of the genus *Dipteronia* are deciduous trees found only in temperate woodlands in China. The genus was first described in 1889 by Daniel Oliver, Keeper of the Herbarium at Kew Gardens, based on collections made by Augustine Henry, a medical officer based in China. The feathery leaves of the two species are quite different from the shape of the typical maple leaf, and the round fruits, known as "samaras", are winged all round (*see below*). Both species are listed as Rare in the China Plant Red Data Book.

Dipteronia sinensis (classed as Lower Risk in the 2000 Red List) is the more widespread of the two, occurring mainly in Sichuan and Hupeh. On the stems of this shrub or small tree are pairs of handsome, large, feathery leaves up to 50cm (20in) long, consisting of five leaflet pairs. Single leaflets are 12cm (5cm) long and 3.5cm (1.5in) wide.

BOTANY

BOTANICAL WINGS

Within the botanical family Aceraceae there are only two genera, *Acer* and *Dipteronia*. Both genera bear paired seeds called schizocarps or samaras. All *Acer* species are distinguished by an oval wing of hard membrane that is attached by an arm to each seed, while the two *Dipteronia* species possess a membrane that entirely encircles each seed (*see left*).

Whether the membrane is formed into a wing or circle, it confers a significant advantage in catching the wind in a spinning motion and delivering the seed within a wide radius of the parent tree.

The flowers, which are inconspicuous, are followed by large bunches of decorative, light green fruits that eventually turn red. The samaras are each 2cm (¾in) in diameter. The tree occurs in small stands scattered in mountainous moist forest areas. Mature trees are said to be becoming scarcer because of cutting and poor regeneration.

Dipteronia dyeriana occurs only in southeastern Yunnan and was named after Sir William Turner Thiselton-Dyer, Director of Kew at the turn of the 20th century. This species, Endangered in the 2000 Red List, is threatened by general deforestation, although it does receive protection within the Laojun Mountain Nature Reserve, in Yunnan. Its feathery leaves are smaller, growing to a length of 40cm (16in).

Dipteronia sinensis is in cultivation in Chinese botanic gardens. It was introduced into cultivation in the UK in the early 1900s, where it can be seen at the Royal Botanic Gardens at Kew and Edinburgh, and in the University Botanic Garden at Cambridge. It is also in cultivation at a handful of other arboreta and botanic gardens around the world.

Both of the *Dipteronia* species have excellent ornamental qualities, and both will grow well in a fertile, loamy soil. *Dipteronia sinensis* is usually available from commercial nurseries, but *Dipteronia dyeriana* is currently less readily obtained. Both species may be propagated by seed, cuttings, and layering.

▲ DIPTERONIA SINENSIS *This young specimen will eventually bear small, greenish white flowers in summer, followed by large clusters of red-brown samaras. The species is grown mainly for its large, feathery leaves.*

Red List: Vulnerable

Dorstenia gigas

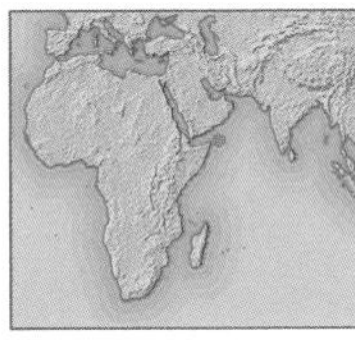

Botanical family
Moraceae

Distribution
Socotra Island, South Yemen

Hardiness
Frost-tender

Preferred conditions
Full sun. Sharply drained, gritty, humus-rich soil with moderate rainwater. A soil comprising pure pumice produces relatively rapid growth.

Like *Dendrosicyos socotrana* (*see p.102*), ***Dorstenia gigas*** is endemic to the Yemeni island of Socotra and is another extraordinary tree of that island's rich and weirdly impressive endemic flora. The species is known locally as *quartab*. As the scientific name might imply, *Dorstenia gigas* is the largest species in the genus, attaining a height of 3.7m (12ft).

Quartab is found only in the Haggier Mountains of north-central Socotra. It has been recorded growing on both limestone and granite formations. In years of average rainfall, populations of the species are largely untouched and unthreatened, since they are confined to vertical cliff faces and areas inaccessible to grazing goats. In drought years, however, whole plants, including the succulent trunk, are sought out, felled, and pulped for use as goat fodder.

In winter, *Dorstenia gigas* produces unusual flowers resembling open-faced figs, often with a dip in the middle. If a flower is fertilized, seeds form in swellings across the centre, eventually to be ejected at least 1m (3ft) from the plant. The seeds may be planted in a mix of granite gravel or pumice. The plant is also propagated by cuttings.

Dorstenia gigas is an odd-looking tree of great horticultural interest and there is a potential risk of overcollection for the market in succulent plants. Conservation measures do not seem to be in place for the species.

STRANGE BLOOMS ► *The flowers first appear as knobby buds; the centre is exposed when the outer edges pull back. Pollen from a male plant is needed to pollinate female flowers.*

DRACAENA DRACO *This massive dragon tree at Icod, near Orotava on the north coast of Tenerife in the Canaries, is an ancient survivor of deforestation. It has long been a popular visitor attraction.*

8 on Red List

Dracaena

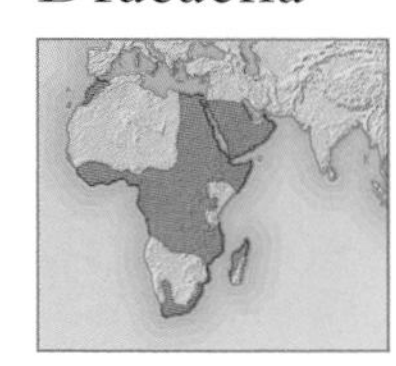

Botanical family
Dracaenaceae

Distribution
Canary and Cape Verde Islands, Morocco, tropical Africa, Arabia

Hardiness
Frost-tender

Preferred conditions
Full sun. Moderately fertile and moist but well-drained soil. Dracaenas require moderate humidity and plentiful water from spring to autumn, but prefer dryness in winter.

The genus *Dracaena* comprises around 40 species. They grow in tropical forest, scrub, and dry, open slopes, and six species are threatened with extinction in the wild. Most dracaenas are native to Africa and Arabia, but in a disjunction, one species is found in South America.

Dracaena draco (Vulnerable in the 2000 Red List) is endemic to islands off the west coast of Africa and now grows in five of the seven Canary Islands, where the total population is reduced to a few hundred trees. The species was also once an important component in the vegetation of relatively arid areas of Madeira and the small neighbouring island of Porto Santo, but today it is reduced to two individuals in the wild. It also occurs in the Cape Verde Islands.

The wild populations of this extraordinary tree have been in decline for a long time. Its red sap, called dragon's blood (*see below*), was a commercially valuable product in the Canary Islands, and felling the trees for extraction of the sap has been one of the main threats to the species. The tree is umbrella-shaped with a stout, silvery grey trunk and evenly spaced branches. Sword-shaped leaves are borne in dense rosettes at the ends of the branches, where the small, greenish-white flowers also appear on branchlets.

MOROCCAN DRACAENA

In 1996, a survey revealed populations of *Dracaena draco* in the Anezi region of the Anti-Atlas Mountains in Morocco. Thousands of individuals exist on steep quartzite cliffs in inaccessible gorges. These populations are likely to represent a distinct variety of the species. The species is protected by local legislation, and is also listed as a species in need of strict protection on the EU Directive on the Conservation of Natural Habitats and of Wild Fauna and Flora.

Two close relatives of *Dracaena draco*, both Endangered in the 2000 Red List, are the Nubian dragon tree, *Dracaena ombet*, which extends from northeast Africa into the Arabian Peninsula, and *Dracaena cinnabari*, the endemic species of Socotra, South Yemen. Exploitation of the Socotran dragon tree for local purposes still takes place. Dragon trees are propagated from seed or from stem cuttings and reach maturity in 30 years. *Dracaena draco* is cultivated in parks and gardens in the Canary Islands, and in botanic gardens elsewhere.

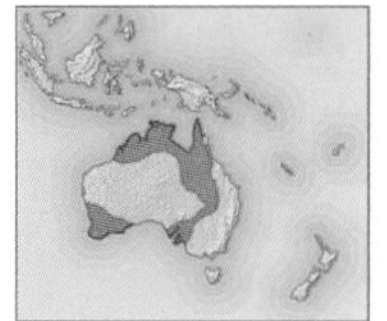

◄ DRACAENA CINCTA *This ornamental Madagascan species and its cultivars are commonly seen in commercial and domestic settings.*

HISTORY

DRAGON'S BLOOD

For many centuries, voyagers have slashed the trunks of *Draceana draco* trees (*see left*) and collected the oozing beads of red sap, called dragon's blood. This product was used by the ancient Egyptians when embalming the dead. In medieval times the dried resin was believed to have magical properties – for example, it was burned to entice lovers to return, and a piece placed in the bed was reputed to cure impotency. Medieval physicians also used the sap medicinally as an external astringent.

Dragon's blood was later used as a stain for mahogany, and 18th-century violin makers used it to impart a lustrous, deep red to the wood of their instruments. In the Middle East the sap of another species, *Dracaena cinnabari*, is still collected. In countries such as the Yemen the resin, also called dragon's blood, remains in use as a traditional medicine and as a dye.

▲ DRYANDRA FORMOSA *has unique orange flowers growing among holly-shaped leaves at the top of foliage-covered stems.*

12 on Red List*

Dryandra

Botanical family
Proteaceae

Distribution
Australia, especially to the north and west

Hardiness
Frost-tender

Preferred conditions
Full sun. Sharply drained, neutral to acid soil of low fertility (or at least low levels of phosphates) is most preferred. Plants may develop iron deficiency if phosphate levels are too high.

The uniquely Australian genus *Dryandra* is named after the Swedish botanist Jonas Dryander, a specialist in Australian plants and librarian to Sir Joseph Banks (*see p.95*). Dryandras grow on sandy heath, on dry, rocky, sandy coasts, and in scrubland. They are closely related to the Australian genus *Banksia*, as well as the true *Protea* genus of South Africa – dryandra flowers, with their silvery interior and golden, furry upper surface, strongly recall *Protea* species. The common names of, for example, ***Dryandra formosa*** – showy dryandra, Australian golden dryandra, and golden emperor – testify to the popularity of the genus with flower arrangers.

There are around 60 species, a number of which are under threat in the wild. All are distinguished by their flowers, in yellow through orange to brown with shades of purple. Some species produce flowers at ground level; these are commonly called honeypot dryandras because of their heavy nectar content.

Most dryandras are well protected from grazing animals by their armoury of spines and tough, unpalatable leaves. However, dryandras are very susceptible to root rots, and where their habitats have been accidentally contaminated by the fungus *Phytophthora* they have joined the list of Australia's most critically endangered plants. One species, *Dryandra montana,* is restricted to fewer than ten plants in the wild and is classed as Endangered in the Red List of 1997.

HONEY POSSUM ► *Native marsupials and birds pollinate dryandras as they deftly ply them for the abundant nectar.*

175*/2 on Red List

Eucalyptus

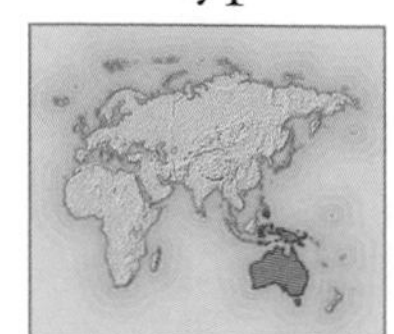

Botanical family
Myrtaceae

Distribution
Australia

Hardiness
Fully hardy – frost-tender

Preferred conditions
Eucalyptus species prefer full sun and fertile, neutral to acid soil that does not dry out. Young trees require shelter from cold, drying winds.

Eucalyptus trees grow in all but the very driest parts of Australia. Of approximately 450 Australian species, over 200 have been planted elsewhere in the world, many of them in gardens. Fast-growing species are raised as plantation crops, although sometimes they grow at the expense of native vegetation. Within Australia, eucalyptus species vary from the massive timber trees of the coastal forests to the stunted but often beautiful "mallee" multi-stemmed forms of the arid interior. Many of the eucalypts have naturally restricted distributions and over 100 are now considered to be threatened with extinction in the wild.

The Queensland western white gum, *Eucalyptus argophloia*, is cited as Vulnerable in the Red List of 1997. An attractive tree with bright, silvery-white bark, it grows to a height of 40m (130ft). The lance-shaped leaves are stalked, alternate, and very firm in texture.

In its natural habitat this species is confined to a small area of woodland or open forest north of Chinchilla in southern Queensland. Much of the land within its range has been cleared for pasture and crops. The small remnants of forest are in an area mainly used for grazing. Some of the scattered populations of this eucalypt are protected in a State Forest Scientific Area; others are in road reserves cared for by local authorities. Protection from grazing on private land, together with the establishment of an additional protected area, should ensure the tree's survival in the wild. The species tolerates saline soils and low-rainfall areas and regenerates readily from seed.

▲ EUCALYPTUS RHODANTHA *has flowers up to 6cm (2½in) in diameter with bright red stamens and prominent yellow anthers.*

▼ EUCALYPTUS EATERS *Koala bears, such as this one at the David Fleays Sanctuary in Queensland, live in eastern Australia. Their diet consists mainly of the leaves and flowers of certain eucalyptus species.*

The Victorian silver gum or Buxton gum, *Eucalyptus crenulata*, is classified as Endangered in the Red List of 1997. Like *Eucalyptus argophloia*, this species is now more abundant in cultivation than in the wild. Trees are grown for shade and reach up to 12m (40ft) in height. The small, toothed leaves are greenish-grey and are often used in cut flower arrangements. The profuse flowers are white to cream and appear in spring.

BUXTON GUM HABITATS

In the wild, the Buxton gum is found only in a small area of Victoria, where there has been extensive clearance of the fertile river flats for agriculture. The silver gum grows on flat or gently sloping sites near rivers or in moist depressions, often as an understorey tree of *Eucalyptus ovata.* The two natural populations of *Eucalyptus crenulata* grow northeast of Melbourne. One population is in the swampy flats of the Acheron River valley. Part of this population lies within the 16.9-ha (42-acre) Buxton Silver Gum Reserve, established in 1978. In addition to this reserve and some trees outside its immediate area, there is a population consisting of around 15 trees over 4ha (10 acres) on the fertile Yarra River flats at Yering. Half of this area has been protected since 1979 within the 125-ha (308-acre) Spadonis Reserve; the other half is on private land.

EUCALYPTUS GUNNII *Juvenile leaves are small and rounded* (top), *later giving way to long, lance-shaped adult foliage* (above).

Despite the Victorian silver gum's protected status, it remains threatened. Edge and "island" erosion associated with small patches of habitat, or changes to drainage or to the fire regime, could contribute to the loss of the last few hundred wild trees. Infestation by the parasitic coarse dodder-laurel, *Cassytha melantha*, is a potential threat to the Buxton population, and grazing and competition from understorey vegetation threaten at Yering. Trees are also attacked by the caterpillar *Uraba lugens* (*see below*). Planting is under way to increase the species population to 1,000 potentially seed-bearing trees over the two sites.

This eucalypt will grow equally well in moist or well-drained soil and is easily propagated from seed. It is available from native plant nurseries, but cultivated varieties of the Buxton gum have been selected for particular forms and no longer represent the natural populations.

Rose gum, ***Eucalyptus rhodantha***, Endangered in the Red List of 1997, is a low-spreading tree with attractive, blue-grey, stalkless leaves, usually in a moderately dense crown. In the wild this species is close to extinction. It is confined to a very small area in Western Australia where its natural habitat is heath scrub on sandy soils. Most of the land within its area of distribution has been cleared for agriculture, but the owners of the site where the rose gum occurs were persuaded to leave the patch of heathland intact. Remaining threats to this attractive plant are collection of the flowers and seed, and damage from roadworks. Also, the increasing isolation of the populations has led to inbreeding. The rose gum is legally protected.

Rose gum grows best in a dry, sunny position. It is moderately tolerant of frosts but may need protection when young. Propagation is by seed and plants flower at 3–4 years. This species is found in botanic gardens in Australia and California, and also in private gardens.

See also Invasive plants, *p.454.*

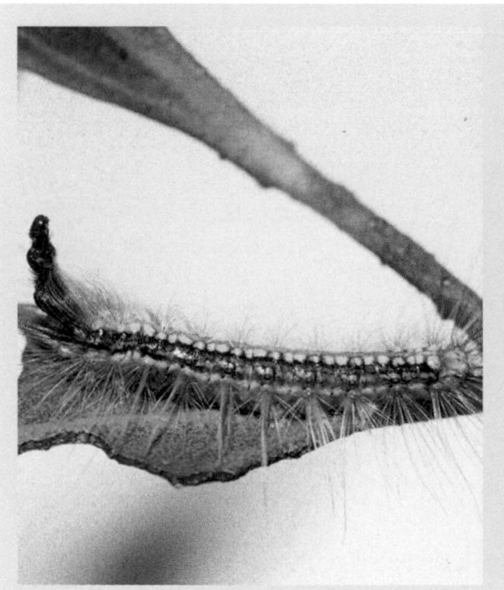

THREAT

GUM-LEAF SKELETONIZER

Marauding insects seldom have a lasting effect on the future prospects of widely distributed tree species, but they pose a significant threat to the survival of small pockets of poorly regenerating species. The reserve of *Eucalyptus crenulata* trees at Yering has been attacked on several occasions by larvae of the gum-leaf skeletonizer (*Uraba lugens, see left*). In 1978, a particularly severe attack by an unidentified lepidopterous leaf miner affected the entire Buxton gum population, with some trees losing as much as 30 per cent of leaf tissue. Such insects are easily controlled with chemical sprays, but resources are often unavailable.

Red List: Vulnerable

Fagus hayatae

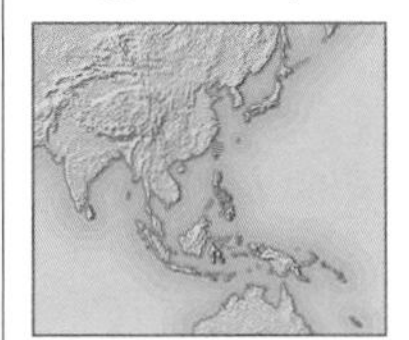

Botanical family
Fagaceae

Distribution
Taiwan

Hardiness
Fully hardy – half-hardy

Preferred conditions
Full sun or partial shade. This tree prefers well-drained soil. As a young plant it requires shade and a sheltered site that provides some protection from wind and direct heavy rainfall.

The Taiwan beech, *Fagus hayatae* is a little-known deciduous tree species found only in the mountains of northern Taiwan at elevations of 1,300–2,000m (4,265–6,560ft). Its natural range is very restricted and estimates suggest that it is confined to an area of less than 1,000 sq km (386 sq miles). Taiwan beech can be found in broadleaved submontane forest in the northern and southern parts of Chatienshan and in smaller populations in Tungshan and Agushan in Ilan County, where the climate is cool and moist.

The trees are slow-growing and populations are largely made up of old individuals. Regeneration is poor since it depends on the appearance of gaps in the existing vegetation, but these are currently being invaded by more aggressive shade-tolerant species such as alpine bamboo and broadleaved species. Habitat clearance and disturbance are also threats to the Taiwan beech, which is morphologically similar to (and may turn out to be synonymous with) *Fagus lucida* from China's central provinces.

The Taiwan beech is propagated by seed. Transplanted young plants are reported to show poor root production and so direct sowing is recommended. The species is in cultivation in the Chia-Yang Botanical Garden in Taiwan, which has national responsibility for its *ex situ* conservation, along with 31 other species of endemic and threatened montane plants of the island. It is also cultivated at Fushan Botanical Garden in Taiwan.

3* on Red List

Forsythia

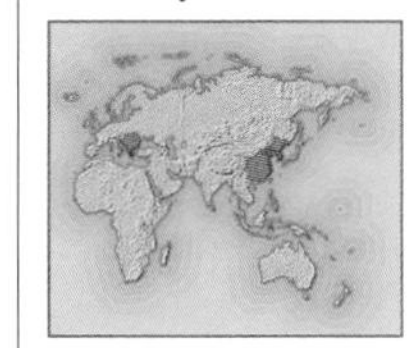

Botanical family
Oleaceae

Distribution
China, Korea, Japan, Europe

Hardiness
Fully hardy

Preferred conditions
Full sun or light, dappled shade. Moderately fertile, moist but well-drained soil produces the best flowering, but plants tolerate all types of soils and survive in dry conditions.

Named after the Scottish horticulturist William Forsyth (1737–1804), the ever-popular genus *Forsythia* is widely familiar as a garden shrub, with many hybrids producing masses of yellow spring flowers. Most garden plants are derived from *Forsythia suspensa*, a species from open woodland in China. In the wild there are around seven species, with one native to Europe and the others to China, Korea, and Japan.

Sadly, the European species, ***Forsythia europaea***, is now very rare in the wild. Distribution of this species is mainly in Albania and Kosovo, possibly extending slightly further east. It grows on serpentine rock in deciduous mixed forests comprising oak, ash, and maple species. During the recent conflicts in the Balkans, rapid and intense forest degradation, caused by fire, pollution, and illegal logging, is thought to have contributed further to losses of this shrub, which is naturally rare.

Forsythia europaea is easy to grow but is considered the least ornamental of the genus and is therefore rarely cultivated outside botanic gardens. Propagation of the plant is by cuttings and seed.

In Japan, two varieties of *Forsythia japonica* are classed as Vulnerable in the Red List of 1997; they are *Forsythia japonica* var. *japonica* and *Forsythia japonica* var. *subintegra*. Another threatened species of the genus is *Forsythia saxatilis*, which is endemic to Mt Kwanak and Mt Pukhan in South Korea and is classed as Endangered in the Red List of 1997.

▼ FORSYTHIA EUROPAEA *is a spreading, deciduous shrub that grows to 2m (6½ft). The flowers appear on the upright branches in March and April, before the leaf buds unfold.*

EUCALYPTUS PAUCIFLORA *SUBSP.* NIPHOPHILA, *called the alpine snow gum, is an especially hardy subspecies, popular with gardeners for the range of colours revealed when the bark is shed.*

Red List: Extinct in the Wild

Franklinia alatamaha

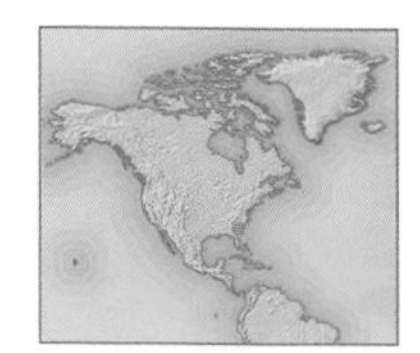

Botanical family
Theaceae

Distribution
USA: Georgia

Hardiness
Fully hardy

Preferred conditions
This plant thrives in full sun, with afternoon shade in hotter areas. It prefers acidic (pH 5–6), moist, well-drained, organically rich soil.

One of the earliest meticulously recorded plant extinctions, the Franklin tree, ***Franklinia alatamaha***, has not been seen growing in the wild since 1803. The last known stand of the trees occupied a very small and distinctive habitat in the acidic sandhill bogs lying at the mouth of the Altamaha River in McIntosh County, Georgia, USA. The Altamaha River corridor remains a priority for plant conservation, continuing to provide critical habitats for such rare species as Radford's dicerandra, *Dicerandra radfordiana*, which grows nowhere else in the world.

Franklinia alatamaha survives as a popular garden tree, despite being quite demanding in its preferred growing conditions. It is a multi-stemmed shrub or small tree with smooth bark that splits to reveal a lighter colour. The leaves are dark green and lustrous, growing up to 20cm (8in) long. In the autumn the foliage turns to shades of orange, red, and purple. The main attraction is its large, creamy white, cup-shaped flowers, which are 6–12cm (2½–5in) across with waxy petals and golden yellow stamens. The flowers have earned the tree a second common name – the "lost camellia" – and are sweet-smelling with a fragrance like orange blossom. The trees have a relatively long flowering period, from July through to the first frosts of autumn.

The Franklin tree is easily raised from seed. Seed capsules should be collected in October to November, allowed to dry and split, and then the seeds should be sown straight away. The tree can also be propagated from stem cuttings, which should be taken from June to August.

Successful growers can register their tree online with the Franklinia Census, monitored by Bartram's Garden in Philadelphia, the historic home of the two explorers credited with saving this tree from extinction (*see right*). The ongoing results of this census, along with further advice for would-be *Franklinia* growers, can be found at the website www.bartramsgarden.org. Pennsylvania has more than 500 plants registered, while North Carolina and New Jersey also have high totals. The UK has around 50 franklinias registered.

▼ RARE BEAUTY *The blossom of this species has been adopted as the State Flower of South Carolina, even though it originally grew in Georgia.*

PEOPLE

FRANKLINIA'S SAVIOURS

The Franklin tree was first observed in its minuscule geographical native range by the botanist John Bartram and his son, William. In 1765 they discovered a small grove of the species growing along the Altamaha River in the state of Georgia in southeast USA. They named the tree in honour of their friend, Benjamin Franklin, one of the drafters and signatories of the US Declaration of Independence. The Bartrams collected seed and took cuttings in order to grow the species in their garden in Philadelphia, Pennsylvania.

On their return, just 38 years later, *Franklinia alatamaha* could no longer be found. But, thanks to the Bartrams' efforts, the species still exists. According to a recent census, more than 2,000 specimens survive in gardens around the globe. Most are in the USA, yet healthy populations exist in Europe, Canada, and New Zealand. Each one is a descendant of the trees propagated by the Bartrams for their garden in Philadelphia.

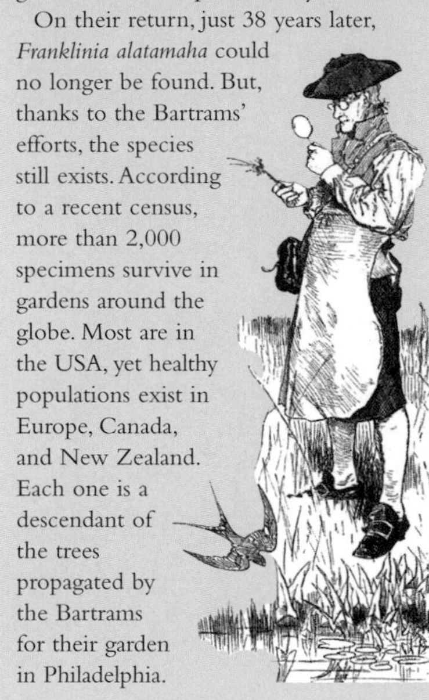

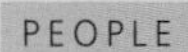

12*/2 on Red List

Hebe

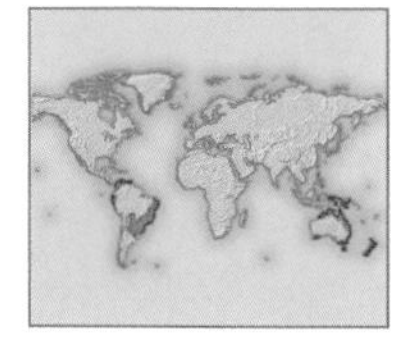

Botanical family
Scrophulariaceae

Distribution
New Zealand, Australia, New Guinea, South America

Hardiness
Fully hardy – half-hardy

Preferred conditions
Sun or partial shade, with shelter from cold, drying winds. Hebes prefer poor or moderately fertile soil that is moist but well-drained.

There are over 100 species in the genus *Hebe* and all but three are endemic to New Zealand – it is that country's largest genus of plants. They occur throughout New Zealand, particularly in river valleys, the alpine and subalpine zones, and in scrub and grassland. Many are widely cultivated around the world, and over 1,000 cultivars are now named. In the wild, some of the species, such as *Hebe armstrongii* and ***Hebe barkeri*** (both Vulnerable in the Red List of 1997), are of conservation concern.

Hebe armstrongii is a moderately hardy "whipcord hebe", so named because its small leaves lie flat against the stems. It grows erect to 60cm (2ft) high, with slender, upright branches. The flowers are white, appearing in long terminal clusters from June to August. The golden-green leaves are tiny and overlap.

In the wild *Hebe armstrongii* is confined to the Castle Hill Area of Canterbury, New Zealand. It is found in the Enys Scientific Reserve and Mount White Station. A major threat to this species is habitat modification by invasive plants, including relatively fast-growing introduced conifers that are causing problems for grassland and shrubland plant communities throughout most of the uplands of South Island. Conservation action for *Hebe armstrongii* includes research and field surveys, control of the invasive plants, and planting of propagated seedlings at the Enys Scientific Reserve.

Also called Barker's koromiko, *Hebe barkeri* has velvety leaves and pale mauve-to-white flowers. Once a significant component of forests in the Chatham Islands, New Zealand, it is now reduced to a handful of populations and scattered isolated individuals. The species is severely browsed by livestock and more recently by possums. Fortunately it is found in various reserves set aside for conservation. Although uncommon in cultivation, the species is easily grown from cuttings and fresh seed.

▼ HEBE BARKERI *is the largest shrub or small tree in its genus. Its overall size and large leaves relative to other hebes may have evolved due to a lack of sunshine and high soil fertility in its Chatham Islands habitat.*

Red List: Vulnerable

Heptacodium miconioides

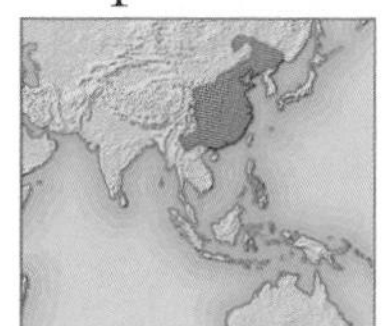

Botanical family
Caprifoliaceae

Distribution
China

Hardiness
Fully hardy

Preferred conditions
The species is best in full sun in areas with long, hot summers that mimic its native habitat, but it also grows well in partial shade. It adapts to most soils as long as they are well drained.

Commonly known as the seven-son flower, *Heptacodium miconioides* is native to three provinces in China – Anhui, Hubei, and Zhejiang – in all of which it is regarded as vulnerable by local botanists. The species typically grows in woods, often in rocky areas. It first reached the attention of the West in 1907, when material of the small tree was collected in China by the great plant hunter Ernest H. Wilson (*see p. 113*).

Heptacodium miconioides grows to a height of 5m (16ft) and bears bold, oval leaves in opposite pairs. The Chinese and American common name of seven-son flower refers to the creamy-white, sweetly scented flowers, which are typically borne in clusters of six flowers encircling a seventh at the centre of the branching flowerhead. But perhaps the finest feature of the plant are the sepals, which expand and turn rosy-red after flowering in a very fine autumn display.

The species is best propagated by seed or by semi-ripe cuttings in midsummer. It flowers at the end of the current year's growth, so pruning should take place during the winter or very early spring.

Red List: Vulnerable (two varieties)

Juglans californica

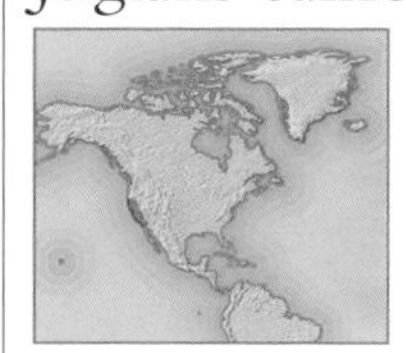

Botanical family
Juglandaceae

Distribution
California, USA

Hardiness
Half-hardy

Preferred conditions
Trees generally occur on moist sites such as north slopes, creek beds, canyon bottoms, and alluvial terraces. Trees grow best in deep, alluvial soils and tolerate full sun and seasonal flooding.

A distinctive species of walnut with attractive, feathery leaves, ***Juglans californica*** is found in two varieties: the Northern California black walnut, *Juglans californica* var. *hindsii,* and the Southern California black walnut, *Juglans californica* var. *californica*, both endemic to that state. The decline of both temperate woodland varieties is mainly attributed to 200 years of grazing, and throughout California the walnut forest has become a much fragmented and declining habitat.

Black walnut trees have considerable cultural significance. Remnant natural stands of walnuts are historically important since they are often associated with Native American encampments. The trees are also ecologically important; rodents, including California ground squirrels and western grey squirrels, eat the nuts, and ground squirrels burrow at the bases of old trees. Raptors such as owls roost and nest in the upper branches.

Currently only two native stands of Northern California black walnut are known, one occurring in Napa County and the other in Contra Costa County. However, the tree is widely naturalized in other parts of California and was formerly cultivated as a rootstock for the common walnut, *Juglans regia*.

The Southern California black walnut is declining throughout its range, which is now limited to the drainage of the Santa Clarita River. The best remaining stands are in the San Jose Hills. Narrow, isolated stands of Southern California walnut sometimes occur in chaparral and occasionally the variety is found in coastal sage scrub. Most Southern California black walnut woodlands are subject to periodic fires and the tree is adapted to regenerate by sprouting from the root crown and trunk after burning.

▲ YOUNG FRUITS Juglans californica *has large fruits for its genus, mid-green when first formed, maturing to deep black. The walnuts inside were eaten by the Chumash people.*

71 on Red List

Ilex

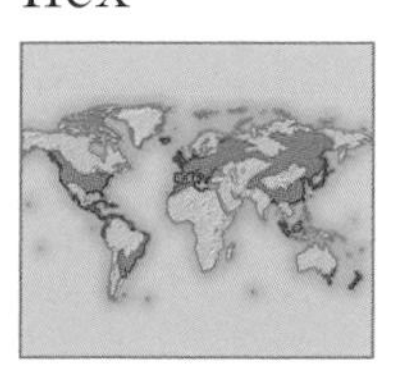

Botanical family
Aquifoliaceae

Distribution
Tropical, subtropical, temperate regions

Hardiness
Fully hardy – frost-tender

Preferred conditions
Full sun or partial shade (variegated hollies have best leaf colour in full sun). Preferred soil is moist, well drained, and humus-rich.

There are around 400 holly species growing in temperate, subtropical, and tropical woodlands around the world. Typically the smooth-barked trees or shrubs have small flowers, fleshy red or black berries, and leathery, shiny leaves. Over 80 *Ilex* taxa are considered to be threatened with extinction in the wild.

The large-fruited holly, *Ilex macrocarpa*, is native to southern China and was introduced into western gardens in 1907. It is included on the 1997 Red List of Japanese Vascular Plants. It is a small to medium-sized deciduous tree with oval, serrated leaves up to 15cm (6in) long. Its large fruits look like small black cherries.

Another threatened member of the genus is a montane forest species, ***Ilex perado* subsp. *platyphylla***, known in Spanish as *naranjero salvaje* and in English as the Canary Island holly. This plant, classed as Vulnerable in the 1997 Red List, is confined to parts of Tenerife and La Gomera in the Canary Islands. It is a species of the "laurisilva" type of forest (*see p. 110*) that occurs mainly on the islands' northern slopes. The populations are small and fragmented, although an important remnant of laurisilva currently receives protection as part of La Gomera's Garajonay National Park, a UNESCO Natural World Heritage Site.

See also Invasive plants, *p.457*.

ILEX PERADO *SUBSP.* PLATYPHYLLA ► *is a parent of the* Ilex x altaclerensis *hybrids now well established in gardens and parks.*

Red List: Extinct in the Wild

Kokia cookei

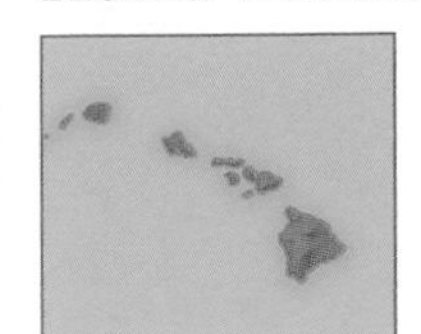

Botanical family
Malvaceae

Distribution
Hawaii

Hardiness
Frost-tender

Preferred conditions
Full sun or partial shade; *Kokia cookei*, like other endangered *Kokia* species, grows within dry forest on soil derived from a lava substrate.

The genus *Kokia* consists of four species, all endemic to the dry tropical forest of Hawaii and either extinct or critically endangered. When ***Kokia cookei*** was first discovered in the 1860s by a Mr R. Meyer, only three small trees of the species were found. They were growing at an altitude of around 200m (650ft) near Mahana, on western Molokai Island. The site formed part of a sheep run and the plants were directly affected by browsing and trampling by domestic and feral animals. Habitat conversion, loss of native pollinators, and seed predation by insect larvae also contributed to the decline. By 1918 the last few specimens were dead.

Some years later, the only known cultivated tree died without producing viable offspring, and the species was thought to be extinct. However, another cultivated specimen was discovered in 1970, and living material from it was grafted onto the Critically Endangered *Kokia kauaiensis*. Currently, the species exists in cultivation at two locations and in managed outplantings at three sites.

Various means of propagation have been used in the attempt to produce more plants of this species, including seeds, cuttings, grafting, tissue culture, and air layering. Land has been allocated to the species, but landowners reluctant to take on the role of managing *Kokia cookei* are in opposition.

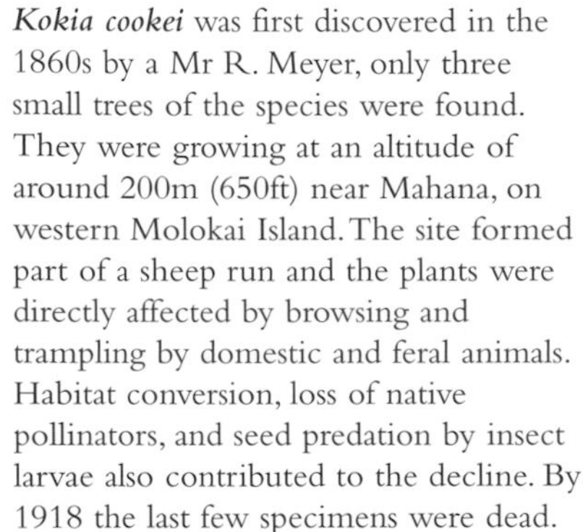

◄ SOLE SURVIVOR *The single tree in cultivation flowered for the first time in 1979 at Waimea Arboretum on Oahu.*

Red List: Lower Risk

Laurelia sempervirens

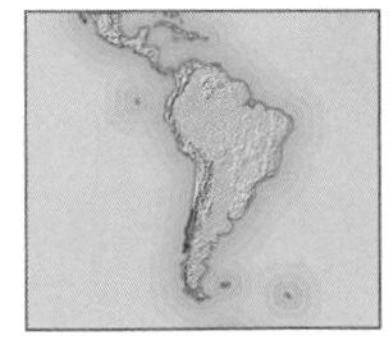

Botanical family
Atherospermataceae/Monimiaceae

Distribution
Chile

Hardiness
Fully hardy (borderline)

Preferred conditions
Full sun or partial shade, sheltered from cold, drying winds. Soil should be moderately fertile, moist but well drained.

Known as Chilean laurel, ***Laurelia sempervirens*** is a handsome evergreen shrub or small to medium-sized tree growing to 15m (50ft). In the wild the species is restricted to central Chile, where it grows in moist seasonal forest in the coastal cordillera, sub-Andean range, and central plains. Its bright green leaves are leathery, serrated, and aromatic. Its bark, leaves, and fruits are used as spices in Latin America.

The tree is at risk of overexploitation for timber in the north of its range. Chile has some of the greatest temperate forests in the world, but improved management and increased protection are needed to ensure that species such as the Chilean laurel do not slide closer to extinction.

▲ AROMATIC SEEDS *of this species are called Peruvian nutmeg, not to be confused with nutmeg from* Myristica fragrans.

Red List: Lower Risk

Laurus azorica

Botanical family
Lauraceae

Distribution
Morocco, the Azores, Madeira, Canary Islands

Hardiness
Half-hardy

Preferred conditions
Full sun or partial shade, sheltered from cold, drying winds. Preferred soils are fertile, moist, but well drained.

The Canary Island laurel, ***Laurus azorica***, is one of the characteristic species and major components of the relict laurel forests, called "laurisilva" (*see below*), that grow in the Canary Islands, the Azores, Madeira, and Morocco. In the Canary Islands this species, known locally as *laurel* or *loro*, is common in the cloud forest zones of Tenerife, La Palma, La Gomera, and El Hierro. Threats to the laurel forests include grazing, fires, and habitat conversion.

This species of laurel grows up to 15m (50ft) in the wild. It has smooth, grey bark and pubescent shoots. The leathery leaves are alternate, elliptic to oval, and up to 15cm (6in) long. They are dark green on the upper surface while the undersurface is paler and hairy. The flowers are greenish-yellow in flattened flowerheads and the fruits are black when ripe. There are only two laurel species, the other being *Laurus nobilis*, the widely planted ornamental sweet bay.

▲ LAURUS AZORICA *differs from the true or bay laurel,* Laurus nobilis, *in having larger and broader leaves.*

HABITAT

LAST RELICS OF A PREHISTORIC FOREST

The laurel forests, also known as "laurisilva", of the Canary Islands occur on the northern slopes of the islands, which are bathed in mist and cloud. They are of great ecological importance both in controlling erosion and in condensing water from the clouds (hence their description as cloud forests).

The broadleaved forests are also of great botanical interest. They contain a high number of rare and endemic species believed to be relics of a flora that covered much of southern Europe and north Africa during the Tertiary period (about 15–40 million years ago).

Laurisilva is dominated by four species of the laurel family: *Laurus azorica* (laurel), *Persea indica* (viñátigo), *Apollonias barbujana* (barbusano), and *Ocotea foetens* (til). Among the many rare species also present are *Arbutus canariensis* (*see p.94*) and *Ilex macrocarpa* (*see p.109*). Clearance for agriculture has been the main threat to the laurel forests; on Tenerife, for example, they have been reduced by more than 90 per cent. However, most of the Canary Island laurel forests are now protected, including those within the Garajonay National Park on La Gomera.

VOLCANIC HABITAT ► *The Canary Islands are volcanic in origin and their laurisilva forest grows in soils derived from volcanic ash and lava. Laurisilva survived in the Canaries, Madeira, and the Azores because of their warm climates – the forests disappeared from much of Europe following the cooling at the end of the Tertiary era.*

43*/4 on Red List

Leucadendron

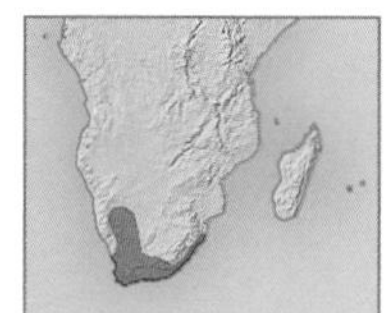

Botanical family
Proteaceae

Distribution
Southern Africa

Hardiness
Frost-tender

Preferred conditions
The genus likes full sun and poor, well-drained, neutral to acid soils. *Leucadendron argenteum* grows on granite clays. *Leucadendron discolor* grows more often on rocky sandstone.

There are over 80 species in the South African genus *Leucadendron*. They are evergreen trees or shrubs with silvery white leaves and solitary terminal flowers. Various species have a precarious future in the wild since their habitat, the remarkable fynbos vegetation of South Africa, faces threats from agricultural conversion, invasive species, and fire.

Leucadendron argenteum (Vulnerable in the 2000 Red List) is a shrub found only in the Western Cape of South Africa, especially on the slopes of Table Mountain. Population counts suggest that there are more than 10,000 mature individuals.

The expansion of Cape Town and the establishment of tree plantations are the main threats to the species. There is also concern about the mixing of gene pools following extensive planting of the species. Regeneration of *Leucadendron argenteum* in the wild seems to be limited.

Leucadendron discolor (Endangered in the 2000 Red List) is a tall shrub, sometimes growing to the size of a tree, that is confined to the western side of the Piketberg Mountains, also in the Western Cape. The total population is between 1,000–5,000 mature individuals, fragmented into three or four major subpopulations and a number of scattered individuals, all of which occur within an area of 20 sq km (8 sq miles). Nearly half of the population occurs within protected areas, where programmes against invasive plant species are in place.

▲ **LEUCODENDRON DISCOLOR** *This male plant bears the red flowerhead that earns the species its common name of "red cone bush". Female plants have yellow flowerheads.*

▲ **LEUCADENDRON ARGENTEUM** *has leaves with a silvery sheen caused by fine hairs that fold flat against the leaves in summer to limit water loss. In winter, the hairs are half-raised.*

Red List: Vulnerable*

Lindera melissifolia

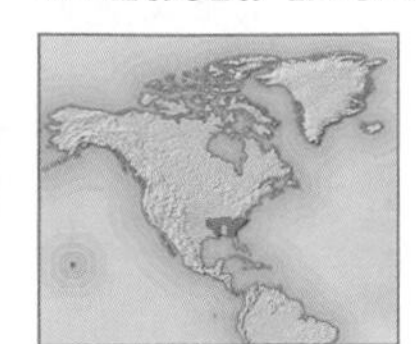

Botanical family
Lauraceae

Distribution
South-east USA

Hardiness
Fully hardy

Preferred conditions
Both full sun and shade are tolerated. This plant occurs on the margins of ponds and swamps in seasonally flooded wetlands and likes fertile, moist but well-drained, acidic soil.

The pondberry, *Lindera melissifolia*, is one of two *Lindera* species native to the USA. The shrub is found only in swampy forests in the southeast of the country. Logging and drainage of the species' habitats have led to its extinction in Alabama, Louisiana, and Florida. It is struggling for survival in six other states.

An additional major problem facing this species is a lack of seedling production. Pondberry is dioecious (both sexes must be present for reproduction) and many colonies are exclusively male. Research is under way to understand more fully the reproductive problems of the species and the reasons for the preponderance of male plants.

Lindera melissifolia flowers in early spring and produces attractive, bright red fruits in late summer. The leaves are aromatic. The species is cultivated at the Mercer Arboretum and Botanic Gardens in Humble, Texas, as part of the Center for Plant Conservation's national collection of endangered plants.

WILDLIFE

CATERPILLAR PLANT

In the USA, pondberry and spicebush shrubs (*Lindera melissifolia* and *Lindera benzoin* respectively), together with sassafras trees (*Sassafras albidum*) and some members of the laurel and magnolia families, are host plants for the caterpillars of the native spicebush swallowtail butterfly (*Papilio troilus*). The caterpillars shelter in folds in host-plant leaves and feed at night. The range of the black-winged butterfly follows that of its host plants, extending from south-eastern Canada to Florida, westward into Oklahoma and central Texas.

Red List: Lower Risk

Liriodendron chinense

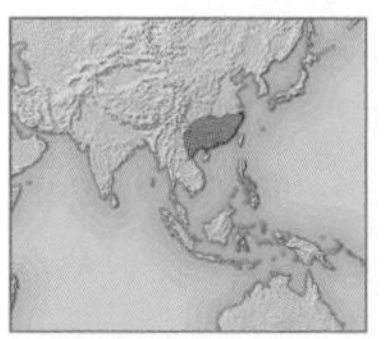

Botanical family
Magnoliaceae

Distribution
Southern China, northern Vietnam

Hardiness
Fully hardy

Preferred conditions
Full sun or partial shade. Moderately fertile, preferably slightly acid, moist, well drained soil.

In the wild, the ornamental and well-known species known as the Chinese tulip tree, ***Liriodendron chinense***, is widely scattered in montane evergreen broadleaved forest in the Yangtze River valley of China, extending farther south into northern Vietnam. The climate where this species grows is mild or cool and humid. Cross-pollination is limited because the mature individuals are scattered in their forest habitats, and therefore regeneration of the species is poor. Extensive logging and clearing of the habitat have affected populations throughout the range. The attractive wood of *Liriodendron chinense* is sought after for use in construction and furniture manufacture.

The tree is a relict species from a once widespread and species-rich genus of the Tertiary era. Fossil remains have been found in Japan, Greenland, Italy, and France. Now only one other species of the genus remains, the North American tulip tree, *Liriodendron tulipifera*.

Discovered at the beginning of the 20th century, the Chinese tulip tree was introduced into cultivation by Ernest Wilson in 1901. The tree differs from its American relative in having smaller, yellow-green flowers. It is fast-growing and thrives in all types of fertile soil, but is much less frequently cultivated than the American species. Propagation is by seed.

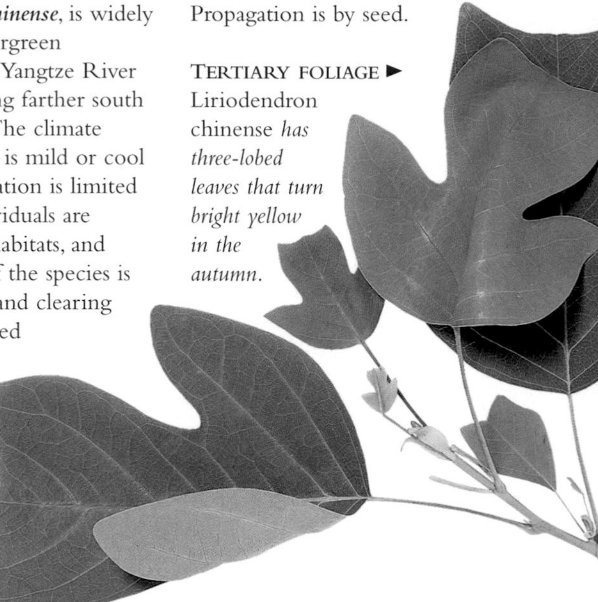

TERTIARY FOLIAGE ► Liriodendron chinense *has three-lobed leaves that turn bright yellow in the autumn.*

Red List: Endangered*

Liquidambar orientalis

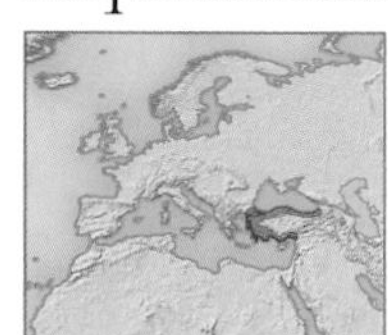

Botanical family
Hamamelidaceae

Distribution
Turkey, Greece: eastern Aegean Islands

Hardiness
Fully hardy (borderline)

Preferred conditions
The species grows in semi-shade and in full sun. Preferred soil is acidic or neutral, moist, and may be sandy, loamy, or even heavy clay. Young plants are susceptible to frost damage.

The Oriental sweet gum or storax, ***Liquidambar orientalis* var. *orientalis***, is a slow-growing, large bush or small tree found in flood plains, marshes, and by the sides of streams in Turkey. Its range in that country was previously far wider than now, and since 1945 forests of *Liquidambar orientalis* have been reduced from 63 sq km (24 sq miles) to 13.5 sq km (5¼ sq miles). Much of that habitat has been converted for agriculture, the trees being cut for firewood. The species was introduced into cultivation about 250 years ago, but it is still relatively rare in UK collections.

Liquidambar trees are the source of a valuable aromatic resin, also called storax, which forms in cavities of the bark and exudes from the trunk. The resin is collected in autumn primarily for the production of a fixative used in the perfume industry. Storax is also used medicinally. It is an ingredient of friar's balsam, a mixture that is inhaled to soothe inflammation of the respiratory tract, and is also used in skin ointments and many forms of treatment.

In the garden, autumn colour is the species' main attraction, along with the orange-red bark, which cracks coarsely into dark brown flakes. Propagation is by seed, sown as early in the year as possible.

Liquidambar orientalis is now the target of a special conservation programme in Turkey. The species is also native to the eastern Aegean Islands of Greece.

▼ FRAGRANT LEAVES *This species has crowded foliage, which in autumn turns to attractive shades of yellow through to red. The leaves are aromatic, especially when bruised.*

29*/45 on Red List

Magnolia

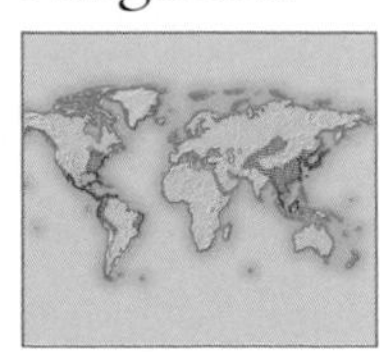

Botanical family
Magnoliaceae

Distribution
Asia; from southern USA to Brazil and western South America

Hardiness
Fully hardy – frost-tender

Preferred conditions
The genus likes moist but well-drained, humus-rich, preferably somewhat acid soil in sun or partial shade, with shelter from strong winds. *Magnolia delavayi* tolerates dry soils.

The genus *Magnolia* is named after a French professor of botany, Pierre Magnol (1638–1715). It is one of the most ancient genera of flowering plants, containing around 120 species; about 25 are grown as hardy plants in cool temperate regions. Magnolias grow in temperate to tropical woodland, scrub, and on riverbanks from the Himalayas to Japan and Malaysia, and from the south-eastern states of the USA to Brazil.

At least 45 magnolia species may be threatened in the wild, and efforts are being made by the Global Tree Specialist Group of IUCN's Species Survival Commission (SSC) to update the data. Assessment of the conservation status and threats to tropical magnolia species, which are less well known and are rarely found in cultivation, has highest priority.

Magnolia delavayi is known to gardeners as the Chinese evergreen magnolia and is on the 2000 Red List as Endangered. It is a temperate species that has been grown for centuries in China. Within Yunnan Province there are specimens associated with temples that are up to 800 years old. Although confined in the wild to its natural habitats in Yunnan and Sichuan provinces, this plant is increasingly cultivated in the West, particularly in its new red forms.

Magnolia delavayi is a large evergreen shrub or tree growing to about 10m (33ft). The bark is pale yellowish-white in colour. The creamy white flowers rarely last for more than two days. The leaves are larger than those of nearly all other hardy trees or shrubs. The species was introduced into cultivation by Ernest Wilson (*see above*) in 1899. It grows well on chalk soils and is often planted against walls. It usually flowers from seed in 3–5 years.

▼ MAGNOLIA CAMPBELLII *VAR.* ALBA *The spectacular spring flowers of this species exemplify the qualities that captivated western gardeners when the genus first became known.*

PEOPLE

PLANT HUNTER

The collector Ernest H. Wilson was born in 1876 in Chipping Camden, England. He was recommended for a visit to China on behalf of Harry Veitch's nursery in order to bring back *Davidia involucrata* (*see p. 101*) for cultivation. Sailing in 1899, the trip lasted three years, during which he found the tree and 400 new plants. This and subsequent trips earned him the sobriquet "Chinese" Wilson. A second trip in 1903 brought many more finds, including *Meconopsis* (poppy) species, rhododendrons, roses, and primulas. In 1906 he moved to the Arnold Arboretum in Boston, USA. Wilson discovered and described more than 3,000 species and introduced over 1,000 plants to the West, including *Magnolia wilsonii* (*see left*).

JAPANESE MAGNOLIA

Six magnolia species are native to Japan. Perhaps the rarest of these is *Magnolia pseudokobus* (Extinct in the Red List of 1997), which has only been recorded from Tokushima Prefecture, Shikoku. Although it is extinct in the wild, the species survives in a few local private gardens.

Much less threatened is *Magnolia stellata*, the star magnolia or shidekobushi, which is widely grown in Japan as an ornamental tree. This species grows on gently sloping land, on hills and valley plains, and also in riverbeds or shallow gorges. It prefers sunny sites with a slight water flow. The occurrence of *Magnolia stellata* as a wild plant has been doubted in the past, but there is growing evidence that the populations in Mie, Gifu, and Aichi prefectures of central Honshu Island are naturally occurring plants.

Introduced into cultivation in the West in 1862, the star magnolia is a fairly slow-growing species that reaches up to 3m (10ft) in height. When the shrub blossoms in March or April it bears a profusion of fragrant white flowers. In the wild the species is threatened by indiscriminate commercial collection and encroaching land development.

Wilson's magnolia, ***Magnolia wilsonii***, is classed as Endangered in the 2000 Red List. This native plant of Sichuan, northern Yunnan, and western Guizhou provinces in China is a deciduous shrub or small tree growing to about 8m (26ft) and is found in forests and thickets at 1,900–3,300m (6,200–10,800ft). Many of the natural populations have been destroyed because the bark is collected for medicinal use, often as a substitute for bark of *Magnolia officinalis* (*see below*). Forest clearance is another factor threatening the survival of Wilson's magnolia. Unfortunately, regeneration of the species in the wild is poor.

Magnolia wilsonii was introduced into Western cultivation by Ernest Wilson in 1908. It is now well established in cultivation and is highly prized for its attractive flowers. *Ex-situ* conservation in botanic gardens and arboreta should contribute to the long-term conservation of this beautiful magnolia species.

Although *Magnolia officinalis* has a very widespread distribution within the broadleaved deciduous forest of China, agricultural conversion, logging, and overharvesting of its bark have caused it to become rare outside cultivation. It is designated as being at Lower Risk in the 2000 Red List.

◄ MAGNOLIA DELAVAYI *The six creamy white inner tepals are cup-shaped, surrounding the unusual, erect, cone-like structure that bears the fleshy seeds.*

Red List: Data Deficient

Malus hupehensis

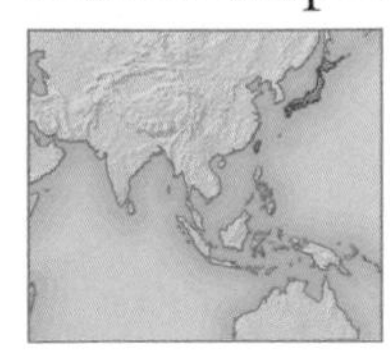

Botanical family
Rosaceae

Distribution
Taiwan, Japan, possibly China

Hardiness
Fully hardy

Preferred conditions
Both full sun and partial shade are tolerated. Moderately fertile, moist but well-drained soil.

Named after Hupeh in China and known to gardeners as the Hupeh crab, ***Malus hupehensis*** is an ornamental species of crab apple that grows to a height of 10m (33ft) and has stiff-spreading branches and oval, sharply serrated leaves. The status of this species in the wild is uncertain. In Taiwan it is found only in a very small area between Taipei and Ilan Counties in the north of the island. The wild population in this area consists of fewer than 50 individuals, occurring in forest growing at altitudes of 1,700–1,900m (5,600–6,200ft). *Malus hupehensis* is also native to Japan and possibly mainland China.

Effective conservation plans for the tree have yet to be drawn up. The species was introduced into Western cultivation by Ernest Wilson in 1900 and is readily obtainable from specialist nurseries.

CHINESE CRAB APPLES ► *The fruits of this rare tree follow the plant's profusion of single, fragrant, white flowers.*

Red List: Endangered*

Mespilus canescens

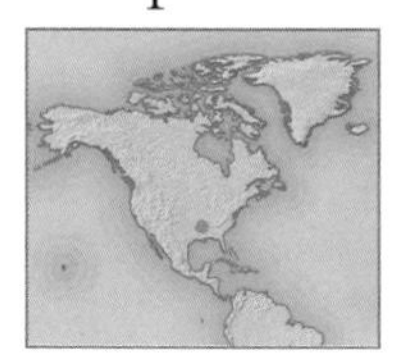

Botanical family
Rosaceae

Distribution
USA: Arkansas

Hardiness
Fully hardy

Preferred conditions
Full sun or light shade. Moderately fertile, moist but well-drained soil.

Stern's medlar, ***Mespilus canescens***, is a deciduous tree native to prairie. Only one other species of the genus is known, *Mespilus germanica*, which grows in southeast Europe and western Asia. Incredibly, Stern's medlar was not known or named by scientists until 1990.

Mespilus canescens is a beautiful, multi-stemmed plant that produces a mass of white blossoms in late April. This floral display is brief, lasting for just a week. The Missouri Botanical Garden is researching propagation of the species.

HABITAT

PRIVATE ACTION

The only wild population of *Mespilus canescens (see left)* occurs in the Mississippi Delta region of Arkansas, USA. Only 25 trees are known. Most of the other woods in the area have been cleared over the years for agriculture. The species receives no protection from federal legislation, but fortunately a private landowner took action to ensure its protection through an agreement with the Arkansas Natural Heritage Commission. Threats to this valuable plant that remain include changes in the water table, chemical runoff from adjacent farmland, and the presence of *Lonicera japonica*, an exotic invasive species.

Red List: Endangered

Michelia wilsonii

Botanical family
Magnoliaceae

Distribution
China

Hardiness
Half-hardy

Preferred conditions
The tree requires reasonably well-drained, fertile, and non-alkaline soil, with constant moisture.

An evergreen relative of the magnolia family, *Michelia wilsonii* is endemic to China and restricted to mountainous areas on the south to west fringe of the Sichuan Basin and Lichuan in western Hubei Province. The tree has a scattered distribution, growing in moist, broadleaved forest at altitudes of 700–1,600m (2,300–5,250ft). Habitat clearance and logging without controls have caused considerable declines in the populations.

Michelia wilsonii is a fast-growing tree that will reach 18m (60ft). The tree is easily grown, though of uncertain hardiness. Yellow, fragrant flowers appear in April to May and are offset by dark grey-green branches bearing shiny foliage, followed by pendulous purple seedpods in August to September.

9 on Red List

Nothofagus

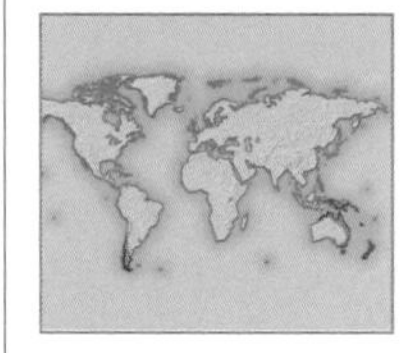

Botanical family
Fagaceae

Distribution
New Guinea to Australia, New Zealand, South America

Hardiness
Fully hardy – frost-tender

Preferred conditions
Grows in full sun, in fertile, moist but well-drained, lime-free soil. Young trees may require shelter from cold winds.

The genus *Nothofagus* consists of about 35 species of trees and shrubs commonly known as southern beech. Species are distributed in the southern hemisphere, generally in moist coastal to montane forest. They provide a useful source for timber and some species have been heavily logged. *Nothofagus* species generally grow well in milder parts of western Europe and North America. As a whole, the genus is considered to have good ornamental qualities, but its species vary in degree of hardiness.

NOTHOFAGUS IN CHILE

Endemic to central Chile, *Nothofagus glauca* occurs on thin or rocky soils up to an altitude of 1,500m (4,900ft) in both the coastal cordillera and the Andes. This species is classed as Vulnerable on the Red List. It grows as a large tree in its native forests, and regeneration of *Nothofagus glauca* is very good in old growth stands, but undisturbed mature forest is increasingly scarce. Almost all of the pure stands in the coastal range, known as Maule forest, have been logged and the land given over to *Pinus radiata* plantations. The Andean populations are presently under conversion also, a response to an increased demand for timber from the woodchip industry. Chile has nearly one-third of the world's virgin temperate forests. Improved management of these globally important temperate forests is a priority to ensure the long-term survival of species such as *Nothofagus glauca*, which is rarely found within protected areas. This species has not yet proved to be a success in cultivation in the West.

Another threatened *Nothofagus* species to be found in Chile is *Nothofagus alessandri*, classed as Endangered in the

SOUTHERN BEECH ► *The distribution of the genus* Nothofagus *extends to the barren, dry lands of Patagonia, shown here, lying at the southernmost tip of South America.*

Red List: Lower Risk

Olearia traversii

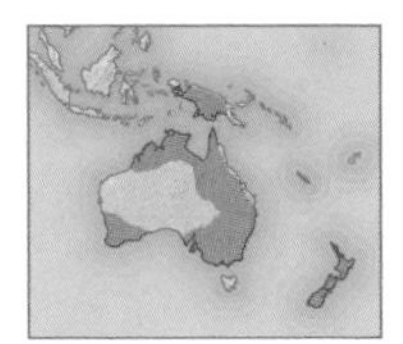

Botanical family
Compositae

Distribution
Australia, New Zealand, New Guinea

Hardiness
Fully hardy (borderline)

Preferred conditions
Full sun, with shelter from cold, drying winds. Soil should be fertile and well drained.

Commonly known as Chatham Island akeake, *Olearia traversii* is one of the largest tree daisies in the world, growing to a height of 15m (50ft) in its temperate forest habitat. The tree has pale, stringy bark and yellow wood, which reportedly smells like curry. The oblong, leathery leaves are green on the upper surface and silky white beneath. The flowerheads consist of tiny, tubular, white flowers and have no petals.

The Chatham Islands, which belong politically to New Zealand and lie about 800km (500 miles) from South Island, have three endemic *Olearia* species of conservation concern. These were once dominant forest species, but forest destruction, burning, and grazing have led to their decline, as well as use of their wood for fence posts and firewood. Over half of the endemic plants of the islands are now threatened with extinction. *Olearia traversii* is now largely confined to sand dunes and paddocks.

Red List of 2000. Like *Nothofagus glauca*, it has suffered devastation as a result of widespread conversion of forests to pine plantations. The species is restricted to a few localities in the Maule region, where *Nothofagus glauca* also occurs.

Two other species of globally threatened nothofagus occur on the botanically rich New Caledonia island archipelago in the Pacific Ocean. The species *Nothofagus discoidea*, Vulnerable in the 2000 Red List, is associated with evergreen rainforest. This type of vegetation covers about 18 per cent of the main New Caledonian island of Grand Terre, but *Nothofagus discoidea* is known only from a few small populations in the southern lowlands.

The second species, *Nothofagus baumannii*, stated to be at Lower Risk in the 2000 Red List, is found in cloud forest and is restricted to three mountain peaks, again in the south of Grand Terre. Fortunately, *Nothofagus baumannii* is protected since it grows within the Rivière Bleue National Park.

Red List: Rare*/CITES II

Orothamnus zeyheri

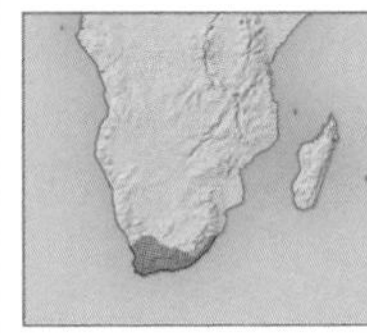

Botanical family
Proteaceae

Distribution
Cape Floral Kingdom, South Africa

Hardiness
Half-hardy

Preferred conditions
In its natural habitat the plant grows in sandy, acidic soils with a high content of peat.

The marsh rose, ***Orothamnus zeyheri***, has been described as "the loveliest of all the proteas". Flowering takes place throughout the year in Western Cape Province, South Africa, although the main flowering period is between March and November. A component of the Cape's fynbos vegetation, the marsh rose is now confined to only a few inaccessible peaks. It is a single-stemmed or scarcely branched shrub that grows to 4m (13ft) in height. The leathery, elliptic leaves are 3–6cm (1–2½in) long with marginal hairs, but it is the flowerheads that attract attention, consisting of red, shiny modified leaves, 5–7cm (2–3in) long, and lemon-yellow flowers.

WILD-COLLECTED FLOWERS

One of the threats to the survival of this species has been wild collection of the beautiful flowers for sale in Cape Town. Collection and sale of the marsh rose from wild populations was banned by law in 1938 and this legislation has generally been effective, although occasionally single blooms have been picked by individuals. The marsh rose was listed in Appendix I of CITES in 1975, presumably as a precaution against international trade, but as the conservation requirements became better understood the species was removed from Appendix I in 1997.

In the wild, the marsh rose grows mainly on steep, south-facing slopes at altitudes of 500–800m (1,650–2,600ft). Like other species of the heath-like fynbos vegetation, the species is fire-adapted, but uncontrolled burning can be a significant threat.

Orothamnus zeyheri is protected in the Kogelberg Nature Reserve, located 90km (56 miles) to the southeast of Cape Town, which is now managed as a biosphere reserve, and also in a reserve near Hermanus, where an outlying population of the species occurs. The protection and partial recovery of *Orothamnus zeyheri*, reduced to an estimated 90 wild plants in 1968, is certainly a botanical conservation success story, but vigilance remains important. All of the Cape's fynbos vegetation is constantly under threat from invasive woody plants such as the Australian wattles (*Acacia* species) that were originally planted as shade and timber trees. In addition to threats to its habitat, the marsh rose is susceptible to damage from fungi, particularly *Phytophthora cinnamomi* and *Pythium* species.

The marsh rose has been cultivated with limited success. Propagation is more by cuttings or grafting than seed, but not without difficulty. Growth conditions in the fynbos are specialized, indicating that the future of the plant probably depends on *in situ* conservation within its habitat.

SEED DISPERSAL ► *This species produces fruit two months after flowering. The plant depends on ants dragging the seeds underground, where they germinate.*

Red List: Rare*

Pimelea physodes

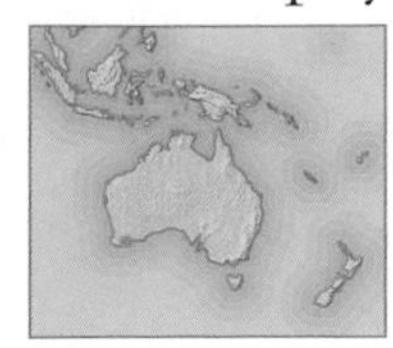

Botanical family
Thymelaeaceae

Distribution
Western Australia

Hardiness
Half-hardy

Preferred conditions
Full sun; well-drained, neutral to acid soil. Plants need regular rainfall while they become established, especially on sandy soils.

In the wild, ***Pimelea physodes***, the Qualup bell, occurs only among the rocks and scrubby vegetation of the Fitzgerald River National Park, Western Australia. Early explorers considered this area to be worthless for farming and settlement but botanically it is one of the most diverse areas in the world.

The Qualup bell was named for its location in Western Australia and its hanging, bell-shaped blooms. The colouring of these bells is magnificent, a combination of bright red and yellow, growing on vivid green and red stems. The flowers are attractive to birds.

Harvesting of the wild flowers of this naturally rare shrub pushed the species to the brink of extinction. A ban on the export of wild-collected flowers was introduced to curb the trade and briefly this species was included in Appendix II of CITES. Fortunately, the Qualup bell is now being grown commercially for the retail florist trade, greatly reducing the threat to the wild population.

Experimentation with propagation of *Pimelea physodes* continues, since the species remains quite difficult to propagate and cultivate. The use of smoke to aid germination, mimicking the effect of bush fires, has been found to be successful with other *Pimelea* species, and vegetative propagation using semi-hardwood cuttings is also possible. However, persistently high summer temperatures can result in plant losses, and plants are also susceptible to botrytis, which can cause leaf drop.

▲ AUSTRALIAN BELLS *The pendent, bell-shaped flowers have beautifully graded coloration. Each stem holds three to five heads and grows up to 60cm (2ft) tall.*

Red List: Vulnerable

Pittosporum dallii

Botanical family
Pittosporaceae

Distribution
New Zealand: South Island

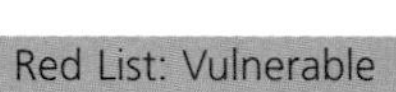

Hardiness
Fully hardy

Preferred conditions
One of the hardiest pittosporum species; thrives in either full sun or partial shade but prefers fertile, moist, but well-drained soil. The species grows from seed, and separated parts of the plant sometimes root in favourable soil.

▲ SERRATED LEAVES *identify this plant as* Pittosporum dallii, *the only New Zealand pittosporum with this leaf characteristic.*

In the wild, Dall's pittosporum, ***Pittosporum dallii,*** has a very restricted distribution, being confined to montane beech forest in the Tasman Mountains of South Island, New Zealand. The tree's habitat appears to be restricted to certain rock types. The species has probably always been rare, but heavy browsing by goats, deer, and possum has contributed to its decline. In the past there has also been a problem with overcollection for horticulture, and road-building projects may yet affect surviving populations.

A small tree that grows to about 6m (20ft), Dall's pittosporum bears many fragrant, creamy-white flowers, although these are rarely seen in cultivation. The bark of the stem and larger branches is light grey, although young wood is reddish-purple in colour. Seedlings are restricted in number at most of the sites where this pittosporum grows, and regeneration is restricted to places that cannot be reached by animals.

Some of the tree's mountain habitat has been included in the Northwest Nelson Forest Park, which offers hope of protection from development. The tree is cultivated to a limited extent in New Zealand and in British gardens.

See also Invasive plants, *p.465.*

25*/5 on Red List

Protea

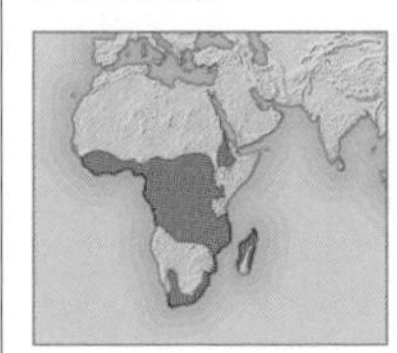

Botanical family
Proteaceae

Distribution
From tropical Africa to South Africa

Hardiness
Half-hardy – frost-tender

Preferred conditions
The genus needs full sun, with day and night temperatures above 15°C (60°F) and 10°C (50°F) respectively. Proteas grow best in well-drained, acidic soils of sand or gravel.

The genus *Protea* was named by Linnaeus after Proteus, the sea-god who was able to change into many different shapes. The genus comprises 115 varying species of evergreen shrubs and, rarely, small trees that grow on rocky hillsides and dry scrub from tropical Africa to South Africa. The flowers, often large, showy, and cone-like, are valued for flower arrangements.

The Swartland sugarbush, *Protea odorata* (Endangered in the Red List of 1997) is an extremely rare South African plant reduced to a handful of wild individuals in Western Cape Province. A native of renosterveld (a special type of lowland fynbos), the species has been almost destroyed by loss of habitat, mainly as a result of agricultural conversion. The rapid spread of an invasive introduced species, the Port Jackson willow, *Acacia saligna*, has also been a serious problem. The decline of

◄ PERFECT PARTNERS *Sugarbirds, also endemic to the Cape, feed on the abundant nectar of proteas, which are pollinated in return. Like humans, they are apparently most attracted by bright, showy species.*

the wild populations of *Protea odorata* has been truly catastrophic. In the early 1980s, over 1,000 plants in five populations were known. By 1996 a survey revealed that just four plants existed in the wild. Due to be reassessed, the species certainly qualifies for the global conservation status of Critically Endangered, even though it was moved from CITES Appendix I to Appendix II control after a 1996 survey identified habitat protection as a more appropriate and urgent conservation strategy than trade control.

Reduced to a handful of plants in the wild, *Protea odorata* is now in cultivation on a small scale in South Africa. Some seedlings from the Kirstenbosch National Botanical Garden and from an agricultural research centre will form the basis of a reintroduction programme at Riverlands Nature Reserve. It is unlikely that the species will become popular in cultivation, but at least *ex-situ* conservation measures can help save *Protea odorata* from extinction.

Another rare species, *Protea lanceolata* (Vulnerable in the 2000 Red List), grows on the southern Cape coast, from De Hoop to Mossel Bay, on calcareous white sands in limestone fynbos and, near Mossel Bay, on gravels. Agricultural activities, particularly planting of cereal crops, and coastal developments have resulted in the decline of what were dense stands in places, and this is likely to continue, especially in the Mossel Bay area. More than half of the sub-populations are threatened by alien invasive plants, particularly acacias. Fortunately, 25 per cent of the known subpopulations are in protected areas.

RENEWAL

REMOVING INVASIVES

The Global Invasive Species Programme has its headquarters in South Africa. Unemployed townspeople are given the work of purging their local mountains of non-native vegetation, thereby helping to ensure the survival of threatened indigenous species. Abseiling down the cliffs to hack out the alien species is an exhausting and dangerous activity, but makes a real difference to the local ecosystems. One million hectares have been cleared, but follow-up applications of herbicide will be needed to prevent regeneration of the alien species.

◄ PROTEA SPECIES *provide spectacular colour among the unique fynbos flora of South Africa's Eastern and Western Cape provinces.*

21*/19 on Red List/1 on CITES

Prunus

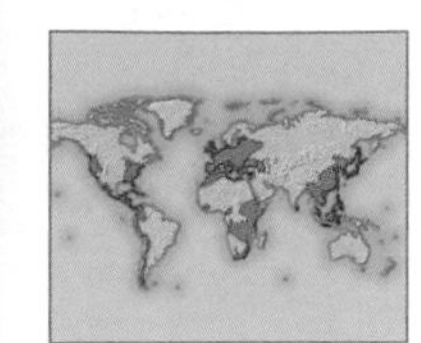

Botanical family
Rosaceae

Distribution
Northern temperate regions, South America, south-east Asia, Africa

Hardiness
Fully hardy – half-hardy

Preferred conditions
Full sun (deciduous species and cultivars) or full sun and partial shade (evergreen species). Preferred soil is moist and well drained, and moderately fertile.

The genus *Prunus* is of worldwide economic importance. It is the source of edible fruits and nuts, such as plums, damsons, almonds, apricots, and cherries, and also provides timber, ornamental plants, and medicinal products. There are over 400 species growing in the wild, mainly in woodland but also in other habitats, such as coastal sands, rocky places, and cliffs. Twenty *Prunus* taxa are recorded as threatened with extinction.

Among the threatened taxa are species that have been heavily exploited, such as ***Prunus africana***, and naturally rare species endemic to islands, such as the three subspecies of ***Prunus lusitanica***, which are threatened by loss of habitat. In situ conservation of the genetic diversity of wild *Prunus* species is considered to be particularly important in Central Asia and west China as a backup for economically important fruit crops.

Prunus africana, sometimes referred to as pygeum or red stinkwood, is classed as Vulnerable in the 2000 Red List. It is a tropical evergreen tree, growing up to 30m (100ft) tall; its small, white flowers set to produce cherry-like fruits. The simple, elliptic, hairless leaves are glossy, dark green on the upper surface, and paler green on the undersurface. When crushed, they smell of almonds. The rough bark is brown to almost black.

Shrinking habitat

Prunus africana is distributed widely in tropical and southern Africa and in Madagascar. Within this broad area it is confined to montane forests, generally at altitudes above 1,500m (5,000ft). These forests provide a habitat for endemic plant and animal species and are of high conservation value. In Cameroon, for example, the fruits are an important food source for birds such as Bannerman's turaco (*see above right*) and the Cameroon mountain greenbul, and primates such as Preuss's guenon. Reduction of the forests by agricultural clearance inevitably has an impact on the prospects of such species. *Prunus africana* is an important tree both for local communities and as a source of international revenue. As well as providing a hard, durable timber, its bark is collected for a variety of local medicinal uses.

The bark is also processed by European companies to produce a drug used in the treatment of a prostate disorder; the bark extract was patented in 1966. Throughout much of its range, and most notably in Cameroon, Equatorial Guinea, and Madagascar, the unsustainable exploitation of the bark of this species for medical purposes, and to a lesser degree its timber, has caused rapid population declines.

▲ PRUNUS AFRICANA *In 1997, the trade in the bark of this species was valued at around £80 million (US$150 million), a reflection of its recently discovered efficacy in treating a prostate disorder. Preserving the species has since become a top conservation priority.*

WILDLIFE

Fruit eater

The habitat of Bannerman's turaco (*Tauraco bannermani*, listed by IUCN as Vulnerable), is restricted to the Bamenda-Banso Highlands of western Cameroon, especially around Mount Oku. The bird feeds on different fruits, including those of *Prunus africana*. However, between 1965 and 1985 the turaco's forest habitat was halved by agricultural clearance.

Two projects, the Birdlife International Kilum Mountain Forest Project and the Ijim Mountain Forest Project, are working with local people to preserve the forest. One initiative is the promotion of bee-keeping – honey commands a good local price, so there is an incentive to preserve the forest that provides honey in quantity.

Commercial production

Cameroon is the major exporter, followed by Madagascar and then Kenya and Tanzania. The tree is generally too rare in other countries to be of commercial use, although it is exploited for local markets. It has been listed in CITES Appendix II since 1994, which should help to ensure the sustainability and legality of international trade. However, monitoring the trade is complicated because of the form in which the products are traded. Within Africa, the species is safeguarded in various protected areas, but bark harvesting has been allowed within some of the protected forests, such as the Parc National de Mantadu in Madagascar.

Fortunately, reafforestation with *Prunus africana* is a distinct possibility in the long term. Fresh seed germinates readily and propagation by cuttings grown in sand without rooting hormones has the potential for success on a large scale.

Prunus in the West

In Europe, one species under threat is *Prunus ramburii*. This plant, Vulnerable in the 2000 Red List, is endemic to the Andalusian sierras (mountain ranges) of southern Spain. *Prunus ramburii* occurs in dry montane scrub. Wild populations are affected by fire, development for tourism, and a lack of pollinators. The fruits are used locally to make an alcoholic drink. A conservation plan for the species is being developed by Cordoba Botanic Garden in Andalusia.

Also in Europe is *Prunus lusitanica*, or Portuguese laurel. The species grows to around 5m (16ft) in cultivation. It has glossy, dark green leaves with red stalks. The scented, cream flowers are produced in slender spikes in June, and are followed by small, black fruits. *Prunus lusitanica* has three subspecies, all of which are threatened with extinction in the wild, although only one of them is currently listed by IUCN. Propagation of the species is by heel cuttings that are best taken in August or September.

Prunus lusitanica* subsp. *azorica is classed as Rare in the Red List of 1997. The clearing of its habitat in the Azores for agriculture and plantations, and the invasion of introduced plants and animals, are the most serious problems.

Prunus lusitanica subsp. *hixa* grows on the island of Madeira. This subspecies is known only from a single location on the north coast, in a small area of the laurisilva forest growing within the

▲ PRUNUS LUSITANICA *SUBSP.* AZORICA *Endemic to the Azores, this tree grows in deep, narrow ravines on the islands, and in stands of undisturbed laurel-juniper forest.*

▲ PRUNUS LUSITANICA *SUBSP.* LUSITANICA *is fully protected by law in France but stands remain vulnerable to fire in the dry summers. A representative group of 20–30 trees may be seen in the Parc National des Pyrénées, south-west of Lourdes on the French-Spanish frontier.*

National Park. There is no regeneration by seed and only one stand of planted specimens has produced seed crops.

Prunus lusitanica* subsp. *lusitanica has the widest distribution of the three subspecies and is widely scattered in moist forest enclaves in the Iberian Peninsula and in humid ravines in the northern Rif Mountains of Morocco. The remaining isolated populations are threatened by wildfire, felling, and declining water availability.

See also Invasive plants, *p.466*.

RENEWAL

PRUNUS AFRICANA

UNSUSTAINABLE HARVEST

For centuries, *Prunus africana* trees have been felled for their timber (*see left*), or simply stripped of their bark, with scant regard for the future of the species. Now this valuable tree is being cultivated on a small scale by rural farmers in Cameroon, where several small commercial plantings have been established. There are also small areas of plantation in Kenya.

As a relatively fast-growing indigenous tree, this protected species is considered to have great potential for reafforestation in afromontane forests. *Ex-situ* conservation in secure field banks will be important as a necessary supplement to legal protection. Ultimately, the goal is to promote a fairly run, sustainable trade in the species.

SEEDLINGS RAISED BY A WOMEN'S COLLECTIVE IN CAMEROON

10*/5 on Red List

Pyrus

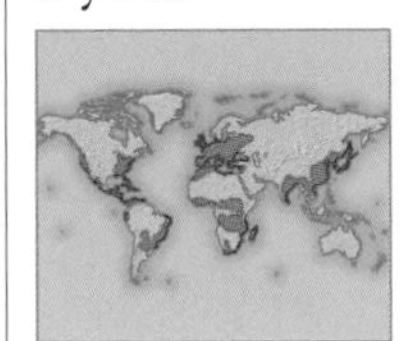

Botanical family
Rosaceae

Distribution
Europe, Asia, North Africa

Hardiness
Fully hardy

Preferred conditions
Full sun; *Pyrus* species adapt well to a wide range of fertile, well-drained soils.

The genus *Pyrus* consists of about 25 species commonly known as the pear. In the wild, species are found in western Europe and across Asia to China and Japan. Turkey, with nine species, is an important centre for wild *Pyrus* diversity. As well as producing edible fruits, a number of species produce timber that is used in the manufacture of furniture and musical instruments. Ornamental forms produce attractive spring blossoms.

Pyrus serikensis, known locally as *yaban armutu* or *zingit*, is endemic to the Pamphylian Plain of Turkey. The plain is mainly composed of alluvium, with occasional small ranges of low hills. As one of the few fertile coastal plains along Turkey's Mediterranean coast, the area has been of exceptional importance to agriculture for centuries. Consequently, little semi-natural vegetation survives. *Pyrus serikensis* is a spiny plant, growing up to 10m (33ft) with white flowers in spring, and round, reddish-brown fruits.

Villagers traditionally left individual *Pyrus serikensis* trees in the fields for shade, but recently these have been felled so that the cotton crop can be sprayed from the air without obstruction. Some trees remain in local woodlands of oak and pistachio and at roadsides, but tourist developments are also affecting these.

The habitat of *Pyrus serikensis* lies within the Tahtalı Dağı Important Plant Area, a site identified as one of Turkey's most valuable areas for wild plant diversity. Protection of this botanically rich area will help to ensure the survival of the attractive pear species.

Another pear found in Turkey and considered of conservation concern is *Pyrus salicifolia*. Fortunately it is also found in neighbouring countries. The weeping form of this species, *Pyrus salicifolia* 'Pendula', is an ornamental often cultivated in parks and gardens.

▼ PYRUS SALICIFOLIA *is native to south-east Europe, the Caucasus, Turkey, and Iran.* Pyrus salicifolia *var.* serrulata, *from Turkey, is classed as Rare in the Red List of 1997.*

70*/54 on Red List

Quercus

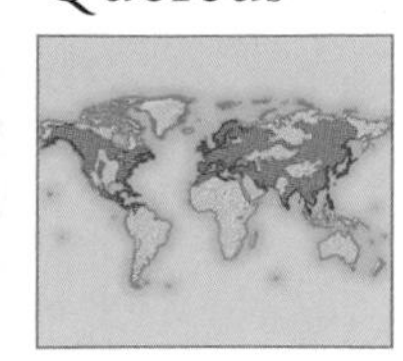

Botanical family
Fagaceae

Distribution
Northern hemisphere extending south to Colombia and Malaysia

Hardiness
Fully hardy – half-hardy

Preferred conditions
Full sun or partial shade (evergreen species prefer full sun). Deep, fertile, well-drained soils, including lime soils. Oaks of borderline hardiness require shelter from frost and cold winds.

The genus *Quercus* has over 500 species growing in woodland and scrub throughout the northern hemisphere. The distribution of oaks extends as far south as Colombia in the Americas, and Malaysia in Asia, with species growing in a wide variety of habitats.

Over the centuries, various oaks have provided humans with timber, cork, tanning agents, dyes, and the acorns traditionally used as fodder for pigs. By long tradition the English oak, ***Quercus robur***, has been fêted as the national tree of England. Unfortunately, over 50 oak species are recorded as threatened with extinction in various parts of the world, and the status of many others remains uncertain and in need of field survey.

QUERCUS ACUTISSIMA
Acorns in cups, in many forms, characterize all oak species.

One of the world's rarest oaks is *Quercus hinckleyi* (Critically Endangered in the 2000 Red List). An evergreen shrub with holly-like leaves, this species is known only from 11 isolated populations in the wild, nine of which are in the Big Bend Ranch State Natural Area, in Texas, USA. The status of the species in neighbouring areas of Mexico is uncertain, and it remains very important to check how many of the trees remain. Most populations of the species consist of fewer than 100 individuals growing in low thickets that cover an area of less than 2ha (5 acres).

Historical climate change is thought to be the main reason that the species has become rare. Hybridization with other *Quercus* species, road construction, overcollection by horticulturists, drought, and grazing are threats today. Although acorn yields are good, all reproduction appears to be vegetative. The tree is listed by the US Endangered Species Act and a recovery plan was devised in 1992.

Mexico has the greatest diversity of oaks of any country in the world, with representative species in ecosystems throughout the country. Of Mexico's approximately 150 species, over half are endemic. At least 30 Mexican oak species are threatened.

Hinton's oak, ***Quercus hintonii***, Critically Endangered in the 2000 Red List, is a notable example. This small, deciduous tree is a species of submontane to montane dry forest that is confined to small areas in Mexico State. *Quercus hintonii* grows to 15m (50ft) in height, with dark bark and foliage that is bright red in early spring. Populations of Hinton's oak have been dramatically reduced as a result of habitat loss from agriculture (cultivation of maize and fruit trees), development of coffee plantations, and road construction. The species is also adversely affected by grazing, which prevents regeneration.

▲ QUERCUS HINTONII, *seen here growing in Nanchititla Ecological Reserve, Mexico. The wood is traditionally used for baking "las finas" bread, which has a characteristic taste imparted by the smoke. The wood was also used for beams and fencing poles, and as firewood.*

PLANTS AND PEOPLE

SUSTAINABLE LIVELIHOOD

The Mediterranean species ***Quercus suber***, or cork oak, provides around 99 per cent of the world's supply of natural cork. Portugal and Spain are the main producers. When the trees reach maturity at 80 or 90 years old, the cork bark is carefully stripped off. Bark may then be removed every nine years or so without damaging the trees, which may live for 200 years.

By far the most important user of cork is the wine industry, but in recent years a shift towards plastic corks and screw-tops has threatened the cork oak forests and the livelihoods of more than 80,000 people.

In the traditional agricultural systems of the cork forests, sheep, goats, and pigs are grazed under the oak trees. Honey is gathered from hives within the forests, and berries and fruits gathered from the forest understorey. The forests are also an important habitat for wildlife, including the Critically Endangered Iberian lynx (*see right*). The lynx has declined dramatically over the past 15 years, and now only an estimated 300 individuals are left in the wild. Past threats have included hunting, but now loss of the animal's habitat is the main concern.

Conservation organizations believe that, should the cork oak forests become no longer economically viable because of falling demand for cork, the forests are certain to be cleared and replaced by less environmentally beneficial forms of land use.

IBERIAN LYNX

A recent study of the distribution and status of Hinton's oak was carried out for the FFI/UNEP-WCMC Global Trees Campaign (*see p.91*). Undertaken by the Sir Harold Hillier Gardens and Arboretum and the University of Puebla, it found that the overall species range covers just 46,000ha (113,660 acres), across three populations that have been highly disturbed. Working with local people, the Global Trees Campaign project developed a conservation strategy that included provision of local training in oak propagation techniques.

Another threatened oak endemic to this part of the world, *Quercus insignis* is remarkable for its egg-sized acorns, which are among the largest found in known oak species. *Quercus insignis* was first described from Veracruz State in Mexico and is now known to be widely distributed through Central America as far south as Panama. The conservation status of this oak is uncertain, but it is rare in the areas where it has been studied.

Quercus insignis grows in cloud forest, the vegetation type that is known locally as *bosque mesofilo de montaña*. This forest type is the most threatened form of forest in Mexico due to the establishment of new coffee plantations. In Veracruz and Jalisco states, forests at the lowest altitude at which *Quercus insignis* grows are being cleared – in this part of Mexico, the best coffee grows at altitudes of 1,200–1,350m (4,000–4,250ft). The Global Trees Campaign project also provides propagation information for this species, which is grown at the Botanic Garden at the University of Puebla and at the Sir Harold Hillier Gardens.

Oglethorpe oak

A species endemic to the USA, the Oglethorpe oak, *Quercus oglethorpensis* (Endangered in the 2000 Red List), has a very restricted distribution in the wild. It is known only from a few counties in the piedmont of north-east Georgia, western South Carolina, and a disjunct population in Mississippi. The oak was first described in 1940, and it is thought that much of its habitat was lost before then. Approximately 1,000 individuals are estimated to exist in the wild.

The most common habitats are roadsides and old fence rows. In Georgia, habitat loss continues to result from clearance for agriculture and forestry; a survey in 1985 failed to locate 11 per cent of the previously identified sites. Poor seed viability and chestnut blight are also reported as threats to the species.

Oglethorpe oak is in cultivation and is quite hardy. In the UK, however, it suffers winter damage because the wood fails to ripen completely in the summer.

QUERCUS ROBUR ► *In 2001, this species was elected as national tree of the USA in a vote organized by the National Arbor Day.*

Red List: Critically Endangered

Ramosmania rodriguesii

Botanical family
Rubiaceae

Distribution
Rodrigues Island, Indian Ocean

Hardiness
Frost-tender

Preferred conditions
Café marron requires a minimum temperature of 19°C (66°F) and a maximum of 23°C (73°F), and relative humidity of 70–80 per cent. The plant needs well-drained, organic compost.

Café marron, ***Ramosmania rodriguesii***, is one of the world's rarest species. Until recently it was reduced to just one surviving individual in the woodland of Rodrigues Island. Its nearest relative, *Ramosmania heterophylla*, the only other member of the genus, was also endemic to the island but is recorded as extinct.

Ramosmania rodriguesii was already scarce when first described by Isaac Bayley Balfour in 1877, and very few plants have been seen since. It was thought to have become extinct but, in 1980, a schoolboy found a single plant growing at the edge of a thicket by the side of a road. A fence was erected around the café marron plant, although continuing damage led to construction of a further three fences.

In 1986, three small cuttings from the attractive, white-flowered shrub were sent to the Royal Botanic Gardens, Kew. Micropropagation using tip cuttings was successful, and cloned seedlings were raised in a tropical glasshouse. Kew seedlings were subsequently returned to Rodrigues Island for planting. Since then, a parent plant has fruited and new plants have grown from seed. The story of café marron shows just how important *ex-situ* conservation can be in saving the world's rarest shrubs from extinction.

EASING THE WAY ► *Pollen failed to travel down the flower tubes (styles), so specialists at Kew removed some of the tubes and fertilized the ovules directly. This resulted in fruiting.*

62*/10 on Red List

Rhododendron

Botanical family
Ericaceae

Distribution
Europe, Asia, Australasia, North America

Hardiness
Fully hardy – frost-tender

Preferred conditions
Species vary in the degree of light they tolerate, from full sun to dappled shade. Generally they prefer moist but well-drained soil that is leafy, humus-rich, and acidic (ideally pH 4.5–5.5).

Of the 500–900 species of the genus *Rhododendron*, at least 30 species are currently thought to be of conservation concern on a global scale. Rhododendrons grow in a range of habitats, from dense forest to alpine tundra, and at a variety of altitudes. Documentation of this extremely popular horticultural genus is incomplete, and a more comprehensive assessment of the status of all the rare and potentially threatened species should be undertaken. Japan has a significant total of threatened rhododendron species; 25 taxa are included on the 1997 Red List of Japanese Vascular Plants.

One threatened Japanese species is ***Rhododendron makinoi***, which is classed as Rare in the Red List of 1997. This plant is native to central Honshu Island, where it grows among ferns and rocks in mountain forest habitats at altitudes of 600–2,300m (2,000–7,500ft). It was named in honour of the Japanese botanist T. Makino, who described it in the early 20th century.

Rhododendron makinoi is a distinctive and elegant foliage plant growing to 2m (6½ft) in height. The striking leaves grow up to 18cm (7in) long. They are narrowly lance-shaped, shiny on the upper surface, and have a thick covering of tan-coloured woolly hairs on the undersurface. At first, the new shoots look like fuzzy white candles as they emerge in June. Gradually, the colour changes from white to tan as the growth expands during July and August. Eventually, the woolly hairs on the upper surface of the leaf rub off, exposing the deep green, shiny surface beneath, but the downy covering remains on the underside of the leaves. Light pink, funnel-shaped, bell-like flowers appear from May to June. The plant is especially hardy.

Along with the species plant, several hybrid forms are in cultivation, including *Rhododendron makinoi* 'Rosa Perle' and 'White Wedding'. Magnificent examples of these hybrids can be seen at the Polly Hill Arboretum on Martha's Vineyard Island, Massachusetts, USA.

◄ RHODODENDRON MAKINOI *is related to* Rhododendron yakushimanum. *Its hardiness makes it well suited to cultivation.*

◄ RHODODENDRON KANEHIRAI, *known as the "Feitsui rhododendron", is named after its habitat, which now lies underwater.*

OTHER RHODODENDRONS

The list of other globally threatened rhododendron species includes five species from Taiwan. Among these is *Rhododendron nakaharae*, an attractive, red-flowered azalea, which is classed as Rare in the Red List of 1997. Another rhododendron from the island is ***Rhododendron kanehirai***, which was classed as Extinct in the Wild on the Red List after the only known habitat of this beautiful shrub was flooded in 1984, a consequence of the construction of a dam. Fortunately, *Rhododendron kanehirai* has been maintained in cultivation for several centuries, so there is a possibility that it might be introduced into a suitable wild habitat in Taiwan. All five of the threatened Taiwanese rhododendron species may be found in cultivation in the island's fine botanic gardens.

See also Invasive plants, *p.467.*

Red List: Rare*

Rosa hirtula

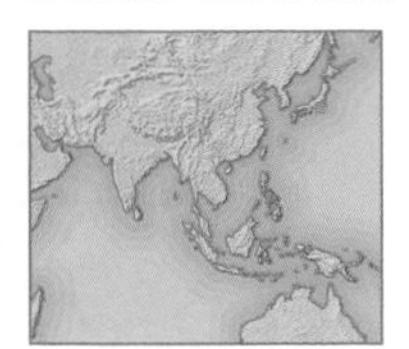

Botanical family
Rosaceae

Distribution
Japan: Hakone Mountains

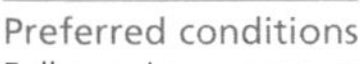
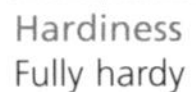

Hardiness
Fully hardy

Preferred conditions
Full sun, in an open, unrestricted site that allows it to spread. Moderately fertile, humus-rich, moist but well-drained soil.

This species of cup-shaped rose grows wild in the mountains of Hakone, Japan. It is included in the 1997 Red List of Japanese Vascular Plants. The Japanese name, *sanshou bara*, or "pepper rose", suggests that its leaves closely resemble those of the Japanese pepper tree. The stems are thorny. The flowers shade from pale pink to deep pink, and are accentuated by beautiful orange-yellow stamens and small leaves.

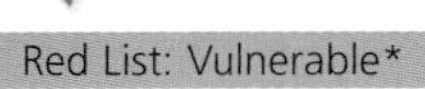

ROSA HIRTULA

See also Invasive plants, pp.467–68.

Red List: Extinct in the Wild

Sophora toromiro

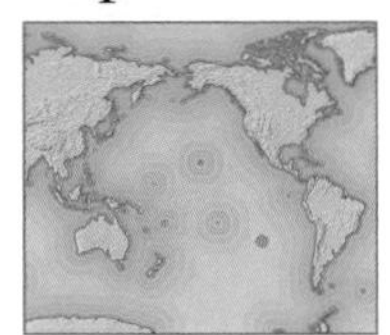

Botanical family
Leguminosae

Distribution
Isla de Pascua (Rapa Nui, or Easter Island)

Hardiness
Frost-tender

Preferred conditions
Full sun, in a sheltered site when growing in cooler latitudes. Soil should be moderately fertile and well drained. Sophoras require long, hot summers to flower at their best.

Commonly known as toromiro, the small tree or shrub ***Sophora toromiro*** is extinct in the wild and is now known only from a few valuable trees that survive in cultivation. The plant is native to the island known locally as Rapa Nui, as Isla de Pascua in neighbouring Chile, and as Easter Island. So far, attempts to reintroduce the species to the island have not been successful. The plant grows to a height of 3m (10ft), with yellow, pea-like flowers and pale green, feathery leaves up to 5cm (2in) long. Its elliptical leaflets are thinly covered by white hairs.

Rapa Nui is now mainly covered by grassland. The last wild toromiro tree grew on the slopes of the Rano Kao crater at the south-west tip of the island. Over 200 years ago, when the species was first discovered by Western botanists, the tree grew in thickets on the volcanic slopes. The main reason for its extinction has been grazing of the seedlings and removal of bark by introduced sheep.

Toromiro was traditionally very important as the only native tree species and source of wood for the people of the island (who, like their language, are also called Rapa Nui). The people used toromiro for making canoes, house frames, and wooden carvings. The traditional Polynesian culture remains strong on Rapa Nui, one of the most isolated places on earth.

Other *Sophora* species facing extinction include Crusoe's mayu-monte, *Sophora fernandeziana*, and Selkirk's mayu-monte, *Sophora masafuerana* (both Vulnerable in the 2000 Red List). These are confined to single islands in the Juan Fernandez group, south-east of Rapa Nui.

Sophora species are readily grown from seed, and *ex-situ* conservation should be able to save other sophoras from following toromiro to near extinction.

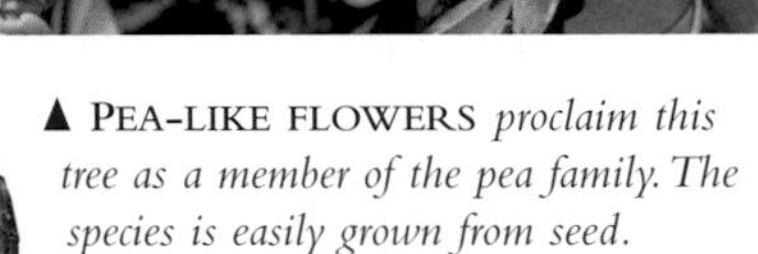

▲ PEA-LIKE FLOWERS *proclaim this tree as a member of the pea family. The species is easily grown from seed.*

RIBBED FIGURE ► *Woodcarvings such as this represent the spirits of long-dead ancestors and were kept as family heirlooms. Early carvings have been attributed to the 14th century.*

Red List: Vulnerable*

Sesbania tomentosa

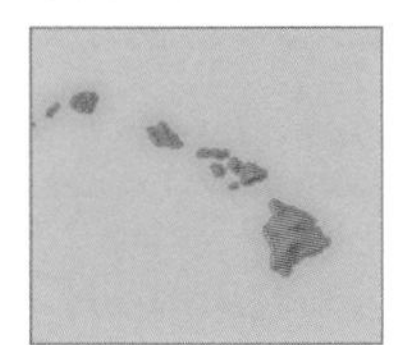

Botanical family
Leguminosae/ Papilionaceae

Distribution
Hawaii

Hardiness
Frost-tender

Preferred conditions
Full sun. The plant prefers moderately fertile, moist but well-drained soil. Sesbanias are found on stream banks as well as dry, sandy areas.

In its native Hawaii the common name for ***Sesbania tomentosa*** is *ohai*. This attractive shrub or small tree was once found among the coastal sand dunes of most of the Hawaiian Islands. Currently it is much reduced in extent and found in only two remote populations on the Hawaiian island of Maui. One population of about 13 plants grows in the Hawaiian National Guard's Kanaio Training Area. In the late 1990s, the Guard erected 3.2km (2 miles) of fencing to prevent grazing, traps were set to prevent rodents from feeding on the fruit and seeds, and weeding carried out to remove encroaching invasive species.

Sesbania tomentosa also occurs within the Kaena Natural Area Reserve, Oahu Island. Fortunately, propagation by commercial nurseries and botanic gardens has been successful. The species may be seen in cultivation at the Amy B.H. Greenwell Botanical Garden, on Hawaii Island, where special emphasis is placed on cultural traditions of land use and the conservation of the native flora.

THREAT

LEISURE PURSUITS

While the extensive sand dunes of the Hawaiian Islands are the natural habitat of ohai and other rare species, some also offer unrestricted access for leisure motorcycling and use of various all-terrain vehicles. The Hawaiian Islands are a major holiday destination, and such activities are set to increase in future. This significant threat to the dunes' flora suggests that some areas must be barred to such sports, perhaps attracting ecotourists in their stead.

▲ COASTAL RARITY Sesbania tomentosa *has large, pea-like flowers and long seed pods. Tiny silver hairs cover the leaves, giving them a whitish appearance. Grazing is the greatest threat to the plant's survival – fencing against goats, deer, and other animals is indispensable.*

11*/22 on Red List

Sorbus

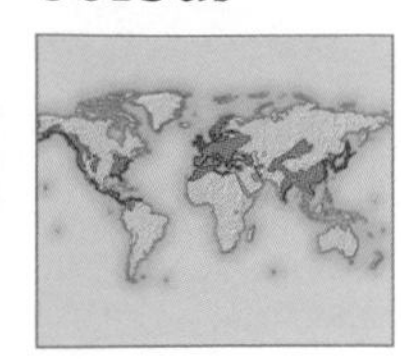

Botanical family
Rosaceae

Distribution
Widely distributed in northern temperate regions

Hardiness
Fully hardy – half-hardy

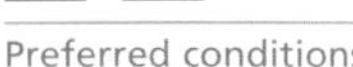

Preferred conditions
Full sun or light dappled shade. Moderately fertile, humus-rich, well-drained soil is best. Species and cultivars with pinnate leaves (divided into leaflets) prefer acidic to neutral soil.

About 200 *Sorbus* species grow in the northern hemisphere, in temperate woodlands, on hills and mountains, and on scree. Ash trees provide timber, the bark has a role in tanning leather, and the edible fruits are used in jellies and brandy. Many ash species are grown as ornamental trees or shrubs, favoured for their attractive leaves, autumn colours, and large clusters of small fruits. Grown from seed, they are valued as pollution-tolerant garden trees.

▲ SORBUS BRISTOLIENSIS *is favoured for its compact, rounded head of branches and its dense clusters of orange-red fruits.*

Eleven of the ash species that grow in the UK are considered to be globally threatened. Two species at risk are endemic to Arran, an island off the west coast of Scotland. The Arran whitebeam, ***Sorbus arranensis,*** and the bastard mountain ash, *Sorbus pseudofennica*, both Vulnerable in the 2000 Red List, are confined to a small area in the north of the island. The Arran whitebeam is now found only in two areas of scattered woodland remnants, mostly on inaccessible, steep slopes, growing on very acidic soils. Reduction of its range is thought to be largely due to clearance of the forest.

Only about 500 plants remain in those areas, most of them being in Arran's Glen Diomhan Nature Reserve, which was established in 1956 to ensure the survival of the two endemic ash species. A fenced enclosure was created to encourage woodland regeneration, but unfortunately it failed to deter grazing animals. Natural regeneration alone cannot save either of the Arran species.

Sorbus arranensis can be seen in cultivation at various botanic gardens and arboreta in the UK. Propagation is by seed, which usually germinates after three months in soil.

Another threatened UK species is the Bristol whitebeam, ***Sorbus bristoliensis*** (Endangered in the 2000 Red List). This ash grows in one of the most important botanical sites in the UK, the gorge cut by the River Avon in the English counties of Gloucester and Somerset. In the 1930s, when the taxonomy of British *Sorbus* species was finally established, two endemic species of the Avon Gorge were distinguished. They are *Sorbus bristoliensis*, with about 100 plants surviving in the wild, and Wilmott's whitebeam, *Sorbus wilmottiana*, Critically Endangered on the 2000 Red List, with about 20 plants growing in the wild. The natural habitat of these rare trees is woodland and scrub on carboniferous limestone. Fortunately, there is some evidence of increases in the populations of these rare ash species.

For all the naturally rare and endemic ash species in the UK, the protection of habitats is the main conservation need. *Ex-situ* conservation provides an important backup. Increasingly, the 11 *Sorbus* species that are under varied threats in the UK are being grown in botanic gardens and arboreta.

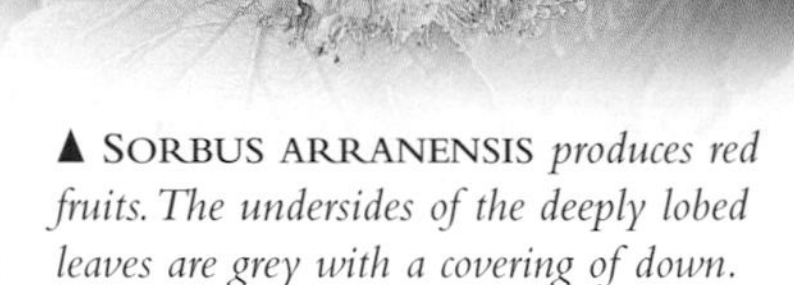

▲ SORBUS ARRANENSIS *produces red fruits. The undersides of the deeply lobed leaves are grey with a covering of down.*

SORBUS IN MADEIRA

The Atlantic island of Madeira has over 60 species of endemic plants, including the endemic species of mountain ash, *Sorbus maderensis* (Critically Endangered on the 2000 Red List). Known locally as *sorveira*, this shrub is perhaps the closest to extinction since it is reduced to a single population of about 30 plants.

The plants grow in a small area of montane woodland on the second highest mountain of the island, Pico do Arieiro, at an altitude of 1,500m (5,000ft). This habitat is within the Ecological Park of Funchal. Recent grazing control has helped to improve the shrub's prospects of regeneration, although fire and plant collectors remain as threats.

Sorbus maderensis is listed as a specially protected species by the EU Directive on the Conservation of Natural Habitats and of Wild Fauna and Flora. An *ex-situ* conservation programme for the species is under way at Westonbirt Arboretum, near Tetbury in Gloucestershire, England.

See also Invasive plants, *p.468.*

◄ THE AVON GORGE *At least 30 different rare and threatened plants grow in this unique, wooded habitat. Among them are two of the UK's rarest* Sorbus *species.*

4 on Red List/3 on CITES II

Swietenia

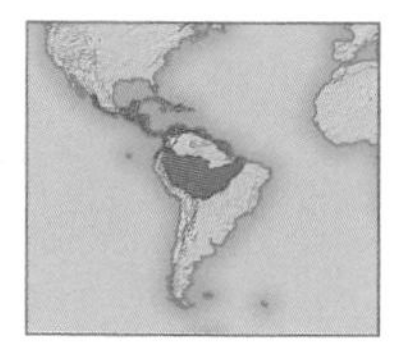

Botanical family
Meliaceae

Distribution
Central and South America

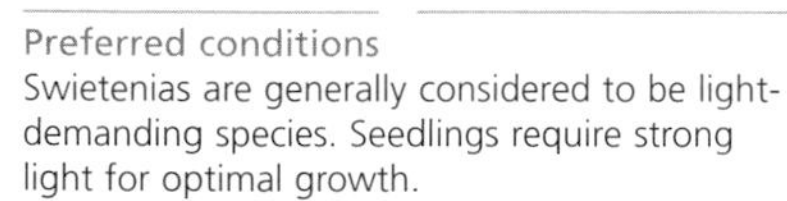

Hardiness
Frost-tender

Preferred conditions
Swietenias are generally considered to be light-demanding species. Seedlings require strong light for optimal growth.

Trees of the genus *Swietenia* grow in the tropical forests of the Caribbean and northern South America. They include three species that produce the true mahogany of international commerce: *Swietenia mahagoni*, ***Swietenia macrophylla***, and the much less used *Swietenia humilis.* Since November 2002 all three have been included on Appendix II of CITES.

Swietenia mahagoni is endemic to the Caribbean and Endangered in the 2000 Red List. This was the first mahogany species to appear on the European market five centuries ago, and from the earliest days of colonial exploration it was used for ships, house fittings, and the furniture of such makers as Chippendale and Hepplewhite in England.

Since the Caribbean mahogany has been widely cultivated, it is difficult to be certain of the species' true natural distribution. The natural stands are now extensively exhausted and the gene pool of the species is much depleted, but it is still found in moist forest on Caribbean islands and in southern Florida, USA, often growing on limestone.

Swietenia mahagoni grows to a height of around 10m (33ft) and has a rounded crown. The leaves are up to 15cm (6in) long, the leaflets are mainly in four pairs, and the white flowers are small. Examples found today are usually weedy trees or bushes. Small quantities of timber from plantations occasionally become available.

Swietenia macrophylla, the big-leaved or Brazilian mahogany, is now the main species in international trade. Vulnerable in the 2000 Red List, this potentially large tree is found in both wet and dry tropical forest, with a patchy distribution ranging from Mexico to Brazil. The trees are notable for their broad buttress roots and are unbranched until about halfway up the trunk. Trees growing today tend to be young and generally attain heights of no more than 30m (100ft) tall.

The Brazilian mahogany remains the most valuable timber of South America. However, in common with other tropical timbers, much of the timber is wasted during logging and processing – there is, at present, little economic incentive to manage natural stands sustainably. Supplies have become locally exhausted, particularly in the northern parts of its range. Knowledge of the northern populations is good, whereas relatively little is known about those in the Amazon. Good stands apparently survive in parts of Brazil and Bolivia.

Regeneration of this species is stochastic, that is, it depends in nature on large-scale disturbance. This pattern of regeneration makes mahogany vulnerable to logging regimes. *Swietenia macrophylla* also grows in timber plantations and reforestation schemes in various parts of the tropics. The largest plantations are in the south and south-east of Asia. The trees are propagated by seed; ripe fruits are collected towards the end of the fruiting season to ensure good germination rates.

SWIETENIA MACROPHYLLA ► *This species can reach over 60m (200ft) in height, but as a result of logging, most of the larger, older trees have disappeared from the forest.*

OTHER THREATENED TROPICAL HARDWOODS

Guaiacum officinale Known as lignum vitae, this eye-catching, blue-flowered, evergreen tree grows to 10m (33ft) in height. It is a slow-growing plant of lowland dry forest, frequently growing in coastland areas. Lignum vitae has been traded for several centuries and overexploitation has taken place throughout the species' range. The hard, oily, and very durable timber was used for propeller shafts and bearings.

Populations are now severely reduced in the Caribbean islands, including the Lesser Antilles, Puerto Rico, Barbados, and the Virgin Islands, as well as Colombia, and extinct or almost extinct in Antigua, Anguilla, and Barbuda. Regeneration is good, but slow. Red List: EN; CITES II.

Intsia bijuga Also called *merbau* or Malacca teak, this species is a slow-growing tree of lowland primary or mature secondary rainforest that reaches to 50m (165ft) in height. It is often found in coastal areas bordering mangrove swamps, rivers, and floodplains. The species prefers rainfall of more than 2,000mm (78in) per year.

Although widespread, extending from Madagascar and India through to Australia and islands of the Pacific, the species has been exploited so intensively for merbau timber that few sizeable natural stands remain. The tree is characteristic of the eastern coastal rainforests of Madagascar, which are among the most threatened in the country. Red List: VU.

Pterocarpus indicus This tropical tree grows to 40m (130ft) in height and mature trees have massive trunks over 3m (10ft) in diameter. It grows naturally in lowland primary and some secondary tropical forest, mainly along tidal creeks and rocky shores. The tree has a natural range extending from southern Burma, through south-east Asia to islands of the Pacific.

Populations have declined because of overexploitation of the timber, called narra. The tree is now probably extinct in Vietnam, Sri Lanka, and Malaysia, and seriously threatened in India, Indonesia, and the Philippines. Heavy exploitation of what are believed to be the largest remaining populations continues in New Guinea.

BUTTRESS ROOTS OF *PTEROCARPUS INDICUS*

Pterocarpus santalinus Known as red sandalwood, this species is restricted mainly to the southern parts of the Eastern Ghats of India, where it grows on rocky ground in dry deciduous forest. The species is in severe decline due to overexploitation of its hard, fragrant, and beautiful wood, which is used for furniture and musical instruments. The dye santalin is also extracted from the timber. The export of red sandalwood is now banned by Indian legislation. Red List: EN; CITES II.

Santalum album This half-parasitic plant, known as sandalwood, is widely scattered in dry deciduous forests, growing in parts of China, India, Indonesia, and the Philippines. Fire, grazing, and most importantly exploitation of the wood (for use in furniture, handicrafts, perfume, and cosmetics) threaten the species.

Currently, cultivation cannot offset the commercial pressure on wild populations. Sustainable management of the wild populations is urgently needed. Control measures attempted by the regional government of East Nusa Tenggara District of Indonesia have lacked community support. Export of the timber is banned from India, but smuggling continues on a formidable scale. Red List VU.

Red List: Vulnerable

Tephrosia pondoensis

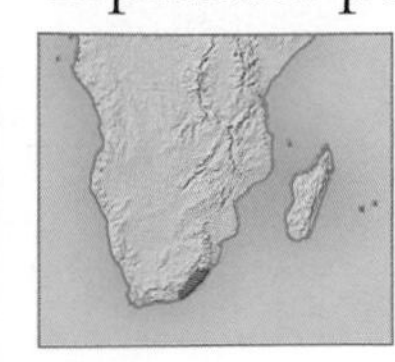

Botanical family
Leguminosae

Distribution
South Africa: Pondoland

Hardiness
Frost-tender

Preferred conditions
Full sun or partial shade. In its generally dry natural habitat the plant grows in sandy soils where there is sufficient ground moisture.

Considered to have good potential as an ornamental due to its pea-like flowers, *Tephrosia pondoensis* is a shrub or small tree endemic to Pondoland in southern KwaZulu-Natal and eastern Transkei in the Eastern Cape, South Africa. The species occurs in small, scattered subpopulations on water courses and moist slopes in sandstone landscapes, on the margins of dry evergreen forest. The plant is called the fish poison pea because, like other *Tephrosia* species, its roots contain rotenone, a piscicide used to control unwanted fish in waterways.

The species' habitat is included in two provincial reserves. *Tephrosia pondoensis* also grows on the margins of a number of demarcated forests in the Transkei, but these forests are no longer well protected and the margins are particularly susceptible to increased disturbance from settlement, cutting for firewood and timber, and browsing by cattle and goats.

Red List: Extinct in the Wild

Trochetiopsis erythroxylon

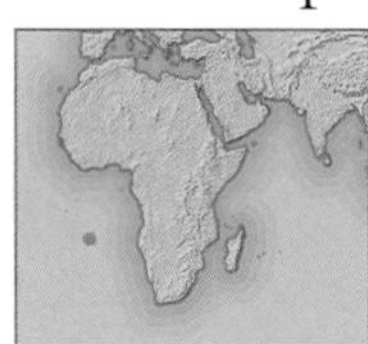

Botanical family
Sterculiaceae

Distribution
St Helena Island

Hardiness
Frost-tender

Preferred conditions
The highlands are cloud-covered for much of the year, with cool temperatures and high relative humidity, and constant exposure to trade winds.

The genus *Trochetiopsis* comprises just three species endemic to St Helena, a remote UK overseas territory in the South Atlantic. ***Trochetiopsis erythroxylon*** and *Trochetiopsis melanoxylon* are classed as Extinct, and *Trochetiopsis ebenus* as Critically Endangered, in the 2000 Red List. The native flora of St Helena consists of about 60 flowering plant and fern species, 50 of which are endemic. Of the endemic plants, six are extinct and 40 threatened with global extinction.

Trochetiopsis erythroxylon, the St Helena redwood, was once common on the island, growing mainly on the central mountain ridge, which now has tiny, isolated remnants of semi-natural forest. The species is a small tree with beautiful flowers and fine, pale green, oval leaves that grow up to 15cm (6in) long. Population declines were rapid as early settlers overexploited the tree for its red timber and bark, while goats reduced the species' chances of regeneration.

The last wild St Helena redwood tree was found in Peak Gut in the 1950s. It is the source of the few existing individuals in cultivation. Inbreeding depression and a impoverished gene pool are manifest in the poor growth and high mortality of cultivated specimens. By contrast, *Trochetiopsis* x *benjaminii*, a hybrid of cultivated forms of this species and *Trochetiopsis ebenus*, is very vigorous and may provide the only chance of survival for this part of the gene pool.

Seed remains viable for several years, and propagation from seed has proved easier than propagation from cuttings.

▼ FLORAL CHANGELING *Initially white, the bell-shaped flower of* Trochetiopsis erythroxylon *gradually changes to pink and then red. The flowers generally grow in pairs.*

4 on Red List

Ulmus

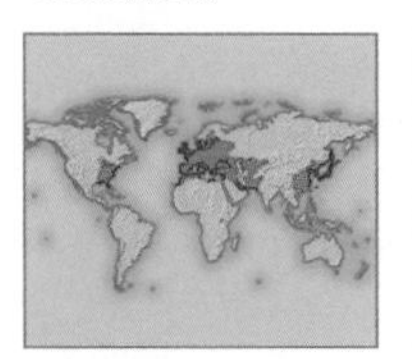

Botanical family
Ulmaceae

Distribution
Most northern temperate regions

Hardiness
Fully hardy

Preferred conditions
Full sun; elm species tend to grow in poor, sandy, acidic to neutral, well-drained soil. If soil is too fertile, trees may be overgrown and leggy.

The genus *Ulmus* consists of about 30 species of deciduous, occasionally evergreen trees, and rarely shrubs, that occur in woodland, thickets, and hedgerows in northern temperate regions. Six species are native to North America and most of the rest are found in China. Elms are an important source of timber and in Europe they have been managed (by pollarding, for example) since Neolithic times. They have also been an important provider of animal fodder.

Ulmus wallichiana, Vulnerable in the 2000 Red List, is a deciduous tree of temperate oak and mixed coniferous forest. Native to Afghanistan, India (Jammu-Kashmir), Nepal, and Pakistan, it grows at altitudes of 1,500–3,000m (5,000–10,000ft). Trees reach up to 30m (100ft) in height, and have a spreading crown. The species provides high-quality fodder for livestock and has been excessively exploited. It is now only found scattered among trees growing in inaccessible sites and protected areas.

In the past, rural villages planted this elm to provide a dependable source of fodder, which they dried for use in the winter. It is unfortunate that this practice is now less common. A wild tree that has been severely hewed cannot reproduce sexually – the coppice-sprouts grow into mature trees but do not produce flowers.

The devastation caused by Dutch elm disease (*see right*) in Europe and America has led to a new use for *Ulmus wallichiana*. Due to its relatively wide gene pool, the species shows a certain degree of resistance to the disease. It is therefore used in breeding programmes to develop elms for urban and landscape planting in temperate areas.

CHINESE ELMS

The three other globally threatened elm species are all native to China and are not yet subject to conservation measures. *Ulmus gaussenii*, Critically Endangered in the 2000 Red List, exists as fewer than 50 trees in a small area of deciduous forest in the Langya Hills of Anhui Province. *Ulmus chenmoui,* Endangered in the 2000 Red List, occurs in a few limestone hills of Anhui and in the province of Jiangsu. *Ulmus elongata,* Vulnerable in the 2000 Red List, is more widespread but is threatened by forest clearance.

Elms can be easily propagated by seed, which is sown fresh since it does not retain viability for long. Vegetative propagation by cuttings, grafting, or layering is also easily accomplished.

▼ ULMUS PROCERA *The English elm was devastated by Dutch elm disease in the late 20th century. The same disease killed tens of millions of American elms after a shipment of infected European timber arrived in the 1930s.*

See also Invasive plants, *p.471*.

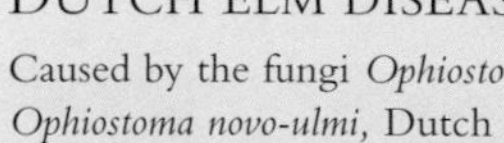

THREAT

DUTCH ELM DISEASE

Caused by the fungi *Ophiostoma ulmi* and *Ophiostoma novo-ulmi*, Dutch elm disease has devastated worldwide elm stocks since 1930. Two species of bark beetle introduce spores of either fungus under the bark, from where the fungus is carried upwards in the sap stream. Once in the crown of a healthy tree, the fungus moves slowly downwards, killing the tree. Elms growing in stands often have roots that are grafted together, and the fungus easily travels from one tree to the next, killing each in turn. Many elms are vegetatively-reproduced genetic clones, making them vulnerable to root-spread infection. The elms planted today tend to be genetically resistant varieties and hybrids that can withstand invasion by the fungi.

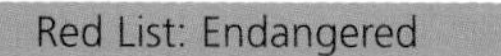

Red List: Endangered

Warburgia salutaris

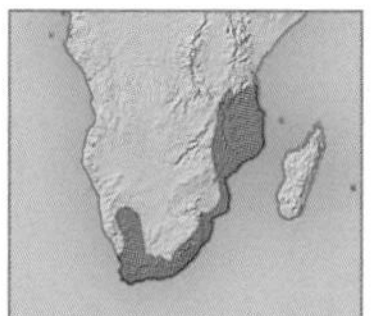

Botanical family
Canellaceae

Distribution
South Africa, Swaziland, Mozambique, Zimbabwe

Hardiness
Half-hardy – frost-tender

Preferred conditions
Full sun; grows in lower forest at altitudes up to 2,000m (6,600ft). Flowers at the beginning of the rains and fruits form late in the rainy season.

The pepperbark tree, *Warburgia salutaris*, has a scattered distribution in southeast Africa. In South Africa it occurs in savannah woodland and coastal forest in northern KwaZulu-Natal; in afromontane forest patches along the Drakensberg Escarpment in Mpumalanga; and on the Soutpansberg and Blouberg ranges in the Northern Province. It is also recorded from afromontane forests in the Eastern Highlands of Zimbabwe and from afromontane and lowland forest patches in Swaziland and Mozambique. The tree is known in local languages as *muranga* (Shona), *isibaha* (Zulu), and *chibaha* (Tsonga).

The bark, stems, and roots of the pepperbark tree all have important roles in indigenous medicine, ranging from the treatment of head and chest ailments to curing people who are bewitched.

▲ PEPPERBARK *has various uses in tribal medicine and magic, but now overcollection is threatening the survival of the species.*

Herbal products from the tree are commonly sold in quantity in urban markets throughout southern Africa. This has led to the near extinction of the species in many of its scattered habitats.

Even where the tree occurs within protected areas, it is difficult to prevent exploitation. Fortunately, there are still large, relatively untouched subpopulations in South Africa's Northern Province, and plants have been reintroduced into two protected areas of KwaZulu-Natal. In addition, a number of projects are under way to provide a cultivated form for use in agroforestry schemes.

Unlisted

Weigela florida

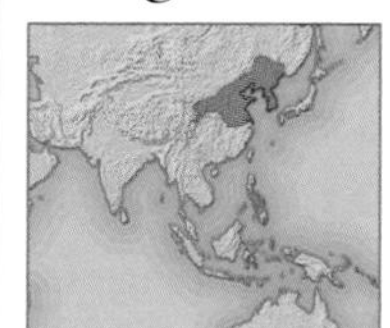

Botanical family
Caprifoliaceae

Distribution
Northern China, Japan, Korea

Hardiness
Fully hardy

Preferred conditions
Full sun or partial shade. Soil should be fertile and well drained, either sandy, loam, or clay. The species tolerates varied ground moisture conditions and soil may be wet, moist, or dry.

Native to Eastern Asia, ***Weigela florida*** is a deciduous shrub that grows to about 2.5m (8ft). It produces graceful, arching stems and medium-green summer foliage. The flowers are funnel-shaped, dark pink with pale pink to nearly white inside, and appear in clusters of 3–5 blooms in late spring.

Weigela florida is often referred to as "old-fashioned weigela" because a great many hybrids with superior flowers are now available. Hybrid flowers range from white through pink and red to purple; the compact cultivar 'Variegata' has white-margined, grey-green leaves.

WEIGELA FLORIDA

3 on Red List

Zelkova

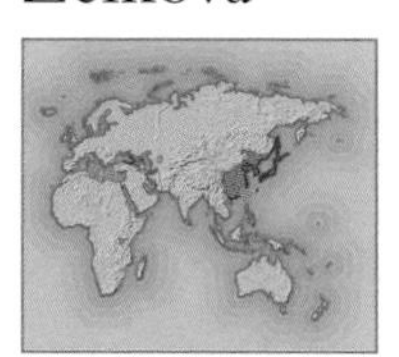

Botanical family
Ulmaceae

Distribution
Italy, Greece, Turkey, Iran, Caucasus, East Asia

Hardiness
Fully hardy

Preferred conditions
Full sun or partial shade; deep, fertile, moist, and well-drained soil. Plants require shelter from cold, drying winds in frost-prone areas.

Belonging to the elm family, the small genus *Zelkova* contains around six species that are native to woodland and shrubby areas of Europe and Asia. All are cultivated and, apart from the unlisted Japanese species ***Zelkova serrata***, all are resistant to Dutch elm disease (*see left*).

Zelkova sicula, Critically Endangered in the 2000 Red List, is the most severely threatened species within the genus. Only about 250 individuals exist in a single, remote population extending 200m (650ft) along the banks of a stream on the northern slopes of the Iblei Mountains in Sicily, Italy. The habitat has been fenced to prevent grazing and other damage. Nevertheless, the future is uncertain for this tree as few of the remaining individuals have produced flowers in recent years and the fruit produced is apparently sterile.

Another Mediterranean island endemic is *Zelkova abelicea*. Vulnerable in the 2000 Red List, this small tree is confined to Crete. Low numbers occur in about 20 alpine localities, especially at the south-east corner of the Omalós Plain. Grazing by goats has a negative effect on the species but most trees regenerate well by suckering. The species makes an elegant, small tree in cultivation.

▲ ZELKOVA SERRATA *is popular for its beautiful yellow and red autumn leaf colour. The smooth, grey bark peels to reveal patches of orange.*

CONIFERS

Conifers have lived on Earth for millions of years. Individual trees represent the oldest, tallest, and most massive living organisms on the planet, and in several countries these extraordinary trees are protected in national parks. Conifers make up 30 per cent of the world's forests and survive in the most hostile environments, from parched deserts to freezing mountainsides, where they stabilize soils. One type of adversity the plants cannot tolerate is acid rain, which has devastated many coniferous forests.

WHAT IS A CONIFER?

Abies numidica *Cedrus deodara* *Cupressus glabra*

With a very few exceptions, all of the 550 or so species of conifers are evergreen trees. In most conifers, the leading shoot is upright and strongly dominant, resulting in the typical straight trunk and the slender, conical outline, produced by whorls of branches in regular tiers. Most conifers have linear, needle-like or scale-like, usually tough and leathery leaves, with parallel veins. The most obvious distinguishing features of conifers, however, are their reproductive structures. Like cycads, they reproduce by means of cones rather than flowers. Conifers are gymnosperms: they bear seeds that are naked, instead of enclosed in an ovary as in the flowering plants (angiosperms). After pollination, the ovule (unfertilized seed) develops into a seed, and as it matures, the individual segments of the cone, called scales, harden and become woody.

▲ **REPRODUCTION CYCLE** *In conifers, ovules usually occur on the upper surface or edge of the scales of a female cone. When the ovules have been fertilized by pollen grains from a male cone, they develop into seeds. As the seeds grow, the scales mature and harden to form the familiar woody cone. When the cone opens, the seeds are released, producing seedlings when they germinate.*

▲ **MEDITERRANEAN FIRS** *Stands of* Abies pinsapo *clothe the hot, dry slopes of El Pinar in southern Spain.*

SURVIVAL STRATEGIES

Absent only from polar regions, conifers have managed to adapt to some of the most extreme conditions. In arid areas, the leathery needles and small, scale-like leaves resist dessication in the face of scorching heat, while on the lofty mountainsides of Canada and northern Asia the trees withstand freezing winds and heavy snowfall. Indeed, the complex chemicals that many conifers produce serve as a form of anti-freeze, enabling them to survive periods of extreme cold. Several species are also well adapted to wildfire: *Araucaria araucana* (*see p.135*) has a thick layer of insulating bark, for example, while the cones of the Monterey cypress (*see p.141*) need heat and smoke to open and shed seed.

HARDWOODS AND SOFTWOODS

Generally speaking, timber from broadleaved trees, such as oak or ash, is classified as hardwood, while the term "softwood" is commonly applied to the timber of coniferous trees. The term "hardwood" does not refer to the density or durability of timber, but rather to its cell structure. In fact, one common softwood, yew (*Taxus baccata*), is among the hardest of all woods. Softwoods such as pine and spruce are highly valued commercially for a variety of industrial and domestic uses, from construction to paper manufacture, and, in general, are faster growing and therefore deemed better suited to managed plantations, since they produce a more rapid return on investment than hardwoods.

◄ **SCOTS PINE** *The graceful* Pinus sylvestris *(Scots pine), a softwood tree, is widely grown commercially for its durable, resin-rich timber.*

SNOWY MOUNTAINS *In Glacier National Park, British Columbia, Canada, stands of conifers, their branches laden with snow, help to regulate snowmelt and stabilize the soil.*

◄ SCALES AND NEEDLES
Conifers have tough, leathery leaves that are scale-like (top) or needle-like (bottom) and well adapted to photosynthesis in severe conditions. The leaf surface is coated with a thick, waxy cuticle that reduces water loss by evaporation.

ANCIENT LINEAGE

Coniferous trees have existed on Earth for many millions of years. Fossil conifer pollen, leaves, and reproductive structures have been found in Permian rocks that date back to the Palaeozoic era, more than 286 million years ago, and fossils that are recognizable as their fellow gymnosperms, ginkgos, date back 270 million years. In the Mesozoic Era (265–248 million years ago) when the early dinosaurs began to evolve – long before the appearance of flowering plants – ancestors of the redwoods, yews, pines, cypresses, and araucarias were dominant elements in both tropical and temperate forests around the world. While their dominance has diminished, conifers continue to be important components of modern-day vegetation – today they make up some 30 per cent of the world's forests. Individual conifers, such as the bristlecone pines (*Pinus longaeva*) of California, Nevada, and Utah, represent the oldest living organisms on the planet.

▲ BRISTLECONE PINE *The oldest trees, growing in the White Mountains of California, are over 4,700 years old and were seedlings when the pyramids of ancient Egypt were being constructed.*

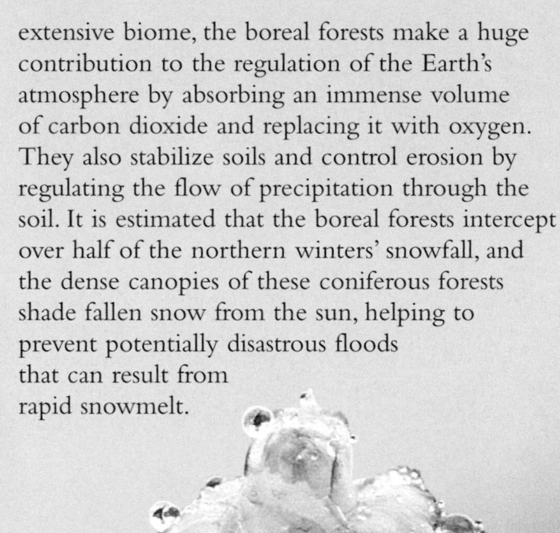

◄ CAUGHT IN AMBER *Resin seeping from the bark of conifers became fossilized over time and hardened into amber, an organic gemstone. Spiders and insects that became trapped in the resin have been preserved over millions of years.*

CONIFER HABITATS

Coniferous forests provide a unique habitat for other plants, especially lichens, mosses, and fungi, and for a diversity of wildlife, from large mammals such as the snow leopard to the tiny dormouse. Many species of owl use conifers as roost sites and birds such as nutcrackers and crossbills have evolved distinctive beaks that are crossed or ridged to extract seed from the cones. The world's most extensive biome, the boreal forests make a huge contribution to the regulation of the Earth's atmosphere by absorbing an immense volume of carbon dioxide and replacing it with oxygen. They also stabilize soils and control erosion by regulating the flow of precipitation through the soil. It is estimated that the boreal forests intercept over half of the northern winters' snowfall, and the dense canopies of these coniferous forests shade fallen snow from the sun, helping to prevent potentially disastrous floods that can result from rapid snowmelt.

▲ BLACK BEAR Ursus americanus, *the American black bear, is a forest-dweller whose diet is 95 per cent plant-based. Bears often climb trees to obtain berries and seeds.*

◄ CROSSBILL *The European crossbill, which inhabits coniferous forests, has evolved a distinctive crossed beak that enables it to crack open cones and extract the nutritious seed.*

COMMERCIAL USES

With their straight stems, conifers are of the highest value to the timber trade and harvesting the wood accounts for about 70 per cent of the world's commercial production of durable, lightweight, resin-rich, and rot-resistant timber. Coniferous timber, generally called softwood, has long been used in all manner of construction and furniture making – industrial and maritime, domestic and religious. Conifers, unlike hardwoods such as oak, are fast-growing and therefore very suitable for managed forests. The wood is also used to make fibreboard and is pulped for paper and textile production; good quality spruce yields paper that can be recycled 6–7 times before it becomes unusable. Conifer bark is used in leather tanning and in oil-spill absorption, and plays an increasingly important horticultural role as a substitute for peat in potting mediums and as mulch material. There is also significant economic value in the sale of conifers as ornamentals in the garden and as Christmas trees. *Pinus pinea* is valued for its highly nutritious nuts – the pine nuts used in pesto – while juniper berries are pressed to give gin its distinctive flavour.

▲ JUNIPER BERRIES *The ripe berries yield an essential oil that is used medicinally.*

▲ LOGGING *In this managed conifer plantation in Clocaenog Forest, Clwyd, North Wales, a forester stacks logs of Sitka spruce* (Picea sitchensis). *Spruce, a fast-growing softwood, is extremely important to the construction and joinery industries throughout Europe and North America.*

CONSERVATION

The International Conifer Conservation Programme (ICCP) is a worldwide scheme for the protection of vulnerable conifers. Coordinated by the Royal Botanic Garden Edinburgh, the programme has established a network of gardens and arboreta that grow some 8,000 threatened plants across 130 sites. Botanists, plant geneticists, and ecologists affiliated with the project undertake habitat surveys, assess the genetic diversity of wild and cultivated conifers, and research propagation methods. An understanding of the way threatened conifers reproduce is essential for producing seed for reforestation and restoration efforts. The ICCP has built upon these studies to establish several local conifer propagation schemes. For example, it was discovered that many cultivated specimens of the endangered *Fitzroya cupressoides* (*see p.142*) in Britain and Ireland were female clones. The ICCP collected seed from wild populations, raised seedlings, and distributed genetically diverse stock of the species.

▲ SPRUCE NURSERY *In the low-altitude forests of northern Bohemia, in the Czech Republic, a replanting scheme has been set up to replace spruces in forests that have been devastated by acid rain.*

LARCH "FLOWERS" *Female reproductive structures of* Larix decidua, *the European larch, are characteristic of other conifer species and differ markedly from their male counterparts. Seen here in close-up, the structures are covered in a sticky resin.*

GARDENER'S GUIDE

- When buying rare conifers, purchase only plants from propagated – not wild-collected – stock. Buying plants responsibly is the most important action a gardener can take to preserve biodiversity in the wild.
- In future there are likely to be more conifers certified as "conservation grade" on the horticultural market, so make a point of seeking these out. These products of local propagation schemes are designed to earn an income for their producers while at the same time taking collection pressure off wild populations.
- Keep meticulous records of the provenance of the plants you purchase. A detailed provenance record increases the conservation value of the specimens growing in your garden and the seed produced by them.
- Find out how you can support research conducted by the International Conifer Conservation Programme (ICCP) and other conifer programmes into ways to ensure the survival of conifer species.
- Support efforts to preserve conifer habitats that are threatened by development or exploitation. Manpower and other resources may be needed in your area.
- Look for forestry products certified as derived from sustainable sources, such as those endorsed by the Forestry Stewardship Council (FSC) or Forests Forever (FF).
- Search out and purchase recycled paper products when buying cards, stationery, and other domestic paper products. Every recycled product you buy will help to reduce the pressure to fell forests for fresh paper supplies.
- Recycle as much of your household waste paper as possible, such as newspapers and cardboard, to contribute feedstock for the recycled paper industry.

31*/15 on Red List

Abies

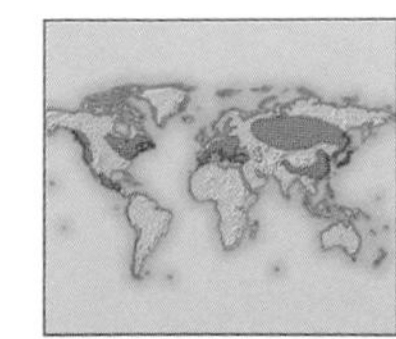

Botanical family
Pinaceae

Distribution
Europe; Asia, south to Taiwan; North America to Guatemala

Hardiness
Fully hardy – frost-hardy

Preferred conditions
Well-drained, neutral to slightly acid soil in sun, with some shade and shelter from cold, dry winds when young.

In common with many members of the family Pinaceae, the firs, or *Abies* species, include several that are globally threatened: there are 32 species on the Red List. The firs are found in northern temperate mixed or coniferous forests, in montane tropical forests, and around the Mediterranean, in forests, woods, or scrub.

Many species of *Abies* have an assured future in cultivation in northern Europe and North America. In gardens, they are often grown as specimens in lawns, and compact species such as ***Abies koreana***, which produces its violet cones on quite young trees, are suited to medium-sized gardens. Many of the larger species can be seen in specialist collections in botanic gardens, pineta, and arboreta associated with old estates and country houses. However, still more needs to be achieved for those plants left in the wild.

▲ ABIES NUMIDICA *Native to the Kabylie Mountains in Algeria, the Algerian fir tolerates drought well, and is sometimes used for hedging in Mediterranean countries.*

▲ ABIES PINSAPO *All three populations of the Spanish fir are found in the Sierras above Malaga. Fire is one of the main causes of its vulnerability, and local authorities close the forests to hikers and picnickers each summer to reduce the risk of accidental fires.*

Mediterranean firs

Many of the threatened species of *Abies* are found in the dwindling forests that surround the Mediterranean, where deforestation, fire, and overgrazing have all taken their toll. Among these, the Greek fir, *Abies cephalonica*, restricted to the mountain tops of southern Greece, is at Lower Risk, Near Threatened, on the 2000 Red List. *Abies nordmanniana* subsp. *equi-trojani*, or Trojan fir, is similarly threatened (Red List 1997). It grows as pure stands in isolated populations on north-facing mountain slopes in western Turkey. ***Abies numidica***, or Algerian fir, from the mountains of northern Algeria, is classed as Vulnerable (2000 Red List).

Abies marocana, or Moroccan fir, is found in the Rif Mountains of northern Morocco. Considered by many to be a subspecies of the Spanish fir, ***Abies pinsapo***, the Moroccan fir differs in its longer, wider needles and shorter female cones. It is listed as Lower Risk, Near Threatened, in the Red List of 1997. The slightly damper climate of the Rif Mountains means that it is probably less at risk from fire than the Spanish fir.

Abies nebrodensis, Nebrodi or Sicilian fir, is Critically Endangered (2000 Red List). The only population, on a high Sicilian mountain, is reduced to 31 trees. This fir probably covered a much wider range in the past, including a more extensive distribution in Sicily, and in the southern Calabrian mountains in mainland Italy.

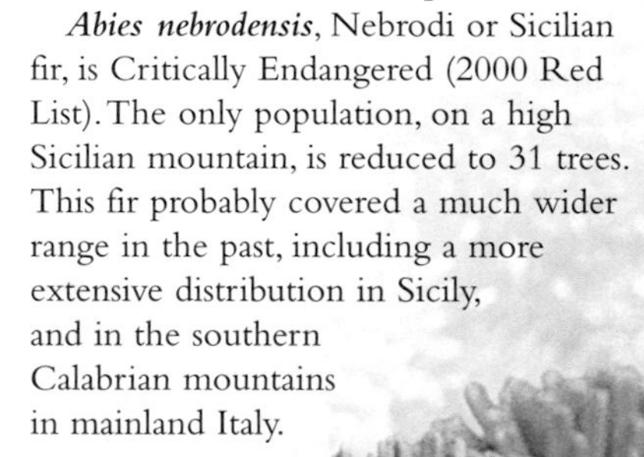

It was recognized as a separate species as recently as 1908, by which time the population had already reached a critical state. A combination of soil erosion through deforestation, climate change, grazing, and competition with its seedlings by beech (*Fagus*) have contributed to its decline. Fire is also a potential hazard. The International Conifer Conservation Programme has a programme to establish more trees in cultivation in Britain. *Abies nebrodensis* is rare in cultivation outside Italy.

Abies pinsapo, hedgehog or Spanish fir, is listed as at Lower Risk, Near Threatened (2000 Red List). It grows only in western Andalucia, southern Spain, where it is restricted to three large populations in the Sierras above Malaga. Two are on limestone rocks in national parks, while the third is on serpentine rocks and currently unprotected. All three populations are at risk from fire. In addition, overgrazing by goats prevents regeneration of seedlings. This species takes 25–35 years to reach maturity, and even then a full seed crop is produced only every 3–5 years.

RENEWAL

A head start

For natural regeneration to occur, enough healthy trees must survive to reproductive maturity to replace those lost by age, disease, or less natural causes. Seedling firs are very vulnerable to competition from other, more vigorous plants and to damage by grazing animals. Raising young plants in the protected conditions of a nursery gives them a head start. Such rescue programmes are in place for *Abies pinsapo* (seen here at Sierra de Grazalema in Spain) and *Abies nebrodensis*. Seedlings are cared for in local nurseries during their most critical years. Once well established, they are re-introduced to the wild.

CONIFERS

◄ **ABIES BRACTEATA** *The remaining populations of the Santa Lucia fir grow in moist canyon bottoms and on very steep, rocky slopes. In these sites, fire, which has been the major cause of its decline and now restricted range, is unlikely to take hold.*

Germination rates are quite low at 50–60 per cent. This is a handsome tree, and is quite widely grown as an ornamental in other parts of Europe.

FIRS IN AMERICA

Abies bracteata, or Santa Lucia fir, is classified as Lower Risk, Conservation Dependent in the 2000 Red List. It occurs only in the Santa Lucia Mountains of California, where it is confined to a narrow coastal strip in Los Padres forest between latitudes 36° and 37° North. Fire has probably been the major cause of this fir's restricted range. Another threat to its long-term survival is seed calcid, insects which normally cause 100 per cent mortality of the annual seed crop. Easily distinguished from all other firs by its elegant, long-bracted cones, the Santa Lucia fir is very ornamental but often difficult to establish in northern Europe because of damage to new growth by late spring frosts.

Abies fraseri, the Fraser or southern balsam fir, is listed as Vulnerable in the 2000 Red List. Found only in the Appalachian Mountains, south-east USA, the Fraser fir is of conservation concern because of its rarity. The remaining stands of this species are of limited commercial value, but their location on the highest peaks and ridges means that they are crucial for watershed protection (*see p.20*). Although they are vulnerable to windfall and lightning strike, by far the most serious threat is the balsam woolly adelgid (*Adelges piceae*), which has spread through the species since 1957, claiming 1.75 million trees by 1970. In the southern Appalachians, the Fraser fir is grown and harvested for the ornamental and Christmas tree markets.

OTHER THREATENED ABIES

Abies forrestii* var. *georgii From the mountains of NW Yunnan, SW Sichuan, and SE Xizang Zizhiqu (Tibet). At high altitudes of 3,000–4,300m (10,000–14,000ft) this species forms pure stands, while at lower levels other conifers including *Picea likiangensis* and *Larix* species, as well as broad-leaved trees, make up the forest. The species is declining due to overcutting and poor regeneration. Planting with commercial timber species in the area is adding to the stresses. It is cultivated in private collections and botanic gardens in Europe. Red List (1997): Vulnerable.

Abies chensiensis This Chinese species is endangered and therefore the Chinese authorities recognize the need to preserve its germplasm.It is protected in nature reserves, but more work is needed on natural regeneration and to prevent seed damage by rats. Red List (1997): Vulnerable.

Abies kawakamii The Kawakami or Taiwan fir is a high-alpine species whose timber is of low commercial value. The greatest threat to this species is probably from fire. Red List (2000): Lower Risk, Near Threatened.

Abies recurvata* var. *recurvata The Min fir is from western China and Xizang Zizhiqu, in Tibet. It is one of several Chinese species that have been described in recent years, some with little detail regarding their status. Red List (1997): Vulnerable.

Abies squamata Also from western China, the flaky fir is unique among *Abies* in having rolls of papery, orange-pink bark hanging from the trunk and branches, similar to that of many of the birches. It is rare in cultivation outside of China. Red List (2000): Vulnerable.

FIRS IN CULTIVATION

As *Abies* are mostly found at high altitudes in the wild, they are relatively cold hardy. Mediterranean species, such as *Abies pinsapo*, grow better in drier areas. Many Chinese, Himalayan, and North American species tolerate cooler, damper conditions. Among those species preferring cooler, wetter conditions are *Abies bracteata* and *Abies fraseri*.

Raising *Abies* species from seed is relatively straightforward, providing the seeds are chilled before sowing for about three weeks below 3°C (27°F). Seed must be sown when fresh; viability drops rapidly in storage. In cultivation, several species are grafted onto reliable stock such as *Abies alba*, the silver fir.

The beautiful violet-blue cones of Abies koreana *are very decorative and, unlike many species of fir, are produced on relatively young trees.*

▲ **ABIES NEBRODENSIS** *is one of the few remaining wild Nebrodi firs in the high Sicilian mountains. Nearby, an intensive rescue programme is under way whereby seedlings are raised in a nursery prior to reintroduction into the wild.*

◄ **ABIES KOREANA** *This hardy fir is found in Korea, and adjacent Russia, where it is at Lower Risk, Near Threatened (2000 Red List). The main wild populations are found above 1,000m (3,280ft) on the island of Cheju, and in Korea at slightly higher altitudes.*

Red List: Lower Risk

Actinostrobus pyramidalis

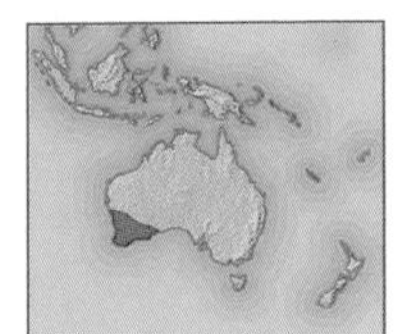

Botanical family
Cupressaceae

Distribution
Western Australia

Hardiness
Half-hardy

Preferred conditions
Well-drained soils and sun or light shade. Drought- and wind-resistant. Tolerates light frost for short periods.

The King George's cypress pine, *Actinostrobus pyramidalis*, occurs in shrublands and sand plains of south-western Western Australia. It is confined to a few locations there, and is listed as being Lower Risk, Near Threatened (2000 Red List). It is threatened because of its narrow natural distribution, and the risk of populations becoming further reduced by fire. *Actinostrobus acuminatus* is in the same category for similar reasons.

Actinostrobus pyramidalis is a compact shrub or small tree that may reach 8m (25ft) tall. It is a dense, conical, evergreen conifer with needle-like juvenile leaves and scale-like adult ones. In frost-free, Mediterranean-type climates, it is sometimes used as a low shelterbelt, or windbreak, and is ornamental when used in a group as a focal point. It tolerates a range of soil types, including dry, sandy, saline, and seasonally wet. It is easily propagated by seed or cuttings.

Red List: Lower Risk

Agathis australis

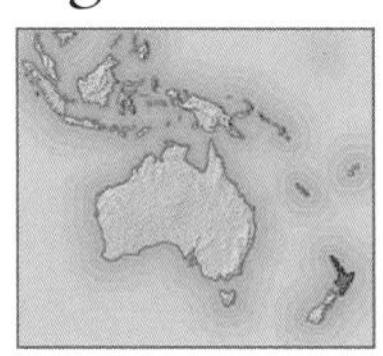

Botanical family
Araucariaceae

Distribution
New Zealand: North Island

Hardiness
Half-hardy

Preferred conditions
Moist but well-drained soil with side shelter from cold and dry winds.

This massive conifer, from the warm-temperate mixed forests of the North Island of New Zealand, is perhaps the most famous of all New Zealand native trees. The kauri is widely grown for its timber in warm temperate and subtropical parts of the world, but the remaining wild populations are classed as Lower Risk, Conservation Dependent (2000 Red List). The main threat has been the historical overexploitation of the trees' valuable timber and resin.

▲ **KAURI CONES** *The kauri's flattened, leathery, almost oblong leaves are markedly different from the needle-like leaves of most conifers. The female cones are spherical, and the male cones are cylindrical.*

PLANTS AND PEOPLE

TIMBER TITAN

Since the colonial times of the 19th century, the light and very durable kauri timber has been highly regarded for a great range of uses, and the resin it produces has also been prized for the manufacture of linoleum, high-quality paints, and varnishes. Towering above the forest canopy to heights of 40m (130ft) or more, kauris can remain unbranched for up to 30m (100ft). It has been said that a single specimen of *Agathis australis* can provide enough timber to build a house and all its furniture. Such high timber values have resulted in an estimated original 1.2 million hectares (3 million acres) of kauri forest being reduced to just 140 hectares (350 acres).

Following cyclones and fire, the kauri is able to regenerate, but, as with all long-lived conifers, regeneration is very slow.

Most of the surviving trees are protected in reserves by the New Zealand Department of Conservation; Waipoua, Mataraua, and Waima reserves make up the largest tracts of forest.

Often storage of large-seeded conifers, such as *Agathis*, is difficult, but seeds of the kauri have been dried and stored at a low temperature for up to six years without loss of viability. Seeds are best sown on the surface of a well-drained seed compost and kept moist. They will germinate within a few months. This species can also be propagated by cuttings, using the tips of leading shoots to ensure an upright, symmetrical habit.

13*/14 on Red List/CITES I

Araucaria

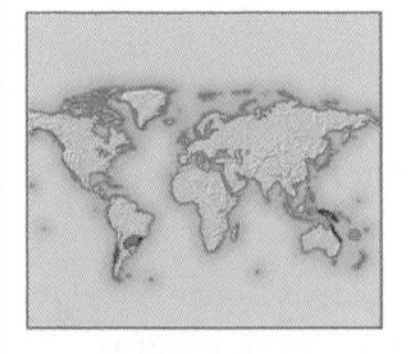

Botanical family
Araucariaceae

Distribution
South America, Australia, Papua New Guinea, New Caledonia

Hardiness
Fully hardy – frost-tender

Preferred conditions
Well-drained, even rocky soils, in sun or part-shade with side shelter from wind.

The monkey puzzles belong to one of the oldest families of conifers, dating back some 70 million years. Fossil and pollen records reveal that during their ancient history they were widespread throughout both hemispheres.

Today they are restricted to the Southern Hemisphere. Most grow in seasonally dry tropical rainforests in South America from Argentina and Chile to Paraguay and Brazil, and in Australia in New South Wales and Norfolk Island.

Some 13 of the world's 18 species occur on the Pacific island of New Caledonia. This botanical paradise has all the classic conservation problems, of which habitat loss due to wildfires and open-cast mining for nickel are the worst. Such pressures have reduced many populations, and 11 of the island's species are listed as threatened. Some species grow extremely slowly and this, coupled with infrequent seed production, severely impairs regeneration. The three species described here are all Vulnerable (2000 Red List).

Araucaria angustifolia, Paraná pine, has suffered continuous decline throughout its natural range (Argentina, Brazil, and Paraguay) due to logging for its valuable timber. The overcollection of its edible seeds has resulted in poor rates of regeneration. About 80 per cent of its original 200,000 sq km (50 million acres) of forest cover has been lost to logging.

Araucaria araucana, monkey puzzle, from Argentina and Chile, is included in CITES Appendix I. During the 20th century, logging dramatically reduced its natural distribution. As a Chilean Natural Monument, felling is now prohibited, but populations are still threatened by illegal logging.

ARAUCARIA ARAUCANA *The distinctive triangular leaves are tough, leathery, and sharply pointed at the tips, and are arranged radially around the shoot.*

▲ **ARAUCARIA ANGUSTIFOLIA** *This ancient and elegant tree forms the primary canopy in some parts of the Brazilian rainforest.*

◄ **ARAUCARIA HETEROPHYLLA** *The symmetrically spaced branches and fan-like arrangement of foliage give the Norfolk Island pine a very distinctive outline.*

Although fire is an important part of the tree's ecology, fierce fires have destroyed large proportions of its ever-dwindling population. In Chile, in December 2000, 45,000 hectares (112,500 acres) of *Araucaria* forest were lost and, in 2003, fire devastated a further 71 per cent of the Malleco National Reserve.

Araucaria heterophylla, Norfolk Island pine, is native to Norfolk Island, Australia. The natural populations have suffered decline since the trees' value as a source of naval timber was spotted by Captain Cook in 1774. Since then, its native habitats have been lost to agricultural use and human habitation. Worldwide, the species is widely cultivated.

ARAUCARIAS IN GARDENS

For over 150 years *Araucaria araucana*, the monkey puzzle, has been grown in northern Europe, particularly in Britain and Ireland, as a specimen tree or sometimes in avenues. It is hardy where winter temperatures do not fall below -20°C (-4°F). Outside frost-free regions, several other species are grown as house or conservatory plants. They all need well-drained soil and tolerate poor, rocky soils. Seed production is normally very good in cultivated trees. Since they were introduced from various sources, these trees are genetically diverse and so have good potential for restoration programmes. Seeds should be collected immediately after they fall and sown on the surface of a freely draining compost. Leave them uncovered and protected from vermin. Cuttings can be rooted from the leading shoot tips of young plants, which will produce upright, symmetrical growth. Young trees resent root disturbance and great care must be taken when transplanting nursery-grown specimens.

PLANTS AND PEOPLE

EDIBLE DELICACY

The starch-rich seeds of *Araucaria araucana* have long formed an important part of the diet of the Mapuche people. Seed collecting by people outside this ethnic group has become widespread. It is estimated that 3,400 tons of *Araucaria angustifolia* seeds are also collected annually for human consumption in Brazil. This commercial collection is greatly impeding the natural regeneration of both the monkey puzzle and Paraná pine.

ARAUCARIA ARAUCANA ► *Seen here in the wild on the volcanic slopes of the Andes, in Conguillo National Park, Chile, it is the only species hardy enough to be grown in the open in northern gardens.*

3 on Red List

Athrotaxis

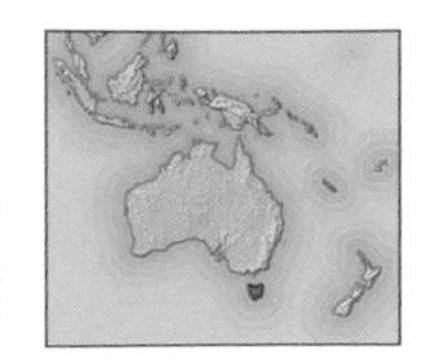

Botanical family
Cupressaceae

Distribution
Australasia: Tasmania

Hardiness
Frost-hardy – fully hardy

Preferred conditions
Humus-rich, moist but well-drained, slightly acid soils in sun; side shelter from dry wind.

The Tasmanian cedars, *Athrotaxis* species, are endemic to Tasmania, where two species are found with an overlapping distribution at altitudes between 800 and 1,300m (2,300–4,300ft). ***Athrotaxis* x *laxifolia*** is a natural hybrid between *Athrotaxis cupressoides* and *Athrotaxis selaginoides* and occurs wherever the parents grow in proximity. All three are capable of reaching ages in excess of 700 years.

They grow in high-rainfall areas, in temperate rainforest and on subalpine moorland, often near glacial lakes where they may form pure stands. In mixed forests, *Athrotaxis* species are associated with eucalyptus, southern beech (*Nothofagus*), and other conifers.

UNDER ATTACK

As with all Tasmanian conifers, these trees are highly susceptible to fire, particularly at the lower altitudes of their natural range. Testimony to this are the extensive stands of dead trees that have been killed by out-of-control camp fires. The Tasmanian Wilderness World Heritage Area has been declared a "fuel-stove-only area" in a desperate attempt to prevent further loss through wood fires.

The two other main causes of tree loss include logging in the past, and, more recently, the fungal disease *Phytophthora*, which is the suspected cause of high-altitude dieback. This root rot often takes advantage of plants stressed by other factors, and is especially prevalent in areas that have been subjected to fires.

About 80–84 per cent of surviving populations are protected within forest reserves. Small stands that exist outside reserves are protected by Tasmanian government policy, which forbids logging. All three species are Vulnerable (2000 Red List).

Athrotaxis cupressoides, the pencil pine, is an important component of subalpine vegetation, where it forms a columnar tree to 15m (50ft) tall in scattered stands. It is extremely sensitive to fire and much of the total population was killed during widespread fires in 1960–61 on Tasmania's Central Plateau. Regeneration is slow due to the grazing of sheep and rabbits.

Athrotaxis selaginoides, King Billy pine, is another extremely fire-sensitive conifer, and about 30 per cent of its total population was destroyed by fire during the 19th century. Its plight was made worse by extensive logging, but today, due to its protected status, only wood from salvaged timber is used. This is found in high-value items, such as the sounding boards in musical instruments (*see also p. 100*).

These ornamental and easily grown conifers are suited to wider use. Both species and the hybrid were introduced into the UK in the mid-19th century and have proved to be very winter-hardy, withstanding temperatures well below freezing. They all thrive in moist, well-drained, acid soils. Although slow-growing, they eventually reach 15m (50ft). Propagation is from freshly sown seed; seed rapidly loses its viability. Cuttings can be rooted using ripened young growth in autumn.

HISTORY

WILLIAM LANNEY

Known facetiously as King Billy, for whom *Athrotaxis selaginoides* is named, William Lanney was the last full-blooded male Tasmanian. In the years following the establishment of a British convict settlement on Tasmania in 1803, almost 5,000 indigenous people were enslaved, hunted, or murdered. In 1830, 75 survivors were removed to Flinders Island by a missionary. Many died on the journey, or later of malnutrition. William Lanney was born there in 1835. At his death, in 1869, he was survived by just two native Tasmanians. After his death, regarded as a human relict, he became a desirable "specimen" and his hands, feet, and skull were stolen by body snatchers.

◄ **ATHROTAXIS X LAXIFOLIA**, *a rare hybrid in the wild, is known as the summit cedar since it is often found on the mountain tops in Tasmania. It is probably the most common* Athrotaxis *in cultivation; it has proved hardy to –15°C (5°F) in the UK.*

Red List: Vulnerable

Austrocedrus chilensis

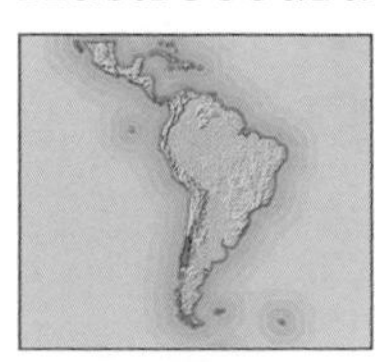

Botanical family
Cupressaceae

Distribution
Southern Chile, southern Argentina

Hardiness
Fully hardy

Preferred conditions
Any well-drained soil of moderate fertility, in full sun, with shelter from dry and cold wind.

The Chilean incense cedar, ***Austrocedrus chilensis***, is the only species in its genus. It is native to the coastal and Andean mountains of Chile, and the eastern Andes of Argentina. It has an extensive distribution both in wetter rainforests and in drier Mediterranean-type vegetation, where it often forms pure stands. Throughout its range, it is often associated with species of the southern beech (*Nothofagus*).

Ancient specimens of *Austrocedrus chilensis* have been dated at 1,000 years of age. These veteran trees occur on rocky outcrops, where the lack of combustible vegetation has spared them from the devastating effects of fire.

Elsewhere, *Austrocedrus* is in decline and is Vulnerable (2000 Red List). In central Chile, most of its forest habitats have been logged, or burnt – or both – and any subsequent seedlings have been grazed by livestock. In Argentina, 85 per cent of its distribution falls outside of protected areas and, there too, most trees have been logged and burnt. The vacant land has been replanted with non-native conifers from North America.

Recent research has established that populations in Argentina, which experience a combination of relatively high rainfall and low altitude, are in decline. This is thought to be caused by the root rot *Phytophthora lateralis*, a fungal disease affecting related species that are threatened in North America.

AUSTROCEDRUS CHILENSIS

CULTIVATION

Although introduced to the UK as early as 1847 by the prolific plant collector Thomas Bridges, *Austrocedrus chilensis* remains uncommon in cultivation. This attractive conifer should be more widely grown for both its ornamental value and its conservation potential. In the past five years, conservation collections have been established in several UK sites, where it grows relatively quickly, in the south particularly. Seedlings from a single tree show remarkable variation in leaf colour, from solid bright green to distinct white markings beneath the leaves.

Propagation is easy by cuttings and spring-sown seed. Seed can be difficult to source; the cones of many wild populations host parasites.

CHILEAN INCENSE CEDAR ► *This specimen of* Austrocedrus chilensis *is seen here on the mountain outcrops of the Parque Nacional Conguillo, from the drier, easterly end of the range. The trees are more tolerant of dry conditions in cultivation.*

PLANTS AND PEOPLE

A PLAGUE OF GOATS

Many species described in this book are under threat from goats. Usually, these are feral goats, introduced by settlers or sailors for meat, milk, and mohair. Exceptionally hardy animals, goats reproduce rapidly, thrive on the toughest grazing, and adapt well to semi-arid environments. But uncontrolled grazing by fast-breeding wild populations spells environmental disaster.

Goats are efficient herbivores, and eat almost all plant parts, including bulbs, roots, and bark; they consume everything below 2m (6ft) in height. Seedlings are an especially succulent food source to a goat, and their consumption prevents natural regeneration. In addition to stripping herb cover, goats disturb the soil with their hooves, causing bare dirt to blow or be washed away. Plants that have not co-evolved with goats are very vulnerable, since they seldom have protective thorns, or an unpleasant flavour to deter grazing.

Red List: Vulnerable

Callitris oblonga

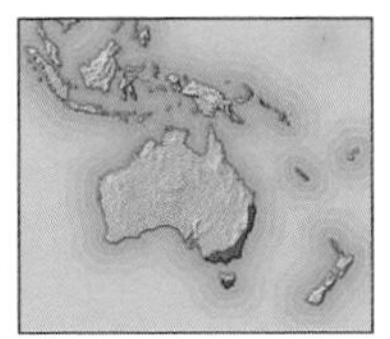

Botanical family
Cupressaceae

Distribution
Australia: New South Wales, Tasmania

Hardiness
Half-hardy – frost-hardy

Preferred conditions
Well-drained, preferably sandy soils in a warm sheltered site in full sun.

The Tasmanian cypress pine, *Callitris oblonga*, found in forest, scrub, and at river margins, is a small tree or shrub, to 8m (25ft). Until recently, it was known only from Tasmania, but the discovery of about 12,000 mature individuals in the tablelands of northern New South Wales has increased its known range. The latter population has been called *Callitris oblonga* subsp. *corangensis* by some authorities.

The Tasmanian population is confined to less than 100 sq km (30 sq miles). Although some small populations are protected in reserves, most are not; some have been killed by fire, or through habitat loss to agriculture. *Callitris oblonga* is Vulnerable (2000 Red List).

Of the five threatened Australian species of *Callitris* grown, *Callitris oblonga* is the most widely cultivated, albeit by conifer specialists. This attractive conifer is hardy in milder, Mediterranean-type climates, where it will grow best given a warm aspect in a sandy, well-drained soil. Propagation is best by seed, which retains its viability over a number of years. Cuttings root very slowly.

2 on Red List

Calocedrus

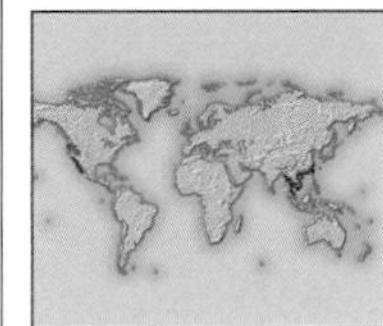

Botanical family
Cupressaceae

Distribution
Myanmar; Thailand; China: Taiwan; Vietnam; western USA

Hardiness
Frost-hardy – fully hardy

Preferred conditions
Any well-drained soil; thrives in sun or partial shade.

The incense cedars, *Calocedrus* species, belong to a genus of only three species, which are represented in both North America and Asia. They are relics of an ancient geological period when the two continents were joined, and are of great interest in floristic studies.

The American *Calocedrus decurrens* is not on the Red List; two Asian species, *Calocedrus formosana* and *Calocedrus macrolepis*, are Endangered, and Vulnerable, respectively (2000 Red List), due to loss of habitat through logging.

Incense cedars form medium-sized trees, 15–25m (50–80ft) tall, and often occur as scattered trees in mixed evergreen forests, at moderate elevations in the mountains. In Yunnan, *Calocedrus macrolepis* is still common in some valleys, but, in southern Vietnam, populations are very small. Some are protected within designated conservation areas.

Neither of the Asian species is commonly grown, but both are good candidates for wider cultivation. Both *Calocedrus formosana* and *Calocedrus macrolepis* make very ornamental garden plants, especially when young. They need a mild, wet climate, where they grow relatively quickly. They are sometimes grown as a plantation tree for their straight-grained wood, which is termite- and insect-resistant; the wood is aromatic and is used for incense.

Red List: Lower Risk

Cathaya argyrophylla

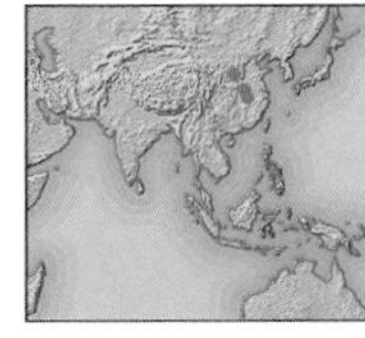

Botanical family
Pinaceae

Distribution
China

Hardiness
Half-hardy

Preferred conditions
Plentiful moisture, moderate humidity, in well-drained, slightly acid soil in sun or partial shade.

This is one of the rare examples of a recently discovered conifer genus. *Cathaya argyrophylla*, the only species known, was first found in the 1950s, but fossil and pollen records were recorded long before its discovery as a living plant. Originally, it was known only from two localities. Today, it is known from 30 or more sites, but, with the exception of one site, the populations in each consist of no more than about 12 individuals.

All of these trees are found on narrow mountain ridges, in relatively inaccessible places away from populous areas. Since most have government protection, they have been listed as Low Risk, Conservation Dependent (2000 Red List). Threats to the long-term future of the slow-growing *Cathaya argyrophylla* are most likely to be competition from faster growing, broad-leaved species.

This conifer has been sought after by horticulturists ever since it was first discovered. It is only since the 1990s that seed has been available from China, and now several botanic gardens and arboreta in Europe and North America have successfully germinated seed. Seed sowing in spring gives the best results; it is difficult to root cuttings.

2*/1 on Red List

Cedrus

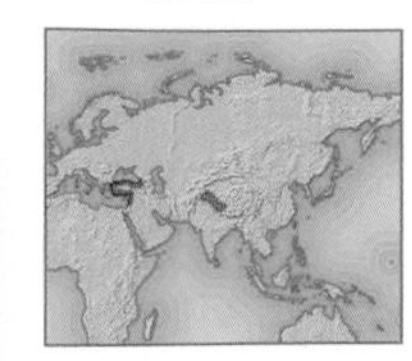

Botanical family
Pinaceae

Distribution
Northern temperate regions of Europe, North Africa, Asia

Hardiness
Fully hardy

Preferred conditions
Well-drained soil (most types) in an open site in sun; tolerates light shade when young.

The genus *Cedrus* contains four species, two of which, *Cedrus brevifolia* and ***Cedrus libani***, appear on the Red List. All species are widely cultivated throughout the temperate world, but the natural populations face various levels of threat. The cedars occur in temperate coniferous and mixed forest, woodland, and scrub and are mostly native to high mountains around the Mediterranean.

The cedar of Lebanon, *Cedrus libani*, was first introduced into Britain in 1638. Its beautiful, horizontally spreading branches make it a highly ornamental feature of landscaped parkland, and many trees persist on country estates to the present day. It is not, however, as attractive as a young tree as the Atlas cedar, *Cedrus atlantica*, from Morocco, or the deodar, ***Cedrus deodara***, from the Himalayas, and is now less commonly planted than either of these two species.

CEDRUS DEODARA *The male cones of the long-needled deodar cedar open in autumn.*

Cedrus brevifolia, Cyprus cedar, from the Troodos Mountains of western Cyprus, is listed as Vulnerable (2000 Red List). This species is probably the least well known of the genus, but is distinguished from the others by its much shorter needles. The greatest threat to this species is its restricted distribution, as it only occurs on high mountains in Cyprus. Were there a forest fire, the entire species could easily be wiped out.

Cedrus libani, found in Lebanon, Syria, and Turkey, is listed as Lower Risk (1997 Red List). In Lebanon, where the cedar forests were much more extensive in ancient times, it is listed as Endangered. The future of the cedar in Syria is also precarious, as in Lebanon, where only five per cent of the original upland forest remains. The species is more plentiful, however, in the Taurus Mountains of Turkey.

The quality of this cedar's timber has been the main cause of its downfall. It has been used for ship building, general construction purposes, and decoration since pre-biblical times, and well before the expansion of the Roman Empire. The earliest record of logging dates back to 4000 BC, and there are many ancient inscriptions that describe this period of logging as lasting some 3,000 years. Deforestation accelerated during the 20th century, forced by the demands for fuel, railways, and the destruction wrought by two World Wars. Of the remaining ten fragments in Lebanon, the Bcharri Grove, the smallest one at 7 hectares (17 acres), was declared by UNESCO to be a World Heritage Site of Cultural Landscape in 1999. Two other remnants have been protected as nature reserves since 1996.

▲ **CEDRUS LIBANI** *in a natural grove beneath the distant, snow-covered mountain Jabal al-Shaykh, is one of only ten remnant populations of the once extensive cedar forests of Lebanon. It has been protected by Lebanese conservation laws since 1992.*

As a result of lobbying by non-governmental organizations (NGOs) and local communities, the first conservation law was ratified by the Lebanese government in 1992. This has paved the way for additional conservation initiatives. Fortunately, the cedars in Lebanon are surrounded by a good deal of national pride, and a number of nurseries there now grow *Cedrus libani*, which is used in afforestation programmes, and is widely planted as an ornamental around the country.

6*/5 on Red List

Cephalotaxus

Botanical family
Cephalotaxaceae

Distribution
Korea, China, Japan, Myanmar, Laos, Vietnam, India

Hardiness
Frost-hardy – fully hardy

Preferred conditions
Moist but well-drained soil in shade, or in sun in cool, damp climates. Needs protection from wind when young.

The plum yews, *Cephalotaxus* species, are a group of very distinctive conifers characterized by their pendent, fleshy, colourful female cones, which are found only in this botanical family.

Plum yews occur in tropical rainforest, and in warm-temperate and temperate mixed forests. Seven of nine species in this genus are threatened. Like many *Cephalotaxus* species, the Taiwanese *Cephalotaxus wilsoniana* is widely distributed, but uncommon through its range; it is Endangered (2000 Red List). ***Cephalotaxus fortunei***, Fortune plum yew, is less threatened and has Lower Risk, Near Threatened status (2000 Red List).

▲ **CEPHALOTAXUS FORTUNEI** *The plum-like fruits, with a single seed surrounded by a fleshy covering, and the yew-like leaves give rise to the common name of plum yew.*

The main threats are habitat loss, poor regeneration due to slow reproductive cycles, slow growth, and long distances between individuals, which can adversely affect pollination. Many species are used locally for timber or firewood. The medicinal properties of their seed oil, and their anti-cancer alkaloids, have led to some overexploitation.

All *Cephalotaxus* prefer moist climates with high levels of shade and will even thrive in clay soils. Once established, they withstand long periods of drought. Unlike many plants, they are known to be highly browse-resistant; for example, they are unpalatable to deer. Propagation is by seed, which takes 10–12 weeks to germinate, or by semi-ripe cuttings in late summer or autumn. Of the two species described, *Cephalotaxus fortunei* is the hardier. *Cephalotaxus wilsoniana* is grown in only a few botanic gardens and needs a warm-temperate climate.

PEOPLE

NAMESAKE

Born in Scotland, Robert Fortune (1812–81) trained at the Royal Botanic Garden Edinburgh. He made several expeditions to China and Japan, often collecting cultivated plants from local nurseries. Travelling in disguise as a Chinese man, he collected tea plants (*Camellia sinensis*) and seedlings for the East India Company's tea plantations in Sikkim and Assam. Fortune introduced many plants to cultivation including *Lilium japonicum*, *Weigela florida*, *Rhododendron fortunei*, and *Sciadopitys verticillata* (*see p. 152*). J.D. Hooker named *Cephalotaxus fortunei* in his honour.

4*/3 on Red List

Chamaecyparis

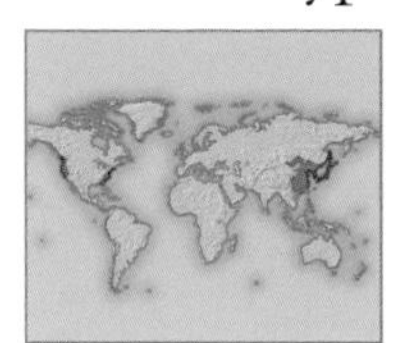

Botanical family
Cupressaceae

Distribution
USA, Japan, China: Taiwan

Hardiness
Fully hardy

Preferred conditions
Mostly moist but well-drained soil in full sun, on steep mountain slopes or valley bottoms.

The false cypresses, *Chamaecyparis* species, are closely allied to the true cypresses, *Cupressus*, and, like them, represent an ancient floristic link between North America and Asia. Most of these elegant conifers form an important component of their native cool-temperate coniferous forests.

The main reason that five members of this economically important genus of conifers are threatened is cutting for their timber. International trade has put great pressure on wild populations. There are also threats from grazing and invasive plants (*see pp.440–471*), which impede regeneration. Recent problems with the fungal root rot, *Phytophthora lateralis*, have exacerbated conservation concerns.

Two Taiwanese species, *Chamaecyparis obtusa* var. *formosana* and *Chamaecyparis formosensis*, from upper montane and subalpine zones, have been threatened by logging; the former is listed as Vulnerable (1997 Red List), the latter as Endangered (2000 Red List). The Atlantic white cedar, *Chamaecyparis thyoides* var. *henryae*, from the forest swamps of south-eastern USA, is at Lower Risk (1997 Red List).

Chamaecyparis lawsoniana, Lawson cypress or Port Orford cedar, is Vulnerable (2000 Red List). Restricted to mountains and canyons from Oregon to California, USA, this is a large forest tree; the tallest known is almost 73m (240ft) tall. Most old-growth forests have been logged, with much of the timber being used for ship building and matches.

▲ **CHAMAECYPARIS LAWSONIANA** *is among the most valuable of North American timber trees. In recent years,* Phytophthora *root rot has begun to kill many of the remaining trees.*

Widely grown as an ornamental and very hardy, the Lawson cypress has given rise to over 200 cultivars, many of them coloured or dwarf forms that are valued for rock gardens and conifer gardens. It is used as a specimen tree and hedge plant, and there is great potential for growing hedges, or even mazes, that have both amenity and conservation value.

In Japan, *Chamaecyparis obtusa* var. *obtusa*, Hinoki cypress, is Vulnerable (1997 Red List) for the same reasons as Lawson cypress. When native populations reached such low levels that logging was banned, timber merchants in Japan turned to America to import the Lawson cypress; its aromatic, white wood is valued for building shrines and temples.

There is an urgent need to establish large commercial populations of all of these species to ease pressure on the remaining stands in the wild. All can be propagated by semi-ripe cuttings and seed, but seed usually shows low levels of germination.

Red List: Lower Risk

Cryptomeria japonica

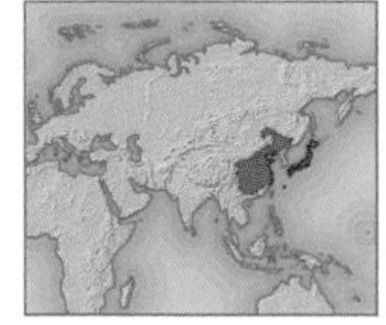

Botanical family
Cupressaceae

Distribution
Japan

Hardiness
Fully hardy

Preferred conditions
Well-drained soil (most types) in sun or part shade, in a sheltered site.

Called the Japanese cedar, or sugi, ***Cryptomeria japonica*** is one of the predominant trees in Japan, where it is found in temperate mixed and coniferous forests. The tallest individuals reach 50–60m (160–200ft), but very few old-growth forests remain. The tree is Lower Risk, Near Threatened (2000 Red List).

Numerous trees still dominate the mountains and hillsides, but most come from the commercial forestry planted after the original *Cryptomeria* forests were destroyed. The species is also grown in Chinese forestry plots. The timber is rot-resistant, easily worked, and has many uses from building to paper making.

The tree has long been planted beside shrines and temples in Japan. Scientists believe that Chinese *Cryptomeria* populations were introduced there for similar purposes many centuries ago. Many are naturalized there, so appear native. The first introductions to Europe and Russia were made in the 1840s.

Cryptomeria japonica is widely grown as an ornamental outside China and Japan, particularly in temperate gardens. Several cultivars are grown, including *Cryptomeria japonica* 'Elegans', which retains its juvenile foliage. *Cryptomeria japonica* 'Nana' is popular for smaller gardens; it reaches 2m (6ft) tall.

Seeds should be soaked in cold water for 12 hours and pre-chilled in open plastic bags for 2–3 months before sowing in spring. Germination rates of 30 per cent are usual. Cuttings taken in spring or summer root readily.

JAPANESE CEDAR ► *This graceful and elegant species has given rise to over 200 cultivars, many of them better known than the species in cultivation.*

Red List: Vulnerable

Cunninghamia konishii

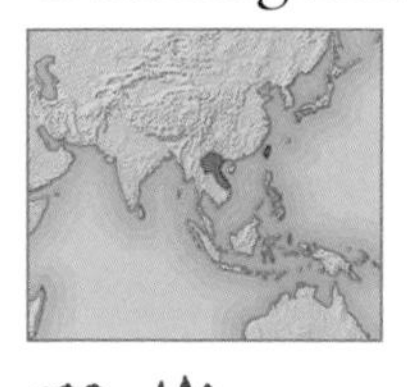

Botanical family
Cupressaceae

Distribution
China: Taiwan, Laos, Vietnam

Hardiness
Fully hardy

Preferred conditions
Any moist but well-drained soil in sun or part shade, with side shelter from wind.

This species, ***Cunninghamia konishii***, is closely related to *Cunninghamia lanceolata*, which is quite well known in cultivation. Some botanists consider it to be a variety of *Cunninghamia lanceolata*. It is considered Vulnerable (2000 Red List).

In Taiwan, it is restricted to the central and northern mountains at 1,300–2,000m (4,225–6,500ft), where it grows with other conifers, such as *Chamaecyparis formosana* and *Chamaecyparis obtusa* var. *formosana*. *Cunninghamia konishii* sometimes forms pure stands in Taiwan.

◄ **CLEARLY CUNNINGHAMIA** *The firm-textured, narrowly lance-shaped leaves are pointed at the tip and have a shiny surface; the leaf arrangement in two ranks along the stem identifies cunninghamias readily.*

This is a rare tree in cultivation, mostly confined to specialist collections. Because of its warm-temperate origins, it is not always tried in places that have traditionally held conifer collections. Plants have been distributed in Britain and Ireland by the International Conifer Conservation Programme during the last ten years and many of these are now over 3m (10ft) tall. High rainfall and mild winters are the most important requirements for this beautiful tree.

RENEWAL

A NEW CASH CROP

In Vietnam, a recent conservation programme has been established to help increase community awareness and to discourage the use of *Cunninghamia konishii* as a building material. In the recent past, much of the area where it grows has been cleared for opium production. The conservation project will concentrate on forest regeneration and a survey of *Cunninghamia's* present distribution. It will also help the local H'Mong people learn propagation and cultivation techniques for agroforestry, so that *Cunninghamia* rather than opium can form the basis of a cash crop.

20*/9 on Red List

Cupressus

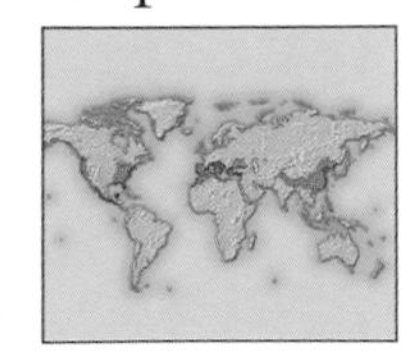

Botanical family
Cupressaceae

Distribution
South-west North and Central America, North Africa to Central China

Hardiness
Half-hardy – fully hardy

Preferred conditions
Well-drained soil, either acid or alkaline, in full sun with shelter from cold, dry winds.

Although the genus *Cupressus* ranges throughout the Northern Hemisphere, these trees are found mostly in small, scattered populations. They grow in scrub and in temperate and subtropical forests, but most cypresses occur in hot, arid, low-rainfall climates, such as those of the south-western USA, Mexico, and North Africa. Those from the Himalaya and China are usually found in dry-summer valleys and on rocky, forested slopes.

In the USA, several species have been reduced to small populations as a result of urbanization, fire, grazing, and cypress canker (*Coryneum* species). Even with national park protection, some are still threatened. This is certainly the case for *Cupressus arizonica* var. *montana*, which is Vulnerable (1997 Red List) but still suffers grazing by cattle. In California, seven threatened cypresses have fewer than 42 populations. These include *Cupressus arizonica* var. *nevadensis*, with nine populations; *Cupressus bakeri*, with eight; *Cupressus goveniana* var. *goveniana* and *Cupressus arizonica* var. *stephensonii*, with two populations each; and *Cupressus guadelupensis* var. *forbesii*, with fewer than five populations. The last two have larger populations in Baja California (Mexico). The endangered *Cupressus goveniana* var. *abramsiana* is restricted to fewer than ten populations, each with no more than 100 immature individuals.

Cupressus macrocarpa, the Monterey cypress, from the chaparral of California, USA, is Vulnerable (2000 Red List). Although grown widely throughout the world as an ornamental, in the wild it is restricted to two natural populations in Monterey County on the California coast. It shares its fragile environment with *Cupressus goveniana* var. *goveniana* and the Monterey pine, *Pinus radiata*.

The Monterey cypress is a dominant component of coastal forests, which form dense communities to a height of about 25m (80ft). Like so many other conifers in California, the Monterey cypress is fire-adapted, and has cones that open to release their seeds when exposed to fire. Historically, the demise of this species is due to a combination of too-frequent fires, forests cleared to make way for housing and golf courses, grazing by livestock, and, perhaps more seriously, the fatal cypress canker. The entire population is protected within reserves.

CUPRESSUS TORULOSA ► *Found in temperate forest in the western Himalaya, where it prefers calcareous soils, this elegant species grows at altitudes of 800–3,000m (2,600–9,750ft).*

MEDITERRANEAN CYPRESSES

In Africa, there are two threatened subspecies of *Cupressus dupreziana*. In Morocco, *Cupressus dupreziana* subsp. *atlantica* is Endangered (2000 Red List), and in Algeria, *Cupressus dupreziana* subsp. *dupreziana* is Critically Endangered (1997 Red List). Although the latter is one of the most drought-resistant plants known, it is unable to regenerate partly because the water table has sunk so low; its numbers have dwindled to fewer than 200 trees through the depredations of logging and goat grazing.

Cupressus sempervirens, Italian cypress, is one of the most widely grown ornamental conifers around the Mediterranean. The most popular form is the narrowly fastigiate cultivar named 'Stricta'. It is precisely because this cypress has been so widely grown for over 500 years that it is difficult to establish its true natural distribution. It is thought not to occur naturally in the eastern Mediterranean. In southern Turkey, it is at Lower Risk, Near Threatened, but is Vulnerable in western Iran (2000 Red List). Found naturally in woodland scrub, the species suffers such conservation threats as urbanization, fire, and grazing, which greatly affects regeneration.

HIMALAYAN CYPRESSES

Several Himalayan species are threatened on account of overexploitation of their timber and for medicines. ***Cupressus gigantea***, Tsangpo cypress, is Vulnerable (2000 Red List) due to the use of its timber in furniture and for its medicinal properties. ***Cupressus cashmeriana***, Kashmir cypress, comes from west-central Bhutan and Arunachal Pradesh in India, where it grows between 2,000 and 3,000m (6,500–9,750ft). It is found in temperate coniferous forests that are heavily influenced by seasonal monsoons. It has been classified as Vulnerable (1997 Red List).

Despite its name, this graceful tree does not occur in Kashmir. It is widely cultivated in that area, however, especially around temples. This planting

▲ CUPRESSUS GIGANTEA *This species is similar to* Cupressus chengiana. *Most of the trees in cultivation have been raised from seed collected in south-east Tibet.*

◄ CUPRESSUS SEMPERVIRENS *This ancient, wind-blasted specimen, seen here in the high, arid White Mountains of Crete, scarcely resembles the more familiar pencil-slim Italian cypress,* Cupressus sempervirens *'Stricta', which is the form that is most widely cultivated throughout the Mediterranean and Italy.*

CUPRESSUS CASHMERIANA ► *This is a particularly fine tree, with unmistakable bluish-grey, weeping foliage. It has been grown mainly under glass in northern Europe, but thrives outdoors in Mediterranean climates.*

has obscured the natural boundaries of its wild distribution. It seems that truly native populations are small, and habitat degradation is the major threat in the long-term future.

Cupressus torulosa, West Himalayan cypress, is listed as Lower Risk, Near Threatened (2000 Red List). It occurs in mixed forests, or in pure stands on mountain slopes and summits. Its fine-textured, termite-resistant timber is ideal for furniture making. It is also aromatic, especially the root wood; the essential oil extracted from this is used as an anti-inflammatory, an antiseptic, and in cosmetics and incense sticks.

CULTIVATION

The cultivation of cypress in Europe dates back before 1500, when the Italian cypress, *Cupressus sempervirens*, was first introduced from the Mediterranean. Since then, this and other species have been widely grown for their garden and landscape value throughout the world.

The Monterey cypress, *Cupressus macrocarpa*, is often used for hedging and as an effective coastal windbreak in mild climates. It is one of the parents of the notorious x *Cupressocyparis leylandii,* the fast-growing Leyland cypress, which has been so widely used for hedges. Most species thrive in hot climates with relatively mild winters, especially in poor, well-drained soils. The Himalayan species also prefer mild winters, but withstand higher rainfall.

All are easily propagated by seed or semi-ripe cuttings. Seedlings grow rapidly and must be transplanted before the roots become spiralled within the pot. Trees with roots that are spiralled when planted out become unstable and prone to wind-rock at maturity.

The most intensive cultivation of cypress, particularly of species from the Americas, was at Rancho Santa Ana Botanic Garden, in California, during the 1930s and 1940s. Here, Carl Wolf and Willis Wagner established experimental plots for the purpose of taxonomic research, and to gain a better understanding of the trees' cultivation needs. Intensive studies were carried out on the devastating effects of the cypress canker.

Although these studies did identify the best species for cultivation in the USA, relatively few of the New World species are grown, mainly due to the problems associated with cypress canker.

▼ **CUPRESSUS MACROCARPA** *Only those trees growing in close proximity to the coast are resistant to the fungal pathogen that causes cypress canker. It is thought that the constant salt spray affords protection by decreasing the fungal spores' viability.*

Red List: Endangered/CITES I

Fitzroya cupressoides

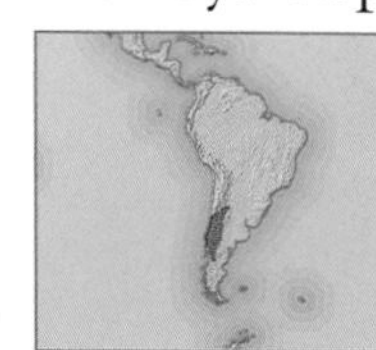

Botanical family
Cupressaceae

Distribution
South-central Chile, Argentina

Hardiness
Frost-hardy – fully hardy

Preferred conditions
Any moderately fertile soil in sun, in a site sheltered from cold and dry winds.

Alerce, ***Fitzroya cupressoides***, is a remarkable tree native to southern Chile and adjacent regions of Argentina. Its entire distribution is within an area of about 360 x 200km (240 x 125 miles), with the most extensive population found in the southern end of its range in the Chilean Andes. The climate is characterized by mild temperatures and annual rainfall between 2,000 and 6,000mm (510–1,525in).

Alerce also occurs in poorly drained lowlands, such as those in Chile's Central Depression, and in freely drained sites in the coastal Cordilleras and Andes, at altitudes of 700–1,500m (2,300–5,000ft). In much of its coastal range it is the dominant species in relatively extensive areas of forest. In the Andes, it is less extensive, and often forms small groves on landslip sites on steep, forested slopes.

RENEWAL

RETURN TO THE WILD

It is generally recognized that if cultivated plant species that are threatened in the wild are going to be of any conservation value, then they should have a broad genetic base. To test this theory, the International Conifer Conservation Programme (ICCP), based at the Royal Botanic Garden Edinburgh, linked up with the Universities of Edinburgh and Valdivia, Chile. They then carried out research into the genetic variability of the historical introduction of *Fitzroya cupressoides* by William Lobb, in 1849.

Leaf samples were taken from 48 trees in British and Irish gardens. Using DNA fingerprinting techniques, it was discovered that all were genetically identical. This suggests that the trees were all female clones derived from a single introduction by Lobb. With this knowledge, the ICCP has sampled seed from a range of wild populations and has distributed plants from these collections, using its network of conservation "Safe Sites" within Britain and Ireland. As a result, today there are more than 400 male and female plants, all genetically unique and of inestimable value for restoration.

It is in the high Chilean Andes that the oldest trees have been found; the oldest is a cut stump with a ring count dating it back 3,600 years. Alerce is one of most famous trees in southern South America due to its incredible timber, which has been especially prized for making roof shingles. Such is the wood's durability that buildings constructed from it in the 17th century still show little sign of weathering today. It is so resistant to rot that logs excavated after having been buried for at least 2,000 years were found to be intact.

Alerce was first introduced to cultivation by William Lobb, one of 22 horticultural collectors employed by the famous nursery of Veitch & Sons, of Chelsea and Exeter (*see p.208*). It is listed as Endangered (2000 Red List).

In cultivation, alerce is mainly restricted to specialist collections, where it has proved to be winter-hardy to –15°C (5°F). It grows best in areas with seasonally high rainfall, but it is tolerant of both well-drained soils and those which tend to be poorly drained.

Propagation is from seed; viability is relatively low. Cuttings taken in autumn root readily, but should be taken from leading shoots; cuttings of sideshoots produce trees with a weeping habit.

◄ **A RARE SIGHT** *Intensive logging has left few surviving mature fitzroyas in this location on the steep slopes of the Chilean Andes, just outside a national park.*

Red List: Endangered

Ginkgo biloba

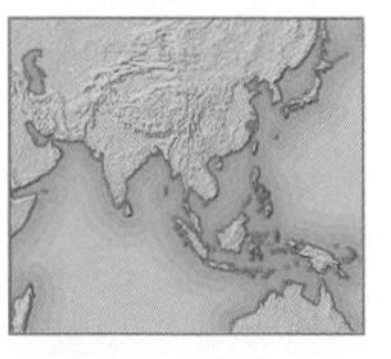

Botanical family
Ginkgoaceae

Distribution
China

Hardiness
Fully hardy

Preferred conditions
Any deep, moderately fertile well-drained soil, in full sun; very tolerant of pollution.

Wild populations of the widely planted ornamental maidenhair tree, ***Ginkgo biloba***, appear to be confined to one mountain in Zhejiang, China. These populations are scattered through broadleaved forests, up to 1,100m (3,600ft). In the wild, *Ginkgo biloba* is considered to be Endangered (2000 Red List). The main threat has been destruction of its forest habitat, and wild populations are now regenerating poorly.

Ginkgo biloba is the only surviving species in the genus and is possibly the most ancient of all living trees. Leaves almost identical to those of ginkgos growing today have been found in Jurassic rocks over 200 million years old. This makes it even older than the dinosaurs. The species has been in cultivation for centuries in China and Japan, where it has been traditionally planted around temples; there, 3,000-year-old trees are found.

Ginkgo was discovered in Japan in 1691, by German physician and botanist Engelbert Kaempfer (*see p.244*), probably in Nagasaki. Kaempfer collected in Japan for the Dutch East India Company, and a tree grown from seed that he collected can still be seen at the botanic garden at Utrecht, Netherlands. The largest trees in the UK are around 250 years old.

The ginkgo is an elegant tree now widely planted along streets and in parks.

PLANTS AND PEOPLE

A HERB FOR ALL ILLS

Ginkgo has been used in Chinese medicine for over 2,000 years, and is now one of the most popular herbal medicines in Europe. It has been used to treat vertigo, tinnitus, and fatigue. It is increasingly recognized as being helpful in improving memory loss and symptoms associated with Alzheimer's disease, as it improves blood circulation and oxygen delivery to the brain. Since it is rich in antioxidants, it helps protect the fragile walls of tiny capillaries, and is also known to reduce blood clotting. Leaves harvested for medicinal use are all derived from cultivated plants, which are grown specifically for the purpose in plantations in France, China, Japan, and Korea.

GINKGO TABLETS

It is tolerant of sulphur dioxide and other urban pollutants.

Ginkgo reaches 30–40m (100–130ft) in height, but the growth rate and shape are variable. The distinctive fan-shaped leaves usually have two lobes, hence the species name, *biloba*. They are pale green, turning darker during the summer and golden in the autumn.

The male reproductive structures are short, thick, yellow catkins that appear in spring. The female structures are borne on separate trees. The resulting fleshy fruits are large and golden-yellow, with an unpleasant smell. Male trees are preferred for street planting.

This tree prefers a sunny site, in moist but well-drained, deep, sandy loam. It is, however, very tough. It was the first tree near Hiroshima to leaf up following the dropping of the atomic bomb in 1945.

BOTANY

THE MISSING LINK?

The ginkgo is called a "living fossil" because fossil specimens, like this Triassic-era mudstone (*right*), reveal that it has changed scarcely at all in more than 100 million years. Fossil records reveal a gradual change from leaves with several lobes about 200 million years ago, to two-lobed ones, more like the living ginkgo, 120 million years ago. For botanists, the tree has been a puzzle; it is so different from any other species that it is classified in a family of its own. It is a gymnosperm (producing naked seeds) like the cycads and conifers, and was once thought to be a missing link between the lower and higher plants, ferns, and conifers.

▲ VARYING FOLIAGE *Ginkgo leaves are carried on both long and short shoots. On mature trees, the leaves are often unlobed compared with younger specimens, which may bear deeply lobed leaves.*

VINTAGE SPECIMEN ► *Older trees develop a broad crown and make spectacular specimen trees in landscaped parks and arboreta. Cultivars with a narrow fastigiate habit are also used in landscaping.*

27*/2 on Red List

Juniperus

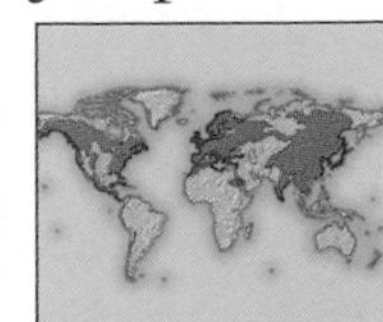

Botanical family
Cupressaceae

Distribution
Worldwide: north temperate to subtropical

Hardiness
Frost-tender – fully hardy

Preferred conditions
Any well-drained soil (preferably alkaline) in full sun or light, dappled shade.

Members of this large genus are found throughout the northern hemisphere. Junipers occur in all types of habitat, from boreal forest to temperate and tropical forest, and in alpine meadows and Mediterranean woodland scrub. There are 27 species of juniper on the Red List, including three cultivated species that are of conservation concern (2000 Red List). All three are highly ornamental.

Juniperus bermudiana, Bermuda juniper, is Critically Endangered, being restricted to the small island of Bermuda in a patch of open, lowland forest. When settlers first arrived, its timber was used for shipbuilding and housing. By the 1950s, the introduction of two scale insects from the mainland USA had defoliated 90 per cent of the forest. Although there was some recovery, the junipers are now fighting for survival against casuarinas and other invasive exotics.

Juniperus cedrus, Canary Island juniper, is endangered on three of the seven Canary Islands. Its valuable timber was harvested in the first 500 years of Spanish occupation, but it is now commercially extinct. This juniper was then further plagued by grazing goats. Felling and grazing were both stopped in 1950. There are several replanting schemes under way, and populations that have survived on cliff edges are expanding with the help of artificial regeneration.

▲ JUNIPERUS RECURVA *VAR.* COXII *The common name of coffin juniper derives from its principal use by the Chinese, who traditionally used the close-grained timber for making coffins.*

HISTORY

PLANT HUNTER

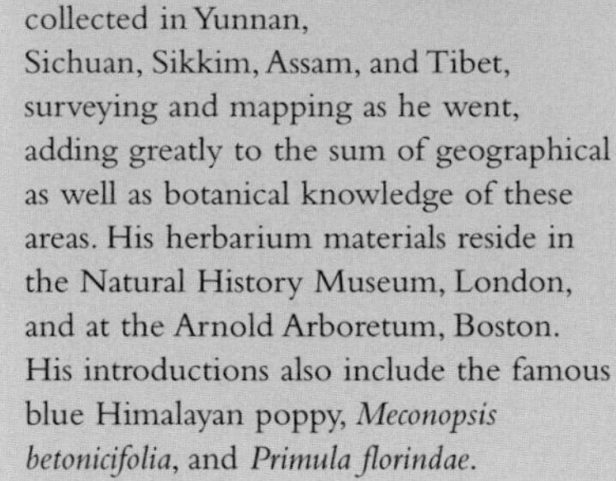

Six years before its introduction in 1920, Frank Kingdon-Ward (1885–1958) discovered *Juniperus recurva* var. *coxii* in Myanmar (then Burma). He also collected in Yunnan, Sichuan, Sikkim, Assam, and Tibet, surveying and mapping as he went, adding greatly to the sum of geographical as well as botanical knowledge of these areas. His herbarium materials reside in the Natural History Museum, London, and at the Arnold Arboretum, Boston. His introductions also include the famous blue Himalayan poppy, *Meconopsis betonicifolia*, and *Primula florindae*.

Juniperus recurva* var. *coxii, coffin or Himalayan weeping juniper, is considered Vulnerable. It occurs along the Himalaya, from Afghanistan to south-west China and Myanmar, in temperate forests, and alpine and montane meadow, at 2,700–4,600m (8,775–15,000ft). In common with many shrubby species that grow in and around the Himalaya, collection for fuel wood is the greatest threat to its survival.

Although there are several replanting schemes under way, efforts are thwarted by erosion, severe winds, and the fact that, at high altitudes, growth rate is very slow. Increased tourism in the area has not helped the plight of this tree.

The Indian Forest Conservation Act of 1980 includes recommendations to halt the progress of conditions that cause continued forest degradation; to enhance conditions under which natural regeneration can occur; and to include and win the support of local people in conservation and management plans.

A tree to 20m (70ft), with grey-green, weeping foliage, *Juniperus recurva* var. *coxii* is a very ornamental species. Reginald Farrer and E.H.M. Cox, well-known plant collectors of the early 20th century, introduced it in 1920. Farrer described it as "one of the most beautiful of all conifers". In Great Britain and Ireland, there are now 80-year-old specimens which are over 13m (43ft) tall.

Since most junipers come from dry habitats, they are particularly tolerant of dry, sandy, or chalky soils and are useful in Mediterranean-style gardens.

Unlisted

Lagarostrobus franklinii

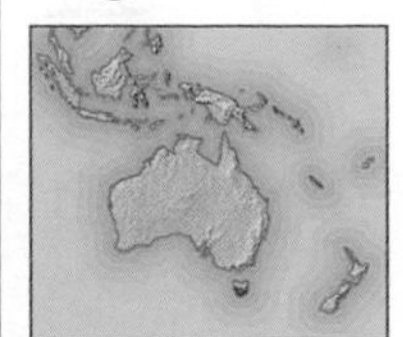

Botanical family
Podocarpaceae

Distribution
Australia, Tasmania

Hardiness
Fully hardy

Preferred conditions
Moist but well-drained soil, in sun or light shade with high rainfall and humidity, and shelter from cold, dry winds.

This remarkable tree, the sole species in the genus, is native to temperate rainforest along the Huon River system, and the adjoining territory near the southern and western coasts of Tasmania. The tree is not listed in the 1997 or 2000 Red Lists.

One of the classic sites where ***Lagarostrobus franklinii*** occurs is on Mount Read, where the trees are almost genetically identical because at present, regeneration only occurs vegetatively by natural layering of low branches.

This Huon pine is yet another example of the extraordinary longevity found in conifers, for this species can live in excess of 3,000 years, making it the oldest living tree in Australasia. Because the wood is extremely resistant to decay, subfossilized specimens have enabled dendrochronologists to construct a continuous tree ring record covering over 12,000 years. This is a world record.

The timber of the Huon pine is widely used in carpentry, and an oil obtained from the wood is used for a wide range of purposes, including in paint, as preservatives, on wounds, to treat toothache, and as insecticide.

Partly as a result of its wide range of uses, the populations of this tree have suffered great losses. It is estimated that up to 15 per cent of its habitat has been destroyed through logging, fire, and flooding for hydroelectric schemes. Some populations have been afforded protection within national parks.

The Huon pine was first introduced to cultivation in the UK as early as 1844, some 34 years after its first discovery by Allan Cunningham, a collector employed by the Royal Botanic Gardens, Kew. In cultivation, it is restricted to specialist collections in areas with wet and mild climates, although it can withstand heavy frosts and snow falls. It is a slow-growing conifer with a weeping habit in youth.

Most of the older trees in cultivation in the UK are male, which may indicate that they are clonal, that is, genetically identical. A much broader genetic base needs to be found if cultivated trees are to have value for conservation purposes.

▲ ELEGANT EVERGREEN *This slender-crowned tree produces cascades of weeping foliage and is a very ornamental specimen for mild areas with warm summers and high rainfall. It is especially attractive in its juvenile years.*

Red List: Lower Risk

Libocedrus plumosa

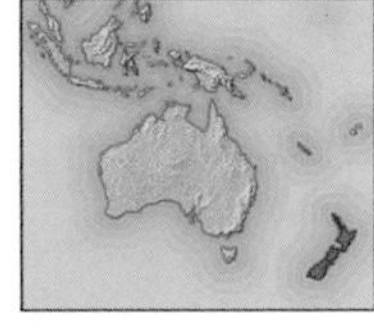

Botanical family
Cupressaceae

Distribution
New Zealand

Hardiness
Frost-hardy

Preferred conditions
Moist but well-drained soil, in sun with shelter from cold, dry winds.

The New Zealand cedar, or Kawaka, from the mixed mountain forests of New Zealand, is found on both North and South Islands, from sea level to 600m (2,000ft). *Libocedrus plumosa*, one of two species native to New Zealand, is protected in national parks, and is Lower Risk, Near Threatened (2000 Red List).

Its highly durable wood is fine-grained with beautiful markings. Logging has destroyed native stands that lie outside protected areas. Other threats include land clearance for agricultural purposes and mining. Regeneration appears to be good in disturbed areas.

This rarely grown species is a good candidate for wider cultivation; it makes a beautiful specimen tree. It has proved reliably frost-hardy, but prefers maritime climates where annual rainfall is relatively high. It is propagated from spring-sown seed, and late-summer or autumn cuttings.

Red List: Critically Endangered

Metasequoia glyptostroboides

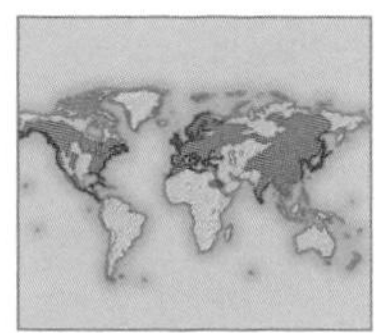

Botanical family
Cupressaceae

Distribution
Western China: Sichuan, Hubei, Hunan

Hardiness
Fully hardy

Preferred conditions
Moist but well-drained soil in sun, or light shade, in areas with moderate rainfall.

The dawn redwood, ***Metasequoia glyptostroboides***, is the sole species in its genus, and one of the few deciduous conifers. It occurs in shady ravines and on damp mountain slopes at altitudes of 750–1,500m (2,450–4,900ft). Soon after its discovery in China in 1941, and in the wake of the revolution of 1949, much of the largest population, about 6,000 trees, was logged.

Laws passed in 1980 prevent cutting for timber or fuel wood, but there is no law against habitat degradation. Although another outlying population was found in Hunan Province, China, in 1986, this ancient species is now Critically Endangered. Natural regeneration seems to be poor and unless further protection is provided, there is little hope for its long-term survival in the wild. The original seed introductions came from just three trees, so the genetic base of this species in cultivation is rather narrow. During the 1980s, it was noted that many cultivated plants showed inbreeding depression. Nonetheless, it is widely planted throughout the temperate world, and is well-represented in botanic gardens and arboreta.

Since the recent introduction of more genetic material from the wild, *Metasequoia* is being considered for trial production as a timber tree. It is fast-growing and very ornamental, and can be easily propagated by seed or cuttings.

A few cultivars have been named, including the bright green 'Emerald Feathers', and 'National', which was selected for its narrow, conical habit.

BOTANY

THE DAWN REDWOOD

In the same year that *Metasequoia* was described as a fossil genus, by S. Miki, a Japanese palaeobotanist, a Chinese botanist found a very unusual deciduous tree in China. Specimens were collected three years later, in 1944, by T. Wang, a forest researcher. A year later, C.H. Wu realized that the living trees belonged to a genus previously unknown to science. Two further populations of 22 and 100 trees were found in 1946. Seeds were sent from China in 1948 to the Arnold Arboretum, Boston, USA, and seeds and plants were then distributed to botanic gardens throughout the world.

▼ TIMBER SOURCE *The long, unbranched trunks are not only majestic, but also a promising source of prime timber. The possibility of commercial forestry rests, in part, on selection and bulking up of good genetic material from the wild.*

Red List: Endangered

Microstrobos fitzgeraldii

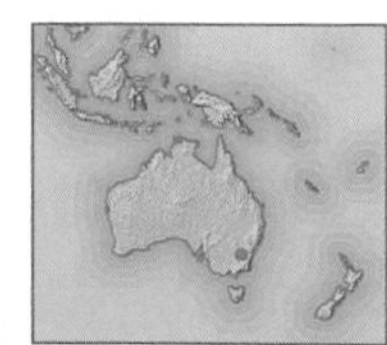

Botanical family
Podocarpaceae

Distribution
Australia: New South Wales

Hardiness
Frost-hardy – fully hardy

Preferred conditions
Reliably moist, but freely draining, soil in sheltered semi-shade.

The dwarf mountain pine, *Microstrobos fitzgeraldii*, occurs only in the Blue Mountains of Australia, where it usually inhabits wet rocks within the spray of waterfalls. It is Endangered (2000 Red List) due to its narrow distribution, and because fires in the Blue Mountain National Park have damaged its habitat.

The species is rarely cultivated, but is sometimes grown in pots by alpine specialists. It does respond well to moisture-retentive soils in semi-shade. It is hardy to frost and snow. It can be propagated from seed, or from cuttings, which root readily.

Red List: Vulnerable/CITES I

Pilgerodendron uviferum

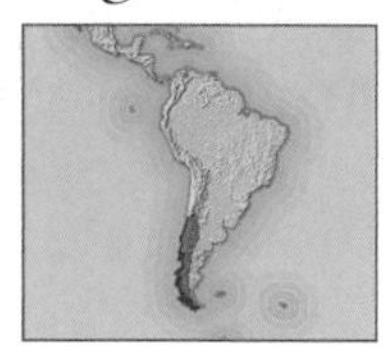

Botanical family
Cupressaceae

Distribution
Southern Chile, Argentina

Hardiness
Frost-hardy – fully hardy

Preferred conditions
Most soils (although usually acidic), including moist and waterlogged ones, in full sun.

With a broad distribution in both coastal and Andean regions, the Chilean cypress, *Pilgerodendron uviferum*, occurs in poorly drained soils. Probably the southernmost conifer in the world, it is often found in its southern, coastal habitats with *Drimys winteri* and *Nothofagus betuloides*. To the north of its range, it occurs with *Fitzroya cupressoides*.

Pilgerodendron uviferum is a narrow, pyramidal, slow-growing tree that can live up to 500 years. It has been heavily exploited for its decay-resistant wood. Although relatively fire-resistant, large-scale forest clearance in colonial times and frequent fires have led to its loss from many parts of its natural range.

Introduced to Europe in the mid-19th century, *Pilgerodendron uviferum* never really found favour, but is easy to grow, withstanding temperatures below -10°C (14°F). It thrives in low or high rainfall, and is one of few conifers that will grow in standing water. Propagation is easy from seed, and from autumn or winter cuttings of the current year's growth.

27*/13 on Red List

Picea

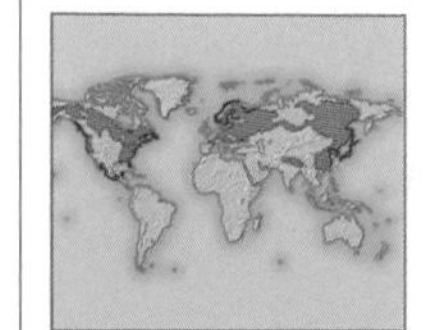

Botanical family
Pinaceae

Distribution
North America, Mexico, Europe, Asia

Hardiness
Frost-hardy – fully hardy

Preferred conditions
Moist but well-drained, neutral to acid soil, in sun; *Picea omorika* tolerates alkaline soils.

The spruces, *Picea* species, are highly prized for their strong, lightweight and fine-grained timber, and, as a result, some species have been overexploited. Even those which once had an extensive distribution, such as *Picea brachytyla*, which is native to central and western China, are experiencing decline due to non-sustainable logging.

Most species are important watershed protectors (*see p.20*) occurring at high elevations on steep slopes. Many are threatened because they have a narrow natural distribution, which makes them even more vulnerable to local human pressures.

Cultivated examples include *Picea morrisonicola*, from the high mountains of central Taiwan, which is Vulnerable (2000 Red List), and two Japanese species, *Picea maximowiczii*, which is also classed as Vulnerable, and ***Picea koyamae***. The two last are restricted to very small populations close to Tokyo, where they inhabit the same forests. In the wild, *Picea koyamae* is represented by only a few hundred individuals, and is Endangered (2000 Red List). In Britain, there are several large, notable stands of these trees with potential for conservation purposes.

PICEA KOYAMAE ► *This was introduced to cultivation in North America and Britain by the famous plant collector Ernest Wilson in 1914.*

THREATENED IN THE WILD

The Brewer or weeping spruce, ***Picea breweriana***, from the Siskiyou mountains of California, and, more rarely, from the coastal ranges of south-west Oregon, USA, is Lower Risk, Near Threatened

◄ PICEA OMORIKA *can be easily identified by its exceptionally narrow habit and dark blue-green needles. It was probably common prior to the last ice age, but is now known only in the Balkans.*

CONIFERS

THREAT

Hard rain falling

Moisture that has been polluted by acid in the atmosphere is known as acid rain, and it damages the environment when it falls. Two common air pollutants that acidify rain are sulphur dioxide and nitrogen oxide. They can be carried over long distances by prevailing winds before falling as acidic rain, snow, fog, or dust. Acid rain is a particular problem in many northern hemisphere conifer belts because the water and soil systems often lack natural alkalinity and cannot neutralize the acid. Acid rain damages leaves and needles, reduces a tree's ability to withstand cold, and inhibits plant germination and reproduction, greatly reducing tree vitality.

(2000 Red List). The key conservation problem is its poor adaptation to fire, although light, short-lived fires may encourage seed germination.

Unlike many of its associated species, the brewer spruce has thin bark and long, weeping branches, which make it particularly susceptible to wildfire. A series of fires in 1987 swept through much of the limited natural range of this conifer. Damage was particularly bad on granite soils, where shallow root systems are extremely prone to fire damage.

Recovery from fire is very slow. Regeneration in open areas left by fire is particularly poor, because the seedlings are not sun-tolerant and suffer moisture stress in the heat and sharply drained mountain soils. The recovery of the brewer spruce from the extensive fires of almost 30 years ago may take decades, if not centuries. In the meantime, more fires may well occur.

Picea omorika, Serbian spruce, from Bosnia-Herzegovina is Vulnerable (2000 Red List). It has a restricted distribution and is said to cover an area of only about 60 hectares (150 acres) of forest, with a population of fewer than 1,000 trees. Here, it occurs in pure stands, or with other conifer and broadleaf species on steep, calcareous, north-facing slopes. Its rarity is due to extensive logging. Fire has also contributed to its demise. A more recent problem is that it hybridizes with commercial plantations of the North American Sitka spruce, *Picea sitchensis*. This could have a devastating effect on its long-term survival if the future generations are unnatural hybrids between *Picea omorika* and *Picea sitchensis*.

Cultivation

Tolerance of extreme exposure to wind and low temperatures make spruces especially suited to shelterbelt plantings. Most, however, are thought to dislike hot weather, or air pollution. All grow in thin, acidic soils; *Picea omorika* tolerates alkaline soils and air pollution.

Most threatened species of spruce are in cultivation, even though some are restricted to specialist collections. In Europe, the Norway spruce, *Picea abies*, is used traditionally for Christmas trees.

All spruces germinate readily from seed, and viability, even after relatively long-term storage, is good, provided that seed is kept in optimum cool, dry conditions.

Of all the threatened spruce species, the American *Picea breweriana*, and the European *Picea omorika*, are the most widely grown; both are outstanding ornamental evergreen trees.

► PICEA BREWERIANA *occurs at high elevations, at 1,200–2,100m (4,000–7,000ft), usually nestling in hollows at canyon heads, in inaccessible regions that experience very heavy snowfall.*

82*/20 on Red List

Pinus

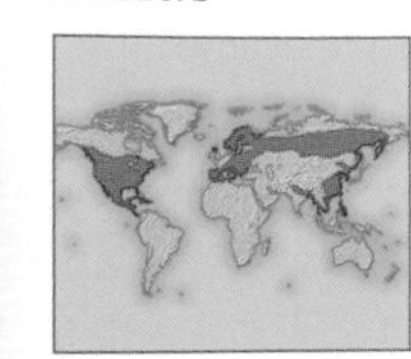

Botanical family
Pinaceae

Distribution
Arctic to Central America; Europe and North Africa to south-east Asia

Hardiness
Frost-hardy – fully hardy

Preferred conditions
Any well-drained soil in sun.

The pines, *Pinus* species, can be found in forests of many types: tropical rainforest, tropical coniferous forest, temperate mixed and coniferous forest, dry Mediterranean forest, and tropical savanna. They are a major component of the boreal coniferous forest, and are an extremely important, and often dominant, component of forests throughout the northern hemisphere.

Ecologically, pine forests are important because they affect the water cycle (*see p.20*) and fire regimes. They even play a role in controlling regional and global climates.

These are just a few reasons for conserving the many threatened species, but, in addition, the human economy depends on pines for timber, pulp, nuts, and resins. This can sometimes be a threat. *Pinus gerardiana*, for example, from the north-western Himalaya, provides an income for over 13,000 indigenous people in Pakistan during years when the nut harvest is good. Uncontrolled harvesting has resulted in virtually no regeneration. It is at Lower Risk, Near Threatened (2000 Red List).

Other threats include narrow natural ranges. In Mexico, *Pinus culminicola*, and *Pinus maximartinezii* are Endangered (2000 Red List). The latter is known only from one location, with between 2,000 and 2,500 individuals; both species have suffered further decline from fire.

Pinus torreyana, which is Vulnerable (2000 Red List), has two varieties restricted to single populations in California. They have been safeguarded by the intensive efforts of local people who have helped establish sanctuaries for their protection, but the threat of fire is still ever-present.

Many pine species are cultivated and widely used for forestry, such as the Cuban endemic, *Pinus caribaea* var. *caribaea*. It is ironic, therefore, that wild populations of such species are still under threat. Other threatened species in this category include *Pinus armandii* var. *mastersiana*, restricted to the mountainous regions of Taiwan; *Pinus ayacahuite* var. *veitchii* from Mexico, threatened by logging; and *Pinus palustris* from south-eastern USA.

PINUS PEUCE

MORE THREATS

Pinus albicaulis, the whitebark pine, ranges from western Canada to north-western USA, and is Vulnerable (2000 Red List). A tree of arid mountain tops, where individuals are known to reach over 880 years of age, this species is well known for its remarkable symbiotic relationship with Clark's nutcracker (*Nucifraga columbiana*, *see box, facing page*).

In Canada, seeds of *Pinus albicaulis* are also an important source of food for bears. In fact, not only does availability of the pine seeds affect bear mortality rates, but it is also said to have a direct correlation with human-bear conflicts. The major threats to this pine are fire and the introduction of the white pine blister rust to its native habitats. This fungus has devastated extensive stands and related pine species.

Pinus aristata, the Rocky Mountain bristlecone pine, from Colorado, New Mexico, and Arizona, USA, is at Lower Risk, Near Threatened (2000 Red List). Closely related to ***Pinus longaeva***, this slow-growing pine is also famed for its longevity; for example, one tree has a cross-dated age of 2,435 years. Such ancient trees grow at the timberline of the Rocky Mountains at 2,300–3,650m (7,500–11,900ft) above sea level, where they experience extremely hot summers and very cold winters. Grazing is the main concern, and most populations now have some sort of protection against this. The new threat of climate change, however, is less easy to solve. Because of its slow growth, this species is often used in small gardens and is a favourite of dwarf conifer enthusiasts.

◄ **PINUS CANARIENSIS** *at Cruz de Tejeda, Gran Canaria, is found on five of seven of the Canary Islands. It is widely grown elsewhere as a timber tree.*

▲ **PINUS LONGAEVA** *Winter in Patriarch Grove, in the White Mountains of California. In such inhospitable, cold, and arid conditions, dead wood can stand for thousands of years.*

Pinus balfouriana, foxtail pine, from California, USA, has Lower Risk, Near Threatened status (2000 Red List). There are two subspecies listed from subalpine altitudes. Both *Pinus balfouriana* subsp. *austrina* and subsp. *balfouriana* are very slow-growing trees, and occur within protected areas, but excessive grazing by wildlife, and fire, are still proving to be problems. White pine blister rust has also afflicted some populations. In cultivation, the foxtail pine is found only in botanic gardens and specialist collections.

Pinus canariensis, Canary Island pine, is a denizen of Mediterranean forests, on volcanic mountains at 600–2,000m (1,950–6,500ft). The species is listed as Rare on the Red List of 1997. In drier locations, it occurs as a widely spaced forest tree, but in wetter areas, such as in the northern cloud belt on Tenerife, it grows in dense forests with a closed canopy. Here, fog-drip that condenses from the needle tips increases the precipitation by a factor of four. Such an increase in water availability also helps sustain the local economy.

Pinus canariensis is the largest native of the Old World, reaching a height of over 60m (200ft). Such trees, however, are rare, as most old-growth forests have been logged out. Although fire is also a great threat, thick bark and a capacity to

resprout from dormant buds makes this pine well able to survive fires.

It is widely grown as a timber species, but in Australia and South Africa, seeding from plantations into native vegetation is causing considerable problems. In the Canary Islands, the leaves are used as a packaging material for bananas, and the glossy, rich-brown cones are often used as Christmas ornaments.

Pinus longaeva, bristlecone pine, from eastern California to southern Nevada, is Vulnerable (2000 Red List). The subalpine habitats of this remarkable species are incredibly hostile. Temperatures on the crest of the mountains range from 70°C (158°F) in the summer to -26°C (-15°F) in winter. Rainfall scarcely exceeds 300mm (12in) per year; very strong winds occur year-round; and electrical storms are frequent. The growing season of six weeks must produce growth and reserves for overwintering. No wonder a pine growing in these conditions, on poor thin soils, has to be highly adapted and armed with effective survival strategies.

The chosen strategy of the bristlecone pine is longevity. Tree-ring counts of living specimens approach some 5,000 years, making this the oldest living, sexually reproducing organism in the world. Some say that the bristlecone pine merely takes a long time to die, preserved by its resinous wood and extreme cold and aridity.

Most populations have full protection from cutting, or gathering of wood samples. The most serious threat is climate change, and it is thought that present rates of regeneration are insufficient to replace the very old and dying trees. The species is widely cultivated, and, due to its slow growth, is often favoured for the smaller garden.

In the wild, *Pinus longaeva* makes a short tree, to about 6m (18ft) tall, with gnarled and little obvious growth; some trees look almost dead. But they have proved immensely useful to dendrochronologists, who date past climatic events from the study of tree rings; during the early part of the 20th century, it was observed that differing widths of growth rings were related to growing conditions. Each year, a tree adds a ring of wood to its trunk, thus producing annual rings, which can be seen in cross-sections of trunk. Cores can be taken from living trees using a slender metal rod. Analysis of the cores helps scientists to construct a chronology of past climatic events. This is highly relevant in environmental studies, especially when assessing such phenomena as climate change.

▼ **PINUS TORREYANA** *may be among the rarest of American pines. They inhabit the canyons that drop down to the California coast, each tree individually sculpted by wind.*

WILDLIFE

FEED AND SEED

Clark's nutcracker, *Nucifraga columbiana*, from the Rocky Mountains of western North America, is one of several bird species with a special relationship with pines. The seeds of the whitebark pine, *Pinus albicaulis*, are the principal food source for this bird, particularly when raising its young. Since the cones of this pine do not open, seed extraction and dispersal are totally dependent on the nutcracker, which uses its specially adapted bill for the purpose. Some seeds are stored in special caches. Dispersal is accomplished when some of these seeds are inadvertently dropped by the bird *en route*, or a seed cache is forgotten.

In Europe, the Macedonian pine, ***Pinus peuce***, and the Arolla pine, *Pinus cembra*, have a similar relationship with another nutcracker, *Nucifraga caryocatactes*. Another bird group, the crossbills, or *Loxia* species, are large finches with crossed bill tips that lever apart conifer cones to extract the seed. Species specialize variously in spruce, firs, larches, and pines.

OTHER THREATENED PINES

Pinus muricata Bishop pine is found in coastal California, USA, and Baja California, Mexico. It is a very variable species in the wild and the focus of much debate among taxonomists. Within its range, it grows in a variety of habitats from swampy terrains and peat bogs, to drier, better drained sites. It belongs to a group of pines with cones that are opened in response to fire and release their seeds onto recently burnt ground beneath the parent tree. With available moisture, the seeds promptly germinate. Grazing goats sometimes destroy young plants. In cultivation, it puts on rapid and bushy growth and is resistant to sea winds, making it ideal for coastal gardens. Red List (2000): Lower Risk, Conservation Dependent.

Pinus peuce Macedonian pine is found in Albania, Serbia, Bulgaria, and northern Greece. This pine occurs on the northern slopes of two Balkan mountain ranges. Seed dispersal is by the nutcracker bird (*see box, above*); seedlings often germinate from uneaten seed caches. Most populations in Bulgaria are protected in national parks. Commonly cultivated due to its shapely habit and tolerance of low temperature; it is one of the most useful pines for landscape planting. It is resistant to white pine rust. Red List (2000): Lower Risk, Near Threatened.

Pinus radiata* var. *radiata Monterey pine is restricted to just three localities in the fog belt on the California coast. Even though this pine has been selected for use in forestry throughout the southern hemisphere, in the wild, it suffers from vandalism. More seriously, 80–90 per cent of some populations are infected with pitch canker. Apart from its use in forestry, the Monterey pine is planted as an ornamental and in coastal windbreaks. Red List (2000): Lower Risk, Conservation Dependent.

Red List: Lower Risk

Platycladus orientalis

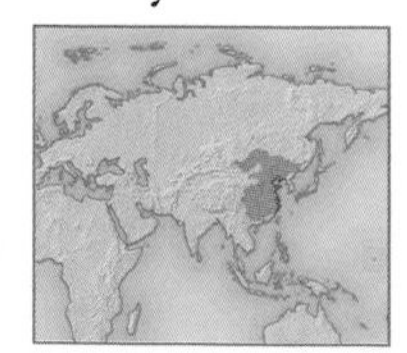

Botanical family
Cupressaceae

Distribution
North-east, central, and south-west China

Hardiness
Frost-hardy – fully hardy

Preferred conditions
Well-drained, even dry soils, in full sun.

Biota, ***Platycladus orientalis***, occurs in mixed forests, many of which have become severely degraded by logging, and by the collection of wood for domestic fires used in cooking. Due to extensive historical cultivation, it is sometimes very difficult to determine which populations are truly native and this causes some difficulty in assessing degree of threat. In 2000, its status was Lower Risk, Near Threatened.

This, one of the most widely grown conifers, is thought to have been introduced to cultivation outside its native country during the early part of the 18th century, if not earlier. In China, it has a long history of cultivation and has gained a place in Chinese folklore. Seeds were placed in the burial casks of the ancient emperors, and trees were planted around their tombs to protect them in the afterlife.

► BIOTA TREES, *with their flattened sprays of scale-like leaves, closely resemble the related cypresses* (Cupressus *species*).

Biota is winter-hardy, but thrives best in hot, sunny climates, where it tolerates periods of summer drought. It is easily raised from seeds or cuttings. Its small stature and neat conical shape make it the perfect conifer for a small garden.

45*/14 on Red List/CITES I & III

Podocarpus

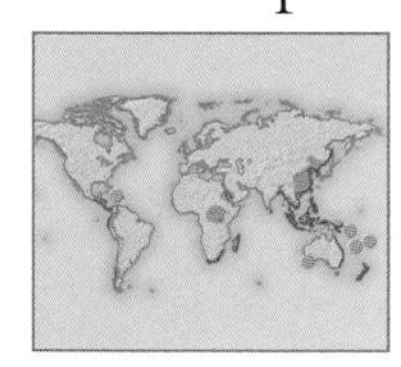

Botanical family
Podocarpaceae

Distribution
South and Central America, Asia, Australasia, South Africa

Hardiness
Fully hardy – frost-tender

Preferred conditions
Leafy, moist but well-drained soil in sun, with shelter from cold or dry wind.

The podocarps, *Podocarpus* species, seldom form extensive stands, but instead are found as individuals scattered through their forest habitats. They usually grow in areas of relatively high rainfall, although some of these forests can be seasonally dry. Most species have a colourful, red to purple, fleshy receptacle, onto which the seeds are set. The nutritious flesh is attractive to birds, which are the main seed distributors. Podocarps are prized for their timber; the wood of the threatened Jamaican endemic, *Podocarpus purdieanus*, for example, is said to have a higher local value than mahogany. On the 2000 Red List, this species was Data Deficient.

The podocarps of South Africa, often known as yellowwood, are much admired for the fine yellow colour of the timber. Yellowwood is commonly used for making butcher's blocks, because the wood is hard, unscented, and does not chip easily.

GLOBAL THREATS

Podocarps are threatened throughout the world by habitat loss, mostly as a result of logging and fire. The natural stands of *Podocarpus costalis*, a native of Taiwan, have been reduced to only a few individuals due to the uprooting of plants for the horticultural trade. It is Endangered (2000 Red List).

PODOCARPUS NUBIGENUS

Few threatened podocarp species are cultivated, and there is a need to widen the range of species in cultivation for conservation purposes. Of those already in cultivation, some are used for forestry, but mostly, they are grown as specimen garden trees. The species most usually seen is the Japanese and Chinese species, *Podocarpus macrophyllus*, often used in topiary. Some species, such as ***Podocarpus salignus*** and *Podocarpus drouynianus*, are used as cut foliage in the florist industry. The most widely grown from temperate South America are *Podocarpus salignus* and ***Podocarpus nubigenus***. Both favour climates with relatively high rainfall and mild winters. The former is the more common, particularly in Ireland, where there are some very large specimens.

Podocarpus nubigenus, Ma-io, from southern Argentina and southern Chile, is Lower Risk, Near Threatened (2000 Red List). It comes from broadleaf and mixed forest. Although not highly threatened, this conifer, like many of the species in the temperate rainforests of South America, is at risk due to massive forest conversion to commercial plantations of Monterey pine (*Pinus radiata*) and eucalyptus.

Traditionally, the wood of Ma-io has been highly sought after thanks to its resistance to decay and its strength; because of these properties, it is often used for internal supports in houses.

Podocarpus salignus, willow-leaved podocarp, is Vulnerable (2000 Red List). It occurs in the temperate broadleaf and mixed forests of southern Chile. It grows at low to medium altitudes in humid forests, especially favouring habitats close to watercourses. Even though natural regeneration is generally very good, logging and replacement plantations of Monterey pine and eucalyptus have eliminated most of the stands of old-growth trees. Unfortunately, *Podocarpus salignus* occurs in few protected areas, and is not helped by the fact that most national parks are located at higher altitudes in the Andes, beyond the altitudinal range of the species.

Propagation of podocarps is generally by cuttings, but seed will germinate freely, if it is not allowed to dry out before sowing. In the case of *Podocarpus nubigenus*, removing the outer seed coat and setting the embryo on top of sphagnum moss in a closed propagation case usually results in good germination.

◄ PODOCARPUS SALIGNUS, *with its slender, shining, blue-green leaves, becomes increasingly elegant as it ages, when the branches begin to weep gracefully. Female trees produce egg-shaped, deep violet-purple fruit in the autumn.*

Red List: Vulnerable

Prumnopitys andina

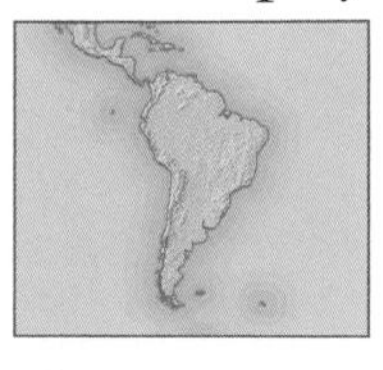

Botanical family
Podocarpaceae

Distribution
Southern Chile, southern Argentina

Hardiness
Fully hardy – frost-tender

Preferred conditions
Moist but well-drained soil in sun, with shelter from cold or dry wind.

The Chilean plum yew, ***Prumnopitys andina***, forms a large, broad-crowned tree, to 15m (50ft) tall. It often occurs on the margins of large rivers, particularly in the Chilean Andes, with one population just across the border in Argentina. Fewer than a dozen known populations remain, most with fewer than 100 individuals. It is listed as Status

PRUMNOPITYS ANDINA ► *There are several populations in the Upper Biobio Valley in southern Chile; one has over 100 mature plum yews. However, in January 2004 a nearby group was lost to flooding, caused by the building of a dam downstream.*

Red List: Endangered*

Pseudolarix amabilis

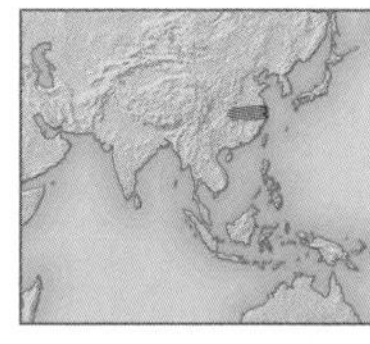

Botanical family
Golden larch

Distribution
China: beside the lower, middle reaches of the Yangtse River

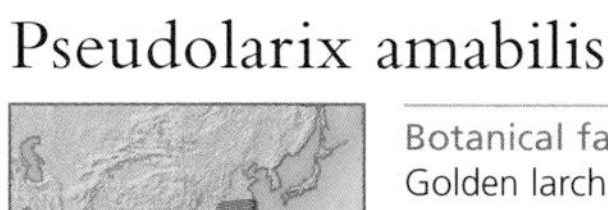

Hardiness
Fully hardy

Preferred conditions
Moist but well-drained soil in sun, sheltered from strong, dry, and cold winds.

The golden larch, ***Pseudolarix amabilis***, an ancient species, was much more widespread 70 million years ago than it is today. Before the Pleistocene Ice Age, its distribution included Europe, most of Asia, and western North America. Now, however, populations remain in China only as

◄ **CHILEAN PLUM YEW** Prumnopitys andina *is closely related to* Podocarpus *and was once included in that genus.*

Not Given (2000 Red List). The timber has been highly prized for fine furniture, and the fleshy, sweet fruits are eaten locally. Indiscriminate logging and wildfires have much reduced populations. Even though female trees produce copious amounts of seeds each year, there is little evidence of regeneration. This is partly due to the fact that the fallen fruits are eaten by domesticated animals.

Collaborative research between the International Conifer Conservation Programme (ICCP) and the Faculty of Silviculture at the University of Valdivia, Chile, is investigating the genetic integrity of the remaining wild populations in order to formulate a conservation policy.

This attractive ornamental species is restricted to specialist collections in arboreta and botanic gardens. Because the seeds are notoriously difficult to germinate, trees in cultivation most likely have been propagated by cuttings, which is a fairly straightforward technique. This implies that they are of limited genetic variability. Another aspect of the ICCP's research is to investigate the problems associated with germinating the Chilean plum yew. A successful outcome for this research will give hope for restoration programmes of this species in Chile.

relict species. Golden larches are listed as Endangered (1997 Red List).

Trees are scattered rather than forming dense forests, and, as with many conifers, good seed production varies from once every three to every five years.

Pseudolarix amabilis is easily propagated from seed or cuttings. Seedlings are best grown 2–3 years before planting them out. Cuttings are best taken from saplings of less than ten years old, when an establishment rate of 70 per cent can be achieved.

The golden larch is well known in cultivation as an attractive, straight-trunked tree with golden-orange autumn foliage and brown cones. It tolerates cold, and should be more widely grown by the gardening public, even though in colder areas young plants can suffer temporary damage from late spring frosts.

◄ **GOLDEN LARCH** *is one of those rare examples of a deciduous conifer, and is considered by many to be the most beautiful. The Latin species name,* amabilis, *means "lovely". Here, the immature female cones cluster at the tips of the branchlets.*

11*/2 on Red List

Pseudotsuga

Botanical family
Pinaceae

Distribution
North America, Japan, China, Vietnam

Hardiness
Fully hardy

Preferred conditions
Any well-drained neutral to slightly acid soil in sun.

Three of the five threatened species of *Pseudotsuga* are in cultivation, albeit rarely. These are huge trees, found in temperate mixed forests, and need a lot of space in cultivation.

Pseudotsuga sinensis var. *sinensis* is native to China and Taiwan, where it is found on limestone peaks and ridges. Its Vulnerable status (2000 Red List) has been given because of continual logging, particularly in China. There, trees are being cut even on relatively inaccessible mountain tops. In Taiwan, its forest habitats are cleared for apple and peach orchards.

Pseudotsuga japonica, from south-east Japan, is Vulnerable (2000 Red List), due to its very localized occurrence. *Pseudotsuga macrocarpa*, from southern California, is Lower Risk, Near Threatened (2000 Red List). Although it has a broad distribution, it is still under pressure from a range of potential threats, including pollution and, ironically, fire suppression. The latter is a problem because the prevention of fire over a long period of time can cause a build-up of large amounts of combustible material, which can then cause more intensive and destructive fires.

▲ **PSEUDOTSUGA MENZIESII** *Douglas fir, from western North America, the most familiar of the genus, is a very valuable timber tree in temperate forestry plantations.*

Red List: Lower Risk

Saxegothaea conspicua

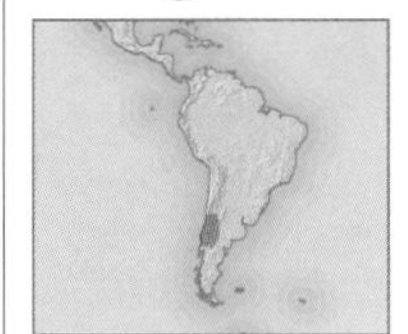

Botanical family
Podocarpaceae

Distribution
Chile, Argentina

Hardiness
Fully hardy

Preferred conditions
Neutral to slightly acid, well-drained soil in sun or partial shade, with some shelter from cold and dry winds.

Prince Albert's yew, ***Saxegothaea conspicua***, is the sole member of this genus. It has a widespread distribution, particularly in Chile, where it is found in the temperate rainforests of the central and southern Andes. Most trees occur at altitudes of 800–1,000m (2,600–3,250ft).

Even though there are still remarkable forests that contain impressive stands of this species, the general threat to the temperate rainforests of South America means that this is still of conservation concern. It is considered to be at Lower Risk, Near Threatened (2000 Red List).

As with all conifers in these forests, the wood of Prince Albert's yew is used for furniture making and construction.

PRINCE ALBERT'S YEW ► *Pale blue-green, bloomed female cones appear at the tips of the shoots in autumn; the male cones cluster at the base of the shoots.*

It was first introduced to horticulture in the 19th century by William Lobb, who worked for the famous nursery firm Veitch & Sons. Although it has never been widely cultivated, Prince Albert's yew is invariably found in specialist tree collections and botanic gardens. Perhaps the largest tree in Europe is a remarkable specimen in Kilmacurragh, Ireland, which is over 20m (70ft) tall.

Saxegothaea conspicua is fairly easy to propagate by cuttings. Seed germinates poorly. Young plants are very difficult to train into single-stemmed trees and the usual form is shrubby, with many leaders.

Red List: Vulnerable

Sciadopitys verticillata

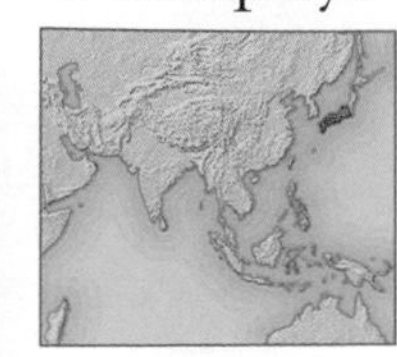

Botanical family
Sciadopityaceae/ Taxodiaceae

Distribution
Japan: South Honshu, Kyushu, Shikoku

Hardiness
Fully hardy

Preferred conditions
Fertile, moist but well-drained soil, preferably neutral to slightly acid, in light shade or sun.

The umbrella pine, ***Sciadopitys verticillata***, occurs in mixed and coniferous cloud forest, at altitudes of 500–1,000m (1,650–3,250ft), where it experiences high levels of humidity and rainfall. Other species associated with it include *Chamaecyparis obtusa* and *Cryptomeria japonica*, which are also threatened. The umbrella pine is classed as Vulnerable (2000 Red List).

This is one of the most ancient conifers, already in decline when other conifers became widespread, in the lower Cretaceous era. The umbrella pine was once sought after for its sweet, spicily scented wood, which was highly valued by boat builders, but it is so slow-growing that it is no longer planted for timber. Other similar, but faster-growing trees, such as incense cedar (*Calocedrus decurrens*) are now grown instead.

Although *Sciadopitys verticillata* was exploited for timber in the past, the situation is now thought to be more controlled. Regeneration studies have shown that seedlings do become established beneath existing forest cover, but gaps in the canopy, especially those with bare, mineral-rich soils, seem to be the preferred sites.

The umbrella pine is sacred in Japan. The ancient tree at Junguji Temple has been designated a natural monument in Kyoto. This is a tree of some 27m (88ft) with a girth of over 4m (12ft). It has been worshipped since the year 1310.

Although largely grown as an ornamental, the species has also been planted for erosion control. It is slow-growing from seed, and often forms plants with several leaders. It can also be propagated from cuttings taken in the summer.

▲ COLLECTOR'S ITEM *The umbrella pine is widely cultivated and likely to be represented in most botanic gardens and arboreta in subtropical and temperate climates.*

SCIADOPITYS VERTICILLATA

Red List: Lower Risk

Sequoia sempervirens

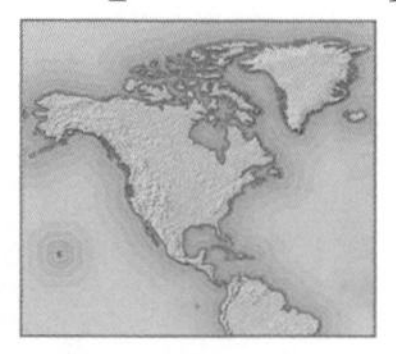

Botanical family
Taxodiaceae

Distribution
USA: south-west Oregon, north-west California

Hardiness
Fully hardy

Preferred conditions
Moist but well-drained soil in sun or dappled shade; light shade in the early years.

The USA is home to many superlative conifers, but perhaps the coastal redwood, ***Sequoia sempervirens***, tops them all. One tree measures 112m (364ft), making it the tallest living thing on earth.

The coastal redwood is capable of living for over 2,000 years. These trees occur in pure stands along a narrow band on the Pacific coast of California and south-west Oregon, where much of the moisture comes in the form of coastal fog.

Sequoia sempervirens is one of the few examples of conifers that are capable of naturally reproducing vegetatively, by regenerating from stump sprouts after major disturbances such as fire. Some shoots that arise directly as root sprouts from old-growth trees are known as "white redwoods". The trees, which generally grow less than 3m (10ft) tall, produce snow-white new growth. They have to obtain all their carbohydrates from the roots of their parent tree, since they lack the chlorophyll needed for photosynthesis.

SEQUOIA SEMPERVIRENS

Although *Sequoia sempervirens* is well protected in several national parks, where it covers an area of over 600,000 hectares (1,482,000 acres), it is still on the Red List. This is because of its past decline due to extensive logging, and ongoing problems with natural regeneration and air pollution. The redwood's current status is Lower Risk, Conservation Dependent (2000 Red List).

Sequoia timber has extraordinary properties apart from its sheer volume. It is an attractive pale terracotta in colour, with a golden shimmer, and easily worked since it is virtually knot-free and straight-grained. It has good termite resistance.

The redwood is exceptionally durable in contact with soil and water, and at one time, California law demanded that wooden uprights beneath buildings all should be made of redwood set in concrete.

The species was first introduced as an ornamental early in the 18th century. It is a very fast-growing conifer, but for best results, it needs to be located in a sheltered valley where there is plenty of available moisture. In colder and more exposed areas, the foliage becomes wind-scorched and brown. It tolerates shade and grows better in its formative years if provided with dappled shade.

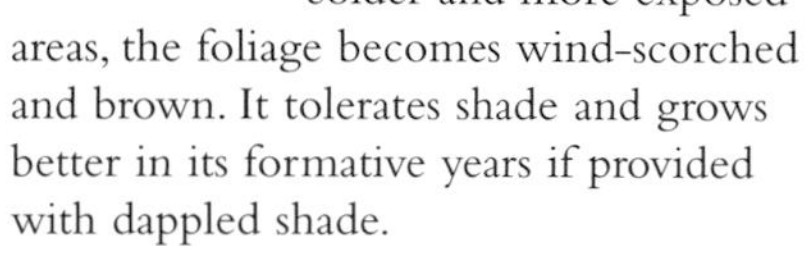

HISTORY

The original Sequoyah

The genus *Sequoia* is named after the celebrated Native American, Sequoyah, who was born circa 1770. He invented the Cherokee syllabary – an alphabet of phonetic symbols for the Cherokee language, which allowed the spoken language to be transcribed into a written form. This gave the Cherokee Nation a high level of literacy and made written communication between eastern and western groups possible. Perhaps more importantly, Sequoyah's work became a vital means of preserving the cultural integrity of the Cherokee Nation, including their traditional beliefs and practices, and complex healing and medicinal knowledge.

SEQUOYAH

Because of its ability to resprout from cut stumps, trees that outgrow their site can be coppiced to produce several new young stems. Stored seed remains viable for many years and best germination results are gained by overnight soaking in water before sowing. The seeds will germinate in about 28 days.

SEQUOIA SIMILARITIES ►
Sequoia sempervirens *is very closely related to* Sequoiadendron giganteum (see p.154). *Both have reddish-brown bark and reach great heights in maturity.*

RENEWAL

Conservation pioneers

The assault on the redwood began in earnest with the Gold Rush. From 1850 on, sawmills processed the timber, and many houses of the period in San Francisco were built from it. At first, the trees were cut with saws and pole axes; but with time, the advent of mechanization led to devastating inroads into the forests.

Conservation pioneers Dr Henry Fairfield Osborn, Dr John C. Merriam, and Madison Grant founded the Save the Redwoods League in 1918, following a trip to devastated forests. Helped by the Garden Clubs of America, they raised funds, matched by the state of California, and began to buy areas of forest in order to preserve them. Today, the Save the Redwoods League protects over 700 honour and memorial groves, including the Children's Forest, and the National Tribute Grove, in honour of servicemen and women who fought in World War II.

Red List: Vulnerable

Sequoiadendron giganteum

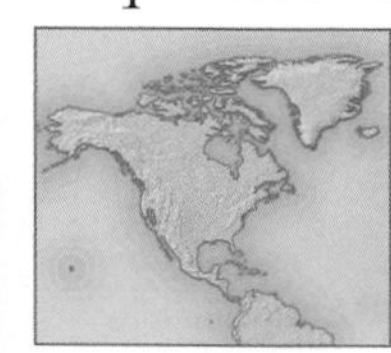

Botanical family
Taxodiaceae

Distribution
USA: California

Hardiness
Fully hardy

Preferred conditions
Moist, but well-drained soil in sun, with side shelter from cold winds when young.

The top contender for the world's biggest tree, ***Sequoiadendron giganteum*** is also one of the most ancient. The largest tree, the General Sherman, stands over 83m (270ft) tall, with a trunk 10m (30ft) in diameter, and has an estimated weight of 6,000 tonnes. The oldest tree has been assessed to be around 3,500 years old.

The Sierra redwood occurs at 900–2,700m (2,900–8,800ft) in 65 isolated groves on the western foothills of the Sierra Nevada. Although still listed as Vulnerable (2000 Red List), all wild populations have full federal protection, an action further endorsed when, on 15th April, 2000, President Clinton laid plans for a 144,000-hectare (355,000-acre) conservation zone.

▲ SIERRA REDWOOD *These gigantic trees develop from a tiny seed, less than 5mm (¼in) long; each tiny cone may produce up to 300 seeds, but only a minuscule percentage germinate and survive to become as tall as their parents.*

The bark of *Sequoiadendron giganteum* is very thick and spongy, making it resistant to fires, which are an important part of the tree's ecology. Ironically, centuries of fire suppression have resulted in denser forests, with the build-up of large amounts of combustible material. Such an accumulation could result in intense fires which would have disastrous consequences for the forest ecosystem. How to manage appropriate fire regimes for the forests is the subject of much ongoing debate.

Sequoiadendron giganteum was first introduced to the UK in 1853, when it was known as *Wellingtonia*, and it soon became very fashionable in Europe to plant this tree either as single specimens, in groves, or even in avenues. Like its close relative, *Sequoia sempervirens*, it prefers sheltered valleys with relatively high rainfall, but it differs in being intolerant of shade and in its inability to resprout from cut stumps. The best form of propagation is by seed, which takes between three and ten months to germinate.

► THE SOFT BARK *of* Sequoiadendron giganteum *is often over 30cm (1ft) thick, and has evolved to be as effective as asbestos in resisting fire. It also contains tannic acid, a natural chemical once used in fire extinguishers.*

PLANTS AND PEOPLE

BIG TREE BATTLES

The discovery of the Big Trees in 1852 inspired visions of large fortunes to be made from the timber. Rapacious logging ensued, resulting in widespread devastation. The Converse Basin grove was reduced to a 1,000-hectare (2,500-acre) wasteland. However, the wood turned out to be brittle, and less than 20 per cent usable, so fortunes did not materialize.

The wanton destruction, however, provoked public outrage and civil action, led by Colonel Stewart, a newspaper editor from Visalia, California. This led, in 1890, to the creation of the Sequoia National Park; all groves have since come into public ownership and are now afforded total protection.

Red List: Vulnerable

Taiwania cryptomerioides

Botanical family
Taxodiaceae

Distribution
China: north-west Yunnan, Taiwan; North Vietnam

Hardiness
Frost-hardy – fully hardy

Preferred conditions
Moist, leafy, but well-drained, soil in a sheltered site; partial shade or sun.

Fossil records show that ***Taiwania cryptomerioides*** was widespread in Eurasia and North America two million years ago, and has remained virtually unchanged since. Today, it occurs only in three areas, and is Vulnerable (2000 Red List). The only population in Vietnam, of about 100 trees discovered in 2001, is threatened by deforestation and fire.

The largest group, some 10,000 trees, was discovered in 2002 in southern Taiwan. Although this population is safely inaccessible, others throughout the natural range have been overexploited for the prized timber. *Taiwania* wood is often used for making coffins because of its extreme durability.

Taiwania has scarcely been cultivated in temperate zones, probably due to the assumption that it is not winter-hardy. Since 1993, however, large numbers of plants have been established in a wide range of sites in Britain and Ireland by the International Conifer Conservation Programme (ICCP), and these have proved winter-hardy to at least –10°C (15°F). The only check to its rapid growth has been slug damage to leading shoots. It is very difficult to propagate from cuttings, but seed germinates easily.

▲ EXTENDING RANGE *Two new, significant populations were discovered in 2001, one in a remote mountain range in Vietnam, the other in southern Taiwan.*

5*/2 on Red List

Taxus

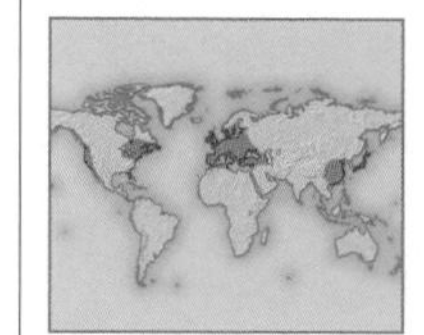

Botanical family
Taxaceae

Distribution
Northern hemisphere: Europe, Asia, North Africa, North America

Hardiness
Fully hardy

Preferred conditions
Well-drained soil in sun or shade.

Because of their generally slow growth rates, yews produce very beautiful and durable wood, and, in many cases, non-sustainable use is the main reason for their demise. In common with many conifers, *Taxus* species are very long-lived, and many examples of trees of 1,000 years or more have been well documented.

The most threatened yew is the Florida yew, *Taxus floridana*, which is listed as Critically Endangered (2000 Red List); it is confined to a 24-km (15-mile) stretch of the Apalachicola River in Florida. One of the largest specimens, at just 6m (20ft) tall, can be seen in Torreya State Park, Florida, USA, where it is protected. It has a similar distribution to another rare member of the same family, *Torreya taxifolia*, and can be distinguished from it by its softer needles and foliage that smells of turpentine.

Taxus brevifolia, the Pacific or western yew, is listed as being at Lower Risk, Near Threatened (2000 Red List). It occurs in western North America, from Alaska to California, with a disjunct population east of the Rockies in Montana and Idaho. This species occurs from sea level to 2,200m (6,500ft), in open or dense forests, moist flats, or deep ravines.

Not surprisingly, there is a rich ethnobotanical history associated with this ancient plant. The wood has always been held in high regard by indigenous peoples of the Pacific Northwest, whose uses for it included bows, harpoons, canoe paddles, tool handles, and ceremonial objects. Its leaves have many reported medical uses. Internationally it is prized by musical instrument makers for lutes and other stringed instruments. Until recently, it was not exploited commercially, but the discovery of taxol (*see box, above*) placed increased pressure on wild populations. Six 100-year-old trees would be needed to provide enough taxol to treat a single patient. Laboratory synthesis should alleviate this threat.

CULTIVATION

As evergreen shrubs or small trees, yews have played an important role in horticulture for centuries. They are amenable to clipping and, as such, have been used for hedges, mazes, and topiary. Yews are easily propagated from seed, cuttings, or by grafting, and tolerate a range of soil and climatic conditions.

PLANTS AND PEOPLE

TAXOL TREE

The drug taxol was first distilled from the bark of Pacific yew in 1962, but did not enter clinical trials until much later. Since then, it has revolutionized the treatment of several forms of cancer, and is one of the most important natural products of the 20th century. In a development that should help reduce the threat to the species' survival, taxol was first synthesized in the laboratory at Florida State University in 1993. Work at Montana State University led to the discovery of a fungus associated with yew that also produces taxol. The use of bio-engineering techniques with the fungi will lead to the production of commercial amounts of taxol.

All species of *Taxus* are poisonous to livestock and humans. People have died from eating the seeds, or drinking tea made from the leaves. Many birds, though, eat the flesh that surrounds the seed, and even swallow the seeds without ill effect. It is thought that most of the seedlings in natural stands of English yew, *Taxus baccata*, are from seed that has passed through the digestive system of birds; it is a feature that seems to help break their dormancy. Otherwise, they take at least two years to germinate.

► TAXUS BREVIFOLIA *in closed forest often grows as a small tree, but in more open sites, it adopts a more shrubby habit, not unlike juniper. One of the largest known trees, at 16m (52ft) tall with a crown 9m (28ft) across, can be seen near Packwood, Washington, USA.*

Red List: Lower Risk

Tetraclinis articulata

Botanical family
Cupressaceae

Distribution
South-east Europe, North Africa

Hardiness
Half-hardy

Preferred conditions
Well-drained soil in a warm, sunny site sheltered from cold and wind.

Despite its overall listing as Lower Risk, Near Threatened (2000 Red List) ***Tetraclinis articulata*** is Critically Endangered on Malta, and Endangered in Spain; in both countries, populations are down to fewer than 50 individuals. It is Vulnerable in northern Morocco and at Lower Risk in northern Algeria.

The wood has been used for centuries in building, including, for example, the roof of Cordoba Cathedral, in Spain. The populations in Spain and Malta have probably been brought to the brink of extinction in the 20th century by urban development.

In northern Morocco, the populations are quite extensive, forming pure stands in some areas. In recent times, the main commercial use has been the root burr, which is harvested after several seasons of severe coppicing to promote the development of burr-wood below ground. This is used for cabinet making, small boxes, and marquetry, and is often sold as thuja burr. A gum, sandarach, is also produced and exported for use in the manufacture of varnishes. Although some plantations do exist, there is concern that wild populations may be exploited if international demand for these products continues.

As this species is only borderline in terms of hardiness, it is not widely grown outside the Mediterranean. Only a few of the mildest gardens in England and Ireland have managed to grow plants of this species. It germinates readily from seed, but establishment from cuttings is more difficult.

TETRACLINIS CONES ► *The blue-grey cones at the tips of the foliage sprays resemble those of the related* Calocedrus.

NATURAL HABITAT ► *In the wild in Spain, this shrubby tree is a component of the typical, rocky scrub known as garrigue that is found around the Mediterranean.*

5*/4 on Red List

Torreya

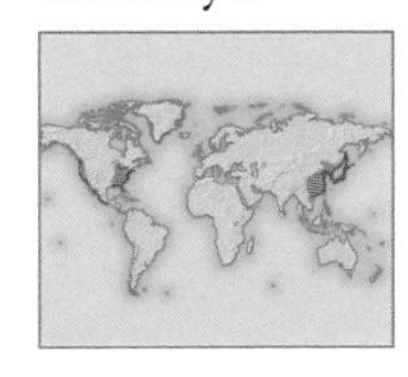

Botanical family
Taxaceae

Distribution
North America, eastern Asia

Hardiness
Frost-hardy – fully hardy

Preferred conditions
Warm, temperate mixed woodland.

The genus *Torreya* includes six closely related species, five of which are threatened. They represent isolated last remnants of a flora that existed in circumpolar regions over 50 million years ago. Their wood is highly prized by cabinet makers, but large trees of any of the species are rarely found now.

Torreya jackii, from eastern China, is listed as Endangered (2000 Red List); natural regeneration is inadequate to sustain populations within its narrow habitat. *Torreya yunnanensis*, from western China, is classed as Data Deficient (2000 Red List), while the third Chinese species, *Torreya grandis* var. *fargesii,* is listed as Vulnerable.

Several species of *Torreya* have been grown as ornamentals: *Torreya nucifera* is probably the best known. This Japanese species is the only one which is not threatened in its native habitat.

Torreya californica is also fairly widely grown in specialist collections in Europe. Also known as the California torreya, or California nutmeg, it is at Lower Risk, Near Threatened (2000 Red List). It is confined to the western slopes of the Sierra Nevada and the central part of the coastal range in California. *Torreya californica* grows along valley bottoms and water courses at 900–2,000m (2,900–6,500ft). This highly aromatic species is the tallest of the genus, with trees as tall as 35m (120ft) recorded, though 20m (70ft) is more common. Before extensive logging greatly reduced the populations, the fine-grained, yellowish-brown timber was used for cabinet making, turning, fencing, and fuel. The species is well adapted to fire; it will sprout from the root, root-crown, and bole even if the top-growth has been burnt.

Torreya taxifolia, the Florida torreya, is Critically Endangered (2000 Red List). The species is now confined to shady ravines along a short stretch of the Apalachicola River in south-west Georgia and north-west Florida. Here, it it is a component of mainly hardwood forest with species of *Magnolia*, beech (*Fagus*) and holly (*Ilex*), and *Taxus floridana*. It is easily distinguished from this species by its sharp, rigid foliage, which smells of tomatoes, and its plum-like fruit.

In common with other torreyas, this species occurs at very low elevations – just 15–30m (50–100ft) above sea level. In the past, it was used for fencing, fuel, and roof shingles, and its populations were thriving until the mid-1950s, when they were hit by a fungal pathogen. By 1962, only non-productive stumps remained in the wild, and, in 1984, the Florida torreya was listed under the US Endangered Species Act.

The population was down to 500 individuals in the late 1990s, all of which are represented in living gene banks in cultivation.

A recovery plan is now under way to re-establish this species in its natural habitat. For now, the continued survival of this species depends on plants held in botanic gardens, which will form the basis of reintroduction programmes.

HISTORY

JOHN TORREY

This eminent scientist (1796–1873) for whom this genus was named, was born in New York, and specialized in the study of American plants. He was mentor to Asa Gray (*see p.197*) and worked with him on *The Flora of North America* (1838–43). His other publications include the *Flora of the Northern and Middle Sections of the United States* (1824) and the *Flora of the State of New York* (1843).

The enormous collection of dried plant specimens he amassed during his lifetime was first left to Columbia College, and it later became the core of the herbarium at the New York Botanical Garden in New York City.

TORREYA CALIFORNICA ► *In the past, large trees were relatively common in the wild. Now the largest known tree, at 14m (46ft), is in a garden in North Carolina.*

7*/1 on Red List

Tsuga

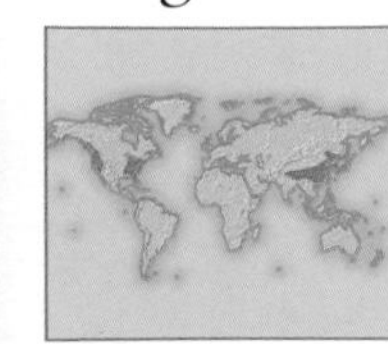

Botanical family
Pinaceae

Distribution
North America, eastern Asia

Hardiness
Fully hardy

Preferred conditions
Leafy, moist but well-drained, slightly alkaline to acid soil, in sun or shade, with side shelter when young.

There are about ten species of *Tsuga* distributed across North America and Asia, the most widespread being eastern hemlock, *Tsuga canadensis*; mountain hemlock, *Tsuga mertensiana*; and western hemlock, *Tsuga heterophylla*.

The Asian species are mostly more limited in their distribution, and these are the ones most at risk. They include *Tsuga chinensis*, from central and western China, and two Japanese hemlocks, *Tsuga diversifolia* and *Tsuga sieboldii* from southern Honshu.

Tsuga forrestii, Lijiang tieshan, from the temperate mixed forests of Yunnan and Sichuan, in China, is Vulnerable (2000 Red List). This species is closely related to *Tsuga chinensis* and is found at subalpine altitudes of 2,700–3,300m (8,800–10,750ft), where it grows with other members of the pine family such as *Abies*, *Pinus*, and *Picea*. *Tsuga forrestii* is not well-known in cultivation.

Although not the major target for logging, *Tsuga* is usually a casualty because it occurs in areas that are felled for forestry. As a result, it suffers habitat degradation, and the few remaining stands are in poor condition with little or no chance of natural regeneration. The widely grown *Tsuga canadensis*, eastern hemlock, is also currently being devastated by hemlock woolly adelgid, both in cultivation and in the wild.

TSUGA CAROLINIANA

Tsuga caroliniana, Carolina hemlock, from the forests of eastern USA, in the Appalachian Mountains of south-west Virginia and Georgia, is at Lower Risk, Near Threatened (2000 Red List).

A medium-sized tree restricted to rocky ledges and cliffs, it is seen most often on south-facing slopes, where it occasionally occurs.

Tsuga caroliniana is not tolerant of fire, and mature populations are only likely to develop in places where fire is absent or rare. The wood can be used for lumber, or pulping, but it is regarded as unimportant in trade. In the past, tannin was extracted from the bark for leather processing.

Tsuga caroliniana is valued for its glossy foliage and graceful habit, and is planted as an ornamental in the USA. Although it was introduced into European gardens in 1886, it remains very rare. In the UK, it is confined to specialist collections in warmer, wetter south-west England and Ireland. This is a very slow-growing tree in cultivation, needing summer heat to grow well.

Hemlocks have been a feature of large gardens and collections in botanic gardens and arboreta for almost 200 years. Their soft, elegant foliage and delicate papery cones are distinctive, but their eventual size makes them unsuitable for small gardens. Most are hardy and some have the advantage of being shade-tolerant when young. Hemlocks are ideal for establishment in wooded areas, where they provide excellent roosts for owls and other birds.

Seed is only produced periodically, about once every three years. The seedlings are very susceptible to fungal diseases that cause "damping off" and may need shading from full sun. While seed germinates in about a month from sowing, seedlings may take two or three years to become established. Cuttings taken from semi-ripe wood in late summer or early autumn usually root in 2–3 months.

▼ TSUGA CAROLINIANA *Although they are not a commercially important source of timber, Carolina hemlocks are an important food source for a range of birds and mammals.*

WILDLIFE

Hemlock communities

Carolina hemlock, *Tsuga caroliniana*, along with the much more common eastern hemlock, *Tsuga canadensis*, is considered essential for the shelter and bedding of white-tailed deer during the winter months. The flaky bark is eaten by beavers, North American porcupines (*see above*), and rabbits. The trees also provide cover and nesting sites for birds, and the seeds are an important food for both birds and mammals. In the Rocky Mountains, the foliage of *Tsuga mertensiana* is a useful food source for browsing animals such as the wild mountain goat.

Red List: Endangered

Widdringtonia cedarbergensis

Botanical family
Cupressaceae

Distribution
South Africa: Western Cape, Cedarberg Mountains

Hardiness
Half-hardy

Preferred conditions
Very well-drained soil and a warm sheltered site in full sun.

The Clanwilliam cedar, ***Widdringtonia cedarbergensis***, is one of a small genus of four species found only in southern Africa, three of which are threatened. Of these, *Widdringtonia cedarbergensis* and *Widdringtonia whytei*, from Mount Mulanje, Malawi, are Endangered, and *Widdringtonia schwarzii*, from Cape Province, is Vulnerable (2000 Red List).

Widdringtonia cedarbergensis occurs in woods and scrub, or fynbos, in the Mediterranean-type climate of South Africa. Historical overexploitation and a subsequent change in fire regimes have brought the Clanwilliam cedar perilously close to extinction. The remaining stands of old mature trees rarely produce seed and there is very little natural generation.

Fortunately, the Western Cape Nature Conservation Organization embarked on a long-term restoration programme in 1987. After establishing a small nursery in the reserve, they planted out about 40,000 seedlings over a nine-year period, of which some 60 per cent have survived.

In addition, the organization has been able to manage the fire regime. Grazing has been controlled, so the future is a little brighter for this elegant tree. The precarious balance between fire and regeneration of the extraordinary fynbos vegetation in this area is very difficult to maintain, though, and financial problems have affected the long-term prospects.

Because of its lack of hardiness, Clanwilliam cedar is not widely grown in Europe or in North America, but it would suit regions with a Mediterranean-type climate and should be more widely cultivated.

The tree is 6–18m (20–60ft) tall, with delicate foliage. Propagation is easiest from seed, which germinates 2–6 weeks after sowing. Seedlings are potted on after a few months and can be planted out in 10–15 months. Cuttings from the current year's growth have the advantage that resultant trees will produce cones more quickly than seed-raised plants.

CLANWILLIAM CEDAR ►
Widdringtonia cedarbergensis *grows in the Cedarberg Mountains in South Africa, where it has been threatened by fire. The genus was named after Edward Widdrington, a Royal Navy captain who studied conifers in the late 18th and early 19th centuries.*

Red List: Critically Endangered

Wollemia nobilis

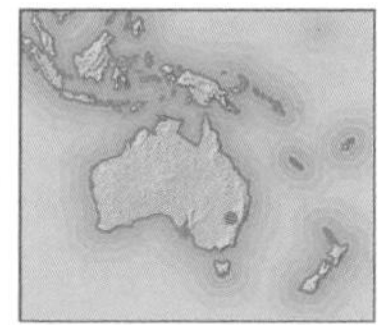

Botanical family
Araucariaceae

Distribution
Australia: New South Wales

Hardiness
Frost-hardy

Preferred conditions
Well-drained, rocky soils, in sun or part shade.

The discovery of the Wollemi pine, *Wollemia nobilis*, in September 1994, in a remote part of the Blue Mountains, only 150km (90 miles) from Sydney, Australia, is being hailed as the botanical find of the century.

The original stand, first discovered by officers with the New South Wales National Parks and Wildlife Service, in what is now the Wollemi National Park, consisted of only 23 mature trees. Today the number of known trees has increased to 100, from three sites. The trees are listed as Critically Endangered (2000 Red List).

This was a significant discovery of a new genus, related to monkey puzzles (*Araucaria*) and Kauri pines (*Agathis*), the only other members of the ancient family Araucariaceae. The fossil records of this group date back over 200 million years, to the time of the dinosaurs.

This remarkable tree remained safe for thousands of years – until its discovery. Now, even though its exact locations are kept secret, there is still a fear that its whereabouts may be discovered. There is concern about risk of introduction of diseases, and of fire from human activity.

Horticultural research has discovered methods for propagation, and this has led to the establishment of an *ex-situ* conservation collection to underpin the conservation of the trees in the wild. The actual techniques for its propagation have yet to be published.

Already, the worldwide demand for this intriguing species has led to an extensive propagation programme. With plans to release plants to the horticultural industry in 2005, it is hoped that this will be a model for how conservation can go hand-in-hand with cultivation.

HERBACEOUS PLANTS

Herbaceous plants are defined as plants lacking woody tissues. If they die back to a resting rootstock and sprout again after a spell of dormancy, they are known as herbaceous perennials. The term herbaceous is also applied to annuals, which live for only one year, and biennials, which make leaf growth in the first year and flower in the second. Non-woody plants of all kinds are found almost everywhere: on alpine and polar tundra, from desert fringe to rainforest, in scrub, in grassland, and in fresh and saltwater wetland, even beneath the surface of lakes and rivers.

WHERE THEY GROW

Herbaceous plants can live where trees and shrubs cannot. Woody growth cannot thrive in places where fertility is too low, such as on sandy soils; where exposure is too extreme, where grazing is too frequent, or where there is simply insufficient moisture. It is in these areas that herbaceous plants dominate, surviving below ground as seed or rootstocks when conditions are too harsh for growth. Even where woody plants do grow, in scrub and forest, non-woody plants can grow among them. If the trees grow so densely that no light can reach the forest floor, then herbaceous plants grow in the canopy where soil collects in branch forks, or on rocky outcrops.

WOODLAND ► *Some herbaceous plants share habitats with woody plants, particularly beneath deciduous trees. Skunk cabbages enjoy damp sheltered woodland.*

◄ COASTAL HABITAT *Where there are no trees, herbaceous communities thrive.* Centranthus *and* Kniphofia *can cope with the salt and exposure at the coast.*

SURVIVAL STRATEGIES

Herbaceous plants are frequently found in intimate association with other plants as a major component of biomes such as forest, tundra, and grassland. As part of the ground flora in deciduous forest, herbaceous perennials and biennials bear the first flowers of late winter and early spring. They are also the first plants to bloom at high altitudes. In both environments the growing season is short and plants have to grow and flower before the canopy closes over (in woodland) or before the first snowfall of autumn (in mountains). Other plants flourish after events such as fires, landslides, or flooding.

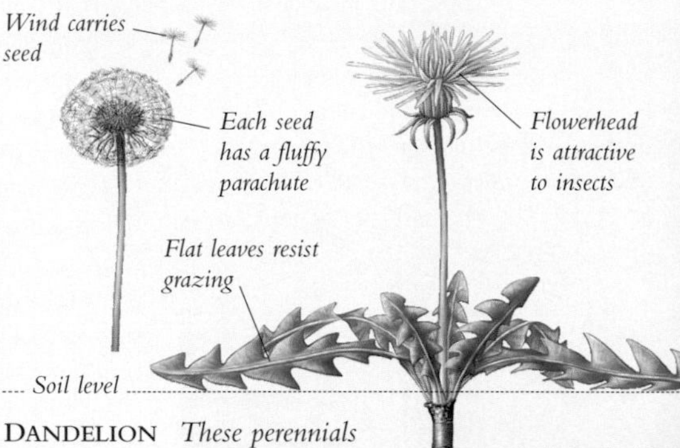

DANDELION *These perennials have a close association with grassland and are well adapted to the conditions there. Wind is an efficient seed distributor on flat, grassy plains.*

LIFE CYCLES AND HABITS

In tropical areas where conditions are relatively uniform, herbaceous perennials can grow all year and are often evergreen. In seasonal climates, perennials can survive harsh periods by retreating underground where they have laid down a food store in their roots. As soon as conditions improve, they spring into growth.

In contrast, herbaceous annuals germinate, grow, flower, set seed, and die within a single growing season. Annuals will rapidly colonize bare ground, where they grow faster than other plants, later releasing plenty of seed to secure a firm foothold for the future. The seed may germinate in the same place the following year, or lie dormant for many years. Seed of the annual field poppy (*Papaver rhoeas*), for example, may last for 50 years in the ground, until disturbance brings the seed to the light and stimulates growth.

▲ HERBACEOUS PERENNIALS *such as peonies survive winter beneath ground and may live for 50 years or more. Some are evergreen and others, known as subshrubs, may develop woody bases.*

▲ BIENNIALS *have a life cycle of two years. This* Echium *can colonize ground rapidly like an annual, but, after a dormant period, enjoys an extended, second growing season.*

▲ ANNUALS, *including sunflowers, grow very fast and have bright flowers to attract pollinating insects quickly. Lots of seed is produced, which germinates rapidly when an opportunity arises.*

POPPY FIELD *The Antelope Valley California Poppy Reserve is dominated by herbaceous plants. Grazing once maintained the landscape, but today controlled burning is used to suppress growth of woody plants.*

VALUABLE RESOURCES

Herbaceous plants provide much of our food, in the form of seeds, oils, leafy vegetables, and root crops such as potatoes and carrots. Seeds from annuals are often full of oils to power rapid germination and can be very nutritious. Leafy vegetables include cabbage, spinach, and salad crops; their fast growth results in leaves that are softer than those of longer-lived plants. Root crops store complex carbohydrates and are staples in the diet of many of the world's peoples.

Many plants naturally yield chemicals to deter insects and animals, which can be put to wider use, such as the insecticide pyrethrin (from *Tanacetum cinerariifolium*). Some defences also inadvertently attract humans, such as the oils in culinary herbs like marjoram (*Origanum* species). Plant chemicals may have medicinal properties also, and herbal remedies are gaining popularity in the West (*see* evening primrose, *below*). In areas where plants are still the primary source of medicines, healers have a vast knowledge of the local flora, built up over generations. Pharmaceutical companies are increasingly tapping into this resource in search of new drugs.

▲ HERBACEOUS PRODUCE *All kinds of food crops are herbaceous. At this stall in Hong Kong, many of the leafy vegetables for sale are from a single, mostly herbaceous family – the brassicas, or cabbage family.*

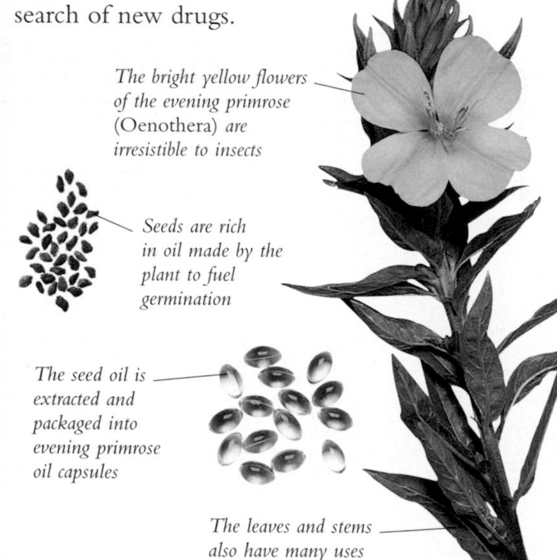

The bright yellow flowers of the evening primrose (Oenothera) *are irresistible to insects*

Seeds are rich in oil made by the plant to fuel germination

The seed oil is extracted and packaged into evening primrose oil capsules

The leaves and stems also have many uses in traditional medicine

The tough vascular structure that helps provide non-woody plants with some support can be turned to use as a source of fibres for textiles. Linen, for example, is woven from perennial flax (*Linum perenne*). Herbaceous plants can also yield dyes such as woad: a blue dye made from *Isatis tinctoria*. It was used by the ancient Britons to stain their skins, according to Julius Caesar.

Herbaceous plants buffer the force of rainfall, and help it to sink into the soil, which reduces erosion and flash flooding. The roots of annual herbs also have an important role in binding and stabilizing disturbed ground. On a grander time scale, herbaceous plants play a part in the formation of soils (podogenesis). Their soft, non-woody growth decomposes rapidly into humus – a property familiar to gardeners who make their own compost. Humus stores soil moisture and helps retain nutrients released during decomposition so they can be used by living plants to fuel new cycles of growth. If decomposition occurs very slowly in cold, wet conditions, without sufficient oxygen, the end product is peat.

WILDLIFE

The fates of many herbaceous plants and their animal pollinators or seed dispersers are intimately intertwined: for example, the endangered Karner Blue butterfly in the USA will only lay its eggs on the blue wild lupin (*Lupinus perennis*). In the last century, intensive farming methods, especially increasing herbicide use, have reduced hedgerow and grassland plant populations. In the UK, plant-diverse meadows have declined by over 95 per cent, as pastures and meadows rich in clovers, vetches, and orchids have been lost. This has contributed to a dramatic decline in over half the UK's native butterfly species, and the extinction of five. Bees are facing a similar future, despite the economic benefits of their role in pollination and honey production. Policies such as leaving verges uncut, and persuading farmers to leave wild areas at field edges, are important in reversing this trend.

◄ BUSY BEE *In the UK, 80 per cent of food crops are pollinated by bees, yet at least one-quarter of bee species are under threat. Similar problems face all countries where agriculture puts pressure on wildlife.*

Woodland plants are important sources of food and cover for insects, birds, and animals. In timber plantations, and in natural woodlands where traditional management has ceased, the canopy can become very dense. This reduces the diversity of wild flowers, which need light. Clearance of woodland glades, controlled burning of grassland and scrub, and reintroduction of herbaceous plants are taking place across the USA and Europe.

GERANIUMS *grow wild in temperate regions worldwide. Try and make room for native species as well as popular cultivars. The flowers of the meadow cranesbill,* Geranium pratense, *from Europe and Asia, are attractive to bees.*

THREATS

Many herbaceous plants are vulnerable to habitat change: urbanization, global warming, and changes in farming practices in particular are all major threats. When grassland is intensively managed, productivity increases but plant diversity plummets. Wetlands, an important habitat for moisture-loving and aquatic plants, have declined by 65 per cent worldwide as land has been drained and natural river systems diverted. Natural woodlands have been replanted with timber crops, or converted to farmland. The problem is exacerbated when invasive species start to dominate the niches that remain for native plants.

▲ GRASSLANDS, *such as the South Downs in the UK, are often converted to farmland. Farming can however be positive: grazing prevents scrub shading out rare plants.*

CONSERVATION

There are designated conservation areas, where herbaceous plants thrive, throughout the world. In addition, many botanic gardens participate in *ex-situ* conservation and reintroduction projects. At Montréal Botanic Garden in Canada, seed and cuttings are being collected from the threatened plants of southern Quebec. In the UK, seed of almost all the native flowering plants has been collected and stored in the Millennium Seed Bank at Wakehurst Place, run by the Royal Botanic Gardens, Kew. Twenty per cent of Britain's coastline and hinterland, which includes chalk and acid grasslands, shingles, and salt marshes, is now run by the UK's conservation-minded National Trust.

▲ WILD FLOWERS *can co-exist with agriculture. Just a narrow strip between a crop and a hedge, untouched by ploughing and herbicides, can dramatically increase biodiversity.*

GARDENER'S GUIDE

- All gardeners have the potential to help conserve herbaceous plants, perhaps more than any other plant group. There are species suitable for any garden, whether large or small, whatever the climate or soil conditions, in any part of the world. The benefits to plants and animals are increased if you can make a space for your local native species.
- Find a reputable wildflower nursery. Some sell plants or seeds of local provenance that they have collected legally and bulked up for use in gardens.
- If space permits, set aside an area of the garden for wild flowers. Get advice from local conservation groups or a good nursery about which valuable local wild flowers will suit the conditions in your garden. Remember never to reintroduce garden plants to true natural areas.
- In smaller gardens, plant a few wild flowers among the other plants – choose species that are important food sources for butterflies and other native wildlife. Local conservation groups or wildflower nurseries can advise on the most valuable species to have. These are often very suitable for boundary areas, such as drives or verges.

BORLOTTI BEANS

- When growing herbaceous crops, try sourcing heritage varieties. They may not crop as heavily or uniformly as modern cultivars, but derive from a richer gene pool, and are usually more attractive to wildlife. The flavour is often very good and the plants ornamental, such as marbled old-fashioned borlotti beans.

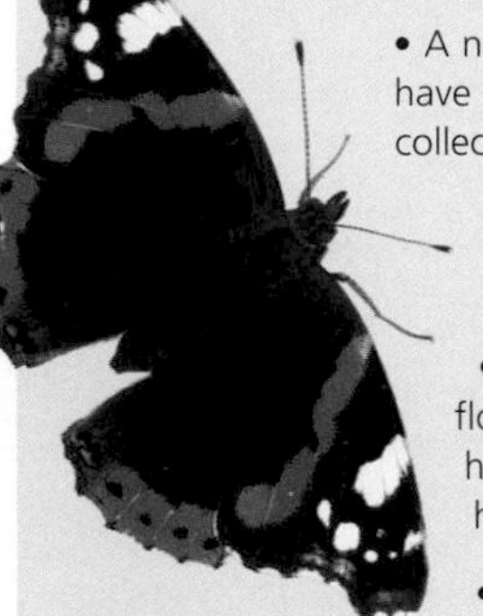

RED ADMIRAL BUTTERFLY

- A number of herbaceous plants have been threatened by over-collection. Make sure the rare or exotic herbs that you want to buy for your garden have been propagated in a nursery.
- If possible, ensure that cut flowers, or raw materials for herbal remedies, have not been harvested from the wild.
- Make sure that the non-native herbs you want to plant have no history of weedy or invasive behaviour in your area.
- A herbaceous plant cannot be properly conserved if its occurrence and ecological niche is not understood. By joining a local group, such as a wildflower society or naturalists' trust, you can learn how to identify and grow herbaceous plants in your area.
- In the 20th century, exploitation of peat for horticultural purposes devastated raised peat bogs. Peat bogs take thousands of years to form and they support a rich, varied, and distinct herbaceous community – many plants of which are now endangered. Avoid using peat in the garden, or at least restrict your use to the minimum, for example using it only in specialist seed composts.
- Join charities such as the WWF or local land conservation organizations. They often buy land with threatened habitats as it becomes available, so that it can be managed with conservation in mind. Volunteer your time and skills to help with their educational and research programmes, or practical conservation projects.

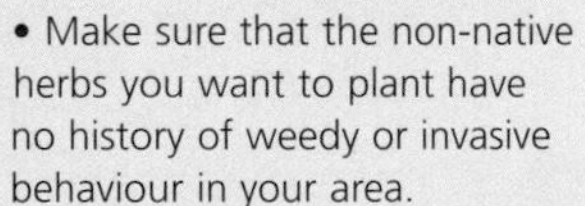

Red List: Vulnerable*

Aciphylla dieffenbachii

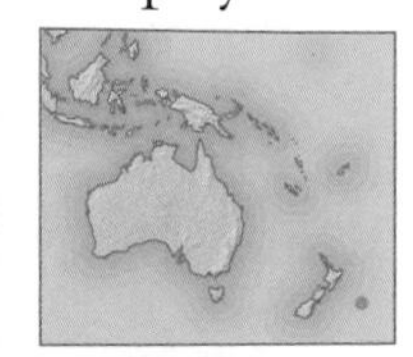

Botanical family
Umbelliferae

Distribution
New Zealand: Chatham Islands

Hardiness
Half-hardy

Preferred conditions
Open site in full sun; moist, but loose, very well-drained soil rich in humus and grit.

Dieffenbach's (or soft) speargrass is one of the spectacular giant "megaherbs" that occur on the coastal cliff and grassland habitats of the temperate south Pacific islands. This handsome species grows in spreading, stout tussocks with blue-green, narrowly segmented leaves. Abundant tiny, cream flowers appear in a large terminal cluster (panicle) that is 60–100cm (24–40in) tall. Unlike other members of the genus, this species is unusual among speargrasses in lacking painfully sharp leaf tips. The species is endemic to the Chatham Island archipelago, 800km (500 miles) east of New Zealand, where it grows on coastal cliffs and open coastal grassland, on volcanic rock.

THREATS TO SURVIVAL

While this speargrass currently occurs on five of the larger Chatham Islands, and also on offshore stacks and islets, like many of the megaherbs, it is threatened by overgrazing by domestic and feral animals, and is now common only on Mangere Island. Ironically, the plant suffers considerable damage from an equally endangered beetle, the speargrass weevil *Hadramphus spinipennis*, of which this speargrass is the principal host.

Additional threats arise from the spread of exotic grasses, which have invaded native grassland swards and which hinder the establishment of speargrass seedlings. Nevertheless, the species has the ability to recover now that efforts are being made to control grazing and restore the natural habitat. Fortunately the species is well established in cultivation, where it is grown for the architectural qualities both of its foliage and its inflorescences.

Like other speargrasses, the species is particularly sensitive to root disturbance. Propagation is best achieved by seed, ideally sown as soon as it is ripe, in a cold frame. As the species is single-sexed (dioecious), both male and female-flowered plants should be planted in the same vicinity in order to maximize seed production.

▲ LUSH FOLIAGE *This impressive plant flowers in the summer and forms large tussocks. The leaves are palatable to grazing animals – one reason why the species is threatened in its wild habitat, where it is restricted mainly to coastal rocky slopes and cliffs.*

Red List: Rare*

Aconitum noveboracense

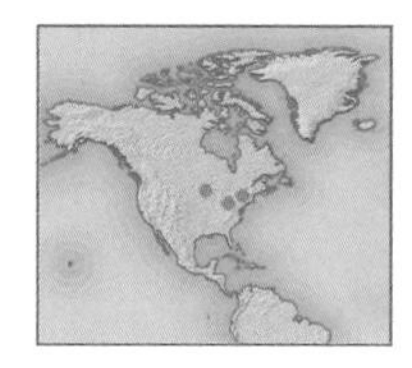

Botanical family
Ranunculaceae

Distribution
USA: Iowa, New York, Ohio, Wisconsin

Hardiness
Fully hardy

Preferred conditions
Grows naturally in humid, temperate woodland; thrives in moist, but well-drained, humus-rich soil, in a semi-shaded position.

Northern monkshood is a showy, clump-forming perennial that will grow to approximately 1m (3ft) in height. It is named after its flowers, which are reminiscent of the cowls of medieval monks. Like other members of the genus *Aconitum*, this species has coarsely lobed leaves.

As a native plant, the species is confined to four north-eastern American states, with the largest number of populations in Iowa and Wisconsin. However, the species remains very rare and endangered in Ohio, where fewer than 120 plants were counted, from just two sites, in 2001. Its rarity in the wild reflects its rather exacting ecological requirements – it favours the high humidity and cool conditions associated with partially shaded cliffs and stony scree slopes. It occurs on both limestone and sandstone. The principal threat to the species arises from habitat destruction, with populations threatened by dam construction, grazing by domestic and wild animals, forestry activities (particularly logging), and urban development. Trampling and collection by botanists and gardeners also threaten some of the more accessible populations.

This monkshood may be propagated by division of its tuberous rootstock, or by seed, which is best sown immediately after ripening in a cold frame.

BOTANY

BUMBLEBEE LURE

The large flowers of monkshoods are adapted for pollination by large insects, notably bumblebees. Experiments have shown that bumblebees find blue flowers especially attractive, and the structure of the monkshood flower also fits the size and shape of mature bees.

When a foraging bee visits the hooded flower, it is forced to delve deep inside in order to reach the nectar, and in so doing it brushes pollen off the flower's anthers. Some of the pollen sticks to the bee's furry body and, later, when the bee visits another flower, the pollen is transferred and the flower is successfully pollinated.

Unlisted

Actaea racemosa

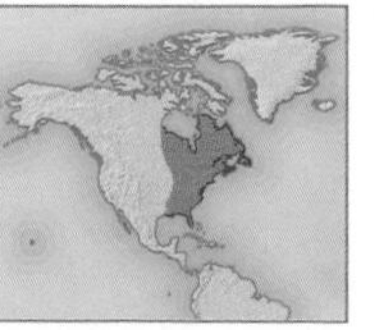

Botanical family
Ranunculaceae

Distribution
Eastern USA, eastern Canada

Hardiness
Fully hardy

Preferred conditions
Partially shady places in temperate, mixed, deciduous forests, especially on mountains; fertile, moist, but well-drained, humus-rich soil.

Black cohosh, or black snakeroot, is a familiar garden plant from North America, and is still commonly referred to by its former scientific name, *Cimicifuga racemosa*. It is a handsome perennial, ideal for partially shaded spots, and grown particularly for its long and slender bottlebrush-like clusters of tiny white flowers, borne in midsummer. Widely available in North America and Europe, it has been awarded the coveted Award of Garden Merit by the Royal Horticultural Society in the UK.

HERBAL REMEDY

Actaea racemosa has become an important medicinal herb, and commercial collection of the species from the wild for therapeutic use is putting the species at considerable risk, at least locally. Black cohosh was used by native Americans for rheumatism, snakebite, and female ailments. Indeed, the species' medicinal qualities have long been valued: the former generic name, *Cimicifuga*, derives from the Latin *cimex*, meaning "bug", and *fugare*, "to drive away", referring to its insect-repellent qualities. Another of its common names, bugbane, refers to this same effect.

This species is also a homeopathic medicine for backache, depression, menstrual problems, osteoarthritis, post-partum depression, and tinnitus. It seems to have broad application, and also treats sore throat, bronchitis, joint pain, fever, insomnia, arthritis, rheumatism, and asthma, as well as enhance the effects of conventional medicines for lowering blood pressure. It should not be taken during pregnancy, and, as with all herbal medicines, only under professional supervision.

The plant is particularly effective in the treatment of the symptoms of menopause. Extracts of this herb have similar effects to those of the female hormone oestrogen, so it is often prescribed as an alternative to hormone replacement therapy.

Medicinal demand is highest in Europe, but large markets also exist in the Far East, and in the USA, where the medicinal market grew five-fold between 1997 and 1998. Wild collection occurs primarily in the states of Illinois, Indiana,

▲ ARCHITECTURAL PLANT *This tall perennial is useful for herbaceous borders, where its dark green, deeply cut leaves contrast with the tall stems, topped by pale flower spikes. The plant exudes a distinctly medicinal smell, as befits its versatility in treating disease.*

Kentucky, Missouri, Ohio, Tennessee, and West Virginia. The American Herbal Products Association (AHPA) estimates that more than nine million plants were harvested between 1997 and 1999. While this species is widely distributed in the eastern United States, and into Ontario and Quebec in Canada, it is regarded as critically endangered or possibly even extinct in Illinois, Iowa, Massachusetts, Michigan, and Mississippi, and considered at risk across most of the rest of its range.

CULTIVATION

Plants in cultivation are grown from division or seed, and the horticultural trade, fortunately, has little or no impact on wild populations. Propagation is best achieved by ripe seed sown fresh in a cold frame or by division in spring.

THREAT

HERBAL MEDICINE

The medicinal properties of black cohosh are found in the resinous rhizomes. When these are lifted, the plant is killed and the impact on wild populations is serious. Demand for wild-collected material is highest because wild plants are believed to have a higher concentration of active compounds, and the species is not yet cultivated in quantities large enough to supply the trade. Black cohosh root is usually taken fresh or dried, and is available in capsule and tablet form.

Red List: Vulnerable*

Alocasia zebrina

Botanical family
Araceae

Distribution
Philippines: Luzon

Hardiness
Frost-tender

Preferred conditions
Tropical, broadleaved forest; an open, humus-rich soil, and partial shade.

Ever since its discovery in the mid-19th century, ***Alocasia zebrina*** has been prized for its architectural stature and bold foliage. Its pale leaf stalks are transversely banded, like a zebra – hence the species name "zebrina". Under good growing conditions, the plant can reach over 2m (6ft) in height.

Today, partly through collection, this species and the related *Alocasia sanderiana* have approached extinction in their native forest in the Philippines archipelago. As a result, both plants are listed on Appendix I of CITES: the species is threatened with extinction and banned from entering international trade unless specimens have their origins in artificially propagated nursery stock.

However, the future of the species looks more assured – at least from the ravages of collectors – since the genus has proved easy to micropropagate within the laboratory. Indeed, another aroid, *Amorphophallus konjac*, was successfully micropropagated as early as 1951. This form of vegetative reproduction does not lend itself to wild restocking programmes, however, because such micropropagated plants are all clones and therefore not genetically diverse enough to sustain viable wild populations in the long term.

SUPPLY AND DEMAND

Nevertheless, this reproductive technique has allowed the trade to satisfy horticultural demand, and this, in turn, has greatly reduced the need for wild-collected material. It is encouraging that, because commercial cultivation of *Alocasia zebrina* was so successful, and demand for its wild plants dramatically reduced, the species, along with *Alocasia sanderiana*, has since been removed from the CITES Appendix.

▲ HANDSOME PERENNIAL *While the species thrives in tropical and subtropical climates, plants in temperate regions are better suited to greenhouses or grown indoors, in an open, humus-rich compost containing peat substitute. In cold conditions, the plant becomes dormant.*

Red List: Vulnerable*

Amorphophallus titanum

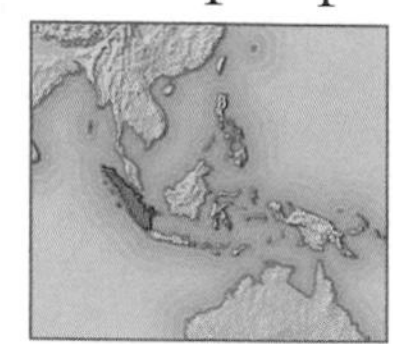

Botanical family
Araceae

Distribution
Indonesia: western Sumatra

Hardiness
Frost-tender

Preferred conditions
Tropical rainforest; shady sites on rich, moist, soil; high humidity and warmth.

Called the corpse flower, *Titan arum* is the media celebrity of the genus *Amorphophallus*, which consists of about 170 tropical species distributed from Africa through Asia to Polynesia, the majority of them Asiatic. First discovered in 1878 by Odoardo Beccari while engaged in botanical study in western Sumatra, it was soon introduced into cultivation. Seed that Beccari sent back to Florence Orto Botanico germinated and was distributed to other gardens.

One seedling was sent to Kew and after 10 years produced a 2.2m (7ft) tall flowerhead and a nauseous scent that came in eye-watering waves and was said to have made ill the artist responsible for illustrating it. The species has now flowered in several botanic gardens around the world, and it always attracts enormous media attention and large numbers of visitors. It flowers only rarely in cultivation; for example, it has flowered only a dozen times or so in cultivation in the USA. The appearance of its singular 1.2–2m (4–6ft) tall flowerhead at a botanic garden is therefore always an event. Notable not only for the stature of its bloom, the well-named corpse flower also produces a revolting stench of putrefaction. In its native habitat, this scent attracts the carrion beetles and sweat bees that succeed in pollinating the numerous inconspicuous female flowers clustered in the lower half of the spike (spadix).

In the wild, the species is threatened by unscrupulous collectors, but the most serious threat is the rapid destruction of its forest habitat in Indonesia.

Titan arum is a difficult species to cultivate. The plant requires heat, shade, and a humus-rich soil, and is perhaps best grown in large, terracotta containers. The tubers are subject to rot, and do not reliably increase in size, neither do they form offsets as readily as other, more common and modestly sized species in the same genus.

Seed is rarely produced in cultivation, because the female flowers open first and have usually faded by the time the male flowers shed their pollen. The seeds need to be cleaned of the pulp of the surrounding berry before planting, and should then be sown immediately, and maintained in warm, moist conditions for germination.

From about the second year onwards, the tubers should be repotted annually when the single leaf dies down, in a new container three times larger than the diameter of the tuber. This plant is very susceptible to attack by nematode worms, so the use of a sterile potting compost is imperative.

BOTANY

Bloom spectacular

Famed for its huge, triffid-like flowerhead and equally impressive, foul smell, *Titan arum* is one of the world's most remarkable plants. The flowering process unwinds over the course of several weeks. First a fleshy flower spike (spadix) shoots up (*see step 1*), wrapped by a bell-shaped bract (spathe). The spathe unfurls over a period of two days (*see step 2*).

After the flower spike fades (*see step 3*), a single huge, umbrella-like leaf emerges from the giant tuber, which can weigh nearly 90kg (200lb). In the wild, this leaf can achieve 6m (20ft) in height, with a 5m (15ft) spread of its compound blade; in cultivation, a more modest 4m (12ft) is usual.

After almost a year's growth, the plant will enter a dormant period lasting several months before either flowering again or merely producing a new leaf.

Unlisted

Amsonia orientalis

Botanical family
Apocynaceae

Distribution
North-eastern Greece, north-western Turkey

Hardiness
Fully hardy

Preferred conditions
Lakesides and marshy grasslands; grows best in partial shade or full sun, in light, moist soil.

Amsonia orientalis is a plant that is much more common in cultivation than it is in the wild. It has, perhaps, received less conservation attention than it deserves, partly because is not endemic to a single nation, but is distributed across two neighbouring countries, Greece and Turkey.

Originally found on the shores of Lake Apolyont (Uluabat) in north-west Anatolia in the early 19th century, the plant has subsequently only ever been found in about seven localities. These sites are in marshes, damp heathlands, and lakesides in north-west Turkey and Thracian Greece.

Today, the largest populations are actually located around the suburbs of Istanbul, perhaps Europe's fastest growing city. Here each of the three existing populations contains fewer than one hundred individual plants, and their habitats are all under intense pressure from urban development. Accordingly, the species is regarded as critically endangered in Turkey.

The plant's status in Greece is less well understood, but it is likely that there are fewer than 2,000 plants currently growing in the wild. Fortunately, the plant has long proved popular with gardeners for its ease of growth, and for its willow-like stems. These grow to 60cm (24in) and are topped by a profusion of flowers in early summer. *Amsonia orientalis* has received the coveted Award of Garden Merit from the Royal Horticultural Society in the UK, and it is now widely grown.

The plant is propagated by seed sown in containers in gentle heat in spring, or by division of established clumps.

▲ **Natural beauty** *This easily grown perennial thrives in full sun, and produces a cloud of pretty, pale blue flowers. It is particularly popular in naturalistic meadow-style plantings, and is suitable for any moist, well-drained, and fertile soil.*

Red List: Rare*

Angelica pachycarpa

Botanical family
Umbelliferae

Distribution
Portugal, Spain

Hardiness
Fully hardy – half-hardy

Preferred conditions
Thrives in temperate coastal grasslands in fertile and well-drained soils in full sun

Glossy-leaved angelica is a relative newcomer to the gardens of North America and Europe, and its introduction to those continents reflects an increasing interest in the architectural qualities of members of the carrot family, Umbelliferae.

This perennial (often grown as a self-seeding biennial) is valued for its exceptionally glossy, almost plastic-like foliage, which forms low domes up to 30cm (12in) tall, as well as lime-green to creamy-white flowerheads borne in abundance. *Angelica pachycarpa* is easy to grow and thrives either in a sunny position in the rock garden or under trees, or as ground cover.

THREAT

OIL DISASTER

On 19 November 2002, the oil tanker *Prestige* split in half and subsequently sank, about 240km (150 miles) off the north-west coast of Spain, carrying 70,000 tonnes of oil. Over the next few weeks, quantities of black, sticky, oil were swept onto the coast of Galicia by onshore winds and currents, causing huge numbers of seabirds to die, and polluting many beaches and coastal habitats.

In the wild, the plant is found mainly along the Galician (Atlantic) coastline of Spain (where the species is however considered Rare), and even more rarely in adjacent parts of Portugal (where the species is considered Endangered). Here the plant is characteristic of wild carrot-fescue grasslands found along the shoreline. Some concern has been expressed for the survival of populations of the plant after the grounding and subsequent breaking up of the oil tanker *Prestige* in 2002, and the release into Spanish waters of its cargo of about 70,000 tonnes of oil (*see box above*).

Angelica pachycarpa is a species best propagated by seed, which benefits from alternating periods of warm (18–22°C/64–72°F) followed by cold (less than 4°C/39°F) temperatures, after which germination takes place.

◄ STATELY BLOOMS *The flowers of* Angelica pachycarpa *are produced in midsummer in flat or slightly domed umbels. These later form fat green seeds.*

GLOSSY FOLIAGE ► *This bushy perennial is a handsome architectural addition to the garden. It has attractive foliage, with large, very leathery leaves.*

Red List: Extinct*

Anthurium leuconeurum

Botanical family
Araceae

Distribution
Mexico

Hardiness
Frost-tender

Preferred conditions
Once found in tropical humid forest, on moist, rich soil.

Anthurium leuconeurum is one of a relatively select group of plants that owes its existence to the efforts of the world's botanic gardens and specialist growers, for the plant is currently unknown in the wild. The species was originally described in 1862 from a collection made by a botanist, Auguste Ghiesbrecht, around 1860 in the Chiapas area of southern Mexico. He introduced it into cultivation in Ghent, Belgium. All plants in cultivation today are derived from that single introduction, and the species has not been rediscovered in the wild since the Ghiesbrecht expedition.

ANTHURIUM CLARINERVIUM ► *Like* Anthurium leuconeurum, *this plant has heart-shaped leaves in a rich, velvety dark green. The colour is a perfect foil for the dramatic veins.*

Different theories exist as to its origins: one suggests that this species may be a narrow endemic, confined to a tiny area and awaiting rediscovery or perhaps already extinct. According to a second theory, the plant originally described may actually be a rare hybrid, perhaps a cross between the related *Anthurium berriozabalense* and *Anthurium clarinervium*, and occurring only very rarely in the wild. Indeed, plants similar to the type drawing have been created in cultivation utilizing these two parents.

Whatever the truth about the plant's origin and status in the wild, its future in cultivation seems assured. It is a very handsome species, grown mainly for its striking, large leaves, as are other anthuriums. These measure up to 40 by 30cm (16 by 12in). They are borne on stems up to 70cm (28in) tall. Conversely, the flowers are rather insignificant. These are typical of the aroid family, and greenish in colour, but with a rather reduced spathe. Since its introduction, *Anthurium leuconeurum* has been extensively used as a parent for a range of hybrids.

In warm climates, grow *Anthurium leuconeurum* in moist, but well-drained, humus-rich soil, in partial shade. The plant requires high humidity and a constant minimum temperature of 16° C (61°F). Propagate by division, in winter or early spring.

RENEWAL

FLOWER TRADE

Anthurium species have been grown commercially in Hawaii for many decades, for the flower trade. In the early days, in the 1930s, they were planted mainly beneath citrus trees, as an adjunct to the tangerine crop. Now some are planted in the shade of trees, but many are grown in special shade-houses. By the 1980s, there were over 200 commercial farms, producing monthly shipments of up to nearly 3 million flowers.

22 on Red List*

Argyranthemum

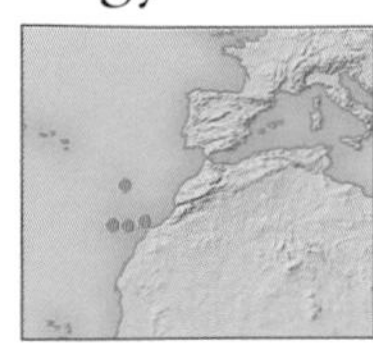

Botanical family
Compositae

Distribution
Spain, Canary Islands, Portugal, Madeira, Salvage Islands

Hardiness
Half-hardy

Preferred conditions
Full sun and free-draining soil on coastal sand dunes, cliffs, moist forests and dry scrub; hardy only to –5°C (23°F).

The genus *Argyranthemum* is of great interest as it provides one of the best examples of an evolutionary process known as adaptive radiation. Studies show that the group originated in the Mediterranean and from there migrated to the island groups that comprise Macronesia (Azores, Canary Isles, Cape Verde Isles, Salvage Islands, and Madeira). As the Mediterranean species spread throughout these scattered islands, they became isolated geographically and reproductively, and subsequently evolved into new species. Species that reached the larger islands usually encountered several different habitats and diversified further as their scattered and secluded populations became more fragmented and adapted to the local environments. With the passage of time, such isolated populations evolved into new species, as Darwin had already ascertained in his famous study of finches on the Galapagos Islands.

◄ ARGYRANTHEMUM FRUTESCENS *A white marguerite daisy, it prefers partial shade and a moist soil, and does well as a pot plant. It flowers all summer. In the wild, it grows in Madeira and the Canary Isles.*

Today, argyranthemums are no longer found in the Mediterranean, because their direct ancestors became extinct when the Pleistocene ice age severely changed that region's climate and altered forever its natural vegetation. Instead, 24 new species are scattered throughout the more climatically stable islands of Macronesia. There, they are found in a wide variety of habitats, each one including a species with a distinct appearance. For example, ***Argyranthemum foeniculaceum***, a glaucous shrub with a distinct candelabra-like habit, grows only on the ancient basaltic inland cliffs of Tenerife, while ***Argyranthemum frutescens***, a low-growing shrub with fleshy green stems, grows at much lower altitudes close to the ocean, but again only on Tenerife. *Argyranthemum callichrysum* is a tall, erect shrub with dark green leaves and is found only in arid, euphorbia-dominated scrub on the island of La Gomera.

Of the 24 *Argyranthemum* species, 22 appear in the 1997 Red List. Those with naturally small distributions are susceptible to extinction from changes in their habitats caused by agricultural practices or tourism. *Argyranthemum broussonetii*, which grows only in the moist laurel forests on La Gomera and Tenerife, is one such species. Fortunately, many species have a degree of protection because they grow in nature reserves.

▲ ARGYRANTHEMUM FOENICULACEUM *Resembling an ox-eye daisy, this pretty species has blue-green foliage and grows to a height of 90cm (3ft).*

Several argyranthemums are cultivated, including all of the species so far mentioned, as well as *Argyranthemum coronopifolium*, *Argyranthemum lemsii*, and *Argyranthemum maderense*. All bear pretty daisy flowers and are long-flowering. However, in colder areas they should be treated as tender perennials or pot plants. Propagate them from stem cuttings in autumn.

ARGYRANTHEMUM MADERENSE *This large-flowered marguerite daisy is actually native to the Canary Islands of Lanzarote and Fuerteventura, and not, as its name suggests, to Madeira.*

35 on Red List*

Arisaema

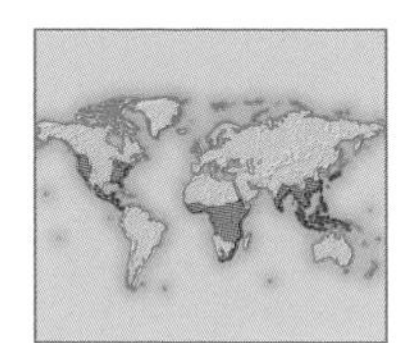

Botanical family
Araceae

Distribution
Temperate and tropical Asia, tropical Africa, North America

Hardiness
Frost-tender – fully hardy

Preferred conditions
Shade and wet, rich soil, in temperate broadleaf and mixed forest, and tropical rain forest.

Botanists estimate that this genus of tuberous aroids, usually known as cobra lilies, contains anywhere from 100 to 200 species. If the conservative number is more accurate, over 25 per cent of the species have been given Red List status. A number of factors combine to make this number so high. Some of the species are exceedingly rare, especially those that occur in the Himalayan forests. Secondly, there is a small, but steady market for the more ornamental species from Asia, and no shortage of local collectors willing to supply the demand. Thirdly, as forest understorey species, cobra lilies are very sensitive to forest disturbance.

Arisaema dracontium, known as green dragon, is found locally throughout the eastern half of the USA, north to Ontario and Quebec in Canada. Although it is not on the 1997 Red List, it is considered threatened or endangered in many of the states and Canadian provinces within its range. The favoured habitat of this species is in low-lying woods on moist soils, where it occurs rather sporadically, but when these habitats are drained for agricultural or other development, the plants die. Green dragon has a greenish, fleshy flower spike (spadix) that extends beyond the less prominent, greenish hood (spathe). The spathe lacks the distinctive purple stripes of its more common relative, jack-in-the-pulpit (*Arisaema triphyllum*). Although green dragon goes dormant in the summer, the soil should not be allowed to dry out.

Arisaema sikokianum is a rare species known only from the southern island of Shikoku in Japan. The tubers produce two leaves with 3–5 leaflets, which are attractively variegated down the centre. This hardy species lends a beautiful spring accent to a shady woodland garden, but will not tolerate standing in water during its winter dormancy. It prefers shade, and a moist but well-drained, fertile soil, for example underneath deciduous trees.

Red berries appear in mid- to late summer on all arisaemas as the spadix withers. The seeds should be sown immediately after cleaning the fruit from around them. They will germinate quickly, grow for a few months, then promptly go dormant until the following spring, when the tubers should be moved to their permanent place in the garden. Two to four years must pass before the plant will flower. The tubers of all species are poisonous if eaten raw.

▲ ARISAEMA DRACONTIUM *Green dragon usually has one large, long-stalked leaf, divided into 7–15 leaflets. It is one of only a few arisaemas native to the USA.*

ARISAEMA SIKOKIANUM *This rare species is one of the more stunning in the genus. The hooded "flower", or spathe, is actually formed from a large bract. The flowers are borne on the central spadix, or flower spike.*

3 on Red List*

Asarum

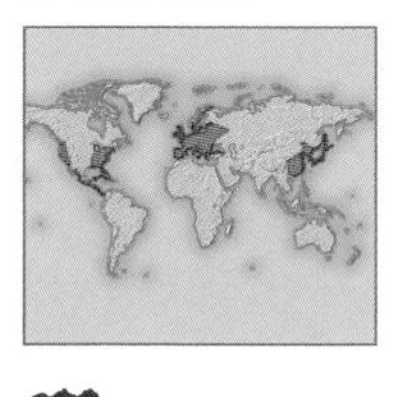

Botanical family
Aristolochiaceae

Distribution
Northern temperate regions, notably north-east Asia

Hardiness
Half-hardy – frost-tender

Preferred conditions
Shaded site, with moist, but well-drained soil.

The genus *Asarum* comprises approximately 70 species with a wide distribution, mainly in the forests of the northern temperate zone. The highest diversity of species is in north-east Asia, notably in Japan. Although not true gingers, some species are often known as wild ginger, and are growing in popularity among gardeners both in North America and in Europe.

However, it is in Japan that the genus has been appreciated by gardeners for the longest period of time, and the art of their cultivation here reaches its zenith. The Japanese cult of *koten engei* – "classical plant" cultivation of native genera – has developed and evolved over centuries. Typically, only native species of the genus are grown, and cultivation focuses on natural variants, individuals collected from wild populations for their subtle differences in leaf shape, markings, or flower colour.

In no other country has the cultivation of native plants and selected forms reached such a peak of development, over such a long period of time. Over the years, many cultivars have been selected, and some of the choicest and rarest change hands for large sums of money. But the fashion for selecting and collecting wild ginger species from wild populations puts considerable pressure on the more local and therefore often more desirable species, and it is thought that no fewer than nine species may be critically endangered in Japan as a result.

WILD GINGER FLOWER ► Asarum caudatum *is native to the west coast of North America. Like other asarums, it is grown for its bizarre flowers. The evergreen foliage is often highly mottled.*

Stamens surround the stigma

Perhaps the most threatened wild ginger is *Asarum minamitanianum*, a clump-forming species prized for its cyclamen-like heart-shaped leaves, and its unusual flowers. The leaves grow up to 10cm (4in) long and are dark green in colour, with yellow-green to grey-green mottling. Its starfish-like flowers are attenuated into three long lobes, each up to 8cm (3in) long, and are liver brown, edged in white. Only known from a few sites on Kyushu Island (at the southern end of Japan's main island), this species has been driven to the verge of extinction in the wild by collection.

Cultivation

Asarum minamitanianum is best grown in frost-free conditions, in an open, humus-rich compost with good drainage. Water plants freely in summer, but keep them barely damp in winter in cool climates.

ASARUM SPECIOSUM ► *This species, Alabama wild ginger, is known in the wild in just three counties in the state of Alabama. It is fairly easy to cultivate.*

Red List: Endangered*

Astelia chathamica

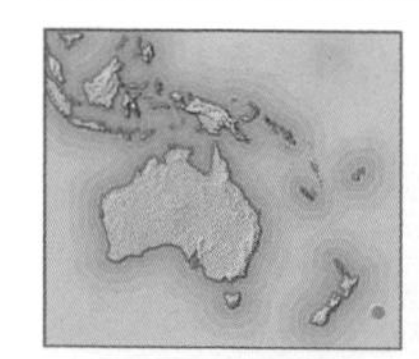

Botanical family
Liliaceae

Distribution
New Zealand: Chatham Islands

Hardiness
Fully hardy – half-hardy

Preferred conditions
Moist but free-draining sites in sun to partial shade in forest, on cliffs, and on stream banks.

Known locally as *kakaha*, or Moriori flax, ***Astelia chathamica*** is endemic to the Chatham Islands, which lie approximately 800km (500 miles) east of New Zealand. Here it grows in a range of moist sites, including cliffs, lake margins, forest floors, and stream banks. Formerly it was widespread, but grazing by domestic and feral cattle, horses, pigs, and sheep has destroyed or damaged much of the natural vegetation cover, and today this species is restricted to isolated or rocky spots in scrub, safe from grazing animals. Locally, efforts are being made to restore lost vegetation, and such efforts may allow the *kakaha* to regain lost territory. For example, a 10-year plan to restore forest on the more isolated Mangere and Rangatira Islands may allow the establishment of "fallback" populations, although currently it is not known whether conditions are moist enough on these islands.

Commonly called silver spears by gardeners, it is undoubtedly the most widely grown of the 25 species in the genus. In recent years, this species has become a favourite among gardeners. By selection and hybridization, the horticultural trade has produced a range of cultivars with leaves in silver, red, and green, that reach up to 3m (10ft) in length. Valued by florists for its striking leaves, *Astelia chathamica* has also become increasingly important as a foliage plant in the floriculture industry.

The plant's dramatic spread in cultivation has largely been due to improvements in propagation, with most material now produced by tissue culture. While such material would be largely unsuitable for restocking programmes in the wild, it does allow gardeners to enjoy growing this species without worrying about the impact of the horticultural trade on wild populations of this *Astelia*.

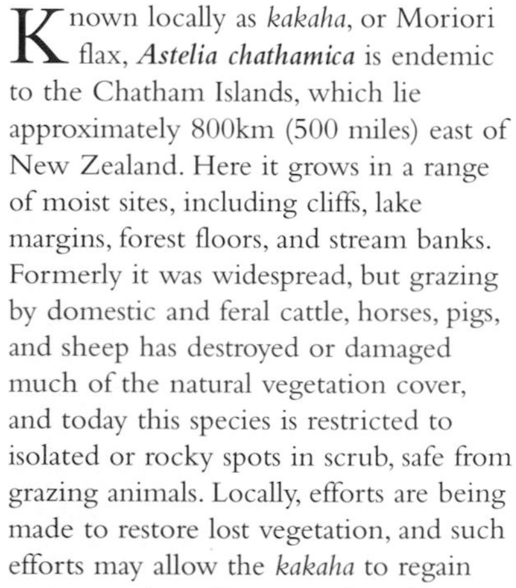

◄ SILVER SPEARS *With its clumps of sedge-like leaves, up to 1.5m (5ft) in length,* Astelia chathamica *is one of the most striking of the Chatham Islands' endemics.*

▲ IT TAKES TWO *Male and female flowers occur on separate plants (dioecious), so* Astelia *plants of the opposite sex are required for successful seed production.*

Red List: Vulnerable*

Azorina vidalii

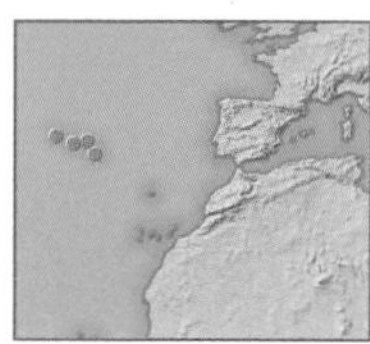

Botanical family
Campanulaceae

Distribution
Portugal: Azores

Hardiness
Half-hardy – frost-tender

Preferred conditions
Full sun to partial shade; rocky, free-draining soils in exposed sites.

The shrubby ***Azorina vidalii*** is the only species in its genus and is restricted in the wild to sea cliffs in the Azores. A relative of campanulas, *Azorina* eventually grows into a shrub up to 2m (6ft) tall, with woody, branched stems and handsome flower spikes. It is given Vulnerable status in the 1997 Red List because it has a very limited distribution. It has also been the victim of overcollection in the past.

Several examples exist of plants from remote islands that have evolved woody stems, and which are consequently quite different from their herbaceous relatives on the mainland. This phenomenon is perhaps best illustrated, however, by the campanula or bellflower family, Campanulaceae. In this family, the mainland representatives are typically small- to medium-sized herbs. Familiar examples include campanula and *Codonopsis*.

In contrast, scattered around the world, on remote islands, are rather unusual woody members of the family Campanulaceae, including the genus *Canarina* from the Canary Islands, *Clermontia* and *Brighamia* from the Hawaiian Islands, and this genus, *Azorina*, from the Azores, which is very closely related to campanulas.

Plants growing on isolated islands tend to evolve a woodier habit than their close relatives on the mainland. This is in spite of the fact that most of the plants that first colonize remote islands are weedy herbaceous species. One of the defining characteristics of weedy species is that they are notoriously well-adapted for long-distance dispersal, and so many survive long-distance travel, arriving ready to colonize remote islands and adapt to the ecological niches that have not yet been filled by other species. In the absence of competition from other plants, some of these weedy colonists eventually evolve into woody species.

NODDING BELLS ► Azorina *flowers are typical of the bellflower family, and are produced in late summer, in pink or white. In colder climates, the plant is best grown in a cool greenhouse.*

64*/32 on Red List

Begonia

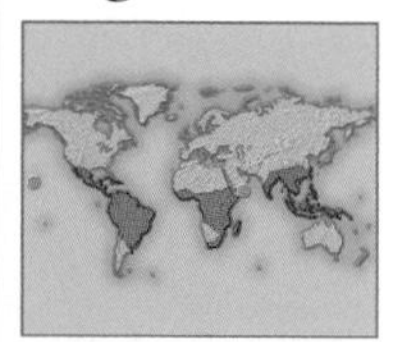

Botanical family
Begoniaceae

Distribution
Tropical and subtropical regions of the world

Hardiness
Frost-tender – half-hardy

Preferred conditions
Rich soil in humid conditions, in tropical or subtropical forests.

With about 1,500 species, *Begonia* is one of the largest of all the genera of flowering plants. As most of these species occur in very small areas of tropical forest – a habitat that has been lost at an alarming and increasing rate in recent years – it is perhaps not surprising that many of them (well over 60) are of conservation concern. Several of the rare or endangered species are found in cultivation.

Begonia dregei is a rare species that grows in subtropical forests along the east coast of South Africa. It was first introduced into cultivation in 1836, and is an important parent of several of the well-known commercial begonia hybrids. In the wild, the populations of this species are highly fragmented, and contain very few individuals, and the

◀ **BEGONIA DREGEI** *The grape-leaf or maple-leaf begonia is a tuberous or semi-tuberous species from South Africa. It has shallowly lobed leaves and small white flowers, and grows best in moist, rich peat.*

plant is listed as Rare in the 1997 Red List. Most plants, however, are protected within nature reserves, so are not threatened with imminent extinction.

Begonia octopetala grows wild only in Peru and is an attractive species that was first cultivated in 1805. Today the species is only rarely cultivated, and the wild populations are also endangered, mainly from habitat destruction. It grows largely in the *lomas* – unusual communities of ephemeral or summer-dormant herbs that grow in the condensation zone created where winter fogs come up against the steep hills adjacent to the west coast of South America.

Most begonias are easily grown in soil or peat-based compost, in humid, shady situations under glass. A few of the smaller species, including *Begonia rajah*, require very high humidity and prefer a terrarium. Tuberous species like *Begonia dregei* and *Begonia octopetala* periodically become dormant, and during that time require very little moisture. In tropical or subtropical climates, several of the larger species and hybrids make excellent garden plants, while in colder climates they are treated as summer bedding. Only one species, ***Begonia grandis***, is reliably frost-hardy.

Begonia rajah was first collected in the vicinity of Mount Lawit on the Malay Peninsula and introduced into cultivation at Singapore Botanic Gardens prior to 1892. From here, cultivated plants were distributed to the UK. Wild plants were not seen again until recently, when a single population was located on cliffs beside a waterfall in an area of logged forest. The species remains rare in cultivation and is known only from this one wild population. The 1997 Red List classifies it as Endangered.

CULTIVATION

MODERN BEDDING

Begonias have become very popular bedding plants, not only in the garden, but also in public places such as parks. A range of bright colours, including red, pink, white, orange, and yellow, is available, offering great scope for eye-catching displays. Some begonias are compact, while others are trailing, and most grow and flower continuously through the summer season. In temperate regions, they need to be brought under cover for the winter. They will grow from cuttings, but most commercial begonia production is from seed. They flourish either in full sun, or partial shade, although they do not compete well with other plants.

▲ **DOUBLE BEGONIAS** *Some of the most popular garden begonias are cultivars with double flowers, and many such forms are now available. These flowers are almost rose-like in appearance, and contrast well with the foliage.*

BEGONIA GRANDIS *Often called hardy begonia, this is a popular perennial for well-drained soil in partial shade and grows to 60cm (24in). Arching, branched stems bear white or pale pink flowers.*

Red List: Vulnerable*

Biarum davisii

Botanical family
Araceae

Distribution
Greece: Crete, Turkey

Hardiness
Half-hardy

Preferred conditions
Full sun, on well-drained soil.

The genus *Biarum* consists of just under two dozen tuberous aroids native to the Mediterranean region, from Portugal and Morocco, east to Afghanistan. They produce strange flask-shaped inflorescences, before the leaves appear. In most species these emit a pungent aroma when the female flowers are fertile. The unpleasant scent is attractive mainly to flies and beetles – the chief pollinators. The flowers of ***Biarum davisii*** are unusual in having a pleasant smell. This species, like most, is autumn-flowering. The small, very narrow leaves appear just after the flowers.

Similar to *Biarum davisii*, the Unlisted *Biarum marmarisense* grows in southern Turkey, on the Marmaris Peninsula. It is somewhat larger than the Cretan species and is also rather easier to cultivate.

Plants should be grown in a soil-based compost, either outdoors in a raised bed, or in a cool greenhouse. They are dormant through the summer and should then be kept dry. Growth resumes soon after flowering in the autumn.

▲ BIARUM DAVISII, *from Crete and Turkey, produces very small flask-shaped, pink-speckled spathes, 5–6cm (2–3in) tall, in the autumn. The spear-shaped leaves appear later. It grows best in a cold greenhouse.*

9* on Red List

Celmisia

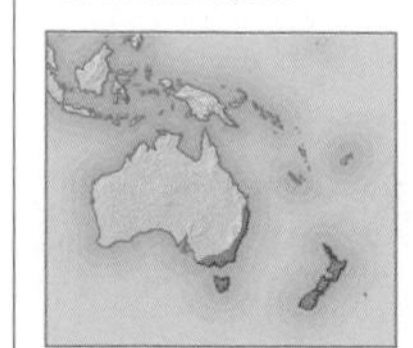

Botanical family
Compositae

Distribution
Australia; New Zealand

Hardiness
Fully hardy

Preferred conditions
Open, well-drained, humus-rich soil in a semi-shaded position.

Snow, or mountain, daisies are a group of about 60 species, largely confined to New Zealand, with a few (three species) in Australia. They are typically alpine plants, occurring on rocks, screes, and in montane grassland, and they have adapted their form in a number of ways in order to survive under extreme conditions. These include a tight, cushion form, and white-felted, and often narrow, leaves. Like many New Zealand alpines, the flowers of all celmisias are white, sometimes tinged pink. Their distribution covers most of New Zealand's mountain ranges.

Mountain daisies are particularly eye-catching when seen flowering *en masse*. Many form rosettes of sword-shaped leaves, ranging in colour from grey-green to silvery grey, or silver. While certain species are relatively widespread, a number are highly localized, in some cases being restricted to single sites or mountains. Reported threats include collection, overgrazing, and the possible longer-term impact of global warming.

Celmisia morganii (Rare) is a handsome, rosette-forming herb with lance-shaped leaves, white-felted beneath, that are up to 45cm (18in) long. Its large, white daisy flowers are approximately 3cm (1in) across. The plant is largely confined to one river gorge in New Zealand's South Island, where a few thousand plants grow on damp rocks, at the base of cliffs, and around waterfalls. A mining tramway was constructed through the gorge, and botanists feared this might damage the population. However, the tramway provided more open ground, which *Celmisia morganii* has colonized vigorously. Unfortunately, the site is now much more accessible, increasing the possible risk of collection by unscrupulous collectors.

▲ CELMISIA SEMICORDATA *The mountain daisy, cotton plant, or* tikumu, *is native to South Island in New Zealand. This large robust perennial has silver-green leaves.*

Celmisia philocremna (Rare*) is an alpine species with tight cushions of grey, glandular-hairy rosettes, topped by large, white, yellow-eyed daisies. Limited to a few populations on rocks and ledges at 900–1,800m (2,950–5,950ft), it is at the top of a creek in the Eyre Mountains of New Zealand's South Island.

Perhaps the most attractive species is ***Celmisia semicordata***, with its conspicuous rosettes up to 1m (3ft) in diameter. It is another species from South Island and grows at 600–1,400m (2,000–4,600ft), in tussocky grassland and subalpine scrub.

HISTORY

GREEK GODS

The generic name *Celmisia* comes from Greek mythology, and commemorates the character Celmis. According to legend, Celmis was born on Crete, and was the son of Anchiale, a nymph. The Mother of the Gods, Rhea, appointed Celmis to guard her newborn son Zeus, and he became a faithful friend and guardian of the young god. Celmis is said to have discovered the art of smelting iron, and he and his four brothers were masters of metallurgy and sorcery. Celmis apparently offended Rhea, and was duly changed into a lump of diamond. The bright flowers of *Celmisia* certainly shine like bright diamonds in their mountain habitat – although far away from Greece.

CULTIVATION

Mountain daisies should be kept moist in the summer, but dry in winter. In wetter climates, it is important to protect them from winter rain. They appreciate cool conditions, and should be grown in a semi-shaded position, or as alpines in pots, troughs, or in a cool greenhouse or frame. Propagation is by seed sown in a cold frame when ripe. They may also be propagated by rooting individual rosettes, or by division. The species hybridize easily, so need to be isolated.

▼ MOUNTAIN DAISY *Rocky mountain grassland is the typical habitat of these plants with pale, showy flowers.*

Red List: Vulnerable*

Clematis addisonii

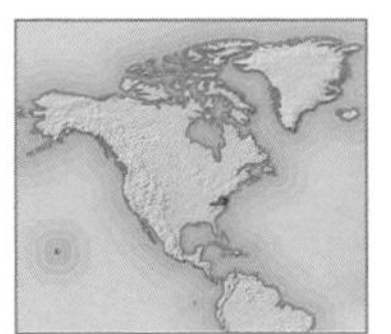

Botanical family
Ranunculaceae

Distribution
USA: Virginia

Hardiness
Fully hardy

Preferred conditions
Fertile, well-drained soil in full sun, in temperate, dry, deciduous woodland.

Addison's leather flower was first found on the banks of the Roanoke River in the state of Virginia, on 29 May 1890, by Addison Brown, after whom it is named. Despite occasional reports from other states, ***Clematis addisonii*** does genuinely appear to be confined to just four counties of western Virginia – Botetourt, Montgomery, Roanoke, and Rockbridge, where approximately 20 populations are known.

In the wild, the plant occurs on rock outcrops and in open woodland, in mixed deciduous-juniper forest, at altitudes of 200–600m (640–1,920ft) on dolomite limestone rock. These "barrens" are notably rich in other rare species, including chestnut lipfern (*Cheilanthes eatonii*), smooth conehead (*Echinacea laevigata*), American gromwell (*Lithospermum latifolium*), plains muhly (*Muhlenbergia cuspidata*), and the goldenrod *Solidago rigida*.

▲ LEATHER FLOWER *This perennial clematis has mostly simple leaves, and bears attractive pink-purple, somewhat leathery, urn-shaped flowers, with creamy-white tips.*

In its natural haunts, Addison's leather flower is considered very rare and imperilled. Here, it is threatened by a range of factors, including excessive grazing, habitat fragmentation, competition with invasive plants, quarrying operations, and road construction. Only in recent years has the plant been introduced into gardens outside the USA, but it is slowly gaining a firm foothold in cultivation among clematis specialists, and those who appreciate its elegant beauty.

Red List: Extinct*

Cosmos atrosanguineus

Botanical family
Compositae

Distribution
Mexico

Hardiness
Frost-tender

Preferred conditions
Moderately fertile, moist but free-draining soil in full sun.

Not only are the flowers of chocolate cosmos an unusual colour, but they also smell of rich, dark chocolate. Unlike the other commonly cultivated cosmos species, the chocolate cosmos is a tuberous-rooted perennial reminiscent of a small dahlia. Formerly found in Mexico, probably near Zimapán, Hidalgo, the plant was first collected in Mexico around 1860 and has only been found on one occasion since. It is now thought to be extinct in the wild, not having been seen in well over 100 years.

All cultivated plants in the UK and North America appear to represent a single clone, which has survived in cultivation since the 1800s. Attempts have been made to reintroduce this species back into the wild in Mexico, but these projects are unlikely to succeed. This is because cultivated plants completely lack genetic variation, since all known plants were propagated from a single, sterile clone.

Also, little or nothing is known about the original habitat of this species, an important prerequisite when attempting reintroduction of a plant into the wild. The best hope for the plant is to maintain it in cultivation by vegetative propagation. Nevertheless, its lack of any genetic variation could prove to be a disadvantage, even to the species in cultivation, because this renders it susceptible to disease.

CULTIVATION

Grow chocolate cosmos in open, free-draining compost and treat it like a dahlia. It forms small tubers, 2–3cm (1in) long; gather them before the threat of frost, and then store in a cool, dry place over winter, ready to replant them in the following spring. Propagation is usually by division of the tubers in the autumn at the time that plants are lifted for storage. An alternative method is to take basal shoot cuttings once the tubers sprout in the spring. Commercially grown plants are often propagated using micropropagation techniques.

▼ DOUBLE CHOCOLATE *Chocolate cosmos is named both for the colour of its flowers, and also for their distinct aroma. The scent is most obvious on a warm evening.*

Red List: Vulnerable*

Datisca cannabina

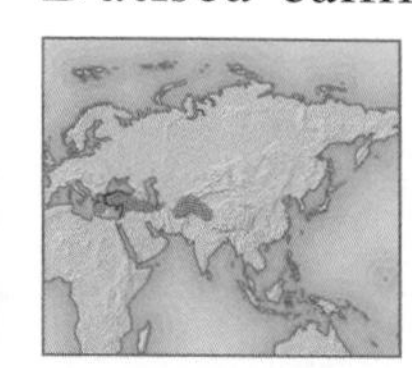

Botanical family
Datiscaceae

Distribution
Crete to the Elburz Mountains of Iran and western Himalaya

Hardiness
Fully hardy

Preferred conditions
Tolerates a wide range of soil types and situations, other than extremely wet or dry ground and positions in full shade; hardy to −10°C (14°F).

Known as false hemp for its elegant, cannabis-like leaves, *Datisca cannabina* is occasionally cultivated in herbaceous borders. It was once more widely grown as a dye plant: a yellow dye for silk was produced from the leaves, stems, and roots.

The interest of this genus lies in the unusual distribution of its two species – this species is from Asia, while *Datisca glomerata* (Unlisted) is from North America – and the fact that it is the closest living relative of the genus *Begonia*. This relationship is beyond doubt, as the two genera share numerous anatomic and genetic features, but is made more interesting by the enormous gulf in outward appearance that separates them, a difference that at first makes it difficult to envision how they could be even remotely related. *Datisca cannabina* is under threat because its populations are sparsely scattered, not well studied, and receive little to no formal protection. In the wild, this species grows along streams at an altitude of 300–1,830m (1,000–6,000ft). It is easily grown in a variety of soils in either full sun or semi-shade and its stems can reach over 2m (6ft) in height. Propagation is usually via division of the rhizomes in spring before growth starts. Seed set is rare in cultivation: the male and female flowers occur on different plants, and few gardeners grow both.

▲ DATISCA CANNABINA *False hemp is a stately ornamental perennial with graceful, arching stems and pinnate, toothed leaves. Both male and female plants bear elegant, pendulous tassels of green or yellowish flowers in summer.*

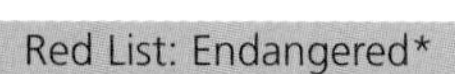

Red List: Endangered*

Delphinium luteum

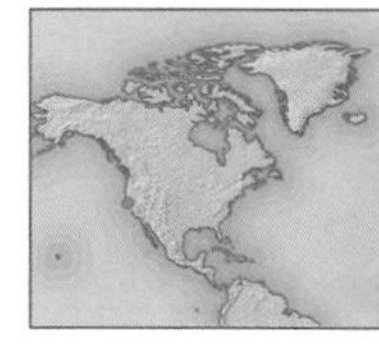

Botanical family
Ranunculaceae

Distribution
California, in coastal areas of Sonoma county

Hardiness
Half-hardy

Preferred conditions
Grows well in moist, shady conditions; just marginally half-hardy, tolerating temperatures below freezing, 0°C (32°C), for short periods only.

With just two small populations that together contain fewer than 50 plants, the yellow larkspur appears to be on the brink of extinction. Although never widely distributed, *Delphinium luteum*'s range within northern California has been reduced in the last 70 years. Its coastal scrub habitat has been badly affected by rock quarrying, residential development, and overgrazing by sheep and deer. The species has also suffered from overcollecting by both botanists and gardeners. Its rarity and narrow distribution leave it particularly vulnerable to random natural disasters, such as fire or a sudden rise in insect predators. Both wild populations occur on private land. At the time of writing, the species is listed as Endangered by the US Fish & Wildlife Service but no beneficial habitat management or protection is in place. There are plans to reintroduce plants grown *ex-situ* to parts of the species' former range (*see p.42*).

This species is rare in cultivation. Although it is protected, wild-collected seed is sometimes offered for sale. This should not be purchased, as it could decrease reproduction in the wild.

THREAT

CALIFORNIA FIRES

Fire is a natural occurrence in California. It can break out when lightning ignites dry vegetation in hot summers, although in recent years human activities have been partly responsible. Fires can cause serious habitat loss or degradation, as well as devastate human communities. Extensive fires in 2003 led to a plan to increase logging in areas of California: supporters claim the scheme will reduce fires and help wildlife, while critics mantain it will increase commercial logging, endanger wildlife habitats, and open up yet more land for grazing.

DELPHINIUM LUTEUM ► *An exception to the towers of blue flowers that are favoured in cultivated delphiniums, the blooms of this species are a soft yellow and borne in short spires. They appear over a long period from early summer.*

Red List: Indeterminate*

Dicentra spectabilis

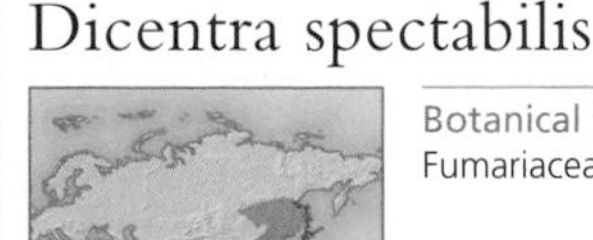

Botanical family
Fumariaceae

Distribution
Eastern Siberia, China Korea, and Japan

Hardiness
Fully hardy

Preferred conditions
Woodland species that likes a cool position in bright or dappled shade; prefers moist, humus-rich soil, ideally acidic but tolerates slightly alkaline conditions; hardy to at least −15°C (5°F).

For centuries, this species has been cherished by Chinese and Japanese gardeners, and it has escaped from cultivation back into the wild on numerous occasions. Wild populations of this beautiful plant are most likely to have been established in moist woodland and mountain habitats throughout much of north-eastern Asia. Indeed, botanists now have difficulty determining exactly where the species is truly native and where it has been introduced.

Dicentra spectabilis was one of the first north-east Asian plants to be introduced to Europe, when plant hunter Robert

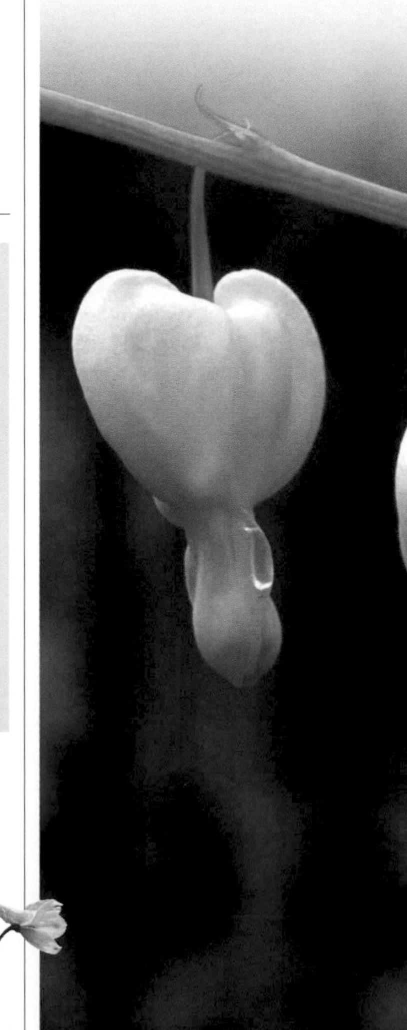

DICENTRA SPECTABILIS ► *The pouched outer petals of the flowers have given this species a number of fanciful common names. Widely known as bleeding heart today, it has also been called lyre flower. Another name, lady in a bath, seems incongruous until the flower is inverted.*

Inner petals protect anthers

Fortune sent plants that had been collected in a Chinese garden back to England in the mid-1800s. Flowering in late spring and early summer, it must surely be one of our most graceful garden plants. Unfortunately its natural beauty has led to the loss of a number of its populations through overcollecting from the wild.

The plant's common name of bleeding heart refers to the resemblance its flowers bear to hearts with drops of blood falling from them. The Chinese have a different interpretation, and suggest that the flower looks like a fish that is holding a smaller fish in its mouth. Bleeding heart is widely available from cultivated stocks. It prefers a partially shady position and a moist, humus rich, slightly acidic soil, and is hardy to −15°C (5°F). Propagation is by seed sown in the spring or by division when plants are either dormant or recently emerged. Root cuttings may also be taken in the spring and summer.

▼ DICENTRA SPECTABILIS 'ALBA' *Both the species and this elegant white form were found in virtually every European and North American garden by the late 19th century. In China, they were already rare in the wild.*

DIERAMA PULCHERRIMUM *The grasslike leaves and slender, graceful stems make this elegant summer-flowering species a rewarding plant for gardeners to grow.*

Red List: Vulnerable*

Dierama pulcherrimum

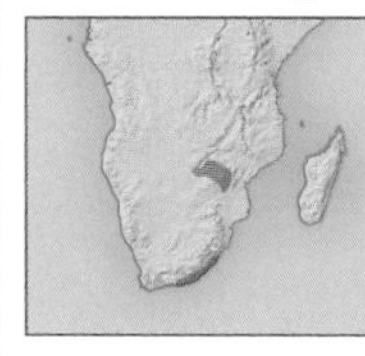

Botanical family
Iridaceae

Distribution
South Africa from the eastern Cape to Transkei

Hardiness
Fully hardy

Preferred conditions
Thrives in an open, sunny, but sheltered position; best in rich, moist soil with plenty of water while in full growth; hardy to −10°C (14°F).

There are 44 species of the genus *Dierama*, the wandflower or angel's fishing rod. They grow wild in an area stretching from the southern Cape of South Africa northwards through southern and eastern Africa to Ethiopia. Despite being so widespread, and found at altitudes ranging from sea level to the mountains, curiously, all are found only in moist grassland habitat.

Of these species, *Dierama pulcherrimum* is perhaps the most commonly found in cultivation. Like the majority of species in its genus, it comes from the Cape. Here, different forms with rich dark purple, rose-magenta, pink, or white flowers occur, and at various times each of these has been introduced into cultivation. The name *pulcherrimum* means "very beautiful", which this species undeniably is. Each summer, plants will produce several arching flower clusters, bearing numerous pendulous flowers at the ends of wire-like stems. In the garden, flowering plants look especially beautiful when grown next to a stream or pond, hence their vernacular name, angel's fishing rod. In the wild, *Dierama pulcherrimum* inhabits a limited area and is Vulnerable in the 1997 Red List. The main threat to the species is from habitat destruction.

Dierama pulcherrimum is easily grown in a moist but well-drained soil that does not dry out in the summer. Although it comes from South Africa, this is naturally a plant of the uplands, growing above 900m (2,950ft) and consequently is hardy to about −10°C (14°F). The species does, however, prefer a sunny, sheltered position. Propagate by seed sown in the autumn. Plants can be divided in the spring, but care must be taken to minimize disturbance to the fleshy roots, which are very brittle. If the roots are accidentally damaged, the plant will take time to re-establish itself.

5 on Red List*

Echinacea

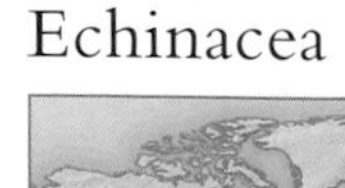

Botanical family
Asteraceae

Distribution
Central and eastern USA

Hardiness
Fully hardy

Preferred conditions
Most species of this hardy genus grow well in a rich, deep, but well-drained soil in full sun, although some grow in poorer soil and most will tolerate some shade.

The botanical name of the coneflower, *Echinacea*, a genus of around 10 species from grassland, dry hillsides, and open woodland in North America, has come to prominence in recent years. In the USA and Europe, extracts from *Echinacea angustifolia* (Unlisted) and the purple coneflower *Echinacea purpurea* (Unlisted) are among the most commonly used herbal products to strengthen the immune system against colds and 'flu. Indeed, demand for the plant has put a huge pressure on the genus. In some areas, such as the Dakota Prairie Grasslands, collection is regulated, but in many areas uncontrolled collection is a major threat. A more longstanding problem, however, is continuing habitat change (*see box*).

ECHINACEA TENNESSEENSIS

Echinacea laevigata (Vulnerable*), the smooth coneflower, is a plant of dry alkaline soils, occurring in open woodland habitats where light levels are relatively high and competition from more vigorous species is low. Traditionally it is typical of oak and cedar barrens, forest margins, and limestone bluffs, where natural fires, thin soils, or grazing herbivores have maintained its preferred disturbed and open habitat. Whilst development, road construction, and collection have led to the decline or loss of populations, changing land use in its native environment remains the most serious threat to the species, even within protected sites. Indeed, in many areas there are no longer fires or grazing animals to disturb the ground, so the plant has established itself along road verges and other rights of way, which are regularly used and maintained. Today around three-quarters of all populations are found along paths, road verges, and rail tracks. Formerly known from around 60 localities in eight states, the species has declined dramatically to about 20 locations, in North Carolina, South Carolina, Georgia, and Virginia. A number have fewer than ten plants each, and *Echinacea laevigata* is on the endangered species list in the USA. It is not as well established in cultivation as many other species, but its large, light pink to purplish flowers borne on stems 1.5m (5ft) high, would look very attractive in gardens.

Rarer still is the Tennessee coneflower, ***Echinacea tennesseensis*** (Endangered*), a species that is known in only about five localities in north-central Tennessee state, where a recent survey located populations ranging between 3,700 and 89,000 plants. Here it occurs in open red cedar forest on stony, shallow soils. Like *Echinacea laevigata*, its open glade habitat has been badly affected by changes in land use. The problem is exacerbated by the plant's poor seed dispersal mechanism and its need for bare ground for germination. Hopefully implementation of a recovery plan, originally drafted in 1983 when the species was placed on the endangered species list in the USA, will ensure a long-term future. Collection for medicinal and horticultural purposes has been identified as a potential threat, but as the species is becoming increasingly available to gardeners, there should be absolutely no reason for this plant to be collected from the wild. One explanation may be that it has been mistaken for a similar species.

▲ ECHINACEA PARADOXA *The paradox referred to in the name is the colour. This coneflower is a yellow-flowered species in an otherwise pink- and purple-flowered genus.*

The rich yellow flowers of ***Echinacea paradoxa*** (Rare*) were once a common feature of the prairie grasslands of Arkansas, Missouri, and Oklahoma, but the majority of these ancient grasslands have since been ploughed up and the yellow coneflower is now very scarce.

Coneflowers are fully hardy perennials, easily cultivated in a deep, fertile well-drained soil in a sunny site. They are easily propagated by seed sown in gentle heat in spring, or can be increased by division in autumn or spring, or by root cuttings in late autumn to early winter.

ANCIENT PRAIRIE GRASSLANDS

OLD-STYLE FARMING, NORTH CAROLINA

WHOLESALE PLOUGHING

THREAT

CHANGING LAND USE

Echinacea is typical of the tall-grass prairie of North America, one of the most threatened ecosystems in the USA. The greatest threats to the survival of the genus still come from destruction and change to this habitat. Following European settlement, much of the prairie was converted to corn and wheat farming and livestock grazing, and wholesale ploughing of ancient grasslands for agriculture continues. The absence of natural fires and the effective extinction of grazing bison in the remaining areas also has a damaging effect, allowing tree seedlings to take root where they would once have been burned or grazed away. With no natural restriction on tree growth, forests will eventually form and encroach on the prairies.

COMMERCIAL CROP *Echinacea is the medicinal herb in highest demand in the USA, and is increasing in popularity in Europe. The trade is worth at least $80 million annually.*

26 on Red List*

Echium

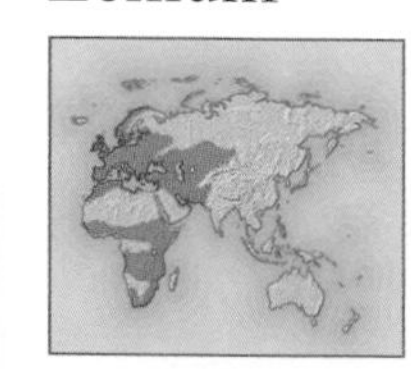

Botanical family
Boraginaceae

Distribution
Western Asia, Europe, and Africa

Hardiness
Frost-tender – fully hardy

Preferred conditions
Require well-drained soils and prefer long summers; sunny position needed by some, others prefer degree of shade from hot sun.

There are about 40 species in the genus *Echium*, spread from western Asia across Europe and Africa, and over 20 are in the Canary Islands. Of these species, most are found nowhere else, and they suffer from the pressures common to island plants, including grazing, habitat loss from development, and possibly horticultural collection.

Echium pininana (Vulnerable in the 1997 Red List) is one of the most spectacular Canary Island species. It produces a single, large rosette of rough-bristled leaves that develops for two or more years before it flowers, sheds abundant seed, and then dies. Flowers are borne in profusion along a single erect, spire-like flowerhead that can reach 3.5m (11ft) under good growing conditions.

The species is known from just a handful of localities in the laurel forests that line the north-eastern coastline of the island of La Palma. Here it occurs in forest clearings and on stony basalt hillsides at 600–1,000m (2,000–3,300ft), within the cloud zone. Surveys in the 1970s highlighted the plight of the species, locating just three populations, probably totalling just a few hundred plants, and it is Endangered in the 1997 Red List. The greatest hope for *Echium pininana* lies in the sustainable management of its laurel forest haunts, unique to the Canary Islands.

BOTANY

BUTTERFLIES

Echiums are often recommended for butterfly gardens. The tall flowering spikes appear in late summer, and are particularly attractive to butterflies. A source of abundant nectar, the flowers range in shade from pink to purple – colours that seem to appeal to butterflies.

The species can survive in a dormant state in extensive soil seed banks, and this seed could be released if appropriate management measures are implemented.

Echium wildpretii (Rare in the 1997 Red List) grows in the sub-alpine regions of Tenerife and La Palma, and is another short-lived perennial with a 60cm (2ft) tall basal tuft of silky haired leaves. Eventually a narrowly pyramidal 2.2m (7ft) spike covered with numerous red flowers arises from this rosette, and the plant dies after setting seed. *Echium wildpretii* grows within a national park, so its future is reasonably secure.

Echium giganteum (Vulnerable in the 1997 Red List) is a shrub within an essentially herbaceous genus; more woody echiums have evolved on the isolated Canary Islands than in mainland habitats. Although at least one population is protected in the Reserva Natural Integral de Pinoleris, the species remains threatened in the wild.

ECHIUMS IN CULTIVATION

The flowering spires of many echiums have earned them the name "tower of jewels" in some places, and several species are cultivated in gardens. Early records indicate that *Echium pininana* was introduced into gardens in Antibes, France, and it is now often found throughout warm temperate parts of the globe. It is a striking species for a moderately fertile, well-drained position in full sun, although it will only tolerate a few degrees of frost, and should be given some protection in frost-prone areas. Propagation is by seed sown in gentle heat in spring or summer. Ironically, *Echium pininana* is very common in the southern UK, France, and in New Zealand, where it was introduced and has since naturalized. In some cases it may be invasive. Investigations by the Royal Botanic Gardens, Kew, into the genetic diversity of such populations have been carried out in the hope that they might provide plants that can be used to restock natural habitats. However, naturalized plants were found to have limited diversity when compared with native plants. Additionally, plants in botanic gardens showed evidence of hybridization. So, the value of such naturalized and botanic garden plants for species restoration remains questionable.

▲ ECHIUM PININANA *can remain in flower for up to three months. Statuesque in habit, and easy to grow, this species has long been a favourite with gardeners.*

Echium wildpretii is not always easy to grow in the open garden, as it needs very good drainage and a sheltered, sunny position. It is only marginally frost-hardy. For best results, it may need to be grown in a frost-free, well-ventilated greenhouse, with protection from winter wet and low temperatures. It requires plentiful water during the growing season, but high levels of atmospheric humidity will rot the leaves. It is easily propagated by seed. Hybrids between *Echium wildpretii* and *Echium pininana* have proved more amenable to cultivation.

See also Invasive Plants, *p.453.*

◄ ECHIUM GIGANTEUM *This is the largest of the woody species found among the island echiums. A native of northern Tenerife, it forms a large shrub with many domed clusters of white flowers in early summer.*

ECHIUM WILDPRETII *This unusual species is perhaps the strangest-looking of the Canary Island echiums. A whole mountainside in full bloom is truly an unforgettable sight.*

Red List: Vulnerable*

Epimedium perralderianum

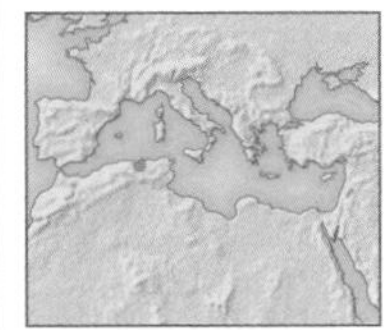

Botanical family
Berberidaceae

Distribution
North Africa, in east Kabylia, Algeria

Hardiness
Fully hardy

Preferred conditions
Best in fertile, moist but well-drained soil; prefers cool, lightly shaded woodland conditions; tolerates sun if soil is reliably moist.

The Algerian species ***Epimedium perralderianum*** currently grows over 3,000km (1,900 miles) away from its nearest relative, the Caucasian species *Epimedium pinnatum* (Unlisted). Remarkably, even though these two species are separated by vast tracts of desert, they are similar in appearance and both grow in woodlands that are dominated by closely related species of cedar, oak, and fir.

Analysis of fossil pollens indicates that these woodlands are very similar to woods that occupied the same sites around 30 million years ago. At that time, this type of woodland stretched continuously from western Asia to south-eastern Europe. Today, however, the woodland and its associated *Epimedium* species are found only in a tiny portion of Algeria, and parts of the Caucasus and Himalaya. This change in distribution resulted from a steady climatic shift during the Pleistocene glaciation, which began roughly 1.5 million years ago. Europe's climate became colder and drier, and this type of woodland began to die out, until the forests became restricted to the western portion of their original range. As Europe's climate changed, North Africa's, by contrast, was becoming wetter and more suitable for woodland development, and was consequently colonized. For a while, new woodland composed of cedar, oak, and fir species prospered there.

PEOPLE

Dr Ernest Cosson

French botanist Dr Ernest Saint-Charles Cosson (1819–89) was born and died in Paris, where his extensive private herbarium is still kept. He worked principally on the flora of areas surrounding Paris, and also on that of North Africa, publishing *Atlas de la flore des environs de Paris* in 1882, and *Compendium Florae Atlanticae*, which is a flora of Algeria, Tunisia, and Morocco, in 1887.

Following the shrinking of Europe's glaciers, North Africa again grew hotter and drier, causing a second woodland retreat, this time to the relatively cool, wet mountains of north-eastern Algeria.

The distribution of epimediums today reflects the dual effects of ancient woodland shrinkage and migration. As a consequence, the limited distribution of *Epimedium perralderianum* in the Babor Massif of Algeria leaves it vulnerable to extinction from habitat degradation.

All cultivated plants of the species come from a single plant collected by French botanist Dr Ernest Cosson at the time of the species' discovery in 1861. This plant was brought to a botanic garden in Paris and distributed to other gardens from there. It is now available from several commercial sources. As this cultivated material represents a single clone, it needs a pollinating partner.

Like most epimediums, this species is easily cultivated in a cool, semi-shaded position and a moist, humus-rich, but well-drained soil. As late frosts may sometimes damage emerging leaves, a sheltered position is preferable. In the autumn, plants should be mulched with leafmould or fine bark chips. Propagation is by division of the creeping rhizome.

▲ EPIMEDIUM PERRALDERIANUM *This evergreen perennial forms a gently spreading clump of glossy leaves that are bronze when they first emerge. Small, spurred, bright yellow flowers appear from mid- to late spring.*

Red List: Rare*

Eryngium alpinum

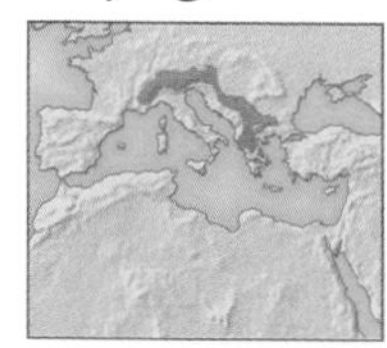

Botanical family
Apiaceae

Distribution
Alps from France to Slovenia, and eastwards to Romania and Serbia

Hardiness
Fully hardy

Preferred conditions
Thrives in open, sunny sub-alpine sites; prefers moderately fertile to poor soils, moist but free-draining; dislikes winter wet; hardy to at least –15°C (5°F).

Flowerheads of an intense metallic-blue make alpine sea holly a striking plant. Like most sea hollies, this species favours open habitats devoid of trees, shrubs, and tall herbaceous plants. The plant is considered an emblem of the Alps, where it commonly grows in areas kept clear of woodland development by traditional hay farming or frequent avalanches. Today, it is threatened by changes in the way that these habitats are managed. A move away from late hay harvests and the extension of early grazing prevent many populations from flowering and setting seed. Commercial picking for the dried flower market is also having a negative effect. Although *Eryngium alpinum* is now protected by law throughout Europe, its populations are slowly declining in size and number.

This rosette-forming perennial has long been in cultivation. Despite its sub-alpine origins and preference for open habitats, it is robust enough in gardens to be grown alongside other tall perennials. It should be given a well-drained soil that does not become excessively dry in summer and is of low to average fertility. The plant is propagated by seed, division, or root cuttings.

▲ **ERYNGIUM ALPINUM** *Flowers appear from midsummer to early autumn. The steely-blue sheen develops most strongly if the plant is positioned in full sun.*

Unlisted

Gentiana lutea

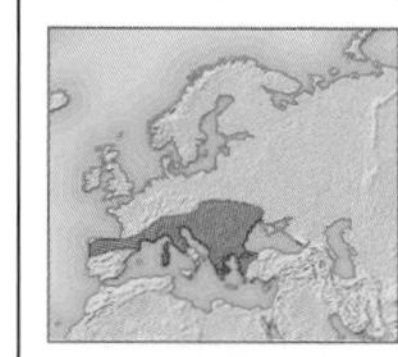

Botanical family
Gentianaceae

Distribution
Southern Europe and western Asia

Hardiness
Fully hardy

Preferred conditions
Mountain locations that are sunny but not too hot; grows best in reliably moist conditions and deep, humus-rich soil.

With stems to at least 90cm (3ft) in height, bold, pleated foliage, and deeply divided, yellow, star-like flowers, *Gentiana lutea* is one of the most distinctive of all gentians.

The yellow gentian is a herbaceous perennial found wild in meadow habitats in the mountainous regions from the Pyrenees through central Europe to western Turkey. Three subspecies are known: whilst *Gentiana lutea* subsp. *lutea* (Unlisted) is found throughout the range of the species, *Gentiana lutea* subsp. *symphyandra* (Unlisted) only occurs in the eastern half of the range, and *Gentiana lutea* subsp. *aurantiaca* (Unlisted) is endemic to a small area of Spain.

Commercial uses

Gentiana lutea is an important plant in herbal medicine and the bitter root has also been widely used as a flavouring. Herbalists have long recommended gentian root as a remedy for digestive disorders, and the plant is still used in herbal preparations. Approximately 300 products currently available in Germany contain derivatives from this species. Extracts are also traditionally included as an aromatic flavouring in alcoholic drinks, most notably gentian bitters, or *enzian schnapps*, and were used to impart a dry flavour to beer before the widespread introduction of hops.

Whilst yellow gentian is cultivated commercially in a few small areas, the majority of the extracts are derived from the roots of plants collected from the wild (*see box*). Sources of plants include Albania, France, Germany, Romania, Spain, and Turkey. *Gentiana lutea* is listed as Vulnerable or Endangered in Albania, Bulgaria, Germany, Portugal, Romania, and Turkey. As a result, the species receives protection through listing on both the European Union Habitats Directive and the Council of Europe Bern Convention. Unfortunately, and despite such formal measures, illegal collection of wild plants is still widespread.

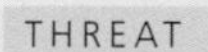

THREAT

Medicinal use

All the commercially available products from the yellow gentian are derived from its roots. Unfortunately, because the plant is a slow-growing perennial, typically with a single long taproot, the collection of roots poses a serious threat to its survival across much of its range. Additionally, the roots are often collected before the plant flowers (when active constituents are said to be at their peak), so the plants cannot set seed. Annual demand is high, and currently estimated at 6,000 tonnes (5,905 tons) of fresh root.

Cultivating gentians

Despite its architectural habit, the yellow gentian is relatively rare in horticulture, with gardeners favouring the smaller-growing, brilliant blue-flowered species of the genus. A tall, fully hardy herbaceous perennial, it makes a stately addition to a border with well-drained, moderately fertile, and humus-rich soil. Site it in sun where summers are cool, or in semi-shade where they are long and hot. Yellow gentian resents disturbance once it is established, so the plant should be raised from seed sown when ripe in a cold frame. Strong plants will produce abundant seed.

GENTIANA LUTEA *is characteristic of pastures in many parts of France and Switzerland. The plant remains untouched by livestock on account of the bitterness of its leaves.*

32 on Red List*

Geranium

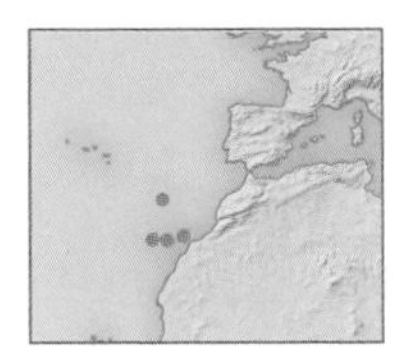

Botanical family
Geraniaceae

Distribution
Canary Islands and Madeira

Hardiness
Half-hardy

Preferred conditions
Moderately fertile soil in sun or shade; will not tolerate very wet conditions, especially waterlogging; species from Canary Islands and Madeira are hardy to –5°C (23°F).

The large and varied genus *Geranium* is found in all but very wet habitats in temperate regions. Two remarkable geraniums that are under threat are ***Geranium reuteri***, also called *Geranium canariense* (Rare★) from the pine and laurel forests of the Canary Islands, and ***Geranium maderense*** (Endangered★) from the laurel forests of Madeira. Both geraniums have suffered population losses through overcollection from the wild, and also through the destruction of their forested native habitats. The species share several features. Both produce beautiful rosettes of large, glossy, aromatic leaves that sit atop a thick stem. In *Geranium reuteri*, this stem can reach 30cm (12in), while in well-grown specimens of *Geranium maderense* it may be 60cm (24in) tall. The Madeiran species is the largest of all the geraniums, and has evolved curious persistent leaf stalks that bend backwards after the blade has fallen off, and in that position prop up the stem. Numerous pink flowers are produced from the rosettes of both species in spring.

GERANIUM MADERENSE
This beautiful but short-lived, evergreen perennial species is best treated as a biennial in the garden.

IN THE GARDEN

Both these species perform best when grown in a humus-rich, gritty soil. As neither is reliably hardy below –5°C (23°F), they are best grown in a cool greenhouse or conservatory in colder regions. In a warm, sheltered microclimate, however, *Geranium reuteri* can be grown outside. This species is also slightly longer lived, and will often flower a second or third year, while *Geranium maderense* usually dies after flowering and setting seed. Both species are easily propagated from seed. Young plants require frequent repotting lest they outgrow their root run.

GERANIUM REUTERI
Easier to cultivate than Geranium maderense, *and also hardier, this species may survive the winter outside, providing it is given a very sheltered position.*

Red List: Endangered*

Gunnera hamiltonii

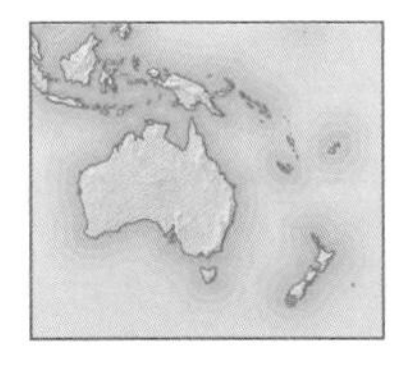

Botanical family
Gunneraceae

Distribution
New Zealand, in Southland and on Stewart Island

Hardiness
Half-hardy

Preferred conditions
Grows best in soil that is deep but not too rich, and permanently moist; prefers position in partial shade and thrives in coastal conditions; hardy to –5°C (23°F).

Diminutive *Gunnera hamiltonii* has the unenviable distinction of being New Zealand's rarest plant. It grows in only two places: near Invercargill on South Island and at Mason Bay on Stewart Island.

Unlike the more familiar large-leafed South American gunneras, this species has tiny, scalloped, bronze-green leaves in rosettes no more than 10cm (4in) in diameter. The plant is adapted to survival in damp, coastal dune slacks, its leaves forming tight-knit swathes that hug the ground and help to stabilize the shifting sands. A drawback associated with its dune habitat is that parts of the plants can be washed away by rough seas. Additionally, the small populations and limited distribution of plants make them vulnerable to threats such as habitat degradation and competition from non-native species. The dune slacks are also fragile environments, which are often difficult to restore when they have been disturbed, for example by farming. Plant populations in both New Zealand sites are plagued by invasive plant species that thrive following human disturbance of the habitat (*see box*).

When conservation work began, an amazing discovery was made. The mainland population was found to consist entirely of female plants, while the Stewart Island plants were all male. Plants from both populations presumably reproduced vegetatively, as no seed set has ever been recorded. The gunnera can cover large areas of sand dune, but each area is effectively a single plant. Plants from both areas have now been brought into cultivation in New Zealand and seed has been produced. As a result, the future of the species seems less precarious, but is still of great concern.

Much less frequently grown than the gigantic *Gunnera manicata* (Unlisted) and *Gunnera tinctoria* (Unlisted), this is a dramatically different species for the garden. It prefers a semi-shaded site in a moist, peaty soil and can be propagated by division in late spring or summer.

THREAT

COMPETING PLANTS

Non-native species frequently pose a threat to the native flora of New Zealand. On Stewart Island, marram grass (*Ammophila breviligulata*) was planted in an attempt to farm the area around Mason's Bay. Now the grass needs constant control to keep it in check. Monterey cypress (*Cupressus macrocarpa*), too, can shade out low-growing plants, and vigorous tree lupins (*Lupinus arboreus, see left*) compete for space with the gunnera and other dune species. The challenge for conservationists is to safeguard vulnerable *Gunnera* populations from threats by other non-native plants.

Red List: Endangered*

Helianthus nuttallii subsp. parishii

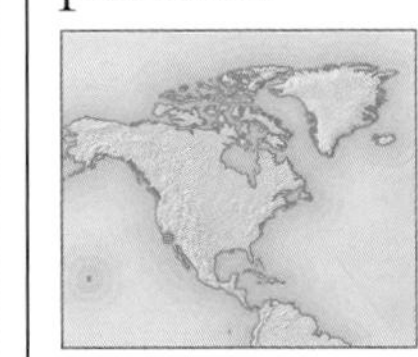

Botanical family
Asteraceae

Distribution
Santa Clara River, Los Angeles County, California, USA

Hardiness
Half-hardy

Preferred conditions
Moist to wet soils in sun; marginally frost-hardy: tolerates below 0°C (32°F) only for short periods.

The Los Angeles sunflower, *Helianthus nuttalii* subsp. *parishii*, is a 3–4m (10–12ft) tall marsh sunflower that once grew throughout the wetlands of Los Angeles County, California. When the marshlands were drained and burned in the 1930s, it was presumed extinct. Then, in 2002, a single population was found in a tiny remnant of privately owned wetland, and aroused much interest among conservationists. Sadly, the land is now threatened by development.

Currently rare in cultivation, this plant is easily grown in a moist position in full sun, and can be propagated by dividing the tuberous roots. The closely related *Helianthus nuttalii* subsp. *nuttalii,* which grows in western North America, is well established in cultivation.

3 on Red List*

Helleborus

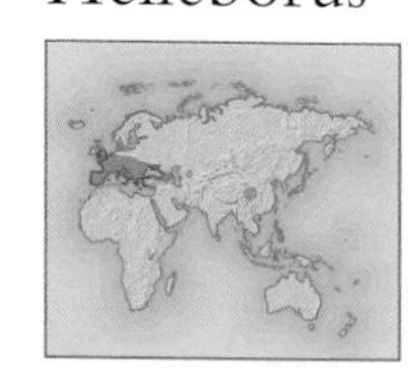

Botanical family
Ranunculaceae

Distribution
Europe to the Caucasus mountains; Himalaya

Hardiness
Frost-tender – fully hardy

Preferred conditions
Semi-shade and moist, humus-rich, neutral to alkaline soil; some prefer acid soil; almost all are hardy to –15°C (5°F).

Hellebores are extremely popular in gardens, and in recent years much has been done to increase the range of form and colour in these prized winter-flowering plants. The species are all restricted to Europe and the Caucasus mountains, with the single exception of the isolated ***Helleborus thibetanus*** (Unlisted), which is a native plant of western China.

Research suggests that hellebores were once widely distributed from Europe to east Asia, but major shifts in climatic conditions and the formation of the Himalaya cut off what is now identified as *Helleborus thibetanus* about 23 million years ago, and left it growing 5,200km (3,231 miles) from the rest of the genus.

◄ HELLEBORUS THIBETANUS *This alpine species has nodding flowers that vary in colour from pink to white and occasionally creamy yellow.*

It is native to north-eastern Xizang (Tibet) and western Sichuan in China, where it inhabits shady woodland banks and glades, flowering in the late winter. Employed as a remedy in traditional Chinese medicine, the plant was much sought after by gardeners but evaded collection for many years, and came into cultivation in the West only late in the 20th century. Unfortunately, its discovery led to many plants being taken from the wild. Some colonies were devastated by local collectors, and plants were traded in both Europe and Japan. Today *Helleborus thibetanus* needs close protection in the wild. It is now reasonably well established in gardens, however, and can be grown from cultivated seed. Although the resulting plants are very slow to reach flowering size, they take the pressure off the vulnerable wild populations. In the garden this species makes a handsome, clump-forming perennial, with divided, bluish-green, coarsely toothed leaves. The nodding flowers are borne on leafy stems. Unlike most other hellebores, the plants are summer-dormant, and die down completely by midsummer.

Helleborus vesicarius (Rare★) also dies down to squat rhizomes in the summer, and produces inflated fruit capsules. It has greenish flowers and delicate-looking leaves, and is found only in a small area of south-eastern Turkey. This species has only recently become established in cultivation, and is threatened by horticultural collection.

▲ HELLEBORUS VERSICARIUS *This beautiful hellebore, with its unusually delicate-looking leaves, is best grown in a bulb frame in most climates.*

Helleborus lividus (Rare★) has the smallest range of all hellebores, being endemic to the island of Majorca and possibly nearby Cabrera. Even here it is now very rare. New building and agriculture have destroyed many of its habitats, and today it is confined to the mountains along the north-east coast in relatively moist sites. It thrives along rivers and grows at the base of rocks, under which its roots find cool, moist earth. Conservationists hope that the rugged terrain and isolation of this mountainous area may protect current plant populations.

HELLEBORUS ORIENTALIS *An elegant garden plant with striking flowers and leathery leaves, this species is fairly easy to grow. It does, however, readily hybridize with several other species.*

THREAT

TRADITIONAL MEDICINE

Helleborus species have long been used in herbal and homeopathic remedies around the world, and *Helleborus thibetanus* is one of the remedies used in traditional Chinese medicine. The active ingredients, which are found in the roots, have been employed by practitioners to treat specific heart conditions. With the continued rise in popularity of complementary medicine, collection of hellebores from the wild has increased.

Helleborus orientalis (Rare*), although distributed through much of the Black Sea region, is threatened by overcollection. It has played an important role in the creation of popular hybrid cultivars, but is rare in cultivation. Most plants that are sold as *Helleborus orientalis* are actually of hybrid origin.

Cultivars of *Helleborus orientalis* are easily grown in a semi-shaded position and well drained, humus-rich, slightly alkaline soil. Propagate by seed or divide in early autumn. *Helleborus lividus* and *Helleborus vesicarius* are not hardy, and the latter needs a warm, dry resting period. Seed is the most reliable means of propagation. When buying hellebores, ensure they are from cultivated stock.

▲ HELLEBORUS LIVIDUS *Unlike most hellebores, this species is not hardy. Keeping it in a frost-free greenhouse during winter flowering may prevent hybridization.*

Red List: Endangered*

Hosta hypoleuca

Botanical family
Hostaceae

Distribution
Japan along the Tenryu River in Aichi and Shizuoka prefectures

Hardiness
Fully hardy

Preferred conditions
Tolerates sun or semi-shade; prefers a fertile, moist but well-drained soil and a position sheltered from cold, drying winds; hardy to –20°C (–4°F).

In the wild this unusual hosta grows on sunny volcanic cliffs. In response to this extreme environment and to protect themselves from the intense solar radiation reflected up from the rocks, the leaves have evolved reflective white undersides. The plant is commonly called the white-backed hosta, and its botanical name is also apt, from the Greek *hypo* for beneath and *leucon* for white.

Hosta hypoleuca is one of the most attractive of the wild hostas. The leaves are large, and in summer it bears bell-shaped white flowers that are suffused with purple. Because of its very limited distribution in the wild it is considered endangered in Japan.

This hosta is not difficult to grow if it is provided with a fertile, moist but well-drained, neutral to slightly acidic soil. It is fully hardy and is one of the few hostas that will tolerate a position in full sun, and is ideally cultivated on a rock wall or slope so that the beautiful white undersurfaces of its leaves can be appreciated from below. Propagation is by division in early spring.

▲ HOSTA HYPOLEUCA *Although it is rare in cultivation, this species of hosta is the parent plant of several more widely grown hybrid cultivars.*

Unlisted

Illecebrum verticillatum

Botanical family
Caryophyllaceae

Distribution
Europe from Spain to Poland; scattered around Mediterranean

Hardiness
Frost-tender

Preferred conditions
Thrives in open, sandy but fertile soils, and in acidic conditions; needs reliable levels of moisture, and will not tolerate complete drought.

Coral necklace, the common name of this trailing plant, is characteristic of shallow-water pools and hollows in western Europe on neutral to acid soils, where livestock are led to water. It is an annual species that favours pools with fluctuating water levels, germinating in spring as water levels drop, growing and flowering rapidly during the warmer summer months, and dying as water levels rise in autumn and winter. The plant's stems hug the ground and this trailing habit, together with the open conditions its seed requires for germination, leaves it unable to withstand competition from more vigorous plants. Sadly, with a decline in rough grazing and the fencing-off of ponds, the species has vanished from many of its old haunts, and it is rare and declining in most of the countries where it occurs naturally. Conversely, the plant has the ability to reappear from buried, dormant seed after a long absence, and can colonize rapidly under suitable conditions. Populations are often found along the rutted tracks, which fill with water in winter, commonly found in managed forests.

◄ ILLECEBRUM VERTICILLATUM *With its combination of trailing coral-red stems, paired sage-green leaves, and an abundance of tiny, chalk-white flowers, the delicate coral necklace is best appreciated at close quarters.*

This delicate annual has potential as an ornamental plant, and at least one major supplier in the UK has offered seed in recent years. In cultivation the plant has done well, often growing considerably larger than its wild siblings. The colour is strongest in full sun, but the plant should not be allowed to dry out. Sow seed *in situ* in spring, once frosts have passed. Plants will self-seed in favourable conditions, or the copious seed can be collected in late summer and autumn for sowing the following year. In gardens the plant may flower until the first frosts.

Red List: Vulnerable*

Iris xanthospuria

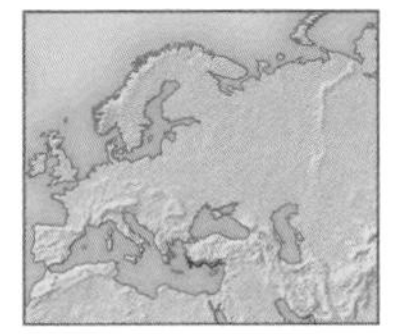

Botanical family
Iridaceae

Distribution
Endemic to southern Turkey

Hardiness
Fully hardy

Preferred conditions
Best in rich, wet soil, but will tolerate well-drained yet moisture retentive soils; position in full sun; hardy to –15°C (5°F).

One of the most localized and arguably the most threatened of all the Turkish irises, *Iris xanthospuria* grows on freshwater coastal marshlands, and in glades in seasonally flooded forests. Limited numbers are also found along ditches in more intensively farmed land. The species is currently known from fewer than 10 localities in the deltas and coastal marshes of southern Turkey, particularly in the south-western corner of the country. It is often closely associated with coastal stands of Oriental sweet gum (*Liquidambar orientalis*), but, like this tree, it has undergone a massive decline over the past 50 years, as sites have been drained, ploughed, and converted to citrus groves. Around three-quarters of the alluvial forests and probably an equivalent extent of marsh habitat have been destroyed. Ironically, efforts to restore gum stands through a programme of tree-planting may be destroying the open habitat preferred by the iris.

This species is rare in cultivation, but worth growing for its long succession of rich golden flowers in early summer. Confined to wet habitats in the wild, it will grow in rich, moist, but well-drained soil in the garden. Sow seed when ripe in a cold frame, or divide rhizomes after flowering.

◄ IRIS XANTHOSPURIA *Once popular in table arrangements in restaurants and hotels in south-western Turkey, this species was not formally described until 1982.*

Red List: Rare*

Isoplexis canariensis

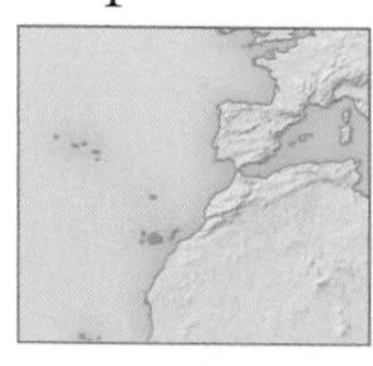

Botanical family
Scrophulariaceae

Distribution
Canary Islands: Tenerife

Hardiness
Half-hardy

Preferred conditions
Grows best in fertile, moist but well-drained soils; prefers warm, sheltered locations in semi-shade; hardy to -4°C (25°F), but prefers frost-free conditions.

The genus *Isoplexis* consists of four species. The name comes from the Greek *isos* meaning equal and *plexis* meaning segment, and refers to the length of the flowers' upper and lower lips, which is the same. It is this characteristic that separates the members of this genus from those of the otherwise quite similar genus *Digitalis,* the familiar foxgloves, to which *Isoplexis* is closely related.

In the wild, all members of the genus are confined to the Canary Islands, the Azores, Cape Verde, and Madeira. Even here they are rare, growing in evergreen forests dominated by trees of the laurel family.

These forests represent a subtropical vegetation type that was widespread throughout Mediterranean Europe and North Africa during the Tertiary period 40–15 million years ago, but is now almost extinct. Long-term climate changes resulting from the southern expansion of the polar ice cap at the end of the Tertiary period are thought to have been responsible for the shrinkage of these forests and their associated species, until they became restricted to the Atlantic islands that are collectively called Macaronesia. These islands were buffered against the climatic change by their oceanic position, and as a consequence maintained much of their subtropical flora. Many of the species native to these islands are found nowhere else in the world. With the arrival of humans on the islands, much of the ancient forest was destroyed (*see box*). Today, little of the original forest remains, and much of what does survive is under considerable pressure.

Isoplexis canariensis, called *cresto de gallo* (cock's comb) on the islands, is the only one of the four *Isoplexis* species that is widely cultivated beyond Macronesia. This species is now extinct on two of the three islands where it was historically recorded, and now only grows wild on Tenerife where forest clearance has been less severe. In these ancient forests it is locally frequent but nevertheless a species of concern, given its restricted global distribution. It is now reliably available from cultivated stocks.

In cultivation all four species require a heavy compost containing a dense, moist loam, with added grit to provide aeration. Plants should be sited in an open but sheltered position, with direct sun for no more than a couple of hours a day. Balanced liquid feeds should be given monthly throughout the growing season to promote healthy leaf growth. *Isoplexis* is tolerant of cold periods, but if grown in frost-prone areas it is best to over-winter plants in a conservatory or frost-free greenhouse and provide a night temperature of 5–8°C (41–46°F). Propagation is usually by seed. This should germinate within 14 days and is best sown in spring. Seedlings should be potted on as soon as they are large enough to handle, and plants may even flower the first year.

ISOPLEXIS CANARIENSIS ► *This bushy evergreen grows up to 1.2m (4ft). It bears beautifully showy flowers over a long period in summer, and can bloom from mid-spring into autumn if grown in a conservatory.*

HISTORY

ISLANDS TRANSFORMED

Until the 15th century these Atlantic islands were uninhabited and pristine. The first settlers brought livestock and probably used fire to clear large areas of the dense evergreen forest that covered much of the islands. Agriculture was slowly established and grapes, oranges, pineapples, tobacco, and tea were all in cultivation. Now, farmers rear cattle for milk and meat, and plantations of the non-native conifer, *Cryptomeria*, provide shelter for the animals. Although emigration has resulted in a limited decline in farming, the growth in tourism has imposed new pressures on the environment.

Red List: Rare*

Kirengeshoma palmata

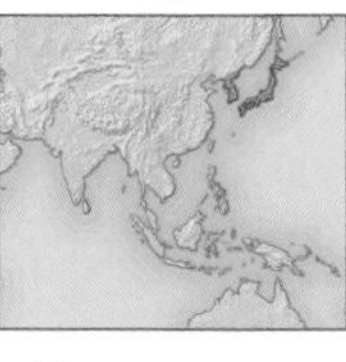

Botanical family
Hydrangeaceae

Distribution
Japan and possibly Korea

Hardiness
Fully hardy

Preferred conditions
Thrives in sheltered and partially shaded sites, and needs neutral to acidic soil, ideally enriched with leafmould. Hardy to at least -15°C (5°F).

The genus *Kirengeshoma* is usually held to consist of two species: ***Kirengeshoma palmata*** from Japan with stems to 2m (6ft) and flowers that barely open, and *Kirengeshoma koreana* (Unlisted) from Korea with stiffer stems to 2.2m (7ft) and more open flowers. However, some botanists recognize just one variable species, *Kirengeshoma palmata*, found in both Japan and Korea. Clearly this is an important distinction, as the genus may contain either one species that is rare or two that are even rarer. Further botanical research is needed to determine the correct nomenclature of the Korean and Japanese populations and their conservation status and priorities.

Kirengeshoma palmata is well established in horticulture, and is easily cultivated in a moist, humus-rich, well-drained soil in a semi-shaded position. This species looks particularly striking in a woodland setting where its tall, elegant stems, palmately lobed leaves, and tubular yellow flowers can be set off against a background of darker green foliage. The flowers are produced in late summer and early autumn. *Kirengeshoma palmata* can be propagated from seed, sown fresh in containers in a cold frame in autumn or spring, although germination can be erratic. Stratifying the seed (*see p.488*) may improve germination rates. Propagate by division of the rhizomes in spring or autumn.

▲ KIRENGESHOMA PALMATA *This elegant, clump-forming perennial is a good choice for a woodland garden or for a shady border with shelter from wind.*

20*/1 species in Red List

Kniphofia

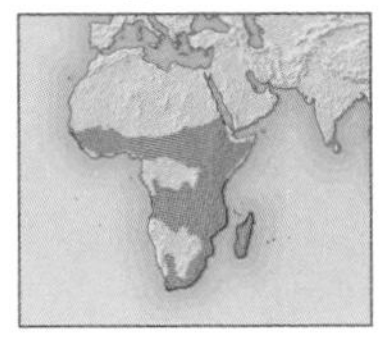

Botanical family
Asphodelaceae

Distribution
South and east Africa, Madagascar, Yemen

Hardiness
Frost-tender – fully hardy

Preferred conditions
Deep, fertile, moist but well-drained soils; site in full sun or partial shade and many tolerate coastal conditions; hardiness varies according to species.

Named in honour of Johannes Hieronymus Kniphof (1704–63), who was Professor of Medicine at Erfurt University in Germany, the 65 species of *Kniphofia* are mostly found in southern and eastern Africa, with one species on Madagascar and another in Yemen. They grow in mountainous areas, damp places, marshes, rough grassland, or rocky places, and several species are threatened.

Kniphofia citrina (Indeterminate on the 1997 Red List) and ***Kniphofia sarmentosa*** (Rare*) are two of the better known, having long been cultivated in European gardens. Both grow wild in the Cape provinces of South Africa, and

▲ KNIPHOFIA CITRINA *Flowering in late summer, this species has pure light yellow blooms opening from green buds, and is relatively small, growing to just 60cm (24in).*

both produce the typical inflorescences from which the genus gets its vernacular name of red hot poker or torch lily. *Kniphofia citrina* grows in dense grass from sea level to 600m (2,000ft) in altitude, and *Kniphofia sarmentosa* by mountain streams. Little is known about the biology of either species, and both are of concern to conservationists due to their limited natural distribution.

Most red hot pokers are easily grown in a fertile, moist but well-drained soil that does not dry out in summer. Full sun suits most species. Propagate by seed or by division in spring or summer.

KNIPHOFIA SARMENTOSA *This tall species has the advantage of a very long flowering season. Its flowers are typical of the genus, with the flower colour fading to yellow towards the base.*

Unlisted

Lepidium oleraceum

Botanical family
Brassicaceae

Distribution
New Zealand

Hardiness
Unknown

Preferred conditions
Tolerates a range of soil types as long as they are fertile and well-drained and will tolerate drought; thrives in open sites and in coastal conditions; likely to be frost tender.

Early in October 1769, the great botanist Sir Joseph Banks and two passengers landed at Poverty Bay, New Zealand as part of Cook's world voyage aboard the HMS *Endeavour*. The team collected almost 40 botanical specimens, of which one of the most significant – at least to the crew of the *Endeavour* – was the New Zealand cress, *Lepidium oleraceum*. Banks suspected that the plant might be rich in vitamin C, and the plant duly proved an invaluable treatment for scurvy, hence its common name, Cook's scurvy-grass.

This is one of a number of closely related cresses found in the coastal zone of New Zealand, all relatives of the unlisted garden cress (*Lepidium sativum*). The plant is a low, spreading perennial, with a stout taproot and a profusion of small white flowers. In the wild, it is confined to fertile, disturbed habitats immediately above the high water mark, typically in the highly enriched conditions associated with shoreline seabird and seal colonies.

During Cook's time the species must have been at least locally abundant, as his crew were able to collect it by the boatload. There is no doubt that it has since undergone a considerable decline in the wild, and it is now regarded as threatened by the national government. The reasons for this decline appear to be complex, but are thought to include habitat destruction, declining seabird and seal populations, browsing by livestock, and competition from more vigorous perennials. In recent years, some populations have been damaged or destroyed – at least temporarily – by severe storms, and ultimately global warming and a rise in sea level may pose the most serious threat to the species. The New Zealand Department of Conservation approved a Coastal Cresses (Nau) Recovery Plan in 1998, "to ensure that viable populations of all extant coastal cress species are restored and self-sustaining in the wild", focusing attention on this little-known plant.

Lepidium oleraceum has proved easy to grow outside its natural habitat. It is suitable for a well-drained, fertile soil in full sun, and although salt, drought, and wind tolerant, its hardiness is not established. Plants are easily raised from seed sown in spring in a cold frame.

◄ COOK'S SCURVY-GRASS *This species is rich in antioxidants, and is worth growing as a culinary herb for the hot, cress-like flavour of its crisp, fleshy leaves.*

Red List: Endangered*

Lotus berthelotii

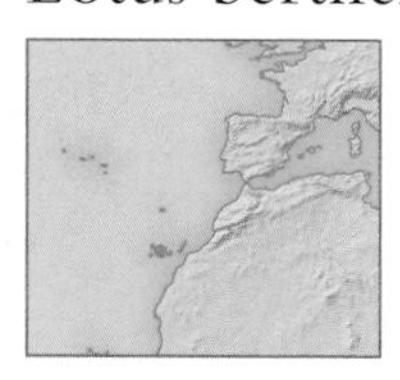

Botanical family
Papilionaceae

Distribution
Canary Islands and Cape Verde Islands

Hardiness
Half-hardy

Preferred conditions
Full sun and moderately fertile, well-drained soils; may survive occasional drops in temperature to 0°C (32°F), but grows best in frost-free locations.

Most of the 100 or so *Lotus* species have small, yellow, pea-like flowers that are pollinated by bees. However, the Cape Verde Islands and Tenerife in the Canary Islands boast the remarkable, orange-red flowered *Lotus berthelotii*.

The clusters of flowers up to 4cm (1½in) in length are distinctive not only for their large size, but also for their brilliant red to orange-red colour and their strongly recurved standard and long beak. The unusual shape and colour of the flower suggest that the plant is bird-pollinated – possibly by sunbirds – yet no such birds remain in the Canaries.

This beautiful species has an unusual sprawling or scrambling habit, with trailing stems from a woody base bearing distinctive silver-grey leaves with linear leaflets. It has been recorded on cliffs at altitudes of up to 1,000m (3,300ft), and the trailing habit is an adaptation to this position, enabling the plant to dangle from cliff faces and present its flowers to passing birds that must have once pollinated them.

Lotus berthelotii is either extinct in the wild or only exists as a few individuals on forested cliffs. As early as 1884 the species was regarded as "exceedingly rare", and collection for horticultural purposes combined with habitat degradation are thought to be the chief reasons for its decline. Even without human influence the species' survival may have been hampered by the lack of suitable pollinators. The plants are thought to be incapable of self-fertilization, so seed cannot be produced to replenish depleted populations.

Although extremely rare in nature, this is one of the most garden-worthy members of its genus, firmly established in cultivation and even reported as naturalized in North America. It is sometimes shy to flower, but the fine blue-grey foliage is attractive when cascading down the sides of a container or hanging basket. Grow in a well-drained loam in full sun and protect from frost. Propagation is by cuttings taken in spring or summer.

PICO DE PALOMA ► *The curved flowers have earned this species the common name of parrot's beak or the Spanish* pico de paloma, *meaning dove's beak.*

76*/1 species in Red List

Lupinus

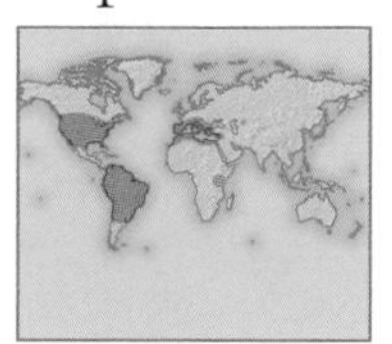

Botanical family
Papilionaceae

Distribution
Southern Europe, north Africa, North America, and South America

Hardiness
Half-hardy – fully hardy

Preferred conditions
Full sun or semi-shade; grow best in moderately fertile to poor, light, sandy soils that are ideally slightly acidic; hardiness varies according to species.

The genus *Lupinus* comprises some 200 species, found in dry grassland and open woodland, on hills, mountains, coastal sands and cliffs, and riverbanks. They have hand-shaped leaves and upright spikes or simple heads (racemes) of pea-like flowers.

The classification of the North American species of *Lupinus* is confused, and it is uncertain at present how many species are recognized. Some authorities

▲ LUPINUS CHAMISSONIS *This showy, shrubby species of lupin remains one of the best-loved flowers of the California dunes and coastline.*

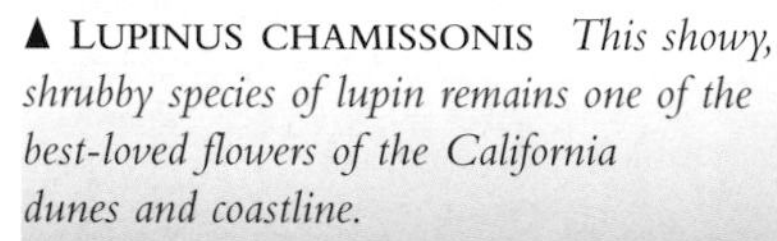

LUPINUS LEPIDUS *The long roots of this lupin penetrate deep into the free-draining volcanic soils of its typical habitat. In dry summers, the silky-haired leaves trap moisture from the frequent mountain mists.*

◄ LUPINUS LEPIDUS *VAR.* LOBBII *This low-growing alpine subspecies was the first plant to colonize the slopes of the volcano Mount St. Helens after it erupted in 1980.*

distinguish a series of closely related species, others just a few. What is certain is that the high montane lupins are very beautiful plants and an important element of the alpine flora. They are excellent nitrogen fixers (*see box*) and also highly attractive to various bees.

Lupinus lepidus (Unlisted) is perhaps the prettiest of the alpine species. It is a plant of screes, rocky moorland, and prairie, in the Rocky Mountains, the Cascade Range, and the Olympic Mountains of Washington State, with many local variants (three on on the 1997 Red List). Plants generally form a woody-based, tufted perennial with numerous leaves, grey-green above and silvery with hairs on the margin and beneath. The short spikes of scented flowers, held just clear of the foliage, are violet to pure blue, generally with a prominent white patch. *Lupinus lepidus* var. *utahensis* (syn. *Lupinus caespitosus* – Unlisted) bears stemless plumes of flowers with a narrow uppermost petal.

Lupinus lepidus* var. *lobbii (also known by the species names *fruticulosus, lyallii,* and *perditorum* – Unlisted) forms dense, prostrate mats of very silvery leaves with short, rounded plumes of violet-blue flowers. Unfortunately, *Lupinus lepidus* is difficult to maintain in cultivation for long. Plants in gardens are very prone to aphid attack, and are generally intolerant of insecticides and disturbance once established. They also dislike excess winter wet, so cultivation within the protection of an alpine house or cold frame is probably the best means of success. Even so, plants tend to die after flowering and seeding only once. Seed germinates readily.

COASTAL LUPINS

In 1817, a Russian ship laid anchor in San Francisco Bay. Among the crew were the French botanist, novelist, and poet Adalbert von Chamisso and the young ship's doctor Johann Friedrich Gustav von Eschscholtz. Chamisso named the genus *Eschscholzia* in honour of his friend the doctor, and the Californian poppy became the state flower. Ten years later, Eschscholtz was able to return the honour with a lupin.

Lupinus chamissonis (Rare*) is a handsome shrub, forming domes to approximately 1.5m (5ft) in height, and is distinctive for its silvery, blue-green leaves and spikes of smoky blue flowers. It is one of a handful of shrubby lupins found along California's coastal zone, occurring in dune scrub and maritime chaparral at the rear of sand dunes in partially unstable, sandy conditions. It is largely confined to dunes in the central Californian coastal dune zone from Los Angeles County to Marin County.

Coastal lupins suffer from habitat degradation through sand extraction and pollution from onshore drilling for oil. Encroachment of invasive, non-native plants, such as marram grass (*Ammophila arenaria*) and the ice-plant (*Carpobrotus edulis*) poses a problem, as does erosion by walkers, horses, and off-road vehicles. Wholesale habitat destruction through industrial and residential development is also a threat: in some areas (such as the northern San Francisco peninsula), as much as 90 per cent of the natural vegetation has been destroyed.

BOTANY

NITROGEN-FIXING ROOTS

Lupins are often found flourishing in infertile and parched vegetation, and were at first thought to be starving the soil of its nutrients. The genus accordingly takes its name *Lupinus* from the Latin for wolf, reflecting a rapacious nature. In reality, like other members of the pea family, lupins improve soil conditions. Bacteria in nodules on the roots convert or "fix" stable, unusable nitrogen from the air into a form that can be absorbed by plants.

See also Invasive plants, p.461.

Red List: Endangered*

Lysimachia minoricensis

Botanical family
Primulaceae

Distribution
Balearic Islands (Minorca)

Hardiness
Half-hardy

Preferred conditions
Damp but well-drained soil that remains reliably moist in summer; prefers locations in full sun, but will tolerate partial shade; hardy to -5°C (23°F).

Minorca has a rich diversity of plant life, and the whole island has been designated a Biosphere Reserve by UNESCO. For one plant this move has come too late: *Lysimachia minoricensis* has the dubious honour of being Minorca's only extinct wild plant, lost during the 20th century. Although the species was described as long ago as 1879, little is known of its existence before extinction. It is thought to have inhabited wet meadows and drain sides near Minorca's southern coastline. This low-growing biennial probably vanished with the loss of grazing animals: these would have kept larger, competing plants in check.

The species is cultivated by botanical institutes across much of Europe, and is gaining popularity among gardeners for its rich, glossy, dark green leaves with silver and grey veins. In summer it bears small cream flowers on stems to 50cm (20in) tall. *Lysimachia minoricensis* is capable of pollinating itself if cross-pollination fails, and one plant can produce over 1,000 seeds with high germination rates. Material from botanic gardens has been reintroduced to the wild, but without success. Its limited genetic diversity suggests that an extremely reduced gene pool was recovered through collection prior to extinction, and this could explain the failure. However, subtle changes in land mangement may be equally responsible.

▲ LYSIMACHIA MINORICENSIS *This is one of the few Mediterranean plants that is extinct in the wild, yet it persists in cultivation in both botanic and private gardens.*

Red List: Rare*

Mandragora officinarum

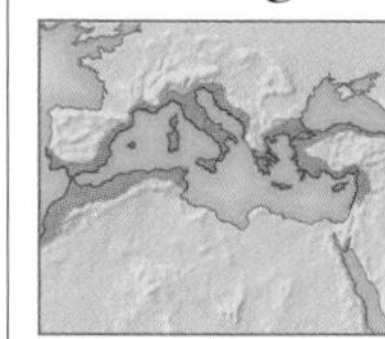

Botanical family
Solanaceae

Distribution
Mediterranean (north Africa, eastern and southern Europe)

Hardiness
Half-hardy

Preferred conditions
Taproots establish best in deep, fertile, moist but well-drained soils; prefers sunny, sheltered locations protected from cold winds and winter wet; hardy to -5°C (23°F).

Mandrake is found in vegetation that typically comprises trees and shrubs with small, hard leaves that withstand summer droughts. Its habitats include open woodland, olive groves, fallow land, and ruins.

For centuries, mandrake has been known as a "hexing herb", a reputation gained partly from its hallucinogenic properties and partly from its vaguely humanoid root. Such is the mandrake's infamy that it can be difficult to distinguish which of its traditional uses have a real basis and which are pure myth (*see box*). For example, one notable use of the plant's narcotic properties was as a pain-reliever for those condemned to die on the cross. In this context, it has been suggested that Christ was given vinegar laced with mandrake – also that this potion put him into a death-like trance and so he did not actually expire.

◄ MANDRAGORA OFFICINARUM *The rosette of wavy leaves surrounds a cluster of flowers. These are greenish white or purple, and are followed by round yellow fruits.*

Many potent alkaloids are found in mandrake roots and, given our current medical knowledge, reports that the herb was effective as a hypnotic, sedative, purgative, and narcotic seem entirely plausible. The physical effects of ingesting mandrake are profound, and include a deadening of pain and a loss of bodily sensation. The plant gained notoriety again more recently as a source of scopolamine, used as a "truth drug" in secret service investigations, and more mundanely to treat motion sickness. Mandrake root is no longer a significant source of alkaloids in western medicine.

COLLECTION AND CULTIVATION

Two millennia of herbal use have caused a great decline in the wild populations of mandrake. As a result of prolonged over-collecting throughout its range, it now has a very scattered distribution around the Mediterranean. No doubt collection of wild plants still occurs, but habitat destruction is probably the greatest single threat to the wild populations today.

Mandrake prefers a warm, sheltered position in a deep, fertile, well-drained soil and is reliably hardy to -5°C (23°F), although it dislikes cold, wet conditions in winter. The plant resents disturbance of its roots, and is often slow to re-establish when transplanted. Propagation is by seed sown in spring, but germination rates are not high. This is an interesting rather than showy plant that is rarely cultivated, although it works well in a dry, sunny border.

HUMANOID MANDRAKE ROOTS

HISTORY

MANDRAKE MYTHS

The mystique of this root goes back over 2,000 years. Pythagoras called it a tiny human, and Theophrastus records symbols and rituals associated with its harvest. It was said that the plant hid by day but shone by night; that the roots protected the bearer from sterility; that it grew only under gallows; and that its shriek when lifted was deadly, and it should only be uprooted by dogs. By the late 16th century, however, its powers were in dispute and in 1597, English herbalist Gerard wrote that he had "digged up, planted and replanted very many".

Red List: Rare*

Mentha requienii

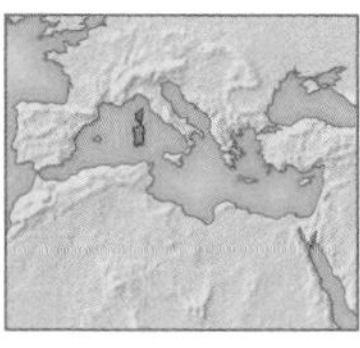

Botanical family
Lamiaceae

Distribution
Corsica, Sardinia, and Isola di Montecristo, Italy

Hardiness
Fully hardy

Preferred conditions
Shady locations – needs shade for at least part of the day; grows best in moist soils of moderate to low fertility; hardy – will recover following temperatures as low as -20°C (-4°F).

When bruised, Corsican mint emits a pleasant, pungent aroma, and the leaves have been used to flavour liqueurs. Some of the plant oils, however, are toxic, and it is not considered safe for human consumption.

This mat-forming evergreen perennial is rare in the wild, and has been overcollected in the past. Nevertheless, it is firmly established in cultivation and has even become naturalized in a few localities in western Europe.

Mentha requienii is a diminutive mint that spreads to form a low carpet of tiny, round leaves. In the summer, minute purple flowers dot the plant's surface and are beautifully set off against its fresh green leaves. Corsican mint is best planted in moist but well-drained, gritty soil and in a position in light shade or with partial sun. It tends to die back if frosted, but usually quickly returns the following spring from the few sprigs that survive the cold, or from buried seeds. Propagate by division of small rooted fragments in spring and summer.

CREEPING MINT ► *Mentha requienii looks especially attractive when it is planted between paving stones, and can even tolerate a moderate amount of trampling.*

Red List: Rare*

Myosotidium hortensia

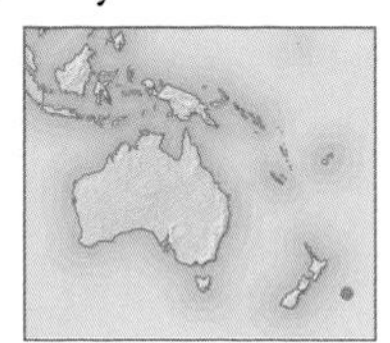

Botanical family
Boraginaceae

Distribution
Chatham Islands, off New Zealand

Hardiness
Half-hardy

Preferred conditions
Humus-rich, moist but well-drained soil; requires a position in semi-shade and protection from cold or drying winds, and thrives in coastal conditions; hardy to 0°C (32°F).

Fifty years ago, anyone visiting the Chatham Islands would almost certainly have been greeted by the sight of *Myosotidium hortensia*, one of the plants endemic to the islands (*see box*). At that time, the Chatham Island forget-me-not formed a lush band of vegetation at the top of many beaches, favouring rock outcrops, boulder beaches, and coastal cliffs. Today, it is found only in a few locations where it has been protected.

This dramatic decline was caused by a combination of factors, including grazing and trampling by pigs and sheep, encroachment by introduced marram-grass (*Ammophila arenaria*), and habitat destruction as a result of coastal development. Even though all of these factors remain problematic, the future of the Chatham Island forget-me-not looks brighter, as several wild populations are now actively managed and protected by the New Zealand Department of Conservation and are increasing. Sadly, despite encouraging signs, the only known white-flowered population was recently the victim of development.

This evergreen perennial is the only species in the genus and an unusual member of the borage family. The large, glossy leaves of *Myosotidium hortensia* resemble those of a hosta, but the flowers are more like those of a forget-me-not (*Myosotis*). *Myosotidium hortensia* is an attractive garden plant, but is rarely seen in cultivation outside the Chatham Islands, owing to the difficulty of cultivating it in most other climates. It requires mild or preferably frost-free winters and cool, damp summers, and grows well in gardens in the western parts of the UK and in North America's Pacific northwest. In suitable areas it can form clumps over 90cm (3ft) in diameter and 60cm (24in) high, with stems branching from a stout, cylindrical rootstock. It should be grown in a position that is bright but never receives full sun, in a well-drained soil to which well-rotted compost has been added. Propagation is usually by seed, as the plants resent disturbance once they are established. Seeds should be pre-chilled at 3–5°C (37–41°F) for two weeks and sown in a cool greenhouse during the spring. They will germinate in one to four months.

HABITAT

CHATHAM ISLANDS

This remote volcanic archipelago consists of two inhabited islands, Chatham and Pitt Islands, and some 40 islets and rock stacks, approximately 800km (500 miles) to the east of New Zealand. They are thought to have been formed around 80 million years ago. Like most oceanic islands, they do not experience extremes of temperature, but they are subject to strong winds and long periods of rain. As a result of the isolation, the wet conditions, and the fertile soils and range of habitat niches, the islands have developed a diverse and unusual flora adapted to maritime conditions, with 47 endemic species. The Chatham Islands were one of the last places in the Pacific to be visited and settled by humans.

▼ CHATHAM ISLAND FORGET-ME-NOT
Flowers appear in early summer, and may be pale or brilliant blue, rarely white. Separate male and female flowers are borne in dense, rounded heads on the same plant.

5 on Red List*

Nuphar

Botanical family
Nymphaceae

Distribution
Temperate and subtropical Europe, Asia, and North America

Hardiness
Half-hardy – fully hardy

Preferred conditions
Grows best in sunny position in open areas of unpolluted, still or sluggish water that is up to 2m (6ft) deep; slightly acidic conditions preferred by some species.

The spherical yellow flowers and large, lily-pad leaves of pond lilies – also called cow lilies and spatterdocks – are characteristic of still or slow-moving water in the temperate and sub-tropical northern hemisphere. These habitats are often under pressure from drainage or water extraction, river traffic, and agricultural or other pollution. Recent studies suggest that there are up to 25 species; five species and subspecies are on the 1997 Red List. Two of these, *Nuphar lutea* subsp. *sagittifolia,* the Cape Fear spatterdock, and *Nuphar lutea* subsp. *ulvacea,* both listed as Vulnerable*, are found in limited areas of coastal freshwater wetland in the south-eastern USA. Large tracts of this habitat have been lost in the last century, but work is being done to secure some of the most significant areas. In Europe, ***Nuphar pumila*** (Unlisted), with isolated and scattered populations in the UK and from north-western Spain to the Russian Federation, is scarce in several countries, and protected in some.

Easily grown, even vigorous, these are plants for deep pools. Divide in spring or propagate by seed sown in wet soil.

NUPHAR PUMILA ► *Hybridization with other more common species and the effects of global warming are the main threats to this relic of glacial flora.*

13 on Red List*

Paeonia

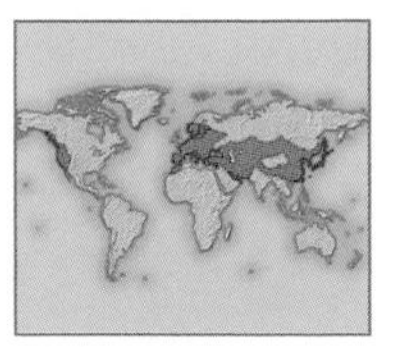

Botanical family
Paeoniaceae

Distribution
Mainland Europe and southern Scandinavia to Japan, western USA

Hardiness
Half-hardy – fully hardy

Preferred conditions
Sun or partial shade; best in deep, humus-rich soil that is moist but well-drained; American species need a dry summer dormancy; hardiness varies according to species.

Peonies are popular garden plants, with numerous species, hybrids, and cultivars in cultivation. While *Paeonia* is not a large genus, with some 35 species, its classification is complex. Botanists divide the genus into three groups: one contains the six shrubby Chinese tree peonies; another the 27 herbaceous, showy-flowered peonies from Europe and Asia, which have long been popular in western gardens; and the last group comprises just two western North American species, very rarely cultivated. Several species, especially the Mediterranean ones, are certainly under threat through development and overcollecting from the wild, but the conservation status of others is less clear.

OLD-WORLD PEONIES

Many European and Asian herbaceous peonies have probably always had very small distributions, particularly those from the Mediterranean region. For example, *Paeonia cambessedesii* (Rare*) is endemic to the Balearic Islands and *Paeonia clusii* (Vulnerable*) is endemic to Crete and Karpathos, where it grows in hills at 1,200–1,500m (3,950–4,950ft).

Today these species are also faced with the twin threats of habitat destruction resulting from agricultural and urban sprawl, and exploitation through overcollecting, underscoring the need for conservation action. Two species that have particularly suffered from overcollecting

▲ PAEONIA CLUSII *was first described by botanist Carolus Clusius (1526–1607), who is credited by some with introducing the genus into European horticulture.*

Red List: Rare*

Origanum amanum

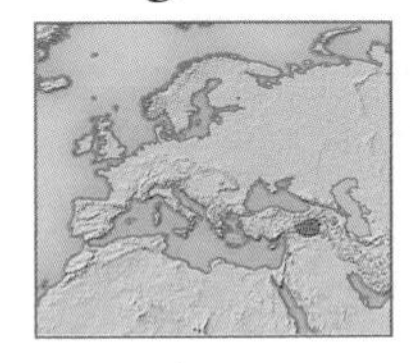

Botanical family
Lamiaceae

Distribution
Southern Turkey (Amanus Mountains)

Hardiness
Fully hardy

Preferred conditions
Best in an open, sunny position with poor to moderately fertile, humus-rich, alkaline soil; well-drained conditions are essential, particularly if winters are wet.

Several of the species of *Origanum* that are cultivated as oregano or marjoram are rare in the wild. The pink-flowered *Origanum amanum* appears to be one of the most at risk. Not only does this species have a limited wild distribution in alkaline rocks on slopes of a single mountain range, but it also appears to hybridize with other *Origanum* species that grow alongside it, and so is at risk from interbreeding and genetic erosion.

Further research is needed in order to determine whether or not the tiny gene pool of this species is being diluted by the genes of its relatives, and hence if the pure species is likely to survive in the wild. If this is an example of natural species turnover, it is unclear what can or should be done to prevent a natural event from taking place. However, if the species is losing habitat as a result of human activities, then there is clearly a need to manage it.

In milder areas this may be grown in an open scree garden, in well-drained, alkaline soil, but as it is marginally frost-hardy an alpine house may be needed.

ORIGANUM AMANUM ► *This shrubby plant grows to 20cm (8in) tall. The fragrant flowers are relatively large compared with most other members of the genus.*

▲ PAEONIA MLOKOSEWITSCHII *Known affectionately as Molly the witch, this species from the Caucasus is a classic garden plant despite the fleeting nature of its spring flowers.*

by botanists and gardeners alike are *Paeonia mlokosewitschii* (Indeterminate★) and *Paeonia cambessedesii*. The latter is now only found in inaccessible parts of its former range. Both of these species are now well established in cultivation, however, and both are easily grown and will thrive in either full sun or partial shade. They are hardy to -10 to -15°C (14 to 5°F), especially if their crowns are protected with a covering of evergreen branches through the winter. In fact, a reasonably low winter temperature is beneficial to these species, as this initiates the period of dormancy during which flower buds develop.

Paeonia clusii is not as hardy, but it is a fine peony for an area with a warm, dry climate that remains above 0°C (32°F).

NEW WORLD PEONIES

While *Paeonia* is primarily an Old World genus, two highly unusual species are found in the western USA. Both flower early in the year, bearing small flowers with strange, incurved petals, before dying down and becoming dormant in summer. Both are very restricted and in need of careful conservation; fortunately some of their localities fall within the boundaries of national parks and reserves, offering some measure of hope for the future.

Paeonia brownii (Unlisted) is a clump-forming, herbaceous perennial with fleshy, glaucous, divided leaves. The solitary, half-nodding flowers are rather hellebore-like and almost spherical, up to 3cm (1¼in) across. They are deep maroon to bronze, with broad overlapping sepals and longer petals with yellow or green margins, hemming in the large boss of stamens. *Paeonia brownii* is distributed from southern British Columbia to Wyoming, California, and Nevada, where it is found in sagebrush, chaparral, and pine forests, in rather dry situations at 900–2,250m (2,950–7,450ft). ***Paeonia californica*** (Unlisted) is even rarer, restricted to a handful of habitats similar to those of *Paeonia brownii* in California, although it does not appear to venture above 1,230m (4,050ft) in altitude. It is a closely allied species, to 60cm (24in) tall, distinguished by its larger, thinner leaves, which are distinctly green beneath, and by its flowers, which bear blackish-red petals with pink margins.

These species are tricky to cultivate, unlike most peonies. The most successful method has proved to be growing them in deep pots within the confines of an alpine house or in a cold frame, where any watering can be carefully regulated once the plants die down in the summer.

FRUIT OPENING

FERTILE AND INFERTILE SEEDS

CULTIVATION

PROPAGATING PEONIES

These long-lived plants prefer a deep soil and do not respond well to any disturbance of their roots. Therefore, although they can be increased by division of large clumps in early autumn, or by root cuttings taken in winter, propagation by seed is preferable.

Peonies are pollinated by insects, especially by beetles. The faded flowers are followed by swollen, dry fruits that split from the tip to reveal the ripe seeds in autumn (*see fruits and seeds of* Paeonia mlokosewitschii, *left*). There are both brilliant red and black seeds: the red seeds are undeveloped and infertile, but their colour attracts birds, which consume and distribute the swollen, black, fertile seeds along with them. The seed should be sown in containers outside, ideally as soon as it ripens. Although peony seed remains viable for many years if properly stored, older seed that has developed dormancy will be much slower to germinate. It is not unusual for germination to take two or even three years.

▲ PAEONIA CAMBESSEDESII *The deep rose-pink, strongly perfumed flowers of this species, borne in spring, can be up to 10cm (4in) across, and the leaves are tinted purple.*

◀ PAEONIA CALIFORNICA *This is the easier of the two North American species to cultivate, although growing it is still fairly challenging. A dry summer dormancy, typical of its native habitat, is essential.*

2 on Red List*

Peltoboykinia

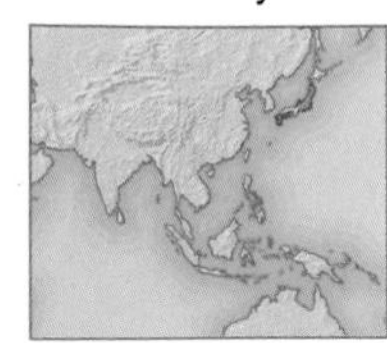

Botanical family
Saxifragaceae

Distribution
Japan (Honshu, Shikoku, and Kyushu Islands)

Hardiness
Fully hardy

Preferred conditions
Damp, moderately fertile, humus-rich soil and cool locations in semi-shade; both species are hardy to -10°C (14°F).

The genus *Peltoboykinia* contains just two species, ***Peltoboykinia watanabei*** (Rare*) and *Peltoboykinia tellemoides* (Rare*), both of which are infrequent members of Japan's forest flora. While similar in appearance, they are classified as distinct species differentiated by the number, shape, and size of their leaf lobes, although some botanists include them in a single species. Both species are deciduous, clump-forming perennials that form a mass of handsome, long-stalked leaves, measuring 10–25cm (4–10in) across. The numerous flowers, appearing in summer, are bell-shaped with narrow, toothed petals, and vary from cream to pale yellow in colour, giving a generous display on established clumps. *Peltoboykinia tellemoides* is the more commonly cultivated of the two species. It grows to about 90cm (3ft) and is almost as wide. *Peltoboykinia watanabei* tends to be slightly smaller, reaching about 60cm (24in).

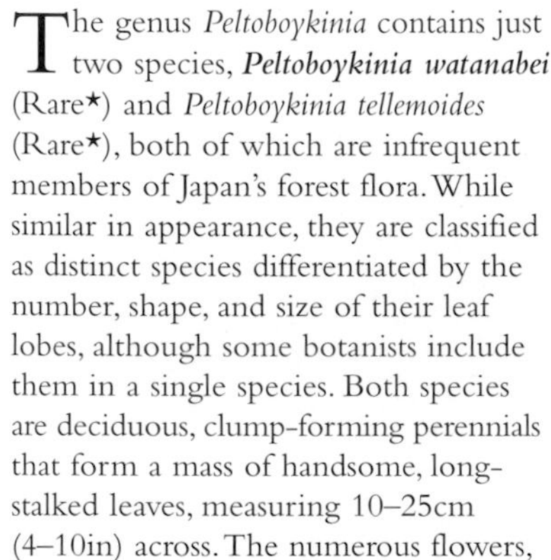

◄ PELTOBOYKINIA WATANABEI *Flowers in branching heads either hang pendulously or face outwards like shuttlecocks. They are held above the leaves on tall stems.*

HABITAT

JAPANESE FORESTS

The low mountains, hills, and plains on the Pacific Ocean side of Honshu, Shikoku, and Kyushu Islands are clothed with broadleaf, mixed, and evergreen forest, with laurel at the coast giving way to oaks and then conifers further inland. The warm waters of the Japan Current (*Kuroshio*) in the Pacific ensure a warm-temperate climate and a long growing season, with tropical typhoons bringing high temperatures and rain in summer. In winter, both temperatures and rainfall levels drop dramatically, and the winters can be even colder than those on the snowy western side of Honshu Island.

Both species are restricted to wooded mountain areas, *Peltoboykinia watanabei* on Shikoku and Kyushu islands, *Peltoboykinia tellemoides* on Honshu. With much land converted to agriculture, remnants of the original forest are now confined to steep mountain slopes and river gorges.

The name *Peltoboykinia* refers to the plants' distinctive, shield-shaped (peltate) leaves and also to their close resemblance to the genus *Boykinia*. In fact, such is the similarity between these plant groups that the two species now classified in *Peltoboykinia* were at one time included within the genus *Boykinia*. This visual resemblance is superficial, however; botanists discovered genetic differences that led to the creation of two separate genera. While these two genera are both classified in the saxifrage family, within the family they are only distantly related. The genera probably inherited similar primitive features from early members of the Saxifragaceae, but have since evolved in relative isolation, which explains the emergence of different genes.

Peltoboykinias are hardy to -10°C (14°F) and make interesting additions to a moist woodland garden or shaded rock garden. They should be cultivated in a fertile, moist but well-drained soil in a semi-shaded position. Propagation is by seed or division, both in spring.

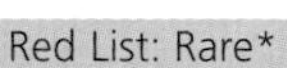

Red List: Rare*

Polemonium pauciflorum

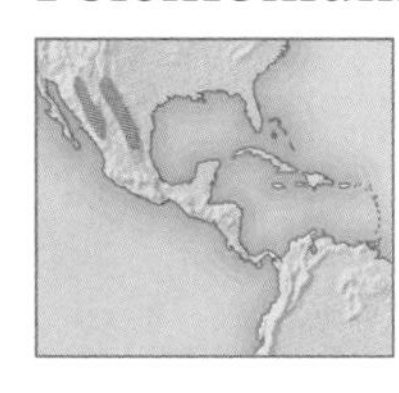

Botanical family
Polemoniaceae

Distribution
Northern Mexico; USA: Texas, Arizona

Hardiness
Half-hardy

Preferred conditions
Position in partial shade is required and plant grows best in a moist, fertile soil; hardy to -5°C (23°F).

Growing on shaded ledges in the Sierra Madre ranges of northern Mexico and neighbouring mountains in the USA, this *Polemonium* has not yet been subject to detailed study. It is unusual within the genus *Polemonium* in having long, tubular flowers. The status of the species is relatively secure, despite its rarity, because of the inaccessibility of its wild populations.

Polemonium pauciflorum is firmly established in cultivation, although it is not frequently grown and is only reliably hardy to -5°C (23°F). It is a short-lived perennial, growing to 50cm (20in) tall with attractive, fern-like foliage, and in summer bears flowers of an intriguing yellow-green hue with reddish tints. It is usually cultivated in a perennial or mixed border in a cool, moist soil in semi-shade. Propagate by seed sown in a cold frame in autumn or spring, or stem cuttings taken in spring.

TUBULAR FLOWERS ► *The horizontal to pendent flowers are borne from early to late summer, either singly or in loose flowerheads.*

Unlisted

Pteridophyllum racemosum

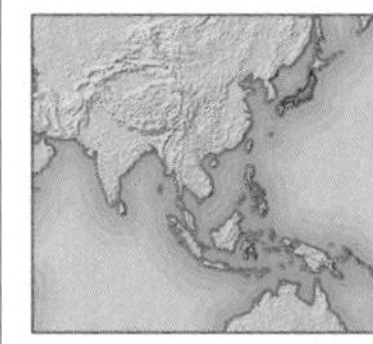

Botanical family
Papaveraceae

Distribution
Japan (Honshu Island)

Hardiness
Fully hardy

Preferred conditions
Moist but well-drained, humus-rich soil that remains reliably moist in summer, and a position in partial shade; requires climate with cool summers; hardy to -34°C (-30°F).

Fern leaf poppy is one of the oddest members of the poppy family. At first glance, the plant has no poppy-like features whatsoever, and as well as being the sole plant in the genus, it is also sometimes placed in a family of its own. The leaf tufts look more like fern fronds, hence the name: *pterido* means like a *Pteris* fern, and *phyllum* means leaf.

This rare Japanese plant is known only from coniferous woodland on the island of Honshu, a habitat under some pressure from development. It is a clump-forming perennial with rosettes of bright green, fern-like leaves, each 10–20cm (4–8in) long, divided into numerous oblong lobes. The delicate, nodding, white flowers are borne in upright flowerheads. Fairly challenging to grow, this is an attractive and unusual plant for the woodland garden. Plants require a cool, moist, acid to neutral, humus-rich soil and dappled shade to succeed. They also grow well in large pots in a shaded cold frame. Summer droughts and slugs and snails are their chief enemies. Plants can be propagated from seed when available, or by cuttings of young shoots or division of the parent plant, both in spring.

▲ FERN LEAF POPPY *The spires of bell-like, four-petalled flowers emerge in summer and can be 25cm (10in) tall on vigorous plants.*

1 on Red List*

Rheum

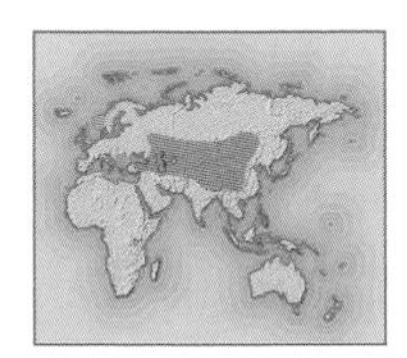

Botanical family
Polygonaceae

Distribution
Eastern Europe to China

Hardiness
Fully hardy

Preferred conditions
Most species prefer a deep, humus-rich, moist soil, some need marshy conditions; a position in semi-shade is ideal; generally hardy to at least -15°C (5°F).

Comprising approximately 50 species, the genus *Rheum* is scattered across the mountainous zones of Asia, and is extending in Europe. Rheums are typically stout perennials with imposing basal leaves, and a number of species have been in cultivation for centuries for their striking, architectural form.

Long before becoming ornamental plants, however, rhubarbs were important medicinal herbs. The roots of the plants – principally from species such as *Rheum palmatum* (Unlisted) – have been widely employed in traditional Chinese and Tibetan medicine for at least 2,000 years, and possibly for as long as 4,500 years. Rhubarb root is used to treat a wide range of ailments, including toothaches, headaches, gastric ulcers, chronic renal failure, and high blood pressure during pregnancy.

Rhubarb is also a familiar culinary plant, grown for the edible leafstalks (petioles). In the mountains of eastern Turkey, the rough petioles of *Rheum ribes* are collected for use as a vegetable, but it is the European *Rheum rhaponticum* (Rare*) and the hybrid *Rheum* x *hybridum*, also called *Rheum* x *cultorum* (Unlisted) that are more widely valued for their culinary properties.

Rheum rhaponticum was introduced into cultivation by the Italian Francisco Crasso, who sent material from Bulgarian populations to the botanic garden of Padua (*see box*) prior to 1608.

RHEUM IN THE WILD *Rhubarbs are found in a range of habitats, from marshy stream banks to rocky slopes. With the exception of the mountain species, most do well in cultivation.*

The plant's importance, first for its medicinal properties and later for its culinary value, was soon established, and it has remained in cultivation ever since. By the 18th century, the species was superseded as an edible by hybrids with similar qualities. Some of these originated from Asia, and they may have included *Rheum palmatum* (Unlisted), *Rheum rhabarbum* (Unlisted), and *Rheum undulatum* (Unlisted). The resulting complex hybrid, *Rheum* x *hybridum*, is now widely grown as a commercial food crop, particularly in the UK, and, since the early 19th century, in the USA.

Rheum rhaponticum in the wild is a rare and highly localized plant growing on damp, north-facing cliffs at about 2,000m (6,600ft) in the Rila Mountains of south-western Bulgaria. More recently, plants identical in appearance have been found growing on Mount Onstadberg in Norway, whilst unconfirmed records suggest that the plant may also have grown in Romania. Aside from its scientific interest, the wild *Rheum rhaponticum* is of particular value as the ancestor of modern culinary rhubarb. It is fully protected, both within Bulgaria and through listing on Appendix I of the Convention on the Conservation of European Wildlife and Natural Habitats (the Bern Convention).

RHUBARB LEAF AND STALK

This species, and others in the genus, are striking architectural perennials, with imposing leaves and tall spires of flowers. They are easily grown in a moist, fertile, humus-rich soil in filtered sun to semi-shade, and are fully hardy. Rhubarbs are best propagated by sowing seeds in a cold frame on ripening in autumn, or by division of the tough rootstock in spring.

◄ **RHEUM RHAPONTICUM** *Although rare in cultivation, this is an interesting species for a garden. Tall plumes of greenish to creamy white flowers are borne above the basal clump of leaves in summer.*

HISTORY

Orto Botanico

The botanic garden of Padua, Italy, is the world's oldest, and dates back to 1545. Originally a garden of "simples" or herbal medicines, it obtained plants from all parts of the globe thanks to the political and commercial power of Venice. It boasts a library, herbarium, and, more recently, several laboratories. The oldest plant in the garden is a palm, *Chamaerops humilis* var. *arborescens*, planted in 1585 and called the "Goethe palm" after the famous German writer who developed his thoughts on evolution, published in *Metamorphosis of Plants*, after studying it.

Red List: Rare*

Romneya coulteri

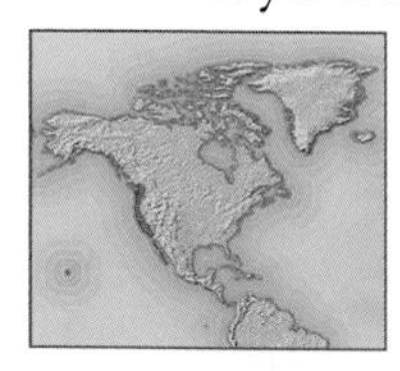

Botanical family
Papaveraceae

Distribution
Western USA: California; Canada: British Columbia

Hardiness
Half-hardy

Preferred conditions
Deep, well-drained, fertile soils, and a sunny site sheltered from cold winds are required. Hardy to -5°C (23°F).

With its white, crepe-textured flowers up to 10cm (4in) across and stems to 2.5m (8ft) in height, the Matilija or tree poppy, *Romneya coulteri*, is a distinctive native Californian flower and one of the most sought-after members of the poppy family. It was introduced into cultivation in 1875, first flowering in the National Botanic Gardens in Dublin, and has remained a favourite in gardens all over the world ever since.

In the wild, the species grows in a few *arroyos* (rocky ravines) in the coastal ranges from Santa Barbara to northern Baja, most notably on the eastern flanks and foothills of the Santa Ana Mountains, where it grows in chaparral, sage scrub, and along dry watercourses at up to 1,200m (3,600ft). The suckering nature of the plant allows it to colonize open, scrubby habitats fairly rapidly, following fire or other disturbances.

Romneya coulteri is under threat from urbanization and road construction, as well as flood-control measures, although significant populations are now within protected areas. Changes in land management – in particular, suppressing natural fires that break out in the coastal scrub – are reducing *Romneya*'s habitats and could threaten the long-term survival of this species in certain localities.

▲ GARDEN FAVOURITE *This dramatic perennial was collected by Dr Thomas Coulter of Newry, Ireland, during botanical expeditions to California and Mexico.*

6 on Red List*

Sagittaria

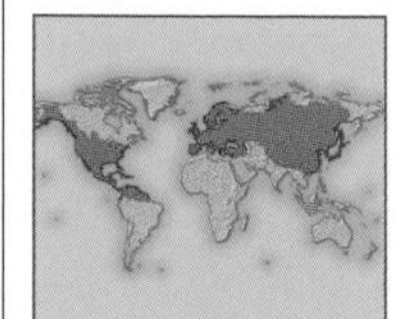

Botanical family
Alismataceae

Distribution
Europe, Asia, and North, Central, South America

Hardiness
Frost-tender – fully hardy

Preferred conditions
Most species prefer wet soil or shallow water in open, sunny conditions; tropical species tend to grow submerged in deeper water.

Noted for their distinctively shaped leaves and branched heads of white flowers, the arrowheads belong to a genus of around 30 aquatic species. The genus is most diverse in the USA, where there are some 25 native species.

Most arrowheads are threatened by water pollution and drainage, and several are very rare. Some are under particular threat due to their particular habitats. Valley or Sanford's arrowhead, *Sagittaria sanfordii* (Vulnerable on the 1997 Red List), is characteristic of seasonal or "vernal" pools on clay grasslands and volcanic slopes in western and central California. Still rarer is the diminutive Kral's water-plantain, *Sagittaria secundifolia* (Endangered on the Red List of 1997), confined to swift-flowing rivers in a tiny area on the border between northern Alabama and Georgia. The most threatened may be the bunched arrowhead, *Sagittaria fasciculata* (Endangered on the 1997 Red List), found in marshy grassland and open woodland. In 1983 there were nearly 30 colonies in north and south Carolina; only five may survive today.

More widely distributed species are also under threat. ***Sagittaria natans***, found from north-eastern Europe to Japan, is of concern in many countries and listed as Critically Endangered* in Japan.

Some of the rarest arrowheads are formally protected through national listing, as well as recovery plans that combine preservation of the last remaining colonies with propagation

SAGITTARIA NATANS *Blooming in summer in shallow lakes, its leaves completely submerged, this hardy species is found as far north as Siberia in the Russian Federation.*

projects to increase the size of populations held in cultivation. Most species may be easily grown in shallow water. Propagate plants by seed sown when ripe in containers of loam-based compost, standing in shallow water, or by division.

Red List: Endangered*

Saintpaulia ionantha

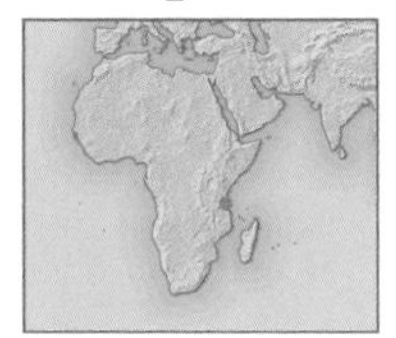

Botanical family
Gesneriaceae

Distribution
Tanzania: coastal plain in Tanga region, also in Udzungwa Mountains

Hardiness
Frost-tender

Preferred conditions
Bright but not sunny position in a humid atmosphere; free-draining soil is essential, and slightly acidic conditions are preferred; grows best at temperatures above 18°C (64°F).

The African violet was chosen by the World Wildlife Fund and the IUCN as the symbol of plant conservation in East Africa, with good reason. It is a strikingly beautiful, instantly recognizable plant, and one that is desperately in need of protection. All 20 species of *Saintpaulia* grow in East Africa, and almost all are threatened. ***Saintpaulia ionantha*** is by far the best known, and the most important horticulturally.

Those who have grown African violets may be surprised to learn that their home is in the wet rainforests of tropical East Africa. Yet in cultivation, plants quickly rot and die if water is allowed to gather beneath their roots or on their fleshy leaves. The answer to this apparent paradox is that in the wild, saintpaulias grow on vertical cliff faces deep within the rainforest, and in this habitat water can continually drain away from their leaves and roots.

◄ POPULAR PARENT *This species has given rise to numerous sports, and is also the species from which almost all of the popular hybrid cultivars are descended.*

Destruction of this limited habitat is the main reason why so many *Saintpaulia* species are threatened. When forests are cleared for much-needed fuel wood, the cliff faces where the African violets grow are exposed to sunlight, which kills these shade-loving plants.

The future indeed looks bleak for *Saintpaulia ionantha* and many of the other wild African violets. Relatively few populations are currently within protected areas, and even within them plants may be threatened by illegal overcollection and forest fires. A number of *ex-situ* conservation projects, such as *Saintpaulia* germplasm collections, have been initiated to preserve the species well into the future. The greatest long-term hope, however, probably lies in creating many more forest preserves – a process that is slowly being achieved. Only time will tell if the African violet has a future in the rainforests of East Africa.

The ideal conditions for African violet cultivation are a position that is light but does not receive direct sunlight, a night-time temperature of 18–20°C (64–70°F), and a daytime temperature of 20–32°C (70–90°F). The atmosphere should be relatively humid, but on no account should African violets be misted, as this will cause them to rot. Plants are best watered from below. Compost should be well drained and aerated and slightly acidic; a mix of equal parts sterilized loam, peat, and perlite is ideal. It is easily propagated by division or by whole-leaf cuttings. Insert the entire leaf stalk into a peat-perlite mix, water infrequently, but provide a humid atmosphere. A new plant will develop from the cut end of the stalk in a few weeks. Rooting hormones are not needed, and delay shoot development.

Red List: Rare*

Salvia albimaculata

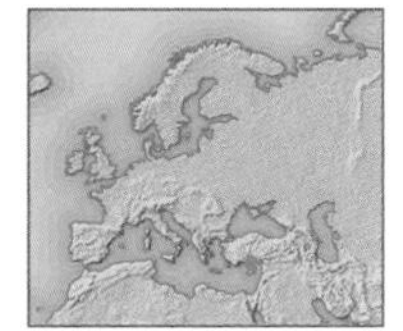

Botanical family
Lamiaceae

Distribution
South-central Turkey

Hardiness
Half-hardy

Preferred conditions
Prefers a light, moist but well-drained soil that is poor to moderately fertile; needs plenty of heat and full sun in summer, and dry winters; hardy to 0°C (32°F).

Comprising some 900 species, the vast genus *Salvia* is distributed mainly throughout Central America and central and south-west Asia. In the latter area, Turkey has the highest level of diversity, with 88 species, of which 45 are endemic.

Few salvia species are as intriguing as the Turkish ***Salvia albimaculata***. It was first collected in 1845, but erroneously identified as the more westerly *Salvia potentillifolia*. It was not until a century later, in 1948, that it was recognized as an altogether more dramatic new species.

Salvia albimaculata forms a low, domed, spreading shrub approximately 30cm (12in) high and up to 90cm (36in) across. The leaves superficially resemble those of the shrubby cinquefoil (*Potentilla fruticosa*), hence the initial mis-identification, but the most dramatic feature of the plant is the large flowers – up to 4cm (1½in) long and a deep inky purple, with a large white spot on the lip. When it was formally described, the species was declared "magnificent" and "very handsome".

In the wild, *Salvia albimaculata* is found in the soft, chalky slopes of the massive canyon country of the Göksu River, in the Ermenek area of south-central Turkey. Here it is relatively abundant, and one of approximately 40 species whose native range is confined to this canyon. The future of this environment, however, is uncertain as proposals for major dam construction on the Göksu River have been put forward.

Introduced into cultivation as recently as 1984, the species is becoming popular among gardeners. It is relatively easy to grow in a temperate, continental climate with high levels of summer sun and dry winters. *Salvia albimaculata* needs well-drained, gritty, loam-based soil and an open, sunny position. In areas with wet winters, this salvia is best grown in an unheated greenhouse, although its size can prove unwieldy. It appears to be short lived, and is best propagated by seed sown in autumn, which germinates erratically in spring, or by semi-ripe cuttings in summer.

WHITE SPOTTED ► *It is the distinctive, brilliant white mark on the lip of the flower that gives the plant its name "albimaculata", meaning spotted with white.*

Red List: Rare*

Scoliopus bigelovii

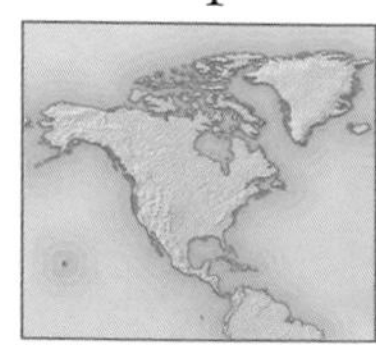

Botanical family
Trilliaceae

Distribution
Western USA: coastal northern California, southern Oregon

Hardiness
Half-hardy

Preferred conditions
Thrives in partial to deep shade, in humus-rich, moist but well-drained soil, with a covering of dry leaves in winter; hardy to -10°C (14°F).

Bigelow's adder's tongue, with shiny, brown-spotted leaves and brownish flowers, is one of two members of a small genus, confined to forests in the western seaboard of the USA and related to *Trillium* (*see pp.200–01*). Like other members of the Trilliaceae family, the floral parts are in whorls of three. The most conspicuous parts of the flowers are the sepals, which have a greenish base colour, finely etched with lines of deep maroon. The three petals are smaller, threadlike, and upright around the three-lobed stigma. Overall, the flower can reach approximately 3.5cm (1½in) across. The plant is not grown commercially, but it is appreciated by enthusiasts who grow it for its particular characteristics and unusual life cycle.

◄ STRONGLY SCENTED *Flowering in the gloom of evergreen forests in winter, this unusual plant attracts pollinating flies largely by scent; the flowers themselves are not particularly showy.*

The brownish flowers have an unpleasant scent, and are adapted to attract pollinating fungus gnats from the closely related families Mycentophilidae and Sciaridae. Initially, the flowers are borne on erect stems to attract these flies, but once pollination has taken place, the stems bend under the weight of the swelling seed pods. The name *Scoliopus* reflects the curving behaviour of the flower stalk. It is derived from the Greek *skolios*, meaning "crooked", and *pous*, "foot". The fruits then ripen in the leaf litter, where they are easily accessible to ants, the principal agents of seed dispersal. The plants flower in early spring, ensuring that the seeds are ripe at the time of greatest ant activity, and each seed bears a fleshy, edible appendage (elaiosome) to attract the insects.

Scoliopus bigelovii was discovered by, and named in honour of, John Milton Bigelow (1804–78), a surgeon and botanist from Ohio, who came upon the plant during a botanical collecting trip in the San Francisco Bay region in the spring of 1854. The plant is confined to the moist shade of coastal forests, and is principally found in redwood forests, below 600m (2,000ft), and, more rarely, in forests of Douglas fir or mixed evergreens.

The species was originally regarded as localized and rare, but is now known to be locally common in the redwood forests of the San Francisco Bay area and in the outer northern coast ranges of northern California. It is often said to be endemic to California, but a few populations are known to extend north into Oregon, home to *Scoliopus hallii*, the other species in the genus. As a consequence of this scattered distribution, *Scoliopus bigelovii* as a whole is not regarded as being at risk, despite being classified as Rare in the Red List of 1997. Individual populations, however, remain at risk from clear-felling operations, sustained browsing by deer, and from urban development.

This small, rhizomatous plant can be cultivated in a moist but well-drained, humus-rich soil, in partial to full shade. It is best propagated from seed, which should be sown in a cold frame immediately on ripening.

REDWOOD FOREST *The adder's tongue is largely confined to stands of mature redwood forest, particularly in the San Francisco Bay area where the atmosphere is moist from sea fogs.*

Red List: Vulnerable*/CITES II

Shortia galacifolia

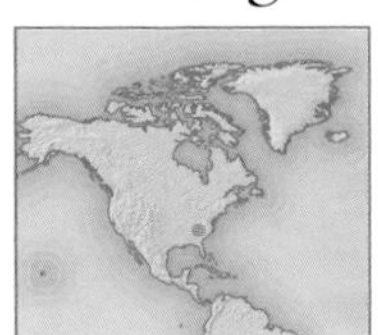

Botanical family
Diapensiaceae

Distribution
USA: Appalachian Mountains of North and South Carolina

Hardiness
Fully hardy

Preferred conditions
A climate with cool, damp summers is preferred. Moist but well-drained soil, humus-rich and leafy, and preferably slightly acidic; in partial to deep shade.

Showy-flowered *Shortia galacifolia*, or Oconee bells, has always been elusive. Endemic to the Appalachian Mountains, it is the only species in its genus to be found in North America; the rest are in Asia. It was first collected in 1787 by French botanist André Michaux, but the specimen in the Paris herbarium remained unnamed, and it was not seen again in the wild until almost 100 years later. When Asa Gray (*see right*) saw this specimen in 1839, he recognized it as a new genus and named it after a friend, the botanist Charles Short.

Michaux's plant was recorded as coming from the "high mountains of Carolina", but Gray's searches for Oconee bells in the wild proved fruitless. In 1877, a few plants were found much lower, on the banks of the Catawba River in North Carolina. Gray renewed his searches, this time in the mountains around Asheville, again to no avail. A second population was found in 1886 in South Carolina's Oconee County, again at a relatively low altitude, by botanist Charles Sargent. Gray had simply been searching too high up. He never saw the species in the wild, but before his death someone brought him a flowering plant.

The rediscovery of *Shortia galacifolia* sparked such immense publicity that collectors flocked to the Carolinas to gather examples. Although a few further populations were located, this almost proved fatal to the species. While collection for the horticultural trade is less of a problem today, there have been recent population losses as a result of habitat disturbance. The woodland favoured by the species has been cleared to make way for roads, farms, and housing, especially second homes. Several dam construction projects have also destroyed wild populations. Plants are abundant in the surviving populations, but such is the scarcity of these that the species is listed as Vulnerable on the Red List of 1997. Conservation efforts by staff at the Botanical Gardens at Asheville have saved at least one population from extinction – they transferred the plants to their native woodland garden, where the shortias continue to thrive.

◄ RARE IN THE WILD *Although extremely attractive, with small, almost circular leaves and white or pale pink, funnel-shaped flowers, this plant is almost as rare in cultivation as in the wild.*

CULTIVATION

This is a plant of temperate mixed and broadleaf forests, found on shaded slopes, creek banks, and rock outcrops in cool and humid ravines. It requires moist but well-drained conditions that are difficult to provide in any but the wettest climates. In regions with suitably moist climates, such as Scotland and the Pacific northwest of North America, it is much easier to grow, but as it is slow-growing, it is not widely cultivated. In areas with naturally high rainfall, grow Oconee bells in soil containing high levels of coarse, acidic leafmould, with grit added for drainage. It is usually grown in full sun or light shade so that its leaves develop deep autumn colour, and it can flower freely. In drier climates a shadier position is recommended, with plenty of water during the summer months. Seed tends to germinate erratically, so propagation is usually by cuttings taken during summer or root division in late winter.

PEOPLE

ASA GREY

Known as the father of American botany, Asa Gray (1810–88) was a leading botanist in the mid-19th century, when the flora of North America was being scientifically recorded. He wrote numerous books, established the Gray Herbarium at Harvard, and was the first director of Missouri Botanical Garden. After studying the similarities between the floras of North America and eastern Asia, Gray became the earliest, leading supporter of Darwin in the USA.

Red List: Rare*

Silene regia

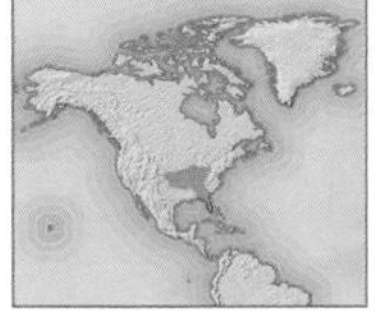

Botanical family
Caryophyllaceae

Distribution
South-eastern USA

Hardiness
Fully hardy

Preferred conditions
Requires a deep, fertile, well-drained soil, ideally neutral to slightly alkaline; thrives in an open position in full sun; hardy to -15°C (5°F).

The name *regia* is derived from the Latin *regius*, meaning royal, and *Silene regia* is one of the most magnificent of the 500 or so species in the genus worldwide. The royal catchfly has stems to 1.5m (5ft), topped by loose heads of vivid red, five-petalled flowers up to 4cm (1½in) across.

The species has a patchy distribution in rocky prairies, open woods, thickets, and glade edges across the south-eastern USA from Florida, Alabama, and Georgia north to Illinois and Ohio, and west into the prairies of Kansas, Missouri, Arkansas, and Oklahoma. Although not nationally protected, it is rare and declining in much of its range. It is listed as possibly Extinct* in Kansas and Tennessee, Endangered* in Oklahoma, Illinois, Kentucky, and Georgia, Vulnerable* in Arkansas, Indiana, Ohio, and Alabama, and Rare* in Missouri. This decline is typical of many grassland species with extensive interconnected "metapopulations", now isolated through habitat loss and fragmentation. Much of the decline is due to the conversion of prairie into agricultural land. Even in surviving areas of grassland, decline continues owing to the suppression of naturally-occurring fires, which once helped to maintain the open habitat suitable for the species. Habitat isolation leaves plants unable to recolonize when conditions improve and research shows that plants in these small, isolated stands produce fewer fertile seeds than those in large colonies.

Today, US conservation organizations such as The Nature Conservancy are working to save the final fragments of original prairie from destruction. In the short term, the reinstatement of fire as a management tool is a top priority, but if the long-term future of *Silene regia* is to be secured, all efforts must be directed at piecing together large blocks of prairie.

Widely grown in the USA, this species is becoming increasingly popular in Europe. It is easy to grow from seed in very well-drained soil in full sun.

CATCHFLY ► *The stems and branches of plants in this genus are covered in minute sticky glands, which trap insects. In the UK, the plant is known as catchfly.*

Red List: Rare*

Stachys candida

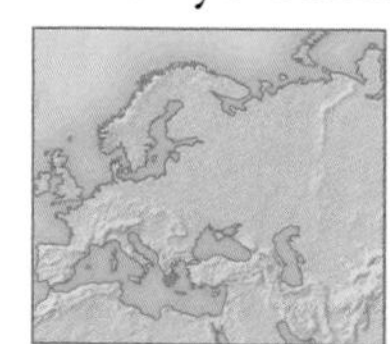

Botanical family
Lamiaceae

Distribution
Southern Greece (Mount Taygetos)

Hardiness
Half-hardy

Preferred conditions
A climate with dry winters and a position in full sun, growing in sharply drained, gritty soil that is moderately fertile; hardy to -5°C (23°F).

◄ MOUNTAIN SPECIES *Like many of the Mediterranean members of its family, the leaves of* Stachys candida *are rich in essential oils with anti-bacterial properties.*

HABITAT

MOUNT TAYGETOS

This mountain massif in the Greek Peloponnese extends for 100km (62 miles), terminating in the rugged Mani Peninsula. Taygetos is one of the botanical hotspots of Greece: some 100 of the species found here are endemic to the country, and 33 of them are wholly or largely confined to this single massif. Famed for its forests of mature black pine (*Pinus nigra*) mixed with broadleaf evergreens and Grecian fir (*Abies cephalonica*), Mount Taygetos has been identified by the World Wide Fund for Nature International as one of its 100 forest hotspots.

With its heavily veined, pink flowers and white-felted leaves, *Stachys candida* is one of several mountain species in the genus that has become a popular rock and alpine garden plant. In the wild, this spreading, shrubby species is confined to limestone rocks and cliffs on Mount Taygetos in Greece (*see box*).

In spite of its exceptional importance to nature conservation and ecotourism, Mount Taygetos remains unprotected. It is particularly vulnerable to fire, and has suffered in recent years from arson, with some fires thought to have been started by those with an eye to speculative housing development. Like many mountain plants native to the Mediterranean region, *Stachys candida* is highly localized, and fire can destroy entire populations.

This low-growing sub-shrub can be cultivated in a very well-drained soil in a sunny situation. It is an ideal plant for a raised rock garden, scree bed, or alpine trough, but because it hates winter wet, it is often grown in an alpine house. Under glass, the species prefers a loam-based compost, with added grit. Propagation is easiest from semi-ripe cuttings taken in summer, or from seed sown in autumn or spring.

Red List: Endangered*

Stylidium coroniforme

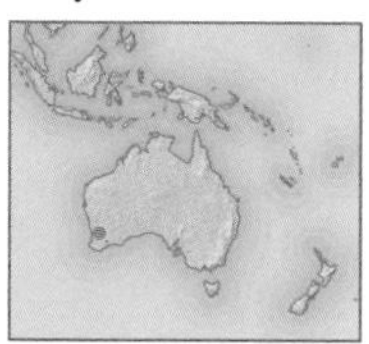

Botanical family
Stylidiaceae

Distribution
Western Australia

Hardiness
Frost-tender

Preferred conditions
Requires a fertile, well-drained soil and an open position in full sun; reliably dry conditions in winter and a minimum temperature of 7°C (45°F) are essential.

Wongan Hills triggerplant is a member of an almost exclusively Australian genus, with one species in New Zealand. The common name, triggerplant, comes from the unusual structure and pollination mechanism of the flowers (*see box*).

Stylidium coroniforme is a low-growing, short-lived perennial, with rosettes of linear, grey-green leaves, and cream-pink flowers in short flowerheads 10–15cm (4–6in) high. It grows in open, bare areas amongst scrub and heathland, on yellow sands overlying laterite, an iron-rich, red soil. Seed persists in the soil for many years, and germinates when low-intensity fire or light disturbance create the open conditions needed by the species.

Individual plants live for 5–15 years, and typically die when their habitat becomes increasingly overgrown.

This species was first discovered in 1964 in the Wongan Hills, in Australia's midwest, and for many years the species was known only from the site of its discovery. By 1980 this population had declined alarmingly to a single wild plant, however. Fortunately, additional sites have since been found in the Wongan Hills. Outlying populations have also been found near Maya, approximately 140km (90 miles) to the north, but plants from this second area may yet prove to be a new species. A survey in 2001 indicated that of the 17 colonies known, 12 supported fewer than 50 mature plants each. Today around 10 populations are known, mostly on public land.

The species is regarded as endangered by the Australian government because of its highly localized distribution, the degree to which its populations are isolated from one another, and the destruction of its preferred habitats. While at least one population lies within a nature reserve, and the bulk of the colonies lie on public land, the species remains under considerable threat from intense fire, overgrazing, and the spread of invasive weeds.

An interim recovery plan for the period 2003–2008 has been drafted by the Western Australian Department of Conservation and Land Management. Activities have included the collection of seed for storage *ex situ*, a reintroduction scheme, and a programme to inform landowners of the presence and value of this rare species on their property.

BOTANY

POLLINATION

The stamens and style of the trigger plant are fused in a bent column, which straightens and snaps when an insect touches it. Young flowers hold pollen in a particular spot, so species can share pollinators without hybridization. In older flowers the style develops and picks up pollen, aiding cross-pollination.

8 on Red List*

Tricyrtis

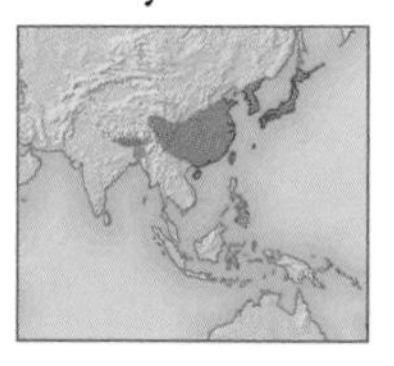

Botanical family
Convallariaceae

Distribution
Eastern Himalayas, south-western China to Japan

Hardiness
Fully hardy

Preferred conditions
Cool, humid, shady conditions and a moist but well-drained, humus-rich soil; most plants are fully hardy, but do best with a covering of snow or dry vegetation in winter.

The flowers of toad lilies bear some resemblance to the related *Scoliopus* (*see p.196*), with two whorls of three petals and three sepals around a column, often of a contrasting colour, carrying the stamens and a style with three forked lobes. The name *Tricyrtis* is from the Greek *tries*, meaning three and *kyrtos*, meaning convex, and refers to the three short spurs at the base of the flower.

JAPANESE SPECIES

There are around 20 species in this Asiatic genus, with over half found only in Japan. They are globally rare, but most of the populations within Japan are reasonably large, with several nationally listed as data deficient. ***Tricyrtis perfoliata*** (Vulnerable in the 1997 Red List) grows abundantly but only in a few mountain gorges on the island of Kyushu, on cliffs flanking the Nanuki River.

This local abundance, however, leaves the toad lilies vulnerable to extinction should anything happen to their limited habitats. These are species of moist woodlands and cliffs, and most are found in very limited areas of forest and rocky gorges. The forests have been much reduced, and are under pressure from urban development and agriculture. The vulnerability of these rare species is

▲ TRICYRTIS OHSUMIENSIS *This spreading, low plant has a thick stem and closely packed, fleshy leaves. Large flowers with petals and sepals that are all the same size are borne in summer.*

▲ TRICYRTIS FLAVA *The rounded leaves have curled tips and purplish spots, and clasp the stem tightly. Flowers are borne singly at the end of each stem from early autumn.*

exacerbated by the fact that they are far more difficult to cultivate than the hybrids generally found in gardens, and so are much rarer in cultivation.

CULTIVATION

Several species make good additions to the garden, with small but interesting flowers from late summer to autumn. ***Tricyrtis hirta*** is the most widely available, although not always a vigorous plant, and hybrids between this and *Tricyrtis formosana* make up the bulk of the toad lilies in cultivation. ***Tricyrtis ohsumiensis*** (Rare in the 1997 Red List) is smaller and less often seen, although its flowers are some of the largest in the genus. ***Tricyrtis flava*** (Rare in the 1997 Red List) is difficult to grow and perhaps the most infrequently seen in cultivation. These perennials may need misting in summer to increase humidity, while in winter, the late-flowering species are prone to frost damage, so need a sheltered position. Those with pendulous stems, such as *Tricyrtis ishiiana* (Vulnerable in the 1997 Red List) and *Tricyrtis perfoliata*, are best on a ledge or slope to appreciate their flowers from below. Those with upright stems, like *Tricyrtis flava* and *Tricyrtis ohsumiensis*, are better on level ground. Propagate by seed sown when ripe, or by division in early spring.

TRICYRTIS PERFOLIATA ► *The flowers are a spectacular sight in autumn, cascading down on their thin, lax stems from vertical cliffs in mountain gorges in Japan.*

TRICYRTIS HIRTA *This is a highly variable species with hairy stems that can reach 60cm (24in). Flowers appear from late summer, with unequal petals and sepals spotted in shades of purple.*

18 on Red List*

Trillium

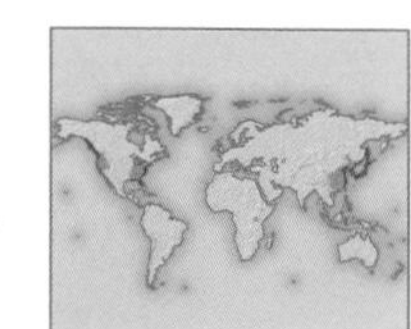

Botanical family
Trilliaceae

Distribution
Temperate North America, north-eastern Asia

Hardiness
Fully hardy

Preferred conditions
Moist but well-drained, deep, humus-rich soil, preferably neutral to acidic; a sheltered location with spring sunshine but shade later in the year.

There are perhaps 50 species in the genus *Trillium*, with the largest number found in the eastern USA. About a half dozen species occur in the western Himalaya and north-eastern Asia, mostly in Japan, and another seven are strictly plants of the western USA.

Of the 15 species included on the Red List of 1997, all but two are from the USA. Two *Trillium* species from the south-eastern USA, *Trillium persistens* (Endangered in the Red List of 1997) and *Trillium reliquum* (Vulnerable in the Red List of 1997), are nationally listed as endangered species, and a number of other species are protected by state laws in at least part of their range.

There is an almost cult-like devotion to trilliums among Japanese gardeners, and this has unfortunately stimulated an export trade in wild-collected rhizomes from the south-eastern USA, where many species occupy restricted ranges. The exact extent of this trade is controversial, and horticulturists and environmentalists have frequently been at odds on just how much of a threat it presents to the continued future of species in the wild.

Complicating this issue is the fact that propagating trilliums from seed is both difficult and slow, and most seed loses viability if it is stored dry for any appreciable time (*see box*).

A much more obvious threat is from logging practices that result in wholesale disturbance of the forest understorey, which shade-loving plants such as trilliums will not survive. Government policies appear to allow timber harvest on public lands leased to timber companies, while many of the species endangered by these practices are protected by other laws on unleased public land. Other threats faced by trilliums in the wild include browsing by grazing animals, deer as well as livestock.

TRILLIUM LUTEUM

TRILLIUM TRAITS

Commonly called wake robin, wood lily, or trinity flower, these plants are easily recognized. The solitary stem, produced in early spring, terminates in three leaf-like bracts below a single, three-petalled flower. The fruit is either berry-like or a capsule, containing seeds partially enclosed by a fleshy, oil-rich appendage (elaiosome). This attracts and provides food for ants, which then disperse the seeds. By midsummer, most trillium plants will have died back to their rhizomes.

The most common species, widely seen in gardens, is *Trillium grandiflorum*. This is one of the most glorious early spring wild flowers found in the eastern deciduous forest of the USA.

Trillium nivale (Rare in the Red List of 1997), the snow trillium or dwarf white trillium, is treated as rare and endangered in a number of states, but is locally abundant in others. It has a limited distribution from West Virginia and Pennsylvania across the south-central Midwest, with sporadic outposts west to Nebraska and north to southern Minnesota and Michigan, a range that essentially marks the extent of the final Pleistocene glaciation 11,000 years ago. It is found on soils derived from limestone, in two distinct habitats, either in crevices or at the base of large rocks, or on the gravel-rich banks of large watercourses. Some evidence suggests that the forms from rocky slopes are more amenable to cultivation.

TRILLIUM NIVALE ► *This is one of the earliest trilliums to flower, often through the snow. The pure white flowers set off by rich bluish-green leaves make it a tempting, but challenging, plant to try in the garden.*

Two species that occur in the south-eastern USA are ***Trillium luteum*** and ***Trillium pusillum***. Of these, *Trillium luteum* is the best known, and often seen in gardens. Its status in the wild is healthier than many, but it is still not common, and is mostly confined to the mountains in south-eastern Kentucky, western Virginia, east Tennessee, western North Carolina and South Carolina, and northern Alabama and Georgia. *Trillium pusillum* (Rare in the Red List of 1997) is protected by law in many states. It has a broad distribution from Virginia west to Kentucky and Arkansas and south to Texas, Mississippi, Alabama, and the Carolinas, but within each state it only occurs in a small area, with scattered populations. The flowers appear in the understorey of moist to swampy woods from mid-spring in the deep south, to early summer in its northernmost and highest outposts.

Most trilliums are suitable for a woodland garden or a moist, shady border. Smaller species, such as *Trillium nivale*, are naturally suited to a shady rock garden and will do better in a site where they are not in competition with other vegetation. Unusually for a trillium, *Trillium nivale* also grows best in a gritty, alkaline soil.

Plants can be grown from seed, but the most usual method of propagation is by division of the rhizomes after flowering, ensuring that each division has a growing point. Cutting out a growing point after the plant has flowered can also stimulate the production of offsets.

BOTANY

PROPAGATION

Propagating trilliums from seed is slow. Seeds must be cleaned, soaked in three per cent hydrogen peroxide solution for ten minutes, and unless harvested while still green, kept moist at around 5°C (41°F) for ten weeks. No growth appears above ground until the second spring. Teams at the Mount Cuba Center for the Study of the Piedmont Flora and the US National Arboretum's Florist and Nursery Crops Laboratory are seeking to perfect tissue culture of trilliums (*left*) and shorten the time to grow plants from seed.

TRILLIUM PUSILLUM *The wavy-margined white flowers of this attractive species age to pink, and the undersides of the dark green leaves are wine-coloured. Many of the scattered populations across the range of the species have distinctive features and are formally recognized as botanical varieties.*

Red List: Rare*

Trollius laxus

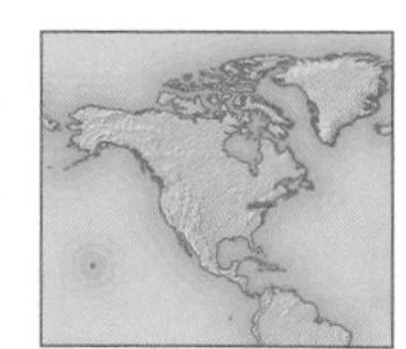

Botanical family
Ranunculaceae

Distribution
Northern USA: north-east Ohio east through north-west Connecticut

Hardiness
Fully hardy

Preferred conditions
A wetland plant thriving in deep, moist, alkaline soil that is fertile and rich in humus; needs a site in at least partial sun, and without competition from more vigorous species.

Once reasonably widespread through much of north-eastern USA, ***Trollius laxus***, spreading globeflower, is now in serious decline, and no longer grows in more than half of its historically documented sites. This is a plant of cold, alkaline wetlands, whose habitats began to be lost as soon as the land was settled, largely due to drainage for agricultural and residential development. Perhaps the biggest threat today is degradation from agricultural runoff.

Two species of globeflower are found in the USA, *Trollius laxus* in the east and the closely related, but more common, *Trollius albiflora* in the west. Both have flowers with spreading petals more reminiscent of a buttercup than the typical globe-shaped blooms of the Asian and European members of the genus. The two American species are very similar and until recently were treated as subspecies of a single species. Their ranges show no overlap, however, and they are unable to cross-fertilize due to differences in their chromosome numbers. A less consistent but more visible difference is that *Trollius laxus* usually has solitary yellow flowers, while *Trollius albiflora* tends to have white flowers in clusters of three or more.

Trollius laxus needs sun for only a few hours a day, but should not be shaded by more vigorous plants. Since it resents disturbance, propagate by seed rather than division. Sow seed as soon as ripe; it may take two years to germinate. Since plants are still taken from the wild, always check that stock is nursery-propagated.

▲ SPREADING GLOBEFLOWER *This species is ideally suited to growing on the damp margin of a pond or in a bog garden, with shade for the hottest part of the day.*

163 on Red List*

Verbascum

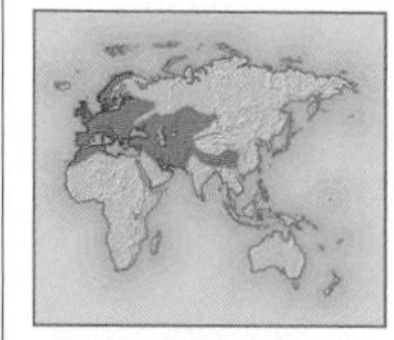

Botanical family
Scrophulariaceae

Distribution
Europe, North Africa, south-west and central Asia

Hardiness
Half-hardy – fully hardy

Preferred conditions
An open site in full sun; gritty, well-drained, alkaline soil of low fertility. The small species dislike winter wet. Most are fully hardy.

Mulleins belong to a genus of some 360 biennials and perennials found in Europe, north Africa, and south-west and central Asia. Turkey is the centre of diversity, with no fewer than 232 species.

The genus is gaining popularity with gardeners, as new species and an ever-increasing range of hybrids reach the market. Among the first species in cultivation were *Verbascum bombyciferum* and ***Verbascum olympicum*** (both Rare on the Red List of 1997). Both are endemic to north-western Turkey, and are largely confined to Uludağ Mountain. This "Great Mountain", the Olympus of mythology, has an immensely rich flora, including over 30 species found nowhere else. It is easily reached from Istanbul, and has long been a magnet for European botanists, who introduced these species into European gardens from their early visits. These two species typify the genus to many people: densely white-felted leaves form an overwintering rosette, 90cm (36in) or more across, from which a flower stem, up to 1.5m (5ft) tall, grows in the second season. The flowerhead is made up of many tightly packed, yellow flowers 3–4cm (1¼–1½in) across. This biennial growth pattern, common to many mulleins, makes them well adapted to growing in scrub and on waste ground: some of the best populations of the very rare *Verbascum olympicum* are to be found around the edges of car parks.

◀ VERBASCUM OLYMPICUM *The flowers, arranged in a branching candelabra, are produced in a plant's second or third year. It then often dies immediately after flowering.*

Very different in form is *Verbascum dumulosum*, an evergreen subshrub no more than 25cm (10in) tall, but spreading to 40cm (16in) or more. In summer it bears a succession of short flowerheads. This species, Rare on the Red List of 1997, is known from a single site on limestone cliffs in south-western Turkey.

Over half of the Turkish verbascums are on the Red List of 1997 but while the majority are highly localized, and several are at risk, most are probably safe. The coarser species are unpalatable to grazing animals, and thrive in areas that may be overgrazed for other plants. Some species are genuinely threatened: the dainty *Verbascum helianthemoides* is confined to a handful of salt steppe sites in central Turkey, where it is under immense threat from ploughing and cultivation of sugar beet and wheat. Similarly, *Verbascum stepporum* (Vulnerable on the 1997 Red List) is known only from a small area in southern Turkey and is threatened by intensive cultivation. It will become increasingly imperilled as the hydroelectric and irrigation schemes of the South-eastern Anatolian Project (GAP) come into operation.

PATIENT PLANTS *Verbascums have adapted to survive harsh conditions for lengthy periods; the seeds of some species have been found to germinate even after a hundred years.*

Unlisted

Veronica spicata *subsp.* spicata

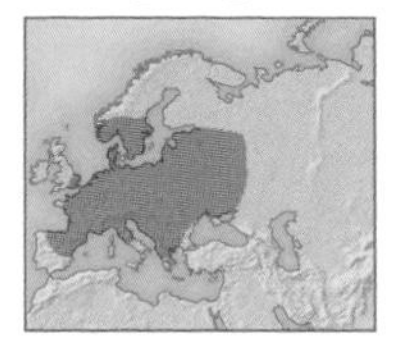

Botanical family
Scrophulariaceae

Distribution
Europe

Hardiness
Fully hardy

Preferred conditions
Thrives in moderately fertile, well-drained soil, preferably alkaline, and an open site in full sun.

The British flora is perhaps the best documented, and among the most actively conserved of any on Earth, yet certain species have nonetheless declined to virtual extinction. One such is *Veronica spicata* subsp. *spicata*, the spiked speedwell, which survives in only a handful of native sites in the Breckland region of East Anglia, and remains in a parlous state overall.

The Brecklands represent one of the most remarkable landscapes in the UK. The area boasts the driest climate in the country, and overall is more continental in climate than any other area. Together with the sandy nature of much of the soil, this has given rise to heath and grassland communities akin to the heaths and steppes of continental Europe. Until the 20th century, Breckland was a wild and desolate place, with "heather in every direction as far as the eye can see". There were "travelling" sands, deposited by ancient glaciers and periodically driven over surrounding country by the wind in "sand floods". The infertile soil made cultivation of arable crops difficult, and land was typically left fallow in the years after cropping. The extensive grass heaths and the abundant open niches provided a home for a range of rare, distinctive plants, including the spiked speedwell, making Breckland one of the top five botanical hotspots in the UK. Continuity of habitat was vital to the survival and spread of the species, which vanished temporarily in unfavourable conditions, only to recolonize later.

Changing landscape

All of this changed in the 20th century. The planting of trees, particularly pines, initially as shelter belts and later in forestry plantations, has altered the landscape beyond recognition. Grass and heathland habitats are a fraction of their original size. Sheep grazing has declined, and rabbits have succumbed to outbreaks of myxomatosis. Low intensity farming, with botanically rich fallow periods, has largely given way to modern intensive cultivation, with irrigation, high usage of fertilizer, and annual cropping.

As a result, the abundance of open habitats has largely been lost. *Veronica spicata*, a perennial with poor competitive ability and mobility, has suffered more than most species: once relatively widespread, it is reduced to a handful of colonies. Only two of any size remain, one on an intensively managed nature reserve, the other on a remote part of Newmarket racecourse, where a long tradition of continual mowing has allowed the largest population to survive. While *Veronica spicata* subsp. *spicata* is not rare at a European level, this does not reduce the need to actively conserve it in places like Breckland. Such disjunct populations often tell scientists much.

This spreading perennial is popular in gardens and easily grown at the front of herbaceous borders or in a rock garden. It is propagated by seed sown in autumn or spring in a cold frame, or by division in spring.

VERONICA SPICATA *Various forms are now widely available in cultivation. Many have blue flowers like the typical wild plant, and there are also pink- and white-flowered selections.*

Red List: Rare*

Xeronema callistemon

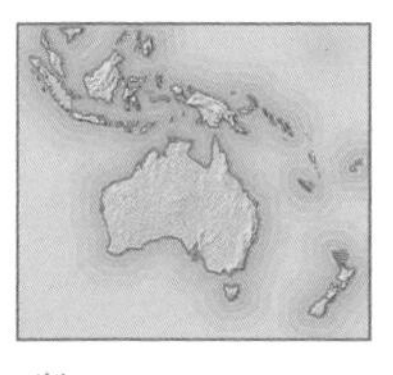

Botanical family
Phormiaceae

Distribution
New Zealand: Poor Knights Islands, Hen and Chickens Islands

Hardiness
Frost-tender

Preferred conditions
Coastal areas with a humus-rich, porous, well-drained soil; position in sun to dappled shade. Plants are likely to be damaged by temperatures below 5°C (41°F).

With red bottle-brush flowers and sword-shaped leaves, Poor Knights lily is one of the most striking of New Zealand's endemic plants. It is largely confined to the Poor Knights Islands, two volcanic islands and numerous stacks about 25km (16 miles) north-east of North Island. An outlying population is found on Taranga, the Hen island in the Hen and Chickens island group, some 55km (170 miles) to the south.

Captain Cook gave the Poor Knights Islands their name when he visited them in 1769 because their profile resembled effigies of crusaders. More significantly, he left pigs with the islanders. These became a useful commodity for barter with mainland tribes until it led to a raid in about 1808, in which many of the islands' inhabitants were massacred. The survivors declared the islands out of bounds and left, never to return. The archipelago has remained uninhabited by humans ever since, and the feral pigs were removed in 1936, in one of the earliest examples of practical island conservation in New Zealand.

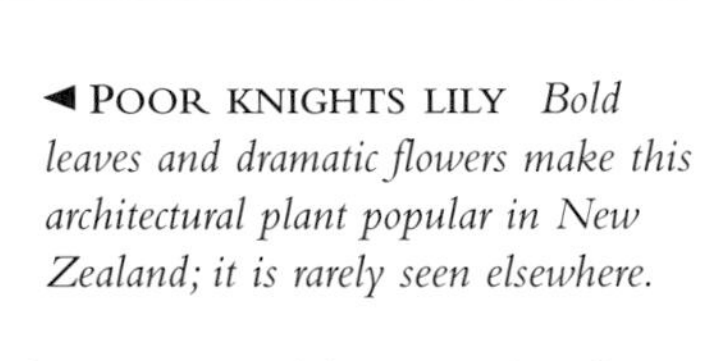

◄ POOR KNIGHTS LILY *Bold leaves and dramatic flowers make this architectural plant popular in New Zealand; it is rarely seen elsewhere.*

The effects of cultivation, fire, forest clearance, and pigs make it unlikely that much original vegetation survived the period of habitation. But the islands suffered little from the introduction of foreign plants and animals. The natural vegetation has reasserted itself, and today the islands have a dense cover of secondary scrub and woodland. They were declared a Strict Nature Reserve in 1975, and the seas were designated New Zealand's second marine reserve in 1981.

Xeronema callistemon was described in the 1920s, and its distribution recorded during the 1930s by eminent botanists Lucy Cranwell Smith and Lucy Moore. It grows to 90cm (36in) on exposed, rocky bluffs, probably always relatively undisturbed, and seems to be as frequent today as it ever was. Strict control of access to the islands should ensure its future, minimizing risks such as fire or the introduction of invasive plants.

This tender perennial needs a humus-rich soil with pumice or other drainage materials added. In very sunny places, it prefers light shade for the hottest part of the day; in frost-prone areas, grow in a greenhouse. Sow seed in an open potting compost under glass, or divide in spring.

CLIMBING PLANTS

Climbers have adopted the successful strategy of using sturdier plants as ladders, so they can reach above other species and ensure their leaves are in sun. Although they are natural competitors, and often winners, plenty are at risk because their preferred habitats, many tropical, are being lost. As a group, climbers boast a range of species that are valuable commercially. Unless action is taken, many useful products and medicines could remain undiscovered. Research into climbers is essential, but they are difficult to study: although they are rooted in the ground, most of their growth occurs high up.

WHAT IS A CLIMBER?

In their early years at least, climbers have no firm supporting tissues of their own and depend on their surroundings to provide a route to the light. Some climbers have no more complex method than to cast their long thin stems over plants or obstacles in their path, but many have developed specialist techniques or equipment to aid their ascent. In addition to having a physical means of climbing, climbers are often extremely sensitive to light. The shoot tips unerringly head up towards the sun, but aerial roots and tendrils often grow away from light, in search of a holdfast on a wall, tree, or shrub that they can latch on to and use to climb.

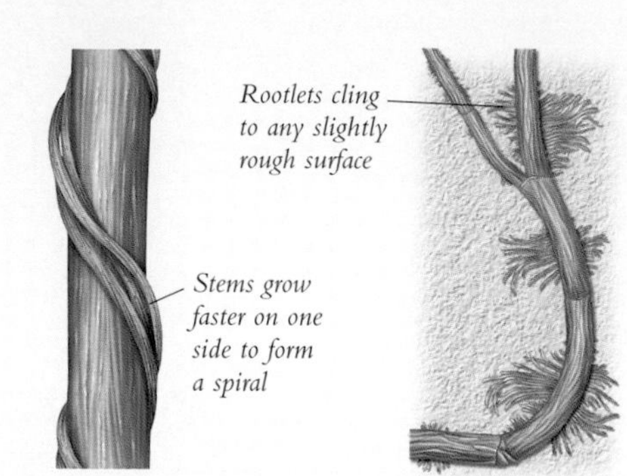

TWINING STEMS *are one of the most common means of climbing, and are employed by wisterias and honeysuckles.*

AERIAL ROOTS *develop in clusters and are used by some climbers, like ivy, to cling to any available support.*

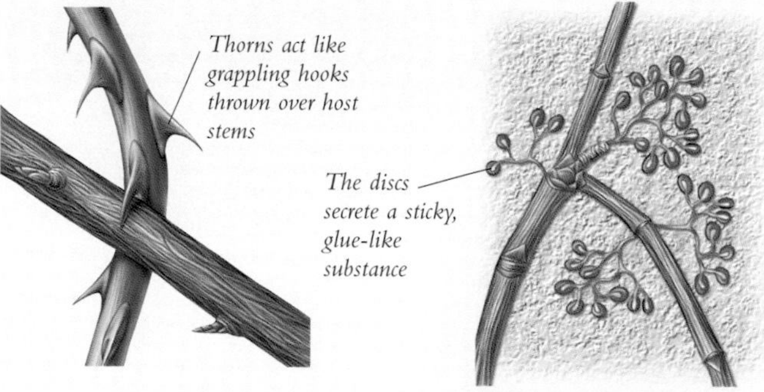

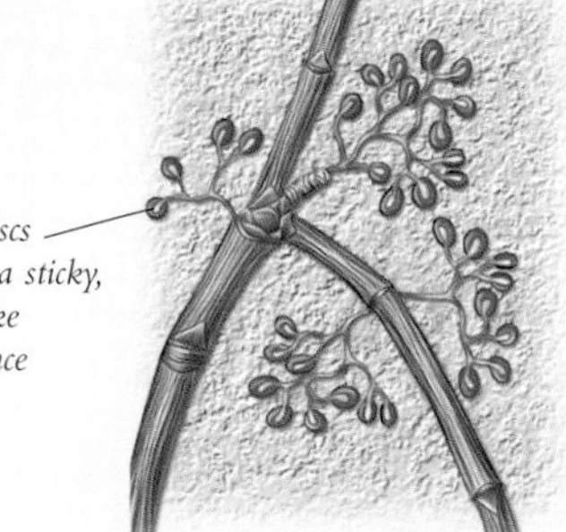

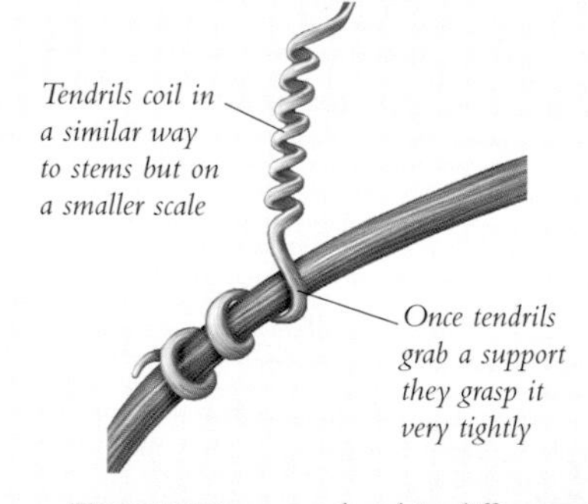

THORNS *Many plants, such as climbing roses, use sharp, downwards-pointing thorns to scramble through the branches of host plants.*

STICKY SUCKERS *are produced on plants such as Virginia creeper as an extension of tendrils. They can adhere tenaciously to the flattest surface.*

TENDRILS *are produced in different places: those of sweet peas at the leaf-tips; passion flowers on main stems. Clematis coil their leaf stalks.*

WHERE CLIMBERS GROW

Climbing plants are found across the world, from tropical to cool temperate zones, in scrub and forests of every type. In scrub, climbers often make use of the clear ground left after a fire to reach the top of burnt trees and shrubs before they resprout. Because climbers need light for germination, in forests they generally inhabit the margins, or the open glades that have been cleared by natural fires, landslips, or fallen trees. They do not form stiff, supporting stems initially, and so they are able to grow very fast, taking advantage of the light that is temporarily available, before the vegetation grows back. By that time, the climber will have reached the top of its support and be basking in the light. In tropical rainforest, for example, woody lianas (twining climbers) are among the first species to take advantage of light gaps caused by fallen trees, and make up about 40 per cent of the canopy.

▲ **HEADING TOWARDS THE LIGHT** *The climber in this small light gap in the dense tropical rainforest of Northern Australia has made a break for the canopy before it closes up.*

SURVIVAL

Climbers and scramblers come in a variety of forms – evergreen or deciduous, annual, herbaceous perennial, and woody plants. They are opportunists that exploit any disturbance by germinating rapidly to take advantage of the vacant niche. Unlike most other plant groups, they do not hold their young stems upright, but concentrate their energies on fast growth, often horizontally along the ground at first. Young climber shoots perform gyratory movements, growing in wide, sweeping motions in order to increase the area in which they might find a support. In their quest for light, the shoots will scramble upwards over trees, shrubs, rocks, or any other obstacle in their path. If its shoots cannot find a support for vertical growth, a climber will sprawl along the ground, eventually mounding itself up over its own stems.

▲ **UPWARDLY MOBILE** *The instinct to climb up cannot be ignored – having found the top of one stem of grass this bindweed has reached across to find a taller stem to continue its ascent.*

COMPETITIVE PERSONALITY *The slender stems of the hedge bindweed grow remarkably fast in all directions and are programmed to twine tightly around any obstacle they touch. The plant also sends out tough rooting stems just beneath the soil surface to colonize new areas.*

CLIMBERS AND WILDLIFE

The energy climbers save by finding other means of support is often spent on producing large, showy blooms and sweet, juicy fruit. These ensure flowers are pollinated and seed is dispersed far and wide. The animals who provide this pollen and seed courier service are rewarded with food. Indeed, some climber and animal species have co-evolved to become dependent exclusively on each other. While such a relationship is working, it is very efficient for both partners, but if a change in the habitat causes the climber to become rare, or forces the animal to move away, the results can be devastating for both species.

◄ MUTUAL BENEFIT *is gained from a close relationship. The hummingbird finds a good source of nectar and the climber is reliably cross-pollinated.*

VALUE TO HUMANS

The effort climbers put into persuading wildlife to pollinate their flowers and disperse their seed is exploited by people: clematis and sweet peas, beans, yams, and grapes are just a few of the major ornamentals and crops in the group.

While much energy is expended attracting useful animals, some climbers can deter herbivores intent on feeding on their soft, lush growth with the plants' natural pesticides. The vines *Lonchocarpus* from South America and *Derris* from south-east Asia, for example, are sources of the popular organic pesticide rotenone – a product that is proving valuable, especially in organic agriculture. The powerful chemicals climbers generate to ward off pests can have other uses. In Ghana, botanists have been working with traditional healers who have a vast knowledge of local plants. Together, they have identified several climbers with medicinal properties including potential treatments for malaria, asthma, hypertension, arthritis, and diabetes.

▲ IDEAL CROP *The purple-podded pea is both ornamental and good to eat, and also fertilizes the soil with nitrogen while it grows.*

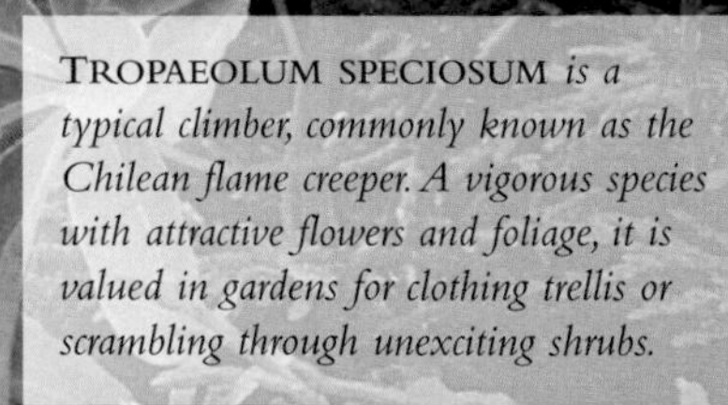

◄ FIBROUS STEMS *The long, flexible stems of climbers, such as* Berberidopsis, *are often woven into baskets, hats, ropes, and many other useful products.*

TROPAEOLUM SPECIOSUM *is a typical climber, commonly known as the Chilean flame creeper. A vigorous species with attractive flowers and foliage, it is valued in gardens for clothing trellis or scrambling through unexciting shrubs.*

CLIMBING PLANTS

INVASIVE VINES

Many climbers are naturally aggressive: the strangler figs (*Ficus* species), for example, have phenomenal appetites for water and nutrients and starve the trees that support them. Although such vigorous growth is tolerated in a climbing plant's natural ecosystem, a climber can become a threat if it is introduced to a different area with a similar climate. The transport of climbing plants worldwide to be grown in gardens or commercially for their fruit, flowers, or foliage has exacerbated this problem. In a new environment, a climber's growth can continue unchecked by the predators and competitors that it would encounter in its native range. A vigorous new climber can steal niches occupied by indigenous species and an invasive climber can smother all other vegetation in its path. Many climbers are among the world's most invasive plants (*see p.440*). The introduced kudzu vine (*Pueraria lobata*) is so invasive in the USA that it is called "the plant that ate the South".

▲ SMOTHERED *Rampant climbers, such as this purple bindweed, can be invasive when introduced to areas that lack the competitive pressures of their native habitats.*

THREATS TO CLIMBERS

Habitat change and loss are the major threats to climbers. Forests and hedgerows in particular are being destroyed at an alarming rate for conversion to agriculture and forestry. Any reduction in habitat area is magnified for climbers because even in ideal conditions, the disturbed areas they prefer make up a tiny proportion of their potential range. Where suitable niches do open up, non-native, weedy plants can snatch them from native vines. On islands, grazing animals introduced by early sailors eat climbers that have no herbivore defences. Plant hunters prefer seed of climbers because it is difficult to extract and move whole plants, so the group has been spared the worst effects of collecting. However, seed removal is damaging to very small populations.

▲ MONOCULTURE *Climbers have little hope of establishing in the crowded, sterile environment of a commercial timber forest.*

NON-CLIMBING CULTIVARS

There is no denying that exuberant and unwieldy climbers can be difficult to grow as a commercial crop or in a small garden. Today, valuable genera are often cross-bred with non-climbing relatives to produce plants that retain their original value, but have no desire to ascend. The resulting cultivars are compact, require no supports, are less weedy, and are easier to harvest. There are now, for example, non-climbing beans, peas, and sweet peas. Important characteristics such as drought tolerance or disease resistance can also be bred into cultivated climbers. However, unattractive or poorly cropping wild relatives with breeding potential for such traits are often ignored, and could become extinct before they are recognized.

GROUND LEVEL ► *This compact bedding sweet pea, 'Knee Hi Mixed', has increased the range of areas in the garden where the pretty, scented flowers can be enjoyed.*

GARDENER'S GUIDE

- Many introductions of climbers collected from the wild for garden cultivation seem to have been in the form of seed, because the unwieldy, living climbers are very difficult to transport. This means the genetic base of cultivated climbers is likely to be wide, and could form a valuable resource for reintroduction programmes for threatened species. Try to maintain this natural variation as much as possible by growing climbing species as well as cultivars, and by propagating plants by seed.
- To avoid uncontrolled collection, governmental agencies throughout the world now have schemes that allow limited seed gathering under licence. Propagation in cultivation then reduces demand for further wild-collected seed. Never purchase seed, or plants grown from seed, that have not been obtained under licence.
- Join the Friends of your local botanic garden or conservation group to find out about native climbers. Indigenous vines often provide food or shelter for wildlife.
- Volunteer for local conservation programmes that aim to propagate climbing plants for reintroduction programmes.
- Try growing "heritage" cultivars of climbing crops or garden plants. They tend to have a richer gene pool than modern cultivars and could prove valuable in the future. They are also often more attractive to insects such as butterflies and bees.
- Keep garden climbers such as ivies and clematis bushy to provide a home for wildlife.
- Turn the vigorous growth of climbers to your advantage by using them to hide an unsightly wall, shed, or tank, or grow them on a trellis to form a screen. Aerial roots or suckers will stick to almost anything but twiners need a support that they can grip. Thorny climbers may need tying to a support.

BIRDS NEST IN CLIMBERS

17 on Red List*

Aristolochia

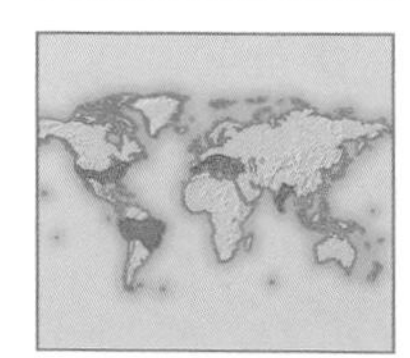

Botanical family
Aristolochiaceae

Distribution
Widespread; especially moist woodland

Hardiness
Fully hardy – frost-tender

Preferred conditions
Fertile, well-drained soil, in sun or partial shade; protection from excessive winter wet.

Occurring both in the tropics and temperate areas, *Aristolochia*, known as Dutchman's pipe, is a large genus of 300–350 species that includes many climbing vines, and some herbaceous and, rarely, shrubby species from damp woodlands worldwide. They vary from fully hardy to frost-tender, depending on whether their native habitat is temperate or tropical.

Their remarkable flowers have long been prized by gardeners and can exceed 30cm (12in) in length in some species. The climbers are vigorous, and easily grown in a fertile soil enriched with organic matter, in sun or filtered light. They are best propagated by seed or softwood cuttings. At the present time only a small selection of Dutchman's pipes are in cultivation, such as ***Aristolochia gigantea*** (Unlisted), but an increasing range is becoming available. Among them are several species that deserve more recognition, including *Aristolochia pfeiferi*, described in 1983, currently only known from a single area of central Panama and Endangered on the Red List of 1997, and *Aristolochia helix,* found in Thailand in 1984 and included as Indeterminate on the Red List of 1997.

There is concern about the decline of *Aristolochia* in the wild and its effect on wildlife. Despite being highly poisonous, the foliage of a number of species is an important food source for the caterpillars of spectacular swallowtail and birdwing butterflies. In some cases, the caterpillars benefit from the plant's defenses by taking up toxic aristolochic acid from the leaves and storing it in their bodies. This makes them very unpalatable, and predators learn to avoid them.

◄ FLY TRAP *The complex flower structure, said to look like a Dutch pipe, creates a trap. Flies are lured inside by a combination of scent and coloration that resembles carrion, dung, or fungus – ideal food sources or egg-laying sites. The confined flies pollinate the flower before escaping.*

The endangered Richmond birdwing butterfly, native to north-eastern Australia, used to be seen in thousands, but numbers have declined over the last century. This butterfly relies on the Unlisted *Aristolochia praevenosa* for food but, as the bush and rainforest in the area has been cleared, the climber has become rare. To make matters worse, the South American ornamental *Aristolochia elegans* (Unlisted) has escaped from gardens and become widely naturalized. The butterflies perceive it as a suitable egg-laying site, but it is toxic to their caterpillars.

Today, conservation of the Richmond birdwing butterfly is being aided by the selection of easily grown, drought-resistant clones of the native *Aristolochia praevenosa,* which are planted out to boost natural populations. Gardeners are also encouraged to grow native Dutchman's pipes, rather than the previously popular *Aristolochia elegans.*

The chemicals in aristolochias can be poisonous to humans. Although they have a historical use in folk medicine, a clue to which lies in the old English name of birthwort, today the importation and use of dried roots is banned in many countries.

! *Aristolochia* vines can be aggressive. Native species are always safe to plant (*see pp.52–53*).

ARISTOLOCHIA GIGANTEA *The giant Dutchman's pipe comes from tropical forests in Panama and is popular in cultivation because of its large, inoffensively scented flowers.*

Red List: Endangered*

Berberidopsis corallina

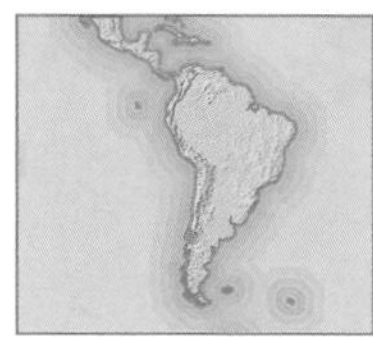

Botanical family
Flacourtiaceae

Distribution
Chile

Hardiness
Half-hardy

Preferred conditions
Humus-rich, slightly acid, moist but well-drained soil, in a partially shaded and sheltered position.

The Chilean coral plant, locally called the coralillo, fascinated the gardening public when it was first displayed. Like many Chilean forest species, *Berberidopsis corallina* favours a sheltered, partially shaded site, and prefers an open, moisture-retentive, slightly acid soil, enriched with organic matter. It is ideally propagated by seed, otherwise by semi-ripe cuttings or layering.

The coralillo was discovered by Richard Pearce, one of three commercial collectors working for the great horticultural dynasty Veitch & Sons. In the late 1800s the nursery on the King's Road in Chelsea, London, was perhaps the most influential in the world in introducing exotic garden and greenhouse plants into cultivation. Visitors to the nursery were intrigued by the coralillo's holly-like leaves, rich red, waxy flowers, and elegant twining habit to an impressive 6m (20ft), and it has remained very popular ever since. Pearce discovered the species during his travels in coastal Chile, in a humid, forested ravine a short distance south of the town of Concepción. It was subsequently found to be confined to a narrow strip of temperate rainforest that grows at low altitudes along 350km (220 miles) of the coastal mountain ranges (cordilleras). Within this small area it grows with other well-known garden plants such as *Drimys winteri*, *Embothrium coccineum*, *Eucryphia cordifolia* (included as Lower Risk on the 2000 Red List), *Lapageria rosea* (*see p.213*), and *Mitraria coccinea.*

ELEGANT CLIMBER ► *The first plant was introduced into gardens from the temperate rainforests of Chile, where it inhabits shady, humid valleys. It will thrive in other temperate and subtropical climates if it is kept moist, and out of drying winds and strong sun.*

The relative accessibility of the coastal forest has made the coralillo very vulnerable. Much of the habitat has been felled for charcoal production, or burnt and replaced with fast-growing non-native timber species such as pine and eucalyptus. As a result the climber is listed as Endangered on the Red List of 1997. It has disappeared from much of its former range, and is now confined to inaccessible ravines that have escaped the ravages of modern forestry operations. Few of these sites receive formal protection within national parks or other monitored areas. Today, the Chilean coral plant is almost certainly more common in gardens than in the wild.

DNA-sequence analyses of leaf material from gardens in the UK by the Royal Botanic Garden Edinburgh has shown that all plants are descendants of Pearce's original introduction of 1860. If he had taken a cutting, all the UK specimens would be clones and useless for conservation purposes; however, genetic variation within the population shows that he must have collected seed, no doubt germinated in Veitch's nursery. This variation provides hope that plants cultivated in gardens could be used in reintroduction programmes in the future, and highlights the value of seed propagation where possible.

3*/12 on Red List

Bomarea

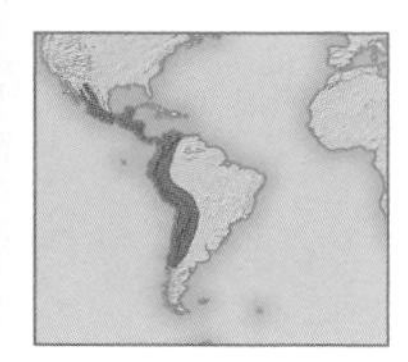

Botanical family
Alstroemeriaceae

Distribution
Central and South America

Hardiness
Half-hardy – frost-tender

Preferred conditions
Fertile, moist but well-drained soil, in full sun, ideally in a humid atmosphere.

These twining climbers have long been noted by gardeners for their clusters of rich red or orange flowers, often dramatically speckled within. They are closely related to the familiar garden plant *Alstroemeria*, which is widely grown for cut flowers.

The 120 species of *Bomarea* are found only in the New World, from Mexico south to Argentina, and are at their most diverse and abundant in Colombia. Most grow in Andean cloud forest (*see also p.67*), which forms when hot, moist air from the Amazon basin is pushed up the slopes of the Andes. As the air rises and cools, the water it holds condenses to form an ever-present mist. The climate is mostly temperate, with a relative humidity of over 90 per cent and the result is a lush, epiphyte-laden forest, which can soar to 25m (80ft) in height. These naturally isolated, humid mountain forests are now severely fragmented after clearance to create farmland. Plants from this most threatened of habitats, always rare and highly localized, are under increasing threat from the extensive destruction.

▲ RAINFOREST FLOWERS *The flowers of this Ecuadorean* Bomarea *are easily seen in the gloom of the understorey. Although these tuberous, generally deciduous climbers grow in a range of habitats, they are most characteristic of the moist temperate forest that clothes the Andes.*

Fortunately the future looks brighter in cultivation, with a range of *Bomarea* species becoming available to gardeners in temperate and tropical climates. They are easily grown in moist but well-drained soil with added organic matter, and like plenty of water in growth, but to be kept just moist during dormancy. The roots of all species must be protected from frost so *Bomarea* plants are best grown under cover in cold climates. Propagate these climbers by seed, sown in spring.

! *Bomarea multiflora* is an invasive species in New Zealand. It is monitored in the wild, and its sale, distribution, and propagation are now banned.

18 on Red List*

Calystegia

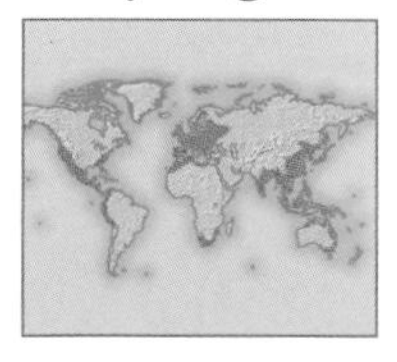

Botanical family
Convolvulaceae

Distribution
North, South America, Europe, east Asia, South Africa, Australasia

Hardiness
Fully hardy – frost-tender

Preferred conditions
Moist, well-drained soil in full sun, although many conditions are tolerated and the genus is fairly adaptable.

To most horticulturalists, members of the genus *Calystegia* are pernicious weeds. They were introduced into gardens for their undeniably pretty flowers, but proved difficult to remove, spreading by tough, far-reaching roots and trailing stems. Yet, in a genus of approximately 25 species worldwide, no fewer than 18 species or subspecies are listed as threatened. Of these, 14 occur in California, which is the centre of diversity for the bindweeds. Plants from this area include the Piute bindweed, included as Rare on the Red List of 1997 (*see below*), and Stebbins' morning glory (*Calystegia stebbinsii*), confined to the chaparral of central California and Endangered on the Red List of 1997.

Like many plants from Mediterranean climates, *Calystegia stebbinsii* is fire-adapted and found in greatest abundance after a burn. Unfortunately, land in California is in great demand for housing, which has directly destroyed a third of all colonies, and has made essential conservation burning very difficult at the remaining sites.

It is not just in California where bindweeds are threatened. In Australia the Norfolk Island *Calystegia affinis*, Rare on the Red List of 1997, is confined to Norfolk and Lord Howe Islands. The sites on Lord Howe Island are all protected but competition from aggressive, non-native species is still a problem. This endangered bindweed is available in cultivation.

! Even rare bindweeds can become aggressive in areas other than their native habitat: plant them responsibly.

◄ CALYSTEGIA LONGIPES *The Piute bindweed is found in scrub in Arizona, California, Nevada, and Utah, where the native American Piute tribe once lived. The area is under pressure from developers and this pretty bindweed is now rare.*

HISTORY

AN ADMIRER

Prince Charles Edward, also known as Bonnie Prince Charlie, was famously impressed by the delicate beauty of bindweed flowers. In 1745 the Prince collected seed of the pink and white candy-striped sea bindweed (*Calystegia soldanella*, Unlisted) whilst waiting to sail from France to the Hebridean island of Eriskay, west of Scotland. Over 250 years later the plant survives on the beach where he landed, and nowhere else in the Hebrides.

Red List: Vulnerable*

Canarina canariensis

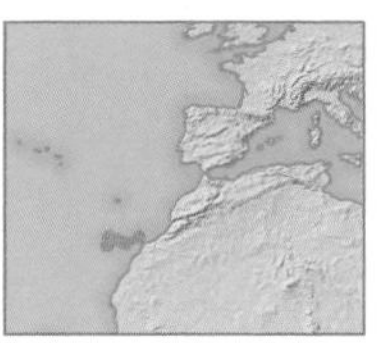

Botanical family
Campanulaceae

Distribution
Canary Islands

Hardiness
Frost-tender

Preferred conditions
Fertile, humus-rich, free-draining, sandy soil in partial shade.

The Canary Island bellflower, ***Canarina canariensis***, known locally as the *bicácaro*, is restricted to the islands of Tenerife, Gran Canaria, La Palma, and Gomera. It scrambles among the tree heather and myrtle heath that grows between the lowland ancient laurel forest (*laurisilva, see p.102*), and the forests of *Pinus canariensis* (*see p.148*) found at higher altitudes.

The Canary Island bellflower grows in one of the most biodiverse temperate regions in the world. The Macronesian Islands of the Azores, Canary Islands, and Madeira are special because they support the last remnant of the forests and scrub that used to blanket the Mediterranean region during the Tertiary period, which ended 1.8 million years ago. The advancing chill of the ice age drove back the ancient species and today they survive only on these warm islands. As many as 70 per cent of all species are endemic, including 520 plants, and this important area is fully protected by the European Union Habitats Directive.

Popular in gardens, the bellflower is a herbaceous vine that is easily grown against a wall or through shrubs, or under cover in frost-prone climates. When growth commences in autumn, the bellflower likes plenty of water, but prefers to be kept dry when dormancy starts in late spring. Propagation is preferable by seed sown as soon as ripe at 15–22°C (59–72°F), or by basal cuttings in late winter or early spring.

▲ CANARY ISLAND BELLFLOWER *This spectacular, easily grown climber scrambles up to 3m (10ft), and is noted among gardeners for its large, bell-shaped flowers borne in abundance during the winter months.*

Red List: Endangered

Clianthus puniceus

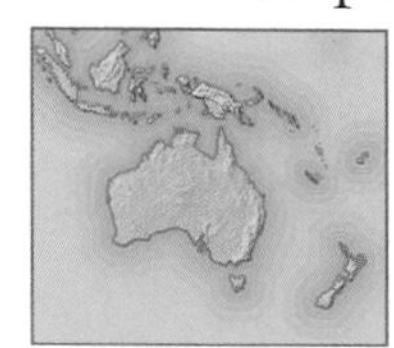

Botanical family
Leguminosae

Distribution
New Zealand: east side of North Island

Hardiness
Half-hardy

Preferred conditions
Fertile, well-drained soil, in a sheltered spot in full sun, and plenty of water in summer.

▲ CLIANTHUS PUNICEUS 'ALBUS'
Forming a smaller, more compact plant than the red type, this form has dramatic white flowers, often tinged with green. It is deservedly popular, being easy to grow and widely available from nurseries.

The New Zealand parrot's-bill, also known as the glory pea, lobster claw, and kaka-beak, was first discovered by the famous and intrepid botanists Joseph Banks (*see p.87*) and Daniel Solander during Captain James Cook's 1769 expedition to New Zealand. By 1830 the climber was popular in the gardens of European settlers in New Zealand and, shortly afterwards, it was brought before the English gardening public, and on sale for the sum of £5 – an extravagant purchase at the time. The parrot's-bill remains one of the most popular native New Zealand garden plants both in its home country, and in temperate climates around the world. In addition to the bright red species, plants with white or pink flowers are available.

TROUBLE IN THE WILD

In spite of its popularity with gardeners, the species remains in a parlous state in the wild. The glory pea is currently known only from a handful of sites on New Zealand's North Island in the Te Urewera National Park and the shores of the East Cape coastline. In some cases, populations support just a handful of individuals and it is thought there may be as few as 200 plants left in the wild. Its range is limited because it germinates and survives only where there is little competition: in disturbed habitats created by tree falls, landslips, or burns. Suitable areas have been in decline because of the expansion of pastoral farming, and any seedlings that gain a foothold often suffer from overgrazing by feral goats and deer. Since 1993 the New Zealand Department of Conservation has undertaken a species recovery programme, aimed at maintaining the range and genetic diversity of plants in the wild. Initiatives include excluding grazing animals with fencing, artificially creating habitats, and establishing new populations using cultivated, but more genetically diverse, stock.

Fortunately, parrot's-bill is easily grown. It prefers a fertile, well-drained soil in full sun, in a spot sheltered from cold, drying winds. The species is often grown against a wall, where it can reach 3m (10ft) in height. Although plants are not fully hardy, they can withstand temperatures down to –5°C (23°F); frost-damaged plants normally regrow in spring. Propagate by seed sown in spring at 13–18°C (55–64°F), or root semi-ripe cuttings in summer.

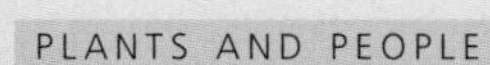

PLANTS AND PEOPLE

NEW ZEALAND NAMESAKE

Clianthus puniceus has the distinction of being the only ornamental plant cultivated in New Zealand prior to the arrival of Europeans. The Maori were attracted by the bright flowers, red being the sacred colour of the Gods. They called the plant *kowhai ngutu-kaka*, or kaka-beak, because the flowers resembled the curved beaks of the native parrot, the kaka.

The kaka parrots are themselves increasingly rare: in common with many New Zealand natives, they are threatened by logging and farming activities that restrict their range. They are also vulnerable to attacks by predators, such as rats, introduced to New Zealand by the European settlers.

CLIANTHUS PUNICEUS *For hundreds of years this evergreen climber has been a favourite in gardens in its native New Zealand and worldwide. It can reach 4m (12ft) and is covered with spectacular flowers from spring to early summer.*

Red List: Rare*

Codonopsis affinis

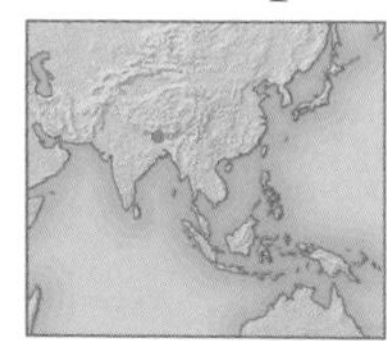

Botanical family
Campanulaceae

Distribution
India: Sikkim

Hardiness
Fully hardy

Preferred conditions
Moist, humus-rich, but well-drained soil; full sun to light shade, in a protected position.

A rarity even when it was first discovered in the mountain scrub of the Himalaya in 1858, ***Codonopsis affinis*** is listed as Rare on the Red List of 1997. It grows at 1,800–3,000m (6,000–10,000ft) in India, although recent records offer hope that the species may also occur in Bhutan and north-east Nepal. Nevertheless, in an effort to boost dwindling natural populations, botanic gardens in the area are aiming to cultivate large numbers to be planted out in the species' native habitat. The project has a good chance of success because *Codonopsis affinis* is easily grown, although it is unknown how plants will survive transplanting into the wild.

The genus *Codonopsis* contains about 40 species of twining and scandant herbaceous perennials, found on rocky slopes from the Himalaya to Japan. Many species, including *Codonopsis affinis*, are gaining popularity among gardeners who admire their flowers. They are delicate, bell- or star-shaped, often have intricately veined or chequered petals, and are borne in abundance during summer and early autumn. Coming from high in the Himalaya, all species are fully hardy. However, some have brittle shoots and benefit from being grown through shrubs or twiggy supports, in a sheltered spot out of damaging winds. They prefer light, moist but well-drained, fertile soil with added organic matter, in sun or light shade, with the roots in shade. Propagation is best by seed, sown in a cold frame in autumn.

THREAT

MEDICINAL TUBERS

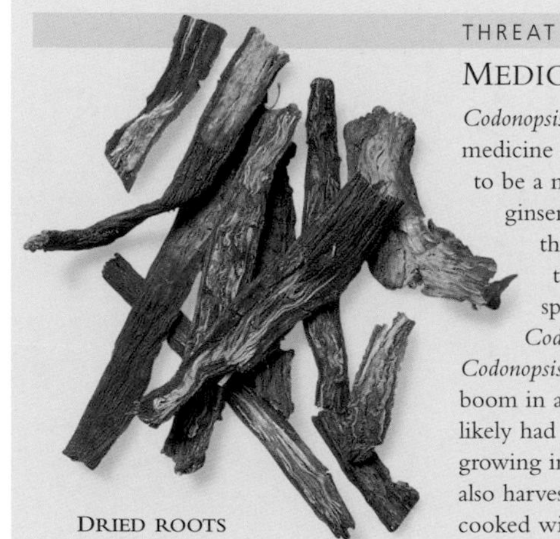

DRIED ROOTS

Codonopsis pilulosa is used in traditional Chinese medicine as an energy tonic, and is considered to be a milder, cheaper alternative to true ginseng. The roots are harvested in their third year, dried, and sold under the traditional name of *dang shen*. Other species are sometimes used, including *Codonopsis tangshen* (*chuan dang*), and *Codonopsis tubulosa* (white *dang shen*). The recent boom in alternative and traditional medicines has likely had an impact on *Codonopsis* populations growing in more accessible areas. The tubers are also harvested for culinary purposes and can be cooked with rice, grilled, baked, or used in soups.

DELICATE CLIMBER ► *This elegantly understated herbaceous climber can twine up to 2m (6ft), and is clothed in large, pungent, heart-shaped leaves, and distinctive, delicate white and purple, bell-shaped flowers.*

1 on Red List*

Holboellia

Botanical family
Lardizabalaceae

Distribution
China, Vietnam

Hardiness
Half-hardy

Preferred conditions
Fertile, well-drained soil, in full sun or partial shade, and a sheltered position protected from drying winds.

The nine species of *Holboellia* are scattered across the Himalaya and China where they can be found climbing in trees or over rocks in moist, temperate forests and thickets. In the past few decades, the area has become more easily accessible and a number of new species have been found. Although these discoveries are still poorly researched, it is thought that they are highly restricted in the wild. The recently described *Holboellia chapaensis,* for example, survives only in dense forest and forest margins in the valleys of south-west Guangxi and south-east Yunnan in China, and accordingly has been classified as Rare on the Red List of 1997. In the future, the range of these rarer species may be extended into gardens.

ORNAMENTAL CLIMBERS

Gardeners' appreciation of the twining evergreens from the genus *Holboellia* is steadily growing. The climbers are closely related to *Akebia* and *Stauntonia*, which are already popular ornamentals, and have similarly handsome foliage, sweet-smelling flowers, and oblong, purple fruits. They make ideal climbers for screening, are very vigorous and grow to 10m (30ft) quite rapidly, if given the right conditions. At present, only species common in the wild, such as ***Holboellia coriacea***, *Holboellia fargesii,* and *Holboellia latifolia* are usually cultivated. There is little doubt that these, and the rarer species, will receive wider recognition among gardeners.

▲ HOLBOELLIA CORIACEA *Grown mainly for its dark, evergreen foliage, this is a widely available, vigorous species, climbing to 7m (22ft), or more in favourable conditions.*

Red List: Rare*

Hoya macgillivrayi

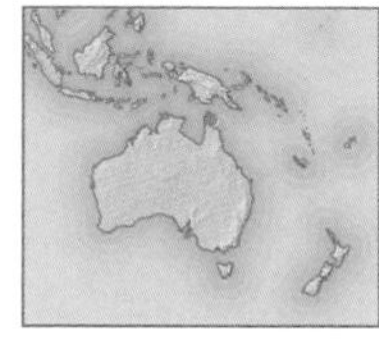

Botanical family
Asclepiadaceae

Distribution
Australia: Queensland

Hardiness
Frost-tender

Preferred conditions
Open, friable, yet moisture-retentive soil in a sheltered, lightly shaded position and a warm humid atmosphere.

The red hoya vine, one of seven *Hoya* species found within Australia, is endemic to the Iron Range and McIlwraith Ranges of the Cape York Peninsula, Queensland. It has been recognized as Rare on the Red List of 1997, and locally by the Queensland government, because of its narrow distribution. However, it is not thought to be endangered because its wide availability in cultivation ensures that collection from the wild poses no threat, and also because the dense forests it inhabits are well protected.

The forests are among the most southerly tropical rainforests, and contain many endemic plants. The flora is notable for the presence of the last remnants of the vegetation that once covered Gondwanaland. This supercontinent existed when South America, Africa, Antarctica, India, and Australia were joined, about 130 million years ago. Living fossils from this time have survived in south-east Asia and the South Pacific because the area's warm, damp climate has remained consistent over time. The area is also home to rare birds, including the red-bellied pitta, and marsupials, such as the spotted cuscus. The importance of the Cape York Peninsula as an unspoilt refuge for plants and animals is such that much of the area has been declared a national park.

◄ WAX FLOWER *The red hoya vine has long been prized by gardeners, particularly in its native Australia where a number of cultivars have been selected.*

Also known as wax flower, *Hoya* is a large genus of milky-sapped, evergreen climbers, and some shrubs and epiphytes. They grow in forests, on cliffs, and near stream margins in the warmer regions of south-east Asia and Australia. Many are cultivated as ornamental plants but few are as dramatic as the red hoya vine, *Hoya macgillivrayi*. It is a strong-growing climber, with fleshy, lustrous, evergreen leaves and clusters of large, cupped flowers that are strongly night-scented.

In warm gardens, it is spectacular when allowed to twine through shrubs and trees or over arches and pergolas, but in cooler areas, it needs the protection of a warm greenhouse. The red hoya vine is an easy to grow, vigorous species that prefers moist but well-drained compost, ideally enriched with leaf mould, pulverised bark, perlite, and charcoal. Container-grown plants perform better when they are root-bound. Propagate by seed if available, sown at 19–24°C (66–75°F) in spring, or by semi-ripe cuttings in late summer.

Red List: Vulnerable*

Ipomoea tenuissima

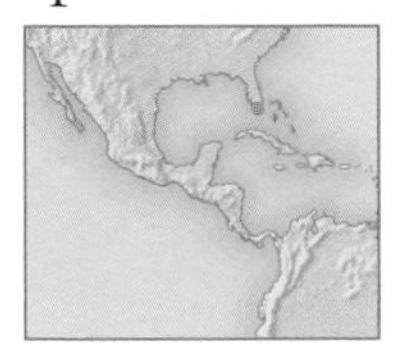

Botanical family
Convolvulaceae

Distribution
Florida

Hardiness
Frost-tender

Preferred conditions
Well-drained soil and full sun, with plenty of moisture while it is in growth.

The morning glories comprise a large genus, of 500–600 species, found mainly in the tropics, extending north to the Mediterranean. Many are popular garden plants grown as perennials in their native climates or as annuals in temperate areas. The native origins of some Unlisted morning glories, such as *Ipomoea purpurea* and *Ipomoea quamoclit*, have been obscured by years of cultivation. Indeed *Ipomoea indica* (*right*) is now classed as invasive.

Not all morning glories are so widespread. In Florida, in an area of open pine forest and seasonally flooded marsh crossed with dry limestone ridges known as the rocklands, lives *Ipomoea tenuissima* – the rocklands morning glory. It is a delicate, twining climber with arrowhead-shaped leaves characteristic of the morning glories, and rich pink, funnel-shaped flowers. The rocklands are found in Miami-Dade County in the extreme south-east of Florida, an area under constant pressure from housing developers. Building work not only destroys wild habitats but also affects those remaining, because naturally occurring brush fires, which do not harm plants but encourage their survival, are suppressed. Endemic species have declined including *Ipomoea tenuissima*. It may soon be responsibly introduced to local gardeners, but it is unknown in cultivation at present.

IPOMOEA INDICA

See also Invasive plants, p.458.

THREAT

SWEET POTATO PEST

Ipomoea includes a number of species of considerable economic importance, such as the sweet potato, *Ipomoea batatas* (*see below*). Unfortunately sweet potato plants imported to Florida in 1993 came with an unwelcome stow-away – the leaf-eating tortoise beetle (*Chelymorpha cribraria*). Within less than a decade it began to infest local species, and has been found on *Ipomoea tenuissima* in the wild. The beetle poses a potential threat to all morning glories as sweet potatoes become more popular and are grown in new regions.

IPOMOEA BATATAS

Unlisted

Lapageria rosea

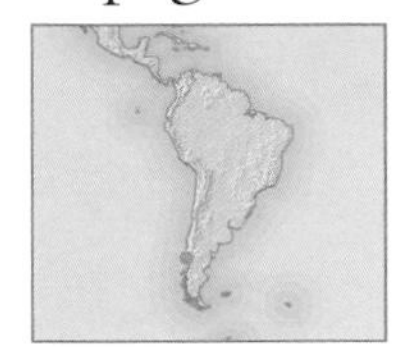

Botanical family
Liliaceae

Distribution
Chile

Hardiness
Half-hardy

Preferred conditions
Moist, open, organic-rich, acid soil, that neither compacts nor becomes waterlogged; humid, shaded or semi-shaded conditions.

PEOPLE

JOSEPHINE DE LA PAGERIE

Lapageria was named for Empress Josephine, Napoleon Bonaparte's wife, whose maiden name was de la Pageric. She loved flowers and had plants sent to her from all over the world, starting a fashion for large and exotic blooms and, most famously, roses. The beautiful gardens she created at the the Château de Malmaison, near Paris, are still admired today.

The Chilean bellflower, known in its native country as the copihue, is widely regarded as the doyen of temperate climbers by botanists and gardeners alike, prized for its large, rich pink-red, waxy bells set off by lustrous bottle-green, evergreen leaves. This and other Chilean endemics, with their extravagant flowers and lush foliage, have long been prized in temperate gardens. They give the appearance of being tropical and exotic, yet have the ability to survive in cooler climates if given a sheltered position and some protection in the coldest months of winter.

In the wild, the Chilean bellflower occurs in the temperate rainforests of southern Chile's coastal mountains, extending east into the foothills of the Andes. Its wiry stems climb into the canopy of its favoured forests of Chilean hazelnut, *Guevina avellana*, and a local avocado, *Persea lingue*.

The Chilean bellflower is declining in numbers as a result of wholesale forest clearance. Today the native forests of lowland southern Chile have largely disappeared, and have been replaced by pasture. This land is often poor quality, infested with gorse (*Ulex europeaus*), and far less productive than the original forest. Plantations of fast-growing timber trees, such as Monterey pine and eucalyptus, are popular replacements for native forest, and in some regions as little as four per cent of the original cover remains. Few of these areas receive protection within national parks and other reserves.

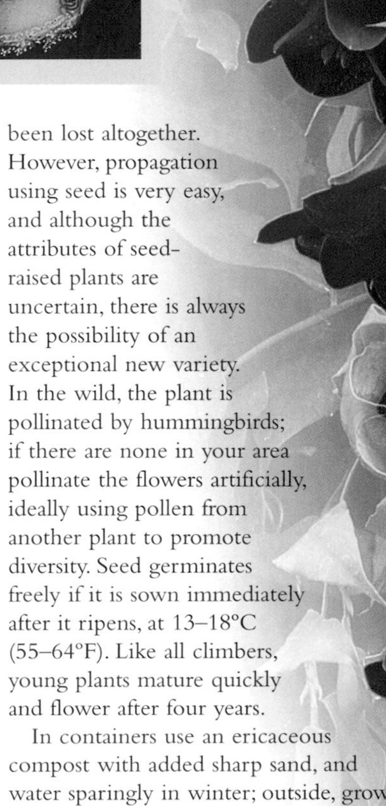

◄ LAPAGERIA ROSEA *VAR.* ALBIFLORA *Plants with unusual colouring, such as this white variety, are naturally rare and often collected. Ensure any plants you purchase are from cultivated stock.*

A number of impressive cultivars of *Lapageria rosea* exist. Most are raised in gardens but some are selected from wild populations with the result that a few colour variants, especially albino plants, are now very rare in the wild. Collection continues because the only sure way of propagating true colours is by layering, or by taking short cuttings in autumn or winter. Unfortunately both processes are difficult and slow, so all named varieties are rare and many have been lost altogether. However, propagation using seed is very easy, and although the attributes of seed-raised plants are uncertain, there is always the possibility of an exceptional new variety. In the wild, the plant is pollinated by hummingbirds; if there are none in your area pollinate the flowers artificially, ideally using pollen from another plant to promote diversity. Seed germinates freely if it is sown immediately after it ripens, at 13–18°C (55–64°F). Like all climbers, young plants mature quickly and flower after four years.

In containers use an ericaceous compost with added sharp sand, and water sparingly in winter; outside, grow in slightly acid, moist but well-drained, humus-rich soil. Like most forest plants, the Chilean bellflower prefers humid, shaded or semi-shaded conditions.

PRIDE OF CHILE ► *The national flower of Chile is this woody, twining vine. It climbs to 5m (15ft) and in summer and autumn bears masses of 10cm (4in) bellflowers.*

24 on Red List*

Lathyrus

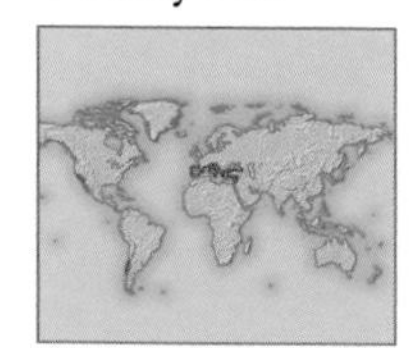

Botanical family
Leguminosae

Distribution
Europe, Middle East, western Asia, North and South America

Hardiness
Fully hardy – frost-tender

Preferred conditions
Well-drained, fertile, humus-rich soil in a warm, sunny position.

This genus of annual and perennial peas is found mostly in areas with Mediterranean climates in the Americas, the Middle East, and across Europe, especially in Turkey. Some are clump-forming, but the majority are tendril climbers that are happiest scrambling through shrubs. With regular deadheading, the flowers will continue all summer and many species make fine garden plants, such as the popular hardy perennial,

▲ LATHYRUS UNDULATUS *This Unlisted perennial pea has only recently become known among gardeners. It is floriferous and easy to grow and propagate from seed.*

Lathyrus latifolius, with its handsome, blue-green leaves and long stems of pink or purple flowers.

However, there is no doubt that the annual sweet pea, ***Lathyrus odoratus,*** is the most popular species in the genus. Many sweet pea cultivars exist, but the origin of the species has been lost through intensive cultivation. It is thought that the small, richly coloured and powerfully scented 'Cupani', sent from Sicily to England by Franciscus Cupani in 1699, most closely resembles the wild plant.

Sweet peas usually have a colour range of mauve, pink, and white so interest has been piqued by reports of a newly described relative with vibrant red and chromium yellow blooms: the Belen pea, ***Lathyrus belinensis***, from the extreme south-west of Turkey. The Belen pea is an annual that scrambles through scrub in orchid-rich clay pasture and often along adjacent road verges. It is highly

LATHYRUS ► ODORATUS *Sweet peas are prized for their sweetly scented flowers which are ideal for cutting, and make them one of the most loved plants of the traditional English cottage garden.*

threatened by pasture reclamation and conversion to arable production.

Fortunately the Belen pea has proved relatively easy to cultivate, and seed is available from a number of nurseries. The species appears to respond favourably to the same basic growing techniques as the sweet pea and other annual *Lathyrus*. In cooler climates, seed can be sown in autumn or early spring in pots, and the young peas planted out in well-manured ground in late spring. Autumn-sown plants are larger, have more flowers, and are more likely to set viable seed during a short, cool summer, but they must be protected from frost and are vulnerable to damping off. In areas with long, hot summers the Belen pea will grow easily outdoors, and self-seed for the following year without intervention. Alternatively, seed can be collected when it is ripe and stored in a cool dry place where, under normal conditions, it should maintain its viability for a few years. In the sweet pea stronghold of the temperate gardens of north-west Europe, the lower light and heat intensity often cause flowers of the Belen pea to fail to take on their natural vibrancy. Hence breeding programmes are underway to introduce its hot colours to cooler-growing species.

Perennial Pea

As well as annuals such as the Belen pea, Turkey is home to a number of perennial peas, commonly called everlasting peas. Of these, one of the most handsome is ***Lathyrus undulatus***, noted for its ease of growth and abundance of bright, magenta-pink flowers. In the wild, it grows in dry scrub, heath, and forest-margin habitats on acid soils in the Istanbul region. Because of this narrow distribution, it is classified as Rare on the Red List of 1997. The bulk of the known populations are under considerable threat from the inevitable pressures of urban expansion because they lie in the vicinity of Istanbul, a huge city of up to 15 million inhabitants. The climate of this region of north-west Turkey, influenced by both the Black and Mediterranean Seas, has a number of localized habitats which are particularly rich in rare plants, with no fewer than 40 species largely or wholly confined to the area. Of these perhaps the best known among gardeners is the rose of Sharon (*Hypericum calycinum*), which is widely used in landscaping schemes in temperate areas where it is invaluable for its freely spreading nature and masses of flowers. In the wild, the rose of Sharon often grows with *Lathyrus undulatus*. The combination of the everlasting pea with its hot magenta blooms, scrambling through the shrubby, golden-flowered *Hypericum*, creates a spectacular splash of colour in the garden, just as it does in the wild. *Lathyrus undulatus* prefers a sunny or semi-shaded spot on fertile loam, and flowers in the second season after sowing. Unlike the better-known *Lathyrus latifolius,* this species is more compact, to just 1m (3ft) in height. Like all *Lathyrus* species, *Lathyrus undulatus* is easily propagated from seed, which germinates best if it is soaked or chipped before sowing. Everlasting peas grow rapidly, particularly if they are given a boost with plenty of moisture and well-rotted organic matter added to the soil.

See also Invasive plants, *p.459.*

LATHYRUS BELINENSIS *In spite of its vibrant colours and its visibility along the verges of Turkey's main southern coastal road, the Belen pea was virtually unknown until 1988.*

Red List: Endangered*

Mandevilla campanulata

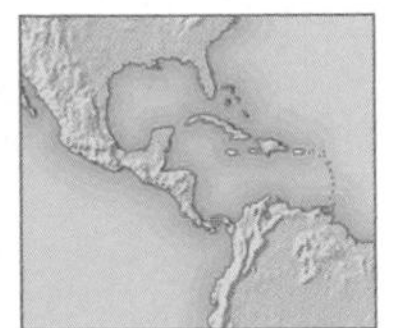

Botanical family
Apocynaceae

Distribution
Panama

Hardiness
Frost-tender

Preferred conditions
Neutral or slightly acid soil, in partial shade, with protection from drying winds.

Notable for its pale yellow, tubular flowers and lustrous green leaves, the recently discovered *Mandevilla campanulata* is a good candidate for tropical and subtropical gardens, and greenhouses in temperate regions. The species was found in 1973 in dense tropical rainforest on the Santa Rita Ridge of the Colón province in central Panama, when rough trackways were cut to allow access for logging vehicles. The tracks also increased access for botanists to this remote area, yet sadly the forestry operations that have allowed people to document and admire this climber are also likely to result in its decline and possible extinction. The forests of the Santa Rita Ridge have many endemic species but are currently unprotected and huge swathes are being irretrievably altered or destroyed by logging, plantations of non-native species, and urbanization. *Mandevilla campanulata* is the rarest of the five *Mandevilla* species found in Panama. There is, however, a good chance it will find sanctuary in gardens because the yellow colour of its flowers offers an alternative to the widely grown varieties which are typically pink or white. Given the vigorous nature of many climbers it is very likely that its introduction into cultivation can be successful.

MANDEVILLAS IN THE GARDEN

Mandevilla campanulata is one of approximately 120 species of mainly woody, twining vines from tropical forest habitats in central and south America. The genus is very ornamental, and the flowers often sweetly scented. A number of forms such as *Mandevilla* x *amoena*, *Mandevilla boliviensis*, *Mandevilla sanderi*, and their cultivars are popular in gardens in their native South America, and throughout the tropics. They have long been grown in greenhouses and conservatories in cooler climates.

Although the unusual yellow *Mandevilla campanulata* is not known in cultivation at the present time, it is likely to respond well to conditions favoured by the other species and varieties. They demand a neutral to acid soil, preferably enriched with humus, and need some protection from strong midday sun and drying winds. Where winter temperatures drop below 5°C (41°F), give *Mandevilla* plants the protection of a warm greenhouse or conservatory with bright, filtered light.

2 on Red List*

Monstera

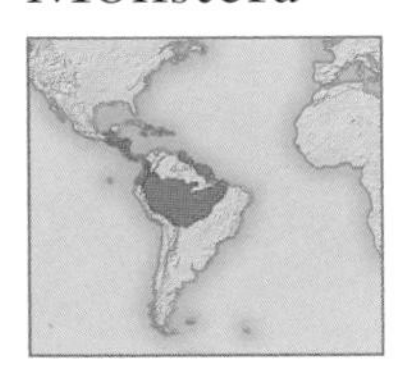

Botanical family
Araceae

Distribution
Central and South America

Hardiness
Frost-tender

Preferred conditions
Humus-rich, moist but well-drained soil in full or partial shade.

Native to the rainforests of Central and South America, the genus *Monstera* is part of the arum family, the Araceae. Known variously as ceriman, Mexican breadfruit, and the Swiss-cheese plant, ***Monstera deliciosa*** (Unlisted) is the best-known member of the genus. In gardens it is widely used as a foliage plant because of its large, lobed leaves. It is the only member of the genus to have edible fruits; they taste like pineapples and are grown commercially.

In contrast, the genus as a whole remains something of a mystery, both taxonomically and ecologically. A review published in 1977 suggested there were just 22 species, yet later work has so far identified 60 species, with 22 of those found in Costa Rica alone. This case illustrates the frustration of botanists and conservationists, who have little hope of describing and documenting the diverse and abundant flora of the tropical forests before it is destroyed by human activity.

Many *Monstera* species have been found in the swamp forests along Costa Rica's Atlantic shoreline, including the endemic *Monstera costaricensis* (Rare★), an impressive epiphyte that climbs many meters in height, tightly pressed to the trunks of trees. The area was part of a larger sweep of lush, tropical forest that ran along the Atlantic coast, from southern Nicaragua to Panama. It was full of huge canopy trees, steadied by vast buttress roots, and adorned with a rich array of epiphytic plants. Catalyzed by government incentives, the area has been heavily settled in recent years. Some of the forest has been logged and converted to cattle pasture and banana plantations. Relatively few large blocks of forest remain untouched and little is protected, even in Costa Rica, a country famous for its national parks and enlightened conservation policies. The Atlantic swamp forests are now among the most threatened habitats in Central America.

MONSTERA DELICIOSA

Plants in the tropical genus *Monstera* are frost-tender and prefer partial shade and moderate to high humidity levels. They are best grown in containers in humus-rich soil, with plenty of water. The rarer species, such as *Monstera costaricensis,* are not yet known in cultivation but with their bold, evergreen leaves could prove to be a valuable addition to the houseplant trade.

❗ These climbers can be aggressive: do not plant in tropical areas outside of their native habitats (*see box below*).

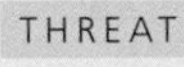

THREAT

RAMPANT CLIMBER

Millions of Swiss-cheese plants have been sold as houseplants since they were first reported in cultivation in 1752. Containers stem their growth so that people remain unaware of the size these climbers can reach. When introduced to the Old World tropics, they can become highly aggressive. There are reports of this species establishing in forests as far apart as Africa and New Zealand. In the Seychelles, cheese plants, together with related species such as philodendrons, are an increasing problem as they invade secondary forests and plantations. There is a considerable risk they will push into the primary, high biodiversity forests. Control by mechanical measures has thus far proved unsuccessful.

Red List: Vulnerable

Mutisia magnifica

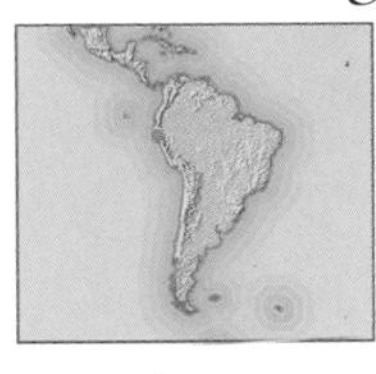

Botanical family
Compositae

Distribution
Ecuador

Hardiness
Half-hardy

Preferred conditions
Fertile, moist, yet perfectly drained soil in a sunny site.

More and more species of *Mutisia* are grown in gardens, where they are prized for their brilliantly coloured flowers. The genus was named after the 19th-century Spanish naturalist José Celestino Mutis and contains about 60 species scattered through South America. They are most abundant in the Andes, where they scramble among forest and scrub by means of tendrils on their leaves.

Mutisia magnifica is a recent discovery from the Uritusinga Hills of south-west Ecuador. When it was found, it was already threatened by forest clearance for charcoal production, and is considered Vulnerable on the 2000 Red List. As its Latin name suggests, it has magnificent flowers and will undoubtedly be well received if it proves possible to cultivate, although it is not currently available. Like other species, it is likely to detest root disturbance, and prefer fertile, moist, well-drained soil. Grow *Mutisia* plants as annuals in cool climates through a support that will let their tendrils twine. Propagation is by seed, sown when ripe, or by stem-tip cuttings in summer.

MUTISIA MAGNIFICA

45 on Red List*

Passiflora

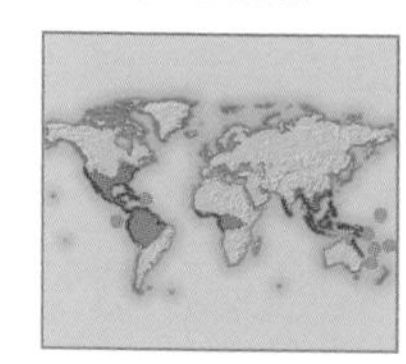

Botanical family
Passifloraceae

Distribution
Sub-tropics and tropics worldwide

Hardiness
Half-hardy – frost-tender

Preferred conditions
Moderately fertile, moist but well-drained soil in full sun or partial shade with shelter from drying winds.

The passion flowers belong to a large genus numbering as many as 600 species. They are found in the tropics and subtropics in the Americas, Africa, Asia, and Australia. Most are vigorous, climbing up to 20m (70ft) by means of tendrils into the forest canopy. Their unmistakable exotic flowers make them very popular with gardeners. Even species that are rare in the wild, such as Colinvaux's passion flower, *Passiflora colinvauxii*, from Santa Cruz in the Galapagos archipelago, have gained a firm foothold in cultivation. In warm climates, passion flowers are very easily grown and require no specialist care. Despite their tropical origins some species, particularly the blue passion flower, ***Passiflora caerulea*** (Unlisted), can thrive outdoors in temperate climates in a sheltered spot. If seed is available, sow it in spring. Otherwise take semi-ripe cuttings in summer.

Many passion flowers have close relationships with animals; their flowers have evolved into a huge variety of shapes and stunning colours to attract a range of pollinating creatures, such as moths, butterflies, bees, hummingbirds, and bats. Their fruits are sweet and distinctly flavoured to encourage wide distribution by animals, including humans.

HELICONID BUTTERFLIES

HUMMINGBIRDS

Some passion flowers are intimately connected with hummingbirds. ***Passiflora citrina,*** Rare on the Red List of 1997, is a good example. In probing for nectar, the birds trigger the style, or female part of the flower, to move towards the bird's head and beak where pollen from other plants is likely to be caught, increasing the chances of pollination.

In Colombia the survival of *Passiflora parritae*, identified as Endangered on the Red List of 1997, is locked to the survival of a hummingbird – the only animal that can pollinate its flowers. The plant is one of the world's rarest and most magnificent passion flowers, with flowers up to 25cm (10in) across. Its sole pollinator is the sword-billed hummingbird, *Ensifera ensifera*, the only bird on earth to boast a bill that is longer than its body length, excluding the tail. Its beak is a remarkable 10cm (4in) in length, just long enough to reach into the giant passion flower, but so large that the bird has to hold its head vertically whilst at rest to reduce neck strain. It is only found at 2,500m (8,200ft) and conservationists fear that global warming is forcing hummingbird populations to move to higher altitudes, leaving *Passiflora parritae* without a suitable pollinator. The only known specimen in cultivation is in the Strybing Arboretum, San Francisco. Unfortunately attempts at propagation have so far proved largely unsuccessful.

▲ PASSIFLORA CITRINA *A rare yellow-flowered species, this small, elegant vine was recently discovered in the moist pine woodlands that clothe the hills of central and western Honduras and eastern Guatemala.*

BUTTERFLIES

Although passion flowers are packed with delicious nectar, their foliage is poisonous to most herbivores. There is an exception – the leaves are the primary food source for the caterpillars of over 70 species of beautiful Heliconid butterflies. The caterpillars store the chemicals safely within their bodies to make them unpalatable to predators. The passion flowers, however, do not give up their leaves willingly – evidence suggests the plants disguise themselves by varying their leaf shapes. As an extra precaution, the leaves often develop raised yellow spots and tightly coiled tendrils, which look like butterfly eggs and larvae respectively. Butterflies searching for an egg-laying site move on to find an unpopulated plant.

See also Invasive plants, *p.465.*

PASSIFLORA CAERULEA *In 1609 an Augustinian friar described the passion flower as "stupendously marvellous". This blue passion flower is common in cultivation today.*

Red List: Indeterminate*

Schisandra grandiflora

Botanical family
Schisandraceae

Distribution
China and India: the Himalaya

Hardiness
Fully hardy

Preferred conditions
Slightly acid, moist but well-drained soil, in partial shade.

The genus *Schisandra* is generally grown for its leathery leaves and pretty flowers. ***Schisandra grandiflora***, Indeterminate on the Red List of 1997, is no exception. This deciduous, twining vine can climb up to 6m (20ft). In the wild it is found in south-west China and northern India, most notably in the Himalaya where it favours forest and scrub up to an altitude of 3,300m (10,800ft). The vine is able to cope with the variety of conditions over this wide altitudinal range in its natural habitat, and so easily adapts to cultivation. *Schisandra grandiflora* has been widely distributed among gardeners and is now grown in many areas including North America, Europe, and New Zealand.

Despite the success of *Schisandra grandiflora* in gardens, scientists know relatively little of its status in the wild and more information is urgently needed. Plants of the Himalaya are increasingly under threat as the area opens up to tourism, with tracks and trekkers carving through virgin habitats. There is also increasing pressure from grazing animals and urban expansion.

Schisandra grandiflora, and other species in the genus, favour rich, well-drained yet moisture-retentive, slightly acidic soil. Ideally they prefer light shade, or full sun if the soil is reliably moist and the roots are in shade. Unlike many climbers, *Schisandra grandiflora* is not rampant, so its semi-twining stems may require tying in. Plants are single-sexed (dioecious), with either male or female flowers. Both male and female plants need to be grown to set seed. Seed is best sown as soon as it ripens in autumn. Otherwise, take semi-ripe cuttings in summer.

▲ HIMALAYAN BEAUTY *The creamy white or pinkish, fragrant flowers are followed by bunches of berries, red as sealing wax, if male and female plants are grown together. The fruits are edible, with a pleasantly tart taste, and can be eaten both raw and cooked.*

Red List: Extinct

Streblorrhiza speciosa

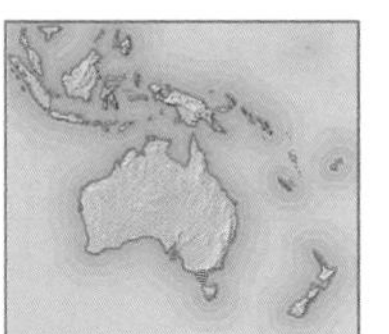

Botanical family
Leguminosae

Distribution
Australia: Philip Island

Hardiness
Half-hardy

Preferred conditions
Once found in moist but well-drained soil in sun or part shade.

When Captain Cook landed on the precipitous, volcanic Philip Island in 1774, it was clothed in a mosaic of dense forest and scrub. In 1788 goats and pigs were introduced to provide meat for the newly inhabited Norfolk Island, situated some 10km (6m) to the north. The stock was later supplemented by rabbits. The animals began eating their way through the lush vegetation noted by Cook. By the time the Austrian botanist Ferdinand Bauer found the Philip Island glory pea, *Streblorrhiza speciosa*, in 1806 it was already doomed. Bauer thought the climber would be popular with gardeners and collected seed for cultivation. He was just in time – by 1830 the levels of grazing were so intense that the woodland and scrub was restricted to inaccessible gullies.

In Europe, the Philip Island glory pea was much admired and widely grown in greenhouses. Sadly, it failed to live up to expectations by flowering only sporadically when confined to a pot, and it is thought to have died out in cultivation within little more than 50 years. Efforts to trace survivors have failed, and the species is classified as completely Extinct on the Red List of 1997. Its loss is particularly tragic because this distinctive genus was represented by just the one species.

THREAT

AN ISLAND TRANSFORMED

This botanical illustration (*left*) is all that is left of the glory pea. Other Philip Island species have fared little better: of the three endemics, *Hibiscus insularis* is the only survivor, albeit Endangered on the Red List of 1997, and the grass *Agropyron kingianum* was last seen in 1912.

Decades of overgrazing turned the island into little more than a colourful desert. Indeed the vegetation was so sparse that the goats and pigs died from starvation. However, since the eradication of rabbits in 1986 there is hope that much of the natural plant cover can be restored. The first step is under way: thickets of introduced African olives are being thinned to open up niches for native shrubs.

Red List: Vulnerable

Strongylodon macrobotrys

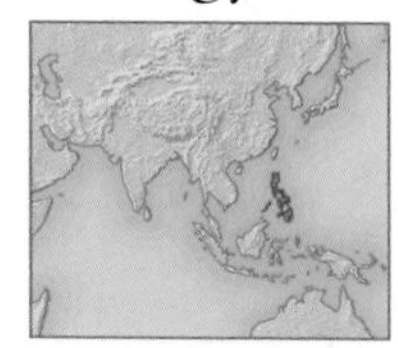

Botanical family
Leguminosae

Distribution
Philippines

Hardiness
Frost-tender

Preferred conditions
Humid conditions in light shade or full sun, and moist, fertile, neutral to acid soils.

Few, if any, of the 8,000 native plants of the Philippines can match the jade vine for sheer magnificence. Its twining stems can reach to 20m (70ft) into the forest canopy, where they produce pendent clusters of flowers to nearly 1m (3ft) in length, each bearing up to 100 luminous turquoise, pea-like blooms. ***Strongylodon macrobotrys*** is undoubtedly the best known, and most spectacular, member of a genus which numbers about 20 species, including some shrubs, from south-east Asia and the Pacific Islands. *Strongylodon macrobotrys* is confined to lowland tropical evergreen forest, up to altitudes of 1,000m (3,300ft), in the Philippines, and is found particularly on the island of Luzon.

THREATENED HABITAT

The Philippines have several native plants in high demand as ornamental and exotic species, such as *Nepenthes,* pitcher plants (*see pp.436–37*), the stag's-horn fern, *Platycerium grande* (*see p.322*), *Cyathea* tree ferns (*see p.316*), and the orchid *Vanda sanderiana* (*see p.422*). Removal for horticultural purposes poses a considerable threat to these small, portable plants but the jade vine, with its sprawling, climbing habit is far from easy to transport.

However, *Strongylodon macrobotrys* is highly vulnerable to the expansion of towns, illegal squatters, selective logging by local people, and, most disastrously, the resumption of large-scale commercial logging. Even protected areas are affected, such as the Palanan Wilderness Area in north-east Luzon – a stronghold for the species, and noted by the International Union for the Conservation of Nature (IUCN) as a Centre of Plant Diversity because of its exceptional botanical richness. The Philippine rainforests are among the most threatened on earth, and it is estimated that up to 80 per cent has already been lost. Indeed, the nation has been forced to start importing timber to supplement declining stocks.

The sheer vigour and size of the jade vine means that in cultivation it needs a sturdy support, outside in warm climates, or in a large greenhouse in temperate areas with a minimum temperature of 15°C (59°F). Indoors the species should be watered copiously during periods of rapid growth and given a general liquid fertilizer application every two to three weeks. Propagate the jade vine by seed or semi-ripe cuttings. Sow seed as soon as it ripens because it is very short-lived, losing its viability within one to two weeks.

EXTRAVAGANT FLOWERS ►
Discovered in 1854 by botanists of the US Wilkes Exploring Expedition in the forests of Mount Makiling on the Philippine island of Luzon, jade vine rapidly became a popular plant for conservatories in temperate regions.

Red List: Endangered*

Tecomanthe speciosa

Botanical family
Bignoniaceae

Distribution
New Zealand: Great Island

Hardiness
Frost-tender

Preferred conditions
Rich, moist soil, with some shade from midday sun; ample rainfall in the growing season.

Just one vine of this handsome, strong-growing, twining climber was discovered in 1945 on Great Island. It was growing at relatively low altitudes, on damp streamside soils, ascending 10m (30ft) into the natural canopy of scrubby *Leptospermum ericoides* trees. Sadly, Great Island, one of the Three Kings group off New Zealand's North Island, has a long history of degradation, first by Maori settlements and later by grazing goats introduced by the Europeans in 1889. As a result, the extensive tracts of mixed coastal forest that are thought to have covered large areas of the island are mostly gone. In 1946, the New Zealand government finally eradicated the feral goats, and within just a few years the populations of many of the rarer Three Kings endemics had begun to expand. Unfortunately, to date the sole plant of *Tecomanthe speciosa* has steadfastly refused to produce fruit or seedlings in the wild, although there are six plants nearby produced by natural layering.

GROWING POPULARITY

The history of *Tecomanthe speciosa* in cultivation is more heartening. Its lustrous, evergreen foliage, pendent clusters of large, cream, tubular flowers, and plight in the wild, have captured gardeners' hearts. Fortunately the single clone is self-fertile and, although the wild plant seems unable, its offspring set copious seed and are easy to grow. The species has been successfully cultivated and distributed within New Zealand and is now becoming widely established in gardens throughout the subtropics, and under cover in cooler climates.

A vigorous climber, *Tecomanthe speciosa* needs a stout arch or pergola for support. In the wild, it enjoys rich, moist but free-draining soil, and copious water during the growing season. If you can provide similar conditions, it will thrive. Plant it in semi-shade or in sun with shade around the roots. It will tolerate coastal winds, but is unlikely to withstand any but the lightest of frosts.

Propagation is easy from seed; fruits take between eight and ten months to ripen and the seed can then be sown at 18–21°C (64–70°F) in spring. Alternatively, take semi-ripe cuttings and root them with bottom heat in summer.

4 on Red List*

Tropaeolum

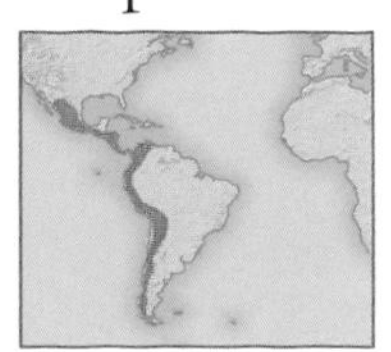

Botanical family
Tropaeoleaceae

Distribution
Central and South America

Hardiness
Half-hardy – frost-tender

Preferred conditions
Moist but well-drained soil in full sun.

Gardeners are already familiar with this genus of annual and perennial climbing, bushy, or trailing plants with their fleshy, lobed leaves and stunning flowers. In the wild, *Tropaeolum* species grow in the Andes of South America, from Patagonia and southern Chile to Colombia, extending into Mexico. They occur in a range of mountain habitats including moist woods, dry scrub, and scree, where they use their tendril-like leaf stalks to scramble up or over any obstacles. Andean ecosystems are on the verge of disaster as global warming threatens to shift habitats upward, and as deforestation continues apace. Luckily, with their brightly coloured flowers and often spectacular foliage, *Tropaeolum* plants are becoming widely distributed through the worldwide gardening community. Available species occasionally include *Tropaeolum looseri*, considered Rare on the Red List of 1997.

◀ TROPAEOLUM TUBEROSUM *Grown as a food crop in its native South America, this perennial climber, which can reach 4m (10ft), is also a popular and widely available ornamental with many cultivars.*

Some widespread, Unlisted species are grown for food. ***Tropaeolum tuberosum*** is cultivated for its edible tubers in the Andes. *Tropaeolum majus,* commonly called nasturtium or Indian cress, is grown in Peru for its edible flowers and leaves. Outside its native range, Indian cress is used as an ornamental annual but is gaining popularity as a salad crop. It has abundant, brilliant red, orange, and yellow flowers and lush foliage throughout the summer.

In areas where naturalization is unlikely, such as cold climates, grow *Tropaeolum* in moist, well-drained soil, in full sun, with plenty of water in the growing season. Sow seed as soon as it is ripe, but germination can be erratic. Alternatively separate tubers in autumn, or root stem-tip cuttings in summer.

! Climbers in the genus *Tropaeolum* can be aggressive, so use them with care in frost-free climates. In New Zealand, the Chilean Flame Creeper, *Tropaeolum speciosum*, was introduced for its handsome red flowers, but escaped into the wild and is now a serious pest, listed on the New Zealand National Pest Plant Accord which prohibits its propagation, distribution, and sale.

1 on Red List*

Vitis

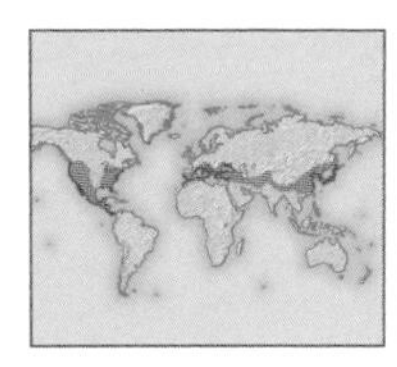

Botanical family
Vitaceae

Distribution
North America, Europe, Asia

Hardiness
Fully hardy

Preferred conditions
Well-drained, neutral to alkaline, humus-rich soils, in full sun or partial shade.

The large genus *Vitis* comprises over 60 species of climbers worldwide, as well as many commercial varieties grown for grapes and wine, and ornamental plants with bright autumn foliage and decorative, if unpalatable, fruit. The greatest diversity of species is in North America and east Asia.

The species we recognize as the domestic grape – *Vitis vinifera* – is the sole Mediterranean representative of the genus and is invaluable for its sugar-rich berries that can be eaten fresh, dried and stored as raisins, or fermented into wine.

Naturally, the wild grape grows in forest, gorge, and streamside habitats, where conditions are humid and mild. Millennia of cultivation have blurred its original distribution, but it is assumed to be indigenous to the Mediterranean and Black Sea shores, east to the southern Caspian forests.

Wild grapes produce male and female flowers on different vines, so only female plants bear the small, often bitter fruit. Since grape cultivation began – the earliest evidence comes from 5,000-year-old sites in the eastern Mediterranean – people have selected large fruit size, sweetness, seedlessness, and self-fertility. Today there are fixed varieties, increased only by hardwood cuttings or grafting.

◀ WILD GRAPE Vitis *species are found in most habitats except extreme deserts and tundra. They form an invaluable genetic stockpile from which useful characteristics can be selected and passed to cultivated grapes via breeding programmes.*

CULTIVATION

CROP AT RISK

Phylloxera is a disease caused by yellow aphids that feed on the roots and leaves of grapevines, causing plants to weaken and die. Symptoms begin to show within five years of the initial infestation, and vines become uneconomic within ten years.

The disease originated in North America and spread rapidly to most wine-growing regions during the mid-1800s with the worldwide transport of plants. In Australia it wiped out vines in Victoria in a decade. The French wine industry faced collapse after *Phylloxera* was introduced in 1860 and devastated 25 million acres of vineyards in 25 years. The saviour was found in wild American grapes that had evolved resistance to the disease. Grapevines were grafted onto New World rootstocks, halting the spread of *Phylloxera*.

Thus, most of the world's grape varieties are based on this single, highly bred, Mediterranean species. It has been transported across the world, coming into contact with new diseases and growing conditions. Whereas local varieties build up resistance and tolerance to their local pests and climate, the imported grape has none. Breeding programmes worldwide aim to find desirable traits in the many North American and Chinese species and transfer them to the domesticated grape. Several wild species have been selected for trial; of these there is great hope for the Callosa grape, *Vitis shuttleworthii*. This American species is locally abundant in parts of central and southern Florida, growing in mixed woods, pine woodlands, and hammocks – dense hardwood stands that form on raised ground. It has relatively large fruit, high levels of resistance to fungal diseases, moderate resistance to the nematodes that spread fanleaf virus, and its Florida provenance brings hope of future cultivars that will thrive in sub-tropical and tropical areas. This gives the Callosa grape immense potential value.

While it grows within a number of protected sites, including the Everglades National Park, the species remains at threat in other areas due to agricultural and urban land development and has accordingly been classified as Vulnerable on the Red List of 1997.

BULBOUS PLANTS

Adapted to life in regions of alternating climatic extremes, bulbous plants are the survivors of the botanical world. Whether true bulbs or bulb look-alikes, such as corms, rhizomes, and tubers, these plants have one thing in common – specialized food-storage organs that enable them to survive a long dormant period, often spent underground. In adverse conditions, especially prolonged drought, many bulbs shed their leaves and live off stored-up nutrients. When favourable conditions return, they rush to produce leaves and flowers, which gives them an edge over other plants.

WHAT IS A BULB?

In bulbous plants, unlike other perennials, new growth always begins below ground, fuelled by stores of sugars and starches. For this reason, botanists call them geophytes, or "earth plants". Different types of bulbous plants store food in different places. In true bulbs, food is stored in a number of small, fleshy "scale" leaves, which form concentric circles around the growing tip. In corms, food is stored in the stem, which swells at the base into a solid mass of storage tissue. In rhizomes, as in corms, food is stored in the stem, which grows more or less horizontally. Tubers are either roots or the underground ends of stems that are swollen by food-storage tissue.

IRIS

LILY

DAFFODIL

GLADIOLUS

ERYTHRONIUM

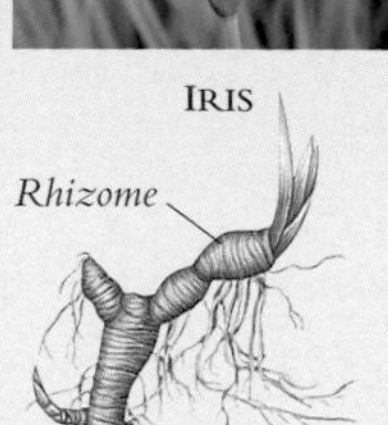

RHIZOME *On iris rhizomes, roots grow from the lower surface, while shoots develop from the ends.*

Basal plate where new roots form

BULB SCALES *In bulbs such as the Turk's cap lily, the fleshy "scales" overlap loosely.*

Thin, papery tunic

BULB IN TUNIC *Narcissi are bulbs that are encased in a papery tunic; onions are also bulbs of this type.*

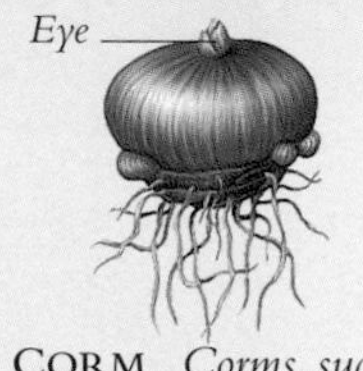

CORM *Corms, such as gladioli, have a top growing point or "eye" and a basal plate from which roots grow.*

Starchy, swollen roots

TUBER *Root and stem tubers occur in many shapes. These are pointed, like an erythonium bulb.*

WHERE WILD BULBS GROW

Geophytes come from almost every corner of the globe, but many of the ornamental species are found in sunny Mediterranean-climate regions with hot, dry summers and cool, wet winters. Particularly rich in bulbous flora are the western Cape of South Africa and the Mediterranean region proper, stretching east to central Asia. From the western Cape come such subtropical bulbous plants as agapanthus, gladioli, and ixias, which cannot tolerate freezing temperatures. Daffodils, crocuses, grape hyacinths, tulips, and fritillaries are among the popular hardy bulbs that come from the Mediterranean region and central Asia. In hot, dry conditions, these bulbous plants lie low, only sending out shoots when they detect sufficiently high moisture levels, and only displaying flowers when pollinators are plentiful. Bulbous plants such as the trilliums of North America and the bluebells of western Europe, native to deciduous forests, produce their foliage and flowers before the trees are in full leaf, when the highest levels of rain and sunlight reach the woodland floor.

▲ ISLAND CROCUS *In its native Corsica, this delicate, scented bulb flowers in spring, when temperatures increase.*

◄ PANCRATIUM TORTUOSUM *Native to sandy and rocky habitats and sometimes called the "desert daffodil", this flowering bulb has characteristic twisted leaves.*

SURVIVAL STRATEGIES

Much of the charm of flowering bulbs lies in their predictable emergence and equally predictable retreat. When adverse weather arrives, many go dormant, shedding their foliage and living off nutrients stored during the previous growing season. Although they appear to be resting, the plants continue to develop even after they seek refuge underground. Their final development is completed only after certain environmental requirements are met and the plants are stimulated to put out roots, leaves, and flowers, starting the cycle anew.

▲ TROPICAL LILY *When the summer rains begin, the vlei lilies (see p.229) bloom in Kgalagadi Transfrontier Park, South Africa.*

TULIPA KAUFMANNIANA ► *Native to Kazakhstan, Uzbekistan, Tajikistan, and Kyrgyzstan in central Asia, these exquisite lily-flowered tulips grow wild on open hillsides, or in rocky or sandy soils where drainage is good. In summer, when the bulbs are baked by the sun, buds are formed.*

WHY BULBS ARE IN TROUBLE

Like so many plants, bulbous species are disappearing as their wild habitats are transformed into farms or human settlements, or taken over by invasive species introduced from other lands. South Africa's incomparably diverse Cape Floral Kingdom includes the richest bulbous flora in the world, but many Cape species have such a tiny range that ploughing a field or building a single house can wipe out the entire world population. To make matters worse, the flora is under great threat from invasive species, including acacias from Australia, and European pines. The same strategy that has allowed bulbous plants worldwide to survive and evolve for millennia has contributed to the recent decline of some species.

▲ **SPARAXIS ELEGANS** *A spectacular member of the iris family, this plant grows in the Cape region of South Africa.*

Bulbs are easy to dig up from the wild and transport around the world when they are in their resting phase and can survive without light, food, or water. The desire of gardeners around the world to grow these beautiful plants has inadvertently encouraged the trade in wild-collected bulbs. Collection from the wild is particularly damaging to bulb populations because entire plants are removed, eliminating the possibility of future vegetative growth. The plants are also dug up when the flowers are still visible and identifiable, before they have shed their seeds. With no seeds left in the soil, populations have no means of replenishing themselves.

Turkey is endowed with a rich bulb flora, including species found nowhere else, and has long been an important source of flowering bulbs for the international market. During the 1970s and 1980s, the scale of the trade escalated. By 1986, more than 60 million bulbs collected by villagers were traded; over half of these were snowdrops. Turkey was not alone, however, and devastating depletions of woodland bulbs also took place in Europe, North America, and Japan. Fortunately, the resulting outcry led to national and international efforts to control the trade in wild bulbs.

THE MODERN BULB TRADE

By the end of the 17th century, Dutch traders were exporting bulbs to several countries, and today Holland remains the hub of the international bulb trade, claiming 65 per cent of the market share. The major countries importing Dutch bulbs are the UK, Germany, and other European nations, as well as Japan, Canada, Taiwan, China, Australia, Mexico, Russia, South Korea, and Costa Rica. According to the International Flower Bulb Centre, based in Hillegom, Holland, about 7 billion bulbs are exported from the country annually. The tulip remains Holland's number-one bulb export, followed by lilies, irises, daffodils, crocuses, gladioli, grape hyacinths, and hyacinths.

In an attempt to reduce pressure on wild populations, three bulb genera have been listed in Appendix II of the Convention on International Trade in Endangered Species (*see p.483*). CITES export permits and government approval are

▲ **BLUEBELLS IN WOODLAND CLEARING** *Native to the deciduous forests of western Europe, bluebells produce their leaves and flowers in spring when light levels are still high, and before the trees overhead are in full leaf. It is illegal to collect bluebell bulbs from woodlands in the UK.*

LEUCOJUM VERNUM ► *The exquisite snowflake is related and similar to the snowdrop* (Galanthus *species*). *Emerging from freezing, snow-covered earth, the bulb has spent the greater part of the year underground, conserving its energy and food supplies.*

▲ BULBS FOR SALE *Amsterdam's famous floating flower market is one of the city's major tourist attractions. Here, responsible gardeners can choose from a huge array of flowering bulbs.*

▲ COMMERCIAL PRODUCTION *In the "Bloembollenstreek" or bulb region of the Dutch town of Lisse, narcissi and other bulbs are cultivated on a massive scale.*

required before all species of cyclamens, snowdrops, and sternbergias can be commercially traded. In 1996 the convention came into force in Turkey, long the epicentre of wild-bulb collection. Today *Cyclamen mirabile* is banned from export in Turkey, and the country's quota system for wild-collected species is considered among the best in the world, capable of protecting its native bulb populations.

The Netherlands, the world's largest bulb producer and exporter, has also taken some serious measures to help bring the trade in wild-collected bulbs to sustainable levels. In 1990 the Dutch bulb industry began phasing in a labelling programme, and, since 1995, all bulbs exported by Dutch firms have been required to bear the labels "from wild sources" or "from cultivated stock". This helps ensure that wild-harvested bulbs shipped to and exported from the Netherlands are clearly labelled.

PROPAGATION PROJECT

For more than a decade, a small project in the mountains of Turkey has proved that cultivating native bulbs in villages close to wild populations can both protect the species and provide much-needed income for the local people. In 1991, the Indigenous Bulb Propagation Project was jointly established by Fauna and Flora International and the Turkish Society for the Protection of Nature. The first trial site for the project was selected in 1993 – the village of Dumlugoze in the central Taurus Mountains. In a scheme to propagate the giant snowdrop artificially, village families were given horticultural training and supplied with bulbs too small to be sold commercially. The first harvest, in 1996, was a great success, and in 1999 demand for the "conservation-grade" bulbs far outstripped supply. By 2000, the amount offered for the villagers' propagated bulbs was about 60 per cent higher than that offered for wild-collected stock.

▲ PROPAGATED SNOWDROPS *Villagers, seen here with trays of snowdrops, are propagating bulbs as part of larger efforts to reduce the trade in wild-collected plants.*

GARDENER'S GUIDE

- When buying bulbs, inquire about their origins, and buy only those that have been either propagated or harvested from the wild under a quota system that can be verified.
- To support bulb propagation by local peoples, buy "conservation grade" bulbs, such as the giant snowdrops being produced by Turkish villagers (*see above*).
- Look for Dutch bulbs, such as alliums (*see below*), labelled "from cultivated stock". Be aware that dealers from other countries are not obliged to comply with this labelling programme.
- Do not hesitate to ask retailers to clarify the origin of the bulbs they sell. Be wary of bulbs marked "nursery grown", for example, as they may have been collected from the wild and grown on in a nursery. Give preference to bulbs labelled "nursery propagated" or "from cultivated stock".
- Consult publications such as *The Good Bulb Guide* (*see p.478*) for lists of companies that have pledged never knowingly to sell wild-collected bulbs.
- Avoid bulbs being sold very cheaply: plants collected from the wild are often less expensive than cultivated specimens.

More than 100 on Red List*

Allium

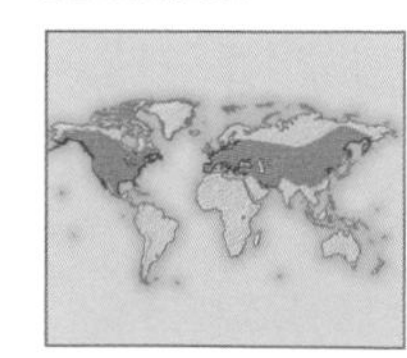

Botanical family
Alliaceae/Amaryllidaceae

Distribution
Mainly North America, Asia, the Himalayas, and the Mediterranean

Hardiness
Fully hardy

Preferred conditions
Most species need full sun and fertile, well-drained soil; some are dormant in dry summers.

The 500 or so *Allium* species are found mainly in the northern hemisphere, in a wide variety of habitats. Alliums will grow in grasslands and savannas; alpine meadows; coniferous forest; Mediterranean forest and scrub; Californian chaparral and woodland; and Central Asian and North American deserts. The edible species, including onions, garlic, chives, leeks, and shallots, have long been of huge economic importance. Ornamental alliums became truly popular as garden plants only in the 20th century.

Many species have narrow ranges of distribution and appear on regional lists of threatened plants, and are little-known in horticulture. The status of the more commercially important ornamental species in the wild is unknown, but there does not appear to be a significant threat from trade in material from the wild, since the ornamental species are readily increased from seeds or offset bulbs and are well established in cultivation. To ensure that valuable rarities are conserved, several germplasm (genetic material) collections have now been established for the *Allium* genus in Europe and the USA.

Allium circinnatum is listed as Rare in the Red List of 1997. It grows in the stony scrub of hillsides on the Greek island of Crete but, being unusual, has attracted the curiosity of collectors and alpine enthusiasts. It has two to five, spirally coiled prostrate leaves that are very hairy. In spring, a hairy flower stalk bears several small, white flowers that are striped pink. It is best cultivated in pots.

Allium flavum var. *pilosum* from Turkey is named for its hairy stems and leaves. While this variety is listed as Rare in Red List of 1997, the species (***Allium flavum***) grows widely in Eastern Europe, on rocky scree and arid steppes up to 2,100m (7,000ft).

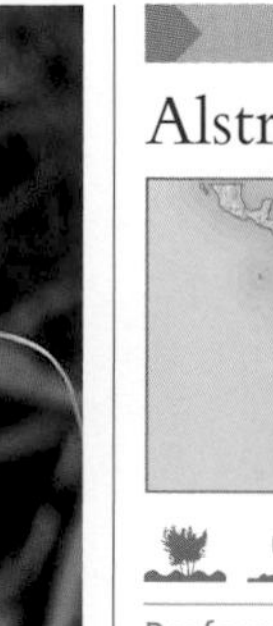

ALLIUM FLAVUM ► *This species is widely cultivated for its dense flowerheads and non-seeding habit. In North America, specialist plant breeding has produced dozens of cultivars in hues other than yellow.*

Allium stipitatum (Vulnerable in the Red List of 1997) occurs from the rocky steppes of Afghanistan and Pakistan to the Tien Shan range of Central Asia. It is rare through most of its range, and has suffered from collection and overgrazing to some degree. The broad, sometimes hairy leaves appear before the flowers.

Allium munzii is found only in the seasonal spring pools and moist grasslands of Riverside County, California. Thirteen small populations are documented, but are threatened by agriculture, mining, off-road vehicle use, and suppression of natural fire cycles. Its showy flowers appear only after winters with sufficient rainfall; in extreme drought, not even leaves are produced. Listed as Endangered in the Red List of 1997, it makes an attractive perennial for wet areas in Californian gardens.

ALLIUM STIPITATUM *This drumstick allium is fairly easy to grow in gardens, soon spreading into a clump. It gives late-spring and early-summer colour, is sweetly fragrant, and makes very good cut flowers.*

Unlisted

Alstroemeria pelegrina

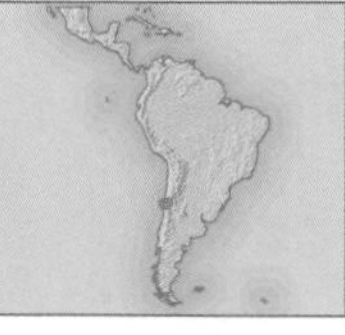

Botanical family
Alstroemeriaceae

Distribution
Coastal lowlands of Coquimbo Province, Chile

Hardiness
Frost-tender

Preferred conditions
Full sun to light shade; sandy, well-drained soil; tolerates a salty, coastal environment; best in a Mediterranean climate with mild winters and summers that are not excessively hot.

Known as the Inca or Peruvian lily, or the pelegrina in Chile, this species has played an important part in the creation of the numerous alstroemeria hybrids now cultivated worldwide as commercial cut flowers. For example, *Alstroemeria pelegrina* 'Alba' has white flowers flushed with green.

In the wild, however, the species is restricted to desert and scrub in the Mediterranean-type climate that occurs in the vicinity of Las Molles in Chile. The plant is usually found close to the shore, growing among rocks, and is vulnerable due to its limited distribution. Coastal development is the chief threat to the continued existence of the species in its narrow native range.

▲ TOUGH BEAUTY *Occurring close to the Pacific Ocean in its native habitat, the Inca lily has developed a high tolerance of salt.*

This beautiful dwarf Inca lily makes a delightful garden perennial that provides spectacular spring and summer colour when grown in a mild Mediterranean-type climate. Like all alstroemerias, the plant produces separate leafy and flowering stems from a network of underground stems (rhizomes) and tuberous roots.

When the foliage begins to yellow in midsummer, allow the plants to dry out. The seed capsules open explosively when dry and ripe, propelling seeds with some force from the plant. The seeds germinate readily within two months, but should be soaked in water for one or two days (changing the water after the first day) to achieve greater speed of germination and uniformity of seedlings.

Unlisted

Anemone blanda

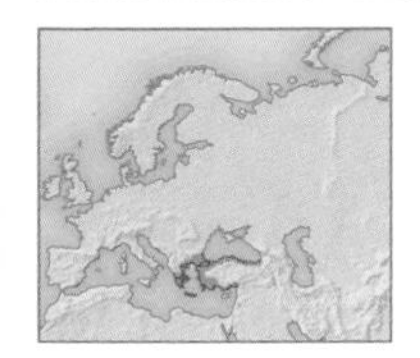

Botanical family
Ranunculaceae

Distribution
Turkey and Greece

Hardiness
Hardy

Preferred conditions
Light to partial shade; fertile, well-drained soil.

Tubers of ***Anemone blanda***, which is not listed by IUCN, have been continually harvested and exported from Turkey in unsustainable numbers for the last 30 years. Millions are estimated to be collected from the country's woodland and scrub annually, and very many of those kept in storage are in a deplorable condition. Sadly, wild windflowers continue to enter the trade, partly due to a perception that they are abundant and partly because, unlike some of Turkey's *Galanthus* species, they are not protected by CITES. Gardeners can play their part in ending the unsustainable harvesting by insisting that the seeds, tubers, and plants of this species that they purchase come from nursery propagation programmes.

Windflowers are typically blue-flowered, and there are pink, white, and purple forms. Buying named cultivars is one way to be sure that material is not wild-collected. 'Radar' is one popular cultivar with red flowers; 'White Splendour' has large, pure white blooms. Windflowers are useful in the rock garden, low border, or woodland garden. The tubers should be soaked overnight in tepid water before planting in autumn. Well-adapted plants spread by seed, which is best sown as soon as ripe.

SPRING FLOWERING ► *Windflowers are beloved for their display of colour in early spring in mild winter zones, and a little later in colder areas.*

Red List: Rare*

Arum dioscoridis

Botanical family
Araceae

Distribution
Southern Turkey, Cyprus, and the Middle East

Hardiness
Half-hardy

Preferred conditions
Full sun or partial shade; well-drained, moderately fertile soil. Mild winters and dry summers are preferred, although this species is able to tolerate summer moisture.

A particularly variable species, ***Arum dioscoridis*** grows in Mediterranean woodland, and some of the regional variants are rarely seen. *Arum dioscoridis* var. *luschanii* from Turkey is cited as Rare in the Red List of 1997, along with four other species from Turkey and Greece, classified as either Vulnerable or Rare. Rarity coupled with tuber collection for processing as a starch source (the latter perhaps contributing to the former) are the main conservation concerns.

Arum dioscoridis is a suitable species for dry woodland rock gardens. In areas of mild temperatures, such as coastal California, the tubers can be planted in full sun. Plant tubers shallowly, about 5cm (2in) below the surface. The unmottled, arrowhead-shaped leaves appear in winter, followed in spring by the nearly stemless spathe, or modified leaf. This is pale yellowish-green, usually spotted purple, although in some forms the purple colour suffuses the spathe almost entirely. The spathe surrounds the tall flower spike (spadix).

This arum can be container grown but be warned that the spadix has a fulsome odour, the means by which the plant attracts the flies that pollinate it in the wild. This odour will taint the greenhouse while the plants are in flower. Nonetheless, the species is enjoyed by bulb connoisseurs, and *Arum dioscoridis* is offered by a number of commercial rare-bulb dealers.

Cultivated plants do not reliably form seed, but the plant is easy to grow when seed is available. The seeds should be separated from the fleshy fruit and planted as soon as they are ripe. Young plants should be kept growing in shade until the leaves begin to yellow, at which time they should be allowed to rest.

▲ ARUM DIOSCORIDIS *Typically of this genus, the flower spike (spadix) is gender-zoned, with the female reproductive organs at the bottom end and the male ones at the tip.*

Unlisted

Babiana stricta

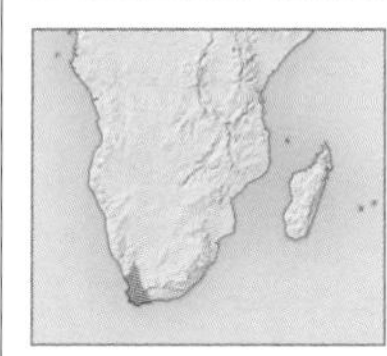

Botanical family
Iridaceae

Distribution
South Africa: Northern Cape and Western Cape provinces

Hardiness
Frost-tender

Preferred conditions
Full sun; rich, well-drained, loamy soil; best in mild, Mediterranean climates, but may adapt to summer rainfall if not excessive; not well-adapted to excessive heat.

The genus *Babiana* contains some 80 veld-dwelling species, of which more than half are restricted to South Africa's Cape region; most of the others are found to the north in Namaqualand. Apart from one species in tropical southern Africa, babianas are not found outside the winter rainfall zone. Many are of restricted distribution, and most are further imperilled by the fact that their habitat – the rich, clay soils of renosterveld – is intensely managed for agriculture. ***Babiana stricta*** and *Babiana rubrocyanea* (sometimes treated as a variety of *stricta*) are the most widely grown species. The former is quite variable and currently does not have any Red List status. Fourteen other *Babiana* species are Red Listed.

Babiana stricta is one of the taller of the baboon flowers, some forms reaching a height of 45cm (18in). It flowers in spring in the northern hemisphere, and flowers remain in good condition for nearly a week. A number of clonally propagated, named cultivars are available, including 'Blue Gem' (violet with purple spots) and 'Purple Sensation' (purple and white). *Babiana stricta* is easily grown from seed in spring (or autumn in zones with mild winters), and seedlings flower in two to three years.

◄ BABIANA STRICTA *Flower colour varies through all shades of blue to purple, white and red. Some forms are fragrant. The pleated leaves are often covered with soft hairs.*

WILDLIFE

WHAT'S IN A NAME

The genus name *Babiana* derives from the Afrikaans word *babiaans*, or baboon, and the plant is so named because these large monkeys, found only in Africa, eat the small corms. Baboons are omnivorous and forage not only for rhizomes and tubers of numerous species, but also for fruit, scorpions, and a variety of insects.

Red List: Rare*

Brodiaea californica

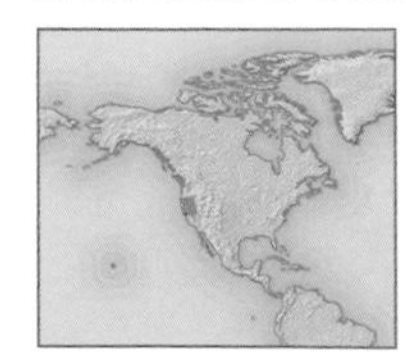

Botanical family
Themidaceae

Distribution
USA: northern California and southern Oregon

Hardiness
Half-hardy

Preferred conditions
Full sun and well-drained, loamy soil; to survive the winter this plant needs dryness in late summer and early autumn.

The bulbous plants of the genus *Brodiaea* are found only in the far west of North America. Most of the species are native to the California chaparral, but a few grow in Oregon to the north and in Mexico. Eight or nine *Brodiaea* species (depending on whether *Brodiaea leptandra* is accepted as a distinct species) appear on the Red List of 1997.

The chief causes for conservation concern are the narrow native ranges of many of the species, coupled with their loss of habitat to development and agriculture. In addition, particularly in the case of ***Brodiaea californica***, there has undoubtedly been some collection of bulbs from the wild for horticultural use, although it is hard to determine the extent to which bulb collecting has contributed to the species' rarity.

Of the numerous native Californian brodiaeas, not all of which are threatened, *Brodiaea californica* is a favourite for inclusion in the garden. It produces the tallest plants and the largest flowers of the genus, and a single plant may carry as many as 15 blooms. Available colour forms of *Brodiaea californica* range from deep to pale blue, through various shades of pink, to nearly white. *Brodiaea californica* var. *leptandra* is typically violet-coloured, with smaller flowers than those of the typical variety, *Brodiaea californica* var. *californica*. The narrow, linear leaves have a tendency to die back on the appearance of the flowers in late spring or early summer.

CULTIVATION

Brodiaea californica has been successfully grown in some gardens outside its natural range, since the corms can tolerate some summer rainfall when dormant, as long as drainage is swift. However, *Brodiaea* bulbs will rot if subjected to very humid summers. The bulbs are reliably hardy in areas where winter temperatures remain above -7°C (20°F), and probably in colder zones too if they are protected. The black, wedge-shaped seeds germinate after 6–8 weeks. Selected forms of all *Brodiaea* species are propagated vegetatively by digging up the parent corms and collecting the numerous offset cormels they produce, and then planting them individually.

▲ **DAYLIGHT OPENING** *The flowers of* Brodiaea californica, *which are the largest of any* Brodiaea *species, close up at night and reopen the following morning. Each funnel-shaped bloom has three fertile stamens and three sterile stamens (called staminodes).*

Red List: Vulnerable*

Brunsvigia litoralis

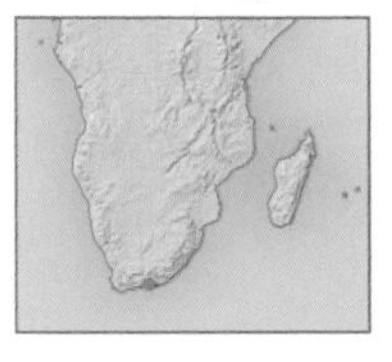

Botanical family
Amaryllidaceae/Liliaceae

Distribution
South Africa: Eastern Cape Province; from Knysna to Port Elizabeth

Hardiness
Frost-tender

Preferred conditions
Full sun, and dry conditions in winter. Soil should be sandy and well-drained.

One of about 20 species in a genus exclusively from South Africa, *Brunsvigia litoralis* is a plant of the coastal veld. It has been adversely affected by development along South Africa's eastern coast, which borders on the Indian Ocean, and secondarily by bulb and plant collection for the indigenous medicinal plant trade (*see below*).

The plant bears large, red, tubular or funnel-shaped flowers in summer after the leaves dry off. The flowers are pollinated by sunbirds. When the seeds are mature, the entire spherical head of dry, capsular fruits detaches from the plant and rolls along the rocks and beach sand in the manner of a tumbleweed, dispersing the seeds as it goes.

Brunsvigia litoralis is reportedly self-sterile and must be cross-pollinated to set seed. Hybrids may be created not only with other *Brunsvigia* species but also with the related genera *Amaryllis* and *Nerine*. The fleshy seeds should be planted as soon as they are ripe, and are best sown by merely pressing them into sand or any well-aerated propagation medium. Seedlings may be transferred to small, individual pots after the first leaf appears. Plants require six years to flower.

In pots, *Brunsvigia litoralis* bulbs should be planted with their shoulders exposed, and are best left undisturbed to encourage flowering. While exposure to high summer daytime temperatures causes no problems, summer nights that remain hot and humid are inimical to good growth and flowering. The bulbs will rot if irrigated during the winter.

TRADITION

AFRICAN REMEDIES

Traditional healers of Southern Africa use more than 1,000 different plants in their *muti* (medicines). In the case of *Brunsvigia* species, the Zulus use them for treating respiratory ailments, and to facilitate enemas, while the Sotho people use extracts from the plants to treat back pain and infertility. The herbalists grind the plants into powders. How these are administered depends on the type and purpose of the herbs – the powders may be taken with water, boiled and drunk as decoctions, smoked, rubbed into cuts in the skin, or licked from the fingers. Herb collectors previously sold only to the herbalists, but selling directly to the public has led to unsustainable overharvesting.

Red List: Endangered*

Caliphruria subedentata

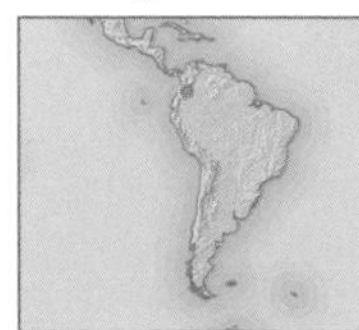

Botanical family
Amaryllidaceae

Distribution
Western Colombia

Hardiness
Frost-tender

Preferred conditions
Moist, fertile soil in deep shade. Whether grown as a houseplant or in a tropical or subtropical garden, it requires high levels of humidity.

Called the dwarf Amazon lily, this is a small bulbous plant with evergreen, hosta-like leaves and delicate, white, funnel-shaped flowers. It is closely related to the true Amazon lilies of the genus *Eucharis*, but differs in the shape of its flower. Never common, it occurs only in the shade of mountain Andean cloud forest in western Colombia. Very little of this habitat survives intact, and few populations of the lily remain today.

The requirement of deep shade prevents this plant from colonizing disturbed, secondary forest. Its spread in the wild is also limited by a scarcity of seed, since only a few form in each seed capsule. Of the two other Colombian endemic *Caliphruria* species, at least one, *Caliphruria tenera*, is presumed extinct.

The bulbs usually bloom in late spring or summer, but flowering may occur at any time of the year after a brief dry period. The bulbs offset generously and form sizable clumps in time. They may be propagated by division, although disturbance frequently inhibits flowering for several years. Individual plants are self-sterile. Seed formation rarely occurs in cultivation, possibly because all plants in horticulture are genetic clones.

33 on Red List*

Calochortus

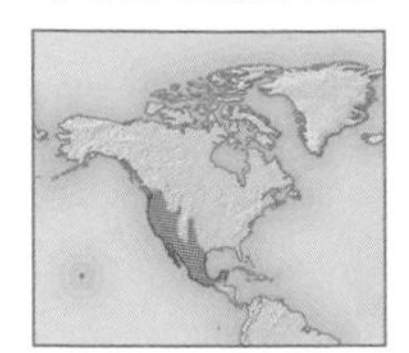

Botanical family
Liliaceae

Distribution
Western North America, Mexico, and Guatemala

Hardiness
Half-hardy

Preferred conditions
Full sun or partial shade in well-drained, loamy soil; species generally require abundant moisture in winter and spring, but a very dry summer.

The mariposa lilies, usually considered quintessentially western mountain plants, are most diverse in California, in the coastal ranges and the Sierra Nevada. There are between 75 and 100 species. Many are local endemics known from one or a few populations of restricted range. There are also a number of Mexican species, including several from the tropical zone, but these are not very well known. The lilies grow in California chaparral and woodland, temperate savanna and prairie, North American desert, alpine and montane meadows, temperate broadleaved, mixed, and coniferous forest, and subtropical mixed forest. Bulbs of some species are edible.

The rare species found in California's coastal hills have been endangered by rapid residential development, especially around Los Angeles, San Diego, and San Francisco. There has also been some collection of bulbs for the speciality market, but it is unclear how much impact this has had.

◄ CALOCHORTUS AMOENUS *This species, called the Sierra globe tulip, is one of the more dependable horticultural performers of the fairy-lily type of calochortus.*

◄ CALOCHORTUS WEEDII *VAR.* INTERMEDIUS *Native to much of San Diego County, California, this plant (Rare in the 1997 Red List) has one of the largest flowers in the genus, sometimes up to 8cm (3in) in diameter.*

Two species, *Calochortus indecorus* and *Calochortus monanthus*, are considered extinct, the latter with some uncertainty. However, the habitats of a few very rare California species have been restored and the species successfully reintroduced. The exact relationships of the mariposa lilies have been controversial; some botanists have allied them with the true tulips (*Tulipa* species), others with the toad lilies (*Tricyrtis* species). Botanists have also advocated recognizing a separate family, the Calochortaceae. Although *Calochortus* is a distinctive lineage within the Liliaceae family, DNA sequence analyses support the retention of the genus within Liliaceae.

Other mariposa lilies

Calochortus amoenus is listed as Vulnerable in the Red List of 1997. The plant is found in the foothills of the southern Sierra Nevada of California, from Madera to Kern Counties, at heights of 300–1,500m (1,000–5,000ft). It colonizes north- and west-facing slopes, sometimes on river canyon walls and is often found in the shade among grasses.

The Mt Diablo fairy lantern, *Calochortus pulchellus* (Vulnerable according to the Red List of 1997), is endemic to Mt Diablo and the surrounding foothills of Contra Costa and Solano Counties in northern California. The flowers appear in late spring. Their sepals are usually greenish and enfold the petals. Very similar, and also rated Vulnerable in the same Red List, is *Calochortus amabilis*, which ranges into Napa and Sonoma Counties. Its flowers are a richer yellow, often with mahogany markings on the petals.

HABITAT

Desert denizen

The arid interior of southern California and southern Nevada is the native range of the alkali mariposa lily, ***Calochortus striatus*** (listed as Vulnerable in California and Endangered in Nevada in the Red List of 1997). This rare endemic of moist alkaline areas occurs on calcareous sandy soil in seasonally moist alkali meadows, desert washes, seeps, and depressions. The purple-pink flowers are very distinctive, differing from those of all other mariposa lilies by the dark purple veins on the petals. The plant is threatened by change in the water table (which eliminates its specialized habitats), urbanization, road construction, and disturbance by grazing. It is a desert gem that should be introduced into more desert gardens.

Cultivation

There has recently been an increased interest in growing mariposa lilies. They may be propagated by bulbils and larger offsets. Many gardeners outside the plants' native range have succeeded in cultivating them in containers, with successful flowering in spring and early summer, and then providing the dry period that so many of the species require. Seed germinates readily if sown fresh; seedlings are best not transplanted until their second year.

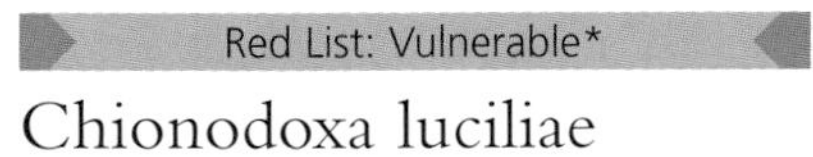

Red List: Vulnerable*

Chionodoxa luciliae

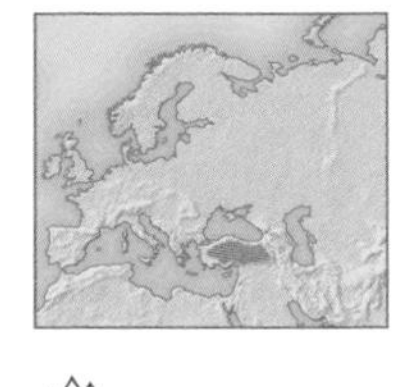

Botanical family
Hyancinthaceae/Liliaceae

Distribution
Turkey

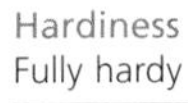

Hardiness
Fully hardy

Preferred conditions
Rich, well-drained soil; must have soil moisture through winter and spring; best left undisturbed until crowded.

Named "glory of the snow", ***Chionodoxa luciliae*** is a popular bulbous plant that grows in alpine meadows and flowers in late winter or early spring. The flowers are typically blue. In the past the bulbs have been collected in wholesale quantities for the international bulb trade, but now the plant is one of the "strictly protected flora species" listed by the Convention on the Conservation of European Wildlife and Natural Habitats, instituted by the Council of Europe.

As the plant's common name suggests, its flowers are often among the earliest to appear when the snows of late winter are still on the ground. The genus *Chionodoxa* is closely related to that of *Scilla* (*see p.253*) and will hybridize with at least some of its species. The taxonomic merger of the two genera has at times been proposed.

Glory of the snow's performance improves markedly with each year after planting. Established bulbs form sizable clumps, and each bulb produces more than one flower stem. The bulbs produce many offsets, which can be separated and planted, and the seed germinates quickly when fresh. Seed may be sown in trays or directly in the garden. Selected cultivars of *Chionodoxa luciliae* include 'Alba' (white) and the lovely pale 'Pink Giant'.

GLORY OF THE SNOW ► *Flowering of the beautiful bulb* Chionodoxa luciliae *starts in late winter and continues into early spring. Pink and white cultivars are also available.*

Red List: Rare*

Clinanthus variegatus

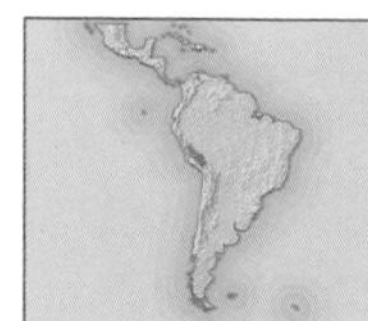

Botanical family
Amaryllidaceae

Distribution
Peru: Andes

Hardiness
Half-hardy

Preferred conditions
Full sun, or partial shade in hot climates; soil should be loamy and well-drained. Flowering may only occur after a particularly heavy rainy season (often coinciding with an El Niño event).

This species, which is still known in horticulture by its previous name of *Stenomesson variegatum*, is one of about 30 in the genus and is found primarily in Peru. It grows at heights of 2,000–3,000m (6,000–10,000ft) in valleys of seasonally dry scrub lying between Andean mountains. It is frequently found on steep slopes, often in the company of cacti. The plant's habitat is not yet under significant human pressure since it is not easily turned to agriculture. Natural rarity and small population size are the greatest threats to this species.

Clinanthus variegatus has large orange and yellow flowers that are probably pollinated by hummingbirds, although no reports of pollination have been published. Pollen carried from another individual is required to set seed.

Growing Clinanthus

In cultivation, the species prospers best in a mild, temperate climate, such as in coastal California, Australia, and South Africa's Cape region. Though very tolerant of high temperatures, the plants are shy to flower in warm, tropical climates. The seeds are flat, black, and winged. Fresh seed should be air-dried for a day, then sown on any well-drained medium and covered shallowly. Germination begins in several weeks and may continue for several months. Pot up seedlings singly when one leaf has formed.

TRADITION

Inca Artefacts

Motifs recognizable as *Clinanthus variegatus* appear on pre-Columbian Inca drinking vessels called *keros*. During ceremonies such as the Festival of the Sun, the people would use two *keros*, one to toast their Inca (king) and his followers, and one to toast the land and the gods. *Chicha*, a fermented drink, was often used. The ceremonial use of *keros* persists among Peruvian peasants even today.

Unlisted

Clivia miniata

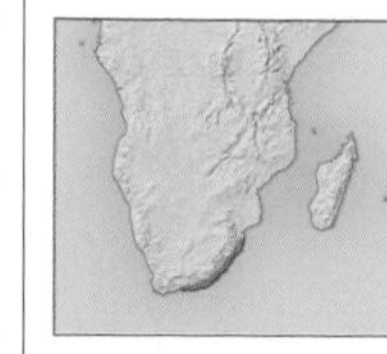

Botanical family
Amaryllidaceae

Distribution
South Africa: Eastern Cape Province to northern KwaZulu-Natal

Hardiness
Frost-tender

Preferred conditions
Partial shade and mild temperatures; rich, fast-draining soil high in organic matter; plants need to stay relatively dry in winter.

Though technically not a bulbous genus, *Clivia* is a member of the amaryllis family, and is often covered in treatises on bulbous plants. The five species are evergreen perennials with thick underground stems (rhizomes) and overlapping leaf bases from which a network of succulent roots emerges.

Until the recent discovery of a clivia species in the winter-rainfall region of South Africa's Western Cape, all were thought to be plants of the summer-rainfall regions of that country and nearby Swaziland. They are part of the berry-fruited tribe of the amaryllis family, which is endemic to Africa. Although no clivia is on IUCN's Red List, all of the species are protected by law in South Africa.

Clivia miniata is a mountain-dwelling species, growing in the understorey of Afro-montane subtropical forests, sometimes wedged into cracks between rocks, occasionally growing on other plants as an epiphyte. It can establish a foothold on sheer cliff if there is shade along with pockets of humus to provide moisture and nutrients.

Clivia miniata is not exactly rare, but neither is it common in the wild. Bulb collection for traditional uses is a major threat (*see above*). Collection for horticulture, widespread in the past, may have lessened since many improved forms are now abundantly available in the trade in America, Europe, Asia, and South Africa. However, the plants take five years to reach flowering size, making them expensive, and stories still circulate of entire populations in the Transkei region being taken for use in landscaping. Deforestation also contributes to the risk of loss, as this exposes the plants to lethal levels of sunlight, drought, and habitat erosion.

Cultivation

Despite their reputation as connoisseur plants, *Clivia* species are remarkably tough and reasonably adaptable. Coastal Californians can grow them outdoors with the ease of garden weeds, as can gardeners in similar climates in Australia, New Zealand, parts of South America, and the Mediterranean region. Otherwise, it is possible for anyone to enjoy their spectacular blooms indoors. *Clivia miniata* does not prosper in high temperatures, especially during the night, and in tropical or subtropical areas the flower stem will frequently fail to elongate adequately (if the plants flower at all), trapping the flowers in between the leaf bases. Cool night temperatures for at least two months in the autumn seem essential to promote proper flower development in the spring. The whitish-brown seeds are large, fleshy and best planted immediately as they do not store well. They should be sown half-covered in a good, friable germination medium. The first roots formed can be surprisingly thick, and seedlings may soon need to be re-potted individually.

TRADITION

Unhealthy Collecting

Its versatility is the bane of *Clivia miniata*, which is threatened by collection for medicinal and ceremonial use by indigenous African peoples. Large quantities are consumed in KwaZulu-Natal province, where the species is most abundant, and where collection for ethnobotanical use by native people is legal. Infusions are prescribed for snakebite and to facilitate childbirth. The plants also serve as charms, for example, to ward off lightning strikes. Consequently, mounds of discarded leaves and excavation holes are almost always observed at population sites of this species.

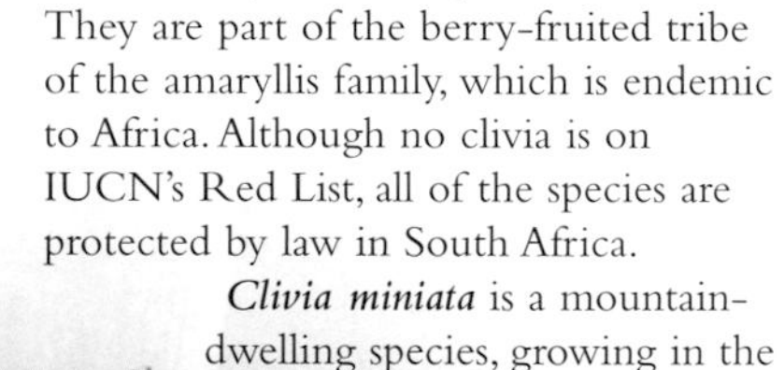

FUNNEL-SHAPED FLOWERS
Clivia miniata *is the only member of its genus to bear fairly wide-spreading, funnel-shaped or rounded flowers. The remaining four* Clivia *species are quite different, having tubular flowers.*

Red List: Vulnerable*

Colchicum corsicum

Botanical family
Colchicaceae/Liliaceae

Distribution
Corsica (France)

Hardiness
Fully hardy

Preferred conditions
Full sun, or light shade in hot climates; soil should be fertile and well-drained.

The genus *Colchicum* contains about 50 species, ranging from Europe through Central Asia to North Africa. All but one bear pink or white flowers that superficially resemble those of the genus *Crocus* (*see p.230*). The two genera are easily differentiated. *Colchicum* flowers have six stamens, whereas *Crocus* flowers have three, and the leaves of *Colchicum* species never have the white line down the middle that characterizes all crocuses.

Allied botanically to ***Colchicum autumnale*** and *Colchicum alpinum*, the Corsican autumn crocus (*Colchicum corsicum*) is a diminutive plant restricted to montane meadow on the island of Corsica in the eastern Mediterranean. Its conservation status derives from its extreme rarity. Protected by French law, it is one of 13 *Colchicum* species named on the IUCN Red List of 1997.

While containing both spring- and autumn-flowering species, the genus *Colchicum* is most valuable in horticulture for the autumn-blooming species, which also happen to be the more easily grown of the two seasonal groups. *Colchicum corsicum* is an easy plant in cultivation, providing the gardener with generous quantities of bulb offsets for separation and planting. The species is a favourite autumn crocus of rock gardeners. The best forms flower in a deep cerise pink.

COLCHICUM AUTUMNALE ► *This species, commonly known as meadow saffron, is a close relative of the rare and threatened* Colchicum corsicum *but has a much wider native range in Europe.*

12 on Red List*

Crinum

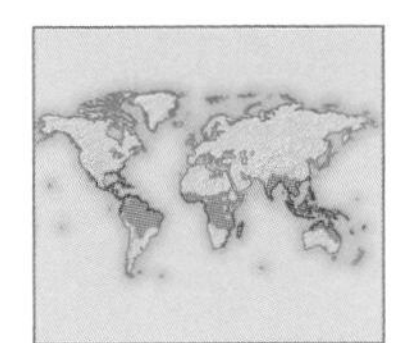

Botanical family
Amaryllidaceae

Distribution
Tropical regions, especially Africa

Hardiness
Frost-tender

Preferred conditions
Depending on species, full sun to full shade. Favoured soils range from sandy and well-drained to moist, loamy, and fertile.

All the other members of its tribe of genera in the amaryllis family are native to Africa, but *Crinum* is the only one to escape that continent. New World crinum lilies are often difficult to distinguish from each other, and probably originated from a single introduction, perhaps when the continents were closer together. Crinums grow in tropical and subtropical grassland, savanna, and shrubland; flooded grassland and savanna; and tropical rainforest. African tropical species were among the first plants to be taken to the Americas by settlers. Indeed, for years botanists mistakenly believed that several African species were native American bulbs from the Amazon. Some *Crinum* species (including all American, Asian, and Madagascan species) have radially symmetrical flowers. Others have bilaterally symmetrical, trumpet-shaped or bell-shaped flowers.

▲ BOTSWANA LILIES *African* Crinum *species, in particular, are threatened by the collection of the bulbs for the indigenous medicinal plant trade. This one, however, is growing freely and prolifically in the protected environment of Chobe National Park in northern Botswana.*

CRINUM SPECIES

Crinum baumii is found in Namibia and Zimbabwe and is cited as Rare in the Red List of 1997. This dwarf species is biologically interesting because it appears to be more closely related to the genera *Ammocharis* and *Cybistetes* than to the rest of *Crinum*, according to genetic analysis. Scarce in its habitat, it is threatened by industrial development in Namibia and collection for the medicinal plant trade. *Crinum baumii* is easy to grow, hybridizes readily with other *Crinum* species, and imparts its compact size to the hybrids.

Crinum jagus, the St Christopher lily, is an evergreen species found in the understorey of tropical rainforest in Equatorial Africa. Its flowers have a fragrance reminiscent of vanilla. It is valued partly because it is not plagued by the foliar fungi affecting many crinums in cultivation. Unlisted by IUCN, it is nevertheless vulnerable in its habitat.

Crinum mauritianum is the only crinum cited as Endangered in the Red List of 1997. Some botanists believe that the Mauritian species is the same as another from Madagascar; others hold that it is a distinct species. The plant's conservation status arises from the ongoing wholesale destruction of its coastal wetlands habitat in Mauritius, coupled with a high degree of invasion by exotic plant species, which are preventing the regeneration of large tracts of native vegetation.

PEOPLE

PLANT BREEDER

Crinums have fascinated plant breeders for centuries, including the famous American Luther Burbank (1849–1926), who produced some hybrids that are still grown today. However, Burbank's primary concern in horticulture was to improve the quality of food plants. His breeding experiments gave rise to more than 800 new varieties, including nuts, grains, fruits, and vegetables, as well as a remarkable range of ornamental flowers.

GROWING CRINUMS

Crinums are easily grown from their large, fleshy seeds. These are best sown half-exposed in moist growing medium and partially shaded. Overwatering should be avoided. Seeds germinate in a few weeks, and usually produce their first leaf in two months.

HABITAT

SEASONAL LAKES

Created by summer rains, shallow, seasonal lakes (called vleis) in the Eastern Cape of South Africa are the habitat of the vlei lily, *Crinum campanulatum* (Rare on the 1997 Red List). During the winter dry season, they are pans of dry, cracked mud, but burst into bloom when the summer rains begin. Thousands of bell-shaped, white and burgundy flowers open together, transforming the barren landscape. As the vlei dries up, the fleshy seeds, which float on the water, touch the mud of the lake bed and germinate. The pans are fragile environments but receive no protection.

CRINUM ASIATICUM ► *Called the poison bulb, this species from tropical South-east Asia is probably the largest bulbous species in the world, capable of achieving a height of 2m (6ft).*

27 on Red List*

Crocus

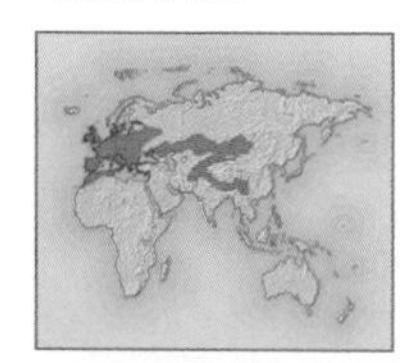

Botanical family
Iridaceae

Distribution
From Portugal eastwards to western China

Hardiness
Hardy – half-hardy

Preferred conditions
From full sun to partial shade, in well-drained soil; species vary in the degree of moisture and organic content they require in soil.

Many of the 80 to 90 species within the *Crocus* genus are rare in nature, of restricted geographical range, or known from few populations. The genus contains species that flower either in spring or in late summer and autumn. *Crocus* species grow in montane meadow; temperate broadleaf, mixed, and coniferous forest; temperate steppe; and Mediterranean woodland.

It is unclear to what extent the past export of wild-collected bulbs has affected the conservation status of crocuses, but the majority of the species listed by IUCN are classified as Rare or Vulnerable, rather than Endangered. On the whole, apart from limited demands from alpine-garden enthusiasts, the market for species crocuses is not large. Most gardeners are far more familiar with the easy-going hybrid stocks created by Dutch breeders, all of which have much more dependable horticultural natures, flowering each year and increasing in numbers over time.

Identification clues

Crocus species are first identified by the white line that runs down the centre of their narrow leaves. That characteristic easily distinguishes crocuses from the superficially similar members of the genus *Colchicum*, alongside the fact that crocus flowers have only three stamens. It should also be noted that there is no *Colchichum* species with flowers of blue or purple, colours that are frequently seen in *Crocus* species. When distinguishing between individual *Crocus* species, two very important features are the form of the corm's coat (the tunic) and the degree to which the style (the connecting tube between the female stigma and ovary) is branched.

▲ CROCUS CYPRIUS *Native to the Troodos Mountains in Cyprus* (see below), *this crocus's alpine habitat supplies it with its very particular moisture requirements. If these can be copied, it can be successfully cultivated in cold frames and cool greenhouses.*

Crocus species

Crocus goulimyi is designated as Rare in the Red List of 1997. First described as late as 1955, this crocus from southern Greece is an autumn-flowering species known only from several populations in the Peloponnese region. Though it is restricted in distribution, its populations are quite large. Its discoverer recorded it as "one of the finest crocuses that I have ever seen in this country". The plant produces offsets rapidly and is very amenable to cultivation, producing its pale lavender, long-tubed flowers before the leaves emerge.

Two subspecies of *Crocus kotschyanus*, subsp. *cappadocicus* and subsp. *hakkariensis*, are cited as Rare in the Red List of 1997. These are found in Turkey only. The species ranges from the Caucasus to Lebanon, where it is found in mountain meadows, scrub, and in stony scree along a broad gradient of altitude. It occurs in zones of high rainfall and in meadows of sufficient elevation that the soil never fully dries out during the summer. Typically the plant produces pale lilac or white flowers with yellow spots near the throat, and a style with only a few branches. This crocus is very adaptable to moist climates, and can be grown in the open in many areas, and even naturalized in the lawn.

Crocus olivieri is a broadly ranging and popularly cultivated crocus that nevertheless has two rare subspecies of conservation concern: both are found in Turkey only. *Crocus olivieri* subsp. *balansae* is cited as Rare in the Red List of 1997, while subsp. *istanbulensis* is noted in the same list as Vulnerable. The spring-flowering species is distributed through the southern Balkan states, Greece, Turkey, and the Aegean Islands. Its flowers range from yellow through orange, sometimes with brown or bronze markings of

▲ CROCUS GOULIMYI *This autumn-flowering crocus, with purple to white blooms, thrives in dry summers and can cope with alkaline soils. In cold winter areas, cultivation under glass is advisable.*

HABITAT

Mountain dweller

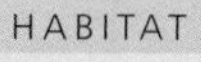

The open scree slopes in the alpine heights of the Troodos Mountains, which lie to the south-west of Cyprus, are the only habitat of ***Crocus cyprius***, one of only two crocus species rated as Endangered in the Red List of 1997. (The other is *Crocus hartmannianus*, and both are native to the island.) The small, pale lilac flowers of *Crocus cyprius*, often marked with a deeper blue blotch on the tepals, appear in early spring at the edges of the mountains' retreating snowfields. This habitat provides two essential requirements for this crocus to prosper: moisture in winter and spring followed by a dry summer.

BULBOUS PLANTS

various sorts. The leaves are sometimes hairy. *Crocus olivieri* is accommodating in cultivation and does very well on alkaline soils.

Crocus speciosus is a widespread, autumn-flowering species ranging throughout the Crimea, the Caucasus, Turkey, and Iran. Two of the three recognized subspecies, *Crocus speciosus* subsp. *ilgazensis* and subsp. *xantholaimos*, are of much more restricted distribution in northern Turkey, earning them the status of Rare in the Red List of 1997. The species is a woodland denizen for the most part, occurring within clearings in broadleaved as well as coniferous forest, but it is also found in open alpine meadows. The flowers are fragrant, lilac blue with darker veins, and sometimes spotted as well. *Crocus speciosus* subsp. *xantholaimos* has a broad yellow area at the throat.

A number of long-established cultivars of *Crocus speciosus* are still available in the horticultural trade. These include 'Albus' (white-flowered); 'Aitchisonii' (large, almost solid lilac flowers); 'Artabir' (pale lilac flowers with very conspicuous, darker veins); 'Cassiope' (large, blue-lavender flowers with a yellow throat); 'Oxonian' (rich, bluish-purple flowers), and 'R.D. Trotter' (large, white flowers). *Crocus speciosus* is a rewarding and undemanding species for interest late in the year. It is adaptable to sun or light shade, sand or clay soils, and it is indifferent to extremes of soil acidity or alkalinity. However, it will not thrive if subjected to excessive summer moisture when the corms are dormant.

CROCUS SPECIOSUS ► *The cup-shaped blooms of this elegant crocus are mauve to violet-blue. The easiest and most abundant-flowering of all the autumn crocuses, it makes a stunning display planted en masse.*

PROPAGATING CROCUSES

Crocus seed germinates rapidly when sown in a sandy medium that is kept moist. It is best sown thinly and the seedlings maintained in the germination container until the second year, whereupon the small corms can be planted in the garden. From seed, plants take three to four years to reach flowering size. Some species produce many small cormels that are genetically identical to the parent plant and can be used for propagation. Crocus corms can also be separated every three or four years in early summer, but if left undisturbed and fertilized, they form sizable clumps that produce a mass of flowers. Mice, voles, and squirrels may feed on the corms.

CROCUS OLIVIERI *SUBSP.* BALANSAE 'ZWANENBURG'
The bronze tinge present in the flower of the rare Turkish subspecies Crocus olivieri *subsp.* balansae *appears more dominantly in this popular cultivar, in which the bronze coloration is suffused throughout most of the flower.*

TRADITION

PRECIOUS STIGMAS

Saffron is by far the most expensive spice in the world, yet its rich, pungent, earthy, somewhat bitter flavour finds it a ready market with food connoisseurs. Saffron is also used as an orange-yellow dye. The source of the spice is *Crocus sativa*, specifically the three long, red stigmas found in each flower. About 80,000 crocus flowers yield about 500g (1lb) of stigmas after they have been dried and lightly toasted. The growing of the crocuses, the hand-picking of their stigmas, and the labour-intensive drying and toasting process all contribute to the spice's high cost. Saffron is thought to have been first used in Mesopotamia around 3300BC. It is now produced in several countries, including Spain, Greece, Turkey, Egypt, Iran, and Kashmir.

CROCUS SATIVA IN NORTHERN GREECE

SORTING THE VALUABLE STIGMAS

Red List: Indeterminate*

Curcuma albiflora

Botanical family
Zingiberaceae

Distribution
Sri Lanka

Hardiness
Frost-tender

Preferred conditions
Partial shade, and moist, fertile soil; the plant requires plenty of moisture during the growing season, and dryness during winter dormancy.

From both a culinary and medicinal perspective, *Curcuma* is an exceedingly important genus in the ginger family. The aromatic underground stems (rhizomes) of several species figure in the traditional medicine of several Asian countries as treatments for many different ailments. Many of the species are very showy garden ornamentals as well. Of the 65 or so species in the genus, only three are currently on the Red List of 1997, although this may reflect our lack of knowledge about Asian tropical botany more than conservation reality. *Curcuma albiflora* is known only from the tropical rainforest of Sri Lanka, where it is considered threatened by its rarity, coupled with habitat destruction and collection for medicinal use.

▲ CURCUMA LONGA *This common relative of* Curcuma albiflora *is the source of the spice turmeric. It contains high levels of curcumin, a highly beneficial phenolic compound that western medicine is beginning to recognize.*

The showy, long-lasting part of the "flower" of *Curcuma albiflora*, like that of most *Curcuma* species, is actually a series of coloured modified leaves (bracts) that surround the individual blossoms. The blooms themselves are white to slightly pink in this species. The flowerhead arises from the rhizome at the base of the plant. This species is not one of the taller-growing "hidden gingers", and the low-blooming flowers, which appear in early summer, are readily visible in the garden. With protection against frost, the rhizomes should survive temperatures falling to -4°C (25°F) in winter.

17 on Red List*

Cyrtanthus

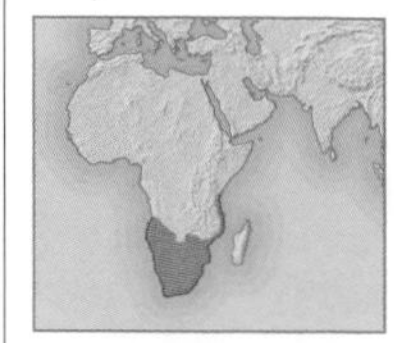

Botanical family
Amaryllidaceae

Distribution
Southern Africa, especially South Africa

Hardiness
Half-hardy – frost-tender

Preferred conditions
Full sun to partial shade; well-drained, sandy, or gritty soil; most prefer intermittent dryness in summer and continuous dryness in winter.

This species-rich African genus of the amaryllis family grows in tropical and subtropical grassland, savanna, and shrubland; fynbos, desert, and scrub; and tropical dry forest. It is also the most diverse genus of the family in terms of floral form, probably the result of selection of specific pollinators, which are known in some cases to include sunbirds, butterflies, and moths. Many of the species have narrow native ranges, while some of the more wide-ranging species show regional variation. Some species react to bushfire by flowering immediately afterwards; some flower only at that time. This irregular flowering habit reduces their reproductive fitness, especially in fire-controlled areas.

The *Cyrtanthus* genus contains both evergreen species (typically found in moist habitats), and deciduous ones (usually found in seasonally dry areas). ***Cyrtanthus elatus***, familiar as Knysna, Scarborough or George lily, is evergreen and grows in mountains of South Africa's southern Cape, a region that receives year-round precipitation. Most of its populations now occur in protected areas, and this lily is not considered threatened, although local collection of bulbs persists.

One of a handful of species from the winter rainfall region of Western Cape Province, *Cyrtanthus carneus,* or fire lily, is cited as Vulnerable in the Red List of 1997. It is a robust, clump-forming species with broad, strap-like, twisted leaves. It is found on sands in limited areas along the coast, where development both for agriculture and tourism is the chief threat to its continued existence.

Another evergreen species is *Cyrtanthus herrei* (Rare in the Red List of 1997), known only from a few populations in Namibia and Northern Cape Province. The large, tri-coloured flowers make it much sought after by bulb collectors, and it flowers more easily in cultivation than *Cyrtanthus carneus.*

Cyrtanthus sanguineus is one of the few truly tropical species, ranging widely from South Africa's Eastern Cape and Natal northwards through East Africa to Kenya. (Three Kenyan subspecies are rated Rare or Indeterminate in the 1997 Red List.) It is a dwarf species, flowering in summer and autumn. The regional colour forms of the species are increasingly rare in the wild, and the species overall is threatened by habitat destruction. However, it is perhaps the easiest of all *Cyrtanthus* species to grow and get to flower in cultivation.

Cyrtanthus have been hybridized for many years, and remain coveted bulbs for collectors. Most are best grown in containers. The seed germinates readily when fresh – some species take as little as ten days, others need several weeks. Seedlings are best kept partially shaded during their first season of growth.

CYRTANTHUS ELATUS ► *Formerly called* Vallota speciosa, *this is the best known* Cyrtanthus *species, and the one that has been most successfully commercialized.*

Red List: Vulnerable*

Daubenya aurea

Botanical family
Hyacinthaceae/Liliaceae

Distribution
South Africa: Northern Cape Province

Hardiness
Frost-tender

Preferred conditions
Full sun, in well-drained, sandy soil; dry summers are an essential condition. In its habitat, night temperatures are cool, about 10–15°C (50–60°F).

The genus *Daubenya* was once thought to consist of a single species, *Daubenya aurea* (the pincushion flower), which inhabits desert and scrub. However, geneticists studying variations in the DNA sequences of plants have recently revealed that other South African genera of the hyacinth family, mostly consisting of a single species, are in fact very closely related to *Daubenya aurea.* The differences among the plants, all characteristics of the flowers, are adaptations to specific pollinating agents. This pattern is often observed in bulbous plants of the Cape region, including those of the genus *Moraea* (*see p.249*).

The pollinating agent in the case of *Daubenya aurea* is the money beetle, a type of scarab. This insect is normally attracted to members of the sunflower family. In order to attract the beetle, the pincushion flower has developed a daisy-like flowerhead by enlargement of the outer tepals of those flowers located on the edge of its condensed flower stem. Both red and yellow forms of the species occur in the wild; the red form is called ***Daubenya aurea* var. *coccinea*.**

The plant's extremely restricted range in Northern Cape Province is the main factor in its conservation status. It is regarded as a highly desirable species by fanciers of bulbous plants, and, in the

◄ RED RARITY Daubenya aurea *var.* coccinea *is one of only two varieties of this rare species. The other,* Daubenya aurea, *has yellow flowers.*

Red List: Rare*

Dietes bicolor

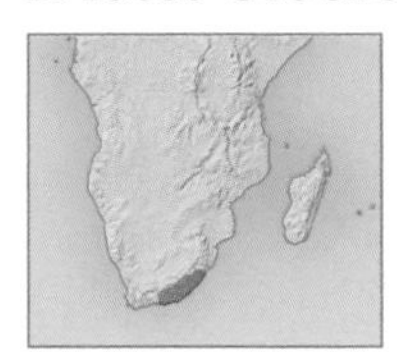

Botanical family
Iridaceae

Distribution
South Africa: Eastern Cape Province

Hardiness
Half-hardy

Preferred conditions
Full to partial shade; full sun only where temperatures do not get excessively high (plants require more water in full sun); rich, moist soil is best, but the plant accommodates most soils.

Of the six species within *Dietes*, an evergreen, rhizomatous genus, five are strictly African and one, in a strange and still unexplained disjunction, occurs on Lord Howe Island, lying between the east coast of Australia and New Zealand. All but one of the African species are found in regions of either year-round or summer rainfall. The genus is thought to be closely related to *Iris*, a genus strictly of the northern hemisphere.

Dietes bicolor, or yellow African iris, is found only in the understorey of warm, temperate forest in the Eastern Cape, where it generally favours the banks of streams and rivers that flow throughout the year. Its range has no doubt contracted due to logging and agriculture, as has a great deal of the original forest in that region of South Africa.

The plant is fast-growing, and for that reason is ideal for massing in shady areas where quick cover is needed. Although it occurs in the wild near watercourses, it is surprisingly drought-resistant. It can also take a fair amount of frost, and is not seriously damaged until temperatures fall below -7°C (20°F) for a sustained period. The plant forms a large, upright clump with an erect fan of leaves, and makes a fine vertical accent in the garden. Propagation by division of rhizomes is done in autumn. The seed germinates readily if sown in spring or autumn in a moist seedling mix. Seedlings can flower in as little as one year after germination.

◄ ONE-DAY BLOOMS *Individual flowers last only a single day but appear successively for a long period of time in spring, and sporadically thoughout the summer.*

Unlisted

Eremurus spectabilis

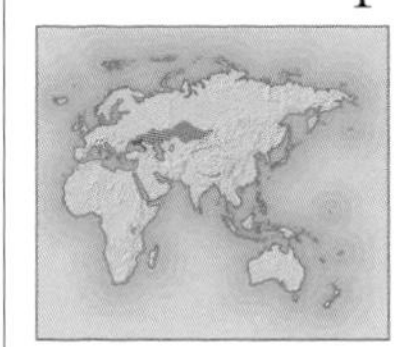

Botanical family
Asphodelaceae/Liliaceae

Distribution
Turkey, Lebanon, Iraq, Iran, western Pakistan

Hardiness
Hardy

Preferred conditions
Full sun; fertile, well-drained, loamy soil; the tall but slender spikes are easily damaged by strong winds and benefit from a sheltered site.

Although ***Eremurus spectabilis*** is unlisted by IUCN or CITES, the European Council has recommended that it be protected. The plant is used in folk medicine throughout its range in the temperate steppe of Turkey and the Causcasus, and overcollection of the bulbs has had a serious impact on the health of wild populations. In a bid to relieve the pressure of wild-harvesting, institutions in Turkey have conducted research on a small scale into means of mass-producing the plant.

Eremurus plants should be left undisturbed in the garden for at least several years. Mature plants with many stems may then be divided to increase the stock. The starfish-like crowns with wide-spreading roots, once individually separated with a sharp knife, may be planted in a hole 15cm (6in) deep. Keeping the growth bud at soil level, sit the crown on an 8cm (3in) mound of coarse sand to help prevent rot. The tubers are brittle and should be handled with care when transplanting. The seed ripens in autumn and should be sown immediately. *Eremurus* requires three to five years to reach flowering size.

▼ SLENDER SPIKES *The shape and orange colouring of* Eremurus spectabilis' *striking flower spikes account for the plant's common name of foxtail lily.*

past, some unscrupulous collectors have reportedly smuggled significant numbers of *Daubenya aurea* out of South Africa. Pressure on the wild populations should be eased somewhat by the achievements of Hadeco, a commercial bulb grower in South Africa, which is now propagating selected forms of the species through tissue culture (*see below*).

The pincushion flower is readily managed in containers in most areas: clay pots are recommended. The plant can be grown in the ground only where its need for dry summers can be adequately met. Another consideration for successful culture is maintenance of the cool night temperatures, 10–15°C (50–60°F), of its native habitat. The prostrate leaves are susceptible to attack by snails and slugs. Seed, if available, germinates easily when sown on a sandy mix and maintained in bright light and cool night temperatures.

RENEWAL

Tissue culture

While some rare plants can be propagated easily by seed or by cuttings, in other cases producing new plants is made very difficult either by natural reproductive limitations or by severe scarcity of plant material. In tissue culture, small pieces of plant tissue (usually taken from a shoot tip, leaf, lateral bud, stem, or root, or from a bulb or corm, without harming the parent plant) are grown into plants on a nutrient medium in a sterile container. For success, some species require specialist techniques and painstaking research into their individual growing needs. Once rooted plants are produced, they are moved from the culture tube to soil in a process called acclimatization. Plants from tissue culture are genetic clones of the parent plant. They increase the stock, allowing further research in which selected genetic lines may be propagated for the species' survival.

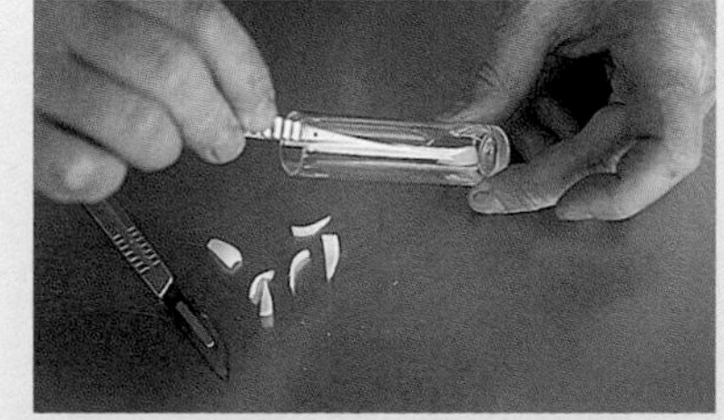

PIECE OF BULB IS PLACED IN TEST TUBE

NEW BULBS GERMINATE IN FLASKS

10 on Red List*

Erythronium

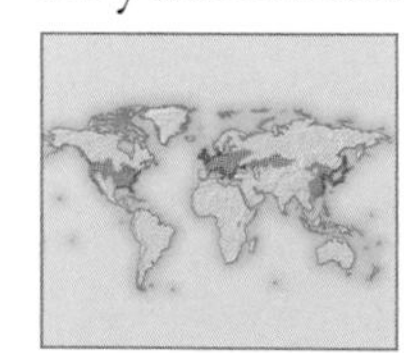

Botanical family
Liliaceae

Distribution
Western North America, Europe, Asia (one or two species)

Hardiness
Fully hardy

Preferred conditions
Full sun to partial shade, depending on species. Rich soil; some species prefer moist but well-drained soil, others can tolerate dryness.

This genus of delightful, late-winter and early-spring wild flowers, called trout lilies, comprises about 20 species. They grow in temperate broadleaved, mixed, and coniferous forest, as well as in temperate prairie and alpine meadows. Their main threat is loss of habitat caused by logging (*see right*), and there is some collection for the international bulb trade. All trout lilies are diminutive plants that produce ground-hugging, usually mottled, leaves from small corms, and one to several flowers with turned-back petals borne on leafless stems.

Erythronium grandiflorum is found chiefly in the Cascade Range of the north-western USA, but also in the Rocky Mountains. Two subspecies appear as Rare on the Red List of 1997. Subspecies *nudipetalum*, with red anthers, is found only in the Bear Valley area in the southern Salmon River Mountains of central Idaho. Subspecies *candidum*, with white flowers, is known only from Washington, Idaho, and Wyoming. The main threats to the rare subspecies are logging, grazing, and hydrological changes brought about by mining. In cultivation the plant prefers cool, damp conditions, and it should never be allowed to dry out.

Other trout lilies

Erythronium mesochoreum grows not in forests but in the prairie. It is considered endangered in the state of Illinois, even though it has yet to receive Red List recognition. It is still sometimes referred to as a subspecies or variety of the widely distributed *Erythronium albidum*, but most botanists view it as a distinct species. It is also found in Kansas, Iowa, and southwards to Texas.

Erythronium propullans, or the Minnesota dwarf trout lily, is cited as Endangered in the Red List of 1997. Known only from that state, and perhaps neighbouring Ontario, this diminutive plant has been observed growing in only 14 populations in the Minnesota woods. It flowers from mid- to late spring, retreating into dormancy by early summer. It rarely produces seed. Instead, the corm sends out an underground creeping stem (stolon) that runs just below the surface, then descends deeper into the soil and produces a new corm at its tip. Botanists report that only about one-tenth of any one population sets fruit annually, and stolons represent the chief means of increase. One study even suggested that most of the seed formed in the wild is the result of hybridization with the broadly distributed *Erythronium albidum*.

An investigation into the genetics of *Erythronium propullans* did in fact indicate that each population of this trout lily consists of a few clones, and that the species is a recent derivative of *Erythronium albidum*. The flowers vary from white to pink to light violet, and the fruit, when produced, nods (the fruit of all other *Erythronium* species consists of an erect capsule). It is also rare to find flowers consisting of the correct number of six tepals. One conclusion is that this rare species is suffering from severe inbreeding. The plants are protected by federal law, and anyone caught disturbing the species faces a huge fine. The species has been successfully tissue-cultured (*see p.233*).

For this, as for all trout lily species, the seed should be collected on ripening and sown immediately on a medium that retains moisture well. The seed is likely to take several months to germinate, and at least three years must transpire before the plants begin to flower.

THREAT

Responsible logging

Rarely abundant wherever they occur, many *Erythronium* species have been negatively affected by "clear-cut" logging practices in the forests and woodlands of the western states of the USA. Clear-cut logging involves felling and removal of all trees within given sectors of forest, clearing the way for heavy vehicles and exposing the ground and understorey plants to harsh weather conditions and erosion. There is now increasing pressure on companies to practise ecosystem-based management, which in old-growth forests spares the biggest trees and allows cutting only of trees below an agreed trunk diameter. With the larger trees remaining, the understorey is partially shielded, aiding regeneration.

▲ ERYTHRONIUM MESOCHOREUM
This drought-tolerant plant is native to the black-soil, tall-grass prairie, a vegetation type that has been reduced to a fraction of its former extent.

ERYTHRONIUM GRANDIFLORUM
Also called the yellow avalanche lily, this erythronium has the largest flowers of the genus, which attracts the attention of gardeners, although the plant is notoriously difficult to grow. Native tribes of western America used the corms as a food source.

Red List: Vulnerable*

Eucharis amazonica

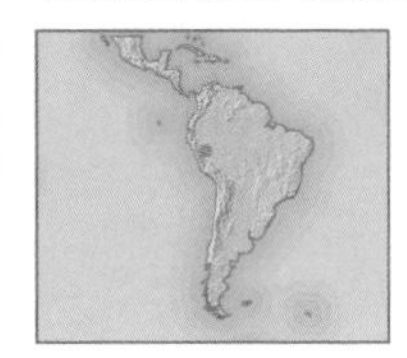

Botanical family
Amaryllidaceae

Distribution
Peru: Huanuco Department

Hardiness
Frost-tender

Preferred conditions
Deep shade and moist, fertile soil; this plant's habitat is one of high humidity.

Known as the Amazon lily, ***Eucharis amazonica*** is the most widely grown member of its genus, much admired for its dark green, hosta-like leaves and large, sweetly fragrant, white flowers. In the wild, however, it is of restricted distribution, known only from a handful of populations in the tropical rainforest that clothes the lower slopes of the Peruvian eastern Andes. The rarity of the Amazon lily is partly explained by the fact that it has a peculiar number of chromosomes for the genus, and is never known to set seed. It is thought that all material of this species, both in its native habitat and in cultivation, may represent a single plant that has been clonally propagated. The fully-fertile species that appears most closely related to it is *Eucharis moorei*, which grows on both sides of the Ecuadorean Andes, and the Amazon lily may have been derived from that species. Deforestation is otherwise the greatest threat to the continued existence of the Amazon lily in its Peruvian habitat.

The Amazon lily may be grown outdoors in subtropical and tropical regions. As long as the bulbs do not freeze, it may persist in warm, temperate climates outdoors. Elsewhere, it may be used as an indoor or outdoor container plant. A short period of drought can sometimes induce flowering, which may occur twice a year, in late winter or spring, and again in summer or autumn. The bulbs resent frequent disturbance, and will form a sizable clump in time. Propagation is by division or bulb scale cuttings.

FRAGRANT FLOWERS ► *Up to six flowers are borne on a leafless stalk in early summer. Very few populations of this prized horticultural plant survive in the wild.*

Red List: Extinct* (but rediscovered)

Eucrosia mirabilis

Botanical family
Amaryllidaceae

Distribution
Southern Ecuador

Hardiness
Frost-tender

Preferred conditions
Partial shade; well-drained sandy soil. The plant requires a long period of dormancy during which bulbs must be dry and safe from frost.

This species, which was originally described as having been collected in Peru, was known for over a century from just a few herbarium specimens and the plate that accompanied its description (it was then named *Callipsyche mirabile*). Recently, a seed house in the UK began offering seed of an amaryllid labelled as *Stenomesson* 'White Parasol', collected in Ecuador. When the first of these bulbs flowered in collectors' greenhouses it became apparent that this plant was indeed a member of the genus *Eucrosia*. Populations were then documented in dry, open mesquite and acacia-dominated woodlands of the inter-Andean valleys in southern Ecuador. This left little doubt that the long-lost plant ***Eucrosia mirabilis*** had been rediscovered. This rare species is one of nine known from Peru and Ecuador, the majority found in the latter country. It may be that the original collections were incorrectly reported as coming from Peru in a bid to mislead rival plant hunters.

Related to more common species such as *Eucrosia aurantiaca* and *Eucrosia eucrosioides*, this plant forms large bulbs in time and rarely offsets. The bulbs are best planted half-exposed above the soil in any good potting medium half-mixed with coarse sand. Watering should be sparing while the tall flower stem develops. If pollinated with pollen from a different plant, seed capsules will form, containing numerous flat, blackish-brown, winged seeds. These germinate readily when ripe, sown on any well-drained germination medium and kept warm and partially shaded. First-year seedlings will typically produce one round leaf, and enter dormancy after a few months.

LENGTHY STAMENS ► *A tall flower stem bears 20–40 flowers with long stamens that greatly exceed the length of the tepals.*

4 on Red List*

Freesia

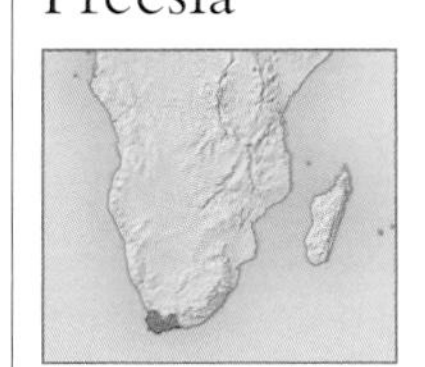

Botanical family
Iridaceae

Distribution
South Africa: Cape region

Hardiness
Frost-tender

Preferred conditions
Full sun; well-drained soil. Plants benefit from strong light levels. Corms need to be dry in summer, as in their habitat.

The genus *Freesia* consists of about 14 species, originating from the desert and scrub (karoo) of the Cape region of South Africa, and the western Cape with its winter rainfall. From this relatively small genetic base has arisen a large assortment of multi-coloured hybrids, including doubles, that were developed in Europe. In most species the flowers are arranged on one side only of the flower stems. A number of the species are powerfully fragrant, sometimes more so than the hybrids, and may be pollinated by bees.

Freesia speciosa is highly scented and is among the larger-flowered species. It occurs sporadically from Montagu to Calitzdorp. The fragile ecology of the karoo scrubs of the Cape has been adversely affected by agriculture and mining. The species is cited as Vulnerable in the Red List of 1997. The abiding reason to grow this species is its dizzying perfume, which can fill a room. ***Freesia sparrmannii***, which IUCN lists as Rare in its native habitat, is another attractive freesia that rewards growers with its white flowers.

Several small corms of these freesias should be potted up together for best effect. In mild winter zones, corms should be planted in early autumn for winter and spring bloom. In mild Mediterranean climates, the corms can stay in the ground. Where summers are wet, and/or winter lows drop below freezing, the corms should be lifted when the foliage dies down in late spring, and stored dry. Freesia seed germinates in 4–6 weeks at cool temperatures of 13–15°C (55–60°F).

FREESIA SPARRMANNII ► *Propagated bulbs of this small freesia, which is well-suited to pot culture, are more easily obtained from nurseries than those of other Red Listed freesias.*

45 on Red List*

Fritillaria

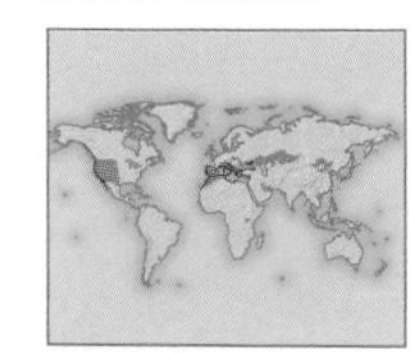

Botanical family
Liliaceae

Distribution
Western USA, central Asia, and the Mediterranean

Hardiness
Fully hardy – frost-tender

Preferred conditions
Full sun to partial shade, best lightly shaded in hot climates; moist, well-drained, loamy soil in winter and spring, but relatively dry in summer.

Of the perhaps 100 *Fritillaria* species found in temperate steppe, montane meadow, and Mediterranean woodland throughout the world, nearly half are now of concern from a conservation perspective. The majority of these beautiful plants, closely related to the true lilies, are rare, and those of greatest concern are not widely cultivated. Although wild collection of *Fritillaria imperialis*, the much coveted crown imperial, has been documented in Turkey, its status in its fairly wide distribution through central Asia to the western Himalayas is not known. However, it is not thought to be endangered. The UN has funded a pilot project to promote the cultivation of the species in Turkey and thereby reduce the pressure on wild populations. Likewise, the popular chequered lily, ***Fritillaria meleagris***, widely distributed throughout meadows in Europe, is now protected in Poland and is increasingly rare in the UK, but has not yet been deemed threatened. However, many of the species native to California, never widespread, are now considered endangered. Habitat loss is the prime cause, although horticultural collection has affected some of the showier species. Thirteen *Fritillaria* species are listed in the California Native Plant Society's Inventory of Rare and Endangered Vascular Plants of California.

▲ FRITILLARIA PALLIDIFLORA *The broadly bell-shaped flowers of this species are faintly foul-smelling. The flowers are often marked inside by brown-red chequering, although on the outside the cream-yellow petals are plain.*

FRITILLARIA MELEAGRIS ► *This species, known as the snake's head fritillary for its mottled and spotted flowers, produces a single or pair of purple or pinkish-purple blooms on each stem; the white form also visible here is* Fritillaria meleagris *f.* 'Alba'.

OTHER FRITILLARIES

Fritillaria gentneri, called Gentner's fritillary or Gentner's mission bells, is known from around 60 localities in south-west Oregon only, scattered across approximately 7,700sq km (3,000sq miles). Many of the populations consist of only one plant, and a reasonably accurate census suggests that a total number of 600 individuals remain. The plant is Endangered according to the Red List of 1997.

Gentner's fritillary is a species of successional grasslands and chaparral lying at the edges of dry, fairly open woodlands. The plant is adapted to and dependent on periodic fires to suppress woody plants from encroaching on its habitat, but for years natural fire cycles have been suppressed by land managers. Habitat loss from agricultural and residential development has also been a factor in its rarity, along with competition from exotic invasive plants.

In spring, the blue-green stems of *Fritillaria gentneri* erupt with showy, reddish-purple, bell-shaped flowers, overlaid by a yellow chequered pattern. Fortunately, this photogenic appearance has resulted in significant public effort to reverse the plant's slide towards extinction. A collection of the species is maintained for conservation purposes at the Berry Botanic Garden in Portland, Oregon. Beautiful as they are, the flowers are short-lived, which limits the plant's appeal as a garden plant, although the species is notable for its many fanciers among UK horticulturists.

Fritillaria pallidiflora is an Asian species noted as Vulnerable in the Red List of 1997. It is native to eastern Siberia and adjacent China, reaching elevations in excess of 2,700m (9,000ft) in the Tien Shan and Alatau Mountains. This tough and dependable fritillary is found in steppe or the margins of coniferous forest, where it is vulnerable to both habitat disturbance and overcollecting for medicinal use, the latter particularly in China. Plants can grow to a height of 45cm (18in). The nodding flowers are pale yellow with reddish spots and appear in late spring. Full sun is preferred in climates where springtime temperatures are moderate, and partial shade where spring warms up early and quickly. One of the more adaptable fritillaries, it grows well in many areas of the USA and Europe.

FRITILLARIA PLURIFLORA *This attractive species is native to California and is described as Rare in the Red List of 1997. It grows in heavy adobe soils of the hot, summer-dry inland valleys.*

GROWING FRITILLARIES

Fritillaries grown from seed may require as much as five years to reach flowering size.

Germination times vary, but are fairly lengthy, and can average six months or more. Seed of hardy species can be sown in containers in the autumn and left to overwinter outdoors or in a frame, or else stored and then sown in spring. Seedlings typically send a "dropper", part of the cotyledon, deep into the soil upon germination, so sowing seed in deeper containers is advisable.

Fritillary bulbs are fleshy and scaly; they are quite fragile and should be handled carefully. Never allow them to dry out. A number of fritillary species produce offsets generously, and these, as well as propagation from individual bulb scales, allow selected forms to be increased by cloning. Removal of too many scales will, however, weaken and reduce the viability of the bulb from which they are harvested.

REDISCOVERY

FRITILLARIA MICHAILOVSKYI

First described in 1914, this attractive species, cited as Vulnerable in the 1997 Red List, was not found again in north-east Turkey until 1965. Material was given to the Dutch bulb industry for propagation purposes, and the exact location of the population was never revealed. The plant's easy cultivation, striking flowers, and dwarf habit quickly made it something of a sensation, and it is now offered by many bulb dealers. This enthusiasm may have stimulated some collection from the wild, but export of collected bulbs from Turkey is now under much greater control than previously. Michael's fritillary has proved adaptable to many different regions, blooms reliably, and increases in the garden. Good drainage is essential, and excess summer rainfall is not conducive to best growth.

5 on Red List*/all on CITES II

Galanthus

Botanical family
Amaryllidaceae

Distribution
Europe and Asia Minor; Pyrenees eastward to the Caucasus and Iran

Hardiness
Fully hardy – frost-tender

Preferred conditions
Full sun to partial shade, depending on species; fertile, well-drained, neutral to slightly alkaline soil that needs to be moist during active growth.

The genus *Galanthus* grows in temperate and Mediterranean woodland, scrub, and alpine meadows and is the most heavily wild-collected genus of bulbous plants in the world. Species populations have also been diminished throughout their range by loss of habitat to logging, agriculture, and other types of development. All *Galanthus* species are now included in CITES Appendix II, thus export of bulbs from their countries of origin is intensively controlled, in theory at least.

The huge demand for snowdrop bulbs is partly fuelled by the cult-like devotion of so-called "galanthophiles" to collecting every possible variant of any given species. Fortunately, some of these enthusiasts will no longer buy bulbs collected from habitat. The propagation of *Galanthus elwesii* bulbs in Turkey (*see box, above*) shows how the demand can be met in a positive and sustainable way.

▲ GALANTHUS PLICATUS *This species is a marvellous garden snowdrop, easy of culture, generous in flower, and sturdy.* Galanthus plicatus *has been used extensively for breeding purposes, since its desirable qualities seem to be inherited by its hybrids.*

Snowdrop species

Galanthus alpinus, Alpine snowdrop, occurs in the Caucasus Mountains of Armenia and the Republic of Georgia. It is cited as Indeterminate in the Red List of 1997. At the western and eastern margins of its distribution it is known only from a few small populations, and is therefore of concern. It is found within and at the edge of deciduous forest, and also within scrub and in clearings, usually on neutral to alkaline soils. It is a sought-after rock garden plant.

Galanthus elwesii, the giant snowdrop, also Red Listed as Indeterminate, has been grown for well over a century. It is extremely variable in nature, with many named varieties recognized. The leaves are a waxy, greyish-green and can be quite broad. The species is found around the eastern islands of the Aegean Sea, in Turkey (*see box, right*), Greece, Bulgaria, southern Ukraine, and the eastern part of the former Yugoslavia. It is quite cosmopolitan in habitat, occurring in oak or pine woodland, oak scrub, subalpine meadows, and forest clearings. This diversity of habitats make it difficult to generalize about its preferred garden conditions, which vary by variety. It is the most tolerant of all snowdrop species to drier conditions.

Other snowdrops

Galanthus plicatus has been grown in Europe for nearly 500 years, but is restricted in the wild to a region roughly corresponding to the western borders of the Black Sea, including the Crimea, Romania, and north-west Turkey. Large populations are rare, but it readily spreads by seed in the garden. Two subspecies are recognized. Subspecies *plicatus* is the more common, occurring within or at the edges of forest in Romania, northern Turkey, and the Crimea. Subspecies *byzantinus*, which is cited as Vulnerable in the 1997 Red List, is restricted to north-west Turkey, in beech and oak woods and scrub, most commonly near water. The two subspecies differ in the number of green spots on the inner tepals – one in subsp. *plicatus*, two in subsp. *byzantinus*.

Galanthus reginae-olgae is found mainly in southern Greece, most prominently in the Peloponnese. It has also been collected on the islands of Corfu, Sicily, and the southern portion of what used to be Yugoslavia. It is typically an autumn-flowering species, but a spring-blooming variety, subsp. *vernalis*, has been recognized. Although not yet of special conservation concern, the desirability of this species and its fairly restricted range, renders it potentially vulnerable.

RENEWAL

Indigenous propagation

It is well known that nursery-grown stock of well-established cultivars of giant snowdrop (*Galanthus elwesii*) performs much more reliably in the garden than wild-collected bulbs, which are often carelessly treated before export. Yet as many as 175 million wild-collected bulbs are estimated to have been shipped from Turkey to the Netherlands for distribution elsewhere in Europe in the late 1980s and early 1990s alone. The majority of these bulbs probably died. Since that time, a propagation and production programme for this species has been launched in Turkey with the assistance of international conservation groups. In time, this should help relieve pressure on wild populations and provide a sustainable economic benefit for participating Turkish villages.

Snowdrops may be grown from seed sown fresh or after a few weeks of storage. Viability drops sharply after a year of cool, dry storage. After removing the elaisome (covering), the seed should be sown thinly, lightly covered, and the containers placed in shade. Seedlings may remain in place for a year unless crowded, since two-year-old plants are easier to transplant. Three to four years usually pass before first flowering occurs. Vegetative propagation of snowdrops is done by plant division or bulb cutting.

▼ GALANTHUS REGINAE-OLGAE *Large flowers and ornamental foliage make this a valuable garden snowdrop. The blue-green leaves, which emerge after the flowers have faded, have a distinctive waxy silver stripe running down the middle.*

65*/1 on Red List

Gladiolus

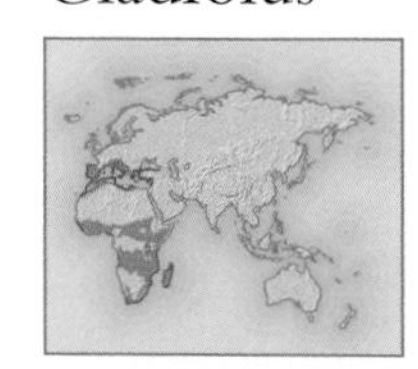

Botanical family
Iridaceae

Distribution
Southern Europe, Middle East, Africa: from the tropics to South Africa

Hardiness
Fully hardy – frost-tender

Preferred conditions
Full sun to partial shade, depending on the species; some prefer well-drained, sandy soil, while others thrive in moist, fertile soil.

Within the iris family, *Gladiolus* is the second largest genus after *Iris* itself. More than 250 species are recognized, of which more than half are restricted to southern Africa, with the vast majority of these found in the Cape. *Gladiolus* species grow in veld, fynbos, tropical and subtropical grassland, savanna, and shrubland. With one exception, all species are deciduous, and the underground corm is typically renewed annually. The genus is marked by its spiky flower stem and winged seed. The diversity of flower form among the many South African species is extraordinary, and reflects the enormous range of pollinating agents to which they have adapted. The flowers of some species have a delightful fragrance. In common with those of other large genera of bulbous plants in the Cape, many *Gladiolus* species are known only from one or a few populations, or occur in areas that are under severe pressure from agricultural, industrial, or residential development.

Many of these small-range species are not cultivated by gardeners, who could provide sanctuary for the genetic material of rare species. At least one *Gladiolus* species (*Gladiolus felicis*) is believed to be extinct, but, perhaps surprisingly, this is one of the hardy European species. Indeed, several species from Europe that have contributed hardiness to their hybrids are now known from a very limited number of populations, due to many centuries of human disturbance to their habitats. In South Africa, many of the Cape species have been used in the development of thousands of cultivated varieties of garden gladioli. The similarity in shape of the many hybrids gives no indication of the great diversity of flower types found among the wild species in South Africa. The corms of just one species, *Gladiolus permeabilis* var. *edulis*, like those of *Babiana* species (*see p.225*), are eaten by baboons in South Africa, and they also feature in the diets of some of the indigenous peoples of the region.

▲ GLADIOLUS PALUSTRIS *This moisture-loving European native makes a delightful accompaniment to a waterside garden, but is equally at home in a moist border. Though it is broadly distributed, local populations in many countries have disappeared.*

HABITAT

MEDITERRANEAN HOME

Free-draining rocky hillsides and areas of scrub, such as this stony patch on Spain's Costa del Sol, are the natural home of several Mediterranean *Gladiolus* species, including *Gladiolus illyricus* (*see left*). This type of habitat is under constant pressure, especially along the coast where land is sought for holiday resorts, building, and agriculture. Although not threatened as a species, *Gladiolus illyricus* has suffered genetic erosion due to the disturbances caused by intensive development.

Magenta-flowered *Gladiolus illyricus* is sometimes thought to be a subspecies of *Gladiolus communis*, another Mediterranean native. Both have proved hardy in temperate areas of western Europe.

Gladiolus alatus, called *kaloentjies* in Afrikaans, is a brilliantly orange-flowered species with a powerful but pleasant fragrance in late winter and spring. The species is broadly distributed, but many areas of its habitat have been transformed by agriculture. *Gladiolus alatus* var. *algoensis*, a variety native to the western Cape, is cited as Endangered in the Red List of 1997. Gardeners fortunate enough to live in regions of the world with a climate similar to that of its habitat, with winter rainfall and summer dryness, can plant it directly in the ground. Gardeners living elsewhere must enjoy it in pots.

OTHER GLADIOLI

Gladiolus palustris, or sword lily, is a native of wetlands throughout Europe. Considered endangered in many individual countries, it received the conservation status of Indeterminate in the Red List of 1997. Sword lily is quite possibly the hardiest of all *Gladiolus* species. Flowering in late spring, it is of intermediate height and bears purple-red flowers. It is a native of moist meadows, and quickly disappears when such areas are drained. There have been some *ex-situ* propagation programmes and re-establishment of the species in protected areas in Switzerland. The species is an excellent summer-flowering bulbous plant for use around ponds.

Gladiolus sempervirens, cliff gladiolus, is notable for the fact that it is the only evergreen species in this large genus, but its summer flowers are also beautiful, large and red with a few white stripes on the lower tepals. Designated Rare in the Red List of 1997, it is found only in surface springs on sandstone slopes of South Africa's south-eastern Cape. It spreads by means of creeping underground stems (stolons), and has proved a very adaptable garden border plant in areas with mild winters. Cliff gladiolus was an important parent of the numerous complex *Gladiolus* hybrids grown throughout the world and would be the basis of any attempt to breed evergreen *Gladiolus* hybrids. In its Cape habitat the cliff gladiolus is pollinated by the large butterfly *Aeropetes tulbaghia*, known as the mountain beauty. Strongly attracted to the colour red, this butterfly is also responsible for pollinating the red-flowering species of a number of unrelated Cape genera, including *Disa* and *Crassula*.

One plant related to the cliff gladiolus, and even more beautiful in flower, is *Gladiolus cardinalis*. Called waterfall gladiolus or New Year's flower, it is a summer-flowering species of wet slopes and waterfalls in the south-west Cape. This plant has not been Red Listed.

PROPAGATION

Gladiolus species produce abundant seed that germinates readily when fresh, and can be stored dry and cold for a year or more. In most cases, seedlings require 2–3 years to flower, although several species produce cormels that are genetically identical to the parent and may reach flowering size in less time.

▲ GLADIOLUS SEMPERVIRENS *The red flowers of this South African evergreen secured it an important role in breeding programmes. Its rhizomatous habit makes it a useful border plant in areas with mild winters.*

GLADIOLUS ALATUS *This species belongs to a group of Cape gladioli known to South Africans as "turkey chicks". They make a conspicuous group on account of their large flowers in relation to their small stature.*

2 on Red List*

Griffinia

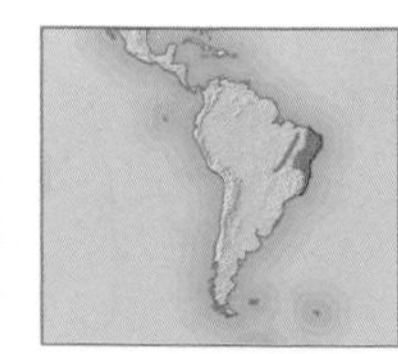

Botanical family
Amaryllidaceae

Distribution
Brazil: the eastern third of the country

Hardiness
Frost-tender

Preferred conditions
As plants of the rainforest understorey, griffinias like deep shade, warm temperatures, high humidity, and moist, acidic, humus-rich soil.

The genus *Griffinia* consists of 12–15 species classified in two subgenera: *Hyline* and *Griffinia*. The two species of subgenus *Hyline* are found in seasonally dry tropical forests and are not in cultivation. Subgenus *Griffinia* consists of 10–13 species found in the understorey of the Atlantic tropical rainforest of eastern Brazil, and its range extends from the north-eastern state of Pernambuco southwards to the states of Rio de Janeiro and São Paulo. Griffinias have also been collected infrequently where this forest type extends inland along rivers in the state of Minas Gerais. All the rainforest griffinias have leaves like a hosta or Amazon lily (*see Eucharis, p.235*) and a few species have attractive white spots on the leaves. The plants produce 5–20 flowers, mostly lilac blue in colour, rarely white. Griffinias are rare in the wild; populations are not very large, and seed production is low, with no more than three seeds forming within each capsule.

Griffinia hyacinthina is the largest-flowered species of subgenus *Griffinia*, and is found only in the states of Rio de Janeiro and São Paulo, in the understorey of primary Atlantic rainforest. It is cited as Vulnerable in the Red List of 1997. The main threat to this species has been the destruction of the Atlantic rainforest, of which less than ten per cent survives.

In cultivation in the northern hemisphere, *Griffinia hyacinthina* may enter a brief period of rest in winter, when the leaves briefly die back. The bulbs should be watered lightly until new growth emerges. The species can brighten a shady site in mild climates or serve as a houseplant in temperate zones. Propagation is by plant division, bulb scale cuttings, or by seed.

◄ GRIFFINIA LIBONIANA *The lilac-blue of the flowers of this species is rare within the Amaryllis family. Some populations of* Griffinia liboniana *have petals that are attractively spotted with silver-white.*

Griffinia liboniana is one of several smaller-flowered species in the genus. It is found in southern Bahia state in the understorey of coastal rainforest and has been collected rarely along interior river forests of Minas Gerais state. Destruction of its habitat is a threat throughout its range. In the Red List of 1997 it is listed as Extinct, but it is actually Endangered. It has broad leaves with white spotting.

Griffinia liboniana is grown and propagated in the same way as *Griffinia hyacinthina*. The bulbs may enter a brief, leafless rest time in winter, especially if exposed to cool temperatures, when they should be watered only lightly. The species makes a fine perennial for a shady site in the subtropical and tropical garden, or a container plant in temperate climates.

12 on Red List*

Haemanthus

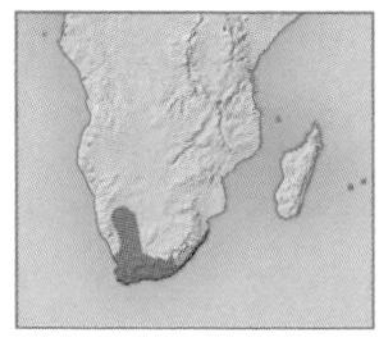

Botanical family
Amaryllidaceae

Distribution
Southern Africa; half of all species in the Cape provinces of South Africa

Hardiness
Frost-tender

Preferred conditions
Full sun; well-drained, sandy soil. Abundant moisture during active growth, followed by dryness in summer.

Natives of the South African fynbos, *Haemanthus* species produce individually small but numerous brightly coloured flowers. The modified leaves lying at each flower base are often just as colourful, and together these form a conspicuous, brush-like attractant to pollinators. Red-flowered species are pollinated by sunbirds and the Cape's endemic mountain beauty butterfly (*Aeropetes tulbaghia*, *see also p.238*), while pink-flowered species are thought to be visited by bees. The berry fruits, mostly red, orange, or pink, are dispersed by animals. One species, *Haemanthus tristis*, produces dun-coloured, leathery fruit that resembles antelope droppings, possibly an evolved means of escaping predation by fruit-eating animals. Suppression of fire has reduced the reproductive success of some species, while land development has adversely affected species with coastal distributions.

Haemanthus amarylloides, one of several species also known as the paintbrush lily, bears pink flowers in autumn, held on a burgundy stem. It is found in the north-

▲ HAEMANTHUS CANICULATUS
The burnt vegetation beneath these flower stems confirms that the species is one of the first to regenerate following a bushfire.

west of Western Cape Province, from Namaqualand to Clanwilliam. Two of its three subspecies are cited as Vulnerable in the Red List of 1997. The bulbs are probably best grown in containers, except where conditions can closely match those of its habitat.

Haemanthus caniculatus is a blood lily with a restricted distribution in swampy coastal flats of the south-west Cape. It is designated as Vulnerable in the 1997 Red List. This species flowers most abundantly after a fire. The red flower stem is spotted near its base, and there are 5–7 brilliant red modified leaves underlying red flowers.

The fleshy seeds of all paintbrush lilies have a high water content and are short-lived. They should be sown as soon as they are ripe in a moist medium, very shallowly covered or left partially exposed. The seedling bulbs may be left in the germination container during the first season's dormant period in summer. Alternatively, several species can be propagated by leaf cuttings.

Red List: Indeterminate*

Hedychium rubrum

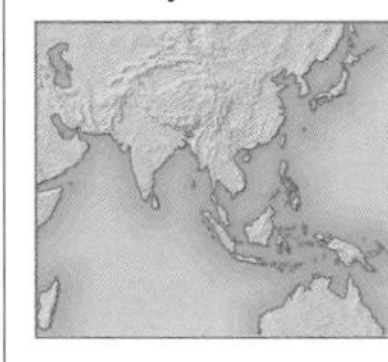

Botanical family
Zingiberaceae

Distribution
India: Meghalaya State

Hardiness
Frost-tender

Preferred conditions
Partial shade in a site with protection against cold wind. Moist, rich soil, containing plenty of organic matter.

The majority of *Hedychium* species, or butterfly gingers, are natives of China and India. A number are fairly well represented in subtropical horticulture, and two have become naturalized weeds in many tropical countries (*see p.456*). Seven species are on the Red List of 1997, and one, *Hedychium marginatum* from India, is considered Extinct. Butterfly gingers are plants from the understorey or margins of tropical and subtropical rainforest (a few are epiphytes), and the main threat they face is habitat disturbance by humans.

Hardest hit are the Indian species, since development continues to gnaw away at their habitat in the remaining tropical forests of the Himalayan foothills. ***Hedychium rubrum*** is known only from rainforest and swamp margins in Meghalaya state, and is emblematic of all of the threatened species of butterfly ginger in southern Asia. It has very large and showy, bright red flowers. Recent taxonomic work places this species under synonymy with *Hedychium greenii*, which extends its distribution to Bhutan and India's Nagaland Province.

Hedychium rubrum may be grown outdoors in mild climates where winter lows rarely dip below -4°C (25°F). In heavy rather than partial shade, the stems tend to lengthen and flowering is inhibited. In colder climates, plants should be lifted after the first frost and the rhizomes stored in cool, dry, dark conditions. Water should be gradually withheld as temperatures begin to fall.

Hedychium species are easily raised from seed sown and lightly covered in any friable germination medium and kept in bright shade at 24–27°C (75–80°F). However, the relative lack of available seed makes rhizome division a more common propagation strategy.

SUMMER BLOOMER ► *The cane-like stems bear slender leaves throughout their length and quickly achieve a height of up to 2m (6ft) during the warm months. Numerous red flowers appear in mid- to late summer.*

5 on Red List*

Hippeastrum

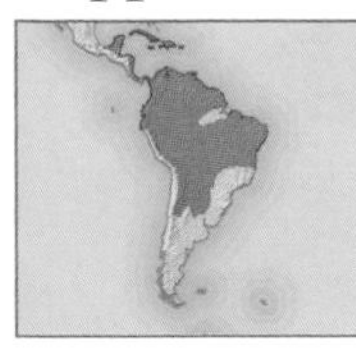

Botanical family
Amaryllidaceae

Distribution
Brazil, Argentina, Peru, Bolivia, Central America, and the West Indies

Hardiness
Frost-tender

Preferred conditions
Partial shade in fertile, well-drained soil; some species like dry conditions in winter, others prefer wetter conditions all year round.

The common name "amaryllis" refers to the genus *Hippeastrum* and should not be confused with the South African genus *Amaryllis*. Although the subject of 200 years of breeding history, the genus *Hippeastrum* is still not well understood taxonomically. It is estimated to consist of 50 species, most of which are robust plants with large flowers.

▲ HIPPEASTRUM RETICULATUM *Unlike the rest of the genus, this amaryllis flowers in the autumn. It grows in the understorey of rainforest. Some populations have a white stripe running down the centre of the leaves.*

They grow in tropical rainforest and dry forest; tropical and subtropical grassland, savanna, and shrubland; and desert and scrub. The two main centres of diversity are in eastern Brazil and the Peruvian and Bolivian Andes. Hummingbirds are the primary pollinators of the plants, most of which have red or pink flowers. Many of the species seem to intergrade, which can make their identification somewhat arbitrary. Destruction of habitat is the gravest threat to the continued existence of many of the species, some of which have restricted native ranges. The largest number of species are found in the tropics at middle altitudes, 1,000–1,500m (3,300–5,000ft), but a few grow at higher altitudes.

HIPPEASTRUM CALYPTRATUM ► *This epiphyte of the coastal rainforest of Brazil has sour-smelling flowers that attract pollinating bats. It also grows on the ground in decomposed leaf litter.*

HIPPEASTRUM SPECIES

Hippeastrum brasilianum is one of a handful of white, heavily perfumed, trumpet-shaped amaryllis species, and is much sought after by enthusiasts. It is restricted to Espiritu Santo in Brazil, a state that has suffered unparalleled disruption of natural habitats since the 16th century. Unlisted by IUCN, it is probably extinct in the wild. In the 1980s, an attempt to find the species near where it had originally been described turned up only plants in cultivation. This plant, which in cultivation produces elegant, fragrant white flowers in spring, resembles the Easter lily (*Lilium longiflorum*). Its strap-shaped leaves are coated with a waxy bloom that seems to render them resistant to the leaf fungi that plague other amaryllis species.

Hippeastrum calyptratum, the green amaryllis, is currently unlisted by IUCN but is a truly vulnerable species. It is one of three epiphytic species in the genus. Within the protected remnants of coastal Atlantic rainforest in Brazil's Rio de Janeiro state (*see below*) it is still possible to see the species festooning the lower canopy of rainforest trees. Plants are also found on the ground, established in decomposed leaf litter. The green flowers produce a foetid fragrance that attracts pollinating bats. A sought-after collector's item, the species can be finicky in cultivation. Most gardeners do not realize that it is an epiphyte, and its succulent roots soon rot if planted in standard potting composts.

Hippeastrum papilio is unlisted by IUCN but is both rare and vulnerable. It created a stir when first introduced into the horticultural trade, and bulbs still fetch a fairly high price. For many years the plant's exact origins were unknown, but at least one wild population was recently discovered growing epiphytically among trees of the Atlantic coastal rainforest in the state of Rio Grande do Sul, in southern Brazil. Little of the original rainforest of the southern states of Brazil has been left intact, but much of what remains is zealously protected. If the existing populations of this species occur within the forest reserves, their prospects of survival are good.

All hippeastrums may be grown from seed, which should be sown shallowly, placed in bright shade, and kept moist. The seedlings should be transplanted into small pots when one leaf has developed. Vegetative propagation is via offset bulbs or bulb scaling.

HIPPEASTRUM PAPILIO 'BUTTERFLY' ► *This is a popular cultivar of the butterfly amaryllis. The parent species is evergreen and produces its boldly patterned flowers twice a year, in spring and autumn.*

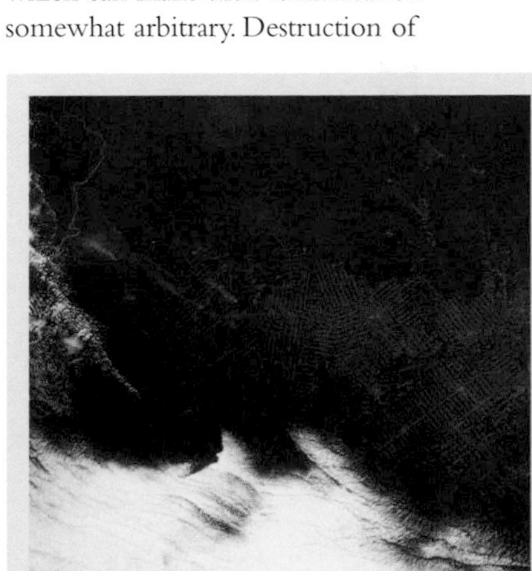

HABITAT

DISAPPEARING RAINFOREST

The loss of tropical rainforest from the coasts of well-populated regions is graphically illustrated by this satellite photograph of Rio de Janeiro state, Brazil. Swathes of mature tropical rainforest have been felled to make way for agriculture and town development. Some remnants of coastal rainforest are now protected by law, but expanding populations in relatively new settlements exert pressure to extend land conversion far into the hinterland.

Unlisted

Hymenocallis speciosa

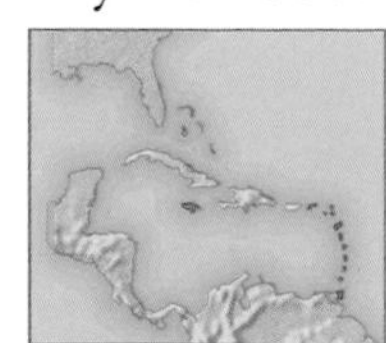

Botanical family
Amaryllidaceae

Distribution
West Indies

Hardiness
Frost-tender

Preferred conditions
Shade, warmth, and high humidity; moist but well-drained soil with rich organic content.

Although not among the six species of its genus on the Red List of 1997, ***Hymenocallis speciosa*** is one of the rarest. It has not been seen in the wild in decades. This *Hymenocallis* is one of the few that are adapted to the understorey of tropical rainforest, and, like all rainforest plants of limited distribution, it is threatened by the rampant deforestation in the tropics. Its genus is most diverse in Mexico, but there are at least 12 species native to south-eastern USA. Another group, including *Hymenocallis speciosa*, is found in the West Indies, but the origins of these are uncertain. Very few species are found in South America, though the related genera *Ismene* and *Leptochiton* are endemic to the Andean region of Peru, Ecuador, and Bolivia.

Hymenocallis speciosa is widely cultivated in the tropics, the plants all deriving from a few early introductions before the species was destroyed in its habitat. Sadly, much of the cultivated material carries viruses. Nevertheless, this lovely spider lily makes a fine evergreen accent in the shady tropical garden. It is also an excellent houseplant, although it can attract spider mites if the air is too dry.

The seed often begins to germinate before it is even sown. It should be half-covered in loose, friable, moist medium, and placed in warm shade. Cultivated material may not produce seed since this species appears not to be self-fertile.

Given the degree to which once-verdant Caribbean islands have been depleted of their trees, it is no wonder this species is now of uncertain origin.

▲ SPIDER-LIKE PETALS *The fragrant flowers are adapted for pollination by hawkmoths and appear for only a short period in late summer and early autumn.*

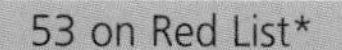

53 on Red List*

Iris

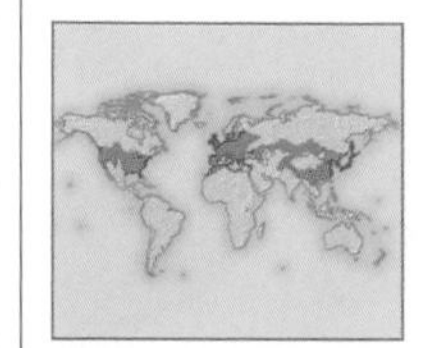

Botanical family
Iridaceae

Distribution
Northern hemisphere, especially Asia and the Mediterranean

Hardiness
Fully hardy – frost-tender

Preferred conditions
Full sun to partial shade, depending on species; well-drained, sandy to gritty soil. Most species prefer dry conditions in summer.

The genus *Iris* is the largest within its botanical family and is found in temperate broadleaf, mixed, and coniferous forest; temperate prairie, savanna, and steppe; flooded grassland and savanna, peat bog, and marshland; Mediterranean forest, woodland, and scrub; and alpine meadow. Sadly, for each geographical area in which the genus occurs there is at least one species that has been Red Listed. As is the case for most Red-Listed rare plants, *Iris* species are threatened most of all by human activity. At least three *Iris* species are extinct: *Iris antilibanotica* and *Iris damascena* from Syria, and *Iris westii* from Lebanon. Many of the threatened species are native plants of the Middle East.

standard

fall

IRIS SUSIANA

BOTANY

THE IRIS FLOWER

Irises grow in a wide range of climatic conditions and habitats, yet their blooms share characteristics that make the iris flower one of the most recognizable in the world. Tepals are modified both into erect or sub-erect "standards" and horizontally displaced or drooping "falls", and the stigma are formed as three stigmatic "crests". Further elaborations include the hairy "beards" found on the falls of certain species. All of these distinguishing characteristics evolved in order to attract specific pollinating insects, and the format has clearly proven successful – *Iris* species have populated every temperate continent in the northern hemisphere.

Iris acutiloba, named as Endangered on the Red List of 1997, is one of the aril or Oncocyclus irises, named for the fleshy covering around the seed. It occurs in only a few populations on the mountain steppes of the Caucasus, and is severely endangered in its native habitats, partly through habitat disturbance for agriculture, and partly through collection for the speciality horticultural trade. The flowers, borne singly on the stem, are white with a striking pattern of blackish-brown veins and bars. As one of the more northern members of its group, *Iris acutiloba* is able to withstand temperatures lower than -12°C (10°F) when protected from wind.

IRIS HISTRIOIDES *Among the earliest of irises to flower, this species is one of the parents of the widely grown Reticulata hybrids. It needs to be kept fairly dry during the summer.*

Native to the steppes of Iran, ***Iris susiana,*** also called the Chalcedonian iris, is considered the easiest Oncocyclus iris to cultivate. It has the rating of Endangered in the Red List of 1997. Its large flowers are a unique lavender-grey in colour, with heavy veins of deep purple-black. This sombre colouring inspired another common name, that of mourning iris.

OTHER IRIS SPECIES

Iris histrioides, cited as Rare in the Red List of 1997, grows in the steppes of central Turkey. It is a member of the Reticulata group of bulbous irises, named for the net-like tunic (covering) that surrounds the bulb. This plant produces a stem 10cm (4in) tall bearing a single, fragrant, large, blue flower with yellow-crested, white-striped falls in early spring. It is propagated easily from the numerous offsets that it produces. *Iris* 'Katharine Hodgkin' is a sky-blue hybrid of this species and *Iris winogradowii*; 'George' is a hybrid of *Iris histrioides* and *Iris reticulata* that has deep purple blooms. The Pacific Coast irises are a distinct group of 10–12 species endemic to western USA. Of these, ***Iris munzii*** is found only in the Sierra Nevada foothills in Tulare County, California, in moist, lightly shaded pastures and along streams. This iris is listed as Endangered on the Red List of 1997. It is thought that the species may have been distributed more widely during the Pleistocene period (one million years ago), when moist, cool conditions were more widespread in California. This species has been used to a large extent in the breeding of the Pacific Coast iris hybrids. *Iris* 'Sierra Sapphire' is a lovely example of the species bred by the late Lee Lenz, an acknowledged authority on the Pacific Coast irises. Munz's iris has little tolerance of heat and prefers cool, moist conditions similar to those of its Californian habitat.

PROPAGATION

All irises with rhizomes (underground stems) can be propagated by division of the rhizomes, although each division must have at least one visible growing point. Although some *Iris* species are self-pollinating, others must be artificially pollinated if they are to set seed. Iris seed is liable to germinate irregularly. Seed of hardy species can be sown in an outdoor bed when ripe and allowed to overwinter, or it may also be stored at 2–4°C (35–40°F) until spring. Depending on species, flowering will occur anything from one to three or more years after germination.

▲ IRIS MUNZII *The endangered Munz's iris grows slowly but attains the greatest height and forms the largest flowers of any iris native to the American Pacific Coast.*

▲ IRIS ACUTILOBA *This iris, as with all members of its group of species with rhizomes (underground or surface stems), has strongly marked bars of deep coloration on its upstanding and descending outer petals. It is endangered in the wild.*

5 on Red List*

Ismene

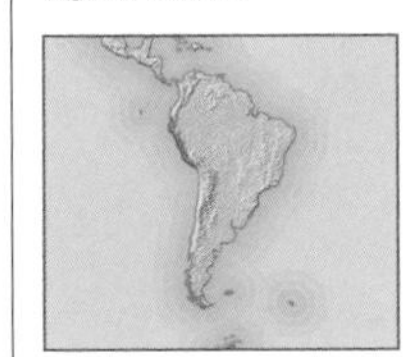

Botanical family
Amaryllidaceae

Distribution
Bolivia, Ecuador, and Peru: Andes regions

Hardiness
Frost-tender

Preferred conditions
Full sun, or partial shade in the hottest areas; fertile, well-drained soil, with plenty of moisture required during the active growth period.

This genus consists of 8–10 species distributed only in desert and scrub of the central Andes Mountains in South America. Three subgroups within the genus suggest that it diversified rapidly as the Andes rose to their present height ten million years ago. *Ismene* species are closely related to the spider lilies (*Hymenocallis* species, *see opposite*). Their flowers differ from one another but all have stems formed by leaves furling tightly around each other at their bases (pseudostems).

Wild-collection of the species has been modest except in the case of *Ismene amancaes*, which was historically plentiful but is no longer. Now, as even marginal lands are developed for agricultural use, the surviving populations of all Ismene species face their greatest threat.

ISMENE SPECIES

Ismene amancaes, called amancay in Peru, is the only yellow *Ismene* species. Cited as Endangered in the Red List of 1997, it is now protected by law from wild-collection for the historic Fiesta de Amancaes (*see below*), which caused a disastrous decline in its numbers.

Genetic analysis of *Ismene amancaes* plants in cultivation revealed some variation in their chromosome numbers, suggesting that some degree of selection for horticultural use may have taken place. In members of the amaryllis family, increases in chromosome number are often correlated with increased vigour and greater numbers of flowers. However, despite the beauty of *Ismene amancaes*, it is less often seen in cultivation than ***Ismene narcissiflora***, or the hybrid between those two species, *Ismene* 'Sulphur Queen'.

▲ ISMENE NARCISSIFLORA *produces flowerheads of up to five strongly scented flowers in summer. The white flowers sometimes have green-striped tubes.*

Ismene narcissiflora, called basket flower, summer daffodil, or Peruvian daffodil, is rated as Rare in the Red List of 1997. More robust and more commonly cultivated than most ismenes, it is found at middling altitudes in valleys between Andean peaks. The most important threat to its existence is its natural scarcity in the wild.

Ismene species are grown as summer-flowering bulbs in temperate climates, after which they are lifted in autumn and stored over the winter. They are also grown in containers. In climates where the temperatures never drop below freezing, bulbs may be left in the ground during the winter, although they must be kept dry. *Ismene* species do not normally produce their fleshy seeds in cultivation.

HISTORY

THE FIESTA DE AMANCAES

Ismene amancaes (*see right*) used to grow so profusely on the sandhills (lomas) around Lima, in Peru, that its blooms were said to paint the hills gold. In colonial times, the Fiesta de Amancaes celebrated this event every year. The roots of the festival lie in ancient Inca culture, and amancay flowers appear on Inca drinking vessels (*keros*). Unfortunately, over the centuries picking of the flowers for the fiesta and collection of both flower stems and bulbs for local use have devastated the population so that only a few healthy colonies remain. The plant has also suffered loss of habitat as a growing human population has spread from Lima into the surrounding hillsides.

AMANCAY BLOOMS TURNED HILLSIDES GOLD

23 on Red List*

Ixia

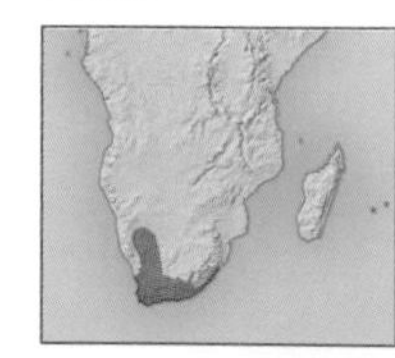

Botanical family
Iridaceae

Distribution
South Africa: Eastern Cape, Western Cape, and Namaqualand

Hardiness
Half-hardy

Preferred conditions
Full sun; well-drained, sandy to loamy soil; best in Mediterranean-type climates in mild winters, but adapts to summer rainfall with good drainage.

There are about 50 species in the genus *Ixia*, all deciduous perennials with corms. Known as corn lilies, most are native to deserts (karoo), the fynbos, or veld in the winter rainfall region of the south-western Cape, South Africa, but a few extend into the summer rainfall region of the eastern Cape. Four of the 23 species on the Red List of 1997 are Endangered. Ixia flower colours and patterning are greatly diversified to attract different pollinating insects. The various species have been hybridized extensively, and some of the hybrids have become cut-flower crops.

Ixia maculata (Vulnerable in the Red List of 1997) has striking, long-lasting, dense spikes of yellow to orange flowers with purple-brown centres in early spring. It is an important constituent of the many *Ixia* hybrids. Once found from the north-west to south-west Cape, it is now unknown in that region.

The otherworldly jade-turquoise flowers of the green ixia, ***Ixia viridiflora*** (Vulnerable in the Red List of 1997), have made it something of a holy grail among bulb fanciers. Unfortunately, it is not an easy plant to cultivate.

Corn lilies are easy to grow in areas that mimic their native climates. Plant them in autumn for spring flowering; water well in winter, then keep dry after the foliage dies back until the autumn. In temperate regions, plant in spring for summer bloom; lift in autumn, and store until the spring, or else treat as annuals.

◄ IXIA VIRIDIFLORA *has large flowers (5cm/3in across) for a corn lily. It is found on rocky slopes in the north-west and south-west Cape.*

Red List: Indeterminate*

Kaempferia rotunda

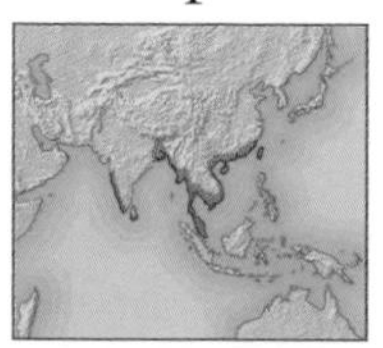

Botanical family
Zingiberaceae

Distribution
Southern China, India, Thailand, Malaysia

Hardiness
Frost-tender

Preferred conditions
Partial shade; moist, fertile soil; rhizomes survive best when dry in winter, with temperatures dropping no lower than -7°C (20°F).

The tropical or Asian crocus, although widespread in tropical rainforests in Asia, is under threat due to habitat destruction and collection for medicinal use. Its exact status is undetermined, and is complicated by the fact that the plant is cultivated widely within its native range, making "wild" populations of the species difficult to distinguish.

The showy "lip" of the flower (*see below*) consists of two modified, sterile stamens. The leaves, mottled with wine-coloured backs and ornate silver patterns on their upper surfaces, remain showy after the flowers have faded. The leaves are longer than those of most other kaempferias, which have prostrate leaves. They grow about 60cm (2ft) tall, forming a pseudostem out of their bases until the autumn cold triggers dieback.

Several cultivars, such as 'Frost', 'Grande', and 'Raven', are available, with differing leaf markings; all look splendid in a shady border.

Seed, if produced, germinates readily on any good seed compost kept moist, shaded, and warm. The rhizomes can also be divided.

PEOPLE

ENGELBERT KAEMPFER

Kaempferia rotunda is named after German polymath and intrepid traveller Engelbert Kaempfer (1651–1716). He studied botany, as well as philosophy, history, medicine, and languages at Cracow, Königsberg, and Uppsala universities. After working for the Persian court in the Middle East and Russia, he joined the Dutch East India Company and went on botanical research expeditions to Ceylon, India, Java, and Siam (now Thailand). In 1690 he reached Japan, where he stayed for four years. He left behind an important body of work, including botanical data and drawings, and ground-breaking cultural histories of Japan. Kaempfer is also thought to have brought the gingko tree from China to Europe.

▲ EARLY ARRIVAL *Fragrant flowers appear in spring, before the leaves, on short stems near the ground. Individual flowers last for only two or three days, but the plants bloom continuously for about a month.*

29 on Red List*

Lachenalia

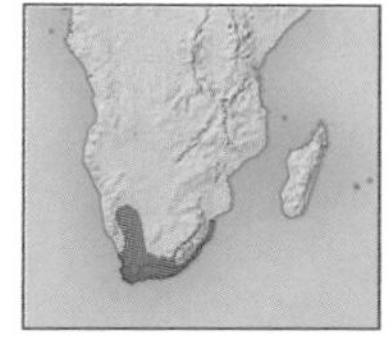

Botanical family
Hyacinthaceae/ Liliaceae

Distribution
Namibia, South Africa: most species in Western Cape Province

Hardiness
Half-hardy

Preferred conditions
Full sun; sandy, well-drained, relatively infertile soil; moderate temperatures; adaptable outdoors only in climates similar to plants' native habitats.

Many of the hundred or so cowslips of South Africa's Cape are restricted to one or a few local populations. Some species grow in huge colonies on thin layers of organic material over rock sheets. Their veld and fynbos habitats are under constant threat from drought and fire, agriculture and other development, and invasive exotic plants, especially from Western Australia. One *Lachenalia* species now exists only in cultivation. *Lachenalia moniliformis* is found only on sandy flats in Western Cape Province and is designated Endangered in the Red List of 1997. The leaves have fleshy bands of reddish-pink and the flowers are marked uniquely blue and pink, with brownish-red spots at the tips and long, yellow stamens. *Lachenalia viridiflora* (also Endangered on the 1997 Red List) is found on granite outcrops in the Western Cape. The colour of the urn-shaped blooms, described as a rich jade-green to almost turquoise, tipped with a narrow band of purple, has made it a collector's item.

Cape cowslips can be propagated by division from aerial bulbils (in some species); leaf cuttings (of broadleaved species); or seed. Seed germinates readily and seedlings need 2–3 years to flower.

LACHENALIA ALOIDES ► *Along with* Lachenalia bulbifera, *this has the largest flowers and is probably the most widely available species. Among this species' many cultivars is 'Quadricolor'.*

RENEWAL

HAPPY ENDING

Until the recent discovery of a solitary population in the Western Cape, *Lachenalia mathewsii* was thought to be extinct in the wild. It has bright yellow flowers with green tips. This species can be seen along with other rare Cape bulbs at the Kay Bergh Bulb House (*see above*) at Kirstenbosch National Botanical Garden. The bulb has been extensively propagated and distributed worldwide.

Red List: Endangered *

Leontochir ovallei

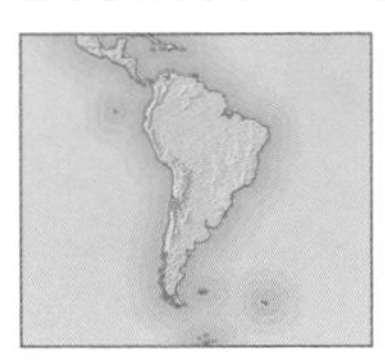

Botanical family
Alstroemeriaceae

Distribution
Chile: Copiapo and Huasco provinces; coastal lowlands

Hardiness
Frost-tender

Preferred conditions
Full sun with good drainage; grows most easily in Mediterranean-type climates with winter rainfall and dry summers. Best situated where the prostrate stems can clamber over rocks.

Garra de león, or lion's claw, is named after its succulent, claw-like leaves, borne on stems that end in large, round, deep scarlet flowerheads. It is one of Chile's most beautiful wild flowers. The thick, prostrate stems typically snake across rocky places in desert and scrub. The stones that cover the soil increase moisture retention, helping the young plants to establish and send out networks of underground shoots and tuberous roots.

Only a few small populations of fairly restricted distribution are known. Although coastal development has probably affected the species to some extent in the past, the chief threat is collection of the showy and long-lasting flowers. Chilean academic horticulturists are currently developing protocols for cut-flower production.

Despite much horticultural interest in the plant, the lack of seed sources has inhibited wider cultivation. If seeds can be obtained, their fleshy jackets, or arils, should be removed before sowing. Soaking the seed for a few days in water before sowing, and changing the water each day, promotes speedy and more uniform germination.

Red List: Vulnerable *

Leucocoryne purpurea

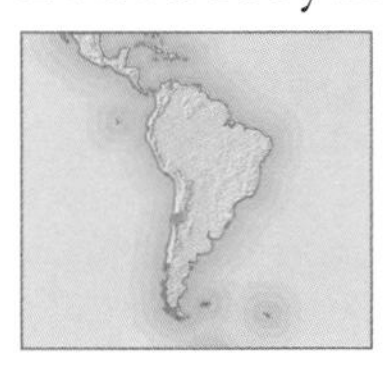

Botanical family
Alliaceae/Amaryllidaceae

Distribution
Chile: Coquimbo Province; coastal lowlands

Hardiness
Half-hardy

Preferred conditions
Full sun; well-drained, sandy soil. The species' native habitat has dry summers and winter rainfall; temperatures are cool and remain above freezing.

This is one of the most beautiful of the dozen or so cormous species of this genus. It is endemic to coastal Chile, inhabiting desert scrub in a Mediterranean-type climate. In spring, after a favourable winter, colonies flower along the Pan American Highway, displaying much variation of colour and pattern. Some populations have been reduced by the development of large resorts along the coast. There are also persistent, but unproven, rumours that high numbers of corms are being exported for the horticultural trade in Europe and Asia. To protect this and other native bulbous plant species, Chilean researchers are developing them as horticultural crops, and breeding interspecific hybrids.

Leucocoryne purpurea is propagated from offset corms or its small, wedge-shaped black seeds, which, if fresh, should germinate readily at 18–24°C (65–75°F). Plants grown from seed usually flower after three years. The corms can rot unless they remain thoroughly dry after the grass-like leaves die back in late spring or early summer.

▲ FLOWERS WITHOUT LEAVES *The wiry flower stems are leafless and each bears between three and eight long-lasting blooms. The leaves resemble a grass and wither before the flowers open. This species can be grown in a cool greenhouse in colder climates for a crop of cut flowers.*

6 on Red List*

Leucojum

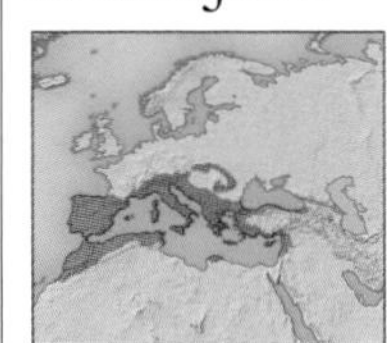

Botanical family
Amaryllidaceae

Distribution
Central Europe and the Mediterranean

Hardiness
Fully hardy – half-hardy

Preferred conditions
Full sun to light shade, in moist but well-drained soil with good humus content; temperatures need to remain above -9°C (15°F) except when the plant is sited in a protective microclimate.

Called snowflakes or snowbells, plants of this genus include two very widely distributed species, ***Leucojum aestivum*** and ***Leucojum vernum***, and about seven more restricted species that occupy narrower ranges in the Mediterranean region. Their habitats vary from forest, woodland, and scrub to marshland and montane meadow. The genus is closely related to the snowdrop genus, *Galanthus*. The less widely distributed (and more rarely cultivated) *Leucojum* species are considered to be the most endangered, more by habitat destruction than by collection for the horticultural trade.

Snowflake species

Leucojum aestivum, or summer snowflake, is the most widely distributed species, found from south-east England, Ireland, and Central Europe east to the Caucasus. The 1997 Red Listing of subspecies *pulchellum* varies from Indeterminate to Endangered in various European countries. Drainage of meadows and loss of woodlands for agriculture, as well as collection for the horticultural trade, has reduced or eliminated many populations of this species, but protection of some populations has actually led to an increase in their size. In 1996, the Council of Europe placed *Leucojum aestivum* on its list of plants subject to exploitation and commerce recommended for protection.

Leucojum vernum, the spring snowflake, ranges across most of southern Europe, where it is found in wet meadows. Subsp. *carparticum* is designated as Vulnerable in the Red List of 1997.

LEUCOJUM VERNUM ► *This threatened species is found in wet meadows across southern Europe. It naturalizes well in lawns when the conditions are to its liking.*

▲ LEUCOJUM AESTIVUM *To reduce the unsustainable demand for wild-collected bulbs of this beautiful species, gardeners can instead grow the robust, nursery-produced cultivar 'Gravetye Giant', which flowers abundantly.*

Drainage of wetlands for agricultural use, as well as collection for the horticultural trade, has rendered the species vulnerable in its natural habitats. It appears on a number of national Red Lists and, like *Leucojum aestivum*, was placed by the Council of Europe on its list of plants subject to exploitation and commerce recommended for protection. *Leucojum vernum* is smaller in stature than summer snowflake, and its elegant, slightly fragrant, bell-shaped, white flowers may be spotted either green or yellow.

Leucojum seed is best sown ripe in well-drained compost in a cold frame. Older seed requires 2–3 months' cold stratification after which it should germinate in 2–4 weeks at 10°C (50°F). If sown thinly, seedlings can be left to grow undisturbed in the pots for the first year and then given an occasional weak liquid fertilization to prevent nutritional deficiencies. Pot bulbs when dormant, and plant out when 3–4 years old.

19 on Red List*

Lilium

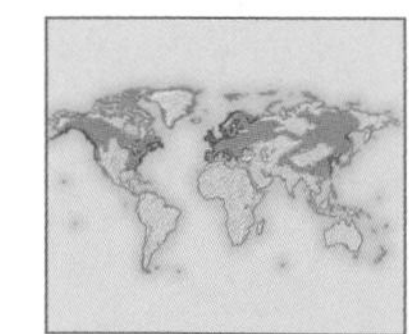

Botanical family
Liliaceae

Distribution
Northern hemisphere: North America, Europe, and Asia to Philippines

Hardiness
Half-hardy

Preferred conditions
Light shade; rich, moist, fertile soil; species vary greatly in the acidity of soil that they prefer – *Lilium rubellum* favours particularly acid soil and tolerates extremely wet conditions.

Although four species are classified as Endangered by human activities on the 1997 Red List, most threatened lily species are naturally rare. Lilies occur in temperate broadleaved, mixed, and coniferous forest, northern coniferous forest, and montane meadow. The North American Lily Society has created a Species Lily Preservation Group dedicated to collecting information about *Lilium* species, as well as conserving the germplasm of lilies in cultivation and in the wild.

Although not currently on the Red List, the Canada lily, ***Lilium canadense***, is a protected plant in at least three states of the USA (Indiana, New York, and Tennessee) and in several eastern provinces of Canada. It is considered "potentially threatened" in several other states. This lily of rich "intervale" meadows at the edges of forest or natural, low-lying, moist meadows, has seen its habitat shrink over the past century as open lands have given way to agricultural, residential, and industrial development. The lily is broadly distributed and many regional variants have been noted in flower colour and sometimes also in ecology (populations at the western fringes of its range tend to be more drought-tolerant). It is a tall-growing species capable of producing as many as 20 nodding, bell-shaped flowers, which are typically pale yellow with dark reddish-black spots.

Lilium iridollae, the pot-of-gold lily, is very restricted in its USA distribution, being found only in south-western Alabama and the adjacent Florida Panhandle, where it grows in bogs and swamps. It is related to the Canada lily, and likewise reproduces by creeping underground stems (stolons). In early to midsummer, up to 8 "turk's cap" flowers are borne on stems 0.9–1.5m (3–5ft) tall. The blooms are a rich yellow with brown spots. A rare component of a few southern wetlands, it should make an excellent species for gardeners in the deep south of the USA and in similar climates elsewhere, where many other lilies will not grow. The species faces threats from unrestricted grazing of livestock as well as draining of wetlands for development. For the best results, sandy soils must be modified with abundant organic matter.

Although without official Red List status, *Lilium kesselringianum* is a rare species that is considered threatened in many nations of the Caucasus through which it ranges. It is related to the better known *Lilium monadelphum* and looks quite similar. It produces 8–10 yellow and brown-spotted, pleasantly fragrant flowers on stems 0.6–1.2m (2–4ft) tall. The bulbs are among the largest in the genus, weighing up to 1.8kg (4lb).

Lilium occidentale, the Eureka lily, inhabits the drier fringes of sphagnum-moss bogs in northern California and adjacent Oregon. Classed as Endangered in the Red List of 1997, it is under threat from logging and other types of habitat

BOTANY

Seed germination

Lily seeds germinate in one of two ways – below ground (hypogeal) or above ground (epigeal). The illustration (*see right*) of dicot seed germination (although lilies are actually monocots) highlights the basic differences. Within the categories of hypogeal and epigeal, germination is further subdivided into immediate versus delayed emergence. Hypogeal seeds tend to be chiefly delayed, while most epigeal types are immediate. It is important to know the category of any given lily species, since it will determine when best to plant the seed. Hypogeal seeds should be planted in late summer or autumn, epigeal in early spring. Of the lily species discussed on these pages, all but the tropical *Lilium philippinense* germinate in the hypogeal manner.

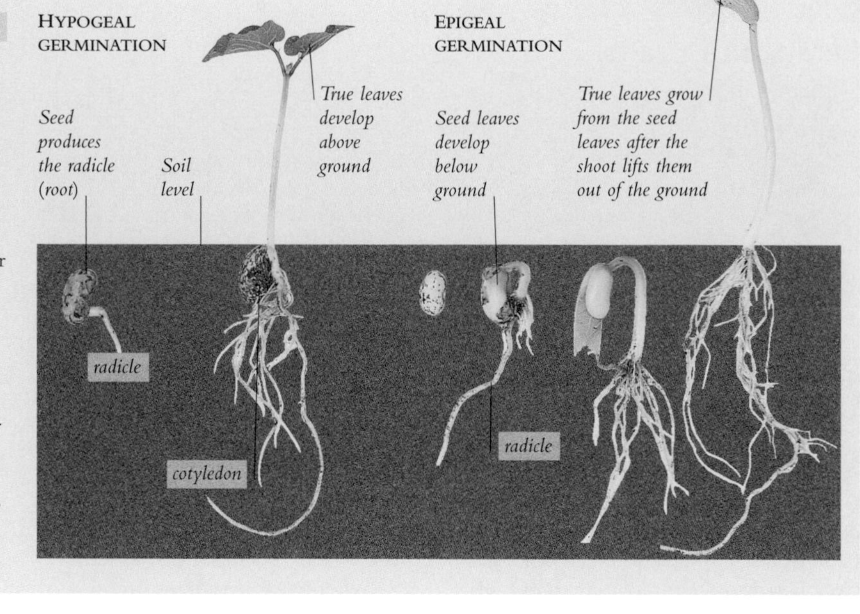

HISTORY

Revered yet rare

Apart from the tulip, there is probably no other genus of bulbous plant that has figured in human history as significantly as *Lilium*. Recognized for their classical beauty, lilies have long been horticultural treasures, but they have also had symbolic importance. For the Romans the lily was a symbol of hope, while Christianity adopted the flower to represent innocence and chastity. The lily is also associated with the Virgin Mary and with resurrection (links that led to lilies being carved on the gravestones of women in the Victorian era). The lily was adopted in Spain by the royal House of Bourbon as its heraldic emblem, while the species *Lilium chalcedonicum* was known as the Red Martagon of Constantinople (Istanbul). Despite the cultural importance of the lily, nearly one-fifth of the Earth's hundred or so *Lilium* species are threatened by human activity.

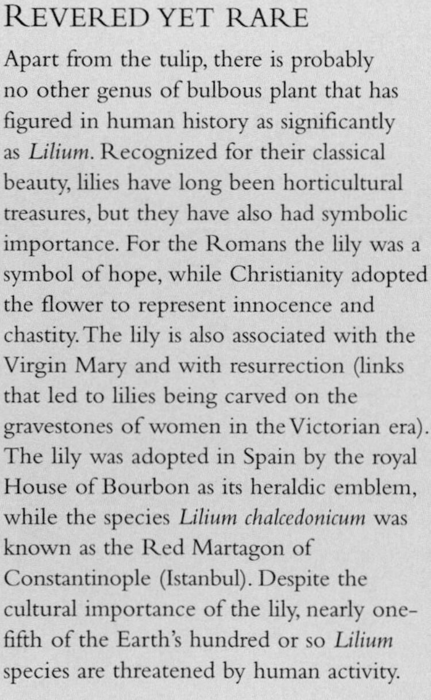

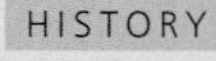

disturbance. The number of known populations of this species has halved. In the 1960s, an entire population in a state park was completely destroyed by the construction of a public toilet facility. The lily is now both a federally and state protected plant. An *ex-situ* preservation and wild reintroduction programme for this species has been successfully initiated at Berry Botanic Garden in Portland, Oregon, under the US Center for Plant Conservation. Each 0.6–2m (2–6ft) tall stem produces up to 15 nodding orange flowers with greenish-yellow throats. Eureka lily is best grown in climates similar to the moderate Pacific coastal climate where it occurs.

Although unlisted by IUCN, *Lilium philippinense*, the world's most southerly growing lily species, is clearly considered threatened by biologists in its country of origin, the Philippines. Its exact status – whether vulnerable or endangered – is not clear, but the reasons for concern are overcollection of the bulbs for ornamental use, as well as loss of habitat for agriculture and other human endeavours. Found only in Luzon province, the species grows on grassy mountain slopes at elevations of 1,500–2,000m (5,000–6,500ft), where it rarely, if ever, encounters freezing temperatures. Closely related to the Formosa lily, *Lilium formosanum*, it grows remarkably quickly from seed and is capable of achieving flowering size in less than a year. In summer, up to several dozen long, white trumpets are produced on stems 90cm (3ft) tall. The tepals are flushed green or brown on their outside surfaces. This species will do well outdoors as long as the bulbs are not exposed to freezing temperatures.

The maiden lily, ***Lilium rubellum***, is found only in Honshu province, Japan, at elevations of 900–1,800m (3,000–6,000ft) on mountain slopes among grass and brush. This elegant lily, classed as Rare in the Red List of 1997, is fairly dwarf, growing to a height of only 30–45cm (1–1½ft). Although rare in nature, the species has been much used in hybridization. It has played an important role in the development of Leslie Woodriff's 'Little Fairies' strain and Norma Pfeiffer's 'Magic Pink' strain. Another hybrid, 'Rosario', is considered to be one of the best to come from *Lilium rubellum*.

Lilium rubellum can be grown in water gardens, as this lily has been observed growing in Japan with its roots submerged. It is one of the earlier lily species in the garden, blooming in late spring, but may be damaged in climates colder than that of its habitat, and where late spring frosts are not uncommon.

▼ LILIUM OCCIDENTALE *Populations of this exceedingly rare example of a turk's cap type of lily are under severe pressure from wild-collecting for cut flowers, as well as bulb collection for garden cultivation.*

◄ LILIUM CANADENSE *Native to eastern North America, this lily is very adaptable to gardens within its native range, but can be short-lived in other climates. Reproducing by creeping underground stems, it usually produces a sizable clump when established.*

CONSERVATION

LILIUM PARDALINUM

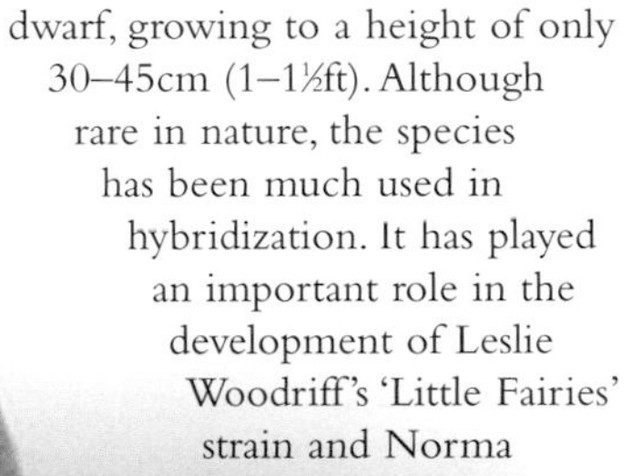

The leopard lily, with as many as ten red and orange flowers dotted purple, borne in early summer, is found in wet meadows and along streams throughout California and into southern Oregon. Subspecies *pitkinense,* the Pitkin marsh lily, is cited as Endangered in the Red List of 1997. It is known from only three populations in California, all of them on privately owned land without federal protection. The owner of one property has not allowed biologists to examine the population on his land for over 25 years. The second site was nearly destroyed by residential development 35 years ago. The third known population, once home to many Pitkin marsh lilies, now consists of only two plants. Fortunately for the future of the species, owners of the latter two sites accepted voluntary protection agreements with The Nature Conservancy.

PROPAGATION

Lily seed should never be germinated in a saturated medium. Lilies generally germinate at cool to moderate temperatures of 13–21°C (55–70°F), and must be monitored for fungal infection and aphid infestation. Vegetatively, lilies may be increased by division of plants and by scale propagation.

▲ LILIUM RUBELLUM *Usually three sweetly scented, trumpet flowers are produced on a stem, although there have been reports of as many as ten per stem in cultivation. The trumpets are shorter than those of many lilies.*

Unlisted

Lycoris aurea

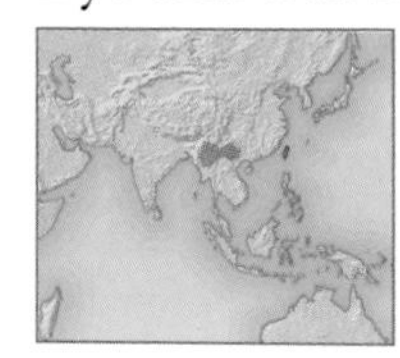

Botanical family
Amaryllidaceae

Distribution
Japan, southern China, Vietnam, Taiwan, and Myanmar

Hardiness
Frost-tender

Preferred conditions
Partial shade; *Lycoris* species thrive in moist, fertile, well-drained soil.

The genus *Lycoris* represents the only true incursion of its botanical family into eastern Asia during its known evolutionary history. The genus is known in China by the Mandarin name *hu di xiao*, which translates wonderfully as "suddenly the earth smiles". *Lycoris* species grow in temperate broadleaf and mixed forest. The conservation status of most of them in the wild is a matter of speculation, since the rarest species grow in areas either poorly explored botanically, including Vietnam, or largely closed to outsiders, such as Myanmar.

There is also confusion in the classification of the yellow-flowered species, in particular, and it is difficult to determine whether they represent one variable species or a number of distinct species. Furthermore, a significant number of additional species have been described in the last two decades. It may seem arbitrary to present ***Lycoris aurea*** as an example of a threatened *Lycoris* taxa, but the degree to which China's forests, both southern warm temperate and subtropical, have been disturbed makes the predicament of this species as good an example as any.

LYCORIS RADIATA ► *Also called the red spider lily, this species from Japan has wavy-margined tepals curved back at the tips and long, protruding stamens. Its hardiness is similar to that of* Lycoris aurea.

Lycoris aurea and its relatives are much appreciated in the lower south-eastern parts of the USA for the fact that their flowers arrive in the garden as a surprise in late summer and early autumn. Large drifts of the bulbous plants are effective below high-branched trees or at the edge of a woodland garden.

Where well adapted, *Lycoris aurea* may naturalize locally. Despite a period of apparent inactivity, the bulbs should never be allowed to become completely dry. The species is not hardy without protection where winter temperatures regularly fall below -4°C (25°F). The round, black-coated seeds may be planted in any good germination mix and transplanted into small pots after the first leaf develops, then set in the garden the following summer. About three years must pass before flowering. Large clumps may be divided in early summer after the leaves die back, taking care to avoid damaging the fleshy roots. Popular alternative species include *Lycoris albiflora* (white-flowering), *Lycoris sanguinea* (red), and *Lycoris squamigera* (pale rose-red).

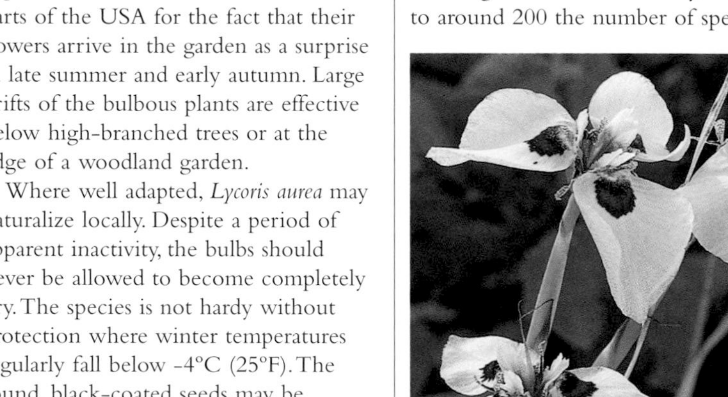

◄ LYCORIS AUREA *This species' common name, golden spider lily, refers to the long flower stamens. The leaves persist through winter and then disappear, after which the flowers unfurl in late spring or early summer.*

52 on Red List*

Moraea

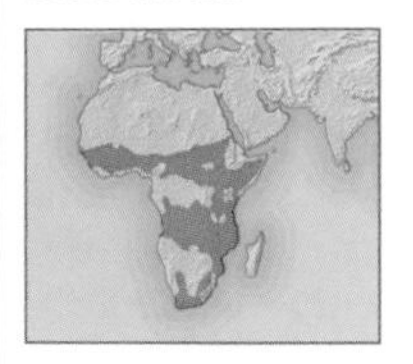

Botanical family
Iridaceae

Distribution
Africa: sub-Sahara and South Africa

Hardiness
Half-hardy – frost-tender

Preferred conditions
Full sun; well-drained, moderately fertile, loamy, or clay soil; a warm site, sheltered from midday sun and winter rainfall.

The taxonomy of the genus *Moraea* was revolutionized recently by the inclusion within it of about half a dozen smaller genera of the iris family. This raised to around 200 the number of species in

▲ MORAEA ARISTATA *The blue spots at the base of each tepal identify this as a true peacock iris. The plant is close to extinction in its native habitat in South Africa, where only one population survives.*

Red List: Vulnerable*

Monocostus uniflorus

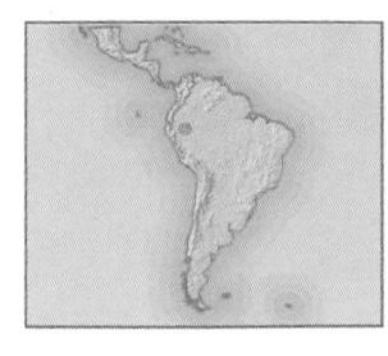

Botanical family
Costaceae/Zingiberaceae

Distribution
Peru: in the lower Huallaga River valley

Hardiness
Frost-tender

Preferred conditions
Bright shade; moist, fertile soil with high acidity and humus content.

The genus *Monocostus* contains only a single species, *Monocostus uniflorus*, or lemon ginger, which grows in tropical rainforest and is classed as Vulnerable in the Red List of 1997. The genus is one of four in the spiral ginger family (Costaceae), which is distinguished from the true ginger family (Zingiberaceae) by its spiralled stems. The showiest part of the flower, like that of all the spiral gingers, is a lip-like, modified sterile stamen. *Monocostus uniflorus* is the smallest in stature of the entire Costaceae family, reaching to a maximum of only 60cm (2ft) in height while many spiral gingers grow to more than 2m (6ft) tall. Restricted to a relatively small area of eastern Peru, lemon ginger is threatened primarily by habitat destruction.

Alongside lemon ginger on the Red List of 1997 are 16 species of the related genus *Costus*, all natives of Central America and northern South America. In the garden, however, *Monocostus uniflorus* provides a delicate contrast to the often coarse appearance of the many *Costus* species. Its lovely lemon-yellow flowers, 5cm (2in) across, are produced singly at the internodes between the slender stems and the spiralled, evergreen leaves. The flowers have a delicate and pleasant fragrance and appear from spring to autumn, and even year-round in tropical climates. Lemon ginger makes a good container plant and has also been used successfully in hanging baskets. It will flower very successfully on a windowsill as long as humidity levels around the plant are kept high. The plants should receive regular feeding for best growth. Lemon ginger may be propagated by division of the rhizomes or by stem cuttings.

Red List: Rare*

Muscari latifolium

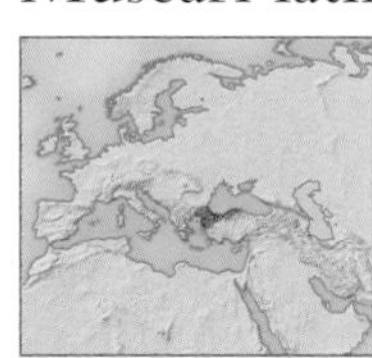

Botanical family
Hyacinthaceae/Liliaceae

Distribution
North-west Turkey

Hardiness
Fully hardy

Preferred conditions
Full sun to light shade; well-drained soil. Once the leaves have died back, dryness is required.

The 30 species of the genus *Muscari* range from Europe, through the Mediterranean region, Asia Minor, and the Caucasus. ***Muscari latifolium*** is a grape hyacinth that grows in the temperate coniferous forest of north-west Turkey. Since being named by Dutch bulb growers as "flower bulb of the year" in 1996, this species has become very popular.

Muscari latifolium is not common in the wild and gardeners should buy cultivated material only, since pressure

the genus. Support for this treatment came from genetic analysis of DNA sequences from the various plant groupings, which indicated that these formerly separated genera were no more than members of *Moraea* that had adapted to attract specific pollinators. Unusually, the flower shapes of these groupings tended to be less elaborate than those of the ancestral form, which bears a resemblance to those of irises. The inclusion of the genera *Galaxia*, *Gynandiris*, *Hexaglottis*, and *Homeria* within *Moraea* raises the total of *Moraea* species on the Red List of 1997 from 31 to 52 species.

Moraea species are strictly a component of open vegetation and are never found in woods or forests. Their habitats include veld, fynbos, tropical and subtropical grassland, savanna, and shrubland. Many of the species are endangered because in nature they grow in very few populations, and inhabit sites that are very likely to become subject to agricultural or coastal development.

◄ MORAEA TULBAGHENSIS *The rich clay soils of the renosterveld in South Africa's western Cape are so valuable for agriculture that species such as* Moraea tulbaghensis *are difficult to find except on small patches of original vegetation that have been preserved.*

A plant that has been offered for years by bulb dealers, ***Moraea aristata*** is one of the true peacock irises (*see right*). It is close to extinction in its habitat and is cited as Endangered in the Red List of 1997. This *Moraea* is known only from one population on the Cape Peninsula in South Africa. It is a beautiful species with white to very light blue flowers. which are marked by an "eye" of turquoise and blackish-blue at the base of the broad outer tepals.

Moraea tulbaghensis, also known under its synonym, *Moraea neopavonia*, is restricted to clay renosterveld sites in the north-west and south-west of South Africa's Cape. It is classed as Vulnerable in the Red List of 1997. The flowers are a brilliant orange in colour and marked with large blue spots at the base of the outer tepals. Two or three flowers are usually open at the same time on the 45cm (18in) stem, making the species a useful cut flower.

Moraea villosa is found on rocky granite and clay hills and flats in the winter rainfall area of South Africa's western Cape. **Subspecies *villosa*** can be seen in the renosterveld clay soils in company with *Moraea tulbaghensis*. The species is variable with flowers ranging from purple, lilac, sometimes pink or orange to cream or white. On the tepals yellow nectar guides are circled by dark, broad bands. *Moraea villosa* subsp. *elandsmontana* is listed as Rare on the Red List (1997).

CULTIVATION

Moraeas from South Africa's western Cape grow most successfully in climates with winter rainfall and dry summers similar to that of their natural habitat. In other climates they need to be grown in containers, protected from frosts during their active growth in winter and spring, and then dried off for their dormancy period in summer. Species from the summer rainfall area of the eastern Cape adapt more easily to subtropical and warm temperate climates that have summer rainfall. As a group, *Moraea* species prefer heavier soils than many other bulbous plants indigenous to the Cape, and they benefit from regular feeding. Most moraeas produce many cormels (small, secondary corms growing from the parent corm), which can be planted to increase the stock of plants. Seed of species from winter-rainfall areas should be sown in late summer, and that of summer-rainfall species in spring.

◄ MORAEA VILLOSA *SUBSP.* VILLOSA *This subspecies shows a remarkable diversity of colour types, many of which have been used by rare bulb enthusiasts to produce a range of garden hybrids. The species is particularly rich in shades of blue.*

PLANT NAME

MORAEA NAMESAKE

The common name of "peacock iris" derives from the species within the genus that have dark blue spots on their large, outer tepals. These spots have an iridescent quality and are reminiscent of the blue "eyes" on peacocks' tail feathers. It is the *Moraea* species possessing the spots on their tepals that have attracted the greatest attention in horticulture.

from the bulb trade for wild-collected bulbs would be a significant threat. Unlike many other *Muscari* species, it produces a very broad leaf, often just one a season. Its flower cluster also has a peculiar feature; the flowers near the apex are sterile and light blue, while those below are fully fertile and a rich, dark, bluish-purple. This two-toned appearance has given rise to the common name "bicolor muscari".

Muscari latifolium flowers from mid- to late spring, depending on climatic conditions. In the garden, it makes an impressive show in large drifts under trees or at the front of the border, and its intense colouring makes it a good foil for deep pink tulips. It can be grown in containers, and is easy to force for winter bloom. Once the leaf dies back in summer, it requires no watering beyond whatever rain falls during the summer. Bulbs should be planted in autumn. Seed, which is produced freely on the plants, germinates readily when fresh. It then takes about three years before plants start to flower.

TWO-TONED FLOWERS ► *The combination of its broad leaf and flower in two tones of blue makes* Muscari latifolium *a choice subject in the early spring bulb garden. It grows to an approximate height of 25cm (10in).*

15 on Red List*

Narcissus

Botanical family
Amaryllidaceae

Distribution
Mediterranean, Atlas Mts, Balkans, northern Europe, China, and Japan

Hardiness
Half-hardy

Preferred conditions
Variable in requirements depending on source; most prefer full sun to light shade and moist, well-drained, loamy soil; southernmost variants need to bake in the summer.

After the genus *Tulipa*, *Narcissus* is the most important genus of garden bulbs in the western world. Relatively speaking, only a handful of its 50–70 species have been developed into mass-market garden bulbs of the temperate zone. *Narcissus* species are found in Mediterranean forest, woodland, and scrub; temperate broadleaf and mixed forest; and montane meadow. Many have narrow distributions, and disturbance of their habitats has taken a considerable toll on the number and size of those populations. In Spain and Portugal, where the greatest diversity of the genus is concentrated, areas continue to be developed for residential and commercial use without regard to their populations of rare *Narcissus*; in some cases, the only known population has been threatened. While some degree of collection for the horticultural trade has undoubtedly occurred, its impact on the conservation status of the genus is not known. Larger-flowered species are probably robbed of their reproductive ability in some areas by the collection of stems for cut flowers.

▲ NARCISSUS BULBOCODIUM *This species is recognizable by its funnel-shaped trumpet and curved stamens. Its tepals are small relative to the size of the trumpet.*

▲ NARCISSUS TRIANDRUS *is a popular miniature species, undemanding to grow, which has been exploited by breeders. Flower colour varies from deep yellow to white.*

Purely on the basis of its economic value, germplasm (genetic material) of *Narcissus* species deserves to be collected officially, but it appears that no conservation collection has yet been formally established. Narcissi readily hybridize where they come into contact, and some, formerly classed as species, are now recognized as natural hybrids.

Narcissus bulbocodium, the hoop-petticoat daffodil, is a delightful miniature species. In the wild it has many variants in colour, size, and shape of trumpet, sometimes within a single population. It ranges from Portugal through Spain and France, occurring in scrub and rocky screes, rarely in meadows on the northernmost edges of its range. In 1996 the Council of Europe placed this species on its list of plants subject to exploitation and commerce recommended for protection. It is not yet listed by the IUCN.

OTHER NARCISSUS SPECIES

Narcissus cyclamineus, the cyclamen narcissus, is another species without Red List status at present, but is recognized as threatened by the European Council. It grows wild in Portugal and Spain, in moist mountain meadows and along streams. It is threatened not only by wild collection but by development of its often picturesque habitats for holiday homes and resorts. It is a desirable miniature in the garden, naturalizing in grass or in light shade under trees. It has been bred with larger daffodils to create the so-called "cyclamineus hybrids".

Narcissus pseudonarcissus is the most common yellow trumpet daffodil in cultivation. Native to western Europe, from Britain in the north, through France and south to Spain and Portugal, it is rich in regional variation. Most of the restricted subspecies, the majority of them native to Spain and Portugal, face the twin threat of habitat destruction and local overcollection for garden use. The subspecies *nevadensis* and *primigenius* are classed in the 1997 Red List as Endangered and Rare respectively.

Commonly called angel's tears, ***Narcissus triandrus*** is native to Spain, Portugal, and north-west France. The French subspecies *copax* is classed in the 1997 Red List as Vulnerable. Biologically, it is highly interesting in that it is the only known species in the amaryllis family to exhibit the syndrome known as tristyly – botanists subdivide the species into three different types based on the length of the style (elongated female flower part between the ovary and the stigma). The threatened subspecies *copax* has the largest trumpet found in the species.

▲ NARCISSUS CYCLAMINEUS *The tepals, turning sharply back from the frilly-edged, tubular trumpet, recall the habit of cyclamen flowers, hence the species' name.*

Narcissi are readily grown from seed, and most species require three years of growth before flowering. Seed is best sown fresh, and takes 1–2 months to germinate. The seedlings, if sown thinly, can stay put until they are two years old, at which time they may be planted in the garden or potted individually.

NARCISSUS PSEUDONARCISSUS *William Wordsworth's poem,* The Daffodils, *was inspired by the sight of this species, here at Ullswater in England. Two Spanish subspecies are Red Listed.*

Unlisted

Neomarica caerulea

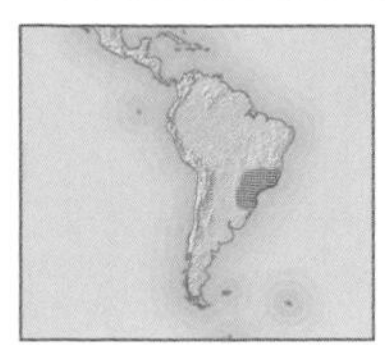

Botanical family
Iridaceae

Distribution
South-eastern Brazil

Hardiness
Frost-tender

Preferred conditions
Partial shade; moist but well-drained, acid soil with a high humus content.

Found only in Brazil, in the understorey of the Atlantic tropical rainforest, the range of this bulbous plant, commonly called "twelve apostles", has been greatly reduced by the fragmentation of its once-extensive coastal forest habitat. ***Neomarica caerulea*** may have once ranged from the state of Rio de Janeiro southwards at least to Santa Catarina State, but is now rare in the wild. However, the species is widely cultivated throughout the world, and thus, ironically, is much better represented in gardens than in its habitat. It is one of perhaps a dozen species in the genus *Neomarica*, all of which are at least vulnerable to extinction in Brazil, although none is Red Listed by IUCN.

The unusual common name arises from the claim that the plant will not flower until 12 leaves have formed. The more ubiquitous common name "walking iris" refers to the species' vegetative means of reproducing itself. New plants form at the nodes of the older flowerheads and gradually their weight causes the stems to bend down towards the soil, whereupon the new growths root and become established. *Neomarica caerulea* forms fans of upright, bluish-green, sword-shaped leaves that grow to a height of 90cm (3ft). From between the leaves arise branched flower stems in early summer, which periodically produce large, blue-to-purple, iris-like flowers that each last a single day. The stigmatic crests have white and yellow markings. In the garden, this makes an eye-catching plant when in flower, while the leaves provide an attractive vertical accent year round.

In regions subject to freezing, this plant can be grown in containers and moved indoors during winter. It is propagated by seed in spring, by division of the rhizomes, or by potting the plants that develop on the old flower stems.

▲ NEOMARICA CAERULEA *has eye-catching flowers that last a single day. Rare in the wild though widely cultivated, this species suits a partially shaded tropical garden.*

5 on Red List*

Nerine

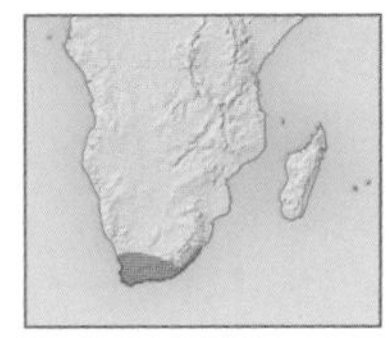

Botanical family
Amaryllidaceae

Distribution
South Africa: Eastern and Western Cape provinces

Hardiness
Fully hardy – half-hardy

Preferred conditions
Full sun to part shade, depending on species; moist, well-drained, sometimes sandy soil.

The nearly two dozen species of the genus *Nerine* are concentrated in the moist grasslands of South Africa's eastern Cape, and in the mountains to the south-west. Their habitats run from near desert conditions, veld, and fynbos to the marshes near the coast. Nerines have been cultivated and hybridized for the garden since the 17th century, with the predominant bloodline derived from *Nerine sarniensis*. A number of species are naturally rare in the wild, and these are the most vulnerable to loss, as native South African vegetation comes increasingly under pressure from population growth and development.

Nerine gracilis has the distinction of being among the rarest in the genus. It is classed as Rare on the Red List of 1997. It was once fairly common in the summer-rainfall, moist grassland lying around exposed limestone slopes in the South African provinces of Mpumalanga and Gauteng, but over-grazing has heavily degraded its habitat. The plant has narrow, deciduous leaves that wither before the flowerheads of star-shaped, pink blooms appear in summer. This species is easy to propagate and readily offsets bulblets. It can withstand a few degrees of frost during the winter, but is best protected wherever temperatures often drop below -4°C (25°F). Plants should be maintained in dry and cool conditions during the winter.

Nerine masonorum is another species cited as Rare on the Red List of 1997. In cultivation it should be watered sparingly, if at all, in the winter. It is an excellent small bulb for the subtropical rock garden, or in containers anywhere.

Nerines are easily grown from their small, fleshy, green seeds. These should be sown as soon as ripe, covered shallowly, and kept moist, moderately warm, and slightly shaded.

NERINE MASONORUM ► *Nearly evergreen, this plant has small, thread-like leaves and grows mainly during spring and summer. The numerous clusters of small flowers appear in summer.*

Unlisted

Pamianthe peruviana

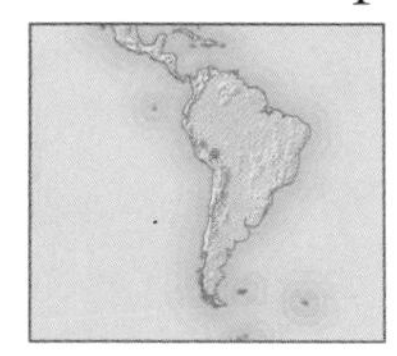

Botanical family
Amaryllidaceae

Distribution
Peru and Bolivia

Hardiness
Frost-tender

Preferred conditions
Full sun to part shade; moderately fertile, moist, but well-drained soil.

A native plant of the mountain tropical rainforest of the western slopes of the Andes, ***Pamianthe peruviana*** grows as an epiphyte (a plant that grows non-parasitically on another plant, without rooting in the soil). That the species actually grew on the branches of trees was not known until *Pamianthe peruviana* was recognized as synonymous with *Pamianthe cardenasii* from Bolivia, a more southerly outpost for the species that has since been completely deforested. A close relative, the more recently described *Pamianthe parviflora*, classed as Vulnerable on the 2000 Red List and known only from a single herbarium specimen, is likely to be extinct in Ecuador. Genetic analysis confirms *Pamianthe* as an unusual genus, being an early branch of the lineage that also gave rise to the genera *Clinanthus*, *Hymenocallis*, and *Ismene*.

Pamianthe peruviana is evergreen, and the elongated bulb has a long aerial neck formed from the overlapping leaf bases.

◄ PAMIANTHE PERUVIANA *produces its large, fragrant flowers, which resemble those of* Pancratium *species, in spring. Pollen from another plant is required to set seed.*

The plant is best grown as a container specimen, preferably in a clay pot, in a mix suitable for epiphytic plants. This can be either a high-quality, peat-based medium mixed with an equal part of medium orchid mix, or else seedling-grade orchid mix by itself. The bulb should be planted with just its basal plate below the surface, and the plant should be staked until it produces new roots and becomes stabilized in the container.

The plant's fruit is a large, slightly woody capsule filled with numerous brown, flattened, winged seeds that will germinate in 6–8 weeks. Seedlings can be moved to small clay pots when several leaves have formed. This plant resents disturbance and established specimens should be repotted only if absolutely necessary – that is, when the mix has broken down to the point where the succulent roots are threatened by rot.

HABITAT

UNDER SIEGE

The cloud forest growing at middle elevations on the western slopes of the Andes has been subjected to so much pressure from development in Peru and Bolivia that very little of it now remains intact. It has been many years since *Pamianthe peruviana* was last encountered in the wild, although it may yet be found in some inaccessible fragment of forest.

Red List: Rare*

Pancratium canariense

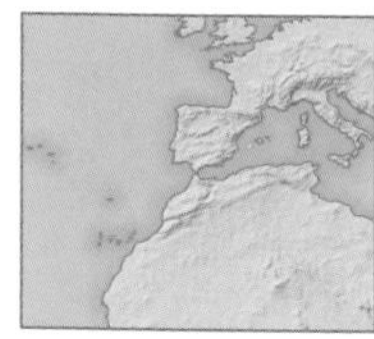

Botanical family
Amaryllidaceae

Distribution
Canary Islands

Hardiness
Frost-tender

Preferred conditions
Full sun or part shade; well-drained, sandy soil; best suited to a Mediterranean-type climate.

A wide-ranging genus that is still poorly known taxonomically, *Pancratium* includes species in Africa, Asia, and Europe. ***Pancratium canariense***, known to Spaniards by the rather poetic name of *lágrimas de la virgen* ("virgin's tears"), has the most western distribution and is found growing in Mediterranean scrub on just some of the Canary Islands. Its rarity makes it vulnerable to disturbance by agricultural or industrial development, but now the species has government protection. Its flowers are very similar to those of the American spider lilies (*Hymenocallis*), to which the genus would seem to be only distantly related. *Pancratium maritimum*, a southern European mainland species, is found on the dunes of the Mediterranean Sea. A single species, *Pancratium tenuifolium*, grows in Africa, and *Pancratium zeylanicum* is native to India and Sri Lanka.

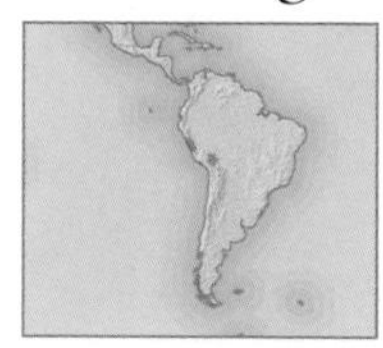

▲ ELEPHANT HAWKMOTH *This specific pollinator plays a key role in the reproduction of* Pancratium canariense, *which has flowers adapted to attract and support it.*

Pancratium canariense is a deciduous bulbous plant best suited to a Mediterranean-type climate, although it can be grown successfully elsewhere in containers if kept dry during the summer months. The plant flowers in autumn, at the same time as the bluish-green leaves emerge from the bulbs. The leaves remain in place during the winter, then die back in spring. The black, D-shaped seeds germinate readily when sown fresh, and seedlings usually flower in their third year.

◄ POWERFUL FRAGRANCE *The 10–12 highly fragrant flowers, borne on a flattened flower stalk, are only part of* Pancratium canariense*'s attraction, since the bluish-green leaves persist through winter.*

Red List: Endangered*

Petronymphe decora

Botanical family
Themidaceae

Distribution
Mexico: Guerrero State

Hardiness
Frost-tender

Preferred conditions
Full sun to partial shade; well-drained, sandy soil; corms require dryness in winter.

The botanical family Themidaceae was once treated as a part of the onion or amaryllis family, but is now known to be more closely related to the hyacinths. Of this family of genera, found variously in Mexico or south-west USA, *Petronymphe* is the rarest genus and consists of only one species. *Petronymphe decora*, or the rock nymph, was restricted to remote rock outcrops in Mexico, where it often grew in cliffs, but the few populations were descended upon by bulb collectors and the plant, once locally abundant, is now seldom seen. Rock nymphs share the lax, wiry stems of the other members of their group. The greenish-yellow flowers are unusual but hardly showy; thus it is something of a mystery why they proved so attractive to collectors. The plant's present scarcity in Mexico is an unfortunate example of how the desire for novelty in horticulture can have disastrous results.

Rock nymphs grow from small corms that must be kept dry in winter. They are best grown in pots or even baskets, where the long, three-sided leaves and wire-like flower stalk can hang down unimpeded. The greenish-yellow, tubular flowers are more of a curiosity than a dazzling display. The small, black, wedge-shaped seeds germinate readily when fresh. Seedlings need 2–3 years before they will flower. Except for the few areas where the plants can remain outside in winter, the small corms are best left in their containers, placed in a cool, dark spot safe from freezing temperatures.

Red List: Rare*

Paramongaia weberbaueri

Botanical family
Amaryllidaceae

Distribution
Peru: Lima, Ancash; Bolivia: La Paz

Hardiness
Frost-tender

Preferred conditions
Full sun, or light shade in very hot climates; well-drained, sandy or loamy soil; demands a very long season of dry dormancy, which in the northern hemisphere would be in the summer.

The largest-flowered member of the amaryllis family, ***Paramongaia weberbaueri***, called cojomaria in South America, superficially resembles a giant yellow daffodil. However, the trumpet of the cojomaria consists of the fused stalks of stamens, rather than tissue from the outer part of the flower. Only one *Paramongaia* species is known, and that from no more than three populations. In Peru, the species has been encountered on the coast near Lima (where the plant community benefits from moisture from fog and the cloud layer), and also at higher elevations in the department of Ancash. The population in Bolivia was originally called *Paramongaia superba* and described as a separate species.

Paramongaia weberbaueri is most closely related to the genus *Clinanthus* (*see p.228*), especially to two species which were originally described as the genus *Callithaumia*. In these two *Clinanthus* species, as in cojomaria, the free portions of the stamen stalks are attached to the trumpet below its rim. The rarity of cojomaria is exacerbated by the collection of the flower stems (which each bear a solitary flower) by local people for sale along the roadsides near its habitat. Like the amancay, *Ismene amancaes* (*see p.243*), which has a somewhat similar flower, cojomaria may only flower profusely in years of exceptional rainfall.

In cultivation, cojomaria is best grown in containers, except in warm-climate areas such as California. Success with this Andean gem, which has evolved in a very specific habitat, requires a subtle understanding of its lifecycle (*see box, above*). Each plant requires pollen from a different plant for it to be able to set seed. The flattened, winged, black seeds germinate readily when fresh, in 6–8 weeks. Seedlings may be carefully transplanted to small pots when one leaf has developed.

CULTIVATION

LIFECYCLE LESSONS

Cojomaria never freezes in its arid, tropical maritime habitat, and it is rarely exposed to temperatures greater than 27°C (81°F). Its season of active growth is little more than 3–4 months in duration. Irrigation should commence on the emergence of the tips of the blue-green, succulent leaves from the apex of the bulb. The 1cm (½in) wide, strap-shaped leaves elongate to about 30cm (1ft) in length. If a bulb is going to flower, the swollen bud will be visible between the innermost leaves. When the leaves begin to yellow, water should be withheld. Bulbs should then be placed in a dry, shady area until new growth appears the following spring. Transplanting, which can inhibit flowering, should be done during the dormant period, taking care to minimize damage to the thick roots. The dormant bulbs may rot if overwatered.

▲ RARE ENDEMIC *There are only three known wild populations of cojomaria, found in the western foothills of the Andes. The intensely fragrant flowers, sometimes almost 20cm (8in) across, are often found for sale in large quantities along roadsides or in local markets in Peru.*

Red List: Endangered

Phaedranassa tunguraguae

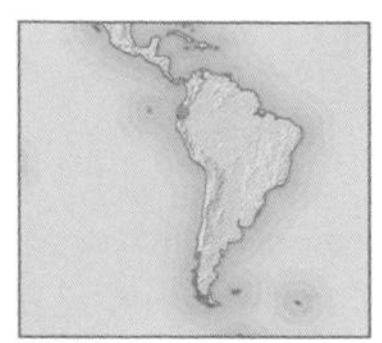

Botanical family
Amaryllidaceae

Distribution
Ecuador: Tunguragua Province

Hardiness
Frost-tender

Preferred conditions
Part shade in fertile, well-drained soil; seedlings benefit from heavier shade than mature bulbs.

◄ TUBULAR BLOOMS *The green-tipped flowers of this Andean native nod gracefully from their stalks, and the dark green, elliptical leaves can make an appealing show in their own right.*

All but two of the *Phaedranassa* species are found only in Ecuador, and most are narrowly distributed. In fact, the majority are known from only single populations. Most are found at higher elevations in scrub between Andean peaks, but a few have colonized steep slopes, landslides, and road cuts in middle-elevation rainforest. ***Phaedranassa tunguraguae***, as the name suggests, is found only on the lower slopes of the Tungurагua volcano in the valley of the Pastaza river. It is one of the larger-flowered species of its genus. The chief conservation threats to this plant are its rarity and disturbance of its habitat.

This species is easier to cultivate than those that are native to higher altitudes. A short, leafless, dry period, preceded by the yellowing of the leaves, is the usual pre-condition of flowering, and alternate periods of drought and regular irrigation may result in flowers several times during the year. The flattened, winged, blackish-brown seeds germinate readily if sown fairly fresh.

Red List: Vulnerable*

Phycella ignea

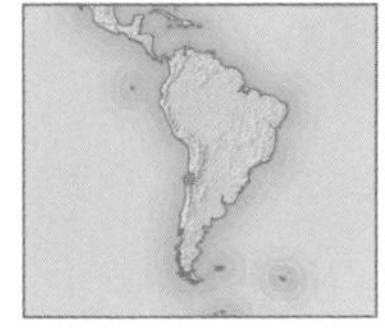

Botanical family
Amaryllidaceae

Distribution
Chile

Hardiness
Frost-tender

Preferred conditions
Well-drained sandy soil in full sun; best in Mediterranean climate; dry in summer.

One of the beautiful amaryllids endemic to Chile, ***Phycella ignea*** flowers in spring, just as the winter rainy season is drawing to an end. There are only a handful of populations of this species all living within the vegetation type known as matorral, which occurs on the coast of central Chile, backed by the Andes mountains. Plants include deciduous shrubs, cacti, thorny species, and one tree, *Jubaea chilensis*. Matorral is under threat from development for agriculture and industry.

Phycella ignea is a beautiful, spring-flowering bulb that can be grown in Mediterranean-type climates where the soil never drops below freezing in winter. The bulbs need to remain dry throughout their period of dormancy in summer or they may rot. In the plant's native habitat, the leaves emerge with the winter rains, and the fiery flowers come later, not appearing until the rains begin to taper off.

▲ NAKED STEMS *The brilliant red flowers of* Phycella ignea *are held well above the low-growing leaves where they can easily attract potential pollinators.*

2 on Red List*

Rhodophiala

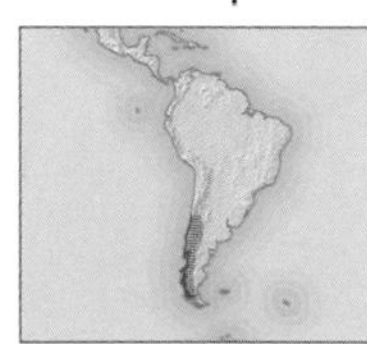

Botanical family
Amaryllidaceae

Distribution
Chile: throughout; Argentina: western regions

Hardiness
Half-hardy – frost-tender

Preferred conditions
Full sun; well-drained sandy or loamy soil; suited to a Mediterranean-type climate; bulbs are best if they remain dry in summer.

The genus *Rhodophiala* consists of perhaps a dozen species native to Chile primarily, with two in Argentina, and another in both countries. It is found in matorral (*see Phycella ignea, above right*); temperate montane meadow; and subtropical grassland, savanna, and shrubland. Several of the species are rare, especially some of those of the high Andes, and their habitats are quite fragile. Some of the lowland species that grow near the coast in central Chile have become sources of local pride, a factor that can only benefit their conservation. However, it is rumoured that large numbers of bulbs have been exported to Europe and Asia in the recent past.

Rhodophiala bagnoldii, the yellow dwarf amaryllis, is a coastal species that grows in enormous drifts close to the shore on dunes and sand flats. After a particularly wet winter, literally thousands of the bright yellow flowers can be seen in flower at once. Despite this, the species is classed as Vulnerable in the Red List of 1997. A few populations are protected in reserves, but the rest could be threatened by agriculture or resort developments near the coast. New roads are planned for some of the less developed regions of Chile's Pacific coast, and in the future these may affect the health of this beautiful bulb.

▲ RHODOPHIALA BAGNOLDII *can be difficult to get to flower in cultivation. In its native South American habitats it is seldom subjected to extremes of heat or cold.*

Rhodophiala phycelloides, the red dwarf amaryllis, is a native of lower elevations and has intensely scarlet flowers. Although unlisted by IUCN, it is certainly vulnerable. It is found close to the coast and also in the coastal range of north-central Chile, inhabiting desert-like vegetation that receives only winter rainfall. Fortunately, a few populations are contained within national parks.

Rhodophiala's black, flattened, D-shaped seeds germinate readily in 6–8 weeks when fresh. Oddly, many Chilean species set seed after the rainy season. The seeds may retain viability for a much longer period than those of other members of the amaryllis family.

Red List: Rare*

Scilla latifolia

Botanical family
Hyacinthaceae/Liliaceae

Distribution
Canary Islands (Spain), Morocco

Hardiness
Half-hardy

Preferred conditions
Full sun to partial shade; well-drained, sandy soil; best suited to Mediterranean climates with mild winters.

One of the larger squill species, ***Scilla latifolia*** grows in the Canaries' Mediterranean woodland and scrub. Its broad, dark-green leaves are often edged in red. Like much of the Canary Islands flora, this plant is vulnerable to disturbance of its habitat by land development, as well as by exotic invasive plants. The presence of this species in Morocco is somewhat controversial and reports of the species from that country may in fact be the related *Scilla maritimum*.

Scilla latifolia is, like the more common and better known *Scilla peruviana*, one of the larger species in the genus, producing fairly large bulbs and broad leaves. It is only hardy where winters are mild, since its season of active growth is winter and spring. It needs a dry summer to succeed, but can be grown in containers anywhere.

◄ LILAC SPIKES *The flower stems, which have a rosy tinge at the tip, appear in late spring; the lilac or pink blooms are loosely distributed along the tip. The fruits that form later turn yellow-orange as they ripen.*

HABITAT

UNIQUE FLORA

Home to 500 endemic vascular plants, the Canary Islands also have the highest proportion of endangered and vulnerable plants in Europe. Some 35 per cent of plants on the islands are non-indigenous species, although not all are invasive. Among the species currently invading is fountain grass, *Pennisetum setaceum*.

Red List: Indeterminate*

Siphonochilus aethiopicus

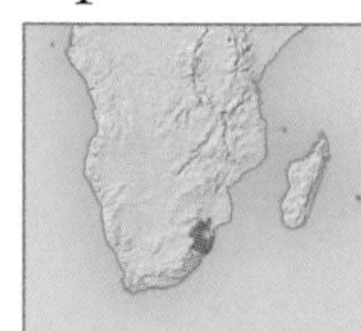

Botanical family
Zingiberaceae

Distribution
South Africa: KwaZulu-Natal and Mpumalanga provinces; Swaziland

Hardiness
Frost-tender

Preferred conditions
Partial shade; rich, fertile, moist soil; bulbs must stay dry in winter or they may rot.

Also called African wild ginger, this species is an understorey herb of tropical savanna and seasonally dry forest closely allied to the genus *Kaempferia* (within which it was once classified). It differs from that genus in that male and female flowers are borne on separate plants. Interestingly, female plants are often smaller than male ones.

Siphonochilus aethiopicus is one of the most prized medicinal plants of South Africa's *muthi*, or traditional healers (*see p.226*). The demand for material of the plant is so intense, and the price it fetches in the South African herbal market so high, that it has been over-collected to the brink of extinction. Micropropagation by tissue culture has been applied in an *ex-situ* programme that has saved wild ginger from complete extinction, but native herb collectors still need to be convinced that material should be cultivated for sale in the markets, rather than harvested from the dwindling wild populations.

Wild ginger demands a soil liberally furnished with organic matter and plenty of water during its active period of growth. Seed may take up to a year to germinate, thus rhizome division when the plants are dormant is a much simpler way to propagate the species. Care should be taken to damage as few of the succulent roots as possible when dividing the plants.

▲ ORCHID-LIKE FLOWERS *Wild ginger produces large but short-lived flowers, similar to those of some orchids, at ground level. The flowers are pleasantly fragrant and usually appear before the leaves in mid- to late spring.*

9 on Red List*

Sparaxis

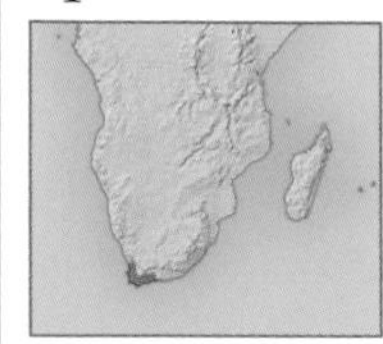

Botanical family
Iridaceae

Distribution
South Africa: Northern Cape and Western Cape provinces

Hardiness
Half-hardy

Preferred conditions
Full sun; fertile, well-drained soils; best in Mediterranean climates, but will also adapt to summer-rainfall areas with mild climates.

The genus *Sparaxis* consists of about 15 species of delicate cormous plants, commonly known as wand flowers. They are native plants of the South African veld. The renosterveld clay soils that harbour most of the species have been intensively claimed by agriculture, while the coastal sand veld inhabited by the remainder is continuously under pressure for resort development. One species, *Sparaxis roxburghii*, is classed as Extinct in the Red List of 1997, but was recently documented.

The genus is quite variable in flower shape, including both radially and bilaterally symmetrical flowers, an example of the way that many Cape species have evolved to attract specific pollinating insects. Bees, long-tongued flies, and scarab beetles have all been implicated as *Sparaxis* pollinators. The brightly coloured, elegant flowers, which all bloom in the late winter and spring in South Africa, have been in cultivation for over 200 years.

Sparaxis elegans is found on clay flatlands of South Africa's north-west Cape. It is classed as Vulnerable in the Red List of 1997. The radially symmetrical flowers are usually salmon pink in colour, with a sharply contrasting eye of purple and yellow; however, white forms occur in nature. The species flowers from late spring to early summer in the northern hemisphere. A cultivar called 'Zwanenburg', with maroon flowers and yellow eyes, has been assigned to this species, but *Sparaxis elegans* has been widely hybridized for many years so its parentage is uncertain.

Four subspecies of ***Sparaxis grandiflora*** are recognized, all varying in flower colours that encompass shades of yellow, purple, and white. The rarest of these, *Sparaxis grandiflora* subsp. *grandiflora*, is classed as Rare in the Red List of 1997. It has remarkable plum-coloured flowers, and is found only to the north-east of Cape Town. The flowers of this species can be 8cm (3in) wide.

Another important species in the breeding history of its genus, *Sparaxis tricolor* in its typical form has an orange-red flower with a yellow centre rimmed in black. It is restricted to an area in the north-western Cape. Known as the harlequin flower, it appears as Vulnerable on the 1997 Red List. Many colours are sold under this species name, but their robustness suggests that they are of hybrid origin. 'Fire King' is a selection with blooms of an intense red.

◄ SPARAXIS ELEGANS *The dark, contrasting eye in the centre of the flowers easily marks the species' influence in the many wand flower hybrids that have been created over the years.*

CULTIVATION

Wand flowers are best planted in masses, in borders or mixed with spring annuals. Since they are fairly low-growing plants, they should be situated where they will not be overshadowed by taller plants. In areas where winter temperatures drop below -4°C (25°F), the plants require protection and are ideally placed in containers that can be brought into shelter. The corms can be lifted and stored until the danger of freezing temperatures is past, but this is likely to compromise the performance of the plants. Whether grown in the ground or in pots, the corms are best left undisturbed for several years, after which time the clumps can be divided.

Seed should be sown in autumn in mild Mediterranean climates, and in spring anywhere else. The seedlings frequently yield interesting variants, and will often flower one year after germination. Vegetative propagation, by the small corms and cormels readily produced by parent plants, is an easier method but of course produces genetic clones rather than the variety achievable by reproduction from seed.

SPARAXIS GRANDIFLORA ► *This large-flowered wand flower, found on clay soils in south-western South Africa, is the parent of a number of garden hybrids.*

2 on Red List*/all on CITES II

Sternbergia

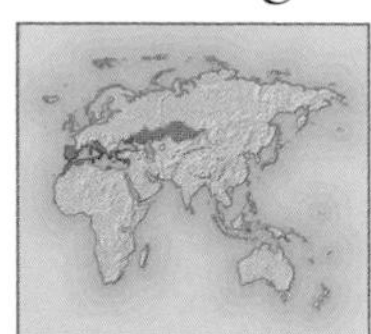

Botanical family
Amaryllidaceae

Distribution
Mediterranean, Balkans, Middle East, Central Asia

Hardiness
Half-hardy

Preferred conditions
Full sun; fertile, well-drained, neutral or slightly alkaline soil; bulbs must be dry in summer.

Only two genera of the amaryllis family appear on CITES' Appendix II: *Galanthus* and *Sternbergia* (*see p.236 and p.255*). Large numbers of bulbs of several *Sternbergia* species have entered the bulb trade in past years, wild-collected from Mediterranean forest, woodland, and scrub, and temperate steppe. The greatest number of species grow in Turkey, which traditionally has had the largest export trade in wild-collected bulbs.

Recently, Turkey has forbidden export of the rarest *Sternbergia* species. Conservation agencies have supported research and pilot projects to develop production of bulbs and replace the wholesale digging of plants from their habitat. Also, many gardening societies are urging their members to purchase only nursery-produced material. Habitat disturbance by logging and agriculture continues to adversely affect *Sternbergia* species.

There are eight spring- or autumn-flowering sternbergias. These are dwarf plants, and all but one have yellow flowers. Superficially they resemble crocuses. Most "hide" their ovary inside the neck of the bulb until just before the seeds are ripe, whereupon the seed capsule emerges on a short stalk.

Sternbergia candida, or white winter daffodil, was only discovered in 1979, in southern Turkey. Overcollection caused it to be classed as Endangered in the Red List of 1997. In the garden the plant readily offsets once it is established, in time producing a mound of white flowers that brighten the last days of winter in warm, temperate climates.

◄ STERNBERGIA CANDIDA *is the only white-flowering and fragrant species of the genus. The flowers usually appear in late winter with the grey, strap-shaped leaves. The species does poorly in pots.*

Sternbergia clusiana, the yellow autumn crocus, produces the largest flowers in the genus. The leaves, appearing soon after, persist through the winter, but are susceptible to damage in severe winters. Native to Turkey and Israel eastwards to Iran, the species grows in lower mountain meadows. Now protected by law from wild-collection, its gravest threat may be the expansion of agriculture.

▲ STERNBERGIA LUTEA *is an easy species to grow and flowers freely in a sunny spot in the garden, increasing in time by offsets. The leaves, which persist in winter, may be damaged by freezing temperatures. Unlike most sternbergias, the species grows well in containers.*

Sternbergia lutea, also called the yellow autumn crocus, is the best-known and most widely grown species of the genus. It is broadly distributed throughout the Mediterranean region, from Spain through Turkey eastwards to Iran and central Russia. Once collected by the millions, the harvesting of bulbs is now controlled by local and international law. This species differs from some others in the genus in that the flowers are held above the ground by short flower stems.

All sternbergias are easily propagated by offset bulbs. After the leaves die back, these can be separated in the dormant period of late spring to early autumn. Seed should be sown in late summer and kept moist, preferably outside or in a cold frame, so that the seeds' requirement of exposure to cold can be met.

Red List: Extinct* (but rediscovered)

Tecophilaea cyanocrocus

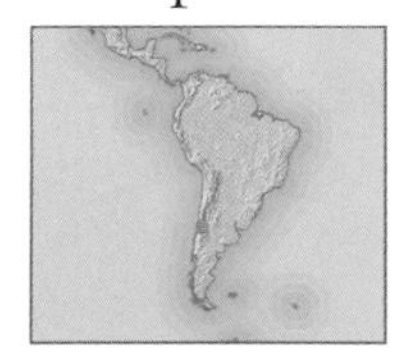

Botanical family
Tecophilaeaceae

Distribution
Chile: Cordillera of Santiago at 3,000m (10,000ft) elevation

Hardiness
Half-hardy

Preferred conditions
Full sun; well-drained, sandy or loamy, slightly acid soil. Moderate temperatures thoughout its growing season, and dryness during dormancy.

The Chilean blue crocus could serve as an emblem for endangered bulb species. It is also a fine example of how horticulture can help to preserve an endangered species. ***Tecophilaea cyanocrocus*** was believed to be extinct in the wild, eliminated by a combination of unrestricted grazing on the grassy, alpine meadows it used to inhabit, and over-collection of the corms. For many years all known living material of this species was cultivated. Finally, a consortium – comprising the Royal Botanic Gardens, Kew; the UK Alpine Garden Society; and the Corporación Nacional Forestal, Chile (CONAF) – was formed to develop a re-introduction scheme for the plant in a protected zone within its former habitat. The conservation genetics group at Kew was set the task of increasing the genetic diversity of their stock of the species by developing seedling populations that incorporated as many different genotypes as possible. Then, in 2001, the species was rediscovered in mountains to the south of the city of Santiago. Maria Teresa Eyzaguirre, of the private R.A. Philippi Foundation for Nature Studies, found two populations, the larger of which occupied an area of no more than 1,000sq m (¼ acre). Interestingly, the flower colour of the population resembled that of the cultivated variety 'Leichtlinii', which no one had ever suspected of occurring in the wild. Several plants with pure white flowers were also observed in a much smaller nearby population.

The Chilean blue crocus is only spectacularly successful in the outdoor garden where its native conditions can be approximated. Horticulturists have succeeded in New Zealand (particularly), on England's southern coast, and in California. The growers often maintain the plant in pots of gritty soil in a cool greenhouse.

The Chilean blue crocus dislikes extremes of temperature. In its habitat fairly high in the Andes, the corms are kept dry and protected from exposure to severely cold temperatures by a winter-long blanket of snow. Neither is the species exposed to hot, humid summers, and it does not tolerate them. The plant can be increased by cormels, which are produced fairly liberally, or by seed. Seed may not develop without hand pollination, and 3–4 years must transpire before corms achieve flowering size. Two cultivars have been recognized in the species: 'Leichtlinii', with lighter blue flowers and a broader central white eye; and 'Violacea', a purple form.

HIGHLY PRIZED BLUES ► *Much coveted for the range of blue coloration in its fragrant flowers,* Tecophilaea cyanocrocus *flowers in spring.*

39 on Red List*

Tulipa

Botanical family
Liliaceae

Distribution
Europe, western and central Asia, and northern Africa

Hardiness
Fully hardy

Preferred conditions
Most species prefer full sun and fertile, well-drained, loamy soil; some may be short-lived if they do not receive a dry period in late summer.

◀ TULIPA KAUFMANNIANA *has the common name of "waterlily tulip", a reference to the way that the flowers open very widely. The outside of the flower has a pink tinge.*

The total of species in the genus *Tulipa* is estimated at somewhere between 100 and 150, reflecting the still uncertain taxonomy of what is unarguably the most important genus of ornamental bulbous plants on Earth, at least in terms of its economic value in the horticultural trade. Tulips grow in temperate broadleaf and mixed forest; montane meadow; temperate steppe; and Mediterranean scrub. They are, of course, historically infamous for the speculative fever they inspired in

▲ TULIPA FOSTERIANA *has large flowers that appear in spring. These open widely in full sun, revealing a dark, almost black eye edged in yellow. It is a parent of some hybrids, including the Darwin types.*

17th-century Holland, when entire fortunes were mortgaged in order to possess a few prized bulbs (*see box, opposite*). Today, over a third of the species within the genus are of conservation concern. Five of the Red Listed species are believed to be Extinct in the wild, while 11 are classed as Endangered. Several factors are responsible for this crisis: overcollection of bulbs from the wild, destruction of habitat, and grazing, coupled with the natural rarity of a number of species. Some European countries, if not yet providing outright protection, have at least placed the species occurring within their borders on national lists of priority species of special concern. The United Nations is currently funding a programme in Kyrgyzstan to develop *ex-situ* production of four native species: *Tulipa greigii*, ***Tulipa kaufmanniana***, *Tulipa kolpakowskiana*, and *Tulipa ostrowskiana*.

TULIP SPECIES

Tulipa fosteriana, the emperor tulip, is rare in its native habitat in eastern Uzbekistan and Tajikistan. It is classed as Endangered in the 1997 Red List. Never abundant in the wild, the populations have suffered from overcollection and habitat disturbance. A number of cultivars have been bred over the years, including 'Madame Lefeber' (scarlet), also known as 'Red Emperor'; 'Purissima' (white); and 'Candela' (yellow). Bulbs of this species are best lifted each year and given a warm, dry rest period.

Tulipa greigii, the Greig tulip, is classed as Indeterminate by IUCN, since its exact conservation status in its native Kyrgyzstan and Kazakhstan is not known. However, the species is presumed threatened by agriculture, chiefly grazing, as well as by collection for horticulture. This species has large, bell-shaped, orange-red flowers, the tepals of which taper towards their tips; they appear in mid-spring. The leaves bear purple-brown, lengthwise bars. A very large number of cultivars are referred to as "Greigii Hybrids", but many resemble the species to a large degree. Some well-known forms include 'Red Riding Hood' (red with a black base); 'Oriental Splendour' (red with a yellow edge); and 'Cape Cod' (pink with a yellow edge). All Greigii hybrids tend to have some mottling or striping on the leaves. The species requires a warm and dry period after it ripens.

Another species listed as Indeterminate by IUCN, *Tulipa kaufmanniana* is a compact plant, adaptable to bedding and rock gardens. In the wild, it occurs on rocky mountain slopes near the melting snow. Its conservation status in Kyrgyzstan and Kazakhstan is not known, but it may be assumed to be subject to the same threats as other tulips in the region. This species has been an important parent of early-flowering hybrids, particularly with *Tulipa greigii* and *Tulipa fosteriana*. The

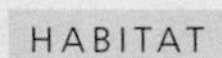

HABITAT

MOUNTAIN HOME

Over half of Tajikistan rises above 3,000m (10,000ft), and many species of tulip grow in profusion in the steppes and on the dry mountain slopes (*Tulipa kaufmanniana* is seen here). The mountain tulips depend on water from melting snow in the spring but are adapted to baking heat in summer and dryness in winter, when the bulbs lie under deep snow. In cultivation they thrive in fast-draining soil in the glasshouse, where the dry conditions may be duplicated.

TULIPA DOERFLERI *A native of Crete, this tulip flowers in profusion in a few meadows of the Plakias region but is classed as Vulnerable in the Red List of 1997 because of its restricted distribution.*

flower is long and white with a yellow tint, and is one of the earliest species to flower in spring. Numerous forms have been given cultivar names, a few of which include 'Shakespeare' (light red with a pink edge); 'Stresa' (yellow and red with yellow spots); and 'Heart's Delight' (red outside, white turning pink with age inside). This tulip needs a dry period after ripening, and it may be safest to lift the bulbs in areas where the summers are very wet.

Tulipa praestans, the multiflora tulip, is native to Tajikistan, where it is subject to overcollection and habitat disturbance. In the Red List of 1997 it is classed as Vulnerable. This species is easy to cultivate, preferring a humus-rich soil. It is one of the most dependable tulip species for successful naturalization in the garden, returning year after year without lifting. It typically bears up to four cup-shaped, orange-red flowers in mid-spring. Notable cultivars include 'Fusilier' (bright red, with up to five flowers); 'Unicum' (yellow-edged flowers, variegated leaves); 'Van Tubergen' (vermilion); and 'Zwanenburg' (brilliant red). This tulip is effective massed in beds, in pots, or in the rock garden.

Tulipa sprengeri is confirmed as Extinct in its native Turkey in the Red List of 1997. This tulip is one of the last to flower, in late sping to early summer. It is also prized for the fact that it has proved very amenable to naturalization in the garden, returning annually without lifting in a variety of climates, and even self-seeding in favourable situations. The plant also adapts to slightly shaded conditions. One of the taller-growing tulip species, it produces bright red flowers on stems up to 45cm (18in) tall. The Royal Botanic Gardens, Kew, is investigating the possibility of re-establishing this species in its native Turkey, using material in cultivation at the London gardens.

◄ TULIPA TARDA *This low-growing tulip can produce as many as five flowers per plant, each on a short stem, giving the effect of a tight floral cluster. A native of subalpine meadows in Central Asia, it is one of the most easily cultivated species tulips, and the bulbs can remain in the ground in most climatic regions.*

Tulipa tarda, the dasystemon tulip, is native to subalpine meadows in the Tien Shan mountains of Central Asia. It is cited as Rare in the Red List of 1997. Introduced into western cultivation in the late 16th century, *Tulipa tarda* spreads by underground creeping stems (stolons), eventually forming clumps. Best planted in large groups for a dramatic spring show, the tulip's diminutive size makes it suitable for the rock garden, or for any border where a blaze of bright yellow is desired.

PROPAGATION

Seed from any tulip species can be sown in any well-drained, preferably sandy germination medium, either when it is ripe or in spring, after a winter storage period. Overwatering should be avoided. Tulip seed germinates within 6–8 weeks. The first leaf is grass-like, and a small bulb will have formed under the surface by the time it appears. Seedlings are best left in their germination containers until they are two years old, with protection from excess moisture that could rot the bulbs during their dormancy. The two-year-old bulbs may then be planted out in the garden.

TULIPA PRAESTANS ► *Like* Tulipa tarda, *this species is appreciated by gardeners because it bears more than a single flower per plant. The flower lacks the blotches usually present at the base of the petals in tulip species.*

HISTORY

TULIP MANIA

In the early 17th century, Amsterdam merchants began a trade in tulip bulbs from Turkey. The rare bulbs commanded such high prices that they became a status symbol among the rich. Connoisseurs favoured the flamed or feathered tulips over the plain ones; in 1623, a single bulb of 'Semper Augustus', the most strikingly marked of the red-and-white flamed tulips, sold for 1,000 florins, when the average annual income was about 150 florins.

Part of the tulip's allure was that it could change colour, seemingly at whim. A plain flower might, in the following spring, produce petals striped in intricate patterns. Unknown to the tulip lovers, this was the work of a virus transmitted by aphids. The virus weakened the bulbs and slowed their production of offsets, which further increased the bulbs' rarity and value.

Tulip mania reached its height in 1634–37, when speculators were staking their fortunes on rare bulbs in the hope of vast profits. The subsequent crash was a disaster for many people, but cultivation of tulips continued. The engraving (*see right*) from a 1699 Dutch gardening book shows how tulips still dominated in the garden.

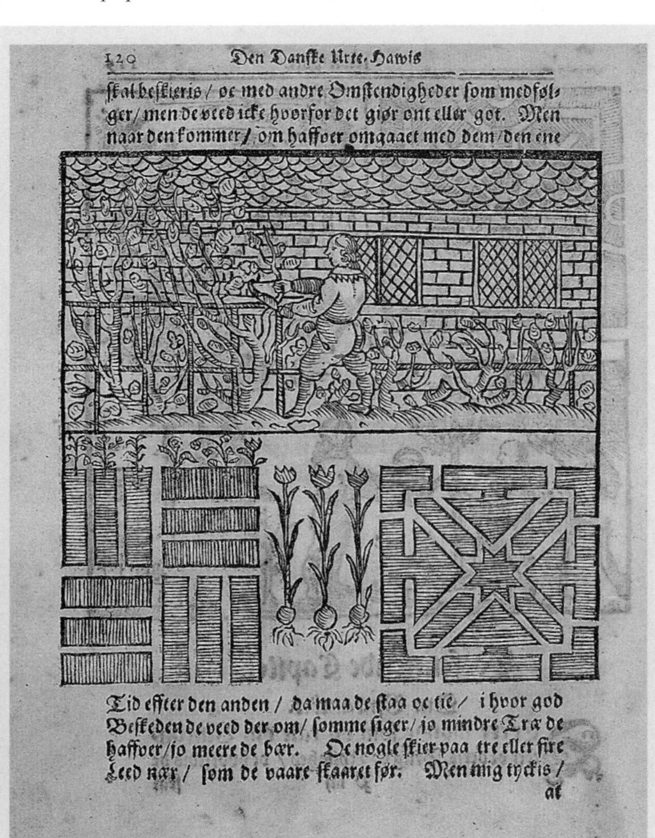

120 Den Danske Urte-Hawis

skal beskieris / oc med andre Omstendigheder som medfølger / men de veed icke hvorfor det giør ont eller got. Men naar den kommer / om haffver omgaaet med dem / den ene

Tid effter den anden / da maa de staa oc tie / i hvor god Beskeden de veed der om / somme siger / jo mindre Træ de haffver / jo meere de bær. Oc nogle skier paa tre eller fire Leed nær / som de vaare skaaret før. Men mig tyckis / at

Red List: Rare*

Watsonia latifolia

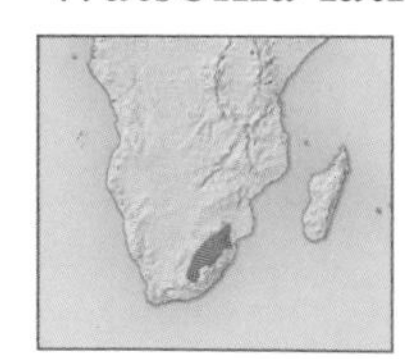

Botanical family
Iridaceae

Distribution
South Africa: KwaZulu-Natal, Mpumalanga, Limpopo; Swaziland

Hardiness
Frost-tender

Preferred conditions
Full sun; well-drained sandy or loamy soil. This species prefers a mineral soil, that is, soil without a great deal of organic matter incorporated.

A genus of the iris family native to the tropical grasslands of southern Africa, *Watsonia* contains over 50 species, centred in South Africa's south-western Cape but extending into the eastern, summer-rainfall region of the country and northwards to Swaziland. Most watsonias are deciduous perennials with corms, but a few species are evergreen. Some of the Cape species are conspicuous by the large size of the populations that they form, especially those species that grow in marshy areas; in flower, they can be a spectacular sight. Eighteen species are on the Red List of 1997; one of these, *Watsonia distans*, is considered to be Extinct.

Watsonia latifolia, one of the more northern species, is very rare in nature, and is a plant protected by law in Swaziland. It is found in open grassland and at the base of granite outcrops, but never in great profusion. The flowers are pollinated by sunbirds.

In the garden, the flowers provide a most unusual shade of red that combines well with the dark blue flowers of plants such as agapanthus. The leaves are quite broad for the genus. The plant is only suitable for outdoor cultivation in regions where there is little or no frost. In marginal zones, it is best to position plants close to a warm, south-facing wall, and to protect the roots with a heavy layer of mulch. *Watsonia latifolia* also lends itself to being grown in containers in the greenhouse.

This species produces several new corms (cormels) each year, which can be separated from the parent corm (*see above*) if the plants are lifted from the ground for winter storage. Seed from this species is sown in the autumn and germinates well, but unevenly; seedlings require 3–4 years to flower. Plants may also be divided in spring.

PROPAGATION

COLLECTING CORMELS

Planting offset cormels is an easy way to propagate new plants. To encourage offsets, corms should be shallowly planted in spring. In summer, flowerheads should be removed to ensure that the plant's energy is directed to production of cormels, rather than seeds. When the foliage dies down, carefully lift the corms with a fork. Cormels will have formed around their bases. These can separated from the parent and planted next spring.

▲ LATE SUMMER STAR *Like several species of its genus,* Watsonia latifolia *is a striking, tall-growing plant with sword-shaped leaves. It produces its long, branched spikes of curved, tubular, mahogany-red flowers in late summer.*

Red List: Endangered*

Worsleya rayneri

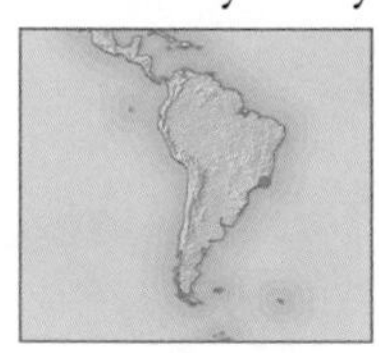

Botanical family
Amaryllidaceae

Distribution
Brazil: Rio de Janeiro state

Hardiness
Frost-tender

Preferred conditions
Full sun; porous, acid soil; best where temperatures do not exceed 27°C (80°F). The plant requires reduced moisture and a cool temperature in winter.

One of the most spectacular plants in the amaryllis family, and something of a legend in horticulture, the blue amaryllis is known from two populations within the tropical shrubland in the mountains to the north of Rio de Janeiro. ***Worsleya rayneri*** is the only species within the genus, and is most closely related to *Griffinia* (*see p.240*) rather than *Hippeastrum* (*see p.241*): hybrids between *Worsleya* and *Hippeastrum* have never been successful. Some botanists believe that the species represents a relic of the earliest lineage of the amaryllis family in the Americas.

Worsleya rayneri clings to domes of granite that rise to elevations of over 2,000m (6,500ft). The plants are bathed by frequent fogs, even during the drier months of the year. During the summer, violent thunderstorms are a frequent occurrence in this habitat. Exposed to extremes of neither heat nor cold, the blue amaryllis is found in small flat areas on the mountain summits, or wedged into crevices on the most precipitous slopes, held in place by its thick roots. This evergreen plant forms a large bulb with an elongated aerial neck, from which arch the two-ranked, sickle-shaped leaves, green with red margins.

In Brazil the blue amaryllis is protected by federal law, but export of plants is not yet prohibited by CITES. Although the two populations have been subjected to sporadic bulb collection for export (many of which probably died in the hands of inexperienced growers), their relative inaccessibility has provided some measure of protection. For this endangered plant, a far greater threat comes from frequent fires, not all of them natural in origin, partly fuelled by an introduced African grass species.

The blue amaryllis is almost legendary for the difficulty encountered in getting the plant to flower in cultivation. It will grow in containers, although it resents disturbance of its velamen-covered roots. (Velamen is a water-absorbing tissue seen on the outside of the roots of epiphytes, as well as *Worsleya*.) The roots also dictate that a very porous mix must be used when planting. Some horticulturists have reported success in growing *Worsleya* exclusively in crushed rock; others vouch for orchid media or a mixture of the latter with high-quality, peat-based potting mix. Wide but shallow clay pots should be used. In their habitat, the plants are exposed to winter temperatures as low as 4°C (39°F), thus a cool, dry period during the corresponding time of the year in the northern hemisphere is recommended. The plants are self-sterile, so pollen from a different individual is necessary to set seed. The black, D-shaped seeds readily germinate within 6–8 weeks when fresh, and the young seedlings may be grown on together for the first year in one large pot of seedling orchid mix.

▲ THE BLUE AMARYLLIS *At the close of winter, which in the high-elevation habitat of* Worsleya rayneri *is a dry season with moderate temperatures, the plant produces a flower stalk bearing 4–6 large, funnel-shaped flowers of an arresting lilac or amethyst blue.*

Red List: Rare*

Zantedeschia pentlandii

Botanical family
Araceae

Distribution
South Africa: Mpumalanga Province

Hardiness
Frost-tender

Preferred conditions
Full sun; fertile, moist, but well-drained soil; requires dryness in winter.

Within the genus *Zantedeschia* are six species with tuberous rhizomes that grow in subtropical grassland, swamps, and lake margins in southern and eastern Africa. Three species from South Africa are Red Listed.

Older residents of the Tonteldoos and Roossenekal areas in the Transvaal region of South Africa remember how, as recently as the 1950s, the mountain summers were brightened by a profusion of yellow flowers, a sight that is now rare to non-existent. The flowers were ***Zantedeschia pentlandii***, the yellow calla lily or golden arum, which still occurs along streams in grassland and in fertile soil accumulations around the rocky outcrops. However, during the 1960s the populations were devastated by traders in the cut stems, which are extraordinarily long-lasting. Nearly extirpated in its habitat at that time, the yellow calla lily has since suffered further losses through agricultural development and predation by porcupines.

Yellow calla lily flowers in late spring and early summer. In cultivation, the rhizomes should be planted several centimetres deep in pots or in the ground. Coming from a cool habitat at middle elevations (up to 1,900m–6,250ft), this species does not favour areas with hot nights in spring and summer, and the plants may enter an early dormancy when exposed to hot, tropical or subtropical climates. If this occurs, the rhizomes should be kept dry and cool until new growth is initiated by cooler temperatures in autumn. Propagation is best achieved by division of rhizomes.

In contrast, *Zantedeschia aethiopica* is a native of wetlands and, in some parts of the world where it has escaped from gardens, is proving invasive (*see p.471*). In northern temperate climates, however, it is a coveted species for water gardens, and is the parent of several fine cultivars.

▲ ZANTEDESCHIA PENTLANDII *While a native of grassland areas, this showy lily often establishes in pockets of soil trapped in rocky outcrops, where moisture is more abundant and the grassy plants are less competitive.*

▼ ZANTEDESCHIA AETHIOPICA 'GREEN GODDESS' *Like its parent, this popular cultivar may be grown in shallow water, 30cm (12in) deep, or at the water margins. The species should be grown with caution, however, because it has proven invasive in some areas.*

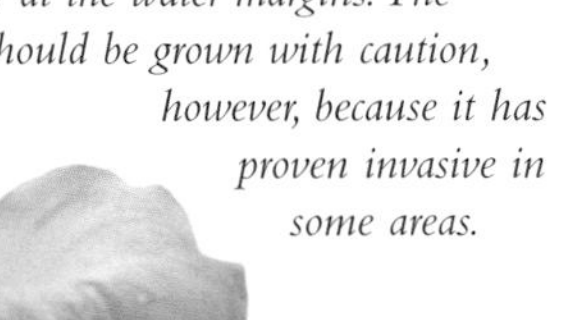

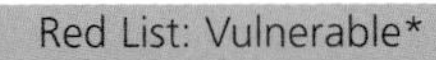

Red List: Vulnerable*

Zephyranthes simpsonii

Botanical family
Amaryllidaceae

Distribution
Florida, USA: subtropical areas

Hardiness
Frost-tender

Preferred conditions
Full sun; moist, sandy soil.

Simpson's zephyr lily is one of two members of this large genus that are found in south-east USA. It grows in subtropical coniferous forest and savanna. This lily has twice as many chromosomes as its close relative *Zephyranthes atamasca* var. *treatiae* (not Red Listed), and has a relatively southerly distribution in the state of Florida. Whereas its relative's flowers are wide-spreading, the rosy-tinged white flowers of Simpson's zephyr lily are more narrowly cup-shaped. The linear leaves are fairly inconspicuous. The gravest threats to this plant, one of the rain lilies, are its restricted range and the development pressures in Florida.

Simpson's zephyr lily is easily grown in gardens within its fairly narrow native range, and is probably hardy wherever winter temperatures do not frequently fall below -4°C (25°F). It is an excellent addition to the Florida native plant garden, and will naturalize into a lawn or in the border if conditions are favourable. The black, D-shaped seeds germinate readily when fresh, and plants may flower in one year from sowing.

REGENERATION

FIRE, THEN FERTILITY

Unlike most rain lilies, which flower sporadically throughout the warm months after a rain storm, *Zephyranthes simpsonii* strictly flowers in spring, most heavily after a fire has swept the pineland margins or prairies where it is usually found. In Florida, controlled burning is carried out in 13,500ha (33,000 acres) of state park each year to promote the survival of fire-dependent plant species.

GRASSES AND BAMBOOS

Grasses comprise one of the largest of all plant families, and one of the most familiar. They often play a major role in shaping the landscape as well as providing food for both wildlife and humans. Various species supply the raw materials that go into bread, sugar, and alcohol, or are used to create grazing for domesticated animals. The swathes of grassland that cover parts of the world began to take their present form millions of years ago, their evolution coinciding with that of many mammals, so that today they are still of enormous ecological, and economic, importance.

WHAT ARE GRASSES AND BAMBOOS?

The term grasses is often applied to a range of slender-leaved, grass-like plants, which also include bamboos, sedges, rushes, bulrushes, and restios. Bamboos are the most closely related to true grasses and belong to the same family, the Poaceae (or Gramineae). All are monocots, with seedlings that produce only one seed leaf. The flowers, often graceful and delicate, are generally inconspicuous, and pollination and seed dispersal is usually by wind. Typical grasses have a fibrous root system and grow rapidly from rhizomes or surface shoots. They will quickly reshoot after the removal of their foliage, an adaptation that allows them to survive constant trampling and grazing by animals. Bamboos are mostly shrub-sized, but some species can attain heights of 35m (120ft).

▲ TRUE GRASSES, *such as this highly ornamental* Stipa gigantea, *have flat leaves and hollow, cylindrical stems.*

▲ SEDGES *are distinguished by their triangular leaves and flower stems. Like this saw-sedge, they inhabit damp sites.*

▲ BAMBOOS *are true grasses but differ in that they have woody, segmented stems, called culms, and narrow, divided leaves.*

WHERE DO THEY GROW?

Grasses have evolved to prosper in a wide range of habitats, from deserts to woodlands, to mountain and alpine sites, and dominate certain types of vegetation such as steppe, savanna, and prairie. The largest areas of natural grassland occur between the climate zones occupied by forest and desert, for example, in the American midwest and parts of central Asia. These fertile grassland soils have, however, been extensively converted to cultivation, especially of grain crops. Reeds and sedges grow in and around wetlands, and may cover large areas of saturated ground. Bamboos typically grow in tropical and subtropical forests, usually in the understorey, where they may be locally dominant or occur as scattered clumps. Management for harvesting them as a crop has resulted in bamboo forests in some regions, notably in China. Bamboos need a warm, wet growing season, but some can withstand cold, even regular freezing, conditions and snow. Restios inhabit dry shrublands, especially in the Cape region of South Africa.

◄ WETLAND *Bulrushes grow in shallow water at the edges of rivers, lakes, and marshes, as here in the Arkansas National Wildlife Refuge in Texas.*

BAMBOO FOREST ► *Bamboos growing in dense thickets usually form the understorey in tropical or subtropical forests. Many species have a clump-forming habit.*

DECEPTIVE BEAUTY *The evening sun lights up a stand of* Phragmites australis *growing along the Rio Grande River in a wildlife refuge in New Mexico, USA. The species is one of the grasses that has become a serious invasive weed.*

SUPPORTING ROLE

Grasses evolved alongside grazing by herbivorous mammals. Even now the main value in nature of grasses and grasslands is as food and as a habitat for a vast diversity of animals. Pollen studies suggest that grass became increasingly abundant some 24–33 million years ago, during the Oligocene epoch, and that large open grassland habitats first appeared 5–25 million years ago, during the Miocene, about the time that herbivorous mammals were diversifying and their populations increasing. Even in the comparatively recent past some grazing mammals, perhaps most famously the American bison (buffalo), roamed grassland habitats in huge numbers. It was partly their grazing pressure that maintained the open prairie habitat. The savannas of East Africa, with their herds of zebra, wildebeest, and gazelles, are an indication of the close co-dependence of herbivorous mammals and grasses. Grasses also support very rare species, such as Przewalski's horse (*see above right*) and the volcano rabbit. Today, many wild grasslands have been replaced by pasture for domestic animals and by fields of cultivated crops.

◄ GIANT PANDA *Just a few species of bamboo are the staple diet of this rare creature, the international symbol of conservation.*

▲ PRZEWALSKI'S HORSES *Wild horses have evolved to feed mainly on grasses, which in turn are adapted to withstand the pressures of repeated grazing.*

▲ MARRAM GRASS *Some grasses, notably marram, can thrive in shifting sands, such as coastal dunes, which they also help to stabilize.*

SYNCHRONIZED FLOWERING

Many species of bamboo flower only at fairly long intervals – anything from 10 to 120 years – and they do so synchronously. Since flowering usually results in the death of the plant, such simultaneous flowering can lead to the sudden disappearance of a species from an entire region, followed by a slow repopulation from seed. This kind of cyclical flowering is rare among flowering plants, and botanists are not sure why it has evolved. One theory is that if masses of seeds are produced at once, a good proportion will survive the activities of seed eaters. Another is that this strategy produces masses of pollen, increasing the chance that some of it will reach a distant, genetically distinct population, thus lessening the possibility of inbreeding.

▲ ABOVE THE CANOPY *The flowering spikes of these tropical bamboos rise above the surrounding forest foliage, aiding the dispersal of pollen and seed by wind.*

A DIVERSITY OF USES

People make daily use of grasses and their relatives in a multitude of ways. Reeds and rushes are used for thatching and to weave mats and screens. Grass is grown as grazing pasture for domestic animals, as lawns to decorate gardens, and as playing surfaces for sports ranging from golf to football. In some cases, grasses may be planted to help reduce erosion or as pioneer species to reclaim polluted soil and slag-heaps. But perhaps their main value to people is as the major grain crops – most notably rice, wheat, barley, maize (sweet corn), rye, millet, and sorghum.

Bamboo, too, is put to a wide variety of uses, and is central to the lives of many millions of people throughout the tropics, most notably in China and south-east Asia, but also in Africa and South America. This is largely because of the unique properties of bamboo culms, which are light (usually hollow), smooth, straight, and very strong. The stems' diameters vary with their age. Bamboo culms are used in constructing buildings and furniture, and also as natural scaffolding, even around modern high-rise buildings in cities, especially in China. Large stems make ideal pipes for irrigating fields, while narrower ones are used for making furniture and traditional musical instruments such as flutes. Many everyday domestic items are made from bamboo, including mats, chairs, rakes, baskets, cooking utensils such as sieves, cups, bowls, chopsticks, and even toothpicks. It is also used to make a form of paper.

Moreover, bamboo is an important food source for millions of people who eat the shoots, especially in China, and for many wild animals. Giant pandas eat it almost exclusively, but it is also part of the diet of others such as the red panda. In South America, spectacled bears and tapirs eat it, as do eastern mountain gorillas in Africa. Several lemur species inhabit the bamboo forests of Madagascar.

▲ HARVEST TIME *Grain forms the staple diet for people over most of the world. Here farmers in India separate the grain from the chaff by winnowing.*

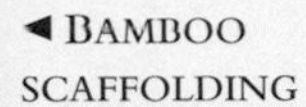

◄ BAMBOO SCAFFOLDING *Bamboo poles are light yet strong, making ideal scaffolding. They are used in many parts of eastern Asia, notably in China, as here in Shanghai.*

CROP ACTION PLAN

The many local varieties of crop plants that now exist around the world form an important genetic resource that it is important to preserve. Known as "landraces", they provide raw material for use by plant breeders. Local varieties are often more resistant to a region's pests than introduced "improved" varieties; they may also compete more effectively with weeds. New imported varieties may offer increased yields, but they often also require greater use of weedkillers and pesticides. In 1996, the Global Plan of Action for the Conservation and Sustainable Utilization of Plant Genetic Resources of Food and Agriculture was adopted at a UN Food and Agriculture Organization conference. The main aim is to create an efficient worldwide system for the conservation of crop plants, as well as their wild relatives. Apart from their use in plant breeding, the wild species form a reservoir of genetic adaptability needed to protect the food supply from potentially harmful environmental change. Genes providing resistance to pests and diseases have been obtained from the wild relatives of crops and the loss of these grasses will diminish future crop development.

EAR OF WHEAT

BAMBOO GROVE *Some bamboos, such as certain species of* Phyllostachys, *can reach a height of up to 30m (100ft). The hard, pole-like culms are harvested and used for a wide range of purposes.*

GARDENER'S GUIDE

- Grasses are increasingly being grown in wildlife-friendly gardens, and are often at their most attractive when mixed with wildflowers. This kind of meadow or prairie garden should be maintained by periodic mowing or burning.
- Wetland species, such as tall reeds, bulrushes, and the larger sedges, are adapted to life around ponds. Bamboos are also suited to moist waterside soils, but their fast-growing, hard rhizomes can pierce even pond linings.
- Avoid planting rampant bamboos, such as low-growing species of *Pleioblastus*, *Sasa*, and *Sasaella*. These can easily take over the garden if not held in check by barriers such as concrete slabs, thick plastic, or large sunken containers.
- A number of grasses and grass-like plants have become invasive and a threat to the biodiversity of native habitats (*see pp.52–3*). Become aware of the species that are problematic in your area and do not plant them.

CAREX ELATA 'AUREA'

GRASSES AND BAMBOOS

Red List: Extinct*

Bromus interruptus

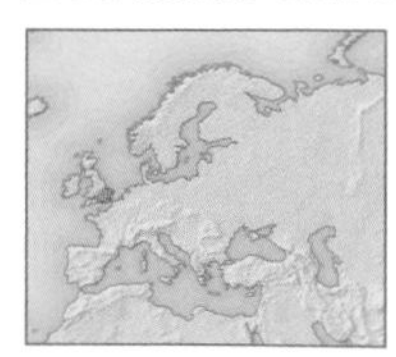

Botanical family
Gramineae

Distribution
England: southern counties

Hardiness
Fully hardy

Preferred conditions
Open sites, with full sunlight; favours disturbed soil, with average moisture.

Rare species often cling perilously to a particular, rather specialized habitat, and sometimes disappear mysteriously even from apparently suitable sites. This is what seems to have happened to the interrupted brome, ***Bromus interruptus***, an attractive grass endemic to southern England, now assumed to be extinct in the wild. But like many grasses, this species is easily overlooked, and it is possible that it may be rediscovered, especially now that more farmers are leaving wild borders around arable fields. If such marginal lands are not sprayed with herbicides, a wide range of wildlife may flourish.

Interrupted brome was known from several sites in southern England, and was last recorded by a farm track in Cambridgeshire in 1972. Fortunately it still survives in cultivation, and Kew's Millennium Seed Bank has stored seeds, so re-introduction remains a possibility.

BROMUS INTERRUPTUS ► *The drooping panicles of this rare grass show the rather loosely clustered spikelets, which give the grass its common name.*

Spikelets are arranged in a rather narrow panicle

HABITAT

PRODUCTIVITY VICTIM

The favoured habitat of interrupted brome was rough field margins on arable land. In crops of sainfoin, clover, or other grasses, it grew as a weed. These sites are often colonized by bright flowers such as poppies (*Papaver*) and ox-eye daisies (*Leucanthemum vulgare*), and are also important habitats for a range of useful insects, including butterflies and pollinating bees. Such habitats are rarer than they were several decades ago, due to improvements in the efficiency and productivity of farming practices.

Red List: Rare*

Calamagrostis foliosa

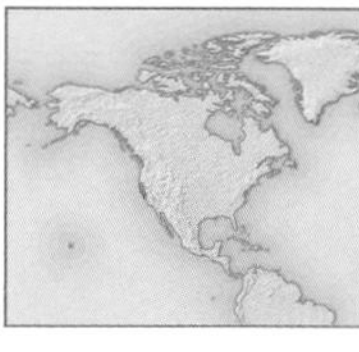

Botanical family
Gramineae

Distribution
California, USA: Del Norte, Humboldt, and Mendocino Counties

Hardiness
Fully hardy – half-hardy

Preferred conditions
Light shade and well-drained soil that retains moderate moisture, in areas such as rocky coastal headlands, riverbanks, and steep slopes.

The Mendocino or leafy reedgrass (*Calamagrostis foliosa*) is a rare small-reed species found in the wild only in California, where it grows on rocky bluffs near the coast, and alongside some rivers. The majority of the remaining wild populations are in the King Range National Conservation Area. Bordered by the Mattole River and Whale Gulch, about 370km (230 miles) north of San Francisco, the King Range National Conservation Area preserves a dramatic 56km (35-mile) section of California's coast. Here the mountain range rises abruptly from close to the sea, and tops out at 1,245m (4,087ft) at King Peak, only 5km (3 miles) inland. This area is known as the "Lost Coast" as the mountains make it rather difficult to reach, so it has escaped much of the development seen elsewhere.

ISOLATED BEAUTY

The landscape here is spectacular, with the rugged coast and nearby mountains clothed in places by old forests of Douglas fir (*Pseudotsuga menziesii*). While the eastern slopes are rather steep, the coastal side is even more precipitous in places. On this western side, there are many very steep cliffs and unstable scree slopes, and many streams draining directly into the sea along rocky gullies. The unstable geology has produced many soil slips and rockslides, that are interspersed with narrow beaches.

In the south, the main vegetation is the old-growth Douglas fir forest and coastal chaparral, while in the north the landscape is dominated by grassland. The mix of habitats found in the King Range National Conservation Area, including chaparral, coastal prairie, forest, shoreline, estuaries, and rocky gullies, attracts many species of birds.

The whole area has a mild climate, but also experiences very high rainfall – the highest rainfall in California – with some sites receiving more than 5,000mm (200in) each year, mainly between October and April. It is also subject to thick sea fogs, especially in the mornings. By contrast, summers are hot and dry, and these dry summers make the overall climate too dry to support forests of coastal redwoods (*Sequoia sempervirens*).

IN THE GARDEN

Fortunately, although under threat in the wild, Mendocino reedgrass is relatively easy to grow and propagate, either from seed or by division of the plants, and it has gained some popularity as a garden species. It is semi-evergreen, with blue-green leaves and attractive pale, arching panicles of flowers.

CULTIVATION

GARDEN GRASSES

Calamagrostis species are commonly known as small-reeds or reedgrasses, and grow by lake or river margins. Several make pretty garden plants and most, like *Calamagrostis foliosa*, are easy to cultivate. Reedgrasses are well named as they bear a resemblance to true reeds, being mostly rather tall, and with rather compact, softly hairy inflorescences, often tinged purplish or pink. Like reeds, they also tend to grow in damp or wet conditions.

The feather reed grass (*Calamagrostis acutiflora*) is a beautiful plant, especially the cultivar 'Karl Foerster'. This has yellowish-green foliage and sends up erect, golden plumes in early summer. The fall-blooming reed (*Calamagrostis brachytricha*) is native to eastern Asia. It is a useful herbaceous perennial, growing to 90cm (3ft), and preferring sunny or partly shaded sites. Its pink flowerheads turn a bronze colour with age. Like many of the reedgrasses, it is also often used in dried flower arrangements.

126*/1 on Red List

Carex

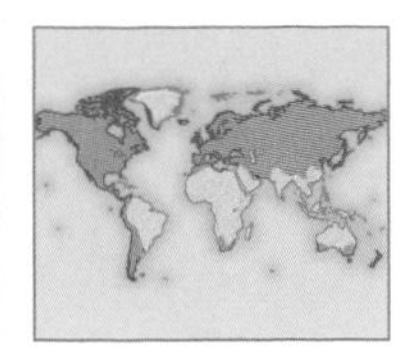

Botanical family
Cyperaceae

Distribution
Worldwide, especially temperate and Arctic regions

Hardiness
Fully hardy (most species)

Preferred conditions
Temperate climates; sun or partial shade; wet or damp soil; some species can withstand drought.

In the wild, sedges grow mainly in wet and marshy habitats, and reach their greatest diversity in temperate and Arctic areas. There are some 1,000 *Carex* species worldwide, and these are among the most difficult plants to identify, as there are many, often superficially similar species. Precise naming requires inspection of detailed features, notably the ripe fruit.

Sedges are increasingly popular as garden plants, for example near water features, and although most are happiest in wet conditions, several can withstand drought and are suitable for drier sites, such as rock gardens. The following rare sedges are among those that have revealed garden potential.

White sedge (*Carex albida*), listed as Endangered in the Red List of 1997, persists in a small area of Sonoma County, California, USA. There are fewer than 1,000 of this evergreen, which is at risk from projects to treat waste water.

Baltzell's sedge (*Carex baltzellii*), listed as Vulnerable in the Red List of 1997, is another rare American sedge; it is found in Alabama, Florida, Georgia, and Mississippi. Its greatest threat is from the clearance of much of its forest habitat. This fairly recently introduced garden plant has very attractive glaucous foliage.

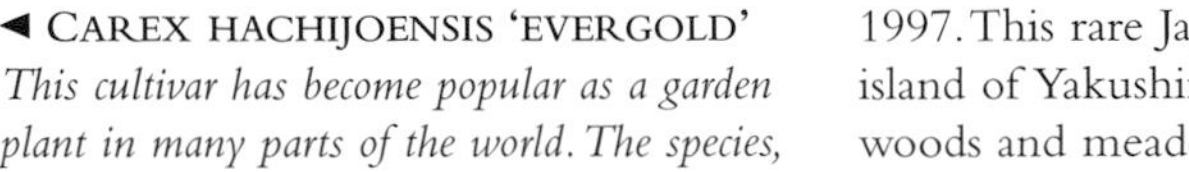

◀ CAREX HACHIJOENSIS 'EVERGOLD' *This cultivar has become popular as a garden plant in many parts of the world. The species, however, is endemic to just a single small island in Japan – Hachijo Island.*

Carex inopinata (Endangered in the Red List of 1997) grows only in a conservation area near Christchurch, New Zealand, on limestone outcrops. It has narrow, bright green leaves, grows to only about 3cm (1¼in), and is good in a rock garden.

Carex hachijoensis (Rare in the Red List of 1997) is from Japan and is a handsome semi-evergreen. The cultivar 'Evergold' produces pretty, yellow-and-green variegated foliage and forms a compact shape, making it very suitable as a border plant.

Morrow's sedge (*Carex morrowii* var. *laxa*) is listed as Vulnerable in the Red List of 1997. This rare Japanese sedge, from the island of Yakushima, is found in wet woods and meadows on the lower mountain slopes. Variegated garden varieties are available.

Carex oshimensis (Rare in the Red List of 1997), again from Japan, is a rare sedge native to dry woods and rocky sites, on Oshima and some other islands. It is a hardy species, and grows well under cultivation.

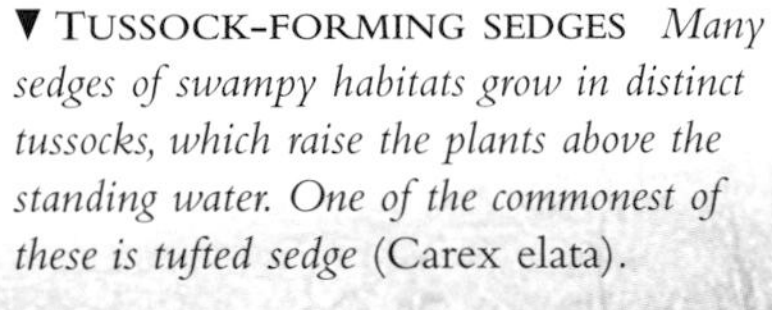

▼ TUSSOCK-FORMING SEDGES *Many sedges of swampy habitats grow in distinct tussocks, which raise the plants above the standing water. One of the commonest of these is tufted sedge* (Carex elata).

6 on Red List*

Chusquea

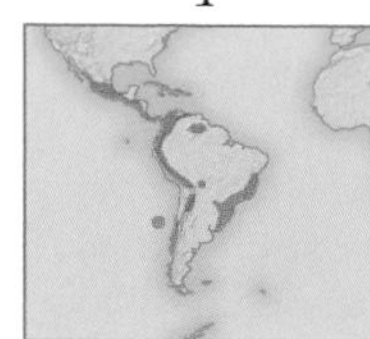

Botanical family
Gramineae

Distribution
Central and South America

Hardiness
Half-hardy – fully hardy

Preferred conditions
Moist, fertile soil; light shade to full sun in an extremely wide range of habitats.

Latin America boasts 20 genera and around 430 species of woody bamboo. One of the most important genera is *Chusquea*, with well over 150 named species, and many more as yet unnamed. Although some chusqueas grow in lowland forest sites, these attractive bamboos are found mainly in montane cloud forests, from Mexico and Cuba in the north, south to Venezuela, Brazil, Argentina, and Chile.

Bamboos typically have hollow stems, but in most species of *Chusquea* the jointed stems, or culms, are solid. Some are clump-forming, while others have a vine-like habit and clamber up trees.

Six species of *Chusquea* are on the Red List of 1997, threatened mainly by loss of habitat, due to logging, and by livestock grazing.

THREATENED SPECIES

Chusquea aperta (Vulnerable in the Red List of 1997) is a rare bamboo that grows in just a few localities in Oaxaca and Veracruz states in Mexico, where it prefers altitudes of 1,750–2,750m (5,775–9,075ft). It has arching, jointed stems, or culms, giving it a rather graceful appearance.

Chusquea bilimekii (Vulnerable in the Red List of 1997) is another Mexican species, found growing at elevations of

▲ CHUSQUEA POHLII *This cloud-forest bamboo from Costa Rica and Panama has arching, jointed stems, or culms, reaching to about 15m (50ft) in length, from which the main branches grow out at right angles.*

2,200-3,000m (7,260–9,900ft). This bamboo is now known in the wild in just three localities. One of these sites is near the Cofre de Perote National Park, and comprises about 100 plants. Local loggers are felling the conifers within the habitat, and livestock may have access.

Chusquea fernandeziana (Vulnerable in the Red List of 1997) is a little-known species described in 1873 and found on the Juan Fernandez Islands, about 700km (435 miles) off the coast of Chile.

Chusquea latifolia (Vulnerable in the Red List of 1997) clambers along the ground for as much as 20m (70ft) and its twining, jointed stems, or culms, may be as long as 40m (130ft). This bamboo from Columbia is unusual in having only two branches on either side of the main branch buds. (Most chusqueas have many branches – in some cases as many as 80 or more.) Large-leaved *Chusquea latifolia* has a network of thin, light, and strong aerial canes that overgrow the host trees and shrubs, and this habit allows the foliage to be distributed over a wide area. The species inhabits mountain cloud forest at altitudes of 1,600–2,950m (5,280–9,735ft).

HABITAT DIFFERENCES

Chusquea longiligulata (Vulnerable in the Red List of 1997) grows in certain montane forests in Costa Rica. It is found in the Talamanca Range between 1,500m (4,950ft) and 2,000m (6,600ft).

As the species name implies, this *Chusquea* has a long elongation at the top of the leaf sheath, measuring 7cm (2¾in). The bamboo reaches a height of 10m (30ft), the leaves are ascending and stiff, and the branches are centred on a usually dormant bud.

Chusquea pohlii (Vulnerable in the Red List of 1997) is from Costa Rica and Panama, where the species inhabits the typical cloud-forest zone at 1,500–2,600m (4,950–8,580ft). An interesting feature of the branch bud is its circular central bud, unlike the triangular shape more common to the genus. The main branch grows at an angle of about 90 degrees from the jointed stem, or culm, and the side branches grow away from the main branch – a branching habit typical of the clambering chusqueas.

CHUSQUEA CULEOU ► *This clump-forming bamboo is native to Chile and Argentina. It grows well in temperate climates and enjoys full sunshine. The jointed stems, or culms, reach a maximum height of around 7m (20ft) and grow close together in a rather dense cluster.*

HABITAT

CORDILLERA

The cordilleras of Central America, and those of the spine-like Andean range of South America, offer a huge variety of microclimates and habitats. Such mountains are the stronghold of the genus *Chusquea*. Here, at altitudes of up to about 3,000m (9,900ft), these bamboos flourish amid the clouds and cool mists, which clothe the slopes, especially during the afternoons. The high altitude and steep slopes have saved some of these wonderful forests from degradation through clearance and grazing, and the remaining cloud forests are some of the most fascinating of all the habitats in Latin America. Trees including oaks (*Quercus*) and conifers often dominate the forests, with bamboos usually growing under their shelter. Most of these bamboos flower profusely in 20–40 years, after which the parent plants die.

Red List: Endangered*

Cortaderia turbaria

Botanical family
Gramineae

Distribution
New Zealand: Chatham Island and Pitt Island

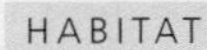

Hardiness
Fully hardy

Preferred conditions
Open sites on limestone covered by sand, in full sunlight; wet or moist soil.

This very large grass produces cream-coloured, rather silky, drooping flowerheads in late spring and summer. Its leaves are long and curved, with sharp margins and fine, silky hairs. In the wild, it is known only from Chatham Island and nearby Pitt Island, about 800km (500 miles) east of New Zealand. Of all the islands in the country, Chatham has one of the most endemic floras.

Attempts are being made to protect the remaining small populations of this grass, and off-site cultivation has proved successful. One conservation problem it faces is hybridization with an introduced close relative, the New Zealand toetoe, *Cortaderia toetoe*.

There are about 25 species of *Cortaderia*, mostly South American, the best known of which is pampas grass (*Cortaderia selloana*). *Cortaderia richardii* is also a popular garden plant. It produces a large mound of arching leaves and tall flower plumes, to 3m (10ft).

See also Invasive plants, *p.450*.

HABITAT

LAKE HURO

The preferred habitats of *Cortaderia turbaria* are the edges of Chatham Island's lakes, lagoons, banks of streams, and similar wet sites. Lake Huro, one of many lakes on Chatham Island, has suitable conditions, with grass- and sedge-rich communities, but its margins have been badly affected by grazing sheep and pigs and other disturbances.

Populations of *Cortaderia turbaria* are now widely dispersed and fragmented, which has made them more vulnerable to sudden changes in ecology, and to competition by introduced species. A fungal wilt disease has also affected this rare grass.

6 on Red List*

Cymbopogon

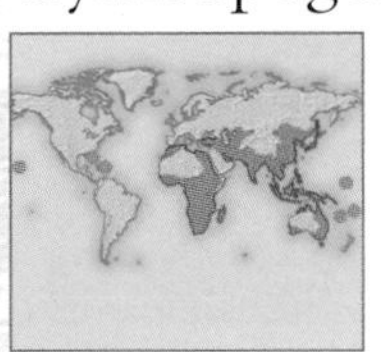

Botanical family
Gramineae

Distribution
South-east Asia, India, Sri Lanka

Hardiness
Frost-tender

Preferred conditions
Full sun, or partial shade, well-drained soil; prefers a hot, humid climate.

The 60 or so species of *Cymbopogon* are native to the Old World tropics, mainly India, Sri Lanka, and south-east Asia. Many of these species yield aromatic oils, and are used traditionally in cooking, for making scents, and also in medicine.

The true lemon grass is ***Cymbopogon citratus*** (Unlisted) from India and Sri Lanka, which is now widely cultivated elsewhere, including in Florida, USA. Lemon grass oil is used in a wide range of products, including air fresheners, perfumes, hair oils, and soaps. In Thai food, the leaves are chopped and cooked with many dishes, to which they add a delicate citrus flavour. The Unlisted *Cymbopogon flexuosus,* from India and south-east Asia, yields Malabar oil. *Cymbopogon microstachys* (East Indian lemon grass), listed as Rare in the Red List of 1997, is found in mixed forests. The species is used to flavour food and tea, and the oil, which is a barctericide, is used in medicine and to repel insects.

Citronella, another famous insect repellent oil, is extracted from *Cymbopogon nardus* (Unlisted).

In warm-temperate regions, lemon grass should be kept in a frost-free pot during winter.

CYMBOPOGON CITRATUS ► *Lemon grass is commonly used to add a distinctive flavour to food in south and south-east Asia, notably Thailand.*

Unlisted

Cyperus papyrus

Botanical family
Cyperaceae

Distribution
Madagascar; tropical Africa

Hardiness
Frost-tender – half-hardy

Preferred conditions
Shallow water, or saturated, fertile soil; grows best at the margins of lakes, ponds, and rivers; full sun or partial shade.

Papyrus is one of the most beautiful of all waterside plants. It grows in dense colonies in shallow water alongside marshes and rivers, notably in northern Africa, particularly in the Sudan, and most famously in Egypt, although it has disappeared from many of its historical sites there. Where the habitat is suitable, stands of papyrus fringe the water, and colonies of the plant also grow up as small islands, surrounded by open water. Papyrus is now quite widespread in the river basins of tropical Africa, such as the Nile, Congo, Niger, and Zambezi, and it is also found in Madagascar. Papyrus also has been introduced widely in warm areas, and is now common in the Mediterranean region and also in south-west Asia.

It is widely cultivated in frost-free regions and is a popular garden plant, gracing many a large ornamental pond or lake, to which its considerable size is best suited. The tall stems grow from a network of woody rhizomes, and usually reach a height of around 3m (10ft), creating clumps, rather like a loose reedbed. But it is the flower clusters that make papyrus especially attractive. The branched flowerheads (umbels) have a soft, rather feathery appearance, and are up to 30cm (12in) across. To thrive, papyrus needs to grow in standing water or in permanently wet soil; in ideal conditions it spreads quickly.

HISTORY

Egyptian motifs

Papyrus features in many ancient Egyptian illustrations, such as the one shown left, where it can be seen with lotus, among other plants and symbols. Lotus was regarded as a sacred flower, and was the symbol of Upper Egypt, while papyrus was the symbol of Lower Egypt. The species of papyrus illustrated here is *Cyperus papyrus* subsp. *hadidii*, and the white lotus is *Nymphaea lotus* var. *aegyptiaca*. Although three species of lotus have been identified in Egyptian art, blue lotus (*Nymphaea caerulea*) is depicted most frequently. The third species of lotus found in Egyptian illustrations, pink lotus (*Nelumbo nucifera*), was introduced to Egypt from Persia.

Egyptian papyrus

The papyrus of Egyptian fame was a particular subspecies, *Cyperus papyrus* subsp. *hadidii*, which grew in the upper Nile. Its natural habitats were backwater marshes, and also seasonally inundated basins on alluvial soils. However, these were exactly the places prioritized for draining and conversion into fertile agricultural land, and planting with crops.

It is clear from travellers' reports from the Middle Ages onwards that true Egyptian papyrus was steadily and rapidly receding, due mainly to drainage and irrigation schemes, as well as the switch to other papermaking processes. *Cyperus papyrus* subsp. *hadidii* seems to have vanished by about the 1820s, although other forms were introduced, for example, to Cairo and Alexandria. In 1968, *Cyperus papyrus* subsp. *hadidii* was eventually found again – there was a small population growing near Umm Risha Lake, in a freshwater marsh, apparently the sole surviving native stand. It is now listed as Endangered in the Red List of 1997.

Papyrus in papermaking

The Egyptian tradition of papermaking from papyrus has a long history, going back some 5,000 years. Indeed the word "paper" itself derives from "papyrus".

In the 10th century, Arabs introduced their new papermaking technology, based on rags and wood pulp, and this method gradually replaced the use of papyrus in Egypt.

Recently, papyrus has been re-introduced and cultivated on the Nile, on an island between Cairo and Giza. Here, too, traditional papyrus papermaking is undertaken, following

PLANTS AND PEOPLE

Making paper

Papyrus stems are harvested by hand, using a sickle or similar tool. The outer green layers are removed from the stems, and the pith inside is sliced into strips. These strips are then hammered and crushed to break the fibres, and soaked in water for about three days. This process makes the fibres transparent and more flexible. The resulting spongy mass is removed from the water, rolled out flat, and left to dry.

Finally, the dried, fibrous material is cut into strips. These are spread out in alternating horizontal and vertical overlapping layers before being pressed and dried further to produce paper. Variations in wetting and drying times produce different shades in the final paper.

as far as possible the methods used by ancient Egyptians. Such displays mainly serve as tourist attractions.

The ancient Egyptians utilized papyrus in many other ways besides papermaking. They used the fibres for weaving mats, making fences, baskets, and rope, and even sandals. The rhizomes were also eaten as food and included in medicines. Boats and rafts were constructed by binding bundles of stems tightly together, taking advantage of the buoyant nature of the papyrus stems.

▼ CYPERUS PAPYRUS *The tall stems of papyrus grow up to about 4m (12ft) tall, and are triangular in cross-section. The stem shape makes them resistant to breaking when forced to bend by the wind.*

Red List: Endangered*

Deschampsia cespitosa subsp. alpina

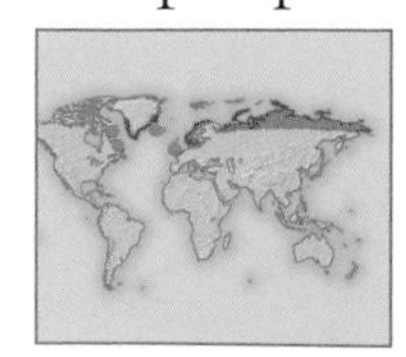

Botanical family
Gramineae

Distribution
Canada, Greenland, Iceland, Britain, Norway, Sweden, Russia

Hardiness
Fully hardy

Preferred conditions
Moist but well-drained soil of any type in light to full shade; prefers rocky sites.

Alpine (tufted) hairgrass (*Deschampsia cespitosa* subsp. *alpina*) is a plant of alpine and mountain meadows – its most typical habitat being bare, stony sites with shallow soil. Although it has quite a wide distribution on three continents, it is nowhere common and is regarded as Endangered over much of its range.

This grass is adapted to withstand the rigours of a harsh climate, notably by its low-growing, clump-forming habit, but also by its ability to produce sexual and viviparous flowers. Instead of eventually forming seeds, the flowerheads may develop vegetatively into tiny plantlets, which drop off, blow away, and become independent plants. This is a good reproductive strategy in exposed sites, where pollination may be unreliable and seeds may find it hard to germinate.

Red List: Extinct

Dissanthelium californicum

Botanical family
Gramineae

Distribution
San Clemente, Santa Catalina, and Guadalupe Islands

Hardiness
Half-hardy

Preferred conditions
Unknown, but thought to have grown in a dry, sunny, open habitat, on poor, moist soil.

The last known sites for this rare grass were three islands off the coast of southern California and Mexico, where it grew in coastal sage-scrub communities, below about 500m (1,640ft). It was last reliably recorded in its California sites in 1912, and is now widely held to be extinct in the wild. *Dissanthelium californicum* is considered also to have been wiped out completely in the Mexican site.

The prime cause of extinction was grazing pressure from introduced, feral goats. Grasses are particularly vulnerable to grazing, and this species soon disappeared. This grass is reported to be available from a few sources, so there is the chance that in the future it could be reintroduced to suitable sites.

54 on Red List*

Festuca

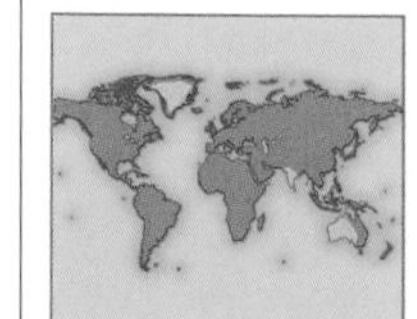

Botanical family
Gramineae

Distribution
Worldwide

Hardiness
Fully hardy (most species)

Preferred conditions
Open sites, fertile, acid to neutral soils (most species); native to cool, humid areas; tolerant of drought conditions.

The genus *Festuca* (fescue) contains about 300 species of annual or perennial grasses, about one-sixth of which are on the Red List of 1997. Fescues are found worldwide, especially in cool and temperate climates, in a range of habitats from high mountain and alpine sites to lowland meadows, steppes, and savannas. Some, such as the common species red fescue, ***Festuca rubra*** (Unlisted), tolerate trampling and are widely used in turf for sports fields and in parks and gardens. Those planted in lawns tend to be the low-growing, fine-leaved species naturally inhabiting upland areas, such as sheep's fescue, *Festuca ovina* (Unlisted) and red fescue. From these and other species, many cultivated varieties have been bred, creating forms well suited to a variety of modern lawns and gardens.

Under different conditions, such as in fertile pastures, fescues grow taller, and many species, notably tall fescue, *Festuca arundinacea* (Unlisted), are sown as a forage crop for cattle and horses. This fescue has a deep root system, penetrating the soil to a depth of 2m (6ft), and therefore remains green for a long time, even during dry summer months. It also has a long growing

◄ FESTUCA RUBRA *Red fescue is one of the most common* Festuca *species, with a wide distribution in many continents. It is also frequently introduced, and finds a use in gardens, often as a constituent of lawns. It thrives under mowing, and tolerates a wide range of conditions.*

FESTUCA COSTATA *The spiky tussocks of this unlisted species are a distinctive feature of this hillside habitat, in the Drakensberg Mountains of KwaZulu Natal, South Africa. With the fescue are the reddish tufts of another grass,* Themeda triandra.

season, from early spring through to late autumn. It is very tolerant of trampling, and is therefore well suited to grazing by horses. Tall fescue is a very important cool-season grass in the USA.

Fescues with shallower roots can survive short periods of drought by ceasing growth and becoming brown and semi-dormant, until rain or irrigation returns. Regrowth from the often extensive rhizomes is rapid, and fescues quickly grow back when suitable conditions prevail.

California fescue, *Festuca californica* (Unlisted), is an elegant grass often planted in gardens in North America, as is blue bunchgrass, or Idaho fescue, *Festuca idahoensis* (Unlisted).

Of particular value in the garden are blue fescues, such as *Festuca amethystina* (Unlisted) and *Festuca ovina* var. *glauca* (Unlisted). Like many other fescues, they develop attractive, semi-evergreen tussocks, and their colouring looks good in rock gardens and borders. Blue fescues are often planted as ground cover, or as edging for beds. Many different cultivars are now available, with foliage colours ranging from blue-grey and bright, electric blue to golden yellow. They are mostly relatively drought-tolerant, and thrive in well-drained rock gardens.

One rare fescue is frequently cultivated in gardens, mainly for its unusual growth habit. This is the hedgehog fescue or porcupine grass (*Festuca punctoria*), listed as Rare in the Red List of 1997. Endemic to Turkey, this species grows above the treeline in the Uludağ mountains. As its name suggests, it has a compact, cushion shape and is also rather spiky. It thrives best on well-drained, poor soil and likes full sun.

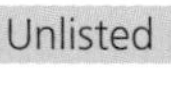

Hierochloe odorata

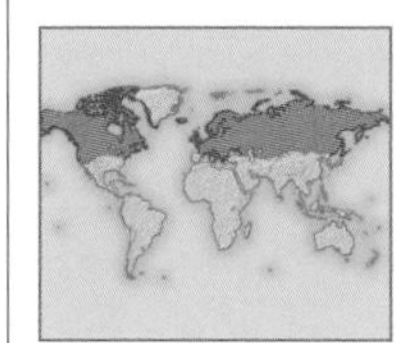

Botanical family
Gramineae

Distribution
Northern North America, northern Europe, northern Asia

Hardiness
Fully hardy

Preferred conditions
Rich damp or wet soils, such as those on the edges of rivers and lakes; cool climate.

Holy grass (***Hierochloe odorata***) is a fascinating species, with a long history of use and tradition, partly bound up with myth and legend. Its distribution is also interesting, for it is a "circumpolar" plant, that is, one that is found in northern North America and northern Europe and Asia, with its distribution centred on the North Pole. It is possible that holy grass was spread by early native inhabitants of North America as they crossed between Alaska and north-east Asia. Alternatively, this unusual distribution may be quite natural, dating back to a period before the loss of the land connection between the continents.

Holy grass has been associated with people for many thousands of years, to the extent that some botanists doubt whether there are many truly wild populations left. It is a hardy grass, and a perennial, but what makes it particularly attractive is its delicate, aromatic smell, and it is this quality that almost certainly explains its close association with people, and also its religious significance.

MULTIPURPOSE PLANT

In Europe and in North America, holy grass has long been regarded as sacred, its dried leaves used as a form of incense, and also woven into braids and baskets. In northern Europe, this grass was strewn in front of church doors on saints' days. Today, it is also used as a flavouring and as a tea for treating sore throat and coughs. Many native American tribes used holy grass for ritual purification, as a perfume, and also in their traditional medicines. It was thought to offer spiritual protection. As a medicine it has been used as a hair wash, a remedy for colds, and also to relieve pain, congestion, and fever.

The vanilla-like scent of holy grass evokes the aroma of newly mown hay. Its main constituent is the chemical coumarin, which is present in varying amounts in other grasses, such as sweet vernal grass (*Anthoxanthum odoratum*), but holy grass has a particularly high content, making it one of the most aromatic of all grasses. In fact, coumarin has been found to have anti-coagulant properties, rather like aspirin, although in large doses it can cause liver damage.

The preferred habitat of holy grass is rich, damp soil in cool climates, typical sites being riversides, lake margins, and wet, peaty grassland. It reproduces mainly vegetatively, from creeping rhizomes, and is best cultivated in moist soil.

▲ HIEROCHLOE ODORATA *This species is relatively common in northern Scandinavia, but is rather rare in other parts of Europe. However, in 1946 it was discovered by two botanists, Desmond Meikle and Norman Carrothers, near Lough Neagh in Northern Ireland. It also grows in Scotland, where it is confined to four counties.*

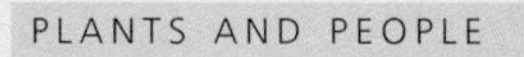

HOLY GRASS AND BISON

Like many grasses, holy grass can withstand a certain amount of grazing, and in one of Europe's finest and least altered forests, the Białowieza Forest in Poland, it forms part of the diet of one of Europe's largest and rarest mammals, the bison. This connection has earned holy grass its Polish name Zubrovka, from "zubr", Polish for "bison". In Białowieza, holy grass grows in small patches in damp forest glades and grassy clearings, some of which are favoured by the bison as feeding sites. Holy grass is also used as a flavouring in some types of Polish vodka. The spirit is distilled from rye grain, diluted with water, and filtered, and the process also involves infusion with extract of holy ("bison") grass.

Red List: Extinct

Hubbardia heptaneuron

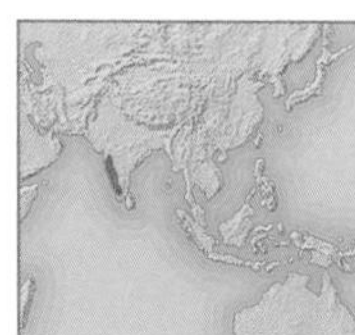

Botanical family
Gramineae

Distribution
India: Maharashtra

Hardiness
Frost-tender

Preferred conditions
Damp, rocky soil; shade; thrives under constant spraying water.

This grass is fascinating for a number of reasons, and not least because it disappeared for a while. Unlike most grasses, it is restricted to a very particular microclimate in its specialized habitat of rocks moistened by the spray of waterfalls. It was first collected in 1919 and described as a new species in 1951, but was only collected twice and so remained something of a mystery. The collection site was near the Gersoppa Falls on the Sharavati River on the border of India's Maharashtra and Karnataka states (then known respectively as Bombay and Mysore). The water supply through the falls was reduced after the construction of a dam, and the grass disappeared, presumed extinct, until its exciting recent rediscovery in a new locality – Tillari ghat in Kolhapur district. It grows at three sites, again on wet rocks.

HABITAT

GERSOPPA FALLS

Gersoppa Falls, also knows as Jog Falls, are some of the finest waterfalls in India. Here the Sharavati River plunges 253m (830ft) into a chasm, splitting as it does so into four separate falls: Raja, Ranee, Roarer, and Rocket. Rocks and vegetation around the falls are shrouded in a highly humid atmosphere of permanent mist and spray. The effect of the waterfall is all the more stunning as it is set amid dense forests dominated by teak, although the water flow has been reduced somewhat by damming the river in recent years. Visitors to the falls can walk down a path into the bottom of the gorge to admire the pools below.

Unlisted

Leymus condensatus

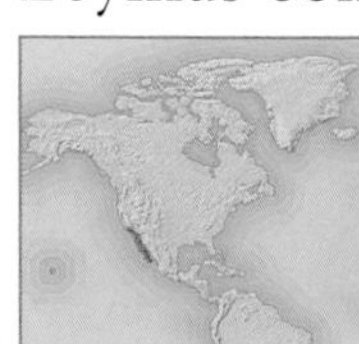

Botanical family
Gramineae

Distribution
USA: California and the Channel Islands; Mexico: Baja California

Hardiness
Half-hardy

Preferred conditions
Full sun or shade; dry or moist soil: withstands drought, but also appreciates regular rainfall.

Giant wild rye (***Leymus condensatus***) is a striking, robust grass found wild in California, USA, and also in Mexico (Baja California), where its typical habitat is chaparral and open woodland, as well as sand dunes, coastal bluffs, and rocky slopes. These are mainly open, sunny sites, and the grass is fairly tolerant of drought and prefers a sandy soil. In some areas it is threatened by grazing, as on the Californian Channel Islands, where goats are a problem.

This ryegrass has found favour with gardeners for a number of reasons, and it is now one of California's most popular native grass species. It is easy to grow and propagate, and tolerates a wide range of conditions. It prefers moist soil, but can also survive periods of drought. Although it spreads well under favourable conditions, it is not as invasive as some other members of this genus, and can be easily controlled. Giant wild rye also is suitable for use as a soil-stabilizer in halting erosion.

This majestic, evergreen grass reaches a height of up to 3m (10ft) when fully grown, and bears tight, rather rigid flowerheads. A popular, grey-blue garden form is the cultivar 'Canyon Prince', which was selected from specimens from Prince Island near San Miguel Island in the Santa Barbara Channel.

LEYMUS CONDENSATUS ►
Open sunny sites, such as California coastal chaparral, are favoured by giant wild rye. This sturdy grass sends up long stalks in the summer, topped by pointed, tight clusters of flowers.

11 on Red List*

Muhlenbergia

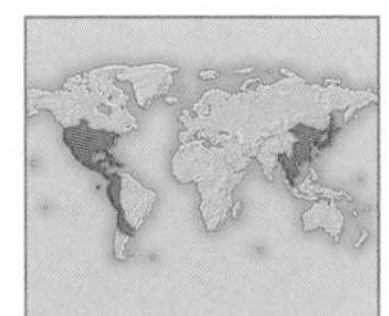

Botanical family
Gramineae

Distribution
USA, Mexico, Central and South America, eastern Asia

Hardiness
Half-hardy

Preferred conditions
Full sun, any soil; tolerant of drought but needs water during the summer months.

Muhly grasses are found mainly in North, Central, and South America, with a few in central and eastern Asia; in all there are around 200 species. They are most concentrated in the south-western USA and in Mexico, where most species are adapted to grow in dry, open habitats such as deserts and semi-deserts, or in open woodland and chaparral. Some spread as pioneers into recently disturbed land.

INFLUENTIAL BOTANIST

This unusual genus is named after a famous German-American, G.H.E. Muhlenberg (1753–1815), who was a Lutheran minister in Pennsylvania. Keen and entirely self-taught, he came to be regarded as the American Linnaeus and was responsible for the names of about 150 plants.

Of the ten muhly grass species listed as being of conservation concern, all but one are found in the south-western United States or in Mexico.

Box Canyon muhly, south-western muhly, or weeping muhly (*Muhlenbergia dubioides*) is endemic to Arizona, where it is found in Pima, Santa Cruz, and Cochise counties. Its status is Vulnerable (Red List of 1997), and it is known only from six sites on riverbanks or dry, semi-desert grassland.

WILDLIFE

VOLCANO RABBIT

The life of one of the world's rarest mammals is intricately connected with a species of muhly grass. The volcano rabbit (*Romerolagus diazi*) is known from only four volcanoes in Mexico – Iztacohuat, Pelado, Popocatepetl, and Tlaloc. One of the smallest and most specialized of all rabbits, it lives in open pine forests at an altitude of 2,800–4,250m (9,250–14,000ft), a habitat locally known as *zacaton*. The ground vegetation here is rich in grasses, including unlisted *Muhlenbergia macroura*, fescues, and other grasses, and it is these that form the main diet of the rabbit.

Sycamore Canyon muhly (*Muhlenbergia xerophila*), which is listed as Rare in the Red List of 1997, is also known only from six locations, all in Arizona.

Bamboo muhly (*Muhlenbergia dumosa*) is Unlisted, and grows in Arizona, and also in Mexico. Its common name refers to its bamboo-like growth, with stems reaching 0.9–2m (3-6ft), and bright green leaves.

Another popular garden muhly grass is the unlisted pink muhly (*Muhlenbergia capillaris*), which grows to about 90cm (3ft) tall. It bears pink flower clusters set against dark green leaves.

Blue muhly grass (*Muhlenbergia lindheimeri*) has attractive, blue-grey foliage and is suitable for poor, sandy soils. It is one of the largest species, and does not appear in the Red List.

◄ **MUHLENBERGIA TORREYI** *Ring muhly (Unlisted) grows in the White Sands National Monument, New Mexico. This challenging habitat is dominated by wave-like dunes of white gypsum sand, and lies in the north of the Chihuahuan Desert.*

2 on Red List*

Olmeca

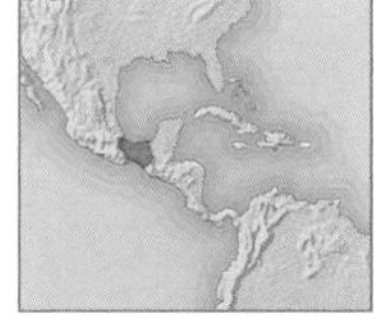

Botanical family
Gramineae

Distribution
Mexico

Hardiness
Half-hardy

Preferred conditions
Moist soils; requires a large area for growth of spreading rhizomes.

There are only two species in this Mexican bamboo genus, which was discovered in 1952, and described in 1981. The species names reflect one of the main differences between the two species: *Olmeca recta* has an erect leaf blade, while *Olmeca reflexa* has a blade that reflexes downward. Both are listed as Indeterminate in the Red List of 1997. *Olmeca recta* grows on acidic volcanic soil in the Tuxtlas Mountains; *Olmeca reflexa* is more widespread and is found on alkaline soils in the states of Chiapas, Oaxaca, and Veracruz.

This is an unusual bamboo genus in that the plant produces a large, fleshy fruit, with a seed about 2.5cm (1in) in diameter. Such a large fruit is almost unknown among the other New World species, but some Asian bamboos develop them. Also noteworthy is olmeca's habit of growth: the plants put out rhizomes which spread for several metres before they send up vertical, jointed stems, or culms. Branching at the nodes also produces long stems, which help these bamboos to clamber among rocks and forest trees. Lower internodes may store water – yet another unusual feature.

Local people use the culms as poles for building and for decoration. However, efforts are being made to cultivate the bamboo and take the pressure off the stands of wild bamboo.

HISTORY

THE OLMEC CULTURE

Olmeca takes its name from an ancient people who inhabited this region of Central America between about 1,500BC and 300AD. This time saw the beginning of agriculture and human settlements, and a flowering of the arts. The Olmecs represented one of the most important pre-Hispanic cultures, and they created characteristic artifacts, including sculpted faces combining human and animal features, especially those of the jaguar, an animal they worshipped. Most impressive were huge heads carved from basalt, and weighing some 40 tonnes. The Olmecs used a calendar, developed a hieroglyphic script, and greatly influenced later cultures, such as the Maya and Aztecs.

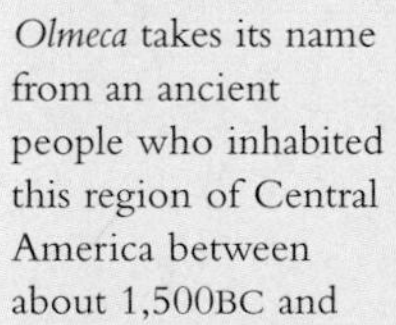

Unlisted

Panicum sonorum

Botanical family
Gramineae

Distribution
USA: Arizona; Mexico; Honduras: El Salvador

Hardiness
Frost-tender

Preferred conditions
Full sun, and dry or moist, well-drained, fertile soils.

Although not on the Red List as a wild species, the ancient cultivated form of this millet, or panic, grass is of conservation concern. The history of its cultivation as an early crop species goes back some 1,500 years, when *Panicum sonorum* was first domesticated in the Sonoran Desert region, in present-day south-west USA, and northern Mexico.

Until recently, this species was thought to be extinct, but has been rediscovered, and its seed collected from farmers in the Sierra Madre. The seeds contain more protein than soybeans, and are rich in the amino acid lysine. *Panicum sonorum* therefore still has great potential as a crop plant, and the US Department of Agriculture has established breeding programmes aimed at increasing the seed size so it can be harvested mechanically. This rare species could play a crucial role in future global crop production.

CULTIVATION

WIDESPREAD CROP

The common name "millet" covers several different grass genera and species, notably common millet (*Panicum miliaceum*), which is widely cultivated in many temperate regions, and little millet (*Panicum sumatrense*), which is grown mainly in India, Nepal, Pakistan, Sri Lanka, eastern Indonesia, and western Myanmar. Other grasses called millet include foxtail millet (*Setaria italica*), which is grown especially in China; teff (*Eragrostis teff*), found mainly in Ethiopia; and finger millet (*Eleusine coracana*), a staple food in eastern Africa, India, and Nepal. To ripen fully, millet needs full sun and warmth, as on this hillside in Nepal.

1 on Red List*

Phyllostachys

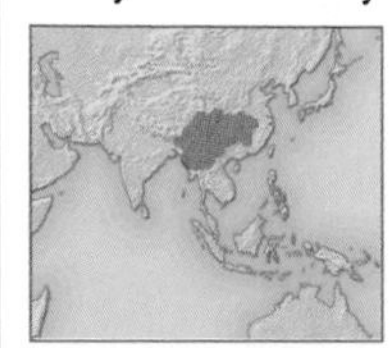

Botanical family
Gramineae

Distribution
China, India, Myanmar

Hardiness
Fully hardy – half-hardy

Preferred conditions
Temperate to subtropical; most species prefer cool to cold winters and hot, rainy summers.

The genus *Phyllostachys* contains about 60 species of bamboo, mostly from China or adjacent parts of eastern Asia, but also from India. Both in natural and in managed stands, they spread rapidly by rhizomes and form thickets.

As with most substantially sized bamboos, the culms (stems) of *Phyllostachys* species are used in India and China for a wide range of purposes, and provide a substitute for wood from trees. In north-east India, these bamboos grow in hedgerows and small plantations, and the leaves are an important source of animal fodder in the winter. In China, bio-product companies are now producing new medicines from bamboo, including some that reduce blood lipids or that contain anti-cancer agents, using the active chemicals in bamboo leaves, flavonoids.

PHYLLOSTACHYS VIRIDI-GLAUCESCENS

Some species, most notably Unlisted ***Phyllostachys pubescens***, are grown and tended as semi-natural "gardened" forests in China, and cropped on a regular rotation to provide a valuable harvest of poles, as well as of edible young shoots.

Phyllostachys viridi-glaucescens (Unlisted) is common but not extensively cultivated across east China, even though it produces reasonably edible shoots. Its stems are also used for small items such as tool handles.

While many bamboos grow tall and strong in their native habitats, some species also grow well in temperate gardens, but usually remain smaller, with weaker jointed stems.

TWO CONFUSED SPECIES

The forests of north-east India and Burma contain many bamboos about which very little is known, and some of these are likely to be threatened as long as the few remaining areas of natural forests continue to dwindle. *Phyllostachys mannii* (Unlisted), for example, is threatened, at least in its native India, although it is considered by some to be the same species as *Phyllostachys decora* (Unlisted), which grows in south-west China (Yunnan Province) and is very hardy and widely cultivated there.

Phyllostachys decora grows from sea level to 2,000m (6,600ft), and is adapted to a wide temperature range of –20°C (–4°F) to 45°C (113°F), in areas with annual rainfall varying from 800mm (32in) to 4,000mm (150in).

Very close to the locality in which *Phyllostachys mannii* was first found in 1889 by Gustav Mann in Meghalaya State, north-east India, lies Cherrapunji, the village with the highest rainfall anywhere in the world, and this bamboo certainly thrives in the rain. It has been known in cultivation for most of the last century.

As with most *Phyllostachys*, it likes cool winters but prefers summer temperatures above 30°C (86°F) and a site in full sun. *Phyllostachys* in the garden requires a good water supply. Also, like most *Phyllostachys* species, it does have rather rampant rhizomes when growing strongly, and may require a physical root barrier in warmer areas. In the cultivated plants, its striped shoots with purple edges and broad leaf blades are quite distinctive.

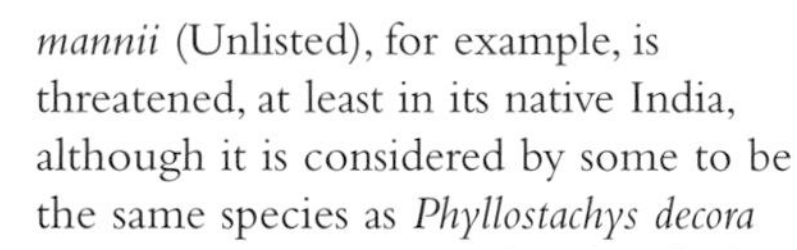

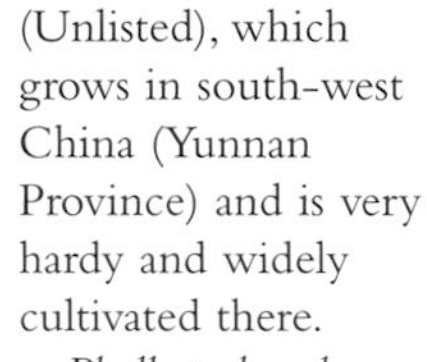

▲ PHYLLOSTACHYS LOGGING *In several regions of China,* Phyllostachys pubescens *is cultivated on hillside sites, as here in Zhejiang Province in the foothills of the Tianmu Shan Range, south-west of Shanghai. The harvested culms are transported by truck for processing.*

PHYLLOSTACHYS PUBESCENS ► *This impressive bamboo sends up straight, hollow, jointed stems, or culms, which reach a height of up to about 30m (100ft) when mature. Forest-like formations dominated by this species can be seen in several sites in China.*

PLANTS AND PEOPLE

INDISPENSABLE BAMBOO

Bamboos are essential to the lives of many people, especially in the tropics and subtropics, and the range of products derived from them is extensive: it includes musical instruments, baskets, bowls, cups, and even combs and toothpicks. Large jointed stems, or culms, are used as sturdy yet light scaffolding and for constructing houses, furniture, and many household items. Split culms are employed as rain gutters, and hollow stems as water pipes. In some parts of rural China, the culms are split and used to make window blinds. Many species bear edible shoots, and bamboo is also included in traditional medicines. Bamboo is used as a fuel, to make charcoal, and in the paper industry.

Unlisted

Pleuropogon hooverianus

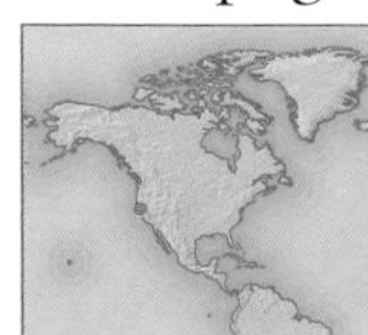

Botanical family
Gramineae

Distribution
California: Marin, Medocino, and Sonoma counties

Hardiness
Half-hardy

Preferred conditions
Full sun or partial shade in moist or wet soils; temporary pools, spring-flooded meadows, and redwood groves.

North coast semaphore grass (*Pleuropogon hooverianus*) is the rarest of five *Pleuropogon* species, four of which occur in western North America, with one-fifth in Asia. It is endemic to California, where it is found in scattered sites along the north and central coastal regions. North coast semaphore grass grows mainly in meadows and freshwater marshes, around temporary pools, and in mixed evergreen forest, its favoured habitat being marshy areas by groves of coastal redwoods (*Sequoia sempervirens*). This rather specialized habitat is under threat from developers. Recently it also seems to have declined in all its known locations; protection is difficult as most of these are on privately owned land.

GARDEN PLANTS

This relatively tall and stately perennial grass produces succulent stems and long, flat, ribbon-like leaves, each with a wide blade. The flowerheads are about 35cm (14in) long, with spaced spikelets drooping from the main flower stalk.

Semaphore grasses make an attractive addition to the garden, and can be used to good effect at the margin of ponds and similar wet sites.

◄ PLEUROPOGON HOOVERIANUS *This grass is rare but not on the Red List. It is from the California coast and is a beautiful plant that reaches 1.5m (5ft) tall and bears graceful, widely spaced flower spikelets.*

Unlisted

Qiongzhuea tumidissinoda

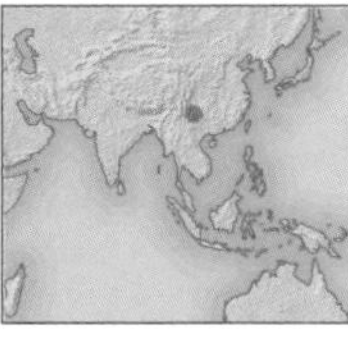

Botanical family
Gramineae

Distribution
China: Yunnan Province

Hardiness
Fully hardy – half-hardy

Preferred conditions
Deep, fertile, preferably acid to neutral soil, in sun or shade, with some shelter; it may need a physical root barrier to control its spreading habit.

This most unusual bamboo, with its prominent nodes, is commonly known as walking stick bamboo, or *qiong zhu*. The latter, and also its generic name, come from the Chinese name, *zhu* meaning "bamboo".

Qiongzhuea tumidissinoda has been known in China at least since the Han Dynasty, some 1,200 years ago and has often appeared in Chinese art. Walking sticks made from this species were exported from China along the Silk Roads to India, Persia, and beyond, yet the plants and their location in China remained a secret, and the species was only described and named scientifically in 1980.

QIONGZHUEA TUMIDISSINODA

Walking stick bamboo grows in south-west China in broadleaved, evergreen forest in the mountains between the north-east edge of the Yunnan Plateau and the Sichuan Basin, from 1,600–2,900m (5,280–9,570ft). It is adapted to withstand a temperature range of –10–29°C (14–84°F), with annual rainfall of 1,100–1,400mm (43–55in) and a soil pH of 4.5 to 5.5.

One threat to the species is overharvesting, especially of its young, pink shoots. These are crisp and delicious, and large quantities are eaten fresh, or they are dried and exported.

Deforestation of much of its natural habitat in Yunnan and Sichuan provinces has also led to the decline of the walking stick bamboo, although protection of the remaining stands in China should ensure its survival. A reserve in Yi Liang County, Yunnan Province, was established in 1984 to protect this and other bamboos.

17 on Red List*

Restio

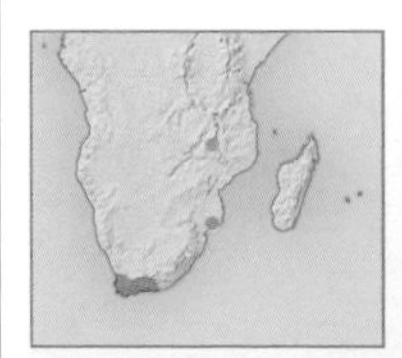

Botanical family
Restionaceae

Distribution
South Africa; mainly Cape region; tropical Africa; Madagascar

Hardiness
Half-hardy

Preferred conditions
Mediterranean climate; survives occasional droughts.

Restio is one of the main genera in the Cape reed family, even though all members of this family are commonly referred to as restios, including *Calopsis* and *Thamnochortus*. There are about 90 *Restio* species, and 40 of these are endemic to the Cape region of South Africa. Some species are also found scattered elsewhere in tropical Africa and in Madagascar. Restios take the ecological place of grasses in the Cape flora, and their tufted, rush- or reed-like habit lends a particular feel to the fynbos community. Sadly, several restios are threatened, as is indeed the whole fynbos habitat itself. Restios have traditionally been used for thatching, and also for making brooms.

Restio dispar is not listed on the Red List. It is found only in the mountains in the south-west Cape region along streams from near sea level to about 1,200m (4,000ft). It is well suited to the garden, but damp soils and frequent frosts damage the plant and stunt growth. Like all Cape reeds, plants are either male or female, and the female flowers of this species are a striking, orange-brown colour.

Green grass reed (*Restio festuciformis*) is listed as Vulnerable in the Red List of 1997, and is well adapted to fire, the seeds germinating much better after exposure to smoke. It grows in tufts, to about 50cm (20in) tall, and produces shiny yellow flowerheads and bright green stems, which turn a golden yellow.

Restio quadratus is a tall, unlisted species, growing to 2m (6ft) and spreading almost as wide. Its stems, which are square in section, are erect and slightly curved, with clusters of sterile branches at the nodes. The young growth produces bright green foliage, with a soft, almost fluffy, appearance. It is best cultivated in full sunlight, and on well-drained soil.

Restio seeds, which are rather large, develop in capsules that split open when ripe. Germination rates can be improved by treating the seeds with smoke, or an "instant smoke" seed primer. This genus has become quite popular for garden plants, which, once established, require little maintenance. They should be watered regularly for at least two months after planting. After about three years, most species will start to die back, and at this stage the dead stems should be removed close to the base of the plant.

The roots of restios are sensitive to frost, especially if the soil is wet, and in regions subject to winter frosts they are probably best grown in containers that can be shifted to a more protected site during the cold weather.

HABITAT

"FINE BUSH"

Restios are prominent in the special habitat known as "fynbos", which is unique to the Cape floral kingdom of South Africa. From the Dutch word meaning "fine bush", fynbos is heath-like, with a shrubby mixture of restios and other plants including proteas, ericas, and pelargoniums. Its plants tend to have tough, leathery leaves or very narrow ones – an adaptation to the frequent drought and occasional fires. Fynbos, which covers a strip from mountain slopes to the coastal plains, is remarkably rich in species, of which about 70 per cent are endemic to this small area.

Red List: Rare*

Sesleria hungarica

Botanical family
Gramineae

Distribution
Hungary: Bükk Mountains; Poland; Slovakia

Hardiness
Fully hardy

Preferred conditions
Full sun or partial shade; well-drained, moist, loamy soil.

This grass is endemic to the Carpathian Mountains, in Poland, Slovakia, and Hungary, mainly in alpine and montane meadows. It also grows on open limestone sites, such as rocky ledges in the central mountains of Hungary. In the Bükk and Pilis mountain ranges in Hungary, it may also be found in a rather special mountain beech community. Here the trees form a low forest, to a height of only 12–15m (40–50ft), over shallow soils on steep slopes. *Sesleria hungarica* grows here alongside *Allium victorialis*, the reedgrass *Calamagrostis varia*, *Phyteuma spicatum*, and the beautiful lady's slipper orchid (*Cypripedium calceolus*).

The Bükk Range is the highest in Hungary, reaching 959m (3,165ft), and it also includes the highest limestone plateau in the country, with a fascinating associated fauna and flora. Some of these mountains are in the protected Bükk National Park, so the future of this rare grass is probably secure.

HABITAT

HUNGARIAN FIELD

Eastern Europe still retains many habitats under traditional agriculture, such as meadows and pastures. These often have a rich flora of mixed grass species and herbs with attractive flowers, many of which have become rare in Europe as a whole. Unlike many of these species, however, *Sesleria hungarica* does not thrive under grazing, and is restricted to high-altitude mountain sites. There it grows with another grass, *Festuca pallens*, and wildflowers such as mountain campanulas and saxifrages (*Saxifraga*).

Red List: Vulnerable*

Tripsacum floridanum

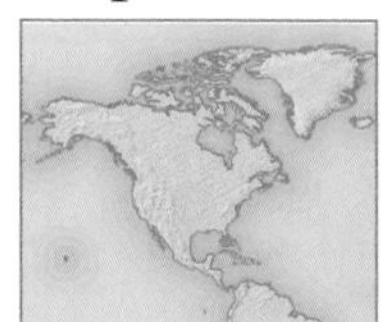

Botanical family
Gramineae

Distribution
Florida, USA: Dade, Collier, and Monroe counties; Cuba

Hardiness
Half-hardy

Preferred conditions
Full sun to full shade; variable soil conditions; very tolerant of drought.

Florida, or dwarf, gama grass (*Tripsacum floridanum*) is a large clump-forming grass, which grows to a height of about 1.2m (4ft). It has long, narrow leaves and tall flowering stems. This fine grass has a highly restricted natural distribution, being known only from about 20 sites in the pinelands of southern Florida, although it possibly also still occurs in Cuba. In the wild, Florida gama grass is threatened by habitat disturbance and invasion by exotic species, as well as construction of houses. The remaining sites are now being managed by burning, (*see box below*) and competing species are being removed.

In addition to its value as an unusual garden ornamental, this gama grass has also played a role in crop improvement. *Tripsacum* is closely related to *Zea* (corn or maize), and the two genera are interfertile. Crop scientists have therefore used genetic traits of Florida gama grass to help develop varieties of corn that are resistant to leaf blight.

HABITAT

Pine Rocklands

The pine forests of southern Florida are an important and distinctive habitat, under threat in many areas, primarily from urban development. Florida gama grass is one of many species restricted to such sites, known locally as "pine rocklands". These have a special flora, containing about 30 plants endemic to this small region. Like many of the forest-floor species in these subtropical communities, Florida gama grass is well adapted to occasional fires, regenerating quickly after burning.

▲ TYPHA MINIMA *This small, dainty bulrush from Europe and Asia is frequently grown in gardens, for example, in small ponds and containers. It is threatened in parts of its natural range.*

Unlisted

Typha

Botanical family
Typhaceae

Distribution
Worldwide

Hardiness
Fully hardy – half-hardy

Preferred conditions
Swamps, pond margins, riversides; some species are also coastal.

There are about a dozen species of *Typha*, commonly known as bulrush, reedmace, or cat-tail, and this genus has a wide global distribution. The most widespread species are southern cat-tail (*Typha domingensis*), *Typha orientalis*, and common bulrush or reedmace (*Typha latifolia*). Southern cat-tail occurs throughout the warmer parts of the world, and *Typha orientalis* mainly in tropical and subtropical south-east Asia and Australasia, while common bulrush is the primary species in most of the Northern Hemisphere.

The smallest species is ***Typha minima***, an interesting plant from a conservation viewpoint. Although it is rare and threatened in one part of the globe, it is extremely abundant in other parts of its range. For example, the species is listed as Endangered on the upper Rhine and in Switzerland (where it is the subject of a reintroduction programme), and has become extinct in the Czech Republic in the past few decades. As a result the species is listed on Appendix I of the Convention on the Conservation of European Wildlife and Natural Habitats (the "Bern Convention"), together with another species, *Typha shuttleworthii*.

Outside Europe, *Typha minima* has a natural distribution that extends through Turkey, Caucasus, Iran, west Pakistan, Afghanistan, and Turkmenistan to Siberia, Mongolia, China, and Japan. Here the species grows along riversides and in swamps, and may be locally very abundant. Outside its native range, the species is even regarded as a potential pest.

◄ SEED DISPERSAL *The flowerhead of the bulrush is a dense, cylindrical spike. The female flowers release large numbers of tiny, fluffy seeds, which are dispersed by the wind.*

1 on Red List*

Zizania

Botanical family
Gramineae

Distribution
North America and Asia

Hardiness
Fully hardy – half-hardy

Preferred conditions
Full sun, in shallow but not stagnant water, such as at the edge of a lake or flooded area.

Wild rice (*Zizania*) is an aquatic grass with edible grains. Two unlisted species – annual wild rice (***Zizania aquatica***) and perennial wild rice (*Zizania latifolia*) – have quite a long history of use as a food. Annual wild rice is harvested mainly as a grain, most famously by Native American people; the young shoots of perennial wild rice are used in Chinese cuisine. In recent years, wild rice has become very popular

ZIZANIA AQUATICA ► *Tall plants of annual wild rice emerge from the water, spreading their large flowerheads. This traditional grain is increasingly available commercially, although its aquatic habitat makes harvesting a challenge.*

as an alternative to common, cultivated rice and is sometimes sold as a mixture with it. The grains are long, slender, and dark, and have a characteristic pleasant, nutty flavour. They have a high protein content and are low in fat, and thus provide a healthy addition to any diet.

WILD RICE

Annual wild rice is widespread in wetlands in the northern, central, and eastern areas of North America, and its grain is harvested commercially. It is sometimes regarded as two species: southern wild rice (*Zizania aquatica*) and northern wild rice (*Zizania palustris*). *Zizania aquatica* var. *brevis*, an estuarine form of southern wild rice, grows exclusively along the St Lawrence River in Canada, in the freshwater marshes in the region of Quebec City. It grows to a height of up to 2.7m (9ft).

In natural stands, annual wild rice often covers large areas, around shallow lakes and alongside streams and rivers, especially those subject to flooding. It needs slow-flowing, clean water 15–150cm (½–5ft) in depth, and a constant or slightly declining water level through the season. Harvesting, usually from boats (often canoes), may last for several weeks. The stalks are pulled into the boat with a stick, and gently beaten to release the mature kernels. The seed is then processed by curing and drying, the hulls removed, and the grains gathered and packaged. Plenty of seed remains in the water to provide a reserve for the following year's growth.

Wild rice is a beautiful grass, well suited to planting around the margin of a garden lake or pond. It will grow to 3m (10ft) tall under optimal conditions, but will also thrive in a container, where a height of about 90cm (3ft) is more typical. The brownish-green flowerheads (panicles) are borne in summer, and turn an attractive yellow in autumn.

Texas wild rice (*Zizania texana*), which is listed as Endangered in the Red List of 1997, is found wild in the headwaters of the San Marcos River in Texas. It is endemic to the state and is federally protected. Unlike annual wild rice, this species is perennial, and most of its growth is submerged in running water, with just the flowering stems raised above the surface. In the 1930s, when this species was investigated and named, cattle were seen feeding on the grass, by sticking their heads deep under the flowing water.

Contemporary threats to Texas wild rice include pollution of the water, control of the flow, recreational use, and competition from other species. Efforts to cultivate it outside its natural habitat have proved difficult. It is thought to have potential as a genetic resource, and efforts have been made to combine some of its properties with other wild rice species.

ALPINE PLANTS

Alpine plants thrive in some of the bleakest and most remote regions of the world. They can resist extreme conditions no lowland plant would tolerate: high light intensities; exposed and very windy positions; large fluctuations in daily temperature; short summers; and bitterly cold winters. Many species flourish in rocky habitats with minimal soil. Although alpine plants belong to a range of diverse families, they have evolved many strategies in common to cope with their harsh environment. The majority are small shrubs, herbaceous perennials, or bulbs, but very few are annuals.

ALPINE HABITATS

High peaks covered in snow are often considered to be the typical alpine habitat. Many plants do live in such areas, on screes (slopes covered with loose stones), moraines (mounds of rocky debris deposited by glaciers), or cliffs, and among boulders. However, alpines inhabit a variety of other places. Alpine meadows, moorland, bogs, marshes, and scrub or woodland can all provide suitable habitats, in polar as well as montane regions. Some alpine plants have a specific niche, but others are quite tolerant and will even venture onto lowland cliffs, or limestone pavement.

MOUNTAINS ► *The Olympic Mountains in Washington, USA are typical temperate peaks that provide a range of habitats for alpine plants. Some plants are very specialized and will only grow at a specific altitude facing a certain direction; others can survive a wider range of conditions.*

HIGH IN THE TROPICS ► *Alpines on very high tropical peaks, such as Mount Kenya, endure surprisingly cold temperatures and extremely strong sunlight.*

◄ POLAR REGIONS *At high latitudes alpines are subject to some of the most difficult conditions, such as snowfall in August in arctic Alaska.*

ALPINE ZONES

Alpines are generally defined as plants that grow above the tree line. Some species are inflexible in their habitat, and can only grow in tundra, where there is little competition from larger plants. Others are more adaptable and may straddle the tree line, or grow in the coniferous forests below. The tree line is a fairly well-defined area where the available resources will no longer support large plants because conditions are too extreme. A tree line exists not only at high altitudes, but also at high latitudes, where the environment is similarly harsh and cannot support tree growth. Alexander von Humboldt (*see also p.28*) first observed that the further the distance from the equator, the lower the tree line. In the Himalaya, the tundra begins around 4,250m (14,000ft), but in polar regions, alpines grow at sea level.

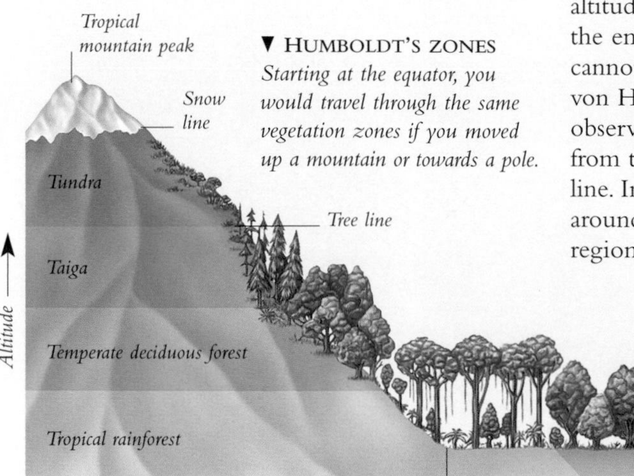

▼ HUMBOLDT'S ZONES *Starting at the equator, you would travel through the same vegetation zones if you moved up a mountain or towards a pole.*

ALPINE MEADOWS *in bloom are among the most amazing sights in nature. In Mount Rainier National Park in the USA, plants burst into life as soon as the snow melts, making the most of the short growing season.*

SURVIVAL STRATEGIES

Alpine plants have evolved to cope with the very extreme conditions imposed by high altitudes or latitudes. Many are blanketed in a deep layer of snow during the winter months, longer in hollows or gullies, and in some years the snow may not melt at all. Beneath the snow no growth can occur and the plants suffer from drought because all the water is frozen. Yet the snow is also essential for survival, providing a layer of insulation against the winter weather, and many alpines have evolved to prefer a dry winter dormancy. Once the snow melts, plants are exposed to drying winds, as well as the sun's rays, which are very intense in the thin mountain air.

HAIRS ▲ *In wet regions, such as where this* Meconopsis *grows in Tibet and western China, long hairs repel water.*

Like desert plants, alpines have evolved features to reduce water loss and protect them against the sun's ultraviolet radiation. To survive, many species bear tiny, closely packed leaves. These are often fleshy or waxy, and have hairs or scales, and fewer leaf pores (stomata) on exposed surfaces. Alpines hug the ground offering minimal resistance to wind that would otherwise dry out and damage them. In addition, a large root system anchors plants and helps to seek out any moisture.

▲ DWARF HABIT *This willow,* Salix reticulata, *here in the Alaskan tundra but also widespread in Europe, hugs the ground so wind passes over it without causing damage.*

◄ LARGE FLOWERS *Although their leaves are often tiny to reduce water loss, flowers of alpines, such as those of* Sempervivum, *are usually relatively large to attract pollinating insects.*

THREATS

Mountains can act like islands, often isolating alpines from adjoining regions. Many species are found on just one side of one peak, often at a specific altitude. Plants with restricted distributions or inflexible habitat requirements are extremely vulnerable to any modification to the environment. A change in the weather, an over-zealous collector, or a thoughtless tourist are all that is needed to cause damage to a population or, in some cases, a whole species.

One problem that affects all alpines, even in the most remote areas, is global warming. As temperatures rise, the tree line and the snow line move upwards and outwards (*see* Humboldt's zones, *p.280*). This has the potential to squeeze alpine habitats, or even eliminate them in some areas.

Inaccessibility protects many alpines but in some places increasingly large numbers of tourists and trekkers are taking advantage of roads or tracks that provide a gateway to mountain environments. Mountains close to large conurbations, such as the Cairngorms in Scotland, Mount Rainier in the American Cascades, and Mont Blanc in France, are in danger of being trampled to death.

Alpine plants also have a large following of people who admire their compact habit and large flowers, and enjoy the challenge of cultivating them. There are good arguments for controlled collection to feed this market, and to provide plants for botanic gardens. However, alpines have been, and continue to be, the focus of selfish plant-hunters acting illegally. Although they remove only a tiny cutting or small amount of seed, the cumulative effect over just a few years can be devastating. Commercial exploitation is also a problem and much money is paid by dealers for wild-collected plants for the horticultural market. The current hotspot for this trade is western China and, although laws to prevent such activities exist, they are difficult to enforce.

▲ CONTROLLED COLLECTION *of desirable species to bring them into cultivation removes pressure on wild populations, and provides a genetic reserve if a species is subsequently overcollected from the wild.*

▲ TOURISM *is increasing in mountainous regions. Although skiing, walking, cycling, and pony trekking (here in the Himalaya) seem harmless, delicate ecosystems are put at risk.*

CONSERVATION

In an age where few areas of true wilderness remain, inhospitable peaks and polar regions are increasingly appreciated and their remoteness gives them some protection. Many reserves, national parks, and even a few world heritage sites have been set up in mountainous regions around the world. Although they often cover only a small portion of a mountain range, they are making a major contribution to the conservation of plants and animals. Controlling access to these areas, preventing the removal of plants and the hunting of animals, and educating people about the alpine environment are all important.

There is a large interest in the cultivation of alpine plants from around the world, especially in western Europe and North America. The largest society, The Alpine Garden Society (UK), has members in many countries. Its prime aims are to promote an awareness of the problems facing alpine species, and an interest in alpine plants by studying them in the wild, and in cultivation. Propagation and distribution of rare alpine species are encouraged among members, and some of the species they grow, although scarce in the wild today, are common in collections. A limited amount of seed from the wild is sometimes required by a society to introduce rare alpines into cultivation, but a strict conservation code is adhered to.

◄ PLANT SHOW *The Alpine Garden Society (UK) is involved in the conservation of alpines by encouraging people to grow and distribute them. At shows, such as this one in London, specimens are judged by compact growth and abundant flowers – signs of healthy plants that have been kept in ideal conditions.*

▲ MOUNTAIN RESERVES *are being created worldwide, and the importance of habitats such as Da Xue Shan, the Big Snow Mountain, in Yunnan is now being recognized.*

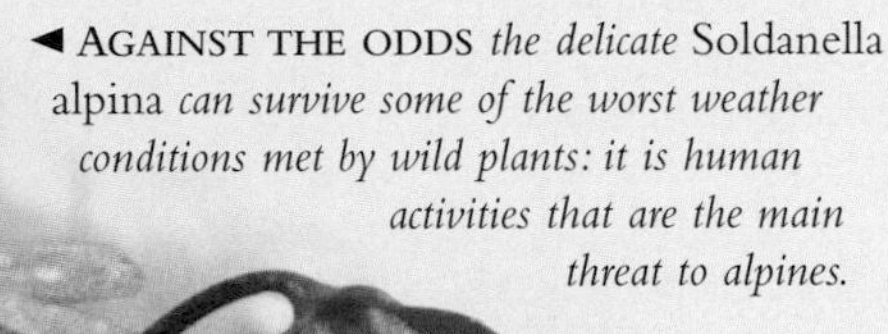

◄ AGAINST THE ODDS *the delicate* Soldanella alpina *can survive some of the worst weather conditions met by wild plants: it is human activities that are the main threat to alpines.*

GARDENER'S GUIDE

- Purchase only from reputable suppliers. There are numerous specialist alpine nurseries in Europe, North America, and Japan supplying a large range of alpine plants, the vast majority of which are grown and propagated from cultivated stock. There should be informative labels giving details of the plant and its origin.
- Never buy an alpine plant that looks like it may have been dug up from the wild – do not purchase supplies that are not accompanied by information on the origins of the plants. Avoid unusually cheap specimens.
- Some alpines may be grown from cuttings or seed collected from the wild: ascertain if they were gained legally. Ask to see import papers and collection licences.
- Be especially vigilant if purchasing plants mentioned in this book, especially those sold on the internet.
- The Alpine Garden Society, the Scottish Rock Garden Club, and the North American Rock Garden Society promote awareness of alpines. They run seed distribution schemes, which can be an excellent way of finding rare species without endangering wild plants.
- Keep meticulous records on plant provenance and propagation – this greatly enhances the conservation value of plants in cultivation.
- Growing alpines is an important part of their conservation but caring for them can be complicated. Make sure you can offer appropriate conditions before you buy. Cultivation information is often given on labels, or ask at reputable nurseries.
- Alpines from temperate regions including Europe, North America, the Andes, and New Zealand – such as gentians and campanulas – prefer a dry, dormant winter, plenty of water in spring, and a relatively dry summer.
- Plants from the mountainous regions of the Middle East, such as *dionysias,* often need very dry, dormant periods in summer and winter.
- Alpines from the monsoon-affected Himalaya and mountains of western China, including *Meconopsis*, and many primulas, prefer a cool, very wet, humid summer and fairly dry conditions for the rest of the year.

DRABA MOLLISSIMA

2 on Red List*/1 on CITES II

Adonis

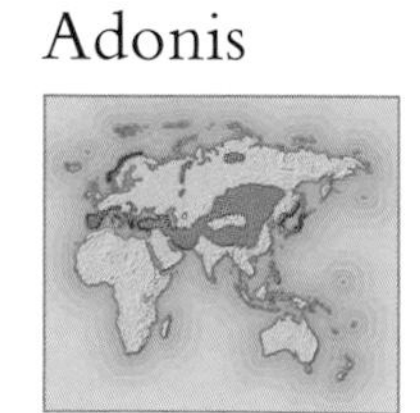

Botanical family
Ranunculaceae

Distribution
Europe, Asia

Hardiness
Fully hardy

Preferred conditions
Open ground in sun or partial shade, where some moisture is held in the soil through the hot summer months.

▲ ADONIS ANNUA *In Greek myth, when Adonis was killed by a wild boar, his lover, the goddess Venus, changed his blood into a red bloom that grew annually where he fell.*

The genus *Adonis* comprises about 20 species of typically red-, white-, or yellow-flowered annuals and perennial herbs found in Europe and Asia, as far east as Japan. Although the genus as a whole is widespread, some species are increasingly at risk, including commonly cultivated alpines, such as ***Adonis vernalis*** and ***Adonis annua***, which are locally threatened, but do not appear on the Red Lists. The species *Adonis cyllenea* is found on just one mountain in Greece and so is classified as Endangered on the Red List of 1997, but it is nevertheless also fairly well established in gardens.

The yellow pheasant's eye, *Adonis vernalis*, does not have a limited distribution – it occurs in Europe and spreads east to western Siberia – but its preferred habitats of fertile, dry, steppe grasslands and open woodlands are under threat. Vast areas of steppe have already been ploughed up to create farmland with the loss of their exceptional wildlife. In addition, *Adonis vernalis* is threatened by the flourishing medicinal plant trade. It is used in preparations to treat circulation, heart, and nervous disorders, and for hyperactive thyroid glands. Worldwide demand for the herb is high but because it grows slowly in cultivation, wild plants are collected, mostly in Bulgaria, Hungary, Romania, Russia, and Ukraine. Fortunately it is now listed on CITES Appendix II, protected by the European Union, and by national laws in some of its native countries. A few areas have adopted sustainable harvesting methods: only the aerial parts of the plant are collected leaving the roots to regenerate the next season.

Adonis species are all fully hardy, thrive on well-drained, moderately fertile soil in full sun, and are ideal for a rock garden. Propagation by seed is preferable because established plants dislike disturbance and are unlikely to survive transplanting. Seed should always come from nursery-raised stock. Sow perennial seed as soon as it is ripe, and grow on individual plants for two seasons before planting out. Sow annual seed *in situ*, in autumn or spring.

ADONIS VERNALIS ► *The yellow pheasant's eye is well established in cultivation, and much prized by gardeners.*

▲ ANDROSACE CYLINDRICA *This rare plant from the western Pyrenees attracts collectors with its neat, rounded cushion, and handsome dark green foliage, which can be completely obscured by the flowers.*

13 on Red List*

Androsace

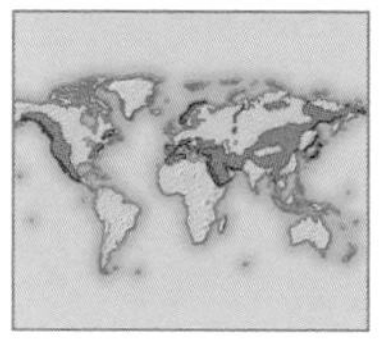

Botanical family
Primulaceae

Distribution
Northern hemisphere: mountainous regions

Hardiness
Fully hardy

Preferred conditions
An open sunny site in moist but well-drained soil, or rock crevices.

There are 153 species of rock jasmine, many highly prized by alpine plant enthusiasts. They are excellent for an alpine house, but some of the more robust species will thrive in a trough or rock garden. Their desirability in the garden means that in some areas they are at great risk from overcollection, although many species are already widely cultivated and further collections from the wild are generally unnecessary.

In China, androsaces are monitored carefully and occasionally collections of a small amount of seed from wild plants are permitted, which brings plants into cultivation and takes some pressure off wild populations. This strategy has brought the startling biennial ***Androsace bulleyana***, from north Yunnan and south-west Sichuan, into cultivation and reduced collection of wild plants. The species is not on the Red Lists.

Unfortunately, even if seed is obtained legitimately, some plants prove tricky to grow. *Androsace delavayi*, which is unlisted, seldom flowers in gardens so profusely as it does in the wild, yet exquisite forms from north-west Yunnan continue to captivate collectors. Even though the plants have proved difficult to cultivate, their seed is constantly being sought.

European species are well protected by law, although it is not always observed by unscrupulous collectors. In Spain ***Androsace cylindrica***, Rare on the Red List of 1997, grows in the canyons of the Ordessa National Park, on cliffs and in open woodland at 1,300–2,500m (4,300–8,200ft). Although it is widely grown, it often hybridizes in cultivation and lack of pristine material may prompt further collections. In Switzerland, a law prohibiting the picking or uprooting of any *Androsace* species protects *Androsace brevis*, also Rare on the Red List of 1997. Ideally, protection needs to be extended to afford all androsaces safety across the whole of their range.

Species outside Europe and China are often poorly known. *Androsace bryomorpha*, Indeterminate on the Red List of 1997, is thought to be restricted to a small area of Tajikistan. It is unusual in having relatively large white flowers. A small reserve has been set up to defend it, ostensibly from local goats, but also from collectors. Despite this, seed has come into cultivation in recent years, almost certainly from an illegal source.

◄ ANDROSACE BULLEYANA *This adaptable plant often forms extensive colonies in the wild, on rocky slopes, sunny banks, and in woodland clearings in China.*

30 on Red List*

Aquilegia

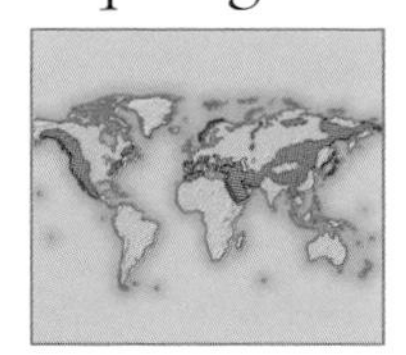

Botanical family
Ranunculaceae

Distribution
Northern hemisphere: mountainous regions

Hardiness
Fully hardy

Preferred conditions
Well-drained soil in full sun or partial shade.

Columbines have been cultivated in gardens for hundreds of years. There are about 70 species; some are large, lowland herbs, but many are neat alpines from mountainous regions throughout the northern hemisphere.

Of the tiny alpine forms the smallest and most charming is the North American *Aquilegia jonesii* from the Rocky Mountains in Wyoming. Although it is not difficult to grow, it can be shy to flower. For this reason seed is constantly being re-collected from the wild, straining the already scattered populations, and it is locally threatened, although not on the Red Lists.

The yellow columbine, *Aquilegia chrysantha*, is easy to cultivate and is a parent of many garden hybrids, but grows wild only in isolated, mountainous areas in Colorado, New Mexico, and Texas. This is probably because a widespread population retreated from the increasing temperatures at the end of the last ice age, becoming stranded at higher, cooler altitudes. Botanists are worried that the yellow columbine, having been affected by climate before, will be pushed to ever higher altitudes by global warming and eventually forced into extinction. So far it remains unlisted on the Red Lists.

Many alpine aquilegias are reasonably easy to grow in moist but well-drained soil in full sun or partial shade. The very compact, high alpines require moist, sharply drained soil in full sun, and some species need expert care, preferring a scree bed or an alpine house, dry winters, and cool summer temperatures.

▼ AQUILEGIA JONESII *Preferring exposed limestone screes in the wild, this alpine will also thrive in a raised bed or stone trough.*

AQUILEGIA CHRYSANTHA *Yellow columbines are threatened in the wild by global warming, but are widely grown for their airy foliage and golden flowers.*

Unlisted

Calceolaria uniflora

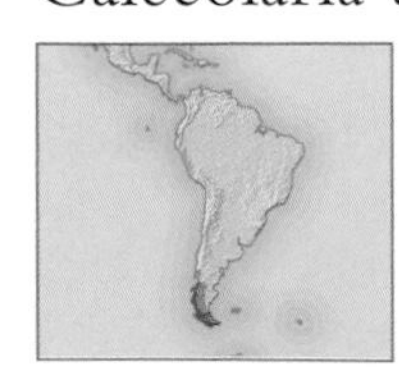

Botanical family
Scrophulariaceae

Distribution
Argentina, Tierra del Fuego

Hardiness
Fully hardy

Preferred conditions
Moist but well-drained, very gritty, acid soil in partial shade. Dislikes winter wet.

For many years this extraordinary and striking plant was known in cultivation under the name *Calceolaria darwinii* and, although ***Calceolaria uniflora*** is now considered correct, it is still occasionally sold under its old name. *Calceolaria uniflora* is popular among alpine plant enthusiasts. The plants form mats or low cushions of dark green oblong- to diamond-shaped, often slightly toothed leaves. The flowers are typical of the genus, but taken to spectacular extremes in *Calceolaria uniflora*, which produces bizarre, solitary flowers in early summer that are huge compared to the rosettes of leaves. The flowers are apparently pollinated by birds, which are said to peck at the prominent horizontal white band on the lower lip, although no-one knows why.

The species is endemic to Argentina and Tierra del Fuego and may have been encountered by Charles Darwin, who explored this region extensively and wrote about his travels in "The Voyage of the Beagle". Many plants from the area have become severely restricted since Darwin's time through overgrazing. *Calceolaria uniflora*, however, is still fairly common, particularly in some reserves that keep out grazing animals, and the exposed, extremely windy, peat moorlands, steppe, and clifftops it inhabits are inhospitable enough to offer some protection. In the wild, the species is immensely variable and many varieties have been recognized; some botanists go so far as to name these as separate species, but most maintain that there is too great an overlap of characteristics to warrant such treatment.

In cultivation *Calceolaria uniflora* can be quite tricky to maintain, which has led to a large number of re-introductions of seed from the wild over the years. The species is best grown in an alpine house in very sharply drained compost with a top-dressing of grit. Although other species and cultivars in the genus are widely available, and easy to grow, they do not have quite the same impact.

UNUSUAL FLOWERS ► *These incredible South American flowers need careful nurturing to thrive in cultivation.*

3 on Red List*

Callianthemum

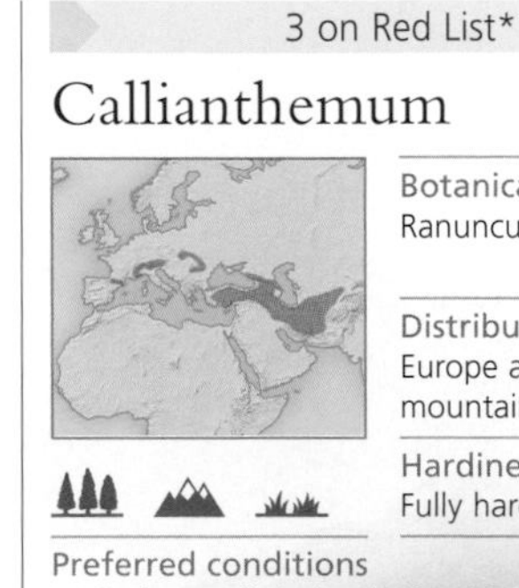

Botanical family
Ranunculaceae

Distribution
Europe and Asia: mountainous regions

Hardiness
Fully hardy

Preferred conditions
Moist but well-drained gritty humus in full sun. Dislikes hot, dry conditions.

There are about ten species of these pretty herbaceous alpine perennials, distributed through the mountains of Europe and Asia. They have flowers resembling buttercups (*Ranunculus*, *see p.304*), to which they are closely related. The species have narrow, scattered distributions and, as mountain environments become more popular for leisure and tourism, their habitats are increasingly accessible and are being quickly degraded by paths, tracks,

▲ CALLIANTHEMUM ANEMONOIDES *comes into flower as the snows melt in spring. Flowering time is so early that the low tufts of glaucous, lobed leaves are barely unfurled and are completely obscured by the blooms. It is available from specialist nurseries.*

142 on Red List*

Campanula

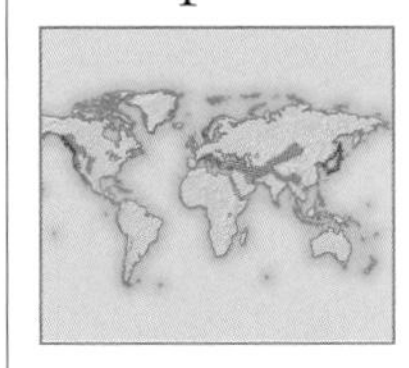

Botanical family
Campanulaceae

Distribution
Northern hemisphere: mountainous regions

Hardiness
Fully hardy

Preferred conditions
Moist but well-drained soil in sun or partial shade; high alpines need sharply drained soil in sun, or a rock crevice, and dislike winter wet.

Campanulas, also known as bellflowers or harebells, are found in the wild in many parts of the world and are very popular garden plants. There are in excess of 300 species ranging from the easy-to-grow lowland annuals and large, leafy perennials, to the tiny, jewel-like alpines, which present more of a challenge in cultivation.

Many of the alpine campanulas have suffered from overcollection in the past but most are now protected, especially in the Alps. Almost all are available as seed from cultivated sources and so there is no need for any further plants to be removed from the wild. However, some are difficult to grow and it is these species which, despite protection, continue to attract collectors. ***Campanula morettiana***, Rare on the Red List of 1997, is such a plant. It is endemic to the Dolomites where it inhabits limestone cliffs. At one time it was more common in collections than it is today, probably because stocks were being repeatedly replenished from the wild. It is rarely long-lived, and does not always produce seed outside its natural habitat. Today it is well protected from collection and is reproducing successfully in the wild. The related *Campanula raineri*, also Rare on

▲ CAMPANULA PIPERI *The profusion of flowers seen on this alpine bellflower in the wild, here on Mount Townsend, USA, can seldom be matched in cultivation.*

and ski slopes. Overcollection and grazing animals also threaten some *Callianthemum* species.

Callianthemum anemonoides is one such species, although it is unlisted on the Red Lists. The plant is a native of Austria, in the north-west Alps, where it inhabits coniferous woodland up to 2,100m (6,900ft). The sumptuous flowers are up to 3.5cm (1½in) across – large compared to the plant. Each of the white petals has a conspicuous nectary at the base that together produce a delicate orange ring around the centre of the flower. *Callianthemum kernerianum,* Vulnerable on the Red List of 1997, is a very similar species, endemic to the Italian Alps, particularly on and around Mount Baldo, up to 2,200m (7,200ft). It is distinguished from *Callianthemum anemonoides* by its smaller stature, not exceeding 8cm (4in) tall, and red flower stalks. Both plants are much sought after by collectors, making them very vulnerable to exploitation.

Callianthemums are choice plants for a rock garden, raised bed, or an alpine house, requiring well-drained yet moist, gritty humus. They enjoy full sun but greatly resent a hot, dry atmosphere and so dislike being grown under glass. Plants tend to be short lived, and seed germination can be both slow and erratic; the best chance of success is from fresh seed sown the moment it is ripe. Species that are popular with enthusiasts, such as ***Callianthemum coriandrifolium***, which is not on the Red Lists, are available as seed, exchanged among growers. Young plants are also sometimes available from specialist nurseries.

▲ CALLIANTHEMUM CORIANDRIFOLIUM *Scattered populations of this pretty species can be found in Europe. The flowers are borne in late spring, and have delicate blue-green foliage.*

Red List: Rare*

Clematis marmoraria

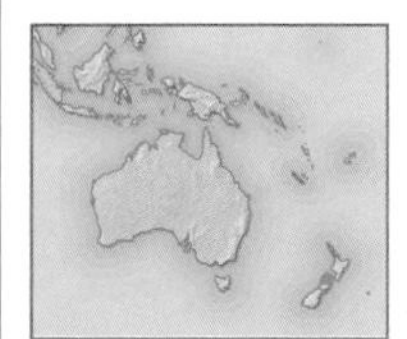

Botanical family
Ranunculaceae

Distribution
New Zealand: South Island

Hardiness
Fully hardy

Preferred conditions
Sunny, sheltered site, in well-drained gritty loam wedged between rocks; dislikes winter wet.

The tiny New Zealand ***Clematis marmoraria*** ranks as the smallest in a familiar genus of 300 species found worldwide. Discovered in 1973, it grows on a few remote crags in mountains to the north-west of Nelson on the South Island of New Zealand. A delightful plant, it immediately appealed to alpine gardeners and today it is far more numerous in cultivation than it is in its wild retreats, where it is considered Rare on the Red List of 1997.

Together with other New Zealand species, notably *Clematis paniculata* and *Clematis petriei,* it has been used to breed a range of non-climbing hybrids. *Clematis marmoraria* is rarely more than 10cm (4in) tall, its stems spreading close to the ground, covered in shiny, leathery, leaves, and flowers up to 5cm (2in) across. The plants are either male or female. Although females bear smaller flowers, they produce silky seedheads maturing from white to tawny yellow. Both sexes need to be grown side by side to produce viable seed, which should be sown fresh. Plants thrive in a cold frame, or outdoors in a trough or raised bed, if protected from winter wet.

▲ TINY CLEMATIS *Unlike familiar garden species, this evergreen alpine does not climb, but bears masses of typical clematis flowers.*

the Red List of 1997, is found in the south-eastern Alps, and also tempts collectors, who claim it to be one of the most splendid species. It has larger, open funnel-shaped flowers of pure mid-blue, set against a tuft of bright green leaves. The plant is certainly best appreciated in the wild because it suffers from the same fickleness in cultivation as *Campanula morettiana.*

In the USA some of the most vulnerable alpine bellflowers are being well protected. ***Campanula piperi***, also known as Piper's harebell or the Olympic bellflower, is an alpine gem from the Olympic Mountains of Washington State. In its more accessible haunts, particularly the popular ski resort Hurricane Ridge, over-collecting has severely reduced its numbers and it is classified as Vulnerable on the Red List of 1997, although it is still locally plentiful on the less visited Mount Townsend. Today all wild collection is prohibited, and unnecessary because it is available in cultivation, requiring a well-drained scree or raised bed. The closely related Castle Crags harebell, *Campanula shetleri*, is very localized, and is also recognized as Vulnerable on the Red List of 1997. It inhabits granite crevices above 1,700m (5,600ft) in the mountains of northern California, in the Shasta-Trinity National Forest. The forest is carefully managed to maintain wildlife habitats and to ensure that none of its endemic plants, such as this pretty white-flowered alpine campanula, become threatened or endangered.

The alpine campanulas are too small for the open garden, but are excellent plants for raised beds and troughs, on a sunny bank, or in crevices in a wall. Some species are best grown in an alpine house and prefer fairly poor, loam-based compost with up to one-third by volume of added grit. Although alpine campanulas are rarely long-lived, replacement plants are easily raised from seed. Sow seed thinly in pots in a cold frame in winter, and it will germinate in early spring, or propagate by division.

◄ CAMPANULA MORETTIANA *The flowers are up to 3cm (1¼in) long, large in comparison to the grey, hairy leaves.*

16 on Red List*

Corydalis

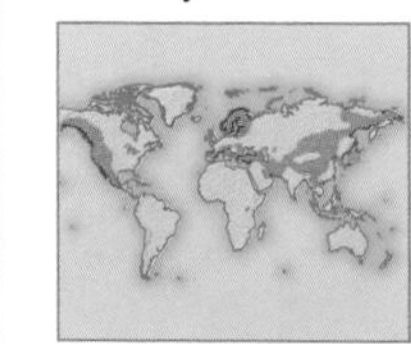

Botanical family
Fumariaceae

Distribution
North America, Europe, Asia

Hardiness
Fully hardy – half-hardy

Preferred conditions
Well- to sharply drained soil in full sun; some species dislike winter wet.

There are as many as 300 species of *Corydalis* scattered mainly across northern temperate zones. These include coarse lowland annuals and perennials, but perhaps most in demand are the small alpine herbs. Sixteen species are listed on the Red List of 1997, and many others are locally threatened. One species that is now widely available in the horticultural trade and so saved from any further collection is the charming ***Corydalis flexuosa,*** which grows wild in China in the famous Wolong Panda Reserve in western Sichuan. It was an instant success when it was first grown in the 1980s, and today various cultivars are available, including 'Balang Mist', 'Blue Panda', 'China Blue', 'Père David', and 'Purple Leaf'. All are excellent plants for the open garden.

In recent years many more species have been introduced into cultivation, which has taken some collection pressure off of wild plants. Collection causes particular damage to *Corydalis* populations because, rather than taking seed, collectors often dig up whole plants, which disturbs the habitat as well as reducing plant numbers. They do this because seed is dispersed rapidly and consequently easily missed. Also, plants are so well camouflaged they cannot be found unless they are in full flower, when seed is still unripe. *Corydalis benecincta*, for example, has fleshy, grey-mottled leaves, which lie flat on the ground, merging perfectly into the scree. It can be easily seen, and therefore collected, only when the pale purple-pink, strongly scented flowers emerge. Although it is restricted to a small area, it remains unlisted on the Red Lists because it is difficult to reach, growing wild in north-west Yunnan at 4,000–5,300m (13,000–17,400ft).

On the high, bleak screes and moraines of the Himalaya, particularly in west and central Nepal, live two very beautiful and closely related species: the yellow-flowered ***Corydalis megacalyx*** and *Corydalis latiflora,* which has greyish-blue leaves. In summer their highly and sweetly scented flowers are visited by various species of bee. Also found on high screes is *Corydalis melanochlora* from the rugged mountains bordering east Tibet, above 4,000m (13,000ft). Plants form a mass of rather fleshy, ferny, blue-grey or blue-green leaves that nestle close to the ground. In summer dense clusters of azure blue, white-spurred flowers appear, produced just clear of the foliage. Despite numerous attempts over the years, none of these high alpines have yet survived well in cultivation. This leaves potential for re-collection, but they remain unlisted on the Red Lists.

The alpine *Corydalis* have differing requirements, although many widely available species prefer moist but well-drained soil in full sun. Propagation is most successful using seed sown as soon after gathering as possible. The high alpines require a well-ventilated alpine house to thrive. Plants are unlikely to grow in the garden unless they have cool, moist summers, where temperatures do not rise much above 24°C (75°F), and dry winters with abundant snow.

▲ CORYDALIS MEGACALYX *is a desirable species that forms mounds of green, finely-divided leaves that are covered in flowers in spring. It is closely related to the grey-blue flowered* Corydalis latiflora *from the same region. Unfortunately neither has proved easy in cultivation.*

◄ CORYDALIS FLEXUOSA *has proved to be a popular and easy-to-grow species.*

9 on Red List*/All on Cites II

Cyclamen

Botanical family
Myrsinaceae

Distribution
Mediterranean and central Europe

Hardiness
Fully hardy – frost-tender

Preferred conditions
Moist but well-drained soil in full sun. Many require hot, dry summers during dormancy.

The genus *Cyclamen* has been grown for more than three hundred years, and all known species of these tuberous perennials are in cultivation. One can be found in bloom in autumn, winter, spring, or summer, making the genus particularly popular in gardens. In the wild, cyclamen are found throughout the Mediterranean with the exception of Spain and Egypt. The characteristic heart- or kidney-shaped leaves are often patterned or have a metallic sheen, and the elegant, often sweetly scented, jewel-like flowers come in shades of pink, purple, red, and, more rarely, white, their petals bent back through 180 degrees.

Even relatively rare species like *Cyclamen cilicium*, Vulnerable on the Red List of 1997, are easily raised from seed. They will flower in the second year, and occasionally naturalize in gardens. This ease of cultivation and year-round appeal has made cyclamen highly vulnerable to exploitation by gardeners who are often unaware of overcollection. Huge numbers were exported from Turkey, in particular, to supply the European market. By 1985, over 6 million cyclamen tubers were being exported to Holland and Germany annually and

CYCLAMEN GRAECUM *grows in scattered populations in Greece, Crete, Turkey, and Cyprus. The flowers appear in autumn, just before the leaves, which are strikingly marked with light green and silver.*

▲ CYCLAMEN PSEUDOIBERICUM *is found wild in south-eastern Turkey in woodlands and shrublands at 500–1,500m (1,640–4,920ft) but is widely available in cultivation.*

many of them perished owing to misidentification and inappropriate care. Eventually, it was the plight of *Cyclamen mirabile,* considered Vulnerable on the Red List of 1997, that shook people awake. Measures were taken and today nine species have Red List status, and all are on CITES Appendix II, which tightly restricts their export from any signatory country in which they are native. Only non-dormant, artificially propagated plants of *Cyclamen persicum*, the florist's cyclamen, are exempt from the regulations. In Turkey such measures have reduced the numbers uprooted, and legitimate commercial production of native species, such as ***Cyclamen pseudoibericum***, classified as Indeterminate on the Red List of 1997, is alleviating pressure on wild populations. Today many species are locally common, although road building, commercial development, and grazing by goats limit their numbers.

Cyclamen mirabile, the catalyst for conservation action, is still restricted but sizeable colonies exist in south-western Anatolia between 330–1,600m (1,100–5,300ft), growing in pine and oak woodlands and maquis. The high demand for these beautiful plants is understandable. The round leaves are deep green, patterned in grey-green, cream, or silver, often overlaid in young leaves with pink or carmine. The small, pink or white, coconut-fragrant flowers, have narrow, twisted petals, which are distinctly toothed at the top.

All the countries bordering the Mediterranean have their own native species. Some, including ***Cyclamen graeceum*** are widespread and not on the Red Lists; others are restricted to just one island. In Cyprus the dainty, winter-flowering *Cyclamen cyprium*, also unlisted on the Red Lists, grows scattered among shrubs and rocks, or in pine woodland, occasionally venturing into ancient vineyards. One or two species extend into central Europe, while others are found farther east, as far as northern Iran where *Cyclamen elegans* is endemic to the Caspian forests in the north, but is locally common and not included in the Red Lists. *Cyclamen somalense*, an unusual species from a single mountain in northern Somalia, was only discovered in 1986. Although this African species is undoubtedly rare, it also remains unlisted on the Red Lists.

Most cyclamen prefer humus-rich, gritty loam, which is moisture retentive but very well drained. They are excellent in the open garden, where they will often naturalize, or in damp climates where they can be grown in raised beds. In colder areas, less hardy species are best grown in pots in an alpine house or cold frame where extreme frosts and excessive rain can be excluded. These Mediterranean plants prefer wet winters and hot, dry summers with little water during dormancy. The Libyan species *Cyclamen rohlfsianum* (Vulnerable on the Red List of 1997) and the unlisted North African *Cyclamen africanum*, require a completely dry dormancy.

OTHER THREATENED CYCLAMEN

Cyclamen alpinum Endemic to south-west Turkey at 400–1,670m (1,300–5,500ft), growing in woodland among tree roots, or in open places and screes. It has dull, green, heart-shaped leaves patterned with grey, silver, or cream; deep pink to carmine purple, primrose-scented flowers appear in early spring. Often sold as *Cyclamen trochopteranthum*. Red List (1997): Vulnerable.

Cyclamen colchicum This species is restricted to the western Caucasus mountains at 300–800m, (1,000–2,600ft) in woodland. Its conservation status is uncertain at present. Currently, it is scarce in cultivation but is notable for its sweetly lilac-scented, pale to carmine-pink flowers. Red List: Unlisted.

Cyclamen intaminatum A dainty species from west and north-west Turkey at 700–1,100m (2,300–3,600ft). Its leaves are plain or marked; the small flowers are white or pink, with faint grey pencil-veins. This close relative of *Cyclamen mirabile* is available in cultivation. Red List (1997): Vulnerable.

Cyclamen libanoticum Without question one of the most endangered species in the wild, it is found among trees and rocks at 750–1,400m (2,460–4,600ft) in Lebanon. Bluish-green, heart-shaped leaves are marbled with grey. Spring flowers open white, then blush pink and have a magenta base. Widely cultivated. Red List (1997): Indeterminate.

8 on Red List*

Daphne

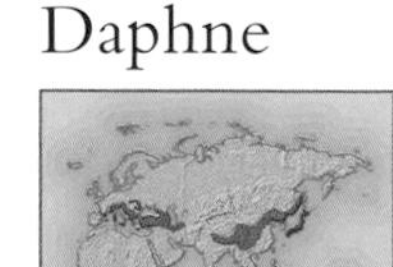

Botanical family
Thymelaeaceae

Distribution
Europe, Asia: mountainous regions

Hardiness
Fully hardy – half-hardy

Preferred conditions
Reliably moist but well-drained, gritty, humus-rich soil, in full sun.

Of all the genera of alpine shrubs, **Daphne** is by far the most popular. Daphnes have a neat habit, profuse blooms, and delightful scent. The smaller species and cultivars are ideal for raised beds and troughs, as well as for pots in an alpine house or cold frame. They are also excellent for alpine plant shows. Their popularity has placed several species in jeopardy from over-zealous collectors. ***Daphne petraea***, for example, from the Lake Como area in the Italian Alps is in high demand for its shiny, deep green, leathery leaves and strongly scented flowers, but it is Rare on the Red List of 1997.

Daphne arbuscula is confined to a few square kilometres in the Carpathian Mountains in Slovakia, on ledges and crevices at 700–1,300m (2,300–4,300ft). Although it is classified as Vulnerable on the Red List of 1997, several populations are found within the boundaries of nature reserves, and it is protected by law. Collection from the wild is limited because plants are fairly widely available from cultivated stock. Cuttings and grafted plants are sold, but all are apparently from a single clone because cultivated plants have never set seed. Not all daphnes are so amenable to cultivation. *Daphne glomerata* is a beautiful species from north-eastern Turkey, yet few plants exist in gardens today, having proved difficult to cultivate. Attempts have been made to introduce it from the wild on numerous occasions over the years. The species is not listed on the Red Lists.

◄ DAPHNE GENKWA *In areas with a continental climate, for example the eastern USA, this plant's spectacular spring display of blooms always excites comments.*

NEW DISCOVERIES

The number of daphnes available to gardeners was dramatically increased when the first of the Asian species was introduced by the famous plant collector George Forrest in 1906. *Daphne aurantiaca,* an unlisted species from Yulongxue Shan (Jade Dragon Snow Mountain) in China has been firmly established in cultivation ever since. However, not all of the introductions have been so smooth. Several attempts were made in the 1990s to introduce seed from a cousin of *Daphne aurantiaca* named ***Daphne calcicola*** but all resulted in failure. Enthusiasts for the genus were sorely disappointed because this unusual species, from high in the mountains of western China, is a beauty. It forms dense, low, grey-green mounds and the bright yellow, scented flowers are borne in the utmost profusion in early summer. Eventually the Chinese authorities authorized limited collection of the species, which is unlisted on the Red Lists. Grafting proved to be the key to success in cultivation and *Daphne calcicola* now exists in several collections.

DAPHNE PETRAEA

The odd-ball of the genus is ***Daphne genkwa***, indeed, it was formerly referred to as a lilac because of its striking flowers, large by *Daphne* standards. It is a very beautiful plant, native to north and central China and Taiwan where it inhabits grassy hills and plains. Unlike the high alpines, it is an excellent plant for the open garden in areas where the climate allows time for its wood to fully ripen before winter. It is not on the Red Lists.

Daphnes thrive in a well-drained, yet moisture-retentive, humus-rich soil with plenty of well-rotted leafmould. They dislike acidic conditions and greatly resent disturbance once they are established. Most of the small alpine species rarely set seed in cultivation but, if available, it should be sown as soon as possible in a well-drained seed compost. A cold period is generally needed for germination. Cuttings of semi-mature wood can be taken in midsummer. When attempting propagation, keep in mind that many choice cultivars are grafted onto a rootstock. Daphnes are occasionally taken from the wild; all plants should be bought from a responsible source.

◄ DAPHNE CALCICOLA *This special plant is often confused with the closely related but more commonly grown* Daphne aurantiaca.

Red List: Vulnerable*

Degenia velebitica

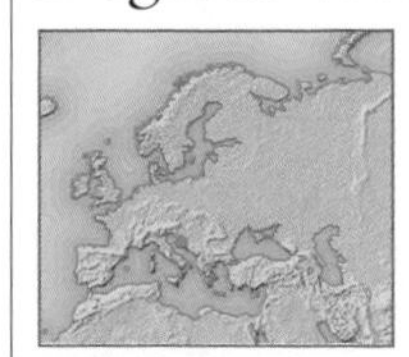

Botanical family
Brassicaceae/Cruciferae

Distribution
Croatia: Velebit Mountains

Hardiness
Fully hardy

Preferred conditions
Moist but well-drained soil in full sun; dislikes excessive winter wet.

This handsome alpine plant is readily raised and maintained in cultivation. ***Degenia velebitica*** is a hummock-forming species with tightly packed, narrow leaves covered with silvery-grey hairs. The bright yellow flowers have four petals arranged in a distinctive cross-shape. If pollination is successful, the flower stalk begins to elongate while the oval fruits develop.

The Velebit Mountains that rise between Croatia and Slovenia are one of the last European strongholds for wild wolves, brown bears, and lynx. They also harbour many rare and very restricted plants, such as *Degenia velebitica*, which is found only on limestone rocks and screes at 1,200–1,300m (3,950–4,250ft). Although not specifically protected at the present time, several wild populations are being carefully watched. At Ljubljana University Botanical Garden in Slovenia *Degenia velebitica* is cultivated and propagated. Plants are then distributed with the intention of relieving any collection pressure on wild populations.

The species is easily grown from seed, which is generally set in abundance. A dislike of winter wet means that it is a plant for a sunny well-drained site, or for a pot in an alpine house.

▲ ONE OF A KIND *The single species in the genus stands out as a decorative alpine for raised beds and scree gardens. It is a member of a large, widely cultivated family.*

75 on Red List*

Dianthus

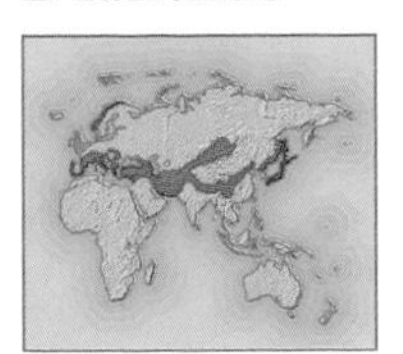

Botanical family
Caryophyllaceae

Distribution
Europe, Asia, Africa

Hardiness
Fully hardy

Preferred conditions
Full sun and sharply drained soil; some species dislike winter wet.

The pinks and carnations are an immensely popular group of plants, and a familiar sight in bouquets and buttonholes. Together they make up the genus *Dianthus*, which contains over 300 species distributed in Europe, Asia, and Africa with a concentration of species in the Mediterranean region. Although many lovely plants are available, among the finest are the tiny, intensely coloured, and often clove-scented alpines.

Dianthus callizonus, Rare on the Red List of 1997, has been declared a "Monument of Nature" in its home country of Romania. It is an interesting species restricted to limestone rocks and screes in the Carpathian Mountains at 1,650–2,200m (5,400–7,200ft). In cultivation, it is known to be a challenge, requiring well-drained gritty compost, a cool root-run, and careful watering. Even then it is rarely long-lived in gardens, which is unfortunate because it is arguably the prettiest of the alpine species. It forms low mats of narrow, glossy, deep green leaves. From these arise stems only 5–10cm (2–4in) high, bearing lavender-pink or carmine flowers, large in comparison at up to 5cm (2in) across.

An impressive example of a cushion-forming pink is the hedgehog pink, *Dianthus erinaceus.* This species can grow up to 50cm (20in) across in the wild, the surface of the hard hummock covered with dense, sharply spine-tipped, deep green leaves. In summer, small, purplish-pink or crimson-pink flowers are borne just above the surface. There are two varieties, *Dianthus erinaceus* var. *erinaceus* and *Dianthus erinaceus* var. *alpinus*, both endemic to Turkey. In cultivation they require a very sunny, sheltered position in sharply drained gritty compost, with protection from winter wet. They can be reluctant to flower freely in cool climates, but will make an impressive mound of foliage. The species is Rare on the Red List of 1997 but is available in cultivation, and *Dianthus erinaceus* var. *alpinus* can often be found in specialist alpine nurseries.

An excellent, but often overlooked, plant for a block of tufa is ***Dianthus simulans***, another cushion-forming pink, which produces dense mounds of grey, spine-tipped leaves. The summer flowers are rose-red and carried on short stems just above the cushion surface, which is often entirely hidden by a blanket of flowers. Not yet widely available from nursery suppliers, the species is currently difficult for gardeners to obtain.

Alpine pinks are neat, hardy, evergreen mat- or cushion-forming plants. Many are excellent grown at the edge of a border, in troughs, or in raised beds. A few of the rarer species require careful cultivation, and occasionally an alpine house to protect them from winter wet.

DIANTHUS SIMULANS ► *This alpine is endemic to the Rhodope Mountains that straddle the border between Greece and Bulgaria.*

Unlisted

Diapensia

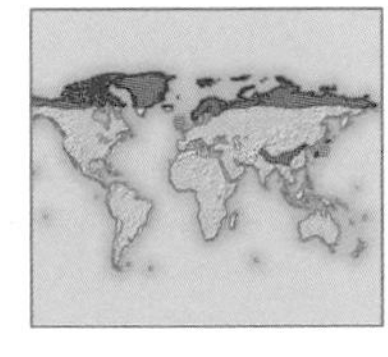

Botanical family
Diapensiaceae

Distribution
Circumpolar and western China

Hardiness
Fully hardy

Preferred conditions
Very well-drained soil in full sun; dislikes hot, dry summers and excessive winter wet.

Regarded as potentially choice plants for enthusiasts, the genus *Diapensia* is found in native habitats similar to those of many plants that thrive in scree or troughs in the garden. Sadly, diapensias have never proved easy to maintain, even by alpine specialists.

Diapensias are found mainly in the Arctic Circle, but with a southward extension into the mountains of western China. They are woody perennials that form small hummocks or mats and the genus is characterized by its very crowded shoots of tiny, leathery leaves and delicate, funnel-shaped flowers with five broad, spreading petals. Of the four species in the genus, ***Diapensia lapponica*** is a typical example and is the most widely distributed, both in collections and in the wild. It inhabits bleak Arctic moorlands, and is found south as far as Japan, New York State, and the Ural Mountains of Russia. There is a single colony in the highlands of Scotland, near Fort William, which is very well protected.

In the Asian outpost of the genus, *Diapensia purpurea* is found only in the high mountains of western China and the neighbouring regions of Tibet at 4,000–5,500m (13,100–18,050ft). It is a plant of rocky moorland slopes, on acid soils, where it forms spreading carpets of tiny, deep green, elliptical leaves, covered in summer with numerous, almost stemless, rose or rose-purple blooms. In the wild the species can form very large plants, more than 1m (3ft) across, which appear to be extremely old. A somewhat similar relative is *Diapensia bulleyana,* also found in western China, which has cream or pale yellow flowers.

Diapensia purpurea was introduced recently into cultivation, but, given that it is adapted to such a high, harsh environment, early indications are that it will be tricky to maintain in gardens. Responsible buyers can help ensure that this will not lead to illegal collections of more plants. When available, *Diapensia* seed needs to be sown as soon as possible because it quickly loses viability.

◄ DIAPENSIA LAPPONICA *A plant of dry rocky slopes, summits, and cliff ledges, it forms deep green hummocks that burst into flower in early summer.*

4 on Red List*

Dionysia

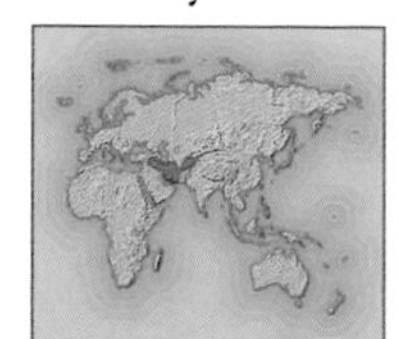

Botanical family
Primulaceae

Distribution
Middle East, central Asia

Hardiness
Fully hardy

Preferred conditions
Very sharply drained soil, or rock crevices; cool damp summers and cold, dry winters.

▲ DIONYSIA MICHAUXII *This unusual species is endemic to the south-west of Iran. Like most dionysias, specific water and temperature requirements have to be met if it is to thrive.*

Once considered very difficult to grow, the genus *Dionysia* now appears in many specialist collections of alpine plants. However, the true species may soon be lost from cultivation because of gardeners' preference for cultivars that are often easier to grow.

The genus *Dionysia* contains about 50 species that are confined mainly to Iran and Afghanistan, with outlying species in south-east Turkey, northern Iraq, Tajikistan, and Oman. They are close cousins of the primulas, with similar flowers that are very attractive to butterflies. Flowers are yellow, pink, or violet and borne on or above cushions of tiny leaf rosettes. In the wild nearly all the species are found on shaded or part-shaded rocks, particularly limestone cliffs.

While one or two species, particularly *Dionysia tapetodes*, have a wide distribution, and are not considered to be threatened, most are confined to one or a single range of mountains. ***Dionysia freitagii***, which forms large, dark green sticky cushions, is restricted to a very limited area in the mountains of northern Afghanistan, close to the city of Mazār-e Sharīf, and the related *Dionysia viscidula*, a very similar but slighter plant, is found in only one gorge in north-west Afghanistan, at around 1,400m (4,600ft). Their limited range makes these species extremely vulnerable to habitat change and collection. This is an increasing problem as Afghanistan becomes more accessible because species have no formal protection and are not on the Red Lists. In contrast, all the Iranian species are in the Iranian Red Data Book, although they also lack an international classification on the Red Lists. They include gems such as ***Dionysia michauxii***, a bright yellow-flowered species, and *Dionysia janthina*, a rare and relatively little-known species that forms pretty, silver-grey, softly hairy cushions. There is no doubt that the whole genus *Dionysia* should be given CITES protection to try and reduce the impact on all wild populations from unlawful collecting.

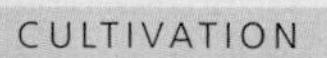

▼ DIONYSIA FREITAGII *Afghan species have become more vulnerable to collectors as the country has become more open to foreign visitors.*

To thrive in damp climates, plants need the protection of a bright and very well-ventilated alpine house. A gritty, extremely well-drained compost is essential and watering should always be done from the base rather than the top. The plants require ample moisture when they are in growth and just enough water in the dormant winter season to save them from drying out completely.

Plants can be raised from seed. Most *Dionysia* species, like primulas, come in two different forms. Each plant is pin- or thrum-eyed; that is, it bears either long- or short-styled flowers, but never both on the same plant. This is a device to prevent inbreeding, so one form must be crossed with the other for successful seed production. This means that any viable collection must contain both forms of the same species. Alternatively, cuttings can be taken of strong non-flowered shoots, although this requires both skill and patience.

OTHER THREATENED DIONYSIA

Dionysia afghanica Restricted to a single locality in north-west Afghanistan, this rare species inhabits a shaded limestone cliff at 1,400m (4,600ft). The plants form neat, grey cushions of crammed leaf-rosettes and bear violet-pink flowers with a distinctive dark eye. The species is available as a single clone in cultivation. Red List: Unlisted.

Dionysia bryoides Endemic to scattered localities in south-west Iran, on partly sunny or shady limestone cliffs at 1,800–2,800m (5,900–9,200ft), where it forms small, deep green cushions of tight rosettes. The solitary, stalkless flowers are pale to deep pink, or violet, with a white centre. Red List: Unlisted.

Dionysia lamingtonii Rediscovered only recently on very few limestone cliffs at 2,850–2,950m (9,350–9,600ft) in western Iran. Plants form very neat, dome-like, grey-green cushions. The solitary yellow flowers completely cover the plant. Red List: Unlisted.

Dionysia microphylla. This species forms hard cushions up to 15cm (6in) across, of closely overlapping scale-like leaves. The pale to deep violet flowers are borne well clear of the foliage. It is endemic to north-west Afghanistan where it inhabits sloping or vertical limestone rocks, sunny or shaded, at 1,200–1,400m (3,900–4,600ft). Red List: Unlisted.

Dionysia mira This species illustrates how closely related *Dionysia* is to *Primula*. It forms tufts of leaves, dusted with white farina beneath. The small yellow flowers are borne one whorl above another, like a candelabra primula. It is the only species found south of the Persian Gulf, in the mountains of northern Oman, where it inhabits limestone rocks or cliffs at 1,500–1,900m (4,900–6,200ft). Plants are protected in the wild and rather rare in cultivation. Red List (1997): Indeterminate.

CULTIVATION

TUFA TROUBLES

Many alpines, including species of *Dionysia*, thrive in tufa but it is a limited resource. Tufa rock is formed when underground water dissolves calcium from limestone. When it eventually reaches the Earth's surface, the water deposits the dissolved minerals as calcium carbonate, or tufa. This process can happen quite quickly, engulfing plants, or indeed anything else in its path. When the organic matter rots away a soft, sponge-like rock remains, ideal for alpines requiring sharp drainage. Demand for tufa is high and has resulted in some natural tufa formations being destroyed. Fortunately artificial substitutes are available.

Red List: Vulnerable*

Douglasia nivalis

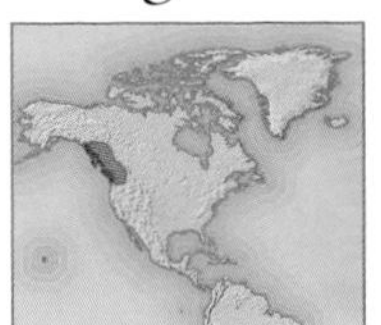

Botanical family
Primulaceae

Distribution
North America: north-west mountains

Hardiness
Fully hardy

Preferred conditions
Moist but very well-drained, gravelly soil in partial shade.

A small, low-growing plant with numerous, lax rosettes of variable grey-green leaves covered with starry hairs, ***Douglasia nivalis*** is a colourful species found in scattered localities in the mountains of north-western North America, from British Columbia to Washington State. In some places, as in the Wenatchee Mountains in Washington State, it has some protection. However, it is a desirable species and could become a victim of collectors. The plant grows in stony ground, among rocks, on ledges, and on stabilized screes. It is sometimes called *Androsace nivalis*.

▲ ABUNDANT BLOOMS *The flowers, just 1cm (½in) across, are borne on short stalks during the late spring and early summer.*

Unlisted

Eriophyton wallichii

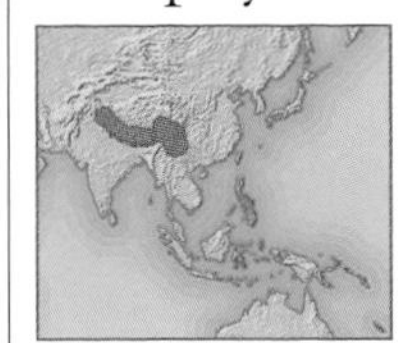

Botanical family
Labiatae/Lamiaceae

Distribution
Himalaya, western China

Hardiness
Fully hardy

Preferred conditions
Cool damp summers, dry, frozen winters, and moist but well-drained soil in partial shade.

Often described as a high-altitude deadnettle, this delightful alpine is related to the true deadnettles (*Lamium* species, *see p.297*). ***Eriophyton wallichii*** is the only species in the genus. It has very hairy leaves and two-lipped flowers that are usually wine-red or purple, although some rare yellow forms exist, which are inevitably prized by collectors. The species inhabits the Himalaya and the mountains of western China, growing at high altitudes on screes and stabilized moraines at 4,300–5,400m (14,100–17,700ft). These mountains are affected by the monsoon and many plants have devised ways to shield their flowers from excessive rains. Some have nodding flowers, which act like an umbrella, some hide their flowers beneath leaves or enlarged bracts, and others achieve the same result by warding off moisture with repellent silky or waxy surfaces. *Eriophyton wallichii* has pairs of toothed, silky grey leaves that fend off the summer rains like a mammal's coat and in winter the plant is protected by a deep layer of snow.

The conditions required by plants used to this climate often make them particularly difficult to grow, even with the protection of an alpine house or cold frame. *Eriophyton wallichii* is no exception and, despite repeated introductions of seed from the wild, very few plants exist in cultivation and even fewer have ever flowered. Even though just a little seed is taken each time, this can affect populations over years, particularly those in well-known locations.

▼ ALPINE NETTLE *This tufted perennial reaches no more than 15cm (6in) in height. Its nettle-like flowers are attractive to bumblebees.*

Unlisted

Draba mollissima

Botanical family
Brassicaceae/Cruciferae

Distribution
Caucasus mountains

Hardiness
Fully hardy

Preferred conditions
Gritty, sharply drained soil or rock crevices in full sun; dislikes winter wet.

This attractive plant is one of the most distinguished members of a large genus of around 300 species that includes a host of popular, cushion-forming alpines. ***Draba mollissima*** has soft yellow flowers that are borne in small clusters, just clear of the cushion. It is endemic to the Caucasus mountains where it inhabits drier cliffs and ledges.

▲ DOWNY LEAVES *Even when it is not in flower, this alpine forms attractive soft, white-haired domes of crammed leaf-rosettes.*

The species is readily raised from seed and is also available from specialist nurseries. However, it is a tricky species to maintain well in cultivation, being prone to partial die-back during the winter months. It needs the protection of an alpine house to thrive, with ample ventilation, a sharply drained compost, and a top-dressing of grit.

Unlisted

Eritichium nanum

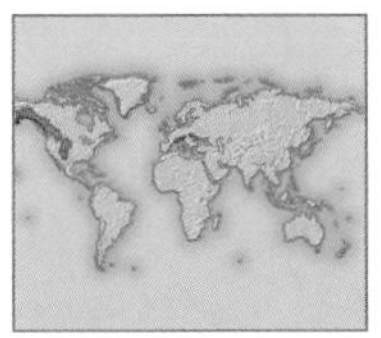

Botanical family
Boraginaceae

Distribution
European Alps, North American Rockies

Hardiness
Fully hardy

Preferred conditions
Moist but sharply drained soil in full sun.

Known as the king of the Alps in Europe, ***Eritichium nanum*** is one of the most beautiful of all high alpine plants, forming soft, low, blue-green cushions, covered in small, yellow-eyed, intensely blue flowers held just clear of the foliage. A variable species, compact in the central and southern Alps and the Carpathians, it is more spreading (*as seen here*) in the Rockies of North America.

▲ ERITICHIUM NANUM

Plants grow at 1,700–2,500m (5,600–8,200ft). Cultivation is difficult and seed has been re-introduced many times. *Eritchium nanum* is now protected in some European localities.

42 on Red List*

Gentiana

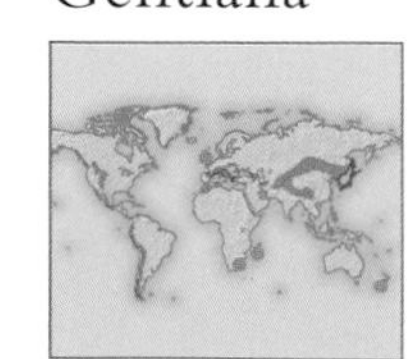

Botanical family
Gentianaceae

Distribution
Widespread

Hardiness
Fully hardy

Preferred conditions
Reliably moist but well-drained soil in full sun, or partial shade in hot, dry areas.

High mountains and deep blue flowers are what spring to mind when gentians are mentioned. However, the genus includes over 300 species, not only in alpine habitats worldwide, but also in some woodlands. Species with yellow, pink, red, white, or purple flowers, as well as various shades and intensities of blue, can be found.

Gentiana depressa is one of the most spectacular species in the wild, where it grows on open, rocky slopes and in alpine meadows at 3,300–4,300m (10,850–14,100ft) in central Nepal, east to Bhutan, and neighbouring regions of southern Tibet. Although not specifically protected, and not on the Red Lists, this gentian does, in places, fall within the boundaries of national parks. The species is rare in cultivation, although it is relatively easy to grow with the protection of a cold frame. It forms flat mats or low cushions of pale, blue-green rosettes with closely overlapping leaves. The flowers are up to 3cm (1¼in) long and appear in the late summer, opening only in bright sun.

Gentiana oschtenica, from the Caucasus and western Asia, is said to resemble a yellow-flowered spring gentian. In the wild it is an uncommon plant, although it does not appear on the Red Lists. Seed from wild plants, which are native to the western Caucasus mountains at about 3,000m (9,850ft), has been collected on a number of occasions in recent years but the species remains rare in cultivation, thriving only in a humus-rich compost with the protection of an alpine house. Germination can be erratic, and the seedlings vary in vigour and the number of flowers they will produce.

Many *Gentiana* species are easily grown in gardens in cool climates – the trumpet gentian, *Gentiana acaulis*, the spring gentian, *Gentiana verna*, and their cultivars are widely available. Gentians prefer climates with cool, damp summers and moist but well-drained soil in full sun. If summers are hot and dry, it is important to provide some shade from hot sun. Species are best propagated by seed sown as soon as it is ripe. Alternatively, divide plants in spring.

▲ GENTIANA VERNA *SUBSP.* BALCANICA *This subspecies of the widely grown spring gentian is native to the Balkans, spreading east as far as the Caucasus.*

▼ GENTIANA DEPRESSA *is one of the most beautiful of the Himalayan gentians and receives some protection in the wild within the national park around Mount Everest.*

Unlisted

Haastia pulvinaris

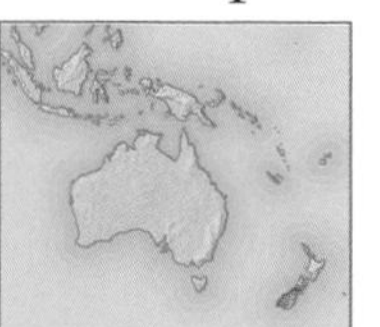

Botanical family
Asteraceae/Compositae

Distribution
New Zealand: South Island

Hardiness
Fully hardy

Preferred conditions
High altitude scree and rocky sites, sharp drainage, and full sun; dislikes winter wet.

As indicted by its common name, the vegetable sheep is one of the strangest-looking cushion plants found anywhere in the world. ***Haastia pulvinaris*** forms spectacular, low, yellowish to buff-coloured domes. In the wild they can reach an impressive size – the largest plant was recorded as more than 2m (6ft) across. The mounds are composed of thousands of very tightly-packed, furry leaf-rosettes, with all but the tips of the tiny, individual leaves hidden by their companions. The small, yellowish-brown flowerheads are held tightly in the centre of the rosettes. They are somewhat inconspicuous, but, taken as a whole, the appearance of the plant is spectacular.

Haastia is a genus of three species, endemic to the South Island of New Zealand. The vegetable sheep are restricted to high rocky ridges and consolidated screes on the drier mountain ranges, at 1,300–2,500m (4,250–8,200ft). Without doubt it is an environment increasingly vulnerable to road development, skiing, and trampling by hikers. In the past the species was much collected but it has proved to be extremely challenging to grow. Specimens any larger than 15cm (6in) are

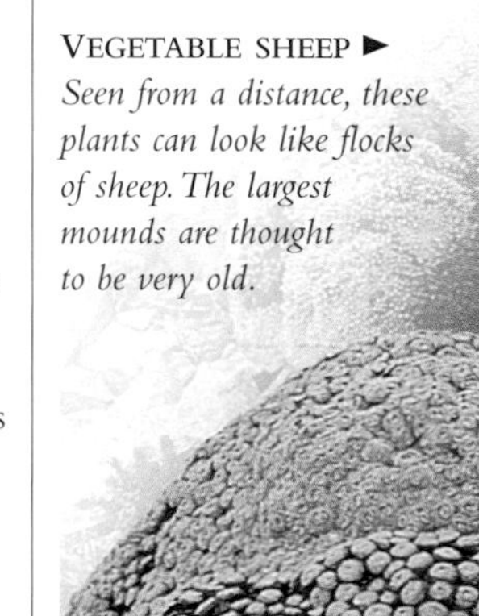

VEGETABLE SHEEP ► *Seen from a distance, these plants can look like flocks of sheep. The largest mounds are thought to be very old.*

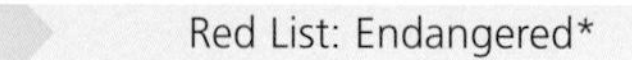

Red List: Endangered*

Hymenoxys lapidicola

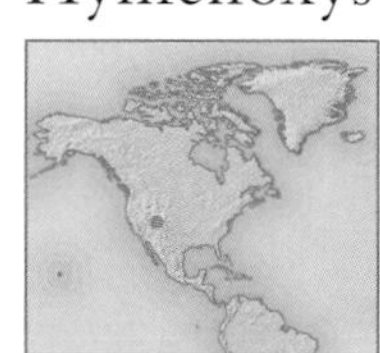

Botanical family
Asteraceae/Compositae

Distribution
USA: Utah

Hardiness
Fully hardy

Preferred conditions
A sunny, dry, position in rock crevices, or poor, well-drained soil; dislikes winter wet.

Meaning "growing in stone", *lapidicola* is an appropriate name for this local endemic. Discovered in 1986, ***Hymenoxys lapidicola***, known as the stone rubberweed, is rare and restricted to inaccessible sandstone cliff terraces, at about 2,500m (8,200ft), on Blue Mountain in the Dinosaur National Monument in Utah, USA. The plants form neat, cushions up to 10cm (3in) across, of narrow leaves in congested rosettes. In summer, flowerheads are borne in the centre of each rosette.

▲ STONE RUBBERWEED *Grown for its cushions of deep-green leaves and summer flowers, this plant is new to cultivation.*

unheard of, quite a contrast to the monsters in the wild. It is slow-growing from seed and young plants are very vulnerable to winter wet. Indeed, it requires a very tightly controlled watering regime throughout the year, resenting too much or too little. *Haastia pulvinaris* is a good example of a plant best appreciated in its native habitat, its glory somewhat diminished by removal to an unnatural setting.

HABITAT

SANDSTONE FORMATIONS

The sandstone terraces where *Hymenoxys lapidicola* grows are part of a geological feature known as the Colorado Plateau, which includes the famous Grand Canyon National Park. This astonishing landscape was formed at least 500 million years ago when the area was covered by shallow seas. A build-up of heavy layers of sediment caused the Earth's crust to sink until heat and pressure deep underground formed sandstone. The continent began to rise again about 10 million years ago, exposing the stone to the elements, which have shaped the formations seen today.

They are solitary, bright yellow, and rather ragged, similar to those of groundsel. The species is now available in cultivation, its small stature making it ideal for raised beds or rock crevices in the open garden, or an alpine house in areas with very damp winters. It prefers very well-drained soil and a position in full sun. Propagation is best by seed sown as soon as it is ripe or in spring.

Unlisted

Incarvillea

Botanical family
Bignoniaceae

Distribution
China: Himalaya

Hardiness
Fully hardy – half-hardy

Preferred conditions
Fertile, moist but well-drained soil in full sun, but with some shade in the hottest part of the day; dislikes excessive winter wet.

Incarvilleas, with their lush leaves and exotic, trumpet-shaped flowers, are often called "hardy gloxinias", but they are not related. Several are popular with gardeners and, although these alpines are from the Himalaya and the mountains of central Asia and western China, many of the larger species are excellent in open borders. Such plants include the robust *Incarvillea delavayi*, the more compact *Incarvillea mairei*, and their cultivars, which mostly have pink or purple flowers with contrasting yellow throats, and fresh green foliage. They are widely available and easy to grow in fertile, moist but well-drained soil, in full sun. Plants dislike very wet winters. Sow seed in spring or autumn, keeping autumn-sown seedlings frost-free; most species of *Incarvillea* take about three years to flower from seed.

CULTIVATION MYSTERY

Similar to *Incarvillea delavayi* in general stature is the lesser-known ***Incarvillea lutea,*** a perennial to 60cm (2ft) tall. It produces attractive bright green foliage and, in summer, erect spikes of yellow flowers borne on sparsely-leaved stems that emerge from the centre of each tuft of leaves. Numerous attempts to cultivate this beautiful plant have been made over the past hundred years.

▼ **INCARVILLEA MAIREI** *VAR.* **GRANDIFLORA** *is a pink, early summer-flowering species that is popular in gardens.*

▲ **INCARVILLEA LUTEA** *The stunning flowers open in summer, at first primrose-yellow then gradually reddening with age.*

Sadly, and somewhat mysteriously, it has not proved easy to maintain. The species is painfully slow from seed, and can take eight or more years to flower, although most plants perish before reaching this age. The reasons for this are not properly understood, especially as so many incarvilleas are a success in gardens.

In the wild *Incarvillea lutea* is a localized species from north-west Yunnan. It favours rocky places at 2,800–3,600m (9,200–11,800ft) in the open, or in the dappled shade of trees and shrubs. These same conditions are preferred by the easily grown *Incarvillea delavayi*, making the cultivation problem all the more mysterious.

Incarvillea longiracemosa was formerly included as a subspecies of *Incarvillea lutea*, but is a generally taller plant, and the flowers are paler yellow and slightly smaller. It is found in damp habitats in an isolated region of south-east Tibet, at 3,650–4,570m (11,900–14–900ft). *Incarvillea longiracemosa* has, so far, fared rather better in cultivation than *Incarvillea lutea*.

Unlisted

Iris afghanica

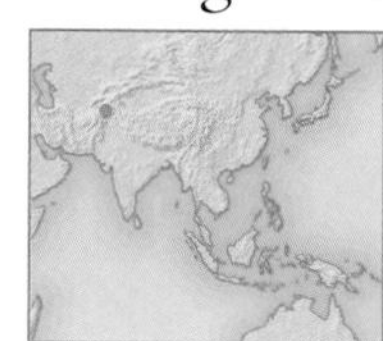

Botanical family
Iridaceae

Distribution
Afghanistan: central and eastern Hindu Kush

Hardiness
Half-hardy

Preferred conditions
Well-drained, neutral to slightly acid soil, in full sun; dislikes damp during dormancy.

One of the most beautiful irises found in central Asia, ***Iris afghanica*** is endemic to the central and eastern Hindu Kush in Afghanistan. If it did not grow in such a politically sensitive part of the world, *Iris afghanica* would certainly be highly vulnerable to over-zealous plant collectors. Indeed, it was collected a number of times in the 1960s and 1970s, not long after its discovery, but has been somewhat inaccessible in the wild since then.

A viral infection of the initial stock plants meant that at first it did not prove particularly easy to maintain. However, seed-raised plants have been more successful in recent years and it is available from specialist nurseries. In the wild, this iris is found in a handful of localities on rocky and grassy mountain slopes at 1,500–3,300m (4,900–10,800ft), where it grows in exposed and sunny sites. Plants form tufts of narrow, bluish-green, slightly scimitar-shaped leaves, up to 35cm (14in) tall in flower, but often less. In late spring and early summer erect stalks carry one or two flowers, each creamy-yellow with contrasting heavy purple-brown veining on the "falls", which have a beard and solid purple signal patch in the centre.

Like most alpine bulbs, this iris is slightly fussy, requiring a summer baking and no water once the leaves begin to wither. During growth it prefers neutral to acid, well-drained soil in full sun. The species can also be grown successfully in deep containers in loam-based, gritty potting compost.

◄ AFGHAN BEAUTY Iris *is a very popular genus, and this species is particularly attractive. It is one of the bearded irises, so called because they have small "beards" of hair in the centre of each downward pointing petal, or "fall". The erect petals are called "standards".*

HABITAT

HINDU KUSH

This vast mountain range lies between Afghanistan and Pakistan. The high alpine zone is at an altitude of 3,000–4,000m (9,800–13,100ft). Here, almost all water comes from snow melt, and temperatures are never warm. Yet plants and animals thrive and there are many endemic species. This is partly because the area has so far remained undiscovered by outsiders and it is sparsely populated by local people. Improvements to roads that run through the mountains may open up the Hindu Kush to farmers, tourists, and plant collectors and the area is unlikely to remain a true wilderness.

Red List: Vulnerable*

Jancaea heldreichii

Botanical family
Gesneriaceae

Distribution
Greece: Mount Olympus

Hardiness
Fully hardy

Preferred conditions
Clefts and crevices in rock or tufa, in partial shade, with plenty of water during growth.

The single species in the genus, ***Jancaea heldreichii*** is an excellent choice for growing in tufa. The plant is very delicate, and similarities can be seen to the related African violets (*Saintpaulia* species). Lilac-blue, bell-shaped flowers arise on arching stalks above rosettes of oval leaves. The foliage is covered with dense silvery-white hairs that give it a soft, silky appearance.

In the wild, *Jancaea* is rare, confined to part-shaded limestone crevices and ledges at 1,000–2,400m (3,300–7,900ft) on Mount Olympus in Greece, an area with a rich flora and quite a few endemic species. At one time *Jancaea* was over-exploited by plant collectors who removed seeds, and even whole plants, but today it receives protection within the national park that protects the mythical seat of the Gods.

Despite their silvery appearance – usually a sign of drought tolerance – plants require ample water during the growing season and like to be kept moist during the dormant winter months. Although *Jancaea* can be grown in pots in a well-ventilated alpine house, plants are fully hardy and grow much better when established on large blocks of tufa or in vertical crevices in a rock garden.

Red List: Vulnerable*

Laccopetalum giganteum

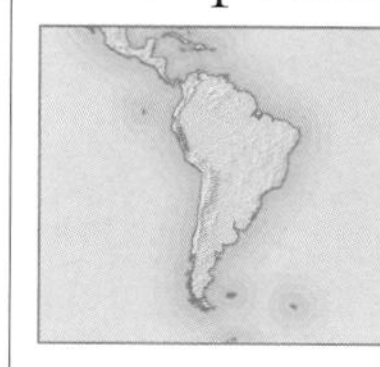

Botanical family
Ranunculaceae

Distribution
Northern Peru: Andes

Hardiness
Half-hardy

Preferred conditions
Cool temperatures and well-drained soil, but with plenty of water when in growth.

At one time included with the buttercups and later with anemones, this species now belongs to a genus of its own. Unfortunately, *Laccopetalum* has not proved as adaptable to growing in gardens as its former relatives. The species has proved difficult to maintain, requiring very specialist conditions, such as those provided by the Arctic bench in the alpine house at the Royal Botanic Gardens, Kew (*see box, right*).

This plant's failure in cultivation is disappointing because it is without doubt an extraordinary species. More handsome than beautiful, plants form clumps of large leaves, up to 50cm (20in) tall, unstalked at the base, and with a distinctly leathery texture. Each stout, leafy stem bears a solitary, almost round flower, which can be up to 15cm (6in) across. It is yellowish-green overall, partly enveloped by green sepals and has a central mass of stamens.

▲ GIANT BUTTERCUP *is the common name of this single species in the genus. It is best appreciated in the wild, in the high, damp alpine meadows of Peru.*

In the wild the species is very localized, growing only in a narrow altitudinal range of 4,100–4,200m (13,500–13,800ft) in the Cajamarca area of the northern Peruvian Andes. The people of the region rely on agriculture, livestock, mining, and tourism, all of which have the potential to affect local plant species. In addition, ***Laccopetalum giganteum*** has been the focus of various plant hunters over the years and is listed as Vulnerable on the Red List of 1997.

ALPINE PLANTS

They prefer a position in semi-shade. *Jancaea* is readily raised from seed, which should be sown in sterile conditions, much the same as those used for germinating fern spores. Plants are available for sale only from specialist nurseries, occasionally under the former spelling, *Jankaea*.

▼ **TINY TREASURE** *This popular alpine is tiny, just 8cm (3in) tall, and is ideal for tucking into rock crevices.*

CULTIVATION

ALPINE HOUSES

Alpine plants inhabit exposed positions at high altitudes or latitudes. In winter they are dormant under a dry, insulating layer of snow, then water from the spring thaw fuels a furious spell of growth and reproduction in the cool summer before the snow returns. In addition, the Arctic summer has 24-hour daylight, and thin mountain air lets through intense sunlight and radiation. These conditions, needed by some alpines to thrive, are recreated in an alpine house. At the Royal Botanic Gardens, Kew, the alpine house has a moat which cools air as it enters, and watering is carefully regulated. There is even a chilled and specially lit "Arctic bench" where temperature and light levels can be further manipulated.

15 on Red list*

Lamium

Botanical family
Labiatae/Lamiaceae

Distribution
Europe, Asia, North Africa

Hardiness
Fully hardy

Preferred conditions
Sharply drained soil in full sun or partial shade; dislikes excessive winter wet.

The 40 species of deadnettles include an array of easy and decorative garden plants, grown as much for their attractive leaves as for their flowers. The alpine species are mainly restricted to the mountains of Turkey, although the popular species *Lamium garganicum* is found also in Italy, Greece, and Iraq and is not on the Red Lists.

Of the alpine deadnettles, *Lamium armenum*, also unlisted, is one of the most delightful plants, restricted to a handful of mountains in Turkey at 1,950–2,700m (6,400–8,850ft), where it grows on scree, coping with the constant movement by means of long, stabilizing roots. It is a mat-forming plant, with pairs of green or grey-green leaves and pink, nettle-like flowers that sit perkily upright. The flowers are up to 5cm (2in) long, large for a plant with a maximum height of just 15cm (6in). *Lamium microphyllum*, Rare on the Red List of 1997, is another desirable western Turkish endemic. The species forms slow-creeping tufts of kidney-shaped or oval, dark green to brownish-green leaves that are just 1cm (½in) long – the species name means "tiny leaves". The flowers appear in late spring and early summer and are delicate rose, with purple streaks and blotches in the throat, and a purple "stag's-horn" upper lip. Probably the smallest of all deadnettles, this species dwells on high screes in Honaz Dağ (Mount Honaz) National Park in Turkey, where it is found just below the summit, at 1,800–1,900m (5,900–6,200ft).

▲ LAMIUM GARGANICUM *SUBSP.* STRIATUM *is a widely cultivated alpine deadnettle, and bears abundant flowers above mats of heart-shaped leaves in early summer.*

In cultivation, alpine deadnettles are attractive, but often short-lived and not as compact as wild plants. They can be grown outside on a raised bed if they are given protection from excessive winter rain, but are best grown in an alpine house where watering can be carefully regulated. Alpine deadnettles have occasionally been known to self-seed; otherwise sow seed in containers in a cold frame in autumn or spring. Alternatively take cuttings of strong young shoots in early summer.

See also Invasive plants, p.458.

Unlisted

Leontopodium monocephalum

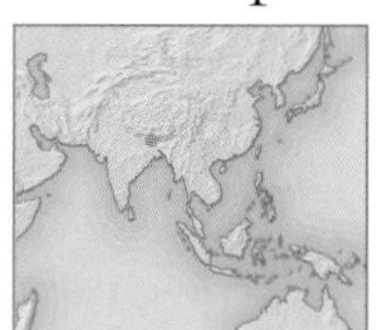

Botanical family
Asteraceae/Compositae

Distribution
Central and eastern Nepal

Hardiness
Fully hardy

Preferred conditions
Sharply drained, slightly alkaline soil in full sun with protection from winter wet.

Edelweiss are generally easily grown in a rock garden, raised bed, or alpine house. They have small flowers, surrounded by large, furry, petal-like leaves (bracts). There are a number of popular and readily available species, such as *Leontopodium alpinum* from the mountains of Europe. Edelweiss prefer very sharply drained, neutral to alkaline soil in full sun, with some protection during wet winters.

Leontopodium monocephalum, sometimes called *Leontopodium evax*, has been introduced as seed on a number of occasions, but it has not proved as easy to cultivate or to maintain as its relatives. This high-altitude edelweiss is native to the Himalaya where it inhabits moraines and screes at 4,300–5,200m (14,100–17,100ft). Plants form small mats, to 20cm (8in) across, of numerous yellowish- or greyish-green, hairy leaves. The summer flowers are citrus-yellow, fading to yellowish-brown, a very unusual colour as most edelweiss are snow-white. Most of this species' known localities are very remote and only accessible with extreme difficulty, and this puts off all but the most committed collectors. More formal protection is provided in the Everest region, where it grows within the boundaries of the national park, although tourism is fast increasing in the area.

▼ **FURRY FLOWERS** *This desirable edelweiss,* Leontopodium monocephalum, *has small flowers surrounded by furry, petal-like leaves, called bracts.*

2 on Red List*/1 on CITES III

Meconopsis

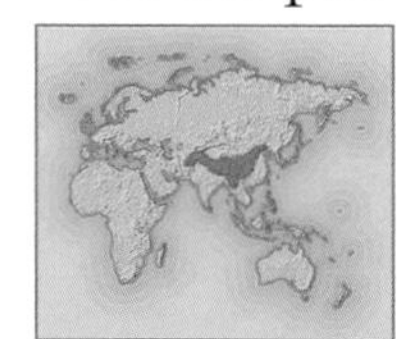

Botanical family
Papaveraceae

Distribution
Himalaya, Burma, China, western Europe

Hardiness
Fully hardy

Preferred conditions
Humus-rich, leafy, moist but well-drained, slightly acid soil in partial shade. Plants dislike cold, drying winds, and young foliage may be damaged by frost.

Sometimes known as Himalayan poppies, the genus *Meconopsis* includes many species and cultivars widely grown in gardens. They are closely related to the common poppy (*Papaver*) and have similar, papery flowers in sumptuous colours. There is one species, *Meconopsis cambrica*, that grows in western Europe, but all other species are found in the Himalaya and western China in alpine meadows, woodlands, and screes. In their native habitats they encounter cool, damp summers and they thrive in similar climates. Sow seed thinly as soon as it is ripe, and keep it moist and in the light. ***Meconopsis betonicifolia***, *Meconopsis grandis,* and their hybrids are very widely grown, but there are many other superb species. Just two species appear on the Red List of 1997, *Meconopsis aculeata* (Endangered) and *Meconopsis latifolia* (Vulnerable): although other Himalayan poppies are unlisted at present, plants should be purchased responsibly so no species become threatened as a result of collection.

▼ MECONOPSIS REGIA *This stately, widely available species can grow up to 2m (6ft) tall. It comes from the rocky slopes and alpine meadows of central Nepal.*

PEOPLE

PLANT HUNTERS

The famous plant-hunting team Frank Ludlow and George Sherriff, here in Tibet, introduced species from the Himalaya into cultivation. Many plants from the region bear their mark, with the species names "*ludlowii*" and "*sherriffii*" cropping up often. *Meconopsis sherriffii* is one of their discoveries, made in Bhutan in 1936. The intrepid pair found it confined to screes and cliffs and areas of dwarf alpine scrub at 4,200–4,600m, (13,800–15,100ft). The solitary pink flowers, borne in early summer, were described by them as "rose-pink like the first flush of dawn on the snows". It is rare in cultivation.

Meconopsis delavayi is an unusual species with fleshy, bluish-green leaves and deep violet-blue, nodding flowers borne on slender stalks. In the wild, this species grows only in north-west Yunnan where it was discovered by Abbé Delavay, a missionary who found it on Yulongxue Shan, the Jade Dragon Snow Mountain, in 1884. It has turned out to be almost entirely limited to this mountain at 3,100–4,400m (10,200–14,400ft), growing on damp limestone screes and in sloping wet meadows, where it is vulnerable to collectors in its more accessible locations. In the garden it prefers a partially shaded trough, with its roots wedged between slithers of rock, in humus-rich compost. Sow seed in pots in a cold frame in late winter, and leave young plants well alone until just before the second season, then separate the small tubers prior to renewed growth.

MECONOPSIS PUNICEA

Another Himalayan gem is *Meconopsis horridula*. Although many nurseries and gardens claim to grow this species, plants invariably prove to be the allied *Meconopsis prattii*, a much easier-to-grow species from the mountains of western China. Unfortunately the true *Meconopsis horridula* has proved fiendishly difficult to maintain outside its natural habitat. In the wild, it inhabits high mountain slopes, among boulders, on stabilized screes and moraines, or in alpine meadows up to almost 6,000m (20,000ft), making it one of the highest-altitude plants recorded. It should come as no surprise that the plant finds it difficult to adapt to life in gardens. Despite numerous introductions of seed it is very rare in cultivation. People continue to try and grow *Meconopsis horridula* for its amazing flowers that are a clear, semi-transparent blue, often with hints or suffusions of pink or purple.

Meconopsis prattii is a worthwhile plant to grow under its own name. It is a larger plant than *Meconopsis horridula*, up to 80cm (32in) tall, but is otherwise similar, covered in yellow to purple spines with deep blue or reddish-blue flowers. Like the common poppies, Himalayan poppies have attractive seedheads; in this case heavily adorned with sharp bristles.

Some *Meconopsis* species are not the classic blue colour, but still have the clear hues the genus is famous for. *Meconopsis punicea* has bright red, nodding flowers from summer to autumn and is quite readily available. *Meconopsis regia* is listed on CITES Appendix III and is rare in cultivation at the present time. The species bears soft yellow or red flowers in late spring, but is attractive all year with evergreen gold or silver hairy leaves.

FREE SPIRIT

The genus contains plants other than *Meconopsis horridula* that are not suited to cultivation except perhaps by an expert, and then only after establishing that plants or seed are from a legitimate source. A charming, tantalizing example is *Meconopsis bella*. 'Bella' means beautiful, an apt description for this tiny, delicate plant, not more than 15cm (6in) tall in flower and often far less. The species has deeply cupped, half-nodding summer flowers, which vary in colour from clear, pale blue to pink or purplish-blue, with contrasting golden anthers in the centre. However, the plant has mostly defied cultivation and, although it has been introduced into gardens numerous times, it has only ever paid a fleeting visit – proving impossible to maintain for more than a few years. Another Himalayan species, it is found in high, remote places where it inhabits shady, moist banks, low, shrubby areas, and cliff ledges. In the wild it is scarce, but unlikely to become endangered because the areas where it grows are so remote and inaccessible.

▲ MECONOPSIS DELAVAYI *is available from specialists, but it is important to check that plants are nursery-propagated.*

MECONOPSIS BETONICIFOLIA
This widely grown poppy can reach up to 1.2m (4ft). It has large, blue-green leaves and, in summer, striking flowers up to 10cm (4in) across.

Unlisted

Mimulus naiandinus

Botanical family
Scrophulariaceae

Distribution
Chile: Andes

Hardiness
Half-hardy

Preferred conditions
Fertile, humus-rich, very moist soil in sun or light dappled shade.

Although widely sold under the name 'Andean Nymph' by nurseries and even supermarkets in the 1980s and 1990s, this monkey flower is now recognized as a distinct species and named ***Mimulus naiandinus***. In 1973 the plant was discovered by John Watson and Martin Cheese, who gathered seed in the Chilean Andes, where it grows in damp, marshy places and along stream margins.

Mimulus naiandinus is a tufted perennial, at first growing upright then becoming more spreading, often rooting at the nodes. The plant does not reach more than 20cm (8in) in height, with pale green, triangular leaves, and typical, two-lipped mimulus flowers, which are borne over a long period in summer. It prefers a damp pocket in a rock garden or border and can be grown as an annual in frost-prone areas.

▲ MONKEY FLOWER *Not many plants become familiar in the garden before they are recognized as a species but this monkey flower is an exception. It is easy to raise from seed.*

Unlisted

Notothlaspi rosulatum

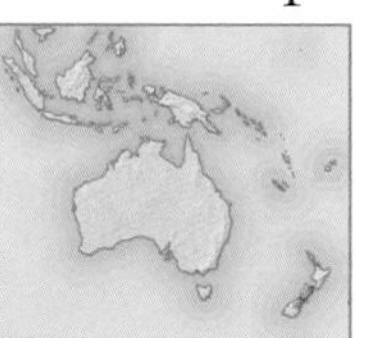

Botanical family
Compositae/ Brassicaceae

Distribution
New Zealand: South Island

Hardiness
Half-hardy

Preferred conditions
Rock crevices or deep, gritty, sharply drained poor soil in full sun.

The few species in the genus *Notothlaspi* are endemic to New Zealand's South Island. The alpine gem ***Notothlaspi rosulatum*** grows on the eastern side of the Southern Alps, from Marlborough to Canterbury, where it is an inhabitant of high, bleak, consolidated screes up to 1,850m (6,100ft). This highly attractive species has been subject to overcollecting and, when it is not in flower and difficult to spot, to trampling. The plant is a biennial and flowers and fruits only once before it dies. During its first year it forms a neat rosette of numerous, pointed and toothed, rather fleshy leaves. They are sea-green, often flushed with bronze or purple and are densely hairy at first. The flower spike erupts from the centre of the rosette in the second season or, if conditions are unfavorable, occasionally in the third, and is densely packed with creamy-white, or rarely buff-coloured, four-petalled, fragrant flowers. These are followed in turn by flattened seed pods. The species is very vulnerable to the vagaries of the weather and in inclement seasons little seed may be produced.

In cultivation *Notothlaspi* plants need deep, gritty, well-drained compost, plenty of light, and the protection of an alpine house. Overhead watering is disastrous, especially during winter dormancy, and all watering needs to be regulated with care. Seed should be sown as soon as it is ripe, and should be carefully spaced because seedlings resent disturbance.

▼ LIFE AMONG BOULDERS *An extensive root system and soft stems help this little plant survive among rocks where there seems to be no soil.*

Unlisted

Nototriche compacta

Botanical family
Malvaceae

Distribution
South America: Andes

Hardiness
Half-hardy

Preferred conditions
Moist but sharply drained soil in full sun; dislikes winter wet.

One of the most exciting and extraordinary plants to see in the wild is ***Nototriche compacta***, which belongs to a genus of almost 100 species confined to the Andes. It grows at 3,100–4,200m (10,200–13,800ft) in rocky scrub and grassland, where plants form silvery-white cushions, and bear small, pale lilac-blue or white mallow flowers in summer. The stamens are bunched together in the centre of each bloom.

These alpine gems are considered challenging to grow: seed readily germinates in cultivation, but plants lose their compactness and, although they will flower, the attention-grabbing quality of plants in the wild cannot be recaptured. Although some gardeners have managed to cultivate this species in the open garden, it is probably best left to experts or, if possible, enjoyed in the wild. The best chance of success comes from ensuring that there is plenty of air movement around plants and placing a protective glass overhead during winter to prevent excessive wet.

▲ SOUTH AMERICAN ALPINE *In the high Andes this plant forms dense cushions, up to 10cm (4in) across, of oak-like, downy leaves. Cultivation requires special conditions.*

Unlisted

Paraquilegia anemonoides

Botanical family
Ranunculaceae

Distribution
Himalaya and western China

Hardiness
Fully hardy

Preferred conditions
Poor, sharply drained, alkaline soil in full sun. Cool summers and cold, dry winters.

Without doubt ***Paraquilegia anemonoides*** is one of the most beautiful of all alpines and a much-cherished plant in cultivation. In the wild the species is predominantly a plant of rock crevices and ledges on cliffs, occasionally venturing onto rough, stabilized screes in areas where it is unlikely to be grazed. It has attractive, finely dissected, ferny-looking foliage that forms neat grey or blue-green tussocks up to 40cm (16in) across. The flowers may be flat to deeply cupped, ascending to nodding, in shades from white to delicate lavender, lilac, and pale or deep violet blue. Colonies of mixed colours can be found in the wild, but usually the colour distribution is regional, or even local.

The variable flowers and foliage are the source of some confusion over how many species of *Paraquilegia* exist – indeed the genus has divided botanists for many years. On one side are those who believe there is a series of about 10 closely related species, while on the other side are botanists who recognize just two. *Paraquilegia anemonoides*, in the broad sense, is spread throughout the mountains of central Asia, from the Tien Shan through the drier rainshadow areas of the Himalaya to the mountains of western China. This wide distribution and the general inaccessibility of its preferred habitats means that the species as a whole is not threatened. However, local populations are under considerable pressure from plant enthusiasts.

The species is readily germinated from seed, and although it is challenging to get the plants to flower as prolifically as they would in the wild it is by no means impossible. In areas with cool summers and cold, dry winters, plants can be grown outside in poor, sharply drained soil in full sun. Elsewhere they are excellent plants for scree beds, troughs, tufa, or an alpine house in gritty, humus-rich loam. They need protection during wet winters, and where summers are hot plants appreciate partial shade.

TRANSLUCENT BLOOMS ► *The flowers and leaves of this widespread alpine species are immensely variable across its range. This plant grows in north-west Yunnan, in China.*

Red List: Rare*

Physoplexis comosa

Botanical family
Campanulaceae

Distribution
Austria and Slovenia: Alps

Hardiness
Fully hardy

Preferred conditions
Gritty, sharply drained soil, preferably slightly alkaline and not too fertile, in partial shade; dislikes wet winters.

Commonly known as the devil's claw, this is one of the most extraordinary-looking alpines. Now the single species in its own genus, it was formerly named *Phyteuma comosum* – a member of a very popular alpine genus that gave it wide recognition. Unsurprisingly, after its name change, it remains a much sought-after plant and has been the target of many collectors over the years. It is a small plant, not more than 10cm (4in) tall, with fleshy, kidney-shaped, sharply toothed leaves that form a lax rosette in which the flowers nestle in late summer. The flowers are pink and bottle-shaped, tapering to a dark violet tip through which protrudes a long, arching style, giving the appearance of a sharp claw.

In the wild, *Physoplexis comosa* is native to rock crevices in scattered localities in the Alps, in south and south-east Austria, the neighbouring parts of Slovenia, and the Dolomites, up to about 2,000m (6,600ft). In some of its native countries it is protected by law and partly falls within the boundaries of various national parks.

Perhaps more curious than beautiful, ***Physoplexis comosa*** has long been cherished among alpine growers as an excellent plant for troughs, rock crevices, or blocks of tufa. It is readily raised from seed or by division of established clumps, or cuttings taken in late spring. Slugs can devour entire plants.

DEVIL'S CLAW ► *This small, tufted, deciduous perennial bears its clusters of curious flowers in late summer.*

49 on Red List*

Primula

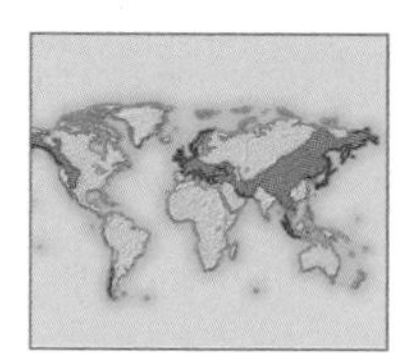

Botanical family
Primulaceae

Distribution
Northern hemisphere: mountainous regions

Hardiness
Fully hardy – half-hardy

Preferred conditions
Moist but sharply drained, gritty, humus-rich, and occasionally slightly alkaline soil. Species tolerate full sun and partial or deep shade.

Primulas include the familiar primrose, *Primula vulgaris*, the oxlip, *Primula elatior*, the cowslip, *Primula veris*, and a host of other fine garden species and innumerable showy hybrids. With about 430 species, *Primula* is without doubt an important genus and contains a great variety of plants: some like deep shade and moist soil and a few are houseplants. The high-alpine species prefer sharply drained conditions in full sun, and some need the protection of an alpine house. Most are easily propagated by seed sown in early spring or division after flowering.

The alpine species are found across mountainous regions in the northern hemisphere, with a few species south of the equator. Many of the finest primulas come from the Himalaya, including the yellow-flowered ***Primula aureata***, which is found above the tree line, colonizing steep, earthy banks and shady rock crevices, often close to streams or waterfalls. This species has neat green leaves heavily marked with white meal, and flowers soon after the snows melt.

Primula aureata is a target for collectors, being accessible on two important trekkers' trails, although it is locally common and not on the Red Lists.

▲ PRIMULA SUFFRUTESCENS *The only primula endemic to California is found in Yosemite National Park.*

Also from the Himalaya, and the mountains of south-west China and south-east Tibet, is the attractive ***Primula dryadifolia***, which is not on the Red Lists. Plants grow on rocky and stony slopes, cliff ledges, and moraines, often on limestone. Here they form dense mats or hummocks of woody stems, clothed with old leaves, with fresh, lax rosettes borne at the shoot tips. The leaves are deep green above, white beneath, and in summer rose to rose-crimson, dark- or yellow-eyed flowers are borne on short stalks. The species was first introduced into cultivation from Yunnan by George Forrest in 1911. Unfortunately, it is not normally a plant for the open garden, requiring humus-rich compost and a shaded cold frame to survive.

◄ PRIMULA DRYADIFOLIA *This beautiful and distinctive species from the Himalaya is a challenge for gardeners.*

Primula sherriffae, also unlisted, is found in south-east Bhutan and north-east India. It is one of the most unusual members of the genus, with extraordinary long-tubed lavender flowers, presumably an adaptation to long-tongued pollinating moths. It is a tender plant, and rare in cultivation.

Species are still being introduced from Asia. In 1994 the unlisted species ***Primula bracteata*** was brought into cultivation using seed legally collected by the Alpine Garden Society. It grows on dry limestone and sandstone cliffs in north-west Yunnan and south-west Sichuan, where it bears pink, primrose-shaped blooms. It is closely related to the well-known *Primula forrestii,* introduced by George Forrest in 1906.

The Himalaya and western China are not the only stronghold of the genus, although undeniably an important one. In Europe, the rather unusual *Primula palinuri*, listed as Vulnerable on the Red List of 1997, is found in coastal habitats in western Italy, centred on Capo Palinuro, where it grows in vertical crevices in sandstone or limestone rocks facing north, north-west, or west. Although some land has been protected from building, especially on Capo Palinuro, *Primula palinuri* is extremely vulnerable to development, and is included in a list of plants that require complete legal protection in Italy. The

Unlisted

Pulsatilla occidentalis

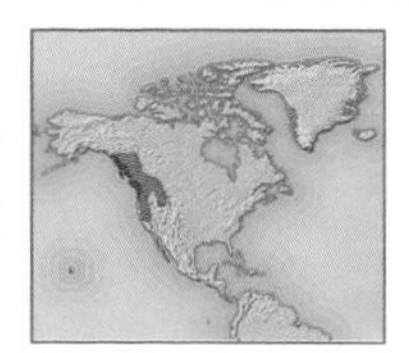

Botanical family
Ranunculaceae

Distribution
Western North America

Hardiness
Fully hardy

Preferred conditions
Moist but sharply drained soil in full sun; dislikes winter wet.

Some plants prove to be inexplicably tricky to cultivate and ***Pulsatilla occidentalis*** is one of these. It has tantalized and frustrated gardeners over the years, although the other pasque flowers are generally easy to grow. The species is native to high rocky slopes and meadows in western North America, from British Columbia to California, east as far as Montana. It is by no means rare, although certainly vulnerable in some places, including Mount Adams and Mount Rainier, in Washington. These areas are increasingly accessible to people, who undoubtedly damage the fragile habitat of this lovely alpine. The carrot-like, delicate leaves and flower buds emerge covered in golden hairs in the spring, as soon as the snow has melted. The flowers open cream, fading to white as the petals unfurl into a broad cup shape, with yellow stamens in the centre. In the wild, *Pulsatilla occidentalis* is found on slopes up to about 3,000m (9,800ft), flowering in late spring or even early summer at the highest altitudes.

Despite numerous introductions of seed over the years, it is not available to gardeners. Cultivated specimens are very poor relations to the splendid plants in the wild and this species is undoubtedly best appreciated in its natural habitat. However, many pulsatillas, although sometimes very slow from seed, flourish in the garden. These include *Pulsatilla alpina*, with white flowers and fluffy seed heads, and the widely available common pasque flower, *Pulsatilla vulgaris*, available in many colours. Grow them in well-drained soil in sun and avoid disturbance. Sow seed as soon as it is ripe.

AT HOME IN THE WILD ► *Despite forming large clumps in the wild, this American pasque flower seems to dislike cultivation.*

CULTIVATION

AURICULA THEATRES

Hybrids and cultivars of the European alpine species *Primula auricula* and *Primula hirsuta* were at the centre of a craze that reached its height with auricula theatres – large wooden display platforms with a black background. These were used to display collections of prize auriculas to full advantage. One still survives in Calke Abbey, Derbyshire, in the UK. Today auriculas are still important horticulturally. They are divided into three groups: alpine auriculas that are excellent for a rock garden; show auriculas that need alpine house conditions to survive; and border auriculas for the open garden.

species is a leafy, tufted perennial with rosettes of rather fleshy, bright green, unevenly toothed leaves up to 20cm (8in) long. The funnel-shaped, yellow flowers are relatively small, but are borne in large clusters. The large, petal-like leaves (bracts) at the base of each flower stalk are a distinctive feature.

California also has an important alpine primula, the only species native to the state. ***Primula suffrutescens*** is an unusual species, listed as Rare on the Red List of 1997. It is found only in a small area covering a distance of some 250km (155 miles) in Yosemite National Park, close to Lake Tahoe and Mount Whitney. Plants grow in granite crevices and screes at 3,300–4,000m (10,800–13,100ft), often close to melting snow near the summits. This primula forms woody mats, with living rosettes of dark green, wedge-shaped leaves covered in sticky glands only at the shoot tips. The rose-pink, red, or purple flowers, have a yellow 'eye' and are borne in clusters of up to nine. Recent DNA-sequence studies show its isolated position within the genus is not just geographical. The plant is now considered to be more closely related to the allied genus *Dodecatheon* than to *Primula* itself. Although it has been in cultivation since 1884 it has never been commonly grown and has proved to be unwilling to flower.

PRIMULA BRACTEATA *Although this rare species from China prefers to be grown with the protection of an alpine house, it is proving very popular with specialist growers.*

PRIMULA AUREATA ► *Much sought after in cultivation, this species has been the focus of plant hunters in Nepal in the past. A rare and very localized primula, it grows in damp valleys, high in the Himalaya, north of Kathmandu.*

Unlisted

Ranunculus weberbaueri

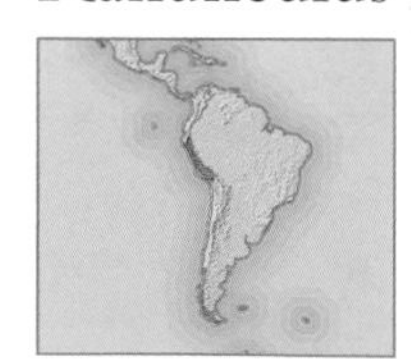

Botanical family
Ranunculaceae

Distribution
Peru: Andes

Hardiness
Half-hardy

Preferred conditions
Moist to almost boggy soil in partial shade, and a cool, humid climate.

A highly unusual buttercup of great distinction, ***Ranunculus weberbaueri*** grows in wet grassy and rocky places, and cliff ledges, and is found only in the Peruvian Andes at 3,400–4,300m (11,150–14,100ft). Apparently unusual in the wild, *Ranunculus weberbaueri* has made several entrances into cultivation in the past, both via uprooted plants and from seed. Unfortunately, neither method has induced it to survive outside its natural habitat. The greatest success has been recorded by the Royal Botanic Garden Edinburgh, where the plants are grown in a moist, gritty loam with added humus, in a partly shaded, well-ventilated cold frame.

This is a plant that will never be easy for the amateur gardener and today is very rarely seen. A trek to the Peruvian Andes would be required to admire its alpine perfection. Its survival in the wild is almost certainly due to its remoteness and general inaccessibility.

◄ UNUSUAL BUTTERCUP *The large flowers, up to 5cm (2in) across, arise from a rosette of thick, kidney-shaped leaves, which are silky with hairs beneath.*

HABITAT

FAR FROM THE LAWN

Although buttercups are familiar in lawns and meadows, there are also many species from high alpine habitats. Here they grow in harsh environments, such as this scree slope above the Fox Glacier, Westland, New Zealand. To survive in the shifting scree, plants need widely spreading root systems to anchor them. Requiring full sun and sharply drained soil, many of these large-flowered, high-altitude buttercups are sold in specialist nurseries.

1 on Red List*

Raoulia

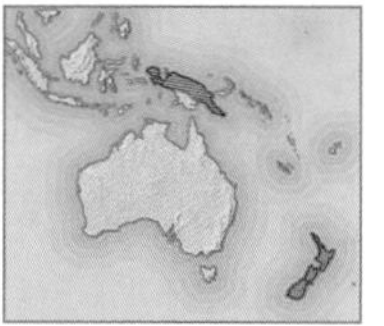

Botanical family
Asteraceae/Compositae

Distribution
New Zealand and New Guinea

Hardiness
Fully hardy

Preferred conditions
Cool, damp summers and dry winters, gritty, humus-rich, moist but sharply drained soil in full sun, or partial shade in warmer areas.

The hummock-forming species of *Raoulia* are among the most spectacular cushion-forming plants to be found anywhere in the world. They are commonly called vegetable sheep, because the white hummocks grow in groups that look like flocks. The genus *Raoulia* contains 20 species confined to the mountains of New Zealand and New Guinea. In New Zealand, road building, skiing, and overcollection of seed have put their fragile communities at risk, and they receive local protection. One species, *Raoulia cinerea*, is listed as Endangered on the Red List of 1997.

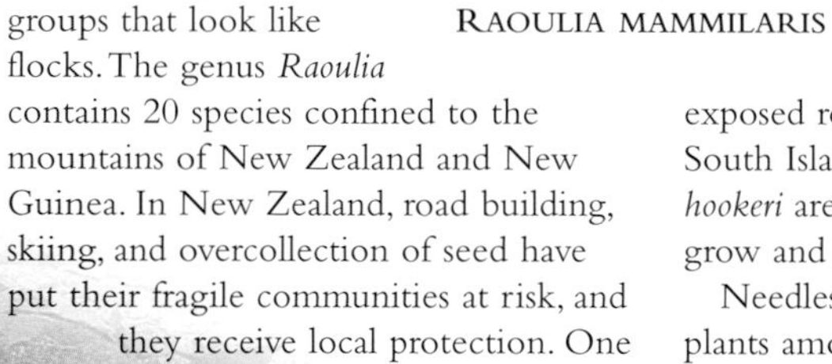

RAOULIA MAMMILARIS

Raoulia eximia, found in rocky places in the mountains of New Zealand's South Island, is one of the most challenging alpines to cultivate. The species is painfully slow-growing from seed and so the large wild plants, up to 2m (6ft) across, must be extremely old. ***Raoulia mammilaris*** is a localized and protected species from exposed rocks and stabilized screes on South Island. *Raoulia australis* and *Raoulia hookeri* are less spectacular but easier to grow and more widely available.

Needless to say, raoulias are popular plants among collectors of unusual alpines, although they are difficult to cultivate successfully. The species need the protection of an alpine house, where they are best grown in a mix of equal parts loam, leaf-mould, and sharp sand, with a top-dressing of grit. Seed is difficult to germinate, but cuttings can be rooted in summer.

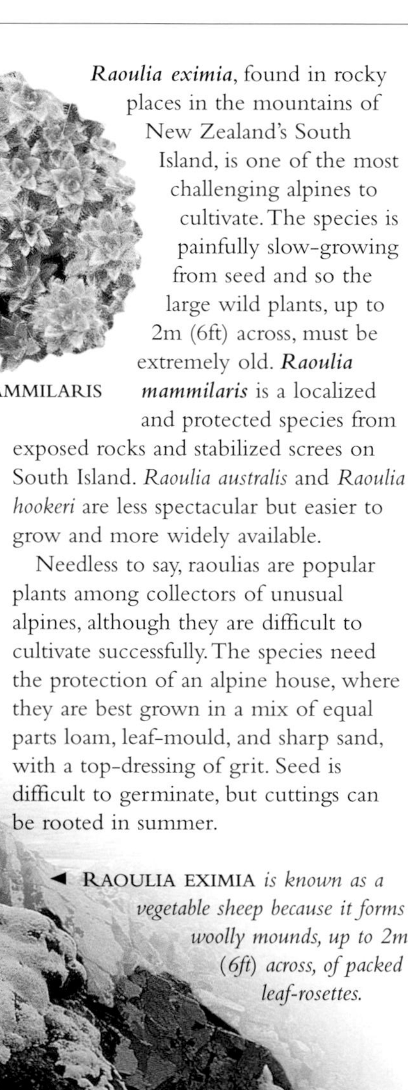

◄ RAOULIA EXIMIA *is known as a vegetable sheep because it forms woolly mounds, up to 2m (6ft) across, of packed leaf-rosettes.*

62*/10 on Red List

Rhododendron

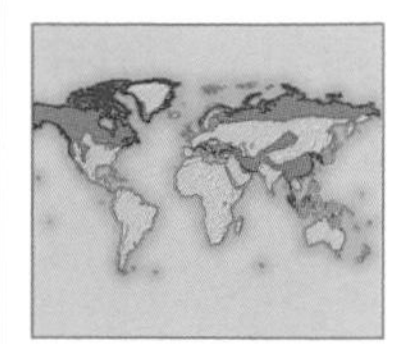

Botanical family
Ericaceae

Distribution
Widespread

Hardiness
Fully hardy – frost-tender

Preferred conditions
Acid, moist but well-drained, humus-rich soil in full sun in cool climates, or partial shade where summers are very hot.

Rhododendrons and their thousands of hybrids and cultivars are deservedly popular. Although those most familiar to gardeners are large shrubs and trees, there are many dwarf alpine species from mountains and tundra in the northern hemisphere, with a concentration of species in the Himalaya.

Rhododendron lowndesii is a refined, deciduous, carpet-forming alpine shrub endemic to central and western Nepal. Here it inhabits rock ledges and slopes at 3,000–4,600m (9,850–15,100ft), in moist, mossy places with peaty soil, often in shade or semi-shade. The delicate, flat flowers are pale yellow, and lightly spotted or flushed with red. Its inaccessibility prevents collection from being major threat; it is also difficult to maintain in cultivation. However, in the wild and in gardens it often hybridizes with the larger, and more common, unlisted *Rhododendron lepidotum*. *Rhododendron lowndesii* is classified as Rare on the Red List of 1997.

SMALLEST RHODODENDRON

Rhododendron ludlowii is a tiny, spreading or creeping evergreen shrub with yellow, bowl-shaped flowers flecked with reddish-brown. It grows in the south-eastern corner of Tibet, at 4,000–4,300m (13,100–14,100ft) on mossy slopes among other dwarf shrubs. The species does not appear on the Red Lists and is quite well protected by its remoteness.

The very high alpine rhododendrons are rare in cultivation but there are many species suitable for the garden, such as the alpenrose, ***Rhododendron ferrugineum***. This unlisted species carpets the mountains of central Europe with its pink blooms in early summer.

Alpine rhododendrons can be grown in full sun in cool climates, but otherwise prefer partial shade. The mat-forming species from very high altitudes need protection within the controlled conditions of an alpine house. All require acid soil conditions; in areas with neutral or alkaline soil, grow plants in containers or raised beds filled with lime-free (ericaceous) compost. Sow seed, if available, in a cold frame, in lime-free propagating compost as soon as ripe. Alternatively root semi-ripe cuttings in autumn.

Rhododendron ferrugineum
Known as the alpenrose, this compact rhododendron is found in the mountains of Europe (seen here in the Pyrenees) and is widely available in nurseries.

64 on Red List*

Saxifraga

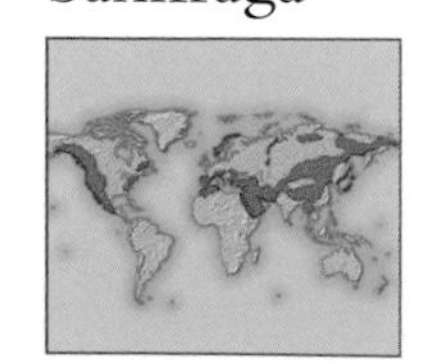

Botanical family
Saxifragaceae

Distribution
Northern hemisphere: mountainous regions

Hardiness
Fully hardy

Preferred conditions
Full sun and well- to sharply drained, neutral or slightly alkaline soil.

Saxifrages appeal not only to alpine enthusiasts, but also to many gardeners. There are over 400 species of these pretty alpines, which are mostly evergreen perennials, from mountainous areas all over the northern hemisphere. The plants range from those suitable for a border or rock garden, to species that need specialist care in a scree bed or alpine house. Saxifrages are popular with collectors because such variety can be found in the genus. Many species, such as ***Saxifraga burseriana*** and ***Saxifraga sancta***, although at risk from habitat loss, are rarely threatened by collection because they are already widely available. However, some rare and prestigious species are still a focus for collectors.

◄ SAXIFRAGA SANCTA *is a neat species from north-eastern Greece. Popular in its own right, it is one of the parents of* Saxifraga x apiculata *and its cultivars.*

Few alpine plants have been more eagerly awaited by saxifrage enthusiasts than *Saxifraga columnaris*, a stunning gem from the Caucasus mountains that was discovered in 1892, but seed was not collected until the 1990s. The species is Indeterminate on the Red List of 1997, but grows in such remote places that climate change seems the only threat. *Saxifraga columnaris* forms dense, silvery-grey cushions of small lime-encrusted rosettes. Venerable specimens of up to 50cm (20in) across have been recorded in the wild but plants are much smaller in cultivation. It is still rare in collections and precious plants are grown carefully in alpine houses. If propagation makes it more common in cultivation, it could be tried in sunny raised beds and troughs, with some protection from winter rain.

The ancient king, *Saxifraga florulenta,* is an extraordinary species, thought to be the last remnant of a once-common group. It is one of the few monocarpic saxifrages – that is, it flowers once, sets seed, then dies. In the first years it forms symmetrical, deep green rosettes. Flowering is quite an event when it happens – in midsummer a spike emerges from the centre of the rosette, reaching up to 45cm (18in) in length, bearing numerous small, flesh-coloured flowers. *Saxifraga florulenta* grows on shady, granite cliffs at 2,000–3,250m (6,550–10,650ft) in the Maritime Alps. The few populations that are known straddle the French–Italian border and are heavily protected by both countries. Flowering is never prolific and regeneration is very poor, so the long-term survival of this interesting species is highly uncertain and it is listed as Rare on the Red List of 1997. The species was probably never common, but in the 20th century overcollecting certainly did much harm. Today, very few plants exist in collections and there are even fewer records of it ever flowering in cultivation. On no account can seed be gathered from the wild, so it is very unlikely this extraordinary species will ever be available to gardeners again.

◄ SAXIFRAGA FLORULENTA *is known as "the ancient king".*

▼ SAXIFRAGA BURSERIANA *Found wild in the Alps, and often cultivated in gardens, this species forms tightly packed, grey-green cushions and flowers in spring.*

▲ STARRY FLOWERS *Hidden beneath the flowers is a dense, very neat, grey-green cushion, attractive even after flowering.*

Red List: Vulnerable*

Trachelium asperuloides

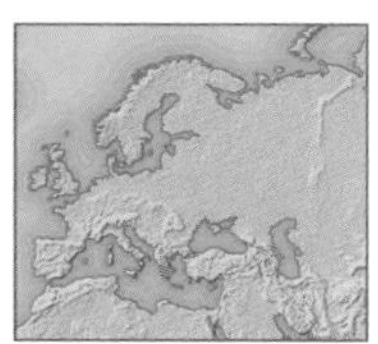

Botanical family
Campanulaceae

Distribution
Greece: northern Peloponnese

Hardiness
Fully hardy

Preferred conditions
Sharply drained, alkaline soil in full sun; dislikes winter wet.

This Mediterranean genus is already represented in gardens in the form of *Trachelium caeruleum* and its cultivars, herbaceous perennials that have become popular cut flowers in recent years. In common with the handful of other *Trachelium* species, they bear starry, narrow-petalled flowers that are highly attractive to butterflies.

The most distinctive and the only widely grown alpine species is ***Trachelium asperuloides***. In late summer the plant is covered in lilac flowers, borne in such large numbers as to conceal the dense cushion of leaves beneath. The species is rare in the wild, endemic to limestone cliffs in the northern Peloponnese in Greece, where it inhabits the upper valley of the river Styx, famous in Greek mythology as the entrance to Hades. In cultivation it is not very long-lived, and does not flower as profusely as in the wild. Plants are best grown in an alpine house in deep, gritty, well-drained compost. Sow seed, when available, in late winter; germination takes place in the early spring. Seedlings need to be pricked out and potted on as soon as they are large enough to handle.

71 on Red List*

Viola

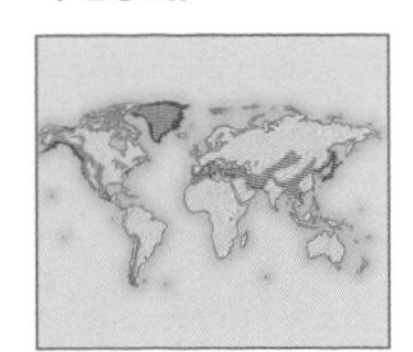

Botanical family
Violaceae

Distribution
Widespread: mountainous regions

Hardiness
Fully hardy – half-hardy

Preferred conditions
Sharply drained soil, not too rich, in full sun or partial shade; dislikes winter wet.

The genus *Viola*, which contains the much-loved pansies and violets, is a large group of about 500 species. They have a cosmopolitan distribution, although the majority of species are found in the cooler regions of the northern hemisphere. Many are excellent plants for the open garden – even some of the alpine species if they are sited with care. However, the high alpines are better grown in troughs and raised beds, or in blocks of tufa. A few species need alpine-house conditions and specialist care if they are to thrive, and these are considered among experts to be challenging and prestigious to grow.

The rare North American *Viola flettii* is one such desirable plant, with metallic leaves and pretty flowers in summer. This species requires skill in cultivation. It is endemic to the Olympic Mountains in Washington State, at 1,500–1,800m (4,900–5,900ft). Overcollection has severely reduced numbers in its accessible locations, especially on Hurricane Ridge, where further disruption is caused by a popular ski and leisure resort. As a consequence, *Viola flettii* is now carefully protected within the Olympic National Park, and is Vulnerable on the Red List of 1997.

The species ***Viola delphinantha***, *Viola kosaninii*, and ***Viola cazorlensis*** form a unique trio, being the only subshrub violas in Europe. They have a charm that has endeared them to alpine gardeners over many years, although they are not very easy to maintain in cultivation. As a consequence they have been repeatedly re-introduced from the wild as seed, which germinates easily. They generally need alpine-house conditions but can be grown outdoors on a block of tufa with protection from winter wet. They are all considered Rare on the Red List of 1997.

◄ VIOLA CAZORLENSIS *Endemic to Sierra de Cazorla, Spain, this viola is protected within the boundaries of a national park. It is available from specialist nurseries.*

VIOLA DELPHINANTHA ► *This localized plant inhabits limestone rock crevices, and occasionally coarse, stabilized screes. In several of its haunts it falls within national parks. Cultivation is best left to experts.*

Of the three, *Viola delphinantha* and *Viola kosaninii* are very closely related in both habit and habitat. They have thin, ascending stems arising from a woody base, deep green, linear leaves, and summer flowers. *Viola delphinantha*, from northern Greece and southern Bulgaria, has rose-lilac to reddish-violet flowers with a long, slender, downcurved spur whose length greatly exceeds the petals. *Viola kosaninii*, from eastern Albania and the neighbouring parts of the Macedonian Republic, differs in having paler, lilac-pink flowers with neatly notched lower petals. The third species, *Viola cazorlensis*, is endemic to the Sierra Cazorla in south-eastern Spain. It is larger in stature than its cousins and has flowers intermediate between the two – rich purplish-pink with the lower petals slightly notched at the tip. This species is easier to maintain in cultivation and can thrive outdoors if planted in a rock crevice or tufa block, shaded from the midday sun in summer. Plants are generally raised from cuttings.

Unlisted

Weldenia candida

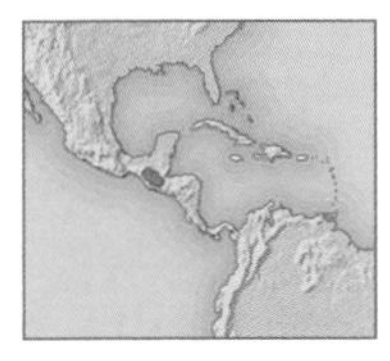

Botanical family
Commelinaceae

Distribution
Guatemala and Mexico

Hardiness
Half-hardy

Preferred conditions
Moderately fertile, sharply drained soil in full sun; dry winter dormancy.

The only species in the genus, the tropical alpine ***Weldenia candida*** was introduced into cultivation at the Royal Botanic Gardens, Kew, in 1893, not long after its discovery in a volcanic crater in Guatemala. Since then the species has been found in several other volcanic, mountainous areas in Guatemala and adjacent areas of Mexico.

Weldenia candida is a sturdy herbaceous perennial with fleshy, tuberous roots. It produces a tuft of foliage at ground level, directly from the crown of the plant, of strap-shaped leaves with somewhat wavy margins. The flowers are borne in long succession in late spring and early summer in the centre of each leaf tuft. They are rather like three-petalled crocus flowers, and are pure white, with prominent bright yellow stamens within. The flowers open only in sunshine, remaining closed during dull weather. *Weldenia* is a member of the family Commelinaceae, which also includes the hardy genus *Tradescantia*. In mild climates it can be cultivated outside with good results in gritty, very well-drained soil, particularly in raised beds. Within its natural habitat, temperatures as low as –6°C (21°F) have been recorded, but it is not reliably hardy in temperate gardens. In areas with cold, damp winters it is best as an unusual and showy subject for an alpine house, or well-ventilated glasshouse. Under cover choose a bright, sunny position and use extra-deep pots filled with a gritty, humus-rich compost. Plants can be watered freely during summer but require a gradual drying out as they begin to die down, preferring to be kept only very slightly moist during their winter dormant season. Unfortunately, fruits are rarely borne on cultivated plants, but, if available, seed should be sown as soon as it is ripe. The plants steadily increase from a single tuft to a closely packed clump, which can be lifted and separated into one-shoot pieces. Each shoot should be replanted in its new position immediately. Division is best carried out when the plants die down in autumn, or when growth commences in spring. However, perhaps the most reliable means of increase in the absence of seed is by cuttings of the thickest and healthiest roots taken in the autumn.

◄ VOLCANO FLOWER *This is an unusual alpine from high in the volcanic region of Central America. It is fairly widely available from specialist nurseries.*

FERNS

The 12,000 species of fern found throughout the world show tremendous diversity in size and shape, ranging from minuscule filmy ferns that are a few cells thick, to species of tree-like dimensions; some have typical fern-like fronds, while others bear little resemblance to a typical fern. Despite their physical diversity, virtually all ferns share certain characteristics – the production of free-floating spores, instead of seeds as in conifers and flowering plants, and a life cycle comprising an alternation of two very distinct generations that live independently of each other.

FERN DIVERSITY

The ferns belong to a group of plants known as pteridophytes, which includes the so-called fern allies: the clubmosses, horsetails, spike mosses, and whisk ferns. These are primitive plants that share a fern-like life cycle.

Ferns and their allies exploit a wide range of habitats and adopt a diversity of lifestyles. Most are terrestrial, evergreen, or deciduous perennials; a few form a woody trunk, like the rough tree fern, *Cyathea australis*, a native to Norfolk Island, where it reaches 28m (90ft) in height. Other ferns, such as the giant *Drynaria rigidula*, are epiphytic or rock-dwelling, while *Lygodium* species climb to reach the light of the forest canopy. At the other end of the spectrum, the tiny fairy moss, *Azolla filiculoides*, lives in water as a free-floating fern.

CLUBMOSSES, *such as the ground pine,* Lycopodium obscurum, *favour moist temperate and tropical environments.*

HORSETAILS *of modern times are much smaller than their giant ancestors, which grew in vast swamps 350 million years ago.*

WHISK FERNS, *such as this* Psilotum nudum *are the only living vascular plants that are lacking both true roots and leaves.*

SPIKE MOSSES *are related to clubmosses. This* Selaginella lepidophylla *is a desert dweller, but most live in humid, shady forests.*

FERN HABITATS

Ferns are most commonly found in damp or humid, sheltered, and shaded environments and occur on the forest floor throughout temperate zones, but are most abundant in the tropics and subtropics. Here, they range from low-altitude rainforest to high-altitude cloud forest. All ferns rely on a film of free water to complete their life cycle, and this requirement makes them less common in arid areas than in other habitats. There are, however, many ferns that are well adapted to life in dry places; *Selaginella lepidophylla*, for example, a resurrection fern, exists in arid areas of Texas and New Mexico, USA, as a ball of curled brown fronds until seasonal rainfall transforms it into a lush, green rosette.

◄ CYATHEA AUSTRALIS, *or the rough tree fern, experiences severe conditions, including frost and snow, in its native mountain forest habitats in south-east Australia.*

FERN FOSSILS

Ferns and their allies were among the first vascular land plants to evolve, and were the dominant plants for millions of years. The earliest fossils of ferns and their allies date back some 400 million years, long before the evolution of conifers and flowering plants. During their life on Earth, their ecological success has fluctuated. They were very prolific and dominated the vast swamps of the Carboniferous period (*see pp. 26–27*). In fact, fossil ferns form a large proportion of the 7 billion tonnes of coal that were laid down then. However, at the end of this period the climate changed and these vast swamps dried up. Ferns and fern allies were ill suited to the new drier, cooler climate, and many became extinct. Even today, ferns are most abundant in the warm, humid environments of the tropics.

▲ FOSSIL FRONDS *in coal measures look very like their modern counterparts.*

TREE FERNS *dwell most frequently in forests, thriving in a sheltered environment. They often form a significant component of the forest understorey, as here in the eucalypt forest of the Yarra Ranges National Park, in Victoria, Australia.*

LIFE CYCLE OF A FERN

All ferns alternate between two distinct kinds of plants in their life cycle, and this phenomenon is known as "alternation of generations". The most familiar stage of the life cycle is the sporophyte generation – this is the mature plant with roots, stems, and fronds. Typically beneath the fronds are clusters of spore cases, or sporangia, all clumped in an organ known as a sorus (the plural is sori).

Inside each sporangia are hundreds or thousands of minute spores which, once released, can float long distances on air currents. If a spore lands on a suitably moist substrate, it will germinate into a tiny, often heart-shaped plant, called a prothallus. This represents the gametophyte generation. The gametophyte plant is rarely more than a few millimetres in size, and is most often seen as a small green blob growing on a moist surface.

Spores contain only half the number of chromosomes of the adult fern, but the gametophyte produces male and female sex cells. After fusion of male and female cells, that is, after fertilization, a new sporophyte grows, with a full complement of chromosomes. This is the adult fern, with roots, stems, and fronds.

◄ AT MATURITY *a fern frond produces many sporangia in sori, usually on the underside of the fronds; the patterns and alignment of sori are characteristic of each species and are useful in classifying and identifying ferns.*

LIFE ON LAND

The ferns, although well adapted to life on land during the adult part of their life cycle, depend on water to reproduce. The spores germinate only in the presence of moisture. A film of water prevents desiccation of the delicate prothallus, and provides the medium through which sperm swim to fertilize the eggs. This is why most ferns occur naturally in damp places. Conversely, many arid-habitat ferns use vegetative means to reproduce.

A few ferns occur only in fresh water, rooted in the bottom of shallow lakes or free-floating; this group frequently become invasive in foreign climates, often having escaped from garden ponds.

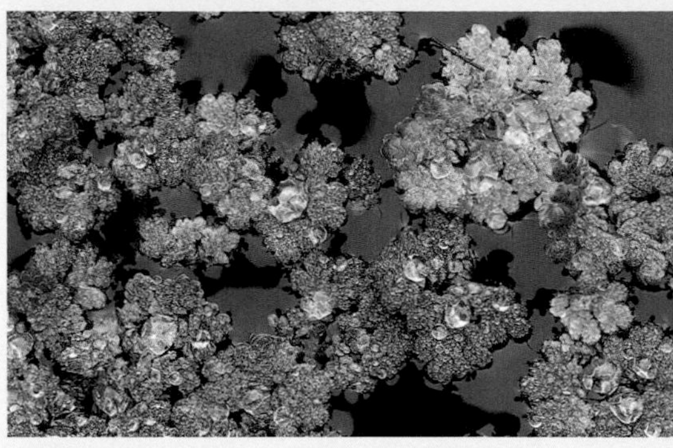

▲ AZOLLA FILICULOIDES *Fairy moss is a freshwater aquatic fern exported from tropical America, and grown for its delicate appearance and fiery autumn colour. It is now a pest in waterways in southern Britain and other countries.*

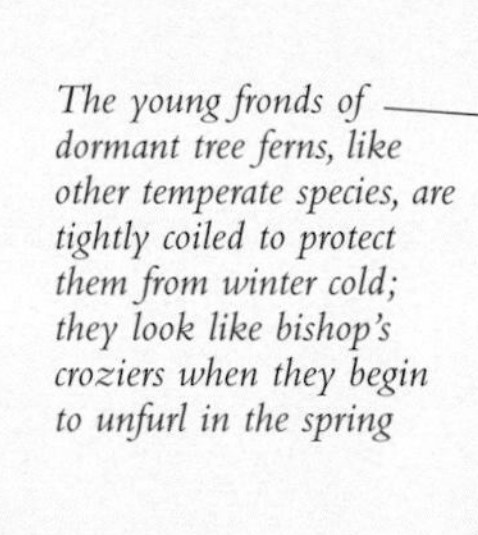

The young fronds of dormant tree ferns, like other temperate species, are tightly coiled to protect them from winter cold; they look like bishop's croziers when they begin to unfurl in the spring

COLLECTING FRENZY

▲ FERNS *like* Hymenophyllum wilsonii *have fronds a few cells thick, so thin that they are translucent.*

Perhaps one of the most illuminating examples of the effects of overcollection happened in Britain in the 19th century. The fashion for ferneries and grottoes in the gardens of the 1840s and 1850s was sparked in part by Thomas Moore's handbooks on ferns, and a general passion for field botany. The delicacy and intricacy of ferns was well suited to Victorian tastes, and ferns not only decorated gardens, but were also reproduced on all manner of household items, from fireplace tiles to antimacassars. As the century progressed, enthusiasm turned to mania – pteridomania – and unscrupulous collectors supplied the demand by the wholesale stripping of vast tracts of countryside in search of the unusual. They collected many mutations with crisped, curled, or finely divided fronds, relatively few of which are still known in cultivation. By the time the fashion peaked in the 1870s, many species had been extirpated. Those that survived in the wild, such as *Woodsia ilvensis*, have not recovered, and in fact many are still declining. Even those in cultivation in the greenhouses at the great estates were lost with the decline of the estates between the two World Wars.

◄ WARDIAN CASES, *sealed glass cases for transporting plants, were invented in the 19th century, and proved very useful for the cultivation of delicate ferns that needed a reliably humid atmosphere.*

CONSERVATION

The distribution patterns and conservation status of ferns are relatively poorly known compared with many conifers and flowering plants. The majority of ferns are tropical, but sampling in the tropics is uneven, leaving great gaps in our knowledge. Listings in the Red List are far from complete; each species may take several years to assess fully. Ferns are often particularly vulnerable on tropical islands. For example, Mauritius has 250 species and varieties of fern, 48 of which have not been seen in recent years and are presumed extinct; 17 fern species are found nowhere other than Mauritius. As is true for many tropical island plants, one of the greatest threats is from introduced plant species, such as Chinese guava and privet. Other threats that are common include habitat loss due to logging, land clearance for agriculture, human settlement, and urbanization including the building of access roads for tourism.

▲ WOODSIA ILVENSIS, *Britain's rarest fern, is a small, evergreen fern that inhabits niches in the moist rocks in the uplands of northern England, North Wales, and Scotland. One Victorian fern handbook claims that "Their rarity rather than their beauty invests these ferns with interest for the cultivator."*

GARDENER'S GUIDE

- Never collect fern plants from their wild habitats.
- If purchasing exotic ferns for landscaping, especially CITES-protected tree ferns, the *Cyathea* or *Dicksonia* species, ask to see evidence of their provenance, and check that they have been collected under licence, or, preferably, grown in cultivation. Tree fern spores germinate readily and small plants raised from spores establish easily in gardens.
- Many ferns are easily grown from spores. Germinating fern spores is much like germinating seeds of flowering plants, but because the spores are so much smaller, special care is needed. Spores can be collected from mature fronds of living cultivated plants, or you can join one of the many fern societies around the world, many of which have active spore exchange programmes. For example, the British Pteridological Society Spore Exchange 2003 lists spores of over 640 available species.
- Growing ferns such as the Boston fern, *Nephrolepis exaltata* (*above right*), which is produced by the million in commercial cultivation, presents no threat at all to ferns in the wild. This species has given rise to the largest known group of decorative fern cultivars. Most of these are not hybrids but sports or mutations, a common phenomenon in the ferns.

FASHION VICTIM *Many* Dicksonia *tree ferns from temperate regions of New Zealand and Australia are now fashionable for landscaping in European gardens. Wild-collected specimens have been put on sale without fronds or roots – and with no chance of survival.*

14 on Red List*

Adiantum

Botanical family
Adiantaceae/Pteridaceae

Distribution
Worldwide

Hardiness
Fully hardy – frost-tender

Preferred conditions
Moderately fertile, humus-rich, moist but well-drained soil in partial shade.

The maidenhair ferns are very familiar, both in the wild and as garden or house plants. They are admired for their elegant, evergreen or deciduous fronds, which unfurl a delicate purplish-pink. The fronds are made up of leaflets (pinnules) with a peculiar, oblong or diamond shape. Spore cases are borne on the undersides of fertile fronds along the margins of the leaf segments, and their arrangement is often used to identify species.

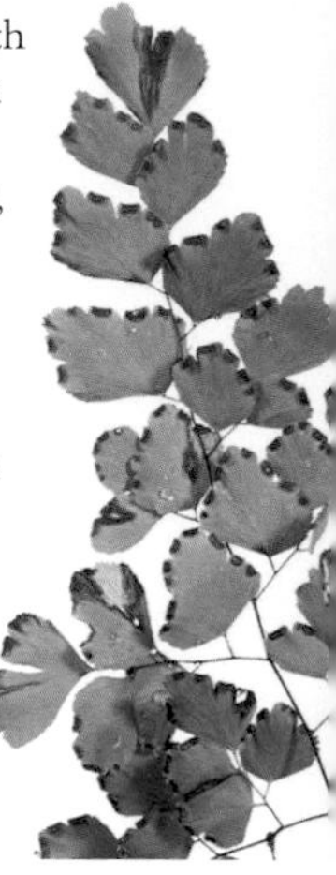

ADIANTUM SPORES

There are over 200 maidenhairs worldwide, both in temperate climates and more abundantly in the tropics and subtropics, especially in the Americas. They are at their most numerous in the moist, mountain forests of the Andes. Maidenhairs vary in size from *Adiantum reniforme* var. *sinense* – just 5cm (2in) in height – to the giant maidenhair, ***Adiantum formosum,*** which can spread to 2m (6ft) or more.

NORTHERN EXPOSURE

One of the most widespread maidenhairs, and a familiar house plant, is *Adiantum cappillus-veneris.* The species is found throughout tropical and temperate regions, but is deciduous in cooler areas. In Canada, where it is known as the southern maidenhair fern, it reaches its northern limit – more than 1,000km (620 miles) further north than the nearest populations in the main range of the species. The fern's survival in Canada requires the warm, humid microclimates created by hot springs in south-eastern British Columbia, where it was first recorded in 1888. Then it was fairly abundant, but the development of a spa resort around the springs has reduced it to only two locations. Both populations

◄ PERFECT FOR MAIDENHAIRS *These delicate ferns are often found in shady rock crevices where the atmosphere is warm and humid. They thrive by this waterfall in the Grand Canyon National Park, Arizona.*

THREAT

THREE GORGES DAM

The Yangtze River is the third largest river in the world, powerful enough to have carved out the Three Gorges, where the Chinese government is building a 2km (1¼-mile) long dam. The Three Gorges Dam project, the biggest of its kind, will create a 600km (370-mile) long reservoir by 2009. Benefits include hydroelectric power, a passage inland for ocean-going ships, and prevention of the floods that have claimed more than a million lives in the 20th century alone. Unfortunately, it will also displace 1.2 million people, flood 400-year-old archeological sites, drown fertile farmland, and destroy the habitat of the rare *Adiantum reniforme* var. *sinense*, as well as many other endemic species.

seem to be sterile because they die down in winter before spores are produced; they can only increase via spreading underground stems (rhizomes). Although *Adiantum cappillus-veneris* is not on the Red List, it is protected under the Canadian Species at Risk Act (SARA).

The primitive species *Adiantum reniforme* was apparently widespread until the last ice age drove it from most of its range. Today there are three isolated groups found where the ice sheets reached their southern limit: in Kenya, the Canary Islands, and China. The Chinese population, *Adiantum reniforme* var. *sinense,* is endemic to warm, humid sites in Sichuan and Wanxian. Plants are only 5–20cm (2–8in) tall, with striking, dark green, kidney-shaped leaves, each patterned with one to three concentric, circular lines. Much of their native habitat is under threat as a result of the Three Gorges Dam (*see box*), and *Adiantum reniforme* var. *sinense* is Endangered on the Red List of 1997.

MAIDENHAIRS IN TROUBLE

Other threatened maidenhairs include *Adiantum vivesii,* which is classified as Indeterminate on the 1997 Red List, and as Endangered by the US Endangered Species Act. This species is known from about 1,000 individuals found on private land in northern Puerto Rico. Plants grow in shaded hollows at the base of limestone cliffs and this very narrow distribution, and potential interest to specialist growers, threatens its survival. *Adiantum viridimontanum,* the Green Mountain maidenhair fern, is found in a total of 21 populations; seven in Vermont and fourteen in Quebec. It is naturally restricted by its preference for mineral-rich serpentine soils, and the outcrops of serpentine rock from which these soils are made. Many of the Canadian plants grow in old or active asbestos mines where further mining could disturb them, and across the whole range, road construction and widening schemes, logging, and overcollection all pose a threat. *Adiantum viridimontanum* is classified as Endangered on the Red List of 1997.

◀ ADIANTUM FORMOSUM *The giant maidenhair fern from eastern Australia and New Zealand is popular in cultivation for its ornamental foliage and elegant habit. It does not appear on the Red List.*

Red List: Endangered*

Angiopteris chauliodonta

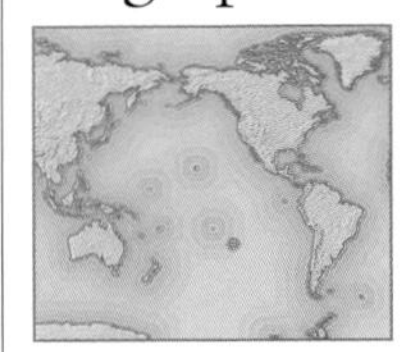

Botanical family
Marattiaceae

Distribution
Pitcairn Islands

Hardiness
Frost-tender

Preferred conditions
Warm, humid atmosphere; moist but well-drained soil with plenty of organic matter; deep or partial shade.

Endemic to Pitcairn and its neighbouring islands in the Pacific, this striking terrestrial fern grows to 3m (10ft) tall and produces fronds up to 4m (13ft) long. *Angiopteris* is an ancient genus that has changed little since the time of the dinosaurs, although the scars left by old frond-bases indicate that these ferns used to produce much larger fronds than they do today.

Small populations of *Angiopteris chauliodonta* are found scattered over the islands. They seem quite healthy and are reproducing both sexually via spores, and vegetatively from the bases of the fronds. However, whether the young ferns will survive to maturity is uncertain because they tend to grow on steep, unstable slopes that are prone to landslips.

See also Invasive plants, *p.444.*

HABITAT

PARADISE LOST

Pitcairn Island lies in the South Pacific, midway between Peru and New Zealand, with only three uninhabited volcanic islands nearby. Pitcairn is also volcanic, with a total area of just 47 sq km (18 sq miles) and a tropical climate. The mutineers of the *Bounty*, led by Fletcher Christian, landed on Pitcairn in 1790 with their Polynesian companions, after ridding themselves of their captain, William Bligh. Since the island was settled, much of the natural vegetation has been cleared for fuel and farming. This has led to soil erosion, and invasive species have gained a firm foothold. In addition, native plants are suffering from a critically small gene pool. Neighbouring Henderson Island has remained virtually untouched and is a World Heritage Site.

39*/1 on Red List

Asplenium

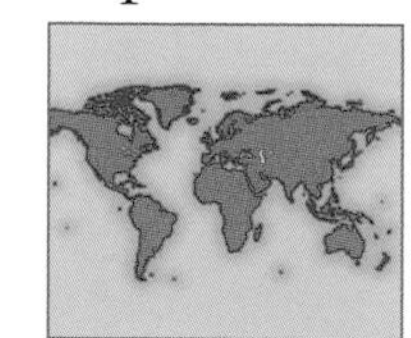

Botanical family
Aspleniaceae

Distribution
Worldwide

Hardiness
Fully hardy – frost-tender

Preferred conditions
Humus-rich, moist but well-drained soil with grit; partial shade.

There are about 700 widespread and diverse aspleniums, or spleenworts, from every continent except Antarctica. They vary from hardy plants, such as the Unlisted ***Asplenium ceterach***, to tender tropical species like ***Asplenium nidus***, also Unlisted. Aspleniums are occasionally sold under the synonyms *Ceterach* or *Phyllitis*, and are grown in both temperate and tropical climates. Many make excellent container plants indoors or out.

Another popular garden fern, especially in its native Australia and New Zealand, is the unlisted hen-and-chicken fern, or mother spleenwort, ***Asplenium bulbiferum.*** This unusual species bears tiny plantlets on the ends of its fronds, which quickly grow if they fall off, or the mother frond touches the soil.

ASPLENIUM BULBIFERUM

Another of the most admired species is the widespread bird's nest fern, *Asplenium nidus*, often grown as a compact house plant in temperate regions, although in its native tropical forests it can grow to 1.2m (4ft) across. In the wild, this handsome spleenwort is epiphytic and grows high in the rainforest trees. The glossy, evergreen, undivided fronds overlap to form a bucket-like "nest" in the centre of the plant to catch falling leaves, which decompose into rich compost just as they would on the forest floor. Besides providing valuable nutrients for the fern, the leaf-filled nest is also a home for ants, termites, and even other plants – undoubtedly an important component of the forest ecosystem.

Other aspleniums depend on habitats rather than creating them, including the central European species *Asplenium adulterinum* subsp. *adulterinum,* listed as Rare on the Red List of 1997, and the Unlisted *Asplenium cuneifolium*. They are endemic to areas of serpentine soils with high levels of heavy metals and low levels of important nutrients – conditions that very few plants can tolerate. Sadly, almost half of the sites that suit these specialist species are threatened.

In the USA, the small fern *Asplenium fragile* var. *insulare* is endemic to the islands of Maui and Hawaii. Several factors threaten the native plants of Hawaii: grazing by feral sheep and goats; invasive alien plants, particularly fountain grass (*Pennisetum setaceum*); and military operations that churn up land and can spark destructive fires. *Asplenium fragile* var. *insulare* is almost certainly adversely affected, although little is actually known about how it lives and what kind of conditions it requires. What is certain is that, although Unlisted, its numbers have fallen to fewer than 300 plants.

ASPLENIUM NIDUS ► *In the wild a large bird's nest fern and its accumulated soil, water, and even ant and termite nests, can weigh up to 200kg (440lb) – a large and important resource for plants and animals in the canopy.*

ATLANTIC DIVIDE

Asplenium scolopendrium **var.** ***scolopendrium,*** sometimes called *Phyllitis scolopendrium* and commonly known as the hart's tongue fern, is widespread in Europe. with over 200 named forms in the UK alone. It is seen on limestone walls, in ravines, and in shady, damp places. However, the North American variety *Asplenium scolopendrium* var. *americanum* has been recognized since the 1880s as one of the rarest ferns. In 1989, it was deemed Threatened in its Entire Range by the US Fish and Wildlife Service, and it is listed as Rare on the Red List of 1997.

The species requires cool, heavily shaded, damp places on magnesium-rich limestone in caves, ravines, overgrown abandoned quarries, or rocky outcrops in dense forest. This specific habitat requirement has taken its toll because suitable sites have been altered by logging, building developments, fresh quarrying, and hiking trails. The main impact of these activities is to thin the shade from trees, allowing too much light, heat, and air movement into the understorey for the fern to survive. Insect infestations that cause shade trees to lose their leaves have also been shown to adversely affect fern numbers.

Encouraging people to grow the American hart's tongue is important to its recovery. *Asplenium scolopendrium* var. *americanum* prefers limy soil, with added crushed shells, and is reputed to be easier to grow than the European variety.

ASPLENIUM SCOLOPENDRIUM ► *The hart's tongue fern is found growing wild in many habitats in the UK and Europe. This damp fissure, or grike, in limestone pavement is an ideal environment.*

ASPLENIUM CETERACH ► *Known as the rusty-back fern, this Unlisted species, from mountains in Europe and Asia, has smaller, tougher leaves than tropical ferns. These enable it to survive in dry, rocky habitats.*

THREAT

INVASIVES IN HAWAII

Invasive non-native plants are a great threat to the native species of Hawaii. One of the most damaging is fountain grass (*Pennisetum setaceum*) – an aggressive plant that takes advantage of the blank canvas left after a burn. The grass thrives in the black, volcanic soils of Hawaii and is itself very flammable, so fires take hold more frequently and burn more intensely than they used to on the island, and the native plants cannot adapt. *Asplenium fragile* var. *insulare* has been affected by fountain grass, as has the endemic fern *Diplazium molokaiense*. Numbers of this fern have dwindled over the past 25 years: it was Rare on the Red List of 1997, and now there is only one individual left on Earth.

17 on Red List*

Botrychium

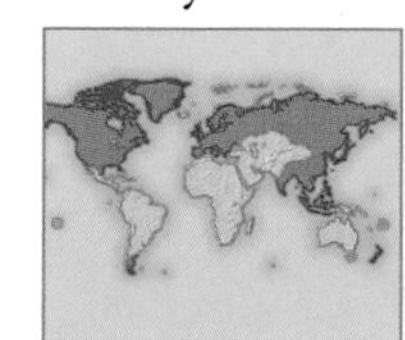

Botanical family
Ophioglossaceae

Distribution
Widespread; mainly northern hemisphere

Hardiness
Fully hardy

Preferred conditions
Rich, moist but well-drained, slightly acid soil with plenty of leaf mould; deep shade.

The grape ferns, or moonworts, are principally found in North America and eastern Asia, although there are scattered species elsewhere. Many are extremely rare, but *Botrychium simplex* is widespread throughout North America, southern Canada, and Europe, and ***Botrychium lunaria*** can be found across America, Asia, and Australia. Moonworts are difficult to grow in gardens because many species are extremely sensitive to soil conditions. The only exception is the North American rattlesnake fern (*Botrychium virginianum*). This species is Unlisted and easy to grow, but it is best to check that plants for sale have not been removed from the wild.

Botrychium ferns do not have typical fiddleheads – their fronds emerge in early spring fully formed, needing only to unfold and expand. The genus has another curious feature: like all ferns, its life cycle consists of two generations (*see p.290*). One is a usually short-lived gametophyte generation that exists while fertilization takes place, and the other is a long-lived sporophyte generation – the fern – which produces spores. But moonworts are unusual because they live underground for up to seven years as gametophytes, during which time they depend on fungal partners for food. Eventually an egg is fertilized and two fronds emerge: one photosynthetic frond with half-moon leaflets, and one frond bearing grape-like spore cases.

Most species, for example the Unlisted blunt-lobe grapefern, *Botrychium oneidense*, from North America, live in acid leaf-litter in shady woodlands and swamps. Like many forest-dwelling species, this one is threatened by invasive plants, changes in the canopy cover, and habitat disturbance, in addition to exotic earthworms (*see box*). Other species, such as *Botrychium pallidum*, the pale moonwort, also from North America, live in sand dunes, open meadows, and fields. Traditionally, grazing animals prevented scrub gaining a foothold in these areas, but increasingly livestock are reared indoors rather than sent out to pasture. The species is listed as Vulnerable on the Red List of 1997 and artificial disturbance may be needed to keep its habitats open.

WILDLIFE

WANDERING WORMS

Many indigenous North American earthworms are thought to have perished under the glaciers of the last ice age, 11,000 years ago. Although worms from the south have been migrating north since the ice sheet retreated, it is an extremely slow process and today much of North America still lacks indigenous earthworms. When Europeans settled in the New World, however, they accidentally brought hardy worms with them, and native forests which evolved without worms have since been invaded by the Eurasian species. Moonworts thrive in the moist, acid leaf-litter of the forest floor, but in places the introduced worms have devoured this litter layer and reduced its thickness by as much as half. As a consequence, many *Botrychium* species are threatened, including the goblin moonwort (*Botrychium mormo*), which is classified as Rare on the Red List of 1997.

▼ BOTRYCHIUM LUNARIA *This Unlisted moonwort is found on three continents but is tricky to spot. The fronds, which emerge in spring, are only 5–20cm (2–8in) tall.*

207*/6 on Red List/CITES II

Cyathea

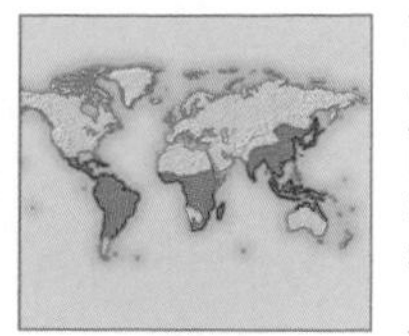

Botanical family
Cyatheaceae

Distribution
South America, Africa, south-east Asia

Hardiness
Half-hardy – frost-tender

Preferred conditions
Fertile, moist but well-drained soil; partial shade, or full sun in damp conditions.

The genus *Cyathea* consists of over 600 species of tree fern, mostly from cloud forests in the tropics and subtropics of the southern hemisphere. Some species are widespread, but others are naturally rare endemics, often found on a single island or mountain. Their trunks are not woody like those of trees, but are erect versions of the spreading underground stems (rhizomes) of other ferns. These narrow trunks, which can be up to 20m (70ft) in height, are covered with fibrous roots that take moisture from the humid atmosphere of the cloud forest. Although some cyatheas are common, many have very restricted distributions, including *Cyathea australis* subsp. *norfolkensis*, which is endemic to Norfolk Island in the South Pacific and is classified as Vulnerable on the Red List of 1997. Over 200 other species of *Cyathea* appear on the 1997 List, and their future is threatened by habitat destruction and collection for the horticultural trade. All *Cyathea* species are included in Appendix II of CITES.

TREE FERN TRADE

Species of *Cyathea* that are native to Australia are the most widely cultivated. The half-hardy rough tree fern, *Cyathea australis*, listed as Vulnerable on the Red List of 1997, should be purchased with care, but the tender scaly tree fern, *Cyathea cooperi*, is Unlisted. Considered fairly difficult to cultivate, these tree ferns can be grown successfully in moist, fertile, well-drained soil in shade, or sun if they are misted on very hot days. The main commercial value of the species, however, lies not in live plants, but in the trunks, which are used as a growing medium for epiphytes, especially orchids.

Most of the cyatheas that go for export, whether live plants or trunks, are harvested from the wild. In Australia and New Zealand, collection is responsible and well regulated, but this is not always the case in other exporting countries, and the provenance of material should be carefully established before purchase.

CYATHEA COOPERI ► *This fast-growing tree fern from Australia is popular with gardeners. The trunk reaches up to 5m (15ft) in height and is topped with impressive fronds to 4m (12ft) in length.*

FERNS

2 on Red List*/ CITES II

Dicksonia

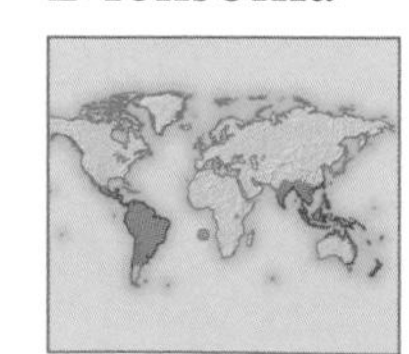

Botanical family
Dicksoniaceae

Distribution
South-east Asia, Australasia, South America

Hardiness
Half-hardy – frost-tender

Preferred conditions
Humus-rich, acid soil in partial or full shade and a humid atmosphere.

The most popular and widely available tree ferns are found in the genus *Dicksonia*, which consists of around 25 species distributed throughout the southern hemisphere. Perhaps the most familiar species to gardeners, at least in cooler climates, is *Dicksonia antarctica*, commonly known as the soft, or woolly, tree fern, which is Unlisted. In recent years it has become a very fashionable plant in European gardens, and is one of very few tree ferns that can survive the winter outdoors in northern temperate regions. The species is half-hardy and so needs a sheltered position with some extra insulation during very cold spells. Although the fern is evergreen in mild climates, it loses its leaves if there is a frost.

The thick, trunk-like stem (rhizome) can reach 15m (50ft) in the wild, although often far less in cultivation, and is usually erect, or occasionally creeping, and covered with fibrous roots and the remains of old leaf bases. The fiddleheads unfurl from the top of the trunk and spread to form a wide, glossy, dark green crown of fronds that can be up to 4m (12ft) long on mature plants.

The soft tree fern is native to Queensland, New South Wales, South Australia, and Tasmania, where it grows in moist gullies and cool temperate forests. Despite its current popularity, *Dicksonia antarctica* is not considered to be under threat in the wild at present, and indeed under Australian law, ferns for horticulture must be harvested in accordance with approved management plans. There has, nevertheless, been some local concern in Tasmania about the high level of harvesting: it is estimated that 80,000 to 90,000 Tasmanian ferns are exported annually. Generally speaking, however, the trade in wild plants is considered to be sustainable.

In New Zealand, the half-hardy, native species *Dicksonia squarrosa* and *Dicksonia fibrosa* are not currently threatened, and grow in abundance beneath plantation forests of pine. Their extraction for export, which takes place just before logging, is fully integrated into the commercial forest management system. Under New Zealand law, poaching is illegal and tree ferns can only be exported with a licence to prove that they have been harvested sustainably. Nevertheless, very large specimens, which can be hundreds of years old, can still tempt collectors.

In the early years of CITES, all species of ferns in the Dicksoniaceae family were included on Appendix II, but following a review they were removed from the listing with the exception of their Asian relative *Cibotium barometz*, which is used in traditional medicine, and the few South American *Dicksonia* species.

South American tree ferns

The reason for the continued protection of South American tree ferns is *Dicksonia sellowiana*, locally called *samambaiaçu*, and a native of the Brazilian Atlantic coastal rainforest. Although unclassified on the international Red List, this species is now considered Endangered on the national Red List of Brazil. This fern requires careful monitoring because the trunks are used to make a very popular orchid growth medium, an industry that is drastically reducing the number of

FOREST OF FERNS ► *This "prehistoric" scene is actually a tropical rainforest in Queensland, Australia.* Dicksonia *trunks are covered in fibrous roots to collect water from the damp atmosphere of the understorey, a climate that can be recreated in the garden by spraying their trunks with water.*

HABITAT

Land of giants

The island of St Helena in the South Atlantic is home to over 50 extraordinary endemic plants, including "cabbage trees" – huge members of the daisy family. They grow in the last surviving thickets of native plants, found high on the central ridge of the island. In these important areas the vegetation is dominated by the tree fern *Dicksonia arborescens*, one of 14 endemic ferns on St Helena. The decline of the lowland forest began with the introduction of goats in the early 16th century, and many species have been lost forever. Today, conservation projects aim to preserve what remains, and a management plan for the tree fern thicket is being implemented by a small team of islanders.

wild plants. The flourishing trade in cultivated orchids for export in Brazil, both as live plants and cut flowers, has exacerbated the problem, as have commercial and private growers in other countries who buy a product called "xaxim", unaware that it is made of tree fern. Projects are underway to stop the use of xaxim. One such initiative involves the popular Brazilian energy-giving drink, green coconut milk. Once the milk is extracted, many tonnes of fibrous coconut husks remain. These are usually thrown into landfill sites where they take years to decay, but increasingly the husks are being processed for use as an alternative to tree fern medium. This material, known as coir, is used in garden composts and makes an excellent substitute for peat-based products, and is also made into biodegradable pots.

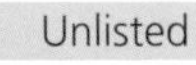

Unlisted

Gymnocarpium robertianum

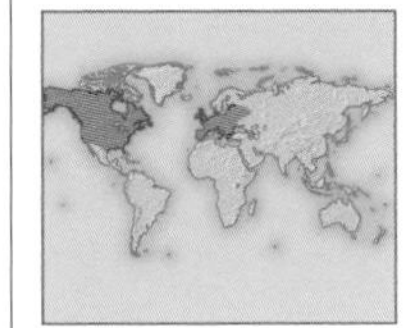

Botanical family
Woodsiaceae

Distribution
North America; Europe

Hardiness
Fully hardy

Preferred conditions
Leafy, slightly alkaline, moist soil in deep shade.

The limestone oak fern, or limestone polypody, ***Gymnocarpium robertianum,*** has proved invaluable to gardeners for providing ground cover in damp, dark places, often on walls or among rocks, where precious few plants can be persuaded to grow. The fern survives in deep shade by tilting its fronds at right angles to any available light to absorb every last ray. The fronds arise from underground stems (rhizomes) in spring, and this is a good time to propagate oak ferns by division, although it is preferable to sow spores when they are ripe, at 15°C (59°F). The closely related common oak fern, *Gymnocarpium dryopteris*, is also popular, and widely available from nurseries.

In the wild, *Gymnocarpium robertianum* grows throughout Europe, the Caucasus, and North America, in damp, mossy places with soils rich in calcium and organic matter. It is also found on limestone pavement, rocky outcrops, and in North American white cedar swamps. But although geographically widespread, the limestone oak fern is locally rare and its specific habitat requirements leave wild populations at high risk if suitable areas are disturbed by logging, or by activities such as hiking and climbing.

▲ GYMNOCARPIUM ROBERTIANUM *The northern white cedar swamps of North America, with their moist, limestone soils, are among the most important habitats of the limestone oak fern.*

Unlisted

Hymenophyllum tunbridgense

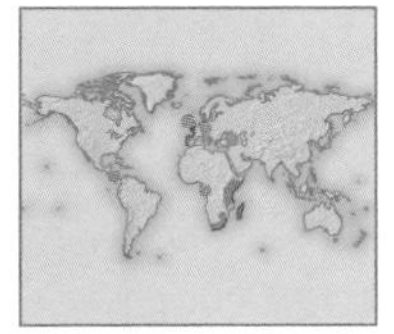

Botanical family
Hymenophyllaceae

Distribution
Azores, western Europe, north-east Turkey, North America

Hardiness
Fully hardy

Preferred conditions
Very damp, humid, shady conditions on rocks or forest floors.

The Tunbridge filmy fern grows in damp, sheltered places on rocks and in woodlands, in isolated populations in the Azores, western Europe, north-east Turkey, and one site in North America. In the UK, it inhabits the wet, temperate forests of the west coast, from Cornwall to Scotland, one of the most important areas in Europe for spore-producing plants – the ferns, mosses, liverworts, and hornworts. However, the Tunbridge filmy fern has been extinct in the UK town of Tunbridge Wells since at least 1875. Reintroduction efforts across the fern's former range have been hampered because it produces green spores that cannot be stored. The micropropagation laboratory at the Royal Botanic Gardens, Kew, has developed a technique that ensures rapid germination and growth to assist the species recovery programme. Attempts are also being made to grow the Tunbridge filmy fern in the Francis Rose Reserve, which was set up specifically to protect spore-producing plants at Wakehurst Place in Sussex, a garden run by Kew.

TUNBRIDGE FILMY FERN ► *This rare fern needs constant moisture, otherwise its transparent fronds, just one cell thick, dry out.*

42*/1 on Red List

Isoetes

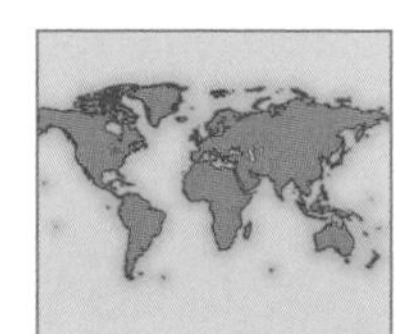

Botanical family
Isoetaceae

Distribution
Worldwide

Hardiness
Fully hardy – frost-tender

Preferred conditions
Shallow, fresh water, including seasonally flooded areas.

The long, tapering leaves of the quillworts look very different from those of ferns. Indeed, they are not true ferns, but are part of an ancient group of fern allies called the lycophytes, to which the club mosses (*Lycopodium, see right*) also belong. Plants in the genus *Isoetes* are thought to be the closest living relatives to the giant lycophyte trees that dominated the coal-forming swamps of the Carboniferous period, 354–290 million years ago. Their ancient origins are apparent during reproduction, when two kinds of spore are produced. The male spore, the microspore, eventually produces sperm which must swim to fertilize an egg produced by a female spore, or megaspore. This is a primitive method of reproduction that requires plentiful moisture, so quillworts grow submerged in shallow, fresh water, surviving as dormant spores if they dry out. There is one exception – the land quillwort ***Isoetes histrix*** (Unlisted).

Quillworts are found worldwide, but many, such as the South African *Isoetes wormaldii,* which grows in seasonal ponds, are under threat. Earthworms apparently eat and disperse dormant spores, but often deposit them away from suitable habitats. This, combined with fluctuating water levels and long-term drought, has reduced the only population to just five plants, and the species is Endangered on the Red List of 1997.

▲ ISOETES HISTRIX *This is the only land-dwelling species of quillwort, mainly found on the Atlantic coast of Europe where it is often submerged during winter. In spring, when floods subside and water levels drop, it bursts into life.*

8 on Red List*

Lycopodium

Botanical family
Lycopodiaceae

Distribution
The Americas, Africa, Europe, Asia

Hardiness
Fully hardy – frost-tender

Preferred conditions
Very well-drained, but permanently moist and humid conditions in partial shade.

The club mosses are found in many parts of the world, in a wide range of habitats, from Arctic and alpine tundra, to tropical and temperate rainforests, where they are most abundant. The club mosses, like the quillworts (*Isoetes, see left*), belong to an ancient group of fern allies. Club mosses have slender stems clothed with tiny leaves, rather than large, fern-like fronds. Their stems repeatedly fork, and terminate in a cone-like structure (strobilus) in which spores are produced. This growth habit follows a simple, primitive pattern, similar to that seen in fossils of some of the earliest plants to colonize land. Indeed, the first club moss appeared in the Devonian period, 418–354 million years ago.

Today many species of club moss are at risk from habitat destruction. Cool-growing terrestrial species, such as ***Lycopodium annotinum***, are under increasing threat as the heaths, wetlands, and forests they favour are altered by human activity. The warm-growing epiphytes face the same problems as other rainforest plants: widespread logging and clearance for agriculture.

Although some species, such as the epiphyte *Lycopodium phlegmaria,* are found across a wide range and are not considered threatened, others have restricted distributions which leave them vulnerable. *Lycopodium nutans*, for example, has been placed on the US list of endangered plants. It is endemic to the Hawaiian island of Oahu, where it is at risk from invasive plants and grazing animals.

The cool-growing species can provide unusual ground cover in moist, shady spots, although the plants resent transplantation and so can be difficult to establish. The tender epiphytes are more commonly grown, either under glass in temperate climates, or on trees in tropical and subtropical areas. If they are to thrive they need misting daily during hot, dry periods. Propagation from spores is very difficult, so divide terrestrial species, or layer epiphytes, at any time of year.

LYCOPODIUM ANNOTINUM ▼ *The stiff club moss is a terrestrial species with a northern distribution, and is not considered threatened. Here, the cone-like structures that are typical of club mosses ascend between alpine bearberry plants and lichens in the Denali National Park, Alaska.*

PLANTS AND PEOPLE

SPORES AND SAFE SEX

Considerable quantities of several species of club moss are collected and dried for a variety of applications in the cosmetics, fireworks, medicinal, and other industries. Gathering plants may damage natural populations, but at the same time it could also promote sustainable management of the forests where the club mosses grow. One surprising use of the talc-like spores is as a dusting agent for rolled condoms to prevent them from sticking. However, recent studies suggest that the spores are allergenic, so manufacturers are switching to cornstarch instead.

8 on Red List*

Marsilea

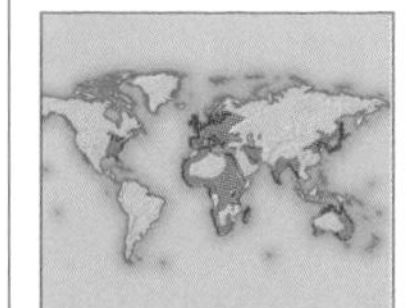

Botanical family
Marsileaceae

Distribution
Widespread

Hardiness
Fully hardy – frost-tender

Preferred conditions
Muddy pond and river margins, in full sun.

There are 65 species of water clover, scattered across all the continents except Antarctica. They are aquatic ferns which grow at the margins of rivers or lakes, spreading with tough, rooting stems (rhizomes) to form large colonies.

The destruction of wetlands by alteration of natural drainage patterns is a worldwide conservation concern, and a major threat to water clovers. *Marsilea villosa*, locally called *`ihi`ihi*, for example, is endemic to the Hawaiian islands, where it is restricted to lowland areas subject to periodic flooding. The species needs water for successful spore release and fertilization, but young ferns can establish only when the water level subsides. The natural flooding cycles on the islands have been changed to suit human needs, and *Marsilea villosa* has suffered. Just 11 populations remain, and the plant is classified as Endangered on the Red List of 1997. If its natural habitats can be restored, the fern will, hopefully, return because all species of *Marsilea* produce drought-resistant spore cases (sporocarps), which can germinate even after 100 years of dormancy.

Common species, such as the unlisted ***Marsilea mutica***, are very adaptable. They can be fully aquatic, amphibious, or even grow on land if the water level drops.

❗ Water clovers can be weedy outside their native ranges, so ensure that plants do not escape from the garden.

▲ MARSILEA MUTICA *This Australian water clover is popular in garden ponds, where it provides shade and shelter for wildlife.*

Unlisted

Pilularia globulifera

Botanical family
Marsileaceae

Distribution
Western Europe

Hardiness
Fully hardy

Preferred conditions
Shallow pond margins on acidic soils.

Pillwort, ***Pilularia globulifera***, may look like a grass, but its true allegiance can be identified when its thread-like leaves unfurl from the coiled fiddleheads that are so distinctive of ferns.

This water fern is an opportunist, favouring the margins of acid pools on clay or peat heathland, particularly the bare mud of new pools or gravel pits, or areas disturbed by watering livestock. When conditions are favourable, plants increase rapidly, but disappear again as the vegetation fills out. Pillwort is increasingly grown on the margins of garden pools: on suitable acid soils, it will thrive and is excellent for masking the edge of plastic pond liners.

The species is confined to western Europe, and is most abundant in the UK, although it is rapidly declining as grazing of heathland and common land becomes increasingly rare. One of the strongholds of the species is the New Forest in the south of the UK, where the ancient practice of grazing cattle and horses on open heath and in woodland continues. Here, pillwort is common, especially at the disturbed margins of ponds and streams where livestock go to drink. However, the New Forest, although protected, is a very popular beauty spot and its environment is under pressure.

▲ MIRACLE PILL *The pill-like spore cases can lie dormant and wait patiently for ideal growth conditions. On the Isle of Wight, off the south coast of the UK, clearance of clay pools resurrected plants after 100 years.*

2 on Red List*

Platycerium

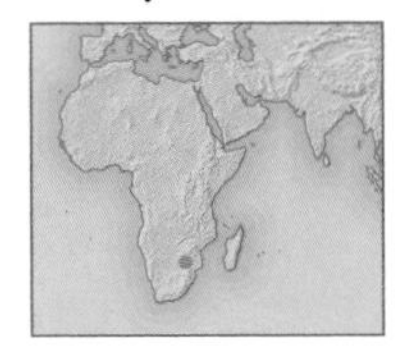

Botanical family
Polypodiaceae

Distribution
Africa, Asia, Australia, South America

Hardiness
Frost-tender

Preferred conditions
Very well-drained yet moist soil, in partial shade, in a moist, tropical or warm-temperate climate.

There are 16 species of *Platycerium* ferns, growing mainly in temperate and tropical rainforest. Staghorns are distinctive, and unusual in producing two kinds of fronds: large, leathery fertile fronds that are forked into antler-like shapes, and rounded, sterile nest leaves. The nest leaves grow close to the host tree, and persist even when they are dead, brown, and papery, to protect the ferns' roots and spreading stems (rhizomes). Like the nests formed by *Asplenium* ferns (*see p.314*), some staghorn nests catch leaf litter, which decomposes to provide the fern with nutrients and a home to various invertebrates in the forest canopy.

Most staghorns are in cultivation due to their ease of growth and propagation, but this wide availability comes at a price: generations of plant hunters have almost denuded the rainforests of their spectacular native staghorns. The situation is not helped by the general threats common to all rainforests, including clearance, logging operations, and invasive non-native plants.

DANGER IN THE WILD

Platycerium grande is a magnificent species from the Philippines. Known as the giant staghorn, it is one of the showiest in the genus, which made it a target for collectors. On the Red List of 1997 it was classified as Endangered and possibly Extinct.

Ridley's staghorn fern, *Platycerium ridleyi,* is another highly prized species that was once found throughout south-east Asia, but was deemed Extinct on the Red List of 1997. In the wild, several plants often grew together on the same host tree, high in the canopy, up to 25m (80ft) from the forest floor. Although this staghorn is found in cultivation, it is seldom as spectacular as it was in the wild. The staghorn fronds are held erect above an enclosed chamber formed by the nest leaves. The chamber was often host to an ant colony, and it is thought that the colony's waste provided the fern with essential nutrients.

◄ FERN FAVOURITES Platycerium superbum *and* P. bifurcatum *are popular in gardens, but rare in the wild. They survive in the rainforests of Fraser Island, a World Heritage Site off the Queensland coast.*

CULTIVATION

Staghorns are striking epiphytes, popular in gardens and under glass in cooler climates. Species such as ***Platycerium bifurcatum*** and ***Platycerium superbum***, both Unlisted, can be grown in humid, partially shaded sites, on trees as they would grow in the wild, or in hanging baskets. In containers they need a very well-drained medium – an equal mixture of coarse leafmould, charcoal, sphagnum moss, and loam is ideal. Spores are produced in large patches on the undersides of the horn-shaped fronds. As with most tropical ferns, using spores for propagation is straightforward as long as proper care is taken. Germination is most successful in a constantly humid atmosphere at a temperature of about 21°C (70°F). Once germination and fertilization have taken place, the tiny ferns quickly take on their distinctive shape. Alternatively, plantlets may form on the end of the roots or rhizomes and these can be detached and grown on.

▲ PLATYCERIUM GRANDE *In humid, tropical climates, the giant staghorn can be grown in the garden, where it forms a crown up to 1.2m (4ft) across, with horn-shaped fronds, up to 1.8m (6ft) long, hanging below.*

Fertile, staghorn frond

Nest leaf

◄ YOUNG STAGHORN *The striking form of staghorn ferns is apparent as soon as they germinate. In the past, this ease of identification made them an easy target for collectors, who could take small, young plants from the wild without attracting attention.*

Unlisted

Salvinia natans

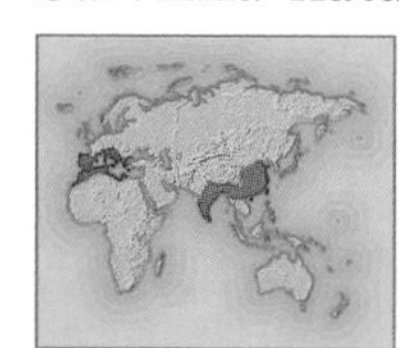

Botanical family
Salviniaceae

Distribution
Southern Europe, North Africa, Asia

Hardiness
Frost-tender

Preferred conditions
Floating on still water in full sun.

This free-floating aquatic fern is widespread and prone to population explosions in Asia, but in Europe the species was scattered by the glaciers of the last ice age. It is now found only in the wetlands of central and eastern Europe, including the Danube Delta in Romania and the Po Basin in Italy. Here the species receives protection on Appendix I of the Bern Convention on the Conservation of European Wildlife and Natural Habitats, which places an obligation on member states to take reasonable and adequate measures to protect the species and its habitat. Protection may prove unnecessary if global warming makes more areas in Europe suitable for the growth of this tender fern. For the moment, pollution and changes to drainage in European wetlands has curtailed the spread of *Salvinia natans*, but in ideal conditions this fern can be vigorous: never put excess plants in natural water courses and do not grow it outside its natural range.

See also Invasive plants, *p.468.*

BOTANY

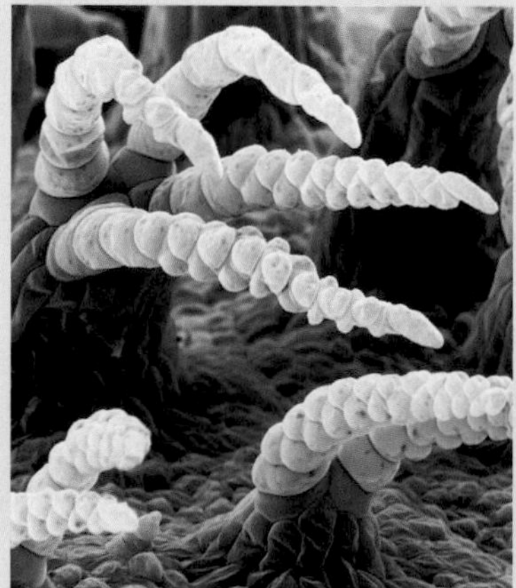

SPECIALIZED HAIRS

The buoyancy of *Salvinia* ferns is achieved by tiny specialized hairs (trichomes) on the leaf surface that are exactly the right size and shape to capture tiny air bubbles. The hairs form a water-repellent coating on the floating leaves to ensure that they bob back to the surface if they are temporarily submerged. The trichomes of the invasive *Salvinia molesta* branch into four, then join at the tip in a characteristic whisk shape.

7*/3 on Red List

Trichomanes

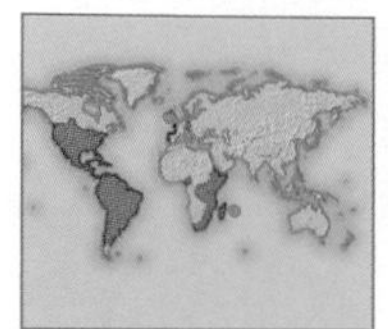

Common name
Hymenophylliaceae

Distribution
USA, South America, Europe, Madagascar

Hardiness
Frost-tender

Preferred conditions
Very moist, shady conditions on rocks and in forests.

This genus suffered greatly during the Victorian fern-collecting craze – pteridomania – when filmy ferns were all the rage. Filmy ferns are so-called because of their translucent fronds, which are only one cell thick. They are difficult to keep in cultivation, requiring shade and perpetual humidity if they are not to dry out and shrivel up, although some species will revive when water becomes more plentiful again. The Victorian collectors got round this problem by keeping their filmy ferns in special glass cases to retain moisture and banish the pollution of the industrial age, to which these delicate ferns are very sensitive. There is a reconstruction of a Victorian fern collection, including a filmy fern case, at the Chelsea Physic Garden in London.

INFERTILE FERNS

Trichomanes speciosum, also called the Killarney fern, is one of the rarest ferns in Europe, where it survives in scattered populations after its numbers were decimated by pteridomania. Habitat destruction continues to threaten remaining plants and the fern is now protected throughout Europe by the European Union Habitats Directive and the Bern Convention, and is classified as Rare on the Red List of 1997. Although the Killarney fern often forms large mats of non-fertile fronds, fertile plants are very rare throughout Europe and especially at the northerly limits of its range. *Trichomanes boschianum* also rarely forms fertile plants, but remains unlisted. This species occurs sporadically in the areas of the eastern USA that remained unglaciated during the last ice age, and at one site in Mexico. *Trichomanes petersii*, Peter's filmy fern, is also found in the south-eastern USA, and south as far as Costa Rica, and is not on the Red List. Unlike many filmy ferns, it will not revive if the fronds dry out, which makes it virtually impossible to grow unless specialist, cool but very humid conditions can be maintained constantly.

In contrast, in New Zealand, and on the Stewart and Chatham Islands, the endemic kidney fern, ***Trichomanes reniforme***, known locally as *raureng*, grows extensively in forested areas, on the forest floor, on rocky outcrops, or as an epiphyte. This fern forms large mats of distinctive and unusual fronds that are slightly sturdier than those of many filmy ferns – a few cells thick rather than just one. In dry conditions, the fronds roll up to reduce water loss. The fern also appears in gardens in New Zealand, where cultivation of native, rather than alien, species is strongly encouraged and very popular.

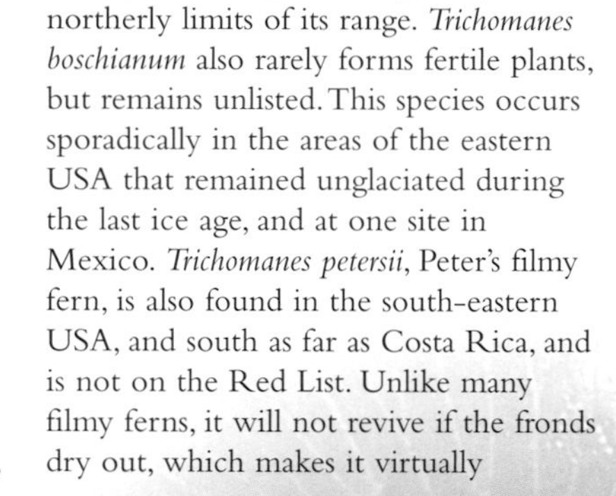

◄ TRICHOMANES SPECIOSUM *The Killarney fern grows in shady, perpetually humid rock crevices across Europe and Macronesia. It is frost-sensitive and the UK is the northerly limit of its distribution. There, it occurs only in the far west and in the Republic of Ireland, where the Gulf Stream keeps the climate warm and moist.*

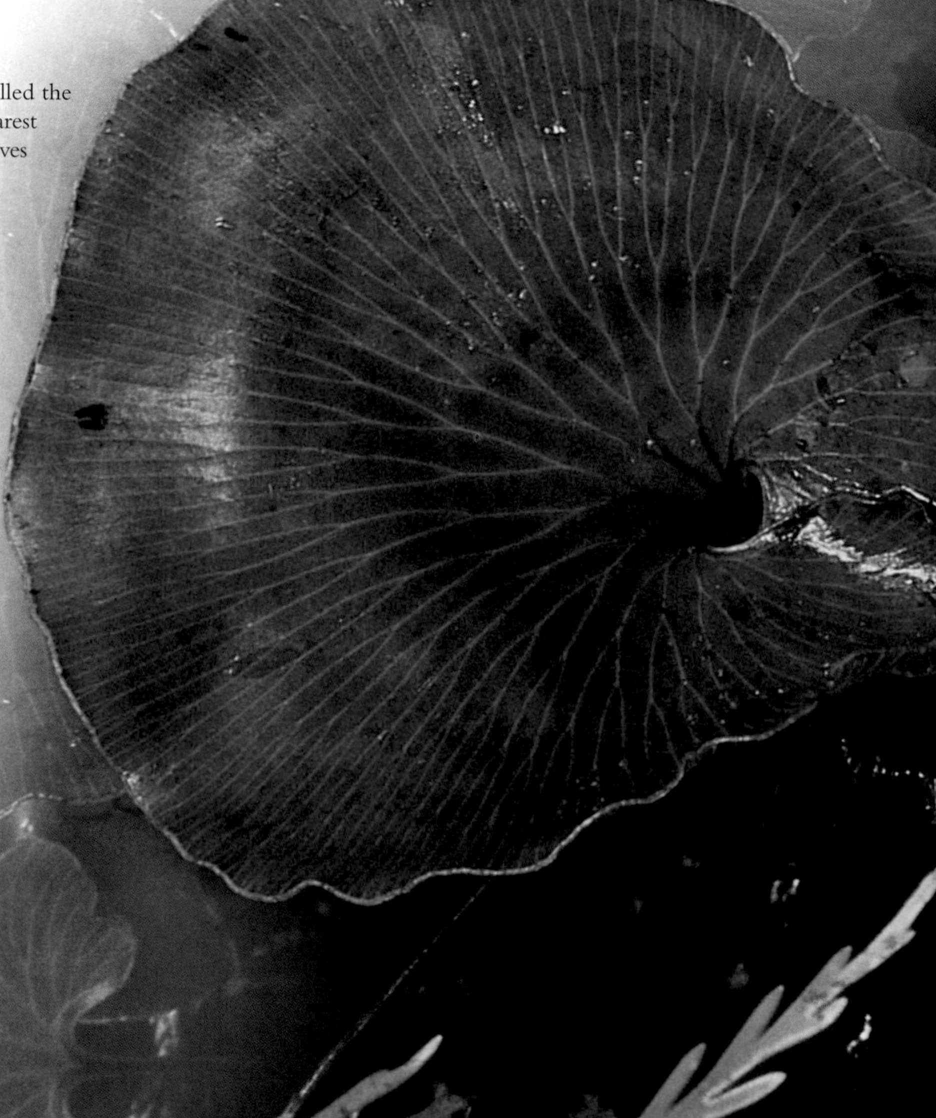

TRICHOMANES RENIFORME ► *The kidney-leaved fern is found only in areas of very high humidity and rainfall in New Zealand. This healthy clump is growing close to Lake Kaniere, in Westland.*

CACTI AND SUCCULENTS

Succulents are the camels of the plant kingdom, adapted to drought and found for the most part in dry, mountainous, windy, or cold habitats, where moisture is finite, fleeting, or frozen. There are some 10,000 species of succulent, many geographically restricted and under threat. The cacti, the best-known succulent family, comprise some 1,500 species; with just one possible exception, all are native to the New World. Several genera and species of cacti and succulents are listed in CITES Appendix I or Appendix II; the entire cactus family (Cactaceae) is included in Appendix II.

WHERE THEY GROW

Succulents show a great degree of specialization in appearance and habitat. At one extreme are cactus species adapted to the desert of northern Chile, where the only moisture is fog from the Pacific Ocean, and at the other are the epiphytic cacti growing in the canopy of rainforests.

▲ TRUE DESERT *Southern Africa is home to 40 per cent of the world's succulent flora. Aloes, such as this* Aloe asperifolia *in the Namib Desert, are characteristic of the landscape.*

While succulents occupy just about every habitat on Earth, two regions, southern Africa and Mexico, are renowned for their remarkable diversity of succulent species. In southern Africa, the highest concentration of rare succulent species is found along South Africa's west coast, and the Namaqualand Desert accounts for one-tenth of the world's succulent flora. The cactus family is widely represented from the open, hot plains of southern Canada to the rocky outcrops of Patagonia's windswept South American landscape, but the diversity is highest in Mexico. One-third of all cacti are native to Mexico and over 75 per cent of Mexican species do not exist in the wild outside the country. Within Mexico, there are unparalleled pockets of cactus diversity and endemism. In particular, the Chihuahuan Desert, one of Mexico's most famous deserts, contains more than half of the country's cactus species, many of which are rare and threatened.

WHAT ARE CACTI?

Perhaps the most extreme examples of leafless succulents are those of the Cactaceae. Cacti are similar to other succulents with one major distinguishing feature: cacti possess areoles, organs from which their spines and flowers arise. It is thought that the spines of many cactus species are modified leaves, which evolved to counter desiccation and protect plants from thirsty grazers. Spines also create favourable microclimates for dew to condense and trickle down the stem to the ground to be absorbed by the roots.

Tubercle

Areole

◄ CACTUS FORM *Areoles, the modified, pad-like buds that produce spines and flowers, are usually arranged along ridges or at the tops of bumps (tubercles). In some cacti, the areoles are crowded together into a dense woolly area (cephalium) at the top of the plant, where all the flowers are produced.*

SURVIVAL STRATEGIES

In arid or semi-arid habitats, plants develop features such as a waxy skin to minimize moisture loss and shallow, radiating roots that absorb rainfall from a wide area. Stem succulents, including cacti, have evolved to have fewer and smaller leaves, or none at all, and store water in their swollen stems. Leaf succulents, such as aloes, retain water in their characteristic fleshy leaves. In a third group of succulents, caudiciforms, the main water reservoir is an enlarged stem base, or caudex.

Many succulents minimize water loss by opening their pores (stomata) to absorb carbon dioxide only at night, when evaporation rates are low. Unable to photosynthesize in the dark, the plants convert the gas into an organic acid, then metabolize it back into carbon dioxide the next day for photosynthesis, a process called crassulacean acid metabolism (CAM).

▲ DISGUISE *Aptly named living stones,* Lithops karasmontana *blends into the stony South African desert, hidden from grazing animals.*

Spines deter grazing animals

Evaporation is limited by the waxy skin

Water is stored in fleshy tissues

A shallow root system absorbs water quickly

SUCCULENT BOTANY ► *Physical adaptations allow succulents to tolerate high temperatures and survive where rainfall is intermittent or infrequent.*

SAGUARO CACTUS *The largest of the cacti,* Carnegiea gigantea *reaches 15m (50ft). It is found in the Sonoran Desert of south-eastern California, southern Arizona, and north-western Mexico, sometimes in dense, forest-like stands.*

HISTORY OF TRADE

Succulents have aroused the curiosity of botanists, growers, and collectors for centuries. South African succulent species and American cacti rapidly became European favourites, in part due to their proximity and accessibility to traders. Early Dutch explorers returning from Indonesia brought plants from South Africa, and Spanish sailors to the New World brought back cacti and agaves, which have since naturalized in Spain, Italy, and France. These imports were welcomed by specialist growers and by entrepreneurs who discovered edible qualities or otherwise practical uses for the plants, but some have become invasive weeds.

Today, dozens of succulent and cactus societies worldwide form one of the largest horticultural networks, and the plants are no longer an obscure obsession. Millions of nursery-propagated plants enter international trade each year, destined primarily for North America, Asia, and Europe.

◄ EXPLORERS *Intrigued by the bizarre forms of succulents, plant collectors brought them home and popularized them in the European horticultural plant market.*

◄ ALOE VERA *The juices of this plant are an effective treatment for sunburn, and the species is used in Chinese and Indian herbal medicine. It is widely cultivated: here a strip of harvested plants await processing.*

ECONOMIC VALUE

The importance of succulents has not been limited to their horticultural value. Spanish settlers in the New World saw the economic potential of prickly pears (*Opuntia* and *Nopalea* species) for their part in producing a valuable red dye. These cacti host the cochineal insect (*Dactylopius coccus*); when dried and crushed they yield a red pigment, at one time Mexico's third most lucrative export.

A number of succulent species are harvested or grown as cash crops. Rural people still support their families both by selling raw plant materials to various industries and by making crafts either for sale to tourists locally or for export to specialist shops overseas.

In Mexico, agaves are cultivated on plantations or harvested from the wild to make the traditional spirits tequila and mescal, whose exports pump tens of millions of dollars into the Mexican economy each year. Many people in South Africa owe their livelihood to *Aloe ferox*, one of nearly 200 *Aloe* species native to southern Africa. Harvesters tap the leaves for a bitter yellow sap, which is sold to the pharmaceutical industry and also used in traditional medicines. Chilean artisans make musical instruments called rainsticks from the woody remains of two columnar cacti, *Eulychnia* and *Echinopsis*, native to Chile's northern desert. Rainsticks are sold to tourists and exported to markets in North America, Europe, and Asia, where they fetch a handsome price.

▲ POTENT PLANT *The hearts of agaves have been harvested and used to make the popular drink tequila for centuries.*

SUPPORTING WILDLIFE

In communities of drought-tolerant (xerophytic) plants, succulents are often the dominant type of vegetation. As such, they serve key ecological functions. The saguaro cactus is the Sonoran Desert's botanical skyscraper. Native woodpeckers nest in them, and other species use them as a refuge from the desert heat and predators. Nectar from the flowers of the spindly and spiny ocotillo (*Fouquieria*), native to the deserts of the south-western USA and Mexico, provides a critical source of nutrition to migrating hummingbirds. The tubular flowers of ocotillos are perfectly shaped to fit the elongated beaks of hummingbirds, hinting that the plant and the bird co-evolved to benefit one another.

◄ SECURE SITE *This Anna's hummingbird* (Calypte anna) *in the Arizona desert has chosen a nesting site protected by the thorns of an ocotillo.*

▲ SEED DISTRIBUTION *The fruits and seeds of* Ferocactus *are eaten by rodents, birds, mule deer, and javelina, a pig-like animal.*

POLLINATION

Succulents need to maximize their chances of reproductive success for flowering to be worth the precious energy and water it consumes. Showy flowers that attract pollinators may collapse when the sky becomes cloudy, to keep their pollen dry. Succulents also emit a dizzying array of scents to entice insects, birds, and bats. The flowers of stapeliads have an odour similar to rotting meat, which flies, the primary pollinators, cannot resist.

Some succulents flower to coincide with specific times of the year or natural events. The *Peniocereus greggii* cactus unfurls its flower on just one night a year, with a seductive scent detectable a mile away. Some species have even formed exclusive relationships with pollinators: only yucca moths can pollinate yuccas, which in turn are the sole source of food for the moth larvae as they develop in the flowers.

THREATS

Shrinking habitat is the greatest threat, particularly where development, agriculture, and mining have taken a heavy toll. Succulents may be the only forage available to livestock in arid climates with little vegetation; relentless grazing does irreversible damage, pushing rare species closer to extinction. Introduced plants can displace native species, and climate change may also affect succulents unable to adapt to rapidly changing conditions.

Unscrupulous collectors are now supplying a new market. Urban populations in the dry south-western USA have risen sharply, and residents and businesses are encouraged, even required, to conserve water. The shift to gardening with xerophytic plants, which use a fraction of the water needed by lawns, has led to a surge in demand for succulents and indiscriminate harvesting of wild plants from deserts.

In addition, some succulents face the shortcomings of their own biology, such as special habitat requirements, or late reproductive maturity, as well as low growth rates and seed output.

▲ INVASIVE SPECIES *Originally introduced as fodder or live fencing, prickly pear* (Opuntia) *quickly spread as an invasive weed in Australia and Madagascar, where efforts are still under way to bring it under effective control.*

CONOPHYTUM ECTYPUM *VAR.* ECTYPUM *Development, overgrazing by elephants and goats, agriculture, and mining are damaging southern Africa's succulents, particularly in Namibia and South Africa.*

CONSERVATION EFFORTS

Habitat protection, harvest and trade controls, and cultivation all mitigate pressures. Governments are developing strategies to protect species in their natural habitat. Where wild populations are an important source of local income, the solution may be local enforcement of sustainable harvest; in other cases, CITES regulations (*see p.37*) and the threat of severe sanctions are effective. Botanical reserves offer the only sanctuary to some species. The Cactus and Succulent Specialist Group of the IUCN (*see p.36*) has published an action plan that identifies vulnerable plants and habitats and the course of action needed. The International Organization for Succulent Plant Study (IOS) has a code of conduct and promotes conservation and study.

◄ HOTSPOTS *International and national laws restrict the collection and trade of plants, such as these yuccas and* Ferocactus *in the Chihuahuan Desert, but more effective protection is needed.*

GARDENER'S GUIDE

- Gardeners can support the conservation of cacti and succulents by exercising care when purchasing plants. Be prepared to ask specific questions.
- Ask if a plant was obtained from the wild or cultivated in a nursery. If a plant appears damaged or stressed, there is a strong possibility that it has been removed from the wild. Such plants do not always transplant well, and may have been illegally or unsustainably collected.
- Avoid purchasing plants or seeds over the internet if the reputation of the vendor cannot be independently verified. The internet has become a conduit for plants of questionable legality.
- Beware of unusually high or low prices. High prices may indicate rarity and a lack of propagated specimens. Very low prices may be a sign of an inferior specimen collected from the wild, and damaged in the process.
- It is not always wrong to purchase wild succulents; some are salvaged from habitat that is about to be developed or may be harvested with the landowner's written permission. Always ask to see the paperwork authorizing legal plant collection.
- Become a member of a local cactus and succulent society to learn more about what you can do for conservation.

43 on Red List*/CITES I and II

Agave

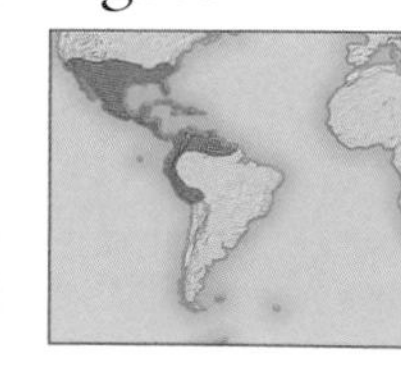

Botanical family
Agavaceae

Distribution
Southern USA to north-eastern South America

Hardiness
Fully hardy – frost-tender

Preferred conditions
Full sun; well-drained, slightly alkaline soil. Hardiness varies according to species; some will be damaged below 0°C (32°F).

Century plants or agaves superficially resemble their African counterparts in the genus *Aloe* (*see pp.332–33*), but there are significant differences in their flowering and reproduction (*see box*). Agaves are found in diverse habitats from dry and temperate forest to grassland, savanna, shrubland, desert, and scrub.

BOTANY

VERSATILE TECHNIQUE

Virtually all agaves are monocarpic, dying after a single flowering. The plant puts all its energy into a massive flowering pole-like stalk, topped by dense clusters of flowers, usually yellowish green and almost always strongly scented. Wildlife feeds on the copious nectar, and night-foraging fruit bats, such as *Leptonycteris curasoae*, are important pollinators. Most agaves produce small bulbils on the stalks, and after the flowers are shed, plantlets develop and drop to the ground where they rapidly take root and develop into genetically identical offspring (clones).

LONG-NOSED BAT FEEDING ON AGAVE

AGAVE SHAWII, MEXICO

A number of *Agave* species are important commercial crops, used for their fibres (*see box*) and sap or, more recently, sold as garden and landscaping plants.

PRECIOUS LIQUID

Agave sap has been used to make liquor for millennia. When the plants approach flowering maturity, the leaves are cut off and the "hearts" or *pinas*, weighing up to 80kg (176lb), are harvested and crushed. Pulque, produced from the fermented juice, must be drunk soon after it is made; the species most used for this are *Agave salmiana* and *Agave mapisaga*.

But the best-known drink is tequila, distilled from the fermented sap of *Agave tequilana*. Tequila can only be distilled in certain areas of Mexico, where the hills, covered with dense *Agave tequilana* plantations, have a characteristically blue colour. Shortages have led some resourceful farmers to make mescal from the *pinas* of other species, notably the popular horticultural plant *Agave americana*. This is leading to serious loss of wild plants in some areas.

COMMONLY CULTIVATED

Agaves are popular in both landscape and domestic planting. Plants will rapidly fill an exposed bed, and many make fine container plants, growing more slowly than in the ground. Some bear striking "bud prints", made by one leaf on another during development. The plants' popularity has led to overcollecting.

One of the most widely cultivated agaves in the world is ***Agave victoriae-reginae***, Queen Victoria's agave. This species has never been plentiful in its natural range of Coahuila, Durango, and Nuevo León in Mexico. Its desirability has put pressure on existing populations, as plants are collected illegally for the lucrative international trade. Although listed as Endangered in the Red List of 1997 and in CITES Appendix II, this species will not become extinct in cultivation; even a conservative estimate is that millions of artificially propagated plants are distributed globally. The flowering stem does not produce any bulbils, but many forms produce offsets that can be removed and planted, and germination of seed is almost 100 per cent successful.

Many agaves have bluish leaves, but few are bluer than the blue agave, ***Agave guiengola***, from the Isthmus of Tehuantepec, Oaxaca, Mexico. The few leaves are an even blue and armed with small, hard, brownish-purple teeth. Classed as Vulnerable in the Red List of 1997, it grows easily from seed or offsets, and thrives in filtered sunlight, surviving down to −5°C (23°F). The flower stem is up to 2m (6ft) tall, but the plant needs a pot at least 90cm (3ft) across to reach flowering size.

◄ AGAVE PARVIFLORA *SUBSP.* PARVIFLORA *grows to just a handspan across. Prized for its white bands and curling filaments, it produces dense greenish flowers and lives on for some years after flowering.*

SMALL IS BEAUTIFUL

Agave growers prefer small, manageable species, and several of these are overcollected, including ***Agave parviflora* subsp. *parviflora***. Listed as Rare in the Red List of 1997, and in CITES Appendix I, it comes from Arizona, USA, and northern Sonora, Mexico. It is self-fertile, producing seed without pollination by another plant. This ability can ensure the survival of a species, at least in cultivation; in the wild it can lead to inbreeding problems. Seedlings grow rapidly if given space; basal sprouts also appear after flowering. This hardy species easily survives temperatures as low as −13°C (9°F), especially if kept fairly dry.

The gypsum agave, *Agave gypsophila* grows in the Mexican states of Guerrero, Jalisco, and Colima. The bluish-green leaves have beautifully wavy edges. They are brittle and break easily yet are still succulent enough to ensure the plant can endure periods of drought. Some forms produce clumps, but not all, so the temptation to collect from the wild is very strong. In the wild it has been found in soils over gypsum – hence its name – and limestone, but it is does not need a special soil mixture. It will be damaged below 5°C (41°F).

Agave nizandensis, Endangered in the Red List of 1997, will happily grow to flowering maturity as a houseplant. The small rosettes have a few leaves, each with a light green central stripe. Strong suckers sprout from the base. The flowering stalk is over 90cm (3ft) tall and carries small clusters of dull yellowish flowers. Temperatures of below –5°C (23°F) will damage the leaves, and below –8°C (18°F) plants will die; they also succumb easily to overwatering.

Growing agaves

Many agaves are easy to grow, if given good drainage. Buy plants or seed that are artificially propagated. Most seed germinates readily if sown fresh, scattered evenly on a tray of well-drained soil and covered with a thin layer of grit. Water the tray from the bottom, standing it in a shallow saucer of water with a fungicide added. Bulbils or basal offsets can also be removed from the parent and planted.

▲ AGAVE GUIENGOLA *The leaves of this agave are unusually soft, and weather damage will leave permanent scars, but it is suited to growing in a container under cover.*

◄ AGAVE VICTORIAE-REGINAE *is prized for its densely packed, ball-shaped rosettes of leaves, although its appearance is variable. The robust flowering pole is up to 4m (12ft) tall. This species is not as easy to grow as its popularity might imply. It does not tolerate overwatering, and will often suddenly succumb to rot.*

COMMERCIAL USES

Sisal production

Agave leaves contain tough fibres that were worked into cord and cloth by the Maya and Aztecs before the Spanish invasion. Subsequently, large plantations were established to provide material for export to make brushes, sacks, and rope around the world. Sisal takes its name from a port in Yucatán, Mexico, from where the fibres were shipped. Obtained from the leaves of ***Agave sisalana***, sisal is still preferred to synthetic materials for many uses, because of its strength, resistance to stretching, and durability when exposed to heat, sunlight, and saltwater. With the recent drive to return to natural products where possible, there is a renewed demand for sisal in other applications where it once had to play second fiddle to synthetics.

Agave sisalana grows very easily in mild, temperate, and even tropical climates, and commercial plantations have spread the species beyond its natural range. In the 19th century it was introduced to East Africa as a crop, and this region is now one of the main centres of production. The species has even become an invasive alien pest plant in many parts of the world, because it reproduces very effectively from bulbils, which strike root easily and grow vigorously.

SISAL PLANTATION

TRIMMED SISAL PLANTS, KENYA

SISAL FIBRES DRYING, MADAGASCAR

OTHER THREATENED AGAVES

Agave arizonica With fewer than 100 wild plants reported in central Arizona, USA, a single catastrophe could easily extinguish the Arizona agave, which is part of the National Collection of Endangered Plants. Widely regarded as a natural hybrid of *Agave chrysantha* and *Agave toumeyana* subsp. *bella*, it should be written *Agave* x *arizonica*. A small plant, it spreads by basal offsets. The stiff leaves are edged with neat, irregular teeth and tipped by pungent spines; the flowering pole of up to 4m (12ft) bears small clusters of yellowish flowers. It prefers filtered sunlight and is easy to cultivate. Red List (1997): Endangered; CITES I.

Agave dasylirioides Probably a very ancient agave with a naturally restricted range near Tepotzlán in Morelos, Mexico. The ball-shaped rosettes often dangle, seemingly precariously, from near-vertical cliffs. The multitude of deep green leaves are thin, flattish, pliable, hardly succulent, and very finely saw-toothed. The unbranched inflorescence is densely packed with greenish-yellow flowers and the upper half is bent downwards. Unusually, the plants do not die after flowering; they also do not form offsets. Red List (1997): Vulnerable

Agave grijalvensis Little is known about the threats to this fairly large agave from Chiapas, Mexico; until 1990 it was confused with *Agave kewensis*. The leaves are comparatively few, generally sword-shaped, thick, and light greenish yellow. Offsets are not produced, so seed is the only means of reproduction. The flowering pole can reach over 5m (15ft) and carries dense clusters of flowers on short side branches. It grows over subsoils containing calcium-rich hardpans. Red List (1997): Rare (as *Agave kewensis*).

Red List: Near Threatened/CITES II

Alluaudia procera

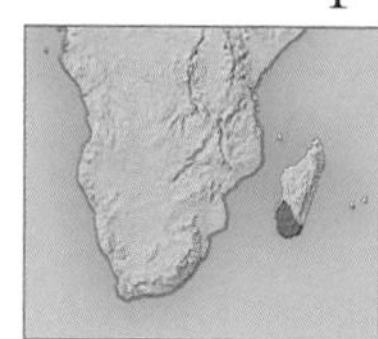

Botanical family
Didiereaceae

Distribution
Southern and south-western Madagascar

Hardiness
Half-hardy

Preferred conditions
Full sun and dry conditions; hardy to 0°C (32°F); can take light frosts with only some tip damage.

The four genera of the Didiereaceae family are all endemic to Madagascar, confined to the island's south and south-west. Their closest relatives are the Cactaceae, and they are sometimes known as the cacti of the Old World; they can be grafted onto cacti, usually *Pereskiopsis*, the most primitive genus.

The genetic relationships within the Didiereaceae are unusual; large jumps in the number of chromosomes suggest past species extinctions within the family. All species are under threat from felling and clearance of the "spiny forest" of cactus-like succulents for agriculture. Another threat has been commercial collecting for succulent plant collectors. To address this problem, the entire family is listed on CITES Appendix II, which helped stop the import of thousands of wild-collected specimens (claimed to be nursery propagated) into Europe during the 1980s.

▼ TINY, ROUND LEAVES *sprout thickly along the spiny stems after rain; they fall at the onset of winter drought.*

WILDLIFE

INTREPID LEAPER

Dense stands of this succulent tree species dominate the spiny forest of southern and south-western Madagascar. The stems are heavily spined but do not deter the lemurs, such as this ring-tailed lemur (*Lemur catta*), that leap from tree to tree.

Alluaudia is the largest genus, with six species. The octopus tree or Madagascan ocotillo ***Alluaudia procera*** is classed as Near Threatened in the 2000 Red List, but it is often felled for house building and charcoal production, both for local use and for sale in the cities. Fortunately, Tsimanampetsotsa Strict Nature Reserve protects *Alluaudia procera* in its natural habitat. This species can be propagated from herbaceous stem cuttings or seed, and is commonly cultivated. Plants grow rapidly, reaching 18m (60ft) in nature, and may reach 3m (10ft) in a few years if grown in large pots; they can, however, easily be maintained at smaller sizes by pruning.

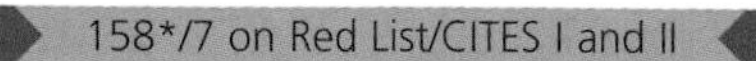

158*/7 on Red List/CITES I and II

Aloe

Botanical family
Asphodelaceae/Aloaceae

Distribution
Africa, Arabian Peninsula, and Indian Ocean islands

Hardiness
Frost-tender

Preferred conditions
Full sun; temperatures above 0°C (32°F); well-drained soil. Some species are particularly sensitive to excessive rainfall, especially in summer.

The genus *Aloe* is arguably the best-known succulent of the southern African flora, varying from trees over 20m (70ft) tall to miniatures the size of small grass, such as *Aloe bowiea*. The flowers are more consistent, being usually bright in colour, pencil-shaped, and in cylindrical or cone-shaped flowerheads. There are over 550 species, found from southern to south-central and north-eastern parts of Africa, the Arabian Peninsula, and on some of the Indian Ocean islands, in habitats ranging from dry forest, grassland, savanna, shrubland, and fynbos to African desert and scrub.

Some species have medicinal and cosmetic uses, particularly the South African ***Aloe ferox*** and the Arabian *Aloe vera*. Most other species have a horticultural appeal. It is an indictment against unscrupulous collectors that many aloes have become endangered due to their desire to grow as many of these species as possible. Wild-collected plants are often illegally taken to unsuitable areas where they almost invariably decline and die a slow death. Such activities are often conducted with no respect for the environmental needs of the plants. Further threats that have contributed to the decline of aloes include the establishment of commercial forests that lead to the destruction of their natural habitats, urbanization, industrial development, road building, and the introduction into their natural habitats of alien invasive plants. More recently, natural predators, once restricted to particular species of aloe, have been transported to populations where they were unknown previously, causing devastation.

Because of these threats, all species except the commercial crop *Aloe vera* are included on CITES: a few are on Appendix I, the rest on Appendix II.

MEDICINAL USES

ALOE FEROX

The bitter aloe is single-stemmed and tree-like, with spine-tipped, thorny-edged leaves. The old, dried leaves skirt the stem, protecting it against fire, desiccation, and scavengers. These plants are a common feature of the dry South African interior, where the copious nectar of their bright flowers provides food for a variety of birds. The leaf sap is tapped from natural populations and used fresh as a treatment for burns and other wounds or exported as a dried purgative called Cape aloes.

▲ ALOE ERINACEA *The charm of this small- to medium-sized aloe lies not only in its stout, prominently prickled leaves but also in the spear-shaped heads of beautiful red, cigar-shaped flowers that yellow with age.*

SOUTHERN AFRICAN ALOES

The greatest concentration of aloes is in southern Africa, where more than 120 are widespread. The beauty of ***Aloe erinacea***, the small hedgehog aloe or *krimpvarkie-aalwyn*, has made it a precious trophy to collect from the wild. Mining and general habitat decline threaten it so it was classed as Endangered on the Red List of 1997. It occurs in a very arid part of southern Namibia and cannot tolerate excessive rainfall, especially in the summer. It is not easy in cultivation; plants removed from their natural habitat die a quick death. Seed germinates rapidly, but seedling survival is poor in the wild: in cultivation, plants should be given a well-drained, sandy soil mixture and never overwatered.

The majestic bastard quiver tree, ***Aloe pillansii***, is threatened by illegal collecting, mining activities, and

HABITAT DESTRUCTION

ENDANGERING ALOES

Aloes on Madagascar have been damaged by a number of recent and historical human activities. The plants themselves are collected for the horticultural trade or damaged by overgrazing and the burning of vegetation for agricultural purposes, while the woody species that grow in association with them are cut for charcoal production. Clearance for plantations of often invasive commercial crops, such as prickly pear (*Opuntia ficus-indica, see p.464*) or, as here, sisal (*Agave sisalana*), is also a problem.

degradation of its habitat in southern Namibia and Richtersveld, South Africa. It is classed as Critically Endangered on the 2000 Red List.

MADAGASCAN ALOES

Most of the aloes found on the islands off the African east coast are concentrated in Madagascar and threatened by habitat degradation (*see box*).

Aloe parallelifolia from central Madagascar is listed as Endangered in the Red List of 1997. A small shrub with several very slender stems, it is known as the parallel-leaved aloe, and has almost cylindrical leaves that curve up to lie nearly parallel to the stem, and sparse reddish-yellow flowers. Propagation is best from cuttings, which strike root readily; very little is done from seed.

Aloe helenae, classed as Critically Endangered on 2000 Red List, is from south-eastern Madagascar. It is a typical single-stemmed species and can be grown from seed, but it is more difficult in cultivation than the similar South African *Aloe ferox* and *Aloe marlothii*.

The most striking and unusual species is *Aloe suzannae*, classed as Critically Endangered in 2000 Red List. It has been brought to the brink of extinction as its south-eastern Madagascan habitat is transformed by harvesting for charcoal production. The plants have long, stout, blunt-tipped leaves with fairly harmless prickles, which vary in orientation from nearly horizontal to gracefully curving up from the stem. *Aloe suzannae* can grow quite tall, up to 4m (12ft), carrying an exceptionally tall and unbranched flowerhead densely covered in small flowers. Unlike many aloes, flowering occurs only after many years, and is highly unpredictable and erratic in cultivation. Plants are propagated almost exclusively from seed.

ALOE PILLANSII *These tall plants leave a lasting impression, with their smooth stems crowned with rosettes of leaves and yellow, cigar-shaped flowers opening from almost round buds.*

▲ ALOE HELENAE *Curved leaves tapering to the ground give this plant a distinct cone shape from a distance. The short, cylindrical clusters of beautifully shaped flowers resemble bottle brushes tinted green, yellow, and red.*

OTHER THREATENED ALOES

Aloe calcairophila The miniature, strikingly beautiful chalk aloe from central Madagascar is easy to identify because its leaves remain in a single row, even into maturity. The leaf edges have prominent but harmless white teeth, and the small, urn-shaped flowers are white. This species is exceptionally easy in cultivation; the preferred method of propagation is through removal of the numerous offsets that sprout from the base of the plants. Seed will also germinate readily if sown in a well-drained, friable soil mixture that is kept moist. Red List (1997): Endangered; CITES I.

Aloe descoingsii subsp. *augustina* This Madagascan miniature aloe from St Augustin, in the south-west, is popular among collectors as a windowsill plant. The small, narrow leaves are slightly to strongly curved downwards and densely adorned with small white spots on raised, sometimes bristly prominences, giving the plants a distinctly warty appearance. The flowers look like small, bright red, almost triangular, lanterns; their mouths are more widely open than in *Aloe descoingsii* subsp. *descoingsii*. Even the stalks bearing the flowers tend to have a deep pink tinge. Although rather slow-growing, the species is easy in cultivation. The most popular method of propagation is rooting the basal sprouts, which are easily removed from a parent plant with a sharp knife. Seed also germinates readily. Plants will grow into a fairly large, ball-shaped clump in time. Red List: Unlisted; CITES I.

Aloe fragilis This species from Cape Manambato, Madagascar, is another of the miniature aloes so popular among collectors. It is a clump-forming species that will develop into small- to medium-sized, semi-spherical mounds or large mats consisting of literally hundreds of rosettes. These rosettes can be removed and each one will produce its own offsets. The dark green leaves are densely spotted with small white flecks. The leaf margins have a bony ridge armed with hard, white teeth. The flowerhead, a sparsely flowered raceme, is usually unbranched, with fairly large red flowers that turn creamy white towards the mouth. It can be confused with two of its Madagascan neighbours, *Aloe guillaumetii* and *Aloe rauhii*, but it has harder leaves with a more glossy sheen. Though it seems to grow in a very specialized habitat in Madagascar, this species is quite easy to cultivate in a well-drained soil mixture. Red List: Unlisted; CITES I.

Red List: Endangered/CITES II

Arthrocereus glaziovii

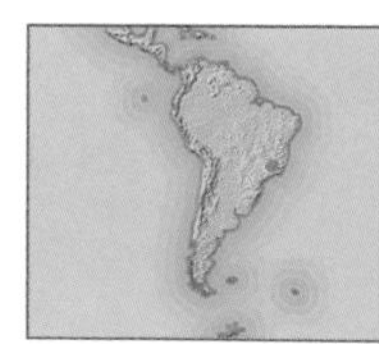

Botanical family
Cactaceae

Distribution
Brazil: Minas Gerais, south and west of the capital, Belo Horizonte

Hardiness
Frost-tender

Preferred conditions
Full sun; temperature range of 10–30°C (50–86°F); well-drained acid soil and water.

There are just four species in this genus, which is endemic to Brazil, and three of them are endemic to the mountainous central and southern parts of the state of Minas Gerais. The short, cylindrical, ribbed stems are densely covered in fine spines, and the nocturnal tubular flowers are sweetly scented and assumed to be adapted for pollination by hawk moths.

All species are rare and endangered, but none more so than ***Arthrocereus glaziovii***. In the wild it is apparently restricted to rocks very rich in iron, upon which little else can grow. This specialization towards a habitat with little competition from other plants has ultimately been its downfall; over the past century it has become endangered by the wholesale removal of much of its habitat to provide iron ore for the many steel works in the region of the city of Belo Horizonte. Its historical range has been estimated at 672 sq km (250 sq miles), but the rocks it favours are much less extensive than even this relatively small area suggests, and many of its former habitats have been completely eliminated. Today a handful of surviving localities is known, some comprising a few hundred plants at most. One is on a mountain ridge overlooking the suburbs of Belo Horizonte. This has survived because it is too close to the city for the iron ore to be extracted safely, although its southern side has been hollowed out by the miners. The best preserved habitat is on the upper reaches of a mountain at the western edge of the range, called the Serra da Piedade. It is a place of religious significance, left relatively undisturbed as a site of pilgrimage. On this mountain, as on some others, the plants are very dwarf, with shortly jointed stems closely hugging the rock surface. Other forms are known, however, some having more erect, elongated, cylindrical stems.

◄ ARTHROCEREUS GLAZIOVII *This variable species is known by several synonyms and the stem forms differ across populations, but all bear similar flowers.*

CULTIVATION TIPS

Soon after it was discovered in the 1880s the species was introduced to cultivation in Europe, and appears to be relatively easy to grow. *Arthrocereus* species are only available from specialist cactus nurseries or through exchange with enthusiasts, and none is common in cultivation. It is reasonable to assume that material will have come from artificially propagated sources and is not wild-collected. The plants are small and the most common form, with short, jointed stems, is easily propagated by cuttings taken as whole stem-segments separated at the joints and placed on a sandy medium until rooted. Seed is rarely available, and will be produced only if two genetically distinct plants flower simultaneously and can be cross-pollinated.

Plants need a minimum winter temperature of 10°C (50°F) and should be cultivated in a sunny greenhouse that is ventilated in summer. It grows on iron-rich rocks in its native habitat, but does not require any additions to its potting medium, which should be gritty, slightly acidic, and very free-draining. Avoid watering with hard (alkaline) tap water; collect rainwater instead, and water regularly during sunny weather from mid-spring to early autumn, allowing the compost to dry out between waterings. At other times of the year, water the plants very sparingly or merely spray them with a fine mist.

Red List: Vulnerable/Cites I

Ariocarpus bravoanus *subsp.* bravoanus

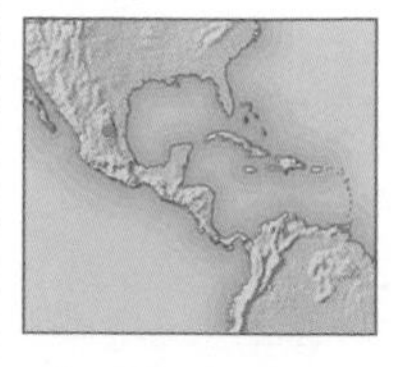

Botanical family
Cactaceae

Distribution
Mexico: Southern Chihuahuan Desert, San Luis Potosí

Hardiness
Frost-tender

Preferred conditions
Well-drained soil with water in the growing season; dry conditions at other times; minimum temperature of 10°C (50°F).

Like all members of the genus, this agave-like cactus, known as *chaute*, is of great interest to collectors. Discovered in the early 1990s, it was known as a population of only a few hundred in the southern extreme of the Chihuahuan Desert. Soon found by unscrupulous collectors, more than 75 per cent of the plants were removed. Fortunately, a few years ago a second population of several thousand individuals was found nearby, but due to its extremely restricted endemism collecting remains a problem: even the collection of the seeds frequently destroys plants.

▲ ARIOCARPUS BRAVOANUS *SUBSP.* BRAVOANUS *This plant was already on the European black market by 1996. Illegal harvesting is still the most serious threat.*

Red List: Rare*/CITES II

Avonia papyracea *subsp.* papyracea

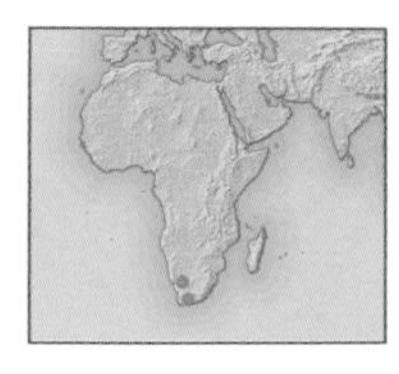

Botanical family
Portulacaceae

Distribution
South Africa: widespread in the Great and Little Karoo

Hardiness
Frost-tender

Preferred conditions
Full sun; minimum winter temperature of 7°C (45°F); well-drained, gritty soil; moderate quantities of water when in growth, virtually none when dormant.

Plants in the genus *Avonia* were, until recently, included in *Anacampseros*. These genera are in the Portulacaceae, a family largely of the southern hemisphere, which includes popular garden plants such as the moss rose *Portulaca grandiflora*. Species of *Anacampseros* are widespread in the arid, desert-like areas and dry savannas of southern Africa, while *Avonia* shares part of this distribution range, but also occurs farther afield, reaching Somalia. The centre of species diversity for these genera is in the western parts of southern Africa.

Avonia papyracea subsp. *papyracea* is popular among collectors. It has a thickened taproot, which supports a small number of short, often curved, stems. These and the small, kidney-shaped leaves are densely covered with layers of white scales that are larger than the tiny leaves, leaving the plants looking like clusters of small, white fingers sprawling on the ground. The flowers are white and arise from among the scales.

Avonia papyracea subsp. *papyracea* is also one of the most widely cultivated species, and is reasonably easy in cultivation if given a well-drained, gritty soil mixture. It is common practice with a plant of this genus to expose the thickened taproot above ground level in order to create the impression of a miniature bonsai plant, but the technique serves little purpose with this particular species, because the root will not support the stems in an erect position.

◄ SPRAWLING PLANT *Most horticulturists and collectors would identify this species as typical of the genus* Anacampseros, *but it has recently been included in* Avonia.

Red List: Rare*

Beschorneria wrightii

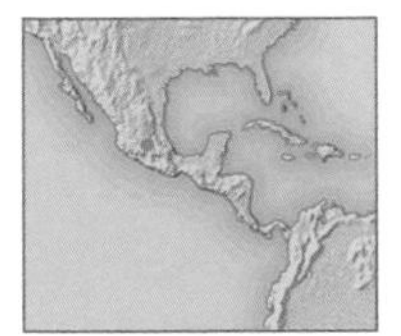

Botanical family
Agavaceae

Distribution
Central Mexico

Hardiness
Frost-tender

Preferred conditions
Reliably moist, well-drained, fertile soil; dislikes extreme conditions and needs some shade from the hottest sun.

The genus *Beschorneria* has seven species distributed from Tamaulipas state, Mexico, to Honduras. The silvery *Beschorneria wrightii* is endemic to Mexico, where it is called *ahuimo plateado*, and its distribution is restricted to a small area of the state of Mexico.

The leaves of this plant are smooth, tough, pliable, and succulent, with a finely toothed margin, and arranged in the form of a rosette at the end of a short stem. The flowers, arranged in lax, branching flowerheads, are pendulous, tube shaped, and downy, with green tepals and yellow apexes. In their natural habitat, flowers are seen from early to mid-spring; they have abundant nectar and are visited by hummingbirds. The plants grow in temperate pine and oak forests, at 2,000–2,200m (6,600–7,250ft) in steep places and cracks in volcanic rocks. This habitat has been altered by the extraction of timber and the introduction of goats.

Beschorneria wrightii is a rare species in nature and was listed in the Red List of 1997. However, the Mexican Official Norm (NOM, which controls the protection of Mexican species under some degree of threat) does treat it as an endangered species, under the PR, or special protection, category.

This species is rarely cultivated, and remains scarce in Mexican botanical gardens. It was known from gardens in Europe and was first described by Sir Joseph Hooker in England in 1901. It has also been grown in Germany and Italy. It is found in some gardens in the USA, especially in California. The species develops well outdoors in a temperate climate with temperatures of 10–20°C (50–68°F), on average. It needs well-drained, nutrient-rich soils and cannot tolerate long periods of drought.

BESCHORNERIA YUCCOIDES

▲ **BESCHORNERIA YUCCOIDES** *is one of this species' cultivated relatives, abundant in gardens with a Mediterranean climate in Europe and the USA. It is propagated by seed, but sometimes produces leafy bulbils that develop into vigorous plants once transplanted.*

106 on Red List*

Ceropegia

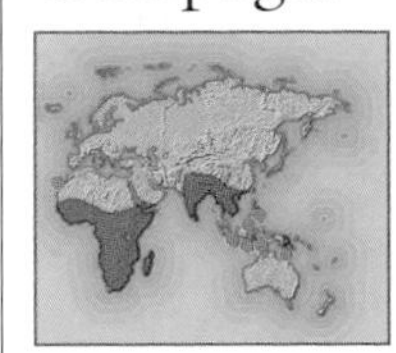

Botanical family
Apocynaceae/
Asclepiadoideae

Distribution
Africa, India, China, and north-west Australia

Hardiness
Frost-tender

Preferred conditions
Wide range of soils in open situations; dry conditions when not in growth, and a minimum temperature of 10°C (50°F).

The genus *Ceropegia* comprises around 180 species and constitutes by far the most popular group within the milkweed family. Habitats range from tropical rainforest and dry forest to shrubland, desert scrub, and grassland.

Ceropegias are erect herbs or twiners, with common names such as chandelier plant and lantern flower. The majority of plants have succulent storage roots (root tubers) combined with stem or leaf succulence. Some even show a stick-like habit, like the Canary Island endemics *Ceropegia dichotoma* and *Ceropegia fusca*, and many of the stem-succulent twiners are leafless. Most of the species are rare, have a scattered distribution, and are difficult to spot in the wild. Large, easily accessible populations are nearly unknown. These conditions make extinction less likely, and not a single species is so far known to be extinct in the wild due to loss of habitat, overgrazing, or collecting. For this reason, *Ceropegia* has been recently deleted from Appendix II of CITES. Nevertheless, the demand for bizarre plant shapes from succulent lovers caused overcollecting of some stem-succulents endemic to Madagascar, such as *Ceropegia dimorpha* and *Ceropegia simoneae*, classed as Endangered in the IUCN Status Survey and Conservation Plan (1997).

Growing Ceropegias

The tropical and subtropical origin of these plants limits their cultivation outside such regions to greenhouse and indoor situations. Only the string of hearts, ***Ceropegia linearis*** **subsp.** ***woodii***, is very widespread as a houseplant. However, they are easily cultivated plants, growing on nearly all kinds of soils, although they prefer well-drained ones like loose garden loam with some leafmould incorporated. The main threats are attacks from mealy bugs and red spiders, and fungi destroying the roots (especially the tubers). One advantage of ceropegias is that, unlike succulent milkweeds, they flower frequently in cultivation.

Since the pollen and styles are located at the bottom of the flower tubes, it is impossible to pollinate their flowers artificially without destroying the corolla, a procedure that the flowers do not usually survive. Instead, micropropagation through *in vitro* tissue culture is used: this works quite well in asclepiads, and in particular in *Ceropegia*.

▲ CEROPEGIA LINEARIS *SUBSP.* WOODII *This species from Africa is a popular pendent houseplant with stems growing to more than 90cm (3ft). They often produce stem tubers at the nodes, making propagation easy.*

BOTANY

Pitfall Flower

Ceropegia juncea's peculiar lantern flower fascinates both horticulturists and biologists studying aspects of flower morphology and pollination biology. The specialized and complex pitfall flowers attract little flies, which are trapped in the flower tube by stiff hairs that prevent them from escaping. The flies pollinate the flowers with pollinia (parcels of pollen) they have brought in from a flower they have previously visited. After pollination, the hairs wither and the flies can crawl out of the now horizontal flowers.

5*/4 on Red List/CITES II

Cipocereus

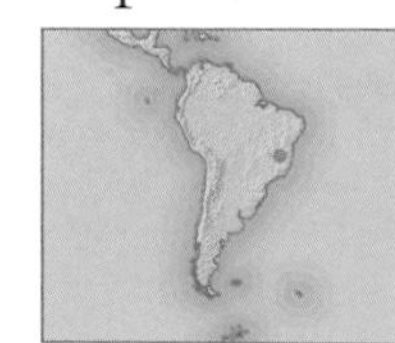

Botanical family
Cactaceae

Distribution
South-eastern Brazil: Minas Gerais

Hardiness
Frost-tender

Preferred conditions
10–30°C (50–86°F) in free-draining, sandy compost with addition of some organic matter; needs regular watering with acidic water during summer, sparingly at other times.

This genus comprises just five species of columnar cacti with blue, waxy, grape-like fruits; all are threatened to some degree. They grow in the rocky subtropical uplands of the east Brazilian highlands known as the *campo rupestre*, an area largely devoid of trees and on very low-nutrient subsoils. Local names include *rabo-de-raposa*, *quiabo-da-lapa*, and *quiabo-do-inferno*. They are naturally rare, some having extremely limited ranges. Only five populations of ***Cipocereus bradei,*** classed as Endangered in the 2000 Red List, are known, restricted to small rocky habitats, none of which are within any reserve or protected area. Plants at some sites are subjected to excessive habitat burning to encourage regrowth of grass for livestock grazing. Seedlings are sometimes seen in cultivation and can even be found in unnamed groups of cacti in garden centres. *Cipocereus bradei* rarely thrives in cultivation beyond the seedling stage; it needs sun, a well-drained, acidic growing medium, and either rainwater or de-ionised water.

▲ CIPOCEREUS LANIFLORUS *The stems, flowers, and fruits of this beautiful cactus are all covered in sky-blue wax, highlighted by the golden-brown spines.*

CONSERVATION EFFORTS

Both ***Cipocereus laniflorus*** and *Cipocereus pusilliflorus* are known from single mountain sites; the latter is classed as Critically Endangered on the 2000 Red list, with fewer than 10 plants found in a recent count. Both are cultivated as part of conservation initiatives at the Royal Botanic Gardens, Kew, but are very rarely seen outside such collections.

Cipocereus laniflorus is classed as Endangered on the 2000 Red List. Its range falls within a strictly controlled private nature reserve, which is officially recognized by the Brazilian government. This reserve is accessible to tourists, mainly nature lovers and mountaineers, but the habitats of the cactus are hard to find, and collecting is prohibited without a permit for scientific purposes. Brazilian scientists are studying the ecology of the threatened plant and animal species in the area, and the reserve's owners, an order of Catholic priests, are well aware of the biological importance of the site. *Cipocereus laniflorus* appears to be an ancient relict species, now isolated from the other members of its genus. It seems relatively straightforward to cultivate in a desert plant glasshouse in botanic gardens, but any availability for general cultivation is strictly controlled by Brazilian authorities.

◄ CIPOCEREUS BRADEI *The seed of this species is regularly collected for the nursery trade, where seedlings of this almost spineless, blue, waxy-stemmed species are popular.*

1 on Red List/CITES II

Coleocephalocereus

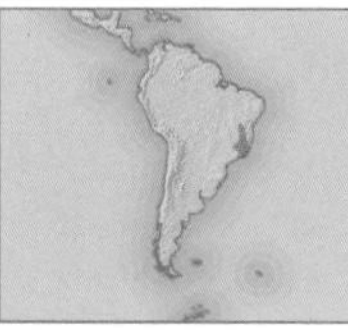

Botanical family
Cactaceae

Distribution
East Brazil: Bahia, Minas Gerais, Espírito Santo, Rio de Janeiro

Hardiness
Frost-tender

Preferred conditions
Minimum temperatures and water requirements vary; all need a hot, sunny site.

The six species of this Brazilian genus inhabit rock outcrops in forests (*see box*). The ribbed stems can be short and squat, or elongated and sprawling or erect on the rock surface to which they cling. Flowers and fruits develop in the cephalia, masses of wool and bristles atop the stems. The short tubular flowers are either nocturnal, whitish, smelling of garlic or mouldy vegetables, and visited by bats, or day-flowering, greenish-yellow or purplish, and visited by hummingbirds. A tiny pore in the club-shaped, pinkish red fruits allows ants to enter and carry away the seeds.

Coleocephalocereus fluminensis* subsp. *decumbens is a prostrate variant of one of the two most widespread species, and is listed as Endangered in the 2000 Red List. The forest surrounding the small areas of rock where it grows has been destroyed, leaving little hope that a protected area will be created, and the largest outcrop is partly within a town, where the cactus is probably seen as a nuisance or even as a danger to children. This species needs a minimum of 20°C (68°F), and has nocturnal flowers.

Coleocephalocereus purpureus, the rarest species, is classed as Critically Endangered on the 2000 Red List. Its short, cylindrical, shallowly ribbed stems bear very long, reddish spines and bright magenta flowers. It is grown from seed when this is available, and once into their second year seedlings grow rapidly given a minimum of 15°C (59°F) and plenty of light. Just five sites are known, likely having fewer than 2,500 plants. All the sites are in thorn scrub (*caatinga*) in the dry valley of the Jequitinhonha River in Minas Gerais; the most important is beside a road, and harvest by unscrupulous collectors is all too easy. The land is privately owned and has no protected areas, despite the fact that the valley is home to many unique plants.

HABITAT

DISTURBED ON ROCKS

Coleocephalocereus aureus grows in full sun on rock outcrops in habitats from dry scrub to humid or seasonally dry forest. Many forests are now highly disturbed or destroyed; this can reduce pollinators, such as bats, and grasses planted for livestock on cleared land can invade the rock face.

▼ COLEOCEPHALOCEREUS FLUMINENSIS SUBSP. DECUMBENS *is currently known in only three clustered localities in north-eastern Minas Gerais state.*

28 on Red List*

Conophytum

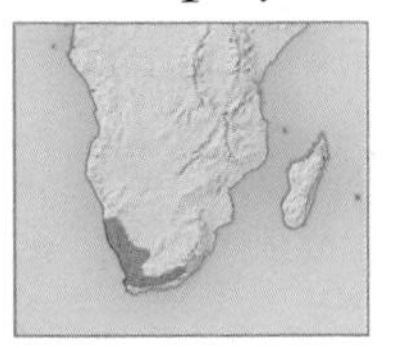

Botanical family
Aizoaceae

Distribution
South Africa, especially the winter-rainfall area in the southern corner

Hardiness
Frost-tender

Preferred conditions
Slightly acid, open soil containing sand and leaf mould; best at 5–25°C (41–77°F), with protection from hot sun; plenty of water in autumn and winter but dry conditions during their natural resting period.

The Aizoaceae, also called mesembs, have succulent leaves and a fruit capsule that can open and close when wetted. The family contains over 1,800 species, from miniature globular plants to small trees. They use camouflage and economize on water to survive the threats of grazing animals and drought. The vast majority and greatest diversity occur in southern African desert scrubs. A few are found in African grasslands and savannas, and Mediterranean scrubs and in fynbos on other continents.

Conophytum and *Lithops* (*see p. 345*) are miniature mesembs, reduced to two very succulent leaves fused at the base. The plant body is below the ground, with only the leaf tips visible at the surface. These resemble surrounding pebbles and are almost translucent to allow sunlight through for growth, giving them their generic common name of window-plants. Other common names across the species include *knopies* (buttons), *toontjies* (little toes), *ogies* (small eyes), *waterblasies* (water blisters), cone plants, and dumplings.

Almost without exception *Conophytum* species have a resting phase in summer, when they are covered by dry leaf sheaths and look dead. With autumn rains, the new leaves become water-filled and firm; many species start flowering. From late winter to spring the plants are in active growth before the summer rest.

Some species have large ranges but are thinly scattered, such as *Conophytum friedrichae, Conophytum angelicae* subsp. *angelicae,* and ***Conophytum pageae***, found in the Northern Cape Province, South Africa and in the diamond-mining area, or Sperrgebiet, of Namibia. Mining and other activities are degrading the habitat of this beautiful and widely cultivated species, which is classed as Vulnerable on the Red List of 1997. In time, plants form large mounds of small bodies with yellowish, creamy, or pinkish flowers opening at night.

▲ CONOPHYTUM HERREANTHUS SUBSP. HERREANTHUS *With large leaves that are triangular in cross-section, this does not have the typical appearance of the genus.*

Other species, such as *Conophytum burgeri* and *Conophytum klinghardtense* subsp. *baradii,* have limited ranges. One of these is the unusual ***Conophytum herreanthus* subsp. *herreanthus***, which is classed as Rare (possibly extinct) in the Red List of 1997. It has a highly localized range near Steinkopf, in Northern Cape Province, South Africa.

OTHER THREATENED CONOPHYTUMS

Conophytum angelicae* subsp. *angelicae This species grows from Northern Cape Province, South Africa to southern Namibia. It is threatened by mining, agriculture, and illegal collecting and is locally listed as critically rare. Plants form clumps of slightly cone-shaped, purplish brown to greenish-grey bodies. The flowers, usually a reddish orange, open at night. Plants can be grown from seed or cuttings. Unlisted.

Conophytum burgeri Unlike most of the genus, Burger's onion grows above ground. It is found near Aggeneys in the Northern Cape Province of South Africa, but is widely cultivated. The bright cherry-pink to dull greenish-grey plant bodies are round or like miniature inverted conical volcanoes, and purplish-pink flowers open during the day in autumn. It can be grown from seed, but is slow, and collecting from the wild is the main threat. Red List (1997): Vulnerable.

Conophytum friedrichae This highly sought-after plant from Northern Cape Province, South Africa, and southern Namibia is locally listed as endangered. The reddish-brown plant bodies are bilobed and resemble miniature loaves of bread with a deep cut in the middle. At first sight, it could be mistaken for *Lithops*. The white, creamy, or pinkish flowers open during the day. Easy to grow from seed, this tolerates high light levels and should be grown in full sun. Unlisted.

Conophytum kling-hardtense* subsp. *baradii Plants form sparse clumps in a small area in southern Namibia. The plants resemble tiny sailing ships. The greyish-green leaves are bilobed and keeled, and the white or creamy flowers open during the day or night. Easy to grow from seed, this does well with sufficient watering. Red List (1997): Endangered.

◄ CONOPHYTUM PAGEAE 'SUBRISUM' *This cultivar's beautiful yellow flowers arise between the fused leaves, usually in autumn and early winter. It can flower as early as midsummer.*

Red List: Endangered/CITES II

Coryphantha maiz-tablasensis

Botanical family
Cactaceae

Distribution
Mexico: Chihuahuan Desert, San Luis Potosí

Hardiness
Frost-tender

Preferred conditions
Deep soil with good drainage and reliably dry conditions in winter; temperature above 10°C (50°F); some shade from the hottest sun.

This North American desert cactus has been found in only five scattered localities in three municipalities of San Luis Potosí, Mexico, at the south-eastern edge of the Chihuahuan Desert region. It has been estimated that it occupies a total area of less than 50 sq km (19 sq miles), in desert scrub and grassland. It almost always grows in association with the endangered cactus *Turbinicarpus lophophoroides* (*see p.354*).

Commercial collecting has reduced several of the populations, and the harvesting of individual plants is a continuing problem. However, the destruction of its habitat is the primary threat to this species. Parts have been severely affected by agricultural development and by road construction and mining operations.

In its natural habitat this plant grows to 3cm (1¼in) tall. In a greenhouse, this plant can grow to 9cm (3½in).

CORYPHANTHA MAIZ-TABLASENSIS ►
In summer, the new growth at the top of the body produces a creamy white flower that has narrow, silky petals.

Unlisted

Cyphostemma juttae

Botanical family
Vitaceae

Distribution
Desert areas of north-western Namibia

Hardiness
Frost-tender

Preferred conditions
Full sun and dry air; well-drained soil; temperature above 5°C (41°F); reliably dry conditions when it is not in leaf.

The Vitaceae, the plant family to which *Cyphostemma* belongs, consists mainly of climbers with tendrils. It contains the grapevine, which is economically important for the production of wine, grapes, and raisins. There are two genera of succulent plants in the family, namely *Cyphostemma* with about 150 species, and *Cissus* with about 350. Most of these are also climbers, but *Cyphostemma*, in particular, includes small trees. Some of these have succulent green stems; in others the axis of the stem and root (the caudex) swells into a fat, trunk-like structure, which is not green. The *botterboom* or butter bush, *Cyphostemma juttae*, is one of the latter species. The caudices are fairly smooth and have a light yellow, papery, peeling surface. The dull green leaves are also somewhat succulent; they are quite large, with indentations along the edges, and are borne in clusters at the tips of the stems. The light yellowish flowers are tiny, but after fertilization large, red fruits develop. They resemble grapes, but are poisonous.

This species is threatened by general habitat degradation and overcollecting of plants and seeds. *Cyphostemma juttae* is surprisingly easy in cultivation. It is very reliable from seed, which spontaneously germinates wherever it is dropped in a garden in an appropriate climate. Even though the species comes from an exceptionally dry part of the world, it can tolerate large amounts of water, received as rainfall or as irrigation while it is in leaf.

CYPHOSTEMMA JUTTAE
These extraordinary looking plants, which resemble massive blobs of molten candle wax, can grow to a height of 2m (6ft) in the right conditions.

10*/5 on Red List/CITES I

Discocactus

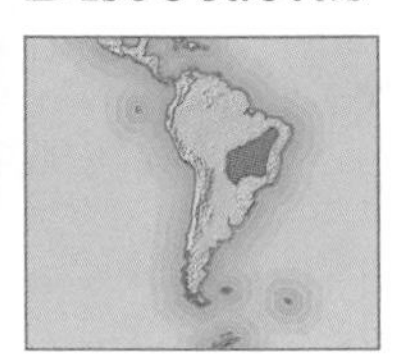

Botanical family
Cactaceae

Distribution
Brazil, north-eastern Paraguay, and eastern Bolivia

Hardiness
Frost-tender

Preferred conditions
Free-draining soil; plenty of light; minimum temperature of 10°C (50°F); frequent but light watering.

▲ DISCOCACTUS BAHIENSIS *The broad, very flat stem is unusual even within this dwarf genus, giving the plant the appearance of having been pressed into the ground.*

The seven dwarf, ground-hugging species of *Discocactus* grow in habitats of savanna, rocky fields, and thorn scrub (*see box*). Recent pressure has come from habitat loss as savanna is converted for growing crops, especially soya beans, but these species are also highly sought-after by collectors, and as a result all are in CITES Appendix I. Common names for the plants are *roseta-do-diabo*, *coroa-do-diabo*, *coroa-de-frade*, and *cabeça-de-frade*.

Discocactus bahiensis, known in Bahia state as *frade-de-cavalo*, is listed as Rare in the Red List of 1997. It seems to have a fairly extensive range in north-eastern Brazil, but may be found in much less than 500 sq km (200 sq miles) of it, with most populations having only a few hundred plants. Some sites are accessible by road and have been visited by collectors. Part of its range was lost in the 1970s by the Represa de Sobradinho, a lake formed by damming the São Francisco River, obliterating the floodplain. Its preference for barren, flat habitats is a problem because these are ideal areas for buildings or livestock corrals. One of the healthiest populations was near a village in northern Bahia, but the ground, with nearly all the plants of the species, was scraped into an embankment for a new road across the floodplain to the city of Juazeiro. No protected areas are known to hold the plant and any stock found in cultivation, where it can be successfully grown, is regarded as an *ex situ* reserve.

The very dwarf ***Discocactus horstii*** is listed as Endangered in the Red List of 1997. It is probably Brazil's most remarkable cactus, having a dark purplish-brown, ribbed stem with many tiny, claw-like spines, closely pressed to its surface. These spines absorb water into the stem, like a wick. The entire plant appears to be pressed into white quartz gravel and only becomes conspicuous when it flowers. It seems to have no unique common name, but it is well known to the residents of the little town of Grao Mogol in northern Minas Gerais, Brazil, because of the frequent visits of Europeans seeking its whereabouts. It was described in the early 1970s, and demand for it caused the nearly catastrophic plunder of its wild habitat. For some years it was thought to be virtually extinct in the wild, but recently a much larger population has been found. Both sites are now protected within a state nature reserve, which is securely guarded, and fortunately the plant is not easy to spot. Seed – which is distributed by ants in the wild – is often offered by specialist suppliers, but few seedlings survive to maturity unless grafted when young. *Discocactus horstii* can be brought into flower in summer in a well-ventilated, sunny glasshouse.

DISCOCACTUS HORSTII *Hawkmoth-pollinated flowers form in the woolly, bristly cap and bloom for a single night.*

HABITAT

ON STONY GROUND

Discocactus are found in gravelly and rocky areas in the central Brazilian savannas (*cerrados*), especially their eastern margins. Here the habitat changes to rocky, treeless upland areas of the eastern highlands (*campo rupestre*) and dry north-eastern thorn scrub (*caatinga*, seen here with ***Discocactus zehntner***). They can survive fire, a regular feature of the savanna habitats. Their low-growing form exploits the cooler air that is drawn in near ground level as the fire passes.

Red List: Critically Endangered

Duvaliandra dioscoridis

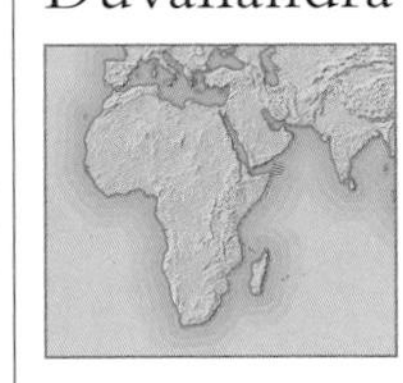

Botanical family
Apocynaceae/ Asclepiadoideae

Distribution
Yemen: Sokotra Island

Hardiness
Frost-tender

Preferred conditions
Free-draining soil; tolerant of a range of conditions; temperature range of 22–28°C (72–82°F) in full sun.

The genus *Duvaliandra* has just one known species, *Duvaliandra dioscoridis*. It is part of the stapeliads, a group of Old World stem succulents. In addition, it is also a close relative of *White-sloanea*, a Somalian endemic that is similarly threatened. *Duvaliandra* is an endemic of Sokotra (politically a part of Yemen). An island in the Indian Ocean, it was designated by WWF as one of the "Global 200" priority natural habitats because of its unique flora, which is often succulent, and about one-third of which is made up of endemic plants. *Duvaliandra dioscoridis*, together with *Aloe perryi*, is the focus of a British research programme looking into the protection and development of Sokotran plants that are under extreme threat. As few as 58 individual plants have been located in the wild by the researchers, underlining the very critical conservation status of this species.

The distribution is restricted to a single population in the Haggeher Mountains in the north-east of the island. Here, these low-growing, clump-forming stem succulents grow freely in crevices of exposed, weathering red granite.

They are endangered for reasons typical of plants in areas used for agriculture all over the tropics and subtropics: habitat destruction by grazing and the collection of building or burning materials or, as in the case of *Duvaliandra*, harvesting by the local people for use as food. As the economic and geographical isolation of Sokotra diminishes, there is a danger that these pressures will increase.

For many years, all *Duvaliandra dioscoridis* plants in cultivation were clones of the first specimens collected, in 1967. The species is self-sterile and, therefore, all attempts to produce seeds from artifically pollinated flowers failed. Then in 2000, a newly collected plant was brought into cultivation in Spain. Fertile seeds have been generated since then, making it possible to introduce new plants to suitable Sokotran habitats. However, further research is needed to develop successful conservation strategies.

Despite its rarity in the wild, this is one of the most easily cultivated stapeliad species and cloned material has long been in cultivation in European greenhouses. The species grows on a wide variety of subsoils, and it is tolerant of both excessive water and cold winter temperatures. Early and easily produced flowers, which are large and showy, make this one of the most attractive stem-succulent asclepiads in cultivation. The single disadvantage, however, is that the strong carrion scent of the flower perfectly mimics the smell of dead mice.

◄ DUVALIANDRA DIOSCORIDIS *The name of the species is taken from "Dioscoridis Insula", the ancient name for Sokotra.*

Red List: Critically Endangered/CITES II

Echinocactus grusonii

Botanical family
Cactaceae

Distribution
Mexico: Querétaro and Hidalgo

Hardiness
Half-hardy

Preferred conditions
Warm position in full sun; fertile, well-drained soil; tolerates spells as low as –4°C (25°F) if fairly dry, but is best above 10°C (50°F).

▲ ECHINOCACTUS GRUSONII *VAR.* ALBA *Where the climate allows it to be grown outdoors, this makes a dramatic landscape plant. Here it is seen on the island of Lanzarote, in beds of the local porous lava pebbles, called* picon.

This handsome, golden-spined giant barrel cactus has been known for a little over 100 years. The common names of ***Echinocactus grusonii*** reflect its habit: known as golden ball cactus or mother-in-law's cushion, it is spherical when young, but becomes barrel shaped with maturity, reaching up to 1.5m (5ft). It is also known as golden barrel cactus or *barril de oro*. This is currently one of the world's most widely cultivated cacti; however, humans have brought this beautiful species close to extinction in its natural habitat of the Mexican desert.

After it was discovered in 1889, thousands of specimens of various sizes were exported to different countries, where they were used as ornamentals for the house or garden. Concerns were already being raised about over-collecting and changing land use in the area by the end of the 19th century. More recently, the building of the Zimapán dam has led to a loss of natural habitat (*see box*). Currently, very few individuals of this species survive in their original habitat and, sadly, illegal collecting of plants continues.

CULTIVATION

The main interest of the plant to gardeners is its strong, deeply ribbed form, emphasized by its golden spines. Flowers are not produced until the plants are quite large, and are less impressive than those of many other cacti, being yellowish, toothed cups. This cactus is fairly easy to propagate and grow, if slow-growing; seedlings are as small as rice grains at one year old, and only need potting after two. In the wild, *Echinocactus grusonii* grows on tufa, a sedimentary calcium carbonate rock. Its main cultivation need is a growing medium with excellent drainage.

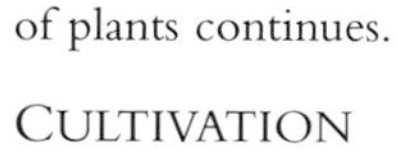

DISTINCTIVE SHAPE

HABITAT

LOST VALLEY

Echinocactus grusonii originates in a small locality on the rolling hills and cliffs of the Moctezuma River canyon, between the Mexican states of Querétaro and Hidalgo. In 1995, the Zimapán dam was built across the river at a deep, narrow gorge to provide hydroelectric power. Many plants were destroyed by the construction of the Zimapán dam, and it had a catastrophic effect on the already devastated wild populations of the golden barrel cactus; the permanent flooding of the valley system for many miles upstream wiped out most of the species' limited natural habitat.

ECHINOCACTUS GRUSONII
Older, larger plants of this species will sometimes produce offsets at the base. They can form dramatic multi-headed clusters, which spread to cover a substantial area.

Red List: Endangered/CITES II

Espostoopsis dybowskii

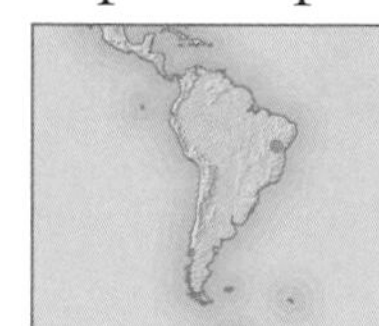

Botanical family
Cactaceae

Distribution
North-eastern Brazil: Bahia

Hardiness
Frost-tender

Preferred conditions
Free-draining, neutral to acidic soils; warm, sunny situation; minimum temperature of 5°C (41°F).

This remarkable Brazilian cactus mimics, and may be related to, the Andean genus *Espostoa*, hence the name *Espostoopsis*. A genus of a single species, it deserves special attention as the loss of the species would also mean the end of an entire genus. It occurs on smooth hills of gneiss and pure quartz outcrops in thorn scrub (*caatinga*), forming a shrub or small tree with many branches. The woolly structures in which flowers and fruits develop (cephalia) are sunk into the branch sides. These may also serve as safe landing platforms amid the spines for bats, believed to pollinate the flowers.

Classified as Endangered due to its fragmented distribution, it may be a relict that has declined in the past due to natural forces. Today humans cause its decline; its apparent abundance at its known sites should not lull us into assuming it is safe. A natural or man-made disaster could eliminate one of the four sites, and with it a key part of the genetic variation of the species. The plants at the northern site, near Juazeiro, are robust, reaching 4m (12ft) and bearing pinkish-red fruits; the southern plants, 400km (250 miles) away, are smaller with more exposed, brownish fruits. It is imperative to establish protected areas to preserve both forms, most urgently the northern population, which straddles a main highway and is being encroached upon by buildings. Two of the southern populations are also close to villages or towns and local people have been setting fire to one.

This cactus is found in cultivation, sometimes under the older name of *Austrocephalocereus dybowskii*, and may be grown on a very sunny windowsill. A well-drained mineral potting compost and non-alkaline water are crucial. Seed and seedlings are sometimes available. The seed may be derived from wild sources, which are unlikely to be affected by occasional harvesting of fruits.

▼ ESPOSTOOPSIS DYBOWSKII *The dense white wool and spines protect the plant from sun and heat and reduce water loss.*

326*/22 on Red List/CITES I and II

Euphorbia

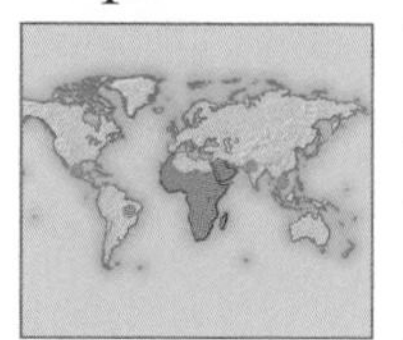

Botanical family
Euphorbiaceae

Distribution
Tropics and subtropics, mostly Africa and islands off the African coast

Hardiness
Frost-tender

Preferred conditions
Range of conditions; hot and mostly dry; active growth ceases when rainfall is reduced.

This very large genus includes about 650 succulent species, from small herbs to large trees, restricted to tropical and subtropical areas, and many more semi-succulent species. Collection from the wild has been a particular threat, and all the succulents in the genus are listed in CITES Appendix II, with some Madagascan species in Appendix I.

Euphorbia cap-saintmariensis grows on limestone rocks within the Reserve Speciale de Cap Sainte Marie, in Madagascar, an area of great importance for its unique succulent flora. This species has erect or creeping stems with rosettes of green to reddish-green leaves up to 2.5cm (1in) long. The inflorescences are pale yellow to olive green. One of the main threats to this species has been overcollection for the international horticultural trade. Thousands were imported into Europe in the mid-1980s, and the species was upgraded to CITES Appendix I. Some species are protected within the Reserve Speciale d'Ankarana in Madagascar. One is ***Euphorbia neohumbertii***, a shrubby species that has spiny, five-angled green stems to 40cm (16in) in height with prominent greyish leaf scars. The green leaves grow to 10cm (4in) and red flower structures are borne at the stem tips. It also grows at la Montagne des Français, an important site for succulents that is not yet protected. Also in the Reserve Speciale d'Ankarana is ***Euphorbia pachypodioides***. The succulent stems of this species grow to 50cm (19in) and show old leaf scars. Bluish-green leaves are arranged at the stem tips, from which reddish-purple flowers on peduncles up to 15cm (6in) long also arise.

Another site considered a priority for conservation because of its exceptionally rich flora and fauna is the Col d'Itremo in the Itremo Massif region. The vegetation of this part of Madagascar is regularly burnt, which may have a detrimental effect on the succulent flora. One endemic *Euphorbia* species found here is *Euphorbia quartzicola*, which is restricted to areas over quartz bedrock. Along with habitat destruction, collecting for international horticultural trade has been a major threat to this species, although it is considered to be difficult in cultivation, and it was transferred from CITES Appendix II to Appendix I in 1989.

▲ EUPHORBIA NEOHUMBERTII *Like many succulent euphorbias, this superficially resembles a cactus, but leaves and flowers arise between the spine clusters, not from them.*

EUPHORBIA PACHYPODIOIDES

THREAT

ISLAND LIFE

The euphorbias favoured by collectors are often from Madagascar and the Canary Islands. The latter is home to *Euphorbia canariensis*, seen here. Species restricted to islands tend to be naturally rare and this rarity appeals to collectors, thus placing extra pressure on the wild populations. Madagascar is also affected by charcoal production and agricultural practices, such as the burning of vegetation, overgrazing, and the growing of invasive crops.

Red List: Rare*

Fenestraria rhopalophylla *subsp.* aurantiaca

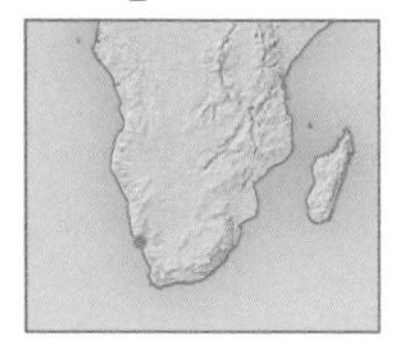

Botanical family
Aizoaceae

Distribution
Southern Namibia and north-western South Africa

Hardiness
Half-hardy

Preferred conditions
Well-drained soil and dry conditions when not in growth; tolerates brief drops to –4°C (25°F) but does best with a minimum of 10°C (50°F).

Each flattened, exposed leaf tip of this charming little species has a "window" over at least half of the area, giving it the common name window plant or *vensterplant*. Mining activities and illegal collecting have had a serious impact on these plants. They are propagated by seed or by division of the small clumps. Seed germinates reasonably well if sown on a very sandy soil mixture that is well drained but kept moist. Plants obtained by division are prone to rot: keep the soil fairly dry after planting and make sure that the soft, thickened roots are not damaged.

WINDOW PLANT *With quite large, yellow flowers and white blooms, most of the soft, elongated, club-shaped leaf of this species is borne underground.*

3 on Red List*

Furcraea

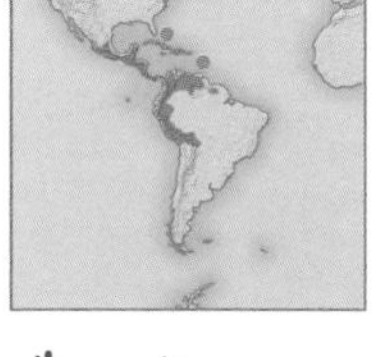

Botanical family
Agavaceae

Distribution
North and South America: Nayarit, Mexico to Bolivia

Hardiness
Half-hardy

Preferred conditions
Rocky, well-drained soils; temperatures above 0°C (32°F), ideally above 10°C (50°F); can withstand drought.

These 25 species of rosetted or tree-like succulents, related to agaves (*see pp.330–31*), are threatened by habitat loss through farming and development.

Macdougall's hemp or *pita del istmo*, ***Furcraea macdougallii***, was part of the tropical deciduous forest in the south-central region of Mexico. Restricted to Oaxaca and Puebla, on sandy and limy soils, it is probably extinct in nature and was last collected in 1961. In spite of this, it has not been included in the Red List. It survives in botanical gardens and in small towns near its original habitat, grown for decoration and as fences, and in many gardens in the southern USA. When dry, the leaves remain attached to the stem for a long time. Downy and whitish flowers to 4cm (1½in) across appear from early autumn when plants are 40 to 60 years old. The species is easily propagated from bulbils: a plant in the Jardín Botánico del Instituto de Biología in Mexico produced 15,000 bulbils but no fruits with viable seeds.

Furcraea niquivilensis, known as *pita de Niquivil*, *maguey de Niquivil*, or Niquivil hemp, is found only on steep, volcanic slopes in the south-eastern mountains of Chiapas state, Mexico, although it is also possible that it grows in the Departamento de San Marcos in Guatemala. It has stems 1–3m (3–10ft) tall ending in rosettes of erect, lance-shaped leaves, with toothed edges that taper to a sharp tip. The teeth are larger than in *Furcraea macdougallii*, as are the lemon-scented downy, red-tinged, greenish flowers, which appear in late spring, and the abundant bulbils, which are the largest of the genus at 5–11cm (2–4½in). This species has apparently almost disappeared from its natural habitat; it has been seen only in cultivation in Motozintla de Mendoza, where the indigenous Mame people use the fibres to weave baskets and cords, but only on a small scale today. It probably still survives in pine and oak and cloud forests near where it is cultivated. It is propagated from bulbils; unlike other species, fruit and seed production are also abundant. There is no record that it is grown in botanic gardens except at the Jardín Botánico del Instituto de Biología. It is not in the 2000 Red List, nor protected by Mexican legislation: its protection is urgent.

▲ FURCRAEA MACDOUGALLII
The leaves of this species have a high saponin content and they have been used macerated in water as a useful substitute for soap. Because of their chemical content, they also intoxicate fish when the leaves are thrown into rivers.

BOTANY

FLOWERING HABIT

Although there are many similarities with agaves, *Furcraea* species differ in that they have flowers that face down, not up. Like agaves, however, these plants are monocarpic, flowering once and then dying. They produce an amazing inflorescence, the tallest recorded at 13m (43ft), with flowers and conical bulbils developing on the short branchlets. In many species, production of viable fruits and seeds from this spike is very low, but the thousands of bulbils that grow on the spike drop off to develop into new plants.

5*/24 on Red List

Hoodia

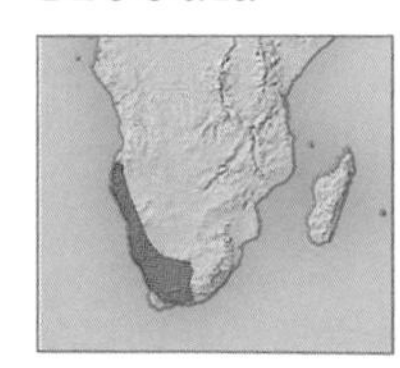

Botanical family
Apocynaceae/Asclepiadaceae

Distribution
Southern Africa: South Africa to south Angola

Hardiness
Half-hardy – frost-tender

Preferred conditions
Reliably hot, sunny site; good drainage and protection from winter wet are essential; cannot be grown outside all year in most climates.

The desert and scrub drylands of south-western Africa, principally the southern Namib and Karoo deserts, are the home of the thorny, cactus-like stem succulents of the genus *Hoodia*. Despite their appearance, this genus is related to neither the cacti nor the many species of cactoid euphorbias (*see p.342*), but rather to the stapeliads, a group of Old World stem succulents that contains over 400 species, and is part of the milkweed family, although the latex of stapeliads is characteristically clear, rather than milky.

Hoodias are impressive perennials that can reach 2m (6ft) in height. They are decorated with star-shaped to shallowly bell-shaped flowers 1–17cm (½–6½in) in diameter. These flowers are usually mud-coloured and foetid-smelling to attract flies for pollination, giving the plants the common name of carrion flowers.

There are 13 known species of *Hoodia*: some are rare, others are fairly common. Only one subspecies, ***Hoodia pilifera* subsp. *pillansii***, is so far classified as Vulnerable in the 2000 Red List, but hoodias are under pressure due to overgrazing and illegal collecting at most of their natural stands. This situation may worsen rapidly, since the value of the stem tissue of ***Hoodia gordonii*** as an appetite supressant has recently been identified as having potential for commercial exploitation (*see box*).

▲ HOODIA PILIFERA *SUBSP.* PILLANSII *When the flowers open en masse, they offer a fantastic sight in their open desert habitats.*

Although the pharmaceutical company involved has a cultivation project for *Hoodia gordonii* in Chile, the state of the natural populations must be observed with special care in the future.

Fortunately, *Hoodia* plants easily set fruits, and these are among the most seed-rich found among the milkweeds, containing up to 500 seeds each.

Hoodias are attractive plants for amateur growers, but they are not easy, and are extremely sensitive to overwatering. The plants demand both high temperatures throughout the year and high light levels in order to induce flowering. Hoodias grafted onto *Ceropegia linearis* (*see p.335*) are much hardier and more floriferous than those growing on their own roots, and grafts that have been done well can live for many years. However, experience in grafting is required for success.

THERAPEUTIC USE

DIET AID

The appetite-suppressing effects of *Hoodia gordonii* were described in 1937 by a Dutch anthropologist studying the San bushmen of the Kalahari Desert. He noticed that they chewed the stems of several *Hoodia* species before and during their nomadic hunts in the hostile environment. More recently, chemists isolated the active substance, P57, and obtained a patent on it. Pfizer, which owns the patent, agreed to share 6 per cent of the profits with the San, a tribe of about 100,000 people. However, weight-loss capsules claimed to contain *Hoodia* stem extract are already appearing on the market from other suppliers.

KALAHARI BUSHMEN STALKING GAME

HOODIA GORDONII *Also called Queen of the Namib or African hats, this species grows to a height of about 80cm (32in). It is found in the sands of the Kalahari Desert, mostly in Botswana.*

Unlisted

Hoya heuschkeliana

Botanical family
Apocynaceae-Asclepiadaceae

Distribution
Philippines: southern Luzon

Hardiness
Frost-tender

Preferred conditions
Moist, humid conditions at 20–25°C (68–77°F); indirect or filtered light; climbing support.

Hoyas, the wax flowers or wax plants of the tropical rainforest, are a widely known group of milkweeds grown by enthusiasts all over the world. With more than 200 species, this is the largest genus among the many epiphytic Asian asclepiads besides *Dischidia*. The two can be distinguished by their leaves: the rather stout and twining *Hoya* plants usually have simple sub-succulent leaves, while some leaves of *Dischidia* are formed into pitchers and the plants are often inhabited by ants. The flowers of hoyas, which are often large and showy, are also conspicuously star-shaped, or sometimes broadly bell-shaped. A unique exception is ***Hoya heuschkeliana***, whose corolla is round to urn-shaped. Such flowers could be easily mistaken for those belonging to *Dischidia*, in which such urn-shaped to tubular flowers are typical. Nevertheless, *Hoya heuschkeliana* has been proven to be a true *Hoya*, offering a good example of convergent evolution, where the two genera have evolved similar flowers to suit the local conditions. Like many other species of *Hoya*, *Hoya heuschkeliana* is of Philippine origin. This species is found only in the forested areas at the southernmost tip of Luzon, an area that is dominated by Mount Balusan, which is still an active volcano. It was first described in 1996 and since then it has been brought into cultivation. Many cuttings have been distributed to German and British *Hoya* collections. It is very easy to grow, and its survival in cultivation seems to be safe. However, its fate in its natural habitat is far less assured. Not only does the species have a naturally restricted range, but the single locality where it is known to grow forms part of Mount Balusan.

Hoya heuschkeliana is an excellent plant in cultivation. It roots easily from cuttings, has a well-branched form, and flowers for almost the entire year if conditions are suitably warm and humid. It can be grown in a pot with soil, in a hanging basket, or fixed onto moss poles. The only feature that might diminish its appeal in the eyes of growers is that the flowers are somewhat small and hidden among the dense foliage.

◄ HOYA HEUSCHKELIANA *The unusual flowers are usually formed on the lower side of the stem and partly hidden.*

Unlisted

Lithops karasmontana *subsp.* bella

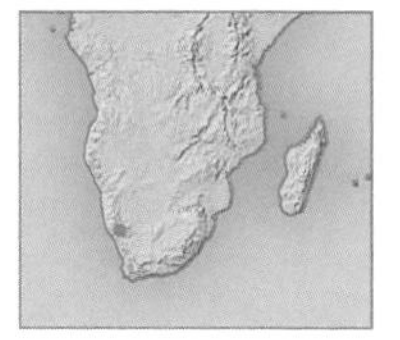

Botanical family
Aizoaceae

Distribution
Southern Namibia: Karas between Aus and Witputs

Hardiness
Frost-tender

Preferred conditions
Firm, well-drained soil; minimum of 12°C (54°F); plenty of water from early summer to late autumn but almost none at other times; protection from hot sun.

Lithops are miniature mesembs (*see* Conophytum, *p.337, for detail*) and are among the best examples of plants that mimic their environment for protection. The only parts of the plant visible above the ground are the leaf tips. These closely resemble pebbles, giving this genus the common names of stone plants or living stones. *Lithops* are restricted to southern Africa, particularly South Africa, where most species occur in summer-rainfall areas that receive very little precipitation in winter. However, this species is native to Namibia. Most flower during the rainy summer season.

◄ LITHOPS SALICOLA *Taking its name from the saltpans where it often grows, this is a rewarding species that flowers in autumn.*

Lithops lesliei and numerous other species of mesemb are highly sought-after as curiosity plants for cultivation all over the world. ***Lithops karasmontana* subsp. *bella***, like many of its relatives, has bright white flowers that attract insects to its pollen sources. After flowering and successful fertilization, seed capsules are formed, which at first sight look rather uninteresting and decidedly woody. However, when they are wetted, the capsules open within seconds to release a few seeds. If the moisture is withdrawn, the capsules close again, and when the next rain arrives, the cycle is repeated. The seeds can last for many years, so these plants are not reliant on the rapid germination of fresh seed. This subspecies is very popular and suffers from illegal collecting.

As a result of their small stature and the fact that they often occur naturally in small cracks among rocks and in other cramped spaces, these plants are well adapted to culture in fairly small pots. In nature they often occur in locally shaded positions and care should be taken not to allow containers to overheat through exposure to intense sunlight.

Most, if not all, *Lithops* species prefer a firm, even hard, soil for successful cultivation. They do not appreciate overly sandy or humus-rich soils, or mixes containing leaf mould. Propagation from seed is by far the most stimulating way of obtaining a crop of *Lithops* plants. Seed will germinate readily if sown fresh on a well-drained mixture containing equal parts sharp, gritty sand and sifted garden soil, with little organic material in it. Because *Lithops* occurs in summer-rainfall areas, watering should be severely limited in the winter. This dry period is the time when the old bodies shrivel, to be replaced by a set of fresh leaves in spring. With the onset of spring and hot summer months, drench the containers properly when water is given, about every 10 to 14 days; this is better than providing less water more frequently.

▼ LITHOPS KARASMONTANA *SUBSP.* BELLA *This attractively mottled plant readily forms clusters of multiple heads, or offsets.*

BOTANY

UNDERCOVER AGENTS

Lithops have no spines to deter browsers and grazers seeking their water stores, and so rely entirely on camouflage for defence. *Lithops dorothea* (Vulnerable in Red List of 1997) is typical of mesembs that hide themselves against the stones among which they grow. The thickly succulent leaves swell and even raise themselves slightly above the ground after rain, only to contract back into the soil when the next dry period inevitably arrives.

126*/24 on Red List/CITES I and II

Mammillaria

Botanical family
Cactaceae

Distribution
Southern USA to northern South America

Hardiness
Frost-tender

Preferred conditions
7–30°C (45–86°F); gritty soil of low to moderate fertility; dry air; protection from winter wet and hot sun.

One of the largest cactus genera, *Mammillaria* is among the most popular for the variety of form, colour, and spine pattern, and beautiful flowers. These cacti are popular houseplants and are also usually found in enthusiasts' collections. The 170 to 180 species of this genus occur from the southern USA to northern South America and the West Indies, but this is a primarily Mexican genus. Plants are found in a variety of habitats, from extremely arid lands to tropical dry oak and pine forests, and grassland: all discussed here are North American desert denizens.

OVERCOLLECTION

Many *Mammillaria* species can be found over extensive areas, but a substantial number have extremely small ranges, and these are usually threatened by the destruction of their natural habitat and by illegal collecting. Many species readily form offsets: buy only nursery-propagated plants.

Mammillaria albiflora, considered by some botanists as a variety of *Mammillaria herrerae*, is classed as Critically Endangered in the 2000 Red List and in CITES Appendix II. It has only been reported from an area 5 sq km (2 sq miles) in size in the semi-desert of Guanajuato, Mexico. The area is heavily disturbed, and is known to both commercial and amateur collectors; populations continue to decline as a result of illegal collecting. Another current threat is urban expansion, but a part of the area inhabited by this species has been set aside for conservation.

Perhaps the most extraordinary species is *Mammillaria luethyi*. Plant histologist Norman Boke first saw a single specimen growing in a coffee can in a hotel in Coahuila in 1952. Photographs reached professional and amateur botanists who speculated about its identity: it is not typical of *Mammillaria* in appearance. It was only almost half a century later, in 1996, that this mysterious, minute plant was discovered in its natural habitat, in limestone slabs somewhere in the northern segment of the Chihuahuan Desert. The plant proved to be an astonishing new species of *Mammillaria*, which differs from other members of the genus in several features of its form, notably its spine structure; it has many tiny spines clustered on top of long, upright, cylindrical bodies. The species is reported to be extremely rare, with no more than 200 plants growing in an area of as much as 200 sq m (240 sq yds). Its discoverers, Jonas Lüthy and George Hinton, have decided not to disclose the locality. The rarer a species is, the more at risk it is; with this species already Endangered in the 2000 Red List and in CITES Appendix II, there is a real danger that the whole population could be wiped out very rapidly if collectors succeed in their intensive searches for it.

Two species endemic to the Tehuacán-Cuicatlán desert valley in Puebla and Oaxaca, Mexico, are ***Mammillaria pectinifera*** and the closely related ***Mammillaria solisioides***, both classed as Endangered in the Red List of 1997, and in CITES Appendix I. *Mammillaria pectinifera* has been repeatedly collected since it was discovered in the 1880s.

◄ MAMMILLARIA ALBIFLORA *Although just 3cm (1¼in) across, this ball-shaped cactus bears spectacular large white flowers, which have attracted the attention of collectors.*

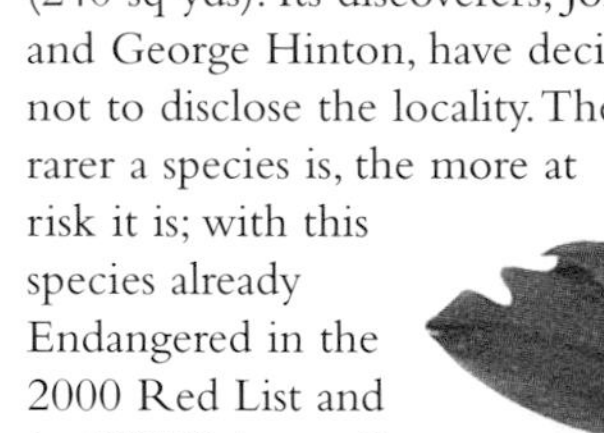

▲ MAMMILLARIA SOLISIOIDES *Some species, such as this one, release their seed slowly. This may help small populations to stay stable, but it threatens their survival if their habitat is disturbed in any way.*

MAMMILLARIA GUELZOWIANA *shows why these cacti have become so popular. It is due to the silky white hairs covering their stem, and an extravagant display of large, glamorous flowers.*

It is endemic to calcareous hills in Puebla, Mexico, where five distinct localities have been recorded; there is evidence that some of these have been severely affected by collectors. The species is threatened by both collection and the destruction of its natural habitat through urban expansion and agricultural activities. Four populations of *Mammillaria solisioides* have been confirmed in Oaxaca and Puebla, in calcareous areas with semi-arid tropical thorn forest and grassland. Numbers are reported to be very low, and the primary threat to this species is illegal collecting. In addition, one of the Oaxaca populations is also threatened by urban expansion, as it is very close to the city of Huajuapan.

One of the most restricted species is *Mammillaria sanchez-mejoradae*. The minute plants of this species are found mainly in the crevices of calcareous rock formations in the north-eastern Chihuahuan Desert in Nuevo León, Mexico. It has been reported from an area of less than 1 sq km (½ sq mile). Since it was first discovered in 1986, the location has become very well known to collectors, and the population is reported to have been reduced by 75 per cent. The consequence is that according to a recent estimate, (2001), there are fewer than 500 individuals left in the population. The species, which is already classed as Critically Endangered in the 2000 Red List and in CITES Appendix II, is expected to continue declining due to commercial collecting activities. There are reports that illegally collected plants of this species are already to be found in private collections in Europe.

CULTIVATION

RESPONSIBLE GROWING

Because many species have small ranges, being endemic to a very specific type of terrain within a small area, they are extremely vulnerable to overcollection. The future of these species depends upon the attitude of collectors. If they persist, many rare species will end up as no more than isolated examples in cultivation, unable to reproduce and evolve in their natural habitat. *Mammillarias* form offsets readily, and artifically propagated plants, often grafted, are widely available.

The beautiful ***Mammillaria schwarzii*** was discovered more than half a century ago. It found its way into private collections as seeds were presumably distributed by its discoverer, Fritz Schwartz. However, the place of origin of this plant remained undisclosed for several decades. Then it was rediscovered in 1987 on a nearly vertical volcanic rock face in the vicinity of San Felipe in Guanajuato, Mexico. Unfortunately, the locality became known to collectors, and the population has since declined by about 90 per cent. In 2001 it was estimated that fewer than 1,000 individuals remain, and the species is listed as Critically Endangered in the 2000 Red List and in CITES Appendix II.

MAMMILLARIA SCHWARZII *has been widely propagated around the world, but illegal collection is still a threat.*

THREATS IN THE HABITAT

Not all threats are from collectors. *Mammillaria crinita* subsp. *scheinvariana* was ironically only formally described after it was extinct in the wild. This plant was discovered in 1994 in Querétaro state, Mexico, by biologist Rafael Ortega. It was found as a result of a plant rescue operation during the filling of the Zimapán hydroelectric dam (*see* Echinocactus grusonii *p.340 for detail*). The species was found only in one area, no more than 50m (160ft) each way, on a steep wall of the Moctezuma river gorge, which was entirely submerged after the construction of the dam. No individuals are known to have survived in the wild, and the plant is listed as Extinct in the Wild in the Red List of 1997 and in CITES Appendix II. The plants collected were taken to the Charco del Ingenio Botanical Garden for propagation. However, with the demise of the garden, the current situation of these plants is unknown.

Mammillaria guelzowiana is restricted to a small area in the vicinity of the Río Nazas Valley in Durango, Mexico, which lies at the western margin of the Chihuahuan Desert. It is classed as Critically Endangered in the 2000 Red List and in CITES Appendix II. Although it has been collected illegally, this is not thought to represent a current threat to the species. However, it has been reported that an intense freeze in 1997 killed 95 per cent of the plants of one of the known populations, reducing it to about 500 individuals. Fortunately, this species is fairly easy to cultivate, and efforts to propagate it are in progress.

▲ MAMMILLARIA PECTINIFERA
Although it is difficult to grow, this attractive plant, with a depressed stem just 3cm (1¼in) across, is often found in private collections.

Red List: Vulnerable*

Manfreda nanchititlensis

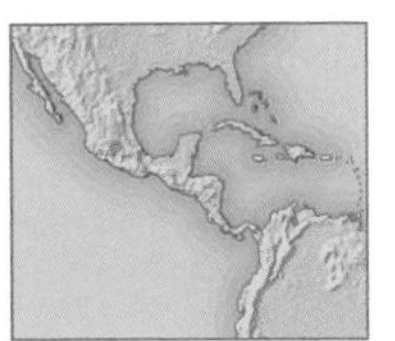

Botanical family
Agavaceae

Distribution
Mexico state

Hardiness
Frost-tender

Preferred conditions
Warm climate; tolerates temperatures down to 5°C (41°F); needs shade during the hottest months of the year.

The genus *Manfreda* is distributed mainly in North America, and contains 28 species, 27 of which grow in Mexico. *Manfreda nanchititlensis* is one of 22 species endemic to the country, and is found only in the state of Mexico, where it is called *amole de Nanchititla* or Matuda's huaco. This herbaceous species, included in *Agave* (*see pp.330–31*) by some, has grass-like leaves with many dark spots. A spike-like flowerhead on a 1.1-m (3½-ft) stem bears purple flowers with fine, thread-like, dramatically long stamens and style that extend far beyond the mouth of the flower. Flowers appear from late autumn to midwinter, the seed capsules in the following months. Plants can be propagated by seed or by separating suckers growing from the fleshy roots emerging from the rhizome.

Wild populations of *Manfreda nanchititlensis* are small. The species is found in forests of oak, sometimes with pine, on volcanic soils, at 1,500–2,000m (4,950–6,600ft), and occasionally in open and sunny places. The forests where it grows are being transformed into agricultural land. It is not included in the international lists of threatened species, but it is included in the PR category, under special protection, in the Mexican Official Norm (NOM), a national regulatory code.

HABITAT

DISAPPEARING FOREST

Clearance of pine and oak forests for cattle farming leads to soil erosion, reducing the chances that the forest will regenerate. Plants of ornamental value are taken by collectors, and pine trees may be left weakened and prone to disease by careless resin extraction. The loss of this habitat could severely affect many species that are unable to colonize other areas.

16*/7 on Red List/all on CITES

Melocactus

Botanical family
Cactaceae

Distribution
Mexico, Caribbean, Central to South America, eastern Brazil

Hardiness
Frost-tender

Preferred conditions
Gritty, free-draining soil; sunny position, and temperature above 16°C (61°F); regular water when in growth, and infrequent rest of year.

This widespread genus occurs on rock outcrops, sand, and gravel. Some species have fragmented ranges associated with particular kinds of rock. Some are threatened by quarrying, tourism, and farming. A few have been overcollected. Four species are in CITES Appendix I. All species here are from eastern Brazil.

Melocactus azureus and *Melocactus pachyacanthus* subsp. *pachyacanthus* are each known from just five small areas in northern Bahia, on flat, low-lying limestone outcrops, which are cleared by farmers or damaged by livestock when the forest is cut down for grazing land. Mining is also a threat. The worst affected is ***Melocactus pachyacanthus*** (Endangered in the 2000 Red List and in CITES Appendix II): the most important site is in a protected area at Gruta dos Brejões. Such threats have taken their toll at two sites of *Melocactus azureus* (Vulnerable in 2000 Red List and in CITES Appendix II). Unless protected areas are created, the only hope is an *ex-situ* seed bank. *Melocactus deinacanthus* (Critically Endangered in the 2000 Red List and in CITES Appendix I) is not closely related to the rest of its genus and is important for the genus's genetic diversity. It is armed with long spines, and was formerly collected. Threats are now its restricted range and proximity to agriculture and roads; the site where it was first collected in southern Bahia has become a quarry. Newly found sites nearby include numerous plants, but are also threatened by human activities. A protected area is urgently needed for this species, which is very rare even in *ex-situ* collections.

Melocactus glaucescens (Critically Endangered in the 2000 Red List and in CITES Appendix I) grows in fine, white quartz sand in the Chapada Diamantina highlands. Such habitats are rare; once known in just two linked sites, the species had a precarious outlook due to fires set by farmers. These sites are now a reserve, and three more have been found in a fairly remote area, but they are small and unprotected.

◄ MELOCACTUS GLAUCESCENS *This beautiful and highly sought-after species is threatened not only by human activity and scarcity of its preferred habitat, but also by natural hybridization with other species.*

MELOCACTUS PACHYACANTHUS *The largest population has a few thousand plants in a dense stand on limestone pavement. Most other habitats have disappeared or are beyond recovery.*

One of two species adapted to survive fire, *Melocactus conoideus* (Critically Endangered in the 2000 Red List and in CITES Appendix I) has an unusual depressed hemispherical stem. It is known from a single site on a ridge above the city of Vitória da Conquista, in a deposit of coarse quartz gravel that is quarried for construction. Horticultural demand, once a threat, is now met from artificially propagated sources. Recently Brazilian cactus enthusiasts, aided by a grant from British enthusiasts, gained permission to create an official protected area, and are prepared to reinforce the population with seedlings. The restricted range means that this species will always depend to some extent on conservation.

The remarkable *Melocactus paucispinus* has a wide but fragmented distribution in the East Brazilian Highlands, with fewer than 100 individuals in some of its seven or eight populations (Endangered in 2000 Red List and CITES Appendix I). It was heavily collected after its discovery. Low-growing and disc-shaped, it is easily confused with the unrelated *Discocactus* (*see p.339*) when not in flower or fruit, with similar very reduced, claw-like, and recurved spines.

All these species are relatively easy to cultivate in a light, warm greenhouse. They develop slowly from seed, which is the only practical means of propagation; today, the seed is usually obtained from plants of cultivated origin.

BOTANY

ANIMAL PARTNERSHIPS

The cephalium of dense bristles, in which flowers and fruits develop, is distinctive and cap-like in this genus, known as Turk's cap cacti, *cabeça-de-frade* or *coroa-de-frade*. The small, bright magenta flowers are almost immersed, as are the fruits at first, but the flowers are highly attractive to hummingbirds, which visit them for nectar and act as pollinators. The club-shaped, berry-like fruits are eaten by birds and lizards, which disperse the tiny seeds.

Red List: Critically Endangered/CITES

Micranthocereus streckeri

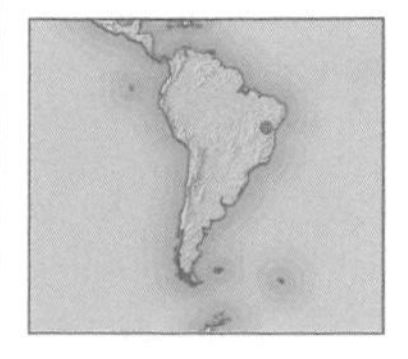

Botanical family
Cactaceae

Distribution
Brazil: central Bahia

Hardiness
Frost-tender

Preferred conditions
Free-draining soil and an open, sunny position, with temperatures above 10°C (50°F), preferably higher.

This beautiful species is known from a single locality in the Chapada Diamantina region of the east Brazilian highlands of Bahia, growing in the tropical savanna habitat of the *campo rupestre*. It was discovered as recently as 1985 and immediately introduced into cultivation in Europe, although it has remained rare and can be seen only in specialist collections. Like all cacti, it is in CITES Appendix II. The Red List status of the species reflects its extreme rarity and the threat posed by its situation above a main highway, which makes collection of plants or harvest of seeds easy for anyone with the right information. Another problem is the apparent tendency of the species to hybridize with adjacent plants of its widespread relative, *Micranthocereus purpureus*. A number of specimens seen at the only locality appeared to show the influence of this species. In total, fewer than 50 pure individuals have been observed, although the mountainous nature of the area has made a thorough survey of potential habitats difficult.

Micranthocereus streckeri is characterized by densely spined, shortly clustering stems bearing sunken cephalia, from

FLOWERING HABIT ► *The woolly, yellowish cephalia develop on the side of the plant, rather than on the top as in many other cacti. Tiny magenta flowers appear in winter, followed by club-shaped pink fruits.*

Red List: Critically Endangered/CITES II

Opuntia chaffeyi

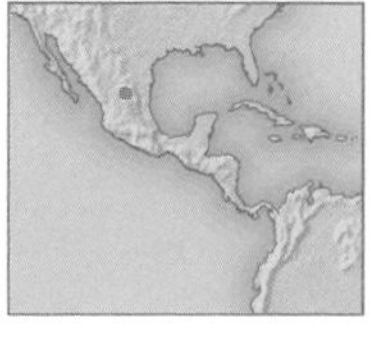

Botanical family
Cactaceae

Distribution
Mexico: Zacatecas

Hardiness
Frost-tender

Preferred conditions
Sunny situation; well drained soil poor in nutrients.

Little is known about *Opuntia chaffeyi*, although the species was originally described as long ago as 1913. It is an unusual species of *Opuntia*, in that it has which flowers and fruits emerge. At the base of the plant is a dense nest of elongated spines to protect the young stems and rootstock from excessive heat reflected from the rock surface.

In cultivation, the species requires a heated, very sunny, well-ventilated greenhouse and a quick-draining, gritty potting compost. It should be watered with rainwater or de-ionised tap water, regularly but sparingly during spring and summer. In winter a temperature of 10–15°C (50–59°F) and little or no watering is the best regime, but on sunny days a light spraying over with a fine mist is beneficial. This species is seldom available, except from specialist sources. Seedlings or cuttings are usually from material maintained in cultivation.

annual stems that die back each year to massive tuberous roots below the ground. So far, the species has only been recorded from three isolated sites. These are in desiccated desert lakes in the Chihuahuan Desert within the state of Zacatecas in central Mexico, at an altitude of around 1,600m (5,300ft).

Recent observations have revealed that the estimated number of individuals in the known populations is critically low. Due to its naturally extremely restricted distribution and the low population numbers, the most serious threat to this plant is the destruction of its habitat through agricultural development.

See also Invasive plants, *p.464*.

9 on Red List*/all on CITES

Pachypodium

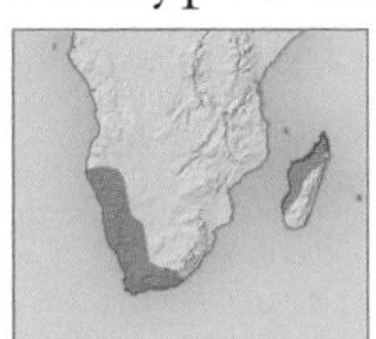

Botanical family
Apocynaceae

Distribution
Western Madagascar, South Africa, and Namibia

Hardiness
Frost-tender

Preferred conditions
Warmth throughout the year; reliably dry conditions when not in leaf.

Most species of *Pachypodium* are part of Madagascar's succulent flora; the rest, such as ***Pachypodium namaquanum*** (Lower Risk), are found in southern Africa. Collecting is a major threat: *Pachypodium baronii*, *Pachypodium ambongense*, and *Pachypodium decaryi* are in CITES Appendix I; all other species are in Appendix II.

The red-flowered *Pachypodium baronii* has two subspecies. The flask-shaped stems of *Pachypodium baronii* subsp. *baronii* grow to 3m (10ft), with spiny, grey-green branches. It is endemic to north-western Madagascar, on bare, rocky areas. The more threatened *Pachypodium baronii* subsp. *windsorii* has spherical stems up to 10cm (4in) in diameter, with leaves at the end of cylindrical, short-spined branches. This is confined to the extreme north of Madagascar, at La Montagne des Français, an area of succulent richness and diversity. By 1983 the population at the first known location was virtually extinct because of collecting for international trade. Both subspecies are in CITES Appendix I.

The variable *Pachypodium rosulatum* has two forms and four varieties. This yellow-flowered succulent has greyish-coloured, flask- or bottle-shaped stems, and young branches covered in hairs and with pairs of spines 5–10mm (¼–½in) long. It is quite widespread on a variety of soil types in western Madagascar.

The peculiar-looking *Pachypodium brevicaule* appeals to succulent plant enthusiasts. It has a flattened body up to 60cm (24in) in diameter with silvery-grey, papery bark, small, sparse, hairy leaves, and lemon-yellow flowers. This Madagascan species occurs at high elevations, in particular in the Itremo Massif. It grows at some of the most important sites for succulent plants, but is not in any protected areas. Populations have been stripped for the horticultural market, with tens of thousands of plants exported to Germany as houseplants in 1985 and 1986. They were declared to be nursery propagated, but investigations revealed that all were wild-collected. Some propagation is done in botanic gardens and in nurseries, but this is considered a difficult species to cultivate.

Pachypodiums can be grafted, but are usually grown from imported seed; vegetative propagation is rarely possible.

PACHYPODIUM NAMAQUANUM
The extraordinary, tree-like elephant's trunk or halfmens *(half-man) grows in the Namib Desert. It reaches 2m (6ft) in height and eventually attains great girth.*

11 on Red List*/CITES I and II

Pediocactus

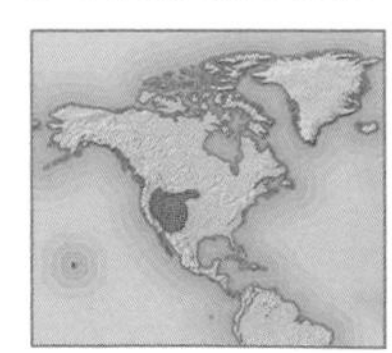

Botanical family
Cactaceae

Distribution
USA: south-western deserts

Hardiness
Frost-tender

Preferred conditions
Full sun and very well-drained soil; dry conditions when not in growth; tolerates temperatures down to 2°C (36°F).

This genus of seven small, globe-shaped cacti is native to the USA. These attractive and naturally rare plants have been popular with enthusiasts, and overcollection has been one of the main threats. They have specialized soil needs, and habitat destruction is another threat. Road building, mining, and trampling by livestock can destroy fragile populations.

The yellow-flowered *Pediocactus bradyi*, the Brady pincushion cactus, grows to about 6cm (2½in) tall and 5cm (2in) across. There is one relatively large population in the Glen Canyon National Recreation Area, and other smaller populations scattered on public and private land. The total population is estimated to be around 10,000 plants. This species is Endangered in the Red List of 1997 and in CITES Appendix I.

The Knowlton cactus, ***Pediocactus knowltonii***, grows to around 5cm (2in) tall and 3cm (1¼in) across, with pink flowers. It is Endangered in the Red List of 1997 and in CITES Appendix I. In 1979 only about 1,000 individuals were known. Their decline was a result of collection, habitat loss, and a misguided rescue operation in 1960, in which thousands of plants were taken from a proposed dam site. The one viable population is on a hill in northern New Mexico; fortunately, this site is owned by The Nature Conservancy.

▲ PEDIOCACTUS KNOWLTONII *grows in pinyon-juniper woodland on gravelly soils. The recovery plan has included efforts to re-introduce plants into parts of its habitat.*

Red List: Rare*/CITES I

Pelecyphora aselliformis

Botanical family
Cactaceae

Distribution
Mexico: San Luis Potosí

Hardiness
Frost-tender

Preferred conditions
Full sun, well-drained soil; minimum temperature of 10°C (50°F); dry conditions when not in active growth.

This small, very attractive cactus has been extremely highly valued in the horticultural market. Its common names *peyotillo*, *peotillo*, and *peoti* refer to the fact that it contains minute amounts of the alkaloids found in peyote (*Lophophora williamsii*), and the names woodlouse or hatchet cactus refer to the distinctive shape of the growing points (tubercles).

This species only occurs in an estimated 720 sq km (278 sq miles), at an altitude of 1,410–1,850m (4,650–6,100ft), in the Mexican state of San Luis Potosí. The actual area occupied within this range is much smaller, as the populations occur in a discontinuous, patchy pattern. It is likely that most of the species' range was originally covered by grassland and has probably been grazed for centuries; it is not clear how this has affected populations. Commercial collecting has been a major threat over the years, and in several of the better-known localities, most of the plants have been removed. Road construction and limestone extraction have also damaged some populations.

FORM AND FLOWERS *These slow-growing plants eventually form groups through offsets. They produce pink flowers, 2–4cm (¾–1½in) across, which appear in spring. Each flower remains open for two days.*

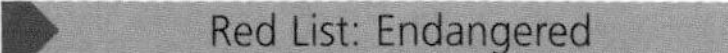

10*/2 on Red List/all on CITES II

Pilosocereus

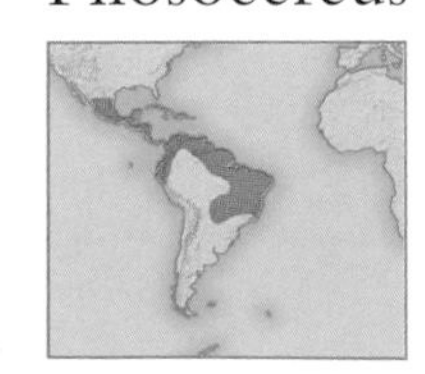

Botanical family
Cactaceae

Distribution
Mexico to Brazil: Bahia and Minas Gerais

Hardiness
Frost-tender

Preferred conditions
Well-drained soil; sunny sites with a winter minimum temperature of 10°C (50°F).

These beautiful columnar cacti form shrubs or small trees with ribbed branches. Before flowering, the spine clusters on the fertile part of the stem develop tufts of woolly hairs, in which the flowers and then fruits are produced. *Pilosocereus* grow on rock outcrops in dry tropical and subtropical forest, and clearance, mining, and building development are all threats.

Pilosocereus fulvilanatus or *quiabo-da-lapa* (Vulnerable in the 2000 Red List) grows to 3m (10ft) with blue, waxy branches and brownish-black spines. There are two subspecies, the more northern subsp. *fulvilanatus*, and the isolated southern subsp. *rosae*. The latter is found on exposed rocks in full sun much farther south in a different river drainage system, and is notable for its slender stems with numerous ribs. It is the more threatened, due to its very limited distribution and location next to a former industrial site, converted into a country club in the 1990s. Access is now restricted due to the private nature of the site, so it is difficult to determine numbers. It was known to be represented by a very small population of a few dozen specimens before the conversion.

A few living specimens of ***Pilosocereus azulensis*** (Critically Endangered in the 2000 Red List) have been found in the region of Pedra Azul, Minas Gerais; it is possible that it also grows south of Vitória da Conquista, Bahia. Further study and protection of the species and its habitat are needed.

Brazilian *Pilosocereus* are relatively easy to grow in a warm, sunny position, but need space. Well-drained, gritty compost and non-alkaline water in spring and summer are essential. The plant can be raised from seed if it is available.

▼ PILOSOCEREUS AZULENSIS *This species is known from one small remnant of dry forest vegetation, which is being cleared for agriculture and use of the wood in charcoal production.*

Red List: Endangered

Pseudolithos migiurtinus

Botanical family
Apocynaceae

Distribution
Ethiopia, Somalia

Hardiness
Frost-tender

Preferred conditions
Full sun with a minimum of 18°C (64°F); well-drained soil; plenty of water when in full growth but dry conditions at other times of year.

Some of the most extreme succulents imaginable, the plants of *Pseudolithos* are the "living stones" among succulent-stemmed plants. They face some of the same environmental pressures as those of the unrelated *Lithops* genus (*see p.345*) found in southern Africa, though the species in *Lithops* are leaf succulents. Plants in *Pseudolithos* usually consist of a single (rarely branched), leafless, and spherical stem that could sit in the palm of the hand. Their surface, which provides protection against excess water loss from transpiration, completes the almost perfect adaptation of the genus to its arid environment; it is also greyish and tessellated, offering some camouflage from herbivores. *Pseudolithos* are found only in rather remote areas of Somalia and Ethiopia, so it is no wonder that they were first described as late as 1959; since then just three additional species have been recorded. Their habitat remains difficult to reach, offering some protection against collecting.

The most widespread and well-known species is ***Pseudolithos migiurtinus*** from Somalia. Called carrion flower, and *dinah* in Somali, these distinctive plants are fairly widely cultivated in greenhouses in Europe and the USA, where seed is regularly produced. As typical "carrion flowers", the showy little flowers emit a strong, foetid scent in order to attract the houseflies and blowflies that are essential as pollinators. Otherwise, this genus is not widely cultivated and is regarded as challenging. *Pseudolithos migiurtinus* is easily raised from seed, and it is the most popular and least difficult species. Mistakes, however, are fatal, because the usually single-stemmed plants die completely if rot sets in. Usually, rotting appears overnight, and the plant turns to pulp within hours. It is important to water carefully, ideally avoiding direct contact of the water with the stems and the root collar. Like many stapeliads, *Pseudolithos* is best watered several times a week in summer, and sparsely or not at all in winter.

To reduce the danger of complete loss of the plants through rotting, grafting on to root or rooted stem tubers of *Ceropegia linearis* is recommended – this is almost essential for some species. Grafted plants will also flower more freely, but as with most stapeliads, flowering in *Pseudolithos* can vary considerably between different individuals within a single species.

THREAT

OVERGRAZING

Grazing is a major cause of desertification in arid areas, and in this harsh desert habitat, where they are intensively sought after by browsing animals, *Pseudolithos* are probably highly threatened. *Pseudolithos migiurtinus* is rare, and general destruction of habitat through overgrazing and the harvesting of material for burning is a considerable threat. *Pseudolithos* are very bitter and so are safe from being taken for consumption by local people, unlike the sweet succulent *White-sloanea crassa*, also from this area.

FOETID FLOWERS ► *The flowers smell not of carrion, as in many plants from this family, but of excrement. This scent is just as effective in attracting pollinators.*

CITES II

Rhipsalis paradoxa *subsp.* septentrionalis

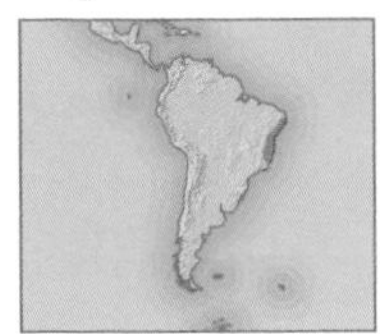

Botanical family
Cactaceae

Distribution
Brazil: Minas Gerais and Espírito Santo to eastern Pernambuco

Hardiness
Frost-tender

Preferred conditions
Moist, slightly acidic conditions and shady locations; minimum winter temperature of 10°C (50°F).

Like most species of *Rhipsalis*, this plant is epiphytic: it takes root in the branches of rainforest trees and hangs to 2–3m (6–10ft), but takes no nourishment from its host tree. It has curiously angled, segmented stems, the angles alternating with one another, and it lacks spines. The relatively insignificant yellowish-white flowers are followed by white, berry-like fruits.

◄ TRAILING STEMS *Epiphytic cacti are widely cultivated as houseplants.* Rhipsalis *species are grown for the decorative twisted effect of their stems, which hang like a curtain. The insignificant flowers of this species appear in late spring, and are followed by tiny fruits.*

The subspecies *septentrionalis* (meaning "northern") is a recently described geographical variant, with more slender stems than the rest of the species. It is found in the southern parts of Brazil's Atlantic Forest, from Rio de Janeiro to the state of Santa Catarina. This northern subspecies is more threatened than southern populations: it is known from just eight localities or sightings, and some of these reports almost certainly relate to patches of forest that no longer exist.

Rhipsalis paradoxa is easily cultivated in a humid, partially shaded conservatory in which the minimum temperature is not allowed to fall below 10°C (50°F). It requires an open potting mix that is rich in organic matter, and likes to be slightly moist during much of the year, but never saturated. Spraying with non-alkaline water (ideally rainwater) is beneficial at any time of year, especially in summer, when adequate ventilation is essential. This plant is easily propagated from cuttings in spring or summer and is available from specialist nurseries or hobbyists. Preservation in cultivation is essential for this species, because there is no current prospect of a secure future in its native habitat.

THREAT

DISAPPEARING FOREST

The Atlantic Forest along the coast of Brazil has been comprehensively altered by human intervention, including agricultural developments, such as this banana plantation on cleared forest land. Nearly three-quarters of the plants here are found nowhere else on Earth, but today less than seven per cent of the forest's original coverage survives. Given this frightening statistic, the future of this *Rhipsalis* does not look bright, despite its considerable range over some 2,000 km (1,250 miles): none of its known sites is within a protected area.

Red List: Endangered/CITES II

Schlumbergera kautskyi

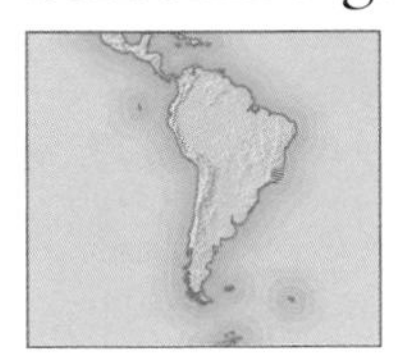

Botanical family
Cactaceae

Distribution
South-eastern Brazil: Espírito Santo

Hardiness
Frost-tender

Preferred conditions
Bright, not too sunny location; minimum 15°C (59°F); moist air; plenty of water in summer but less when in flower.

This is the smallest-flowered of the Christmas cacti. Like the other five species in this genus, it grows as an epiphyte or lithophyte, perching on trees or rocks. The other species are found in the Atlantic Forest of Rio de Janeiro and adjacent parts of São Paulo, with its high mountain peaks and coastal sugar-loaf inselbergs. ***Schlumbergera kautskyi*** is the most geographically isolated of the species, outside the range of the others. It is Endangered by virtue of being known from only two small mountain localities, both in an area where expensive holiday homes and residential developments are causing significant change. Most of the Atlantic Forest in this region was destroyed prior to the discovery of the species in 1986, leaving only small fragments on the mountains where it grows. It is now probably too late for these habitats in central Espírito Santo to become effective nature reserves. The plant's future in the wild may rest on the hope that developments do not reach the last unaffected, steepest slopes.

These cacti are not difficult to grow, given a well-drained, organic potting compost. They appreciate good light, but not full sun, all day long and benefit from cooler conditions during hot summer weather, when they can be placed outdoors in the partial shade of trees. In winter, when flowers develop, the plants should be kept in good light and a minimum of 15°C (59°F). Watering should be less frequent, and it is important to keep the plant in humid air away from draughts and sudden fluctuations in temperature. Propagation by cuttings (detached stem-segments) is straightforward and is sometimes made even easier by the production of aerial roots on the stem to be separated.

◄ DELICATE BLOOMS *The funnel-shaped magenta flowers of* Schlumbergera kautskyi *are the smallest in the genus, only half as large as on the species usually grown as houseplants. They are followed by elongated, yellowish fruits.*

▼ SCHLUMBERGERA TRUNCATA *This widely cultivated species resembles* Schlumbergera kautskyi, *with its flattened stem-segments and sharply pointed marginal teeth, but it has larger flowers and round, pink fruits.*

18*/19 on Red List/all on CITES I

Turbinicarpus

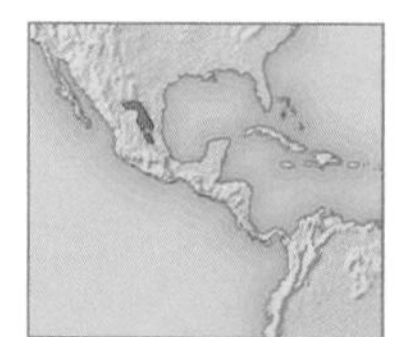

Botanical family
Cactaceae

Distribution
Mexico: Chihuahuan Desert

Hardiness
Frost-tender

Preferred conditions
Sharply drained, moderately fertile soil and full sun; plenty of water when in growth, but almost none at other times.

This genus comprises about 24 species of usually small cacti. The entire genus is endemic to the Chihuahuan Desert of Mexico, primarily scattered in the south-eastern portion. Most species have extremely small ranges, and are commonly restricted to one or a few small localities. Although these plants are rare in cultivation, they are very sought-after among hobbyists. The great demand has resulted in the plundering of whole populations. In fact, the most dramatic cases of depredation of cacti in Mexico are of members of this genus. Due to the heavy pressure from illegal collectors, the entire genus is in CITES Appendix I.

The locations of two species remained unknown for decades after their initial discovery. The only known population of ***Turbinicarpus gielsdorfianus*** (Critically Endangered in the 2000 Red List), is dependent on a specific soil type in an area of less than 1 sq km (½ sq mile) in San Luis Potosí. Unfortunately, its rediscovery in 1988 revived the interest of collectors, and the population has been reduced to 20 per cent of its original size. It is now estimated to stand at about 4,000 mature individuals. Since the single small locality of ***Turbinicarpus saueri*** in southern Tamaulipas state was rediscovered in 1998, collectors have removed half of the plants. The species was recently estimated to comprise fewer than 200 individuals, and is Critically Endangered in the 2000 Red List. Illegal collecting is clearly a major threat. The demand for wild plants persists despite the fact that both species are easy to grow and propagate, and cultivated stock has been available for decades.

Turbinicarpus schmiedickeanus subsp. *gracilis* is also limited to a single locality, in southern Nuevo León. It grows among calcareous rocks surrounded by xerophytic vegetation. Recent estimates suggest that the total population is just 10,000 individuals. Illegal collecting is a serious threat, although habitat destruction caused by overgrazing and soil erosion also take their toll. The species is listed as Near Threatened on the 2000 Red List, but this subspecies, like many others, is Critically Endangered.

Another narrowly endemic species is ***Turbinicarpus ysabelae***, found in a single locality in southern Tamaulipas. According to a recent census, the population contains fewer than 250 plants. It is highly threatened by illegal collecting, and is Critically Endangered in the 2000 Red List.

▲ TURBINICARPUS YSABELAE *The single known population of this species grows in a limestone area of about 2,000 sq m (2,400 sq yd). The habitat is adjacent to an expanding urban area, and is also much visited by cactus collectors.*

TURBINICARPUS SAUERI *This small plant is only known to grow in a single area of less than 1 sq km (½ sq mile) in high limestone hills.*

HABITAT

CHIHUAHUAN DESERT

This relatively high-altitude region, typically 1,000–1,500m (3,300–4,950ft), has hot summers with fairly low rainfall, and dry, cold winters with frequent night frosts. The soil is typically rich in calcium, derived from limestone beds. It is a scrub desert, with no large cacti, only small species, here *Turbinicarpus gielsdorfianus.*

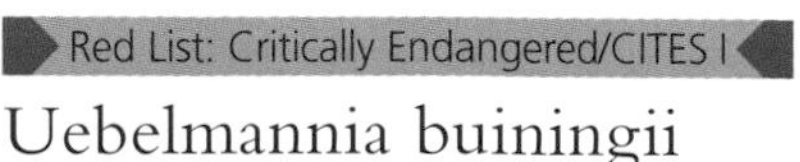

Red List: Critically Endangered/CITES I

Uebelmannia buiningii

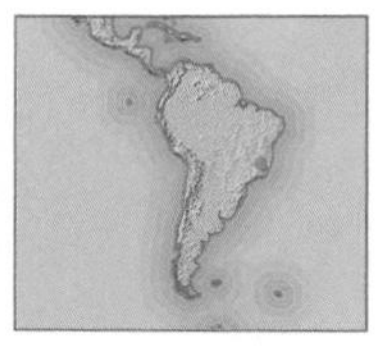

Botanical family
Cactaceae

Distribution
South-east Brazil: east-central Minas Gerais

Hardiness
Frost-tender

Preferred conditions
Soilless quartz gravel or on rocks; sunny conditions with high humidity; winter minimum of 10°C (50°F).

Unlike its relative, *Uebelmannia gummifera*, this species, known as *coroa-de-frade* or *cabeça de frade*, was never particularly abundant. It has a small, spherical, ribbed, purplish stem with a woolly apex from which the bright yellow flowers emerge and spread their slender petals. Repeated collecting for the horticultural trade and the depredations of goats have caused the species to decline sharply, and it is now nearly extinct in the wild. Just a few dozen plants persist in its unique habitat, the upland quartz rocks and gravels in the *campo rupestre* of the Serra Negra region in Minas Gerais, Brazil. In recognition of these problems, the three species of the genus were added to CITES Appendix I in 1993. This may have helped control international trade in wild-taken plants, although the illegal export of seed has continued.

CULTIVATION

These are not the easiest of cacti to grow and remain rare in cultivation. They need nutrient-poor, freely draining, mineral soils and very pure, non-alkaline water, not tap water. Conditions should be sunny but not baking, with a winter minimum of 10°C (50°F) and regular spraying to maintain a level of humidity that mimics their cloud forest habitat. The easiest way to cultivate them is as grafted specimens. Just one nursery is currently offering this species in Europe from artificially propagated seeds; other suppliers of seed may be obtaining their stock from the few remaining plants in habitat, so always enquire about the source of the seed.

▲ HARD TO GROW *This species is difficult in cultivation and may not survive even when grown as a grafted specimen.*

10 on Red List*

Yucca

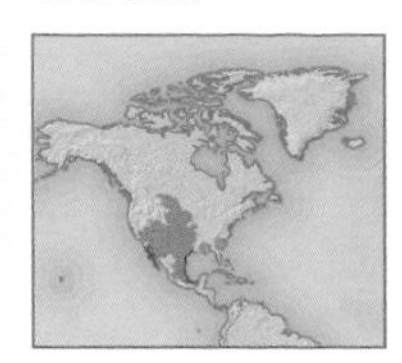

Botanical family
Agavaceae

Distribution
North America through Central America and the West Indies

Hardiness
Frost-tender – fully hardy

Preferred conditions
Sandy, well-drained soils in warm, fairly dry climates; winter minimum temperatures vary by species.

Comprising approximately 50 species, the genus *Yucca* is distributed across mainly desert environments in North America and south to Guatemala. Plants are in high demand for landscaping in the USA. Ten *Yucca* species are included in the Red List of 1997, and some are protected nationally, but none are protected under CITES. Some yuccas produce bulbils or suckers, but many can only be propagated by seed. They do not set seed without their unique pollinators or careful hand-pollination, making it all the more important that their natural habitat be protected.

Yucca brevifolia, a tree growing to 10m (30ft), is endemic to higher altitudes of the Mojave Desert in the south-western USA. It exists as distinct stands in scattered localities, which implies that perhaps the present numbers are only a remnant of a once much larger distribution.

Yucca queretaroensis is commonly called *estoquillo*, *izote estoquillo*, or Queretaro's yucca. It is endemic to Guanajuato, Hidalgo, and Queretaro in Mexico. It is a tree-like species, growing to 5m (15ft) tall, and has a rosette of more than 500 linear leaves with finely serrate edges and a slender thorn at the apex. Flowers are bell-shaped, whitish, and borne from late spring in compact, branching flowerheads up to 90cm (36in) long. This species belongs to the xerophytic scrub of the Moctezuma River basin, growing on pronounced slopes, on sandy and rocky soils, at 1,000–1,300m (3,300–4,300ft). The leaves are used to thatch roofs, and the flowers are edible. The species is not included in the international lists of threatened species, but the Mexican government gives it special protection (PR), because its habitat is being altered. This species is known to grow in some botanical gardens in Mexico, but is rarely cultivated although the large number of leaves in the rosettes and on the stem make it very decorative. It is found in dry climates with average temperatures of 18–25°C (64–77°F), although it can survive in slightly cooler climates if it is grown in sandy, well-drained soil.

YUCCA BREVIFOLIA *The Joshua tree is said to have been named by the Mormons, who likened its form to the prophet Joshua with his arms raised. Weevil damage provokes the distinctive branching. It is most abundant in the Joshua Tree National Park, California.*

WILDLIFE

POLLINATION SPECIALISTS

Yucca pollination is very specialized, carried out solely by *Tegeticula* yucca moths, mostly *Tegeticula yuccasella*. These moths lay their eggs in the flowers, and the emerging larvae eat seeds at the tip of the seed capsule, leaving the rest to develop normally. Anything that affects the moth populations affects the yuccas.

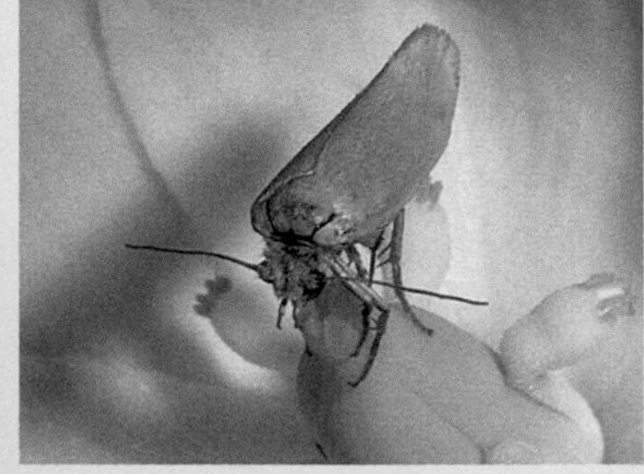

PALMS AND CYCADS

Evergreen trees or shrubs with arching, divided leaves, palms and cycads look somewhat alike but botanically are not closely related. Palms are angiosperms that produce panicles of small flowers. Cycads are gymnosperms, ancient seed plants that reproduce via cones, not flowers. Although they dominated the planet's vegetation during the Jurassic, the Age of Dinosaurs, cycads are now one of the world's most imperilled plant groups. Only 11 genera survive, mostly in small, disjunct populations on several continents. According to the IUCN's recently published Cycad Action Plan, 52 per cent of the cycads that remain are threatened with extinction.

HOW AND WHERE THEY GROW

The most familiar form taken by palms and cycads is a tall, single trunk topped by an impressive crown of divided leaves. However, while most palms do have a prominent, solitary trunk – ranging from pencil-thin to extremely stout – numerous palms form clumps, some have trunks that hug the ground, and some are climbers, like the rattans. Cycads, too, range in size, from giants with lofty, columnar trunks to small species with subterranean stems. Both palms and cycads are most commonly found in tropical and subtropical forest, but both groups also contain savanna plants, capable of surviving periods of drought and sporadic fires, and there are palms that grow in oases and wadis in the blistering Sahara Desert, and on the slopes of the Himalayas.

◄ WHERE CYCADS GROW *Cycads are found on every continent except Europe and Antarctica. Australia, Mexico, and South Africa are centres of diversity, containing more than half of the world's species.*

REPRODUCTION

The largest inflorescence and the biggest seed in the plant kingdom are produced by palms. The largest inflorescence is produced by *Corypha umbraculifera*, rising some 10m (33ft) from the top of the plant and carrying millions of flowers. Most palms have both male and female flowers on the same plant (known as monoecious) but in others, the males and females are on separate plants (dioecious). All cycad plants are either male or female, and the cones of each sex are usually quite different in size and shape and, to some extent, colour. The female cones of most cycads are larger and bulkier than the males, and those of large cycads can be impressive, weighing more than 20kg (44lb). The seeds of most palm and cycad species are recalcitrant: the viability of ripe seed decreases rapidly, and they cannot be dried and stored at low temperatures in seed banks (*see p.44*) without killing the embryo. Cycad seeds are slow to germinate, and the plants take a long time to grow. This not only makes propagation time-consuming, but also leads to the poaching of mature plants.

◄ PALM FLOWERS IN BUD *Linnaeus called palms "the prince of plants" for their elegant, lofty form and magnificent inflorescences.*

PALM SEED ► *The seeds of the oil palm* (right) *are of great economic importance, but the biggest palm seeds are those of the double coconut,* Lodoicea maldavica (*see p.374*).

▲ CYCADS CONES *Male cones of* Encephalatos (*see p.370*), *here of* E. frederici-guillemii, *are striking, but female cones can be spectacular; those of* E. transvenosus *can weigh in at 45kg (99lb).*

PALMS IN THE MALDIVES
Many of the world's most endangered palms belong to small or even monotypic genera, consisting only of a single species, and are endemic to oceanic islands and other remote areas.

▲ COCOS NUCIFERA *Source of food, oil, fuel, and building materials, the coconut is one of the plants most useful to humankind.*

◄ WEAVING RAFFIA *The raffia palm* (Raphia farinifera) *is native to Madagascar. However, stocks there have been decimated by overexploitation.*

USE AND EXPLOITATION

Palms rank below only the grasses and legumes in importance to humankind. They are essential to the economies and the ecology of tropical regions. They put food on the table for millions of people daily, from coconut (*Cocus nucifera*) to dates (*Phoenix dactylifera*) and heart-of-palm, or palm-cabbage, which is taken from many species. The oil palm produces cheap cooking oil that is used around the world. Palms also furnish materials for furniture, fibres, building, clothing, fuel, waxes, wines, and many other products. They are also prized for their beauty, both outdoors in tropical and subtropical landscapes and as indoor potted specimens in cooler climes.

Cycad stems, roots, and leaves have long been harvested for magical and medicinal use. Several cultures have even developed methods of preparing food from cycads, which contain several powerful toxins. Cycads are of great interest to scientists, gardeners, and the general public because they provide a glimpse of ancient ways of life in the plant world. Cycad enthusiasts pay extraordinary sums for the new species and forms introduced by specialist nurseries. Both indoor growers and landscapers appreciate not only their striking forms but also their slow growth rates and predictable dimensions.

The passion for possessing cycads is a major cause of their demise. The enormous prices fetched by large specimens fuel a flourishing international trade in plants – not always legal. *Encephalartos* species in Africa, for example, can sell for thousands of dollars. Populations of a number of cycad species have been decimated by rampant poaching of mature plants. Poachers often seek out female plants, skewing the natural sex ratio of wild populations and reducing even further their capacity for regeneration.

As a result, all cycads are protected by CITES against overexploitation due to international trade, although enforcing such regulation remains an ongoing struggle.

COMBATING THREATS

Along with many other groups of plants, palms and cycads have suffered at the hands of humans. The two main threats to palms are over-exploitation and habitat destruction. Conversion of land to human settlements and farms, such as oil palm plantations in south-east Asia, as well as large-scale timber operations have had devastating impacts. The use of rattan has caused the populations of numerous species to decline drastically. Demand for seed of rare palms by nurseries and palm enthusiasts is a growing problem, because the plight of these species can be aggravated by even limited seed collection.

Around the world, cycad populations are wiped out to make way for crops. In some countries, they are killed when forests are clear-cut for timber. While CITES regulations have reduced poaching to some extent, cycad theft and smuggling is still a huge and profitable enterprise.

Botanic gardens and government agencies are struggling to find ways to propagate the most threatened palms and cycads, and efforts are under way to grow other species that are at risk and reintroduce them to the wild. The recalcitrance of palm and cycad seed complicates such conservation efforts, because it means that these plants can only be conserved as growing specimens, and it is difficult to find enough space in botanic garden collections to preserve the genetic diversity present in wild populations of most threatened species. Complicating matters even more, species in the same genus can cross when grown in close proximity, and hybrid seed has little conservation value.

For example, for more than a decade, Lowveld National Botanical Garden has been working to propagate South African species of *Encephalartos* for conservation purposes. Seeds are collected from every known locality and the resulting representatives of each population (about 100 of each are needed to make a viable resource) must be planted in separate orchards to minimize the mixing of gene pools. Surplus specimens are sold to the public to reduce pressure on wild plants.

▲ OIL-PALM FARMING *Much habitat destruction is caused by the conversion of land to oil-palm plantations (here in the Cameroons).*

▲ GARDEN USE *The current craze for "exotic" styling in gardens exacerbates the pressure on wild cycads, already vulnerable to poaching to satisfy the demands of specialist collectors.*

◄ HEARTS-OF-PALM *Many palms have been forced to the brink of extinction by the demand for this delicacy, which involves harvesting the apical bud, or growing tip, of the plant.*

FROM AN ANCIENT WORLD *Cycads have seen the dinosaurs come and go; their 300-million-year history and primitive appearance contribute to the passion they inspire as subjects for cultivation.*

SUCCESS STORIES

In 2001, the U.S. Fish and Wildlife Service arrested members of a smuggling ring who had sent protected cycads, valued at half a billion dollars, to the USA from South Africa, Australia, and Zimbabwe. South African conservation authorities are now putting microchips into cycads growing in the wild to discourage theft and enable them to trace stolen plants.

While cultivation has been a mixed blessing for one rare cycad, it has been an unmitigated boon to an associated butterfly once feared extinct. Once abundant in the pre-development pinelands and hammocks of its native Florida, *Zamia integrifolia*, the Florida coontie, is now uncommon in the wild, threatened by the poaching of large plants for the landscaping industry. As this cycad has become more common in gardens, however, so has the rare Atala butterfly, *Eumaeus atala*. The cycad is the preferred food of Atala caterpillars.

EUMAEUS ATALA

GARDENER'S GUIDE

- Gardeners can help conserve palms and cycads by being careful not to support the illegal trade in wild plants and seeds. Buy only from responsible nursery suppliers, and ask questions about the origin of any plant you would like to purchase. Buy plants that have been propagated in nurseries from cultivated specimens, and ask to be shown copies of export papers.
- Five signs of a poached cycad are: (1) The plant is a rare species not ordinarily found in cultivation; (2) the plant is not fully rooted and looks as if it has been dug up from the wild; (3) the plant carries scars from its natural habitat, such as blackening from veld files, holes chewed by porcupines, and irregular growth patterns due to periods of drought; (4) the supplier can't tell you anything about the plant's origin; (5) the plant is surprisingly inexpensive – an especially suspicious sign if it is a good-sized specimen.
- Because palms and cycads tend to hybridize readily, it is difficult for gardeners to maintain accurate records about the origins of plants. However, it is possible to support efforts of botanical gardens and government agencies to preserve their habitats, propagate the plants, and reintroduce them into the wild.
- One consequence of the passion for cycads is that significant collections can be found around the world. For some threatened species, this could be an important genetic resource. At the very least, as many cycads as possible should be propagated to help satisfy the demand. If you don't have enough space to do this on your own, offer to help pollinate and propagate the cycads at your local botanic garden. The gardens that are pollinating their plants to increase seed production are doing this largely with volunteer help.

Red List: Critically Endangered

Attalea crassispatha

Botanical family
Arecaceae/Palmae

Distribution
Southern Haiti

Hardiness
Frost-tender

Preferred conditions
Fertile, moist but well-drained soil, in sun, but with shade when young.

The prospects for *Attalea crassispatha,* or carossier palm, are bleak. It is endemic to Haiti, where it is found in deciduous forest. There are fewer than 50 mature palms of this species in the wild, and it is Critically Endangered.

The carossier palm survives as mature trees in pastures and home sites, but there is no natural regeneration. Children eat the immature seeds, which taste like coconut, and seedlings are grazed by farm animals. Fortunately, there are individuals of this rare species in cultivation outside Haiti.

Botanists are especially interested in the carossier palm because it is the only Caribbean representative of the genus *Attalea*; the remaining 30 or so species are found in Mexico, and in Central and South America.

Attalea crassispatha is a large palm that reaches 20m (65ft) or so, with feathery, deep green leaves and a stout trunk. In cultivation, it grows moderately quickly given a fertile soil and abundant moisture. Some shade is beneficial for the two or three years after germination, but later, the palm needs full sun. It is drought-tolerant at maturity.

Like all palms, carossier undergoes a prolonged "establishment phase", during which time the palm's stem is formed underground. During this period, the palm appears to be merely a tuft of leaves rising from the ground, but once the stem achieves the mature diameter, the palm begins building an above-ground stem.

RENEWAL

TREES FOR RENT

At the Caribbean Botanical Gardens for Conservation conference, held at Fairchild Tropical Garden in May of 2002, the tropical ecologist, Joel Timyan, spoke of a novel means that he had devised to conserve *Attalea crassispatha.*

He decided to modify a customary practice in Haiti – the renting of mango trees to harvest the fruit – and apply it to the seeds of three particularly free-fruiting carossier palms.

Conservation contracts were drawn up to rent these trees from their owners. This will ensure a seed harvest that can be raised in cultivation in order to restore populations in the wild.

The tree rental fee, which is usually paid at the beginning of the school year, often means that the tree's owner can afford to send a child to school.

This is a technique that could be applied to other species that are threatened by their use as a valuable food source in their native environments.

Red List: Critically Endangered/CITES II

Beccariophoenix madagascariensis

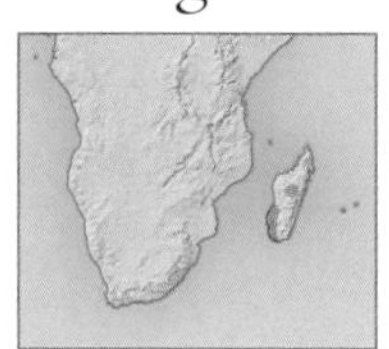

Botanical family
Arecaceae/Palmae

Distribution
Madagascar

Hardiness
Half-hardy

Preferred conditions
Well-drained soil with ample water in open sun; tolerates partial shade and is moderately drought-tolerant.

Madagascar has a rich palm flora with over 170 species, nearly all of which are endemic to this biodiversity hotspot. Of the endemic species, 86 are listed as Extinct, Critically Endangered, or Endangered, and the remainder are listed under lower categories of threat.

The manarano palm, ***Beccariophoenix madagascariensis***, occurs at 900–1,000m (2,925–3,250ft), often as an emergent above the canopy of tropical rainforest. It was classified as Critically Endangered in the 2000 Red List. The IUCN/SSC Palm Specialist Group recommends the establishment of an *ex-situ* breeding programme in Madagascan botanic gardens. There are only two known populations, one in Mantady and the other in the south-west of the island.

Overexploitation for timber, food, and fibre are among the reasons that this palm is now listed as Critically Endangered, but even in the early 1900s, it had been overused for the manufacture of hats. The overcollection of seed for the horticultural trade is also proving to be a problem.

▲ ROBUST AND HANDSOME *This palm makes a particularly attractive specimen as a young plant, but should be bought only from nursery-propagated stock.*

Named after Dr Odoardo Beccari, a 20th-century Florentine collector of palms, it is robust and handsome, is without thorns, and has large, feathery, rich green leaves. It is especially valued in cultivation for landscaping in tropical and subtropical gardens, where it is known to tolerate light, short-lived frosts. The edible fruit is similar to, but not as well-flavoured as, the true date palm, *Phoenix dactylifera.*

Seed germinates within 3–5 months if sown as soon as it ripens. Although, as is the case with many palms, growth is slow in the early years, it grows fairly quickly once it becomes established.

Red List: Endangered

Bentinckia nicobarica

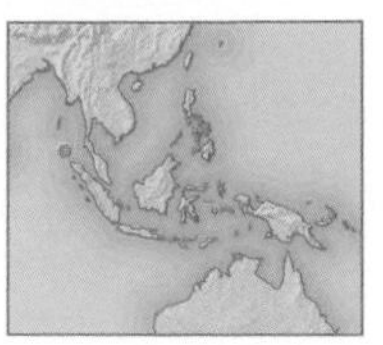

Botanical family
Arecaceae/Palmae

Distribution
India: Nicobar Islands

Hardiness
Frost-tender

Preferred conditions
Plentiful moisture, constant humidity, and a humus-rich soil; sun for mature plants, but shade when young.

This palm, *Bentinckia nicobarica*, is endemic to the evergreen tropical rainforests in the coastal lowlands of Great Nicobar, an island in the Andaman Sea, to the north of Sumatera. It is listed as Endangered in the 2000 Red List and has been threatened in the past by habitat loss and overharvesting of the edible palm heart.

There are plans to make the Nicobar Islands a major tourist destination, to establish Great Nicobar as a free-trade port, and to increase the military presence on the islands. In addition, road development and the promotion of cash crops (rubber and cashews) threaten the surviving forest habitat, despite the high levels of statutory protection of the Nicobar Islands' rainforests as a whole. Important decisions with regard to how to control these developments and conserve resources are yet to be made.

As a result of its rarity and isolation, little is known about the evolutionary biology, ecology, or current wild status of *Bentinckia nicobarica.*

Growing to about 15m (50ft) tall, it has a highly ribbed and narrow trunk that is topped by a pale green crown-shaft and elegantly arching leaves. The large inflorescences that are held beneath the leaves give rise to clear scarlet fruits.

Bentinckia nicobarica is sensitive to the cold, but thrives in tropical conditions. Ideal for parks and large gardens, it makes a handsome specimen or avenue tree. It was among the top four fastest-growing tropical palms in trials at the Fairchild Tropical Garden, in Miami, Florida, USA.

In cultivation, it responds well to a consistently moist but well-drained, moderately fertile soil and should be given a sheltered position in sun or dappled shade.

The only other species in the genus, *Bentinckia condapanna*, is found in Kerala and in the hills of Tamil Nadu, in India, more than 2,000km (1,250 miles) away.

Bentinckia condapanna is listed as Vulnerable in the Red List of 1997. It is given this listing because it suffers from land clearance and consumption of the palm heart by elephants. It is known as Lord Bentinck's palm, after William Cavendish Bentinck, Governor of Madras, India, 1803–1807.

CITES II

Bowenia

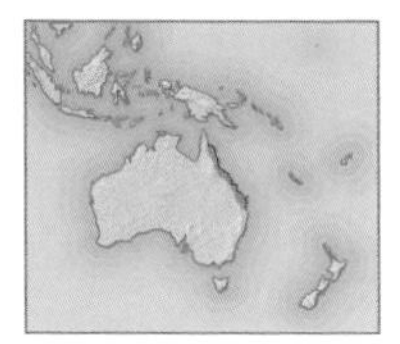

Botanical family
Zamiaceae

Distribution
Australia: Queensland

Hardiness
Frost-tender

Preferred conditions
Fertile, free-draining soil with ample moisture, in shade.

The name Byfield fern refers to one of the two species described in this genus, *Bowenia serrulata,* and highlights the remarkable resemblance between ferns and these small cycads, which grow in the understorey of closed- or open-canopy tropical rainforest.

Deforestation for commercial forestry plantations has destroyed large tracts of habitat for one species, ***Bowenia spectabilis***, but it survives remarkably well in plantations and second-growth forests. Plant collectors have had little impact on the bowenias, partly because plants that are removed from the wild do poorly in cultivation; any damage to the stem leads to rotting and death.

Both species of *Bowenia* are still quite abundant in nature, but are included in CITES Appendix II because of concern about the impact of trade on all cycad species, and because customs officers often find it difficult to distinguish them from other, more threatened, cycads.

Bowenias are decorative plants with underground stems and fine, glossy green leaves. Fresh leaves are popular for flower arrangements and were even used as a garnish until it was discovered that cycad leaves and seeds are toxic when eaten.

Attractive plants when grown well, bowenias have a reputation for being difficult to grow. Frost-tender and suitable only for tropical and subtropical gardens, both species must have excellent drainage and shade. Plants rot rapidly when grown in poorly drained soil and may become dormant at low temperatures, or if there is insufficient water or fertilizer. They respond well to mulching and regular feeding.

Bowenia species are grown either from seed sown fresh, or by division of the tuberous stem. The cut stem sections are treated with sulphur to prevent infection. Plants from seed, which germinates in 12–18 months, are more easily grown on, and a good-sized specimen can be had in as little as five years.

▲ BOWENIA SPECTABILIS *At first glance, with its slender leaf stalks and glossy, neatly divided leaves,* Bowenia spectabilis *resembles a fern.*

PEOPLE

WHO NAMED BOWENIA?

The genus *Bowenia* was first described in 1863 by Joseph Dalton Hooker (1817–1911). At this time, he was Assistant Director to his father, Sir William Jackson Hooker, at the Royal Botanic Gardens, Kew. It was the elder Hooker who proposed the name *Bowenia*, in honour of the Governor of Queensland, Australia, Sir George Ferguson Bowen.

Although better known for his plant collections in Sikkim and the Himalayas, as a young man, J.D. Hooker collected plants in South America, New Zealand, Australia, and South Africa, when on the four-year voyage of HMS *Erebus* to the Antarctic. He was a friend and regular correspondent of Charles Darwin, and a supporter of Darwin's then revolutionary theories of evolution.

When Hooker took over as Director of Kew on his father's death in 1865, he continued his father's developments there and it was under his directorship that Kew first became a world centre for botanical research, as it still is today.

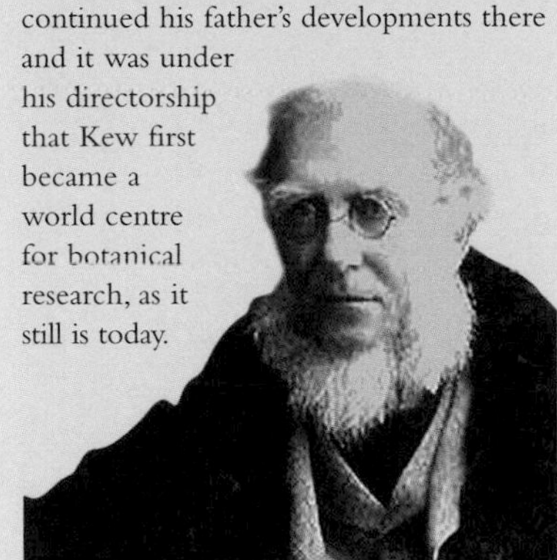

Red List: Endangered

Brahea edulis

Botanical family
Arecaceae/Palmae

Distribution
Mexico: Isla Guadalupe

Hardiness
Half-hardy

Preferred conditions
Well-drained soil; needs shade when young, but thrives in sun when mature.

Although the Guadalupe palm, ***Brahea edulis***, is widespread in cultivation, in nature, it is listed as Endangered in the 2000 Red List. It is found only in the grasslands and dry, deciduous forests of the island of Guadalupe, off the western coast of Mexico.

Like many islands, Guadalupe has suffered from the introduction of exotic and invasive species. In this instance, goats were introduced to the island in 1830, so that passing ships would be able to provide themselves with fresh meat.

Unfortunately, the goats quickly multiplied on the island and ate every plant within reach, including seedling palms. Ships no longer stop at Guadalupe, but the goats remain. Adult palms on Guadalupe survive, but their seedlings are quickly devoured by the plague of goats. Eventually, as the adult palms succumb to disease or age, this species will be extinct in the wild.

When this occurs, *Brahea edulis* will have the dubious distinction of being the first palm known to have become extinct in modern times. The solution for *Brahea edulis* is simple: remove the goats from the island. The conservation of *Hyophorbe lagenicaulis* (*see p.372*) shows that island populations can recover once grazing animals, such as goats and other predators, are removed.

The Guadalupe palm grows well in mild, Mediterranean-type climates, such as that of southern California and adjacent western Mexico. Beautiful specimens grow at Huntington Botanical Gardens in San Marino, California. In cooler areas, young specimens would make attractive plants for pots in the greenhouse or conservatory.

◄ EDIBLE SEEDS *The specific name "edulis" means edible, and refers to the seeds that are contained in fruits, 3cm (1.5in) across, which ripen to glossy chestnut-black during summer. The fruits take about a year to ripen following pollination, and fresh seed takes a further 4–8 months to germinate.*

4*/2 on Red List

Butia

Botanical family
Arecaceae/Palmae

Distribution
Brazil, Uruguay

Hardiness
Half-hardy – frost-tender

Preferred conditions
Well-drained soil in sun or partial shade; drought- and wind-resistant.

The jelly palm, or wine palm, ***Butia capitata***, is native to the grasslands and woodlands of Uruguay and Brazil, where it remains abundant in the wild, often in populations of several thousands.

It is the most commonly grown of the genus, a graceful and slow-growing tree valued for its blue, blue-grey, or grey-green leaves and for its round, yellow or orange, edible fruits. One of the toughest of palms, it is tolerant of short spells at temperatures of –10°C (14°F).

In contrast, other *Butia* species, which all come from Brazil, are experiencing a range of threats. The most serious is habitat loss, but *Butia microspadix* is threatened simply because it is very rare. *Butia campicola* is also rare, and *Butia eriospatha,* from the Atlantic forests, *Butia microspadix,* and *Butia purpurascens,* from dry scrub in Goiás, Brazil, are described as Vulnerable in the 2000 Red List.

▲ BUTIA CAPITATA *Unlike other members of the genus, the jelly palm is abundant in the wild and commonly grown. It can lend tropical style to gardens that endure short-lived frosts.*

12*/17 on Red List/all on CITES I

Ceratozamia

Botanical family
Zamiaceae

Distribution
Mexico, Guatemala, and Honduras

Hardiness
Frost-tender

Preferred conditions
Fertile, humus-rich soil; light, dappled shade to deep shade; moderate to high humidity, in a sheltered site.

Most species of *Ceratozamia* are found in mountain forests at altitudes of 800–1,800m (2,600–5,850ft). These are often in cloud forests at the higher end of their altitudinal range. However, some of the more widespread species also occur at lower elevations. All species are under threat, often due to overcollection for the horticultural trade, sometimes as a result of forest clearance for banana and coffee plantations.

▲ CERATOZAMIA HILDAE *The slender stems and arrangement of narrow, thin-textured leaves give a bamboo-like appearance to this elegant, rapidly maturing cycad, which may produce cones in as little as five years.*

As a result of the high level of threat, this largely Mexican genus has been listed in CITES Appendix I since 1985 to prevent further trade in wild plants. Restrictions associated with CITES listing do appear to have reduced the trade in cycads collected from the wild. While several thousand plants were exported from Mexico between 1982 and 1985, no further exports have been recorded since 1985.

RED LIST SPECIES

In the 2000 Red List, *Ceratozamia euryphyllidia*, *Ceratozamia zaragozae*, *Ceratozamia kuesteriana*, and *Ceratozamia norstogii* are listed as Critically Endangered; *Ceratozamia sabatoi* is listed as Endangered; ***Ceratozamia latifolia***, *Ceratozamia microstrobila*, and *Ceratozamia*

Red List: Endangered*

Calyptronoma rivalis

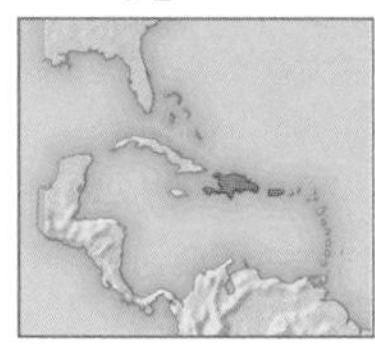

Botanical family
Arecaceae/Palmae

Distribution
Puerto Rico

Hardiness
Frost-tender

Preferred conditions
Reliably moist, acid soil in full sun.

The manac palm was described as Endangered in the Red List of 1997, and was under threat of extinction. In Puerto Rico, it is restricted to stream sides in damp, limestone forest in the north-western part of the island, where there are just three natural populations comprising an estimated 250 palms.

This is an elegant palm that can reach 15–18m (50–60ft) in height, with leaves 3m (10ft) in length. A tree with exceptional form, it would make a very ornamental asset in any tropical garden.

This species suffered habitat loss due to road building, clearance for pasture, fire, and flooding. Recently, the Puerto Rican Department of Natural Resources has undertaken a number of projects to reintroduce the plant.

Red List: Critically Endangered

Carpoxylon macrospermum

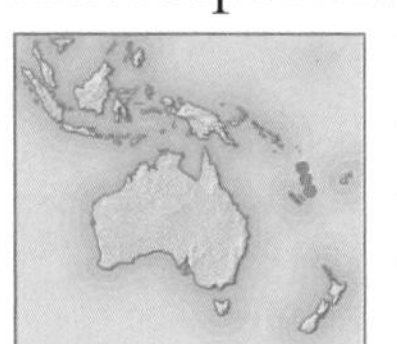

Botanical family
Arecaceae/Palmae

Distribution
Vanuatu

Hardiness
Frost-tender

Preferred conditions
Reliably moist, humus-rich soil in sheltered sites in partial shade.

This extraordinary palm is the only one in its genus (monotypic). It was discovered on the island of Aneityum in 1859, but was considered extinct until its rediscovery in Espiritu Santo Island, Vanuatu, in 1987.

Now considered Critically Endangered on the 2000 Red List, it grows wild on valley sides, by rivers, and in coastal forest. A survey executed in 1994 found about 150 reproductively mature trees in the whole of Vanuatu, with fewer than 40 mature trees in the wild. It is the subject of a conservation plan managed by the Vanuatu Forestry Department.

This tall and graceful tree is one of the world's most beautiful palms. It needs warm, moist conditions and is unlikely to tolerate cold.

Now increasingly seen in tropical collections, the plant grows quickly as long as it has sufficient moisture.

Seed from cultivated trees is marketed by the Carpoxylon Palm Conservation Committee. The funds generated are used to support the conservation of the wild populations (*see below*).

PLANTS AND PEOPLE

CONSERVATION INCENTIVE

The people of Vanuatu value *Carpoxylon* for food, medicine, contraception, and many other uses. The Foundation for the Peoples of the South Pacific (a non-profit, non-governmental organization) devised a Profitable Environmental Protection Project, with the objective of developing profitable enterprise as a tool for conservation. This led to the formation of the Island Palm Products Company, which now markets *Carpoxylon* seeds to retailers in Australia and the USA. Such commercial activities create local economic incentives, raise awareness of the need to conserve, and earn profits for *in-situ* conservation. This increases the survival chances of this palm, ensuring its cultivation throughout tropical zones.

THREAT

LAND CLEARANCE

Forest clearance for agriculture (here for a banana tree plantation) is a common cause of threat to cycads. Coffee cultivation, too, is a threat. Many species of *Ceratozamia* grow in the forest understorey, which is partially destroyed when shade-tolerant coffee is planted beneath the forest canopy. The introduction of sun-tolerant varieties of coffee has, ironically, brought little relief, as this has led to the opening up of forest areas and the complete destruction of cycad habitat. Even if the cycads are not destroyed during forest clearance, most of them do not survive out in the open, where they are exposed to scorching sun, drier atmospheres, and degradation of moisture-retaining soil humus.

miqueliana are listed as Vulnerable. The bamboo cycad, *Ceratozamia hildae*, is also listed as Endangered. This unusual cycad from tropical dry forest has an underground stem and clustering leaflets along the leaf stalk (rachis).

It was first imported into the USA for commercial horticulture in 1960. Its popularity, due to its compact size and unusual leaves, led to thousands of plants being taken from the wild, leaving a few scattered populations with a very high risk of extinction.

Ceratozamia hildae is an attractive container plant, but also grows well in open ground given well-drained, neutral or alkaline soil and shelter in shade or dappled sunlight. Mature plants are relatively hard to obtain, but the species is easily propagated from fresh seed even though seed viability is often quite low.

HABITAT

PLANT PARADISE

Cloud forests occur in mountains at altitudes of 1,000–3,000m (3,300–9,850ft), usually associated with rainforests growing at lower altitudes. The environment is characterized by lush, sheltered, cool conditions and high humidity resulting from being almost constantly wrapped in cloud. In the understorey, plants grow in leafy, humus-rich soils, are sheltered from wind and sun, and are adapted to shade or dappled sunlight. Plants from such specialized habitats need to be provided with similar conditions to ensure success in cultivation.

Ceratozamia kuesteriana is a small cycad, with bronze-coloured new leaves and long, narrow leaflets. It was discovered in 1841, but was not found in any collections until more than 100 years later. Since it first appeared in collections, however, it has declined dramatically due to the mass removal of plants from its tropical coniferous forest habitats.

While CITES protection prevents harvesting of wild populations, *Ceratozamia kuesteriana* occurs only in restricted areas, and is extremely vulnerable to habitat destruction. Clearing of forests to make way for agriculture is affecting large areas of Mexico and has greatly increased the risk of extinction for this species.

Ceratozamia kuesteriana is a much sought-after species. Its size and growth form make it an ideal container plant, but it grows well in the ground as well as in a pot, and in sun or shade. Plants are not yet widely available but can be propagated from fresh seed.

Ceratozamia latifolia is still relatively secure in the understorey of tropical rainforests on the steep mountains of eastern Mexico, but significant numbers of plants have been removed by plant collectors and its survival is further threatened by land clearance for plantation crops.

Due partly to the large number of plants removed from the wild, *Ceratozamia latifolia* is one of the more available species in cultivation. It thrives in cooler tropical and subtropical climates, preferring well-drained soils and shade. It is easy to grow and is propagated by fresh seed or division of basal suckers.

Ceratozamia miqueliana occurs in scattered groups in tropical rainforest across a wide area. It is a beautiful plant with broad, lime-green leaflets and long, arching leaves. *Ceratozamia miqueliana* experiences similar problems to those of many other Mexican cycads; too many plants have been taken from the wild by plant collectors, leaving behind small populations that may struggle to survive.

Cycads are particularly vulnerable to population collapse because this group of plants is single-sexed (either male or female) and changes in sex ratio, asynchronous coning (when male and female cones ripen at widely different times), large distances between male and female plants, and the local extinction of beetle pollinators can all lead to reproductive failure.

When plant collectors remove adult plants from a population, especially when they target female plants, they set in motion a series of changes that have far greater consequences than the simple loss of a few mature plants.

Ceratozamia miqueliana thrives in tropical and subtropical climates. It grows best in shade and prefers an organic mulch and light fertilizer applications. It needs sheltered conditions and grows well in containers indoors and under glass. It is easily propagated from seed.

▼ CERATOZAMIA LATIFOLIA

Found only in cloud forests in the mountains of San Luis Potosí and Hidalgo in eastern Mexico, where the dominant trees are various species of oak (Quercus), *this attractive small cycad has been overexploited for the horticultural trade.*

57 on Red List*

Chamaedorea

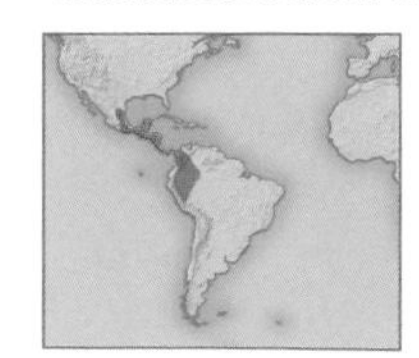

Botanical family
Arecaceae/Palmae

Distribution
Mexico, Central and South America

Hardiness
Frost-tender

Preferred conditions
Moist but well-drained, neutral to acid, humus-rich soil, in shade or dappled shade.

There are approximately 100 species of *Chamaedorea*. They occur mainly in the understorey of tropical rainforest.

Many species have suffered from forest clearance for agriculture. Several are threatened by overcollection for the horticultural trade, since most species in the genus are relatively small and elegant, easily grown, and highly valued for interiors and small gardens.

Chamaedorea stolonifera makes a highly attractive container plant with stolons tending to spill gracefully over the edge of the pot. It is easily propagated by division in early spring or early autumn.

A clustering palm, *Chamaedorea stolonifera* grows to 1–2m (3–6ft) tall, its stolons creeping over the soil and through cracks in rocks to form colonies up to 10m (33ft) across in the wild. It is perhaps the most extreme example of the stoloniferous habit in palms. The stems bear several attractively divided leaves at the upturned tips and have been likened to snakes rearing their heads. Cultivated since Victorian times by collectors who prized the unusual, *Chamaedorea stolonifera* is seldom grown now, and male plants are unknown in cultivation. It is, therefore, propagated by division and renewed commercial interest has resulted in overexploitation of wild plants for propagative material.

Once thought to be extinct, the species was rediscovered in the 1980s. The threat from commercial growers and land clearance for coffee plantations leaves this species listed as Vulnerable in the Red List of 1997.

Chamaedorea metallica, metallic palm, is one of the most durable for indoor use because it tolerates drought and low light levels and is not susceptible to mites.

The metallic palm has long been a collector's palm, and is threatened. Collecting from the wild has pushed it close to extinction in the rainforests of the Atlantic slopes and lowlands of southern Mexico. This palm is one of the most distinctive of its genus. The solitary stems are slender and ringed with swollen leaf-base scars. The leaf blades are either entire or variously divided and cup upwards towards the stem tips. Its most attractive feature, however, is the metallic sheen on the deep blue-green leaves. It is commonly propagated by seed. Plants typically reach 1–2m (3–6ft) in cultivation.

CHAMAEDOREA ELEGANS *The familiar parlour palm is commonly grown as a shade-tolerant house plant.*

Chamaedorea tuerckheimii is difficult to grow and usually seen only in botanical collections, where its needs can be met. Otherwise, these tiny palms are best appreciated, and best preserved, in the wild.

Chamaedorea tuerckheimii may, in fact, be the world's smallest palm; its distinctively pleated leaves are only 15cm (6in) long. It occurs in the cloud forests of Guatemala and Veracruz.

The greatest threat is from plant collectors, who dig up mature plants and sell them locally and internationally.

These palms are as delicate as they are beautiful, and sadly, once removed from their humid mountain home, they succumb to low humidity, poor water quality, and infestation by mites.

HISTORY

A VOGUE FOR PALMS

Palms, being mainly tropical in origin, are nearly all frost-tender plants. The central growing point amid the crown of leaves is especially vulnerable to cold, and if that is frost-damaged, the whole trunk dies.

Early experiments in growing palms outdoors in northern climes were doomed to failure. But as glasshouse technology proceeded apace in the 19th century, cultivating palms became highly fashionable, and specialist structures were built to house them. The vogue swept across Europe and palm houses were built both by wealthy individuals and botanic gardens; the Palm House at the Royal Botanic Gardens, Kew, finished in 1848, is a famous example.

A palm house has a much higher roof than most other glasshouses, and frequently includes an aerial walkway or balcony to afford a close-up view of the palms.

So amenable are they to cultivation, palms later became a focal point of public winter gardens, many of which had a palm court. By the turn of the 20th century, palms had become valued as house plants, for which their adaptability to pot cultivation and tolerance of shade makes them suitable.

POTSDAM PALM HOUSE *An 1833 painting by Karl Blechen displays a living collection during the period when building palm houses was the height of fashion.*

7*/2 on Red List

Coccothrinax

Botanical family
Arecaceae/Palmae

Distribution
USA: Florida, to the Caribbean

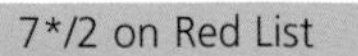

Hardiness
Frost-tender

Preferred conditions
Fertile, well-drained soil, in a sunny site with plenty of root space and room to grow.

Silver palms of the genus *Coccothrinax* are found in sand dunes, scrub, and pine forests of Florida in the USA, the Bahamas, Cuba, the Dominican Republic, Barbados, and Lesser Antilles. There are an estimated 30–50 species in the genus. Some, like ***Coccothrinax barbadensis***, the Barbados silver palm, are supremely elegant and widely cultivated. In some of its natural habitats, however, *Coccothrinax* has been brought almost to the brink of extinction by overcollection and the clearance of coastal forest and scrub for development.

Most thatch palms are easily grown, given sun and good drainage, and while many are slow-growing in the wild, they respond well to fertilizer and irrigation in cultivation. Although some, such as *Coccothrinax crinita* and *Coccothrinax barbadensis*, may survive several degrees of frost for brief periods, in general they are best suited to tropical gardens.

The slow growth rate of several of these palms, however, such as the widely grown silver thatch palm *Coccothrinax fragrans*, slows further if confined in a pot. So, in cool climates, they can be grown in greenhouses or conservatories.

Red List: Endangered

Copernicia ekmanii

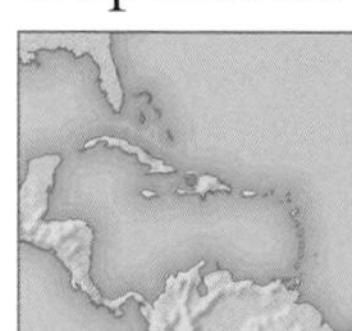

Botanical family
Arecaceae/Palmae

Distribution
Northern Haiti

Hardiness
Frost-tender

Preferred conditions
Full sun, moderately fertile soils with excellent drainage; tolerates periodic drought and alkaline soils.

The coastal thorn forests on the rocky shores of northern Haiti are home to this striking Caribbean palm. ***Copernicia ekmanii***, called locally *jamm de pay*, was thought to be almost extinct in the wild, until a team of botanists confirmed the palm's survival in 1996.

Copernicia ekmanii is listed as Endangered as it is confined to a small area of Haiti, a nation with considerable challenges to conservation. Although protected to some extent by the harsh and extremely inaccessible conditions of

COCCOTHRINAX BORHIDIANA ► *Beautiful as a juvenile plant, when the crown of overlapping fans of leaves can be viewed at close quarters, this palm needs space to spread.*

Cuba is especially rich in thatch palms, with many *Coccothrinax* species confined to specific rock and soil types, such as limestone or serpentine rocks. Such species are vulnerable because their areas of occupancy are so small.

Coccothrinax borhidiana, named after Hungarian botanist Attila Borhidi (1932–), is confined to a small region of scrubland on the north coast of Cuba. It is highly threatened by construction and habitat destruction, and is Critically Endangered. A palm of extraordinary appearance, prized for its tight crown of green leaves and skirt of dried ones, *Coccothrinax borhidiana* is elegant, but slow-growing in cultivation. It needs sun and plenty of root space, even when young. It occurs on limestone and tolerates alkaline soils.

Coccothrinax crinita, the old man palm, another Cuban native, is known for its curiously fibrous or "hairy" trunk. The fibres are remnants of leaf bases that clothe the stem, but these palms often go "bald" as the fibres slough off with age.

This palm is confined to small populations in western Cuba, where it is threatened by habitat destruction and overharvesting for the traditional local use as thatch and fibre. Although it is very rarely found growing in the wild, it is widely cultivated in southern Florida.

◄ COCCOTHRINAX BARBADENSIS *This lonely Barbados silver palm owes its survival to its unique location in a cemetery. Here it is protected from the land clearance that has destroyed much of its natural habitat on the island of Barbados.*

its home range: even there, grazing by goats, charcoal-making, and harvesting for thatch continue to threaten it.

Copernicia ekmanii is cultivated, but not widely, although it is a handsome garden ornamental that is well suited to cultivation. It is tolerant of alkaline soil and periodic droughts, and, as an adult, can withstand brief and occasional cold spells. It is not fast-growing, but its almost spherical crown of silver-green leaves is worth the wait.

Copernicia ekmanii is related to the more familiar carnauba wax palm, *Copernicia prunifera*, a native of South America. This species is grown in large-scale plantations in Brazil for the industrial production of carnauba wax. The wax is collected from the leaves and is used as a polish, and in cosmetics, such as lipsticks.

SILVERY LEAVES ► *This is an especially ornamental palm on account of the silvery cast to the long-stalked leaves, which shimmer in the sunlight as the fronds twist in the wind.*

Red List: Extinct in the Wild

Corypha taliera

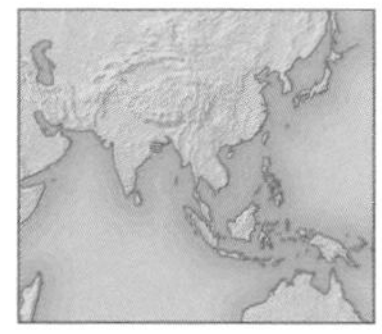

Botanical family
Arecaceae/Palmae

Distribution
India

Hardiness
Frost-tender

Preferred conditions
Moist but well-drained soil in sun; slow-growing but needs plenty of space.

The genus *Corypha* contains some of the world's most spectacular palms, including the extraordinary talipot palm, *Corypha umbraculifera*, with a huge inflorescence that is 6m (20ft) long. Like all *Corypha* species, it dies after flowering. All are large palms – much too big for small gardens.

Corypha taliera is a magnificent tree. It deserves to be grown where its stature can be appreciated and the enormous inflorescence can be seen from a distance. Sadly, this palm, which was endemic to Bengal in northern India, has long been thought to be near total extinction with only a few specimens in cultivation, at Howrah Botanic Garden, and Fairchild Tropical Garden, among other places. Wild specimens had been recorded in one village in Bengal, and a single tree was found in the former scrub jungle at Dhaka University by Professor M Salar Khan. This tree is now sheltered within an enclosure at the residential quarters of the university's Pro Vice-Chancellor.

Sadly, this species is now extinct in the wild. The future of *Corypha taliera* will depend on careful genetic management and, in the long term, on reintroduction into secure forest reserves.

▼ MAJESTIC PALM *The crown of this species soars to 25m (75ft) and the huge leaves can reach 5m (15ft) across.*

Red List: Extinct in the Wild

Cryosophila williamsii

Common name
Arecaceae/Palmae

Distribution
Honduras

Hardiness
Frost-tender

Preferred conditions
Moist soil in shade with shelter from strong wind.

This frost-tender palm, *Cryosophila williamsii,* is listed as Extinct in the Wild in the 2000 Red List. However, a small but growing colony is surviving in a protected forest reserve in north-eastern Honduras. This small patch of rainforest is its last stronghold.

This species is still threatened with extinction because of deforestation caused by slash-and-burn subsistence agriculture, which has devastated the region's rainforest. Only 5 sq km (1,250 acres) of primary forest remain.

The trunk of this species has upward-growing roots that function as spines, a feature that is characteristic of the genus. Its graceful, evenly divided leaves and small stature, up to 7m (22ft) tall, make this palm an attractive ornamental for tropical gardens. Unfortunately, if deforestation continues, gardens in the tropics may be the only place where future generations can see this palm.

It needs 50 per cent shade, moisture, and high humidity, and tolerates limestone soils, but it has been found to be very difficult to cultivate.

21*/35 on Red List/CITES II

Cycas

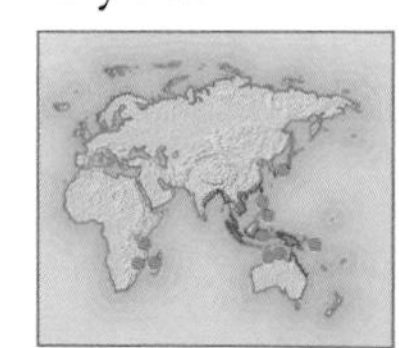

Botanical family
Cycadaceae

Distribution
Australia, Pacific Islands, SE Asia to Madagascar, east coast of Africa

Hardiness
Half-hardy – frost-tender

Preferred conditions
Very light shade to full sun, excellent drainage in leafy, acid soil; good air movement.

This ancient genus, commonly known as sago palm, has a fairly wide distribution from China and Japan, west through south-east Asia to India, and south through the Pacific Islands to Australia. One species is found on Madagascar and along the east coast of Africa. They are found in open habitats ranging from arid savanna and tropical grasslands, through temperate prairie and steppes, to the shade of temperate broadleaved or mixed forest and humidity of tropical rainforest.

CYCAS TAIWANIANA

Their elegant form is valued by landscapers all over the world, for garden use and interior decoration. In many cases, collection from the wild for the horticultural trade has been a major cause of decline, along with habitat destruction, as is the case with ***Cycas taiwaniana*** from China. Although known in cultivation for hundreds of years in that country, it is now endangered by these dual threats; more than half of its natural habitat has been cleared.

The "blue" cycads are highly valued by growers. The blue colour is caused by a waxy bloom on the leaves which, in some species, occurs only on young leaves. Blue cycads occur in drier areas and most are species of *Cycas* and *Encephalartos*, although *Macrozamia macdonnellii* (*see p.376*) from the dry Australian interior also has striking blue leaves. The demand for blue plants has led to the decline of many populations.

High light levels and good drainage are crucial to success with the blue-leaved species; forest cycads prefer more shade. Most cycads respond well to fertilizer and additional moisture in dry periods. Many thrive in containers, so are ideal for conservatories or greenhouses.

They are propagated by fresh seed, which does not remain viable for long. Seeds of several desirable species remain hard to obtain, but seed banks in botanical gardens should make seeds and seedlings more widely available.

International assistance

Because so many cycads are declining in the wild, the IUCN/Species Survival Commission Cycad Specialist Group has been coordinating efforts to include all threatened species in gene banks in botanical gardens. Nong Nooch Tropical Garden in Thailand is one of the gardens in this network that has established gene banks for south-east Asian cycads, including *Cycas condaoensis*. The aims of the group are to ensure that species do not die out altogether, and to assist with the conservation of cycads in habitat.

Cycas condaoensis is one of several species listed as Vulnerable that are cultivated in Vietnam. This grassland and shrubland cycad is not well known outside its native area, but is desirable and likely to become popular. Despite its moderate size and suitability as a landscape plant, *Cycas condaoensis* has not been taken from the wild in large numbers. It is threatened because it occurs only in a small area on the Côn Đao Islands, so human disturbance or environmental disaster could affect survival in the wild.

Cycas hoabhinensis, another threatened cycad, is also popular in Vietnamese gardens; because of its compact stature, it is ideal for containers. It is declining because plant collectors have depleted more accessible colonies in its forest habitat. The threat to wild populations is likely to worsen as the demand for plants from outside Vietnam grows. The obvious solution is to propagate from seed, but there is currently no programme set up for this.

Cycas panzhihuaensis, on the other hand, is becoming more widely available due to the mass propagation effort in China (*see box, above*), but is still hard to obtain. It exists in several large colonies in the wild, but its future is uncertain. Strip and pit mining for iron and titanium have destroyed large areas of its grassland habitat, and it has also been collected for sale to gardeners. The dark green foliage and bright yellow, citrus-scented male cones make it an attractive ornamental.

CYCAS RUMPHII MALE CONE

Cycas revoluta, from Japan and China, is widely proclaimed as a model for conservation through cultivation. Every year, millions of plants of this most commonly cultivated cycad are traded from cultivated stock grown in nurseries worldwide. While seeds are still harvested from wild plants, this poses little threat to the vigorous populations in the wild.

Cycas rumphii is another cycad at Lower Risk, but for different reasons. It belongs to a group of cycads that bear seeds with a spongy layer that enables them to float. Consequently, it is found mainly on island coastlines. Plants have also been dispersed by people using the seeds for food. Although island habitats are being cleared for agriculture and logging, *Cycas rumphii* is vigorous and appears less vulnerable to habitat clearing than other *Cycas* species.

Cycads in Australia

Australia is renowned for the large numbers of cycads that occur in the wild, and is the only country where cycads have sometimes been treated as weeds. Farmers have cleared thousands of cycads because of the toxic effects they have on cattle that consume them (*see box, facing page*). This has added to the decline of some rare species, such as the Endangered *Cycas megacarpa*.

Cycas megacarpa is one of relatively few threatened Australian cycads, but it shows how some of the less common species can become threatened as pressures intensify. In addition to removal by farmers and habitat clearance, species such as *Cycas megacarpa*, with its erect, narrow trunk and crown of strongly arching leaves, are at risk from collecting to satisfy demand for Australian cycads among plant collectors.

Cycas megacarpa is drought-resistant and is a fine plant for large containers, as well as landscaping, especially in "water-wise", or "xeriscape", gardens, designed with water conservation in mind. It tolerates a wide range of conditions, including short-lived light frosts. Seeds are becoming more readily available.

Cycas cairnsiana, the Mount Surprise cycad, falls into the Lower Risk category of threat. This is a blue-leaved cycad, in which the blue colour is retained even on older leaves. The intense blue leaves and relatively short stem, at 2.5m (8ft), make this beautiful cycad an ideal landscape plant.

It is still quite rare in cultivation, but large numbers of seeds are produced in wild populations, so the potential for a substantial increase in plants propagated in cultivation is good. Seedlings can be hard to establish, but they will grow at a reasonable rate if fertilized frequently. *Cycas cairnsiana* is still abundant in the wild, but collecting to satisfy the plant trade is regarded as a real threat.

RENEWAL

Conservation success

Named after the mining town, Panzihua, in China, the Dukou cycad, *Cycas panzihuanensis*, is threatened by both mining and overcollection for the local horticultural trade. It is protected by a local initiative undertaken by a mining company and local villagers. The mining company has helped to establish a temporary reserve near one of the largest wild colonies, and a wall has been erected to protect the plants from the frequent fires that are started by local villagers. From this reserve, large numbers of seeds have been collected from the wild and propagated in a special nursery in Panzhihua to supply the nursery trade and for re-introduction to the wild.

BOTANY

Fire-adapted cycads

In some of the regions where *Cycas* species grow, wildfires are common, such as Queensland, Australia, where ***Cycas media*** resprouts after fire (*see left*), and many cycads are adapted to resist it. Although all living tissues above the ground may be burned and blackened, apparently beyond hope of resurrection, the persistent leaf bases that clothe the trunk afford some protection to the growing tip. In some cycads, there is a subterranean stem that is protected by an insulating layer of soil. So new growth usually emerges rapidly after a fire and plants recover unharmed. Several cycads, notably some species of *Encephalartos* (*see p.370*), actually need fire to stimulate the production of cones, and several *Cycas* species are known to produce seed in abundance after burning.

CYCAS REVOLUTA CONE

CYCAS REVOLUTA *Seed, leaves, and male cones are still harvested from wild plants for propagation and decoration, but this poses no threat to the vigorous populations in the wild.*

PLANTS AND PEOPLE

CYCAD TOXINS

The fruits or stems of many cycads are used as food, despite the toxins they contain, which affect both the liver and nervous system. On the island of Guam, consumption of cycad starch has declined due to its link with degenerative disease. In Australia, farmers cleared cycads to stop cattle eating the leaves, which cause paralysis of the back legs, known as "staggers". Native people who eat cycad products treat them before use, by heat, leaching, or fermentation; it is thought that Australian Aboriginal people knew of these processes some 13,000 years ago.

◄ SPECTACULAR INFLORESCENCES *Sprays of dark purple fruits extend from inflorescences that reach lengths of 1m (3ft) or even longer.*

Red List: Endangered*

Dictyosperma album

Botanical family
Arecaceae/Palmae

Distribution
Mascarene Islands

Hardiness
Frost-tender

Preferred conditions
Moist but well-drained, fertile soils, in full sun or light, dappled shade.

Three varieties of ***Dictyosperma album***, hurricane or princess palm, were listed as Endangered on the Red List of 1997. Endemic to the islands of Mauritius, Rodrigues, and La Réunion, the species itself is threatened with extinction in the wild as a result of past habitat loss, and by animal predation on the fruits and seedlings. *Dictyosperma album* var. *aureum*, from Mauritius and Rodrigues, is also threatened, while *Dictyosperma album* var. *conjugatum* lives in the wild as a single tree protected on Round Island, just north of Mauritius.

This tall, elegant palm tolerates salt and wind and grows rapidly in fertile, well-drained soils in warm, frost-free climates. Populations of *Dictyosperma album* are being managed by the Mauritian Wildlife Foundation and the National Parks Service of Mauritius.

123*/54 on Red List/CITES II

Dypsis

Botanical family
Arecaceae/Palmae

Distribution
Madagascar

Hardiness
Frost-tender

Preferred conditions
Moist but well-drained fertile soil; sun or dappled/part-day shade.

Of this largely Madagascan genus, 54 species are listed by IUCN as experiencing threat, and most of these are classified as Vulnerable, Endangered, or Critically Endangered. The natural habitats of *Dypsis* species are mostly forests, but these range from the white-sand coastal forest and lowland rainforest to moist mountain forest. A few species, such as *Dypsis madagascariensis,* are found in palm grassland and scrub.

These palms suffer land clearance for agriculture or settlements, overgrazing by domestic animals, and forest fires. Among the Critically Endangered, *Dypsis ifanadinae* survives as only 50 individual plants due to continuous clearance of its lowland tropical rainforest habitat in Madagascar. Similar numbers of *Dypsis intermedia* and *Dypsis interrupta* are left and there are fewer than 20 individuals of *Dypsis mangorensis.* This last is from rainforest on the north-eastern coast of Madagascar, in the Mananara Nord Biosphere Reserve, which is part of a pilot project aiming to integrate biodiversity conservation, buffer zone development, and participation of local people in management decisions.

Among the commonly cultivated species, the drought-tolerant ***Dypsis decaryi*** is easy to identify by the crown of strongly arching leaves neatly aligned in three ranks. This palm reaches 6m (20ft) or more in height, at maturity. The leaves develop reins (strips of leaflet margins) that hang from the leaves. It bears copious flowers and fruit.

Dypsis decaryi occurs in Madagascar, in the Anosyennes Mountains, where humid forest grades into drier spiny forest. Andohahela National Park contains the only protected site for *Dypsis decaryi,* which is described as Vulnerable in the 2000 Red List. The entire wild population numbers about 1,000 plants, which are threatened by fires and by overcollecting of seed for the horticultural trade. Education projects aimed at local people and tourists have helped raise awareness about the need for its protection.

In cultivation, *Dypsis decaryi* is robust, preferring soil that is well drained and irrigated, although it withstands a range of soil types and light, short-lived frost. Like most cultivated *Dypsis* species, it responds well to adequate moisture and fertilizer. It is, however, susceptible to Lethal Yellowing Disease.

DYPSIS LUTESCENS

Dypsis lutescens, until recently known as *Chrysalidocarpus lutescens*, is one of the most easy palms to grow. Graceful and fast-growing, it has gold-flushed fronds and looks good planted as single specimens or as clumps. Despite its narrow distribution in the wild, it is one of most widely grown of palms in the warmer regions of the world. That such a commonly cultivated palm should be threatened in the wild seems to defy logical explanation, but *Dypsis lutescens* is one of several threatened Madagascan palms listed in CITES Appendix II.

▲ DYPSIS DECARYI *is prized throughout the tropics and subtropics for its architectural form and durability as a landscape plant.*

Dypsis lutescens is only found in the wild along the eastern coast of Madagascar, in a narrow strip of coastal forest, where it grows on white sand. The human population of the island of Madagascar is straining natural ecosystems to breaking point. *Dypsis lutescens* is not listed as being under threat at present, but could rapidly become so if its coastal habitat is disturbed.

12*/9 on Red List/CITES II

Dioon

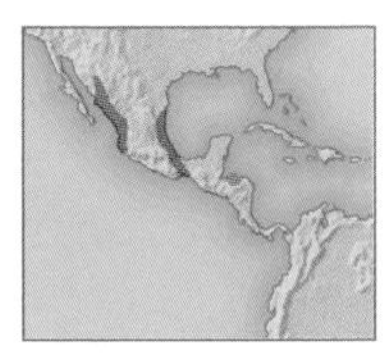

Botanical family
Zamiaceae

Distribution
Mexico, one species in Honduras

Hardiness
Frost-tender

Preferred conditions
Full sun in soils with perfect drainage; moist but well-drained soil in partial shade for *Dioon spinulosum* and *Dioon rzedowskii.*

The vast majority of *Dioon* species are on the 2000 Red List. They are threatened by land clearance for crops, such as coffee, which has caused large-scale loss of habitat; and, on a smaller scale, to losses due to subsistence farming. Overcollection for the horticultural trade has also been an increasing problem.

Adding to the losses are the collection of seeds and wholesale removal of plants from the wild by collectors, as well as the practice of decapitating plants. This had traditionally been done to obtain the large, edible seeds, since it was easier to decapitate than to climb. Recently, the demand for seeds and plants of several species has been so great that a market has developed for the lopped-off crowns, which are sold as established plants.

Some species can recover from such lopping, while others, like ***Dioon spinulosum***, simply rot at the wound and die. This is sadly ironic, as this species grows rapidly from seed and is one of the most trouble-free cycads to cultivate. It occurs in rainforest in Oaxaca, Vera Cruz, in Mexico, and is listed as Vulnerable on the 2000 Red List. It thrives in filtered sun and partial shade. Although it prefers neutral to alkaline soils, it tolerates a range of soil types, and thrives in filtered sun and partial shade.

Dioon edule, the virgin palm, or palmita, is included in CITES Appendix II, and described as at low risk. The scientific name, *edule*, derives from the widespread practice of eating *Dioon* seeds. A related species from Honduras is called "teosinte", the same name given to maize, which is an indication of the value of cycads as food in some areas.

▲ DIOON EDULE *Subsistence farmers from near Monte Oscuro, Mexico, transport recently lopped-off crowns of* Dioon edule *to market.*

In each case, local people have devised ways to eliminate the powerful toxins found in cycad seeds. However, eating the seeds is not the primary threat. The real threats are land clearance for agricultural crops, overcollection, and decapitation.

As one solution, conservation authorities have set up a nursery run by subsistence farmers near Monte Oscuro in Mexico. The local community propagates seedlings of these decorative, medium-sized cycads from wild-collected seed. The villagers understand that they need to conserve the natural habitat as a seed source, and that they also need to re-introduce nursery-produced plants into the wild to compensate for seed removal. Gardeners can assist conservation efforts by purchasing plants from sustainable-use nurseries, such as this one.

DIOON SPINULOSUM

Dioon edule is a popular cycad that is widely grown in a range of climates. There are a number of variants with different growth requirements, of which *Dioon edule* var. *edule* is most adaptable. Plants grow easily from seed, which needs to be stored for 6–8 months before it will germinate.

Dioon merolae is a medium-sized cycad with distinctive flat leaves from tropical dry forest of Chiapas and Oaxaca in Mexico. It suffers from many of the same problems as other species, especially habitat destruction and the practice of lopping crowns for decoration. The stress this causes to decapitated plants, and the absence of reproduction during the time it takes to produce a new crown, has resulted in declining numbers in the wild, and *Dioon merolae* is listed as Vulnerable in the 2000 Red List. There are still some populations where seed production is good and these populations need to be protected.

Various research and conservation organizations in Mexico, together with the botanic garden in Xalapa (Jardín Botánico), have formed partnerships with international agencies such as GTZ-Germany, MAB-UNESCO, and Fauna & Flora International to promote the sustainable use of cycads. Populations of *Dioon merolae* in the buffer zone of La Sepultura biosphere reserve are included in this programme.

Dioon merolae is a handsome plant that is easily propagated from seed and grows relatively quickly in well-drained soils with regular fertilizer. Despite its potential as a landscape plant, it is scarce in cultivation and gardeners should be careful not to buy plants that may have been collected from the wild.

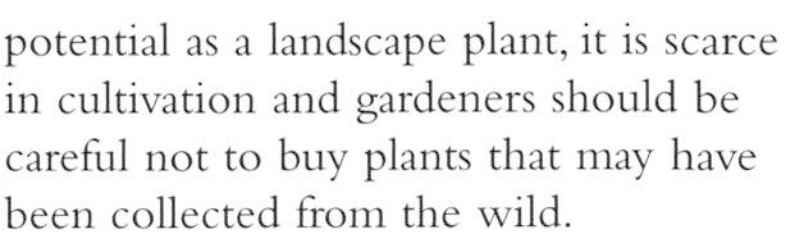

Dioon rzedowskii occurs mostly on limestone cliffs, in rocky crevices with accumulations of humus. It has been saved from many forms of human disturbance by being inaccessible in this steep, inhospitable habitat. It does however, typically grow in small groups that decline quickly when disturbed. Despite the terrain, people have begun to cultivate the surrounding lands, which increases the risk to *Dioon rzedowskii*. It is now listed as Vulnerable.

The dense crown of many arching leaves makes this an attractive, medium-sized ornamental for gardens. In nature, the trunks of old plants may be more than 5m (15ft) long, and these large plants tend to develop a reclining habit in which the lower part of the stem grows along the ground.

Dioon rzedowskii tolerates a range of conditions in cultivation, including sun or partial shade, but favours neutral to alkaline soils. It is easy to grow and plants propagated from seeds reach a good size in relatively few years. A shortage of seeds in the horticultural trade means that *Dioon rzedowskii* may be difficult to obtain. It prefers neutral to alkaline conditions but tolerates a range of soils, in partial shade or filtered sun.

DIOON MEROLAE ► *In the mountains of Chiapas, the species occurs at 900–1,200m (2,925–3,900ft), often in association with oaks* (Quercus) *and pines* (Pinus).

PEOPLE

THE LONDON CONNECTION

The genus *Dioon* was first described in 1843 by John Lindley (1799–1865). He spelled it *Dion*, not the now accepted *Dioon*, leading to a long dispute over the original spelling. The name comes from the Greek *di-*, meaning two, and *oon*, meaning egg, referring to the paired ovules found in all cycad genera. As a botanist, horticulturist, and administrator, Lindley was crucial in helping to establish the Royal Botanic Gardens, Kew. The Royal Horticultural Society's Lindley Library is named for him.

56*/46 on Red List/CITES I

Encephalartos

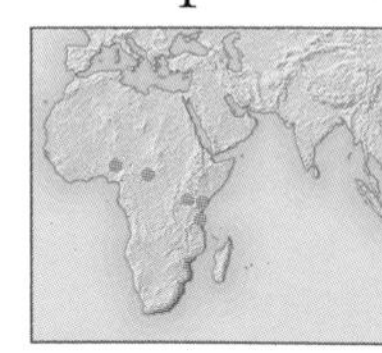

Botanical family
Zamiaceae

Distribution
Pockets in West, East, and Central Africa, Eastern Cape

Hardiness
Frost-tender – half-hardy

Preferred conditions
Sun or partial shade; well-drained soil, with shelter from wind.

This cycad grows in pockets in West, Central, and East Africa and sweeps up the Eastern Cape as far as Swaziland and Mozambique, in diverse habitats from grassland to forest, and coastal scrub to rocky mountain sites. Two-thirds of *Encephalartos* are Rare or threatened.

Some, like ***Encephalartos villosus*** and ***Encephalartos longifolius***, are at Lower Risk, but are listed on CITES Appendix I to prohibit trade in wild-collected plants, as customs officials cannot easily distinguish these from other species, especially when traders cut off the leaves.

Many *Encephalartos* make very fine landscape plants. They are easy to cultivate, preferring well-drained soils in full sun or partial shade. ***Encephalartos lehmannii***, *Encephalartos longifolius*, and ***Encephalartos transvenosus*** need full sun.

While several South African cycads have suffered habitat loss, in many cases it is the combination of beauty, rarity, and garden desirability that has led to the major threat – unscrupulous traders. With high levels of illegal harvesting, gardeners should be sure to obtain plants only from reputable sources.

Microchip protection

There are an estimated one million cycads in private gardens in South Africa, many of them wild-collected. The Critically Endangered ***Encephalartos latifrons*** has declined due to habitat loss and overcollection; there are fewer than 100 plants in nature. Law enforcement officers long struggled to prosecute theft since they could not prove where plants came from. Conservation officials have now inserted microchips into the stems of all remaining plants so that illegally harvested plants can be easily identified.

Encephalartos altensteinii is among the most popular cycads for landscaping. As many as one-third of all the wild plants have been removed by collectors. This cycad hit the headlines in 1995, when nearly 400 – some ten per cent of the remaining wild population – were dug out by a collector. The plants were confiscated by authorities and one of the largest re-introduction programmes ever was launched to return them to the wild.

Species loss has repercussions for fauna, too. *Encephalartos altensteinii* is host to more insect species than any other cycad. No fewer than 11 cycad-specific insects are associated with it, including pollinator weevils; beetles that feed on the cones and stem; and moths that use cycad poison to advantage by storing it in their bodies to protect against predators.

Encephalartos kisambo, the Voi cycad, is named for the town of the same name near the Maungu Hills, where it occurs. The area is rich in plants and birds and is a biodiversity "hotspot". Because the Voi cycad occurs in such a small area, any environmental or human disturbance has a devastating effect on this Endangered species. One of the main threats is deforestation by the charcoal industry, but plant collecting has also reduced the incidence of plants in the wild. Ironically, the illegal harvest of a large number of seeds and seedlings from this species has resulted in thousands of plants in cultivation.

The crown of *Encephalartos kisambo* grows to 7m (22ft) across and needs plenty of space, which makes this cycad suitable only for large gardens. This decreases the likelihood that demand will increase significantly.

Encephalartos lehmannii, karoo cycad, occurs on land that is mostly too dry for crops, but goats continually defoliate natural stands. Defoliation induces plants to produce repeated flushes of new leaves. This, in turn, attracts leopard magpie moths, whose caterpillars emerge *en masse* to feed on young leaves. The repeated defoliation reduces coning and eventually kills the plants.

RENEWAL

Return to the wild

In nature, overcollection has left male and female plants of *Encephalartos latifrons* too far apart for successful pollination and the beetle pollinators appear to be extinct. But several farmers are artificially pollinating plants on their land to produce seed. New research on pollinators aims to find out whether different beetles will be able to fertilize the populations. Here, seedlings are being returned to the wild. Since 1915, Kirstenbosch National Botanical Garden, in South Africa, has built up a collection of *Encephalartos latifrons* and supplied seedlings to gardens all over the world. These plants will be a valuable resource once scientists have scanned their DNA to assess their suitability for re-introduction programmes.

HISTORY

Cycad saviour

The mist-shrouded hill of Modjadji is home to the sacred Rain Queen, a matriarchal figure for the Balobedu tribe. It is also home to the world's most spectacular cycad forest, which covers the hillside in an almost pure stand of canopy plants, 13m (42ft) tall, with an understorey of seedlings and juveniles. The Rain Queen's presence has protected the cycads over the past 300 years.

The conservation of *Encephalartos transvenosus* has been further secured by the establishment of reserves at Modjadji and two other sites. A cycad nursery has also been set up near Modjadji, which has supplied thousands of cultivated plants to collectors.

▲ ENCEPHALARTOS LEHMANNII
The karoo cycad occurs in an arid but starkly beautiful landscape that is characterized by unusual rock formations and a variety of succulent plants.

Because *Encephalartos lehmannii* is a desirable blue-leaved cycad, many have been removed from the wild to satisfy the demand for garden plants. It is listed as Near Threatened in the 2000 Red List. The commercial value of some species, especially blue-leaved ones, is so great that criminal syndicates are involved in the trade, even though *Encephalartos lehmannii* is easy to propagate. Plants produce multiple suckers that can be removed and transplanted, and they also grow well from seed.

Encephalartos longifolius was discovered by the Swedish botanist, Carl Thunberg (1743–1828), as he explored the south-east coast of South Africa in 1772. He noted that the Khoi people made bread from its starch-rich stems during famines, but this has never threatened the wild populations of the plant.

The main threat is collectors, who have systematically depleted populations near roads and towns. Prolific seed production and ready germination has reduced some of the pressure to remove plants from the wild. Small plants are now more available.

ENCEPHALARTOS LONGIFOLIUS *produces up to 400 seeds per cone, and large numbers of seedlings have been propagated in cultivation.*

Encephalartos paucidentatus is another species that has declined substantially due to collecting. The Ida Doyer Nature Reserve (now part of the Songimvelo Reserve in Mpumalanga, South Africa) was set up to help protect this cycad. Conservation officials monitor plants in the relatively small area. The National Botanical Institute in South Africa has conservation collections of threatened species like this. Because cycad seeds cannot be stored for long periods, it is not possible to include them in seed banks, such as the Millennium Seed Bank run by the Royal Botanic Gardens, Kew. Instead, gene banks have to be maintained as living plants, which means that about 20 plants from each population need to be grown in specially managed collections.

Encephalartos woodii has come perilously close to total extinction; it is extinct in the wild and is now known only in cultivation. A single male plant was discovered by John Medley Wood in the Ngoye Forest, South Africa, in 1895. Several offsets of this plant were taken to Durban Botanic Gardens between 1903 and 1916, and the last stems were removed from the wild in 1916.

Offsets from these plants have in turn been distributed to botanical gardens all over the world. All 500 plants in collections are male, making *Encephalartos woodii* the best known bachelor in the plant world. Attempts are being made to stimulate a change of gender (*see p.49*), but, as yet, with no success in producing a female companion.

One bright spot for cycad conservation is that most species grow well from seeds, which are more readily available than mature plants.

◄ **ENCEPHALARTOS VILLOSUS** *As rare cycads command high prices, the name "poor man's cycad" is ironic; but compared to other species, this one is abundant. It is listed in CITES because customs officials cannot easily distinguish it from more threatened cycads.*

▼ **ENCEPHALARTOS ALTENSTEINII** *Here, a confiscated consignment of illegally harvested plants has been replanted in habitat in the Eastern Cape, South Africa; many plants survived the ordeal, but it is too soon to tell whether they can still reproduce successfully.*

All on Red List

Hyophorbe

Botanical family
Arecaceae/Palmae

Distribution
Mascarene Islands

Hardiness
Frost-tender

Preferred conditions
Moist but well-drained soil, in full sun or light shade.

The bottle palms, *Hyophorbe* species, are endemic to the Mascarenes, an island group that lies some 700km (450 miles) east of Madagascar in the Indian Ocean. They occur in forest from sea level to altitudes of 700m (2,300ft), on both volcanic and limestone soils.

The genus *Hyophorbe* includes some of the world's most threatened palms and – paradoxically – some of the world's most popular garden palms. All five of these single-trunked, pinnate-leaved palms are threatened with extinction in the wild as a result of habitat loss, and due to the introduction of non-native grazing animals and pests and diseases.

The most commonly cultivated is ***Hyophorbe lagenicaulis***, a relatively quick-growing palm endemic to Round Island, north of Mauritius. Only five mature plants of this Critically Endangered palm are known in the wild, but seed has been collected. Seedlings have been raised in nurseries set up by conservation partners including Fauna & Flora International; Royal Botanic Gardens, Kew; the government of Mauritius; and the Mauritian Wildlife Foundation. Grazing animals have been removed from Round Island.

Hyophorbe verschaffeltii, which is Critically Endangered, is endemic to Rodrigues, and has a rather slimmer trunk than *Hyophorbe lagenicaulis.* Of the three other species, only *Hyophorbe indica* is normally seen in cultivation. It is listed as Endangered and comes from La Réunion. The two remaining species are also at the brink of extinction. *Hyophorbe vaughanii,* from Mauritius, survives with only three trees known in the wild. *Hyophorbe amaricaulis* is known only as a single tree growing in the Curepipe Botanic Gardens, Mauritius.

Bottle palms are excellent garden palms, being neither too large nor too demanding and, while not particularly cold-tolerant, they do tolerate exposure, wind, and salt spray. All grow relatively quickly and look best in group plantings that accentuate their architectural form.

◄ HYOPHORBE LAGENICAULIS *With neatly arching leaves emerging from a smooth, swollen trunk and prominent crown-shaft, the bottle palm is good for medium-sized gardens; it reaches about 6m (20ft) in height.*

Red List: Vulnerable

Jubaea chilensis

Botanical family
Arecaceae/Palmae

Distribution
Chile

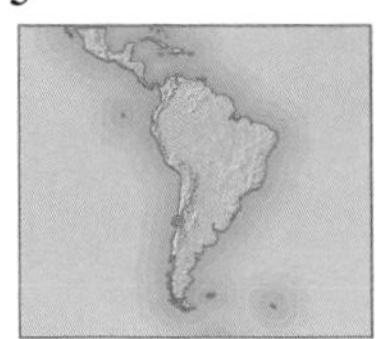

Hardiness
Half-hardy

Preferred conditions
Well-drained soil in full sun with low atmospheric humidity.

The Chilean wine palm, or honey palm, is the only species of the genus *Jubaea*, and this Vulnerable species is becoming increasingly rare in the wild. ***Jubaea chilensis*** is threatened by habitat loss, uncontrolled wild fires, and cutting of trunks for the production of wine. The seeds – called pygmy coconuts or coquitos – were a valued part of the indigenous people's diets.

The palm is endemic to the Chilean matorral, the dry, scrubby woodland that extends along a strip of central Chilean coast that is about 100km (60 miles) wide. This ecoregion has been severely affected by fire, mining, logging, rubbish dumps, urbanization, and the pollution of air, water, and soil. Unfortunately, this is the least-protected region in Chile. Natural regeneration of *Jubaea chilensis* is insufficient to outstrip these environmentally destructive practices.

Jubaea is a magnificent ornamental palm that reaches about 25m (80ft) in height. It is able to withstand lower temperatures than most palms, making it a popular choice for the cool subtropics and warm temperate regions. It grows best in a Mediterranean-type climate that closely matches that of Chile.

However, experimental plantings in the UK, France, and the northern Pacific coast of the USA suggest that this palm is hardier than previously suspected. A huge specimen graces the Temperate House, Royal Botanic Gardens, Kew.

Provide this palm wih full sun, well-drained soil, and low humidity. It is impressive as a street or avenue tree, where the extraordinary columnar trunks can be appreciated.

HISTORY

WHAT'S IN A NAME?

When Alexander von Humboldt found the Chilean wine palm, he named it in honour of Juba (*below*), king of Numidia from 25 to19BC. As a baby, in 46BC, King Juba had been captured as a war trophy by Julius Caesar, and was brought up and educated in Italy. So scholarly and respected did he become that he was restored to kingship and married one of the daughters of Antony and Cleopatra. He is thought to have been an influential teacher of Pliny the Elder in botany and zoology.

▼ CHILEAN WINE PALM *In its native matorral,* Jubaea chilensis *clothes the ravines and ridges of the coastal hills from sea level to altitudes of 600m (2,000ft).*

Red List: Endangered

Kentiopsis oliviformis

Botanical family
Arecaceae/Palmae

Distribution
New Caledonia

Hardiness
Frost-tender

Preferred conditions
Well-drained soil, including alkaline, in partial shade, with moderate humidity.

An impressive palm with striking leaf scars and a large crown of stiffly ascending leaves, ***Kentiopsis oliviformis*** is a collector's favourite that is becoming more common in the gardens and landscapes of the subtropics and tropics.

It is native to a small area of the island of New Caledonia, which has one of the world's most threatened island floras. Almost 77 per cent of its plants are strictly endemic, occurring nowhere else in the world. The New Caledonian flora includes 31 endemic palms, in 15 genera that are not found anywhere else.

The primary threats to this species are cattle, which eat the seedlings, and loss of habitat from land clearance. As yet, no measures have been taken to protect *Kentiopsis oliviformis* from these threats to its survival and it is listed as Endangered in the 2000 Red List.

In the wild, *Kentiopsis oliviformis* emerges above a moderately moist rainforest canopy. At 30m (100ft) high, it is the tallest palm in New Caledonia. This magnificent palm tolerates a wide variety of conditions, including the seasonally dry, alkaline soils of Florida in the USA. It does not tolerate cold, and for best growth needs a semi-shaded situation.

Kentiopsis oliviformis has great potential for use in the horticultural trade as it is both drought-tolerant and adaptable in terms of soil pH and type. This palm is slow-growing in nature, although growth rate varies according to watering and fertilizer regimes; it typically reaches a height of 8m (25ft) in cultivation.

▲ COLLECTOR'S FAVOURITE *From a young age,* Kentiopsis oliviformis *grows an elegant crown of upright, arching leaves.*

Red List: all Endangered

Latania

Botanical family
Arecaceae/Palmae

Distribution
Mascarene Islands

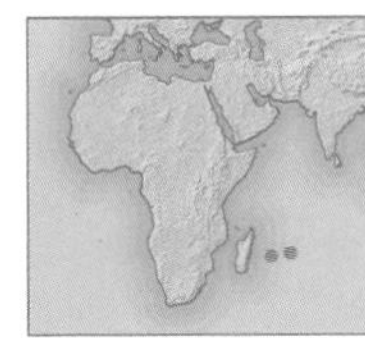

Hardiness
Frost-tender

Preferred conditions
Any well-drained soil in full sun.

All three species of the *Latania* palms are threatened with extinction. The most commonly cultivated is ***Latania loddigesii***, which is endemic to the offshore islands of Mauritius, Round Island, and Gunner's Quoin. The other two species are more rare in cultivation. *Latania lontaroides* from La Réunion has about 100 plants left in the wild, and *Latania verschaffeltii,* from Rodrigues, has about 500 wild trees. All three species are listed as Endangered in the 2000 Red List.

Latanias are spectacular-looking fan palms, with blue- or grey-green leaves, and they make excellent garden plants for warm climates. They are robust trees that are handsome planted as dramatic clumps, or as single specimens, to display the straight trunks and angular leaf canopy to great effect. All three species are able to tolerate drought (*Latania loddigesii* is particularly drought- and salt-tolerant) and will grow in a variety of soil types.

▲ LATANIA LODDIGESII *bears tiny white or pale yellow-green flowers in summer, in branching panicles that reach 1.5m (5ft) or more in length.*

Red List: Lower Risk/CITES II

Lepidozamia

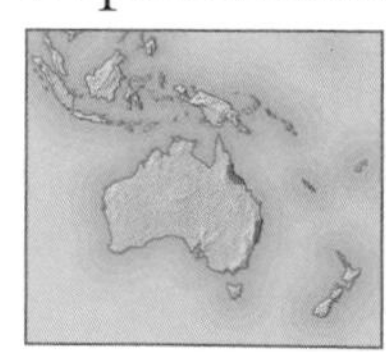

Botanical family
Zamiaceae

Distribution
Australia: Queensland, New South Wales

Hardiness
Half-hardy – frost-tender

Preferred conditions
Well-drained, fertile, reliably moist soils in sun or shade.

The only two known species of *Lepidozamia* occur in large colonies in damp, shaded gullies in the tropical rainforests along Australia's eastern coast. In the past, Aboriginal people harvested and ate the large seeds of both species, but with no apparent effect on the abundance of wild populations.

As with other peoples who used cycad seeds in their diet, Australian Aboriginal people learned to overcome the powerful toxins contained in the seeds. In this case, they used heat in order to denature the toxins.

In many parts of the world, cycad populations have declined to the point where beetle pollinators have become extinct. In contrast, thousands of beetles flock to visit the female cones of ***Lepidozamia peroffskyana***.

Lepidozamia species are listed as being at Lower Risk on the 2000 Red List, but are included in CITES Appendix II due to concern about the impact of trade, and because of the difficulty customs officials have in distinguishing them from other cycads that are more threatened.

Both species make attractive specimens in tropical and subtropical gardens and can be grown in containers. Their natural habitat in deep shade on the forest floor lends them a tolerance of low light conditions and they grow well indoors. The tropical *Lepidozamia hopei* does well in warm, humid conditions and is less suitable for open landscapes than the hardier *Lepidozamia peroffskyana*, which suits warm to temperate climates. Mulch and feed them regularly and water in dry periods. Both grow readily from seed, if it is sown fresh; it has a limited viability.

LEPIDOZAMIA PEROFFSKYANA ► *One of the outstanding features of this species is the tall male cone, which opens in spirals to reveal the mature pollen sacs; it remains for some time in this very ornamental state.*

Red List: Vulnerable

Livistona carinensis

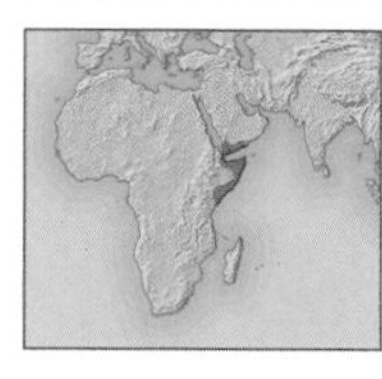

Botanical family
Arecaceae/Palmae

Distribution
Somalia, Djibouti, and Yemen

Hardiness
Frost-tender

Preferred conditions
Well-drained soil in full sun.

Found only in very arid areas, in oases, and wadi mountain valleys (micro-habitats with a perennial water source, which are also the locations sought by a growing human population in need of land for agriculture), *Livistona carinensis*, or bankoualé palm, is known to occur only in three widely scattered populations, in Djbouti, Somalia, and Wadi Hajar in eastern Yemen.

This is an arid, semi-desert region that borders the Red Sea and the Gulf of Oman. It has low rainfall, with yearly averages of 100–200mm (4–8in) and is one of the world's harshest desert regions. *Livistona carinensis* occurs well outside the main centre of *Livistona* distribution in south-east Asia, and is an intriguing bio-geographic curiosity that is thought to be the last survivor of a greater *Livistona* diversity that once flourished in the wetter palaeoclimate of Africa and the Arabian Peninsula.

Livistona carinensis survives in scattered populations. With increasing demands on water for irrigation, the water table in many areas is becoming too low to support *Livistona carinensis*, and little or no regeneration occurs. In addition, until recently, this palm tree was highly sought after for roof timbers, because it is regarded as termite-resistant.

Like other species in the genus, *Livistona carinensis* is adaptable and proves easy to grow in the arid tropics and subtropics. It may be able to tolerate colder temperatures than most *Livistona* species. While this tree is relatively slow-growing, it is very tough and is tolerant of both salt and drought.

Since its introduction to cultivation by Peter Bally, it has proved tolerant of both arid and humid climates. There are successful plantings in Miami, USA, and Mombasa, Kenya.

Red List: Vulnerable

Lodoicea maldivica

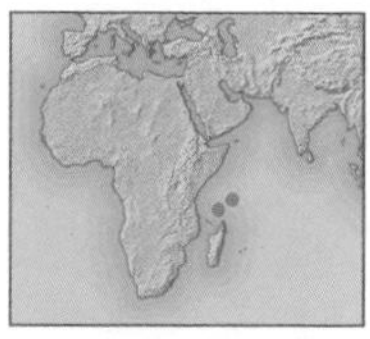

Botanical family
Arecaceae/Palmae

Distribution
Seychelles

Hardiness
Frost-tender

Preferred conditions
Fertile, moist, but well-drained neutral to acid soil; sun.

The single species in the genus *Lodoicea* is one of the world's most legendary and mysterious plants. The huge leaves, with leafstalks growing to 4m (12ft) long, can be up to 6m (20ft) in length, and 4m (12ft) across. The massive seeds of *Lodoicea maldivica*, or the *coco de mer*, are the largest in the plant kingdom, and defy evolutionary and ecological explanations. Having first been described on the basis of seeds found floating near the Maldives (hence the specific name), the only remaining wild populations occur on two islands in the Seychelles.

The World Heritage Site of the Vallée de Mai, in the Praslin National Park, has been managed by the Seychelles Islands Foundation since 1989. The other main population is protected in the Curieuse National Park. These populations are susceptible to fire and encroachment by invasive plants. They are managed by the Seychelles government. Legislation ensures sustainable harvesting of the fruit, while strategic firebreaks prevent fires sweeping through the forest stands.

Lodoicea is a spectacular palm for humid tropical gardens. There, seeds should be germinated in the ground, in very deep, rich soil. The *coco de mer* grows best with high levels of moisture and humidity and neutral to acid soil. It is a dioecious species (male and female flowers are borne on separate plants) and fruits are only produced when male and female trees are grown together. It usually takes 25–50 years for a plant to reach maturity. In cultivation, fruiting trees can be seen at the Foster Botanic Garden, Hawaii; in the Victoria Botanical Garden, Seychelles; and at the Royal Botanic Garden, Peradeniya, Sri Lanka.

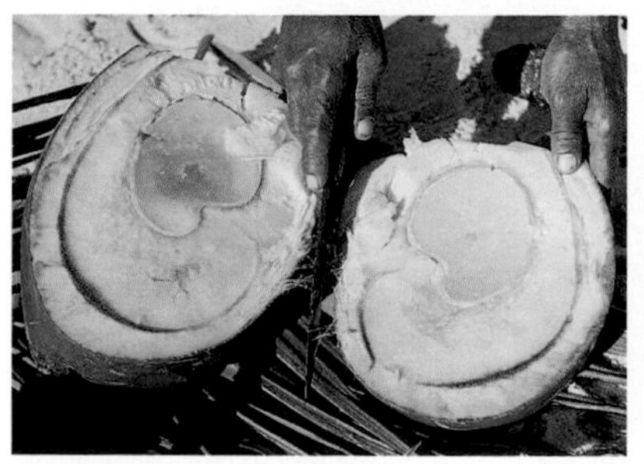

▲ LODOICEA NUTS *can take 5–7 years to mature, and contain two very large seeds. The creamy white flesh and the jelly inside have long been considered great delicacies.*

INTACT FOREST *This is the largest intact forest of the endemic palm,* coco de mer, *seen here in habitat at Vallée de Mai, Seychelles, which was declared a World Heritage Site in 1983.*

RENEWAL

GROWING *COCO DE MER*

In 2003, the Victoria Botanical Garden in the Seychelles gave the gift of a seed of *coco de mer* to the Royal Botanic Garden Edinburgh. Under the conditions imposed by the Seychelles government, trade in nuts has been closely controlled since 1995, so this was a very special gift. The two-lobed seeds can weigh as much as 30kg (66lb), and are the largest and heaviest plant seeds known. The suggestive form has given rise to most of the legends associated with *coco de mer* and, not surprisingly, it is believed to possess aphrodisiac properties. Much prized, the seeds have commanded high prices for centuries. This palm has proved difficult to raise in cultivation in cool temperate climates. However, in Edinburgh, after storage for a mere four months in warm, dark, humid conditions in the tropical propagation house, the seed began to germinate in August 2003.

1) To prompt germination, the seed was kept moist, in a closed case in dark conditions at a temperature of 24°C (75°F).
2) The cotyledon stalk, a root-like structure, emerged from between the lobes.
3) The germinating seed was transferred to a large pot, twice the diameter of the seed, with the cotyledon stalk facing downwards into specially prepared, open-textured, free-draining potting compost.

4) One of the secrets of successful propagation is to provide a pot that is deep enough for the emerging root. The root will penetrate to depths of 4m (12ft), before it branches. Bottomless pots are stacked one above the other to gain necessary depth.
5) The seed, now buried to half its depth, was watered thoroughly. The first leaf is expected to take a further 12 months to emerge.

11*/13 on Red List/CITES II

Macrozamia

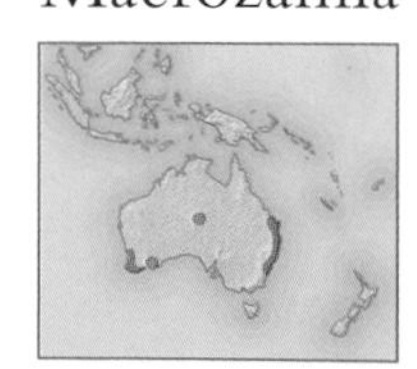

Botanical family
Zamiaceae

Distribution
Australia

Hardiness
Half-hardy – frost-tender

Preferred conditions
Well-drained soils in full sun to partial shade; *Macrozamia macdonnellii* does not tolerate humid conditions.

◄ MACROZAMIA RIEDLEI *This fairly low-growing species seldom forms a trunk. It grows wild in both forest and scrub in Western Australia, and so tolerates both full sun and partial, dappled shade. It thrives in warm temperate zones and will tolerate a sharp but short-lived frost.*

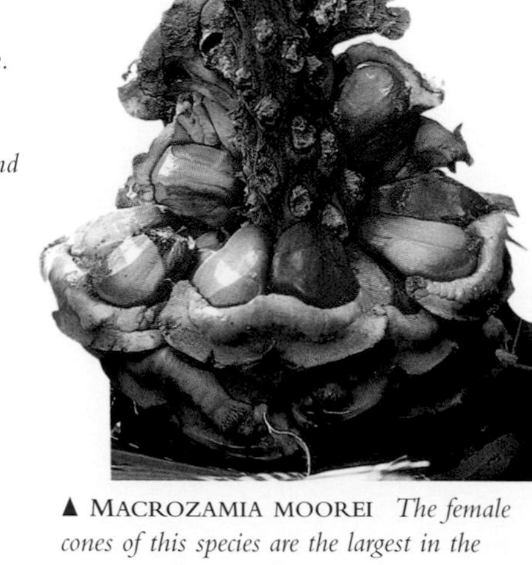

▲ MACROZAMIA MOOREI *The female cones of this species are the largest in the genus. Seed-grown plants are slow-growing, taking 50 years to produce cones, and 100 years to reach 2m (6ft) in height.*

Macrozamias are a characteristic feature of some Australian landscapes and can be a dominant component of the vegetation. They occur in a wide range of habitats, from desert and scrub, through tropical and subtropical grassland and savanna, to temperate broadleaf forest. Most *Macrozamia* species have short or underground stems, but several species have robust, upright trunks, such as the spectacular ***Macrozamia moorei***.

Scientists recognize two groups of macrozamias: one group consists of small, inconspicuous plants that tend to be scattered in the landscape; the second group consists of larger, showier plants, which may occur in dense populations.

All *Macrozamia* species appear in CITES Appendix II, and in the 2000 Red List some 13 species are listed as Endangered or Vulnerable. Although macrozamias are often common plants, plant collecting has contributed to a substantial decline in some species, and land development has destroyed the habitat of many more. Farmers used to eradicate macrozamias to prevent stock poisoning, but this practice has now been stopped.

Even common species can contribute to conservation objectives. For example, *Macrozamia moorei*, which is an attractive landscape ornamental, is quite common in the wild, and Australian authorities have allowed the removal and export of several hundred plants based on the principle of sustainable use. This allows trade in wild-collected plants, provided that there is no negative effect on population survival. The supply of large common cycads to the landscape trade reduces demand for rare species.

The species best suited to garden cultivation are those with dense crowns of long, arching leaves, including ***Macrozamia communis***, *Macrozamia macdonnellii*, *Macrozamia moorei*, and ***Macrozamia riedlei***. Most of these species are best planted in the ground because mature plants are quite large, but young specimens can make attractive container plants. *Macrozamia macdonnellii* occurs in the arid centre of Australia and does not tolerate humid conditions.

Most species grow well from seed, which should be sown fresh, since it has a short viability period.

▼ MACROZAMIA COMMUNIS *This magnificent species is seen here in its natural habitat forming large colonies in the understorey of the tall sclerophyll forest of Bateman's Bay, New South Wales, Australia.*

Red List: Critically Endangered/CITES I

Microcycas calocoma

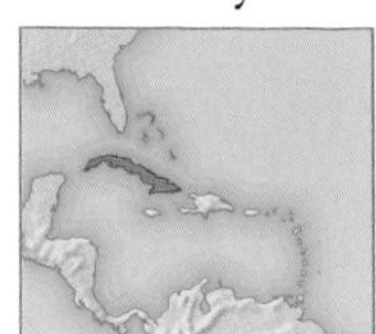

Botanical family
Zamiaceae

Distribution
Cuba

Hardiness
Frost-tender

Preferred conditions
Well-drained soil; full sun.

The scientific name *Microcycas,* meaning small cycad, is not at all appropriate for this tall and beautiful plant. The strong resemblance to palms is reflected in the local Cuban name, *palma corcho*, or cork palm.

The only species in this genus, cork palm is restricted to the dry tropical forests on the island of Cuba, where it is Critically Endangered; only a few hundred plants survive in the wild.

Although they are protected from collectors, populations have declined in the past because of the effects of land clearance for agriculture. Seed set is also poor, and scientists suspect that the beetle pollinators are extinct. Cuban botanists are trying to raise funds to study the pollination mechanisms of this unusual plant.

Most of the plants in cultivation originate from Fairchild Tropical Botanic Garden or the Montgomery Botanical Center in Miami, USA. Both gardens hand-pollinate cultivated plants in their collections, and seeds and seedlings have been distributed to gardens and collections all over the world.

Microcycas calocoma is extremely sensitive to cold weather, but grows well in tropical and subtropical gardens with regular mulching and fertilizing. It is, however, difficult to obtain because of the shortage of plants in cultivation. Nevertheless, it can be propagated readily from seed. However, the seed, which has a short viability, should be sown within one or two months of collection.

Red List: Lower Risk

Phoenix theophrasti

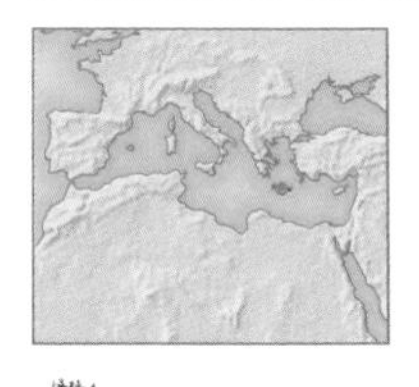

Botanical family
Arecaceae/Palmae

Distribution
Crete, Turkey

Hardiness
Frost-tender

Preferred conditions
Well-drained, preferably sandy, soils in full sun; some shade when young.

The Cretan date palm, ***Phoenix theophrasti***, was thought to be endemic to Crete until the discovery of a few small Turkish populations. In both regions, the palms are at risk.

This palm is well known in botanical literature, having appeared there since the time of the philosopher Theophrastus (about 372–286 BC), after whom it was named. *Phoenix theophrasti* was probably more widespread on Crete in the past; it once appeared on Roman coins that were minted at Ierápetra, where it is no longer found.

On Crete, there are eight populations; the largest is at Vai, with a few thousand trees that are protected under Greek law.

Phoenix theophrasti is usually found on sandy soils close to the sea and is under threat because of land drainage and tourism. In particular, car parking and camping in the palm groves at Vai has prevented natural regeneration.

Camp fires, which have been a problem in the past, are now prohibited. The trees in the grove at Vai, a tourist attraction in its own right, have had the dry lower fronds removed to further reduce the risk of fire. This, however, has tended to reduce regeneration because of direct damage to the sideshoots and due to trampling of seedlings by passersby.

The wild populations grow in areas with a high water table. Pumping to provide fresh water, for human consumption or local enterprises, can lower the water table and kill the trees.

The trees reach about 10m (30ft) in height, and the main stem usually has a few shorter sideshoots from the base, each topped by a dense head of narrow fronds. The Cretan date palm would make an unusual specimen for gardens with a Mediterranean-type climate. Trees can be propagated by seed and removal of sideshoots when small.

▲ THINNED THICKETS *Under natural conditions* Phoenix theophrasti *forms dense, impenetrable thickets. Thinning of the thickets has allowed access to the groves and reduced regeneration.*

▼ PRITCHARDIA PACIFICA *The Fiji fan palm hails from Tonga, but was introduced to Fiji before European colonization.*

16*/17 on Red List

Pritchardia

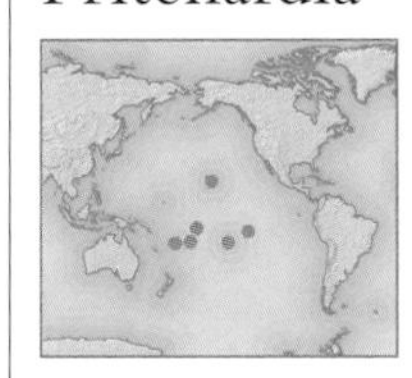

Botanical family
Arecaceae/Palmae

Distribution
Pacific Islands

Hardiness
Frost-tender

Preferred conditions
Moist but well-drained soils; sun or part shade with high humidity and wind shelter.

The lo'ulu palms include some 28 species in the genus *Pritchardia* and are found in the Pacific archipelagos of Hawaii, Fiji, Samoa, Cook Islands, Tonga, and the Tuamotus. Of these, 23 species are endemic to the Hawaiian Archipelago. The most commonly cultivated species in Hawaii are ***Pritchardia pacifica*** (Unlisted) and *Pritchardia thurstonii* from Fiji, classed as Lower Risk on the 2000 Red List.

Many Hawaiian species are Critically Endangered, such as *Pritchardia alymer-robinsonii,* surviving as two wild trees on the island of Ni'ihau, and *Pritchardia munroi*, with one wild tree on Molokai.

Only two Hawaiian species survive as reproducing wild populations, and these have been protected from predators such as goats and rats. They are ***Pritchardia remota*** on Nihoa, classed as Endangered on the 2000 Red List, and, from Huelo Rock, Molokai, *Pritchardia hillebrandii*, which is Unlisted.

The lo'ulu palms are quick-growing, attractive trees for tropical gardens. They do not tolerate persistent drought or cold, even though some, such as *Pritchardia munroi,* are from very dry habitats. Some, notably *Pritchardia thurstonii*, have inflorescences that can reach up to 3m (10ft) in length.

► PRITCHARDIA REMOTA *bears large, flamboyant inflorescences and produces large quantities of viable seed.*

Red List: Indeterminate*

Pseudophoenix sargentii

Botanical family
Arecaceae/Palmae

Distribution
Florida, Cuba, Mexico, Central America

Hardiness
Frost-tender

Preferred conditions
Well-drained soil in sun; some shade when young; tolerates drought, salt spray, and occasional frost.

Found in the wild in coastal scrub and seasonally dry forest, *Pseudophoenix sargentii,* the buccaneer palm, or Sargent's cherry palm, was first described from plants discovered in 1886 on Elliott Key, a small coral-rock island off the coast of Miami, Florida, USA. Shortly thereafter, Elliott Key was plundered for palms to use in Miami landscaping projects, and large areas of the island were cleared for agriculture. Elliott Key is now part of Biscayne National Park, but because fewer than 50 adult palms exist here, *Pseudophoenix sargentii* was listed on the Red List of 1997. However there is not much data available on the severity of the risk. The Florida Department of Agriculture and Consumer Services also lists this plant as Endangered in Florida.

Elsewhere, the palm is also threatened. In the Bahamas and Yucatán Peninsula of Mexico, trees are weakened by beach-front developments. In eastern Cuba and some small islands off the coast of Hispaniola, and Dominica, the future of the palm is more secure. Because these are remote areas, they have not yet been subjected to pressure from developers.

This palm is of great interest to botanists because it represents an isolated evolutionary lineage within the palm family. It does not appear to be related to other Caribbean palms.

Pseudophoenix sargentii is a slow-growing palm, but so durable and elegant that it is worth the effort to grow it. The graceful, feathery leaves and bright red fruit that follow the flowers are an asset to any garden, and many different kinds of indigenous bees are attracted to its greenish-yellow flowers. *Pseudophoenix sargentii* responds well to regular irrigation and fertilizer.

Red List: Indeterminate*

Roystonea regia

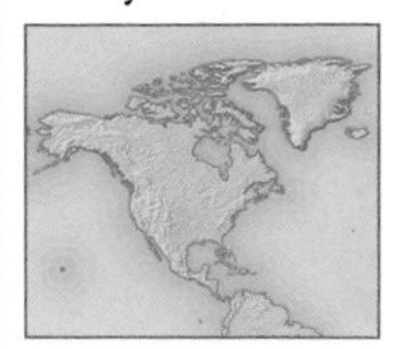

Botanical family
Arecaceae/Palmae

Distribution
USA: Florida

Hardiness
Frost-tender

Preferred conditions
Fertile, well-drained but consistently moist soil in sun; half-hardy when mature; tolerates storms.

The Florida royal palm is one of the most widely cultivated palms. Tall and stately, with a bright green crown-shaft and a crown of massive leaves, it is grown as a specimen or in avenues.

The Floridian and Cuban royal palms were once considered separate species: *Roystonea elata* and *Roystonea regia*. Both are now classified as *Roystonea regia*. In the wild, this is limited to five populations

Red List: Endangered

Sabal bermudana

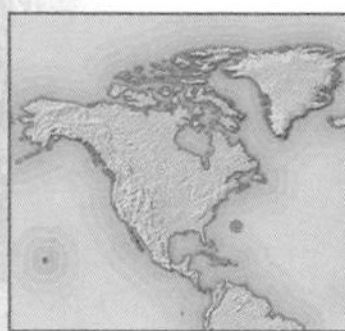

Botanical family
Arecaceae/Palmae

Distribution
Bermuda

Hardiness
Half-hardy

Preferred conditions
Well-drained, preferably alkaline soils in partial shade.

Sabal palms are widespread in the Caribbean and Central America, often growing in dry, open habitats. The Endangered ***Sabal bermudana*** is one of three endemic trees in the Bermudas, with wild populations surviving in reserves such as Paget Marsh, which is a residual cedar and palmetto hammock.

All species are similar in looks and cultural needs. *Sabal bermudana* is a slow-growing palm, to 6m (20ft) tall, and more robust than the widely cultivated cabbage palmetto, ***Sabal palmetto***.

The sabal palms grow best in partial shade, on moist but well-drained, calcareous soils. They can tolerate relatively low temperatures.

SABAL BERMUDANA ▼ *is considered one of the hardiest palms in Europe. Extremely large specimens can be seen in Mediterranean gardens.*

◄ SABAL PALMETTO *covers vast expanses of the state of Florida. It is harvested from the wild on a very small scale for hearts-of-palm, but large numbers are transplanted from the wild for use in landscaping.*

Red List: Near Threatened/CITES I

Stangeria eriopus

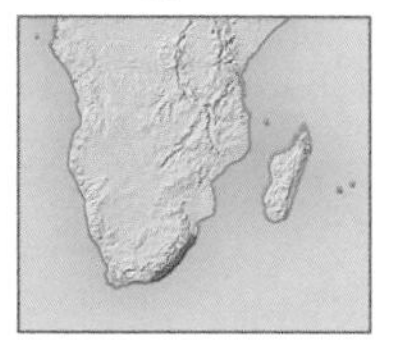

Botanical family
Stangeriaceae

Distribution
South Africa

Hardiness
Half-hardy

Preferred conditions
Sandy, acid, well-drained soil; grassland forms prefer sun, forest forms thrive in shade.

This small cycad, with broad, papery leaves and a swollen, subterranean stem, was first mistaken for a fern on account of its distinctly fern-like, divided leaves; it looks quite unlike other cycads. *Stangeria eriopus* is found in tropical and subtropical grassland and savanna, and in

in the Everglades and Big Cypress Swamp. It is listed as Endangered in Florida by the Florida Department of Agriculture and Consumer Services.

Roystonea regia grows on moist hammocks (ground raised above the surrounding wet land) in dense forests, amid shrubs, trees, ferns, and epiphytes. Although these populations are now protected, they have been threatened by drainage for construction and road building, which lowers the water table and renders land more vulnerable to fire. It is now confined to the south-west tip of the state, but in the 18th century *Roystonea regia* was sighted by William Bartram along the St Johns River, far north of its current range.

In the past, wild seedlings and juvenile plants were taken for use as ornamentals. Seed germinates readily, and if it is sown in pots, the resulting seedlings transplant better than specimens that have been removed from the open ground.

shrubland and evergreen forest on the eastern coast of South Africa.

It is less popular with plant collectors than other South African cycads. The rapid decline of *Stangeria eriopus* in recent years is largely due to overcollection for medicinal uses.

There are two recognized forms: a forest form, which prefers shaded situations and produces relatively long, lush leaves; and a grassland form, which prefers sunny situations and produces shorter, more leathery leaflets.

Tolerant of light, short-lived frost, *Stangeria eriopus* is a moderately hardy species suitable for subtropical and warm temperate gardens. It is easy to grow in a sheltered site in shade or sun, if mulched, regularly watered, and given slow-release fertilizers. Seed sown fresh takes about 12 months to germinate.

THREAT

MEDICINAL PLANT

The fleshy stems of *Stangeria eriopus* are widely traded at traditional medicinal plant markets (*left*). Traditional healers use the tuberous stems for a variety of ailments and one study showed that more than 3,000 plants were being traded every month from markets around the city of Durban alone.

Demand is increasing, possibly due to the search for cures for HIV-Aids, and populations of this cycad around urban centres have been completely wiped out. In response, the Durban Botanic Gardens and the National Botanical Institute of South Africa have set up a project that will survey and map the remaining populations, and develop a living gene bank.

The ultimate aim is to propagate and grow plants in cultivation in order to supply them to traditional healers and to re-instate natural populations in the wild.

Red List: Lower Risk

Veitchia metiti

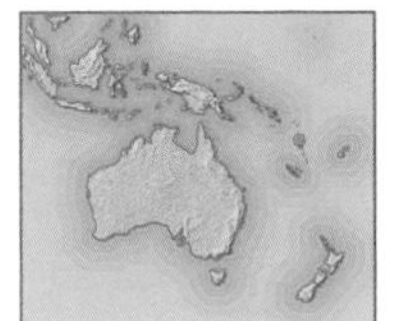

Botanical family
Arecaceae/Palmae

Distribution
Vanuatu: Banks Islands

Hardiness
Frost-tender

Preferred conditions
Well-drained but moist soil in partial shade or in full sun in reliably moist soils.

This palm was first collected from the tropical rainforests of the New Hebrides (now Vanuatu) in 1908, and named in 1920 by Odoardo Beccari, before becoming lost to science until just a few years ago.

The genus name commemorates James Veitch and his son, John Gould Veitch, members of the family of famous English nurserymen, who sponsored many Victorian plant collectors.

Veitchia metiti was known only from that first collection for 76 years, until it was rediscovered on Vanua Lava, in the Banks Islands of Vanuatu, by a team of botanists studying the palms of that remote chain in the western Pacific.

This beautiful palm has been introduced into cultivation, where its strikingly graceful leaves clustered at the top of notably slender stems make it a handsome addition to a tropical garden. Species in this genus typically bear showy flower panicles beneath the foliage, which are often followed by clusters of colourful fruit.

Like the more widely cultivated *Veitchia arecina* (formerly known as *Veitchia montgomeryana*), which is now Endangered in the wild, *Veitchia metiti* is fast-growing. It thrives in partial shade as a juvenile, but tolerates full sun when mature, as long as abundant soil moisture is available. Little is known about its hardiness, but it is likely to be sensitive to the cold. It is propagated by fresh seed, preferably sown as soon as it is ripe.

The outlook for *Veitchia metiti*, in common with many island palms, is unclear. It is classified as being at Lower Risk. However, according to the 2000 Red List, the survival of *Veitchia metiti* is dependent on conservation efforts. As long as its native islands remain undisturbed by humans or natural disasters, it will flourish. A cataclysmic cyclone, or a new logging concession could be devastating; forest clearance would spell doom.

The same applies to other species which extend into the Fijian islands. Three were included in the Red List of 1997: *Veitchia pedionoma* (from Vanua Levu, Fiji), *Veitchia petiolata* (also from Vanua Levu), and *Veitchia simulans* (from Taveuni, Fiji) are classified as Vulnerable.

Red List: Lower Risk

Verschaffeltia splendida

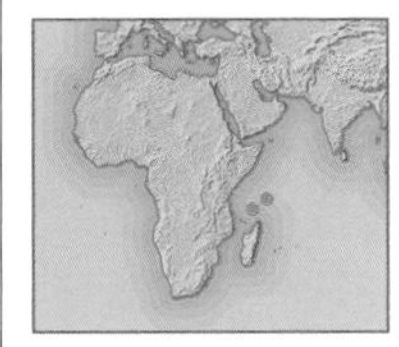

Botanical family
Arecaceae/Palmae

Distribution
Seychelles

Hardiness
Frost-tender

Preferred conditions
Well-drained, neutral to acid soil; high humidity; shelter from strong or dry winds.

The single species in the genus *Verschaffeltia* is a palm that is native to small patches of mist-wrapped, humid rainforest on the oldest islands of the Seychelles group. Mature palms grow to more than 20m (70ft) tall, and emerge above the tree canopy. In the wild, ***Verschaffeltia splendida*** is restricted to the eroding hillsides and stream beds that slice through these ancient, crumbling granite islands. It is listed as Lower Risk in the 2000 Red List, but its survival depends on conservation efforts.

Verschaffeltia splendida, known as the Seychelles stilt palm, has found its way into cultivation as an ornamental because of its bizarre, aerial stilt roots. The stilt roots appear during the first year after germination, and are probably an adaptation to life on uneven and shifting ground. These, combined with the black spines on the trunk and leaf stalks, make this unusual palm one of the most prized in cultivation.

WILDLIFE

MUTUAL BENEFITS

The Seychelles black parrot (*Coracopsis nigra barklyi*) and *Verschaffeltia splendida* have a mutually dependent relationship. The tree's fruits form an irreplaceable part of the black parrot's diet, while the bird is the only known vector (carrier) for the dispersal of the tree's large, ridged seeds. Without each other, both tree and bird populations would suffer.

This species needs frequent misting when grown under glass. It is very intolerant of dry winds. When grown in a glasshouse, or in a forest understorey, its leaves are paddle-shaped and undivided. With exposure to wind, the leaves usually become irregularly tattered.

The palms of the Seychelles are extraordinary in that all six endemic species are the only members of their genus (monotypic). As well as the *Verschaffeltia*, *Deckenia nobilis*, *Lodoicea maldavica*, *Nephrosperma vanhoutteanum*, *Phoenicophorium borsigiana*, and *Roscheria melanochates* can be found under the protective umbrella of the Vallée de Mai Nature Reserve, Seychelles.

▲ STILT PALM *Emerging above the tree canopy, this palm produces an inflorescence that may reach 2m (6ft) long; the clustered olive-green fruits that follow are about 2.5cm (1in) across.*

Red List: Lower Risk

Washingtonia filifera

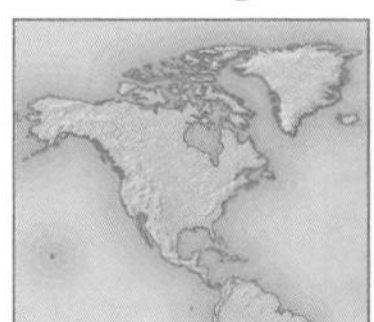

Botanical family
Arecaceae/Palmae

Distribution
USA: SE California, western Arizona; Mexico: Baja California

Hardiness
Frost-tolerant

Preferred conditions
Sun and fertile, well-drained soil; tolerates drought and frost.

The California fan palm may be one of the most widely cultivated of palms. In the Red List of 1997, several populations of this species were deemed Lower Risk, and it was recommended that these natural stands unaffected by human impacts be given protection.

Washingtonia filifera is found in desert and semi-arid regions, usually in gorges and canyons, which often channel rainfall into palm oases. In Arizona, some populations of *Washingtonia filifera* are listed as Endangered; there are probably around 25,000 palms to be found around 116 seeps, springs, and streams in the Sonoran Desert. Wild trees are threatened by flash floods and are killed by vandals who set fire to the "skirts" of dead, dry fans.

The Californian fan palm is traded internationally as an ornamental plant. In gardens it is valued for its tall straight trunk, shaggy with the remnants of the dead fans, and topped by a rounded crown of grey-green leaves. It is a splendid avenue tree, although in such public spaces, the dry skirts are often removed to avoid fire hazards.

▼ SHAGGY TRUNKS *Usually seen in a neatly shorn state in cultivation, the California fan palm is truly magnificent in its natural condition in the dry desert habitat of the south-western USA.*

Red List: Lower Risk

Wodyetia bifurcata

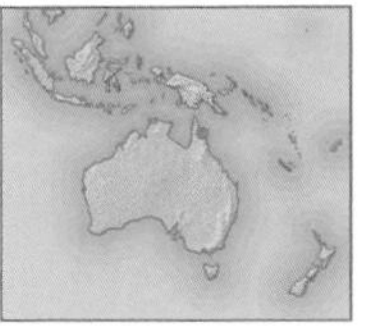

Botanical family
Arecaceae/Palmae

Distribution
Australia: Queensland

Hardiness
Frost-tender

Preferred conditions
Well-drained soil in full sun; tolerates short-term drought, wind, and short-lived frost.

Since it was named in 1983, *Wodyetia bifurcata*, the foxtail palm, has become a popular subtropical and tropical landscape plant. Mature trees have trunks up to 15m (50ft) tall, topped by a slender, blue-green crown-shaft, and spectacular leaves shaped like a fox's brush. It is a tough and easily cultivated palm that grows rapidly and tolerates a wide range of growing conditions.

Found only in Cape Melville National Park, Queensland, Australia, in woodland, scrub, and gravelly hillsides, this species has been threatened by theft of wild seeds for the horticultural trade.

Although still classed as Lower Risk in the 2000 Red List, *Wodyetia* is highly fecund, producing huge quantities of seed. Thousands of plants have now been raised in cultivation all over the world.

HABITAT

AUSTRALIAN HOTSPOT

Queensland's natural habitats are exceptionally diverse. They range widely from reefs and mangroves to wetlands and rainforests. A wide array of forest types are found there, from cloud forest, as illustrated here, to seasonally dry forest. These native habitats provide a home to the greatest biodiversity in Australia, with more than 8,000 species of native plants. Of this extraordinary flora, 13 per cent is rare and threatened, and receives strict protection.

30 on Red List/all on CITES II

Zamia

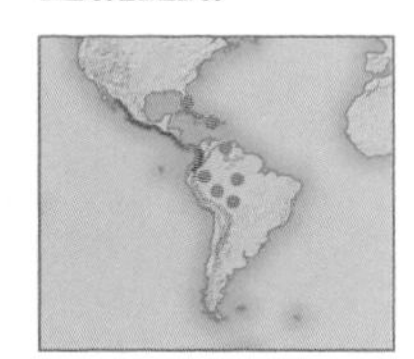

Botanical family
Zamiaceae

Distribution
USA: Georgia and Florida; Central and South America

Hardiness
Frost-tender

Preferred conditions
Well-drained soil in sun, or in shade with shelter and moderate to high humidity.

These cycads occur from Georgia and Florida, USA, and Mexico, south to Bolivia and Peru. *Zamia* species are found in a range of habitats, from open environments such as grassland, savanna, scrub, and open woodland, to more shaded and enclosed conditions, as found in tropical and mountain rainforest.

Their natural habitat provides clues to their needs in cultivation. Those from open habitats need full sun and well-drained soil, while rainforest species prefer shade or dappled sunlight with higher levels of humidity.

Many species are valued as garden and landscape plants for subtropical and tropical gardens; a few, like ***Zamia furfuracea*** and *Zamia pumila,* thrive in warm-temperate zones. *Zamia* species are sculptural plants with handsome leaves and cones; several, such as *Zamia lacandona, Zamia purpurea,* and *Zamia splendens,* have young foliage that is attractively coloured on emergence.

Over half of the known species of *Zamia* experience a degree of threat, with land clearance and habitat loss being the most common causes. Large areas of rainforest have been cleared across Central and South America, leading to the decline of many species.

Inevitably, with a genus that is so attractive and adaptable to cultivation, overcollection for the horticultural trade has taken a huge toll. Thousands of plants of *Zamia pumila* have been removed from tropical dry forest, and although it is classed as Near Threatened on the 2000 Red List in most of its range, it has become locally extinct in Cuba.

Zamia furfuracea, which is Vulnerable in the wild, is, ironically, one of the most common cycads in cultivation. This popular cycad was once exported from Mexico at a rate of several tons per month, resulting in the rapid decline of wild populations. Fortunately, it grows well in cultivation and produces seed within a few years, so that cultivated specimens are now readily available.

Conservation authorities in Mexico and elsewhere have set up community nurseries to provide a legal, sustainable income for villagers. Buying plants from them benefits cycad conservation directly, as well as providing an incentive to preserve intact habitat. Seed-raised plants reach maturity within three years.

Due to its abundance in collections, *Zamia furfuracea* has also been an important plant for scientific study. Until as recently as 1986, all cycads were thought to be wind-pollinated. However, studies on cultivated *Zamia furfuracea* in Fairchild Tropical Garden in Miami, USA, showed that beetles were the main pollinators. In this case, the pollinators were cycad-specific beetles from Mexico that had been introduced accidentally together with the cycads. Since then, beetles have been shown to pollinate cycads in Mexico, Africa, and Australia.

VICTIMS AND SURVIVORS

Zamia lacandona has survived the destruction of rainforest in Mexico. Its underground tubers remain untouched during logging operations, and the plants are able to resprout. There are even reports of plants surviving in cornfields, where they seem able to adapt to the higher light levels. However, continuing land clearance remains a threat to wild populations and it is classed as Endangered in the 2000 Red List.

Zamia lacandona is propagated by seed, but cone production is irregular in cultivation. It is difficult to produce seedlings in quantity, so plants are scarce.

Zamia obliqua, from tropical rainforest in Colombia and Panama, is listed as being Near Threatened, but suffers the recurring threat of land clearance and habitat loss common to cycads across Asia and the Americas. Unlike cycads with tuberous subterranean stems, which can survive logging and ploughing, the tall, single-stemmed cycads die out quickly when their habitat is disturbed. For this reason, *Zamia obliqua* responds badly to land clearance.

Zamia vazquezii has the ideal profile for commercial plant collectors: small, easy to dig up, and in high demand. The fern-like leaves make this a very handsome cycad for indoor use. It is also an attractive border plant. As a result, wild populations have been devastated.

Overcollection has been the main threat to this Critically Endangered species. Forest clearance is adding to the pressure on the small number of plants that have survived.

Ironically, unlike many larger cycads, the species grows quickly from seed. In cultivation, given open, well-drained soils and light shade, *Zamia vazquezii* reaches maturity in 4–5 years and plants can produce large quantities of seed, so it should be possible to produce enough cultivated plants to satisfy demand.

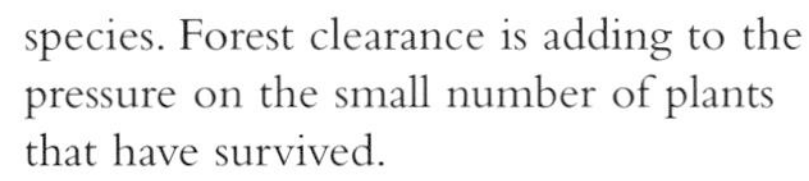

HISTORY

STAPLE STARCH

Most peoples who use cycads as a food source usually do so only in hard times. The Seminole peoples of Florida, however, used stem starch from *Zamia integrifolia* as a staple food. In this engraving of 1591, they are shown transporting crops for storage in a public granary. The starch was ground and leached before cooking to remove the toxins, and was known as coontie. White settlers referred to it as Florida arrowroot. In the 1800s, there was industrial extraction of Florida arrowroot, for use as laundry starch.

Several other species, including *Zamia chigua, Zamia pumila,* and *Zamia obliqua,* were used to make flour for tortillas. *Zamia lacandona* is a staple of the indigenous people of the tropical Lacandona rainforest in Mexico.

► ZAMIA FURFURACEA *in a nursery run by villagers near Cienaga del Sur, Venezuela. Plants propagated locally are used to re-establish colonies that have been plundered in the wild.*

OTHER THREATENED ZAMIAS

Zamia pumila A dwarf cycad that has the distinction of being the first species of *Zamia* to be described, in 1659. Large populations in the wild are chiefly threatened by forest clearance, but overcollection has also reduced numbers. Because *Zamia pumila* propagation in cultivation has been so successful, it is probably no longer economical to harvest from wild populations. Between 1995 and 2000, at least 40,000 artificially propagated plants were exported from places where the species does not occur naturally. Red List: Near Threatened.

Zamia pygmaea The Least cycad is from grassland, savanna, and scrub in Cuba but little is known about its ecology and conservation. Original descriptions did not say where plants were collected. Few plants have been found in western Cuba. Cuban botanists and staff from the botanic garden in Havana are gathering information on indigenous cycads. As this becomes available, it will be possible to determine this diminutive species' status and conservation needs. It is scarce in cultivation, but grows quickly and several individuals and organizations are propagating plants. Red List: Lower Risk (Data Deficient); CITES II.

Zamia variegata With yellow or cream flecks on the dark green leaflets, this endangered cycad has obvious appeal to collectors, but the main threat has been loss of habitat in Guatemala and Mexico over the past 30 years. There may be only 500 plants left in the wild, and while there may be undiscovered populations, forest clearing is taking place so rapidly that they may be destroyed before they are found. Red List: Endangered.

ORCHIDS AND BROMELIADS

Both orchids and bromeliads form their own botanical families – the Orchidaceae and the Bromeliaceae – but share exotic good looks and unusual growth habits. These are some of the most specialized plants on Earth, many occupying narrow niches and often relying on fascinating relationships with other plants, fungi, or animals for their growth, habitat, or reproduction. They have glamour, mystery, and romance in abundance, and are two of the most eagerly collected, and fiercely guarded, plant groups on the planet.

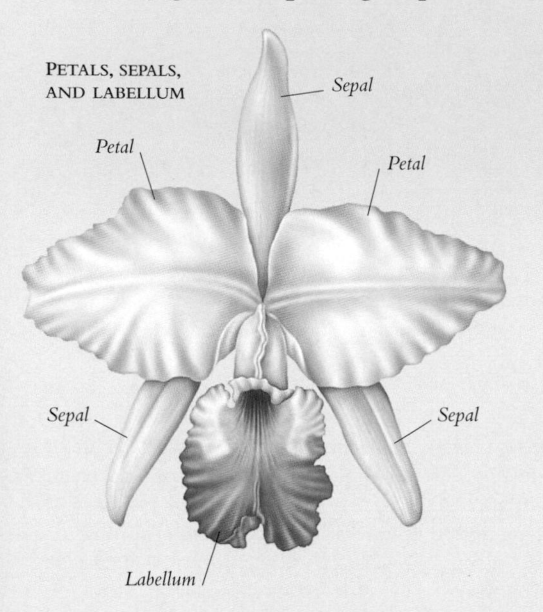

FLOWERS

With 20,000 to 30,000 species, the orchids are arguably the largest and the most diverse flowering plant family – roughly one in every ten known flowering plants is an orchid. Despite their diversity, the spectacular flowers show an unmistakeable family resemblance and all are bilaterally symmetrical. An orchid flower has three sepals alternating with three petals, the distinctive lower petal modified into a large lip or labellum that serves as a landing pad for pollinators.

Bromeliads are a smaller family, comprising around 2,700 species, but also have showy flowers. Many flower only once: from the centre of each rosette of lush foliage emerges a large flower stalk (scape), surrounded by brilliantly colourful petal-like leaves (bracts).

WHERE THEY GROW

Orchids belong to a far-flung family, growing in all terrestrial ecosystems that support flowering plant life, from Greenland to Tierra del Fuego at the tip of South America, but the greatest abundance and diversity of species are found in the tropics. Ecuador, Panama, Costa Rica, and Mexico are particularly rich in species, while Africa has the least diverse tropical orchid flora.

Bromeliads are less widespread – with the exception of one West African species, they grow only in the subtropical and tropical regions of the Americas, extending northwards from Chile and Argentina to the state of Virginia in the USA.

Both families can be found in very diverse habitats within their range, from cool mountain-tops to arid desert, but are at their most abundant in the global habitats that are most at risk – moist, tropical forests. As many thousands of acres of tropical forest are destroyed annually for timber or for grazing cattle, countless orchid and bromeliad populations are lost. Individual species of orchids and bromeliads often occupy a very specialized environmental niche, and this puts them at particular risk from habitat loss. An estimated three-quarters of tropical orchid species occur in three or fewer sites, particularly in cloud forests, and in these specialized environments, bromeliads also thrive but they, too, are at risk.

◄ COSTA RICA *The Monteverde Cloud Forest Reserve in Costa Rica is a centre of diversity and refuge for Bromeliads. This* Guzmania *is a typical "tank" bromeliad, storing water in its rosette of leaves and using its roots as an anchor.*

ORCHID FLOWERS *Blooming in every shade and countless combinations of colours, orchid flowers are exquisite. The cypripediums, or lady's slipper orchids, are highly prized by enthusiasts and have spectacular, soft, pouch-like lips (labellums), hence their common name.*

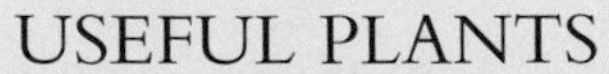

HOW THEY GROW

Some orchids and bromeliads grow in the ground (terrestrials) but most grow on trees (epiphytes) or on rocks (lithophytes). There are even underground orchids in Australia, which can only be seen when their flowers peep from cracks in the soil. Epiphytic orchids grow high up on trees in tropical forests where light is plentiful – very little light reaches the forest floor. They have fleshy aerial roots, which can absorb moisture and nutrients from the air. Some tree-dwelling orchids are leafless and use their green fleshy roots as both leaves and roots – a remarkable feat in the plant kingdom. Epiphytic bromeliads grow not only on trees and shrubs but even on cacti. Some "tank" bromeliads live on nutrients from the organic debris that lands in their rosette of leaves, or tank, and this rosette also stores their water supply. Tillandsias are called "air plants" because they are able to take moisture and nutrition directly from the atmosphere.

▲ **EPIPHYTES,** *such as this air plant, should not be confused with parasites – they do not harm their host. Growing on tree branches, they can enjoy more light than on the ground, and they are also out of the reach of grazing animals.*

▲ **LITHOPHYTIC,** *or saxicolous, orchids and bromeliads grow on or among rocks. There are few nutrients and little water, and exposed rock can get very hot, but specialists, such as this Australian rock lily orchid, have no competitors.*

▲ **TERRESTRIAL** *orchids and bromeliads grow in soil. All temperate orchids are terrestrial, living in grasslands, or marshy areas. Most terrestrial bromeliads, such as this* Guzmania, *grow in the shady forest understorey.*

USEFUL PLANTS

It is not surprising that the orchid family, which has such a wide distribution, has been the source of food and medicines for countless cultures around the globe. In commercial terms, the horticultural and cut-flower trade is the most valuable – dendrobiums grown in Hawaii are alone worth over $10 million per year. Today, the only orchid-based product that rivals plant production for the flower market is vanilla, which comes from the vanilla vine, a climbing orchid (*see p.423*).

Bromeliads also boast an important crop – the pineapple. The fruit was first described in 1483 by Columbus, who said it looked like a pine cone, hence its common name. Today over 10 million tonnes of pineapples are produced each year around the world. Pineapples also produce an important enzyme, bromelain, which is used to tenderize meat, as well as in the leather and pharmaceutical industries.

PINEAPPLE ► *The fruit of each flower on the bromeliad flower spike swells, eventually fusing into a single fruit around the stem – the core of the pineapple. Cultivated fruits are sterile – they have no seeds.*

CONSERVATION

The major threat to orchids and bromeliads is habitat loss (*see p.382*). Although some species can adapt to changing conditions (*see below*), many cannot, especially the large numbers living in small, niche environments that are particularly susceptible to habitat change.

In addition to their narrow distributions, orchids and bromeliads are also at risk because of their beauty – a number of species are threatened by overcollection from the wild. A few hours of activity by unscrupulous collectors can wipe out an entire population, or even an entire species. All orchids are listed in CITES Appendix II, and those particularly at risk, such as the plants of the *Paphiopedilum* genus, the tropical slipper orchids, are included in Appendix I. Certain bromeliads are also at risk from overcollection, especially the air plants, *Tillandsia*, seven of which are on CITES Appendix II, affording them some protection from unsustainable international trade. Today, propagation of rare and new species for trade is encouraged, especially in countries where the species are native, to reduce demand for plants collected from the wild. Few reserves have been established specifically for orchid and bromeliad preservation, but those that do exist may hold 50 to 60 per cent of the world's orchid species.

▲ **BROMELIAD DENIZEN** *The water-filled tanks are home to small animals, including this frog, which was found on a bromeliad in Monteverde Cloud Forest.*

◄ **LIFELINE** *A few bromeliads, such as these seen growing on telephone wires in Tobago, in the West Indies, cling tenaciously to life by adapting to surprising new niches as their tropical forest environment is destroyed.*

PROPAGATION

Orchids produce enormous quantities of seed and so, theoretically, all threatened species could be saved if they could be propagated artificially. For decades, however, orchids were often impossible to propagate by seed, since they depend on fungal partners for the nutrients needed for germination and growth. The breakthrough in orchid propagation came when American plant physiologist Lewis Knudson (1884–1958) produced a medium containing various nutrients that proved successful. The medium, Knudson C (KC), is now widely used by orchid growers and enthusiasts.

Propagating bromeliads from seed also presents a challenge. Seed loses its viability very quickly and often requires very specific conditions to germinate. Bromeliads can also be very slow to mature from seed – taking between two and fifteen years to flower. However, they do produce offsets. Today both orchids and bromeliads are artificially increased using micropropagation techniques.

▲ **MICROPROPAGATION** *Small pieces of plant tissue or seed, with suitable nutrients, can be used to propagate a plant in sterile conditions. This technique is invaluable for increasing numbers of endangered orchids and bromeliads.*

GARDENER'S GUIDE

Novice and expert growers alike can play an important role in the conservation of orchids and bromeliads. Some threatened species that have been obtained legally, and are being nurtured in private collections and propagated by networks of conservation-minded growers, provide an important stockpile of plants that could, in future, be used to restore the endangered plants to the wild.

- Make some room in your collection for different species, especially the less showy and charismatic ones.
- Do not collect plants from the wild and avoid purchasing jungle plants or other species that have been wild-collected. Buy orchids that have been certifiably propagated in nurseries.
- Promote genetic diversity by purchasing specimens that have been propagated by seed. Flasked seedlings of threatened species are often available.
- Propagate the species in your collection by seed, producing seed capsules via cross-pollination. Work with a group of enthusiasts to develop a propagation plan, sharing plants and pollen with other growers, keeping records, and cross-pollinating as many different plants as possible.
- If you grow rare orchid and bromeliad species for exhibition, guard against artificial selection by resisting the pressure to produce the unnatural flower shapes favoured by show judges.
- Volunteer to assist propagation and preservation work at botanic gardens and other conservation groups.
- Join an orchid or bromeliad society and help support or establish an active conservation committee.
- Support national and international efforts to preserve the native habitats of orchids and bromeliads. Many international organizations, for example the WWF, are working hard to save threatened environments, especially tropical forests, around the world.

TILLANDSIA DEIRIANA

10*/9 on Red List

Aechmea

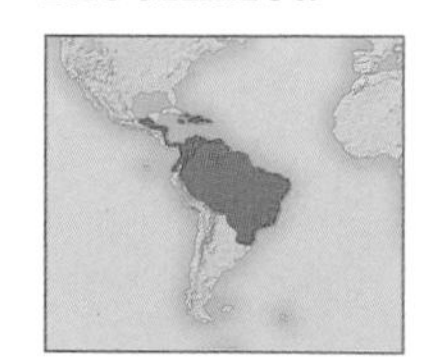

Botanical family
Bromeliaceae

Distribution
Southern Mexico, Central and South America, West Indies

Hardiness
Frost-tender

Preferred conditions
Bright, filtered light and low to moderate humidity; open, well-drained soil.

The genus *Aechmea* includes about 200 species of mostly epiphytic plants that naturally find footholds in the niches and crevices of rainforest trees. A few aechmeas live on rocks, but they are rarely found as terrestrials on the forest floor. Species occur in moist, or seasonally dry, subtropical, and tropical rainforest of several types – coastal, lowland, and montane.

These plants are perhaps most familiar to gardeners as urn plants, as in the silver vase plant, *Aechmea fasciata*, which is commonly grown as a house plant in cool temperate regions.

▲ AECHMEA ORLANDIANA *has long been valued in cultivation, not only for its red and yellow flowerheads, but for the variously maroon-spotted or banded leaves.*

Aechmeas are valued for their vivid flower spikes, often with colourful, long-lasting bracts, which may be followed by coloured berries. Many fruit without cross-pollination. In the wild, they are pollinated by hummingbirds, bees, or, in many Amazonian species, by ants.

Most *Aechmea* species have spiny, toothed leaves arranged around a central urn, which retains water. This "tank" is often home to amphibians and other creatures, such as the bromeliad crab, *Metopaulias depressus*, which raises its broods within its shelter.

Although some aechmeas are tough and drought-tolerant and survive in degraded environments, others have been threatened by habitat loss due to land clearance for agriculture, settlements, mining, and timber extraction, which is ongoing in the Amazonian forest. Several species, such as the Vulnerable *Aechmea biflora,* and the Endangered *Aechmea manzanesiana* (2000 Red List), both from Ecuador, have been intensively collected for the horticultural trade.

Botanists in Ecuador have been involved in surveying and recording the endemic flora as a vital first step in assessing the condition of often rare and little-known species. The *Libro Rojo de las Plantas del Ecuador* (Red Book of Endemic Plants of Ecuador) lists several of the *Aechmea* species described here.

Aechmea allenii, from Panama, is listed as Vulnerable (Red List of 1997). Most coastal rainforest in Panama has been cleared for agriculture, leaving a fragmented habitat. *Aechmea allenii* has also suffered from overcollection. This small, beautiful plant, seldom exceeding 70cm (28in) tall, produces red- or pink-bracted spikes of white or mauve flowers, followed by purple berries. It has been in cultivation for many years and its cultural needs are well known. It is propagated by seed, which is produced in abundance.

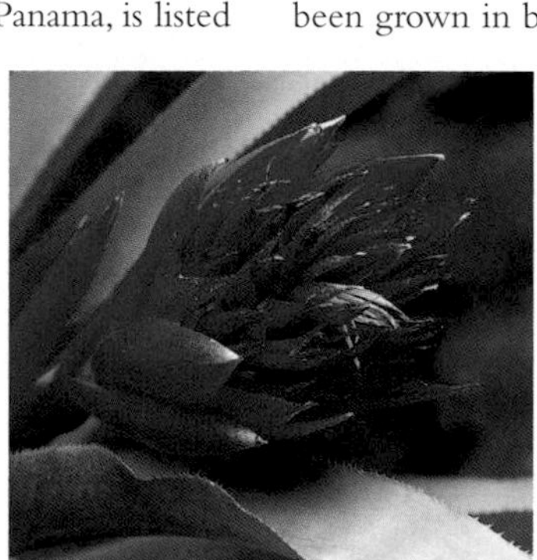

AECHMEA ALLENII

Aechmea dichlamydea var. *dichlamydea*, found in tropical rainforest in Tobago, is Indeterminate (Red List of 1997). Currently a focus of eco-tourism in Tobago, with populations protected there in a national park, this is one of the few blue-flowered bromeliads, and it attracts all types of pollinators. An additional concern for *Aechmea dichlamydea* var. *dichlamydea* is that its habitat could be damaged or destroyed by a hurricane.

Aechmea mariae-reginae, the Queen Mary bromeliad, is a rare species that dwells at the tops of tall trees in coastal rainforests on the Caribbean side of Costa Rica. It has been threatened by collectors – the plant is highly coveted for its extraordinary flowerhead, a white-woolly cone over 50cm (20in) in length, with bright pink tresses, up to 20cm (8in) long, hanging from the base. The flowers are pollinated by hummingbirds.

Aechmea mariae-reginae bears male and female flowers on different plants, and both are needed to produce seed. As in many species, female plants are less common, and many males are required to ensure pollination. This has exacerbated the threat from over-collection. The species has, however, been grown in botanic gardens for many years, and the easiest way to propagate it is by cuttings. Although this results in a limited gene pool, it can go some way towards fulfilling the desires of collectors.

Aechmea orlandiana var. *orlandiana*, from the tropical Atlantic forest of Brazil, is thought to be extinct in the wild. This beautiful species once occurred near Itapemirim, in the state of Espirito Santo, but has not been found there since 1940. The species and the many cultivars that derive from it, however, are quite common in cultivation, and easy to grow.

Aechmeas should be grown in a free-draining mix of peat and coarse bark, with bright filtered light and good air circulation. Propagation by seed requires indirect light, high humidity, and a sterile mix of equal parts peat and sand. Germination usually occurs within four weeks. After about six months, the crowded seedlings can be pricked out, and plants can be fully mature within 2–3 years depending upon the species. Aechmeas flower best if kept fairly dry, and watered only when the growing mix is dry to the touch.

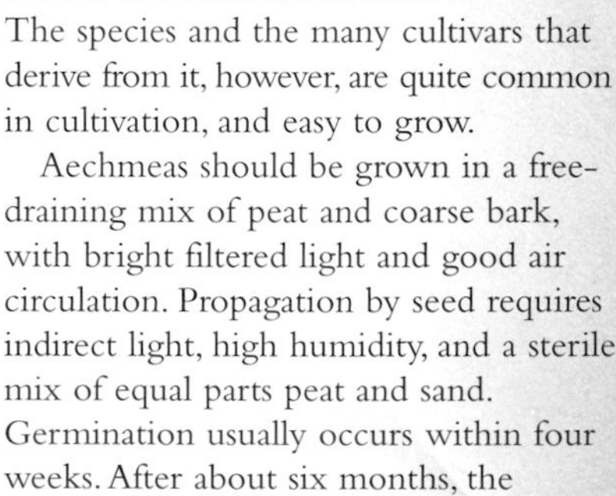

AECHMEA MARIAE-REGINAE ► *The local people of the Costa Rican Caribbean coast hold this exquisite flower in high esteem, and use the plants for ceremonial purposes, to decorate their churches.*

HABITAT

A BUTTERFLY'S WING BEAT

Pressures of overpopulation have led to the expansion of Lita, Ecuador, in the heart of the Chocó-Darién rainforest, over the last 15 years. Since then, roads, urban areas, and pasture have replaced the rainforest. The degraded habitat reflects more solar energy than virgin forest, and consequently the local weather pattern now includes a dry season. The destruction of the Brazilian Amazon rainforest still escalates as well. Rainforest plays a part in climate control, and computer modelling suggests that its loss not only will cause significant reduction in local rainfall, but also that the effects will extend to the midwestern USA, causing a drop in summer rainfall, when it is needed most for agriculture.

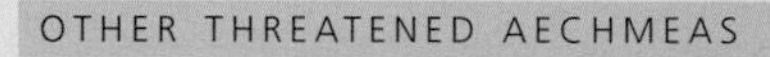

OTHER THREATENED AECHMEAS

Aechmea abbreviata This species is listed in the Red Book of Endemic Plants of Ecuador. It is found in lowland rainforests of the Amazon Basin that are rapidly being replaced by pasture. Because of its rarity, this plant is only available from botanical gardens and a few reputable nurseries. Unlisted.

Aechmea aripensis Under threat because of its limited distribution, it is found only in cloud forest in Trinidad and northern Venezuela. It is in cultivation in specialist nurseries, valued for the red bracts and blue flowers that give an impressive display in tropical gardens. Unlisted.

Aechmea kentii Once abundant, these compact, easily grown plants were over-collected from the wild, and their forest habitats have been reduced to fragmented remnants. It is now in the Red Book of Endemic Plants of Ecuador. It is popular in cultivation and is propagated in specialist nurseries. Red List (2000): Endangered.

Aechmea pittieri Threatened by loss and fragmentation of its habitat in the degraded coastal forests of Costa Rica, this plant survives marginally due mainly to its tolerance of drought and strong sun. The stiff and well-armoured leaves, which form an urn, have earned this plant the nickname of *el tiburon* or "shark's teeth". Easily grown and widely available. Unlisted.

Aechmea warasii Endemic to tropical Atlantic rainforest of Espirito Santo, Brazil, this species is known only from a small locality and is highly threatened with extinction. Flowers open violet and turn a deep red the following day, and are pollinated by hummingbirds. The species has been in cultivation for over 30 years, and is available from established nurseries. A highly adaptable and easy to grow, shade-loving species. Unlisted.

Red List: Endangered*/CITES II

Aeranthes arachnites

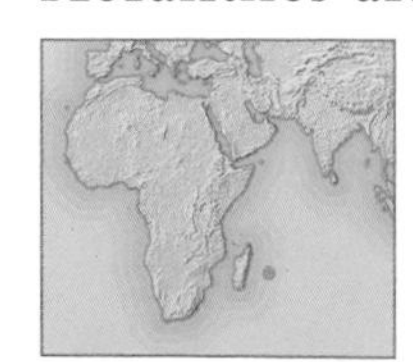

Botanical family
Orchidaceae

Distribution
Mascarene Islands

Hardiness
Frost-tender

Preferred conditions
Epiphytic on trees in humid semi-shade, with buoyant air movement.

This rare orchid, ***Aeranthes arachnites,*** is classified as Endangered in the Red List of 1997. It is one of a genus of some 40 species that are native to the islands off the eastern coast of Africa. It occurs as an epiphyte on trees in the shade of the rainforests of the Mascarene Islands, where it is threatened by habitat loss. This is partly due to clearance for agriculture, but the main pressure is the use of timber in construction, and for fuelwood and charcoal.

Although reforestation must be the prime aim, reducing fuel consumption will help conserve remaining forest. To this end, the WWF and the Association to Save the Environment designed a more fuel-efficient cooking stove that can reduce wood or charcoal consumption by 30–50 per cent. This not only reduces pressure on the forests, but also releases more of impoverished families' income to provide food and clothing for their children.

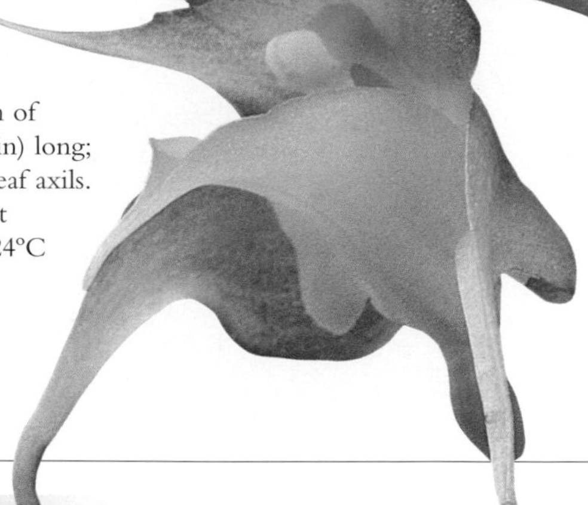

► AERANTHES ARACHNITES *is an elegant orchid, valued in cultivation for the many delicate-stemmed spikes of translucent flowers that open one by one over a period of several months between spring and autumn.*

Aeranthes arachnites produces a fan of leathery leaves, each to 30cm (12in) long; the flower spikes arise from the leaf axils. Plants need filtered light, constant humidity, and a minimum of 19–24°C (57–75°F). They must be fed and watered regularly during growth and flowering, and kept almost dry when at rest. Grow in a mix of coarse bark and coir.

Red List: Endangered*/CITES II

Barkeria lindleyana

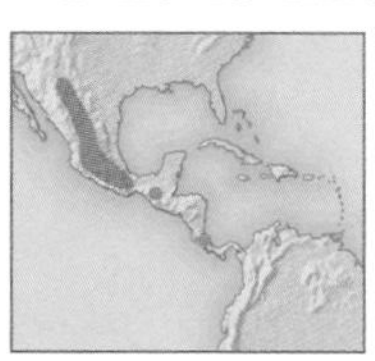

Botanical family
Orchidaceae

Distribution
Mexico, Costa Rica

Hardiness
Frost-tender

Preferred conditions
Bright light, fairly high humidity, ample moisture.

Eight of fifteen *Barkeria* species, rainforest-dwelling epiphytic orchids, appear on the Red List of 1997. This beautiful species, ***Barkeria lindleyana***, occurs from Mexico to Costa Rica in montane tropical forests that experience a winter dry season. In San José, Costa Rica, it often colonizes spontaneously in gardens. The species is named for John Lindley (*see p.369*), and the genus for George Barker, a 19th-century Birmingham (UK) orchid collector. It is listed as Endangered on the Red List of 1997, threatened by habitat loss and over-collection for horticulture.

Barkeria lindleyana has been coveted in cultivation for its handsome, rosy-pink flowers, which appear in autumn, arising from the end of the pseudobulb. Each arching stem, up to 60cm (2ft) long, may carry up to 20 blooms. The long, narrow leaves are spaced along thin, reed-like, clustered pseudobulbs; the species is leafless in winter, a response to the dry season in the wild.

CULTIVATION

Grow *Barkeria lindleyana* in a bright, sunny position, with fairly high humidity and good air circulation. It can be grown epiphytically, or in pots of free-draining compost if care is taken that the roots do not remain wet for extended periods. Feed regularly during growth; plants need a dry rest in winter when watering should be restricted to once a week.

◄ BARKERIA LINDLEYANA *was introduced to Britain from Costa Rica in 1842. The individual flowers can be up to 8cm (3in) across, and last for several months. In the wild, blooms vary in colour from white through palest lilac and rosy pink to deep purple.*

The lip of the bloom has blotches of deep pink and white

Sepals and petals are a clear rose-pink

CONSERVATION

COSTA RICA RAINFOREST

The small Central American country of Costa Rica is famously foresighted in its conservation policies, and has won many environmental awards for the protection of its rainforests.

About a quarter of its land area receives some form of statutory protection. The largest expanse of rainforest on the Caribbean rim lies along the San Juan River – home to a huge diversity of flora and fauna, including sloths and jaguars. This, however, is a small remnant of lowland tropical forest that once covered most of central America. Much of the forest outside Costa Rica's reserves and national parks has been cleared for ranching and agriculture. Even with a ban on logging some 60 native trees, illegal clearance still goes on, and despite the government's best efforts, less than 20 per cent of the original forest remains.

Unlisted

Billbergia

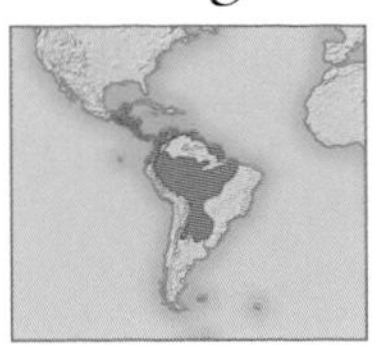

Botanical family
Bromeliaceae

Distribution
Southern Mexico to Bolivia, Brazil, and northern Argentina

Hardiness
Half-hardy – frost-tender

Preferred conditions
Open, well-drained soil; bright, indirect light or partial shade.

Found in tropical deciduous forest and rainforest, from sea-level to 1,700m (5,500ft), billbergias do not appear on the Red List, although several are threatened by habitat loss. The species described here, natives of the Atlantic rainforest of Brazil, are put at risk by habitat destruction. Most species are epiphytic or rock-dwelling perennials; a few are terrestrial. They are pollinated by hummingbirds or bats.

Several, such as ***Billbergia pyramidalis*** and ***Billbergia nutans***, are common in cultivation; they increase readily and are easy to care for. Higher-altitude species, such as *Billbergia sanderiana*, tolerate temperatures at, or just below, freezing. Others, especially Ecuadorian species, are frost-tender. Billbergias are tolerant of poor light and neglect, but the flowers, with large colourful bracts, are always short-lived. This has affected their popularity, although many species have striking, banded or coloured foliage.

Billbergia horrida is a true epiphyte. Rainforest destruction in the state of Espirito Santo, Brazil, has brought it to the brink of extinction. It produces an upright spike of green, blue-tipped flowers, which open in the evening. The flowers are fragrant – an unusual trait in billbergias. Studies suggest that bats are the primary pollinators.

Billbergia horrida var. *tigrina* is widely available in the nursery trade. It has stiffly upright, dark-banded silvery leaves edged with strong, black spines.

◄ BILLBERGIA PYRAMIDALIS *is often grown as a houseplant in cool climates; the leaves are sometimes flushed purple, and in many individuals, the orange-pink flowers are tipped with kingfisher-blue at maturity.*

► BILLBERGIA NUTANS *has been called the friendship plant on account of its ease of propagation; rosettes twisted from the parent plant root readily and were often given away as tokens of friendship.*

Billbergia leptopoda from Minas Gerais and Espirito Santo, Brazil, grows either as an epiphyte or a terrestrial. This small species is valued in cultivation for its dark green, yellow-spotted foliage. Strong light increases the decorative yellow spotting. The leaves appear to have been shaped with a curling iron, hence its common name, permanent wave plant. The erect flowerhead barely tops the leaf rosette. It has bright red floral bracts and yellow petals with blue tips.

Billbergia sanderiana, an epiphyte from south-eastern Brazil, was once widely distributed, but is now confined to the few national parks. It is a small plant with green leaves edged with coarse black spines; the pendent flowerhead has striking pink petal-like leaves (bracts) that enclose jade green flowers with pale blue tips. Sadly the flowerhead falls quickly, especially in warm conditions.

The species described above are ideal as house plants. They need well-drained soil, kept on the dry side; plants rot if kept too wet. They prefer indirect light, and thrive on neglect. Most species produce offsets after flowering, and form colonies that are attractive even when the plants are not in flower. The offsets can be used for propagation.

Red List: Rare*/CITES II

Broughtonia sanguinea

Botanical family
Orchidaceae

Distribution
West Indies

Hardiness
Frost-tender

Preferred conditions
High light, with shade from the hottest sun; high humidity; warm, moist conditions.

Endemic to the islands of the West Indies, the members of this genus of five species have become threatened both by loss of habitat to agriculture, and overcollection for the horticultural trade. *Broughtonia negrilensis* and ***Broughtonia sanguinea***, both epiphytes, grow in low-altitude rainforest in Jamaica, and are listed as Vulnerable in the Red List of 1997.

Broughtonia sanguinea bears attractive and highly coloured flowers and is commonly grown in orchid collections in Florida, USA. This species has been highly sought after by plant breeders due to the brightness of the flower colour, its compact habit, and its ease of cultivation. It has been extensively used in hybridization, especially with *Cattleya* species, to produce plants known as cattleytonias, with intensely coloured blooms. This means that plants should be readily available from specialist orchid nurseries as stock that has been artificially propagated in cultivation.

Broughtonia sanguinea produces a compact cluster of egg-shaped pseudobulbs, with two narrow, leathery, dark green leaves. The slender, semi-pendulous flowerheads, to about 30cm (12in) in length, arise from the tip of the pseudobulb between autumn and spring.

Growing Broughtonia

Broughtonia sanguinea has proved vigorous in cultivation. It requires warm conditions at a minimum of about 20°C (68°F) with high light levels and high humidity at all times. It is best grown on a raft, or cork slab, or in pots of free-draining material such as coarse bark. Plants should be watered moderately but frequently, and given fertilizer regularly during the growing and flowering season. As the pseudobulbs mature, water should be withheld gradually and the plants dried off for their dormant period.

◄ BROUGHTONIA SANGUINEA *Noted for its neat, compact habit and the distinctive brilliance of the flowers – the Latin name* sanguinea *means "blood red" – this species was first introduced from Jamaica to the UK in 1793.*

74*/4 on Red List/CITES II

Bulbophyllum

Botanical family
Orchidaceae

Distribution
Pantropical, especially south-east Asia, New Guinea, Australasia

Hardiness
Frost-tender

Preferred conditions
Epiphytic; dappled sun or bright, filtered light; high humidity and plentiful moisture.

This, the largest genus in the orchid family, is found throughout the world's tropical zones. There are over 3,000 known species of *Bulbophyllum* found in a diversity of tropical and subtropical habitats, from dry forest to high-altitude rainforest and cloud forest.

Some 76 species appear on the Red Lists of 1997 and 2000, threatened by forest clearance for plantation crops, as in *Bulbophyllum filiforme*, from Cameroon, (Critically Endangered, 2000 Red List), or for human settlement, smallholder farming, and timber, as in *Bulbophyllum modicum* from Cameroon and Equatorial Guinea (Endangered on the 2000 Red List).

Two of the world's smallest orchids belong in this genus: *Bulbophyllum minutissimum*, with pseudobulbs the size of a pinhead, and *Bulbophyllum pygmaeum*, which is best viewed with a hand lens. Others, like *Bulbophyllum lobbii*, have large and spectacular blooms, while ***Bulbophyllum beccarii*** has the largest leaves of any known Asian orchid; they are cupped to capture leaf litter falling from the canopy.

This epiphytic genus includes some of the most bizarre orchids, tipped with mobile tufts, spider-like filaments, and sinister, flesh-like appendages; sometimes the whole flower is designed to quiver in the breeze to attract pollinating insects, which are mostly flies.

Formidable fragrances

Among the most notable attributes of this genus are its floral fragrances, designed to attract pollinators. They have been described variously as smelling of urine, blood, dung, or rotting meat, with *Bulbophyllum beccarii*'s scent compared to the rotting flesh of dead elephants.

Bulbophyllum beccarii is Unlisted, but is rare and threatened by habitat loss to farming and logging. Found in lowland rainforests, from sea level to 600m (2,000ft) in Brunei, Kalimantan, Sabah, and Sarawak, it is usually associated with dipterocarp trees. These resinous trees, of the family Dipterocarpaceae, form a dominant component of tropical forests. In south-eastern Asia, they are the main timber-yielding family of trees.

Bulbophyllum beccarii has been highly sought after by collectors as far back as 1880, which may also have contributed to its rapid decline in the wild. The vile-smelling flowers appear from spring to summer.

Bulbophyllum rothschildianum, from north-eastern India, has also been endangered by overcollection, but is now readily available as cultivated stock. Although in flower for only a short period, this spectacular species, with large and interesting flowers, would make an outstanding addition to a collection.

Bulbophyllum species are best grown on fibrous slabs, or in a highly free-draining epiphyte compost. They need bright light and plentiful water when in growth, with regular light feeding. Gradually withdraw water to allow a rest period during summer, but do not allow to dry out completely; water sparingly in winter. These species need a temperature of about 15–20°C (60–68°F).

HISTORY

Odoardo Beccari

Bulbophyllum beccarii (below) was named by the Florentine botanist Odoardo Beccari (1843–1920). It is a remarkable species, with some of the largest leaves of any Asian orchid, but is now rare and restricted due to over-collection, farming, and logging in the lowland rainforests of Malaysia. Beccari discovered the species, along with the amazing *Amorphophallus titanum* (*see p.166*), during his travels in south-east Asia, described in *Wanderings in the great forests of Borneo*. This book remains an important and vivid account of the diversity of this ecosystem before it was disturbed by modern development.

BULBOPHYLLUM ROTHSCHILDIANUM ► *bears large, foul-smelling flowers in an inflorescence up to 30cm (12in) long; it blooms from autumn to spring.*

▲ BULBOPHYLLUM MACRANTHUM *This species, from the tropical forests of Burma, is much sought after for its fleshy-textured flowers that emerge a rich burgundy then fade over time to dull violet with deep violet speckles.*

62 on Red List*/CITES II

Caladenia

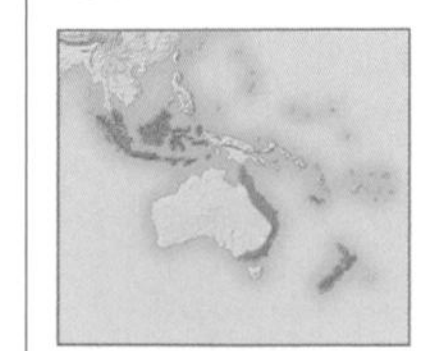

Botanical family
Orchidaceae

Distribution
Australia, south-east Asia, south-west Pacific

Hardiness
Frost-tender

Preferred conditions
Sandy, leafy, acid soil; part shade, or filtered light; constant moisture when in growth, and a dry period after flowering.

There are over 150 species of spider orchid, occurring mostly in Australia. They have long, tentacle-like, drooping petals, giving a spider-like appearance. *Caladenia* species occur in temperate Australia, from Western Australia to Victoria and New South Wales to Queensland, through to south-eastern Asia and the south-west Pacific.

They are found in a range of habitats, from shrubland to woodland and dense forest; one species is semi-aquatic. These are terrestrial orchids, producing growth from a tiny tuber in the cooler, damper autumn or winter, before flowering in spring. Some flower before the appearance of the single leaf.

Caladenia includes some of the rarest of Australian orchids, with several species at risk because of their very restricted range; a few, such as *Caladenia pumila*, are Extinct on the Red List of 1997. Others are Endangered or Vulnerable due to human pressures.

The coastal areas of the far south-west of Australia have become prime tourist destinations, and land has been cleared for holiday homes. Settlement brings

22 on Red List*/CITES II

Calanthe

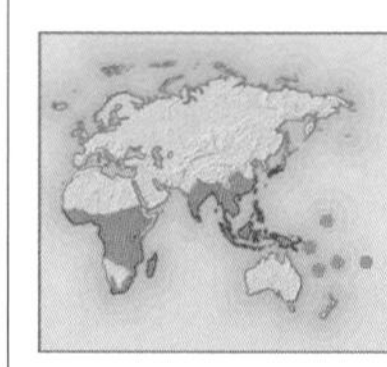

Botanical family
Orchidaceae

Distribution
Africa, Madagascar, Asia, Australia, and Polynesia

Hardiness
Frost-tender

Preferred conditions
Rich, open, and freely draining soils with plentiful organic matter; semi-shade and moderate humidity in warm conditions.

This is a widespread genus of about 150 species of deciduous and evergreen terrestrial orchids, with some 22 species appearing on the Red List of 1997, experiencing varying degrees of threat. One species, *Calanthe whiteana*, from Sikkim, is Extinct. *Calanthe* species are found in tropical and temperate regions in an arc that transcribes the globe from Africa to the south Pacific, north to Japan and south to Australia, with a centre of distribution in south-

with it competing weed species and land disturbance by feral animals; there are also problems with grazing by rabbits.

Many of these orchids bear graceful flowers, and are coveted by collectors. *Caladenia excelsa*, for example, has flowers to 30cm (12in) across, the slender filaments being a further 40cm (16in) long. It is Vulnerable (Red List of 1997) in the wild, and this (and other species) should be bought only with a guarantee that plants are from cultivated stock. *Caladenia caesarea*, the dwarf spider orchid, has mustard-yellow flowers, with a lip striped dark brown. This orchid is Vulnerable on the Red List of 1997, because its home in the granitic, coastal heathlands of south-western Australia is pressured by land clearance for tourism.

Caladenia huegelii, the grand spider orchid, is a spectacular plant with large, red-streaked green flowers, and a fringed lip. It is Vulnerable (Red List of 1997) due to loss of its woodland habitat to housing and farming, and is restricted to a few hundred plants in the wild.

Caladenia rosella grows in woodland, on dry slopes in gravelly soils, often in open areas. Habitat loss has resulted in local extinction in New South Wales, and it is Endangered in Victoria (Red List of 1997). Grown in botanic gardens, it is included in an Approved Recovery Plan (2001) by the NSW Parks and Wildlife Service.

Caladenias can be grown in a mix of leafmould and coarse sand. They are best re-potted annually when the plant dies back, including some of the old soil to ensure the transfer of its obligate fungal partner. Seeds sprinkled at the base of the plant at leaf emergence germinate within six weeks and can be transplanted once dormant. Seeds will also germinate on a nutrient-agar medium. Seed-grown plants take about three years to flower.

▲ CALADENIA FLAVA, *the cowslip orchid of Western Australia, is not threatened in the wild, where it occurs on sandy soils. The yellow, crimson-spotted flowers give rise to the name* flava, *which means pure yellow.*

eastern Asia. The plants grow in tropical to subtropical rainforest, shrubland, and grassland, often in deep accumulations of organic matter in shaded sites.

Calanthe species are characterized by their attractively pleated leaves, making them decorative foliage plants when not in flower. They are cultivated mainly, however, for the grace and beauty of their flowers and were among the most popular of winter-flowering orchids in Victorian times. Most are easy to grow and flower freely. Some have found favour with collectors and plant breeders, and a number of hybrids are available for the home orchid enthusiast.

Calanthe ceciliae is an evergreen species found on the forest floor in the hills of Sumatra, Java, and Malaysia, where it is now rare on account of local over-collection for the horticultural trade. On a flower stalk some 45cm (18in) tall, it produces several orange-suffused, violet flowers with a white lip and violet spur.

Calanthe vestita is locally rare and threatened by habitat loss to farming and urbanization, and also by overcollection in parts of Burma, Thailand, Sulawesi, and Java. It grows in the understorey of lower montane forest, up to altitudes of 700m (2,300ft). This deciduous terrestrial is very occasionally seen growing epiphytically. The flowerhead, to 60cm (2ft) tall, bears up to 12 long-lasting flowers.

Calanthe species require warm, semi-shaded conditions, constant moisture, and moderate humidity during the growing season. They are grown in free-draining mixes such as leafmould, granulated composted bark, and charcoal. Good flowering depends on regular monthly feeding when in leafy growth, so that the pseudobulbs are as large as possible before a dry resting period. Commercially available hybrids are especially rewarding and easy to grow.

► CALANTHE VESTITA *has been important in the history of orchid growing, as the parent of one of the first orchid hybrids,* Calanthe *Veitchii, produced in the 1850s.*

Unlisted

Canistropsis

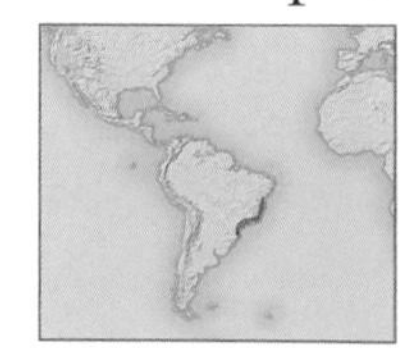

Botanical family
Bromeliaceae

Distribution
Brazil

Hardiness
Half-hardy – frost-tender

Preferred conditions
Well-drained soil in partial shade; withstands drought but not direct sunlight; tolerates temperatures at or just above freezing.

This genus was separated from *Nidularium* and *Canistrum* in 1997, by Elton Leme, a modern-day Brazilian taxonomist. Small- to medium-sized, epiphytic or terrestrial bromeliads from tropical rainforest, they have suffered mainly from destruction of their favoured habitats.

Canistropsis species are valued in cultivation for the colourful bracts that form a cup around a flowerhead of white or greenish-white flowers, which produce juicy berries with many purplish-brown seeds. Their compact size and tolerance of drought and dry atmospheres, along with their preference for shaded conditions make them particularly good house plants, although they prove remarkably hardy for plants of tropical origin.

Canistropsis billbergioides, an inhabitant of the lower levels of the forest canopy, was once extremely common in the Atlantic rainforest of south-eastern Brazil. It is now found only within the protected confines of the several national parks in its original habitat.

This is a variable species, and leaf colour ranges from green to lavender. The flowerheads can be yellow, orange, or purplish-red and, in cultivated plants, often have glossy bracts. Bracts in wild plants are often clothed in grey scales.

The most ornamental individuals of this and other wild species have, over the years, been selected for commercial cultivation, and are increased by seed or offsets. European nurseries produce thousands of young plants from seed annually for sale in garden centres and supermarkets. These cultivated plants are, unfortunately, highly unlikely to be a suitable basis for any recovery plan.

Canistropsis elata may be extinct in the wild. Today, the town of Mambucaba, Rio de Janeiro, Brazil, completely covers the locality in the Atlantic rainforest where this species once grew. Its native forest is now flooded, and it is unlikely that this species (formerly *Nidularium microps*) has survived. It is, however, self-fertile, and produces seed which has been used to increase stocks in cultivation.

Canistropsis elata has adapted readily to cultivation as a houseplant. The inflorescence is held high above the central rosette of purple leaves (*elata* means tall). The floral bracts are wine-red, and enclose bluish-white flowers that are slightly fragrant. It increases naturally by means of stolons. Other *Canistropsis* species can be increased from seed or offsets; they enjoy similar conditions to *Canistrum* (*below*).

CANISTROPSIS BILLBERGIOIDES 'PERSIMMON' ► *A selection of the species produced in great numbers in commercial cultivation, this cultivar is notable for the glossy golden bracts that surround the tiny, greenish-white flowers.*

Unlisted

Canistrum

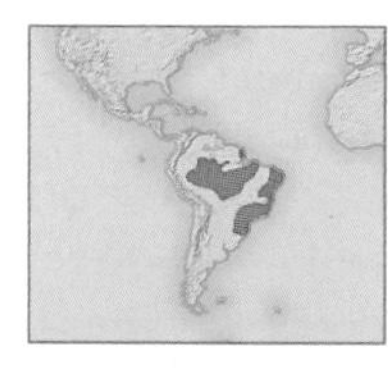

Botanical family
Bromeliaceae

Distribution
Brazil

Hardiness
Half-hardy – frost-tender

Preferred conditions
Well-drained soil in partial shade; withstands drought but not direct sunlight; tolerates temperatures at or just above freezing.

Like related *Canistropis* species, members of this genus are Unlisted, but experience severe threats due to the combination of a restricted distribution and loss of habitat. Their plight is made worse by the fact that they do not occur at all in areas of forest that are protected. *Canistrum camacaense* is from the wet, lowland tropical rainforest in southern Bahia, Brazil, where it grows as an epiphyte or terrestrial in the diffuse light of the forest understorey. Its future is uncertain, as it is not found in the legally protected forest in this region near Una, Brazil, and its own habitat is being rapidly destroyed.

It has been in cultivation for many years prior to the escalating loss of its habitat, and is usually available from specialist mail-order nurseries. It forms a rosette over 1m (3ft) across, with long creeping stolons emerging from it, which can be detached for propagation. The floral bracts are iron-red, in contrast to the showy yellow flowers.

Canistrum fosterianum grows in the rainforest near the city of Salvador, Bahia, Brazil, but again not in protected areas, and so is threatened with extinction. It grows epiphytically in small trees, or as a terrestrial in sandy soils in coastal tropical rainforests. This species forms an attractive, tubular rosette of dark green leaves marked with rust-coloured cross-banding. From the centre grows a tulip-shaped rosette of red-pink floral bracts, with tiny white flowers.

▲ CANISTRUM AURANTIACUM *was first introduced to Europe in 1873; named from the Latin,* canistrum, *meaning basket, the inflorescence was thought to resemble a basket of flowers.*

Canistrum fosterianum has been in cultivation for 50 years and has parented many hybrid offspring, which are difficult to distinguish from the original type species collected by Mulford Foster.

Canistrum montanum occurs in the Atlantic rainforest in a mountainous region of Bahia, Brazil, where it enjoys the constant cool humidity of cloud and mist. A beautiful species of the forest canopy, it is now found mainly in the few trees that have been left to provide shade for cacao plantations. It is not found in protected areas, and is threatened with extinction.

Canistrum seidelianum also grows in the cacao-growing uplands of Bahia, Brazil. With the species' native forests all but destroyed, the offspring of the original collections, made over 20 years ago, may be the last of the type left. The leaves are banded with dark purple, and form a tight rosette enclosing pale pink petal-like leaves (bracts) and yellow flowers. There are only two clones of this rare plant known in cultivation.

Canistrum species need good drainage, good ventilation, and filtered light, and can be grown epiphytically on bark slabs, or in pots containing a mix of equal parts granulated bark, leafmould, and coarse grit. Propagation is by seed, cuttings, or separation of stolons.

11 on Red List*/CITES I and II

Cattleya

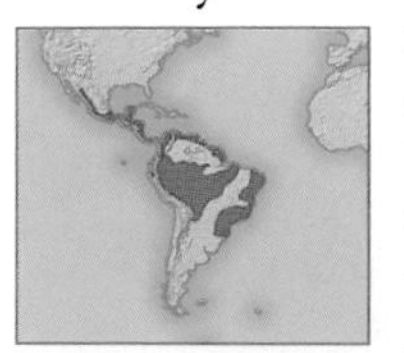

Botanical family
Orchidaceae

Distribution
Central and South America

Hardiness
Frost-tender

Preferred conditions
Warm temperatures and plenty of light; moderate humidity; free-draining soil.

Few orchids have so captured the imagination of gardeners and horticulturists – nor is that surprising. Cattleyas possess shapely and brilliantly coloured, sumptuously textured flowers, with billowing petals in a dazzling palette, from magenta to shades of lilac, rose, pink, and ruby red; from pastel lemon to the most intense gold, orange, and fiery vermilion; and from green to pure white. Their glistening lips, often extravagantly large and heavily fringed, are suffused and veined with vivid hues. Their sensual perfume can fill a room.

A SURPRISE PACKAGE

The chance discovery of cattleyas is one of the strangest sagas in orchid history. The first plants came to England as packing material around ferns and mosses collected by naturalist William Swainson in the uncharted tropical jungles of northern Brazil. He had no idea what these packing plants were, and neither did those who received them in England, including ardent plant collector William Cattley, of Barnet, London. Cattley potted up some of the strange-looking specimens out of curiosity.

In 1818, they produced spectacular purple flowers with a darker lip, and the orchid world was never to be the same again. John Lindley (*see p.369*) recognized them as something new and beautiful – indeed, a new orchid genus. He described it in botanical journals and immortalized Cattley by naming the new plant *Cattleya labiata* in his honour.

Unfortunately the horticultural world was to lose *Cattleya labiata* for the next 71 years, despite the efforts of major orchid-growing companies in England, France, and Belgium, who sent collectors to Brazil from the 1830s through the 1880s to search for it. Swainson had omitted to tell anyone where he collected the originals, and had since disappeared into the wilds of New Zealand. Speculation that he collected the species near Rio de Janeiro sent plant

WILD CATTLEYAS ► *Seen here in the Venezuelan rainforest, they grow epiphytically, cascading from their host tree. When seen in the natural state, it is not difficult to understand the cattleyas' ability to incite collectors to possess them.*

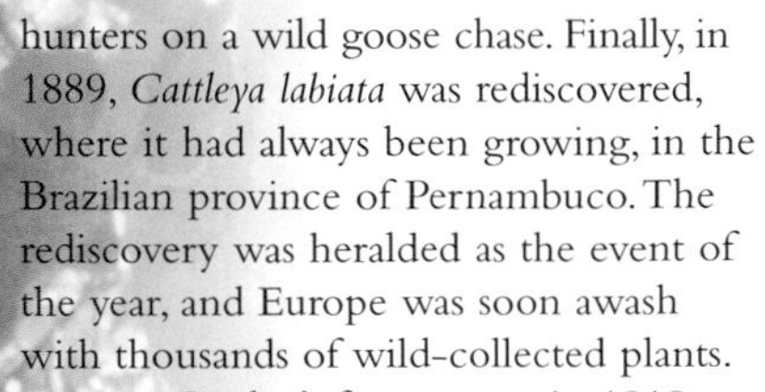

hunters on a wild goose chase. Finally, in 1889, *Cattleya labiata* was rediscovered, where it had always been growing, in the Brazilian province of Pernambuco. The rediscovery was heralded as the event of the year, and Europe was soon awash with thousands of wild-collected plants.

Since Cattley's first success in 1818, collectors have scoured South American jungles in search of new cattleyas, with one record of an order for 5,000 plants of one species alone. Their astonishing beauty, ease of culture, and diversity of forms stoked demand from 19th-century European orchid collectors to the point where large areas of rainforest were denuded, and many once common and widespread species even now survive in critically low numbers.

THREATENED CATTLEYAS

The genus comprises 70 species of epiphytes, or occasionally rock-dwelling plants, from Central and South America. They occur in coastal lowland forest to mountain forest at altitudes of 2,000m (7,000ft). Most are found growing in sunny positions on upper branches of rainforest trees. Habitat loss continues, and over-collection remains a serious threat.

Cattleya bowringiana is listed as Rare on the Red List of 1997, due to the demands of the horticultural trade. The species is now restricted to Central America, from Honduras to Guatemala, where it grows on tree branches near fast-flowing streams. It gained collectors' attention because of the remarkable number of flowers, up to 20, produced at any one time.

Cattleya porphyroglossa is also blighted by overcollection, as well as habitat loss for farming in its natural range in Brazil,

▲ CATTLEYA BOWRINGIANA *is extravagantly lovely, producing crystalline-textured, rosy-purple flowers with garnet-flushed lips, in spikes of up to 20, opening from autumn to winter.*

near Rio de Janeiro and Espirito Santo. Although Rare on the Red List of 1997, it is easy for collectors to find, growing on low trees in coastal swamps.

Cattleya trianae is the spectacular Christmas cattleya, its winter flowering giving it its common name in the northern hemisphere. Highly sought after by 19th-century plant collectors, it has long been depleted over much of its natural range. Rare in the wild, this rainforest epiphyte from Colombia is now listed in CITES Appendix I. Fortunately, *Cattleya trianae*, and the many fine hybrids derived from it, are now commercially propagated in abundance.

CATTLEYA TRIANAE

Cattleyas need good light, excellent ventilation, and a minimum temperature of 15–19°C (57–66°F). Provide a free-draining potting mix of composted bark or similar material, with plenty of water and regular fertilizer during the growing season. *Cattleya bowringiana* tolerates lower temperatures if dry.

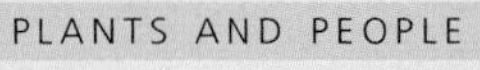

PLANTS AND PEOPLE

THE CORSAGE ORCHID

For centuries orchids, and cattleyas in particular, have symbolized passion. They are often known as the corsage orchids, after the small arrangements that are pinned on a woman's dress, or placed on her wrist, as a token of admiration. Author John Updike left no doubt about the erotic association of orchid corsages at American high school proms. As he looked back on the formal dances of his youth in the late 1940s, he remembered the corsages he pinned on the gowns of smooth-shouldered girls. "What heats, what shadowy, intense glandular incubations came packaged with the five-dollar purchase of a baby orchid in a transparent plastic box", he wrote.

Red List: Rare*/CITES II

Chiloglottis longiclavata

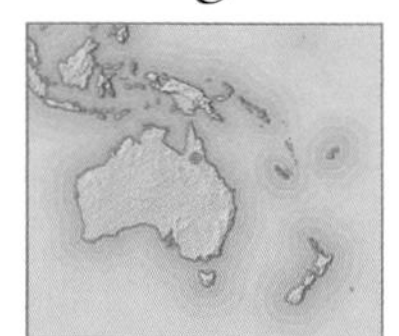

Botanical family
Orchidaceae

Distribution
Australia: Queensland

Hardiness
Half-hardy – frost-tender

Preferred conditions
Well-drained, gravelly soil; a well-lit site with shade from hot sun and good air movement; plentiful moisture while in growth followed by a dry rest period.

Restricted to an area from the Atherton Tableland, near Cairns, to the Herberton Ranges in Queensland, Australia, this tuberous, terrestrial orchid, *Chiloglottis longiclavata*, is listed as Rare in the Red List of 1997. It occurs at elevations of 400–700m (1,300–2,300ft) in gravelly loams, forming colonies in woodland, and in gullies in open forest.

The flowers are a fascinating example of co-evolution. They use a method of sexual deception to attract the male of a thynnid wasp species by offering sexual attractants known as pheromones, and visual cues, including a lip with distinctive protuberances that resemble the form of the female wasp.

Chiloglottis longiclavata sprouts from a sub-surface tuber and can produce underground offset tubers, enabling the species to form localized colonies. The plant produces a basal rosette of leaves with distinctive wavy margins. The flowers are produced singly from the rosette from autumn to early winter.

Chiloglottis longiclavata responds well to cultivation in a specialized terrestrial orchid mix that is free draining and contains good amounts of well-composted organic material, such as leafmould. The shallow root system is best accommodated in trays or pans. The plants need regular watering during the growing season but once the leaves and flowers have died back, tubers must be dried off and kept cool and dry until growth recommences. Plants can be repotted when in a dormant period. They are best grown under brightly lit, shade-house conditions with good air movement; in cool temperate zones, they may be grown in an alpine house.

Red List: Rare*/CITES II

Coelogyne cristata

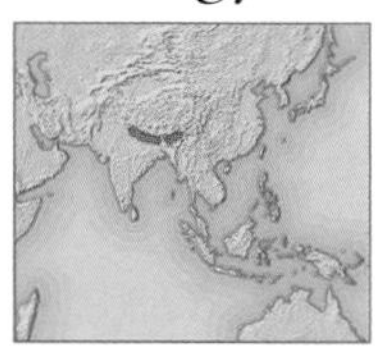

Botanical family
Orchidaceae

Distribution
Himalaya: Nepal to northern India

Hardiness
Frost-tender

Preferred conditions
Niches in trees and rocks; bright indirect light and good air circulation; ample supply of moisture while in growth followed by a cool, dry rest period.

Since it was introduced to cultivation in Europe from Nepal in 1837, ***Coelogyne cristata*** has become threatened in the wild as a result of overcollection for the horticultural trade across its entire range in the foothills of the Himalaya from Nepal to northern India.

► COELOGYNE CRISTATA *is widely grown by orchid enthusiasts and readily available from specialist nurseries from commercially propagated stock. An elegant specimen for a pot or a basket, it may also be grown on a slab of bark.*

This remarkably beautiful species is listed as Rare in the Red List of 1997.

It occurs as an epiphyte on trees, and occasionally on rocks, at moderate to high elevations in the Himalaya. The plant has pairs of narrow leaves that arise from the top of the short, rounded pseudobulbs, which are produced at regular intervals along a creeping rhizome. The pendulous inflorescence is 20cm (8in) long, with up to ten highly scented, pure white flowers of an exquisite crystalline texture.

A cool-growing orchid, it prefers a minimum temperature of 10–13°C (50–55°F) with good air movement, abundant moisture, and regular fertilizer when in growth. It should be potted in a coarse, free-draining mix.

16 on Red List*/CITES II

Corybas

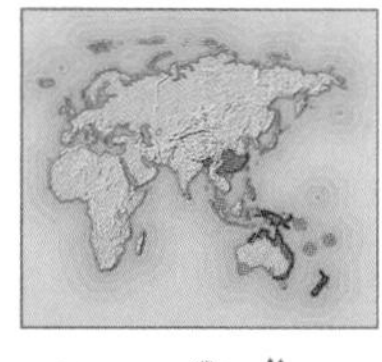

Botanical family
Orchidaceae

Distribution
Indo-Malaysia, Australia (including Tasmania), and New Zealand

Hardiness
Frost-tender

Preferred conditions
Acidic, humus-rich soil; cool temperatures and high humidity; ample moisture while in growth.

The helmet orchids comprise 100 herbaceous, tuberous, terrestrial and semi-terrestrial species, some 16 of which are variously listed as Rare, Endangered, or Vulnerable in the Red List of 1997. These diminutive orchids occur in deeply shaded areas on the floor of humid forest and woodland and in wetland areas, in northern India, southern China, New Guinea, south Pacific, Australia, and New Zealand. The hooded, usually helmet-shaped flowers are curious, rather than beautiful, but are often admired for their complex structure and glistening, deliquescent surfaces. They are pollinated by gnats, which supposedly confuse the flowers for mushrooms or toadstools. The plants produce squat, solitary flowers from the base of the leaves between winter and early spring. Once the flowers are pollinated, the stalk elongates to ensure effective seed dispersal.

◄ CORYBAS OBLONGUS *is seen here nestling in a moist, mossy substrate that helps maintain a humid atmosphere. The hooded dorsal sepal suggests the shape that gave rise to the common name, helmet orchid.*

Corybas species are best grown in shallow pans of a gritty, open mix with granulated bark and leafmould. Plants need to be kept moist and cool during the winter growing season, with a free flow of humid air and bright light.

After flowering in spring, the plants should be allowed to dry out, and any re-potting of the tiny tubers should be done in summer. In suitable conditions, the plants will multiply to produce large numbers of daughter plants each year. If left undisturbed, *Corybas* species will clump up rapidly to form a fascinating floral display.

Unlisted

Cryptanthus

Botanical family
Bromeliaceae

Distribution
Eastern Brazil

Hardiness
Frost-tender

Preferred conditions
Warmth and moderate to high humidity coupled with a humus-rich soil.

These terrestrial bromeliads, also known as earth stars or starfish plants, are native to the dry tropical forests and rainforests in the mountains of eastern Brazil, to altitudes of 1,600m (5,000ft). Most are found growing in deep shade on the forest floor. Although Unlisted, *Cryptanthus* species are at risk, or have even become extinct in the wild, because of loss of natural habitat.

The generic name comes from the Greek *cryptos*, which means hidden or covered, and *anthos*, flower; the flowers are hidden among the bracts and are ornamentally insignificant. Most species produce a central, flattened rosette of sometimes wavy, usually finely serrated leaves. These are grey and scaly beneath, while the upper surface is usually longitudinally striped or banded in bright colours.

Cryptanthus bivittatus, once native to eastern Brazil, is commonly grown as a house plant, yet is now unknown in the wild. The dark green leaves have two lighter green, longitudinal stripes (*bivittatus* means two-striped).

▼ CRYPTANTHUS ZONATUS *Another casualty of lost habitats, this species, known as zebra plant, features horizontal silver banding on a background that varies from dark green to chocolate brown or dull red.*

▲ **CRYPTANTHUS BIVITTATUS** *has been used as a parent to the many hybrids now in commercial cultivation.*

Cryptanthus bivittatus var. *atropurpureus* has red-flushed leaves, striped with red; they turn purple in the sun.

Cryptanthus lacerdae, silver star, is also known only in cultivation and is extinct in the wild. Smaller than *Cryptanthus bivittatus*, it has densely toothed, dark green leaves with silver-white margins, and two broad, longitudinal silver stripes. The undersides of the leaves are clothed in dense white scales.

Earth stars have been grown in Europe since the early 1800s, and are well established in cultivation. They are widely available from propagated stock, and many brilliantly coloured cultivars have been bred and selected. With leaves in all shades of green, variously striped and banded with pink, red, purple, ivory, silver, and coppery shades, these make desirable foliage plants for the home, conservatory, or terrarium.

These plants should be grown in a humus-rich substrate, in a fairly humid atmosphere, in shade or bright, indirect light. Provide temperatures above 15°C (60°F). Propagate by cuttings, or remove offsets; young plants may fall off the mother plant, and will need re-rooting.

7 on Red List*/CITES II

Cymbidium

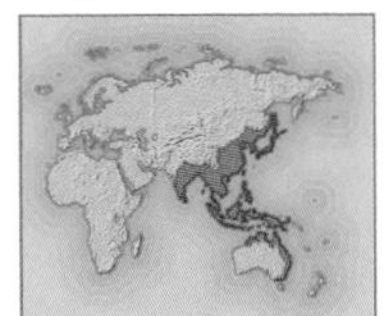

Botanical family
Orchidaceae

Distribution
India to Japan, Australasia

Hardiness
Frost-tender

Preferred conditions
Depending on the species, niches in trees and rocks or sites with free-draining, moisture-retentive soil; good light levels but shade from hot direct sun.

There are about 50 species in the genus *Cymbidium*, and seven are on the Red List of 1997, facing threats that leave them Vulnerable, Endangered, or Rare, by dint of loss of habitat and overcollection for horticulture or traditional medicine. Many *Cymbidium* species naturally have small and inconspicuous flowers, but have been successfully hybridized to produce larger blooms. In fact, they are among the most highly hybridized of orchids. Since the first cross was bred at Veitch's nurseries in the 1880s (*Cymbidium eburneum* x *Cymbidium lowianum*), thousands more have followed. It is one of the best known of orchid genera to florists.

Commercial cultivation of the genus is a huge industry; cymbidiums are easy to grow, tolerate cool growing conditions (4°C/40°F), and have long-lasting flowers and decorative foliage. Desirable forms, such as those with albino or variegated foliage, command high prices. The demand has fuelled a lucrative trade in wild plants that has also led to the decimation of species in accessible habitats.

Cymbidium species are found from India through south-east Asia to Japan, south to eastern and northern Australia. They grow as epiphytes, rock-dwellers, or terrestrials in temperate or tropical rainforest, and in dry forest and woodland, occurring at altitudes from sea level to 3,000m (9,850ft) or more. *Cymbidium canaliculatum* (Unlisted) is one of the world's few arid-zone epiphytic orchids, and can be found in desert-like areas of northern Australia, with roots penetrating the cooler, moister conditions in the rotting heartwood of a host tree.

Cymbidium devonianum is threatened in the wild due to excessive collecting in its range, from Nepal and north-east India to north-east Thailand. Found at elevations up to 1,500m (4,900ft), this compact, epiphytic or rock-dwelling species lives in mixed mountain forests. It produces a pendent inflorescence, to 45cm (18in) long, of 15–35 green flowers evenly placed along the stem, in late spring. A beautiful, distinctive species, it benefits from regular applications of half-strength fertilizer.

CYMBIDIUM DEVONIANUM

Although Unlisted, *Cymbidium tigrinum* is restricted due to overcollection in its native range from Myanmar and north-east India. It grows on rocks or in rock crevices, often in almost full sun. It occurs at elevations of 1,500–2,700m (4,900–8,850ft) and often experiences frost in winter. The olive-green autumn flowers have lined, spotted lips, margined with purple.

Cymbidium wenshanense is a rare but Unlisted epiphytic species, recently discovered in Yunnan and Vietnam. In spring, a stalk up to 40cm (16in) long bears 3–7 large, scented, flowers; the petals do not open fully.

Cymbidium wenshanense is typical of several orchids native to China that have been iconic in Chinese horticulture for over 1,000 years. In fact, the genus has a long history of over 2,500 years in cultivation in China, and there is written evidence that the plants were much admired by Confucius in about 500BC. The elegant, grassy foliage and delicate form of the scented blooms have been immortalized in Chinese brush painting through the centuries.

▲ **CYMBIDIUM TRACYANUM** *in the forest of Doi Inthanon National Park, Thailand, is typical of delicate species threatened by the acquisitive interest of modern collectors.*

Cymbidiums grow best in bright filtered light, in a free-draining mix that retains moisture, such as composted bark with charcoal and perlite. Keep moist in summer, feeding regularly, then water sparingly in winter.

PLANTS AND PEOPLE

CONFUCIAN FAVOURITE

The earliest records of orchids in the history of humankind are found in Chinese documents from 1000BC onwards, and are thought to refer to *Cymbidium ensifolium*. Cymbidiums were known as *lan* and are often mentioned in the poetry and song of the early dynasties. Confucius, the Chinese philosopher, was under the spell of cymbidiums, calling them "the king of fragrant plants". He said that "acquaintance with good men is like entering a room full of *lan*".

8* on Red List/all on CITES II

Cypripedium

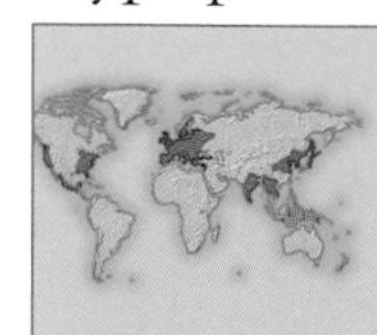

Botanical family
Orchidaceae

Distribution
Europe, Asia, and the Americas

Hardiness
Fully hardy – half-hardy

Preferred conditions
Damp, leafy, humus-rich, acid or slightly alkaline soils, depending on species; cool sheltered site, in dappled or partial shade.

The slipper orchids, *Cypripedium* species, are a widespread genus of about 45 species. They range across the Northern Hemisphere from the Americas to Europe and Asia, with a concentration of species from the foothills of the Himalaya. They are terrestrial orchids that inhabit the woodland and forest floor, meadows, marshes, and bogs, occurring over a wide altitudinal range, including high elevations that experience considerable snow and frost. Some species occur in subtropical conditions in Mexico.

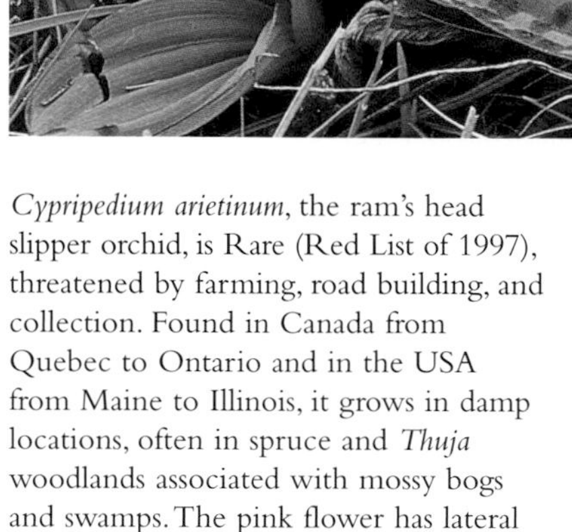

CYPRIPEDIUM REGINAE

The slipper orchids have a perennial over-wintering underground rhizome or rootstock, which gives rise to an annually deciduous, leafy stem with the leaves often pleated or attractively ribbed. The lip of the flower is enlarged to form a slipper-shaped pouch, hence the common name. The pouch invites small bees into the opening of the lip, channelling them into a one-way system that ensures pollination as they leave. Flowering occurs between spring and summer.

BEAUTIFUL VICTIMS

These desirable plants have been prime targets for acquisitive collectors for centuries, and depletion of species has been most clearly seen near centres of population. In the UK, *Cypripedium calceolus* dwindled in northern woods to a single wild individual, and several Chinese species were rendered extinct in the wild within a couple of years of their discovery.

A number of species have also come under threat from loss of habitat via changes in agriculture resulting in land drainage and loss of meadowland, and to forestry.

Standard horticultural techniques have proved of little use in propagating these exquisite plants. The Sainsbury Orchid Conservation Project, founded in 1983 in association with the Royal Botanic Gardens, Kew (*see right*), was set up to discover suitable propagation methods, in glasshouse and laboratory, and then to re-establish plants in quantity at safe sites in the wild.

The dust-like seeds of these (and many other) orchids need a symbiotic (mutually beneficial) association with a soil-dwelling fungus – a mycorrhizal fungus – in order to germinate. The research of the Sainsbury project has refined methods of propagation, including sowing on sterilized agar plates inoculated with suitable mycorrhiza. Several thousand *Cypripedium calceolus* seedlings, and of other genera, have been established in the wild as a result.

Due to the highly restricted nature of most species, gardeners should only buy plants certified as being from artificially propagated sources to discourage collection from the wild. Alternatively, and for improved reliability of growth and flowering, choose one of the many hybrids available from commercial nurseries.

Cypripedium acaule, the moccasin flower or pink lady's slipper orchid, is widespread but becoming less so, due to habitat loss and overcollection. It grows in habitats ranging from seasonally wet bogs to dry conifer woodlands, from Canada south to Tennessee and west to Minnesota and adjacent states in the USA. Unlike many other slipper orchids, it prefers acidic soils.

CYPRIPEDIUM MARGARITACEUM ►
This species from Yunnan and Sichuan often forms small colonies in leaf litter in the wild, in coniferous woodland. It is threatened by overcollection for medicinal use.

Cypripedium arietinum, the ram's head slipper orchid, is Rare (Red List of 1997), threatened by farming, road building, and collection. Found in Canada from Quebec to Ontario and in the USA from Maine to Illinois, it grows in damp locations, often in spruce and *Thuja* woodlands associated with mossy bogs and swamps. The pink flower has lateral sepals that sweep downwards and outwards like a ram's horns.

Cypripedium irapeanum is Endangered and Vulnerable (Red List of 1997) in Mexico and Guatemala respectively due to farming and forestry. It favours pine-oak forest, and is usually associated with volcanic areas. This golden-flowered species reaches 1m (3ft) in height, but is difficult to cultivate due to its dependence on mycorrhiza.

Cypripedium kentuckiense, the Kentucky slipper orchid, is Endangered from Alabama to Texas (Red List of 1997) by habitat loss and overcollection. This species blooms in hues of cream through amber in late spring. Its cultivation needs are not well understood.

Cypripedium reginae, the showy lady's slipper orchid, one of the finest of the slipper orchids, is Unlisted, but threatened because coveted by enthusiasts; it has been relentlessly overcollected. It is also threatened by habitat loss to agriculture. This moisture-loving species occurs from Quebec to Minnesota, in moist or seasonally wet sites associated with swamp margins, bogs, and moist prairies. Spectacular in flower in early summer, it can grow in colonies of several thousand, with stems to 85cm (34in) tall, bearing up to four pink-lipped, white flowers.

Most cypripediums require specialized cultivation related to the conditions in which they are found in the wild. Plants from artificially propagated sources are becoming available from nurseries.

RENEWAL

BACK FROM THE BRINK

Cypripedium calceolus is one of the best known of the slipper orchids; records date back to 1541. It is found in woodlands, often of a highly calcareous nature. The species is common and widespread from Spain to Siberia, China, and Japan, but in the UK it declined until there was just one wild plant.

Its conservation is assured due to the Sainsbury Orchid Conservation Project and Royal Botanic Gardens, Kew. The project's aim is to re-establish terrestrial British and European orchids at safe sites in the wild, raising them from seed to maintain genetic diversity. Seed is difficult to germinate, but when symbiotic mycorrhizal fungi were added, germination was successful. The first wild flowering of artificially raised *Cypripedium calceolus* was reported in 2000.

▲ **CYPRIPEDIUM KENTUCKIENSE** *is very similar in appearance to* Cypripedium calceolus, *but produces larger flowers.*

HISTORY

THE LEGEND OF VENUS

According to Roman myth, Venus, the goddess of love and beauty, was out hunting with her lover, Adonis, when a thunderstorm forced them to flee for cover. As was her wont, the goddess took full advantage of their enforced intimacy, but failed to notice she had lost her slipper. When the storm had passed, a mortal saw Venus's slipper and reached to pick it up. Suddenly, it was transformed into a ravishing flower with a slipper-shaped petal.

In 1737, when Swedish botanist Carolus Linnaeus was contemplating a genus name for a European slipper orchid, he choose *Cypripedium*, alluding to Cyprus, the mythological birthplace of the goddess of love, and *pedilum*, or slipper. More than a century later, classical mythology once again came to the service of science when Ernst Hugo Heinrich Pfitzer named *Paphiopedilum*, *Cypripedium*'s sister genus of tropical slipper orchids, in honour of Paphos, an alternative name for Aphrodite, the Greek goddess known as Venus to the Romans. Paphos is the city in Cyprus close to the site where Aphrodite is said to have been born from the ocean.

CYPRIPEDIUM ACAULE ► *Native Americans once used this species medicinally to treat irritability and insomnia. It has been in cultivation since the 1700s, and although it grows and flowers in garden conditions, it often dies after flowering.*

5 on Red List*/CITES II

Dactylorhiza

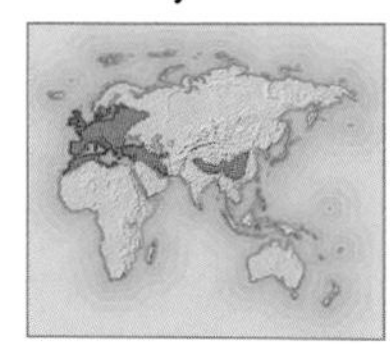

Botanical family
Orchidaceae

Distribution
Europe, North Africa, Asia, North America

Hardiness
Fully hardy

Preferred conditions
Damp, leafy, humus-rich soil in partial shade or dappled shade; some sun in reliably moist soils, but protection from hot midday sun.

The spotted, or marsh, orchids make up a genus of about 33 species of tuberous perennials, producing erect spikes of flowers in white and shades of red, pink, violet, or purple. They are found in marshes, meadows, heaths, damp streambanks, and dune slacks throughout Europe, temperate Asia, and North America. Many species are of concern to conservationists, because of the loss of suitable habitat resulting from land drainage for agriculture and changes in grassland management.

A number of species, including the robust marsh orchid, *Dactylorhiza elata*, common spotted orchid, ***Dactylorhiza fuchsii***, and southern marsh orchid, *Dactylorhiza praetermissa*, are fairly well known in cultivation, where they are often grown in naturalistic meadow plantings, or in rock gardens. These species are now readily available from commercially propagated stock; some commercial producers use techniques developed by the Sainsbury Orchid Conservation Project.

The Sainsbury research project, undertaken in association with the Royal Botanic Gardens, Kew, developed techniques of symbiotic mycorrhizal propagation in the laboratory with *Dactylorhiza praetermissa* and *Dactylorhiza fuchsii*. The resulting seedlings of both species have been naturalized successfully in several protected sites in the UK, as part of a joint conservation effort with the National Trust and English Nature.

These successful experiments in propagation and re-establishment are part of ongoing studies in the ecological requirements of orchids, and involve a range of conservation organizations. These include many local Naturalists' Trusts, whose members contribute a great deal of local expertise regarding suitable sites for re-introduction. Some groups have also trained members to pollinate those orchids that are at the extremes of their natural range, where natural pollinators may not be present. Group members may also be active in providing security against theft by collectors.

Populations of *Dactylorhiza fuchsii* subsp. *sooana* in Poland and Slovakia are described as Vulnerable in the Red List

DACTYLORHIZA MACULATA ►
The heath spotted orchid is found on acid soils, on heathland, moorland, and in bogs and light woodland.

▼ DACTYLORHIZA FUCHSII
The common spotted orchid is seen here in its typical habitat in calcareous (limestone) soils in grassland.

of 1997. *Dactylorhiza chuhensis*, which occurs at elevations of up to 2,300m (7,550ft) in the mountains of Turkey, is classed as Endangered in the Red List of 1997, its status probably the result of local overcollection for medicinal use.

Dactylorhiza species are usually sturdy plants, with flower spikes up to 30cm (12in) tall, arising from a basal cluster of lance-shaped green leaves, marked with brown to dark purplish-violet spots. The spikes are often densely flowered.

Most *Dactylorhiza* species are hardy and are best grown in partial shade, or in dappled, part-day sun, in loamy, leafy soil enriched with granulated bark and leafmould. The plants are in growth during spring and summer, when they need ample moisture. These orchids are sometimes grown in pots, using a terrestrial orchid mix comprising equal parts of leafmould, grit, and granulated bark.

Propagation of established colonies in cultivation is by division of the tubers in early spring.

66 on Red List*/CITES I and II

Dendrobium

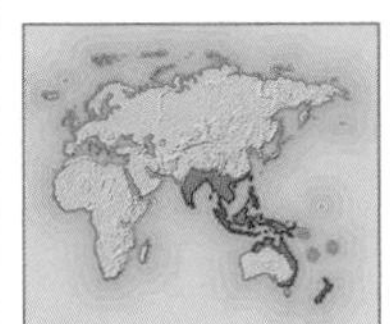

Botanical family
Orchidaceae

Distribution
India, Japan, South-east Asia, and New Guinea to Australia and Oceania

Hardiness
Frost-tender

Preferred conditions
Good light; excellent air circulation; plentiful moisture when in growth, with a dry dormant period; temperatures vary according to geographical origin.

With about 1,200 species ranging across India, Japan, south-eastern Asia, and New Guinea to Australia and Oceania, *Dendrobium* is one of the largest genera of orchids. *Dendrobium* species occur from near sea level, sometimes almost within reach of sea spray, to mountain altitudes of 3,000m (9,850ft), and from humid rainforest to the semi-arid regions of eastern and northern Australia. They range in size from diminutive plants less than a few centimetres across, to substantial growers of 1.5m (5ft) or more in height, and are often epiphytes, although a number of species are rock-dwelling.

Dendrobium bigibbum, the Cooktown orchid, is found near Cooktown in far north Queensland, Australia, but ranges from north-eastern Australia to southern New Guinea. It is listed as Vulnerable in the Red List of 1997 and is under threat because of extensive overcollection and habitat loss to farming.

This species grows as an epiphyte on a range of tree species in warm, moist lowland sites with a pronounced seasonal fluctuation in moisture and temperature. It is one of the most attractive of the genus, and has been the foundation for much hybridization and development of orchids for the Asian cut-flower market.

The inflorescence is some 40cm (16in) long, and arises from the upper leaf axils in autumn, and the mauve to purple flowers are very long lasting. This is an easily grown species that has given rise to numerous cultivars, many of which make highly desirable additions to the orchid greenhouse.

Dendrobium cruentum is one of the few orchid species to be given the highest protection against trade in jungle-collected plants by CITES. Introduced to cultivation in 1884, *Dendrobium cruentum* had suffered from rampant over-collection for international and local trade for over a century from its native habitat, the humid lowlands of Thailand and Myanmar. Then, in 1996, the species was upgraded from CITES Appendix II to Appendix I in an effort to save the last populations in the wild. This orchid is particularly desirable for two major reasons: its flowers are strikingly coloured and it is willing to produce flowers throughout the year. Borne singly or in pairs, the pale green to cream blooms have a white lip with brilliant blood-red marks and red side lobes. They are about 7.5cm (3in) across and exceptionally long-lived, with a light but distinctive fragrance.

Fortunately, *Dendrobium cruentum* is easy to germinate from seed, and artificially propagated plants are widely available from orchid nurseries.

◄ DENDROBIUM BIGIBBUM *produces a spectacular display of up to 20 flowers per spike in spring. Such prolific blooming makes heavy energy demands on the plant, but it responds well to regular feeding and a prolonged dry rest after flowering.*

Dendrobium moorei is listed as Vulnerable in the Red List of 1997 because of its very restricted distribution and loss of habitat to farming. It has also suffered the impact of overgrazing by feral animals on the World Heritage site of Lord Howe Island, off the east coast of Australia, where it grows at 100–850m (330–2,800ft) elevations as an epiphyte, a rock-dweller, or even occasionally a terrestrial plant. It is valued by orchid growers for its pure white flowers and ease of culture.

In cultivation, most *Dendrobium* species need moderate to strong light, good air circulation, abundant moisture when in growth, and a period of dry dormancy. They prefer free-draining mixes of bark or similar media, and regular applications of fertilizer during the growing season. Since they come from such a wide range of habitats and altitudes, there are species to suit almost every temperature regime, from cool to warm.

RENEWAL

MICROPROPAGATION

In eastern Asia, several *Dendrobium* species are used to prepare Shi-hu, which is taken as a tonic or strengthening medicine, or as an aphrodisiac. However, the demand for these medicines has threatened wild populations. Today, some of this pressure is relieved by cultivating orchids on a large scale using micropropagation techniques, and making them available to the herbal medicine industry. Many thousands of new orchid plants are produced using small pieces of tissue grown on an artificial medium in strictly controlled, sterile conditions. It is a technique that has helped orchid, and many other plants under threat from overcollection for medicinal or horticultural purposes.

21 on Red List*/CITES II

Disa

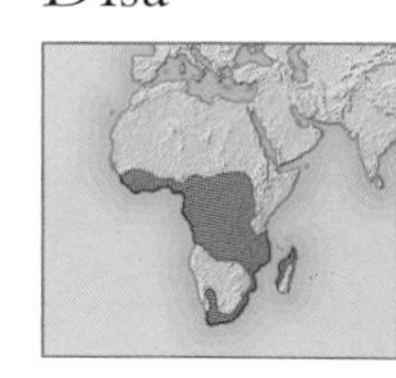

Botanical family
Orchidaceae

Distribution
Central and eastern Africa, South Africa, Madagascar

Hardiness
Frost-tender

Preferred conditions
Free drainage, in cool, humid conditions in semi-shade or filtered light; moisture while in growth.

This genus of about 125 species of terrestrial orchids occurs throughout Africa, especially in south-central Africa and South Africa. Those most amenable to cultivation occur in damp, streamside sites. These include *Disa cardinalis*, *Disa racemosa*, *Disa tripetaloides,* and ***Disa uniflora***, all Unlisted, which have been used as parents of the finest, widely available hybrids.

Disa species historically have been threatened by overcollection and habitat loss to agriculture, and urbanization continues to pose problems.

Disa barbata, for example, was first found at several sites near Cape Town, South Africa, in the early 19th century. All the original locations are now used for housing and only a single population on one reserve is now known. This beautiful species grows in deep, seasonally moist, sandy heath dominated by reed-like restios. It is now being propagated by tissue culture for the re-introduction of plants to the wild, and, eventually, for potential sale to growers.

In cultivation, free drainage is essential, with abundant soft water during growth, which should be applied so that it does not touch the foliage. Water should be reduced but not withheld after flowering.

► DISA UNIFLORA *is one of the most striking orchids of the genus, valued for its long flowering period and the size and colour intensity of its blooms, which vary from white through to yellow, pink, and red. It is one of the parents of the innumerable hybrids that are among the easiest of the genus to grow.*

13 on Red List*/CITES II

Diuris

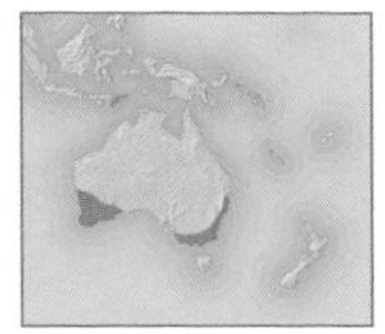

Botanical family
Orchidaceae

Distribution
Australia, Timor

Hardiness
Frost-tender

Preferred conditions
Humus-rich, acidic, free-draining soil; good light; ample moisture in growth, and a dry rest after flowering.

There are about 55 species of donkey orchids, comprising some of the most colourful and easily grown terrestrial species endemic to Australia, with one species in Timor. The flowers employ floral mimicry to attract small, pollinating bees, often resembling in colour and form the pea flowers of leguminous plants that share their habitats.

Diuris species are found in swamps, grassland, and temperate and subtropical woodlands and scrub. All are deciduous, with leaves dying back to spherical or spaghetti-like underground tubers. In the wild, they bloom mostly in spring and become dormant in summer.

Diuris fragrantissima, the scented donkey orchid or sunshine diuris, is Endangered in the Red List of 1997, and is close to extinction in the wild. It is restricted to a population of about five plants in a grassland remnant in Victoria, south-east Australia. It has suffered habitat loss with change of land use for housing and agriculture, competition from introduced weeds, grazing, and illegal collection for the horticultural trade.

National and regional governments in Australia have formulated an exceptional series of recovery plans for Australian habitats and their endemics, among them one for *Diuris fragrantissima*. The plan involves protection of the existing site, coupled with re-introductions there and in other safe sites. Research into cultivation and propagation techniques will help ensure viable collections of mature plants in the wild and in cultivation. Control of weeds, pests, and predators is essential; hand-pollination may be necessary.

Diuris micrantha, one of the smallest of the genus, is Endangered in the Red List of 1997, and restricted to just a few sites near Perth, Western Australia. It favours seasonally wet places, occurring among rushes in full sun, and is threatened by drainage and urbanization. Seed has been placed in cryogenic storage at Kings Park and Botanic Garden in Perth. Plants have been raised for re-introduction to the wild. Research has shown that this species can be seed-raised in the laboratory with or without symbiotic mycorrhiza, flowering within 2–3 years.

DIURIS LONGIFOLIA

Diuris species need a well-drained, acidic mix, rich in organic matter, in sun or semi-shade, with good ventilation and ample water during the winter growing season preceding flowering. Plants need a dry rest after flowering. Growth resumes as temperatures fall in autumn.

Red List: Vulnerable*/CITES II

Dracula vampira

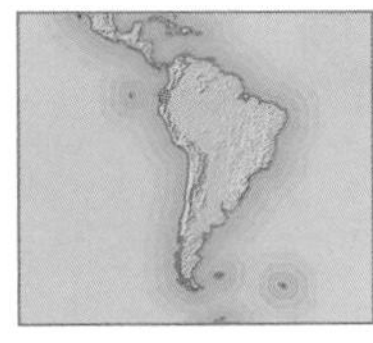

Botanical family
Orchidaceae

Distribution
Ecuador

Hardiness
Frost-tender

Preferred conditions
Damp, shady, airy conditions; free drainage with ample moisture while in growth.

The genus *Dracula* consists of some 90 species of epiphytic orchids, occurring in Andean forests from Central America to Peru, including about 20 that are Rare, Vulnerable, or Endangered on the Red List of 1997. Collectors still seek Ecuadorian ***Dracula vampira*** because of its remarkable flowers, which are the closest yet to the mythical "black orchid". Known as the "black chimaera" because of the intense coloration of the flowers, it is fairly widely available as cultivated stock.

The species is restricted to the forested slopes of a single mountain, at altitudes of up to 2,000m (6,550ft). It is an epiphyte that usually grows lodged in humus-rich niches between the branches of trees. On a plant of 30cm (12in) tall, an inflorescence of up to 20cm (8in) long will bear several flowers, each up to 30cm (12in) long and 10–15cm (4–6in) wide.

DRACULA VAMPIRA

The large, sinister-looking flowers are basically green, but the green is often obscured by dark purple-black stripes and suffusions. In good conditions, flowers may be borne almost year-round. The plants need moist, shady conditions in a free-draining mix of composted bark or similar, preferably in a basket, with a minimum of 10°C (50°F). Apply dilute fertilizer regularly.

PLANTS AND PEOPLE

BLACK ORCHIDS IN MYTH

Do black orchids really exist? Those with dark markings, such as *Dracula roezlii*, *Coelogyne pandurata*, or *Ophrys muscifera*, are almost black. Perhaps the closest to true black is found in the highly bred *Paphiopedilum* Black Velvet 'Candor Neat'. For hundreds of years, hybridizers have sought to create black flowers of all types, but black orchids have a particularly irresistible allure for many.

Writers too cannot resist the exotic appeal. In Rex Stout's 1941 novella, *Black Orchids*, detective Nero Wolfe leaves his comfortable New York brownstone in pursuit of a truly black orchid at a flower show, and ends up pursuing a murderer at the exhibition. Black orchids also play a leading role in the comic strip *Brenda Starr*, in which the red-headed American reporter dashes around the world in pursuit of a story. But Brenda's true love, handsome Basil St. John, suffers from a rare disease, treatable only with an elixir extracted from a rare black jungle orchid.

55 on Red List*/CITES II

Epidendrum

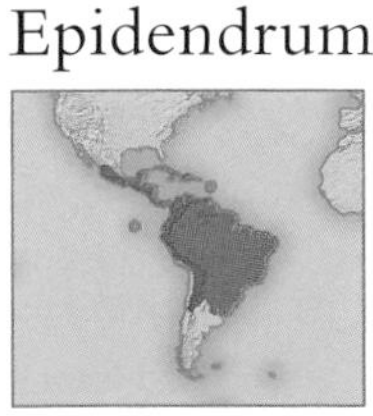

Botanical family
Orchidaceae

Distribution
Tropical North, Central, and South America

Hardiness
Frost-tender

Preferred conditions
Free drainage; bright filtered light; most need moisture all year; minimum temperatures vary according to origin.

There are 500–750 species of *Epidendrum*, although some species are being re-classified by botanists into *Encyclia* and other related genera. Epidendrums are found from lowland tropics to cloud forests at altitudes of 1,000m (3,250ft), as epiphytes, terrestrials, and rock-dwellers.

Formerly known as *Nanodes medusae*, the epiphytic ***Epidendrum medusae*** is listed as Endangered on the Red List of 1997 and is found only in Ecuador. The remarkable form of the flowers has made it a prime target for unscrupulous collectors. The plant has pendulous stems densely clothed in leaves that are dull green, but sometimes suffused with purple. The flowers appear in the spring, with yellow-green petals suffused red-brown, and a lip of fleshy, deep maroon that is one of the more spectacular found in the orchid family, both in size and colour and in the incredible filament-like fringe of elaborate, tentacle-like hairs.

Epidendrum medusae has adapted readily to cultivation, and this highly desirable species with its intriguing flowers is now widely available as artificially propagated plants from specialist orchid nurseries. It is best grown epiphytically on a log, or in a basket suspended in the greenhouse, so that the flowers can be displayed to best effect.

Epidendrum ilense, discovered in 1977 in Ecuador, was listed as Extinct just 20 years later in the Red List of 1997. This epiphytic species, native to the tropical forests on the coastal plains of Ecuador, was under threat even as it was discovered, as the trees were being felled for their timber. The original location was destroyed within a few years of discovery of the plants. The species survives in cultivation, where it is noted for the year-round production of delicate white blooms with a distinctive fringed lip and a pronounced citrus fragrance that is noticeable at nightfall. This elegant species is now widely propagated from artificial sources and has borne a variety of beautiful hybrids.

In general, *Epidendrum* species require bright filtered light, good air movement, and constant moisture throughout the year (more sparing in winter), with light but regular feeding. Montane species need a minimum of 10°C (50°F); tropical species do best at minimum 19°C (66°F).

◄ EPIDENDRUM MEDUSAE, *formerly* Nanodes medusae, *has a mass of fleshy filaments fringing the lip, which gave rise to the species name; it refers to the writhing serpent locks of the Gorgon of Greek myth, Medusa.*

12*/32 on Red List

Guzmania

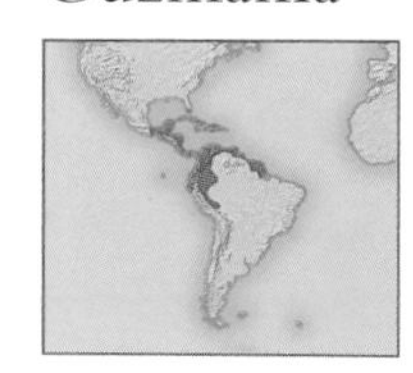

Botanical family
Bromeliaceae

Distribution
Central and north-west South America, Caribbean

Hardiness
Frost-tender

Preferred conditions
Warm, shady conditions with moderate to high humidity and an open, well-drained, slightly acidic soil.

Over 120 species of evergreen epiphytic bromeliads make up the genus *Guzmania*, named for the Spanish naturalist Anastasio Guzman (died 1802). Long valued in horticulture for the funnel-shaped rosettes of glossy leaves that cup conspicuously colourful stems, vivid floral bracts, and white or yellow flowers, many guzmanias have become severely threatened by the loss of their rainforest and cloud forest habitats.

The deforestation of Ecuador's Pacific coastal rainforest is especially critical. In the 1980s, Alwyn H. Gentry, a field biologist at Missouri Botanical Garden, reported that less than 0.1 per cent remained. By 2004, losses had increased by a factor of 100. Gentry's peers at the Missouri Botanical Garden estimate that 90 per cent of all bromeliads (approximately 450) in Ecuador face extinction within ten years.

The drier climate and disruption of seasonal rainfall patterns induced by rainforest loss accelerates the process of extinction. With few exceptions, these bromeliads are unable to adapt to the higher light level and exposure in secondary forest environments. Even if reforestation began today, the original topsoil has long since washed away, leaving no suitable substrate to permit regrowth of the plants.

▲ GUZMANIA WITTMACKII *is the most important pollen parent of cultivated bromeliads today; some 20 million plants of cultivars carrying its genes are produced commercially each year throughout the world.*

Guzmania alborosea, with a magnificent inflorescence of white flowers enclosed by pink sepals and floral bracts, is pollinated by hummingbirds. It was once common in the coastal Pacific rainforest of Ecuador, at elevations of 800–1,200m (2,600–3,900ft). In the northern Chóco-Darién forests, a compact form is found in very wet conditions, in dense, small-tree forests with an annual rainfall of 8,000mm (312in). Further south, near Pinas, in south-central Ecuador, a much larger form, up to 1m (3ft) across, was once common. The southern forest is even more damaged than the Chóco-Darién forests. Here there are fewer than six forest remnants, each less than 100 sq km (30 sq miles) in extent. *Guzmania alborosea* is rare in cultivation and only occasionally available as artificially propagated stock: make sure plants are of cultivated provenance before buying. Propagation is from seed or cuttings.

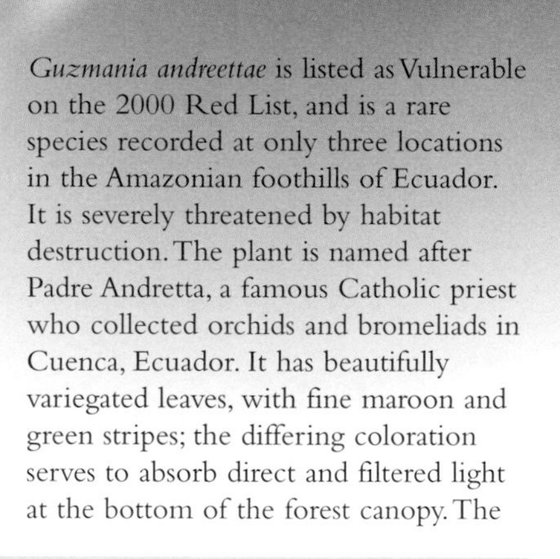

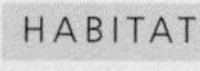

Guzmania andreettae is listed as Vulnerable on the 2000 Red List, and is a rare species recorded at only three locations in the Amazonian foothills of Ecuador. It is severely threatened by habitat destruction. The plant is named after Padre Andretta, a famous Catholic priest who collected orchids and bromeliads in Cuenca, Ecuador. It has beautifully variegated leaves, with fine maroon and green stripes; the differing coloration serves to absorb direct and filtered light at the bottom of the forest canopy. The cone-shaped inflorescence is handsomely marked, but the small, white, night-blooming flowers are attractive mainly to pollinating moths and flies.

Guzmania andreettae is rarely seen in cultivation, but is grown in botanical gardens. It can be propagated by cuttings, but to produce seed a second clone is needed for cross-pollination.

Guzmania wittmackii (Unlisted) is found in the tropical rainforest of southern Colombia and western Ecuador. It is the primary pollen plant for numerous bromeliad cultivars, including 'Orangeade', which was introduced by Henry De Meyer of Belgium in the mid-1960s, and began an explosion of *Guzmania* cultivars. This bromeliad is famous for its long primary bracts, ranging from soft pink through bright yellow-orange with many shades of red found in between. Flowers can be white or yellow. The species is easy to grow where humidity is adequate.

Guzmania blassii, from the wet and humid rainforests of western Costa Rica, is uncommon. Although first described as recently as the 1980s, this is yet another example of a plant threatened in the wild that has made considerable economic contribution to horticulture; many commercial cultivars have been bred from it. Its virtues as a parent of hybrids include its ability to thrive at relatively low temperatures of 5°C (40°F). The striking inflorescence,

HABITAT

SAN RAFAEL WATERFALL

Water gathered in the cold upper reaches of the Andes cascades down the eastern side of the Cordilleras Real, eventually joining the Amazon as it flows down to the Atlantic Ocean. On its route, in the San Rafael Cloud Forest Reserve, the river encounters a fault line, creating the spectacular San Rafael waterfall – plunging some 145m (475ft), it is the largest waterfall in Ecuador. In this extraordinary habitat, the cool, moist atmosphere of the cloud forest becomes almost saturated with moisture. This, coupled with the shelter of trees and undergrowth, which sustain an equable temperature, and the bright light that filters through the canopy, create perfect conditions for growth, and orchids, bromeliads, and other epiphytes thrive here in abundance – for the time being.

▲ **BIRD POLLINATION**
Many bromeliads, here Guzmania nicaraguensis, *are pollinated by hummingbirds.*

with bright red bracts and yellow sepals, lasts up to three months.

Propagation is by offsets; two different clones are needed to produce seed. The plant is cultivated as a species mainly in botanic gardens.

Guzmania donnell-smithii is a native of tropical rainforest at low to moderate elevations in the mountains of Costa Rica. It is a curious example of a plant that has been able to exploit new niches when its native habitat is under threat.

GUZMANIA LINGULATA

Seeds of *Guzmania* species have a pappus, or crown hair, like the seeds of a dandelion, to aid wind dispersal. With the destruction of the rainforest from Costa Rica south to Ecuador continuing at an alarming rate, seed dispersal by wind now occurs over greater distances, because the forest canopy, which prevented the seeds from spreading widely, has now gone. *Guzmania donnell-smithii* has recently extended its range as far south as northern Ecuador.

This species is a fairly recent discovery and seed has been distributed by the Bromeliad Society International. The plant is self-fertile, producing copious seeds without cross-pollination. Easy to grow, this small plant has bright red bracts and yellow sepals that last over a month; it has proved a popular pot plant and has been propagated commercially for sale in Europe.

Guzmania rubrolutea is listed as Endangered in the 2000 Red List. Found in 1981 in the Amazonian foothills by the San Rafael waterfall (*see box, left*), this narrowly endemic species needs high humidity and temperatures above 5°C (40°F) to thrive. Habitat lost to agriculture and oil pipelines jeopardize its fate in the wild.

Guzmania species need shade and a minimum of 10–15°C (50–60°F); they will tolerate high temperatures of 33°C (90°F). Humidity of 50–80 per cent is ideal. They prefer slightly acidic growing media, and rainwater or filtered water. Many guzmanias can be grown epiphytically if misted frequently. If growing from seed, indirect light is needed for germination. Never fertilize young plants until there are three sets of leaves, then use weak solutions of fertilizer (wash foliage afterwards). Avoid overcrowding and maintain high humidity with good air circulation. Patience is needed as plants grow slowly and it may take several years for them to flower.

OTHER THREATENED GUZMANIAS

Guzmania harlingii This giant bromeliad forms a rosette up to 1m (3ft) across. It bears large, branched spikes covered in yellow flowers. Native to the Chóco-Darién rainforest of Ecuador. It is threatened in the wild but several large nurseries are growing this plant from seed and it should be widely available in the near future. Red List (2000): Vulnerable.

Guzmania longipetala is found in the Lita region of Ecuador, where severe habitat destruction makes cultivation vital to ensure the survival of this striking species, with its apple-green leaves and orange and yellow flowers. Propagation is by seed; plants are not self-fertile. Cultivars with almost black leaves have recently been introduced. Unlisted.

Guzmania rauhiana Named after the late German botanist, Dr Werner Rauh, this species was discovered after the building of new roads in the Lita region of Ecuador opened up the area to botanists, and unscrupulous collectors. It is found in the Chóco-Darién rainforest, which is severely threatened by habitat destruction. Safety in cultivation is assured, as it is self-fertile and produces abundant seed. Red List (1997): Rare.

Guzmania macropoda is threatened due to its limited distribution and habitat loss. Found only in the Panama Canal zone, this small plant has a tube of leaves covered with a silver frosting; the shape and form suggest that this might be a passive carnivorous plant, digesting insects that fall into the "tank" formed by the leaves. Despite the level of threat, it is easy to cultivate. The plant is self-fertile and produces seed in abundance. Red List (1997): Indeterminate.

Guzmania xanthobractea A large bromeliad, spectacular in bloom, it is native to the west coast tropical rainforest of central Ecuador. Once common and often seen near ridge tops, it is being pushed rapidly towards extinction by the drying out of the entire western forest. Today it is found wild in only a few private reserves in the Mindo area. It is occasionally available from cultivated sources, from specialist nurseries and botanic gardens. Red List (2000): Near Threatened.

LAELIA ANCEPS ► *This Mexican endemic is adaptable and can be found growing in and colonizing a variety of agriculturally managed situations, including coffee plantations, and trees in pastureland, at elevations up to 1,500m (4,900ft). Its adaptability in nature translates into ease of cultivation, which has made this orchid a very popular species with orchid collectors, endangering wild populations.*

PLANTS AND PEOPLE

FLOR DE MUERTO

Every year in early November, Mexicans celebrate the Day of the Dead by setting out a meal for the spirits of their ancestors, who are enticed to the feast with a trail of marigold flowers. For weeks, families prepare items to be sold or used during the fiesta, especially small sweetmeats in the shape of animals, skulls, and fruits. To make these traditional confections, they use the pseudobulbs of two orchids, *Laelia speciosa* and *Laelia autumnalis*, called *flor de muerto*, or flower of the dead. These are ground up into a paste, with a little water. Flour, sugar, lemon juice, and egg whites are added. After maturing for a few days, the dough is placed in a hand-carved wooden mould, and decorated with dye.

13 on Red List*/CITES I and II

Laelia

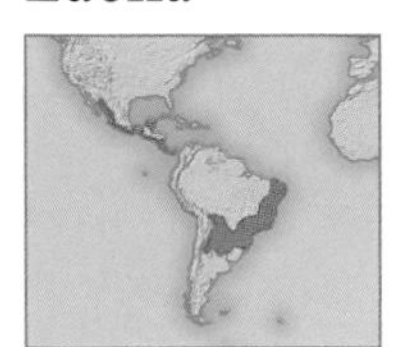

Botanical family
Orchidaceae

Distribution
Mexico, Brazil

Hardiness
Frost-tender

Preferred conditions
Cool, light sites; moisture while in growth and flower, but dryness once new growth matures.

Previously comprising some 60 species, the genus *Laelia* has been revised by botanists and now consists of just a few Central American xerophytic epiphytes (adapted to survive in arid conditions). They live on trees and rocks in cool forests at up to 2,600m (8,300 ft), preferring to perch on oak trees where they receive a lot of light. Laelias have proved popular with gardeners in their native countries and internationally.

The Brazilian *Laelia milleri*, introduced to cultivation in the 1960s, was illegally collected to the point of near extinction, until it was protected under CITES Appendix II. It has been widely propagated from seed by orchid nurseries, which has virtually halted the trade in wild plants, although local collection remains a threat.

In Mexico, laelias are much loved, and have been offered up in religious celebrations for centuries. Some species are on the verge of extinction because they are collected from the wild and sold at incredibly low prices in the markets of Mexico City, Guadalajara, and many other small towns, for ornamental use in homes and gardens.

Laelia gouldiana, a spectacular species once endemic to Mexico's Sierra Madre Orientale, is considered Endangered in the Red List of 1997. It has been found in a semi-wild state around Hidalgo. Presumably grown in gardens from wild collections, these plants had formed healthy clumps on stone fences and mesquite trees. They seem, however, to be divisions of a single, self-sterile clone and set no seed. Attempts to breed genetically different individuals have failed. This clone of *Laelia gouldiana* has been propagated by division, and is in cultivation around the world.

Collection is not the only menace. ***Laelia anceps*** is Endangered in the Red List of 1997, threatened as its forest habitat is burned to make way for agriculture. But its beauty has also led to intense collection pressure and until recently, plants of this, and other species, were exported in huge quantities.

Red List: Vulnerable*/CITES II

Lepanthes calodictyon

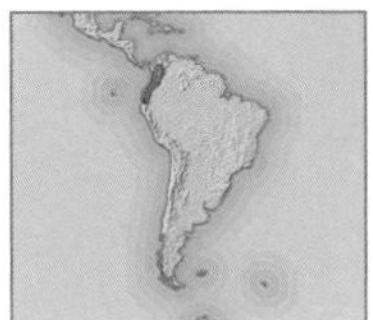

Botanical family
Orchidaceae

Distribution
Ecuador, Colombia

Hardiness
Frost-tender

Preferred conditions
Sheltered, humid, semi-shaded sites with a regular moisture supply.

This rare and unusual orchid species has endured the pressure of collection for the horticultural trade, and is classed as Vulnerable in the Red List of 1997. Found from the Andes through the western cordillera of Colombia and Ecuador, ***Lepanthes calodictyon*** is one of the more highly prized orchids. The plant is epiphytic on trees growing at elevations of 750–1,300m (2,450–4,250ft).

It is valued for its remarkable lime- or emerald-green leaves with a network of rose-pink through lavender and red-brown markings, held on long, slender stems. The single leaf produces a multi-flowered inflorescence, which lies on the leaf surface, with vivid, minute, red and yellow flowers, 5mm (¼in) across. The species varies greatly across its range in terms of leaf size, colour, and pattern.

These fragile orchids are best grown mounted on slabs of bark or a similar material. They need water, warmth, and constant humidity throughout the year, with regular, dilute fertilizer.

▲ LEPANTHES CALODICTYON *is a delicate species and can be challenging to cultivate. The intricacies of its tiny flowers can best be appreciated with a magnifying glass; each has long-tailed petals and a fringed lip.*

Red List: Endangered*/CITES II

Lycaste skinneri

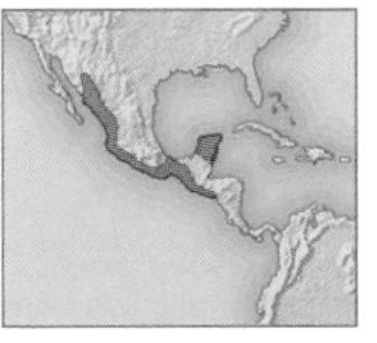

Botanical family
Orchidaceae

Distribution
North and Central America: Mexico to Honduras

Hardiness
Frost-tender

Preferred conditions
Cool, partially shaded site with good air circulation; freely draining soil and ample moisture when in growth; dry rest period.

One of the most beautiful of its genus, and certainly the one with the largest flowers, *Lycaste skinneri* was once widely distributed from Chiapas, Mexico, south through Guatemala into Honduras and El Salvador. Its beauty has made it attractive to collectors, and today the species is Endangered (1997 Red List), partly due to burning of its habitat for agriculture and charcoal production, but largely as a result of extensive over-collection since the early 19th century, when vast numbers of plants were shipped to the UK and Europe.

The flowers, 10–18cm (4–7in) across, are very variable in colour, from pure white through blush to deep pink and lavender. Some 50 distinct varieties have been described. The most highly prized of the many forms is ***Lycaste skinneri* var. *alba***. This pure white form is often singled out as one of the world's most beautiful orchids. Because its large, triangular white sepals resemble a nun's cap, it is commonly called the nun's orchid, or *monja blanca* (white nun) in Guatemala, where it is the national flower.

In nature, *Lycaste skinneri* grows epiphytically in the forks of fairly large trees, usually at an elevation of about 1,650m (5,000 ft) in seasonal cloud forest. Although its survival in the wild is precarious, artificially propagated plants are readily available, and the species has been used to create a number of hybrids.

LYCASTE SKINNERI *VAR.* ALBA

PEOPLE

George Ure Skinner

Lycaste skinneri was collected in the 1840s by plant hunter George Ure Skinner (1804–1867). The son of a Scottish clergyman, Skinner pursued his own vocation – orchids – with a passion. For 34 years he roamed the jungles of Central America, finding nearly 100 new species. About to settle down in Europe with his collections, he embarked on one last expedition, his 40th Atlantic crossing. On the eve of departure from Panama, at the end of this final trip, Skinner died of yellow fever, and never made it home. *Lycaste skinneri*, *Barkeria skinneri*, and *Cattleya skinneri* were all named after him.

Lycaste skinneri is very amenable to cultivation and can be grown in any free-draining epiphyte mix, such as coarse bark. It should be watered freely and regularly during the growing season; then water should be withdrawn gradually as new growth matures to allow a cool, dry rest period. When new roots appear, recommence watering and applications of fertilizer. The species grows best in partial shade where there is good, buoyant air circulation. The plant has a long bloom period, with individual flowers lasting for as long as 6–8 weeks.

50 on Red List*/CITES II

Masdevallia

Botanical family
Orchidaceae

Distribution
Central and South America

Hardiness
Half-hardy – frost-tender

Preferred conditions
Cool, airy, lightly shaded conditions; free drainage but with a constant moisture supply.

This large genus numbers some 350 species, native to the rainforests of Central and South America. Many grow as cloud forest epiphytes at moderate to high elevations, up to 2,500m (8,200ft). Several are rock-dwelling, especially those from higher altitudes.

Some 50 species appear on the Red List of 1997, due in part to the showy flowers in remarkably vibrant colours, which have attracted the attentions of orchid collectors. The flowers are unusual in that the sepals fuse to form a distinctive, tricorne-hat shape, often with slender tails on the sepals; in some species, the lip is greatly reduced and almost invisible at the centre of the flower. There are huge numbers of hybrids available in cultivation.

Masdevallia davisii is rare and restricted to Peru, where the plant has been over-collected in the wild. It is coveted on account of its fragrance and the striking colour of the flowers. It is also a compact and easily cultivated species.

Masdevallia rosea, from Colombia and Ecuador, is now rare, having been highly sought after by collectors on account of its bright, orange-tubed, carmine flowers. Both these species are now available from specialist orchid nurseries from artificially propagated sources.

Most *Masdevallia* species need water all year round, but should not be kept too moist during the darker days of winter. They can be propagated by division.

▲ MASDEVALLIA STUMPFLEI *is a rock-dwelling species found in rocky niches on the slopes of Cordillera Blanca in the Peruvian Andes; this example shows the brilliant colour and the tricorne form that is characteristic of many species in the genus.*

Red List: Rare*/CITES II

Mexipedium xerophyticum

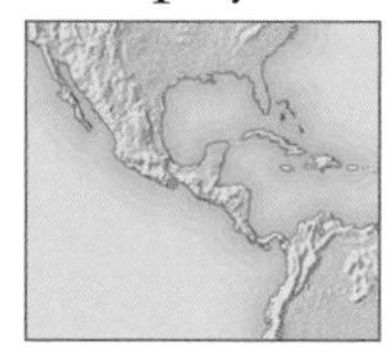

Botanical family
Orchidaceae

Distribution
Mexico

Hardiness
Half-hardy

Preferred conditions
Frost-free conditions that are drier than those needed by most slipper orchids; free drainage.

This recently discovered slipper orchid, ***Mexipedium xerophyticum***, was first collected in 1985, and originally named *Phragmipedium xerophyticum*. It is Rare on the Red List of 1997, with only seven plants in the wild, all in an area of just 1 sq km (0.3 sq miles).

From small divisions of live plants collected in 1988, thousands are grown around the world, representing an *ex-situ* conservation collection. There is still an urgent need to reinforce wild populations, ensuring that there are sufficient plants in several sites to avoid extinction.

In the wild, *Mexipedium xerophyticum* is found only in Oaxaca, Mexico, on karst (limestone) outcrops at an elevation of about 300m (1,000ft), as part of a unique xeric (arid) plant community that is surrounded by rainforest. The exact locality is one of the orchid world's most closely kept secrets.

Mexipedium xerophyticum grows on vertical, treeless, north- or east-facing cliffs, in leaf debris in crevices, or on exposed rock, where it is shielded from the fierce midday sun.

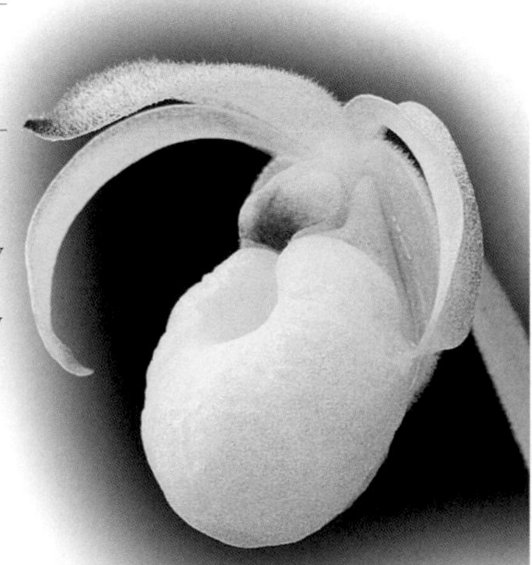

▲ MEXIPEDIUM XEROPHYTICUM *has showy, pink-flushed white flowers, to 2.5cm (1in) long and wide, on branched inflorescences above a fan of tough, light green leaves.*

Red List: Vulnerable*/CITES II

Miltoniopsis warscewiczii

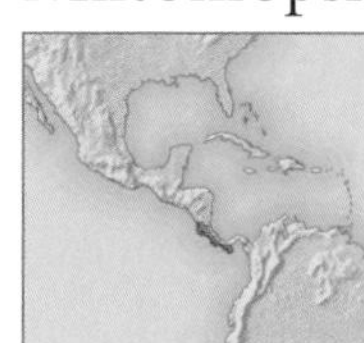

Botanical family
Orchidaceae

Distribution
Costa Rica, Panama, Venezuela

Hardiness
Frost-tender

Preferred conditions
Bright, filtered light and good air movement; moderate humidity.

This species is Endangered in Costa Rica and Vulnerable in Panama (1997 Red List) as a result of over-collection for the horticultural trade. *Miltoniopsis warscewiczii* is an epiphyte, found at altitudes to 2,000m (6,550ft), often forming large clumps at the tops of tall trees in cloud forest.

The species bears an inflorescence 30cm (12in) long, from winter to spring, with 3–6 fragrant, creamy white flowers, blotched rose-purple, 7cm (3in) across. This beautiful orchid has been one of the foundation species for breeding in this genus, resulting in some prize-winning hybrids. Plants from artificially raised sources are available from specialist orchid nurseries.

NOMENCLATURE

HYBRIDS AND GREXES

Breeding from *Miltoniopsis* species (and other orchids) has resulted in many complex hybrids. Partly to keep track of the complexities, a naming system of two or three levels was devised. In *Miltoniopsis* Anjou 'Robert St Patrick', *above*, the genus name is *Miltoniopsis*; Anjou, the so-called grex name, describes a group of hybrids raised from the same two parent plants (species or hybrids); and the cultivar 'Robert St Patrick' describes a clone, one of a set of genetically identical individuals increased by vegetative means from a single selected seedling of the grex.

Unlisted

Nidularium

Botanical family
Bromeliaceae

Distribution
Brazil

Hardiness
Half-hardy – frost-tender

Preferred conditions
Acidic conditions with good drainage; filtered light and moderate to high humidity.

Bird's nest bromeliads generally grow in shady conditions beneath the forest canopy of the Atlantic tropical forest of Brazil and are under threat from loss of habitat to logging, agriculture, and new settlements.

Most *Nidularium* species have attractive foliage and are interesting when not in flower. The leaves are either totally green, or green with red or purple undersides; this coloration is an adaptation to absorb that part of the light spectrum found in the darkest areas beneath the forest canopy. The floral parts of *Nidularium* species can appear to be poorly developed; in fact, the inflorescence is often nest-like; the genus is named after the Latin word *nidus*, meaning nest.

Nidularium apiculatum is endemic to the tropical rainforest of Itatiaia National Park, near Rio de Janeiro, Brazil. Protected there but nevertheless quite rare, it is found in the understorey where it is epiphytic on trees, or on rocks in deep moist shade at elevations of 700–1,200m (2,300–3,900ft). The lime-green leaves form a loose, funnel-form rosette around red flowers with green margins.

Nidularium campos-portoi is an unusual species from the Atlantic rainforest of eastern Brazil, near Rio de Janeiro, where it is threatened by habitat loss. It has light green leaves, and, during flowering, the central leaves turn yellow, then red, from the centre of the plant to the tips of the leaves. The inflorescence rises slightly above the foliage and within the bright yellow-red primary bracts emerge attractive yellow flowers surrounded by white floral bracts. This species is propagated by division and has been in cultivation for many years.

Nidularium fradense is found in the Atlantic rainforest to the north-east of Rio de Janeiro, in the leaf litter of the forest floor, sometimes forming large populations. Threatened because of forest destruction, this species has beautiful pink primary bracts and a symmetrical floral rosette, which is large compared with the size of the plant. Nestled in the centre are small blue flowers.

Nidularium fulgens is named aptly; its name means the "plant that shines". This species, also from the Atlantic rainforest around Rio de Janeiro, has been over-exploited commercially for the last century, and mature specimens have been sold to the general public in Brazil. Today it is protected in the Serra dos Orgaos National Park, but is still threatened by local collectors who sell plants in bloom in the city. *Nidularium fulgens* has lime-green leaves suffused with dark green spots. From the centre of the rosette rises a floral rosette of red, sometimes orange or pink, primary bracts. The flower petals are dark blue with white margins. An inhabitant of the forest floor and the canopy, this species tolerates a range of light and humidity levels and is easily grown.

Nidularium kautskyanum is a jewel-like bromeliad from the cloud forest of Espirito Santo, Brazil. It lives on the forest floor, where coarse quartzite rock provides perfect drainage. The plant needs cool shade, making it especially sensitive to forest clearance; there is no current record of it in any protected areas.

▲ NIDULARIUM INNOCENTII *is a very variable species from Brazil, long cultivated for its vivid red floral bracts. Several varieties, once common in the wild, are now known only in cultivation.*

RENEWAL

SUSTAINABILITY

Nidularium species are part of the diverse vegetation of the Salto Morato Reserve, in Paraná, Brazil. The reserve is a centre for research, and training and education for careers in park management, conservation, and eco-tourism. The programme is funded by the Eco-Development Fund, a joint project of the Inter-American Foundation and the Fundaçao O Boticário de Proteçao á Natureza (the latter an offshoot of O Boticário, a botanical/natural cosmetics company). Other projects provide sustainable employment for local people, including growing organic coffee in the shade of forested corridors that will link isolated forest fragments, forming wildlife corridors for flora and fauna.

Unlisted/CITES II

Odontoglossum crispum

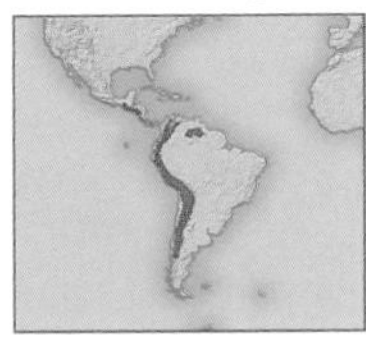

Botanical family
Orchidaceae

Distribution
West Indies, Brazil to Colombia, Peru

Hardiness
Frost-tender

Preferred conditions
Good drainage; cool temperatures and high humidity; bright indirect light.

An orchid often cited as one of the most magnificent, ***Odontoglossum crispum*** was discovered by Karl Theodor Hartweg in 1841, in cloud forest near Bogota, Colombia, at an elevation of about 2,300m (8,000ft). It has gracefully arching flower sprays, up to 1m (3ft) long, bearing 3–4 large, crystalline white flowers. The lip is often yellow, with a few red spots. Once quite common in its native cool, misty, high-altitude forests, it was a spectacular sight in bloom, cascading from tree branches like giant snowflakes. However, it did not remain abundant for long.

The European obsession with orchids probably began in 1836, when *Psychopsis papilio*, the butterfly orchid, was displayed at a show organized by the Royal Horticultural Society in London. When plants of *Odontoglossum crispum* were introduced to Europe in the mid-19th century, they sparked a furore, and by then, orchidmania was out of control.

In 1889, the showiest specimens were fetching in excess of 150 guineas at auction – a small fortune in those days, far exceeding the average yearly salary. Collectors scoured the mountain tops near Bogota. There are records of one shipment of 40,000 specimens, and there were many shipments. The demand was so insatiable that collectors did not bother to remove plants individually from their host trees, but razed entire forests to strip them more quickly. Thousands of plants were taken by mule or canoe to coastal cities, where they were shipped to Europe. Inevitably, many perished en route, but a few survived and they merely stoked further demand for the dazzling flowers.

Today, the mountain-top forests are gone, and the species is hard to find even in the few fragments that remain. The gene pool of those that do survive is impoverished. Hundreds of specimens exist in private collections around the world, but the genetic diversity of these, too, is narrow, having been bred for the past 150 years for ease of cultivation and flowers approaching human ideals of perfection.

◄ ODONTOGLOSSUM CRISPUM *gets its name from its petals and sepals, which have "crisped" margins.*

18 on Red List*/CITES II

Ophrys

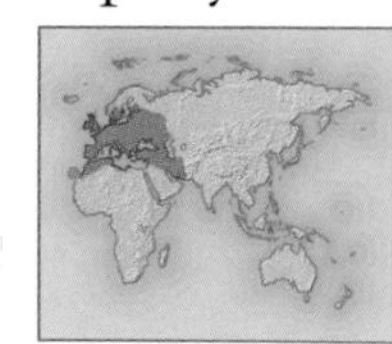

Botanical family
Orchidaceae

Distribution
Europe, western Asia, North Africa, Mediterranean Islands

Hardiness
Frost-hardy – fully hardy

Preferred conditions
Partial shade; free drainage; abundant moisture during growth followed by a dry rest period after flowering.

This is a genus of about 140 species of herbaceous terrestrial orchids that die down each year to a roughly spinning-top-shaped underground tuber. Known collectively as bee, spider, and mirror orchids, *Ophrys* species employ amazingly inventive systems for ensuring their flowers are pollinated.

The flowers have a thick, ample lip that is rich brown, maroon, or yellow in colour. In the centre of the lip is a distinctively shaped zone that is shiny and metallic, bright blue to violet, and marked with lines and patterns. The broad tip of the lip is velvety, and, in many species, the edges are fringed with dense, dark hairs. In other words, the flowers look just like the enticing body of a female bee or wasp. The flowers also exude a scent that simulates the pheromones produced by receptive females.

Each species of *Ophrys* is generally pollinated by its own pollinator species, as in ***Ophrys scolopax*** (*see right*). When the male lands on a flower, it grasps the lip and attempts to copulate with it. In the process, the flower deposits pollinia on the insect's head, to be carried and placed on the next flower it visits. The orchids flower about two weeks before the female insects emerge, prompting some speculation that this process serves not only to fertilize the orchid flowers, but also to give the male insects much needed practice in the art of copulation.

Charles Darwin, famous for his five-year-long voyage on H.M.S. *Beagle* and the resulting theory of evolution, wrote that no subject interested him as much as the orchid. Indeed, Darwin used orchids to illustrate his theory in *On the Origin of Species*, published in 1859. Three years later he published *On Various Contrivances by Which Orchids are Fertilised by Insects, and On the Good Effects of Intercrossing*. It took Darwin's unparalleled powers of observation and deduction to understand the mechanisms by which orchids lure insects to pollinate their flowers.

Shiny, pale violet-blue zone at the base of the lip

OPHRYS INSECTIFERA ► *The fly orchid has a distinctively patterned shield, or speculum (from the Latin word, meaning like a mirror or window pane), on the base of the lip. This feature has given some* Ophrys *species the name of mirror orchids.*

He reportedly studied the curious reproductive rites of ***Ophrys apifera***, and other species, while strolling through the English countryside in Devon.

Darwin also studied a whole range of exotic orchids, and was supplied with them by the great horticulturists and scientists of the day, including James Veitch (of the famous Veitch & Sons nurseries), Dr Lindley (*see p.369*), and Dr Hooker at Kew. Dr Lindley, an early European orchid expert, produced the first significant scientific classification of the family as a whole.

OVERCOLLECTION

Many species in the genus are becoming rare due to unscrupulous overcollection on account of their conspicuous and curiously attractive flowers, and growers are urged to only buy plants from reputable sources where the plants have been artificially propagated.

The major conservation concern for many has been land drainage and other changes in agricultural practices, most especially the different grassland management techniques that have led to the decline of old-fashioned hay meadow habitats.

A few species have long been used in a concoction called salep, a sort of mucilaginous, strengthening medicine that originated in Turkey and Iran.

Ophrys lycia is endemic to Turkey, in calcareous soils at elevations of 400–600m (1,300–1,950ft), where the plant is rare due to habitat loss and over-collection for the production of salep. It is listed as Endangered in the Red List of 1997. One of the largest populations occurs in a cemetery near Agulla, Turkey, where the plants appear to be protected thanks to the tuber gatherers respecting the Islamic tradition that a cemetery is a sacred and untouchable location.

The requirements of this species in cultivation are similar to many others in the genus: a gritty, humus-rich, sharply drained medium with plentiful moisture in the growing season. Most need partial or dappled shade. Bee orchids bloom early in the year and become dormant during the summer months, after flowering, when they need to be kept dry.

If pot grown, late summer or early autumn is the best time to repot the tubers and separate offsets.

OPHRYS SCOLOPAX ► *The woodcock orchid is native to thin, chalky grasslands and scrub around the Mediterranean, including the islands of Cyprus, Corsica, and Sardinia. It is seen here in the embrace of a male* Eucera longicornis, *the long-horned bee.*

▲ OPHRYS APIFERA *The bee orchid is found in grassland, at woodland edges, in hedgebanks, and in scrub. A versatile orchid, it is also seen on dune slacks and in maquis. It has been studied and propagated by the Sainsbury Orchid Conservation Project and re-introduced to several safe sites in the UK.*

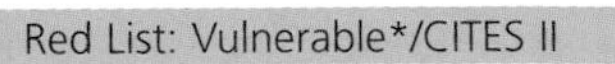

Red List: Vulnerable*/CITES II

Orchis canariensis

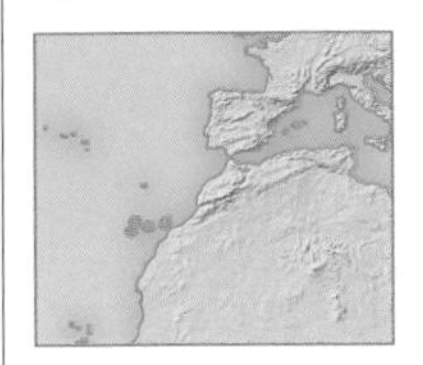

Botanical family
Orchidaceae

Distribution
Canary Islands

Hardiness
Frost-tender

Preferred conditions
Leafy, gritty, well-drained, acidic soils; partial shade; moisture while in growth and a dry period after flowering.

Species of *Orchis* are found from the Mediterranean to Asia, as far east as Japan. The genus includes about 33 species, nine of which appear on the Red List of 1997. ***Orchis canariensis*** is a naturally rare species, endemic to the Canary Islands. It grows in ravines, thickets, and seasonally dry, pine woodland, on undisturbed ground, to elevations of 1,400m (4,600ft). It is listed as Vulnerable in the Red List of 1997 because of habitat disturbance.

Orchis canariensis produces leaves each year from underground tubers, during the winter to spring growing season. In spring, a dense, erect inflorescence of pale pink flowers is produced, each flower with a prominent nectar spur.

In Mediterranean-type climates, this orchid can be grown in a rock garden, or naturalized in open, free-draining, leafy, acidic soils in partial shade. If grown in an unheated glasshouse, use an acidic terrestrial orchid mix and top-dress with pine needles. Ample water is needed during growth, followed by a warm, dry rest after flowering.

▲ ORCHIS CANARIENSIS *is seen here wedged in a shaded rock crevice; in its natural habitat it produces flowers in late winter and early spring. The prominent nectar spur behind the flower is thought to be important in attracting pollinating insects.*

33 on Red List*/CITES I

Paphiopedilum

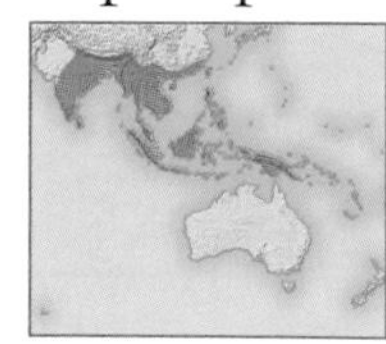

Botanical family
Orchidaceae

Distribution
India, China, south-east Asia to New Guinea

Hardiness
Frost-tender

Preferred conditions
Shade and high humidity in warm conditions; good air circulation; seasonal rainfall.

Containing about 60 species of evergreen orchids known as tropical slipper orchids, the genus *Paphiopedilum* ranges from India eastwards across south China to the Philippines, and throughout south-east Asia to New Guinea and the Solomon Islands. The vast majority of species are endemic to a very narrow range; a few are known only at one or two sites.

Most species are found in lower montane, evergreen or deciduous forests in shade, growing in small colonies with their roots in leaf litter. Paphiopedilums often occur on calcareous soils, and can range up to 2,000m (6,550ft) or more in cloud forests. In many of these areas, rainfall and humidity are high, but the rainfall is seasonal. The plants' thick, leathery leaves are able to cope with this periodic drought, and desiccated plants recover rapidly when rains resume. Although the majority of paphiopedilums are terrestrials of forest floor, grassland, or rocky slopes, some species have assumed an epiphytic existence.

Many species remain threatened in the wild by overcollection, and more recently, by destruction of habitat by logging and forest fires.

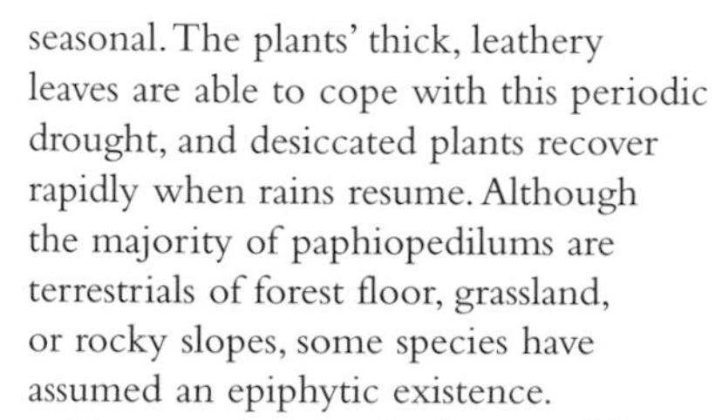

PAPHIOPEDILUM FAIRRIEANUM

Concern for survival

Slipper orchids are named for their pouch-lipped flowers. This characteristic makes them attractive to collectors unfortunately, and a new site is often stripped bare as soon it is discovered. The extremely limited distribution of many species only serves to exacerbate matters. CITES Appendix I listing prohibits international trade in wild specimens; these were among the first plants to be protected on this scale. Artificially propagated species are listed in Appendix II and may be traded commercially. To minimize pressure on wild populations, growers should only purchase species propagated from seed, under licence.

The popularity of slipper orchids in horticulture dates back to the 18th century, when the first species were brought into cultivation. Slipper orchids proved the ideal orchids for growing in the smaller greenhouse, or under lights, as the plants generally need low light levels. Years of cultivation experience have shown that most species can be grown easily from seed under artificial conditions, which makes it even more shocking that wild-collected plants are still sold on the black market.

▼ IN THE WILD *Rooting in moist, leafy or mossy substrates that are typical of the forest floor, the tropical slipper orchids often form small colonies. Enchantingly beautiful, they have inspired an extraordinary desire to possess in unscrupulous collectors.*

CULTIVATION

Collection victim

After its discovery in the early 20th century, examples of *Paphiopedilum delenatii* were introduced to France. All died apart from one in the Lecoufle brothers' collection. This survivor self-pollinated and plants grew and thrived; they even became commonplace. Following decades during which it was feared extinct in the wild, the orchid was rediscovered in Vietnam in the late 1990s, but then became a victim of export trade, as Vietnam sought much needed income from native orchids. Many turned up in Japan, however, where breeders used "wild" pollen on inbred, cultivated plants to produce robust, genetically diverse seedlings. Now mass-produced in the Netherlands, plants are sold for a fraction of the cost of black market wildlings. Another population was found in 2000, but responsible cultivation still may not be able to stem wild collection.

Paphiopedilum delenatii (*see box, above*) was listed as Extinct in the Wild in the Red List of 1997, and although it has recently been rediscovered in Vietnam, it is still critically threatened. It grows on acidic soils on rocky sites and cliff faces at 800–1,500m (2,600–5,000ft). Most specimens of this popular and easy-to-grow species in cultivation are a result of the early propagation success with the original introductions on the part of the noted French orchid nursery of Vacherot and Lecoufle. *Paphiopedilum delenatii* has richly mottled leaves and pink-lipped white flowers in spring.

Paphiopedilum fairrieanum was first introduced into cultivation in 1857. Few plants tantalized the orchid world as much as this. For almost half a century after its introduction to Europe, its place of origin remained a mystery. By December 1904, it had become so rare in cultivation, with only one surviving plant and four seedlings, that a reward of £1,000 was being offered for its rediscovery.

This species is still Endangered in its habitat (1997 Red List) in northern India and Bhutan because of overcollection, the impact of uncontrolled fires, and grazing by goats. Found on rocky, limestone substrates, it survives in protected grasslands and forests up to 2,200m (7,200ft). The flower appears in late winter and is white, veined with rich maroon-purple.

Paphiopedilum rothschildianum is the rarest in nature of all slipper orchids, known only from steep slopes and cliff faces on the lower elevations of Mount Kinabalu, Sabah, Borneo. Prized on account of a 1m (3ft) long inflorescence which bears a succession of flowers 30cm (12in) wide in spring, it is considered Endangered in the wild in the Red List of 1997.

Although it is protected within Mount Kinabalu National Park, continued illegal collection and forest fires still threaten the last populations in the wild.

The enigma of the spectacular, tail-like petals of species such as *Paphiopedilum rothschildianum* has mystified orchid biologists. What could spur the evolution of such unlikely if magnificent petals? The lurid flower colours and curious furry warts and spots found on the floral segments have been cited as clues that these slipper orchids may be pollinated by flies. One of the few detailed pollination studies of the genus found that *Paphiopedilum rothschildianum* is indeed pollinated this way, by a syrphid fly (*Dideopsis aegrota*).

Paphiopedilum sukhakulii is rare and naturally restricted to tropical rainforest in north-east Thailand, where it grows in sandy loams near streams at altitudes of 1,000m (3,300ft). It was named after Prason Sukhakul, manager of the Bangkrabu Nursery, Bangkok, the source of the first European consignments. Few recent introductions have had such an impact on orchid enthusiasts.

Though first seen in cultivation as late as 1964, it has become one of the most common *Paphiopedilum* species in cultivation, and has had a profound impact on hybridization. The species bequeaths to its offspring a set of wide and very horizontal petals covered with dark spots. These traits continue to be expressed through several generations.

The species was first seen in cultivation from a consignment of wild-collected plants, and was noted as new to science when plants flowered in a commercial collection. The leaves are attractively mottled and the single flower, almost 12cm (5in) wide, is produced in late winter to spring on a tall, erect stem.

This attractive species has proved reliably easy in cultivation and is a popular parent species of many modern slipper orchid hybrids. It and its many hybrid offspring are readily available from most reputable orchid nurseries from artificially propagated sources.

PAPHIOPEDILUM ROTHSCHILDIANUM ► *Despite extensive searching for more than a century, this species has been located at only four sites, two now destroyed, on ledges and steep slopes at an altitude of 600–1,200m (2,000–4,000ft) on Mount Kinabalu in Borneo.*

PAPHIOPEDILUM SUKHAKULII *shows the perfect* Paphiopedilum *form, with waxy-textured flowers, notable for the width and bold markings of the lateral petals, and the large, deep, glossy pouch. Flowers are held on an upright stem above rich green leaves, mottled with paler yellow-green. The species is Indeterminate on the Red List of 1997.*

Red List: Indeterminate*/CITES I

Peristeria elata

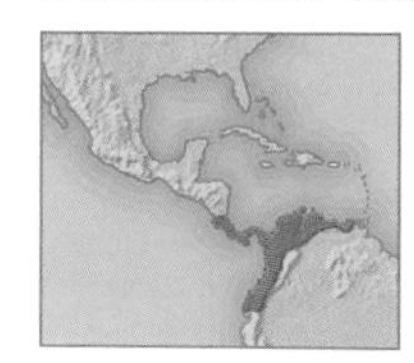

Botanical family
Orchidaceae

Distribution
Costa Rica, El Salvador, Panama, Colombia, Venezuela

Hardiness
Frost-tender

Preferred conditions
Warm conditions with bright, filtered light; moderate to high humidity with good air circulation; ample moisture year-round.

One of the first orchid species to be given protection from international trade under CITES Appendix I, ***Peristeria elata*** is erect and stately, with a slender flower stalk to 1.5m (5ft) tall, bearing 10–15 fragrant flowers. Each has five waxy white petals, with a white, dove-like form inside. The plant's English common name is dove orchid, while in its native Central America, it is known as Espiritu Santo, or Holy Ghost orchid. The blooms are used to decorate churches at Easter.

This revered species is the national flower of Panama, where it was once plentiful, from near sea level to 600m (1,950ft), growing as a terrestrial in grass, or on rocky outcrops, often at the edge of tropical forests. Dove orchids have been picked by indigenous peoples, and by international collectors. The lethal combination of overcollection and habitat destruction led *Peristeria elata* to its CITES listing. However, despite this official protection, and the availability of artificially propagated plants at low cost, wild specimens are still being collected.

In its native habitat, this species grows in surface leaf litter. In cultivation it is best grown in shallow clay pans half-filled with broken crocks for good drainage, topped by a leafy terrestrial mix. After flowering, reduce watering to give a semi-dry rest. Resume watering as new growth appears.

▲ SCENTED ORCHID Peristeria elata *bears strongly scented flowers with five waxy petals cupping inner parts that bear an uncanny resemblance to a dove. It is Indeterminate on the Red List of 1997.*

12 on Red List*/CITES II

Phalaenopsis

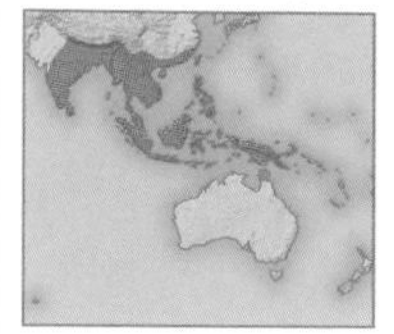

Botanical family
Orchidaceae

Distribution
India, Himalayas, south-east Asia, north-eastern Australia

Hardiness
Frost-tender

Preferred conditions
Well-drained soil, evenly moist but never waterlogged; semi-shade with moderate warmth and humidity.

The genus *Phalaenopsis* consists of about 60 species of mostly epiphytic, occasionally rock-dwelling orchids found throughout tropical Asia, China, India, Nepal, Indonesia, New Guinea, and the extreme north-east of Australia. Plants occur at a range of altitudes, from tropical, low-altitude, steamy rainforests, to higher altitude cloud forest. They are known as moth orchids. The 1997 Red List includes 12 species of *Phalaenopsis*, seven of which are declared Vulnerable and three of which are classed as Endangered. This is mostly a result of overcollection. The beautiful flowers, which, in some species, are produced in large numbers along the stem, are the key attraction.

Phalaenopsis amboinensis is native to the Indonesian archipelago and is named after the island of Ambon, formerly one of the famed Spice Islands. The plants are listed as Rare in the Red List of 1997 as a result of extensive overcollection for horticulture. Recent fires in the region and logging practices have further eroded the habitat of this species, with the result that the plants are now highly threatened in the wild. The flowers are star-shaped, thick-textured, and attractively banded in golden red-brown on a creamy white or pale yellow ground.

PHALAENOPSIS AMBOINENSIS

Phalaenopsis amboinensis has short stems and three to four, overlapping, succulent green leaves and abundant fleshy roots. It is easily grown in semi-shade, with good air circulation. Plants grow best in a very open mix of coarse bark with evenly moist, but not wet, conditions. Temperatures should be maintained between 15–25°C (60–77°F) for maximum growth and flowering. Plants often require a slight drop in temperature for flower initiation.

Phalaenopsis stuartiana is restricted to the Philippines, where it has dramatically declined in the wild as a result of the clearing of its native habitat and overcollection for horticulture – growers are drawn by the species' spectacular flowering habit and attractive foliage. It is listed as Vulnerable in the Red List of 1997.

The species is allied to the pink-flowered *Phalaenopsis schilleriana* (Vulnerable in the Red List of 1997), which is also restricted to the Philippines, and is very similar in appearance and habit.

Phalaenopsis stuartiana is an epiphyte of lowland rainforests, where the plant thrives in warm, humid conditions. It is grown for its large, attractive, rather fleshy leaves, which are mottled with silver-grey above, and purple beneath, making the plants ornamental even when not in flower.

The white flowers, 7cm (3in) across, are not only long lasting, but are produced in abundance on very long, arching, multi-branching inflorescences. Flowers may completely cover the plant.

Phalaenopsis stuartiana adapts readily to cultivation and can form large specimen plants if given warmth and regular light fertilizing. It should only be acquired from specialist orchid nurseries who use artificially propagated sources.

All species require year-round humidity. In addition, since most grow in deep shade, the broad, almost succulent leaves are highly susceptible to scorching in full sun.

The genus is notable for the intensive hybridization carried on by commercial nurseries, which has resulted in many fine horticultural hybrids in a remarkable variety of colours. The period from seed to first flowering is among the shortest in commercially available orchids, at around 18 months. Plants are mass produced in commercial glasshouses for sale as pot plants. Moth orchids need a minimum of 15°C (60°F), although they can tolerate cooler conditions for short spells.

▲ PHALAENOPSIS STUARTIANA *produces pristine white petals of a substantial texture with an upcurved lip marked and mottled with gold, amber, and red-brown; it is sweetly fragrant.*

4 on Red List*/CITES I

Phragmipedium

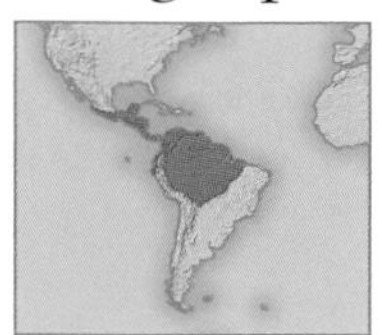

Botanical family
Orchidaceae

Distribution
Mexico, Central and South America

Hardiness
Frost-tender

Preferred conditions
Bright filtered light; moderate humidity with good air circulation; ample moisture when in growth, less moisture in autumn and winter.

The phragmipediums are New World tropical slipper orchids. Most of the 20 or so species are terrestrial or rock-dwelling, but a few are epiphytic tree-dwellers. Among the latter are species such as *Phragmipedium caudatum*, with long, ribbon-like petals resembling those of some *Paphiopedilum* species (*see p.410*).

Phragmipedium species are often found growing between rocks, in the splash zone of waterfalls, and on streambanks. The plants, which lack pseudobulbs, have fibrous roots. Their leaves are leathery and strap-shaped. Upright stems that arise from the centre of the leaves bear one or several flowers with a large, slipper-shaped lip.

Like most tropical slipper orchids, phragmipediums are threatened by habitat loss and over-zealous collectors. All species are listed in Appendix I of CITES, giving them the highest degree of protection from abuses due to international trade.

Phragmipedium besseae is listed as Vulnerable in Ecuador, and thought to be Extinct in parts of Peru in the Red List of 1997. The introduction in 1981 of this species, with its vivid scarlet flowers, rekindled interest in growing the tropical lady slippers, most of which sport drab green, or moss-coloured flowers and had been all but forgotten for decades.

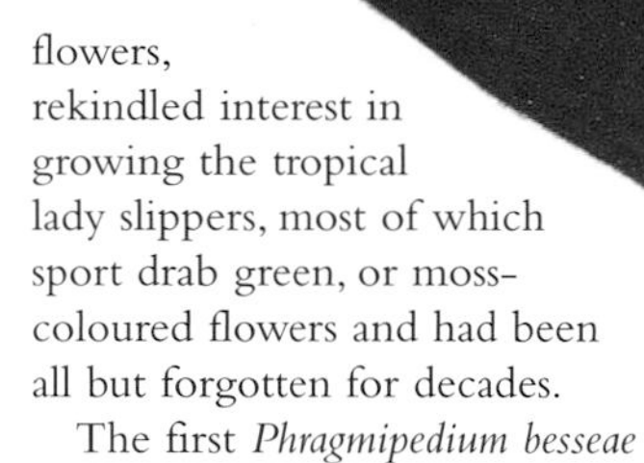

PHRAGMIPEDIUM BESSEAE *VAR.* DALESSANDROI ► *In a genus whose members' colouring is usually described as subtle, this brilliantly coloured species is an outstanding parent for breeding new hybrids.*

The first *Phragmipedium besseae* was discovered in Peru in 1981, and subsequently found in large numbers in Ecuador. This species has been held up by orchid enthusiasts as an example of an orchid species saved by commercialization. When some of the Ecuadorian plants were first brought to the USA and Europe and offered for sale, their prices were astronomical. Some small divisions reportedly sold for US$600 – a recipe for rampant greed and unscrupulous collecting. Then a wholesale orchid nursery in California, USA, developed techniques for mass producing the species from seed, and made plants available quickly and cheaply, thus reducing pressure on the surviving wild populations. The species has since been used to create spectacular hybrids.

These are cool-growing orchids that need good air circulation and moderate humidity; grouping plants on humidity trays is helpful for plants in the home. They need bright, filtered light and are best grown in pots that restrict the roots. A well-drained, open medium that permits drenching when watering is best. Reduce fertilizer in autumn and winter.

57*/22 on Red List

Pitcairnia

Botanical family
Bromeliaceae

Distribution
Central and South America, West Indies, west Africa

Hardiness
Frost-tender

Preferred conditions
Well-drained, humus-rich, acid soil; filtered light or shade; high humidity.

Members of this genus are mainly denizens of tropical Central and South American forests, except for one species that also occurs in Africa – the only bromeliad found outside the Americas – *Pitcairnia feliciana*.

Most of the 260 species are evergreen terrestrial bromeliads; a handful grow as epiphytes. There are 22 species on the 2000 Red List, variously classed as Vulnerable, Endangered, Critically Endangered, and Near Threatened.

◄ BRILLIANT BRACTS *The bright floral bracts are characteristic of pitcairnias, and, in various species, are usually in striking contrast with the bright orange, green, yellow, or scarlet flowers that they surround.*

One of the most beautiful pitcairnias, the epiphyte *Pitcairnia nigra*, was once a common species in the western forests of Ecuador, but the degradation and loss of its habitats has caused this plant to become threatened.

Pitcairnia nigra is a small to medium-sized plant, with long, dark green, spoon-shaped leaves forming an open rosette. From the centre of the leaves arises a bright red spike. Along the length of the inflorescence, black flowers appear in sharp contrast to the glowing red bracts.

This species is fast-growing in good conditions. If allowed to become too dry, it will lose some of its leaves, but it will re-grow. Most species prefer a free-draining growing mix high in organic matter. Propagation is by division or by seed, which is often produced without pollination. Germination of the seed is rapid and bright light is needed to produce consistent growth. Once the young plants have several sets of new leaves, the plants can be given dilute fertilizer. If fed well, they grow quickly and flower within two years.

Red List: Vulnerable*/CITES II

Platanthera praeclara

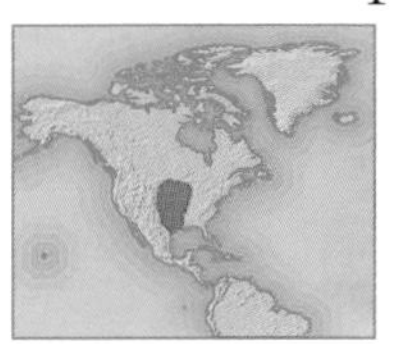

Botanical family
Orchidaceae

Distribution
Central USA

Hardiness
Fully hardy

Preferred conditions
Humus-rich, well-drained but moisture-retentive, slightly alkaline soils; sun or light shade; ample moisture when in growth, followed by a dry summer rest.

The eastern prairie fringed orchid, ***Platanthera praeclara***, is Endangered and Vulnerable in many parts of its range in central USA, and feared Extinct in South Dakota, in the Red List of 1997. The species is threatened by extensive land clearance for farming and ranching.

The plants, which bear flowers with a fringed lip, occur in grassland, hence their common name. They are usually found in boggy dips in sand prairies which are mown for hay – cutting the grass annually may be necessary to ensure continued growth and flowering of the species in nature.

Platanthera praeclara is considered one of the most beautiful in the genus, producing up to 25 sweetly scented, ivory-white flowers, each with a prominent and heavily fringed lip. The pollinators are long-tongued moths that seek nectar from the base of the floral spur.

This deciduous terrestrial sprouts annually from a dormant underground tuber and needs a slightly alkaline, humus-rich medium that mimics its native damp calcareous prairie soils.

Conservation programmes in the USA have successfully grown the species from seed to flowering and trial reintroductions are planned.

▲ SWEET-SCENTED BLOOMS *This is an American representative of a genus of around 85 species found in heaths, woodland, scrub, and grassland throughout temperate regions of the world. Some even occur on the tundra and in Arctic birch and conifer forests.*

Red List: Rare*/CITES II

Pleione formosana

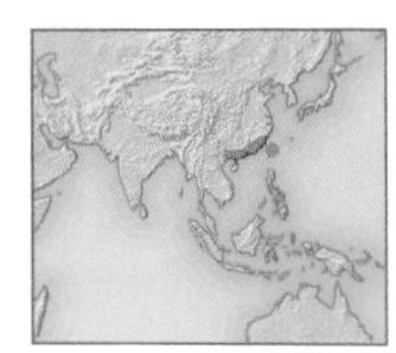

Botanical family
Orchidaceae

Distribution
Taiwan, eastern China

Hardiness
Half-hardy

Preferred conditions
High altitudes; cool temperatures; moisture during the growing season.

This orchid is rare in many locations due to overcollection; the species is a popular houseplant and garden subject in its native regions. It is Rare on the Red List of 1997. A dainty, decorative species, it grows at elevations of 500– 2,500m (1,600–8,000ft), on mossy areas rich in leaf litter, around the base of trees or on rocky outcrops. It produces nearly spherical pseudobulbs up to 3cm (1¼in) long and 3.7cm (1⅜in) wide. The deciduous leaf, produced singly at the top, is pleated and up to 25cm (10in) long and 5cm (2in) wide.

The 25cm (10in) tall bloom appears in spring from the base of the leafless pseudobulb, on a short spike bearing one or two flowers. These are 10cm (4in) across, disproportionately large for the size of the plant, and range in colour from white to magenta pink. While the sepals and petals are long and narrow, the lip is large and flared, fringed at its edge, spotted or streaked with other colours.

Pleione formosana can be grown indoors without extra heat in a shallow container, using a soilless compost or a sphagnum moss substitute. It needs regular watering, but reduce the amount once winter approaches, when the plants should be kept dry. Fertilize at intervals once the new growths appear in spring.

◄ SUMPTUOUS LIP *This beautiful species is well known in cultivation and is the easiest of pleiones to grow. It is the source of many fine hybrids, all with a disproportionately large flower.*

Unlisted/CITES II

Pleurothallis tuerkheimii

Botanical family
Orchidaceae

Distribution
Tropical Central America

Hardiness
Half-hardy

Preferred conditions
On trees in shady locations with a constant source of moisture.

This is one of an astonishing 1,000 or more *Pleurothallis* species, making the genus one of the largest in the orchid family. The plants are highly variable in form, from small tufts on tree trunks to large, bushy species. The genus is popular with orchid enthusiasts because the plants provide growth forms and flowers that are botanically complex and intriguing, while their ease of culture and compact growth makes them ideal for the smaller greenhouse or growing area.

Pleurothallis tuerkheimii is one of the more widespread species and can be locally abundant. It is mostly epiphytic, occurring in woodland and rainforest locations from near sea level up to elevations of 3,000m (10,000ft) or more. The plant produces a single large, leathery leaf above an elongated, pseudobulb-like structure called a ramicaul. The flower spike grows out from the leaf axil and all the pendulous flowers open simultaneously.

As a houseplant, this species can be grown in any free-draining material, such as bark, or can be mounted on tree fern slabs or similar supports. It grows throughout the year and needs frequent watering and regular doses of dilute fertilizer. The plant should be given some shading and may benefit from additional warmth during cold spells, though this may not be strictly necessary.

PLEUROTHALLIS TUERKHEIMII

Red List: Vulnerable*/CITES II

Pterostylis cucullata

Botanical family
Orchidaceae

Distribution
South-eastern Australia

Hardiness
Half-hardy

Preferred conditions
Coastal locations; sandy soils under shrubs; moisture during winter.

This species is becoming restricted as a result of land clearance for agriculture and urban development, and invasion by weeds. It is Vulnerable on the Red List of 1997. Though slower growing than others in the 120-species-strong genus, it is a reliable plant, forming dense colonies over several years if left undisturbed. The dormant tubers sprout in response to a drop in temperature. The oval leaves are produced in a loose basal cluster of 5–7; each leaf is 10cm (4in) long by 3cm (1¼in) wide. The solitary flowers appear on 15cm (6in) long stems in late winter to spring and last up to two weeks.

The flowers have a curious touch-sensitive lip (as do all species in the genus) and pollination is by small fungus gnats, attracted to the flower because it resembles a fungal fruiting structure. Once the gnat alights on the trigger point, the lip catapults the insect into the reproductive sections of the flower.

These plants can be easily grown in the home, in a free-draining compost containing leaf litter. They require constant moisture in the winter season. However, the plants must be given a dry, prolonged rest period of up to six months after flowering. Daughter tubers can be removed from a dormant plant and used to multiply the species. Plants should be bought from reputable specialist nurseries.

LEAFY GREENHOOD ► *With its multi-leaved rosette and distinctive hooded green flowers,* Pterostylis cucullata *has earned itself this descriptive common name.*

Red List: Vulnerable*

Puya raimondii

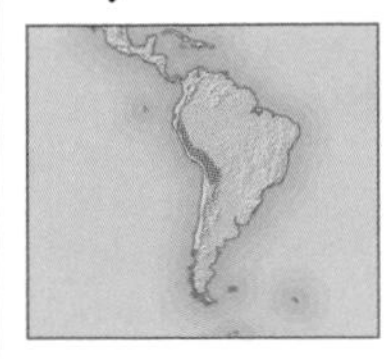

Botanical family
Bromeliaceae

Distribution
Andes Mountains of Peru and Bolivia

Hardiness
Fully hardy

Preferred conditions
Full sun and moisture at elevations of 4,000m (15,000ft) and above.

The largest known bromeliad species, ***Puya raimondii*** is classed as Vulnerable on the Red List of 1997. It occurs in the Andes of South America, and in flower the plant reaches an astonishing 10–12m (30–40ft) with a broad inflorescence, 2.4m (8ft) wide. It flowers and fruits only once, then dies (it is monocarpic), without forming offsets. It reproduces only by seed.

Puya raimondii holds the record for taking the longest time to flower. Legend has it that it takes 150 years to flower, but more recent estimates put the time it takes to reach reproductive maturity at closer to 80–100 years. From August to November 1986, a specimen flowered in the University of California Botanic Garden at Berkeley, where the milder climate sped up the time from seed to flower to a mere 28 years.

A conservation project has been set up in Bolivia, where the species has a wide distribution.

THE THRILL OF DISCOVERY

Puya raimondii was discovered by Antonio Raimondi, a scientist and botanist, who wrote: "The travelling botanist who has the thrill of surprising these strange and admirable plants while they are in flower can do nothing but stop and contemplate ecstatically for some time such a beautiful specimen.". Although it was discovered only in the 19th century, this bromeliad is now considered to be one of the most ancient plant species in the world. It is often described as a "living fossil".

Puya raimondii is a wonderful sight even when not in flower, with its rosette of tough, waxy leaves tipped with sharp spines. The spiky rosette itself grows up to 3m (10ft) tall and offers protection to a variety of birds – nests have been found within the leaves. It often flowers in groups, studding the mountainsides with its impressive spires.

RECORD FLOWER SIZE ► *This bromeliad has the largest known flowerhead and is a truly spectacular sight in bloom, with over 8,000 individual white flowers studding the spike. It generally blooms for about three months, during which time it is pollinated by up to 17 species of hummingbirds.*

Red List: Indeterminate*/CITES I

Renanthera imschootiana

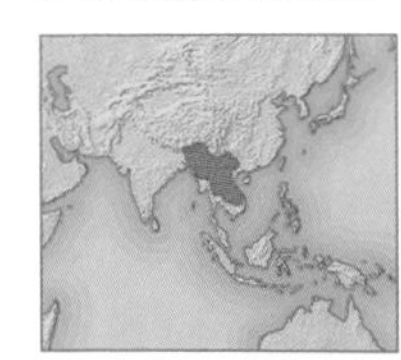

Botanical family
Orchidaceae

Distribution
India, Myanmar to Vietnam, south China

Hardiness
Frost-tender

Preferred conditions
Warm to hot temperatures, with strong light and plenty of humidity.

Intensive overcollection is currently threatening ***Renanthera imschootiana***. It is in demand locally for its spectacularly vivid red flowers, which are well displayed on long, much-branched flower spikes. A scrambling epiphyte, it grows on trees at altitudes of 500–1,500m (1,650–5,000ft). The leafy branches, up to 1m (3ft) long, are somewhat tangled, with regularly spaced, narrow, leathery leaves to 11cm (4½in) long. The flower spike is produced in summer from the axils of the upper leaves. It extends to 45cm (18in) with long-lasting flowers up to 6cm (2½in) across, often covering the plant.

This species makes a fine floral display in a large greenhouse, where it is best grown in baskets suspended from the roof. It needs an open epiphytic compost and abundant watering and fertilizing in the growing season. It is available from specialist orchid nurseries, who use artificially propagated sources.

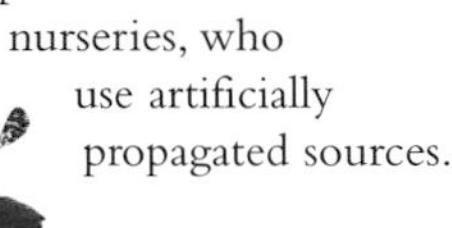

▲ COLOURFUL BLOOMS *This highly desirable orchid is one of 15* Renanthera *species native to the Asian subcontinent. Its bright flowers are produced in great abundance.*

Red List: Indeterminate*/CITES II

Rhynchostylis coelestis

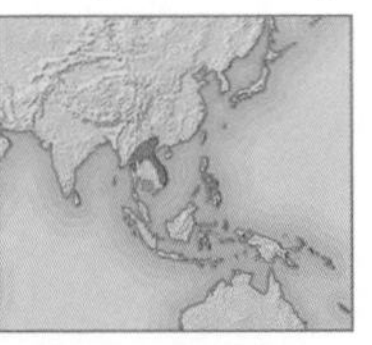

Botanical family
Orchidaceae

Distribution
Mountains of Thailand, Cambodia, Vietnam

Hardiness
Half-hardy

Preferred conditions
Moderate to warm temperatures and strong but indirect light.

One of only four species in the genus, ***Rhynchostylis coelestis*** is at risk because of overcollection for horticultural use. The unusual pale blue colour of the flowers has made it sought-after for gardens, and it is also highly prized by collectors. The plant grows as an epiphyte on large trees in deciduous tropical forests, where it has to withstand an extended dry season. It copes with this seasonal drought thanks to its fleshy leaves, 20cm (8in) long by 2cm (¾in) wide. The erect flower spikes arise from the leaf axils from summer to autumn. They open to form attractive blooms which are also highly fragrant.

This species can be grown as a houseplant in very open compost, preferably in a basket. Care should be taken to ensure that the plant is left dry between waterings, as the crown is susceptible to rot if it stays wet for extended periods. During the growing season it needs abundant moisture and regular fertilizing.

One of the most beautiful species in the genus, *Rhynchostylis coelestis* is now well established in various hybridizing programmes and it has given rise to a range of cultivars bearing flowers with the same attractive shade of blue as the parent plant.

▼ UNUSUAL COLORATION *The beautiful, dense spike of scented flowers borne by this species has given it the common name of the fox-tail orchid.*

1 on Red List*/CITES II

Rhizanthella

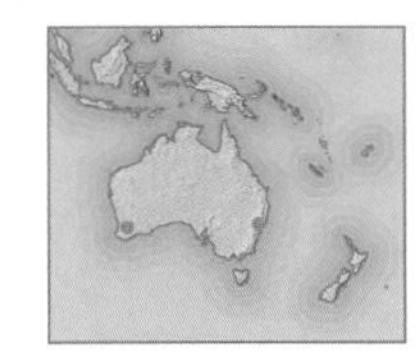

Botanical family
Orchidaceae

Distribution
Western and eastern Australia

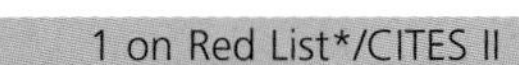

Hardiness
Frost-tender

Preferred conditions
Underground. *Rhizanthella gardneri* needs the presence of a mycorrhizal fungus and the root system of *Melaleuca uncinata* to germinate seed.

While the vast majority of orchids perch on trees or grow in fields or woods, two Australian species live underground, taking a troglodytic existence to extremes of evolutionary development. All parts of the western Australian species, ***Rhizanthella gardneri***, including the flowerhead, are buried in the soil. Surveys confirm that it survives in just a few small reserves, under threat from salt encroachment, drought, and decline in the health of the broom honey-myrtle, its symbiotic partner. It is Vulnerable in the 1997 Red List.

The eastern species, *Rhizanthella slateri*, may flower in or below the leaf litter. Its white and purple translucent bracts surround the small, spirally arranged flowers within. It was rediscovered in June 2002 by a 13-year-old boy bush-walking in New South Wales. Its conservation status is uncertain.

▼ RHIZANTHELLA GARDNERI *has bracts around an opening near the soil surface, so that insects can enter and pollinate the flower in bloom below ground.*

Red List: Vulnerable*/CITES II

Sarcochilus hartmannii

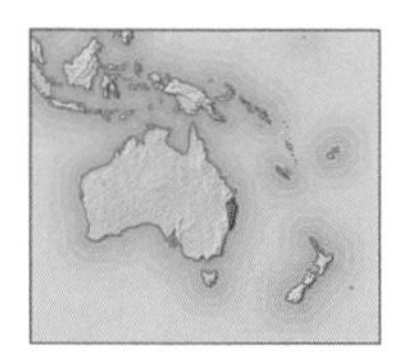

Botanical family
Orchidaceae

Distribution
Australia: Queensland to New South Wales

Hardiness
Half-hardy

Preferred conditions
High light levels, warmth and humidity, with good air circulation.

Hartmann's orchid, or orange blossom orchid, native to the Great Dividing Range of eastern Australia, is diminishing in the wild due to overcollection and land clearing. In fact, its popularity has led to some colour forms being illegally collected to the point of extinction in the wild. The species is listed as Vulnerable on the schedules of the New South Wales Threatened Species Conservation Act.

Whereas most of the 13 species in the genus are epiphytes, four species, including ***Sarcochilus hartmannii***, are rock-dwelling. They are often found growing in rainforest and moist temperate woodlands on boulders, cliff faces, and other exposed sites up to elevations of 1,000m (3,300ft). The species may grow in shady ravines or highly exposed cliff ledges and it tolerates a huge temperature range, from 0°C (32°F) to 38°C (100°F). The plant produces multi-branching stems to 50cm (20in) long with up to ten fleshy to leathery leaves, 20cm (8in) long by 2cm (¾in) wide. The 25cm (10in) flower spikes are arching to erect and produce up to 25 glistening, thick-textured white flowers with red centres that resemble orange blossom in both shape and size.

◄ HARTMANN'S ORCHID *has blooms similar to orange blossoms, hence its other common name. The plants make handsome specimens and form an impressive massed floral display in a short period.*

This species is easy and rewarding to grow in a cool (not cold) greenhouse, if it is provided with very free-draining and open compost and high levels of light, with good air movement and humidity. It should be kept moist throughout the year. Buy from specialist orchid nurseries which use cultivated sources to propagate the plants.

1 on Red List*/CITES II

Sophronitis

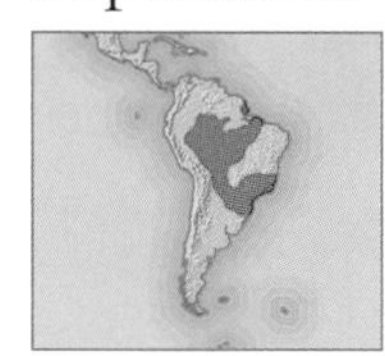

Botanical family
Orchidaceae

Distribution
Eastern Brazil, Bolivia, Paraguay

Hardiness
Half-hardy

Preferred conditions
Good light, free air movement, with moisture in the growing period and a dry spell in winter.

The genus comprises about 70 species, including rock-dwellers that were once included in the genus *Laelia*. It exhibits some of the most vivid, sought-after orchids in cultivation. Concern has arisen because of intense pressure by collectors over more than a century and a half and, more recently, clearing of land for ranching in Brazil, Bolivia, and Paraguay. *Sophronitis mantiqueirae* var. *varonica* is classed as Endangered in the Red List of 1997.

The plants grow as lithophytes on rocky outcrops at moderate to high altitudes, often in exposed and well-lit conditions. They range in size from medium to miniature and produce a single leathery leaf. Many species have highly reduced pseudobulbs and need moist conditions throughout the growing season as well as a distinct dry rest period (except for the species *Sophronitis coccinea*).

In cultivation, plants need free-draining epiphytic compost, high light levels, and good air circulation. The compact size and ease of growth of some species has made the genus popular with growers, especially where space is limited.

Formerly classed with the *Laelia* genus, *Sophronitis jongheana* is a magnificent species. Discovered in 1854, it became highly desirable during the "orchidmania" of the late 19th century. As a result of this early pressure and continued collection for horticulture, it is close to extinction in its native range in Minas Gerais, Brazil, although it is not Red Listed. This species is epiphytic in rainforests at 1,300–1,600m (4,000–5,000ft). The pseudobulbs are slightly compressed, 5cm (2in) long, and have one terminal leaf. The terminal flower spike, 4cm (1½in) long, produces two large and spectacular flowers, 10–15cm (4–6in) across, from late winter to early spring.

Sophronitis jongheana can be grown at home under intermediate conditions (it needs winter heating in cold climates). It should be potted up in a free-draining mix of composted bark or similar medium, watered and fertilized in the growing and flowering season, and allowed to dry out during winter dormancy. Plants are easily propagated from seed, and many fine hybrids have been obtained from this species.

Sophronitis lobata (Unlisted) is in demand for its large, decorative blooms and its ease of culture, and as a result is threatened by overcollection. Plants are now restricted to rock outcrops and occasionally grow as epiphytes on trees at elevations of 200–800m (650–2,500ft) above Rio de Janeiro in Brazil. The club-shaped pseudobulbs, 10–20cm (4–8in) tall, bear a single leathery leaf. The 35cm (14in) flower spike is produced from spring to early summer and carries six or more flowers, each 15cm (6in) across. This species also requires an open compost of bark or similar mix and can be mounted or grown in a basket to ensure good air movement. Reputed to be shy to flower, the plants are best maintained undisturbed for some years.

▲ SOPHRONITIS LOBATA *is Endangered in the wild, although artificially propagated plants are readily available from specialist orchid nurseries.*

2 on Red List*/CITES II

Spathoglottis

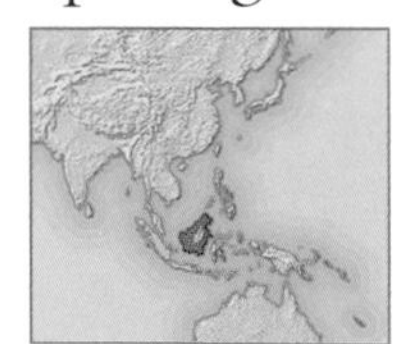

Botanical family
Orchidaceae

Distribution
Borneo

Hardiness
Half-hardy

Preferred conditions
Semi-shade to full sun in tropical or sub-tropical climates; well-drained, humus-rich soil.

These highly decorative orchids form a genus of about 40 species, confined to habitats in south-east Asia and Australia. *Spathoglottis* are reliable as garden plants when grown in tropical to subtropical climates in positions of semi-shade or full sun. They are often used in landscaping too, for their almost perpetual flowering habit as well as their attractive pleated and shapely leaves.

The terrestrial species *Spathoglottis confusa* is locally rare and restricted, due to the loss of its habitat in Borneo (Rare in the Red List of 1997). Land has been claimed for farming and for development from both lowland forest and lower mountain forest at elevations of 100–1,800m (300–5,500ft). This orchid grows to 90cm (3ft) tall and has two or three pleated leaves, each up to 60cm (2ft) long and 5cm (2in) wide. They are produced on slender stalks arising from semi-buried, rounded pseudobulbs. Flowers up to 7cm (3¼in) across open sequentially on slender stalks up to 80cm (32in) long in summer.

Spathoglottis confusa is readily grown in well-drained soil that is rich in

8 on Red List*/CITES II

Spiranthes

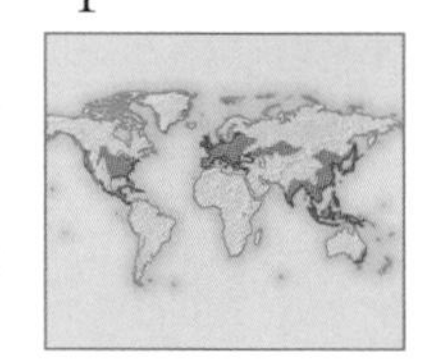

Botanical family
Orchidaceae

Distribution
USA, Europe

Hardiness
Fully hardy

Preferred conditions
Dry, open locations, preferably in limey soil, in indirect sun or light shade.

Eight species of this large genus are included on the Red List of 1997, and three of those are listed as Endangered. All are listed on CITES II.

Spiranthes longilabris is restricted to the south-eastern USA, where the species favours low, wet pine woodlands and prairies, and is becoming uncommon in some locations because of habitat loss. It is listed as Rare in the Red List of 1997. A terrestrial species, it sprouts from a dormant tuber by producing a tuft of 3–5 grass-like leaves which are keeled and rigid. The inflorescence is up to 50cm (20in) tall and is produced in autumn once the leaves of the orchid have died down. Up to 30 flowers emerge in a single rank, arranged in a spiral.

The plant can be grown in a specialist terrestrial mix and needs abundant watering when in growth, less so when the plants are resting after flowering. Its obvious attributes and late flowering make this species an interesting subject for container and garden culture.

Spiranthes spiralis is a widespread species. It grows in the south-eastern USA and in Europe, from the Mediterranean to Scandinavia and from Britain to Turkey. It is one of the hardiest terrestrial orchids and grows in a variety of habitats, but prefers low-lying, wet pine woodlands and short, mown grassland. It is also suffering from habitat loss in some situations.

Spiranthes spiralis sprouts from a distinctive small cluster of tubers during the growing season, from winter through to early summer, when the species dries back to the dormant tubers. The flower spike, up to 50cm (20in) tall, starts growing in autumn, before the leaf rosette, so the plant should not be disturbed during this growth phase. Up to 30 flowers, 10mm (½in) across, are produced on a single spike. The leaf rosette comprises 3–5 grass-like leaves up to 15cm (6in) long and 5mm (¼in) wide. This species is also known as long-lip ladies' tresses or autumn ladies' tresses.

In gardens this species may spread to form clumps. It should be carefully managed to ensure that plants do not naturalize in wild areas outside the garden, as has in fact occurred in some areas of Scandinavia.

SPIRANTHES SPIRALIS ► *Ladies' tresses, the old English name for members of the genus* Spiranthes, *refers to the spiral arrangement of the flowers, like a plait of hair.*

Red List: Vulnerable*/CITES II

Stanhopea tigrina

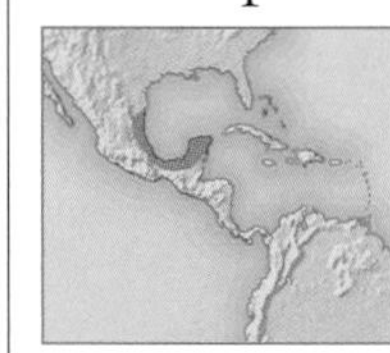

Botanical family
Orchidaceae

Distribution
Mexico

Hardiness
Half-hardy

Preferred conditions
Forest trees with good light and free air circulation.

This species is threatened in its native habitat in Mexico because of over-collection; it is in demand for its large, conspicuously patterned flowers. An epiphytic orchid, ***Stanhopea tigrina*** grows on stout branches in forest areas and produces its flowers on pendent spikes suspended below the branch on which it grows. Its clusters of large, grooved, egg-shaped pseudobulbs are each up to 6cm (2¼in) tall, while the solitary ribbed leaf is 35cm (14in) long and 10cm (4in) wide.

The plant's most remarkable feature is the large, fleshy flowers, produced from summer to autumn on the spikes that emerge from the base of the pseudobulb. They hang below the plant, displaying the pendulous blooms most clearly. The impressive, highly fragrant flowers, which are 12cm (5¼in) or more across, have a heavy, waxy texture and a striking tiger-like coloration – though each flower lasts only a matter of days.

BOTANY

KRAKATOA

The huge eruption in 1883 of the famous volcano Krakatoa is thought to have wiped out all life on three of the four Krakatoa islands. This provided botanists with a rare opportunity to chart the colonization of a virgin landscape. A survey 13 years later, in 1896, revealed a *Spathoglottis plicata* to be one of three orchids among the pioneers to re-colonize the bare ground, together with a *Cymbidium* and an *Arundina*. By 1906 *Spathoglottis* was recorded as being common on the islands. It is thought that the tiny seeds were easily blown across the 40-km (30-mile) stretch of water between these islands and Java, then rapidly across the devastated open landscape.

humus, provided it is given good light, with adequate moisture and regular fertilizing, particularly during the growing season.

The species produces a remarkable display of flowers even indoors when grown as a specimen plant. Because of their flowering habit, the plants must be grown in fibre-lined, slatted hanging baskets, in a free-draining epiphytic compost, so that the flowers can push down through the compost and out of the base of the basket. They require good light, free air movement, and some protection from frosts. They should be kept moist throughout the year with regular, moderate fertilizing. Plants from artificially propagated sources are readily available from most specialist orchid nurseries.

▲ POLLINATOR GUIDES *Two projections at the base of the fleshy lip guide pollinating insects into the flower at the correct angle.*

7 on Red List*/CITES II

Thelymitra

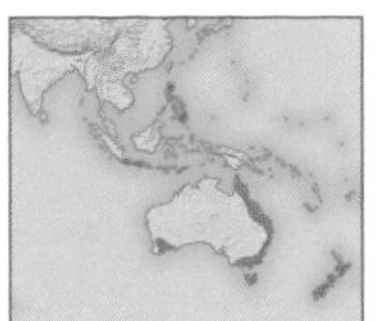

Botanical family
Orchidaceae

Distribution
Australia, New Zealand

Hardiness
Frost-tender

Preferred conditions
Warmth and bright light in the growing season; cool and shade in the dormant season.

This genus of mostly Australian and New Zealand terrestrial orchids has many species noted for their spectacular coloration. This ranges from clear blue to rich purple and can be almost iridescent in ***Thelymitra variegata***. Some species are highly scented, with fragrance ranging from lemon to a remarkable cinnamon scent in *Thelymitra stellata*. Seven species of *Thelymitra* are on the 1997 Red List (two Rare, two Endangered, and three Vulnerable), all as a result of extensive clearing for agriculture or for building, as well as some illegal collecting.

Thelymitra are known as sun orchids due to the flowers' remarkable ability to open and close on a daily cycle, with the opening linked to temperatures above 15–18°C (59–64°F). Because the flowers need to open every day, they have developed a modified lip that resembles the other parts of the perianth.

These orchids dwell in woodland, from mallee scrub to Mediterranean forest, with subtropical species found in savanna woodland. The tuberous plants produce a single leaf from dormant tubers at the onset of rainfall.

Thelymitra apiculata, Rare on the Red List of 1997, grows under remarkably harsh conditions, often in sparse scrub where it is exposed to direct sunlight with little protection. The species is summer dormant and sprouts in response to cooler, moister winter conditions from an underground tuber.

Thelymitras can be grown indoors in free-draining terrestrial orchid compost. They require brightly lit conditions and good air circulation in the growing season. Once flowering has finished, the plants should be gradually dried off and maintained cool and dry during the dormant period.

THELYMITRA VARIEGATA ► *This striking species produces up to 15 flowers in mid- to late winter on a 30cm (12in) stem that is supported by a single, wide, erect leaf. In cultivation, the plants require warm to cool conditions and low humidity, and grow well in a free-draining, sandy loam mix that is kept moist (but not wet). It is Unlisted.*

Because of their need to open daily, sun orchids have modified the third inner petal, or lip, to resemble the other parts of the perianth. This gives the flowers a regular arrangement uncharacteristic of most other orchids.

100*/22 on Red List/CITES II

Tillandsia

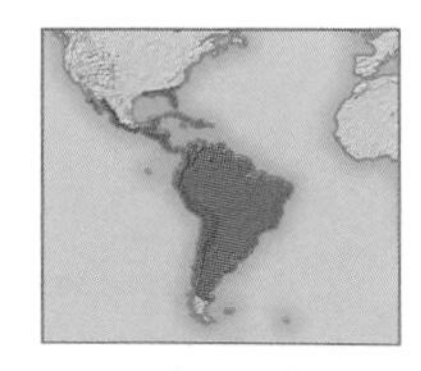

Botanical family
Bromeliaceae

Distribution
North, Central, and South America, Caribbean

Hardiness
Frost-tender – fully hardy

Preferred conditions
Tree branches or rocks; good air circulation; depending on the species, high rainfall to drought conditions; bright or indirect light.

HABITAT

SPANISH MOSS

Tillandsia usenoides, which grows from the southern USA down to Argentina, is also known as Spanish moss. Its Latin species name, meaning "looks like Usnea", a lichen, was bestowed by Linnaeus (*see p.30*). It is most commonly an epiphyte of oaks or cypresses, although it is found on other tree species. Its long, filamentous masses can hang down in lengths of up to 7m (20ft) and provide a home for wildlife, including birds, snakes, and bats. Spanish moss was once prized as a stuffing for mattresses, and it padded the seats of the first Model T Ford cars. Today the US Geological Survey tests samples collected along main roads to monitor levels of pollution absorbed by the plant's scales.

Currently 25 species of *Tillandsia*, the air plants, appear in the 2000 Red List, because commercial collection of the genus in large numbers over many years has badly affected wild stocks. Three species are considered as Near Threatened, and the remainder are classed as Threatened. *Tillandsia's* range extends from Central Baja California to southern Chile on the west coast of the Americas, and on the eastern side species are found from northern Florida to southern Texas, and throughout the Caribbean to southern Brazil and southern Argentina.

This is the largest genus in the bromeliad family, with over 400 species. Most tillandsias are quite small, with decorative foliage. Although the majority come from areas of high rainfall and are smooth and green, the air plants familiar as horticultural curiosities are covered in white scales, known as trichomes. These cup-shaped structures both prevent water loss and soak up moisture and nutrients from the air. The plants can survive for a long time without water, and in times of extreme drought become dormant until reawakened by a shower. This has made many such species popular as house-plants, because they can survive almost any amount of neglect.

Many air plants grow on rocks and dry tree branches. The types with a grey or white coating on their leaves are adapted to bright light as well as to drought conditions; in fact they dislike low light levels. Those with smooth green leaves are from rainforests and need warm, humid, gloomy conditions to thrive. They can also be successfully grown indoors.

Tillandsia cyanea, the pink quill, was once abundant in the tropical rainforests of western Ecuador but is now restricted to rocky outcrops. The species was once exported by the tens of thousands when millions grew in the wild, but now the host trees are almost extinct due to human activities. Today millions of young plants are produced annually from seed and tissue culture in Europe and the USA; in fact more plants are produced in nurseries than grow in the wild. The species is Near Threatened on the 2000 Red List.

TILLANDSIA MATUDAE

◄ TILLANDSIA CYANEA *has spectacular cuttlefish-shaped bracts growing out of the rosette of fine green leaves. The short-lived true flowers are bright purple and appear singly from each notch of the pink bract.*

Tillandsia dyeriana is endemic to a narrow area of mangrove swamps near Guayaquil, Ecuador, a habitat currently threatened by the fast-growing shrimp industry. The species is near extinction in the wild (Critically Endangered on the 2000 Red List), but many plants are now grown from seed in commercial quantities.

A native of southern Mexico and Guatemala, *Tillandsia eizii* is found at lower elevations in wet tropical forests at an altitude of 1,200m (3,600ft). Epiphytic on trees, this slow-growing bromeliad is not Red Listed but is subject to habitat loss due to destruction and predation by squirrels in the few remaining sites. The plant is used in ceremonial rites by the local Maya Indians who live near San Critobal de Las Casas in Chiapas, Mexico.

As the species becomes scarce in the wild, the specimens being brought to market are becoming smaller and poorer in quality. The dual use of the forest for ceremonial and commercial purposes has severely reduced numbers of this once-abundant species.

Tillandsia magnusiana (Endangered on the 1997 Red List) is rapidly disappearing due to both habitat loss and local commercial exploitation. It is native to temperate oak and pine forests at 1,100–1,600m (3,500–5,000ft) in southern Mexico, near the state of Chiapas. A decorative epiphyte, it is highly coveted by the local Maya Indians and during the holiday season is used in ceremonies.

Although not Red Listed, the epiphyte *Tillandsia platyrhachis* is severely threatened due to habitat loss. It only grows on trees in primary forests and is native to the tropical rainforests of eastern Colombia, Peru, and Ecuador. When in bloom, its many large,

TILLANDSIA INSIGNIS ► *Hummingbirds, such as this stunning magenta-throated woodstar, are drawn to the brightly coloured flower bracts of* Tillandsia *species. In April 2004* Tillandsia insignis *was transferred to the genus* Werauhia.

TILLANDSIA FASCICULATA *This variable species grows in a wide range of altitudes in Florida and the Caribbean area to South America. The erect flower stalk carries brightly coloured bracts bearing 5–12 flattened branchlets.*

feathery, pink spikes can be seen from a considerable distance, which helps to attract hummingbirds to the fragrant violet flowers. Through propagation by division or seed, more plants are now becoming available.

The rapid habitat destruction of *Tillandsia umbellata*, if it continues unabated, will cause this tropical rainforest plant to be nearly extinct within ten years. It is classed as Endangered on the 2000 Red List. The small plant is nonetheless a strong candidate for use in horticulture, with its narrow green leaves, large blue flowers with white centres, and sweet fragrance.

In general tillandsias prefer bright but indirect light and good air circulation. They dislike being wet and cold, and when growing them as houseplants it is best to water them early in the morning. Species with white coating need only a light spray every now and again. Weak fertilizer is only necessary during the warm summer months when the light is bright; the foliage should be carefully washed afterwards.

The fruits contain many white hairy seeds with a pappus that aids seed dispersal, like dandelion seeds. To propagate them, sow seed on a well-drained medium in a brightly lit place. Although humidity aids germination, constant watering can cause the young plants to dampen off. Tillandsias grow very slowly, especially from seed. After several years, the young plants may be only 1cm (½in) high and it could take ten years or more for a plant raised from seed to flower.

OTHER THREATENED SPECIES

Tillandsia confertiflora is a native of tropical dry forests in southern Ecuador and northern Peru, where it is scarce as a result of habitat destruction. The leaves are covered with fine grey scales; the mauve flowers are fragrant, especially in the morning. Unlisted.

Tillandsia kautskyi A small, silvery species from a cool and humid climate, this plant multiplies readily by division and has been in the horticultural trade for many years. It was discovered on the farm of Roberto Kautsky, where there was rainforest before agriculture severely degraded the area. CITES II; Red List (1997): Endangered.

Tillandsia moscosoi is found only in a small area of rainforest in the Dominican Republic, where the few remaining trees will probably be cut down soon. Growing at elevations of 1,200m (3,500ft), this species can tolerate near freezing temperatures if kept dry. Unlisted.

Tillandsia wagneriana Cultivated in Europe as a houseplant for many years, this is a spectacular native of northern Peruvian and southern Ecuadorian rainforests, where it is severely threatened with extinction. The plant is grown commercially from seed. Unlisted.

15 on Red List*/CITES I and II

Vanda

Botanical family
Orchidaceae

Distribution
India to south-east Asia, Philippines, Australia

Hardiness
Half-hardy

Preferred conditions
Epiphytic on tree branches; humidity; plenty of sunshine.

While large, sky-blue flowers of any kind are rare and remarkable, large sky-blue orchids are bound to arouse interest. *Vanda*, an epiphytic genus found from north-east India to southern China and Thailand, bears blooms, 8–10cm (4in) across, in varying shades of blue and purple, with darker tessellated markings.

When ***Vanda coerulea*** was discovered in 1837, it was not rare in its native habitat – deciduous tropical forests with a distinct dry season found at elevations of 750–1,220m (2,500–4,000ft). But enormous numbers were collected and sent to Europe, and within a few decades the orchid had been stripped from the small oaks that are its favoured host trees. This was the first orchid to receive legal protection from its native government, India. In the 1970s it was listed on CITES Appendix I, giving it the highest level of protection from unsustainable international trade. But the species never recovered and indeed has continued to decline. It is classed as Rare in the Red List of 1997. Plants are now available from artificially propagated sources, but have been bred to produce blooms of show standard, rather than to maintain the genetic diversity necessary to survive and adapt in the wild.

Vanda sanderiana was until recently thought to be a lone species in the closely related genus *Euanthe*. This beautiful orchid has suffered from extensive overcollection for horticulture in the Philippines, where it grows as an epiphyte in rainforest trees. It is classed as Vulnerable in the Red List of 1997.

Along with *Vanda sanderiana*, *Vanda coerulea* has been the foundation on which nearly all *Vanda* hybrids, and many intergeneric hybrids, have been created. These species were also used to produce the remarkably beautiful hybrids widely used today in the orchid cut-flower trade.

▲ VANDA COERULEA *In the wild, the flowers are produced in autumn and winter on stems 75cm (30in) long. Up to 20 ravishing blooms can occur on a single flower spike.*

10 on Red List*/CITES II

Vanilla

Botanical family
Orchidaceae

Distribution
South America, tropical USA, Africa, Malaysia

Hardiness
Frost-tender

Preferred conditions
Swamps or leaf litter; good light; excellent air circulation.

This genus of around 100 species, mostly climbers, is native to the tropics and subtropics.

The leafy, or oblong-leaved, vanilla orchid, *Vanilla phaeantha*, is naturally limited in distribution, though it is widespread throughout swamps in Florida and the Caribbean islands to Suriname. Classed as Rare in the Red List of 1997, it is one of the largest-

31*/10 on Red List

Vriesea

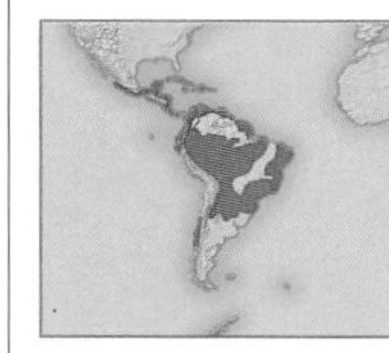

Botanical family
Bromeliaceae

Distribution
Southern USA, Central and South America, West Indies

Hardiness
Frost-tender

Preferred conditions
Tree branches or rocks; temperatures warm or cool, depending on native habitat; partial shade.

Vrieseas can be divided into two groups. The green-leaved types, similar to guzmanias (*see pp. 402–403*), prefer a humid environment; the grey-leaved types are similar to tillandsias (to which they are related, *see pp. 420–421*), and prefer generally drier conditions. The only difference between tillandsias and vrieseas is that the latter have a nectar scale at the base of each petal. Work by botanists may yet reshuffle which plants belong to which genus.

The green-leaved ***Vriesea hieroglyphica*** is an epiphytic species endemic to the Brazilian Atlantic rainforest. It reaches 1.5–2m (4½–6ft) in height and has been called "the king of bromeliads". Commercial exploitation throughout its range is now coupled with rapidly increasing habitat destruction, and although it is Unlisted, this once-common species is now nearly extinct in the wild.

The green-leaved species are usually epiphytic, but are often found growing on rocks (especially the large, watertight tank types in Brazil). Vrieseas with soft green leaves will grow best indoors under glass, given some shade, and with their central cups kept topped up with water in the growing season. They prefer an acidic compost, with a pH range of

◄ VANILLA PLANIFOLIA *The flowers of this species are large, showy, and either yellow or green, but individually they are very short-lived, often flowering for less than 24 hours. In cultivation the plants are shy flowerers until they reach a considerable size.*

flowered orchids in the continental USA. A tall, robust, scrambling vine, it has roots at each leaf node, enabling it to climb up the trunks of trees to the light that is essential if it is to flower.

The leaves are fleshy, green, 10cm (4in) long, and 3.5cm (7in) wide. The flower spikes produce up to 12 flowers, 15cm (6in) wide, which are very short-lived (usually only a matter of hours), in late winter to summer.

The seedlings germinate in rotting leaf litter, stumps, or logs and climb rapidly towards the light. The plant is only suitable for larger greenhouses; it can be mounted on slabs of tree fern fibre or similar and requires good light and air movement, with regular watering.

Some species of vanilla, such as *Vanilla humblotii*, grow in desert conditions, creeping along the ground. Their leaves are reduced to tiny leaf bracts and their roots are stunted, but their fleshy stems retain enough moisture for survival.

CULTIVATION

COMMERCIAL CROP

Today, ***Vanilla planifolia*** is the only orchid to be grown as a commercial crop for its fruit, dried and fermented as vanilla pods, which are in worldwide demand for use as vanilla flavouring and vanilla essence. The species is one of the most primitive orchids, originally cultivated for its flavouring by the Aztecs of Mexico, then introduced into Europe by the Spanish conquistador Cortés in the 16th century. A native of tropical South America, the plant scrambles through tree branches to climb high into the canopy on its long, vine-like stems. Its leaves are thick and fleshy and its roots only occasionally hang down to the ground. Once vanilla orchids have reached a certain size, their flowering and fruiting is almost continuous, if they have plenty of sun. Annual world production of vanilla pods is estimated at around 1,500 tonnes.

5–6, or can be grown on pieces of bark. Slower growing than guzmanias from seed and cuttings, vrieseas require less fertilizing. The leaves should be carefully washed after fertilizer application. The grey-leaved species should be treated exactly like their *Tillandsia* cousins.

Vrieseas often produce seed without fertilization, but it takes a long time for the green-leaved types to flower. They produce large feathery seeds resembling *Tillandsia* seeds. Patience will reward the hobbyist with a truly wonderful plant, however. After the initial year or two, in which the young seedling seems to struggle for existence (an adaptation to survive the dry season), an explosion of growth occurs and the plant doubles in size every six months or so. Those living in cooler environments will find vrieseas easier to grow than guzmanias. They can be grown outdoors under shade cloth, if temperatures do not fall below freezing. Vrieseas thrive on neglect.

▲ VRIESEA HIEROGLYPHICA *Named for the beautiful hieroglyphic-like dark green markings on its lustrous wide leaves, this species bears cream flowers on a tall stalk; they are bat-pollinated and stay open at night.*

Unlisted

Wittrockia cyathiformis

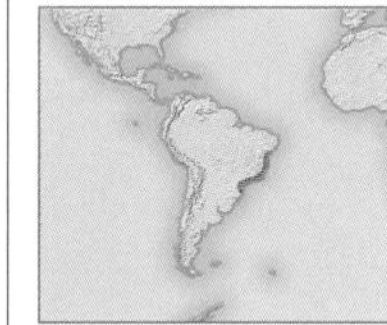

Botanical family
Bromeliaceae

Distribution
Eastern Brazil

Hardiness
Half-hardy

Preferred conditions
On trees or rocks with good air circulation, cool temperatures, and high humidity.

Once wide-ranging in the Atlantic rainforest of eastern Brazil, between the states of Bahia to the north and Paraná to the south, *Wittrockia cyathiformis* is now confined primarily to the protected areas of the Serra dos Orgaos and Bocaina national parks. Found at elevations between 800m and 2,000m (2,500–6,000ft), the species is highly adaptable. It is found growing on rocks, on the forest floor, or as an epiphyte in the trees of the rainforest. However, only seven per cent of its original forest remains, so tropical gardens and greenhouses could be the last refuge for this plant.

Bearing the largest floral "scape" of all the species in the genus – the name refers to the cup-shaped flowerhead – this species has striking, rose-pink bracts surrounding the yellow petals. The 30cm (12in), lime-green leaves are armed with sharp spines, making the plant most suitable for places where it can grow with plenty of space around it.

Wittrockias have been available in the horticultural trade for many years and are readily found in bromeliad nurseries, where they can be bought by mail order. They should be grown under glass in epiphytic compost in bright, filtered light. *Wittrockia cyathiformis* is hardy to 5°C (41°F). Propagation of this magnificent species is by division or seed.

HABITAT

COASTAL FOREST RESERVES

One-third of the remaining Brazilian Atlantic rainforest is now officially under some sort of protection, through national parks as well as biological reserves and ecological stations. The Serra dos Orgaos and Bocaina national parks, for example, created in 1971, are covered in the lush, jungle-like vegetation that once blanketed much of the Atlantic coast north and south of Rio de Janeiro. Most was cleared to grow coffee or for logging Brazilian hardwoods. Initiatives to conserve the biodiversity of plants and animals by protecting their habitats include plans for Brazil's first canopy walkway.

CARNIVOROUS PLANTS

These plants grow in poor soils and supplement their nutrition by capturing prey. Carnivorous plants are primarily insectivorous, trapping insects that range in size from microscopic *Daphnia* to houseflies and wasps. Other animals, such as frogs and even the occasional small mammal, are also accidentally captured by large species. Plants are sometimes precisely adapted to particular prey: for example, different species of *Nepenthes* capture either termites or ants. Carnivorous behaviour has evolved in over 600 carnivorous species in 18 genera spread across eight families. Many of these plants are under pressure, and several are listed on CITES.

TRAPPING MECHANISMS

Carnivorous traits have evolved several times, resulting in a variety of forms. These range from the thumbnail-sized, sticky *Drosera* to the huge pitcher traps of *Nepenthes*, and the bladders of *Utricularia,* which suck in the prey.

Some plants use enzymes to digest the prey and then absorb the nutrients. Some, called para-carnivorous plants, need another organism to break down the prey for them. Others, such as *Nepenthes*, can partly break down the prey, but are aided by bacteria living in the pitchers. The bladders of the genus *Utricularia* may have evolved to house the micro-organisms that live in them and help digest prey.

◄ **MUTUAL BENEFIT** *Assassin bugs* (Permeridea morlothii) *eat the prey trapped by* Roridula dentata. *The bugs' faeces in turn fertilize the plant.*

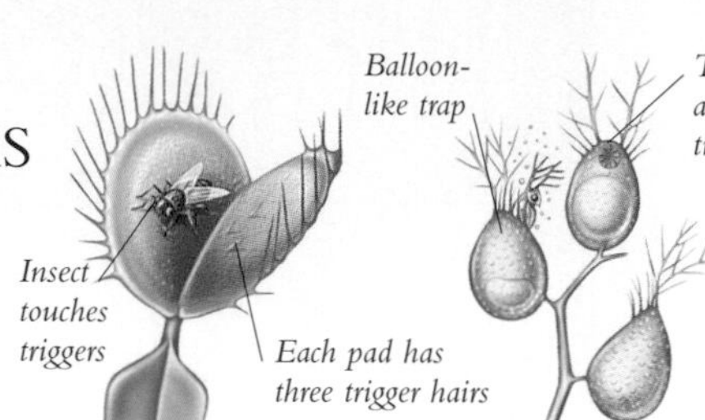

STEEL TRAP *This trap slams shut when the trigger hairs are tripped. It is best known in the popular Venus flytrap* (Dionaea muscipula).

BLADDER TRAP *This is found in* Utricularia *and other bladderworts. The traps have a vacuum inside; when triggered, it sucks in the prey.*

PITFALL *This is typical of pitcher plants such as* Sarracenia. *A trail of nectar guides insects into the mouth of the pitcher, where they fall in.*

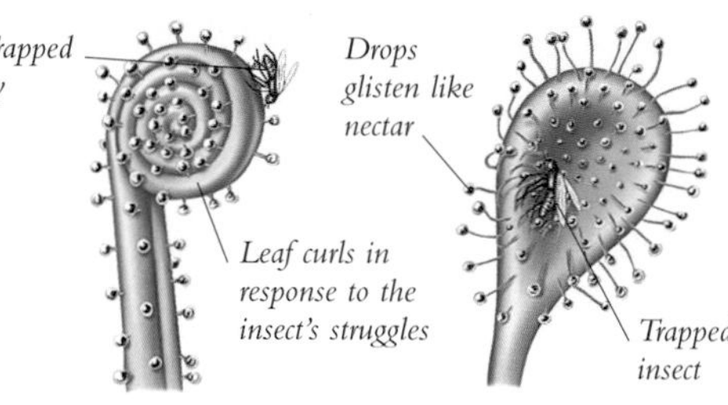

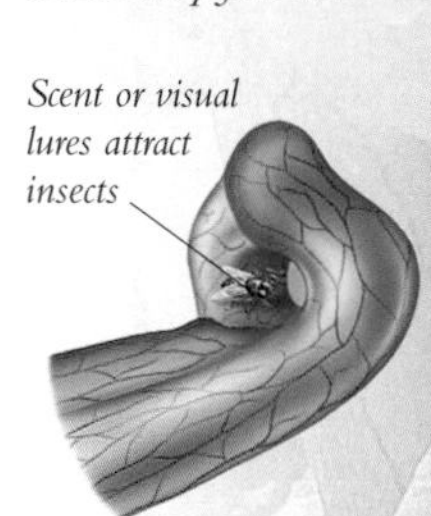

ACTIVE FLYPAPER Drosophyllum lusitanicum *first traps insects on sticky hairs. The leaves then slowly move to hold them more securely.*

PASSIVE FLYPAPER *The sticky droplets secreted at the ends of hairs on species such as* Drosera capensis *trap any insect that lands on them.*

LOBSTER POT *Prey lured into the traps of* Sarracenia psittacina *is funnelled deeper by hairs that allow only one-way movement.*

PEAT BOG ► *A valley bog, such as this in the New Forest in the UK, supports a range of plant types. Sphagnum moss and carnivorous sundews are found around the edges, where nutrient levels are lowest. Bog myrtle and moor grasses grow in the richer soil in the middle.*

◄ **TROPICAL HABITAT** *Most of the nutrients in a lush tropical rainforest are locked up in the plants; the underlying soils are actually nutrient-poor. Carnivorous genera such as* Nepenthes *establish on the ground, while epiphytic genera such as* Utricularia *grow high in the moss-covered trees.*

WHERE THEY GROW

Carnivorous plants evolved to survive in environments with extremely low nutrient levels. The peat bogs of the boreal region are a prime example, but the dry bushland of Western Australia also has these characteristics, although it only receives moisture in winter and is powder dry in summer. As a result, carnivorous plants are found from the cold of Siberia, where *Pinguicula variegata* grows, to the hot, steamy rainforests of Borneo, which are home to *Nepenthes*.

Many temperate genera survive harsh winters by producing resting buds, while some of the sundews and a few bladderworts of Western Australia produce tubers to survive dry seasons, and *Sarracenia* have rhizomes.

Most carnivorous plants are terrestrial. Epiphytic types are rare, being restricted to only a few species of *Nepenthes*, *Pinguicula*, and *Utricularia*; a number of *Utricularia* species are also aquatic. No woody carnivorous plants have been recorded.

COBRA LILY *The sole species of the genus,* Darlingtonia californica *is found on the west coast of North America. The "fangs" hanging from the tip of the hood are covered in nectar to attract insects into a lobster-pot trap.*

HISTORICAL TRADE

The first tropical pitcher plant was described by Étienne de Flacourt, governor of Madagascar, in the 1650s. In the 18th and 19th centuries, colonial trade routes brought a wealth of rare and bizarre plants from exotic lands to a fascinated European public. With technological advances, including glass Wardian cases for transporting plants, and the use of stove houses to recreate tropical conditions, the popularity of carnivorous plants surged during the 19th century. *Nepenthes* were a particular favourite, first introduced into the Royal Botanic Gardens, Kew in the 1780s, and exhibited in 1843 at the Royal Horticultural Society's annual show. New hybrids flooded the market, and by 1879 plants were being propagated by the thousand to satisfy demand in Europe. In the first half of the 20th century economic slumps and war slowed demand, but today the sales of carnivorous plants, and the existence of one international and over 10 regional carnivorous plant societies, are testament to their enduring popularity.

▲ **INFINITE VARIETY** *The differing trap mechanisms found in carnivorous plants have intrigued gardeners and biologists alike for centuries. The enthusiasm for exotic species reached fever pitch among the Victorians.*

HISTORY

CHARLES DARWIN

Carnivorous plants challenge our intuitions about how animals and plants interact, since it is more usual for animals to eat plants than vice versa. Early accounts of carnivorous plants in the tropics were peppered with lurid tales of huge man-eating traps and bladderworts that devoured crocodiles.

Naturalist Charles Darwin (1809–1882) shared the Victorian fascination with carnivorous plants but not the credulity. His book *Insectivorous Plants*, published in 1875, was the first major work on these species. It not only outlines the natural habits of the plants in the wild, but also catalogues in great detail an exhaustive range of experiments carried out upon them with different substances and stimuli. Darwin's experiments began with plants native to the UK, but he considered the Venus flytrap (*Dionaea muscipula*) from North America "one of the most wonderful" plants in the world.

THREATS TO THE PLANTS

The threats faced by carnivorous plants are very similar to those faced by other plants. Habitat loss, pollution, agricultural run-off, and soil erosion take their toll, and changes in land management create openings for invasive species. The remaining populations are vulnerable to overcollection.

Habitat loss is the main threat. Carnivorous plant habitats are not productive, so they are often developed or drained for pasture. In many cases, the decline in carnivorous plants is synonymous with the decline in wetland habitats. This is illustrated by the case of *Sarracenia* pitcher plants, which are endemic to the eastern states of the USA, including Alabama. Alabama has lost over 50 per cent of its wetlands in the last century, and today it is estimated that just 2.5 per cent of the original *Sarracenia* habitat remains. If the water table is lowered or fires are suppressed, woody plants grow up and herbaceous plants are shaded out. The open, nutrient-poor and species-rich habitats are soon lost.

Species that are reduced to ever-smaller habitat fragments have little or no buffer to offset any encroachment from human activity. They are also susceptible to collectors seeking plants from obscure locations or showing new variations.

▲ **DRAINING A RAISED BOG** *Conversion to agriculture is a common reason for drainage of bogs. Ditches like this one in the UK lower the water table, posing a threat to a wide area of habitat.*

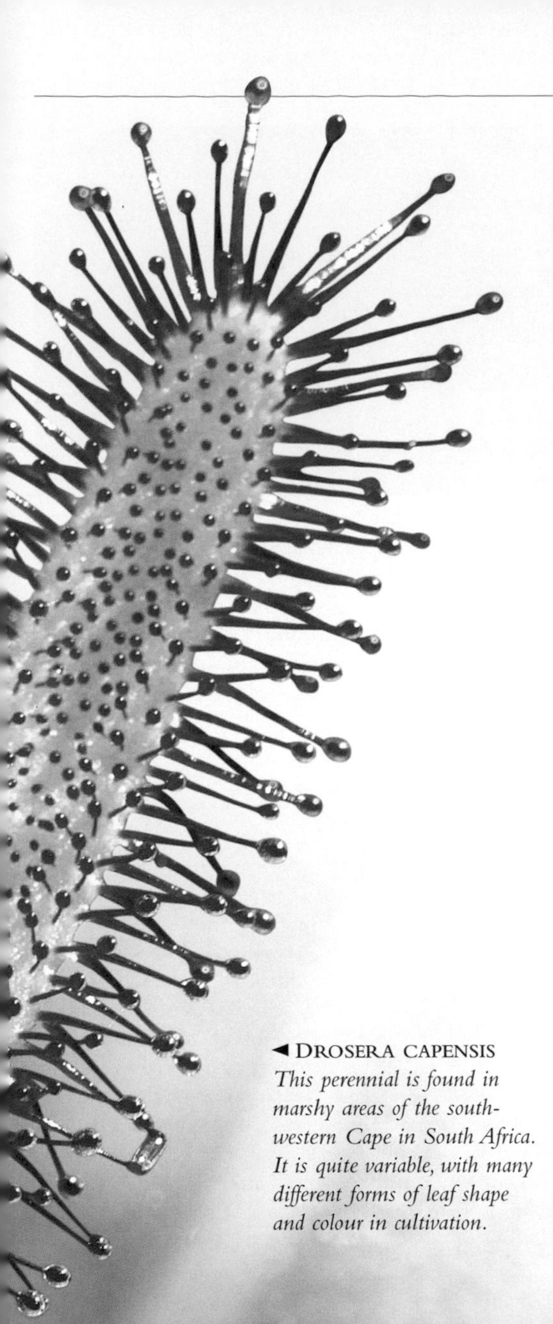

◄ DROSERA CAPENSIS *This perennial is found in marshy areas of the south-western Cape in South Africa. It is quite variable, with many different forms of leaf shape and colour in cultivation.*

▲ SARRACENIA JONESII Sarracenia *species have rhizomes that can persist for several years after land is drained, but they may be out-competed by the surrounding vegetation.*

CONSERVATION EFFORTS

Two major concerns in conserving habitats are controlling invasive woody species and preserving water levels. Trees and shrubs are cut back or burned so that they are vulnerable to disease and eventual death. Soil restoration must be carried out with minimal disturbance. One effective method is to dam any ditches using the organic debris accumulated from clearing the site. This slows the flow of water, reducing erosion, and maintains or restores a high water table, an intrinsic element of these habitats.

The Conservation Program of the Atlanta Botanical Garden (ABG) in Georgia, USA, is a fine example of plant conservation at work. ABG has one of the world's largest carnivorous plant collections, and is active in monitoring, restoring, and conserving habitats. Working with landowners and other organizations over 15 years, ABG has developed simple, cost-effective management and propagation techniques. It conducts many projects with The Nature Conservancy, and is also under contract to the US Fish & Wildlife Service to assist in the recovery of several endangered carnivorous plant species.

GARDENER'S GUIDE

- Millions of carnivorous plants are artificially propagated each year, but plants are still dug from the wild. This is a serious issue when many populations are reduced to tiny fragments of the species' original distribution. To ensure you are not contributing to the demise of these vulnerable populations, it is important to shop carefully.
- Always ask questions about the origins of the material, buying only propagated stock or seed, or wild stock from a sustainable harvesting scheme. If the labelling does not state whether the plant has been wild-collected or artificially propagated, ask the retailer to be more specific. Doing so encourages retailers to put such information on the packaging in the future.
- Ask to see paperwork confirming the credentials of any non-native carnivorous plants. A wildlife agency or the CITES authority in either the country of origin or the country where the nursery outlet is based can provide more information.
- Plants may be wild-collected and then grown in a nursery for a period of time. For this reason, it is important to seek out plants labelled "nursery propagated", not "nursery-grown". Always check that a plant has been collected from the wild. Be especially wary of rare plants, particularly sizable specimens, that are being sold cheaply.
- Support or join a plant exchange scheme with a conservation policy. Ask what steps they take to ensure they and their members are not contributing to illegal collection or the demise of wild populations.

▲ MICROPROPAGATED PLANTS *The Venus flytrap* (Dionaea muscipula) *is so easy and inexpensive to propagate in large numbers that there is no reason to collect wild plants. It is an ideal first plant for the novice. Many vividly coloured cultivars are propagated in this way.*

MODERN TRADE

Today, carnivorous plants are once again popular. They are traded worldwide, both commercially and among enthusiasts or hobbyists.

The hobbyist grower is supported by numerous carnivorous plant societies, which provide a forum for discussion, newsletters, conferences, plant sales, swaps, and exhibitions. The largest of these organizations is the International Carnivorous Plant Society, which encourages its members to contribute to the conservation and study of carnivorous plants around the world.

Commercial production of artifically propagated plants in large-scale nurseries is mainly based in the USA and the Netherlands. These nurseries tend to concentrate on producing perhaps the most popular carnivorous plant, the Venus flytrap (*Dionaea muscipula*). This species is listed in CITES Appendix II (*see p.37*), together with the pitcher plants (*Sarracenia*) and the tropical pitcher plants (*Nepenthes*), limiting trade to seeds and artificially propagated plants. A selection of other carnivorous plant species is listed in CITES Appendix I, affording them more stringent protection.

Unlisted

Aldrovanda vesiculosa

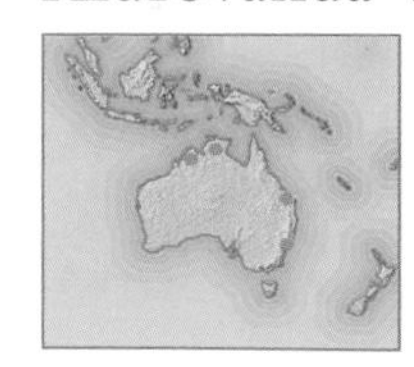

Botanical family
Droseraceae

Distribution
Australia, Timor, Japan, India, Africa, central and eastern Europe

Hardiness
Half-hardy – frost-tender

Preferred conditions
Clear water with pH 6–7, high CO_2, and low in nitrogen and phosphorus in bright light; temperate forms need 16–30°C (61–86°F), tropical forms require 25–32°C (77–90°F).

The only species of *Aldrovanda* is a rare, carnivorous aquatic plant that is found in temperate to subtropical regions, in small pockets scattered over four continents. Called the waterwheel plant, it grows in shallow, standing water of acidic lakes and ponds where there is decomposing bottom litter, in areas with very warm summers. It is threatened by increasing levels of pollution and habitat drainage, especially in Europe. It is now extinct in the wild in Japan, although *ex situ* conservation projects are raising plants for possible re-introduction programmes.

Reddish tropical forms grow all year in warm temperatures, while temperate forms go dormant in winter. They form fleshy buds, or turions, in autumn which sink to the silty bottom, then reshoot when the water warms up in spring. New plants can also be obtained by simply breaking the plants apart.

This species is not easy to cultivate because of the very specific conditions that it requires to survive. Use a large tank with a cup of peat per 5 litres (1.1 gallons) of water. Keep the water clear of algae and tadpoles by changing it whenever necessary. The presence of water fleas will help to maintain a clear tank as well as feed the plants.

▲ SNAPPING SHUT *The traps in the waterwheel plant are similar to, but smaller than, those of their close relative, the Venus flytrap* (Dionaea muscipula). *The traps close in less than a second, faster than on any other plant, and will stay shut for about a week if live prey is caught.*

1 on Red List

Byblis

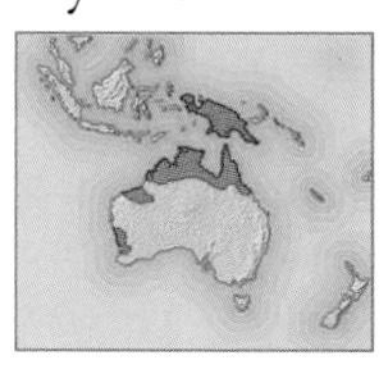

Botanical family
Byblidaceae

Distribution
Australia, Papua New Guinea

Hardiness
Half-hardy – frost-tender

Preferred conditions
Full sun to partial shade; 3–40°C (37–104°F) depending on species; sandy, seasonally dry to moist soils.

Byblis was the name given to a child featured in a Greek myth who, when rejected, wept so copiously that her tears turned into a fountain. The beads of sticky mucilage on the stems of this plant are said to recall those tears, making its leaves glisten and creating the refractive effect that gives byblis its common name of rainbow plant.

The first species was discovered in the 19th century, but more recently several new species, subspecies, and forms have been discovered. *Byblis* are found in peat bogs and marshland: habitats that are shrinking as a result of drainage and land conversion for agriculture.

▲ BYBLIS LINIFLORA *is the easiest species to grow. Its pretty flowers seed prolifically in favourable conditions and the seed does not need smoke treatment to germinate.*

Byblis liniflora is the most commonly cultivated species. A tropical annual, it is found throughout the northern areas of Western Australia, the Northern Territory, Queensland, and south-east Irian Jaya in Indonesia, where it enjoys the warm conditions with a summer monsoon and a drier winter. The delicate stems clothed in sticky hairs are 30cm (12in) tall. One of the newest forms, called 'Darwin Red', was recently discovered in the Northern Territory and has maroon flowers.

Byblis gigantea is perennial and grows in coastal regions in the south-west of Western Australia, around Badgingarra, Eneabba, and Perth. These areas enjoy a Mediterranean-type climate with cool, wet winters and hot, dry summers. The seed often lies dormant for years until a bush fire followed by the rainy season induces rapid germination of seedlings. When the swamps they inhabit dry out in the summer months, the plants become dormant, then rejuvenate when the rain returns in the autumn.

Several species have been discovered in recent years. They are *Byblis aquatica*, so far known at a few sites around Darwin, as well as *Byblis filifolia* and *Byblis rorida*, both of which are distributed throughout the Kimberley region of Western Australia. These populations seem stable.

Seed is available for cultivation, but some seeds, such as those of *Byblis gigantea* and 'Darwin Red', require smoke treatment to break their dormancy before they will germinate. *Byblis liniflora* seed should be dried before storing. Use a soilless compost mixed in equal parts with lime-free sand.

BOTANY

MYSTERIOUS MUCILAGE

Carnivory in plants has evolved in various ways in different groups, and is present in eight separate families. *Byblis* species have two types of gland. Prey get caught on the first, which are hair-like, stalked (*see left*), and covered in a clear, sticky mucilage. The other type of gland, on the surface of the leaf, secretes digestive juices. However, no enzymes or bacteria have been located in these juices and theories as to how digestion can occur without enzymes have yet to be fully worked out. Some theories point to the role of fungi, others to the presence of assassin bugs, which live on the plant, digest the trapped insects, and then secrete a nutritious fluid which may be easier for the plant to digest.

▲ BYBLIS GIGANTEA *This is the largest* Byblis, *and can reach 70cm (28in) in height. Its leaves are coated with sticky mucilage and the flowers range in colour from a beautiful deep pink to violet-blue. It is classed as Critically Endangered on the 2000 Red List.*

Red List: Vulnerable

Cephalotus follicularis

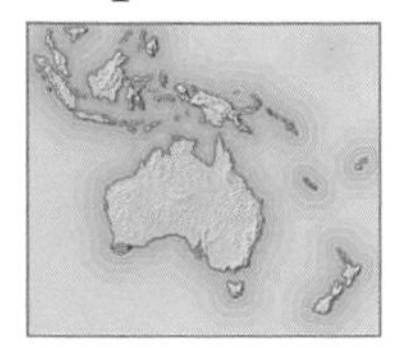

Botanical family
Cephalotaceae

Distribution
Western Australia: Albany

Hardiness
Half-hardy – frost-tender

Preferred conditions
Full sun to partial shade; 3–30°C (37–86°F); wet, acidic, peaty, sandy soils.

A bizarre, evergreen perennial, this is the sole species in the genus. It is native to the coastal peat swamps of the extreme south-west of Australia, ranging from Augusta to Cheyne Beach and Cape Riche, where the Mediterranean-type climate has warm, dry summers and cool, wet winters. This region is under pressure from farming and the expanding human population.

International trade was considered enough of a threat to place the species on Appendix II of CITES. It has since been removed because trade has reduced and cultivated plants are now widely available, if still expensive. A colony of thousands of plants has also been found around a man-made dam near Albany.

In cultivation, this slow-growing species dislikes being waterlogged, so care should be taken with watering. The plants also revert to green coloration in low light. The summer flowers are not impressive, and are best removed to prevent them from weakening the plant. Propagation from seed is possible, but is slow (eight weeks is not unusual) and its viability is limited. Sow seed promptly in a thin layer on a peat or sand and moss mix and keep the container in a damp, cold environment for six months.

This species has a tendency to die back for no apparent reason; however, the plants may reshoot provided that the old growth is removed.

▼ IRRESISTIBLE LURE *Also known as the Albany pitcher plant, this species is best known for small, stout pitchers with distinctly toothed rims, or peristomes. Crawling insects are lured inside by the scent of the nectar.*

Unlisted

Darlingtonia californica

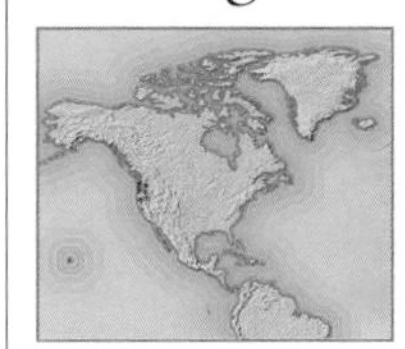

Botanical family
Sarraceniaceae

Distribution
USA: northern California to south-west Oregon and Seattle

Hardiness
Fully hardy

Preferred conditions
Warm to temperate, with cool summer nights; but survives winter temperatures as low as –10°C (14°F); full sun to partial shade (more important in warmer areas); cool, wet soil.

Closely related to the sarracenias (*see p.438*) found in the eastern states of the USA, ***Darlingtonia californica***, also known as the cobra lily, is the only species in its genus. It can be found on mountain slopes as high as 2,600m (9,000ft) down to the Pacific coastal bogs at sea level. It grows on poor soils rich in magnesium (serpentine), as well as iron (ultramafic), often in association with fast-flowing, cool streams.

The plants lure prey into the pitcher mouth with nectar trails along the pitcher's "tongues". Light entering the "windows", or fenestrations, in the back of the pitcher hood confuse the prey and cause them to fall to the bottom of the pitcher, where they drown. Bacteria and other micro-organisms break down the prey and the nutrients, released in fluid, are then re-absorbed by the plant.

CULTIVATION

In the wild, cobra lilies may grow up to 90cm (3ft) tall, but in cultivation they reach a height of only 45cm (18in). It can be difficult to reproduce the cool, wet conditions that this species needs around its roots (the plant is intolerant of high temperatures). The best solution is to use sphagnum moss beds or an open, well-aerated growing medium such as sphagnum moss and perlite. Keep the growing medium cool either next to a stream or in a pot that is regularly flushed from above with cold, soft water. Sow seed thinly, keep damp, and place in strong light at 21–29°C (70–84°F). Reduce the temperature when seedlings have two true leaves. Cobra lilies can also be propagated by division.

▲ DARLINGTONIA IN BLOOM *Flowers appear on long stems in spring and produce spiky seeds. Plants grown from seed take three years to form pitchers 10–13cm (4–5in) tall.*

BOTANY

MISSING PIGMENT

The red coloration in many carnivorous plants is caused by a pigment called anthocyanin. The anthocyanin-free variant of *Darlingtonia californica*, 'Othello', was originally discovered in 1997 at a single location. Such unusual forms are of particular interest to horticulture, but unfortunately also attract unscrupulous collectors. Plants from the site have been carefully hand-pollinated, and the seed collected, in the hope of satisfying demand. The site is also threatened by logging, however, and attempts have been made to buy the seep bog in which the 'Othello' population is found.

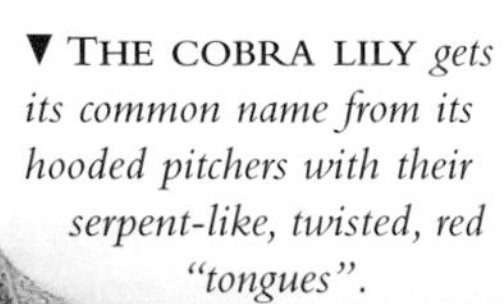

▼ THE COBRA LILY *gets its common name from its hooded pitchers with their serpent-like, twisted, red "tongues".*

Red List: Vulnerable/CITES II

Dionaea muscipula

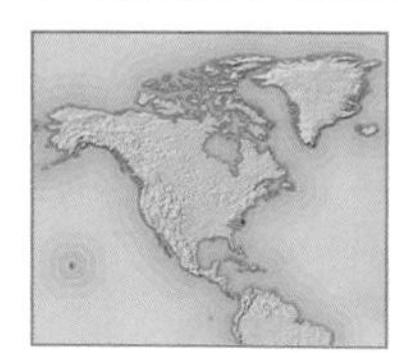

Botanical family
Droseraceae

Distribution
USA: North Carolina, South Carolina

Hardiness
Half-hardy

Preferred conditions
Full sun to partial shade on moist, free-draining, acid soil; warm and wet in summer, but cool and damp in autumn and winter; 2–38°C (36–100°F), survives light frost.

Probably the best known of all the carnivorous plants, the Venus flytrap is endemic to the pine savannas on the coastal plain of both North and South Carolina, USA. Almost all plants in the wild occur primarily in North Carolina, but the flytrap has been transplanted outside its limited native range to New Jersey, California, and a small area of the Florida panhandle, where it thrives.

Dionaea muscipula is the only species in this genus, but it is very variable from seed in colour, trap size, and tooth shape. It attracts its prey with the colour and ultraviolet markings on its trap pads, as well as with nectar. There is a large number of natural and hybridized varieties, with names such as 'Beartrap', 'Dentata', and the fabulous 'Red Dragon', with needle-like teeth and entirely red traps and leaves.

Decline in the wild

Unfortunately, the fame of this species has partly been its downfall. The huge numbers harvested in the past for international trade led to it being listed in CITES and in the USA Endangered Species Act. Overcollection from the wild still occurs, to supply the domestic trade in the USA in particular and to replenish parent stock in nurseries. The plant is also under siege from habitat destruction through land conversion for agriculture and construction, aggressive forestry practices, and the suppression of natural fires that maintain the open savanna landscape.

Thousands of plants are now micro-propagated to perpetuate various cultivars, such as 'Fine Tooth', 'Giant', and 'Royal Red'. These cultivars are recommended to gardeners because purchasing these plants will not endanger species in the wild colonies.

▲ **Venus flytrap in flower** *Gardeners growing these plants for the novelty of their "traps" often pinch out the 30cm (12in) flower stems that later emerge, because this encourages more traps to form. However, the plant is beautiful in flower, and will set seed that can be raised by the amateur, although germination is very slow.*

Darwinian favourite ► *Charles Darwin wrote that the Venus flytrap is "one of the most wonderful plants in the world". It was a favourite experimental subject of his and he spent much time observing the plant's responses to being "fed" a number of different food items, such as cheese. He would have noticed that each trap closes fully only a few times before it blackens, then dies.*

Cultivation

This plant grows best in equal parts of lime-free sand and peat. Keep it warm at 21–38°C (70–100°F) and wet in the growing season, but provide cooler conditions (2–10°C/ 36–50°F) in the dormant season. Many cultivated plants die from being overwatered or from constant use of hard water, found in many public water sources. If possible, use distilled or rainwater. Many plants are thrown away every year in the belief that they are dead, when they are merely in a dormant phase.

Seed can be sown on a peat and sand mix. Sprinkle it thinly and keep in high humidity and good light.

Digesting the prey *A grasshopper approaches an open flytrap (1), attracted by the nectar secreted from glands at the base of the "teeth". As the insect moves into the trap, it must touch one of the six trigger hairs on the leaf lobes twice, or any two of the hairs once in 20 seconds, for the trap to snap shut (2). This fail-safe mechanism prevents the trap from being triggered by debris. The interlocking teeth stop the insect escaping (3) and close to form an airtight seal. Digestive enzymes are then released so the plant can absorb the nutrients.*

13 on Red List*

Drosera

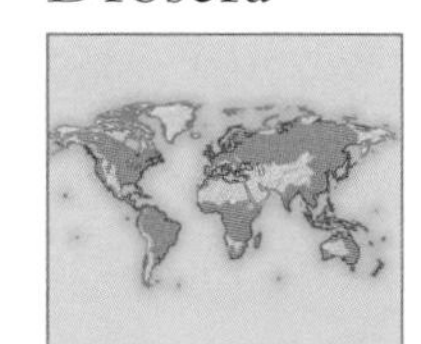

Botanical family
Droseraceae

Distribution
Worldwide

Hardiness
Frost-tender

Preferred conditions
Full sun; acid soil and warm summers to 38°C (100°F), or 35°C (95°F) for tropical species; in winter no less than 2–4°C (36–39°F), or 16°C (61°F) for tropical species.

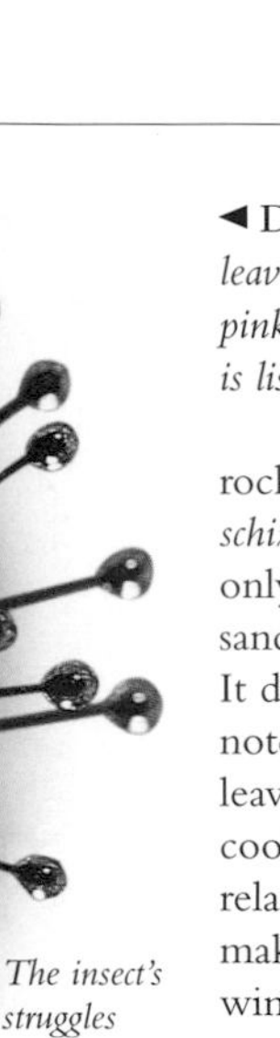

The leaf surface secretes digestive enzymes

The insect's struggles prompt the tentacles to pull it into the centre

◀ DROSERA ROTUNDIFOLIA, *or round-leaved sundew, bears spikes of small, white or pink flowers in summer and early autumn. It is listed as Endangered on the 1997 Red List.*

Known as sundews, the 100 or more species of *Drosera* are found in peat bogs and marshland in all climatic zones of the world, from the arctic regions in Canada to the tropics of Brazil and Australia. Their carnivorous leaves are covered in glands or tentacles that produce the drops of sticky mucilage which makes the plants glisten on a dewy morning or in late afternoon sun, giving them their common name. The species range in size from a few millimetres across to over 30cm (12in) tall, and the leaves are equally variable.

Bog drainage and conversion of land for agriculture, as well as construction and forestry, have led to an overall decline in sundew populations over the last 100 years – as is the case with many species that favour wetland habitats. Some sundews are confined to small areas that are susceptible to continual habitat disturbance and loss, and many plants have been taken from the wild for trade.

Nearly all sundews are readily available in cultivation. They can be divided into different groups according to their size and shape and the growing conditions they require. Most species are evergreen, but some become dormant in winter or summer. All but a few favour full sun. The Cape sundews from South Africa, with thick, fleshy roots, and the tuberous sundews have adapted to cope with the heat by dying down in hot, dry summers and bursting into growth when autumn rains arrive. Pygmy sundews, the smallest species at less than 2cm (¾in) across, have a tight bud to protect the growing point from hot sun. Temperate sundews vary in form from upright to prostrate rosettes; some produce winter buds. The distribution of fork-leaved sundews is limited to the east coast of Australia and one species in New Zealand, in warm-temperate to tropical climates.

The thread-leaved sundew, ***Drosera filiformis***, is a beautiful species growing to 50cm (20in), with erect linear leaves that glisten and shimmer. There are two varieties, *Drosera filiformis* var. *filiformis* from the north-east and mid-Atlantic USA to a small area of the Florida panhandle, and *Drosera filiformis* var. *tracyi* from the Gulf Coast.

Drosera rotundifolia is a frost-hardy rosetted species. Although still common in sphagnum moss bogs across the cool, mountainous regions of North America, Europe, and Asia, it is locally threatened by activities such as bog drainage and peat cutting (*see below*).

Other rosetted sundews comprise a group of closely related tropical species endemic to small areas of rainforest in Queensland, Australia. All three were classed as Vulnerable on the 1997 Red List. *Drosera adelae* has long, lance-shaped leaves and grows by streams in sand by the coast; ***Drosera prolifera*** grows on wet rocks and stream banks. *Drosera schizandra*, the notched sundew, is known only from one peak, where it prefers sand in deep shade by mountain streams. It derives its common name from the notch at the tip of its older, flat, oval leaves. *Drosera adelae* tolerates brighter, cooler, less humid conditions than its relatives, but it cannot take frost. All make lovely plants for a humid windowsill or terrarium.

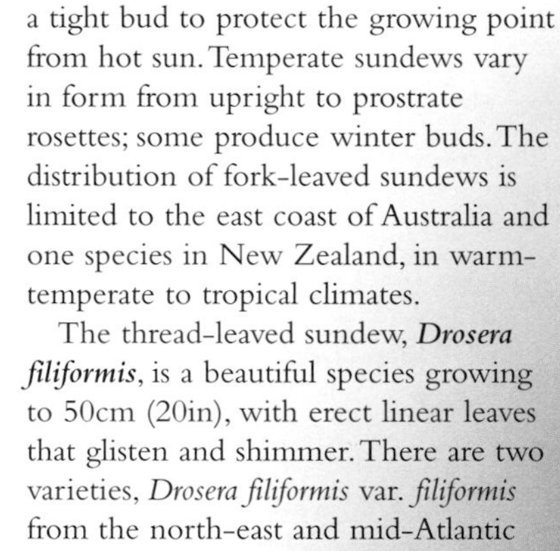

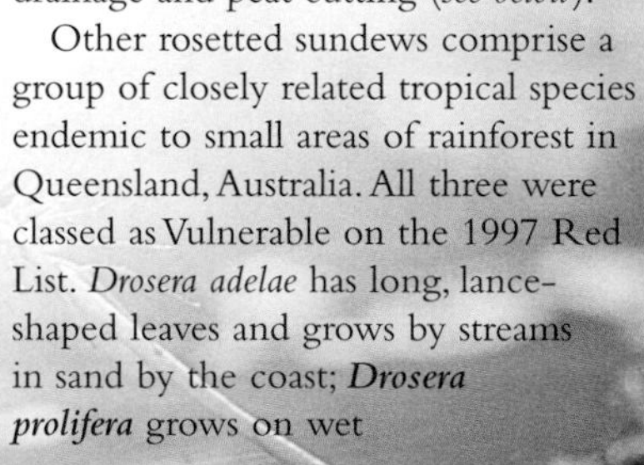

DROSERA FILIFORMIS ▶ *The thread-leaved sundews, or dew-threads, are most at risk in the southern part of their North American range, where acidic bogs and wetland grass savannas are being destroyed.*

HABITAT

PEAT CUTTING

Peat bogs form over millennia, as successive layers of wetland vegetation decay. Healthy bogs are moist, highly acidic, and very low in nutrients, so only a few specialist plants are able to survive, like sundews and many species of sphagnum moss, which carpet the surface of the bogs in brilliant green, ochre yellow, and rusty red. For centuries European farmers cut blocks of peat to use as fuel. In recent decades, peat has been cut for horticultural purposes: sphagnum moss, used to line wire baskets, and sphagnum peat moss, used as a soil conditioner, are both prized for their high water-holding capacity. Once peat is cut, the bog is drained and becomes drier, and its living flora begins to die off.

CULTIVATION

Sundew propagation methods vary, but seed is the preferred way to raise new plants, followed by leaf cuttings. Use an open compost of peat or perlite and sphagnum moss. Some of the temperate species, such as ***Drosera anglica***, *Drosera filiformis*, and *Drosera rotundifolia*, survive the winter by producing a tight dormant bud known as a hibernaculum. Even if the plant itself dies back, the bud will reshoot in spring. This bud can be removed and stored in an airtight bag or container with some moist sphagnum moss for four to five months, or it can be transplanted at this time and kept in a drier medium.

▲ DROSERA PROLIFERA *Unlike its close relatives, this tropical species spreads quickly by producing new plants wherever the tips of its flower stems touch the ground. This gives an easy method of propagation.*

OTHER THREATENED SUNDEWS

Drosera callistos This pygmy sundew has bright orange flowers. It is restricted to a small area east of Perth (Gidgegannup, The Lakes, and Brookton Highway), in Western Australia. It grows in sandy soils where mineral crusts form (laterite soil), but its habitat is being lost to farming and the expanding human population. Unlisted.

Drosera graniticola An erect, tuberous sundew, to 20cm (8in) tall, known only from a handful of plants at three locations, it grows on granite outcrops in one area of the eastern wheat belt of Western Australia, so it is threatened by farming and the encroaching human population. 1997 Red List: Rare.

Drosera hamiltonii A rosetted, tuberous sundew with large, dark pink flowers, it is restricted to a narrow area from Augusta to Albany in Western Australia and is difficult to locate in the peat swamps that it shares with *Cephalotus follicularis* (see p.429). It is becoming rare due to habitat loss from farming. Unlisted.

Drosera macrantha An erect, sprawling tuberous sundew growing to 1.5m (5ft), this relatively common species from the eastern wheat belt of Western Australia has been divided into several distinct subspecies and forms. Often very localized, it must be considered threatened by habitat destruction. Unlisted.

Drosera ordensis A non-tuberous sundew, with golden or red, round traps, it is restricted to two rivers on the border of Western Australia and the Northern Territory. It grows in sand by sandstone outcrops in a relatively untouched area that is hard to reach. This plant is a tropical species from the Petiolaris group, endemic to north-eastern Australia. Unlisted.

Once an insect's soft tissue is dissolved and digested, only the husk, or exoskeleton, remains; the tentacles then revert to their upright positions

▲ DROSERA REGIA *This rare species, known as the king sundew, has leaves that reach to over 30cm (12in) in height and large, deep pink flowers. It is found in only a few colonies in South Africa and is listed as Rare on the 1997 Red List.*

The "dew" contains sugar to attract insects, but once caught, the dew blocks the insect's breathing pores so that it suffocates

◄ DROSERA ANGLICA, *known as the English, or great, sundew, grows in sphagnum moss bogs, often in association with* Drosera rotundifolia. *It is widespread in temperate regions of Japan, North America, and Europe, but is locally threatened by the destruction of its habitat.*

Unlisted

Drosophyllum lusitanicum

Botanical family
Droseraceae

Distribution
Northern Morocco, Portugal, southern Spain

Hardiness
Frost-tender

Preferred conditions
Full sun; free-draining, alkaline, poor soil (rocky or sandy), with a wet and a dry season; moisture from fog or dew.

Called the Portuguese sundew or dewy pine, this species is a close relation of the sundews (*Drosera*) and the Venus flytrap (*Dionaea muscipula*). But unlike those and many other carnivorous plants, it is found on dry hills rather than acidic wetlands. It is not well known in cultivation due to the difficulties in keeping seedling plants alive in the first couple of years. Any root disturbance usually kills the plant.

The only successful way to propagate *Drosophyllum lusitanicum* is from seed. Rub the seeds between sandpaper and soak them, then sow onto vermiculite, and keep damp. Transplant germinated seedlings into the pots where they will remain. Keep them damp for the first two weeks, but thereafter let them dry out between waterings.

▲ NATIVE HABITAT *The dry, alkaline coastal hills where this species grows receive most of their water in the wet Mediterranean winters, although nightly fogs provide welcome moisture during the warm, dry summers.*

◀ STICKY LEAVES *These 20cm (8in) leaves form an upright rosette that looks like a cluster of pine needles.*

Unlisted

Heliamphora

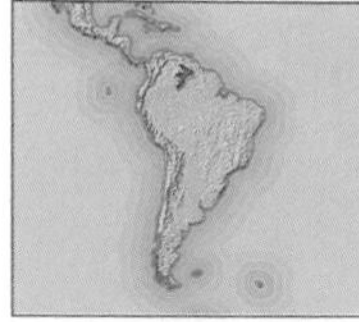

Botanical family
Sarraceniaceae

Distribution
Northern Brazil, Guiana, southern Venezuela

Hardiness
Frost-tender

Preferred conditions
Open, well-drained soil that is low in nutrients; regular, heavy rainfall and high light levels; 3–26°C (37–79°F).

This genus consists of primitive pitcher plants with a simpler structure than their relatives: these pitchers are cone-shaped, with a wide mouth leading to a hood, or nectar spoon. Also known as sun pitcher plants, heliamphoras are most closely related to the pitcher plants found in North America – darlingtonias and sarracenias. Sun pitcher plants are found only on the Guiana Highlands which stretch across three South American countries, in unique mountain habitats called tepuis (*see facing page*). Currently eight species are known, but new species are still being discovered, the most recent being ***Heliamphora folliculata***, ***Heliamphora hispida***, and *Heliamphora chimantensis*.

The populations of *Heliamphora* species are not currently threatened, because their habitats are largely inaccessible, often on summits reached only by helicopter. However, the sites are now being visited more frequently and the plants occasionally collected. Given the extremely fragile nature of these ecosystems and the small population numbers, the increase in levels of disturbance could be devastating to plant colonies. Many tepuis have now been declared national parks, affording these plants some protection.

Heliamphora nutans was the first species to be described, when it was discovered on Mt Roraima, perhaps the most famous tepui of all, in 1839. Today it can be found on the Sierra Paceraima highlands in southern Venezuela. While many of the species discovered subsequently are easier to cultivate and look more spectacular, this is still the most commonly grown.

The smallest species, ***Heliamphora minor***, has stocky pitchers 5–8cm (2–3in) high and can spread to make large, low-growing clusters. A form that occurs on the Chimanta group of tepuis is characterized by the long "hair" inside its pitchers and its rich red colour when grown in full sun. *Heliamphora heterodoxa* was discovered in 1951 on the Mt Ptari-Tepui plateau in the Sierra Paceraima highlands in southern Venezuela. It also grows in warmer temperatures in the lowland savannas, or Gran Sabana, around the mountain. This is an excellent plant for terrariums.

▲ HELIAMPHORA MINOR *has pale flowers borne on tall stems (approx. 25cm/10in long). In cultivation, this species flowers all year round.*

CULTIVATION

All species are available in cultivation as micropropagated plants, including natural varieties and man-made cultivars. In most climates, they must be grown in terrariums or greenhouses to ensure the humidity they require. Water them frequently with cold water that is low in dissolved mineral salts. These plants are slow-growing and can be a little difficult – they resent being transplanted, for instance, and this makes them rather expensive to buy.

Propagation by seed can be slow. The sown seed should be kept in a humid environment at 20–22°C (68–72°F), not in direct sunlight. Many species respond to different light levels: the brighter the light, the redder the pitchers.

▲ HELIAMPHORA NUTANS *The giant pitcher plant forms vast carpets of pitchers 10–15cm (4–6in) tall in the acidic humus that collects in the swamps of Mt Roraima, Venezuela.*

▲ **HELIAMPHORA FOLLICULATA** *This new species comes from Los Testigos Table Mountains in the south of Venezuela. The flower colour varies from white to whitish-pink. It is named after the "bubble" (follicle) formed by the nectar spoon, the most distinct characteristic of this species.*

▲ **TEPUIS PLATEAU** *Large clumps of heliamphoras (here,* Heliamphora folliculata*) can be found in the shallow pools and boggy depressions on the windswept plateaux of the tepuis. The heavy annual rainfall ensures high humidity levels.*

HABITAT

REMOTE TEPUIS

The tabletop mountains, called tepuis, that form the Guiana Highlands were immortalized in Sir Arthur Conan Doyle's famous novel, *The Lost World*, inspired by the 1884 expedition of British botanist Everard Im Thurn to Mt Roraima.

Each tepui has huge cliffs of sandstone rising sheer out of the surrounding tropical jungle or savanna to a height of 1,000–3,000m (3,300–10,000ft). They are constantly shrouded in cloud, and the resulting low temperatures and daily torrential rain create great runoffs, including Angel Falls, the highest waterfall in the world at 979m (3,212ft). The rains wash away nutrients and organic material but also form the shallow marshes and pools scattered over the plateaux. As well as poor, fragile soils, the tepuis have a harsh highland climate with strong winds, intense sunlight, and ultraviolet radiation. The days are only moderately warm and the nights are cold.

Despite the conditions, the isolation of these habitats over 1.8 billion years has created a rich flora with many endemic species, including all heliamphoras and other carnivorous plants such as *Drosera roraimae* and *Utricularia humboldtii*.

▲ **HELIAMPHORA HISPIDA** *This newly discovered species of* Heliamphora *comes from Cerro Neblina, Venezuela, where it can form large, dense clumps of pitchers in the shallow acidic marshes scattered over the mountain tops. The flowers are white, or white and pink, and are borne on stems 50cm (20in) tall.*

CARNIVOROUS PLANTS

19*/59 on Red List/CITES I and II

Nepenthes

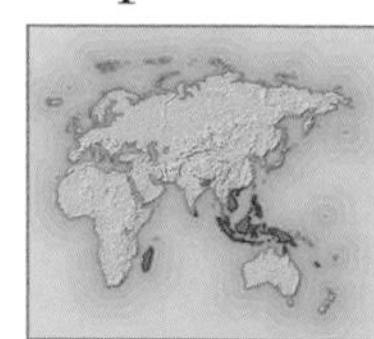

Botanical family
Nepenthaceae

Distribution
Madagascar, Seychelles, Sri Lanka, north-eastern India, south-east Asia

Hardiness
Frost-tender

Preferred conditions
Bright light; 70 per cent humidity; low-nitrogen, moisture-retentive soil; lowland species need 21°C (70°F) at night, 29°C (84°F) by day; highland species 10°C (50°F) at night, 21°C (70°F) by day.

▲ NEPENTHES AMPULLARIA *is remarkable in that the lids do not cover the pitchers, enabling the plant to "feed" on falling leaves as well as insects. It is Unlisted on the Red Lists but is in CITES Appendix II.*

▲ NEPENTHES LOWII *is found in open moss forest in Sabah and Kuching, Borneo, at 1,800–2,600m (6,000–9,000ft). Vulnerable on the 2000 Red List and in CITES II.*

Monkey cups, or tropical pitcher plants, originate from peat bogs, marshlands, and tropical rainforest in south-east Asia, many from Sumatera (Sumatra) and Borneo. They have recently regained popularity, with the discovery of new species in Sumatera and the Philippines, and the availability of propagated material for most species. Over-collecting currently affects only a few species. Of greater concern is habitat destruction, either through logging, particularly in lowland areas, or forest fires, which caused the near-extinction of species like *Nepenthes campanulata* and *Nepenthes clipeata*. Several species, such as *Nepenthes rafflesiana*, tolerate disturbance and thrive in secondary forest, while others (*Nepenthes sumatrana*) are sensitive to the least disturbance. Trampling and collection by tourists along the summit trail of Mt Kinabalu have badly affected colonies of ***Nepenthes villosa***.

The many species vary in their growth habit. Some carpet the ground, like ***Nepenthes ampullaria***, which is common and widespread on lowland heath and peat swamp forest in Borneo, Sumatera, New Guinea, and Thailand. Others, like *Nepenthes inermis*, grow as epiphytes on trees in the high moss and mountain forests of Sumatera Barat and Jami.

TRAP MECHANISMS

Nepenthes trap insects in their pitchers; these traps vary widely, from a simple vase structure to a pitcher with "teeth" around the rim (peristome). Some have evolved to catch a certain type of prey. Most species have two types of pitcher: the lower one catches ground-dwelling insects and the upper pitcher traps flying insects. The prey may be attracted by nectar, colour, or even rotting victims. Pitchers can support mini-ecosystems: the prey they trap in turn attracts other organisms. Frogs may lay their eggs in the pitchers and bats even sleep in them. The lid of ***Nepenthes lowii*** (*left*) may act as a perch for birds or tree shrews, which defecate into the pitcher while feeding on nectar on the lid. These pitchers may also trap leaves.

CULTIVATION

Different *Nepenthes* species must be grown at different temperatures. Highland species are popular in temperate countries; the few species from higher altitudes need cooler conditions. Many tropical pitcher plants grow in open, bright habitats; these may need extra lighting in cultivation. Fogging systems can create the right atmosphere for highland species and fans can help to increase air circulation.

Depending on the species, *Nepenthes* can be grown in a range of media, including perlite, vermiculite, fine bark, peat, lava rock, pumice, sand, or charcoal, as long as the mix is open and free-draining, and kept moist. Propagation is possible from seed, but most species are increased using tissue culture.

Lid prevents rainwater from falling into the pitcher and diluting the digestive fluid

Pitchers are formed from modified leaves and lined with tiny, waxy scales that stop insects gaining a foothold and escaping

► NEPENTHES BURBIDGEAE *can hybridize in the wild with* Nepenthes rajah *to produce plants like this one. These species are sparsely distributed on poor soils in the lower mossy forests around one mountain in northern Borneo at 1,200–1,800m (4,000–6,000ft), near the boundary of a national park. The peristome, or rim, of the pitcher has glands that secrete nectar to lure insects inside it. It is classed as Endangered on the 2000 Red List and is in CITES Appendix II.*

BOTANY

NEPENTHES ARISTOLOCHIOIDES

This bizarre species has pitchers shaped like a Dutch pipe or the flowers of the climber *Aristolochia*; the porthole-like mouth recalls that of *Darlingtonia californica* and *Sarracenia minor*. It is known only from the moss forest of Gunung Tujuh (2,000–2,500m/ 6,600–8,250ft), in the Kerinci-Seblat National Park of Sumatera, where it grows as both a terrestrial and an epiphyte. It may also occur on the Gunung Kerinci where it was collected nearly 50 years ago. Little of its high-altitude habitat has been cleared, but its striking looks and rarity have made it highly sought after and threatened by collection. It is Critically Endangered on the 2000 Red List and is in CITES Appendix II.

◄ NEPENTHES VILLOSA *is found on two mountains in northern Borneo, in cloud moss and subalpine forests, at 2,400–3,200m (8,000–10,500ft) – higher than any other species. It is sought after for its impressively toothed peristome, but is difficult to cultivate. It is listed as Vulnerable on the 2000 Red List and is in CITES Appendix II.*

▼ NEPENTHES RAJAH *Heavily collected for its huge pitcher which holds a litre of liquid and traps rats, lizards, and frogs, this species grows in grassy clearings at 1,500–2,600m (5,000–8,500ft) on two peaks in northern Borneo. It is now protected in a national park. It is Endangered on the 2000 Red List and is in CITES Appendix I.*

OTHER THREATENED NEPENTHES

Nepenthes adnata Known only from east-central Sumatera Barat, where it grows at 600–1,100m (1,950–3,600ft) on moist, mossy, sandstone cliffs in dense shade, this plant has small pitchers; the lower ones have two fringed wings. The small population is often visited by plant collectors. 2000 Red List: Data Deficient; CITES II.

Nepenthes albomarginata A lowland species from heath forest and other open vegetation, including lower ridges and peaks up to 1,100m (3,600ft), in Borneo, peninsular Malaysia, and Sumatera, it may be adapted to attract termites by means of white, mealy hairs around the pitcher mouth. Pitchers range from green through red to near-black. 2000 Red List: Lower Risk; CITES II.

Nepenthes bellii A lowland species, resembling a dwarf *Nepenthes merrilliana* with small, round pitchers. It grows on the north-east coast of Mindanao island, Philippines. 2000 Red List: Endangered; CITES II.

Nepenthes bicalcarata Confined to the lowland peat swamp forests of north-west Borneo up to 950m (3,150ft), the largest species in the genus reaches high into the canopy. Uniquely, it has two sinister, fang-like projections from the lid, which may be a specialized means of capturing ants. 2000 Red List: Vulnerable; CITES II.

Nepenthes campanulata This was thought extinct after its only known site, the 300m (1,000ft) limestone cliffs of Ilas Bungaan, Kalimantan, eastern Borneo, were burnt in the 1982 drought. However, some plants were rediscovered in 1997 at an undisclosed site and remain rare. It is unlike the other *Nepenthes* of Borneo; the bell-shaped pitchers resemble those of *Nepenthes inermis*. 2000 Red List: Extinct; CITES II.

Nepenthes clipeata This species has distinctive pitchers, with rounded bottoms fanning out to a cone. Probably the most endangered species, it is confined to one site in West Kalimantan, on remote granite cliffs at 600– 800m (2,000–2,650ft). The colony was reduced to 2–6 plants by 2001 and it is not protected by any national park system. Without urgent action, it will become extinct in the wild. The only hope lies in availability through tissue culture. 2000 Red List: Critically Endangered; CITES II.

Nepenthes diatas This is abundant in the moss and mountain forests of the Gunung Bandahara massif, Sumatera, at 2,500–3,000m (8,000–9,500ft). Its upper pitchers are tall and thin, the lower ones stout with pronounced teeth. Little of its high-altitude habitat has been cleared. 2000 Red List: Lower Risk; CITES II.

Nepenthes gracilis A still common species found in a wide range of lowland habitats, from Thailand to Sulawesi, up to 750m (2,500ft), it varies in colour. Its habitats have been heavily cleared, particularly with recent forest fires. Red List: Unlisted; CITES II.

Nepenthes inermis This has an unusual upper pitcher, without a rim or lid. To stop prey being washed out of the pitcher, insects are caught on the sticky pitcher wall and washed to the base by rain. Water is tipped out by a pivoting action of the tendril. It is epiphytic and found at 1,500–2,600m (5,000–8,500ft) in the moss and mountain forest of Sumatera Barat and Jami. Its altitude has protected its habitat from being cleared. 2000 Red List: Vulnerable; CITES II.

Nepenthes macrophylla A terrestrial or epiphytic species confined to the moss forest summit of one peak in northern Borneo, at 2,200–2,500m (7,250–8,000ft). Pitchers have a broad lid and well-developed teeth. Its habitat is not protected and it is threatened by collectors and damage by visitors. 2000 Red List: Critically Endangered; CITES II.

Nepenthes merrilliana Probably one of the largest species in the genus with pitchers over 50cm (20in) tall, it grows only in the north-east of Mindanao Island in the Philippines, in lowland forests on mineral-encrusted hills. 2000 Red List: Vulnerable; CITES II.

Nepenthes mikei The distinctive pitchers, both upper and lower, are green but marked with black lines. Known only from two mountains in Sumatera, this species is found among the stunted vegetation of peaks and cliffs at 1,000–2,800m (3,500–9,250ft). It is difficult to distinguish from *Nepenthes angasanensis*, so may be more widespread. Although this species is very localized, little of its habitat has so far been cleared. 2000 Red List: Vulnerable; CITES II.

Nepenthes mira Recently described, with pitchers that resemble those of *Nepenthes rajah*, this is from the highlands of Palawan, Philippines. Red List: Unlisted; CITES II.

Nepenthes mirabilis Found from Indochina to Micronesia and Australia, it is probably the most widespread *Nepenthes* of moist, open habitats up to 1,500m (5,000ft). It is rare in habitats where other species grow, suggesting it is unable to compete. This is a variable species; one of the most bizarre forms, var. *echinostoma*, has a very swollen rim, or peristome. Red List: Unlisted; CITES II.

Nepenthes rafflesiana A widespread species of lowland heath and peat swamp forest up to 1,200m (4,000ft) in Borneo, peninsular Malaysia, and Sumatera. Being adaptable, it is found in secondary habitats that result from human disturbance. This species varies greatly in size, and in colour from dark purple pitchers to yellow ones speckled red. Red List: Unlisted; CITES II.

Nepenthes sibuyanensis This lowland species has pale yellow to slightly red pitchers with some red spots. It is known only from one mountain in Sibuyan, in the Philippines, where it grows in open, grassy areas with *Nepenthes alata* and *Nepenthes argentii*. 2000 Red List: Vulnerable; CITES II.

Nepenthes sumatrana From dense, undisturbed forests in north-west and central Sumatera, this species is mainly found up to 800m (2,650ft), where it is rare and intolerant of disturbance. The spectacular aerial pitchers widen to a large open mouth with a colourful peristome. It is threatened by large-scale habitat destruction. Red List: Unlisted; CITES II.

Nepenthes talangensis This beautiful species has small, round pitchers with a rich yellowish or red hue and is known from a single mountain, Gunung Talang, in Sumatera Barat, where it grows in moss and mountain forest at 1,800–2,500m (5,900–8,250ft). 2000 Red List: Endangered; CITES II.

Nepenthes truncata This lowland species grows only on open mountainsides on Mindanao Island, the Philippines, and has pitchers as voluminous as those of *Nepenthes rajah*. 2000 Red List: Endangered; CITES II.

12 on Red List*

Pinguicula

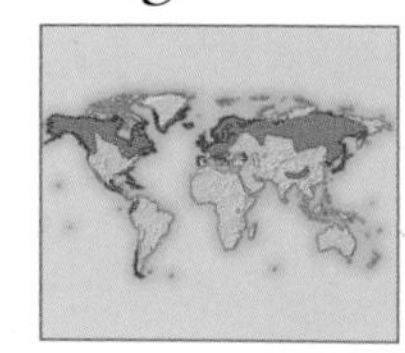

Botanical family
Lentibulariaceae

Distribution
Asia to Japan, Europe, North Africa, North America to the Andes

Hardiness
Half-hardy – frost-tender

Preferred conditions
Full sun to partial shade; light, sandy, usually acidic soil, but some species like alkaline soil; temperate species: 0–29°C (32–84°F); tropical and subtropical species: 2–32°C (36–90°F).

There are over 65 *Pinguicula* species. The name means "little greasy one"; both this and its common name of butterwort derive from the mucilage secreted from thousands of tiny glands covering the leaves. Pinguiculas inhabit peat bogs and marshland in the northern hemisphere, from Arctic glaciers down to Alabama, USA. Some species occur in Mexico, the Caribbean, and South America, where they can tolerate drier conditions and even grow as epiphytes on trees. The usually green leaves form rosettes and the flowers range in colour from white through lurid pink to purple.

Pinguicula crystallina can be found in various habitats from vertical limestone cliffs to wet meadows and sphagnum moss bogs. It is not easy to grow, but in some locations it will survive low temperatures. Cultivated plants need high humidity and an open, sandy compost.

Pinguicula gypsicola grows in Mexico, on wet gypsum cliffs in the shade beside cacti, agaves, and other plants that are adapted to dry conditions (xerophytes). In winter, temperatures can sink below 0°C (32°F) and it almost never rains. The summers are very hot and rain is scarce, but moisture is provided by heavy dews at night. In cultivation, this species should be grown in a free-draining mix of sand and vermiculite and kept completely dry, but with high humidity during winter.

Pinguicula planifolia, or Chapman's butterwort, has a very narrow range along the Gulf Coast, USA, from the Florida panhandle to Louisiana. It is endangered through loss of habitat from land conversion, drainage, and overuse of herbicides, particularly along powerlines where many carnivorous plants survive. It grows in wet, acidic soils together with many other carnivorous plants such as sarracenias and utricularias, in indirect to partial shade provided by taller native plants. In cultivation, the flowers need to be hand-pollinated to set seed.

▲ PINGUICULA CRYSTALLINA *has pale blue to rose-pink flowers. It is found in Cyprus with subcolonies reported in Turkey. One sub-species grows on tufa rock by the sea in Italy.*

▲ PINGUICULA PLANIFOLIA *is easily recognized, with large leaf rosettes that turn deep red to maroon, glistening like raw meat in the sun, in contrast to its delicate flowers.*

Pinguicula vallisneriifolia is an unusual, temperate species from southern Spain. It has grassy leaves up to 20cm (8in) long, and white to violet flowers. New plants arise on creeping stems, or stolons. It prefers wet, humid, shady spots on vertical limestone cliff faces, but can be grown in well-drained, alkaline soil with a little peat. The species was listed as Endangered on the 1997 Red List.

Pinguicula filifolia is a tropical species with narrow, threadlike leaves and violet flowers. It is found in western Cuba where the light is intensified by the white sand in which the plant grows, at swamp edges. It needs high humidity.

Pinguicula ionantha (Vulnerable on the 1997 Red List) has white, violet-centred flowers. It is limited to ditches and ponds in the Florida panhandle, and is federally listed as endangered. The sites are now threatened by drainage schemes.

10*/3 on Red List/CITES I and II

Sarracenia

Botanical family
Sarraceniaceae

Distribution
Canada, south to the Gulf Coast and west to Texas, USA

Hardiness
Frost-tender

Preferred conditions
Good, but indirect, light; acid, infertile soil; waterlogged in growing season; prefers 2–35°C (36–95°F).

There are eight species and various subspecies in this carnivorous genus. Known as pitcher plants or trumpet pitchers, they are popular and easy to grow, but are all threatened in the wild to varying degrees. Species in the south of the range are particularly susceptible because they occur in a restricted number of sites, which are threatened by mismanagement, changes in water levels, and over-collection.

Fortunately, many species can survive periods of unsuitable environmental conditions because they can regrow from rhizomes. In addition, habitat restoration projects on many sites aim to raise the water table to ensure the necessary wet, acidic conditions, and to re-introduce periodic bush fires to clear competing plants. *Sarracenia* populations have responded well to such measures.

All species, except ***Sarracenia psittacina***, catch insects by means of a pitfall trap: an insect loses its footing while following a trail of nectar and falls into the pitcher (*see also p.424*).

13 on Red List*

Utricularia

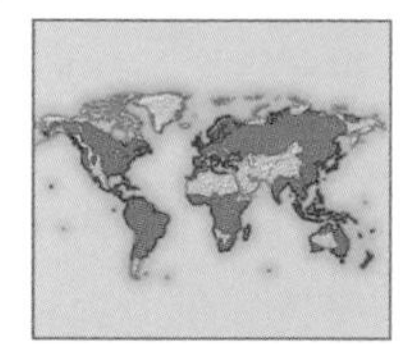

Botanical family
Lentibulariaceae

Distribution
Worldwide

Hardiness
Half-hardy – frost-tender

Preferred conditions
Temperate species: 7–28°C (45–82°F); tropical species: 10–35°C (50–95°F); tuberous species: 4–38°C (39–100°F); North American species: 1–32°C (34–90°F).

This highly adaptable genus has 214 species, distributed across climatic zones from the Arctic to the tropics, except for most arid regions and many islands. Most species occur in South America. The plants are prized for their often beautiful, orchid-like flowers. The common name, bladderwort, arises from the small traps (0.2–12mm across) that occur along the plant's stems and leaves. Populations are in decline, particularly in their wetland habitats, as a result of pollution and habitat loss.

Aquatic species usually form free-floating mats in open water. ***Utricularia inflata***, from the eastern coastal plain of the south-eastern states of the USA, has traps at the end of its flotation tubes.

◄ UTRICULARIA INFLATA *This aquatic supports its flower stems on rosettes of 4 to 10 air-filled tubes that float in acid ponds and ditches.*

This species is threatened by drainage of its habitat, pollution from agriculture, and aggressive forestry practices leading to land conversion or soil runoff into waterways. It can be easily grown in tanks or ponds if the water is acidified by adding one part peat to 5 litres (1 gallon) of water. The plants become dormant in colder temperatures.

Terrestrial species live in permanently or seasonally waterlogged soils. In the eastern wheat belt of Western Australia, *Utricularia menziesii* grows in sandy soils at the edge of swamps and other damp habitats which are under threat from farming and urban expansion. Unusually, it forms hairy tubers to ensure its survival below the hot, dry soil during summer. It can take several years to produce its single, blood-red flowers.

One of the rare, epiphytic species is *Utricularia quelchii*, much sought-after for its beautiful, vermilion flowers and a great plant for terrariums and hanging baskets in acid compost. It is found on Mt Roraima, in tropical Venezuela (*see* Remote Tepuis, *p.435*), a fragile habitat vulnerable to trampling from visitors.

◄ SARRACENIA PURPUREA *has purple or greenish-purple flowers in spring, and is widespread in the eastern USA and Canada. It is easy to grow.*

▲ SARRACENIA PSITTACINA, *unlike other sarracenias, has prostrate pitchers with "lobster-pot" traps. Once an insect is lured into the dome, it is confused by light seeping through clear panels (fenestrations) in the pitcher walls, wanders into the pitcher tube, and becomes stuck.*

Pitcher plants can be grown indoors or in a bog garden without any winter protection. They need a peaty compost and should be kept standing in 2–3cm (¾–1¼in) of rainwater for the growing season. When the plant becomes dormant, remove the old leaves and keep the compost slightly damp. Most species need good light to bring out the best colour in the pitchers. They can be propagated by seed.

Sarracenia leucophylla, the white-topped pitcher plant, is found on the eastern Gulf Coast in pine savanna and seep bogs. It is listed as Vulnerable on the 2000 Red List, since mismanagement of sites has resulted in the suppression of periodic fires, on which its existence depends. Unlike other sarracenias, it produces two flushes of pitchers in one season. But it resembles other red-flowered species in having forms that lack anthocyanins (red pigments) and so bear yellow flowers. One such form is known as "Schnell's Ghost", after Don Schnell, foremost writer on carnivorous plants of the USA and Canada.

Another red-flowered species with yellow forms is *Sarracenia psittacina*, the parrot pitcher, with a unique "lobster-pot" trapping mechanism. It is found in pine savannas and seep bogs along the Gulf Coast and Georgia's Atlantic coastal plain, where it may at times be partially submerged. Giant forms have been found around the Okefenokee Swamp.

Sarracenia purpurea subsp. *purpurea*, known as the northern pitcher plant, is the most widespread, and least threatened, of the pitcher plants from Canada and north-eastern USA. It has been introduced and grows wild in peat bogs in Europe, such as Roscommon in Ireland. Its horizontal pitchers may be suffused red, and pure yellow or green forms also exist. Another subspecies, *Sarracenia purpurea* subsp. *venosa*, or southern pitcher plant, is found on the eastern and Gulf coasts of the USA, on coastal plain savannas and occasionally in seep bogs. It is more variable than its northern cousin, *Sarracenia purpurea* subsp. *purpurea*. The variety *montana* is found in upland areas, while var. *burkii*, with pink flowers, grows on the Gulf Coast.

Sarracenia rubra occurs throughout the south-eastern USA, but was listed as Rare on the 1997 Red List. It is the most difficult species to classify, having five currently recognized subspecies, two of which are on the federal protection list.

One of the rarest forms is *Sarracenia rubra* subsp. *alabamensis*, confined to a handful of bogs in central Alabama and listed on CITES Appendix I. It has stocky pitchers, a wide pitcher mouth opening and wavy lid, and it stands up to 75cm (30in) tall. Known as the Alabama canebrake pitcher plant, it was once more common, with nearly 30 populations known in the 1960s. Change in its habitats, including falling water table levels, as well as mismanagement in the form of suppressing natural bush fires, have reduced the number of sites by almost 60 per cent over the last 40 years. There are now only 11 sites in central Alabama, many privately owned. Half the plants are at one site and only three sites are truly viable.

The federally protected *Sarracenia rubra* subsp. *jonesii*, or mountain sweet pitcher plant, is also listed on CITES Appendix I. It can be found in mountain seep bogs in a small area of the Carolinas. Anthocyanin-free forms are also known. They grow 15–25cm (6–10in) tall with a distinctive bulge. *Sarracenia rubra* subsp. *gulfensis* is largely restricted to the seep bogs of the western Florida panhandle.

SARRACENIA LEUCOPHYLLA ► *Known as the white-topped pitcher plant, this is the most elegant sarracenia species. The pitchers are delicately interlaced by red or green veins.*

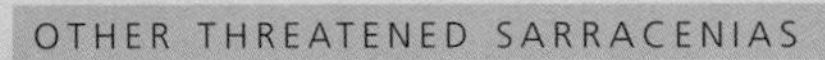

◄ SARRACENIA FLAVA, *or yellow pitcher plant, is one of the best-known species of savanna and seep bogs. It is extremely variable, with seven recognized forms. The reddest varieties are the most sought after by collectors.*

OTHER THREATENED SARRACENIAS

Sarracenia alata The pale pitcher plant has two distinct zones of distribution, often in pine savannas. Its western range stretches from Texas to Louisiana, and its eastern from Mississippi to Alabama. The tall pitchers (up to 75cm/30in) can vary widely in shape and colour. The pitchers often have pointed lids and can be pale green with red veining or a deep red, almost black colour. 2000 Red List: Lower Risk; CITES II.

Sarracenia minor Found in southern North Carolina and Florida, USA, mainly in pine savannas but also in very wet habitats such as the Okefenokee Swamp, where a giant form can be found. Called the hooded pitcher plant, it is thought to be the most primitive species of *Sarracenia* and mostly seems to trap ants. 2000 Red List: Lower Risk; CITES II.

Sarracenia oreophila The green pitcher plant is federally protected, with a very limited distribution in the Appalachian Mountains across Alabama, Georgia, and North Carolina, USA. Many colonies have disappeared from open seep bogs and the remaining 26 populations are isolated. Unlike other species, this plant has a distinct period when it produces only flat, blade-like leaves. It is highly sought-after by collectors. 2000 Red List: Critically Endangered; CITES I.

Sarracenia flava There are several varieties. The most typical, var. *flava*, has deep red veins and is common in the Atlantic coastal plain; red var. *atropurpurea* is rare in the Atlantic coastal plain of the Carolinas and the Florida Panhandle. The pure green var. *maxima*, intensely red-veined var. *ornata*, and copper-lidded var. *cuprea* all occur on the coastal plain of the Carolinas and north-west Florida. All on 1997 Red List; CITES II.

INVASIVE PLANTS

INVASIVE PLANTS

Non-native plants are described as invasive when they can, have, or are likely to spread into native flora or managed plant systems, develop self-sustaining populations, and become dominant or disruptive to those systems. Some harm not only plants but people and animals as well. The native floras of some parts of the world, such as the Hawaiian Islands and New Zealand, are highly susceptible to displacement by invasives.

Many plants in this section should not be grown by the gardener, and cultivating some of them is illegal. Waste from invasive plants should be composted completely, either at municipal facilities or by leaving it to decompose in the sun in special bins or black plastic bags. Composting does not kill all seeds, so these should be removed for disposal in the rubbish (where allowed by law).

(100) Featured in ***One Hundred of the World's Worst Invasive Alien Species***, a list published by IUCN that includes not only land and aquatic plants but also all other life forms.

ACACIA BAILEYANA

Abrus precatorius
Rosary pea

The common name of this woody climber with herbaceous branches derives from the use of its seeds as beads to create necklaces. All parts of the plant are toxic if eaten. The seeds are dispersed by birds and the plant can be very hard to control because it becomes deeply rooted. ECOLOGICAL IMPACT This plant can climb high into trees in open but undisturbed forests. It is essentially pantropical, being found in south-eastern states and Hawaii in the USA, India, several Caribbean islands, Belize, and some South Pacific islands. It probably should not be planted anywhere.

Acacia auriculiformis
Earleaf acacia

This tree can begin to flower after only one year and produces numerous seeds, which are eaten and dispersed by birds. It tolerates varied soil conditions and has been used both for timber and as an ornamental shade tree in tropical areas around the world. ECOLOGICAL IMPACT Bacteria in the root nodules of this tree can alter soil chemistry by "fixing" nitrogen in the soil. The plant is native to New Guinea and parts of Australia but has become naturalized in other places. Tropical areas, especially islands, appear to be vulnerable. It now invades in Florida and the South Pacific.

Acacia baileyana
Cootamundra wattle

Masses of fuzzy yellow flowers are produced by this pretty tree in the winter. The flowers turn into large numbers of seeds that are dispersed by both birds and the explosive opening of the pods. The seeds can remain viable for many years. ECOLOGICAL IMPACT The tree adds nitrogen to the soil via bacteria in its root nodules. It is native to south-east Australia but has become very invasive in other parts of Australia and appears to hybridize with other acacias there. Avoid planting in parts of Australia where it is not native, and in areas where other native acacias are established, such as in Hawaii and South Africa. Bushfire accelerates the germination rate of seeds. Land managers in areas where fire is a part of the ecosystem should take care to monitor growth of new seedlings of this acacia following recent fires.

ABRUS PRECATORIUS

Acacia confusa
Formasan koa

Planted as a street ornamental and reforestation species in warm climates around the world, this tree produces abundant fruit and begins seed production after only two or more years. The germination rate of the seeds may increase after fire. ECOLOGICAL IMPACT This tree adds nitrogen to soil via bacteria in its root nodules. Formosan koa is also believed to have allelopathic properties, meaning it can poison plants that grow beneath or near it. The tree is currently invading in Taiwan and Hawaii, and places with similar climates also may be susceptible to invasion.

Acacia decurrens
Early black wattle

Like most other invasive acacias, this tree is native to Australia. ECOLOGICAL IMPACT Studies in South Africa indicate that the tree consumes water at a higher rate than native plants there, potentially monopolizing local water availability. It also adds nitrogen to the soil through nodules on its roots. There is additional concern that this wattle may hybridize with native acacias and compromise their gene pool. Avoid planting this tree in parts of Australia where it is not native and where it is spreading, and in other places where other native acacias exist, such as California, South Africa, and other countries with a Mediterranean climate.

Acacia farnesiana
Mimosa bush, sweet wattle

One of the most invasive members of this genus, thorny *Acacia farnesiana* begins reproducing after only two years and produces abundant seed each year. In warm climates the tree flowers all year and perfumes are made from the fragrant flowers. Cutting back may not be enough to control it since it regenerates from the cut stump. ECOLOGICAL IMPACT The tree adds nitrogen to the surrounding soil through bacteria in its root nodules. Mimosa bush is native to Latin America, but is already invading in South Africa, Australia, Madagascar, and other places with a warm, dry climate.

Acacia melanoxylon
Blackwood

Like most acacias, blackwood prefers a warm climate, but this tree can grow in a wider climatic range than most. It begins producing seed at only two years of age and reliably produces highly viable seed each year. The seeds can remain viable in the soil for up to 50 years, making eradication of a population a long-term enterprise. It also produces root suckers, increasing the population vegetatively. ECOLOGICAL IMPACT The tree adds nitrogen to the soil through bacteria in its root nodules. It is invading in California, South Africa, New Zealand, and other places with a warm, Mediterranean climate.

Acer negundo

Box elder

The plain-leaved form of box elder grows very vigorously and can cause problems. Although plants are dioecious (either male or female), the females can produce seed without fertilization. Box elder also spreads by root suckers. 'Flamingo', the safe garden form with pink-and-white-splashed leaves, is male and only propagated vegetatively. ECOLOGICAL IMPACT While this tree can invade dry areas, it prefers wetter soil and may do most damage alongside rivers and streams, trapping sediments and obstructing flow. It is invasive in parts of Australia, such as New South Wales, and central Europe, as well as in parts of the USA.

ACER PLATANOIDES

Acer platanoides

Norway maple

An ability to tolerate tough urban conditions has made this species a popular street tree in many parts of the world. It is very tolerant of poor soil and grows rapidly. The flowers do not require fertilization to form viable seeds and the seeds germinate easily, especially when they are fresh. ECOLOGICAL IMPACT In north-east USA the tree has become a serious pest in forests, where it outcompetes its close relative, *Acer saccharum*. Avoid planting in the northeast and middle Atlantic sections of the USA, and in other temperate areas with high summer rainfall.

Acer pseudoplatanus

Sycamore

A native of parts of northern Europe, this tree is invasive in other parts of Europe and the UK (where it was probably introduced early in the 16th century). Sycamore reliably produces seed each year, and for unknown reasons seed production skyrockets in some years. ECOLOGICAL IMPACT This tree can invade open woods and be a problem in wetlands. It has become very invasive in New Zealand, the UK, north-east USA, and parts of temperate South America. It should be monitored when planted in areas with high summer rainfall, and should not be planted near wetlands that experience dry summers.

Acer tataricum *subsp.* ginnala

Amur maple

Native to northern Asia, the amur maple is a shrubby small tree that has seeded in forests and open areas in the New England and upper midwest regions of the USA. A single tree can produce up to 5,000 seeds annually, without the need for cross-pollination, and these seeds germinate readily. Amur maple also resprouts vigorously when cut back, hindering control. ECOLOGICAL IMPACT In grasslands this maple shades out native species, and it is also capable of invading open forests. It is currently invading in the east and upper midwest of the USA, where planting should be avoided. Caution should be exercised when planting in areas with similar climates.

Aegopodium podagraria

Ground elder, goutweed

This herbaceous plant, native to Europe and Asia, appears to spread vegetatively rather than by seed, and it is nearly impossible to eradicate. Its roots are brittle, breaking off easily when the plants are pulled, and it resprouts almost immediately from the fragments. Most herbicides are ineffective against ground elder. ECOLOGICAL IMPACT This weed is locally invasive virtually everywhere it is grown in temperate areas. It should never be grown where the roots, or root fragments caused by disturbance, can escape into open countryside. It can be planted where the roots are completely contained, such as in a heavy planter, but the waste must be disposed of carefully.

Ageratina adenophora

Eupatory, Crofton weed

This member of the herbaceous daisy family can grow to nearly 2m (6ft) and is sometimes included in garden plantings for its structural value. No pollination is needed for seeds to form, and each plant produces up to 10,000 seeds, which are spread by the wind. Roots may also form where stems bend over and touch the ground. Eupatory needs light to germinate, so it is generally found in open areas. ECOLOGICAL IMPACT Livestock may be poisoned by this plant, and it also appears toxic to other plants that grow nearby. It is native to Mexico, but is now invasive in many parts of the tropical and subtropical parts of the world, including north-eastern India, Nigeria, south-east Asia, Yunnan Province in China, several South Pacific islands, South Africa, New Zealand, Australia, and western North America. It should not be planted anywhere.

AILANTHUS ALTISSIMA

Ailanthus altissima

Tree of heaven

Partly because of its wide tolerance of environmental conditions, this tree is one of the most aggressively invasive species known. It can grow in heavily polluted urban sites with minimal soil, alongside streams, under shade, or in the open. It spreads by root suckers, but also produces ample wind-dispersed fruits that germinate easily. After germination, plants can begin to produce flowers and fruits in only six weeks. That the plant is short-lived is little solace, since many thousands of seeds are produced during the life of a single tree. ECOLOGICAL IMPACT This tree forms dense stands and is allelopathic, poisoning plants beneath and around it. Avoid planting it anywhere.

Albizia julibrissin

Silk tree

Sweet-smelling, pink powder-puff-like flowers make this tree a gardener's favourite. The podded fruits appear to be dispersed by small animals, as well as wind. The tree is fast-growing but short-lived and is often found in disturbed areas as well as alongside streams. It resprouts vigorously when cut back. ECOLOGICAL IMPACT This tree has become invasive in temperate areas with summer rainfall. Avoid planting in any warm, temperate climate with high summer rainfall. A notable example is south-east USA, where it currently invades.

AEGOPODIUM PODAGRARIA

Albizia lebbeck

Woman's tongue

Feminists may not be enthused by the common name of this tree species, but its long, podded fruits do somewhat resemble a tongue. When dry, the pods rattle in the wind, suggestive to some people of the scolding of an irate woman. Each tree produces lots of pods, and these may remain on the tree for several months. The tree is fast-growing, can spread by root suckers, and the seeds remain viable for at least four years, so control can be difficult. However, the species is sensitive to drought. Stumps of felled trees are likely to resprout. ECOLOGICAL IMPACT This tree is widely naturalized throughout the tropics and is rapidly expanding its range in some areas. It is already invading Florida, the Caribbean, the South Pacific, and South Africa. On this account it should not be planted anywhere in the tropics, especially in wetter areas.

AMMENOPHILA ARENARIA

Allium triquetrum

Three-cornered garlic

In recent years bulbous plants of the genus *Allium* have left the preserve of the vegetable garden and become popular as ornamentals. Some have giant, bursting heads of brightly coloured flowers, but the three-cornered garlic is more modest, with numerous drooping white flowers on each stalk. Gardeners have recognized that it can get out of control in the garden and there is now concern that it is jumping the garden gate and invading tracts of open countryside. ECOLOGICAL IMPACT In the UK three-cornered garlic is known to naturalize in hedgerows and woodland edges. In Australia the plant has been named as one of 50 species that should never be used in that country's gardens. It is also found in New Zealand. Do not plant it anywhere near open countryside, where it could easily spread, and be careful to contain it in your own garden. There are reports that the seeds have the capacity to remain viable in the soil for more than 18 years.

Ammenophila arenaria

European beach grass

Many Americans probably think of this grass as a native species since it has been planted widely in the USA. It can stabilize dunes but it can also change beach topography by increasing their steepness. It spreads almost entirely by rhizomes (underground stems), which may be fragmented by wave action. ECOLOGICAL IMPACT The grass reduces native plant species, and the number of arthropod species also decreases in some areas. It is native to a wide latitude range, from Scandinavia to northern Africa. This adaptability to climates makes it likely to spread wherever it is introduced. It should not be planted in gardens and should be planted in dune areas only under strictly controlled conditions.

Ampelopsis brevipedunculata

Porcelain berry

This climber's relationship to grape vines is evident in the shallowly lobed leaves and the grape-like berries. The latter turn shiny blue and purple, mottled with white and grey, hence the common name of porcelain berry. The seeds germinate easily, and plants can also reproduce from stem and root fragments. The root system is not easily removed. ECOLOGICAL IMPACT The fast-growing stems form extensive tangles that blanket the ground, trees, and shrubs in woodland areas, reducing the diversity of native species and increasing the vulnerability of trees to wind damage. The plant is native to north-east Asia and invades in the middle Atlantic and north-eastern parts of the USA. In similar temperate areas with high summer rainfall, such as much of Europe, any plantings should be monitored.

AMPELOPSIS BREVIPEDUNCULATA

Angiopteris evecta

Mule's foot fern

A member of an ancient group of ferns that first appeared 400 million years ago, this plant certainly looks prehistoric. It has giant fronds up to nearly 3m (10ft) in height and a canopy that can spread even wider than that. Perhaps being descended from one of the first species to establish on land has also given it the ability to invade in new lands even now. ECOLOGICAL IMPACT Although increasingly rare in its native Australia, the fern is invasive in moist forests and rainforests in Hawaii. It should not be grown there because the tiny spores are dispersed by wind and can travel long distances. Plants growing near any moist forests should be monitored.

Anredera cordifolia

Madeira vine

With closely arranged, fleshy, small leaves on its 3–7m (10–22ft) stems, this climber has an odd appearance. It has small, fragrant, white flowers but it does not set fruit often, reproducing instead from small tubers found in joints between the stems and leaves. Mature stands may comprise 1,500 tubers per square metre. The vine is very hard to kill. ECOLOGICAL IMPACT The climber covers shrubs and small trees in gardens, vacant lots, and open countryside, growing as much as 1m (3ft) a week. The succulent leaves and tubers make the climber very heavy, often damaging the plants it covers. It is native to tropical America but is a pest in parts of New Zealand, Australia, South Africa, and several South Pacific islands. It should not be grown in warm, moist climates.

Araujia sericifera

Cruel plant

This climber's large fruits open to release seeds, which are spread by wind and water. Its sap is toxic and irritates the skin. Moths and butterflies are attracted to the white flowers but become trapped in the oozing sap, where they die. ECOLOGICAL IMPACT Climbing over shrubs and small trees, this plant forms a dense canopy that denies light to plants beneath. The climber invades disturbed ground, as well as wet forests and along rivers

ARAUJIA SERICIFERA

and creeks in open countryside. Native to Peru, it has escaped cultivation in a number of places, including New Zealand, South Africa, western North America, eastern Australia, and Israel.

Arctotheca calendula

Capeweed

A mat of underground stems topped by yellow flowers is formed by this member of the herbaceous daisy family. It can spread from seed and fragments of the plants. Some cultivars are reputedly sterile, but this has yet to be ascertained.
ECOLOGICAL IMPACT The plant grows vigorously and covers native vegetation. Streamside areas, grasslands, and dunes appear to be the most affected. Native to South Africa, the plant now invades in Australia and California. It should be expected to invade in southern Europe (recent reports suggest that it may have begun to invade in coastal Spain), temperate South America, and in other Mediterranean-type regions.

Ardisia crenata

Coral berry

People living in temperate areas may know this only as a pot plant, but in warmer areas it is a shrub that is grown outdoors. Its bright red fruits are attractive to birds, which disperse seed throughout forests. The fruits may be retained on the plant for one year, giving the birds plenty of time to find them. The germination rate of the seeds tends to be very high.
ECOLOGICAL IMPACT Coral berry can achieve densities of up to 100 plants per square metre, replacing many species of native vegetation.

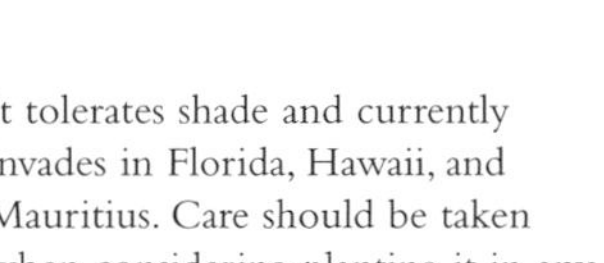

It tolerates shade and currently invades in Florida, Hawaii, and Mauritius. Care should be taken when considering planting it in any wet tropical or subtropical area.

Ardisia elliptica

Shoebutton ardisia

The shoebutton ardisia is a beautiful, low-growing shrub. The purple-black fruits attract birds, which disperse the seeds. The plant produces seeds without fertilization of the flower, and the seeds germinate easily.
ECOLOGICAL IMPACT This shrub can quickly take over an area with its creeping underground stems. It tolerates shade and can dominate a forest understorey in wetter, warm-climate areas. It also increases the nitrogen levels in soils through bacteria in its root nodules, altering the chemistry of soil in which native plants have evolved. Avoid planting in warm areas with ample rainfall, such as Hawaii, Florida, Indonesia, and parts of the Caribbean, where the shrub currently invades.

Arundo donax

Giant reed

Originally from India, the giant reed was introduced very early in the Mediterranean and is now planted widely. It has been used for erosion control, paper, roof thatch, reeds for woodwind instruments (the original Pan pipes may also have been made from its cane), as an ornamental plant, and for many other purposes. The reed rarely produces seed, but can grow rapidly from stem and root fragments to dominate hundreds of acres in huge populations of genetically identical clones. The reed prefers wetter areas and has transformed floodplains and rivers, where it spreads from fragmentation during floods and other disturbances.
ECOLOGICAL IMPACT The giant reed outcompetes native plants and does not provide food or habitat for native animals. In the USA it is believed to be reducing the food sources and habitat of a number of rare animal species. Because of the aggressive nature of this species, enormous care should be taken to prevent the spread of root fragments anywhere that it is grown. It currently invades in southern Africa, parts of North America and Mexico, Caribbean islands, South Pacific islands, Australia, and south-east Asia.

Asparagus asparagoides

Bridal veil

This herbaceous plant is popular in horticulture and floriculture for its feathery foliage. Birds love its fruits and disperse the seeds along streams and into open forests.
ECOLOGICAL IMPACT The plant can smother understorey vegetation and suppress regrowth in the dry, coastal communities it invades. The roots may have fleshy tubers that must be destroyed to prevent the plant from resprouting. Native to South Africa, it is now invasive in southern and western Australia and is spreading in southern California and New Zealand. Southern European countries should be aware of its invasive potential.

ARCTOTHECA CALENDULA

ARUNDO DONAX

Asparagus densiflorus

Asparagus fern

Despite the common name, this herbaceous plant is not a fern but a lily. Also called *Asparagus sprengeri*, it is grown in pots in cool climates, and in the ground in warmer ones. Birds spread the seeds from the red fruits, which should be removed from the garden before they can be eaten. The plant also forms tubers that must be destroyed or the plant will resprout following removal.
ECOLOGICAL IMPACT Asparagus fern creates spiny tangles of foliage in warm climates. It likes well-drained soils and invades both dry and moist forests and coastal areas. Native to South Africa, it is now invasive in Florida, some Caribbean islands, Queensland and New South Wales in Australia, Hawaii, and several South Pacific islands. Southern European countries need to be alert to its invasive potential.

Asparagus scandens

Climbing asparagus

Birds eat the orange-red berries of this climber and disperse the seeds. The roots produce small tubers which can resprout if left in the ground during control attempts.
ECOLOGICAL IMPACT The plant is shade-tolerant and invades forests and disturbed areas. It scrambles up surrounding vegetation on thin, wiry, spiky stems that can grow 4m (12ft) long. Originally from South Africa, climbing asparagus is reported as invasive in New Zealand and Australia. Because other asparagus species are invasive in warm climates, it should be grown with care in southern parts of Europe and the USA, as well as parts of Central and South America.

Azolla filiculoides

Water fern

This aquatic fern is found free-floating on still or slow-moving waterways. It is distinguished by the reddish colour it develops when it is stressed. The fern's fast-growing, invasive nature is now so well known that it is doubtful whether it is ever used in gardens intentionally.
ECOLOGICAL IMPACT Water fern forms dense mats on the surface of waterways, blocking sunlight and threatening the submerged plants growing beneath, as well as the fish that feed on those plants. The fern is invasive in many countries and should never be used. Water gardeners should carefully inspect all purchased plants to ensure that fragments of *Azolla filiculoides* are not introduced into the water when the plants are established.

BAUHINIA VARIEGATA

Bauhinia variegata

Mountain ebony, orchid tree

Most members of the genus *Bauhinia* are not invasive, but this tree is an exception. The flowers are undeniably attractive, and it may be tempting to leave plants to the point that they get out of control.
ECOLOGICAL IMPACT The wind-dispersed seeds travel and germinate easily and dense stands can form, both from seed and from root suckers. In Florida the orchid tree has displaced native vegetation in at least two different plant community types. It is also invasive in parts of the Caribbean, South Africa, and parts of the South Pacific, so it should be avoided in favour of other, less invasive bauhinias. *Bauhinia variegata* is a parent of the hybrid *Bauhinia* x *blakeana*, which is said to be sterile and therefore a good alternative for planting.

Begonia cucullata

Wax begonia

Native to South America, this herbaceous species is commonly used as a bedding plant and in containers in temperate gardens, and as a houseplant. It spreads by seeds that are dispersed by wind and water.
ECOLOGICAL IMPACT Currently it appears to be spreading in Florida and Georgia in the USA, where it has escaped cultivation into disturbed wet forests. It has even been found growing on floating mats of debris and vegetation in streams. It should be monitored when grown near open countryside in a tropical or subtropical location.

Berberis darwinii

Darwin's barberry

A native of Chile and Argentina, this spiny shrub is beginning to become invasive elsewhere. The fruits are eaten by birds, which then widely disperse the seeds.
ECOLOGICAL IMPACT In New Zealand it has been found to invade a number of plant community types, including intact vegetation and dense colonies. Avoid planting in New Zealand, Australia, and in other mild, temperate climates.

Berberis thunbergii

Japanese barberry

The Japanese barberry is a popular shrub used to form thorny hedges, which prevent trespassing. The plant tolerates shade and forms thickets in native forests. It spreads quickly because it can form multiple stems from the root system. Birds greatly enjoy its fruits and spread its seed widely, helping it to become invasive in some parts of the world. Another factor in its spread is that deer avoid its spines, allowing new populations to develop unchecked.
ECOLOGICAL IMPACT The plant has become a problem on some nature reserves in north-east USA, where it can form dense stands. Avoid planting there and in other temperate areas where winter temperatures are cold. Because the seeds require cold chilling before germination, planting may be less problematic in warmer climates.

BERBERIS DARWINII

Bischofia javanica

Bishopwood

This fast-growing tree produces a great many seeds, which germinate easily in warm climates. It can resprout vigorously if the trunk is cut. The tree produces flowers and fruit early. Once an individual becomes established in a forest, it rapidly begins seeding to produce a large population. Formerly popular as a street tree, its use in landscaping is now discouraged.
ECOLOGICAL IMPACT The plant is rapidly spreading and invading native plant communities in southern Florida. Avoid planting in Florida and in the South Pacific region, where it currently invades.

Bocconia frutescens

Plume poppy

Bocconia is an interesting woody member of the usually herbaceous poppy family that grows in dry and moderately moist forests. It has huge, spectacular leaves and the flowers, though not remarkable, are produced on the plant all year long in its preferred warm climate. It produces large numbers of seeds that germinate easily and are dispersed by birds. It will resprout if not completely uprooted and the waxy leaves make it resistant to control by most herbicides.
ECOLOGICAL IMPACT This plant has become a serious problem in Hawaii, especially on the dry, leeward side of the islands. There are indications that it is capable of growing in the wetter areas too. Planting should be avoided in the dry tropics, and carried out with caution in the wet tropics.

BUDDLEJA DAVIDII

Broussonetia papyrifera

Paper mulberry

The ancient Polynesians brought the paper mulberry tree with them when they colonized new islands, using the plant as a source of fibre for tapa cloth. It is therefore difficult to determine where it is native in the South Pacific. Paper mulberry is very fast-growing and it spreads vegetatively though root suckers. It resprouts after cutting back.
ECOLOGICAL IMPACT It has spread in forests and along rivers throughout the south-east USA, where it can form dense thickets and decrease wildlife habitat. Avoid planting here and in other mild areas with high summer rainfall.

Bryophyllum tubiflorum

Mother of millions

Also known as *Kalanchoe delagoensis*, this succulent native of Madagascar has a fascinating method of reproducing: it produces new little plantlets and bulblets on the margins of the leaves, which develop into new plants when they are dislodged from the mother plant. The species is grown as an ornamental in a number of hot, dry locations because of this habit.
ECOLOGICAL IMPACT This plant is spreading rapidly on dry slopes and forests on the Hawaiian Islands, the Virgin Islands, parts of Australia such as south-east Queensland, and South Africa. Areas with similar climates, such as the Mediterranean countries, should monitor open countryside wherever it is present.

Buddleja asiatica

Dogtail

Found along roads and on relatively fresh lava flows in Hawaii and other islands in the South Pacific, this scrambling shrub is very tolerant of poor soils. It also grows in wet meadows and forests. The shrub produces flowers and fruits all year and the small wind-blown seeds germinate easily. In forested areas it may use nearby trees for support, almost like a vine.
ECOLOGICAL IMPACT Avoid planting in the South Pacific, where it is invasive, and in areas with similar warm climates.

Buddleja davidii

Butterfly bush

The butterfly bush is popular with people who enjoy attracting butterflies to their gardens (the tiny purple flowers in lilac-like flowerheads have copious nectar in summer). Unfortunately, this shrub can become invasive in a variety of conditions, from roadsides to alongside rivers and streams. Its tiny wind-dispersed seeds are easily blown great distances.
ECOLOGICAL IMPACT The butterfly bush grows easily in soils low in organic matter. It is a problem in the UK and New Zealand, and in parts of the USA such as the Pacific north-west and the middle Atlantic states.

Buddleja madagascariensis

Smoke bush

The common name "smoke bush" is used for both this shrub and another species, *Cotinus coggygria*. This buddleja produces attractive flowerheads that offer nectar to a variety of pollinators. The fragrant, yellow-orange flowers and the fruits may be found on the plant all year, and the small seeds are easily dispersed by wind. Unlike most buddlejas, the seeds are found in pulpy fruits and may be dispersed by birds. Lax in habit, it can sprawl over other species and in forests it uses surrounding trees for support.
ECOLOGICAL IMPACT In parts of Hawaii it has formed dense thickets that exclude native vegetation. Avoid planting in moist or wet areas in warm climates, especially in Hawaii, Australia, and South Africa, where it already invades.

BUTOMUS UMBELLATUS

Butomus umbellatus

Flowering rush

This aquatic plant grows in water and along the edges of marshes and shorelines. It is not a "true" rush and is distinguished by pink flowers in a flattened flowerhead. It grows to 1.5m (5ft) tall. The flowers have earned it increasing popularity as a pond garden plant. It spreads through seed and small bulblets that are dispersed by water.
ECOLOGICAL IMPACT Native to Eurasia, the plant is invasive in the eastern and upper midwestern parts of North America, and all of the southern provinces of Canada, where it competes with native shoreline vegetation, including willows and cattails. It is very hardy and has the potential to invade

CARDIOSPERMUM GRANDIFLORUM

wetlands in virtually any temperate area. It should not be used in water gardens where there is a risk of escape into natural watercourses.

Caesalpinia decapetala

Mysore thorn, wait-a-bit

Grown as a hedge plant because of its fearsome, impenetrable thorns, this shrub is also cultivated for ornamental purposes. The thorns can make the plant difficult to control by cutting, and can pose a danger to animals. (In Hawaii it was reported that a cow wandered into a patch of Mysore thorn and ended up impaled, remaining imprisoned until death.) Unlike most plants in the pea family, Mysore thorn does not "fix" nitrogen. It flowers and fruits all year long, although the heaviest flowering of bright yellow flowers is generally in spring months. The seeds may be spread by birds and rodents, or on car and truck tyres. Stems that reach the ground have also been reported to root, increasing the spread.
ECOLOGICAL IMPACT While the plant is invasive in Hawaii and other tropical areas, it is less of a problem in Mediterranean-type climates.

Calophyllum antillanum

Alexandrian laurel

This frost-tender shrub is very tolerant of salt water. Its large seeds have been shown to remain viable after rivers have carried them into ocean currents. The plant produces abundant seeds every year, and the seeds appear to maintain viability over a long period of time. In a favourable site they germinate easily.
ECOLOGICAL IMPACT This plant has invaded mangrove and other coastal forests, where it can outcompete native species. Avoid planting this laurel in any warm-climate coastal area.

Calotropis procera

Calotrope, rubber bush

A shrub with rubbery leaves and pretty, small, pink flowers, this plant also has a milky sap that contains a cardiac poison. People and livestock have been injured by it, especially in arid areas where the chemical appears to become more concentrated. The fruits of this member of the milkweed family burst open when ripe and numerous seeds with silky hairs are released into the wind. The plant responds to soil disturbance by spreading rapidly, but there are also reports that it invades areas with less disturbance and that it tolerates poor soil conditions.
ECOLOGICAL IMPACT The plant can be very competitive with native vegetation. It is invasive in Australia and there is concern that it could spread in Hawaii and other islands.

Cardiospermum grandiflorum

Balloon vine

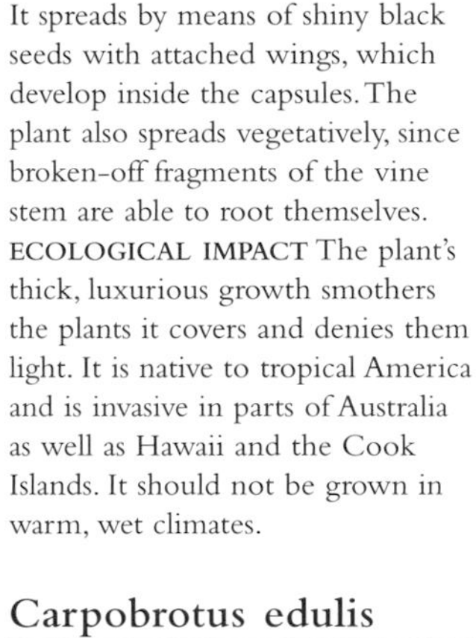

The common name of this climber derives from its puffy fruit capsules. The plant grows quickly, aided by tendrils that attach it to the trees and other structures that it climbs. It spreads by means of shiny black seeds with attached wings, which develop inside the capsules. The plant also spreads vegetatively, since broken-off fragments of the vine stem are able to root themselves.
ECOLOGICAL IMPACT The plant's thick, luxurious growth smothers the plants it covers and denies them light. It is native to tropical America and is invasive in parts of Australia as well as Hawaii and the Cook Islands. It should not be grown in warm, wet climates.

Carpobrotus edulis

Hottentot fig, kaffir fig

This succulent plant is planted on exposed sites to control erosion, although it is unclear how effective it is in that role. The plant has colourful flowers and spreads both vegetatively and by seed. Mammals eat the fruits and disperse the seeds, and their digestion processes also facilitate the seeds' germination.
ECOLOGICAL IMPACT The plant grows into a thick mat, suppressing other vegetation growing in its vicinity. It is native to coastal South Africa and has invaded coastal areas of California. It appears to be able to colonize both dry and moist sites. It has been shown to reduce soil alkalinity. Plantings should be monitored in regions such as southern Europe, Australia, and temperate South America.

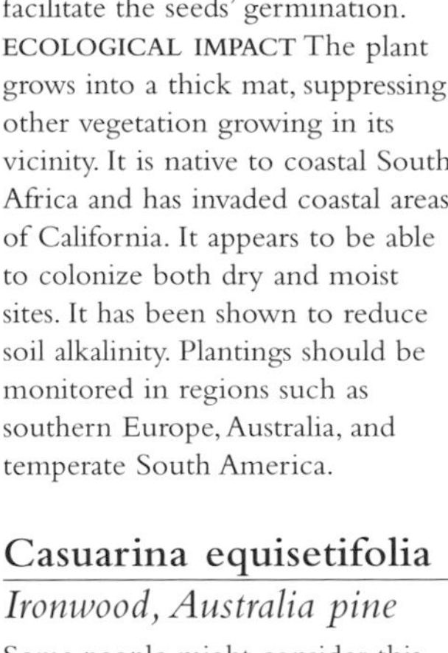

Casuarina equisetifolia

Ironwood, Australia pine

Some people might consider this tree the ultimate invasive plant. It produces many tiny, wind-dispersed seeds, which may be produced without fertilization of the flower, and which can germinate easily without any pre-treatment. The plant can grow to a height of 3m (10ft) in its first year, and it begins to produce seeds after only a year or two. It has been widely planted as a windbreak species around the world. It can change soil chemistry by increasing the nitrogen content via bacteria in its root nodules.
ECOLOGICAL IMPACT Ironwood can form dense stands. The tree produces deep litter that both smothers and poisons nearby vegetation. Planting this species has been made illegal in some areas and its introduction should be avoided nearly everywhere.

CALOTROPIS PROCERA

INVASIVE PLANTS

CELASTRUS ORBICULATUS

Casuarina glauca
Suckering Australian pine

Less salt-tolerant than its close relative, *Casuarina equisetifolia* (*see p.447*), this tree is no less aggressive. Not only does it set thousands of tiny seeds per plant, it also spreads aggressively through root suckers.
ECOLOGICAL IMPACT Like its relative, it can smother and poison other nearby plants, and it also changes soil chemistry. In Florida its elimination of plants contributes to the erosion of beaches, and it also interferes with the nesting of rare sea turtles and American crocodiles. In fact, few animals can benefit from its impenetrable thickets. In landscape settings it has been known to invade water and sewer pipes, causing maintenance problems. Planting this tree should be avoided nearly everywhere.

Cayratia thomsonii
Bushkiller, sorrel vine

The common name of "bushkiller" speaks volumes about this aggressive climber of the grape family, which is native to tropical Asia. It is only just beginning to become a problem in warm, humid parts of Texas and Louisiana in the USA, but it appears to be on its way to becoming a serious pest.
ECOLOGICAL IMPACT The climber grows quickly to blanket shrubs and trees, blocking light and weighing down supporting plants. There are even reports of it coming up through cracks in concrete. It should not be grown in south-eastern USA, and should be monitored carefully when grown elsewhere.

Celastrus orbiculatus
Oriental bittersweet

Also known as the staff vine, this fast-growing climber has yellow leaves in the autumn that underscore yellow-and-red fruits. The plant can be enormous – up to 20m (70ft) in length with a diameter of 10cm (4in). The seeds are spread by birds, although humans using the plant for autumn flower arrangements have also been known to disperse them. Open woods and meadows are vulnerable to its invasion.
ECOLOGICAL IMPACT A powerful climber, it aggressively scrambles over shrubs, blocking their light, and it also twines around their stems, constricting the flow of water and nutrients. Storm damage to trees is made worse by its weight. Cutting only causes it to produce suckers from the roots, making it more vigorous. It is native to parts of Asia, and is very invasive in the north-eastern and middle Atlantic states in the USA, where it should never be planted. Caution should be used when planting in temperate areas with summer rainfall, such as much of Europe.

Celtis australis
Southern nettle tree

Otherwise called the Mediterranean hackberry, this tree is generally regarded as a good, medium-sized shade species in most of the places where it has been grown. However, it has very invasive roots that can lift pavement in a landscape setting. It bears clusters of purple fruit that attract birds and other wildlife, which disperse the seeds. It tolerates both heat and poor soil conditions.
ECOLOGICAL IMPACT The invasive roots may increase the tree's ability to outcompete native species. Southern nettle tree has invaded in parts of Australia, especially around Canberra. It is unusual for a tree to be invasive in only one place, so monitoring for seedlings should be done if the tree is grown elsewhere.

Celtis sinensis
Hackberry, Chinese elm

This hackberry species is a medium-sized and well-shaped tree that is grown in many warm, temperate areas. It has plentiful orange fruit that is relished by birds. It is tolerant of many soil conditions and is able to withstand high winds.
ECOLOGICAL IMPACT This tree is currently invasive only in south-east Queensland, Australia, where it is reported to outcompete native vegetation. Like *Celtis australis* (*see above*), it is invasive in only one region, but gardeners should monitor for seedlings when growing the tree in other places.

Centaurea cyanus
Cornflower

Long-lasting flowers in shades of brilliant blue to lilac distinguish this herbaceous plant, also called the "bluebottle". The other common name of "cornflower" is attributed to its weedy nature in corn and agricultural fields. It is frequently found in wildflower seed mixes – it is a native wildflower in Europe and western Asia. It is also grown for floriculture and the edible petals

CENTRANTHUS RUBER

may be used to decorate food. It is an annual but produces prolific and highly viable seed, and populations can increase rapidly.
ECOLOGICAL IMPACT Away from its native habitat, cornflower readily invades grasslands and disturbed areas. It should not be planted near grasslands in Australia or North America. Once established, it can be difficult to control without the use of herbicides, which may also kill other nearby plants.

Centaurea macrocephala
Big-headed knapweed

The knapweeds are a notoriously invasive group of herbaceous plants, and most have little horticultural value. This species is grown for its showy yellow flowerheads, hence the occasional common name of "yellow fluff". Each flowerhead produces about 200 seeds, which are spread by wind. The flowerheads are sometimes used in dried flower arrangements, and seed may be spread through careless handling. The species can grow to a height of 1.5m (5ft) and produces a deep taproot that makes established plants difficult to eliminate.

CENTAUREA CYANUS

ECOLOGICAL IMPACT Seed dispersal by wind enables plants to establish in meadows near where the species has been deliberately cultivated. Big-headed knapweed is native to the Caucasus Mountains of western Asia and appears to be invading in parts of the Pacific north-west and upper midwest of the USA. This species may spread slowly, but should be monitored to ensure it does not invade open countryside near where it is grown.

Centranthus ruber
Red valerian

Often found growing in perennial beds, this herbaceous plant may have red flowers but more often they are pink and sometimes even white. It produces large numbers of wind-blown seeds. Successful in poor soils in very hot locations, it is more competitive in areas with more fertile soil and some moisture.
ECOLOGICAL IMPACT The seeds establish freely along roadsides and cracks in pavement, as well as in open countryside that is regularly disturbed. The plant is native to the Mediterranean and has become invasive in similarly hot areas, such as Australia, western North America, and higher-altitude locations in the Hawaiian Islands. It should be considered as potentially invasive wherever it is planted, and should be monitored when established near open countryside.

Cestrum diurnum
Day jessamine

In warm climates this shrub can grow so well that it is able to flower and fruit nearly continuously. The flowers are known for their fragrance during the day and the plant has been a popular ornamental shrub or small tree in warm climates. It does not easily tolerate shade or wet soil. Birds eat the black berries and disperse the seeds, enabling the species to rapidly invade native plant communities.
ECOLOGICAL IMPACT The fruits are poisonous to humans as well as livestock. Day jessamine is invasive in parts of southern USA, Hawaii, Puerto Rico, Guam, and American Samoa. Avoid planting in warm areas with high rainfall.

Cestrum nocturnum
Night jessamine

The flowers of this shrub are fragrant at night, making it a popular ornamental, although the scent is so strong that some people find it objectionable. Flowers and fruits are produced all year and the white berries are eaten by birds, which distribute the seeds.
ECOLOGICAL IMPACT The plants form dense, impenetrable thickets in moist and wet forests in tropical

CHRYSANTHEMOIDES MONILIFERA *SUBSP.* MONILIFERA

climates. The shrub resprouts when cut back, making control difficult. Night jessamine is invasive in parts of southern USA, Hawaii, and many other islands in the South Pacific. Avoid planting in any warm area with high rainfall.

Cestrum parqui

Willow-leaved jessamine

Fragrant, yellow-green flowers and dark violet berries make this species a popular shrub. It may be the hardiest within the genus *Cestrum* and grows in relatively cool climates. Birds eat the fruits and disperse the seeds.
ECOLOGICAL IMPACT The fruits are poisonous to humans and livestock. The shrub invades along rivers, where it is very competitive with native flora. It resprouts when cut back, making control difficult. It has become invasive in parts of Australia, New Zealand, Hawaii, and other South Pacific islands. Care should be taken to control it in areas with similar climates.

Chasmanthe floribunda

African cornflag

Bright orange flowers and tolerance of stressful conditions have made this bulbous plant popular with warm-climate gardeners. The seeds are orange and are dispersed by birds. The plant is notorious for its ability to "heavily reseed".
ECOLOGICAL IMPACT The plant colonizes open areas and may exclude native species. Native to South Africa, it is invasive in western USA and southern Australia. Gardeners in these areas, and in regions such as southern Europe and temperate South America, should use it with caution. There is concern that it is becoming invasive in New Zealand.

Chrysanthemoides monilifera *subsp.* monilifera

Boneseed, bitou bush

A native of South Africa, this shrubby member of the daisy family has been used for a number of purposes, from dune stabilization to ornamental use. It grows vigorously and produces large amounts of seed, which is dispersed by birds and small mammals.
ECOLOGICAL IMPACT In southeast Australia, where invasions are most severe, the shrub has been shown to decrease native species abundance. It should not be planted in Australia or New Zealand, or in any area with a Mediterranean-type climate. If the shrub is found invading, it is recommended that plants be pulled immediately while they are young and shallow-rooted. Once they are established, colonies can be very hard to eradicate.

Cinnamomum camphora

Camphor tree

This tree is widely grown in the tropics for various purposes as well as ornamental use as a shade tree. Historically, parts of it were distilled to provide camphor, although that product is now synthesized artificially. The wood was favoured for cabinetry. The tree produces abundant fruits (up to 100,000 seeds per plant) that germinate easily. Birds eat the fruits and disperse the seeds.
ECOLOGICAL IMPACT The tree forms dense thickets and outcompetes native vegetation. It is invasive in parts of Australia, in southern USA, and in Hawaii. It should not be planted in warm areas with abundant rainfall, and it can also invade dry sites.

Citharexylum spinosum

Fiddlewood

In warm climates fiddlewood is commonly used as an ornamental tree, and it is even occasionally used as a street tree. Its shiny green leaves turn bronze in cooler weather and it has fragrant, white flowers. The tree is very fast-growing and begins to produce seeds at an early age. Flowering may be sporadic throughout the year and appears to be correlated with the rainy season. The fruits, which are red when they first appear and then black, are eaten by a variety of bird species that go on to disperse the seed widely.
ECOLOGICAL IMPACT The tree is currently invasive in many parts of southern USA and Hawaii and should not be planted there.

Clematis vitalba

Old man's beard

The dry, one-seeded fruits (achenes) of this climber are amply covered by poufs of white hairs, hence the common name. Another common name for the plant is traveller's joy, but it is hard to imagine how this European native species was ever called that – for most people, an invasion is more likely to elicit despair. Each plant can produce up to 100,000 seeds in a year, which are mainly spread by wind. In addition, the plant spreads vegetatively, since older stems can root when they touch the ground.
ECOLOGICAL IMPACT This tough, woody climber mounts into trees and completely engulfs them – even trees that are 20m (70ft) tall. The stems develop side branches that can grow 10m (30ft) in a single season, which makes them extremely difficult to pull from the trees. The species is a serious pest in New Zealand and is becoming a problem in the Pacific northwest of the USA. It should never be grown in those regions, and gardeners in other mild, temperate climates should use it with great caution.

CLEMATIS VITALBA

Clerodendrum philippinum

Glory bower

Sometimes an invasive plant gains a role in the folk culture where it invades. Probably native to south China, this shrub is such a plant in Hawaii, where the white, fragrant flowers are popular for use in the garlands known as "leis". The shrub produces vigorous root suckers and its seeds are distributed by birds.
ECOLOGICAL IMPACT Glory bower can spread very aggressively along roadsides and into native vegetation. It is invasive in tropical countries ranging from South America to Africa. Despite its widespread popularity for its decorative appearance, great care should be taken that the roots are contained and fruiting is prevented.

CONIUM MACULATUM

Clidemia hirta

Koster's curse

The common name of this shrub should give the reader an idea of how it is regarded by many gardeners. The leaves are attractive, although the flowers are small. Each small plant flowers and fruits continually, producing up to 5,000 seeds in one year. The fruits are spread by a variety of creatures, from birds and pigs to humans. The plant can also spread vegetatively – the lower branches may root when they touch the soil, and dislodged leaves can root in wet areas. Koster's curse tolerates a variety of environmental conditions.
ECOLOGICAL IMPACT In areas of open countryside the shrub forms dense thickets that are virtually impenetrable. This plant is such a pest, and there are so many attractive, noninvasive alternatives available, that there is no reason to grow it anywhere.

Coccinia grandis

Ivy gourd, scarlet gourd

It is unusual for a species of the gourd family to have bright red fruit, and that explains why this climber is often grown in warm-climate gardens. The plant is native to Africa and Asia, and in Asian countries the young fruits are sometimes cooked and eaten in savoury dishes.
ECOLOGICAL IMPACT This fast-growing climber can form a dense canopy that covers trees. It is also known to act as a reservoir for viruses that attack other members of the cucumber and gourd family. Ivy gourd is very invasive in Hawaii and other South Pacific islands, as well as in the Caribbean, parts of southern USA, and Australia. It should not be planted in these regions (it is illegal to grow it in Hawaii and parts of Australia) and it should not be grown in other areas with similar climates.

Colocasia esculenta

Coco-yam, taro

People may grow this herbaceous plant as an ornamental for its striking, large leaves, or for the rootstock, which is used to make a number of starchy foodstuffs. The plant is native to India and is widely cultivated in tropical areas. There are a great many cultivars; it is believed that the native Hawaiians grew about 300 different kinds.
ECOLOGICAL IMPACT This plant is known to have escaped cultivation in some of the places it is grown. Care should be taken to ensure that it does not invade wetlands or wild aquatic systems.

Conium maculatum

Hemlock

The Greek philosopher Socrates died from ingesting this herbaceous plant, and the reddish blotches on its stem are sometimes referred to as "the blood of Socrates". While many people are aware of its poisonous nature, the plant is still occasionally grown in gardens. It should always be handled with great care. Despite growing to a height of 2.5m (8ft), it is an annual.
ECOLOGICAL IMPACT In addition to being poisonous to humans and animals, hemlock is quite an invasive plant and can eliminate native vegetation by forming dense stands along streams and in open forests. The plant is native to Europe and has become widespread in many temperate parts of the world, including North America, Australia, New Zealand, and South America.

Coprosma repens

Looking-glass plant

This multi-stemmed shrub is often used as an ornamental, either as a sheared hedge or as a small specimen tree. The leaves have an almost mirror-like shine, hence the common name. The plant is tolerant of salt spray. Birds eat its orange fruits, and the seeds germinate easily in light shade, especially if there has been some soil disturbance.
ECOLOGICAL IMPACT This plant competes aggressively with some native species, and is invasive in New Zealand and parts of Australia such as New South Wales and Tasmania. There is concern that the plant is also spreading in western USA, especially in California.

CORTADARIA JUBATA

Cortadaria jubata

Jubata grass

One of two commonly grown grasses of the genus *Cortadaria*, jubata grass is less hardy than its relative, *Cortadaria selloana* (*see below*), but it is able to produce seed without fertilization by pollen. Each plant may have 10 or more plumed flowerheads, and each plume can produce up to 100,000 seeds, which are easily dispersed by wind. The seeds readily germinate in open, moderately moist, sandy soils.
ECOLOGICAL IMPACT This plant displaces native species in coastal areas and may also be found in open woodlands. It is known to be suppressing regeneration in sequoia forests in California, USA. Because of its highly fertile nature, jubata grass should not be planted anywhere near open countryside. It currently invades warmer areas on the North American west coast.

Cortadaria selloana

Pampas grass

This grass is less invasive than jubata grass (*see above*) because it is less likely to produce viable seed. Plants produce either all-male or all-female flowers, and the female flowers are not self-fertilizing. Male plants are rarely cultivated because their plumes are not as fluffy as female plumes, so little seed is set.
ECOLOGICAL IMPACT While it appears to be less invasive than jubata grass, pampas grass has been found invading in some parts of California, where it competes with native vegetation in coastal areas. This recent discovery indicates that care should still be taken when planting near open countryside.

Cotoneaster franchetti

Grey cotoneaster

This ornamental shrub is spreading increasingly in some uncultivated areas. The fruits are held on the

plant through winter and birds favour them, spreading seeds far and wide. Grey cotoneaster grows very quickly and appears to outcompete some native vegetation.
ECOLOGICAL IMPACT In the Pacific northwest of the USA and northern California it has become common in open forests and along forest edges. It is also found in Australia, including Tasmania. Plantings in other Mediterranean-type climates should be monitored.

Cotoneaster glaucophyllus

Large-leaf cotoneaster

As its scientific and common names imply, this shrub has larger leaves than most cotoneasters and they are a whitish grey. The plant produces red fruits that are dispersed by birds. It invades along rivers in some areas, where it grows rapidly and outcompetes some native vegetation. It is also very tolerant of dry conditions and may also invade upland areas.
ECOLOGICAL IMPACT This plant is invasive in New Zealand and parts of Australia, such as New South Wales.

COTONEASTER PANNOSUS

Cotoneaster pannosus

Silverleaf cotoneaster

This attractive shrub produces many pink flowers followed by bright red fruits. It retains its leaves in the winter, so it may do well in Mediterranean-type climates where it can photosynthesize – despite the fact that the parts of China to which it is native receive more summer rainfall. The seeds can occur without fertilization of the flower, enabling isolated plants to form viable seed. While birds disperse the fruits, reports indicate that it is not a particularly favoured food. It colonizes wooded areas and is resistant to drought.
ECOLOGICAL IMPACT Avoid planting the species in parts of Australia such as Tasmania, Victoria, and New South Wales, in California, and in other areas with Mediterranean-type climates.

Crassula helmsii

Australian stonecrop

Most gardeners think of the stonecrops as plants for arid gardens, but this is an aquatic plant often found in freshwater aquatic habitats. It has been sold as a pond plant but its rapid growth makes it poorly suited for that purpose. Now that its invasive nature is known, fewer garden suppliers sell it. The plant is sometimes offered as *Tillaea helmsii* or *Tillaea recurva*, but these names are incorrect.
ECOLOGICAL IMPACT Australian stonecrop forms dense mats that clog ponds and ditches and block light to native submerged plants. Care should be taken to ensure that fragments are not spread by boats or fishing equipment. *Crassula helmsii* should never be planted in the UK and it is reported also to be invading in south-east USA. Because of its aggressive growth, planting it should be avoided everywhere.

Crataegus monogyna

Common hawthorn, may

Abundant red fruits that resemble tiny apples are an attraction of this tree. The fruits remain on the branches long after the leaves have dropped. In the USA, most native species shed their fruit relatively early in the season, so this trait may increase the likelihood of birds eating the fruit and distributing the seeds. Common hawthorn can produce seed without fertilization, which may increase its ability to spread from isolated individuals. Because the seeds require cold-chilling prior to germination, the species is of highest concern in colder regions.
ECOLOGICAL IMPACT One study in the USA showed that birds preferred the fruit of the common hawthorn to that of a native American hawthorn, with the effect that dispersal of seed of the American native was decreased. Common hawthorn should not be planted near open countryside in western USA, where it may hybridize with the native hawthorns.

Crocosmia x crocosmiiflora

Montbretia

Gardeners admire the bright orange-red flowers and sword-like leaves of this South African bulbous plant, a relative of the iris. While it may be found in both temperate and warmer climates, it is more invasive in higher temperatures. Spreading by numerous small seeds per plant, it also expands vigorously by rhizomes, especially when the corms attached to the rhizomes are separated from the parent plant as a result of soil disturbance.
ECOLOGICAL IMPACT This plant is

CRATAEGUS MONOGYNA

particularly invasive in moist soils and may spread through ditches into waterways. It invades in Australia, California, and European regions with mild temperatures. Care should be taken to prevent it from entering open countryside.

Cryptocoryne beckettii

Water trumpet

This aquatic plant has interesting flowers and has been one of the most popular aquarium and pond plants for more than 60 years. Unfortunately, it has been found invading in two different river systems in the USA, in Florida and Texas. The Texas population is quite large and appears to be spreading rapidly. In some rivers plant growth has extended from bank to bank.
ECOLOGICAL IMPACT In Florida and Texas the plant is affecting at least one very rare plant species. While its growing preferences may be particular enough to restrict its invasiveness to only certain sites, in warm climates it should never be grown where it could escape into freshwater systems.

Cryptostegia grandiflora

Rubber vine

The common name of this climber refers to the latex found in its sap, which has been used as a source of rubber. The plant begins to produce seeds when only seven months old. These are highly viable and are dispersed by wind and water. Rubber vine is native to the island of Madagascar, but most of its habitat there has been destroyed.
ECOLOGICAL IMPACT This fast-growing climber can smother trees as tall as 15m (50ft). It can dominate both dry and wet forests and in open country it forms impenetrable thickets. It is a pest in Hawaii and other South Pacific islands, as well as Queensland, Australia, the Caribbean, and the island of Mauritius. A related species, *Cryptostegia madagascariensis*, invades in Florida.

Cupaniopsis anacardioides

Carrotwood

The inner bark of this tree, a native of Australia, is orange, hence the common name. Carrotwood produces large numbers of seeds that are dispersed by birds. In Florida, USA, it began spreading almost immediately after introduction as a landscape tree.
ECOLOGICAL IMPACT Carrotwood invades a number of native plant community types, including coastal forests. In some places it is found with as many as 24 separate saplings per square metre and it is capable of outcompeting even other very aggressive invasive species. It is currently invading coastal counties in southern Florida and, given how invasive it is there and how quickly it has become a serious pest, monitoring for seedlings should be carried out in the areas surrounding wherever it is grown.

Cynanchum louiseae

Black swallow wort

This sprawling, climbing shrub is native to southern Europe and is often invasive where it is planted. It grows rapidly and covers native vegetation. It resprouts when cut, has seeds that are dispersed by wind, and has an extensive root system that allows it to spread vegetatively. The dark purple flowers smell rotten and attract flies.
ECOLOGICAL IMPACT Black swallow wort invades a broad range of habitats, from urban areas to woodlands. In north-east USA it can outcompete a related species on which monarch butterflies depend. The monarchs lay their eggs on swallow wort, but the larvae are unable to complete their life-cycle. Currently it is mostly invasive in north-east USA, into Missouri, but, because it can grow so rapidly and spread so easily, care should be taken wherever it is planted.

CROCOSMIA x CROCOSMIIFLORA

Cytisus scoparius

Common or Scotch broom

This broom is one of the most invasive shrubs in areas with Mediterranean-type climates. Its evergreen stems allow it to photosynthesize in the winter when temperatures are mild and there is ample rainfall. The seeds live for 80 years, but it is unclear whether they survive for so long outside the laboratory. The seeds are spread both by forcible ejection from the pod and by ants attracted by the covering of the seed. The ants effectively "plant" the seeds by dragging them to their mounds. ECOLOGICAL IMPACT This broom can change soil chemistry, partly by increasing the nitrogen content via bacteria in its root nodules, especially in the low-nitrogen grasslands it tends to invade. It should not be planted in any area with a Mediterranean-type climate characterized by mild, wet winters.

CYTISUS SCOPARIUS

Cytisus striatus

Portuguese broom

Like other shrubs of the genus *Cytisus*, Portuguese broom is very tolerant of poor soil conditions. It has green stems that enable the plant to photosynthesize in mild winters. The plants themselves are relatively short-lived, even in cultivation, but the seeds may persist in the soil seed bank for many years. ECOLOGICAL IMPACT The plant can change the chemistry of soil by means of bacteria in its root nodules, which increase the soil's nitrogen content. It has been found to be invasive in areas with warm, Mediterranean-type climates, such as central California, USA, and south-central Chile, and care should be taken in other such areas.

Dalbergia sissoo

Indian rosewood

Used as both an ornamental and a timber tree, this fast-growing plant has seeds that are spread by wind, water, and birds. The seeds germinate easily. The tree has become locally widespread in some tropical and subtropical climates. ECOLOGICAL IMPACT The plant changes soil chemistry using bacteria in root nodules to increase its nitrogen content. It is invasive in warm, wet areas, such as south Florida, Hawaii, and Taiwan.

Daphne laureola

Spurge laurel

This attractive, evergreen shrub flowers in winter and produces fruit in the summer. The availability of the fruit during summer, when food can be scarce, entices birds to consume and spread them. Spurge laurel is very shade-tolerant and has been increasing rapidly in forests and oak savannas of the Pacific north-west of North America. ECOLOGICAL IMPACT The shrub has an extensive root system that can outcompete native plants in forests, including woody species. It should not be planted in the north-western region of North America or in any other similar region.

Daphne mezereum

Mezereon

Flowering in early spring, this shrub has fragrant, purple, albeit poisonous, blooms. The seeds can be slow to germinate, although germination rates improve after cold-chilling and scarifying of the coat. Absence of these factors may be inhibiting a greater spread of this plant in wild areas. ECOLOGICAL IMPACT The plant is spreading in moist forests in Ontario, Canada, and parts of north-eastern USA. Care should be taken when planting in cold, moist areas.

Delairea odorata

German ivy, parlour ivy

Also called *Senecio mikanioides*, German ivy is a climbing member of the daisy family, native to South Africa. It may spread mostly by vegetative means since seed viability is believed to be very low. ECOLOGICAL IMPACT Like most climbers, the vine affects native vegetation by climbing over it and intercepting light. Its weight can cause trees and shrubs to collapse. The climber contains alkaloid chemicals that are toxic to many mammals and fish. Studies in California found invaded forests to have 50–95 per cent fewer species than similar but non-invaded forests. The climber should not be planted in Australia, California, or Italy, where it currently invades. It should be avoided in all regions similar to its South African habitat, such as Mediterranean countries and temperate South America.

DIOSCOREA BULBIFERA

Digitalis purpurea

Common foxglove

Foxglove is a popular biennial herbaceous plant, but its abundant production of tiny, wind-dispersed seeds ensures that it does not stay in the garden for long. It can be found along roadsides, in open forests, and in river and stream corridors. The plant probably originated in the western Mediterranean region, but that is difficult to confirm because it can be found in large naturalized populations around the world. ECOLOGICAL IMPACT The foxglove can outcompete native plants, but, perhaps more seriously, the leaves contain a stimulant that poisons people and livestock. Individuals believing they are making tea of the similar-appearing comfrey can fall victim to foxglove, and some people report numbness in their hands after weeding large patches of the plant. Foxglove should not be planted where it could spread into open countryside.

Dioscorea bulbifera

Air potato

"Air potato" describes very well the reproductive method of this climber. Small tubers (bulbils) appear on the stems, which can grow 10m (30ft) up into the tree canopy in just one season. The slightest disturbance can send a shower of the tubers to the ground, where even the smallest can take root and grow new plants. While they are not the finest of yams, cultivars of this species are used as food in most humid tropical areas around the world. ECOLOGICAL IMPACT The plant smothers the trees and shrubs that it covers, colonizing both dry upland areas and moist habitats. It is native to tropical Asia, but was introduced into Africa at an early time and is a noxious weed in Florida, USA. Cultivars are grown around the world, but these do not seem very persistent in the wild. The wild type, which has a more bitter fruit, should not be grown anywhere.

DIGITALIS PURPUREA

ECHIUM PLANTAGINEUM

Dipsacus fullonum, Dipsacus laciniatus
Teasel

The seedhead of this herbaceous plant is commonly used in dried flower arrangements. When used for this purpose all the tiny seeds should be removed to prevent spread. The deep taproot can make it difficult to eradicate. *Dipsacus fullonum* and *Dipsacus laciniatus* are both grown under the common name "teasel" and are similar in appearance and invasiveness.
ECOLOGICAL IMPACT Teasel colonizes a variety of habitats, from grasslands to wet meadows. It is native to Europe but is now common in many temperate areas. Care should be taken so that it does not spread into open countryside.

Echium plantagineum
Viper's bugloss

The common name does not hint at the pretty, bright-purple flowers of this herbaceous plant. The many seeds have rough surfaces that help them attach to passing animals. The seeds may also be spread by water and can survive for up to 10 years in the soil seed bank. Bees make a delicious honey from the flowers. Cattlemen in arid places use dried plants as emergency fodder, although alkaloids in the leaves of fresh plants are capable of poisoning livestock.
ECOLOGICAL IMPACT The plant can grow quite densely, excluding native species. The plant is a serious problem in many parts of Australia and should never be grown there. It is also expanding its range in northern North America. Because it has demonstrated its invasive capability, it should always be grown with care to prevent it from escaping into wild areas.

Egeria densa
Brazilian elodea

Formerly grouped with *Elodea* (*see below*), but now classed in a separate genus, *Egeria densa* is grown for its use in aquaria. Unfortunately, people cleaning their aquaria often dump the whole contents into freshwater systems, and that has allowed this aquatic plant to spread widely. Just a small fragment of stem with two nodes is enough to get it started, and it may become rooted in the mud or survive as free-floating mats of intertwined branches. The thick mats impede boat traffic and create an expensive problem for irrigation projects and hydroelectric utilities, as well as affecting water supplies to urban areas.
ECOLOGICAL IMPACT Thick mats of Brazilian elodea cut off light to other submerged species. The plant is native to parts of tropical America and has escaped aquaria in Chile, the UK, New Zealand, Australia, and several parts of the USA. This species should never be deposited in any freshwater system anywhere. Boats should be checked to ensure they do not spread it.

Eichhornia crassipes (100)
Water hyacinth

While considered to be one of the worst weeds in the world, this aquatic plant is still commonly grown as a pond ornamental. It has large, violet flowers sitting on spongy bladders that help it to float. The plant spreads mostly by rhizomes. Its high transpiration rate may decrease water levels in some lakes and ponds. It also impedes boat traffic. The weed can be found in virtually all types of freshwater, from lakes to ditches.
ECOLOGICAL IMPACT The plant can quickly form a dense colony that shades out all submerged vegetation below. It alters water temperatures, and affects native wildlife as well as vegetation. This species is native to Brazil but should never be planted in warm climates, even in places where it appears confined to a water body. There is concern that more frost-hardy forms are arising, so care should also be taken when growing as an annual in cooler areas.

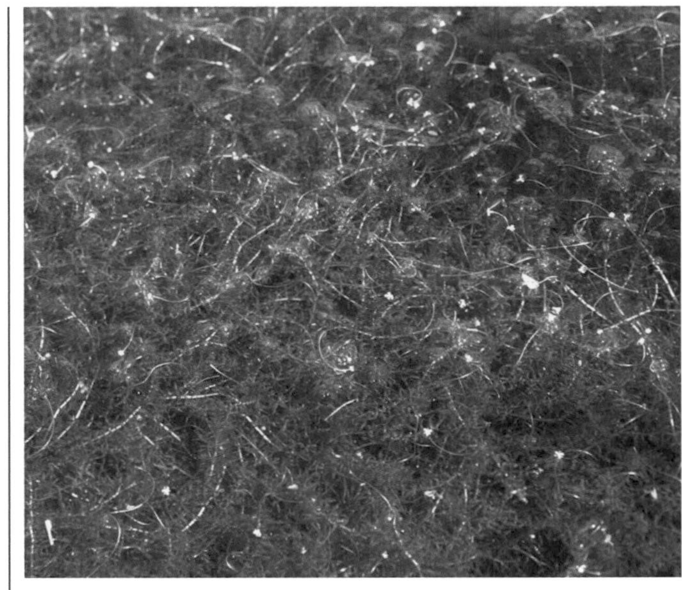
ELODEA CANADENSIS

Elaeagnus angustifolia
Oleaster

In the mid-1900s this shrub was commonly planted to prevent soil erosion and as a wildlife habitat, and it has also been grown as an ornamental. It has silvery scales on its leaves. The prolific quantities of small, silver-yellow fruits are dispersed by birds. The plant is tolerant of many soil types, and resprouts vigorously when cut back.
ECOLOGICAL IMPACT Oleaster can change soil chemistry by releasing nitrogen produced by nodules on its roots. It can grow along riverbanks and may alter the flow of rivers by trapping material in its branches as they hang into the water, thereby enabling the buildup of soil. The plant is a pest in drier parts of western USA, but has also been found to be invasive in many states with higher rainfall, including the states of New England. When the shrub is planted in areas with a dry climate, rivers and wetter areas, in particular, should be checked regularly for invasiveness.

Elaeagnus umbellata
Autumn olive

Widely planted in the 20th century, this shrub was used for erosion control, wildlife enhancement, and landscape enhancement. A number of animals, including the raccoon, fox, skunk, and possum, have been implicated in the dispersal of its seed. A single shrub can produce up to 54,000 seeds, making it a notable and beneficial food source for a wide range of wildlife.
ECOLOGICAL IMPACT Autumn olive invades forests and grasslands, where it can alter soil chemistry through nitrogen produced in, and released from, its root nodules. It can form an impenetrable thorny thicket and shade out native herbaceous species, as well as woody saplings. The species is invasive primarily in eastern USA, from Maine to South Carolina, but also in the midwest, from Oklahoma to Minnesota. Autumn olive should not be cultivated in any area with a comparable temperate climate.

EICHHORNIA CRASSIPES

Elodea canadensis
Canadian pondweed

Used for decades to furnish aquaria in several parts of the world, this aquatic plant has ended up colonizing ponds and slow-moving waterways. The pondweed forms dense mats of vegetation that make waterways nigh impassable to boat traffic. Any boat that has come into contact with the weed should be checked to prevent it from carrying fragments into unaffected areas.
ECOLOGICAL IMPACT The dense mats inhibit native vegetation. While not currently as invasive as some other aquarium and pond plants that have escaped cultivation, Canadian pondweed is invasive in the UK, northern Europe, and several areas in Australia. It should not be introduced into any temperate freshwater ponds that are not completely contained.

EPIPREMNUM PINNATUM 'AUREUM'

Epipremnum pinnatum 'Aureum'
Golden pothos

Also known to botanists as *Scindapsus aureus*, this climber is very popular as a potted plant in temperate areas, where it is not hardy. In warmer climates, pieces of stem root easily when disturbed. The plant does not often produce seeds. Its sap can irritate the skin of some people, so gloves should be worn when uprooting the plant.
ECOLOGICAL IMPACT Golden pothos grows into the tree canopy and has colonized openings in some wet forests. Native to the Solomon Islands, it invades in Hawaii, other South Pacific islands, and Florida. When grown in warm, wet climates, it should be kept away from open countryside, and garden waste should be disposed of appropriately.

Erica lusitanica

Portuguese heath

Most shrubs in the heath family are not invasive, but this one is an exception and it easily dominates a shrub community. It is popular with florists for arrangements. Portuguese heath produces large amounts of tiny seeds that are spread by wind and water. The branches can root where they touch the ground.
ECOLOGICAL IMPACT The plant crowds out native vegetation and is thought to alter soil chemistry. In Australia it can be found in Victoria, New South Wales, and Tasmania; it also invades in northern coastal California and New Zealand.

Eschscholzia californica

California poppy

The California poppy, the official flower of the state of California, USA, is a native herbaceous plant of western America and is now found throughout North America and also in Chile. It is often included in readily available "wildflower" seed mixes, but these should not be used where there is a risk of plants escaping into open countryside.
ECOLOGICAL IMPACT This poppy self-seeds vigorously and, while it often prefers disturbed areas, it can be competitive in grasslands. Because its native region has winter rain and summer drought and it has proved to be invasive in comparable regions, it should not be planted in Australia, South Africa, and temperate South America.

Eucalyptus globulus

Tasmanian blue gum

Many eucalyptus tree species can be invasive, but probably none more so than this one, perhaps because it has been so extensively planted outside its native Australia. It is used mostly as an ornamental, though it has been planted in the past as a timber tree – unfortunately, it has not proved especially useful for that purpose. It has many tiny seeds that are spread by wind and water and germinate easily. The tree matures quickly and regenerates from the stump when cut back. The wood and the leaves have copious oils that increase its flammability. It is not a good species to have growing near houses or in places where lightning strikes are common.
ECOLOGICAL IMPACT The oils may poison plants growing under the trees. They are distributed either by water dripping from the leaves or by water seeping from leaves decomposing on the ground. The tree should not be planted in areas with Mediterranean-type climates. It has also been shown to be invasive in higher-altitude zones in the tropics, so planting in such areas should be carefully monitored.

ESCHSCHOLZIA CALIFORNICA

Euonymus alatus

Winged spindle

Brilliant red autumn colour is the glory of this shrub. It also has curious, corky ridges on the stems and interesting, brightly coloured fruit and seeds. Birds disperse the seeds widely, making the shrub very invasive in many temperate areas with summer rainfall.
ECOLOGICAL IMPACT This plant can form dense thickets in open woodlands, grasslands, and some other forest types – it has been noted to be widely tolerant of many environmental conditions. In the USA it invades the New England and upper midwest states. Because it is spreading fast in these areas, it should not be grown in any place with a similar climate.

Euonymus fortunei

Wintercreeper

Also called creeping euonymus, this shrub can be very aggressive in its vegetative growth, climbing up into nearby trees and shrubs. It is very tolerant of a range of environmental conditions, from shade to sun and varying soil types. The popular variegated form appears to be less invasive. It may not produce seed in all locations, especially if it is not growing vertically, but can spread by means of vigorous vegetative growth. When it does produce seeds, these are spread by birds.
ECOLOGICAL IMPACT The plant can outcompete native species for light, water, and soil nutrients. It should be avoided in eastern USA, and care should be exercised in other areas with similar climates.

Fallopia japonica (100)

Japanese knotweed

This perennial herbaceous plant, also known as *Polyganum cuspidatum*, is now recognized as a serious pest in the open countryside of a great many regions. It appears to respond to disturbance by sending out underground stems (rhizomes) several metres long to start new populations away from the disturbance. The rhizomes are known to regenerate even when buried 1m (3ft) deep. The seeds are windborne, although in invasive populations it is believed that fertile male flowers are rare and most seed develops from pollen of other closely related invasive species, such as *Fallopia sachalinensis* (*see below*).
ECOLOGICAL IMPACT This plant forms dense rhizomatous stands that can become nearly impossible to control. While it invades roadsides, of greater concern are its invasions along rivers, where floods wash fragments of the rhizomes downstream to form dozens of new populations. The plant is currently causing serious problems in the UK, Europe, Australia, New Zealand, and North America. Do not plant it anywhere – once it is established, it is hard to eradicate, even with the aid of chemicals.

Fallopia sachalinensis

Giant knotweed

Like its better-known cousin, *Fallopia japonica* (*see above*), the giant knotweed (also called *Polyganum sachalinensis*) is a native of Japan. It is a perennial herbaceous plant that grows up to 4m (12ft) tall in one season. The leaves are larger and more heart-shaped than those of *Fallopia japonica*, and it is likely that some infestations attributed to Japanese knotweed are actually of this plant. Giant knotweed invades in many of the same regions as its cousin, and it may provide the pollen that enables Japanese knotweed to produce viable seed in the absence of male flowers.
ECOLOGICAL IMPACT This plant forms dense clumps that spread rhizomatously and those rhizomes produce new plants when separated from the mother plant. It is causing

FALLOPIA JAPONICA

FICUS PUMILA

serious problems in the UK, Europe, and many parts of North America. Do not plant it anywhere. Once it is established, it is hard to get rid of, even with chemicals.

Ficus altissima
False banyan

The false banyan tree is a "strangler fig", first growing epiphytically (without rooting into soil) on a tree branch, then growing to overcome the tree. In some tropical areas where it has been introduced it can become problematic. Fig trees are pollinated by tiny wasps, and, if these are absent, seeds are not produced. In southern Florida, USA, where the wasp appears to be present, the seeds are spread by birds to native forests.
ECOLOGICAL IMPACT The tree can kill the plants over which it grows. Avoid planting in southern Florida (it is prohibited in the Miami area), or in any other warm area where the wasp is present.

Ficus microcarpa
Curtain fig, Malay banyan

A popular tree in tropical and subtropical climates, curtain fig can become very invasive in areas where its pollinator, a tiny wasp, has also been introduced Birds eat the tiny fruits and disperse the seeds, which can germinate and grow almost anywhere they land, including on housetops. Curtain fig can cause considerable damage to structures.
ECOLOGICAL IMPACT This tree grows very quickly and can invade native forests. Like the false banyan, *Ficus altissima*, it can strangle a tree with its growth. Avoid planting in Hawaii, parts of Mexico, southern Florida, and the Bahamas, or in any other warm climate where the pollinator has been introduced.

Ficus pumila
Creeping fig

The only thing that prevents this climber from becoming a greater problem is the lack of a specific pollinating wasp species in the places it is grown. Instead of reproducing from seeds in these areas, it spreads vigorously by stem growth from the walls and other structures it is planted to cover. It can climb high into trees and adheres to surfaces by exuding a rubbery substance that cements the aerial roots to whatever it climbs. If left long enough it becomes more of a woody shrub than a climber.
ECOLOGICAL IMPACT This fig does not grow as fast as many other invasive climbers, but if unchecked it creates a dense, smothering mass. It is poisonous to dogs. The fig is native to Asia and it has been found invading in New Zealand, western Australia, and Florida, USA. It should not be grown near open countryside in warm, wet climates, and garden waste should be disposed of appropriately.

FOENICULUM VULGARE

Foeniculum vulgare
Fennel

It may be a culinary delight, but fennel has also spread quickly over acres of grassland and riverbanks in some parts of the world. The tiny seeds of this herbaceous plant blow in the wind and are also spread along roads and paths, from where the plant moves out into surrounding vegetation. The seeds of garden plants can be used for cooking, but it is better to remove them before maturity and buy commercial seeds, to prevent local spread.
ECOLOGICAL IMPACT Fennel can grow into dense clumps that exclude all other vegetation. Areas with climates similar to that of its native Mediterranean region appear to be the most vulnerable, although it has also invaded in eastern North America. It should never be planted near open countryside. If growing for the anise-scented foliage or roots, or as an ornamental, never allow it to go to seed. The deep taproot can make it difficult to remove without herbicides.

Fuchsia magellanica
Hardy fuchsia

Hummingbirds are attracted to the pendulous red-and-pink flowers of this shrub, reminding us that even invasive plants have their charms. The seeds are consumed and spread by birds. The plant tolerates any soil type and can establish easily in wet forests, but in dry areas it needs disturbance to get established.
ECOLOGICAL IMPACT This shrub is capable of forming dense thickets in and along the edges of open woods. Cold, moist areas are the most susceptible, including Ireland, Tasmania, Réunion Island, and the higher, more mountainous parts of Hawaii, where it already invades.

FUCHSIA MAGELLANICA

Genista monspessulana
French broom

While not as hardy a shrub as Scotch broom (*Cytisus scoparius*, *see p.452*), French broom can be just as aggressive in milder climates. It can begin to reproduce by seed after only two years and some estimates indicate that mature stands may produce as many as 100,000 seeds per square metre.
ECOLOGICAL IMPACT The shrub shades out native species, slows reforestation efforts by shading saplings, increases the nitrogen in the soil through bacteria in its root nodules, and may increase fire frequency and intensity in some places. Do not grow in California, many parts of Australia, such as Tasmania, and other places with a Mediterranean-type climate.

Gladiolus caryophyllaceus
Wild gladiolus

In its native South Africa, this bulbous plant is a rarity. It produces small cormlets that scatter when the corm is disturbed, and it also reproduces by seeds. The plant is sometimes sold as *Gladiolus hirsutus*. Other, noninvasive gladiolus species also appear on the market under the name "wild gladiolus".
ECOLOGICAL IMPACT This plant can become a pest away from its native habitat, spreading rapidly in open areas. It currently invades in south-western Australia. Gardeners in areas with a similar climate, such as southern Europe, western North America, and temperate South America, should use it with caution.

Gomphocarpus fruticosus
Milk bush

The puffy fruits of this herbaceous plant are oval in shape, inflated, and silvery-green with soft spines. The plant is noted by gardeners for its attractiveness to breeding butterflies.
ECOLOGICAL IMPACT Milk bush is capable of forming dense thickets in moist sites. The fruits and plants can be poisonous. It is native to South Africa but as an escaped garden shrub it has become a serious weed in Western Australia, and is also spreading along the Mediterranean coast. Areas with similar climates, such as California and temperate South America, should monitor open countryside where it is grown.

GOMPHOCARPUS FRUITICOSUS

Grevillea banksii
Kahili flower

The Proteaceae are known for their striking flowers, and this member is no exception. The shrub is a native of Queensland, Australia, and bears dense plumes of red, pink, or creamy white flowers on the ends of its branches. However, it also has become invasive, producing numerous seeds that are dispersed by the wind. The seeds have a high viability and remain viable for at least a few years.
ECOLOGICAL IMPACT This shrub is allelopathic, which means that it can poison plants that grow under or near it. It is now invasive in many parts of Hawaii and in KwaZulu-Natal in South Africa.

Grevillea robusta

Silky oak

As well as being planted as a beautiful ornamental tree, the silky oak has been widely introduced for forestry purposes. Unlike most invasive plants, it has a long juvenile period and often must attain 10 years of age before beginning to flower and fruit. However, it does fruit freely once it is established in an area. Its wind-dispersed seeds germinate easily but may be short-lived. The tree prefers dry soils. ECOLOGICAL IMPACT This tree produces allelopathic chemicals, which can poison and destroy nearby plants. It is a species of concern in Hawaii and the Caribbean, especially Jamaica.

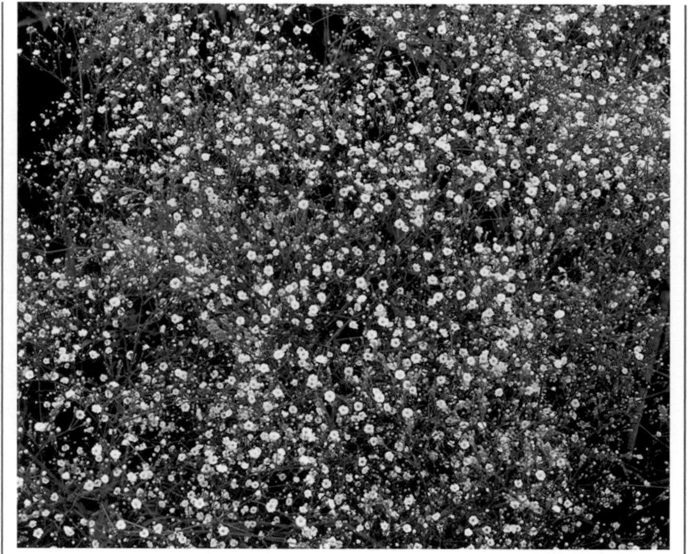

GYPSOPHILA PANICULATA

Gypsophila paniculata

Baby's breath

This distinctive herbaceous plant is very popular because of its much-branched stems of tiny white flowers, which have a frothy appearance prized by florists. But while it is an attractive addition to gardens and floral arrangements, it is also an aggressive seeder that has been declared a noxious weed in several states of the USA. A single plant may produce nearly 14,000 seeds, which are mostly dispersed by the wind and by people collecting the plant for flower arrangements. The seeds germinate easily. ECOLOGICAL IMPACT Baby's breath invades along the shores of lakes and streams, as well as dry areas, and in some places it is colonizing the habitat of rare native plants. It is native to the Mediterranean region eastwards into Asia, and currently it invades only in parts of North America. Despite being a valuable ornamental, its invasive potential should be carefully monitored.

HEDERA HELIX

Hakea sericea

Mountain hakea

A prickly shrub native to dry areas of Australia, this species has been used for hedges in some areas with comparable climates around the world. It is also used for firewood and soil erosion control. ECOLOGICAL IMPACT In South Africa, especially in the fragile Cape region, it has become one of the worst invasive species. Huge tracts of the Cape's native vegetation, known as "fynbos", have been destroyed by mountain hakea, where it appears to be highly competitive with native species. Do not plant it in South Africa and other areas that experience a warm, Mediterranean-type climate.

Hedera helix

Common ivy, English ivy

In some parts of the USA, people get together on Saturdays to participate in programmes called "The No-Ivy League" or "Ivy-OUT", gathering in parks to remove this woody climber and restore urban forests. The purplish-black fruits ripen on upright stems in the spring and are eaten by birds (there is little else available); the seeds are then distributed and germinate in surrounding woods. Plants can also establish from just a small part of the stem separated from the main stem. Common ivy is sometimes used for erosion control, but its shallow roots actually make it ineffective for this purpose. ECOLOGICAL IMPACT Common ivy reduces the diversity of native species on the forest floor, and foresters say that ivy-shrouded trees are the ones that usually sustain the greatest damage in storms. While the species is invasive, it seems that most cultivars are less so. A study in the USA's Pacific north-west found that only four of 400 common ivy cultivars were invading: *Hedera helix* 'Hibernica' (also known as *Hedera hibernica* 'Hibernica'); *Hedera helix* 'Baltica'; *Hedera helix* 'Pittsburgh'; and *Hedera helix* 'Star'. Common ivy is native to parts of the UK, Europe, and Asia, and is widely grown in temperate areas. It is currently invasive along the east and west coasts of the USA, and in Australia, New Zealand, Hawaii, and Brazil.

Hedychium coronarium

White ginger lily

The ornamental gingers are beautiful herbaceous plants. This species has fragrant, white flowers in dense flowerheads at the end of lush, green stems. The plant may not produce seeds frequently but its underground stems (rhizomes) enter open countryside through fragmentation after disturbance. Well-meaning people trying to "spice up" forest areas, and those dumping garden waste, also contribute to its spread. ECOLOGICAL IMPACT The plant spreads voraciously by its rhizomes, overwhelming native plants in wet, warm forests and the sides of streams. White ginger lily is native to the Himalayas, but is extensively found in wet forests throughout the South Pacific islands, including New Zealand. If planted anywhere, it should be carefully contained, so that no rhizomes are able to spread out from the cultivated areas.

Hedychium gardnerianum (100)

Kahili ginger

"Kahili" were staffs crowned by circular plumes of bright feathers, a symbol of royalty in Hawaii. The flowerhead of this herbaceous species, with its dense, 360-degree yellow blooms and red stamens, really does look like a kahili. This ginger is not quite as fragrant as the white ginger lily (*see above*). It spreads quickly by underground stems (rhizomes), and also produces scarlet seeds, dispersed by birds. ECOLOGICAL IMPACT The rhizomes are very aggressive and dominant in wet forests where it invades. The species is native to the Himalayas but is invasive in New Zealand, Hawaii and some other South Pacific islands, the Caribbean, South Africa, and La Réunion. Gardeners will wish to grow such a beautiful species, but extreme care should be taken to prevent seed production and rhizome spread.

Heracleum mantegazzianum

Giant hogweed

With its huge leaves and an even huger flowerhead that attains a height of 3m (10ft), this herbaceous plant is definitely an oddity in temperate regions. For that reason, it has become a popular ornamental. It spreads only by its many seeds, which are wind-dispersed. The seeds are used as seasoning in some Middle Eastern foods, and it is sometimes grown for that purpose. Its deep root can make it difficult to control.

HEDYCHIUM GARDNERIANUM

HERACLEUM MANTEGAZZIANUM

ECOLOGICAL IMPACT The plant seeds freely and forms a canopy that displaces native plants in forests and along streams and rivers. It is a federally listed noxious weed in the USA, and is also invasive in the UK and Europe. Further, chemicals in its sap can cause serious skin burns that may result in permanent scarring, especially when exposed to sunlight. Never allow any parts of this plant to touch the skin.

Hesperis matronalis

Dame's rocket

A large quantity of easily germinated seed is produced by this herbaceous plant. While native to Eurasia, the plant is often included in American wildflower seed mixes, to the extent that many Americans assume it is a native wildflower.
ECOLOGICAL IMPACT Dame's rocket can quickly dominate moist meadows and forests, excluding native plants. The plant is considered to be invasive in much of the USA. It would be better removed from wildflower mixes anywhere, but if used in the garden all fruits should be collected before they mature. In areas where it is listed as a noxious weed, the laws should be observed.

Hiptage benghalensis (100)

Hiptage

This climber beguiles with its attractive and fragrant flowers. Its winged seeds are dispersed by wind, allowing it to spread widely. The IUCN has described it as one of the 100 worst invasive organisms in the world in a list that includes all other invasive life forms. Despite this, little is known about the plant.
ECOLOGICAL IMPACT The climber's fast growth and its habit of growing into tree canopies enable it to smother native vegetation. It has been reported invading in La Réunion, Mauritius, Hawaii, and Florida – it is especially bad in the first two locations. Avoid planting in islands and other regions that have a warm climate.

Hydrilla verticillata

Hydrilla

Sold in the aquarium trade, this aquatic plant has been spread widely by fish enthusiasts dumping the unwanted contents of their tanks into fresh waters. The plant mostly spreads vegetatively, through regeneration of stem fragments and stem tubers. A single node on a fragment is all it takes for a new plant to establish.
ECOLOGICAL IMPACT Hydrilla forms large mats that impair drainage and irrigation channels. Boats cannot move through it, and any swimmer entangled in it may easily become exhausted and drown. Hydrilla can fill the water column and reduce native vegetation, fish, and water quality. It should never be put into any freshwater system anywhere, and boats should be checked to ensure they do not spread fragments into unaffected areas. The plant is invasive in the South Pacific, Australia, and parts of North and Central America.

Hydrocotyle ranunculoides

Floating pennywort

This fleshy aquatic plant is used in aquaria and in ponds. It invades slow-moving waterways, where it quickly develops patches of 15sq m (50sq ft) in a single season, growing up to 20cm (8in) a day. The plant reproduces mostly from stem and root fragments that are carried by the water or have been spread by waterfowl. In Europe it may be labelled as the native *Hydrocotyle vulgaris*, so aquarists should check the identification carefully.
ECOLOGICAL IMPACT The plant can exclude native plants, fish, and invertebrates. A native of North America, this species invades in Central and South America, northern and southern Europe, and western Australia. It should not be put into any freshwater areas that are not completely contained. Boats should be checked to ensure they do not spread fragments.

Hygrophila polysperma

Indian swamp weed

Sometimes called "green hygro", this is another aquatic plant that has found its way from aquaria into freshwater systems. The plant grows from bottom to surface in water 3m (10ft) deep, as well as creeping along the edges of streams and waters. It can rapidly form very dense stands of stems which then fragment and establish new plants elsewhere. One population in Florida is reported to have increased from 0.04 ha (0.1 acre) to over 0.41 ha (1 acre) in one season. The plant can produce seeds, but it is believed that most new populations have grown vegetatively.
ECOLOGICAL IMPACT Indian swamp weed outcompetes native vegetation. It is currently found in fast-moving streams in some parts of south-eastern USA. It should not be put into any freshwater system that is not completely contained. Areas with warmer climates are most vulnerable, but this species may also be tolerant of cold conditions.

HYDROCOTYLE RANUNCULOIDES

Hypericum canariense

Canary Island St. John's wort

The bright yellow flowers of this shrub are very attractive and long-lasting, and it is therefore widely grown as a garden plant. The wort produces many small seeds per plant, and these are spread easily by wind. The shrub has escaped cultivation in parts of California and Hawaii.
ECOLOGICAL IMPACT This shrub rapidly outcompetes native plants and is invasive in Calfornia, Hawaii, and south-western Australia. Avoid planting in these and similar areas with a warm climate. Reports indicate that the shrub can take possession of 90 per cent of a site once it invades, and the population may spread at a rate of 45–90m (150–300ft) per year.

Ilex aquifolium

English holly

People in many temperate areas love this evergreen shrub for its attractive red fruits set against dark green, glossy foliage. Unfortunately, birds also find their fruits attractive and gobble them up, only to deposit the seeds later beneath trees. English holly is very shade-tolerant and is becoming invasive in forested areas, including old-growth forests of the Pacific north-west of the USA.
ECOLOGICAL IMPACT The species adds a tall shrub layer in forests that do not usually have such a structure, threatening the native understorey vegetation. Avoid planting in north-western North America.

Impatiens glandulifera

Policeman's helmet

If law enforcement officers began to wear pointed pink helmets, this herbaceous plant – also known as "Himalayan balsam" – would be aptly named. The flowers hang by a thin stem and have a distinctive pointed hood at the rear. The hood contains nectar, which attracts the pollinating insects. This annual species grows to 3m (10ft) tall in one year and is spread exclusively by seeds, which are dispersed more than 6m (20ft) by an explosive release of up to 800 seeds per plant.
ECOLOGICAL IMPACT This plant invades along streams, in wetlands, and in moist forests. Native to the western Himalaya, it is very invasive in the UK and throughout Europe, and is considered a noxious weed in parts of the USA. It should not be planted near open country with wetlands or streams anywhere.

HESPERIS MATRONALIS

Imperata cylindrica

Cogongrass

A native of south-east Asia, this grass is so invasive that it is currently considered by experts to be one of the world's worst weeds. The plant reproduces both from underground stems (rhizomes) and small, windborne seeds that are dispersed over long distances. A red-tipped garden form sold as "Japanese blood grass" is the most common type currently seen in gardens, and it appears to be less invasive than the wild type. However, the red-tipped form is also more cold-tolerant and there is concern that it may hybridize with the wild type and confer greater hardiness, enabling the more invasive type to spread into colder areas.

ECOLOGICAL IMPACT This grass infests pine woodlands, dunes, wetlands, and grasslands, where it outcompetes native species and increases the risk of fire. Cogongrass has been reported as invasive in 73 countries. The all-green form should never be grown, and Japanese blood grass should not be grown anywhere that cogongrass is found invading.

IMPERATA CYLINDRICA

Ipomoea indica

Blue dawn flower

While a number of the "morning glories" are invasive, this climber appears to be the most problematic of the common garden forms. It has lovely true-blue flowers tinged with purple. It sends out numerous stems that root at the nodes as it moves across soil, spreading from fragments detached from the mother plant but rarely producing seed.

ECOLOGICAL IMPACT In warm climates the climber grows very quickly, invading open areas and smothering the vegetation it covers. It is found in tropical areas around the world but is considered to be invasive in parts of Australia, South Africa, and North America. It should not be grown near open countryside in warm, wet climates, and garden waste should be disposed of appropriately.

Iris pseudacorus

Yellow flag

The bright yellow flowers of this bulbous plant have made it a popular choice for water gardens. In the wild the flat seeds, bobbing along the waves in lakes and ponds, have the appearance of little ships looking for harbour. Once on land, they root on the shoreline. The underground stems (rhizomes) are tough and can withstand many weeks of drought, only to resprout as soon as water arrives.

ECOLOGICAL IMPACT The plant forms dense rhizomatous stands that exclude native shoreline vegetation and make it difficult for waterfowl to move between water and land. It is native to Europe, but is invasive in many parts of the world and is a noxious weed in parts of the USA. It should never be grown where it can get into natural waterways.

IRIS PSEUDACORUS

Jasminum dichotomum

Gold Coast jasmine

Like the Brazilian jasmine (*Jasminum fluminense, see below*), this climber is grown in part for its powerful fragrance at night. It produces small, fleshy berries over a long period of the year, and the seeds contained within are widely dispersed by birds and small mammals.

ECOLOGICAL IMPACT This jasmine can completely cover native vegetation, growing into the canopy of mature forests where it can kill trees and prevent regeneration. It has become very invasive in Florida and Hawaii and should not be grown there. Gardeners should use caution when growing it in other warm, humid climates.

Jasminum fluminense

Brazilian jasmine

Although the common name of this climber indicates that it is from Brazil, it is actually native to Africa and was brought to Brazil by the Portuguese, where it invaded and was first described by botanists. Like the Gold Coast jasmine (*see above*), this species is fragrant in the evening. The seeds are dispersed by birds and small mammals and germinate very easily.

ECOLOGICAL IMPACT This jasmine is an aggressive climber that can destroy mature forests by denying light to the trees. Its weight in the tree canopy increases wind damage. It is a major pest in Florida and Puerto Rico, the Caribbean, Hawaii, and Guam. It should not be grown if there is any possibility of its escaping into natural areas.

Jasminum sambac

Arabian jasmine, pikake

In India the flowers of *Jasminum sambac* are said be sacred to Vishnu, the god that protects against evil, and for many the intense fragrance of Arabian jasmine is a definitive scent of the tropics. In some locations, however, this climbing plant is an invasive species. It flowers and fruits continuously, and birds eat the fruits and disperse the berries. Plants may also spread along underground stems (rhizomes).

ECOLOGICAL IMPACT This climber can invade native plant communities and become a serious pest. It should not be planted in Florida, Hawaii, and the Caribbean.

Jatropha gossypifolia

Bellyache bush

The bellyache bush gets its name from its poisonous tendencies, which appear to be much more lethal than a simple bellyache. This shrub thrives on a variety of soil types and reproduces both by seeds and by regeneration of broken branches. There are concerns that the seeds may remain viable in the soil for up to 15 years, long after the parent plant has been removed.

ECOLOGICAL IMPACT Early in the 1900s the shrub was a favoured ornamental in tropical areas, but it is now notorious for poisoning humans and wildlife, as well as climbing over and smothering native vegetation. Avoid planting it in northern and western Australia, Hawaii, and in areas with similar climates.

LANTANA CAMARA

Koelreuteria paniculata

Golden rain tree

A beautiful, fast-growing species, the golden rain tree has bright yellow flowers followed by interesting, pink, papery fruits. Unfortunately, it is also very invasive, capable of producing thousands of seedlings that require continual control. Golden rain tree tolerates many different soil types, drought, and air pollution – this adaptability makes it capable of invading a number of different plant community types.

ECOLOGICAL IMPACT It is capable of outcompeting native vegetation and is currently invading in Florida and Hawaii.

Lagarosiphon major

Curly pondweed

Although sometimes used in garden ponds as an oxygenating plant, this aquatic species is more commonly found in the aquarium trade. It is somewhat similar to its relatives in the genus *Elodea*, and may be found mislabelled as *Elodea crispa,* a nonexistent species.

ECOLOGICAL IMPACT Curly pondweed grows submerged in slow-moving waterways and rapidly clogs them, blocking light to native submerged plants. It is very invasive in a number of countries, but is especially so in the UK and New Zealand. It should never be planted in ponds and, if used in aquaria, the tanks should be cleaned carefully and never emptied into waterways.

Lamium galeobdolon

Yellow archangel

This fast-growing, herbaceous ground-cover plant has yellow, strongly hooded flowers and, often, variegated leaves. It is very shade-tolerant and once established can thrive in dry soil. Rarely does the plant appear to set seed.

ECOLOGICAL IMPACT The harm in this plant comes from its ability to spread very rapidly through vegetative growth. It can be planted safely in most places, but it should never be grown near open countryside, and garden clippings should never be thrown onto heaps from which it may easily spread.

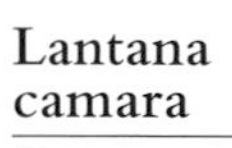

Lantana camara

Lantana

Considered to be a significant problem in at least 50 countries, this shrub probably qualifies as one of the most serious invasive species in many areas of the tropics. Its flowers are attractive, usually showing a mixture of colours from orange to yellow on the same flowerhead. (Its many garden varieties have other colours that tend towards pink.) The flowers attract butterflies and hummingbirds and many animals, from chickens to wild birds, eat the juicy fruits and disperse the seeds.

LEPTOSPERMUM LAEVIGATUM

LATHYRUS LATIFOLIUS

ECOLOGICAL IMPACT Lantana can invade open countryside and form dense thickets, often made impassable by razor-sharp thorns on the stems. The plant is native to Central America, but is now invasive in most parts of the tropics and subtropics, including Australia.

Lathyrus latifolius

Everlasting pea

Bright pink flowers make this a beautiful climber, but the cut flowers do not last as those of the annual sweet pea, and the aggressiveness and persistence of this species soon becomes tiresome. Once established, it is very difficult to eradicate because it has a deep and incredibly tough root system. It is mostly found in disturbed areas. The stems reach only a few metres in length, but a single root crown can produce numerous stems.
ECOLOGICAL IMPACT Because the plant is aggressive and difficult to control, it can harm other species in the areas it invades. It is native to Europe and is extensively naturalized through North America and other temperate climates. It should not be allowed to spread into open countryside. Seedpods should be removed as they appear.

Lathyrus tingitanus

Tangier pea

In addition to being grown as an annual sweet pea, Tangier pea is sometimes used as a forage plant. It is an annual climbing species with bright pink flowers.
ECOLOGICAL IMPACT The dead stems are creating a fire hazard in western Australia, where the plant is spreading rapidly along roadsides and into the bush. It is native to the western Mediterranean and Azores, but is invading in western and southern Australia, western North America, and New Zealand.

Lavandula stoechas

French lavender

A source of lavender honey and oil, this herbaceous plant is one of the showiest of the lavenders. French lavender is drought-tolerant with evergreen leaves and spreads by seed. As its name implies, it is native to the Mediterranean.
ECOLOGICAL IMPACT French lavender is increasingly being found along roadsides and in open countryside, and is thus finding its way onto weed lists. It is unknown whether this species has negative effects on native plants or their habitats. The plant is well-adapted to climates with winter rain and summer drought, such as found in Australia and New Zealand, where it invades, as well as California, temperate South America, and South Africa, where it should be grown with care.

Leonotis leonurus

Lion's ear

A tall, gangly, annual herbaceous plant with odd-looking flowers in rounded, spiny clusters, lion's ear spreads by seeding freely.
ECOLOGICAL IMPACT This plant mostly invades highly disturbed areas, but it can be a problem along streams and rivers, where it forms dense thickets. It is native to tropical Africa, but is found throughout southeast USA, Hawaii, and Australia. It should be monitored where grown to ensure it does not get into open countryside.

Leptospermum laevigatum

Tea tree

An attractive, small tree with aromatic foliage, tea tree has numerous small flowers of pink and red and an interesting, shredded bark. The small seeds are dispersed by wind and do not live long. This species has an extensive root system, and resprouts if the trunk is cut back.
ECOLOGICAL IMPACT The tree can form dense thickets whose shade eliminates other species. It is invasive in South Africa and Hawaii and should be avoided in places with Mediterranean-type climates. Some other leptospermum species have shown tendencies towards invasiveness, and they also should be used with caution.

LEUCANTHEMUM VULGARE

Leucanthemum vulgare

Marguerite, ox-eye daisy

Also known as *Chrysanthemum leucanthemum*, this herbaceous plant is probably the most familiar daisy to many gardeners. It was introduced centuries ago as an ornamental, and has been spread in seed both as an agricultural contaminant and intentionally as a "wildflower" species. It produces a great many seeds that germinate easily, and the new plants can begin reproduction when only a few months old. It also spreads by means of underground stems (rhizomes).
ECOLOGICAL IMPACT This daisy colonizes crops and roadsides, but is also found in native plant communities, such as grasslands, wet meadows, and streamside forests. It outcompetes native plants and carries viruses that can spread to crops. Native to Europe, it is widely invasive in 40 countries, including those of North and South America, Africa, and Australia.

Leycesteria formosa

Himalayan honeysuckle

Many a person, attracted by this shrub's drooping flowerheads of interesting white flowers with red modified leaves (bracts), has planted it only to discover that the seeds start germinating everywhere. Seeds continue to produce long after the plant has been discarded because they can remain viable in the soil for several years. In some parts of the world, birds seem to be attracted to the seeds, dispersing them outside garden boundaries.
ECOLOGICAL IMPACT Plants have been found far outside cultivation, especially in moist areas, where they create dense thickets and exclude native species. The species has been invasive in parts of New Zealand, where it is a significant pest, and in parts of Australia, such as New South Wales. Care should be taken in areas of the USA and Europe with similar climates.

Ligustrum japonicum
Japanese privet

This evergreen shrub, with glossy leaves and small, fragrant flowers, is commonly used for hedging in many parts of the world. The seeds, when freshly dispersed by birds, can germinate immediately at high rates, and the young seedlings grow to reproductive maturity very quickly. The shrub is tolerant of poor or dry soils, but it cannot thrive if its roots are waterlogged. ECOLOGICAL IMPACT This privet forms dense thickets in woodlands and excludes native vegetation. The plant is extremely invasive in south-eastern USA and is considered a major pest in Florida. It is also invasive in New Zealand. Parts of other continents with warm summers, regular rainfall, and mild winters may also be at risk.

Ligustrum lucidum
Chinese privet

Chinese, glossy, or broad-leaf privet is a shrub or tree known for its copious fruit production. Birds favour the bluish-black, one-seeded fruits and distribute the seeds, which can germinate easily. Chinese privet is more tolerant of wet soils than some other privets and may be found in low-lying forests and even bogs. It is quite common for plants sold under this name actually to be *Ligustrum japonicum* (*see above*), so check with a local botanist or horticulturist if you are not sure. ECOLOGICAL IMPACT Like most privets, this species is shade-tolerant and has been found to invade forested systems, excluding native vegetation. Chinese privet is found throughout eastern USA and parts of Australia, such as south-east

LIGUSTRUM LUCIDUM

Queensland. Areas of Europe with significant summer rainfall may also be susceptible to invasion.

Ligustrum obtusifolium
Border privet

Forests and other shady environments suit this deciduous shrub very well, but it is more commonly found along the forest edges and in abandoned fields. In one such field, researchers found plants at a density of 6,082 plants per hectare. The small, white flowers are not as fragrant as those of some privets. The fresh seed can germinate easily and is dispersed by birds. ECOLOGICAL IMPACT The shrub is very invasive in parts of the USA, such as the north-east and midwest. It should be monitored when grown in places outside its native Japan, especially those with summer rainfall, including much of Europe.

Ligustrum sinense
Privet

Not only does this shrub produce fruits that birds disperse to new locations, it can also spread by means of root sprouts. The shrub is very difficult to control once it is established and mature colonies can produce up to 1,300 fruits per square metre – there are accounts of even higher fruit production in Australia. Reports from that country suggest that it can be found in some areas literally "anywhere that birds fly". ECOLOGICAL IMPACT This privet can form thickets that easily exclude native vegetation. In Florida it is believed to be encroaching on a globally endangered plant species. Avoid planting *Ligustrum sinense* in eastern USA, parts of Australia such as New South Wales, and other areas with similar climates.

Lilium formosanum
Taiwan lily

With large, fragrant, white flowers perched on stems up to 1.5m (5ft) tall, this is a gorgeous bulbous plant. However, it grows from seed to flowering in just a few months, and each flower can produce hundreds of tiny seeds that are spread widely by the wind. The plant also spreads from bulblets formed on its bulb, dispersed whenever the soil is disturbed. ECOLOGICAL IMPACT This lily is invading roadsides and native plant communities over wide areas in Australia, from Queensland to Victoria. Caution should be exercised wherever it is grown.

LILIUM FORMOSANUM

Linaria dalmatica
Dalmatian toadflax

The flowers identify this herbaceous plant as a member of the snapdragon family. Dalmatian toadflax is a perennial with a deep, extensive, creeping root system. It is not quite as aggressive as yellow toadflax (*see below*), but it is still able to invade open countryside. ECOLOGICAL IMPACT Once established, the plant competes with native species for water. It spreads through open forest and shrub-dominated communities, as well as disturbed areas. It is considered a problem plant in Russia, North America, and parts of Europe to which it is not native. With many noninvasive and more attractive snapdragons to choose from, there is no need to use this species.

Linaria vulgaris
Yellow toadflax

Although this herbaceous species is relatively small, a mature plant can produce up to 30,000 tiny seeds per year. The plant also spreads through underground stems (rhizomes). In one tested population there were as many as 180 stems per square metre. In another location, a patch that was originally 0.4ha (1 acre) in size expanded to cover 34ha (85 acres) in five years through seeding and rhizomatous growth. Seeds can last for up to 10 years, so control is a difficult and lengthy proposition. ECOLOGICAL IMPACT The ability of yellow toadflax to spread very densely and rapidly makes it a highly effective competitor for water and soil nutrients. The plant is considered a problem in Russia, North America and South America, New Zealand, Australia, South Africa, the Caribbean, and parts of Europe to which it is not native. Many other noninvasive snapdragons can be grown, so there is no need to use this invasive species.

LINARIA VULGARIS

Lonicera x bella
Bell's honeysuckle

Most hybrid plants lack invasive tendencies, perhaps because many of them are sterile and do not produce seeds. This shrub is an exception – it does produce seeds and it also spreads vegetatively. Both of its parents, *Lonicera morrowii* and *Lonicera tartarica* (*see p.461*), are also seriously invasive plants. ECOLOGICAL IMPACT In some parts of the USA, such as the middle Atlantic states and upper midwest, the hybrid has been reported as a pest on natural preserves. It should not be grown in eastern and midwest USA, and care should be taken in places with similar climates, such as in Europe.

LONICERA JAPONICA

Lonicera japonica
Japanese honeysuckle

Children love the Japanese honeysuckle for the sweet nectar they suck from the base of the flower, and adults love it for the fragrance. This climber can grow up to 15m (50ft) a year once it is established, and it may spread both by birds eating the fruit and dispersing the seeds and by vegetative means. The plant's ability to photosynthesize in winter may also help it to spread quickly. It develops an extensive root system, which can make control difficult. ECOLOGICAL IMPACT This plant quickly covers and engulfs native species. It currently invades eastern USA, many South Pacific islands, the Canary Islands, and parts of New Zealand and Australia. Care should be taken when it is planted anywhere with summer rainfall.

Lonicera maackii
Amur honeysuckle

This often tree-like, multi-stemmed shrub is tolerant of many soil types, as well as drought conditions, enabling it to colonize in varied localities. Birds disperse the seeds, although the fruits have been recorded as a poor source of energy. ECOLOGICAL IMPACT The plant has become very invasive in some parts of the USA, especially the middle Atlantic states, where it is forms dense thickets in forest understories. Its persistence in those understories has led to long-term changes in the composition of plant and animal communities in the forests. The shrub is invasive in temperate parts of North America that receive ample precipitation in the summer months. It should be avoided wherever the climate is similar, such as in parts of Europe.

Lonicera morrowii

Morrow's honeysuckle

Like its relatives, this shrubby species has red berries that are attractive to birds. It can easily colonize disturbed areas, as well as open areas in forests, and it appears to tolerate both wet and dry sites. Where its habitat overlaps with that of *Lonicera tatarica* (*see below*), the two species hybridize to produce *Lonicera* x *bella* (*see p.460*).
ECOLOGICAL IMPACT Morrow's honeysuckle leafs out early in the year and retains the leaves until well into the autumn – characteristics that allow it to shade out and exclude other species in the forest openings it invades. The shrub is invasive in the USA through much of the east and midwest. It should be avoided in other temperate areas with summer rainfall, such as Europe and parts of Asia.

Lonicera tatarica

Tartarian honeysuckle

The tartarian honeysuckle is perhaps the most invasive of all the shrub honeysuckles. It reproduces by both fruits and soil layering of branches. Because the seeds require a cold period of up to 90 days before germination, it is likely to remain a problem only in cold areas.
ECOLOGICAL IMPACT Studies have shown that *Lonicera tatarica* is allelopathic, or poisonous, to other plants, such as young pine trees. It currently invades only in nature reserves along the east coast of the USA, but areas with cold winters and ample summer rainfall, such as northern Europe, may be at risk.

LUPINUS ARBOREUS

Lunaria annua

Honesty, satin flower

The American common name for this popular herbaceous species is "moneyplant", and the flat, shiny seed membranes that remain after the seeds have fallen certainly resemble silver coins. However, the attraction of the dried plant to flower arrangers is offset by the invasiveness of the living species.
ECOLOGICAL IMPACT Once established, this plant can spread through open and shady forests, as well as disturbed areas. It does not appear to cause major problems, but in areas where it is well established it can grow densely and compete with native plants. It is native to southeast Europe and can invade forests in temperate areas, so it should not be planted where it could escape into woodland.

LUNARIA ANNUA

Lupinus arboreus

Tree lupin

Also called, more appropriately, the yellow bush lupin, the tree lupin is actually a woody shrub. It is topped by spikes of bright yellow flowers. Places that have little organic matter in the soil are the most likely to be invaded by the tree lupin.
ECOLOGICAL IMPACT By using its root nodules to change atmospheric nitrogen into a form usable by the plant, the tree lupin increases the soil's nitrogen content. With time this affects the composition of native species in the affected habitat. The dense shrubs may also prevent some native species from getting enough light. In British Columbia, Canada, the tree lupin has been found to hybridize with a native lupin species. It is invasive in northern California, the Pacific north-west of the USA, and the UK. Care should be taken when planting it, since it appears to tolerate both frost and drought.

Lygodium japonicum

Japanese climbing fern

Ferns are not usually thought of as climbers, but this species, as well as its relative, *Lygodium microphyllum* (*see below*), certainly is an aggressive climber. It does well in both wet and upland forests and reproduces by spores carried by the wind.
ECOLOGICAL IMPACT This climbing fern can scramble to the top of the forest canopy, where it forms dense, intertwined mats that suffocate trees and shrubs. It can carry forest fires from ground level to the canopy in locations where surface fires, but not canopy fires, are part of the natural ecosystem. It invades both dry and wet forests in south-eastern and southern USA and should not be planted there.

Lygodium microphyllum

Old World climbing fern

Mounting into the tree canopy of wet forests, this climbing fern forms mats up to 1m (3ft) thick. Studies have shown that the leaves have an unusually high rate of photosynthesis, which allows this species to grow very quickly. The spores, appearing on the underside of the leaves, travel long distances in the wind.
ECOLOGICAL IMPACT Like Japanese climbing fern, *Lygodium japonicum* (*see above*), it forms mats that both suffocate trees and allow fires to climb to the tree canopy. The fern is native to Africa, Asia, and Australia, and is considered to be a very serious weed in Florida, where it is smothering tens of thousands of hectares. It should not be planted wherever it is invading.

Lysimachia vulgaris

Yellow loosestrife

Purple loosestrife (*Lythrum salicaria*, *see p.462*) may be notorious, but yellow loosestrife is an even more invasive herbaceous plant than its distant cousin. Yellow loosestrife invades the margins of lakes and marshes. The plant spreads not only by seed but also by underground stems (rhizomes). It may not produce seed early in its life – reports indicate that blooming plants are usually those that have been in an area for a few years. However, once established this perennial species blooms annually. Any young plants should be removed immediately after detection to prevent them from producing seed.
ECOLOGICAL IMPACT Yellow loosestrife is invasive in many parts of North America and has the potential to be so in other temperate areas. It should never be planted where its rhizomes or seeds could escape and colonize a freshwater system.

LYGODIUM JAPONICUM

LYTHRUM SALICARIA

Lythrum salicaria

Purple loosestrife

For decades purple loosestrife was a popular perennial herbaceous plant, with its pink flower spikes and its ability to tolerate wet soils. Unfortunately, it is also a prolific producer of seed – an estimated two million seeds may be produced on a single mature plant. Seeds are spread by wind or water and by mud that clings to feet, animals, and boats. There are claims that some hybrids are sterile, but so far all of those plants have shown to be at least partially fertile. Purple loosestrife is considered a noxious weed in many states of the USA, and it can be found to some degree in all states.
ECOLOGICAL IMPACT In the USA it now dominates wetlands, greatly excluding native plant species and decreasing wildlife food and shelter. Because of its ability to spread, its harm to wetlands and wetland animals, and its tolerance of cold and mild temperatures, it should never be grown – there are many noninvasive alternatives.

Macaranga mappa

Bingabing

A striking tree with enormous leaves, bingabing also has capsule-like fruits with a fleshy seed coat that makes them attractive to birds, ensuring dispersal of the seeds.
ECOLOGICAL IMPACT Dense groves of the tree have invaded the forests around Hilo, Hawaii, where it was once seeded by an aircraft following a fire. It appears to be reproducing actively and spreading from this initial seeding. Commonly grown in wet tropical areas as an ornamental, it should be monitored for its invasive ability following its performance in Hawaii.

Macfadyena unguis-cati

Common cat's claw vine

This fast-growing, evergreen climber, also widely known as *Bignonia unguis-cati*, rapidly covers sheds, fences, and shrubs. It produces numerous wind-dispersed seeds from its bean-like fruits. Assisted by clawed tendrils, it climbs up trees and shrubs in forests and open areas. The stems can root at each leaf node while lying on the ground, so fragments of the stem quickly establish new plants. At about every 50cm (20in) a tuber forms on the roots, and each tuber can send up new climbing stems. The climber tolerates a variety of soils and has very deep, thick roots that are almost impossible to kill without herbicides.
ECOLOGICAL IMPACT The climber forms dense mats on the forest floor that exclude native species. It is native to tropical America and invades in eastern Australia, Hawaii, New Caledonia, and south-eastern North America.

Melaleuca quinquenervia

Paperbark tree

The thick, papery bark, narrow evergreen leaves, and bottlebrush flowers of this tree make it an interesting ornamental. In the past it was used as an afforestation tree in places such as the great Everglades wetlands of Florida, USA. However, it is now a serious threat to such ecosystems and has become one of the world's worst pest trees. The thick bark helps it to tolerate fire, which triggers the release of tiny seeds that germinate within three days. The tree flowers and fruits after only one year and can grow up to 2m (6ft) a year. It also spreads vegetatively, resprouting if cut back.
ECOLOGICAL IMPACT This tree is believed to be allelopathic, poisoning surrounding plants. One study found that paperbark forests have 60–80 per cent fewer small mammals than native forests in the same areas. It has been a problem primarily in Florida, but also on some islands in the South Pacific.

MACARANGA MAPPA

MACFADYENA UNGUIS-CATI

Melia azedarach

Bead tree

Melia has purple flowers and is often used as a street tree in warm climates. It is very fast-growing and can spread rapidly. The flowers do not need pollination to produce fruit and the many seeds produced each year germinate easily. The tree may also form root suckers and spread vegetatively. Few insects seem to have any effect on the tree, which probably increases its ability to survive in varied places.
ECOLOGICAL IMPACT Bead tree has been reported as invading undisturbed forests and stream banks in South Africa, states of south-eastern USA, Latin America, and islands in the South Pacific, where it outcompetes native vegetation. Most of the tree's extensive roots are in the upper part of the soil, so it may have a competitive advantage over native species in absorption of water and nutrients. Its invasive ability should be monitored wherever it is grown.

Merremia tuberosa

Spanish morning glory

Also known as *Ipomoea tuberosa*, this climber is sold by florists for dried flower arrangements. Its main attraction for this purpose is the woody fruit, which resembles a rose (another common name for the species is "wood rose"). It has been suggested that the plant is sometimes spread by people discarding unwanted arrangements that contain seeds. The plant produces lots of fruits and dried seeds can remain viable for many years.
ECOLOGICAL IMPACT Spanish morning glory can quickly climb up into the canopy of forests and smother trees and shrubs, starving them of light. It grows so immense that it can pull them down with its weight. It is native to tropical America, but is now widely distributed in most of the world's tropical areas. It should not be planted near open countryside, and flower arrangements should be disposed of carefully.

Mesembryanthemum crystallinum

Crystalline iceplant

A succulent relative of the Hottentot fig (*Carpobrotus edulis*, *see p.447*), the crystalline iceplant looks somewhat like that plant except that it has smaller leaves and flowers. It is sometimes planted along roadsides and banks for erosion control.
ECOLOGICAL IMPACT The plant can take in and store water from the soil, denying water to native plants. Unusually high levels of nitrate salts can accumulate under it, preventing the establishment of native species. Originally from South Africa, it is invasive in California, so it should be used with caution in the Mediterranean region, Australia, and temperate South America.

Miconia calvescens

Miconia

Despite its beautiful, huge, purple leaves, this tree species should never be planted outside, under any circumstances. Its small fruits are eaten by mostly non-native birds and the seeds have a very high rate of germination.
ECOLOGICAL IMPACT Miconia has become a major pest in the South Pacific, especially in Tahiti, where it has replaced as much as 70 per cent of the native forest. It is spreading rapidly in other places where it has been introduced, such as Hawaii. It forms dense forests and completely excludes light at the understorey level, preventing the survival of native species. Avoid planting except in conservatories in temperate areas.

Mikania micrantha

Mile-a-minute plant

It may not grow at the rate of one mile a minute, but this herbaceous plant does grow very quickly. It reproduces by the seeds, which are dispersed by wind or by attaching to clothing, or by detached stem fragments, which can root when brought into contact with soil.
ECOLOGICAL IMPACT The plant can smother native species in open areas. Native to tropical America, it is invasive throughout the South Pacific, tropical Asia, and is now in Queensland, Australia. It is on the United States Noxious Weed List.

MESEMBRYANTHEMUM CRYSTALLINUM

Mimosa pigra

Cat-claw mimosa

Sometimes called the "giant sensitive plant" because it can fold its leaves in response to stimuli, this aggressive woody shrub produces up to 42,000 seeds per plant and the seeds remain viable for many years. Cat-claw mimosa is invasive in many parts of the world, and it can increase flooding by obstructing water flow in rivers and streams.
ECOLOGICAL IMPACT The plant reduces the numbers of birds and other small animals typically found in unaffected areas. The shrub is currently invasive in Florida (where planting it has been prohibited in some areas), northern Australia, and Vietnam. Avoid planting near wetlands in warm climates.

Miscanthus sinensis

Chinese silver grass

In recent years this grass has become very popular and it has spread outside cultivation in many areas. It does not appear to be increasing rapidly, but it has been found along roadsides and in forest edges and clearings. Although it can reproduce by seed, it mainly spreads by means of underground stems, and small pieces of only 4cm (1½in) can start a new plant. Whether the cultivated forms produce seed is debated, but the species does in Asia so it likely that at least some of the cultivated forms also produce seed.
ECOLOGICAL IMPACT The grass is an active invader of disturbed areas in its native Asian habitat, so it is not surprising that it also colonizes disturbed areas wherever it escapes cultivation. In temperate regions with summer rainfall, such as much of Europe, the grass should not be allowed to become established in areas of open countryside.

Morella faya

Firetree

Also known as *Myrica faya*, the firetree is spreading like wildfire in some parts of the world. In Hawaii, whole forests are dominated by the tree. Each plant produces thousands of tiny seeds that are spread by birds (plants with mostly female flowers may produce as many as 37,000 seeds in a year). The tree resprouts from the cut stump.
ECOLOGICAL IMPACT A study in Hawaii found that, through bacteria in its root nodules, the plant added more than four times as much nitrogen per year to the soil compared to soil of unaffected forests. It is abundant in the Volcano National Park area on Hawaii Island and is found on other islands. It should be grown with care.

Moringa oleifera

Horseradish tree

The horseradish tree has some interesting properties that are of potential importance in developing countries. Its seeds may help bind bacteria in water, helping to make it more drinkable, and the leaves are believed to be very nutritious.
ECOLOGICAL IMPACT The tree has been reported as invasive in islands in the South Pacific and the Caribbean. It flowers in its first year and produces copious amounts of seed. Care should be taken to prevent its escape wherever it is grown in warm and wet climates.

Myosotis scorpioides

Water forget-me-not

This herbaceous perennial spreads with slow-growing, underground creeping stems, but it mostly reproduces by seed. It prefers a waterside habitat, releasing its seeds downstream, where they germinate easily in the wet soil.
ECOLOGICAL IMPACT Colonizing the banks of streams, it is most aggressive in moist areas. It is native to Eurasia and is widespread in North America and parts of Australia. It should not be planted near open countryside.

Myosotis sylvatica

Wood forget-me-not

Charming, small, blue flowers account for the frequent inclusion of this herbaceous plant in mixes of "wildflower" seeds. The species self-sows freely and prefers a drier habitat than does its cousin, *Myosotis scorpioides* (*see above*).
ECOLOGICAL IMPACT The plant spreads rapidly in dry locations. It is native to Eurasia and is widespread in North America and in parts of Australia. It should not be planted near open countryside.

MYOSOTIS SYLVATICA

Myriophyllum aquaticum, Myriophyllum spicatum

Milfoil, parrot's feather

Both of these aquatic plants are used in aquaria and end up in freshwater systems when tanks containing the plants are emptied into them. *Myriophyllum aquaticum* reproduces entirely by regeneration from stem fragments and sections of underground stems. *Myriophyllum spicatum* reproduces from fragments but may occasionally produce seed.
ECOLOGICAL IMPACT The milfoils invade lakes, ponds, and slow-moving waterways, where they compete with native plants, obstruct the passage of boats, and provide good breeding conditions for mosquitoes. These species should not be put into any freshwater system, and are extremely hard to control once established. Boats should be cleaned after travelling an infested waterway.

Nandina domestica

Heavenly bamboo

Not a bamboo at all but a shrub, heavenly bamboo has pretty, finely dissected leaves. Seeds germinate easily without any treatment. It is not one of the worst invasive plants, perhaps because it is somewhat fussy about soil types, but it has been found outside cultivation in some parts of the continental USA and Hawaii, where birds spread the seeds.
ECOLOGICAL IMPACT In Florida it can form dense thickets and is believed to be contributing to the endangerment of at least one native herbaceous species. It is invasive in mild, temperate parts of the USA with summer rainfall. Gardeners in other areas, such as parts of Europe, should also plant this species with care. New cultivars that do not produce fruit are being developed and should be sought at nurseries in preference to their invasive parent.

Nephrolepis cordifolia

Tuberous sword fern

Unlike many sword ferns, this fern can form small, round tubers on the rhizomes when growing in good soils; disturbance of its underground stems and tubers can result in regeneration from either. The plant also reproduces by means of spores. It is sometimes called "Boston fern", but that name correctly belongs to a cultivated variety of another species, *Nephrolepis exaltata* 'Bostoniensis'. No fern that produces any form of tuber can be called a true "Boston fern".
ECOLOGICAL IMPACT The tuberous sword fern colonizes in shady forests and can spread aggressively, forming dense stands that easily displace native ground plants. The species is native to Australia, where it may also be invasive in some situations and circumstances, and is considered a pest species in south-eastern USA. Gardeners in warm, wet regions should plant it with caution.

MYRIOPHYLLUM AQUATICUM

Nephrolepis multiflora
Asian sword fern

Unusually for a sword fern, this plant is able to tolerate dry conditions. It is a common fern in Hawaii, but reports that it is native to Hawaii appear to be erroneous. Like the tuberous sword fern, *Nephrolepis cordifolia* (*see p.463*), it is sometimes incorrectly called a "Boston fern". The Asian sword fern can be distinguished by a line of erect hairs on the upper surface of the "leaf" blade.
ECOLOGICAL IMPACT The fern can be found in Hawaii aggressively spreading along roadsides, paths, and even lava flows. It is native to India and tropical Asia and invades in Hawaii, Puerto Rico, and the Caribbean islands. Gardeners should use it with caution.

Nicotiana glauca
Tree tobacco

Not so much a tree as a shrub, tree tobacco has pendulous, tubular flowers that are sweetly scented. Hummingbirds are attracted to the flowers, a factor that appeals to many gardeners. However, each plant produces up to a million tiny seeds, and dispersal by wind and water enables the plants to reproduce rapidly. The plants are very drought resistant and may be found in very poor soils.
ECOLOGICAL IMPACT Alkaloids in the leaves can make them toxic if consumed. Tree tobacco is invasive in many parts of the world, especially in drier areas, and should not be grown in California, Australia, Mexico, South Africa, Israel, or India. Areas with similar climates should also avoid it.

OPUNTIA FICUS-INDICA

Nymphaea odorata
Fragrant water lily

With their white and pink, many-petalled flowers and floating leaves, fragrant water lilies are among the most beautiful aquatic plants. They spread by seed moving in the water, as well as by underground stems (rhizomes) that break into pieces. One established rhizome can produce a plant that covers 5sq m (16sq ft) of water surface.
ECOLOGICAL IMPACT The plant can grow in very dense mats that decrease light to other aquatic plants beneath, and which also impede boat traffic and swimming activities. It is native to eastern North America, but is invasive in the western part of the continent. Other temperate wetlands, such as those in Europe, are also likely to be vulnerable. This water lily should never be planted in lakes or ponds and it should not be allowed to escape into them. It can be planted in contained garden ponds.

Ochna serrulata
Mickey Mouse plant

As the common name "Mickey Mouse plant" suggests, this shrub has unusual fleshy fruits borne on spreading red sepals that somewhat resemble the head of the famous cartoon mouse. Mainly because of this curious resemblance, the plant is often cultivated in warm climates. It also has bright yellow flowers. The seeds are often spread from the garden by birds into surrounding open countryside and forests.
ECOLOGICAL IMPACT The plant currently invades at lower altitudes on several South Pacific islands and in south-east Queensland, Australia. Areas with warm, wet climates appear to be most at risk.

Opuntia ficus-indica
Prickly pear, Indian fig

This cactus is grown as much for its flat stems (called *nopalitos*) and growth habit as for its fruit. The fruit are spine-covered but taste good, and are known in parts of the world as a "famine food" that can be eaten when other, less drought-resistant species are unavailable. Some cultures value the fruit more highly than that, although eating it can cause gastric distress.
ECOLOGICAL IMPACT Prickly pear can be very invasive and forms dense, impenetrable thickets, notably in dry areas. In some places a moth was introduced to control it, creating controversy among people who enjoy its fruits. Parts of Africa, Madagascar, Australia, the Caribbean islands, and the Mediterranean countries are infested. It should not be grown near open countryside.

ORNITHOGALUM UMBELLATUM

Ornithogalum umbellatum
Star-of-Bethlehem

White, star-shaped flowers are the horticultural attraction of this bulbous plant. It rarely produces seedlings, but the main bulbs produce small bulblets that easily become detached when the plant or soil is disturbed. Each bulblet is capable of producing a new plant.
ECOLOGICAL IMPACT Star-of-Bethlehem infests disturbed areas, crops, and some open woods, where it may affect native vegetation. It is native to parts of Europe and North Africa and is invading in areas of eastern North America. It should be grown with care in temperate areas with similar climates.

Oxalis pes-caprae
Bermuda buttercup, soursob

Most gardeners know the Bermuda buttercup as a garden weed, but many still plant it for the yellow flowers and its ability to spread as a ground cover in dry areas. This herbaceous plant may not reproduce by seed, but it produces numerous small bulblets that are spread by soil disturbance, wind, or water, each one producing a new plant.
ECOLOGICAL IMPACT The plant is reported as invading in dunes and other dry areas, where it both outcompetes native plants and inhibits the germination of their seeds. It is native to the Cape region of South Africa and invades in several Mediterranean countries, western North America, and all states of Australia. It may also be invasive in other semiarid areas. If planted in the garden, all waste should be disposed of in a way that does not spread the bulblets.

Paraserianthes lophantha
Cape wattle, swamp wattle

Also called *Albizia lophantha*, Cape wattle is a small, spreading tree with delicate fern-like foliage. It produces a large number of seeds annually during summer months, and these may be dispersed by ants, water, or human movement. Coastal forests appear to be most vulnerable. Cape

OXALIS PES-CAPRAE

PASSIFLORA TARMINIANA

wattle reproduces at an early age and grows very fast – up to 3m (10ft) a year in good conditions. It vigorously resprouts when cut.
ECOLOGICAL IMPACT In South Africa, cape wattle is considered to be such a serious invader that biological control agents have been introduced to reduce its numbers. While it is native to parts of Australia, it is also invasive in other parts of the country, and in South Africa, New Zealand, and parts of Hawaii. Warmer countries in Europe should plant it with care.

Passiflora suberosa

Corky passion vine

Dry sites are particularly favoured by this climber. Birds eat the purple fruits and disperse the many seeds. The plant is used by butterflies as a source of food for larvae.
ECOLOGICAL IMPACT Corky passion vine is an aggressive weed that can quickly grow into the subcanopy and canopy layers of open forests and shrublands. It is native to the Galapagos Islands, but invades many South Pacific islands, Caribbean islands, La Réunion, and South Africa. In Australia, nurseries in New South Wales have been asked not to sell it because it is so invasive and widespread, which is probably good advice for the other places it invades too.

Passiflora tarminiana

Banana passion flower

If the legend is to be believed, this climber, also called *Passiflora mollissima*, was introduced in Hawaii to beautify an outhouse. With large, pendant, pink flowers and interesting, banana-like fruit, it certainly succeeded in that, but it also spread throughout the islands. Wild pigs are known to spread the fruits in Hawaii, but birds and humans spread them in other locations. The fruits are edible, but are mostly used for making juice.
ECOLOGICAL IMPACT This climber can smother the canopy of trees 10m (30ft) high. It is native to tropical South America and is invasive in Hawaii, New Zealand, South Africa, and Asia. Although it is a beautiful plant, it is so invasive that planting should be strictly avoided in wet, tropical areas.

Paulownia tomentosa

Empress tree, foxglove tree

As one of the common names suggests, this tree has been associated with royalty, perhaps because it sports enormous flowerheads of royal purple blooms in spring, before the leaves emerge. It grows very quickly and begins flowering and fruiting early, producing seeds that germinate easily and are dispersed by wind. Up to 2,000 seeds may be contained in each capsule, and there are potentially hundreds of capsules on mature trees – it has been estimated that one tree can produce 20 million seeds in a year. The tree also forms colonies through root suckering, and it resprouts when cut back. It is able to tolerate poor soil consisting of little organic matter.
ECOLOGICAL IMPACT Despite its attractive flowers and its many uses, from ornamental horticulture to production forestry, this tree should only be grown where it can be closely monitored, because it is highly invasive. The empress tree has invaded deciduous forests in south-eastern USA.

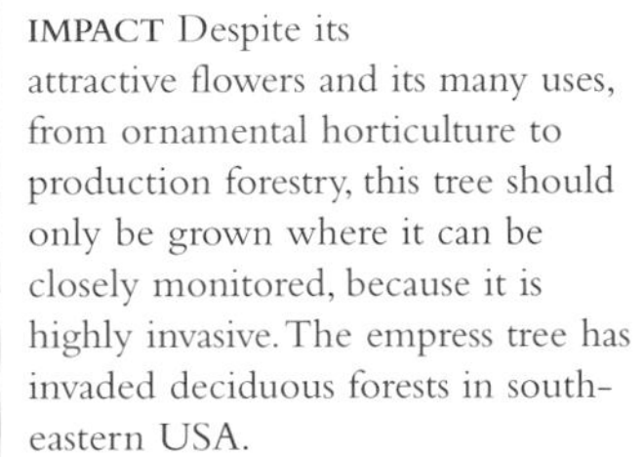

PENNISETUM SETACEUM

Pennisetum setaceum

Fountain grass

One of the first grasses to become popular in horticulture, fountain grass has a dense, clumping form and compact flowerheads that are somewhat pink when young. Unfortunately, it is able to produce seed without pollination and it can grow in a wide range of different environments, from bare lava flows in Hawaii to grasslands in California, at many altitudes.
ECOLOGICAL IMPACT While this grass can outcompete native species, the greatest danger it poses is an increased risk of fire where it invades. As the grass dies, the dry foliage provides tinder that is easily ignited by lightning or careless people. Surrounding plant species that are less well adapted to fire may not recover, and ground-nesting birds and terrestrial animals are also affected. Fountain grass is native to Africa and the Middle East, but is invasive in the warmer parts of North America, Hawaii, Fiji, South Africa, and Australia. Because many other noninvasive grasses are available, there is little reason to plant this one.

Phragmites australis

Common reed, Norfolk reed

Genetic forms (genotypes) of the common reed are found on every continent except Antarctica. Therefore it is difficult to know where the reed is native and where it is introduced, and where both have occurred. This tall grass, also called *Phragmites communis*, has been used for a number of different purposes and is sometimes grown in gardens. In Europe it is still used for roof thatch and this may outweigh some environmental concerns.
ECOLOGICAL IMPACT The reed is extremely invasive in wetlands and along water bodies, where it forms dense stands over underground stems (rhizomes). It can be found in ditches, but it also invades and degrades pristine water systems. In the USA, genetic studies are showing that the most aggressive forms of the reed are not native plants. The reed is also capable of evaporating more water than can be accumulated after rainfall. This high water consumption, and the plant's thick leaf litter, together discourage native species from germinating beneath it. The dense stems also prevent waterfowl from moving between nest and water.

PHRAGMITES AUSTRALIS

Phyllostachys aurea

Golden bamboo

Bamboos are actually treelike tropical or semitropical grasses with hollow stems. The golden bamboo seldom produces seed, but its steel-like underground stems (rhizomes) can soon cause gardeners to regret planting it. Very little can stop their growth and they are notorious for penetrating root barriers intended to contain them. Golden bamboo is very commonly planted in a number of climates, but is most likely to become a problem in areas where there are warm summers, especially when planted in rich soil.
ECOLOGICAL IMPACT Golden bamboo is currently found invading in parts of the southern USA, Hawaii, and Queensland, Australia. If the bamboo must be grown, effective root barriers should be established prior to planting. The bamboo should never be planted anywhere near open countryside that could be invaded by the rhizomes, even when root barriers are used. Special care should be taken when growing the bamboo in areas with warm summers.

Pistia stratioides

Water lettuce, shell flower

A floating perennial, this aquatic plant spreads by means of creeping underground stems (stolons). The linked plants form dense mats on the water surface, and these may fragment to form new populations. The soft, thick leaves form a rosette, which could be the origin of the common name "shell flower". The plant is known to facilitate the breeding of two species of mosquito. It is grown for its foliage in water gardens, but noninvasive alternatives exist.
ECOLOGICAL IMPACT The vast mats affect submerged vegetation, fish, and other animals, and impede boat traffic. This species is invasive in at least 40 countries and is a serious weed in Sri Lanka, Ghana, Indonesia, and Thailand. It should not be grown anywhere.

Pittosporum undulatum

Australian mock orange

In colder climates this tree, also called "cheesewood", appears to behave itself, but in warmer areas it is known for being very invasive, with high germination rates for its orange-red seeds, spread by birds.
ECOLOGICAL IMPACT Some reports suggest that the tree is allelopathic, using chemicals in the leaves or roots to poison other plants. It grows quickly, tolerates shade, and creates a dense understorey that can exclude native plants. It is invading in California, Hawaii and other Pacific islands, Jamaica, South Africa, and other warm places where it is cultivated. It invades parts of Australia, although it is native to other parts of that country. Southern European countries should use it with care.

PISTIA STRATIOIDES

Polygala myrtifolia

Sweet pea bush

A native of Africa, this shrub has dense, alternately arranged, evergreen leaves, and flowerheads of purple blooms that resemble those of the pea family, though it is not actually of that family.
ECOLOGICAL IMPACT The shrub currently invades streamside forests in parts of Australia and New Zealand. Areas with similar mild climates, such as temperate parts of Europe and North and South America, should monitor plantings.

Potentilla recta

Sulphur cinquefoil

A herbaceous perennial, this plant has numerous pale yellow flowers on stems with many leaves, but few branches. Only this potentilla species appears to be invasive.
ECOLOGICAL IMPACT This potentilla is very competitive in native grasslands and open forests. It is very tolerant of drought, and few animals seem willing to browse it. For these reasons sulphur cinquefoil appears to be spreading. It is native to the Mediterranean area but can invade in a range of climates, including some that are much wetter in the growing season and much colder in the winter than its native habitat. The plant is found throughout much of North America. Care should be taken when using this species anywhere, to ensure it does not spread.

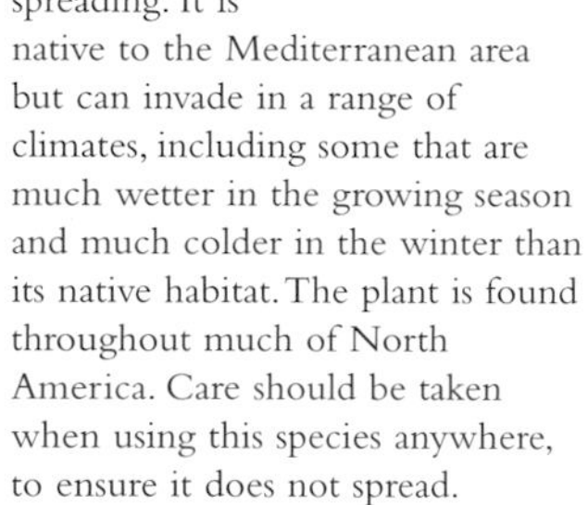

POTENTILLA RECTA

Prunus laurocerasus

Cherry laurel

Birds are attracted to this shrub's purple-black fruits (which also stain pavements where they fall) and they disperse the seeds throughout forests. The shrub has a number of cultivars, but it is the large-leaved wild type that appears to be a problem. It is very shade-tolerant and is commonly encountered in forests near urban areas where temperatures are mild. Fortunately, many of its cultivated varieties appear to be noninvasive.
ECOLOGICAL IMPACT Cherry laurel is invading forests in the Pacific north-west of the USA, as well as the Blue Mountains of Jamaica and parts of Australia, such as New South Wales. The species form should not be grown in areas with mild wet winters.

PRUNUS LAUROCERASUS

Prunus lusitanica

Portugal laurel

Currently less invasive than cherry laurel (*Prunus laurocerasus, see above*), Portugal laurel is a shrub that can be found in forested areas where birds have dispersed the seeds. It is hardier than the cherry laurel, so potentially it can invade a wider latitude range.
ECOLOGICAL IMPACT There is concern that this shrub may be increasing its spread in some parts of the world and, given its tolerance of shady forest understories, it could become more of a problem in forests. It currently invades the Pacific north-west of the USA, as well as the Blue Mountains of Jamaica. The species form should be carefully watched for invasiveness when grown in areas with mild, wet winters.

Psidium cattleyanum

Strawberry guava

At first sight, this shrub is attractive, with its shiny, evergreen leaves, red bark, fragrant flowers, and tasty-looking tart fruits. However, once you have seen it covering a hillside, having obliterated the plants that were there before, it loses its appeal. It can spread by underground stems (rhizomes) and its seeds germinate easily. The flowers do not need to be pollinated to set viable seed.
ECOLOGICAL IMPACT The plant is able to poison nearby plants, as well as outcompete them for water, nutrients, and light. It forms dense thickets in warm areas with high rainfall and tolerates shade, which allows it to invade forests and then alter them through proliferation. This is one of the worst invasive species on several South Pacific islands, where it should not be grown. It also invades moist areas of Florida. If grown in warm, wet climates, extreme caution is needed.

Psoralea pinnata

Blue butterfly bush

This shrubby tree, also called "blue psoralea", has thin branches, needle-like leaves, and fragrant, purple, pea-like flowers in the summer. The plant produces large numbers of seeds that are capable of surviving in the soil for long periods, making it difficult to fully eliminate from an area. Bushfire may stimulate the seeds to germinate.
ECOLOGICAL IMPACT Native to southern Africa, this shrubby tree is invasive in parts of Australia such as Tasmania, Victoria, and Western Australia, as well as New Zealand. Gardeners in western North America and warm-climate countries in southern Europe should take particular care when growing this species.

PUERARIA MONTANA VAR. LOBATA

Pueraria montana *var.* lobata

Kudzu

In the USA they call this climber "the vine that ate the South", a reference to its incredibly fast rate of growth there – up to 30cm (12in) a day. In the South, "kudzu queens" are named for festivals. Kudzu is also sold as a medicinal product. It has a taproot up to 2m (6ft) in length and 20cm (8in) thick, from which up to 30 stems can grow. The hard-coated seeds may persist for years.
ECOLOGICAL IMPACT The climber is found in open areas and at the edges of forests and along roadsides, where it scrambles up trees to reach the light. It covers trees and shrubs with a solid blanket of leaves, girdling their trunks and stems and breaking branches and whole trees with its crushing weight. It is native to Asia but is invasive on several islands of the South Pacific, as well as in south-eastern North America. Kudzu has recently been found in the Pacific north-west of North America, where it appears to have been introduced intentionally. In view of its extreme invasion of south-east USA, it is not wise to grow it anywhere.

Rhamnus cathartica

Common buckthorn

So invasive has this shrub become in the USA that nurseries in parts of that country actively discourage its sale and use. Its fruits are eaten by birds and mammals and may have a laxative effect, facilitating dispersal of the small seeds. The plant is able to tolerate both dry and wet conditions and can be found in open forests, but it also invades grassland. Its relative, *Rhamnus frangula* (alder buckthorn), invades wetlands in temperate areas and should also be used with care.
ECOLOGICAL IMPACT Common buckthorn forms dense stands. It leafs out earlier than most native species in most of the places it invades, giving it a competitive advantage. A native of Eurasia, the shrub is invading most of the northern states in the USA and Canada. Because it is so strongly invasive there, care should be taken when growing it in any comparable temperate climates.

Rhaphiolepis indica

Indian hawthorn

Gardeners tend to regard this tidy, evergreen shrub as very useful for some locations because it is thrifty in its consumption of water.
ECOLOGICAL IMPACT Indian hawthorn has become invasive in at least parts of Australia, notably south-east Queensland, and is of

RHODODENDRON PONTICUM

moderate concern there. When the shrub is grown in areas with similar climates, it should be monitored.

Rhododendron ponticum

Members of the genus *Rhododendron* are rarely invasive, but this shrub is an exception, perhaps because it is capable of spreading vegetatively as well as by seed. Just one flowerhead can produce up to 7,000 seeds. A single plant, with its branches rooting where they touch the soil, can cover 100sq m (1,000sq ft). Combined, the active seed production and rampant vegetative spread make it a formidable plant.
ECOLOGICAL IMPACT Native to south-eastern Europe and western Asia, this shrub is most famous for its invasion of the UK, especially to the west, and Ireland. It forms dense thickets where little else survives, and invades woodlands, as well as heath habitats. Like most rhododendron species, it contains potentially toxic chemicals – these may help repel animals and insects that might otherwise browse it. Feral populations have also been found in New Zealand. There are many other attractive, noninvasive rhododendrons to grow in preference to this species.

Rhodomyrtus tomentosus

Downy rose myrtle

Profuse rose-pink flowers are produced by this evergreen shrub in the spring. Its fruits somewhat

RICINUS COMMUNIS

resemble blueberries and can be used to make jam. The seeds are dispersed by birds and mammals.
ECOLOGICAL IMPACT Seeds fall to the ground under the plants, creating dense thickets when they germinate. The shrub is spreading fast in areas such as Florida, where it is taking over open pine forests and increasing fires. Avoid planting in Florida, Hawaii, and other warm parts of the USA, and in parts of Asia such as Thailand.

Ricinus communis

Castor oil plant

For years children were given the dreaded castor oil for medicinal purposes, and this is indeed the shrub from which it is derived. It also makes an attractive, shrubby tree that has become popular in horticulture in recent years. Some of the most popular cultivated varieties have red leaves. The red, spiky seedpods explode when ripe, scattering the seeds, which may also be dispersed further by rodents and birds. The seeds are very poisonous to humans and the deadly chemical ricin is made from them. The plant tolerates dry sites and poor soils.
ECOLOGICAL IMPACT Castor oil plant can form dense thickets that exclude other species. It was probably native to Africa, but is now found throughout the tropics. It should not be grown in warm climates, even those in which it does not currently invade, such as southern Europe and warm parts of North and South America. In temperate climates it is increasingly grown as an annual, which should be relatively safe.

Robinia pseudoacacia

Black locust, false acacia

Fragrant, pendulous flowers and an interesting rope-like bark distinguish this tree. It is shade-intolerant, very fast-growing, and can be found in forests and grasslands, and along streams. The hard seed coat can make germination difficult, but once a plant is established it can spread 1–3m (3¼–10ft) a year by root suckering. The seeds may remain viable in the soil for up to 10 years, increasing the probability of successful germination.
ECOLOGICAL IMPACT Like many other members of the pea family, this shrub can add nitrogen to the soil, changing the composition of plant communities by driving out nitrogen-sensitive species. Parts of western USA have large infestations of *Robinia pseudoacacia*, as does northern Europe. While the tree is native to a region with high summer rainfall, it also appears to do well in dry climates along rivers or in low-lying areas. Therefore, care should be taken to monitor areas surrounding plantings.

Rosa canina

Dog rose

A native of Europe, dog rose is the most common wild English rose. This shrub spreads from landscape plantings by seed, which is highly viable. While bees pollinate the rose, it is also capable of producing seeds without fertilization.
ECOLOGICAL IMPACT Dog rose can form dense thickets in disturbed areas, open forests, and along streams. In the USA it should not be grown near open countryside in the north-east, midwest, or western states. The shrub is also invasive in southern Australia.

Rosa eglanteria

Eglantine rose, sweet briar

For gardeners this shrub has fragrant leaves, reminiscent of green apples, in addition to its flowers. However, unlike many roses, it is capable of

ROBINIA PSEUDOACACIA

spreading vegetatively by root suckers. Birds and small mammals increase the problem by eating the orange-red fruit and spreading seeds.
ECOLOGICAL IMPACT This rose is so invasive in Australia that it has been highlighted as a species that should not be grown in that country. It has also invaded in many parts of the USA, especially in the west, and in New Zealand. Care should be taken when growing it in any temperate climate.

Rosa multiflora

Multiflora rose

In the USA this shrub is thought to be a highly invasive species. Some of the recorded invasions may be the result of deliberate introductions to provide food and habitat for wildlife, rather than being escapes from ornamental plantings. The flowers of the multiflora rose are small and of the "wild-rose" type. Perhaps the shrub's clusters of red "hips" are its most ornamental trait. One plant can produce hips containing hundreds of thousands of seeds, and the shrub can also spread through the rooting of branches that touch the ground.
ECOLOGICAL IMPACT Open habitats such as grasslands appear to be especially vulnerable to invasion. Currently the multiflora rose primarily invades in midwestern USA. Since it is so invasive there, it should be monitored in other American areas and wherever it is grown in other temperate places.

ROSA MULTIFLORA

ROSA RUGOSA

Rosa rugosa

Hedgehog rose

This rose is an interesting example of a shrub that is invasive after introduction to a new place, but is rare in at least part of its native range – in this case China. It can be found in coastal areas, both in eastern Asia, where it is native, and in places where it invades, such as western North America. It produces numerous small seeds in its persistent "hips" that are dispersed by water and small mammals. The rose also spreads vegetatively.
ECOLOGICAL IMPACT The stems are very thorny and the rose can form dense, impenetrable thickets. Avoid planting in coastal areas, especially in the north-east and western areas of North America, as well as temperate parts of Europe and Australasia. It is very hardy and may be able to spread in colder areas as well.

Ruellia brittoniana

Mexican petunia

A herbaceous species with bright purple flowers, this plant is native to Mexico. It blooms heavily despite heat and humidity, and its many seeds are able to germinate easily. Butterflies favour the flowers.
ECOLOGICAL IMPACT The very traits that make this plant popular with gardeners – its adaptability to heat, drought, and poor soil – make it a persistent weed where it establishes. It is invasive in parts of south-east USA. Because of its competitive nature and its ability to sow seed freely, it should not be used near open countryside.

Salvinia molesta

Giant salvinia, water fern

This highly invasive aquatic plant reproduces by growing new plants from buds on existing plants – these new plants then fragment and float off to begin budding new plants themselves. In favourable conditions the plant can double itself every 2.2 days. Thus, if just one plant is added to a water body, there could be 8,000 plants the first month, 67 million after the second month, and 4,500,000,000 million plants after the fourth month. During times of stress, such as low temperatures or drought, the plants die back to the dormant buds and wait for more favorable times. Mats have been recorded as deep as 1m (3¼ft). Despite this, water fern is still sold for water gardens.
ECOLOGICAL IMPACT The plant can colonize lakes, streams, rivers, wetlands, and virtually any aquatic system. Its dense mats block all sunlight to the submerged plants fed upon by fish, as well as decreasing the oxygen needed by fish. In the USA it is federally listed as a noxious weed. Boats should always be checked very carefully after removal from any water where infestation might be present.

Sapium sebiferum

Chinese tallow tree

The seeds of this tree have a waxy covering that can be used to make candles or soap – hence the common name. It is also one of the few trees to produce autumn colour in warmer-climate areas such as the southern USA, and is therefore a popular ornamental. It spreads aggressively by suckering and also produces highly viable and easily germinated seed that is spread by birds as well as by water.

RUELLIA BRITTONIANA

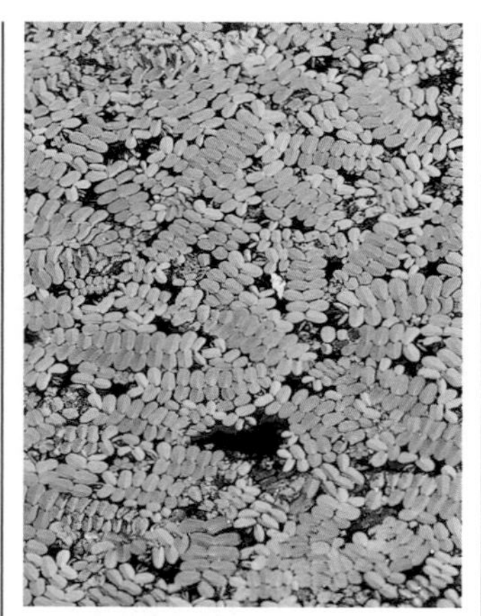

SALVINIA MOLESTA

ECOLOGICAL IMPACT The tree can outcompete native plants in a number of habitats, including coastal prairies and wetlands, as well as upland areas. A native of China, it currently invades parts of Australia, South Africa, and southern and western USA. Gardeners in warmer parts of Europe and the temperate areas of South America should not plant it, or they should at least monitor it carefully if they do.

Schefflera actinophylla

Australian ivy palm

People in temperate areas may know this species as a potted plant found in bank lobbies, but in warmer climates it is an outdoor tree. The fruits, which are attractive to birds, are held above the branches on erect groups of stalks that resemble red octopuses. After a fairly long juvenile period, delaying the onset of fruiting, this plant can become very invasive.
ECOLOGICAL IMPACT This plant can colonize in undisturbed forests, and in Florida, USA, it appears to be shading out an endangered species. It is native to Australia and New Guinea, but invades a number of other South Pacific islands, Christmas Island in the Indian Ocean, and Hawaii, as well as Florida. It should not be grown on subtropical or tropical islands and should be monitored wherever it is grown outdoors.

Schinus terebinthifolius

100

Brazilian pepper tree

The fruits of this plant, actually a shrub, are indeed peppery tasting and are sold as pink peppercorns in gourmet shops. However, in large doses, or to some small animals, they can be intoxicating and even poisonous. The plants begin producing the seeds at a young age and many species of birds and small mammals disperse them.
ECOLOGICAL IMPACT Plants that grow nearby are poisoned, and stands of Brazilian pepper generally lack an understorey. The shrub has invaded in many places around the world, but may be at its worst in Florida and Hawaii. Also affected are many South Pacific islands, Indian Ocean islands, Bermuda, the Bahamas, and Australia. Because it is so invasive and has serious effects in open countryside, it should not be grown as an ornamental.

Senecio elegans

Purple groundsel

The pretty, purple, daisy-like flowers of this low-growing, annual herbaceous plant make it popular with gardeners. Less popular is its ability to seed itself freely.
ECOLOGICAL IMPACT Reports of the plant's invasive ability are fairly recent, so it is unclear how widespread it might be or what impacts it might have. It is native to South Africa but is invading in New Zealand, Australia, and California. Care should be taken to prevent it from escaping gardens, especially in Mediterranean-type climates.

Senna alata

Empress candle plant

With bright yellow flowers held high in erect flowerheads, this makes an attractive, tall shrub. It does not spread vegetatively but is quick to produce fruit, which allows the populations to build up rapidly.
ECOLOGICAL IMPACT A colonizer of open areas, it can form dense thickets and is especially aggressive in areas where water is not limited. Empress candle plant is native to Mexico and South America, but this species has become invasive in parts of the South Pacific, Australia, Africa, and Asia.

Senna pendula

Climbing cassia

Actually a sprawling shrub, climbing cassia produces flowers and fruits throughout the year. This continual seed production has led it to increase its range and abundance quickly after introduction. It does not reproduce vegetatively.
ECOLOGICAL IMPACT The shrub displaces native vegetation in open forests and coastal areas and can climb over adjacent vegetation once established. It is invasive in several parts of the world, including Hawaii, parts of Australia such as Queensland, the Bahamas, and south-western Europe.

Sorbus aucuparia

Mountain ash, rowan

Birds love the mountain ash tree, especially when the orange-red fruits have fermented a little after cold weather sets in – they can be found having raucous parties in the canopies. The seeds are unable to germinate easily, and the plant does not spread vegetatively, both of which prevent it from becoming a more serious problem.
ECOLOGICAL IMPACT The tree is very common in forests within and surrounding cities in the temperate areas where it is grown, including the northern part of North America. The numbers are probably due to the sheer volume of seeds dispersed by birds. Care should be taken when planting mountain ash near forests in temperate climates.

Sparaxis bulbifera

Harlequin flower

While there are other invasive *Sparaxis* species, this bulbous plant, also called *Ixia bulbifera*, is the most common one. It has cream and purple petals. Small corms grow off its main corm, and spread mostly occurs when the small corms become detached from the parent.
ECOLOGICAL IMPACT The plant is invading wetlands in parts of southern and western Australia. All *Sparaxis* species are native to South

SPARAXIS BULBIFERA

SPHAEOPTERIS COOPERI

Africa and are easy to grow in Mediterranean-type climates. Seeds should be removed from cultivated plants to slow dispersal and spread.

Spartium junceum

Spanish broom

Interesting, spiky green branches and yellow, pea-like flowers identify this shrub as one of the group known as "brooms". The green branches photosynthesize well, so this species is essentially leafless. The green stems allow it to grow year-round in places where winter temperatures are mild. The plant is very tolerant of different soil types, it reproduces easily from seed, and it resprouts when cut back.
ECOLOGICAL IMPACT Spanish broom adds nitrogen to soil, which can be a disadvantage for some native species. It is invasive in the west coast of the USA, South Africa, and Australia. It appears to be most invasive in Mediterranean-like climates, where the mild winters with ample rainfall may give it a competitive advantage.

Spathodea campanulata

African tulip tree

The flowers of this tree are a spectacular red-orange and attract birds to pollinate them. However, the tree is very invasive in wet forests and each fruit pod contains up to 500 dry, wind-dispersed seeds. The tree grows very rapidly, up to 2m (6ft) a year, and produces flowers and fruits year-round. The plant may also spread vegetatively through root sprouts.
ECOLOGICAL IMPACT The young plants tolerate shade and can form dense thickets in forest understories. As the name implies, the species is native to tropical Africa and is widely planted throughout the tropics. It is invasive in most of the places it is grown, and is a serious problem in the South Pacific, including tropical Australia. Although it is a beautiful tree, planting it near open countryside should be done with extreme caution.

Sphaeopteris cooperi

Australian tree fern

Also known as *Cyathea cooperi*, this tree fern is perhaps the most common, and also the most hardy, tree fern used in gardens. It can grow up to 12m (40ft) tall, with a spray of huge fern leaves at the top. Like other ferns, it spreads by means of tiny, wind-borne spores that often travel long distances.
ECOLOGICAL IMPACT The fern invades moist forests and rainforests and in some places has dominated high-quality forest. It is native to eastern Australia, but is invading in western Australia, where it is cultivated, and in Hawaii and French Polynesia. It is likely to be invasive in warm, wet climates.

Sphagneticola trilobata

Wedelia

Yellow flowers are produced all year round by this ground-covering herbaceous plant, also known as *Wedelia trilobata*. Seed is rarely set, which might suggest an ideal garden plant. Individuals with mature fruits have been seen, but it appears that spread is more likely to occur by fragments of the stems quickly rooting after plants are transferred, either accidentally or intentionally, into forest habitats.
ECOLOGICAL IMPACT Wedelia grows rapidly and can form a very dense cover, choking out native species and preventing their regeneration. The plant grows in moist or dry forests. It is native to tropical America but has spread throughout the South Pacific islands and warm, humid parts of North America. Garden waste should not be spread near open countryside.

Spiraea japonica

Japanese spiraea

Masses of small, bright pink flowers have made this attractive shrub a popular landscape species. It flowers profusely and the seeds are tiny, wind-dispersed, and germinate easily. A variety of light and soil conditions are tolerated.
ECOLOGICAL IMPACT The plant is fast-growing and displaces native species. It invades open forests in the USA, central Europe, and Australia. Temperate areas with ample summer rainfall appear to be most at risk. There are many garden cultivars, so seek advice as to which ones are less invasive.

Symphoricarpos albus

Snowberry

A modest shrub that was introduced into several countries, snowberry is an ornamental native to western USA. It has small, pinkish flowers in the spring and produces white berries throughout the winter. In the UK it has also been used to enhance game habitat, although birds do not usually eat the fruits.
ECOLOGICAL IMPACT Snowberry is mostly invasive in the UK and some parts of northern Europe, though it has the potential to be invasive following disturbance in other locations. Care should be exercised when planting it near open countryside in any part of the world where it is not native.

Syngonium podophyllum

Goosefoot

In temperate areas, and in gardens where the climate is relatively warm, this climber is used as a potted indoor plant. There are at least 10 cultivated varieties, but there is no information about whether some are less invasive than others. The plant spreads mostly through rapid growth and, if allowed to climb, can develop stems that are several centimetres thick. The climber rarely produces seeds, but dispersed stem fragments are able to root quickly.
ECOLOGICAL IMPACT Goosefoot can smother trees and shrubs and girdle them with its stem. It is native to Central America and invades in many South Pacific islands, Christmas Island, Florida in the USA, and South Africa. It should not be grown near open countryside in warm, wet climates, and garden waste should be disposed of appropriately. If kept under control it can make an effective ground cover plant, but it should not be allowed to climb.

SPATHODEA CAMPANULATA

Tamarix aphylla
Athel

Some people lump this evergreen shrub together with *Tamarix ramosissima* (*see below*), and some name all three as *Tamarix chinensis*. Traditionally they were separate species, and that treatment is observed here. Athel is tolerant of many soils, including alkaline and salty soils. It begins to produce seed when only a year old and is capable of producing hundreds of thousands of seeds. In some of the areas where it is invasive it does not appear to produce seed, instead spreading vegetatively. Athel is less invasive than *Tamarix ramosissima*.
ECOLOGICAL IMPACT Athel has become very invasive in central Australia over the last 30 years, and invades to a lesser degree in south-western USA. It should not be grown in any region with a warm, Mediterranean-type climate.

Tamarix ramosissima 100
Tamarisk

Riverbank ecology in southwest USA has been completely altered by this shrub. Each plant produces hundreds of thousands of tiny seeds, which are very short-lived but germinate easily. If the seeds land in a site with adequate water, such as a riverside or wetland, germination occurs, followed by very rapid growth. Tamarisk has a deep taproot that enables it to survive in very dry places once it is established, and it can transpire huge volumes of water through tiny pores on the leaves.
ECOLOGICAL IMPACT Tamarisk's high consumption of water has resulted in a lowering of the water table in numerous locations, causing native species with shallower roots to die out from the community. Native to central Asia and northern Africa, the plant has been extremely invasive in south-western North America, South Africa, and central Australia. It has the potential to do the most harm in warm, arid regions, but planting should be avoided everywhere.

TAMARIX RAMOSISSIMA

Terminalia catappa
Indian almond

Often grown as a shade tree in warm areas, this species has very large leaves. The almonds are edible and the tree may be planted in some areas as a food source. Indian almond is fast-growing and begins to produce its fruit at a young age. The plants are salt-tolerant, and the seeds are spread by ocean current as well as by various mammals.
ECOLOGICAL IMPACT Indian almond primarily invades coastal areas. It is native to Indonesia, but is now found throughout the South Pacific as well as parts of Florida and the Caribbean.

Tournefortia argentea
Tree heliotrope

A woody member of the forget-me-not family, this shrub is fast-growing and begins to flower and fruit all year round at an early age. The fruits may spread by dropping into and floating on ocean currents.
ECOLOGICAL IMPACT This shrub can crowd coastal areas, although it may have some beneficial uses for wildlife – there are reports that it provides a nesting habitat for seabirds. Tree heliotrope is native to parts of tropical Asia and Australia, but it is invasive in the South Pacific, where it has been used as an ornamental. It should be planted with care on tropical islands and monitored so that any plants that become established can be removed early.

Tradescantia fluminensis
Wandering Jew

Commonly used as a houseplant and in patio container gardens, this herbaceous species has become an important pest in some places, in both agricultural and uncultivated areas. The plant covers the ground in the form of dense mats as much as 60cm (24in) thick. It spreads mostly by vegetative means, with fragments rooting at the nodes and forming new mats.
ECOLOGICAL IMPACT This plant smothers native species and prevents regeneration in moist forests, especially if there is some disturbance. In New Zealand it is a very serious problem, associated with decreased diversity of invertebrate species and a reduction in seedlings of native species. Native to tropical South America, it should not be grown where escape is possible in south-east USA, New Zealand, and New South Wales, Australia, or in other humid tropical areas where it does not yet invade.

Tradescantia spathacea
Moses-in-the-cradle

The modified leaves (bracts) beneath the flowerhead of this herbaceous plant resemble a cradle. That oddity, and the stiff, purplish leaves, have made this species – also called three-men-in-a-boat – a popular potted and ground-cover plant. It grows well, spreading by seeds or by reestablishing itself after being discarded as garden waste.
ECOLOGICAL IMPACT Moses-in-the-cradle invades in dry, shady areas and can even colonize rocky outcrops. The plant is poisonous and can cause skin irritation when handled. It is native to tropical America and is invasive throughout the South Pacific and south-east USA. It should always be carefully monitored when grown anywhere near open countryside.

Tradescantia zebrina
Wandering Jew

This herbaceous plant shares the common name of its cousin, *Tradescantia fluminensis* (*see above*). Purple and green leaves and a hardy constitution make this a popular ground-cover plant in warm areas. It spreads mostly by vegetative means, with fragments rooting at the nodes and forming dense mats.
ECOLOGICAL IMPACT This plant's ability to thrive under difficult conditions also helps to make it invasive – it can be found in dry

TRADESCANTIA ZEBRINA

areas and on steep banks, where its dense carpets smother native plants. It is native to Mexico and is a pest in Queensland, Australia, many South Pacific islands, and the Galapagos Islands. It should always be carefully monitored when grown near open countryside.

Ulmus pumila

Siberian elm

Reproduction of this tree species is either by the wind-blown seeds, which germinate easily, or, more commonly, by vegetative root sprouts. The tree can form dense thickets in a relatively short time because of its rapid growth. It is resistant to Dutch elm disease and is considered to be one of the hardiest of the elms, able to survive both drought and cold temperatures.
ECOLOGICAL IMPACT This elm can quickly invade and dominate grasslands and it has been shown to hybridize with native elms. It is invading in several spots in North America, especially in dry areas like New Mexico, where it is a noxious weed, and the upper midwest. Because it is so adaptable it should be carefully monitored in Europe, Australia, and South Africa.

Verbena bonariensis

Purple top

Nodding heads of tiny flowers have made this a very popular perennial herbaceous species in many places. Unfortunately, it also produces numerous tiny seeds that are very long-lived for small seeds – they have been known to persist for several years in areas where the adult plants have long been controlled. The seeds germinate easily and young plants can begin producing seeds within weeks.
ECOLOGICAL IMPACT Purple top can be very aggressive in grasslands and survives very well in dry sites. It is native to South America and invades in south-east Australia, South Africa, and other areas with dry climates. It has also been found in areas with summer rainfall, such as south-eastern USA. Care should be taken when planting it virtually anywhere, but especially where it may be able to invade grasslands.

VERBENA BONARIENSIS

VINCA MAJOR

Veronica persica

Persian speedwell

Liked for its sky-blue flowers on wiry stems, this herbaceous plant is sometimes used in container gardens. It is an annual species and reproduces by seed. The plants usually die off after the seed has been produced.
ECOLOGICAL IMPACT Persian speedwell forms dense ground cover that can affect the regeneration of native species, but it is more often found in turf and gardens. The plant is invasive in many parts of the USA, Europe, and Australia. Efforts should be made to prevent the widespread distribution of seeds.

Vinca major

Greater periwinkle

A popular evergreen ground-cover herbaceous plant with bluish-purple flowers, greater periwinkle is also an aggressive climber. Rarely does this species, or its relative, the lesser periwinkle (*Vinca minor*), spread by seed. Greater periwinkle appears to be more likely to establish in less disturbed and drier places than *Vinca minor*. In stressful conditions it may die back, then resprout when the weather improves.
ECOLOGICAL IMPACT In areas where the species persists from earlier plantings, or where fragments are deposited, it can smother native plants. It is very shade-tolerant and streamside corridors may be especially vulnerable. It is native throughout Europe to North Africa. Planted areas of either periwinkle species should be contained effectively to prevent spread from their original plantings. Periwinkle appears to grow especially well in Mediterranean climates.

Watsonia meriana *var.* bulbillifera

Bulbil watsonia

Native to South Africa, bulbil watsonia is a bulbous plant that appears to invade only in several parts of Australia. There are actually several watsonias capable of being invasive, but this one is probably the worst. Not only does it have an underground corm (bulb), it also produces bulbils at the bases of shoots that are easily dislodged when the plants are disturbed.
ECOLOGICAL IMPACT It grows along creeks and in open areas and can form dense colonies that exclude native species. It does not spread rapidly, but it can be very difficult to control once it is established, especially when it is along watercourses where herbicide cannot be used. Animals do not graze watsonia, and by eating everything around it they gradually allow it to colonize the pasture. Only Australia is adversely affected at present, but it should be used with great caution when grown in similar climates, such as southern Europe, western North America, or temperate South America.

Wisteria floribunda

Japanese wisteria

One of the few wisterias to have fragrant flowers, this hardy climber grows very quickly. It has creeping underground stems (stolons) that may root, and it also produces large seeds that can be spread by water.
ECOLOGICAL IMPACT The climber overtops trees 20m (70ft) high, forming dense thickets where little else grows. Its stems twine clockwise, forming girdles around tree trunks that may grow to many centimetres thick. The seeds are poisonous to many mammals. This plant is found throughout south-east USA, where it is very invasive. There do not appear to be reports of it invading in other temperate areas, but established plants should be carefully monitored.

Wisteria sinensis

Chinese wisteria

Also called *Wisteria chinensis*, this is a fast-growing, woody climber. Fresh seeds germinate easily, but the young plants are slow to flower. The plant can "fix" nitrogen in its root nodules, which helps it to colonize poor soils. It spreads through the rooting of stems on the ground, as well as by producing seeds that can be spread by water.
ECOLOGICAL IMPACT This climber can quickly cover a forest. Its seeds are poisonous to mammals. It is very invasive throughout south-east USA and should not be grown where it can be dispersed into open countryside. There is concern that it could be invasive in mountainous parts of Hawaii, but it is not currently known to be invasive in other temperate areas, though it should be carefully monitored.

WISTERIA SINENSIS

Zantedeschia aethiopica

Arum lily

Gardeners prize the long-stemmed flowerheads of this bulbous plant, with their surrounding showy white modified leaves (bracts). However, in warm climates birds disperse the seeds and the plants can become a problem, especially in moist places. The plants also spread from fragments of the thick roots, so care should be taken to collect these when plants are disturbed.
ECOLOGICAL IMPACT Plants sometimes wash downriver and reestablish following floods, so streamside infestations should be controlled. All of the plant is poisonous and it has been known to kill children and animals. Native to South Africa, it exists as a garden escape in moist areas in warm climates in many parts of the world. It is becoming an increasingly serious problem in Australia and should not be planted there.

USEFUL ADDRESSES

SPONSORS OF THIS BOOK

Botanic Gardens Conservation International (BGCI)
Descanso House
199 Kew Road
Richmond
Surrey TW9 3BW, UK
Tel: +44 (0)20 8332 5953
E-mail: info@bgci.org
www.bgci.org

AUSTRALIA

Botanic Gardens Trust (Royal Botanic Gardens), Sydney
Mrs Macquaries Road
Sydney
NSW 2000, Australia
Tel: +61 (0)2 9231 8111
E-mail: rbg@rbgsyd.gov.au
www.rbgsyd.gov.au

Royal Botanic Gardens, Melbourne
Birdwood Avenue
South Yarra
Vic 3141, Australia
Tel: +61 (0)3 9252 2300
E-mail: webmaster@rbg.vic.gov.au
www.rbg.vic.gov.au

BELGIUM

National Botanic Garden of Belgium
Domaine de Bouchout
B-1860 Meise, Belgium
Tel: +32 (0)2 269 3905
E-mail: office@br.fgov.be
www.br.fgov.be

BRAZIL

Jardim Botânico do Rio de Janeiro
Rua Pacheco Leao 915
22460-030 Rio de Janeiro, Brazil
Tel: +55 (0)21 294 6012/512 2077
E-mail: jardimbotanico@jbrj.gov.br
www.jbrj.gov.br

CANADA

Royal Botanical Gardens
680 Plains Road West
Hamilton/Burlington
Ontario L7T 4H4, Canada
Tel: +1 (905) 527 1158
E-mail: info@rbg.ca
www.rbg.ca

Jardin Botanique de Montréal
4101 Sherbrooke East
Montreal
Quebec H1X 2B2, Canada
Tel: +1 (514) 872 1400
E-mail: jardin.botanique@ville.montreal.qc.ca
www2.ville.montreal.qc.ca/jardin/jardin

CHINA

Wuhan Botanical Garden
The Chinese Academy of Sciences
Moshan, Wuchang District, Wuhan
Hubei 4300, China
Tel: +86 27 7409218
www.whiob.ac.cn

GERMANY

Botanischer Garten und Botanisches Museum Berlin-Dahlem (BGBM)
Königin-Luise-Strasse 6–8
14191 Berlin, Germany
Tel: +49 (0)30 838 50100
www.bgbm.org

INDONESIA

UPT Balai Pengembangan Kebun Raya-LIPI
Jalan Ir. H. Juanda No. 13
PO Box 309
Bogor
16003 Java, Indonesia
Tel: +62 (0)251 322 187/321 657

MEXICO

Jardín Botánico del Instituto de Biología
Universidad Nacional Autónoma de México (UNAM)
A.P. 70-614, C.P. 04510
Coyoacán
Mexico D.F., Mexico
Tel: +52 (01) 56161297
www.ibiologia.unam.mx

NETHERLANDS

Utrecht University Botanic Garden
Budepestlaan 17, De Uithof
Utrecht, The Netherlands
Tel: +31 30 253 2143
E-mail: botgard@bio.uu.nl
www.botanischetuinen.uu.nl

NEW ZEALAND

Wellington Botanic Garden
Wellington City Council
Glenmore Street
Wellington, New Zealand
Tel: +64 4 801 3071
E-mail: info@wcc.govt.nz
www.wcc.govt.nz/recreation/gardens/botanic

SINGAPORE

Singapore Botanic Gardens
National Parks Board
Cluny Road
Singapore 259 569
Tel: +65 471 9933/34/43
E-mail: nparks_mailbox@nparks.gov.sg
www.nparks.gov.sg

SOUTH AFRICA

Kirstenbosch National Botanical Garden
National Botanical Institute
(renamed the South African Biodiversity Institute)
Rhodes Drive, Newlands
Cape Town, South Africa
Tel: +27 (0)21 799 8783/8620
E-mail: info@nbict.nbi.ac.za
www.nbi.ac.za

SPAIN

Jardín Botánico Canario "Viera y Clavijo"
Apartado de Correos 14
Tafira Alta, Las Palmas
Gran Canaria, Islas Canarias, Spain
Tel: +34 928 353604/353342
www.step.es/jardcan

UK

Royal Botanic Garden Edinburgh
20A Inverleith Row
Edinburgh EH3 5LR, UK
Tel: +44 (0)131 552 7171
E-mail: info@rbge.org.uk
www.rbge.org.uk

Royal Botanic Gardens, Kew
Richmond,
Surrey TW9 3AB, UK
Tel: +44 (0)20 8332 5655
E-mail: info@kew.org
www.rbgkew.org.uk

USA

Brooklyn Botanic Garden
1000 Washington Avenue
Brooklyn
New York 11225, USA
Tel: +1 (718) 623 7200
E-mail: postmaster@bbg.org
www.bbg.org

Chicago Botanic Garden
1000 Lake Cook Road
Glencoe
Illinois 60022, USA
Tel: +1 (847) 835 5440
www.chicagobotanic.org

Missouri Botanical Garden
Gray Summit
St. Louis
Missouri 63166-0299, USA
Tel: +1 (314) 577 9400
www.mobot.org

OTHER BOTANICAL GARDENS AND ARBORETA

The botanical gardens and arboreta listed below are included either by special request of contributors and consultants to this book or because there are references to them in the text.

AUSTRALIA

Kings Park and Botanic Garden
Botanic Gardens and Parks Authority
West Perth, WA 6005, Australia
Tel: +61 (0)8 9480 3600
E-mail: enquiries@bgpa.wa.gov.au
www.kpbg.wa.gov.au

FRANCE

Jardin des Plantes
Service des Cultures MNHN
43 Rue Buffon
F-750005 Paris, France
Tel: +33 43 36 12 33

INDIA

Indian Botanic Garden
Botanical Survey of India
PO Botanic Garden
Howrah 711 103
West Bengal, India
Tel: +91 67 32 31 35

IRELAND

National Botanic Gardens
Glasnevin
Dublin 9, Ireland
Tel: +353 1 8377596/8374388

ITALY

Botanical Garden
University of Florence
"Giardino dei Semplici"
Via P.A. Micheli, 3
50121 Florence, Italy
Tel: +39 (0) 55 2757402
E-mail: ortbot@unifi.it
www.unifi.it/unifi/msn/ortus/obfr_eng.htm

Botanical Garden
University of Padua
Via Orto Botanico, 15
Padua, Veneto, Italy
Tel: +39 (0) 49 656614
www.cbft.unipd.it/pdtour/garden.html

Botanical Garden
University of Pisa
Via Luca Ghini 5
I-56100 Pisa, Italy
Tel: +39 (0) 50 551345

MEXICO

El Charco del Ingenio
San Miguel de Allende
Guanajuato 37700, Mexico
Tel: +52 (415) 154 4715
E-mail: charco@laneta.apc.org
www.laneta.apc.org/charco

Jardín Botánico "Ignacio Rodriguez Alconedo"
Av. San Claudio s/n Col San Manuel Ciudad
Universitaria Edificio No 76
Puebla, Mexico CP 72590
Tel: +52 (12) 244 3938

NEW ZEALAND

Christchurch Botanic Garden
Rolleston Avenue
Christchurch, New Zealand
Tel: +64 (0)3 941 7590
E-mail: LeisureandParks@ccc.govt.nz
www.ccc.govt.nz/parks/BotanicGardens

SOUTH AFRICA

Durban Botanic Gardens
70 St Thomas Road
Durban 4001, South Africa
Tel: +27 (0)31 201 1303
www.durban.gov.za/parks/

Lowveld National Botanic Garden
Nelspruit, South Africa
Tel: +27 (0)13 752 5531
E-mail: curator@glow.co.uza
www.nbi.ac.za/lowveld/mainpage.htm

SPAIN

Jardín Botánico de Córdoba
Avda de Linnao s/n
Apdo. 3048
14080 Cordoba, Spain
Tel +34 (9) 57 200 355
E-mail: jardinbotcord@cod.servicom.es

THAILAND

Nong Nooch Tropical Garden
34/1 Sukhumvit Hwg
Najomtien, Sattahip
Chonburi 20250, Thailand
Tel: +66 38 709358
www.pttaya.com/nong nooch

UK

Cambridge University Botanic Garden
Cory Lodge, Bateman Street
Cambridge CB2 1JF, UK
Tel: +44 (0)1223 336265
E-mail: enquiries@botanic.cam.ac.uk
www.botanic.cam.ac.uk

Chelsea Physic Garden
66 Royal Hospital Road
London SW3 4HS, UK
Tel: +44 (0)20 7352 5646
E-mail: enquiries@chelseaphysicgarden.co.uk
www.chelseaphysicgarden.co.uk

Eden Project
Bodelva
St. Austell
Cornwall PL24 2SG, UK
Tel: +44 (0)1726 811911
E-mail: info@edenproject.com
www.edenproject.com

Royal Botanic Gardens
Wakehurst Place
Ardingly
Near Haywards Heath
West Sussex RH17 6TN, UK
Tel: +44 (0)1444 894066
E-mail: wakehurst@kew.org
www.rbgkew.org.uk

Sir Harold Hillier Gardens and Arboretum
Jermyns Lane
Ampfield, Romsey
Hampshire SO51 0QA, England
Tel: +44 (0)1794 368787
E-mail: info@hilliergardens.org.uk
www.hilliergardens.org.uk

University of Oxford Botanic Garden & Harcourt Arboretum
Rose Lane
Oxford, OX1 4AX, UK
Tel: +44 (0)1865 286690
E-mail: postmaster@botanic-garden.ox.ac.uk
www.botanic-garden.ox.ac.uk

Westonbirt Arboretum
Forestry Commission
The Forestry Authority
Tetbury, Gloucestershire GL8 8QS, UK
Tel: +44 (0) 1666 880 220
E-mail: enquiries@forestry.gov.uk
www.forestry.gov.uk/westonbirt

USA

Amy BH Greenwell Ethnobotanical Garden,
82-6188 Mamalahoe Highway
Captain Cook,
Hawaii 96704, USA
Tel: +1 (808) 323 3318
E-mail: pvandyke@bishopmuseum.org
www.bishopmuseum.org

Arnold Arboretum of Harvard University
125 Arborway
Jamaica Plain
Massachusetts 02130-3500, USA
Tel: +1 (617) 524 1718
E-mail: abweb@anarb.harvard.edu
www.arboretum.harvard.edu

Asheville, The Botanical Gardens at
151 W.T. Weaver Boulevard
Asheville
North Carolina 28804-3414, USA
Tel:+1 (828) 252 5190
E-mail: botgardens@main.nc.us
www.ashevillebotanicalgardens.org

Atlanta Botanical Garden (ABG)
1345 Piedmont Avenue NE
Atlanta
Georgia 30309, USA
Tel: +1 (404) 876 5859
E-mail: info@atlantabotanicalgarden.org
www.atlantabotanicalgarden.org

Bartram's Garden
54th Street and Lindbergh Boulevard
Philadelphia
Pennsylvania 19143, USA
Tel: +1 (215) 729 5281
E-mail: explore@bartramsgarden.org
www.bartramsgarden.org

Berry Botanic Garden
11505 SW Summerville Avenue
Portland, Oregon 97219, USA
Tel: +1 (503) 636 4112
E-mail: register@berrybot.org
www.berrybot.org

Fairchild Tropical Botanic Garden
11935 Old Cutler Road
Miami, Florida 33156, USA
Tel: +1 (305) 667 1651
E-mail: webmaster@fairchildgarden.org
www.fairchildgarden.org

Huntington Botanical Gardens
1151 Oxford Road
San Marino
California 91108, USA
Tel: +1 (626) 405 2100
E-mail: webmaster@huntington.org
www.huntington.org

Lady Bird Johnson Wildflower Center
4801 La Crosse Avenue
Austin
Texas 78739, USA
Tel: +1 (512) 292 4100
www.wildflower.org

Mercer Arboretum & Botanic Gardens
22306 Aldine Westfield Road
Humble
Texas 77338-1071, USA
Tel: +1 (281) 443 8731
E-mail: webmaster@cp4.hctx.net
www.cp4.hctx.net/mercer

Montgomery Botanical Center
11901 Old Cutler Road
Miami
Florida 33156, USA
Tel: +1 (305) 667 3800
E-mail: montgome@fiu.edu
www.montgomerybotanical.org

National Tropical Botanical Garden (NTBG)
3530 Papalina Road
Kalaheo, Kauai
Hawaii 96741, USA
Tel: +1 (808) 332 7324
www.ntbg.org

The New York Botanical Garden
Bronx
New York 10458, USA
Tel: +1 (718) 220 6504
www.nybg.org

Polly Hill Arboretum
809 State Road
West Tisbury
Massachusetts 02575, USA
Tel: +1 (508) 693 9426
E-mail: Postmaster@pollyhillarboretum.org
www.pollyhillarboretum.org

Rancho Santa Ana Botanic Garden
1500 N. College Avenue
Claremont
California 91711-3157, USA
Tel: +1 (909) 625 8767
www.rsabg.org

Regional Parks Botanical Garden
Tilden Regional Park
Berkeley
California 94708-2396, USA
Tel: +1 (415) 841 8732
E-mail: info@nativeplants.org
www.nativeplants.org

Santa Barbara Botanic Garden
1212 Mission Canyon Road
Santa Barbara
California 93105, USA
Tel: +1 (805) 682 4726
E-mail: info@sbbg.org
www.santabarbarabotanicgarden.org

Strybing Arboretum and Botanical Gardens
Strybing Arboretum Society
9th Avenue at Lincoln Way
San Francisco
California 94122, USA
www.strybing.org

United States National Arboretum
3501 New York Avenue, NE
Washington, DC 20002-1958, USA
Tel: +1 (202) 245 2726
California 94122, USA
www.usna.usda.gov/
Floral and Nursery Plants Research Unit
www.usna.usda.gov/Research

University of California Botanical Garden
200 Centennial Drive
5045 Berkeley
California 94720-5045, USA
Tel: +1 (510) 643 2755
E-mail: mrmr@uclink.berkeley.edu
www.botanicalgarden.berkeley.edu

University of Wisconsin-Madison Arboretum
1207 Seminole Highway
Madison
Wisconsin 53711-3726, USA
Tel: +1 (608) 263 7888
http://wiscinfo.doit.wisc.edu/arboretum

Waimea Arboretum and Botanical Garden
59–864 Kamehameha Highway
Haleiwa
Hawaii 96712, USA
Tel: +1 (808) 638 8655

CONSERVATION CONTACTS

Australian Network for Plant Conservation (ANPC)
GPO Box 1777
Canberra,
ACT 2601, Australia
Tel: +61 (0)2 6250 9509
E-mail: anpc@anbg.gov.au
www.anbg.gov.au

Bamboo of the Americas (BOTA)
Sponsored by the American Bamboo Society (ABS)
1541 Sunset Drive
Vista
California 92081-6532, USA
Tel: +1 (760) 726 4038
www.bambooofthea mericas.org

Barbara Everard Trust for Orchid Conservation
(UK-based charitable trust administered by the Orchid Society of Great Britain; aims to help conserve private collections of orchids whose owners can no longer care for them.)
www.orchid-society-gb.org.uk/conservation.html

BG-BASE (UK) Ltd
c/o Royal Botanic Garden Edinburgh, *see p.473*
E-mail: k.walter@rbge.org.uk
http://rbg-web2.rbge.org.uk/bg-base

BioNET-International
Technical Secretariat
Bakeham Lane
Egham
Surrey TW20 9TY, UK
Tel: +44 (0)1491 829036/7/8
E-mail: bionet@bionet-intl.org
www.bionet-intl.org

California Native Plants Society (CNPS)
2707 K Street Suite 1
Sacramento
California 95816.5113, USA
Tel: +1 (916) 447 2677
E-mail: cnps@cnps.org
www.cnps.org

Canadian Botanical Conservation Network
Royal Botanical Gardens
PO Box 399
Hamilton, Ontario L8N 3H8, Canada
Tel: +1 (905) 527 1158 ext 309
E-mail: cbcn@rbg.ca
www. rbg.ca/cbcn/en/

Center for Plant Conservation
P.O. Box 299
St. Louis
Missouri 63166, USA
Tel: +1 (314) 577 9450
E-mail: cpc@mobot.org
www.centerforplantconservation.org

Center for Urban Horticulture
University of Washington
Box 354115
Seattle
Washington 98195-4115, USA
Tel: +1 (206) 543 8616
E-mail: urbhort@u.washington.edu
www.urbanhort.org

CITES (Convention on International Trade in Endangered Species of Wild Flora and Flora)
International Environment House
Chemin des Anémones
CH-1219 Châtelaine,
Geneva, Switzerland
Tel: +41 22 917 8139/40
E-mail: cites@unep.ch
www.cites.org
CITES appendices
www.cites.org/eng/append/appendices.html
CITES-listed species database
www.cites.org/eng/resources/species.html

Conservation International
1919 M Street, NW Suite 600
Washington, DC 20036, USA
Tel: +1 (202) 912-1000
www.conservation.org

Convention on Biological Diversity (CBD)
393 Saint Jacques Street, Suite 300
Montreal
Quebec H2Y 1N9, Canada
Tel: +1 (514) 288 2220
E-mail: secretariat@biodiv.org
www.biodiv.org

Convention on the Conservation of European Wildlife and Natural Habitats (also known as the Bern Convention)
Council of Europe, *see below*
http://conventions.coe.int/Treaty/EN/treaties/html/conservation.104.htm

Council of Europe
Avenue d'Europe
67075 Strasbourg Cedex
Tel: +33 (0) 3 8841 20 00
E-mail: infopoint@coe.int
www.coe.int

Counterpart International
(formerly The Foundation for the Peoples of the South Pacific)
1200 18th Street, NW
Suite 1100
Washington, DC 20036, USA
Tel: +1 (202) 296 9676
E-mail: communications@counterpart.org
www.counterpart.org

English Nature
Northminster House
Peterborough PE1 1UA, UK
Tel: +44 (0)1733 455000
E-mail: enquiries@english-nature.org.uk
www.english-nature.org.uk

ETC group (action group on Erosion, Technology and Concentration, formerly Rural Advancement Foundation International, or RAFI)
478 River Avenue, Suite 200
Winnepeg, MB R3L, Canada
Tel: +1 (204) 453 5259
www.etcgroup.org

European Centre for Nature Conservation (ECNC)
P.O. Box 90154
5000 LG Tilburg, The Netherlands
Tel: +31 13 594 4944
E-mail: ecnc@ecnc.org
www.ecnc.nl

European Union Habitats Directive
DG Environment Unit B.2
200 Rue de la Loi
BU-9 03/201
B - 1049 Brussels, Belgium
E-mail: nature@cec.eu.int

Fauna & Flora International (FFI) UK
Great Eastern House
Tenison Road
Cambridge CB1 2TT, UK
Tel: +44 (0)1223 571000
E-mail: info@fauna-flora.org
www.fauna-flora.org

Fauna & Flora International (FFI) USA
Presidio Building 38, Suite 116
P.O. Box 29156
San Francisco
California 94129-0156, USA
Tel: +1 (415) 561 2910
E-mail: info@fauna-flora.org
www.fauna-flora.org

Food and Agriculture Organization of the United Nations (FAO)
Viale delle Terme di Caracalla
00100 Rome, Italy
Tel: +39 06 57051
E-mail: FAO-HQ@fao.org
www.fao.org

Forestry Stewardship Council (FSC)
FSC International Center
Charles-de-Gaulle 5
53113 Bonn, Germany
Tel: +49 (228) 367 66 0
E-mail: fsc@fsc.org
www.fsc.org

Forests Forever (FF)
50 First Street, Suite 401
San Francisco
California 94105, USA
Tel: +1 (415) 974 3636
E-mail: mail@fsc.org
www.forestsforever.org

Foundation for the Peoples of the South Pacific, *see* Counterpart International, *above*

Global Biodiversity Information Facility (GBIF)
GBIF Secretariat
Universitetsparken 15
DK-2100 Copenhagen ø, Denmark
Tel: +45 35 32 14 70
E-mail: gbif@gbif.org
www.gbif.org

Global Invasive Species Programme (GISP)
The GISP Secretariat
Claremont 7735
Cape Town, South Africa
Tel: +27 21 799 8800
E-mail: gisp@nbict.ac.za
www.gisp.org

Global Partnership for Plant Conservation
c/o Botanic Gardens Conservation International, *see p.473*
www.plants2010.org

Global Trees Campaign (a partnership between FFI and UNEP-WCMC)
Gateway also to the SoundWood and UK Wood Waste campaigns.
www.globaltrees.org

Instituto de Biología
Universidad Nacional Autónoma de México (UNAM)
04510 Mexico City, Mexico
Tel: +52/55 5622 9070/71/92
E-mail: botanica@ibiologia.unam.mx
www.ibiologia.unam.mx

International Conifer Conservation Programme (ICCP)
c/o Royal Botanic Garden Edinburgh, *see p.473*

International Organization for Succulent Plant Study (IOS)
www.iosweb.org

International Plant Genetic Resources Institute (IPGRI)
Via dei Tre Denari 472/a
00057 Maccarese (Fiumicino)
Rome, Italy
Tel: +39 06 6118.1
E-mail: ipgri@cgiar.org
www.ipgri.cgiar.org

IUCN – The World Conservation Union
Rue Mauverney 28
Gland 1196, Switzerland
Tel: +41 22 999 0000
E-mail: mail@iucn.org
www.iucn.org
IUCN Bookshop
www.iucn.org/bookstore
IUCN Red List
www.iucnredlist.org
IUCN-SSC (Species Survival Commission) Plants Programme
www.iucn.org/themes/ssc/plants/plantshome
IUCN-SSC Specialist Groups
www.iucn.org/themes/ssc/sgs/sgs.htm

MAB *see* UNESCO Programme on Man and the Biosphere, *above*

Millennium Seed Bank Project (MSBP)
Wakehurst Place
Ardingly
Haywards Heath
West Sussex RH17 6TN, UK
Tel: +44 (0)1444 894100
E-mail: msbsci@kew.org
www.kew.org/msbp

Mount Cuba Center for the Study of the Piedmont Flora
Box 3570
Barley Mill Road
Greenville
Delaware 19807-4244, USA
Email: jfrett@mtcubacenter.org
www.mtcubacenter.org

National Botanical Institute
(renamed the South African Biodiversity Institute, or SANBI)
Private Bag X7
Claremont 7735
South Africa
Tel: +27 21 799 8800
E-mail: webmaster@nbi.ac.za
www.nbi.ac.za

National Collection of Endangered Plants
see Center for Plant Conservation, *p.475*

The Nature Conservancy
4245 North Fairfax Drive, Suite 100
Arlington
Virginia 22203-1606, USA
Tel: +1 (800) 628 6860
E-mail: comment@tnc.org
www.tnc.org

People and Plants International
A.B. Cunningham, director
84 Watkins Street
Fremantle 6162, Australia
Tel: +61 08 9336 6783
E-mail: peopleplants@bigpond.com
www.rbgkew.org.uk/peopleplants

Planta Europa
c/o Plantlife International, *see below*
E-mail: coordinator@planteuropa.org
www.planteuropa.org

Plant Conservation Alliance (PCA)
Bureau of Land Management
1849 C Street NW, LSB-04
Washington, DC 20240, USA
Tel: +1 (202) 452 0392
E-mail: plant@plantconservation.org
www.nps.gov/plants

Plantlife International
14 Rollestone Street
Salisbury
Wiltshire SP1 1DX, UK
Tel: +44 (0)1722 342730
E-mail: enquiries@plantlife.org.uk
www.plantlife.org.uk

Ramsar Convention on Wetlands
Rue Mauverney 28
CH-1196 Gland, Switzerland
Tel: +41 22 999 0170
E-mail: ramsar@ramsar.org
www.ramsar.org

Sainsbury Orchid Conservation Project
c/o Royal Botanic Gardens, Kew, *see p.473*
www.kew.org/gowild/wildscience/sainsbury

Smithsonian Institution
PO Box 37012
SI Building, Room 153, MRC 010
Washington, DC 20013-7012, USA
Tel: +1 (202) 357 2700
E-mail: info@si.edu
www.si.edu

Society of Conservation Biology
4245 North Fairfax Drive, Suite 400
Arlington, Virginia 22203-1651, USA
Tel: +1 (703) 276 2384
E-mail: info@conbio.org
www.conbio.org

SoundWood, *see* Global Trees Campaign, *p.476*

Southern African Botanical Diversity Network (SABONET)
SABONET Coordinator
c/o National Botanical Institute
(renamed the South African Biodiversity Institute, or SANBI)
Private Bag X101
Pretoria 0001, South Africa
Tel: +27 12 8043200
E-mail: nrn@nbipre.nbi.ac.za
www.sabonet.org

SSC (Species Survival Commission), *see* IUCN – The World Conservation Union, *p.476*

TRAFFIC International (a joint programme of WWF and IUCN)
219a Huntingdon Road
Cambridge CB3 0DL, UK
Tel: +44 (0)1223 277427
E-mail: traffic@trafficint.org
www.traffic.org

UNEP-WCMC (United Nations Environment Programme-World Conservation Monitoring Centre)
219 Huntingdon Road
Cambridge CB3 0DL, UK
Tel: +44 (0)1223 277722
E-mail: info@unep-wcmc.org

UNESCO Programme on Man and the Biosphere (MAB)
7 Place de Fontenoy
75352 Paris 07 SP, France
Tel: +33 (0)1 45 68 10 00
E-mail: mab@unesco.org
www.unesco.org/mab

UNESCO World Heritage Centre
7 Place de Fontenoy
75352 Paris 07 SP, France
Tel: +33 (0)1 45 68 15 71
E-mail: wh-info@unesco.org
whc.unesco.org

United Plant Savers
P.O. Box 400
E. Barre
VT 05649, USA
Tel: +1 (802) 479 9825
E-mail: info@unitedplantsavers.org
www.unitedplantsavers.org

N.I.Vavilov Research Institute of Plant Industry
44 B.Morskaya Street
190000 St Petersburg, Russia
Tel: +7 (812) 315 5093
www.vir.nw.ru

World Conservation Monitoring Centre (WCMC),
see UNEP-WCMC, *above*

World Conservation Union,
see IUCN, *p.476*

World Land Trust
Blyth House
Bridge Street,
Halesworth,
Suffolk IP19 8AB, UK
Tel: +44 0845 054 4422
E-mail: info@worldlandtrust.org
www.worldlandtrust.org

World Resources Institute (WRI)
10 G Street NE, Suite 800
Washington, DC 20002, USA
Tel: +1 (202) 729 7600
partners.wri.org

WWF International (World Wide Fund for Nature)
Avenue du Mont-Blanc
1196 Gland, Switzerland
Tel: +41 22 364 88 36
www.panda.org

WWF-UK
Panda House
Weyside Park
Godalming
Surrey GU7 1XR, UK
Tel: +44 (0)1483 426 444
www.wwf-uk.org

WWF-US
1250 24th Street NW
Washington, DC 20037-1175, USA
Tel: +1 (202) 293 4800
E-mail: via website
www.worldwildlife.org

HORTICULTURAL SOCIETIES

Alpine Garden Society, UK
AGS Centre
Avon Bank
Pershore
Worcestershire WR10 3JP, UK
Tel: +44 (0)1386 554790
E-mail: ags@alpinegardensociety.org
www.alpinegardensociety.org

American Camellia Society
Massee Lane Gardens
100 Massee Lane
Fort Valley
Georgia 31030, USA
Tel: +1 (478) 967 2358
E-mail: ask@camellias-acs.org
www.camellias-acs.org

American Fern Society
E-mail: webmaster@amerfernsoc.org
amerfernsoc.org

American Horticultural Society
7931 East Boulevard Drive
Alexandria
Virginia 22308, USA
Tel: +1 (703) 768 5700
E-mail: sdick@ahs.org
www.ahs.org

American Orchid Society
16700 AOS Lane, Delray Beach
Florida 33446-4351, USA
Tel: +1 (561) 404 2000
E-mail: theaos@aos.org
www.orchidweb.org

British Pteridological Society
E-mail: webmaster@eBPS.org.uk
www.nhm.ac.uk/hosted_sites/bps

Bromeliad Society International (BSI)
E-mail: webmaster@bsi.org
www.bsi.org

International Bulb Society
P.O. Box 336
Sanger
California 93657, USA
E-mail: webmaster@bulbsociety.com
www.bulbsociety.org

International Carnivorous Plant Society
PMB 330
3310 East Yorba Linda Boulevard
Fullerton
California 92831-1709, USA
E-mail: carl@carnivorousplants.org
www.carnivorousplants.org

New England Wild Flower Society
at Garden in the Woods
180 Hemenway Road
Framingham, MA 01701, USA
Email: newfs@newfs.org
www.newfs.org

North American Lily Society (NALS), incorporating the Species Lily Preservation Group (SLPG)
NALS Executive Secretary
PO Box 272
Owatonna, Minnesota 55060, USA
E-mail: gilman@ll.net
www.lilies org

North American Rock Garden Society
www.nargs.org.uk

Royal Horticultural Society
80 Vincent Square
London SW1P 2PE, UK
Tel: +44 (0)20 7834 4333
E-mail: info@rhs.org.uk
www.rhs.org.uk

Scottish Rock Garden Club
E-mail: info@srgc.org.uk
www.srgc.org.uk

PUBLICATIONS

BGjournal, *Cuttings*, and
Roots: BGCI Education Review
All published by Botanic Gardens Conservation International, *see p.473*

Conservation Biology
Published by the Society of Conservation Biology, *see p.477*
Blackwell Science, Inc
350 Main Street
Malden
Maine 02148-5018, USA
Tel: +1 (781) 388 8250
E-mail: journals@blacksci.com
www.jstor.org/journals/08888892.html

Conservation in Practice
Published by the Society of Conservation Biology, *see p.477*
Department of Biology
P.O. Box 351800
University of Washington
Seattle
Washington 98195-1800, USA
Tel: +1 (703) 276 2384

Fauna & Flora and
Oryx, the International Journal of Conservation
Both published by Fauna & Flora International, *see p.476*

The Good Bulb Guide
List of suppliers who pledge to deal only in bulbs that have not been wild-collected, or to inform their customers if stock is wild-collected. Cost: $1.00.
Faith Campbell
8208 Dabney Avenue
Springfield
Virginia 22152, USA

Plant Talk
Published by the National Tropical Botanical Garden, *see p.475*
Editorial/Subscription Office
10 Princeton Court
55 Felsham Road
London SW15 1AZ, UK
E-mail: pt enquiries@dial.pipex.com
www.plant-talk.org

The World Conservation Bookstore
Online mail-order service for more than 2,000 specialist titles including all Red Data listings.
www.iucn.org/bookstore
Also available through Island Press for customers in North America and Canada: www.islandpress.org

OTHER USEFUL WEBSITES

The websites of many of the organizations detailed on the previous pages provide a great wealth of information about global and local plant and animal conservation programmes and associated subjects. The websites listed below are also highly recommended.

www.arkive.org
ARKive, the world's centralized digital library of images of life on Earth

www.bgci.org
Includes the Botanic Gardens Conservation International (BGCI) online database of more than 2,200 botanic gardens worldwide. The database pinpoints which plant species are being cultivated in the individual botanic gardens and in what frequency. *See also* BGCI, *p.473*.

www.biodiversityhotspots.org
Features an interactive map of biodiversity hotspots. *See also* Conservation International, *p.476*, and map, *p.479*.

www.biodiv.org
Offers a gateway to the work of the Biodiversity Convention and provides links to the websites of national biodiversity authorities worldwide. *See also* Convention on Biological Diversity, *p.476*.

www.championtrees.org
Details the largest specimens of tree species, presented by The Earth Restoration and Reforestation Alliance (TERRA).

www.earthtrends.wri.org
Environmental information portal of the World Resources Institute. *See also* World Resources Institute, *p.477*.

www.futureforests.com
Campaigns for greater awareness of global warming and how individuals can take steps to reduce it.

www.globio.info
Website of GLOBIO, a project of the United Nations Environment Programme (UNEP, *see p.477*) that maps present and future impacts of humans on the biosphere. The findings are presented with the aid of maps and some videoclips and animations.

www.panda.org/about_wwf/where_we_work/ecoregions
Presents the WWF's Global 200 project, a ranking of the world's most biologically outstanding terrestrial, freshwater, and marine habitats. *See also* WWF International, *p.477*, and map, *pp.480–81*.

www.plants2010.org
Website of the Global Partnership for Plant Conservation, dedicated to the worldwide implementation of the Global Strategy for Plant Conservation.

www.plantsforlife.net
BGCI resource for children interested in plant conservation: projects, information for schools, and useful links.

www.plant-talk.org
Website associated with *Plant Talk*, the magazine on plant conservation worldwide. *See also Plant Talk, above.*

www.rbgkew.org.uk/peopleplants
People and Plants Online, a partnership of WWF-UK and UNESCO, with the Royal Botanical Gardens, Kew as Associate. *See also* People and Plants International, *p.477*.

www. ukbap.org.uk
Website of the UK Biodiversity Action Plan (UKBAP).

BIODIVERSITY HOTSPOTS

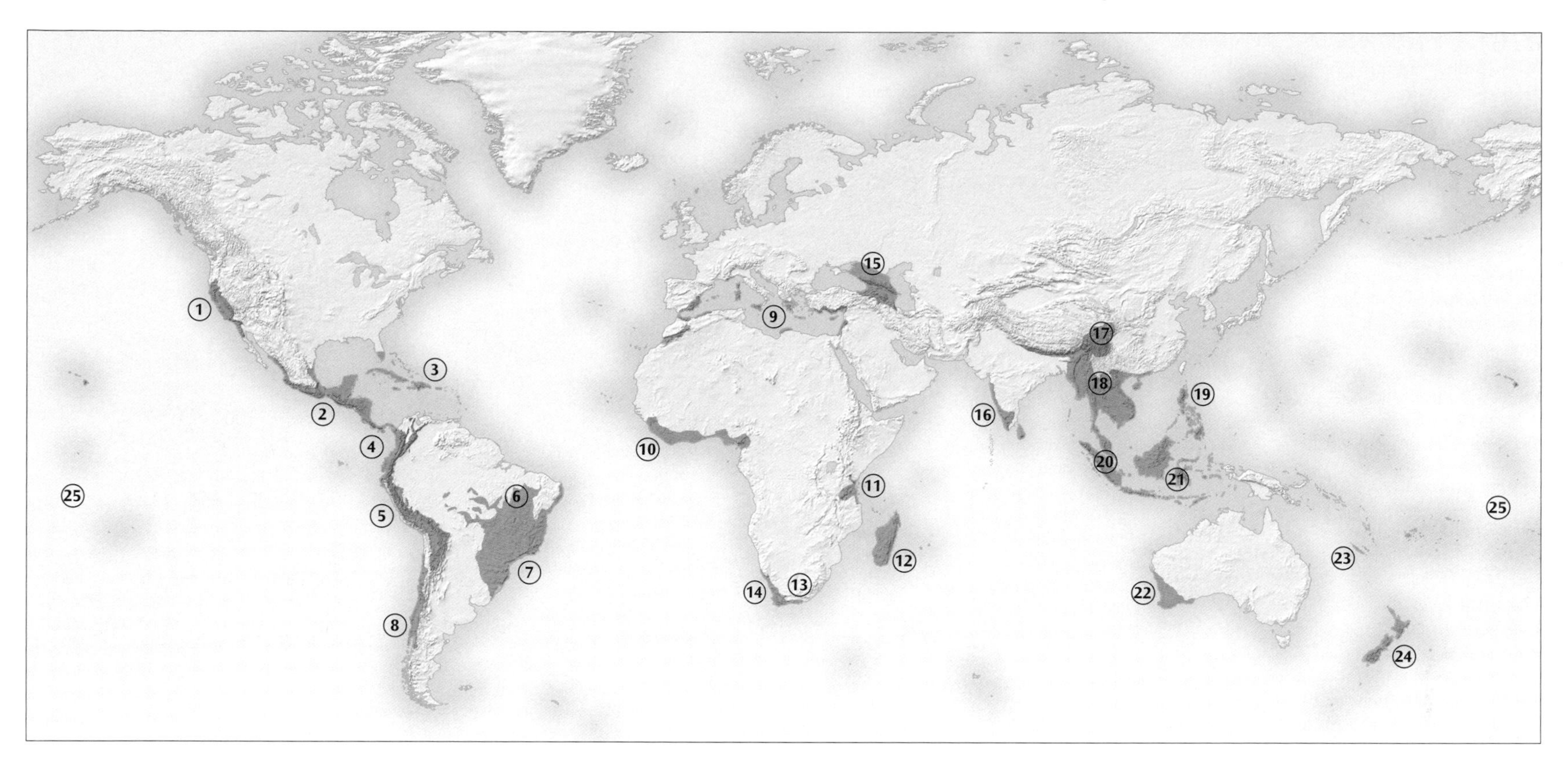

AS DEFINED BY Conservation International, a biodiversity hotspot is a region that supports 1,500 endemic species (0.5 per cent of the global total), but where more than 70 per cent of the habitat has been lost. The 25 hotspots shown here contain 44 per cent of all plant species in only 1.4 per cent of earth's land area.

1 The California Floristic Province stretches for 1,800 km (1,120 miles) along the western coast of North America. The region covers 70 per cent of California, as well as small parts of Oregon and Baja California (Mexico), and outlying islands.

2 The Mesoamerica hotspot spans most of Central America, encompassing all subtropical and tropical ecosystems from central Mexico to the Panama Canal. Mesoamerican forests are critical for the preservation of the biodiversity of the western hemisphere.

3 The Caribbean hotspot includes most of the island groups in the Caribbean Sea as well as the southern tip of Florida, including the Keys and the Everglades.

4 The Chocó-Darién-western Ecuador hotspot extends 1,500 km (930 miles) along the western flank of the Andes. It includes the tropical forests of Panama's Darién Province, the Chocó region of Colombia, coastal Ecuador, and north-western Peru.

5 The Tropical Andes are the richest and most diverse region on Earth, containing 15–17 per cent of the world's flora in only 0.8 per cent of its area. The hotspot covers parts of Venezuela, Chile, Argentina, Colombia, Ecuador, Bolivia, and Peru.

6 The Cerrado spreads across the central Brazilian plateau and is the country's second largest biome after Amazonia. Having a pronounced dry season, it supports a unique array of drought- and fire-adapted plant species.

7 The Atlantic Forest region lies along Brazil's Atlantic coast, extending inland to Paraguay and Argentina. Less than ten per cent remains. Of its 20,000 plant species, 8,000 are found nowhere else.

8 Central Chile is a virtual continental island of 300,000 sq km (115,840 sq miles) with a richly endemic flora. It is bounded by the Pacific Ocean, Andes Mountains, and the Atacama Desert.

9 The Mediterranean Basin extends over 2,362,000 sq km (912,000 sq miles). It spans from Portugal to Jordan, and from northern Italy to the Canary Islands, and includes the sea's many scattered islands.

10 The Guinean Forests hotspot includes all of the lowland forests of West Africa, from Sierra Leone to Cameroon, as well as Bioko, Pagalu, São Tomé, and Principe islands in the Gulf of Guinea.

11 The Coastal Forests hotspot stretches along most of Tanzania's eastern coast and the Eastern Arc Mountains. It contains 13 per cent of Africa's plants, yet occupies only 0.1 per cent of the land area of tropical Africa.

12 The Madagascar and western Indian Ocean islands hotspot is one of the most important and threatened conservation priority areas on Earth. Madagascar alone has ten plant families and five primate families seen nowhere else.

13 The Cape Floristic Region hugs the coastline of the far south-western tip of South Africa. It has 8,200 vascular plant species within 74,000 sq km (2,850 miles), of which six plant families and 5,682 species do not grow elsewhere.

14 The succulent Karoo hotspot runs from the south-western coastal tip of South Africa up the Atlantic coast to southern Namibia. It covers 116,000 sq km (45,000 sq miles) of desert and is home to the richest succulent flora on Earth.

15 The Caucasus hotspot consists of 500,000 sq km (193,000 sq miles) of mountains in Eurasia between the Black Sea and the Caspian Sea. The region has twice the animal diversity of adjacent lands.

16 The western Ghats and Sri Lanka hotspot consists of two blocks of montane forest separated by 400 km (250 miles) of land and sea. Its diverse, endemic fauna and flora are greatly threatened by human activity.

17 The south-west China hotspot covers over 800,000 sq km (309,000 sq miles) of temperate to alpine mountains. It is twice the size of California and includes small parts of Myanmar and eastern India.

18 The Indo-Burma hotspot covers about 2 million sq km (772,000 sq miles) of tropical Asia east of the Indian subcontinent. Its fauna and flora are not fully documented and new species continue to be discovered.

19 The Philippines hotspot includes more than 7,000 islands in the westernmost Pacific Ocean. With high levels of diversity and endemism in the tiny remaining portion of natural habitat, the Philippines are among the top five hotspots for global preservation.

20 The Sundaland hotspot covers the western half of the Indo-Malayan archipelago, an arc of 17,000 equatorial islands between the Asian mainland and Australia. It includes two of the largest islands in the world, Borneo and Sumatera.

21 The Wallacea hotspot lies mostly in Indonesia, between Java and Papua Province in New Guinea. The flora and fauna are so varied, with such high levels of endemism, that protected areas are needed on every island to preserve biodiversity.

22 The south-west Australia hotspot lies on Australia's south-western tip, within the state of Western Australia. This Mediterranean-type hotspot has high diversity and endemism among its plants and reptiles.

23 New Caledonia is one of the world's smallest hotspots, consisting of Grande Terre Island, the smaller Loyalty Islands, and a 1,600-km (995-mile) encircling coral reef in the South Pacific.

24 The New Zealand hotspot covers 270,534 sq km (104,500 sq miles). Like New Caledonia, it was largely isolated from the evolution and spread of placental mammals.

25 Micronesia, Polynesia, and Fiji make up a hotspot of at least 1,415 islands scattered across the southern Pacific Ocean. In this vast expanse of sea, the total land area is only 46,012 sq km (17,750 sq miles).

ECOREGIONS

RECOGNIZING THAT resources available to conservation organizations are limited, the WWF drew up a list of ecological regions – collectively named the Global 200 – that are crucial to the conservation of global biodiversity. The aim of the WWF analysis is to ensure that all important ecosystems are represented in global biodiversity strategy. The Global 200 list includes 43 marine ecoregions in addition to the sites shown here.

TROPICAL AND SUBTROPICAL MOIST BROADLEAF FORESTS

1 Guinean Moist Forests
Benin, Côte d'Ivoire, Ghana, Guinea, Liberia, Sierra Leone, Togo

2 Congolian Coastal Forests
Angola, Cameroon, Democratic Republic of Congo, Equatorial Guinea, Gabon, Nigeria, São Tomé & Príncipe, Republic of Congo

3 Cameroon Highlands Forests
Cameroon, Equatorial Guinea, Nigeria

4 North-eastern Congo Basin Moist Forests
Central African Republic, Democratic Republic of Congo

5 Central Congo Basin Moist Forests
Democratic Republic of Congo

6 Western Congo Basin Moist Forests
Cameroon, Central African Republic, Democratic Republic of Congo, Gabon, Republic of Congo

7 Albertine Rift Montane Forests
Burundi, Democratic Republic of Congo, Rwanda, Tanzania, Uganda

8 East African Coastal Forests
Kenya, Somalia, Tanzania

9 Eastern Arc Montane Forests
Kenya, Tanzania

10 Madagascar Forests and Shrublands
Madagascar

11 Seychelles and Mascarenes Moist Forests
Mauritius, Réunion (France), Seychelles

12 Sulawesi Moist Forests
Indonesia

13 Moluccas Moist Forests
Indonesia

14 Southern New Guinea Lowland Forests
Indonesia, Papua New Guinea

15 New Guinea Montane Forests
Indonesia, Papua New Guinea

16 Solomons-Vanuatu-Bismarck Moist Forests
Papua New Guinea, Solomon Islands, Vanuatu

17 Queensland Tropical Forests
Australia

18 New Caledonia Moist Forests
New Caledonia (France)

19 Lord Howe-Norfolk Islands Forests
Australia

20 South-western Ghats Moist Forests
India

21 Sri Lankan Moist Forests
Sri Lanka

22 Northern Indochina Subtropical Moist Forests
China, Laos, Myanmar, Thailand, Vietnam

23 South-east China-Hainan Moist Forests
China, Vietnam

24 Taiwan Montane Forests
China

25 Annamite Range Moist Forests
Cambodia, Laos, Vietnam

26 Sumateran Islands Lowland and Montane Forests
Indonesia

27 Philippines Moist Forests
Philippines

28 Palawan Moist Forests
Philippines

29 Kayah-Karen/Tenasserim Moist Forests
Malaysia, Myanmar, Thailand

30 Peninsular Malaysian Lowland and Mountain Forests
Indonesia, Malaysia, Singapore, Thailand

31 Borneo Lowland and Montane Forests
Brunei, Indonesia, Malaysia

32 Nansei Shoto Archipelago Forests
Japan

33 Eastern Deccan Plateau Moist Forests
India

34 Naga-Manupuri-Chin Hills Moist Forests
Bangladesh, India, Myanmar

35 Cardamom Mountains Moist Forests
Cambodia, Thailand

36 Western Java Mountain Forests
Indonesia

37 Greater Antillean Moist Forests
Cuba, Dominican Republic, Haiti, Jamaica, Puerto Rico (USA)

38 Talamancan and Isthmian Pacific Forests
Costa Rica, Panama

39 Chocó-Darién Moist Forests
Colombia, Ecuador, Panama

40 Northern Andean Montane Forests
Colombia, Ecuador, Venezuela, Peru

41 Coastal Venezuela Montane Forests
Venezuela

42 Guianan Moist Forests
Brazil, French Guiana (France), Guyana, Suriname, Venezuela

43 Napo Moist Forests
Colombia, Ecuador, Peru

44 Río Negro-Juruá Moist Forests
Brazil, Colombia, Peru, Venezuela

45 Guayanan Highlands Forests
Brazil, Colombia, Guayana, Suriname, Venezuela

46 Central Andean Yungas
Argentina, Bolivia, Peru

47 South-western Amazonian Moist Forests
Bolivia, Brazil, Peru

48 Atlantic Forests
Argentina, Brazil, Paraguay

49 South Pacific Islands Forests
American Samoa (USA), Cook Islands (New Zealand), Fiji, French Polynesia (France), Niue (New Zealand), Samoa, Tonga, Wallis and Futuna Islands (France)

50 Hawaii Moist Forests
Hawaii (USA)

TROPICAL AND SUBTROPICAL DRY BROADLEAF FORESTS

51 Madagascar Dry Forests
Madagascar

52 Nusu Tenggara Dry Forests
Indonesia

53 New Caledonia Dry Forests
New Caledonia (France)

54 Indochina Dry Forests
Cambodia, Laos, Thailand, Vietnam

55 Chhota-Nagpur Dry Forests
India

56 Mexican Dry Forest
Guatemala, Mexico

57 Tumbesian-Andean Valleys Dry Forests
Colombia, Ecuador, Peru

58 Chiquitano Dry Forest
Bolivia, Brazil

59 Atlantic Dry Forests
Brazil

60 Hawaii's Dry Forests
Hawaii (USA)

TROPICAL AND SUBTROPICAL CONIFEROUS FORESTS

61 Sierra Madre Oriental and Occidental Pine-Oak Forests
Mexico, USA

62 Greater Antillean Pine Forests
Cuba, Dominican Republic, Haiti

63 Mesoamerican Pine-Oak Forests
El Salvador, Guatemala, Honduras, Mexico, Nicaragua

TEMPERATE BROADLEAF AND MIXED FORESTS

64 Eastern Australia Temperate Forests
Australia

65 Tasmanian Temperate Rain Forests
Australia

66 New Zealand Temperate Forests
New Zealand

67 Eastern Himalayan Broadleaf and Conifer Forests
Bhutan, China, India, Myanmar, Nepal

68 Western Himalayan Temperate Forests
Afghanistan, India, Nepal, Pakistan

69 Appalachian and Mixed Mesophytic Forests
USA

70 South-west China Temperate Forests
China

71 Russian Far East Temperate Forests
Russia

TEMPERATE CONIFEROUS FORESTS

72 Pacific Temperate Rainforests
Canada, USA

73 Klamath-Siskiyou Coniferous Forests
USA

74 Sierra Nevada Coniferous Forests
USA

75 South-eastern Coniferous and Broadleaf Forests
USA

76 Valdivian Temperate Rainforests and Juan Fernandez Islands
Argentina, Chile

77 European-Mediterranean Montane Mixed Forests
Albania, Algeria, Andorra, Austria, Bosnia and Herzegovina, Bulgaria, Croatia, Czech Republic, France, Germany, Greece, Italy, Liechtenstein, Macedonia, Morocco, Poland, Romania, Russia, Slovakia, Slovenia, Spain, Switzerland, Tunisia, Ukraine, Yugoslavia

78 Caucasus-Anatolian-Hyrcanian Temperate Forests
Armenia, Azerbaijan, Bulgaria, Georgia, Iran, Russia, Turkey, Turkmenistan

79 Altai-Sayan Montane Forests
China, Kazakstan, Mongolia, Russia

80 Hengduan Shan Coniferous Forests
China

BOREAL FORESTS/TAIGA

81 Muskwa/Slave Lake Boreal Forests
Canada

82 Canadian Boreal Forests
Canada

83 Ural Mountains Taiga
Russia

84 Eastern Siberian Taiga
Russia

85 Kamchatka Taiga and Grasslands
Russia

TROPICAL AND SUBTROPICAL GRASSLANDS, SAVANNAS, AND SHRUBLANDS

86 Horn of Africa Acacia Savannas
Eritrea, Ethiopia, Kenya, Somalia, Sudan

87 East African Acacia Savannas
Ethiopia, Kenya, Sudan, Tanzania, Uganda

88 Central and Eastern Miombo Woodlands
Angola, Botswana, Burundi, Democratic Republic of Congo, Malawi, Mozambique, Namibia, Tanzania, Zambia, Zimbabwe

89 Sudanian Savannas
Cameroon, Central African Republic, Chad, Nigeria, Democratic Republic of Congo, Eritrea, Ethiopia, Kenya, Nigeria, Sudan, Uganda

90 Northern Australia and Trans-Fly Savannas
Australia, Indonesia, Papua New Guinea

91 Terai-Duar Savannas and Grasslands
Bangladesh, Bhutan, India, Nepal

92 Llanos Savannas
Colombia, Venezuela

93 Cerrado Woodlands and Savannas
Bolivia, Brazil, Paraguay

TEMPERATE GRASSLANDS, SAVANNAS, AND SHRUBLANDS

94 Northern Prairie
Canada, USA

95 Patagonian Steppe
Argentina, Chile

96 Daurian Steppe
China, Mongolia, Russia

FLOODED GRASSLANDS AND SAVANNAS

97 Sudd-Sahelian Flooded Grasslands and Savannas
Cameroon, Chad, Ethiopia, Mali, Niger, Nigeria, Sudan, Uganda

98 Zambezian Flooded Savannas
Angola, Botswana, Democratic Republic of Congo, Malawi, Mozambique, Namibia, Tanzania, Zambia

99 Rann of Kutch Flooded Grasslands
India, Pakistan

100 Everglades Flooded Grasslands
USA

101 Pantanal Flooded Savannas
Bolivia, Brazil, Paraguay

MONTANE GRASSLANDS AND SHRUBLANDS

102 Ethiopian Highlands
Eritrea, Ethiopia, Sudan

103 Southern Rift Montane Woodlands
Malawi, Mozambique, Tanzania, Zambia

104 East African Moorlands
Democratic Republic of Congo, Kenya, Rwanda, Tanzania, Uganda

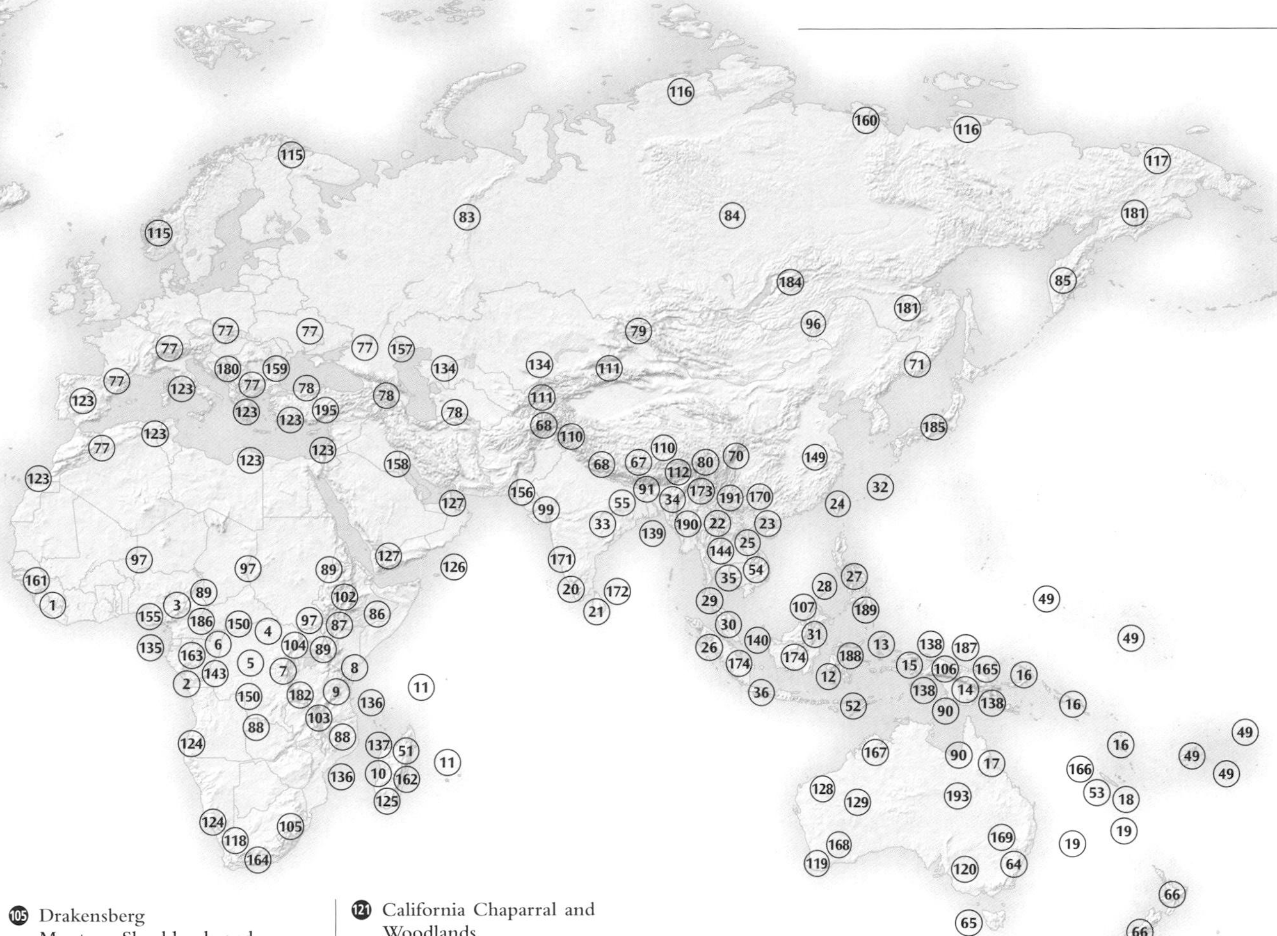

105 Drakensberg Montane Shrublands and Woodlands
Lesotho, South Africa, Swaziland

106 Central Range Subalpine Grasslands
Indonesia, Papua New Guinea

107 Kinabalu Montane Scrub
Malaysia

108 Northern Andean Paramo
Colombia, Ecuador, Peru, Venezuela

109 Central Andean Dry Puna
Argentina, Bolivia, Chile, Peru

110 Tibetan Plateau Steppe
Afghanistan, China, India, Pakistan, Tajikistan

111 Middle Asian Montane Steppe and Woodlands
Afghanistan, China, Kazakhstan, Kyrgyzstan, Tajikistan, Turkmenistan, Uzbekistan

112 Eastern Himalayan Alpine Meadows
Bhutan, China, India, Myanmar, Nepal

TUNDRA

113 Alaskan North Slope Coastal Tundra
Canada, USA

114 Canadian Low Arctic Tundra
Canada

115 Fenno-Scandia Alpine Tundra and Taiga
Finland, Norway, Russia, Sweden

116 Taimyr and Siberian Coastal Tundra
Russia

117 Chukote Coastal Tundra
Russia

MEDITERRANEAN FORESTS, WOODLANDS, AND SCRUB

118 Fynbos
South Africa

119 South-western Australia Forests and Scrub
Australia

120 Southern Australia Mallee and Woodlands
Australia

121 California Chaparral and Woodlands
Mexico, USA

122 Chilean Matorral
Chile

123 Mediterranean Forests, Woodlands, and Scrub
Albania, Algeria, Bosnia and Herzegovina, Bulgaria, Canary Islands (Spain), Croatia, Cyprus, Egypt, France, Gibraltar (United Kingdom), Greece, Iraq, Israel, Italy, Jordan, Lebanon, Libya, Macedonia, Madeira Islands (Portugal), Malta, Monaco, Morocco, Portugal, San Marino, Slovenia, Spain, Syria, Tunisia, Turkey, Western Sahara (Morocco), Yugoslavia

DESERTS AND XERIC SHRUBLANDS

124 Namib-Karoo-Kaokeveld Deserts
Angola, Namibia, South Africa

125 Madagascar Spiny Thicket
Madagascar

126 Socotra Island Desert
Yemen

127 Arabian Highland Woodlands and Shrublands
Oman, Saudi Arabia, United Arab Emirates, Yemen

128 Carnavon Xeric Scrub
Australia

129 Great Sandy-Tanami Deserts
Australia

130 Sonoran-Baja Deserts
Mexico, USA

131 Chihuahuan-Tehuacán Deserts
Mexico, USA

132 Galápagos Islands Scrub
Ecuador

133 Atacama-Sechura Deserts
Chile, Peru

134 Central Asian Deserts
Kazakstan, Kyrgyzstan, Uzbekistan, Turkmenistan

MANGROVES

135 Gulf of Guinea Mangroves
Angola, Cameroon, Democratic Republic of Congo, Equatorial Guinea, Gabon, Ghana, Nigeria

136 East African Mangroves
Kenya, Mozambique, Somalia, Tanzania

137 Madagascar Mangroves
Madagascar

138 New Guinea Mangroves
Indonesia, Papua New Guinea

139 Sundarbans Mangroves
Bangladesh, India

140 Greater Sundas Mangroves
Brunei, Indonesia, Malaysia

141 Guianan-Amazon Mangroves
Brazil, French Guiana (France), Suriname, Trinidad and Tobago, Venezuela

142 Panama Bight Mangroves
Colombia, Ecuador, Panama, Peru

LARGE RIVERS

143 Congo River and Flooded Forests
Angola, Democratic Republic of Congo, Republic of Congo

144 Mekong River
Cambodia, China, Laos, Myanmar, Thailand, Vietnam

145 Colorado River
Mexico, USA

146 Lower Mississippi River
USA

147 Amazon River and Flooded Forests
Brazil, Colombia, Peru

148 Orinoco River and Flooded Forests
Brazil, Colombia, Venezuela

149 Yangtze River and Lakes
China

LARGE RIVER HEADWATERS

150 Congo Basin Piedmont Rivers and Streams
Angola, Cameroon, Central African Republic, Democratic Republic of Congo, Gabon, Republic of Congo, Sudan

151 Mississippi Piedmont Rivers and Streams
USA

152 Upper Amazon Rivers and Streams
Bolivia, Brazil, Colombia, Ecuador, French Guiana (France), Guyana, Peru, Suriname, Venezuela

153 Upper Paraná Rivers and Streams
Argentina, Brazil, Paraguay

154 Brazilian Shield Amazonian Rivers and Streams
Bolivia, Brazil, Paraguay

LARGE RIVER DELTAS

155 Niger River Delta
Nigeria

156 Indus River Delta
India, Pakistan

157 Volga River Delta
Kazakhstan, Russia

158 Mesopotamian Delta and Marshes
Iran, Iraq, Kuwait

159 Danube River Delta
Bulgaria, Moldova, Romania, Ukraine, Yugoslavia

160 Lena River Delta
Russia

SMALL RIVERS

161 Upper Guinea Rivers and Streams
Côte D'Ivoire, Guinea, Liberia, Sierra Leone

162 Madagascar Freshwater
Madagascar

163 Gulf of Guinea Rivers and Streams
Angola, Cameroon, Democratic Republic of Congo, Equatorial Guinea, Gabon, Nigeria, Republic of Congo

164 Cape Rivers and Streams
South Africa

165 New Guinea Rivers and Streams
Indonesia, Papua New Guinea

166 New Caledonia Rivers and Streams
New Caledonia (France)

167 Kimberley Rivers and Streams
Australia

168 South-west Australia Rivers and Streams
Australia

169 Eastern Australia Rivers and Streams
Australia

170 Xi Jiang Rivers and Streams
China, Vietnam

171 Western Ghats Rivers and Streams
India

172 South-western Sri Lanka Rivers and Streams
Sri Lanka

173 Salween River
China, Myanmar, Thailand

174 Sundaland Rivers and Swamps
Brunei, Malaysia, Indonesia, Singapore

175 South-eastern Rivers and Streams
USA

176 Pacific North-west Coastal Rivers and Streams
USA

177 Gulf of Alaska Coastal Rivers and Streams
Canada, USA

178 Guianan Freshwater
Brazil, French Guiana (France), Guyana, Suriname, Venezuela

179 Greater Antillean Freshwater
Cuba, Dominican Republic, Haiti, Puerto Rico (USA)

180 Balkan Rivers and Streams
Albania, Bosnia and Herzegovina, Bulgaria, Croatia, Greece, Macedonia, Turkey, Yugoslavia

181 Russian Far East Rivers and Wetlands
China, Mongolia, Russia

LARGE LAKES

182 Rift Valley Lakes
Burundi, Democratic Republic of Congo, Ethiopia, Kenya, Malawi, Mozambique, Rwanda, Tanzania, Uganda, Zambia

183 High Andean Lakes
Argentina, Bolivia, Chile, Peru

184 Lake Baikal
Russia

185 Lake Biwa
Japan

SMALL LAKES

186 Cameroon Crater Lakes
Cameroon

187 Lakes Kutubu and Sentani
Indonesia, Papua New Guinea

188 Central Sulawesi Lakes
Indonesia

189 Philippines Freshwater
Philippines

190 Lake Inle
Myanmar

191 Yunnan Lakes and Streams
China

192 Mexican Highland Lakes
Mexico

XERIC BASINS

193 Central Australian Freshwater
Australia

194 Chihuahuan Freshwater
Mexico, USA

195 Anatolian Freshwater
Syria, Turkey

THE IUCN RED LISTS

Two complementary sources for the conservation status of plants exist: the printed 1997 IUCN Red List of Threatened Plants (also available at www.unep-wcmc.org/species/plants/red-list.htm), and the IUCN Red List of Threatened Species (available at www.iucnredlist.org), updated annually since 2000. The categories used in the 1997 and 2000 Red Lists are given below. (Plants named as "Unlisted" have not yet been evaluated, but may well be globally threatened.)

CATEGORIES USED IN THE 1997 RED LIST

EXTINCT (EX)

Taxa that are no longer known to exist in the wild after repeated searches of the type localities and other known or likely places.

EXTINCT/ENDANGERED (EX/E)

Taxa possibly extinct in the wild.

ENDANGERED (E)

Taxa in danger of extinction and whose survival is unlikely if the causal factors continue operating. Included are taxa whose numbers have been reduced to a critical level or whose habitats have been so drastically reduced that they are deemed to be in immediate danger of extinction.

VULNERABLE (V)

Taxa believed likely to move into the Endangered category in the near future if the causal factors continue operating. Included are taxa of which most or all the populations are decreasing because of over-exploitation, extensive destruction of habitat or other environmental disturbance; taxa with populations that have been seriously depleted and whose ultimate security is not yet assured; and taxa with populations that are still abundant but are under threat from serious adverse factors throughout their range.

RARE (R)

Taxa with small world populations that are not at present Endangered or Vulnerable but are at risk. These taxa are usually localized within restricted geographic areas or habitats or are thinly scattered over a more extensive range.

INDETERMINATE (I)

Taxa known to be Extinct, Endangered, Vulnerable, or Rare but where there is not enough information to say which of the four categories is appropriate.

CATEGORIES USED IN THE 2000 RED LIST

EXTINCT (EX)

A taxon is Extinct when there is no reasonable doubt that the last individual has died. A taxon is presumed Extinct when exhaustive surveys in known and/or expected habitat, at appropriate times (diurnal, seasonal, annual), throughout its historic range have failed to record an individual. Surveys should be over a time frame appropriate to the taxon's life cycle and life form.

EXTINCT IN THE WILD (EW)

A taxon is Extinct in the Wild when it is known only to survive in cultivation, in captivity or as a naturalized population (or populations) well outside the past range. A taxon is presumed Extinct in the Wild when exhaustive surveys in known and/or expected habitat, at appropriate times (diurnal, seasonal, annual), throughout its historic range have failed to record an individual. Surveys should be over a time frame appropriate to the taxon's life cycle and life form.

CRITICALLY ENDANGERED (CR)

A taxon is Critically Endangered when the best available evidence indicates that it meets any of the criteria for Critically Endangered, and it is therefore considered to be facing an extremely high risk of extinction in the wild.

ENDANGERED (EN)

A taxon is Endangered when the best available evidence indicates that it meets any of the criteria for Endangered, and it is therefore considered to be facing a very high risk of extinction in the wild.

VULNERABLE (VU)

A taxon is Vulnerable when the best available evidence indicates that it meets any of the criteria for Vulnerable, and it is therefore considered to be facing a high risk of extinction in the wild.

LOWER RISK (LR)

In very recent versions of the Red List, this category is superseded (*see below*). A taxon was classed Lower Risk when it did not satisfy the criteria for any of the categories Critically Endangered, Endangered, or Vulnerable. Taxa identified as Lower Risk were placed in three subcategories. The subcategory **Conservation Dependent (LR/cd)**, now discontinued, included taxa that were the focus of targeted conservation programmes which, if terminated, would result in the taxa qualifying for one of the more threatened categories within five years. The subcategory **Near Threatened (LR/nt),** now the separate category **Near Threatened (NT)**, included taxa that, while not Conservation Dependent, were close to qualifying as Vulnerable. The subcategory **Least Concern (LR/lc)**, now the separate category **Least Concern (LC)**, included taxa that did not qualify for Conservation Dependent or Near Threatened status. Widespread and abundant taxa are included in this category.

DATA DEFICIENT (DD)

A taxon is Data Deficient when there is inadequate information to make a direct, or indirect, assessment of its risk of extinction based on its distribution and/or population status. A taxon in this category may be well studied, and its biology well known, but appropriate data on abundance and/or distribution are lacking. Data Deficient is therefore not a category of threat. Listing of taxa in this category indicates that more information is required and acknowledges the possibility that future research will show that threatened classification is appropriate. If the range of a taxon is suspected to be relatively circumscribed, and a considerable period of time has elapsed since the last record of the taxon, threatened status may well be justified.

THE GLOBAL STRATEGY FOR PLANT CONSERVATION

The 1992 Earth Summit in Rio de Janeiro led to the adoption by world leaders of the Convention on Biological Diversity (CBD, *see also p.36*). In 2002, the CBD signatories endorsed the Global Strategy for Plant Conservation (GSPC), the first major global agreement to include clear conservation targets. It was determined that all the targets of the strategy should be met by 2010.

The immediate threat to 100,000 plant species through destruction of natural habitat, unsustainable exploitation, and inappropriate development is addressed in the global strategy by 16 targets, grouped in terms of five broad goals.

UNDERSTANDING AND DOCUMENTING PLANT DIVERSITY

1. A widely accessible working list of known plant species, as a step towards a complete world flora.
2. A preliminary assessment of the conservation status of all known plant species, at national, regional, and international levels.
3. Development of models with protocols for plant conservation and sustainable use, based on research and practical experience.

CONSERVING PLANT DIVERSITY

4. At least ten per cent of each of the world's ecological regions effectively conserved.
5. Protection of 50 per cent of the most important areas for plant diversity assured.
6. At least 30 per cent of production lands managed consistent with the conservation of plant diversity.
7. 60 per cent of the world's threatened species conserved *in situ*.
8. 60 per cent of threatened plant species in accessible *ex situ* collections, preferably in their country of origin, and ten per cent of them included in recovery and restoration programmes.
9. 70 per cent of the genetic diversity of crops and other major socio-economically valuable plant species conserved, and associated local and indigenous knowledge maintained.
10. Management plans in place for at least 100 major alien species that threaten plants, plant communities, and associated habitats and ecosystems.

USING PLANT DIVERSITY SUSTAINABLY

11. No species of wild flora endangered by international trade.
12. 30 per cent of plant-based products derived from sources that are sustainably managed.
13. The decline of plant resources, and associated local and indigenous knowledge, innovations, and practices, that support sustainable livelihoods, local food security, and health care, halted.

PROMOTING EDUCATION AND AWARENESS ABOUT PLANT DIVERSITY

14. The importance of plant diversity and the need for its conservation incorporated into communication, education, and public awareness programmes.

BUILDING CAPACITY FOR THE CONSERVATION OF PLANT DIVERSITY

15. The number of trained people working with appropriate facilities in plant conservation increased, according to national needs, to achieve the targets of this strategy.
16. Networks for plant conservation activities established or strengthened at national, regional, and international levels.

PLANT SPECIES CONTROLLED BY CITES

The Convention on International Trade in Endangered Species of Wild Fauna and Flora (CITES) was established in 1972 to monitor and control international trade in threatened animals and plants. It operates through a system of export and import licences relating to three lists of endangered species, called CITES Appendices I, II, and III. Controls on Appendix I species are particularly strict; those on Appendices II and III are less so. Detailed below are the plant species whose international movements are monitored by CITES. If you want to trade listed plants, you should seek permission from the UK Management Authority. Apply to DEFRA, Global Wildlife Division, 1/16 Temple Quay House, 2 The Square, Temple Quay, Bristol BS1 6EB.

CITES APPENDIX I

International trade is permitted only in artificially propagated plants, although trade in wild specimens for scientific purposes is occasionally allowed. All parts of the plants are controlled, including the seeds. Flasked seedlings and tissue cultures of orchids are exempt from CITES trade restrictions.

Agavaceae
Agave arizonica, Agave parviflora, Nolina interrata

Aloeaceae (Liliaceae)
Aloe albida, Aloe albiflora, Aloe alfredii, Aloe bakeri, Aloe bellatula, Aloe calcairophila, Aloe compressa including vars. *rugosquamosa* and *schistophila, Aloe delphinensis, Aloe descoingsii, Aloe fragilis, Aloe haworthioides* including var. *aurantiaca, Aloe helenae, Aloe laeta* including var. *maniensis, Aloe parallelifolia, Aloe parvula, Aloe pillansii, Aloe polyphylla, Aloe rauhii, Aloe suzannae, Aloe thorncroftii, Aloe versicolor, Aloe vossii*

Apocynaceae
Pachypodium ambongense, Pachypodium baronii, Pachypodium decaryi

Araucariaceae
Araucaria araucana

Asclepiadaceae
Ceropegia chrysantha

Asteraceae (Compositae)
Saussurea costus

Cactaceae
Ariocarpus spp.; *Astrophytum asterias; Aztekium ritteri; Coryphantha werdermannii; Discocactus* spp.; *Echinocereus ferreirianus* var. *lindsayi, Echinocereus schmollii; Escobaria minima, Escobaria sneedii; Mammillaria pectinifera, Mammillaria solisioides; Melocactus conoideus, Melocactus deinacanthus, Melocactus glaucescens, Melocactus paucispinus; Obregonia denegrii; Pachycereus militaris; Pediocactus bradyi, Pediocactus knowltonii, Pediocactus paradinei, Pediocactus peeblesianus, Pediocactus sileri; Pelecyphora* spp.; *Sclerocactus brevihamatus* subsp. *tobuschii, Sclerocactus erectocentrus, Sclerocactus glaucus, Sclerocactus mariposensis, Sclerocactus mesae-verdae, Sclerocactus papyracanthus, Sclerocactus pubispinus, Sclerocactus wrightiae; Strombocactus* spp.; *Turbinicarpus* spp.; *Uebelmannia* spp.

Crassulaceae
Dudleya stolonifera

Cupressaceae
Fitzroya cupressoides; Pilgerodendron uviferum

Cycadaceae
Cycas beddomei

Euphorbiaceae
Euphorbia ambovombensis, Euphorbia capsaintemariensis, Euphorbia cremersii, Euphorbia cylindrifolia, Euphorbia decaryi, Euphorbia francoisii, Euphorbia handiensis, Euphorbia lambii, Euphorbia moratii, Euphorbia parvicyathophora, Euphorbia quarziticola, Euphorbia stygiana, Euphorbia tulearensis

Fouquieriaceae
Fouquieria fasciculata, Fouquieria purpusii

Leguminosae (Papilionaceae)
Dalbergia nigra

Nepenthaceae
Nepenthes khasiana, Nepenthes rajah

Orchidaceae
(All flasked seedlings are excluded from control) *Cattleya trianaei; Cephalanthera cucullata; Cypripedium calceolus; Dendrobium cruentum; Goodyera macrophylla; Laelia jongheana, Laelia lobata; Liparis loeslii; Ophrys argolica, Ophrys lunulata; Orchis scopulorum; Paphiopedilum* spp.; *Peristeria elata; Phragmipedium* spp.; *Renanthera imschootiana; Spiranthes aestivalis; Vanda coerulea*. In addition the European Union (EU) includes all European orchids in Appendix I.

Pinaceae
Abies guatemalensis

Podocarpaceae
Podocarpus parlatorei

Rubiaceae
Balmea stormiae

Sarraceniaceae
Sarracenia alabamensis ssp. *alabamensis, Sarracenia jonesii, Sarracenia oreophila*

Stangeriaceae
Stangeria eriopus

Zamiaceae
Ceratozamia spp.; *Chigua* spp.; *Encephalartos* spp.; *Microcycas calacoma*

Zygophyllaceae
Guaiacum sanctum

CITES APPENDIX II

International trade in wild and artificially propagated plants is allowed subject to licensing. Excepted from licensing control are seeds, flasked seedlings, tissue cultures, and cut flowers of artificially propagated orchids, fruits of artificially propagated vanilla species, fruits of naturalized or propagated orchids, and pads of *Opuntia* subgenus *Opuntia*.

Agavaceae
Agave victoriae-reginae

Aloeaceae (Liliaceae)
Aloe spp. except *Aloe vera* and *Aloe barbadensis*

Amaryllidaceae
Galanthus spp.; *Sternbergia* spp.

Apocynaceae
Pachypodium spp.; *Rauvolfia serpentina*

Araliaceae
Panax quinquefolius, Panax ginseng

Arecaceae (Palmae)
Chrysalidocarpus decipiens; Neodypsis decaryi

Berberidaceae
Podophyllum hexandrum

Bromeliaceae
Tillandsia harrisii, Tillandsia kammii, Tillandsia kautskyi, Tillandsia mauryana, Tillandsia sprengeliana, Tillandsia sucrei, Tillandsia xerographica

Cactaceae
Cactaceae spp. other than those in Appendix I

Caryocaraceae
Caryocar costaricense

Crassulaceae
Dudleya braskiae

Cyatheaceae
Cyatheaceae spp.

Cycadaceae
Cycadaceae spp.

Diapensiaceae
Shortia galacifolia

Dicksoniaceae
Dicksonia spp. (originating in the Americas only); *Cibotium barometz*

Didiereaceae
Didiereaceae spp.

Dioscoreaceae
Dioscorea deltoidea

Droseraceae
Dionaea muscipula

Euphorbiaceae
Euphorbia spp. (succulent spp. only)

Fouquieriaceae
Fouquieria columnaris

Juglandaceae
Oreomunnea pterocarpa

Leguminosae (Papilionaceae)
Pericopsis elata; Platymiscium pleiostachyum; Pterocarpus santalinus (only logs, woodchips, and unprocessed broken material)

Meliaceae
Swietenia humilis, Swietenia mahagoni

Nepenthaceae
Nepenthes spp.

Orchidaceae
Orchidaceae spp.

Orobanchaceae
Cistanche deserticola

Portulacaceae
Anacampseros spp.; *Avonia* spp.; *Lewisia maguirei, Lewisia serrata*

Primulaceae
Cyclamen spp.

Proteaceae
Orothamnus zeyheri; Protea odorata

Ranunculaceae
Adonis vernalis (all parts and derivatives); *Hydrastis scanadensis*

Rosaceae
Prunus africana (except seeds, pollen, tissue cultures, and flasked seedling cultures)

Sarraceniaceae
Sarracenia spp.

Scrophulariaceae
Picrorhiza kurroa

Stangeriaceae
Bowenia spp.

Taxaceae
Taxus wallichiana (except finished pharmaceutical products, seeds, pollen, tissue cultures, and flasked seedling cultures)

Thymeleaceae
Aquilaria malaccensis (except seeds, pollen, tissue cultures, and flasked seedling cultures)

Valerianaceae
Nardostachys grandiflora

Welwitschiaceae
Welwitschia mirabilis

Zamiaceae
Zamiaceae spp.

Zingiberaceae
Hedychium philippinense

Zygophyllaceae
Guaiacum officinale

CITES APPENDIX III

International trade in the following species requires an export permit from the country that listed the species with CITES, or a certificate of origin.

Gnetaceae
Gnetum montanum

Magnoliaceae
Magnolia liliifera var. *obovata*

Meliaceae
Swietenia macrophylla

Papaveraceae
Meconopsis regia

Popocarpaceae
Podocarpus neriifolius

Tetracentraceae
Tetracentron sinense

EUROPEAN UNION ANNEX D

Additional to the three CITES appendices is this listing, adopted by the European Union (EU), of plants for which import notification is required when importing to the EU. The controls apply only to live plants except where an asterisk indicates that notification is also required for all parts and derivatives.

Agavaceae
Calibanus hookerii; Dasylirion longissimum

Araceae
Arisaema dracontium, Arisaema erubescens, Arisaema galeatum, Arisaema jacquemontii, Arisaema nepenthoides, Arisaema sikokianum, Arisaema speciosum, Arisaema thunbergii var. *urashima, Arisaema tortuosum, Arisaema triphyllum; Biarum davisii* subsp. *davisii* and subsp. *armariense, Biarum ditschianum*

Asteraceae (Compositae)
*Arnica montana**; *Orthonna armiana, Orthonna cacloides, Orthonna clavifolia, Orthonna euphorbioides, Orthonna hallii, Orthonna herrei, Orthonna lepidocaulis, Orthonna lobata, Orthonna retrorsa*

Ericaceae
*Arctostaphylos uva-ursi**

Gentianaceae
*Gentiana lutea**

Lycopodiaceae
*Lycopodium clavatum**

Menyanthaceae
*Menyanthes trifoliata**

Parmeliaceae
*Cetraria islandica**

Passifloraceae
Adenia fruticosa, Adenia glauca, Adenia pechuelii, Adenia spinosa

Portulacaceae
Ceraria spp.

Trilliaceae
Trillium catesbaei, Trillium cernuum, Trillium flexipes, Trillium grandiflorum, Trillium luteum, Trillium pusillum, Trillium recurvatum, Trillium rugelii, Trillium sessile, Trillium undulatum

GLOSSARY

A

ACHENE 1) A dry, one-seeded fruit that does not open. 2) Any small, brittle, seed-like fruit; a "naked" seed.

ACID Of soil – with a pH value of less than 7. (Cf. Alkaline and Neutral.)

ADAPTIVE RADIATION The process by which plant and animal species, deriving from common ancestors, evolve specialized characteristics by natural selection that enable them to utilize aspects of their environments, such as varied food sources, and so fill ecological niches.

ADVENTITIOUS Arising from places where growths do not normally occur; e.g., adventitious roots may arise from stems.

AERIAL ROOT A plant root growing above ground to provide anchorage and, on an epiphyte, to absorb atmospheric moisture.

AFROMONTANE Refers to high-altitude areas on the mountains of Africa.

AIR-LAYERING A propagation method in which a cut in an aerial stem is covered by moist sphagnum moss and sealed in a plastic sleeve to induce rooting.

ALIEN SPECIES A plant occurring outside its natural range as a result of intentional or accidental dispersal by human activities.

ALKALI MEADOW A grassland or meadow area where the water table is shallow (1–3m/3–10ft deep) and the soils are alkaline; e.g., parts of the Owens Valley, California, USA.

ALKALINE Of soil – with a pH value of more than 7. (Cf. Acid and Neutral.)

ALLELOPATHY The suppression of the growth of one species by another by the release of toxic substances during, e.g.,, the decomposition of its shed leaves.

ALPINE 1) growing above the tree line. 2) High-altitude plant from above the tree line and usually snow-covered in winter. (Cf. Montane.)

ALPINE GRASSLANDS Grass-dominated vegetation in mountainous regions of the northern hemisphere, including the eastern Himalayas and the Tibetan Plateau.

ALTERNATE Of leaves – occurring successively at different levels on opposite sides of a stem. (Cf. Opposite.)

ANGIOSPERM Flowering plant that bears ovules, later seeds, enclosed in ovaries. (Cf. Gymnosperm.)

ANNUAL A plant that completes its life-cycle – germination, flowering, seeding, dying – in one growing season.

ANTHER The part of a stamen that produces pollen; it is usually borne on a filament.

AQUATIC Any plant that grows in water; it may be free-floating, totally submerged, or rooted on the bottom with leaves and flowers above the water surface.

AREOLE A depressed or raised area bearing spines, branches, or flowers in cacti.

ARIL Coat covering some seeds, often fleshy and brightly coloured.

ARROYO 1) A watercourse in an arid region. 2) A water-carved gulley or channel.

ARTHROPOD An insect or spider with a jointed, segmented body. Arthropod counts are conducted in the assessment of habitats and ecosystems.

ASEXUAL REPRODUCTION A form of reproduction not involving fertilization.

ASYNCHRONOUS CONING The ripening of a conifer's male and female cones at widely different times.

AXIL The upper angle between a part of a plant and the stem that bears it.

AXIS A rachis, stalk, or stem on which organs such as flowers, leaves, or leaflets are arranged.

B

BACK-BULB Of orchids – a dormant, old pseudobulb without leaves.

BAMBOO A treelike, usually tropical or subtropical, woody grass, sometimes with a hollow stem.

BARK The tough covering of woody roots, trunks, and branches.

BASAL At the base of an organ or structure; e.g., a basal sucker emerges from the base of the parent plant.

BASAL PLATE A compressed stem, part of a bulb.

BERRY A fleshy fruit with one to many seeds developed from a single ovary.

BIENNIAL A plant that flowers and dies in the second growing season after germination.

BIODIVERSITY The variability among all forms of living organisms and the ecological complexes of which they are part; this includes diversity within species, between species, and of ecosystems.

BIODIVERSITY HOTSPOT An ecological area that is unusually rich in its species of fauna and flora, many of which may be unique to that area.

BIOMASS 1) The amount of living matter calculated to exist in a given unit area or volume of habitat. 2) The sum total of plant materials, living and dead, plus animal waste in a given area.

BIOME A major region in which distinctive plant and animal communities are well adapted to the physical environment of their area of distribution.

BIOSPHERE RESERVE A combination of protected areas and surrounding lands that is managed in ways that promote both conservation and sustainable use of natural resources. (Cf. UNESCO Biosphere Reserve.)

BLOOM A whitish, powdery or waxy coating on some fruits and plant parts.

BOG A type of peat-accumulating wetland that receives water mainly from rainfall. The peat is mostly derived from sphagnum moss, and bogs are usually found in basins where there is no outlet for water. Bogs are poor in nutrients and are acidic to levels toxic to most plants. (Cf. Seep bog.)

BOG PLANT A plant whose natural habitat is soil that is permanently damp and acidic, or one that thrives in such conditions.

BOLE The trunk of a tree.

BONSAI Production of dwarf trees or shrubs by pruning or root restriction.

BOREAL FOREST Forest composed almost entirely of conifers that occurs in the latitudinal belt 50–60°N across the USA and Eurasia, which experiences short, moist summers and long, cold winters. The four main genera are *Abies*, *Larix*, *Picea*, and *Pinus*. *Abies* and *Picea* dominate in North America; *Pinus* is common in Eurasia.

BOTANIC GARDEN An institution holding documented collections of living plants for the purposes of scientific research, conservation, display, and education.

BOTTOMLAND WOODS Low, wet forests found along the edges of lakes, rivers, and sinkholes that mark a transition between drier upland hardwood woodlands and very wet forests in river floodplains and wetlands. Bottomland forests tolerate frequent floods but not the constant flooding of swampy environments.

BRACT A modified, often protective, leaf at the base of a flower or flower cluster. Bracts may resemble normal leaves, or be small and scale-like, or large and brightly coloured.

BROADLEAVED Describes trees and shrubs that have broad, flat, usually deciduous leaves, in contrast to the narrow, needlelike leaves of conifers.

BUD Immature organ or shoot enclosing an embryonic branch, leaf, inflorescence, or flower.

BUFFER ZONE An area of land surrounding a protected zone that is subject to controls intended to protect the interests of the central zone. E.g., the dependence of local people on materials taken from the zone may be accommodated by involving them in management of the local resources in ways that emphasize sustainability.

BULB A modified underground bud acting as a storage organ and consisting mainly of fleshy, more or less separate or tightly packed, scale leaves on a much reduced stem (basal plate).

BULBIL A small bulb-like organ, often borne in a leaf axil, occasionally on a stem or flowerhead.

BULBLET A small developing bulb produced from the basal plate of a mature bulb outside the tunic.

BURR 1) A prickly, spiny, or hooked fruit, seedhead, or flowerhead. 2) A woody outgrowth on the trunk or root of some trees.

BUTTRESS ROOT Fluted or swollen tree trunk that aids stability in shallow rooting conditions. (Cf. Stilt root.)

C

CAATINGA A thorny, dry, tropical woodland found in northeast Brazil.

CALCAREOUS 1) Consisting of or containing calcium carbonate. 2) Growing on limestone or in soil impregnated with lime.

CALCIPHILE A calcium-loving plant, often found in wetland habitats.

CALICOLE Lime-loving; a plant that thrives in alkaline soil.

CALIFUGE Lime-hating; a plant that will not grow in alkaline soil.

CALYX (pl. calyces) The collective name for sepals, the outer whorl of segments, usually green, that enclose the flower bud.

CAMBIUM A layer of meristematic tissue capable of producing the new cells that increase the girth of stems and roots.

CANOPY The uppermost, spreading, branching layer of a forest.

CARPEL The female part of the flower of flowering plants that contains the ovules; several carpels in a flower are collectively know as a pistil.

CATKIN A cluster of small, petalless flowers.

CAUDEX The swollen stem base of a woody-based plant such as a palm, cycad, or some succulents.

CAUDICIFORM Resembling or possessing a caudex.

CEPHALIUM The woody, flower-bearing, densely spined area at the stem apex of some cacti.

CERRADO A South American savanna type of vegetation with open woodland of short, twisted trees, typically very rich in endemic species.

CHAPARRAL Mediterranean-type vegetation found west of the Sierra Nevada Mts, USA. Coastal sage chaparral consists of low, aromatic, drought-resistant, deciduous shrubs; further to the east, the foothills chaparral consists of a variety of woody shrubs.

CHLOROPHYLL The green plant pigment that is mainly responsible for light absorption and hence photosynthesis in plants.

CLAY DESERT A clay flatland (*see below*) that receives insufficient moisture to support permanent vegetation.

CLAY FLATLAND A vast, highly fertile area where sands and gravels deposited by glaciers are covered by thick clay, laid down over thousands of years when seas covered the area in the wake of the glaciers.

CLIMAX COMMUNITY The final stage in the successional development of an ecological community, at which it remains stable under the prevailing environmental conditions. Climax species are often woody plants.

CLIMBER A plant that climbs or clings by means of modified stems, roots, leaves, or leaf-stalks, using other plants or objects for support.

CLONE 1) A group of genetically identical plants produced by vegetative propagation or asexual reproduction. 2) An individual plant in such a group.

CLOUD FOREST High-growing woodlands that depend more on enveloping clouds than rainfall for their supply of moisture.

COASTAL DESERT A dry coastal region where sea fog is the principal source of moisture.

COASTAL RAINFOREST A coastal area where mild temperatures and high levels of rainfall result in dense, fast-growing trees.

COASTAL SAND VELD An arid coastal type characterized by sand dunes. Plant growth is inhibited by minimal rainfall and strong, salt-laden winds that shift the sand, uproot plants, and deprive them of moisture.

COASTAL WETLAND A low-lying coastal area where continual or regular flooding by either the sea or freshwater rivers results in saline or freshwater marshes.

COLD-CHILLING A propagation technique in which seeds of cold-dwelling species are refrigerated to the same low temperatures that trigger germination in the wild.

COLD DESERT A desert exposed to very cold temperatures in winter.

COLONY Many individuals of a single plant species established in a location to the exclusion of other plants.

COMPOUND Divided into two or more subsidiary parts, e.g., a leaf divided into two or more leaflets.

CONE The densely clustered bracts of conifers and some flowering plants, often developing into a woody, seed-bearing structure, as in the familiar pine cone.

CONIFERS Mostly evergreen trees or shrubs, usually with needlelike leaves and seeds borne naked on the scales of cones.

CONIFEROUS FOREST Evergreen forest that includes broadleaf and needleleaf species.

CONSERVATION The management of human use of the biosphere for the maximum benefit of both the present and future generations.

CORM A bulblike, underground swollen stem or stem base, often surrounded by a papery tunic. A corm is replaced annually by a new corm that develops from a terminal or lateral bud.

CORMEL A small corm developing around a mature corm, usually outside the main corm tunic, as in *Gladiolus*.

CORMLET A small corm arising at the base (and usually within the old tunic) of a mature corm.

COROLLA The interior whorl of the perianth of a flower, comprising several free or fused petals.

CORYMB A broad, flat-topped, or domed inflorescence of stalked flowers or flowerheads arising at different levels on alternate sides of an axis.

COTYLEDON A seed leaf; the first leaf to emerge from the seed after germination, often markedly different from mature leaves. Flowering plants (Angiosperms) are classified into monocotyledons (one) and dicotyledons (two) depending on how many cotyledons are contained in the mature seed. In Gymnosperms they are often produced in whorls.

CRETACEOUS PERIOD The geologic period of 142–65 million years ago, which saw the earliest flowering plants and the ascendancy of the dinosaurs.

CROSS-FERTILIZATION The fertilization of the ovules of a flower as a result of cross-pollination.

CROSS-POLLINATION The transfer of pollen from the anther of a flower on one plant to the stigma of a flower on another plant; the term is often loosely applied to cross-fertilization. (Cf. Self-pollination.)

CROWN 1) The basal part at soil level of a herbaceous plant where roots and stems join and from where new shoots are produced. 2) The upper, branched part of a tree.

CROWNSHAFT The upper section of a palm or cycad trunk that bears leaves and inflorescences.

CULM The jointed, usually hollow, flowering stem of a grass or bamboo.

CULTIVAR A contraction of 'cultivated variety' (abbreviated to cv.); a group (or one among such a group) of cultivated plants clearly distinguished by one or more characteristics when propagated either asexually or sexually. (Cf. Variety.)

CUSHION PLANT A tight, ball-like mass of stems closely covered by leaves or flowers; the flowers often appear on short stalks above the cushion.

CUTTING A section of leaf, stem, or root separated from a plant and used for propagation.

CYME A flat or round-topped, branched inflorescence with each axis ending in a flower, the oldest at the centre and the youngest arising in succession from the axils of bracteoles (secondary bracts).

D

DAMPING OFF Collapse of seedlings and young plants caused by fungi, which rot the bases of stems and roots.

DEAD-HEAD To remove spent flowerheads in order to prolong flowering and prevent self-seeding.

DECIDUOUS Describes plants that shed leaves at the end of the growing season and renew them at the beginning of the next; semi-deciduous plants lose only some of their leaves at the end of the growing season.

DEHISCENCE Term used of fruits (usually capsules) and anthers to describe the process of opening at maturity to release their contents.

DEHISCENT Describes a fruit, usually a capsule, or an anther that splits along definite lines to release seed or pollen.

DESERT A region that has no or very low rainfall, daily extremes of heat and cold, and cold, strong, abrasive winds. (*See also* Coastal desert, Loma, Semi-desert.)

DESERTIFICATION The often rapid removal by wind and rain of fertile, nutrient-rich topsoil from land that has been denuded of plant cover by the activities of humans and grazing animals.

DICOTYLEDON A flowering plant that usually has two cotyledons or seed leaves in the seed; a dicotyledon is also characterized by the (usually) net-veined leaves, the petals and sepals in multiples of two, four, or five, and by the presence of a cambium. (Cf. Monocotyledon.)

DIE-BACK The death of a shoot, beginning at the tip, due to damage or disease.

DIOECIOUS Bearing male and female reproductive organs on separate plants. (Cf. Monoecious.)

DIPTEROCARP A member of a family of tall, mainly evergreen trees native to Africa and south Asia.

DISTRIBUTION The native distribution of a plant encompasses the areas of the world where it naturally grows wild. Also called its native range.

DIVISION The propagation of a plant by splitting it into two or more parts, each with its own section of root system and one or more shoots or dormant buds.

DORMANCY The state of temporary cessation of growth and slowing down of other activities in whole plants, usually during the winter; seed dormancy: non-germination of seed when placed in conditions suitable for germination due to physical, chemical, or other factors inherent in the seed; double (seed) dormancy: non-germination of seeds due to two dormancy factors in the seed.

DROPPER ROOT A deep-growing root produced by seedlings of certain species when they germinate.

DRUPE A fruit consisting of one or several hard seeds (stones) surrounded by a fleshy outer covering.

DRY FOREST A forest growing in an area with a predominantly dry climate.

DUNE SLACK The low-lying depression between two lines of dunes.

DWARF A small or slow-growing variant of a species resulting from hybridization, mutation, or specific cultivation methods. (Cf. Bonsai.)

E

ECOREGION A large unit of land or water containing a distinct assemblage of plant and animal species, having a boundary that approximates the original extent of the natural community prior to major changes in land use.

ECOSYSTEM The interdependent relationships between the community of plants and all other organisms that coexist in a particular environment.

ECOTONE A transitional area between two adjoining ecological communities, often having a high level of biodiversity.

ELAIOSOME A seed appendage, usually rich in oil, that attracts animals (especially ants) as a food source, and so aids dispersal.

ELFIN WOODLAND Low-growing woodland, also called pygmy woodland. On the Colorado Plateau, USA, elfin woodland mainly consists of pinyon pine (*Pinus edulis*) and juniper (*Juniperus osteosperma*).

EMBRYO The part within a seed that has the potential to grow into a new plant. (Cf. Endosperm.)

EMBRYOGENESIS A tissue-culture technique used in plant conservation in which new embryos are propagated from somatic (body) cells that, in normal circumstances, would not be capable of division (self-reproduction).

EMERGENT Describes a forest tree that grows taller than most other tree species in the forest.

ENDEMIC Restricted to a localized area, or peculiar to an area. Also used for a plant that is endemic to a particular locality.

ENDOSPERM The food reserve within a seed that fuels germination. (Cf. Embryo.)

EPIGEAL A type of seed germination in which the seed is pushed above soil level by elongation of the hypocotyl. (Cf. Hypogeal.)

EPIPHYTE A plant that grows on another plant without being parasitic; it obtains moisture and nutrients from the atmosphere without rooting into the soil.

ERICACEOUS 1) Belonging to the Ericaceae. 2) Heath-like or belonging to the genus *Erica*. 3) Describes compost with a pH of 6.5 or less, suitable for growing acid-loving plants.

EVERGREEN Describes plants that retain their foliage for more than one growing season; semi-evergreen plants retain only a small portion of their leaves for more than one season.

EXOTIC A plant that is native to one part of the world but has been introduced elsewhere.

EX SITU CONSERVATION The conservation and maintenance of samples of plants outside their natural habitat, usually in the form of seed, pollen, vegetative propagules, tissue or cell cultures, or individual plants.

F

F1 HYBRID A vigorous and uniform, first-generation offspring, derived from crossing two distinct, pure-bred lines.

FAMILY A category in plant classification, a grouping together of related genera, e.g., the family Rosaceae includes the genera *Rosa*, *Sorbus*, *Rubus*, *Prunus*, and *Pyracantha*.

FARINA A powdery deposit naturally occurring on some leaves and flowers.

FASTIGIATE Describes branches that turn upwards and close to the trunk, as in some conifer species.

FEN A peat-forming wetland habitat that receives water mainly from groundwater seepage. Conditions are less acidic and more fertile than in bogs; fens can become alkaline and support non-acid-loving plants. (Cf. Bog.)

FERAL Describes a domestic animal that has adapted to life in the wild yet remains distinct from true wildlife species.

FERN A flowerless, spore-producing plant consisting of roots, stems, and leaflike fronds. (*See also* Frond.)

FERTILE Of plants – producing viable seed; shoots bearing flowers are said to be fertile shoots as opposed to non-flowering (sterile) shoots.

FERTILIZATION The fusion of a pollen grain nucleus (male) with an ovule (female) to form a fertile seed.

FESCUE A widely cultivated pasture and lawn grass, having stiff, narrow leaves.

FILAMENT The stalk of a stamen that bears the anther.

FLORA All plant life, but especially used in terms of the plants characteristic of a region, period, or particular environment.

FLOWER The reproductive organ of a great many plant genera.

FORB A non-woody, broadleaved, perennial plant other than grass that grows in a grassland habitat.

FOREST-EDGE HABITAT A transitional, buffering habitat between grasslands and forest, typically supporting a high diversity of fauna and flora.

FORMA A variant, abbreviated as f., within a species that is usually distinguished only by minor characteristics. *Clematis montana* f. *grandiflora* is a large-flowered, more vigorous form of *C. montana*; also loosely used for any variant of a species.

FOSSIL RECORD The account of earth's prehistory that has arisen from the dating and analysis of fossils (mineralized relics of plant and animal matter) and the geological contexts in which they were formed.

FRESHWATER WETLAND A wetland at least partly divided from the sea and replenished by freshwater sources such as rivers, e.g., the Everglades in southern Florida, USA.

FROND The leaf-like organ of a fern. Some ferns produce both barren and fertile fronds, the latter bearing spores. 2) Loosely applied to large, usually compound leaves such as those of palms.

FRONTIER FOREST Intact, natural forest.

FROST-TENDER Describes plants that withstand temperatures down to 5°C (41°F). (Cf. Fully hardy, Half-hardy.)

FRUIT The fertilized, ripe ovary of a plant containing one to many seeds, e.g., berries, hips, capsules, and nuts; the term is also used of edible fruits.

FRUIT SET The successful development of fruits after pollination and fertilization.

FULLY HARDY Describes plants that withstand temperatures down to -15°C (5°F). (Cf. Half-hardy, Frost-tender.)

FYNBOS Vegetation of the Mediterranean-type region of the Cape of South Africa. Fynbos comprises 8,500 species, of which 70 per cent grow nowhere else. Bulbous plants and shrubs of the heath family grow beneath proteas and other broadleaved shrubs.

G

GALLERY FOREST A forest growing along a watercourse in a region otherwise devoid of trees.

GAMETOPHYTE The plant body, in species showing alternation of generations, that produces the gametes (germ cells that fuse with other germ cells during fertilization). (Cf. Sporophyte.)

GARRIGUE OR GARRIQUE An exposed habitat of the Mediterranean basin with open cover of heaths and aromatic herbs such as lavender and thyme.

GENE BANK Propagating materials that are collected and stored under conditions that retain viability for long periods. They can include seed, pollen, tissue culture, vegetative propagating material, DNA, and even whole plants grown in plantations.

GENETIC EROSION A permanent reduction in the richness or evenness of genes or combinations of genes to be found in an individual or population. This reduction in genetic resources is detrimental to both short-term viability and evolutionary potential.

GENOTYPE All or part of the genetic constitution of an individual or group. In plant conservation, geneticists may collect and combine as many of a plant's genotypes as possible in order to increase its genetic diversity and survival prospects.

GENUS A category in plant classification (pl. genera) ranked between family and species. A group of related species linked by a range of common characters; e.g., all species of horse chestnut are grouped under the genus *Aesculus*. (*See also* Cultivar, Family, Forma, Hybrid, Species, Subfamily, Subgenus, Subspecies, and Variety.)

GEOPHYTE A plant that survives the winter by having underground buds such as bulbs, corms, or rhizomes.

GERMINATION The physical and chemical changes that take place when a seed starts to grow and develop into a plant.

GERMPLASM The cells of any living organism that carry its hereditary characteristics, and which are distinct from its other cells. The collection of germplasm can be essential in preventing the extinction of highly endangered species.

GLABROUS Smooth and hairless.

GLAUCOUS With a blue-green, blue-grey, grey, or white bloom.

GRAFTING A method of propagation by which the scion (shoot) of one plant is artificially united with the rootstock of another plant so that they eventually function as one plant.

GRASSLANDS (*See* Temperate grasslands.)

GROWING POINT The tip of a shoot from which new extension growth develops.

GREX A collective term, mainly applied to orchids, for all the progeny of an artificial cross from known patents of different taxa. E.g., in *Miltoniopsis* Anjou 'Robert St Patrick', Anjou is the grex name.

GYMNOSPERM A tree or shrub, usually evergreen, that bears naked seeds in cones rather than enclosed in ovaries, e.g., conifers, cycads. (Cf. Angiosperm.)

H

HABITAT The environment in which a plant occurs in the wild.

HALF-HARDY Describes plants that withstand temperatures down to 0°C (32°F). (Cf. Fully hardy, Frost-tender.)

HAMMOCK An area of dense forest with many species of ferns, epiphytes, trees, palms, and shrubs that occurs on land slightly higher than the surrounding terrain, as is common in the swamp forests of Florida, USA.

HARDY Describes plants that can withstand freezing temperatures in winter. (Cf. Fully hardy, Half-hardy, Frost-tender.)

HEEL A cutting made from a side shoot with part of the main stem attached, often preferred for the successful propagation of certain species.

HERBACEOUS A non-woody plant in which the upper parts die down to a rootstock at the end of the growing season. The term is chiefly applied to perennials, although botanically it also applies to annuals and biennials.

HERBICIDE A chemical that kills plants.

HERMAPHRODITIC Describes plant species in which the male and female sex organs are present together in single flowers.

HIBERNACULA A part of a plant that protects the embryo during winter, such as a bulb or a bud.

HIP The closed and ripened, firm or fleshy fruit of the rose, containing many seeds.

HUMUS The chemically complex, organic residue of decayed vegetable matter in soil. Also often used to describe partly decayed matter, such as leaf mould or compost.

HYBRID The offspring of genetically different parents, usually of distinct taxa (*see* Taxon). Hybrids between species of the same genus are known as interspecific hybrids; those between different but usually closely related genera are known as intergeneric hybrids.

HYBRIDIZATION The process by which hybrids are formed.

HYDROPHYTE A plant growing in water or very moist conditions.

HYPOCOTYL The portion of a seed or seedling just below the cotyledons.

HYPOGEAL Type of seed germination in which the seed and the cotyledons remain below the soil surface while the young shoot (plumule) emerges above soil level. (Cf. Epigeal.)

IJK

INBREEDING DEPRESSION A reduction in vigour and fertility of offspring resulting from sexual reproduction within a small, genetically similar (inbred) plant or animal community.

INDEHISCENT Describes a fruit that does not split open to release its seeds. (Cf. Dehiscent.)

INDIGENOUS Originating and growing naturally in a particular region or environment.

INFLORESCENCE A group of flowers borne on a single axis (stem); e.g., a raceme, panicle, and cyme.

IN SITU CONSERVATION 1) The conservation of biological diversity in nature. 2) The conservation of a plant within its natural habitat. (Cf. Ex situ conservation.)

INTERGRADATION The merging of certain plant species through a continuous series of intermediate forms, so that individual plants are difficult to identify positively as one species or another.

INTERNODE The portion of stem between two nodes.

INTERPLANTING A combination of several species planted together for visual effect.

INTERVALE MEADOW A meadow lying at the edge of forest, often having a rich variety of fauna and flora.

INTRODUCTION The establishment of a plant in an area where it has never been known to occur naturally.

INVASIVE Describes a vigorous plant that is able to overwhelm more delicate neighbours if it is not restricted in spread.

IN VITRO Literally, in glass. Refers to biological processes or reactions made to occur, not inside the body of the organism, but in an artificial environment; e.g., in vitro fertilization.

JURASSIC PERIOD The geologic period of 199.5–142 million years ago, which saw the earliest birds and mammals.

KAROO SCRUB Drought-resistant shrublands and lower-altitude xeric communities growing in South Africa's Karoo-Namib biogeographical region, much of which is now contained within the country's Karoo National Park.

KEEL 1) A prominent longitudinal ridge, usually on the underside of an organ such as a leaf, resembling the keel of a boat. 2) Two lower, fused petals of a pea-like flower.

L

LABELLUM Lip, particularly the prominent third petal of iris or orchid flowers. (*See also* Lip.)

LANDRACE A primitive or antique plant variety, often highly adapted to local conditions, that is usually associated with traditional agriculture.

LATERAL A side growth that arises from a shoot or root.

LAURISILVA Broadleaved cloud forest that grows on mist-shrouded slopes in the Canary Islands, the Azores, Madeira, and Morocco. It is dominated by four species of the laurel family, *Laurus azorica*, *Persea indica*, *Apolonnias barbujana*, and *Ocotea foetens*, but many other rare species are also present.

LAX Describes the loose, floppy habit of some plants.

LAYERING A propagation method in which a stem is pegged to the soil while still attached to the parent plant, to induce rooting. (*See also* Air-layering, Mound-layering.)

LEACHING The loss from the topsoil of soluble nutrients by downward drainage.

LEADER 1) The main, usually central, stem of a plant. 2) The terminal shoot of a main branch.

LEAF A plant organ, variable in shape and colour but often flattened and green, borne on a stem, that performs the functions of photosynthesis, respiration, and transpiration.

LEAF CUTTING A method of propagation in which a new plant is grown from a leaf. Only certain species bear leaves that will produce roots in favourable conditions.

LEAFLET One of the subdivisions of a compound leaf.

LEGUME A one-celled fruit of the Leguminosae family that splits along two sides to disperse ripe seed.

LIANA A tropical climbing plant.

LIGNIN A naturally occurring substance that helps to provide strength and structural integrity in plants.

LIGNOTUBER A starchy swelling on an underground root or stem. Fire-resistant plants regrow by drawing resources from lignotubers when their top growth is burned, as do certain plants when their leaves are consumed by animals.

LIGULE A thin, strap-shaped appendage at the base of the blade of a leaf, especially in grasses.

LIME Loosely, a number of compounds of calcium; the amount of lime in soil determines whether it is alkaline, acid, or neutral.

LIP A prominent lower lobe on a flower, formed by one or more fused petals or sepals.

LITHOPHYTE A plant naturally growing on rocks (or in very stony soil), usually obtaining most of its nutrients and water from the atmosphere.

LLANO Savanna consisting of extensive, grassy, treeless plains in Venezuela and elsewhere in South America.

LOAM A term used for soil of medium texture, often easily worked, that contains more or less equal parts of sand, silt, and clay, and is usually rich in humus. If the proportion of one ingredient is high, the term may be qualified as silt-loam, clay-loam, or sandy loam.

LOBE A segment, usually rounded, separated from adjacent segments by clefts extending halfway or less to the centre of the parent organ (such as a leaf).

LOMA A plant community that grows in desert-like conditions in mountains or on steep coastal slopes and depends on fog and/or cloud for moisture.

M

MALLEE Shrubland growing in the Mediterranean conditions of southern and south-western Australia. It includes pungent eucalyptus shrubs and wildflowers.

MANGROVE 1) Trees and shrubs of the genus *Rhizophora* that colonize wetlands, forming dense masses kept in place by tangles of stilt roots. 2) Other tree species with habits similar to those of the true mangroves.

MAQUIS A habitat consisting of dense shrub thickets, particular to Corsican and other Mediterranean coastlines.

MARGINAL AQUATIC A plant that requires permanently moist conditions, from pure mud to water 30–45cm (12–18in) deep.

MARSH An open, freshwater or saltwater wetland habitat subject to frequent or continual flooding, with plants growing on the margin. Dominated by herbaceous plants; emergent plants include cattails, bulrushes, grasses, sedges, and water-tolerant forbs, but no woody plants.

MASS EXTINCTION EPISODE A geologically short period (perhaps a few hundred thousand years) during which 75–95 per cent of species become extinct.

MAT A densely tangled mass of interwoven plant stems, often on the ground or water surface but also found in the tree canopy.

MATORRAL Mediterranean-type vegetation found on the coast of central Chile, backed by the Andes Mountains. Tree species, such as *Jubaea chilensis*, grow with deciduous shrubs, cacti, and thorny species.

Meal A fine dust or powder covering the leaves and other parts of certain plants.

Mediterranean 1) The European region consisting of 1–2 per cent of the world yet containing 20 per cent of plant species. It is characterized by cool, wet winters and warm or hot, dry summers, and poor soils. Vegetation consists of low, fire-adapted shrubs. 2) An area that experiences a climate comparable to that of the Mediterranean region. (*See also* Chaparral, Fynbos, Garigue, Mallee, Maquis, Matorral.)

Medium A compost, growing mixture, or other material in which plants may be propagated or grown.

Megaherb A plant species originating in the sub-Antarctic islands of New Zealand that, in response to harsh growing conditions, low ambient light, and rich soil nutrients has evolved larger leaves and flowers than related species growing in more favourable environments.

Megaspore The female spore that produces an egg for sexual reproduction in in some fern families. (Cf. Microspore.)

Meristem Plant tissue that is able to divide to produce new cells. Shoot or root tips may contain meristematic tissue and may be used for micropropagation.

Mesophyte A plant intermediate between a xerophyte (drought-resistant plant) and a hydrophyte (water-dwelling plant).

Mesozoic era The geologic period of 252–65 million years ago.

Mesquite Extensive woodland thickets, mainly in south-western USA, dominated by spiny, leguminous trees and shrubs of the genus *Prosopis*, especially *P. glandulosa*.

Metapopulation An interconnected association of plant and/or animal populations in habitat "islands" that go extinct locally but migrate and recolonize.

Microhabitat The immediate surroundings of a plant or animal within its habitat.

Micronutrients Chemical elements essential to plants but needed only in very small quantities, also known as trace elements. (Cf. Nutrients.)

Micropropagation The growing of plants from seed or small pieces of tissue under sterile conditions on nutrient-enriched media.

Microspore The male spore that produces sperm for sexual reproduction in some fern families. (Cf. Megaspore.)

Midrib The primary, usually central, vein of a leaf or leaflet.

Miocene epoch The geologic epoch of 24–5 million years ago (within the Tertiary Period) that saw the appearance of the first hominids and the increase and diversification of herbivorous mammals.

Mixed forest Northern forest consisting of both broadleaf and needleleaf species.

Monocarpic Describes a plant that flowers and fruits only once before dying; such plants may take several years to reach flowering size.

Monocotyledon A flowering plant that has only one cotyledon, or seed leaf, in the seed; it is also characterized by narrow, parallel-veined leaves, and parts of the flower in three or multiples of three.

Monoecious Bearing separate male and female reproductive organs on the same plant. (Cf. Dioecious.)

Monopodial Growing indefinitely from a bud on the apex or end of a stem. (Cf. Sympodial.)

Monotypic Having only one component, e.g., a genus containing only one species.

Monsoon forest A type of tropical, deciduous, lowland forest that grows in a hot, humid environment with heavy seasonal rainfall. The trees open their leaves at the onset of the monsoon. The most common species is teak.

Montane Growing on relatively cool mountain slopes just below the tree line.

Moraine A mass of debris, often including highly fertile soil, scoured from the earth's surface by glaciers and deposited in the form of ridges and mounds.

Moss forest A form of cloud forest characterized by an abundance of mosses that cover the tree trunks and the forest floor. Bromeliads, orchids, clusia trees, and tree ferns are often present also.

Mound-layering A propagation method in which the basal section of a stem is earthed up to induce rooting.

Mucilage A gummy secretion present in various parts of plants, especially leaves.

Multi-stemmed Describes shrubs that produce numerous stems at or near their base, rather than branches from a single stem or trunk.

Mycorrhizal Refers to a mutually beneficial association between a soil-dwelling fungus and the roots of a plant.

N

Native 1) Naturally growing wild in a particular area. 2) A plant that naturally grows wild in a particular area. (*See also* Distribution.)

Naturalize To establish and grow as if in the wild.

Nectar A sugary liquid secreted from a nectary, often attractive to pollinating wildlife.

Nectary Glandular tissue usually found in the flower, but sometimes found on the leaves or stems, that secretes nectar.

Needle A stiff, linear leaf of a conifer.

Neutral Of soil – with a pH value of 7, i.e., neither acid nor alkaline.

Niche 1) An area of habitat providing the conditions necessary for an organism to survive. 2) The position of an organism within its ecosystem, dependent upon its behaviour and relationships with other organisms of the ecosystem.

Nitrogen fixing The capability of some plant species, notably members of the Leguminosae, to increase the nitrogen content of the soil through the activity of bacteria present in nodules on their roots. Nitrogen fixing can give such plants a competitive advantage since many rival species cannot tolerate the raised levels of soil nitrogen.

Node The point of the stem from which one or more leaves, shoots, branches, or flowers arise.

Nopalito A flat stem segment of a prickly pear.

Noxious weed A plant that has a poisonous or otherwise harmful effect on the surrounding vegetation.

Nurse shrub A shrub that provides shelter, shade, and moisture for seedlings of other species growing close by, thereby improving their chances of reaching maturity.

Nut A one-seeded, indehiscent fruit with a tough or woody coat, e.g., an acorn. Less specifically, all fruits and seeds with woody or leathery coats.

Nutrients Minerals (mineral ions) used to develop proteins and other compounds required for plant growth. (Cf. Micronutrients.)

Offset A young plant that arises by natural vegetative reproduction, usually at the base of the parent plant; in bulbs, offsets are initially formed within the bulb tunic but later separate out. Also known as offshoots.

Old-growth forest Woodland that has evolved undisturbed over centuries and may contain many ancient trees.

Oligocene epoch The geologic epoch of 33.5–24 million years ago (within the Tertiary Period) during which grasses became increasingly abundant.

Open-pollination Natural pollination. (*See also* Pollination.)

Opposite Describes two leaves or other plant organs at the same level on opposite sides of a stem or other axis. (Cf. Alternate)

Organic Chemically, refers to compounds containing carbon derived from decomposed plant or animal organisms.

Ornamental A plant cultivated for its attractive appearance rather than its use.

Orphan species A plant that is extinct in its native habitat and survives only in cultivated foster habitats.

Ovary The basal part of the pistil of a flower, containing one or more ovules; it may develop into a fruit after fertilization. (*See also* Carpel.)

Ovule The part of the ovary that develops into the seed after pollination and fertilization.

Oxygenating plant An aquatic plant that, by producing oxygen during photosynthesis, contributes to the oxygen dissolved in the surrounding water and so benefits fish and other aquatic wildlife.

PQ

Palaeozoic era The geologic period of 543–252 million years ago.

Pampas Temperate grasslands of southern South America.

Panicle A branched raceme. Loosely applied to freely branched, corymb-like and cyme-like inflorescences.

Pappus An appendage or tuft of appendages that crowns the ovary or fruit in various seed plants and functions in wind dispersal of the fruit.

Paramo A cold, damp, high-altitude mountain region of grassland and shrubland found only in the Andes Mountains.

Parthenocarpic The production of fruit without fertilization having taken place.

Pathogen A micro-organism that causes disease.

Peatland A wetland habitat, evolved in a glacial lake bed, in which organic matter has built up and replaced some of the water because conditions are too cold for decay.

Peduncle The stalk of an inflorescence.

Pendent Hanging downwards.

Perennial Strictly, any plant living for at least three seasons; commonly applied to herbaceous plants and woody perennials (i.e., shrubs and trees).

Perianth The collective term for the calyx and corolla, particularly when they are very similar in form, as in many bulb flowers.

Perianth segment One portion of the perianth, usually resembling a petal and sometimes known as a tepal.

Peristome The fringe or rim, sometimes toothed, of a pitcher plant's urn.

Permian period The geologic period of 290–252 million years ago, which saw the world's only mass extinction of plant species.

Petal A modified leaf, often brightly coloured; one part of the corolla usually of a dicotyledonous flower. (Cf. Tepal.)

Petiole The stalk of a leaf.

pH A measure of alkalinity or acidity, used horticulturally to refer to soils. The scale measures from 1 to 14; pH7 is neutral, above 7 is alkaline, below 7 is acid.

Photosynthesis The production of organic compounds required for growth in plants by a complex process involving chlorophyll, light energy, carbon dioxide, and water.

Phylum The first level of division of living organisms, each of which may contain one or more classes. In turn, classes are divided into orders, and orders into families.

Pineland Forest of pine, oak, and cedar species. Over 1 million acres of pinelands are conserved in the Pinelands National Reserve in southern New Jersey, USA

Pine rocklands A habitat in which stands of pines are interspersed by hammocks of hardwood trees, such as survive in southern Florida, USA.

Pinnate Describes a compound leaf with leaflets (pinnae) arranged alternately or in opposite pairs on a central stalk.

Pistil *See* Carpel.

Pitcher plant A plant with leaves or part leaves that are modified into hollow vessels in which insects and plant matter fall or are trapped and digested, often by means of a liquid secretion.

Pith The soft plant tissue in the central part of a stem.

Pleistocene epoch The geologic period of 1.8 million–10,000 years ago, which saw extensive glaciations of the northern hemisphere and the early evolutionary development of humans.

Plume A flowerhead with the appearance of a tuft of feathers.

Pod An ill-defined term generally applied to any dry, dehiscent fruit; it is particularly used for peas and beans.

Pollen The male cells of a plant, formed in the anther.

Pollination The transfer of pollen from anther to stigmas. (*See also* Cross-pollination, Open-pollination, Self-pollination.)

Pollinator 1) The agent or means by which pollination is carried out (e.g., insects, birds, wind). 2) A plant required to ensure seed set on another self- or partially self-sterile plant.

Polyembryonic Containing more than one embryo in the ovule or seed.

Population A group of individuals of common ancestry that are much more likely to breed with one another than with individuals from other populations.

Prairie Temperate grasslands of North America, including the short-grass prairies of western North America.

Prairie potholes Depressions in flat, formerly glaciated terrain that collect rain and melting snow to form temporary or permanent marshlands, colonized by aquatics, bulrushes, cattails, and sedges.

Primary forest Mature, virgin forest that has evolved over centuries and has never been disturbed by humans.

Propagate To increase plants by seed or by vegetative means.

Prostrate Describes plants or parts of plants that lie on or creep along the ground.

Pseudobulb The thickened, bulb-like stem of a sympodial orchid arising from a (sometimes very short) rhizome.

R

Raceme An inflorescence of stalked flowers radiating from a single, unbranched axis, with the youngest flowers near the tip.

Rachis The main axis of a compound leaf or inflorescence.

Rainforest A forest, usually tropical, with high rainfall, high humidity, and high temperatures all year round.

Ramicaul A slender, upright stem that resembles an elongated pseudobulb, a part of some orchid species.

Recalcitrant seeds Seeds unable to survive the reductions in moisture content or lowering of temperature necessary for long-term storage.

Receptacle The enlarged or elongated tip of the stem from which all parts of a simple flower arise.

Recovery plan *See* Species recovery plan.

Recurved Arched backwards.

Reintroduction The release and management of a plant in an area where it formerly occurred, but where it is now extinct or believed to be extinct.

Relict A group of plants or animals that exists as a remnant of a formerly widely distributed group.

Renosterveld A type of lowland fynbos in South Africa, typically growing on rich, clay soils.

Respiration The release of energy from complex organic molecules as a result of chemical changes.

Resting period *See* Dormancy.

Restoration The process of intentionally altering a site to reproduce the historic, indigenous ecosystem in its natural structure, function, diversity, and dynamics.

Rhizome A specialized, usually horizontally creeping, swollen or slender, underground stem that acts as a storage organ and produces aerial shoots at its apex and along its length.

Rib 1) A ridge, normally vertical, formed on the stem of a cactus. 2) The primary vein on a leaf.

Riparian wetlands Shallow-water habitats restricted to the sides of watercourses that support a variety of aquatic and water-loving species.

Root The part of a plant, normally underground, that anchors it in the soil, and through which water and nutrients are absorbed.

Rootbound Describes a plant with tangled and densely crowded roots due to its being kept too long in a container.

Rootstock The complete underground part of a plant. In propagation, the scion of one species may be grafted onto the rootstock of a related species to promote more or less vigorous growth.

Rosette A cluster of leaves radiating from approximately the same point, often borne at ground level at the base of a very short stem.

Runner A horizontally spreading, usually slender, stem that runs above ground and roots at the nodes to form new plants. Often confused with stolon.

S

Salinization The formation of plant-killing surface crusts of salt resulting from a rise in the water table, e.g., when deep-rooted native plants are replaced by shallow-rooted crops in hot areas.

Saltmarsh A coastal wetland that may experience dramatic, twice-daily fluctuations in water level between high and low tides. The water may be partially salty (brackish) or fully saline. Typically found in estuaries, backwaters, behind barrier beaches, and along bays or inlets. Salt-tolerant grasses and sedges are the dominant vegetation.

Salt steppe A region – e.g. parts of Anatolia, Turkey – characterized by lakes that dry up in the summer. The large areas of exposed lakebed are covered by a salt layer that is thickest at the lake centre. Salt steppes support numerous species adapted to varying levels of soil salinity and moisture availability.

Samara A dry, winged, single seed.

Sand plain Flat terrain covered by sand deposited by outwash during the glacial era and subsequent seas or lakes. The deep sands may be acidic, as in the sand-plain element of South Africa's fynbos. Also known as sand prairie.

Sap The juice of a plant contained in the cells and vascular tissue.

Savanna A flat, dry habitat covered with low shrubs and dotted with small trees, where the ground layer is dominated by grasses. (*See also* Tropical savanna.)

Scale cutting A method of propagating bulbous plants in which new plants are grown from individual scale leaves separated from a parent bulb.

Scandent Ascending or loosely climbing. (*See also* Climber.)

Scape The leafless stem of a solitary flower or inflorescence.

Scarification A propagation technique in which the hard outer covering of certain seeds is abraded or treated chemically before sowing. This increases the rate of water uptake and promotes germination.

Scion The shoot or part of a shoot that is bonded to the rootstock of a second plant by grafting.

Sclerification The hardening of cells (e.g., in leaves) by the formation of a secondary wall and the deposit of lignin.

Scree A slope comprising rock fragments formed by the weathering of rock faces. High-altitude alpines that need excellent drainage are grown in garden scree beds.

Scrub A habitat with poor or dry soil, covered with bushes and small trees.

Secondary forest Trees that grow in the place of virgin, primary forest that has been removed by logging, forest clearance, or a natural agent such as wildfire. Some plant species that establish in secondary forest were never part of the primary forest that grew in the locality.

Seed The ripened, fertilized ovule, containing a dormant embryo capable of developing into an adult plant.

Seedbank 1) An establishment dedicated to the survival of endangered species in which thousands of seeds are stored in conditions calculated to preserve and prolong their viability. The seeds are drawn from as many populations of the endangered species as possible to maximize genetic diversity. 2) The stock of dormant and viable seeds present in the soil.

Seedhead Any fruit that contains ripe seeds.

Seed lead *See* Cotyledon.

Seedling A young plant that has developed from a seed.

Seep bog A low-lying area or hillside saturated by groundwater which has been prevented from flowing downwards by impervious rock and so seeps out laterally, sometimes over a wide area. (Cf. Bog.)

Self-fertile Describes a plant that produces viable seed when it is fertilized with its own pollen. (*See also* Fertilization, Pollination, Self-pollination, and Self-sterile.)

Self-incompatible *See* Self-sterile.

Self-pollination The transfer of pollen from the anthers to the stigma of the same flower, or alternatively to another flower on the same plant. (Cf. Cross-pollination.)

Self-seed To shed fertile seeds that produce seedlings around the parent plant.

Self-sterile Describes a plant unable to produce viable seed after self-fertilization, and requiring a different pollinator in order for fertilization to occur. Also known as "self-incompatible", or incapable of self-fertilization.

Selva A name for tropical rainforest.

Semi-desert A dry region, less arid than true desert, that receives sufficient rainfall to support some life.

Semi-evergreen Describes a plant the retains most or some of its foliage throughout the year.

Semi-ripe wood cutting A cutting taken from semi-mature wood in mid- or late summer, occasionally in early autumn.

Semi-terrestrial Growing partly in the soil and partly epiphytically. (Cf. Epiphyte, and Terrestrial.)

Sepal The outer whorl of the perianth of a flower, usually small and green, but sometimes coloured and petal-like.

Serrated With sharp, forward-pointing teeth on the margin (mainly of leaves).

Serrulated Minutely serrated (mainly of leaves).

Sexual reproduction A form of reproduction involving fertilization, giving rise to seed or spores.

Shrub A woody-stemmed plant, usually branched from or near the base, lacking a single trunk.

Simple Undivided (mainly of leaves). (Cf. Compound.)

Softwood cutting A cutting taken from young, non-woody growth, from spring to early summer.

Spadix A fleshy spike bearing tiny flowers, usually sheathed by a spathe.

Spathe A bract (modified leaf) that encloses a flower.

Speciation The process by which new biological species are formed.

Species A category in plant classification, the lowest principal rank below genus, containing closely related, very similar individuals. Species are capable of interbreeding freely with each other but not with members of other species.

Species recovery plan A comprehensive, practical plan of action to safeguard a species against further loss or deterioration of its remaining genepool.

Speculum A glossy raised area, varying from square- to diamond- or horseshoe-shaped, on the lip of some orchid flowers.

Spike A long flower cluster with individual flowers borne on very short stalks or attached directly to the main stem.

Spine A stiff, sharp-tipped, modified leaf or stem.

Sporangium A body that produces spores on a fern.

Spore The minute, reproductive structure of flowerless plants, such as ferns, fungi, and mosses.

Sporocarp A spore casing.

Sporophyte The form of a plant, in species having alternation of generations, that produces asexual spores (Cf. Gametophyte.)

Spur 1) A modified petal with a hollow, basal projection, often containing nectar. 2) A short branch or branchlet along a main branch on which flowers and fruit are produced.

Stamen The male reproductive organ in a plant, comprising the pollen-producing anther and usually its supporting filament or stalk.

Stem The main axis of a plant, usually above ground, that supports structures such as branches, leaves, flowers, and fruit.

Stem-tip cutting A cutting taken from the soft tip of a non-flowering stem, usually from spring to autumn.

Steppe Temperate grasslands of Eurasia, from the Ukraine to Russia and Mongolia.

Sterile 1) Unable to produce flowers or viable seed. 2) Describes flowers without functional anthers or pistils (*see* Carpel).

Stigma The apical portion of a carpel, usually borne at the tip of a style, which receives pollen prior to fertilization.

Stilt root A stabilizing, obliquely angled root produced from the trunks of trees adapted to shallow or waterlogged soil. (Cf. Buttress root.)

Stolon A horizontally spreading or arching stem, usually above ground, that roots at its tip to produce a new plant. Often confused with runner.

Stomata Microscopic pores in the surface of aerial parts of plants (leaves and stems), allowing transpiration to take place.

Strain A hereditary line of descent. The linear development of a hybrid is known as a strain.

Stratify To expose seed to cold in order to break dormancy, either by refrigeration before sowing, or by sowing outdoors in autumn or winter. Also called pre-chilling.

Strobilus The reproductive organ of a conifer. Male strobili resemble catkins; female ones resemble mature cones in miniature. Some ferns develop spore-producing strobili.

Stylar column Styles fused together.

Style The usually elongated part of a carpel between the ovary and the stigma; it is not always present.

SUBFAMILY A category in plant classification, a division within the family.

SUBGENUS A subdivision within a genus, higher in rank than species.

SUBMONTANE Describes a type of forest found in lowland and hilly sites that consists mainly of deciduous, broadleaved trees.

SUBSHRUB 1) A low-growing plant that is entirely woody. 2) A plant that is woody at the base but has soft, usually herbaceous, growth above.

SUBSPECIES A distinct form or race of a species, higher in rank than varietas (var., *see* Variety) or forma (f., *see* Forma).

SUBTROPICAL Refers to the high-temperature zone located between tropical and temperate regions. Rainfall occurs mainly as heavy downpours during the monsoon season.

SUBTROPICAL MIXED FOREST Mixed broadleaf and needleleaf evergreen forest growing in the subtropics; typical genera are *Araucaria*, *Nothofagus*, and *Podocarpus*.

SUCCESSIONAL CHANGE 1) Alterations in the mix of vegetation as low-growing, light-loving species grow tall and provide conditions for shade-loving, understorey species. 2) Alterations in the composition of an area's vegetation resulting from changes taking place in the environment. E.g., a reduction in the frequency of bush fires can depress reproduction of fire-dependent species and allow relative proliferation of non fire-dependent species.

SUCCULENT A drought-resistant plant with thick, fleshy leaves and/or stems adapted to store water. All cacti are succulents.

SUCKER 1) A shoot rising from an underground bud on the roots or rootstock of a plant. Suckering is one means by which plants reproduce vegetatively; the new plant is a genetic clone of the parent. 2) A sticky pad, sometimes an extension of a tendril, produced in great numbers by some climbing plants to assist ascent.

SUSTAINABLE DEVELOPMENT Development that meets the needs of the present without compromising the ability of future generations to meet their needs.

SWALE A low-lying or depressed and often wet stretch of land.

SWAMP A freshwater or saltwater wetland that is dominated by trees.

SYMBIOSIS A close and mutually beneficial association of two dissimilar animal or plant species.

SYMPODIAL Form of growth of a shoot in which the terminal bud ends in an inflorescence or dies, in which case growth continues by lateral buds. (Cf. Monopodial.)

SYNONYM An alternative name by which a plant is known in botany. In citations, the correct botanical name is given first, followed by its synonym (syn.).

T

TABLETOP MOUNTAIN *See* Tepuis.

TAIGA The Russian name for boreal forest, widely used elsewhere. (*See* Boreal forest.)

TAPROOT The primary, downward-growing root of a plant (especially a tree); also loosely applied to any strong, downward-growing root.

TAXON (pl. taxa.) A group of living organisms at any rank; applied to a group of plants or entities that share distinct, defined characteristics.

TEMPERATE Refers to zones located between the subtropics and the polar circles, which experience distinct seasons without extremes of temperature. Rainfall occurs throughout the year.

TEMPERATE GRASSLANDS Grassy vegetation growing in very fertile soils in semi-arid, continental climates between forest and desert. The grasses are sometimes classed as tall, mixed, or short, with short grasses characteristic of drier climates. Fire occurs periodically in the grasslands. (*See also* Pampas, Prairie, Steppe, Veld). (Cf. Savannas.)

TENDRIL A modified leaf, branch, or stem, usually filiform (long and slender) and capable of attaching itself to a support. (*See also* Climber.)

TEPAL A single segment of a perianth that cannot be distinguished as either a sepal or a petal, as in *Crocus* or *Lilium*. (*See also* Perianth segment.)

TEPUIS The tabletop mountains of the Guyana Highlands in eastern Venezuala, a high plateau environment characterized by deep fissures and vertical cliffs.

TERRARIUM A glass container, often a globe, in which plants are grown, especially plants that require a high humidity level.

TERRESTRIAL Growing in the soil; a land plant. (Cf. Epiphyte and Semi-terrestrial.)

TERTIARY PERIOD The geologic period of 65–1.8 million years ago, which saw the earliest large mammals and hominids.

THORN A sharp outgrowth on a stem.

TISSUE CULTURE The propagation of plant tissue under sterile conditions in artificial media.

TOOTHED With a margin divided into tooth-like segments (mainly of leaves). In plants such as cacti and hollies, frequent spines protect the leaf margin.

TOP-DRESS 1) To apply fertilizers or mulches to the soil surface around plants. 2) To apply material such as stone or grit to the surface of the soil, or compost around a plant, in order to improve drainage and reduce moisture loss.

TRANSLOCATION Movement of substances such as dissolved nutrients or weedkillers within the vascular system (conducting tissues) of a plant.

TRANSPIRATION The loss of water by evaporation from plant leaves and stems.

TREE A woody, perennial plant, usually with a well-defined trunk or stem with a head or crown of branches above.

TRICHOME Any type of outgrowth from the surface tissue of a plant, such as a hair, scale, or prickle.

TRISTYLY The natural occurrence of three different lengths of style in individuals of the same species, resulting in the species being subdivided botanically into three different types based on the length of the style.

TROPICAL Refers to the zone between the Tropics of Cancer and Capricorn, with a hot, humid climate that promotes lush plant growth. Rainfall may occur throughout the year or mainly during the monsoon season. (Cf. Subtropical.)

TROPICAL DRY FOREST Forest with a canopy lower than that of tropical rainforest, typically 30–40m (100–130ft) high. Relatively few lianas are found. Often a habitat for numerous endemic species.

TROPICAL EVERGREEN LOWLAND FOREST Grows in hot, humid areas with constant high rainfall. Trees have shallow roots, buttresses at the base, branching near the crown, and leathery, unlobed leaves.

TROPICAL EVERGREEN MOUNTAIN FOREST Grows in cool and damp areas with high rainfall. Lichens and mosses are common under the shallow-rooted trees.

TROPICAL RAINFOREST Tall-growing forest with a canopy 40–50m (130–160ft) high, with emergent trees exceeding 60m (200ft). This forest type has the richest species diversity of all forests.

TROPICAL SAVANNA Grass-dominated habitats ranging from treeless plains to woodlands with a grassy understorey. They typically experience less rainfall than tropical forests, with a distinct dry season, and are found between equatorial rainforests and subtropical deserts. (*See also* Cerrado and Llano.)

TUBER A swollen, usually underground, organ derived from a stem or a root, used for food storage.

TUBERCLE A small, wart-like projection, as found on the stems of some cacti.

TUFA Porous, moisture-retentive limestone rock, used for the cultivation of alkaline-loving, rock-dwelling alpines.

TUFTED Growing in dense clusters (tufts).

TUNDRA The treeless zone lying between the icecap and the timber line of North America and Eurasia. The subsoil is permanently frozen; mosses and lichen characterize the landscape.

TUNIC The fibrous membrane or papery outer skin of a bulb or corm.

TUNICATE Enclosed in a tunic.

TURION 1) A detached, overwintering, usually fleshy, bud produced by certain water plants. 2) A term sometimes applied to an adventitious shoot or sucker.

TUSSOCK A dense tuft of vegetation, especially of grass.

U

UMBEL A flat or round-topped inflorescence in which numerous stalked flowers are borne from a single terminal point.

UNDERSTOREY VEGETATION Plants that grow beneath the tree canopy of forests. When the tree canopy is very thick the understorey plant species may be adapted to require only minimal light and water.

UNDULATE Describes leaves that have a wavy margin.

UNESCO United Nations Educational, Scientific and Cultural Organization.

UNESCO BIOSPHERE RESERVE An internationally recognized terrestrial or coastal ecosystem nominated to fulfil three functions: to contribute to the conservation of landscapes, ecosystems, species, and genetic variation; to foster sustainable development; and to provide support for the global conservation effort.

UNESCO WORLD HERITAGE LIST A list of cultural and natural sites identified by the World Heritage Committee as being of outstanding value to humanity and deserving of protection. The list comprised 582 cultural properties, 149 natural sites, and 23 mixed sites in 2004, and additional properties are nominated each year.

V

VARIABLE Varying in character from the type; particularly of seed-raised plants that vary in character from the parent.

VARIEGATION Irregular arrangements of pigments, usually the result of either mutation or disease, mainly in leaves.

VARIETY 1) Botanically, a naturally occurring variant (varietas – var.) of a wild species, between the rank of subspecies and forma. 2) Also used commonly but imprecisely used to describe any variant of plant. (Cf. Cultivar.)

VASCULAR PLANT A plant containing food-conducting tissues (the phloem) and water-conducting tissues (the xylem).

VEGETATIVE GROWTH Non-flowering, usually leafy growth.

VEGETATIVE PROPAGATION Non-sexual techniques for increasing plants, by cuttings, division, grafting, or layering.

VEGETATIVE REPRODUCTION Any increase of plant numbers that occurs by non-sexual means, e.g., new shoots that erupt from creeping roots and organs such as rhizomes, and shoots from arching branches, or even detached leaves, that take root in the soil.

VELAMEN A water-absorbing tissue covering the aerial roots of certain plants, including many epiphytes.

VELD Temperate grasslands of South Africa.

VIABILITY The capacity to germinate. Depending on species, some seeds are viable for only a few days while others can germinate after many years in the soil.

VIVIPAROUS 1) Describes a plant that forms plantlets on leaves, flowerheads, or stems. 2) Also applied loosely to plants that produce bulbs or bulbils on these organs.

VLEI South African name for a shallow seasonal lake that fills during the summer rains, then disappears over the winter dry season.

WXYZ

WEEPING Describes a tree or shrub that is pendent (hanging downwards) in habit.

WETLAND A habitat that is more or less saturated with water year-round and supports aquatic and water-loving plants. (*See also* Bog, Fen, Freshwater wetland, Marsh, Peatland, Saltmarsh.)

WET MEADOW A habitat in which lowland meadow this is saturated by ground water gives way to relatively dry upland.

WHORL An arrangement of three or more organs arising from the same point.

WILDLIFE Living animals that have never been domesticated.

WOODY Describes stems or trunks that are hard and thickened rather than soft and pliable. Also called ligneous. (Cf. Herbaceous.)

XERIC Relating to or growing in dry conditions.

XEROPHYTE A plant adapted to survive in arid conditions, either by the reduction of stems and leaves to minimize water loss, or by having water-storage tissue, as in cacti and other succulents.

XYLEM The woody part of plants, consisting of supporting and water-conducting tissue.

YUNGAS Subtropical, montane, deciduous and evergreen forests that flank the eastern slopes and central valleys of the central Andes from northernmost to southernmost Peru. Yungas support one of the richest montane forest ecosystems.

INDEX

Numbers in italics refer to pages where the subject is illustrated.

A

B

C

D

E

F

G

H

K

L

M

N

O

P

Q

R

S

T

PICTURE CREDITS

The publisher would like to thank the following for their kind permission to reproduce their images:

ABBREVIATIONS KEY:
a=above; b=below; c=centre; f=far; l=left; r=right; t=top.

FFI=Fauna & Flora International; FLPA=FLPA - Images of Nature; GPL=Garden Picture Library; OSF=Oxford Scientific Films; SPL=Science Photo Library; Woodfall=Woodfall Wild Images

1 NHPA/Rod Planck; **2-3** RSPCA Photolibrary/P Harcourt Davies/Wild Images; **4-5** Alamy Images/Mark Lewis; **6-7** NHPA/Ernie Janes; **8-9** NHPA/Daniel Zupanc; **10-11** Alamy Images/Jamie Marshall; **12-13** Woodfall/M Biancarelli; **14-15** A-Z Botanical Collection; **16** National Geographic Image Collection/Paul Chesley (cra), SPL/Andrew Syred (ca), SPL/Georgette Douwma (bc), SPL/NASA GSFC (bl); **16-17** Corbis/Douglas Peebles; **18** Corbis/Thomas Wiewandt;Visions of America (tl), Corbis/William Manning (br), SPL/Eye of Science (bc); **18-19** SPL/Martin Dohrn; **19** SPL/Vaughan Fleming (br); **20** Corbis/John MacPherson (tc); **20-21** FLPA/Minden Pictures (cra); **21** SPL/Dr Jeremy Burgess (tr); **22** Corbis/Owen Franken (tr), Corbis/Royalty-Free (c); **22-23** Corbis/Keren Su; **23** Corbis/Enzo & Paolo Ragazzini (tc), Corbis/Tim McGuire (cl); **24** SPL/Peter Menzel (br), Martin Walters (cb); **24-25** National Geographic Image Collection/Panoramic Images/Warren Marr; **25** Alamy Images/Nigel Cattlin/Holt Studios Int.(bc), SPL/George Bernard (bl); **26** DK Images/Natural History Museum (cfl), SPL/Andrew Syred (cra); **27** DK Images/Natural History Museum (ca); **28** FLPA/Albert Visage (tr), OSF/David M Dennis (bl); **28-29** FLPA/F Lanting; **30** akg-images (bl), DK Images/Christine M Douglas (cra); **30-31** Corbis/Susan G Drinker; **32** Corbis/Tony Arruza (bl), Mary Evans Picture Library (tr); **32-33** Still Pictures/Mark Edwards; **33** FFI/ Juan Pablo Moreiras (cra); **34** FFI/ Juan Pablo Moreiras (c), NHPA/Rod Planck (bl), Still Pictures/Hartmut Schwarzbach (tr); **34-35** Getty Images/Rich Frishman; **36** Corbis/Louise Gubb (bc), SPL/Peter Chadwick (ca), **36-37** Alamy Images/Trevor Smithers ARPS; **37** Copyright Trustees of the Royal Botanic Gardens, Kew (tr); **38** Ecoscene/Wayne Lawler (bl); **38-39** Alamy Images/David Copeman; **39** Alamy Images/Nicholas Rigg (tr), Holt Studios Int. (tl); **40** BGCI/Bian Tan (bc), www.bridgeman.co.uk/Ashmolean Museum, University of Oxford, UK (cbl), Corbis/David Lees (ca); **40-41** GPL/John Glover; **41** Corbis/Chinch Gryniewicz/Ecoscene (cr), Corbis/Nik Wheeler (bc); **42** Photo courtesy of Royal Botanical Gardens, Hamilton (ca), University of Wisconsin-Madison Arboretum (bc), (bcl), (bl); **42-43** Corbis/Raymond Gehman; **43** Eden Project, UK/copyright Charles Francis (tc), SPL/Tony Craddock (tr); **44** copyright Trustees of the Royal Botanic Gardens, Kew (br), SPL/Klaus Guldbrandsen (c), Corbis/Antoine Gyori/Sygma (bl); **44-45** Corbis/ Reuters (t); **45** A-Z Botanical Collection (bl); **46** A-Z Botanical Collection (cl); **46-47** Peter Anderson/Designer Philip Van Wyck; **48** Garden World Images (bc), (cfl), (cra); **48-49** Garden World Images; **49** Dr John Donaldson (cfr), San Marcos Growers, Santa Barbara, CA (tc), Luke Sweedman,Kings Park and Botanic Gardens, West Perth, 6005, Western Australia (br); **50** NHPA/David Woodfall (c), Plant Images/C Grey-Wilson (bl); **50-51** Corbis/Wolfgang Kaehler; **51** Corbis/Joseph Sohm/ChromoSohm Inc. (tr); **52** NHPA/Dr Eckart Pott (cfr), Garden World Images (ca); **52-53** OSF/Chris Perrins; **54** Neil Diboll/Prairie Nursery, Westfield, Wisconsin, USA (tr), S Edgar David & Associates Landscape Architects/Rob Cardillo Photography (crb), John Glover (cfr), New England Wild Flower Society/Dorothy Long (bl); **55** Corbis/Joe McDonald (bl), Mt. Cuba Center, Inc (r); **56** Alamy Images/Gay Bumgarner (bc), Alamy/John Glover/GPL (cfr), Jason Hawkes (cfl); **56-57** Corbis/David Aubrey; **57** Alamy Images/Gay Bumgarner (tc), Alamy Images/Gil Hanly (br); **58** Alamy Images/Val Duncan/Kenebec Images (bl), SPL/Susumu Nishinaga (cal); **60-61** NHPA/James Carmichael Jr; **62** Andrew Byfield (bl); **63** Ecoscene/Andrew Brown (cl), (cr), Holt Studios Int./Hew Prendergast (br), Woodfall (bl); **64** Ardea.com (tr), FFI/Juan Pablo Moreiras (c), Holt Studios Int./Alan & Linda Detrick (cfrb), Holt Studios Int./Inga Spence (cfra), NHPA/Kevin Schafer (cra), NHPA/Rod Planck (br), SPL/Martin Bond (crb), Still Pictures/David Woodfall (bl); **64-65** Ecoscene/Andrew Brown; **66** FLPA/Minden Pictures (tr), Still Pictures/Jany Sauvenet (bc), Still Pictures/Klein Hubert (bl); **67** Bruce Coleman Ltd/Gerald S Cubitt (clb), Ecoscene/Andrew Brown (r), OSF/Harold Taylor (tr), OSF/John Brown (tc), OSF/Michael Fogden (cr); Still Pictures/Mark Edwards (tl), Still Pictures/Patryck Vaucoulon (br), Still Pictures/Walter Hodge (bl); **68** FLPA/Steve McCutcheon (bl), SPL/Simon Fraser (bc), (bcr), (br), Woodfall/J Cornish (tr); **69** Ecoscene/Andrew Brown (r), NHPA/John Shaw (tl), (tr), OSF/Konrad Wothe/SAL (br), SPL/Alan Sirulnlkoff (bl), SPL/Kaj R Svensson (cr), SPL/Vaughan Fleming (bc); **70** Forestry Commission (cra), Holt Studios Int. (crb), SPL/Simon Fraser (b), Still Pictures/J P Sylvestre (tr); **71** New England Wild Flower Society/Hal Horwitz (br); **72** Holt Studios Int./Nigel Cattlin (cfra), FLPA/Martin Withers (tr), FLPA/Minden Pictures (cfrb), NHPA/Ann & Steve Toon (c), NHPA/N A Callow (cra), NHPA/Rod Planck (crb), OSF/Dinodia Picture Library (br), Still Pictures/Fred Bruemmer (bl); **72-73** Ecoscene/Andrew Brown; **74** Ecoscene/Andrew Brown (l), Lonely Planet Images/Carol Polich (br), NHPA/ANT Photo Library (tl), NHPA/Jonathan & Angela Scott (c), NHPA/Mirko Stelzner (cl), OSF/Stan Osolinski (tr), Woodfall/N Hicks (bl); **75** Holt Studios Int. (cl), Holt Studios Int./Bob Gibbons (tr), P Cliff Miller, Inc. Landscape Artistry. Lake Bluff, IL (br); **76** OSF/M Wendler (tr); **77** Ecoscene/Andrew Brown (r), Ecoscene/Anthony Cooper (br), Ecoscene/Nick Hawkes (cr), Holt Studios Int./Peter Wilson (clb), FLPA/Fritz Polking (bl), Still Pictures/Paul Springett (tr), Still Pictures/Peter Frischmuth (tc); **78** Ecoscene/Andrew Brown (bl), NHPA/Geoff Bryant (cra), OSF/John McCammon (cfra), OSF/Mike Slater (tr), OSF/Wendy Shattil & Bob Rozinski (br), **78-79** Woodfall; **80** Ecoscene/Wayne Lawler (br), GPL (ca), GPL/Briggite Thomas (tr), FLPA/Jurgen & Christine Sohns (cl); **81** Corbis/Andrew Brown/Ecoscene (tr), Onne van der Wal (cr), NHPA/Dick Roberts (br), NHPA/Nigel J Dennis (bl), (ca), Still Pictures/Ted Schiffman (tc); **82** FLPA/Minden Pictures (c), NHPA/Anthony Bannister (cfrb), OSF/Adam Jones (cfra), OSF/Konrad Wothe (cra), OSF/Richard Packwood (crb), OSF/Tim Shepherd (br), Still Pictures/Roland Seitre (bl), Woodfall/T Mead (tr); **82-83** Holt Studios Int./Hew Prendergast; **84** Holt Studios Int./Hew Prendergast (l), FLPA/Mark Newman (cr), OSF/Martyn Colbeck (cl), OSF/Michael Fogden (tr), Woodfall (tl), Woodfall/T Mead (bl); **85** Corbis/Steve Kaufman (cl), **85** GPL/John Glover (cb), GPL/Zara McCalmont (br), OSF (tr); **86-87** Woodfall/A Watson; **88** Natural Visions/Brian Rogers (clb), Ardea.com/Jean-Paul Ferrero (bl); **88-89** NHPA/David Woodfall; **90** Alamy Images/geogphotos (bl), A-Z Botanical Collection (cla), NHPA/David Woodfall (tr); **91** OSF/Ian West (tr), Martin Walters (tc); **92-93** Woodfall (b); **93** Corbis/Carl & Ann Purcell (tr), Woodfall (tl); **94** Natural Visions/Ian Tait (br), David Graber (bl), Plants of Hawaii/Forest & Kim Starr (tc), NHPA/Geoff Bryant (tr); **95** Australian National Botanic Gardens (bl), Corbis/Pam Gardner/Frank Lane Picture Agency (br), Mary Evans Picture Library (cr), (tr), OSF/Kjell Sandved (tc); **96** A-Z Botanical Collection (tr), www.bridgeman.co.uk/Museo Nacional de Antropologia, Mexico City, Mexico (cl), NHPA/Kevin Schafer (bl), The Wellcome Institute Library, London (cr); **97** Corbis/Ric Ergenbright (b), Missouri Botanical Garden (cl), Photos Horticultural (tr); **98** GPL/Howard Rice (tc), NHPA/Geoff Bryant (tr), OSF/Rob Nunnington (b); **99** Corbis/Andrew Brown/Ecoscene (br), FFI/Juan Pablo Moreiras (tr); **100** FFI (br), FFI/Evan Bowen-Jones (tr), OSF/Bert & Babs Wells (bl); **101** NHPA/E Hanumantha Rao (b), NHPA/Geoff Bryant (tr); **102** Corbis/Peter Johnson (br), Friends of Soqotra/Sue Christie (tc); **102-103** NHPA/Vicente Garcia Canseco; **103** A-Z Botanical Collection (tr), bihrmann.com (br), FLPA/Keith Rushforth (cl); **104** OSF/Mike Slater; **105** Holt Studios Int./Bob Gibbons (bl), NHPA/ANT Photo Library (br), OSF/Bert & Babs Wells (tr); **106** Don Herbison-Evans/Macleay Museum, University of Sydney (bc), NHPA/Ann & Steve Toon (bl), Jerry Pavia Photography Inc (tc); **106-107** Natural Visions; **107** A-Z Botanical Collection (br); **108** The John Bartram Association, Historic Bartram's Garden (cra), The G.R. 'Dick' Roberts Photo Library/Shannel Courtney (br), Martin Gardner (ca); **109** Michael Charters (tr), Photos Horticultural (br); **110** Gerald D Carr, PHD (cl), FLPA/Keith Rushforth (bc), Photos Horticultural (tr); **111** Natural Visions (tr), Garden World Images (br); **112** Corbis/Joe McDonald (cl), DK Images/Andrew Butler (b); **113** NHPA/Andrew Ackerley (bl), Photos Horticultural (tcl), Copyright Trustees of the Royal Botanic Gardens, Kew (tcr), Garden World Images (br); **114** Arkansas Natural Heritage Commission (cl), FLPA/Martin Withers (tr); **114-115** FFI/Juan Pablo Moreiras; **115** Natural Visions (br); **116** OSF/Bert & Babs Wells (cal); **116-117** FLPA/Wendy Dennis (b); **117** Bruce Coleman Ltd/Rita Meyer (tl), Edward Ruiz (cra); **118** Corbis/Kevin Schafer (tr), Eric Crichton Photos (br), Tony Cunningham (bl); **119** Natural Image/Bob Gibbons (br), (tl), Still Pictures/Mark Edwards (bl), (cl); **120** Forest Light/Alan Watson (tr), Nature Picture Library Ltd/John Cancalosi (br), Still Pictures/Romano Cagnoni (bl); **121** A-Z Botanical Collection; **122** Sally and John Perkins (bl), Copyright Trustees of the Royal Botanic Gardens, Kew (tr), Cheng Yu-Pin (c); **123** Alamy Images/Jeff Smith (bl), Corbis/James L Amos (car), The G.R. 'Dick' Roberts Photo Library (tr), Plants of Hawaii/Forest & Kim Starr (br), Roger's Plants Ltd (cl); **124** Ardea.com (bl), Martin Gardner, (tcl), Royal Botanic Garden Edinburgh (tr); **125** Natural Visions (b), FFI/Evan Bowen-Jones (tr); **126** Tim Upson (bl); **126-127** OSF (b); **127** Corbis/Roger Tidman (cl), Tony Cunningham (tc), Photos Horticultural (bc), Garden World Images (br); **128** Martin Gardner (tr); **128-129** Corbis/William Manning; **130** OSF/Judd Cooney (clb), OSF/T C Middleton (tc), OSF/Tony Tilford (bl); **130-131** NHPA/Guy Edwards; **131** NHPA/David Woodfall (tl), OSF/Richard Packwood (tr); **132** Martin Gardner (bl), (tc), **132-133** Garden World Images (cl); **133** S L Jury (br), Royal Botanic Garden Edinburgh/Sid Clarke (tl); **134** Natural Visions (bl), Corbis/Hulton-Deutsch Collection (cl), Holt Studios Int./Silvestre Silva (br), **134-135** NHPA/ANT Photo Library; **135** NHPA/Lutra (tl), Royal Botanic Garden Edinburgh/Debbie White (tr); **136** Andrew Lawson (bl), WL Crowther Library, State Library of Tasmania (tc); **136-137** Martin Gardner; **138** Natural Visions (bc), Copyright Trustees of the Royal Botanic Gardens, Kew (br); **139** OSF/Terry Heathcote (tc), Garden World Images (bcl), (cr); **140** FLPA (cr), OSF/Niall Benvie (bl), Garden World Images (tc); **140-141** A-Z Botanical Collection; **142** DK Images/Natural History Museum (br), Sabina Knees (bl); **143** Natural Visions/Heather Angel (br); **144** Copyright Trustees of the Royal Botanic Gardens, Kew (tc), Garden World Images (bl), (cr); **145** GPL/Howard Rice (b); **146** A-Z Botanical Collection (tr), GPL/John Glover (br); **147** John Game (r), OSF/Richard Packwood (tl); **148** Martin Gardner (bl), Still Pictures/Galen Rowell (tr); **148-149** A-Z Botanical Collection; **149** NHPA/T Kitchin & V Hurst (tr); **150** Martin Gardner (c), FLPA/Keith Rushforth (tc); **150-151** Martin Gardner; **151** Martin Gardner (tl), Royal Botanic Garden Edinburgh/Debbie White (bl), Garden World Images (br); **152** Corbis (br), DK Images/Andrew Butler (bl), OSF/Breck P Kent/AA (cr); **153** Woodfall/D Woodfall (r); **154** Garden World Images (bl); **154-155** Photos Horticultural; **155** Corbis (tr); **156** NHPA/David Middleton (br), SPL (tr), Garden World Images (bl); **157** A-Z Botanical Collection (br), Martin Gardner (cal), (tr), SPL (cb); **158-159** Corbis/David Muench, Natural Visions/Colin Paterson-Jones (cr), NHPA/T Kitchen & V Hurst (tl); **160** Mark A Johnson (c), Garden and Wildlife Matters (cl); **160-161** Corbis/Macduff Everton; **162** Corbis/Michael S Yamashita (tc), **162-163** OSF/Konrad Wothe; **163** Photos Horticultural (bc), (tr), Woodfall (ca); **164** The G.R. 'Dick' Roberts Photo Library (t), Clive Nichols (b); **165** A-Z Botanical Collection (tl), Joseph Dougherty (br); **166** A-Z Botanical Collection (br), Copyright Trustees of the Royal Botanic Gardens, Kew (tc), (tcr), (tr); **167** Natural Visions/Heather Angel (br), A-Z Botanical Collection (cl), Corbis/Antoine Serra/In Visu (tr), Eric Crichton Photos (cr), OSF/Deni Bown (cb); **168** Natural Image/Bob Gibbons (tl), Photos Horticultural (b); **169** Natural Visions/Heather Angel (br), Corbis/Gunter Marx Photography (cb), Rick Mark (cl); **170** GPL/Neil Holmes (tr), FLPA/Francois Merlet (bc), S & O Mathews Photography (cl), **170-171** Garden World Images; **171** Corbis/Cardinale Stephane (tr), OSF/Deni Bown (tl); **172** Holt Studios Int./Peter Wilson (bl), NHPA/ANT Photo Library (br); **173** John Glover (br), Photos Horticultural (cl); **174** Corbis/Kim Kulish (cb), Photos Horticultural (t), Garden World Images (bcr); **174-175** A-Z Botanical Collection; **176** Alamy Images/Jenny Andre (cb), Corbis/Jim Sugar (bc), Corbis/Tom Bean (bl), Jerry Pavia Photography Inc (tc), J S Peterson/plants.usda.gov (cl); **176-177** Corbis/Steve Terrill; **178** Corbis/Michael Boys (cr), Garden and Wildlife Matters/Debi Wager Stock Pics (tc), OSF/Bob Gibbons (bl); **178-179** Ardea.com/Thomas Dressler; **179** Copyright Trustees of the Royal Botanic Gardens, Kew (tr), Garden World Images (br); **180** Garden and Wildlife Matters (tc), OSF/Tom Leach (bc), Garden World Images (bl); **181** Garden and Wildlife Matters (tr), Natural Image/Bob Gibbons (cb), NHPA/Kevin Schafer (cl); **182** GPL/J S Sira (tr), Natural Image/Bob Gibbons (tc), Garden World Images (b); **183** Andrew Byfield (br), GPL/Howard Rice (tr), NHPA/Martin Garwood (cr), Copyright Trustees of the Royal Botanic Gardens, Kew (cl), Garden World Images (bl); **184** Natural Visions/Colin Paterson-Jones (r), Mary Evans Picture Library (bl), NHPA/Brian Hawkes (tc); **185** Jerry Pavia Photography Inc (cl); **186** Natural Visions/Heather Angel (bc), OSF/John McCammon (t), Photos Horticultural (cr); **187** NHPA/David Middleton (tl), OSF/Breck P Kent (cr), Plant Images/C Grey-Wilson (b); **188** Mr Fothergill's Seeds (bl), NHPA/G I Bernard (cl), RSPCA Photolibrary/P Harcourt Davies/Wild Images (t); **189** Ardea.com (b), OSF/Robin Bush (t); **190** Ardea.com/Bob Gibbons (t), Natural Image/Bob Gibbons (bc), Garden World Images (br); **191** Natural Visions (cra), DK Images/Roger Smith (br), (tr), Charles E Jones (bl), Clive Nichols (tl); **192** Corbis/Photowood Inc. (tr), Garden World Images (cr), (cl); **193** A-Z Botanical Collection (bl), Corbis/David Lees (br), Garden World Images (t); **194-195** A-Z Botanical Collection; **195** Natural Visions/Heather Angel (tc), Andrew Byfield (br); **196** Hutchison Library/Edward Parker (b); **197** A-Z Botanical Collection (c), Jerry Pavia Photography Inc (br), Copyright Trustees of the Royal Botanic Gardens, Kew (tr); **198** Natural Visions/Heather Angel (bc), Ardea.com/Bob Gibbons (tc); **198-199** Garden and Wildlife Matters/Martin P Land; **199** Garden World Images (tl), (tr); **200** Center for Conservation and Research of Endangered Wildlife, Cincinatti Zoo (bl), OSF/Adam Jones (tc); **200-201**Garden World Images; **202** Natural Visions/Heather Angel (br), Neil Holmes Photography (bl); **203** Natural Image/Bob Gibbons (tr), Garden World Images (bc); **204** Natural Visions (tr), OSF/Kathie Atkinson (bc); **204-205** NHPA/Jim Bain; **206** Michael and Patricia Fogden (cla), Martin Gardner (bcl), (bl), OSF/Chris Sharp (cl); **206-207** Photos Horticultural; **207** Ardea.com/Ake Lindau (tc), OSF/Terry Heathcote (bc); **208** OSF/Geoff Kidd (bc), Plant Pictures World Wide (tr); **209** Mary Evans Picture Library (cr), OSF/Tom Leach (br), Garden World Images (bl), Still Pictures/Kevin Schafer (tc); **210** Nature Picture Library Ltd/John Cancalosi (bl); **210-211**Garden World Images; **212** Natural Visions (cb), Garden World Images (cr); **213** DK Images/Dave Watts (bl), DK Images/Wallace Collection (c), John Glover (br), Garden World Images (tc); **214** GPL/Howard Rice (br), Photos Horticultural (cl); **215** Eric Crichton Photos (c), Garden and Wildlife Matters/John Feltwell (bl), Copyright Trustees of the Royal Botanic Gardens, Kew/G Lewis (cr); **216** Photos Horticultural (tc); **216-217** OSF/Geoff Kid; **217** Photos Horticultural (tr), Copyright Trustees of the Royal Botanic Gardens, Kew (bc); **219** A-Z Botanical Collection (tc), (bl), NHPA/David Middleton (cr); **220** A-Z Botanical Collection (cb), RSPCA Photolibrary/Hans Christian Heap (bc), Woodfall (cra); **220-221** DK Images/Jacqui Hurst; **222** Alamy Images/David Moore (bl), Natural Visions/Brian Rogers (tl); **222-223** OSF/Konrad Wothe; **223** FFI/Mike Read (cfr), Still Pictures/Joern Sackermann (tl); **224** GPL/J S Sira (b), Andrew Lawson (cr); **225** Natural Visions (cr), DK Images/Roger Smith (tr), Jerry Pavia Photography Inc (bc), Plant Images/C Grey-Wilson (cl); **226** Corbis/David Turnley (cr), Photos Horticultural (bl); **227** Dr Hugh P McDonald (tl), Gary A Monroe (cr), Clive Nichols (br), Garden World Images (cl); **228** Corbis/Roger De La Harpe/Gallo Images (tr), DK Images/British Museum (bl); **229** Corbis/Underwood & Underwood (cr), DK Images/Roger Smith (tr), Hutchison Library/Michael Kahn (bl), Dr Alan Meerow (bc), OSF/Adrian Bailey (c); **230** Garden World Images (tl); **230-231** Natural Visions (bl), GPL/Chris Burrows (b); **231** A-Z Botanical Collection (tr), Corbis/Caroline Penn (br), Eye Ubiquitous/Bob Gibbons (cr); **232** John Manning (br), OSF/Deni Bown (tr), Photos Horticultural (cl); **233** Natural Visions/Colin Paterson-Jones (cl), Holt Studios Int./Nigel Cattlin (r), Connall Oosterbroek (bl), (clb); **234** FLPA/Mark Newman (tr), NHPA/David Middleton (b), Kenneth R Robertson (cr); **235** Garden and Wildlife Matters/Debi Wager (br), Copyright Trustees of the Royal Botanic Gardens, Kew (bl); **236** DK Images/Roger Smith (bl), John Game (cr), GPL/Howard Rice (tr); **237** FFI//Mike Read (tr), GPL/Howard Rice (br), Andrew Lawson (cl); **238** Ardea.com/M Watson (bl), John Manning (br), Garden World Images (tr); **239** Photos Horticultural **240** Natural Visions/Colin Paterson-Jones (c), Dr. Alan Meerow (tc), Garden World Images (br); **241** Corbis (bl), GPL/Chris Burrows (br), Dr Alan Meerow (cl), Garden World Images (tr); **242** Dr Alan Meerow (bl), Photos Horticultural (br), Copyright Trustees of the Royal Botanic Gardens, Kew (t); **243** Dr Alan Meerow (br), (tr), Photos Horticultural (bl), Copyright Trustees of the Royal Botanic Gardens, Kew (cl); **244** British Library, London (cr), National Botanical Institute-Kirstenbosch/Graham Duncan (br), OSF/Deni Bown (bc), (cr); **245** Natural Visions/Heather Angel (bl), Clive Nichols (tr), OSF/Konrad Wothe (br); **246** www.bridgeman.co.uk/Linnean Society, London, UK (bl); **246-247** Corbis/Gary Braasch (b); **247** Garden World Images (tr); **248** GPL/Philippe Bonduel (cr), Plant Images/C Grey-Wilson (cl); **249** Natural Visions/Colin Paterson-Jones (c), (tl), GPL/Howard Rice (b); **250** John Glover (b), (tl); **251** NHPA/Kevin Schafer (br), OSF/Densey Clyne Productions (tl), Photos Horticultural (bl), (cr); **252** NHPA/Stephen Dalton (tc), Royal Botanic Garden Edinburgh/Debbie White (br), Garden World Images (cl); **253** Corbis/Nik Wheeler (crb), GPL/John Glover (bc), Martin Gardner (cl), Dr. Alan Meerow (tl), (tr); **254** Natural Visions/Colin Paterson-Jones (bl), GPL/David Dixon (br), GPL/Jerry Pavia (tr); **255** Plant Images/C Grey-Wilson (br), (tr), Garden World Images (cl); **256** DK Images/Roger Smith (tc), Plant Images/C Grey-Wilson (bl), Garden World Images (br), (cl); **257** www.bridgeman.co.uk/Private Collection/The Stapleton Collection (br), Photos Horticultural (cr); **258** Kellydale Nursery/Tony Palmer (cr), Photos Horticultural (bl); **259** Corbis (br), DK Images/Roger Smith (bl), Connall Oosterbroek (tc); **260** Ardea.com/John Mason (cl), GPL/Neil Holmes (cfl), OSF/Mike Hill (c), OSF/Ronald Toms (bc); **260-261** Corbis/David Muench (bl); **262** Natural Visions/Heather Angel (ca), OSF/Daniel Cox (cfl), Still Pictures/Christine Sourd (tc), Zefa Visual Media/E Koch (bl); **262-263** Photolibrary.com/Botanica/Acevedo Melanie; **263** Natural Visions/Heather Angel (c); **264** Ardea.com/A P Paterson (tr), Ardea.com/Chris Knights (cl), Woodfall/R Revels (cr); **265** Ardea.com/J L Mason (b); **266** Holt Studios Int./Peter Wilson (bl), Diego Ugarte (tc); **266-267** Photos Horticultural; **267** Ardea.com/Don Hadden (cr); **268** Corbis/Archivo Iconografico, S.A (tc), Corbis/Reza/Webistan (bcl); **268-269** NHPA/Martin Harvey; **270** OSF/C E Jeffree (tc); **270-271** Natural Visions/Colin Paterson-Jones (b); **271** Ardea.com/P Morris (br), © Ulster Museum 2004/Reproduced with the kind permission of the Trustees of the National Museums & Galleries of Northern Ireland (tr); **272** Corbis/Hulton-Deutsch Collection (cl); **272-273** Photos Horticultural; **273** Corbis/George D Lepp (tr), Corbis/Scott T Smith (c), DK Images/© CONCULTA-INAH-MEX; Authorized reproduction by the Instituto Nacional de Antropologia e Historie; Museo de Antropologia de Xalapa (crb); **274** Corbis/Craig Lovell (clb), Corbis/Nevada Wier (br), Martin Walters (tr); **274-275** Ardea.com/Pascal Goetgheluck; **276** Bart & Susan Eisenberg (cla), Dr Chris Stapleton (bcl); **276-277** NHPA/Nigel Dennis; **277** Natural Image/Bob Gibbons (crb); **278** Hutchison Library/Nigel Smith (clb); **278-279** OSF/Professor Jack Dermid; **279** NHPA/Stephen Kraseman (tl), OSF/Deni Bown (t); **280** A-Z Botanical Collection (ca), Corbis/Chinch Gryniewicz (c), Corbis/Scott T Smith (cfl); **280-281** OSF/Mark Hamblin; **282** Natural Visions/Heather Angel (clb), (tr), Corbis/Tania Midgley (bl), Garden World Images (bc), Woodfall/N Hicks (tl); **282-283** Corbis/Bryan Knox/Papilio; **283** The Alpine Garden Society (tc), Plant Images/C Grey-Wilson (tr); **284** Garden and Wildlife Matters/John & Irene Palmer (tc), Photos Horticultural (bl), Plant Images/C Grey-Wilson (br), Garden World Images (tr); **285** Alpine Garden Society (bl), A-Z Botanical Collection (r); **286** Laurie Campbell Photography (bl), Photos Horticultural (tr), Plant Images/C Grey-Wilson (br); **287** Ardea.com/Bob Gibbons (tl), A-Z Botanical Collection (b), Plant Images/C Grey-Wilson (tr); **288** Plant Images/C Grey-Wilson (b), (t); **288-289** Plant Images/C Grey-Wilson (b), Plant Images/C Grey-Wilson (t); **290** Ardea.com/Bob Gibbons (c), Plant Images/C Grey-Wilson (bl), (br), (t); **291** Natural Visions/Brian Rogers (b), Garden World Images (t); **292** Plant Images/C Grey-Wilson (br), Garden World Images (bl), (t); **293** Plant Images/C Grey-Wilson (bl), Garden World Images (br), (cl), (cr); **294** OSF/Bob Gibbons (br), Plant Images/C Grey-Wilson (bl), (t); **294-295** Natural Visions; **295** Corbis/Joseph Sohm/Visions of America (bl), GPL/Christopher Fairweather (br), Plant Images/C Grey-Wilson (tr); **296** Corbis/Julian Calder (t), DK Images/Neil Fletcher (bl), Copyright Trustees of the Royal Botanic Gardens, Kew (br); **296-297** Ardea.com; **297** Ardea.com (t), Photos Horticultural (bl), Plant Images/C Grey-Wilson (br); **298** Plant Images/C Grey-Wilson (br), (c), Royal Botanic Garden Edinburgh (t), Garden World Images (bl); **299** GPL/Neil Holmes; **300** Natural Visions/Brian Rogers (br), The G.R. 'Dick' Roberts Photo Library (bl), Photos Horticultural (t); **301** Plant Images/C Grey-Wilson; **302** Corbis/David Muench (c), Neil Holmes Photography (t), Plant Images/C Grey-Wilson (b); **303** Ardea.com (br), Corbis/Patrick Johns (cl), Plant Images/C Grey-Wilson (bl), (t); **304** OSF/Robert Peakes (c), Corbis/Tui De Roy (tr), Plant Images/C Grey-Wilson (b), Garden World Images (tl); **305** OSF/Bob Gibbons; **306** Ardea.com/Bob Gibbons (tr), DK Images/Copyright Trustees of the Royal Botanic Gardens, Kew (br), Garden World Images (bl); **307** Ardea.com/Bob Gibbons (c), GPL/J S Sira (b), Garden World Images (t); **308** NHPA/ANT Photo Library (bl), NHPA/John Shaw (cfl); **308-309** Bruce Coleman Ltd/Gunter Ziesler; **310** FLPA/Tony Wharton (bl), NHPA/Laurie Campbell (tr); **310-311** Corbis/W Perry Conway; **311** NHPA/Alberto Nardi (tc); **312** Corbis/David Muench (l); **313** Ecoscene/Stephen Coyne (tc), OSF/Michael Brooke (cr); **314** FLPA/Jeremy Early (br), FLPA/Minden Pictures (cl), NHPA/Ivan Polunin (tr), OSF/Waina Cheng (bl); **314-315** OSF/Martyn Chillmaid; **316** Ardea.com/John Mason (bl); **318** Tim Upson (bl); **318-319** Corbis/Mark A Johnson; **319** Corbis/Andrew Brown/Ecoscene (br), Kerry S Walter (tr); **320** Natural Visions/Heather Angel (bl); **320-321** Corbis/Charles Mauzy; **321** A-Z Botanical Collection (br), GPL/Howard Rice (tr); **322-323** Ecoscene/Wayne Lawler; **323** A-Z Botanical Collection (tr), OSF/Deni Bown (br); **324** Nature Picture Library Ltd/David Tipling (tcr), SPL/Eye of Science (clb); **324-325** NHPA/Geoff Bryant; **326** RSPCA Photolibrary/Pete Oxford/Wild Images (cfl); **326-327** SPL/William Ervin; **328** Alamy Images/Lightworks Media (cfr), Corbis/Bettmann (cfl), FLPA/Jurgen & Christine Sohns (cfl), SPL/Mark de Fraeye (tc); **328-329** NHPA/Daryl Balfour; **329** Corbis/George H H Huey (cfr), NHPA/Daniel Heuclin (tc); **330** OSF (cfl), RSPCA Photolibrary/Beth Davidow (bl); **331** Natural Visions/Heather Angel (cfr), Ardea.com/Richard Waller (cra), Jerry Pavia Photography Inc (tl); **332** Ardea.com/Alan Weaving (br), Dr. Ulrich Meve (tr), NHPA/Martin Harvey (bl), NHPA/Nigel J Dennis (tc); **332-333** Natural Visions/Colin Paterson-Jones; **333** FLPA/E & D Hosking (tr), NHPA/Martin Harvey (tl); **334** Graham Charles (cl), Héctor M Hernández (cr), OSF/Michael Fogden (bc); **335** A-Z Botanical Collection (br), Garden and Wildlife Matters (cl), Photos Horticultural (tr); **336** Graham Charles (bl), (br), (tr), Dr Nigel P Taylor (tc); **337** Terry Smale (br), (tl); **338** Carlos Gómez-Hinostrosa (tr), Garden World Images (b); **339** Graham Charles (bl), (tc), Dr. Ulrich Meve (br), Garden World Images (cl); **340** Ardea.com (tc), Héctor M Hernández (bl); **342** A-Z Botanical Collection (tr), Graham Charles (bl), Garden World Images (cr); **343** Abisai Josue Garcia Mendoza (br), OSF/Michael Fogden (tr), Jerry Pavia Photography Inc (bcr); **344** Ardea.com/Chris Harvey (bl), Dr. Ulrich Meve (r), (tc); **345** Ardea.com/John Mason (bl), Dr. Ulrich Meve (tc), Photos Horticultural (br); **346** Carlos Gómez-Hinostrosa (tc); **346-347** Garden World Images (tr); **347** Ardea.com/A Greensmith (br), Garden World Images (bc), (c), (tc); **348** Graham Charles (bl), (br), (tc); **348-349** Graham Charles; **349** OSF/Tim Jackson (r); **350** NHPA/Stephen Krasemann (bl); **350-351** Garden World Images; **352** Marlon Marchado (bl), Dr. Ulrich Meve (br); OSF/Mary Plage (cra); **353** Prof. Dr. W Barthlott (cr), (tl), NHPA/Martin Wendler (tr); **354** Carlos Gómez-Hinostrosa (t), Garden World Images (br), William C Weightman (bc), (bl); **355** Ardea.com/Francois Gohier (c), OSF/Michael Fogden (br); **356** Natural Visions/Heather Angel (bc), Garden World Images (cl); **356-357** Corbis/Dallas and John Heaton; **358** Corbis/Chris Hellier (tl), OSF/Edward Parker (bc), Garden World Images (tc); **358-359** Corbis/Douglas Peebles; **359** OSF/Brian Kenney (tr); **360** Martin Gibbons & Tobias W Spanner (c); **361** Dr John Donaldson (tc), Mary Evans Picture Library (cr), GPL/Mel Watson (b); **362** Martin Gibbons & Tobias W Spanner (cr), Garden World Images (t); **363** Ardea.com/Wardene Weisser (br), Nature Picture Library Ltd/Bruce Davidson (tl), OSF/Michael Fogden (bl); **364** www.bridge-man.co.uk/Hamburg Kunsthalle, Hamburg, Germany (bl); **365** A-Z Botanical Collection (cr), Martin Gibbons & Tobias W Spanner (bc), (cl), (tc); **366** Natural Visions (cl), A-Z Botanical Collection (bl), GPL/Howard Rice (crb), Martin Gibbons & Tobias W Spanner (tc); **367** A-Z Botanical Collection (c), Corbis/Penny Tweedie (br), Garden and Wildlife Matters/Debi Wager Stock Pics (cl); **368** A-Z Botanical Collection (bc), Martin Gibbons & Tobias W Spanner (tl), Garden World Images (cr); **369** Bob & Marita Bobick (br), Dr John Donaldson (tc), Garden World Images (c), Getty Images (bl); **370** Dr John Donaldson (tc), Photos Horticultural (cr), Still Pictures/Walter H Hodge (bl); **370-371** Dr John Donaldson (b); **371** Natural Visions (cal), RSPCA

Photolibrary/John Bracegirdle (tr); **372** OSF/Michael Brooke; **373** The Art Archive/Musée du Louvre Paris/Dagli Orti (tc), Martin Gardner (bl), Martin Gibbons & Tobias W Spanner (br), (tr); **374** NHPA/Anthony Bannister (bc), SPL/Dr Jeremy Burgess (c); **374–375** NHPA (t); **375** Royal Botanic Garden Edinburgh (bcl), (bcr), (cb), (cbl), (crb); **376** Natural Visions (tr), Dr John Donaldson (b), Garden and Wildlife Matters/John & Irene Palmer (tc); **377** A-Z Botanical Collection (bl), Martin Gibbons & Tobias W Spanner (br), (tr); **378** Corbis/Tim Thompson (l), Dr John Donaldson (br), Garden and Wildlife Matters (bc); **379** Martin Gibbons & Tobias W Spanner (br), Nature Picture Library Ltd/David Pike (tr); **380** A-Z Botanical Collection (br), NHPA/Daniel Heuclin (l); **381** www.bridgeman.co.uk/Library of Congress, Washington D.C., USA (tr), Corbis/Tony Arruza (br), Dr John Donaldson (bc); **382** FLPA/Minden Pictures (bl); **382–383** Plant Images/C Grey-Wilson; **384** Corbis/Michael & Patricia Fogden (c), Corbis/Richard T Nowitz, RSPCA Photolibrary/John Bracegirdle/Wild Images (bl), RSPCA Photolibrary/Margaret Welby (ca), Garden World Images (cal); **385** Hutchison Library/Edward Parker (tr); **386** Joseph Dougherty (cr), FFI/Juan Pablo Moreiras (tc); **386–387** Garden World Images; **388** Joseph Dougherty (cb), Johan Hermans (tr); **389** A-Z Botanical Collection (ca), Garden and Wildlife Matters/Debi Wagner Stock Pics (tr), Johan Hermans (br); **390** Ch'ien C Lee (bl), Corbis/Hal Horwitz (tc), OSF/Mary Plage (br), Copyright Trustees of the Royal Botanic Gardens, Kew/Rodolfo Pichi Sfurmolili (cl); **391** OSF/Chris Perrins (t); **392** Corbis/Ricardo Azoury (bc), Andrew Steens (cr); **392–393** Ecoscene/Kjell Sandved; **393** Corbis/John-Marshall Mantel (br), GPL/Frank Leather (cr); **394** Natural Visions/Heather Angel (bl), OSF/Deni Bown (c); **394–395** OSF/Kjell Sandved (b); **395** Eric Crichton Photos (tl), Ecoscene/Papilio/Alastair Shay (tr); **396** Nature Picture Library Ltd/David Tipling (bc), Plant Images/C Grey-Wilson (tr); **397** NHPA/Rod Planck (l), Jerry Pavia Photography Inc (tr); **398** Laurie Campbell (tr); **398–399** NHPA/Ernie Janes; **399** Corbis/Joel Creed/Ecoscene (br), NHPA/ANT Photo Library (tr); **400** Photos Horticultural; **401** Johan Hermans (b), (cr), OSF/Bert & Babs Wells (cl); **402** Corbis/Kevin Schafer (bl), Photos Horticultural (tc); **402–403** Michael & Patricia Fogden; **403** OSF/Deni Bown (cl); **404** A-Z Botanical Collection, Hutchison Library/Isabella Tree (br); **405** A-Z Botanical Collection (bc), Johan Hermans (tr); **406** Johan Hermans (bl), OSF/Michael Fogden (t); **407** NHPA/Haroldo Palo Jr (cl), Photos Horticultural (tr); **408** Natural Visions/Brian Rogers (br), Natural Visions/Heather Angel (bl); **409** Holt Studios Int./Bob Gibbons (br), SPL/Claude Nuridsany & Marie Perennou (l); **410** A-Z Botanical Collection (bl), Jerry Pavia Photography Inc (tc), Garden World Images (cl); **411** Garden World Images (tr); **412** Eric Crichton Photos (tr), Garden World Images (bl); **413** Corbis/Hal Horwitz (br), SPL/Dr Morley Read (bc), Garden World Images (tc); **414** Garden World Images (b), (c); **415** Corbis/Galen Rowell; **416** Greg Allikas (br), Eric Crichton Photos (cl), OSF (bl); **417** Eric Crichton Photos (tr), Joseph Dougherty (c); **418** OSF/Dani/Jeske/AA (t), OSF/Richard Manuel (bc); **419** Eric Crichton Photos (bl), OSF/Bert & Babs Wells (r), Garden World Images (tl); **420** OSF/C C Lockwood (t), Photos Horticultural (cr); **420–421** Corbis/Michael & Patricia Fogden (b); **421** FLPA/David Hosking (t); **422** Corbis/Owen Franken (tl), OSF/Nils Reinhard (tr), Garden World Images; **423** Corbis/Joel Creed/Ecoscene (br), DK Images/C Andrew Henley (cl); **424** Natural Image/Bob Gibbons (bl), (cbl), OSF (cfl); **424–425** Corbis/David Muench; **426** Alamy Images/Popperfoto (cbl), Corbis/Bettmann (cfl), NHPA/David Woodfall (cal); **426–427** A-Z Botanical Collection; **427** Corbis/Lynda Richardson (bcr), (cr), Garden World Images (tcr); **428** OSF (tr); Bert & Babs Wells (br), Barry Rice/sarracenia.com (bl), (cl); **429** Ardea.com/Jean-Paul Ferrero (bl), NHPA/ANT Photo Library (br), OSF/Paul Franklin (tr), Barry Rice/sarracenia.com (cr); **430** Corbis/Hal Horwitz (tc), SPL/Dan Suzio (cb), (clb), (crb); **430–431** Getty Images/Steve Hopkin; **432** Corbis/Richard Cummins (bl), ICCE Photolibrary/Andy Purcell (tc), OSF/Breck P Kent (cl); **432–433** NHPA/Laurie Campbell; **433** Photos Horticultural (cr), Barry Rice/sarracenia.com (tc); **434** Ardea.com/M D England (cl), Nature Picture Library Ltd/Neil Nightingale (br), Barry Rice/sarracenia.com (bl), (tr); **435** Corbis/Richard List (tr), Dr. Andreas Wistuba/wistuba.com (b), (cl), (tl); **436** OSF/Zig Leszczynski/AA (cl), RSPCA Photolibrary/Hans Christian Heap (tc), Dr. Andreas Wistuba/wistuba.com (bl); **436–437** Photos Horticultural (b); **437** NHPA/Mark Bowler (tl), Garden World Images (c), (r); **438** OSF/Jim Bockowski/AA (tr), Photos Horticultural (tr), Barry Rice/sarracenia.com (bl), Garden World Images (c); **438–439** Nature Picture Library Ltd/Neil P Lucas (r); **439** GPL/Jerry Pavia (cr), OSF/Philippe Henry (tl); **440–441** OSF/David M Dennis; **442** NHPA/ANT Photo Library (t); **443** FLPA/I Rose (b), FLPA/Keith Rushforth (tr); **444** OSF/David Cayless (b), Photos Horticultural (tr); **445** Natural Visions/Brian Rogers (bl), NHPA/Geoff Bryant (tr), Garden World Images (cr), (tl); **446** Natural Visions (bl), FLPA/Roger Wilmshurst (tr), OSF/Mike Birkhead (t); **447** NHPA/Daniel Heuclin (b), Photos Horticultural (t); **448** FLPA/Roget Tidman (br), Photos Horticultural (tl); **449** FLPA/R Wilmshurst (r), Garden World Images (tl); **450** GPL/Pernilla Bergdahl (br); **451** A-Z Botanical Collection (br), Ecoscene/Frank Blackburn (tr), Garden World Images (cl); **452** Ecoscene/Papilio Photographic/Dennis Johnson (tc), FLPA/E & D Hosking (b), Garden World Images (cr); **453** A-Z Botanical Collection (tc), Ecoscene/Joel Creed (b), OSF/John McCammon (tl), Garden World Images (cr); **454** Ecoscene/Andrew Brown (bl), Garden World Images (br); **455** GPL/Philippe Bonduel (tl), FLPA/David Hosking (tr), FLPA/Rolf Bender (bl), Garden World Images (cr); **456** Natural Visions (bl), Corbis/Douglas Peebles (br); **457** Natural Visions (b), (cr), (tl); **458** Garden and Wildlife Matters (tl), FLPA/Leo Batten (bl); **458–459** OSF; **459** Ecoscene/Chinch Gryniewicz (tr), FLPA/Ian Rose (tl); **460** FLPA/J Tinning (cr), Garden World Images (b); **461** Natural Visions (t), FLPA/Leo Batten (bl), Photos Horticultural (br); **462** Plants of Hawaii/Forest & Kim Starr (bl), FLPA/Francois Merlet (br), Garden World Images (tr); **463** A-Z Botanical Collection (tr), OSF/Bob Gibbons (b); **464** Natural Visions (tl), Natural Visions/Colin Paterson-Jones (b), FLPA/D P Wilson (tr); **465** Garden and Wildlife Matters/John Feltwell (cl), FLPA/P Haynes (br), Garden World Images (tl); **466** A-Z Botanical Collection (tr), FLPA/Peggy Heard (bl); **466–467** Natural Visions; **467** Natural Visions (tr), A-Z Botanical Collection (br); **468** Natural Visions (tc), (tl), Natural Visions/Brian Rogers (br), Photos Horticultural (bc); **469** Ecoscene/Wayne Lawler (tl), NHPA/Elizabeth MacAndrew (br); **471** Garden World Images (tl).

For further information, see www.dkimages.com

ACKNOWLEDGMENTS

This book was sparked by the generosity of Mrs Marjorie Arundel

Editorial assistance:

Joni Blackburn, Joanna Chisholm, Sharon Lucas, Pamela Marmito, Carole McGlynn, Anja Schmidt

Design assistance:

Martin Hendry, Ted Kinsey

Dorling Kindersley would like to thank:

The directors of the botanic gardens that sponsored this book (*see 473*); the library, picture library, and administrative staff at the Royal Botanic Gardens, Kew, the Royal Botanic Garden Edinburgh, BGCI (Botanic Gardens Conservation International), and FFI (Fauna and Flora International), in particular Fiona Bradley, Barbara Bridge, Etelka Leadlay, Chris Loades, Marilyn Ward, and Debbie White ABIPP.

Also, thanks to Terry Hewitt of Holly Gate Cactus Nursery and Dr Julien Shaw, RHS International Orchid Registrar, for their advice.